FORMULATION AND PROCESS DEVELOPMENT STRATEGIES FOR MANUFACTURING BIOPHARMACEUTICALS

FORMULATION AND PROCESS DEVELOPMENT STRATEGIES FOR MANUFACTURING BIOPHARMACEUTICALS

Edited by

Feroz Jameel
Susan Hershenson

WILEY

A JOHN WILEY & SONS, INC., PUBLICATION

Published by John Wiley & Sons, Inc., Hoboken, New Jersey
Published simultaneously in Canada

Library of Congress Cataloging-in-Publication Data:

ISBN 978-0-470-11812-2

10 9 8 7 6 5 4 3 2 1

CONTENTS

FOREWORD ix

PREFACE xiii

CONTRIBUTORS xvii

PART I PREFORMULATION AND DEVELOPMENT OF STABILITY-INDICATING ASSAYS: BIOPHYSICAL CHARACTERIZATION TECHNIQUES 1

1 THE STRUCTURE OF BIOLOGICAL THERAPEUTICS 3
Sheryl Martin-Moe, Tim Osslund, Y. John Wang, Tahir Mahmood, Rohini Deshpande, and Susan Hershenson

2 CHEMICAL INSTABILITY IN PEPTIDE AND PROTEIN PHARMACEUTICALS 41
Elizabeth M. Topp, Lei Zhang, Hong Zhao, Robert W. Payne, Gabriel J. Evans, and Mark Cornell Manning

3 PHYSICAL STABILITY OF PROTEIN PHARMACEUTICALS 69
Byeong S. Chang and Bernice Yeung

4 IMMUNOGENICITY OF THERAPEUTIC PROTEINS 105
Steven J. Swanson

5 PREFORMULATION RESEARCH: ASSESSING PROTEIN SOLUTION BEHAVIOR DURING EARLY DEVELOPMENT 119
Bernardo Perez-Ramirez, Nicholas Guziewicz, and Robert Simler

6 FORMULATION DEVELOPMENT OF PHASE 1–2 BIOPHARMACEUTICALS: AN EFFICIENT AND TIMELY APPROACH 147
Nicholas W. Warne

7 LATE-STAGE FORMULATION DEVELOPMENT AND CHARACTERIZATION OF BIOPHARMACEUTICALS 161
Adeola O. Grillo

8 AN EMPIRICAL PHASE DIAGRAM–HIGH-THROUGHPUT SCREENING APPROACH TO THE CHARACTERIZATION AND FORMULATION OF BIOPHARMACEUTICALS 173
Sangeeta B. Joshi, Akhilesh Bhambhani, Yuhong Zeng, and C. Russell Middaugh

9 FLUORESCENCE AND PHOSPHORESCENCE METHODS TO PROBE PROTEIN STRUCTURE AND STABILITY IN ICE: THE CASE OF AZURIN 207
Giovanni B. Strambini

10 APPLICATIONS OF SEDIMENTATION VELOCITY ANALYTICAL ULTRACENTRIFUGATION 231
Tom Laue

11 FIELD FLOW FRACTIONATION WITH MULTIANGLE LIGHT SCATTERING FOR MEASURING PARTICLE SIZE DISTRIBUTIONS OF VIRUS-LIKE PARTICLES 253
Joyce A. Sweeney and Christopher Hamm

12 LIGHT-SCATTERING TECHNIQUES AND THEIR APPLICATION TO FORMULATION AND AGGREGATION CONCERNS 269
Michael Larkin and Philip Wyatt

PART II DEVELOPMENT OF A FORMULATION FOR LIQUID DOSAGE FORM 307

13 EFFECTIVE APPROACHES TO FORMULATION DEVELOPMENT OF BIOPHARMACEUTICALS 309
Rajiv Nayar and Mitra Mosharraf

14 PREDICTION OF AGGREGATION PROPENSITY FROM PRIMARY SEQUENCE INFORMATION 329
Mark Cornell Manning, Gabriel J. Evans, Cody M. Van Pelt, and Robert W. Payne

15 HIGH-CONCENTRATION ANTIBODY FORMULATIONS 349
Steven J. Shire, Jun Liu, Wolfgang Friess, Susanne Jörg, and Hanns-Christian Mahler

16 DEVELOPMENT OF FORMULATIONS FOR THERAPEUTIC MONOCLONAL ANTIBODIES AND Fc FUSION PROTEINS 383
Sampathkumar Krishnan, Monica M. Pallitto, and Margaret S. Ricci

17 REVERSIBLE SELF-ASSOCIATION OF PHARMACEUTICAL PROTEINS: CHARACTERIZATION AND CASE STUDIES 429
Vikas K. Sharma, Harminder Bajaj, and Devendra S. Kalonia

PART III DEVELOPMENT OF A FORMULATION FOR LYOPHILIZED DOSAGE FORM 457

18 DESIGN OF A FORMULATION FOR FREEZE DRYING 459
Feroz Jameel and Mike J. Pikal

19 PROTEIN CONFORMATION AND REACTIVITY IN AMORPHOUS SOLIDS 493
Lei Zhang, Sandipan Sinha, and Elizabeth M. Topp

20 THE IMPACT OF BUFFER ON SOLID-STATE PROPERTIES AND STABILITY OF FREEZE-DRIED DOSAGE FORMS 507
Evgenyi Y. Shalaev and Larry A. Gatlin

21 STABILIZATION OF LYOPHILIZED PHARMACEUTICALS BY CONTROL OF MOLECULAR MOBILITY: IMPACT OF THERMAL HISTORY 521
Suman Luthra and Michael J. Pikal

22 STRUCTURAL ANALYSIS OF PROTEINS IN DRIED MATRICES 549
Andrea Hawe, Sandipan Sinha, Wolfgang Friess, and Wim Jiskoot

23 THE IMPACT OF FORMULATION AND DRYING PROCESSES ON THE CHARACTERISTICS AND PERFORMANCE OF BIOPHARMACEUTICAL POWDERS 565
Vu L. Truong and Ahmad M. Abdul-Fattah

PART IV MANUFACTURING SCIENCES 587

24 MANUFACTURING FUNDAMENTALS FOR BIOPHARMACEUTICALS 589
Maninder Hora

25 PROTEIN STABILITY DURING BIOPROCESSING 605
Mark Cornell Manning, Gabriel J. Evans, and Robert W. Payne

26 **FREEZING AND THAWING OF PROTEIN SOLUTIONS** **625**
Satish K. Singh and Sandeep Nema

27 **STRATEGIES FOR BULK STORAGE AND SHIPMENT OF PROTEINS** **677**
Feroz Jameel, Chakradhar Padala, and Theodore W. Randolph

28 **DRYING PROCESS METHODS FOR BIOPHARMACEUTICAL PRODUCTS: AN OVERVIEW** **705**
Ahmad M. Abdul-Fattah and Vu L. Truong

29 **SPRAY DRYING OF BIOPHARMACEUTICALS AND VACCINES** **739**
Jim Searles and Govindan Mohan

30 **DEVELOPMENT AND OPTIMIZATION OF THE FREEZE-DRYING PROCESSES** **763**
Feroz Jameel and Jim Searles

31 **CONSIDERATIONS FOR SUCCESSFUL LYOPHILIZATION PROCESS SCALE-UP, TECHNOLOGY TRANSFER, AND ROUTINE PRODUCTION** **797**
Samir U. Sane and Chung C. Hsu

32 **PROCESS ROBUSTNESS IN FREEZE DRYING OF BIOPHARMACEUTICALS** **827**
D. Q. Wang, D. MacLean, and X. Ma

33 **FILLING PROCESSES AND TECHNOLOGIES FOR LIQUID BIOPHARMACEUTICALS** **839**
Ananth Sethuraman, Xiaogang Pan, Bhavya Mehta, and Vinay Radhakrishnan

34 **LEACHABLES AND EXTRACTABLES** **857**
Jim Castner, Pedro Benites, and Michael Bresnick

35 **PRIMARY CONTAINER AND CLOSURE SELECTION FOR BIOPHARMACEUTICALS** **881**
Olivia Henderson

36 **PREFILLED SYRINGES FOR BIOPHARMACEUTICALS** **897**
Robert Swift

37 **IMPACT OF MANUFACTURING PROCESSES ON DRUG PRODUCT STABILITY AND QUALITY** **917**
Nitin Rathore, Rahul S. Rajan, and Erwin Freund

INDEX **941**

FOREWORD

Since the introduction of recombinant therapeutic proteins in the 1980s, dozens of products have been successfully commercialized, and hundreds of new ones are currently in clinical trials. These products provide uniquely effective treatments for numerous human diseases and disorders. They have revolutionized the practice of medicine, saving and improving countless lives. But the most promising protein-based drug will not be of benefit to patients unless it can be manufactured, shipped, stored, and delivered to the patient, while minimizing degradation of the protein. This is a daunting challenge because proteins can readily aggregate, even under solution conditions that greatly favor the native state. Also, proteins are susceptible to numerous pathways of chemical degradation. Adding to the challenge is the potential that even if a small fraction of the protein molecules in a dose is degraded, an immunogenic response may be triggered with the potential to cause adverse effects in patients. Furthermore, the therapeutic protein must be produced at commercial scale using a complicated process that has been developed and documented to consistently result in a high-quality product. Also, the appropriate analytical methods must be developed and validated to ensure that degradation products can be quantified accurately and precisely. Clearly, the successful development of a commercialized therapeutic protein product requires multidisciplinary efforts of experienced, skilled scientists, engineers, and managers, and tremendous expenditure of capital. It is critically important for management to be cognizant of these challenges and to provide the appropriate resources to the development efforts for therapeutic proteins, as well as to establish reasonable timelines for this work, which is so vital to ensuring product quality and protecting patients' safety.

Over the years, as recombinant therapeutic proteins have been developed, the field as a whole has had to learn how to do this properly, and many of the important guiding principles and practical strategies had to be learned "on the job" as products were being developed. There were no established academic or industrial foundations for these efforts, because never before had recombinant proteins been used to treat human diseases and disorders. Fortunately, during this time, many of the leaders in the key disciplines published papers and books describing the continually improving, state-of-the-art approaches to stabilizing proteins, analyzing degradation products, and developing successful formulations.

An example of this type of on-the-job training was the research focusing on developing stable lyophilized formulations of proteins. In the early to mid-1980s, expertise from parenteral sciences and process engineering were applied to formulation and process development for freeze-dried therapeutic proteins. The results of early efforts were often commercially viable freeze-drying cycles and formulations that provided good cake structure but did not stabilize the protein very well. At the same time, researchers from materials sciences, food sciences, and even zoology were working to understand the mechanisms by which various excipients succeeded or failed to stabilize proteins during freezing, drying, and storage in the dried solid. Combined efforts from all of these disciplines gradually led to determination of these mechanisms as well as to the discernment of the key physical properties (and associated analytical methods) that govern long-term storage stability of dried proteins. As a result, we now have fairly straightforward, rational approaches to development of stable freeze-dried protein formulations. But many challenges remain, particularly understanding the quantitative linkages between different degradation pathways (e.g., oxidation, aggregation) and physical properties of the dried formulation (e.g., glassy-state dynamics, protein structure).

Also, during the years of development of therapeutic proteins the types of degradation products that could be studied, and the quality and resolution of analytical methods have vastly improved. These improvements allow for better understanding of causes and pathways for degradation. However, they also lead to more stringent criteria for the definition of a stable protein product. There are still many analytical challenges. For example, size exclusion chromatography (SEC) is the key method used to quantify levels of protein aggregates and monomer. But this method can provide misleading results because aggregates can dissociate or form during SEC and/or adsorb to the column resin. Thus, values obtained from SEC may not actually represent the true aggregate levels in the protein drug container, so there is a continued effort to investigate methods that can be used to corroborate results from SEC. Currently the most promising approach is analytical ultracentrifugation (AUC). But this method has its own challenges in proper sample handling, data analysis, and appropriate training of personnel. The field must continue to strive to improve SEC and AUC methods for aggregate quantification and to explore new methods (e.g., field flow fractionation).

Today we benefit from the numerous advances in the field that have been made over the last few decades. But we also face many new challenges to developing safe and effective therapeutic proteins. For example, monoclonal antibody products that have doses with relatively high protein concentrations (e.g., $\geq$100 mg/mL) can be difficult to manufacture, stabilize sufficiently, and analyze properly. Additionally, the use of prefilled syringes as product containers has recently led to new issues with protein stability that had to be resolved. In general, we must conduct research to gain more fundamental insights into the effects of the various product containers and their component materials on protein stability. Similarly, we must work to understand how various key processes steps (e.g., filling vials or syringes with pumps) affect protein stability, to increase awareness of these issues, and to create effective strategies to investigate these potential problems and to mitigate them.

Another challenge facing many companies is the need to develop consistent approaches for protein formulation studies, characterizing analytical methods, and studying protein stability during various processing steps. This does not mean that there should be "platform" formulations for drug product or platform analytical methods; indeed, any platform approach must be confirmed for each individual molecule, and there are many examples of surprising results. Rather, it is important to incorporate the scientific knowledge that has been gained across the industry in rational approaches that are developed and agreed on by educated and experienced personnel to ensure product quality and safety. As more companies develop global operations, such an approach may have the added benefits of promoting best practices between sites and individual researchers, minimizing unproductive conflicts, and speeding product development.

As has been the case throughout the history of working with recombinant therapeutic proteins, the field will take on current and future challenges and learn how to overcome them. Certainly, with future insights into disease pathologies and creation of new therapeutic protein categories, delivery approaches, and analytical methods, even more challenges will arise. With the strong foundation of excellence in therapeutic protein product development and rational approaches to delineate and solve problems, the field will successfully overcome these barriers, and new medicines will be made available for the benefit of patients.

In this book, experts from around the world provide comprehensive overviews of the many important steps involved in—and the critical insights needed for—the successful development of therapeutic proteins. The book is a state-of-the-art summary of what we have learned together as a field as we have worked to define the theory and practice of proper development of safe and effective medicines based on biotechnology. Moreover, it documents how researchers from numerous companies and universities contribute to furthering our insights and expertise for developing therapeutic proteins. The editors and authors are to be congratulated for their leadership in these efforts and their willingness to continue to communicate openly about where we are as a field and where we are going.

John Carpenter

PREFACE

The unraveling of the human genome, the concomitant explosion of proteomics, and an ever-increasing interest in proteins to treat an expanding range of medical indications have lead to growing interest in the development and production of biomolecules for therapeutic use. The identification of a new candidate drug compound is preceded by substantial scientific efforts and considerable capital investment. In order to realize the value to patients and the healthcare industry, the new drug molecule must be formulated and manufactured in an appropriate dosage form that can be conveniently used by the patient. Understanding the underlying challenges at each step of development and commercialization of the drug product dosage form is central to the successful launch of a biological therapeutic.

In order for proteins to manifest their proper biological and therapeutic effect, their conformational and structural integrity must be maintained at all stages of the development and commercialization process. Biomolecules are generally very sensitive to their microenvironment due to their complex and fragile structures. Once a new biologic has been identified for therapeutic use and product development, the first steps in the development process are determination of the physical and chemical properties of the molecule, identification of the major degradation pathways, and development of stability-indicating analytical methods as well as other biophysical characterization techniques. The information gathered from these early studies is used to identify excipients and conditions that will keep the protein therapeutic molecule in the native conformation and promote long-term product stability. Several chapters in this book discuss the latest biophysical and biochemical characterization techniques, as well as approaches to conducting the early physicochemical characterization studies.

The protein or peptide drug active must then be formulated for preclinical and clinical testing in conditions that preserve the chemical and physical integrity of the molecule, as well as render it in a form suitable for administration to patients. This is generally accomplished by screening the protein under a variety of excipients and conditions and monitoring stability as a function of time, temperature, and other stresses to identify/select the best conditions for further development. Liquid dosage forms may be preferred because of their greater convenience and lower manufacturing costs. However, lyophilized formulations may be required in some cases to attain adequate shelf-stability or where enhanced stability at higher temperatures or other special features are desired. At early stages, lyophilization may also offer a faster or more reliable path to develop an initial clinical formulation. This book contains a number of chapters relating to early formulation development strategies, platform approaches for initial antibody formulations, high-throughput strategies based on statistical design, and design space considerations. Additional chapters focus on the challenges associated with stability and analysis in the development of high concentration antibody formulations, and the impact of high concentrations on manufacturing and dose delivery. Case examples are provided to illustrate these approaches and offer specific applications.

Concurrent with preclinical and clinical testing of the candidate drug compound, the process development group will typically evaluate additional options available for expression, recovery, purification, and characterization of the drug substance for commercial production. Alternative formulations of the drug product for commercial use will also typically be explored. At this stage, the requirements in terms of stability, shelf-life, and ruggedness are typically much greater than for the earlier stages of development. In addition the focus on minimizing cost of goods and increasing throughput and manufacturing ease and consistency are significantly greater at this stage. Robust conditions for storage and shipment of the bulk drug substance must be identified. During subsequent commercial manufacturing, the purified bulk drug substance needs to be processed and prepared for successful fill/finish of final dosage form and, may go through freeze-thawing, formulation, mixing, filtration and filling operations prior to finishing as a lyophilized or liquid dosage form. Although these unit operations have been studied during earlier stages, the stresses generated and the mechanisms of denaturation in a manufacturing setting may be different, depending on scale, equipment and facility. Chapters dedicated to drug product process development discuss in detail, illustrated with case studies, methodology to develop, characterize and "optimize for scalability" all the manufacturing processes relating to drug product prior to their transfer to manufacturing sites. Additionally, these chapters provide guidance on formulation design considerations to stabilize the drug against the stresses that typically arise during large-scale manufacturing and commercialization in the cGMP environment.

There is growing interest in devices to simplify injection, particularly for products that will be sent home with patients for self-administration. This has led to increased interest in more complex container closures, such as prefilled syringes, either as standalone injection devices or as a component of a more complex injection device such as an auto-injector. The more complex primary containers may introduce additional stresses for the protein drug, as well as increased manufacturing challenges. Several chapters address considerations common to all container closures, as well as specific issues related to the more complex primary containers such as prefilled syringes.

Once the commercial formulation and configuration have been recommended and all the process parameters are locked into, the process is transferred to manufacturing. In simple terms technology transfer is referred to as transfer of a new product design from development (internal or external) into an operational environment for validation and robust sustained production. It can be between sites at a single company or from company to company and may involve a scale change or adaptation to a different equipment train. It is very complex operation that demands in-depth understanding of manufacturing challenges associated with the design of the facility, equipment train, scale, and operational procedures, besides development of robust processes and analytical methods. Chapters relating to technology transfer will discuss the manufacturing challenges and requirements and provide guidance to the reader as to when in the development phase these requirements need to be incorporated to mitigate the risk of failures and delays in getting the product to the market.

In recent years the field has evolved rapidly in many dimensions. The dramatic expansion in number and diversity of protein therapeutics, new scientific and technical approaches, the evolving regulatory landscape, and changes in marketing requirements and expectations for patient compliance make it imperative to update the available information. This book provides a comprehensive overview and guide to formulation and process development as well as manufacturing of biopharmaceutical drug product, covering both fundamentals and specialized considerations. Case histories are included to illustrate challenges and successful approaches for each phase as well as various classes of protein therapeutics, along with thoughtful analysis of lessons learned. Contributors have been selected from both industry and academia and have a wide range of experience and expertise in this area. The book will benefit scientists and engineers involved at various stages of product development, commercial production, project management, clinical, regulatory affairs, and quality assurance, and can serve as an introduction and reference for students who are contemplating a career in the biopharmaceutical industry.

Color versions of some of the text illustrations can be found at the following ftp site address:
ftp://ftp.wiley.com/public/sci_tech_med/formulation_biopharmaceutical

FEROZ JAMEEL
SUSAN HERSHENSON

Thousand Oaks, California
La Jolla, California
May 2010

CONTRIBUTORS

Ahmad M. Abdul-Fattah, Biogen Idec, San Jose, California

Harminder Bajaj, Process Development Sciences, Maxygen, Inc., Redwood City, California

Pedro Benites, Lanthens Medical Imaging, North Billerica, Massachusetts

Akhilesh Bhambhani, Macromolecule and Vaccine Stabilization Laboratory, Department of Pharmaceutical Chemistry, University of Kansas, Lawrence, Kansas

Jim Castner, Ph.D., Senior Principal Scientist, Lanthens Medical Imaging, North Billerica, Massachusetts

Byeong Chang, Ph.D, Symyx Technologies, Inc., Camarillo, California

Rohini Deshpande, Scientific Director, Division of Translational Sciences, Amgen Inc., Thousand Oaks, California

Gabriel J. Evans, Legacy BioDesign LLC, Loveland, Colorado

Wolfgang Friess, Professor, Department of Pharmacy, Pharmaceutical Technology and Biopharmacy, Ludwig Maximilians–University Munich, Munich, Germany

Erwin Freund, Scientific Director, Drug Product and Device Development, Amgen Inc., Thousand Oaks, California

Larry A. Gatlin, Global Head Technical Development, Parenteral Center of Emphasis, Pfizer Global Research & Development, Groton/New London Laboratories, Pfizer, Groton, Connecticut and Product Novartis Consultant

Adeolla O. Grillo, Human Genome Sciences, Inc., Rockville, Maryland.

Nicholas Guziewicz, BioFormulations Development, Genzyme Corporation, Framingham, Massachusetts

Christopher Hamm, Research Chemist, Merck & Co., Inc., Rahway, New Jersey

Andrea Hawe, Division of Drug Delivery Technology, Leiden/Amsterdam Center for Drug Research, Leiden University, Leiden, The Netherlands

Olivia Henderson, Principal Scientist I, Protein Pharmaceutical Development, Biogen Idec,

Susan Hershenson, Vice President, Pharmaceutical and Device Development, Genentech Inc., South San Francisco, California

Chung C. Hsu, Director, Genentech Inc., San Francisco, California

Maninder Hora, Vice President, Product Operations & Quality, Facet Biotech, Redwood City, California

Feroz Jameel, Ph.D., Drug Product and Device Development, Amgen Inc., Thousand Oaks, California

Wim Jiskoot, Professor, Division of Drug Delivery Technology, Leiden/Amsterdam Center for Drug Research, Leiden University, Leiden, The Netherlands

Sangeeta B. Joshi, Associate Director, Macromolecule and Vaccine Stabilization Laboratory, Department of Pharmaceutical Chemistry, University of Kansas, Lawrence, Kansas

Devendra S. Kalonia, Associate Professor, Department of Pharmaceutical Sciences, University of Connecticut, Storrs, Connecticut

Sampath Kumar Krishnan, Principal Scientist, Process and Product Development, Amgen Inc., Thousand Oaks, California

Michael Larkin, Research and Development, Wyatt Technology Corporation, San Francisco, California

Tom Laue, Professor, Department of Biochemisty, University of New Hampshire, Durham, New Hampshire

Jun Liu, Late Stage Pharmaceutical and Processing Development Department and Pharmaceutical and Device Development Department, Genentech Inc., San Francisco, California

Suman Luthra, Pfizer Global Research and Development, Pfizer Inc., Groton, Connecticut

X. Ma, Department of Formulation, Freeze-Drying, and Drug Delivery, Global Biological Development, Bayer HealthCare, Berkeley, California

D. MacLean, Department of Formulation, Freeze-Drying, and Drug Delivery, Global Biological Development, Bayer HealthCare, Berkeley, California

Hanns-Christian Mahler, Director, Formulation R&D Biologics, Pharmaceutical and Analytical R&D, F. Hoffmann-La Roche, Switzerland

Tahir Mahmood, Department of Chemistry and Worm Institute of Research Medicine, The Scripps Institute, La Jolla, California

Mark Cornell Manning, Legacy BioDesign LLC, Loveland, Colorado

Sheryl Martin-Moe, Director, Late Stage Pharmaceutical and Processing Development Department, Genentech Inc., San Francisco, California

Susanne Jörg, Pharmaceutical and Analytical Development, Novartis Pharma AG, Basel, Switzerland

Bhavya Mehta, Drug Product and Device Development, Amgen Inc., Thousand Oaks, California

C. Russell Middaugh, Professor, Macromolecule and Vaccine Stabilization Laboratory, Department of Pharmaceutical Chemistry, University of Kansas, Lawrence, Kansas

Govindan (Dan) Mohan, Ph.D., President, Applied Prime Technologies, Cupertino, California

Mitra Mosharraf, HTD Biosystems Inc., Hercules, California

Rajiv Nayar, President, Product Development, HTD Biosystems Inc., Hercules, California

Sandeep Neema, Senior Director, Pharmaceutical Sciences, Global Biologics, Pfizer Inc., Chesterfield, Missouri

Tim Osslund, Principal Scientist, Division of Translational Sciences, Amgen, Inc., Thousand Oaks, California

Chakradhar Padala, Ph.D., Senior Scientist, Drug Product and Device Development, Amgen Inc., Thousand Oaks, California

Monica M. Pallitto, Principal Scientist, Process and Product Development, Amgen Inc., Thousand Oaks, California

Xiaogang Pan, Bayer Technology Services (Asia), Shanghai, China

Robert W. Payne, Legacy BioDesign LLC, Loveland, Colorado

Bernardo Perez-Ramirez, BioFormulations Development, Genzyme Corporation, Framingham, Massachusetts

Micheal J. Pikal, Professor, Department of Pharmaceutical Sciences, School of Pharmacy, University of Connecticut, Storrs, Connecticut

Vinay Radhakrishnan, Group Leader, Pharmaceutical R&D, Global Biologics, Pfizer Inc., Chesterfield, Missouri

Rahul S. Rajan, Principal Scientist, Process and Product Development, Amgen Inc., Thousand Oaks, California

Theodore W. Randolph, Gillespie Professor of Bioengineering, Department of Chemical and Biological Engineering, University of Colorado, Boulder, Colorado

Nitin Rathore, Senior Scientist, Drug Product and Device Development, Amgen Inc., Thousand Oaks, California

Margaret S. Ricci, Director, Process and Product Development, Amgen Inc., Thousand Oaks, California

Samir U. Sane, Group Leader, Genentech Inc., San Francisco, California

Jim Searles, Ph.D., Director of Development, Aktiv-Dry LLC, Boulder, Colorado

Ananth Sethuraman, Senior Scientist, Drug Product and Device Development, Amgen Inc., Thousand Oaks, California

Evgenyi Y. Shalaev, Ph.D., FAAPS, Associate Research Fellow, Parenteral Center of Emphasis, Pfizer Global Research & Development, Groton/New London Laboratories, Pfizer, Groton, Connecticut

Vikas K. Sharma, Early Stage Pharmaceutical Development, Genentech Inc., San Francisco, California

Steven J. Shire, Group Leader, Late Stage Pharmaceutical and Pharmaceutical and Device Development Department, Genentech Inc., San Francisco, California

Robert Simler, BioFormulations Development, Genzyme Corporation, Framingham, Massachusetts

Satish Singh Pharmaceutical Sciences, Global Biologics, Pfizer Inc., Chesterfield, Missouri

Sandipan Sinha, Research Fellow, Department of Pharmaceutical Chemistry, University of Kansas, Lawrence, Kansas

Giovanni B. Strambini, Professor, CNR, Institute of Biophysics, Pisa, Italy

Steven J. Swanson, Ph.D., Executive Director, Clinical Immunology, Amgen, Inc., Thousand Oaks, California

Joyce A. Sweeney, Senior Investigator, Merck & Co., Inc., Rahway, New Jersey

Robert Swift, Senior Principal Engineer, Drug Product and Device Development, Amgen, Inc., Thousand Oaks, California

Elizabeth M. Topp, Professor, Department of Pharmaceutical Chemistry, University of Kansas, Lawrence, Kansas

Vu L. Truong, Vice President, Aridis Pharmaceuticals, San Jose, California

Cody M. Van Pelt, Legacy BioDesign LLC, Loveland, Colorado

D. Q. Wang, Department of Formulation, Freeze-Drying, and Drug Delivery, Global Biological Development, Bayer HealthCare, Berkeley, California

Y. John Wang, Late Stage Pharmaceutical and Processing Development Department, Genentech Inc., San Francisco California

Nicholas W. Warne, Ph.D., Director, Formulations Group, Wyeth BioPharma, Andover, Massachusetts

Philip Wyatt, Wyatt Technology Corporation, Santa Barbara, California

Bernice Yeung, Ph.D, Symyx Technologies, Inc., Camarillo, California

Yuhong Zeng, Macromolecule and Vaccine Stabilization Laboratory, Department of Pharmaceutical Chemistry, University of Kansas, Lawrence, Kansas

Lei Zhang, Department of Pharmaceutical Chemistry, University of Kansas, Lawrence, Kansas

Hong Zhao, Department of Pharmaceutical Chemistry, University of Kansas, Lawrence, Kansas

PART I

PREFORMULATION AND DEVELOPMENT OF STABILITY-INDICATING ASSAYS: BIOPHYSICAL CHARACTERIZATION TECHNIQUES

1

THE STRUCTURE OF BIOLOGICAL THERAPEUTICS

Sheryl Martin-Moe, Tim Osslund, Y. John Wang, Tahir Mahmood, Rohini Deshpande, and Susan Hershenson

1.1. INTRODUCTION

The first synthetic drug, acetylsalicylate, produced in 1895 and patented as aspirin [1] (Bayer in 1900), marks the beginning of the modern pharmaceutical industry. Throughout the early to mid-1900s there was significant emphasis on development of synthetic antibiotics for infectious diseases and small organic molecules, which continues to this day as the mainstay of the traditional (small-molecule) pharmaceutical industry. Advances in understanding the mechanism of reproductive functions and metabolic diseases, such as diabetes and short stature, led to the discovery of polypeptide hormones. The early polypeptide hormone drugs were purified from organs, such as insulin from animal pancreas or growth hormone from cadaver pituitary. The first animal-derived insulin preparation to become commercially available was Iletin, derived from bovine or porcine sources (Eli Lilly in 1923) [2]. Growth hormone, which had to be derived from a human source, was originally produced by special order in hospital laboratories and only commercialized much later in the United States in 1976 [3]. Although these products were breakthroughs in the treatment of diabetes and dwarfism, there were serious limitations with this type of production, including availability of organs and issues with transmission of infectious diseases [4].

Formulation and Process Development Strategies for Manufacturing Biopharmaceuticals, edited by Jameel and Hershenson

During World War II, Edwin Cohn developed the plasma fractionation process whereby blood components such as human serum albumin were produced to supplement blood loss in wounded soldiers [5]. Polyclonal antibody therapeutics derived from human plasma, such as immunoglobulins, were available as treatment as early as the 1940s [6]. In 1968 the first of the blood enzymes, antihemophilic factor VIII was commercialized by Baxter's Hyland Division [7,8]. Although these products represented life-saving breakthroughs for the treatment of patients with chronic conditions such as hemophilia, the production of plasma proteins also suffered from safety and production-scale limitations since they, too, were isolated from natural sources [9,10].

Advances in the synthesis of peptides in solution and by solid phase made it possible to produce peptides on the industrial scale [11]. However, until the advent of recombinant DNA technology, purification from natural or semisynthetic sources, with the attendant limitations of scale and/or concerns about impurities and infectious agents, remained the only means of producing the larger polypeptide therapeutics, such as insulin and human growth hormone.

The biotechnology industry began formally in 1976 with the founding of the first biotechnology company, Genentech. Recognizing the significance of being able to manipulate genes using the newly discovered tools of restriction endonucleases and DNA ligases, Stanley Cohen and Herbert Boyer (a co-founder of Genentech) outlined in a series of papers the foundation of modern biotechnology [12]. Recombinant DNA production systems are integrally related to the structure and function of the protein products. The initial phase of recombinant production started with expression and purification in bacterial systems such as *Escherichia coil*. Although successful for many products, there can be serious challenges with refolding some proteins purified from *E. coli*. In addition, if posttranslational modifications, such as glycosylation, are required for activity, then *E. coli* is not viable as a host since the machinery for this is not present. Subsequently, methods were developed for cloning and expression of DNA sequences in yeast where refolding was not an issue but post-translation modifications were limited, and in insect cell systems using Baculovirus where modifications were closer (but still not identical) to those produced by human cells but where scalability was an issue [13]. It was not until mammalian cell systems such as Chinese hamster ovary (CHO) cells were established that recombinant DNA technology could produce products closely resembling the full range of human protein structures. Mammalian systems enabled production of larger and more complex protein therapeutics because of the ability of these cell systems to fold proteins correctly and add posttranslational modifications such as *N*-linked and *O*-linked glycans essential for the biological activity and stability of many proteins. In addition, efforts were initiated to improve pharmaceutical or pharmacokinetic properties of proteins. Approaches included modification of the sequence, addition of glycosylation sites, and fusions with domains such as the Fc portion of an antibody or chemical modifications such as addition of poly(ethylene glycol) or lipid. Today there are well over a hundred protein therapeutics on the market in the United States alone, representing a wide variety of structural classes; and the natural human proteins and chemically modified derivatives continue to be a fruitful source of new therapeutic products [14].

The first monoclonal antibody therapeutics to be introduced were murine antibodies, based on the work of Kohler and Milstein on the continuous expression of monoclonal antibodies in mouse hybridoma systems [15]. The first monoclonal antibody product produced by this technology was muromonab-CD3 (Orthoclone OKT-3 by Ortho in 1986), for reversal of kidney transplant rejection [16]. Over the following decade methods for increasing the human content of therapeutic antibodies through production of human–murine chimeras, followed by humanized and fully human MAb constructions were developed. Today fully human antibodies are now commercially produced [17]. Currently monoclonal antibodies constitute the most rapidly growing class of human therapeutics and the second largest class of drugs after vaccines. There are 26 approved monoclonal antibody-based biopharmaceutical products, mainly for the treatment of cancer and autoimmune diseases, and there are well over a 100 drug candidates currently under clinical development [18].

Antibodies have served as a natural biomolecular scaffold for various applications. The variable region serves as the antigen-binding site and provides an effective humoral response against foreign substances and invading pathogens. A key advantage of antibody therapeutics is their high level of specificity for the relevant disease targets, which minimizes cross-reactivity and off-target toxicity and can thereby serve to reduce adverse effects compared to other therapeutic approaches. New technologies for generating humanized and fully human monoclonal antibody therapeutics are being developed with the aims of extending the range of targets, increasing the efficiency of production. In addition, the field continues to experiment with modified forms of the basic monoclonal antibody platform to extend the already impressive performance of this dominant class of biological therapeutics [19,20].

The following section introduces some of the major structural classes of recombinant therapeutic proteins. Consideration has been limited to this class and does not include related topics such as synthetically derived peptides, protein vaccines, or diagnostic reagents. Even for therapeutic proteins, the chapter is by no means comprehensive, but is rather intended to provide illustrative examples of each major class. In addition some of the recent trends in the development of new classes of therapeutic proteins are briefly considered. Figure 1.1 illustrates a few examples of the structural range of current therapeutic protein structures, in comparison to small-molecule drugs such as aspirin or an antibiotic.

1.2. NATIVE HUMAN PROTEINS AND ANALOGS

1.2.1. Polypeptide Hormones, the First Recombinant Therapeutics

The first wave of the biotechnology industry in the 1980s targeted replacement of existing, nonrecombinant therapeutic proteins with high-purity, fully human proteins produced using the new recombinant DNA technology. The first human protein to be cloned and expressed was a polypeptide hormone, somatostatin, in 1977, based on the work by Herbert Boyer et al. in laboratories at the University of California, San Francisco and the City of Hope [21]. Soon after, the human polypeptide hormones insulin [22] and growth hormone, or somatotrophin [23], were cloned and expressed

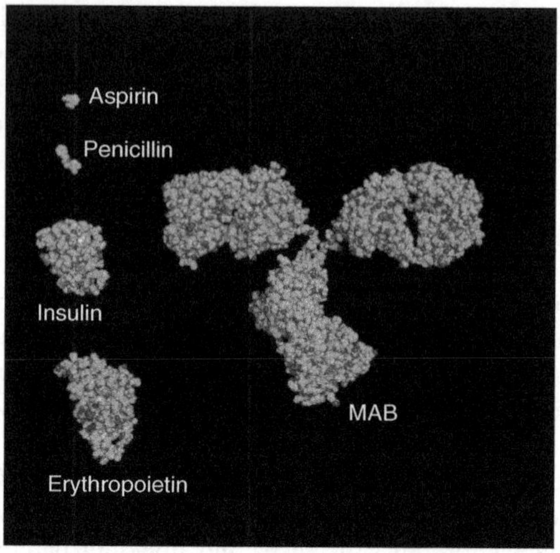

Figure 1.1. Molecular representations comparing the structural complexity of types of parenteral drugs: aspirin (180 Da), penicillin (334 Da), insulin (5808 Da), erythropoietin (36,000 Da), and a monoclonal antibody (MAB) (150,000 Da).

by recombinant DNA methods at Genentech for commercial purposes. To understand the pressing medical need for recombinant sources, consider the cases of insulin and growth hormone.

Prior to the production of recombinant insulin, insulin-dependent diabetics were treated with insulin preparations from either bovine or porcine sources. Human insulin was made from conversion of porcine insulin using a combination of enzymatic and chemical treatment of the porcine product [24]. The three-dimensional conformation is similar for insulin analoges from human, bovine, or porcine sources, in that the A chain forms two antiparallel α helices, and the B chain forms a single α helix followed by a turn and a β strand. The arrangement of the chains buries the disulfide bonds and the aliphatic side chains in the nonpolar core. Insulin has the tendency to self-associate forming dimer, hexamer, and multimers but is complexed with zinc to form the hexamer [25,26]. Although the basic structure of the human, porcine, and bovine insulins is similar, the preparations derived from animal tissues contained many impurities (proinsulin, arginine insulin, and desamidoinsulin), some of which elicited immune responses exacerbated by the dosing frequency. Additionally, seemingly minor sequence differences between human and bovine insulin, in particular, may have elicited antibody responses that reduced biological activity over time, resulting in insulin resistance or altered pharmacokinetic profiles [27].

Recombinant human insulin was licensed to Eli Lilly by Genentech for development, and was the first recombinant human protein therapeutic to be approved by the US Food and Drug Administration (FDA) in 1982. The original Genentech production process involved insertion of the nucleotide sequence coding for the insulin A and B chains into two different *E. coli* cells that were cultured separately at large scale. After purification, the A and B chains were incubated together under appropriate conditions to promote interchain disulfide bonds. Eli Lilly improved on this method by insertion of a proinsulin nucleotide sequence into *E. coli*, allowing a single fermentation and purification process. After purification there is proteolytic excision of the C peptide to produce human insulin [24,28]. Today there are a number of insulin and insulin analog products on the market produced by a variety of host cells and processes; for example, in addition to a number of *E. coli*-derived products, at least one recombinant human insulin is produced in yeast (Novolin by Novo Nordisk, approved in 2005).

Insulin variants have also been designed with specific pharmacokinetic properties. In 1987 the first group to design fast-acting insulin was at Novo Nordisk [29]. The FDA eventually approved three analogs of insulin for fast onset: insulin lispro (Humalog by Lilly, in 1996), insulin aspart (NovoLog by Novo Nordisk in 2000), and insulin glulisine (Apidra by Aventis in 2000). They differ from insulin in Lys-Pro at B28 and B29, Asp at B28, and Lys and Glu at B3 and B29, respectively [16]. These modifications generally disrupt the hexamer association, shifting the equilibrium in favor of the monomer that is absorbed more rapidly [26].

For long-acting insulin, FDA granted approval of insulin glargin (Lantus by Aventis in 2000) and insulin detemir (Levemir by Novo Nordick in 2005) [16]. Insulin glargin contains an extra Gly on the A chain and two Arg residues on the B chain. Insulin detemir is novel as B30 is omitted and a fatty acid (myristate) is attached at B29 Thr [30,31] (see Table 1.1 for current insulin products).

In the case of growth hormone, patients (mainly children) had been treated with drug derived from the pituitary gland of human cadavers. These patients often experienced loss of response to the therapy, which was also linked to induction of neutralizing antibodies attributed to the quality of the preparations [32]. These observations created a drive to introduce high-quality, native human protein therapeutics that would avoid or minimize these adverse reactions and exposure to infectious agents [23]. Genentech's somatrem (Protropin, met-hGH) was approved for treating growth hormone deficiency in children in 1985. The approval process was expedited by the FDA after a number of deaths caused by virus contamination in pituitary somatropin [33]. The plasmid used at that time was designed for *E. coli* cytoplasmic expression, resulting in a protein, somatrem, with an extra methionine at the *N* terminus of human growth hormone. By designing a plasmid with a signal sequence, it was shown that *E. coli* could remove the signal sequence during the secretion process to produce growth hormone without *N*-terminal methionine. On the basis of this sequence difference, Lilly's recombinant somatropin, Humatrope, also received orphan drug exclusivity and received approval in 1987 [16]. By the end of 2008, there were several recombinant somatropins on the market (Table 1.1).

Some of the smaller peptide hormones may be produced by either recombinant DNA technology or chemical synthesis. Nesiritide (Natrecor, commonly known as "brain" or *B*-type natriuretic peptide) has the same 32 amino acid sequence as the endogenous peptide, which is produced by the ventricular myocardium, a single chain with a disulfide bond between cysteines at 10 and 26. In the late 1990s, Scios made

TABLE 1.1. Recombinant Protein Parenteral Drugs Approved That Are Not Antibodies

Common Name	Nonproprietary Name[a]	Trade Name	Main Indication	U.S. Approval Year	Manufacturer or Licensee
		Hormones			
Insulin	Insulin	Humulin®	Diabetes mellitus	1982	Eli Lilly
	Insulin lispro	Humalog®	Diabetes mellitus	1996	Eli Lilly
	Insulin aspart	NovoLog®	Diabetes mellitus	2000	Novo Nordisk
	Insulin glulisine	Apidra®	Diabetes mellitus	2004	Aventis
	Insulin glargine	Lantus®	Diabetes mellitus	2000	Aventis
	Insulin determir	Levemir®	Diabetes mellitus	2005	Novo Nordisk
Growth hormone	Somatrem	Protropin®	Dwarfism	1985	Genentech
	Somatropin	Humatrope®	Dwarfism	1987	Eli Lilly
	Somatropin	Norditropin®	Dwarfism	1987	Novo Nordisk
	Somatropin	Nutropin®	Dwarfism	1993	Genentech
	Somatropin	Tev-Tropin®	Dwarfism	1995	Ferring
	Somatropin	Genotropin®	Dwarfism	1995	Pharmacia/Pfizer
	Somatropin	Saizen/Serostim®	Dwarfism	1996	EMD Serono
	Somatropin	Zorbtive®	Short-bowel syndrome	2003	EMD Serono
	Somatropin	Omnitrope®	Dwarfism	2006	Sandoz
	Somatropin	Valtropin®	Dwarfism	2007	LG Life
	Somatropin	Accretropin®	Dwarfism	2008	Cangene
IGF-1[b]	Mecasermin	Increlex®	Dwarfism	2005	Tercica
Chorionic gonadotropin	Choriogonadotropin alpha	Ovidrel®	Infertility	2000	Serono
Luteinizing hormone	Lutropin alpha	Luveris®	Infertility	2004	Serono
Follicle-stimulating hormone	Follitropin alpha	Gonal-F®	Infertility	1997	Serono

Follicle-stimulating hormone	Follitropin beta	Follistim®	Infertility	1997	Organon
Glucagon	Glucagon	Glucagon®	Hypoglycemia	1998	Eli Lilly
Parathyroid hormone (1–34)	Teriparatide	Forteo®	Osteoporosis	2002	Eli Lilly
Calcitonin	Calcitonin-salmon	Fortical®	Osteoporosis	2006	Unigen
BNP[c]	Nesiritide	Natrecor®	Congestive heart failure	2001	Scios
		Cytokines			
Erythropoietin	Epoetin alfa	Epogen®/Eprex®	Anemia	1989	Amgen/Ortho
Erythropoietin	Epoetin delta	Dynepo®	Anemia	2002 (EU)	Aventis
Erythropoietin	Darbepoetin alfa	Aranesp®	Anemia	2001	Amgen
PEG[d]-Epo	Methoxypolyethylene glycol epoetin beta	MIRCERA®	Anemia	2008	Hoffmann La Roche
IL[e]-2	Aldesleukin	Proleukin®	Renal cell carcinoma	1992	Chrion
G-CSF[f]	Filgrastim	Neupogen®	Neutropenia	1991	Amgen
G-CSF PEG	Pegfilgrastim	Neulasta®	Neutropenia	2002	Amgen
GM-CSF[g]	Sargramostim	Leukine®/Prokine®	Posttransplantation	1991	Immunex
IL-11	Oprelvekin	Neumega®	Thrombocytopenia	1997	Genetic Institute
IL-1Ra	Anakinra	Kineret®	Rheumatoid arthritis	2001	Amgen
Anti-IL-2 receptor	Denileukin diftitox	Ontak®	Cutaneous T-cell lymphoma	1999	Ligand/Eisai
		Interferons			
IFN[h]-α-2b	Interferon alfa-2b	Intron A®	Hairy cell leukemia	1986	Schering Plough
IFN-α-2a	Interferon alfa-2a	Roferon A®	Hairy cell leukemia	1986	Hoffmann La Roche
IFN-α-n3	Interferon alfa-n3	Alferon N®	Genital warts	1989	Interferon Sciences
IFN-α-con-1	Interferon alfacon-1	Infergen®	Hepatitis C	1997	Amgen

(*continued*)

TABLE 1.1. Recombinant Protein Parenteral Drugs Approved That Are Not Antibodies (*Continued*)

Common Name	Nonproprietary Name[a]	Trade Name	Main Indication	U.S. Approval Year	Manufacturer or Licensee
PEG-IFN-α-2b	Peginterferon alfa-2b	Pegintron®	Hepatitis C	2001	Schering Plough
PEG-IFN-α-2a	Peginterferon alfa-2a	PEGasys®	Hepatitis C	2002	Hoffmann La Roche
IFN-β-1b	Interferon beta-1b	Betaseron®	Multiple sclerosis	1993	Chiron
IFN-β-1a	Interferon beta-1a	Avonex®	Multiple sclerosis	1996	Biogen
IFN-β-1a	Interferon beta-1a	Rebif®	Multiple sclerosis	2002	Serono
IFN-γ-1b	Interferon gamma-1b	Actimmune®	Chronic granulomatosis	1990	Genentech/InterMune
		Growth Factors			
KGF[i]	Palifermin	Kepivance®	Mucositis	2004	Amgen
PDGF[j]	Becaplermin	Regranex®	Diabetic foot ulcer	1997	Ethicon/OMJ
BMP-2[k]	Dibotermin-α	INFUSE®	Degenerative disk disease	2002	GI/Medtronic
		Coagulation Factors/Thrombolytic Agents			
FVIII	Antihemophilic factor	Recombinate®	Hemophilia A	1992	Baxter/GI/Wyeth
FVIII	Antihemophilic factor	Kogenate®	Hemophilia A	1993	Bayer/Miles, Inc.
FVIII B deleted	Antihemophilic factor	ReFacto®	Hemophilia A	2000	GI/Wyeth
FVIII	Antihemophilic factor, plasma/albumin-free	ADVATE®	Hemophilia A	2003	Baxter
FVIII B deleted	Antihemophilic factor, plasma/albumin-free	XYNTHA®	Hemophilia A	2008	Wyeth
FVIIa	Coagulation factor VIIa	NovoSeven®	Hemostasis aid for Hemophilia A/B	1999	Novo Nordisk
FIX	Coagulation factor IX	BENEFIX®	Hemophilia B	1997	GI/Wyeth
Hirudin	Lepirudin	Refludan®	Heparin-induced thrombocytopenia	1998	Bayer/Hoechst Marion Roussel

Thrombin	Thrombin, dermal	Recothrom®	Hemostasis aid	2008	ZymoGenetics
Antithrombin III	Antithrombin III	ATryn®	Regulation of hemostasis	2009	GTC Biotherapeutics
Activated protein C	Drotrecogin alfa	Xigris®	Sepsis	2001	Eli Lilly
tPA	Alteplase	Activase®	Myocardial infarction	1987	Genentech
tPA[l]	Reteplase	Retavase®	Myocardial infarction	1996	Boehringer Mannheim
tPA	Tenecteplase	TNKase®	Myocardial infarction	2000	Genentech
		Enzymes			
DNAse	Dornase alpha	Pulmozyme®	Cystic fibrosis	1993	Genentech
β-Glucocerebrosidase	Imiglucerase	Cerezyme®	Gaucher's disease	1994	Genzyme
Asparaginase-PEG	Pegaspargase	Oncaspar®	Acute lymphoblastic leukemia	1994	Enzon
Urate oxidase	Rasburicase	Elitek®	Uric acid management in leukemia	2002	Sanofi-Synthelabo
α-L-Iduronidase	Laronidase	Aldurazyme®	Mucopolysacchariodosis	2003	BioMarin/Genzyme
α-Galactosidase A	Agalsidase beta	Fabrazyme®	Fabry disease	2003	Genzyme
α-Glucosidase (GAA)	Alglucosidase alfa	Myozyme®	Pompe disease	2006	Genzyme
Iduronate-2-sulfatase	Idursulfase	Elaprase®	Hunter syndrome (mucopolysaccharidosis II)	2006	Shire
Arylsulfatase B	Galsulfase	Naglazyme®	Maroteaux–Lamy syndrome (mucopolysaccharidosis VI)	2005	BioMarin
Hyaluronidase	Hyaluronidase	Hylenex®	Adjuvant to increase absorption and dispersion of other injected drugs	2005	Halozyme

(continued)

TABLE 1.1. Recombinant Protein Parenteral Drugs Approved That Are Not Antibodies (*Continued*)

Common Name	Nonproprietary Name[a]	Trade Name	Main Indication	U.S. Approval Year	Manufacturer or Licensee
		(Fc Conjugates)			
TNFR-Fc[m]	Etanercept	Enbrel®	Rheumatoid arthritis, psoriasis	1998	Amgen/Wyeth/ Immunex
LFA-3-Fc[n]	Alefacept	Amevive®	Plaque psoriasis	2003	Biogen
CTLA-4-Fc[o]	Abatacept	Orencia®	Rheumatoid arthritis	2005	BMS
TPO[p] mimetic (Fc peptibody)	Romiplostim	Nplate®	Thrombocytopenia	2008	Amgen

[a]Nonproprietary name: U.S. adopted name. [b]Insulin-like growth factor (IGF); [c]β-type natriuretic peptide (BNP); [d]Pegylated (PEG); [e]Interleukin (IL); [f]Granulocyte-colony stimulating factor (G-CSF); [g]Granulocyte macrophage-colony stimulating factor (GM-CSF); [h]Interferon (IFN); [i]Keratinocyte growth factor (KGF); [j]Platelet derived growth factor (PDGF); [k]Bone morphogenic protein (BMP); [l]Tissue plasminogen activator (tPA); [m]Tumor necrosis factor receptor (TNFR); [n]Lymphocyte function-associated antigen (LFA); [o]Thrombopoietin mimetic (TPO).

Source: FDA [16].

nesiritide by total peptide synthesis for clinical development, then for cost reasons switched to manufacture the peptide using an *E. coli* expression system. The recombinant version was approved by the FDA in 2001 for congestive heart failure [16]. Together with glucagon (29 amino acids) and PTH (1–34), they are the smallest proteins produced by recombinant DNA technology. Salmon calcitonin (32 amino acids) was made either by total chemical synthesis in the Novartis product, or by the rDNA process in Unigene's product. This trend suggests that the cost of recombinant DNA production is very competitive, with only small peptides consisting of up to 10 amino acids produced exclusively by total chemical synthesis. Today there are a wide variety of peptide hormone products on the market for treatment of diverse conditions (see Table 1.1 for approved products in the United States).

1.2.2. Cytokines

Cytokines are proteins secreted from a variety of cells, including inflammatory leukocytes and some nonleukocyte tissues. Unlike hormones, which are produced in distinct glands and secreted into the bloodstream in order to act over long distances, cytokines are produced and act locally where they regulate immune functions, inflammation, and hematopoiesis. They are generally produced in low concentrations in the human body and bind very tightly to their cognate receptors, often in the picomolar range (as compared to hormones that bind in the nanomolar range). The cytokine ligand often causes dimerization of the receptor as either a homo- or heterodimer. The binding of the cytokine to its receptor typically causes a phosphorylation event from an intracellular kinase that is either part of the receptor or is closely associated with the intracellular domain of the receptor, triggering a cascade of intracellular reactions [34].

1.2.2.1. Four-Helix Bundle Cytokines. This family of proteins shares a four-helix bundle motif in which the first two helices are oriented in the opposite direction relative to the second set of helices. This "up–up–down–down" motif requires two long loops between the first and second helix, a small loop between the second and third helix, and another long loop between the third and forth helix [35]. This unusual connectivity between the loops was not apparent when the first cytokine structure of interleukin-2 (IL-2) was published [36] but was later corrected [37]. IL-2 (aldesleukin or Proleukin by Chiron in 1992) is used for treatment of kidney (renal) carcinoma [16].

The four-helix bundle cytokines can be subdivided into two subclasses (Fig. 1.2) based on secondary and tertiary structure, those with long α helices and short crossing angles between the helices, such as granulocyte colony-stimulating factor (GCSF); and those in which the helices are significantly shorter and the crossing angles between helices are relative large, for example, granulocyte macrophage colony-stimulating factor (GMCSF). Often the cytokines with the shorter helices have a small β strand connecting the long loops between helices A and B and C and D. Both classes have yielded therapeutics, including methionyl GCSF or filgrastim (Neupogen by Amgen, in 1991) to treat chemotherapy-induced neutropenia and GMCSF or sargramostim (Leukine originally by Immunex, now Berlex in 1991) used postchemotherapy or bone marrow transplantation to stimulate neutrophil recovery and mobilization of hematopoietic

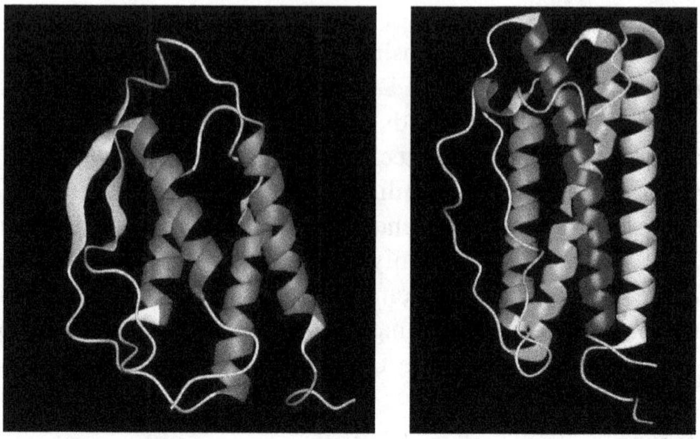

(a) (b)

Figure 1.2. Ribbon structures of (a) GCSF and (b) GMCSF, illustrating the two major subclasses of four-helix bundle cytokines. Note the shorter helices and longer crossing angles of GCSF. See the insert for color representation of this figure.

progenitor cells into the bloodstream [16]. Endogenous human GM-CSF is a glycoprotein consisting of 127 amino acids. Sargramostim differs from that by a substitution of the amino acid leucine for arginine at position 23 to stabilize the protein within the expression system (yeast), and contains both *N*-linked and *O*-linked oligosaccharides composed of mannose and *N*-acetylglucosamine [38]. A modified version of GCSF has been developed in which a single polyethylene glycol (PEG) is conjugated at the amino terminus, distal from the helical bundle and the lower half of the A helix, where much of the receptor binding occurs [39]. Pegylation enhances the circulation time, allowing less frequent dosing than unmodified GCSF [40]. Pegfilgrastim has been commercialized (Neulasta by Amgen in 2002) for treatment of patients undergoing chemotherapy to decrease the incidence of infection by febrile neutropenia [16].

Many members of the cytokine family of proteins are glycosylated in their native states, often including both *N*- and *O*-linked glycans. During the initial phase of the biotechnology industry, therapeutic proteins were produced in bacteria, which are incapable of performing posttranslational glycosylation. This approach was successful for a number of cytokines, including GCSF, growth hormone, and interferons. However, bacterial production was not successful in all cases. For example, erythropoietin produced in bacteria lacks *N*-linked carbohydrates and, as a result, was quickly eliminated from circulation. Therefore, it was necessary to develop more complex manufacturing processes such as expression in CHO cells; CHO cell expression systems were used for commercial production of erythropoietins such as epoietin alpha (Epogen by Amgen in 1989 and Procrit manufactured by Amgen for Johnson & Johnson) for anemia associated with kidney failure, generic Zidovudine-treated HIV, and cancer patients on chemotherapy [16].

Second-generation products have been developed that are indicated for dosing intervals longer than that for recombinant erythropoietin. In one example, two additional *N*-glycosylation sites were added to the molecule, resulting in a novel erythropoiesis-stimulating protein, darbepoietin alfa (Aranesp by Amgen in 2001), used for the treatment of chemotherapy-induced anemia [41]. In a second example, pegylation was employed to modify epoietin beta [42] (Mircera by Roche, in 2007), recently approved in the European Union (EU) for the treatment of anemia associated with chronic renal failure [16]. Both approaches have led to regulatory approval for treatment of anemia with less frequent dosing.

The interferons provide additional examples of helical cytokine structures. Although the structure of interferon alpha has proved to be elusive, a homology model suggests that it falls in the typical four-helix bundle cytokine structure [43]. IFN alpha-2 was the first cytokine to be approved by the FDA in two versions (A and B), concurrently in June 1986. IFN alfa-2B (Intron A from Schering-Plough in 1986 for the treatment of hairy cell leukemia, followed by numerous indications, including genital warts, AIDS-related Kaposi's sarcoma, hepatitis B and C, malignant melanoma, and follicular lymphoma) [16] and IFN alfa-2A (Roferon by Hoffman La Roche in 1986 for hairy cell leukemia, and later hepatitis C and chronic mylegenous leukemia) [16]. PEGINTRON, PEGylated Intron A, was introduced subsequently (Schering-Plough in 2001) [16]. It is chemically conjugated using 12-kD PEG with the greatest proportion ($\approx$48%) derivatized to the His^{34} residue, and retains about 28% *in vitro* antiviral activity relative to Intron A. In clinical studies, PEGINTRON demonstrated a serum half-life of 48–72 h consistent with once-weekly dosing [44]. Roferon has also been PEGylated using branched 40-kD PEG at lysine residues (PEGasys by Hoffmann-La Roche, in 2002). Plasma half-life was increased from 3–8 h to 65 hours, and T_{max} was prolonged from 10 to 80 h. Although it retains only 7% of the *in vitro* antiviral activity, it is efficacious by weekly dosing [45].

The structure of interferon beta has been determined to be a helical bundle topology similar to other cytokines, but with five alpha helices in its bundle [46] instead of four. Interferon beta-1b, which is bacterially produced and therefore nonglycosylated, was approved for treatment of patients with multiple sclerosis (Betaseron by Chiron in 1993). Cysteine 17 is mutated to serine to avoid disulfide scrambling [16]. Interferon beta-1a (Avonex by Biogen, in 1996) is a 166–amino acid glycoprotein made in CHO cells, where it is also glycosylated. The impact of the glycosylation can be seen in the recommended doses and administration schedule for the two products, which are quite different in that Betaseron is administered at 0.25 mg subcutaneously every other day, whereas Avonex is given 0.03 mg intramuscularly once a week [16].

Some of the more diverse structures in the cytokine family include interferon gamma (Actimmune by Intermune in 1999), used for chronic granulomatous disease and delaying progression of severe malignant osteoperosis, and stem cell factor (Ancestim by Amgen, currently licensed to Biovitrum AB) [16]. The core structure of IFN gamma contains six alpha helices that self-associate to form a homodimer in its active state [47]. The native factor contains a transmembrane domain in addition to the four-helix bundle core motif that accounts for the biological activity. It also self associates to form a noncovalent dimer with a head-to-head orientation between the 2 four-helix bundles. Ancestim (Stemgen by Amgen, available in New Zealand), consists of the extracellular domain only [48].

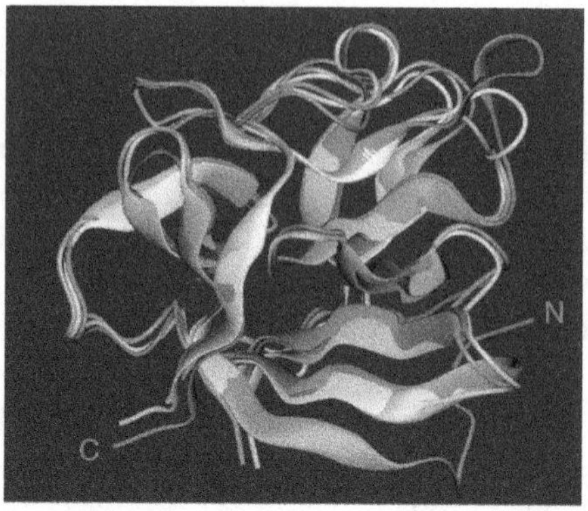

Figure 1.3. Superimposition of the ribbon structures of FGF-1 (green), FGF-2 (blue), and FGF-7/KGF (red). The β trefold is a common structural motif for the FGF family of proteins. See the insert for color representation of this figure.

1.2.2.2. Fibroblast Growth Factor (FGF) Family of Cytokines. In contrast to the four-helix bundle cytokines, this family of proteins is structurally composed of β-trefoil architecture. Twelve antiparallel β strands are folded into three β–β–β (trefoil) units (Fig. 1.3). Twenty-three members of this family have been identified on the basis of sequence homology and structure. Although there is only 40% sequence homology between FGF1 and FGF7/KGF (keratinocyte growth factor), their conformational structures are almost identical [49]. The most significant difference between the two structures is the receptor-binding domain, which is slightly larger in FGF7. Although both FGF1 and FGF2 advanced to clinical trials, the only member of the FGF family to be approved in the United States is KGF (Palifermin by Amgen in 2004, currently licensed to Biovitrum AB), used to alleviate mucositis, a common complication of high-dose chemotherapy and radiation treatments associated with bone marrow transplant [16]. [FGF2 (Fiblast by Kaken Seiyaku) was approved in Japan in 2001 for use in wound healing and is currently (as of 2009) in clinical trials in the United States [50].] Anakinra, an interleukin-1 receptor antagonist (IL-1RA), adopts a similar β-trefoil fold [51], although not related to the FGF family of proteins in sequence, and has been approved for treatment of rheumatoid arthritis (Kineret by Amgen in 2001) [16].

1.2.2.3. Cystine Knot Cytokines. Resolution of the structures of the neurotrophic factors, nerve growth factor (NGF), brain-derived neurotrophic factor (BDNF), and neurotrophin 3 (NT3), revealed a novel protein topology in which two molecules form a homodimeric, biologically active protein [52]. One unusual aspect of this characteristic fold is the position of six cysteines, which was eventually termed the "cystine knot" motif. In this motif two disulfide bonds form an amino acid ring in the peptide backbone, which is penetrated by a third disulfide bond (Fig. 1.4). The cystine knot motif is also shared by human transforming growth factor-β2 (TGFβ), vascular endothelial growth factor (VEGF), and a variety of sequence-unrelated but structurally similar proteins [53]. Although many of the original therapeutic

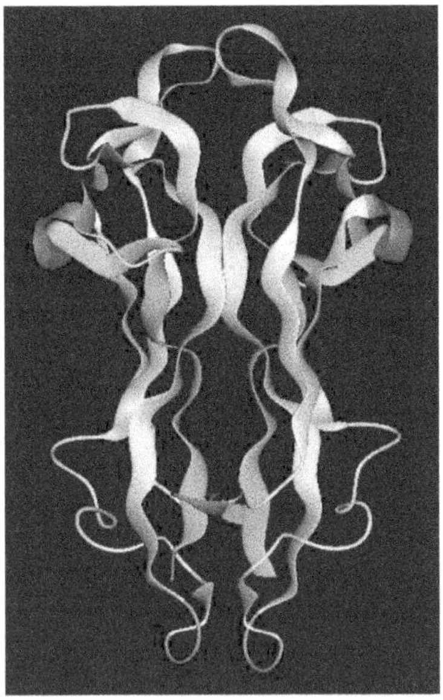

Figure 1.4. Ribbon structure of BDNF, representing the structure of the family of proteins termed *cystine knot cytokines*. See the insert for color representation of this figure.

candidates in this structural class have not been commercialized, a few have made it to the market. Platelet-derived growth factor (Regranex or becaplermin by OMJ in 1997) in carboxymethyl cellulose topical gel was approved for treatment of diabetic foot ulcers [16]. A second use of PDGF is in a calcium phosphate matrix used for the treatment of periodontal bone defects, (GEM 21S by Biomimetics in 2005). Other cases of growth factors incorporated into a matrix are bone morphogenic protein-7 (BMP-7) (OP-1 Putty by Stryker Biotech in 2001) as osteoinductive and conductive bone graft material, and BMP in an absorbable collagen sponge (dibotermin-α as INFUSE® bone graft device by Wyeth in 2004), which is indicated for spinal fusion procedures in skeletally mature patients with degenerative disk disease (all are mentioned in Ref. 54).

1.2.3. Enzymes and Blood Factors

This is a highly varied class of proteins, comprising a number of diverse structures. As with insulin and human growth hormone, a number of products in this category had prerecombinant antecedents. These include serum-derived versions of albumin, blood factors such as factor VIII, RhoGam [Rh(rhesus) factor], Gamimune, and immune globulins. The procurement of a safe blood supply has been difficult until relatively recently, and transmission of hepatitis and HIV viruses has occurred, resulting in infection of patients using blood-derived products. In the past, over 60% of hemophilic patients were likely to be accidentally infected [55]. Today the production of these factors from human plasma is significantly safer with the addition of advanced methods for blood screening for infectious agents and addition of viral inactivation steps to the production process [6]. Although many blood-derived products now exist as recombinants, even today not all members of the very diverse and complex category are available as recombinant products.

Recombinant antihemophilic factor (recombinant factor VIII) is, at 2332 amino acids and 280 kDa, the largest commercial protein made by recombinant technology. In its native form it is a glycoprotein with a domain structure of A1–A2–B–A3–C1–C2, in which the heavy chain is composed of A1–A2–B domains and the light chain comprises the remaining domains [56]. There are 25 Asn glycosylation sites [57], six tyrosine sulfation sites [58], eight disulfide bonds, and three free cysteines [59], and the presence of a metal ion [60]. Recombinant factor VIII (rFVIII) was first cloned by Genentech, licensed to Bayer, and approved by the FDA (Kogenate in 1992). Several rFVIII products are currently on the market, including Kogenate FS (Bayer); ReFacto, an engineered version of factor VIII with a deleted B domain, resulting in a molecule that is 110 kDa, smaller than the full-length protein (Genetics Institute/Wyeth). Xyntha, an albumin-free formulation of ReFacto, and Advate (Baxter) are serum-free factor VIII products [16].

In 1987 Genentech received approval for Alteplase, a recombinant tissue plasminogen activator (r-tPA) to dissolve blood clots in patients with acute myocardial infarction. tPA is a multidomain serine protease that converts plasminogen to plasmin, a required step in blood coagulation [61,62]. Alteplase is a single-polypeptide chain composed of 527 amino acids held together by 17 disulfide bridges with four potential sites for *N*-linked glycosylation. By homology with other proteins, alteplase is divided into five domains. A domain homologous to fibronectin type I finger extends from residues 1 to 43; residues 44–91 are homologous to human epidermal growth factor; the kringle 1 (92–173) and kringle 2 (180–261) regions are homologous to the kringle regions found in plaminogen and prothrombin; the *C*-terminal region of alteplase, comprising residues 276–527, is homologous to the trypsin family of serine proteases and contains a catalytic active site formed by His [322], Asp [371], and Ser [478] [63]. Retavase (Reteplase, by Boehringer Mannheim in 1996, also for dissolving blood clots) is a nonglycosylated deletion mutein of tissue plasminogen activator (tPA), containing the kringle 2 and the protease domains of human tPA. Retavase contains 355 of the 527 amino acids of native tPA (amino acids 1–3 and 176–527) which allows production in *E. coli*. It has a longer half-life than does its parent molecule, tPA, which is attributed in part to lack of glycosylation and resultant inability to be cleared by the mannose receptor pathway [16,64]. Tenecteplase (TNKase by Genentech in 2000) is another r-tPA variant. TNKase is approved for treatment of acute myocardial infarction. It contains a number of amino acid changes (Asn for Thr [102], Glu for Asn [116], and tetraalanine for amino acids 296–299) compared to alteplase, which results in enhanced fibrin selectivity [16].

Following the observation that it was effective in reducing the viscosity of human sputum, DNase was developed as a therapeutic to reduce the concentration of free DNA in the lungs of patients with cystic fibrosis. The original product was isolated from bovine sources; however, this was replaced by the recombinant form dornase alfa,

(Pulmozyme by Genentech in 1993). Dornase alfa is a single-polypeptide chain of 260 amino acids with two *N*-linked glycosylation sites. It is classified as an αβ protein with 2 six-stranded β-pleated sheets packed against each other into a hydrophobic core [16]. The α helices and extensive loop regions flank the two antiparallel β sheets. The conformation and stability are both highly dependent on the binding of calcium. Pulmozyme is delivered by inhalation using a jet nebulizer [65].

Lysosome storage disorders are genetic diseases caused by a deficiency in enzyme activity, and patients have benefited from enzyme replacement therapy. The first products were modified placental proteins that have been largely replaced by recombinant proteins. Gaucher's disease is an example of an autosomal recessive lipid storage disorder caused by the deficiency of the lysosomal enzyme β-glucocerebrosidase. The mannose terminated recombinant product is imiglucerase (Cerezyme, by Genzyme in 1994) and is produced by recombinant DNA technology using CHO cells. Purified imiglucerase is a monomeric glycoprotein of 497 amino acids, containing four *N*-linked glycosylation sites and has a molecular weight of 60,430 Da. Imiglucerase differs from placental glucocerebrosidase by one amino acid at position 495, where histidine is substituted for arginine. Somewhat different from the carbohydrate structures of native placental glucocerebrosidase, the oligosaccharide chains have been modified to terminate in mannose sugars. These mannose-terminated oligosaccharide chains are specifically recognized by endocytic carbohydrate receptors on macrophages, the cells that accumulate lipid in Gaucher's disease. This recognition facilitates the enzyme to enter these cells more efficiently to cleave the intracellular lipid that has accumulated in pathological amounts [16]. This example illustrates molecular modifications that may be introduced to influence biodistribution, thereby enhancing activity [66].

The foregoing examples serve to illustrate a fraction of the highly diverse category of enzymes and blood factors that are currently produced using recombinant DNA technology, including factors that were originally commercialized from human plasma sources and later converted to recombinant DNA production and others whose production was initially enabled by the advent of recombinant DNA technology. The latest breakthrough in this field occurred in February 2009, when the FDA approved recombinant human antithrombin expressed in genetically engineered goats (ATryn by GTC Biotherapeutics) for the prevention of perioperative and peripartum thromboembolic events, in hereditary antithrombin deficient patients [16]. (For a complete listing of enzymes and blood factors approved in the United States, see Table 1.1.) At the same time, there are a number of blood enzymes and factors that continue to be isolated from human serum, presumably due to the complexity of producing these proteins using recombinant technology combined with substantial improvements in the safety of the blood supply [67], such as albumin (by Octapharma in 2006) used to restore blood volume, alpha-1 proteinase inhibitor (Aralast by Baxter in 2002, Zemaira by Aventis Behring in 2003) for genetic emphysema, antihemophilic factor/von Willebrand factor complex (Alphanate, by Grifols Biologics in 2007 and Humate-P by CSL Behring in 2007), for hemophilia, C1 esterase inhibitor (Cinryze by Lev Pharmaceuticals in 2008) for use in hereditary angioedema attacks, immune globulin (by numerous companies) for immune deficient diseases, thrombin (Evithrom by Omrix Biopharmaceuticals in 2007) for an aid in hemostasis, and protein C (Ceprotin by Baxter in 2007) for venous thrombosis and purpura fulminans [16].

1.2.4. Fusion Proteins

In the early 1980s the preferred approach for most biotechnology companies was to make recombinant proteins that were structurally as close as possible to the naturally occurring human form, with the expectation that this would minimize antigenicity. Almost from the beginning, however, the need to overcome stability, pharmacokinetic, or other limitations of the native proteins led to the introduction of various strategies to engineer protein therapeutics to improve their pharmaceutical properties (see earlier discussion). Fusion proteins represent a further evolution in the technology to improve pharmacokinetics. One of the initial purposes for developing fusion protein products was to join two different activities in a single molecule in the hope that the combined functionality would have a synergistic effect. A more widespread application has been to fuse an active binding protein or domain with the Fc domain of an IgG (see Table 1.1, p. 12). The binding domain confers the desired activity, while the Fc domain serves to increase overall molecular size and prolong circulation time. The complete molecule can be expressed recombinantly, which eliminates the need for a chemical modification step such as pegylation. A well-studied fusion protein is etanercept (Enbrel by Immunex/Amgen in 1998 for moderate to severe rheumatoid arthritis, 2002 for psoriatic arthritis, 2003 for ankylosing spondylitis, and 2004 for moderate to severe plaque psoriasis). The fusion protein is bivalent, having two TNFα receptors fused to the Fc region of an IgG1 antibody (see Fig. 1.5). It functions by binding to tumor necrosis factor alpha (TNFα) to interrupt the immunological cascade that triggers rheumatoid arthritis in many patients [16,68].

Peptibodies may be considered a class of Fc fusion proteins in which a biologically active, nonnative peptide is fused with the Fc portion of an antibody (see next section for discussion of Fc properties). The fusion can take place at either the amino or carboxyl terminus or even, in some cases, within the Fc domain. Peptibodies can be expressed and refolded from bacteria, whereas the more complex Fc fusion proteins generally require mammalian cell expression systems. The first peptibody to be approved by was romiplostim (Nplate by Amgen in 2008) [16,69,70]. The molecule was designed to target chronic immune thrombocytopenic purpura and has been demonstrated to be an effective thrombopoiesis-stimulating protein. It is composed of a disulfide-bonded human IgG1 heavy-chain and κ light-chain constant regions with two peptide sequences fused at residue 228 of the heavy chain with a polyglycine linker between the peptide and F_C fragment.

A unique fusion protein is denileukin diftitox (ONTAK by Seragen in 1999), approved for the treatment of patients with persistent or recurrent cutaneous T-cell lymphoma whose malignant cells express the CD25 component of the IL-2 receptor. Denileukin diftitox is a genetically engineered cytotoxic fusion protein consisting of the amino acid sequences for the enzymatically active portion of diphtheria toxin fused to the sequence of human interleukin-2, resulting in a molecule that is cytotoxic for cells bearing the target IL-2 receptor [16].

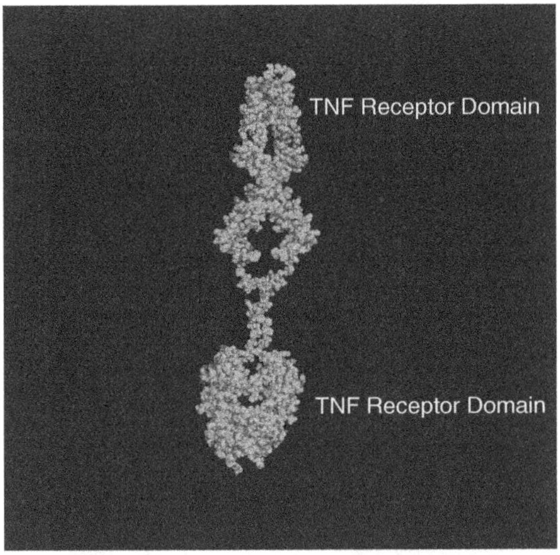

Figure 1.5. Structure of entanercept, a fusion protein with two tumor necrosis factor receptors fused to the Fc of an IgG1.

1.3. ANTIBODIES

Antibodies or immunoglobulin proteins are produced in response to non-self antigens and are a key mechanism for defence against pathogenic organisms and toxins. The distinguishing characteristics of antibodies are the enormous diversity and specificity of antigens that can be recognized. The antibody protein consists of two light (~ 25-kDa) and two heavy (55–65-kDa) chains that adopt β-pleated sheet structures arranged in a symmetric fashion producing an overall Y-shaped quaternary structure. There are five classes of antibodies, IgG, IgM, IgD, IgA, and IgE, which are defined by their particular heavy chains. The main serum antibody, IgG, is the antibody class used almost exclusively in therapeutic antibodies (Fig. 1.6). The two heavy chains are joined at the hinge region by disulfide bonds, and the two light chains are each linked to a heavy chain by a disulfide bond such that the intact molecule is about 160 kDa. Each chain has a variable (V) domain at the amino terminus containing about 110 residues that is involved in antigen binding and a constant (C) domain at the carboxyl terminus that determines the subclass or isotype. Some mouse and human antibodies are glycosylated at a conserved site in the CH2 domain (Asn297) with an *N*-linked fucosyl biantennary complex that can be sialylated. *O*-Linked oligosaccharides have been reported mainly in the hinge region and possibly the Fc region [71]. Glycans can affect binding to antigens and are important for interactions with other components of the immune system, including complement and Fcγ receptors on cytotoxic T lymphocytes.

There are two major variations within the light chain that are designated *kappa* and *lambda*. Most antibodies developed for therapeutic use have the kappa version

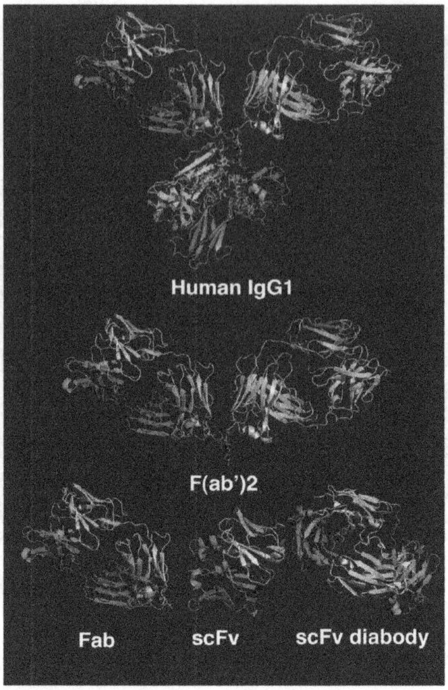

Figure 1.6. Examples of therapeutic antibody constructs. Cartoon structural representations of an intact human IgG1 monoclonal antibody [72] and F(ab′)$_2$ and Fab fragments derived from it, an scFv [72], and an scFv dimer diabody [74]. The heavy chains are shown in magenta and green and light chains, in yellow and wheat. The carbohydrate component of the whole IgG1 molecule is shown as a stick representation and the cystiene residues linking the heavy and light chains are shown in red. The images were produced from PDB files using Pymol (DeLano Scientific, LLC). See the insert for color representation of this figure.

of the light chain; this bias is also found in the structural database, where there are almost 20 times as many kappa structures than lambda. Differences in amino acid sequence between the two light chains lead to specific variations in domain packing, thermodynamics, and structure [78]. The insertion of a single amino acid at the elbow between the variable domain and the constant domain within the light chain of the lambda variation may introduce more flexibility for the overall domain allowing the elbow angle to be significantly larger than the kappa version of the light chain [76].

There are four subclasses of IgG in humans: IgG1, IgG2, IgG3, and IgG4. From a sequence and structural perspective, the major difference between the IgG subclasses resides in the length and flexibility of the hinge region. This region links domains of β-pleated sheets and can be cleaved by proteases to produce large fragments that delineate specific functions. The portion of the constant region between the *C* terminus and the hinge is the Fc (crystallizable) fragment, and the remaining Fab fragment contains the antibody-combining (i.e., antigen-binding) site. Most of the approved therapeutic human(ized) monoclonal Abs are the IgG1 kappa subclass [71]. The IgG1 subtype has a hinge region 15 amino acids in length, with two interchain disulfide bridges. This configuration allows the Fab region to rotate freely on the tethering region of the first disulfide bond. In contrast, the IgG2 hinge region is 12 amino acids in length, with a polyproline helix and four disulfide bonds. This more rigid configuration restricts the flexibility of the Fab [80]. Subclass IgG3 has an extended hinge region of 62 amino acids with up to 11 disulfide bonds. Although this results in a larger molecule [78], it is more susceptible to proteases, resulting in a shorter half-life [79–81]. The hinge region of the IgG4 is three amino acids shorter than the IgG1 hinge and contains a proline substitution for serine, which allows for an intrachain disulfide bond. The intrachain disulfide bond changes the position of the Fab relative to the Fc allowing direct contact of the CH1 domain with the CH2 domain, resulting in a more compact structure relative to the other IgG molecules. In addition, although IgG4 antibodies are expressed as bivalent molecules similar to the other IgGs, IgG4 antibodies have a propensity to swap light and heavy chains [82], creating bispecific structures and half molecules [83].

The human subclasses of IgGs also differ in effector functions, mediated by their Fc domains. The Fc domain recruits cytotoxic effector functions through complement, which is referred to as *complement-dependent cytotoxicity* (CDC) and/or through interactions with Fc-gamma receptors (on gamma globulins) which is referred to as *antibody-dependent cell-mediated cytotoxicity* (ADCC) and *antibody-dependent cell-mediated phagocytosis* (ADCP) [84]. Complement-dependent cytotoxicity results from interaction of cell-bound antibodies with proteins of the complement cascade system starting with binding of complement protein C1q to the Fc domain. This causes a conformational change in C1q that initiates an enzymatic cascade of complement proteins that spreads rapidly and ends in formation of the membrane attack complex that causes lysis of the target [85]. ADCC and ADCP are governed by engagement of the Fc region with a family of Fcγ receptors (FcγRs) [86]. In humans, this protein family comprises FcγRI (CD64); FcγRII (CD32), including isoforms FcγRIIa, FcγRIIb, and FcγRIIc; and FcγRIII (CD16), including isoforms FcγRIIIa and FcγRIIIb. FcγRs are expressed on a variety of immune cells, and formation of the Fc/FcγR complex recruits these cells to sites of bound antigen, typically resulting in signaling and subsequent immune responses such as release of inflammation mediators, B-cell activation, endocytosis, phagocytosis, and cytotoxic attack of the target cell. All FcγRs bind IgG Fc, yet with differing high (FcγRI) and low (FcγRII and FcγRIII) affinities [87]. Furthermore, whereas FcγRI, FcγRIIa/c, and FcγRIIIa are activating receptors characterized by an intracellular immunoreceptor tyrosine-based activation motif (ITAM), FcγRIIb has an inhibition motif (ITIM) and is therefore inhibitory. Engineered Fc variants have been created with improved FcγR affinity and specificity that provide significant enhancements in cytotoxicity [88,89].

Broadly, IgG1 and IgG3 are more competent at inducing cytotoxicity than are IgG2 and IgG4 [18,90]. If the mechanism of drug action requires a cytotoxic event, then IgG1 may be the most desirable subclass with the best half-life. If effector function is not needed or is considered undesirable then IgG2 or IgG4 may be preferred. For example, denosumab, recently filed for postmenopausal osteoporosis and patients undergoing hormone ablation for prostate or breast cancer, is an IgG2, which may provide a safety advantage (Amgen, filed in 2008) [91–93].

The first monoclonal antibodies were murine antibodies obtained by hybridoma technology. The half-life of murine MAbs is typically relatively short, likely because of the inability of the MAbs to interact with the human neonatal Fc receptor [90]. This property of murine MAbs is used to advantage for diagnostics or therapeutics, where it is desirable to clear the MAb quickly after the procedure. For example, two radioisotope-conjugated anti-CD20 MAb preparations, ibritumomab tiuxetan-^{111}In or- ^{90}Y, (Zevalin by IDEC in 2002) and tositumomab- ^{131}I, (Bexxar by SmithKline Beecham in 2003), both for treatment of non-Hodgkin's lymphoma, utilize murine MAbs that are cleared relatively rapidly [16]. There are relatively few commercial murine MAb therapeutics, most likely owing to the potential for human anti-mouse immunogenic responses [94].

Several years after the introduction of murine MAbs, a new class of therapeutic MAbs consisting of mouse/human chimeric antibodies received FDA approval. Chimeric MAbs generally combine the variable regions of a mouse antibody with the effector or constant regions of a human antibody, resulting in antibodies that are approximately 65% human in sequence. Chimeric antibodies generally exhibit reduced immunogenicity and increased plasma half-life compared with those of murine Mabs [90]. Examples of chimeric MAb products include rituximab that targets CD20 (Rituxan by IDEC and Genentech in 1997) for the treatment of patients with B-cell non-Hodgkin's lymphoma, and infliximab, which targets tumor necrosis factor alpha (TNFα) (Remicade by Centocor in 1998 for Crohn's disease and in 1999 for rheumatoid arthritis) [16].

The next step in increasing the human component of MAb therapeutics was to generate humanized MAbs. These antibodies generally combine the hypervariable regions of a mouse (or rat) antibody within the framework of a human antibody molecule. Most of the MAbs on the market today are in this class. Examples include daclizumab, which targets CD25 (Zenapax by Hoffman-La Roche in 1997) [16] for prophylaxis of acute organ rejection in patients receiving renal transplants; anti-HER2/neu antibody, trastuzumab (Herceptin by Genentech in 1998), approved for treatment of metastatic breast cancer in HER2-positive patients [16]; and bevacizumab, which targets VEGF (Avastin by Genentech, in 2004 for treatment of colorectal cancer, in 2006 for advanced lung cancer, and in 2008 for metastatic breast cancer) [16].

Fully human MAbs can now be produced by replacing murine immunoglobulin genes with human immunoglobulin genes, leading to the production of fully human monoclonal antibodies [95]. Examples of fully human MAb products include adalimumab (Humira, by Abbott in 2002), a fully human anti-TNF MAb for treatment of rheumatoid arthritis and subsequently for psoriastic arthritis in 2005, Crohn's disease in 2007, and plaque psoriasis in 2008; and panitumumab (Vectibix by Amgen in 2006) for treatment of colorectal cancer [16].

As with other classes of protein therapeutics, immunogenicity continues to be a safety consideration for antibodies, although the advent of humanized and fully human MAb constructs has led to a reduction in the general level of risk and facilitated their application to an increasingly wide range of therapeutic indications [90,96]. Today, monoclonal antibodies constitute by far the largest and fastest growing category of biological therapeutics, with 26 currently approved in the United States alone (see Table 1.2) and hundreds in various stages of clinical and preclinical development [18]. The majority of the current MAb products are approved for cancer therapy [18]; however, as experience with this class of therapeutics increases, the range of indications is expanding rapidly.

1.3.1. Antibody Conjugates

Attempts to produce antibody conjugates carrying a toxin can be traced to the 1970s [97]. Early attempts included conjugation of diphtheria toxin (a protein) to an antibody, followed a few years later by conjugates of daunomycin and adriamycin, small molecules that were covalently bound to antibodies [98]. Interest was driven by the potential advantages of conjugates, including reduced toxicity due to better localization of the toxin as well as, in some cases, greater efficacy based on a combination of effects from both the targeting moiety and the toxin. Early drug candidates failed for a number of reasons including insufficient information regarding tumor antigens, murine antibody prompted immune responses, short biological half-life of murine antibodies, and issues with linker chemistries. For conjugates containing protein toxins, inefficient internalization of the toxin posed another hurdle [99].

Despite the considerable challenges and early low success rate, interest in this format has persisted, based on its obvious potential advantages, and several antibody conjugates have been approved by the FDA. These include conjugates with isotopes such as Bexxar and Zevalin [100], and the only toxin conjugate, gemtuzumab ozogamicin (Mylotarg by Wyeth in 2000), approved for the treatment of elderly patients suffering from CD33-positive acute myeloid leukemia [16,100]. A humanized IgG4 kappa MAb was selected for this application because the effector functions are not needed, and the IgG4 subclass reportedly has reduced nonspecific binding to normal tissues [101]. The anti-CD33 antibody is conjugated with a cytotoxic antitumor antibiotic, ozogamycin (*N*-acetyl-γ-calicheamicin). The drug load is about four to six molecules of ozogamycin per antibody molecule, with approximately 50% of the antibody conjugated and the remaining 50% unconjugated. The linker contains two cleavable bonds (disulfide and hydrazone) [102].

Today most conjugates are based on humanized or human MAbs, and new linker chemistries have also been developed. A comprehensive review article on antibody–drug conjugates has recently been published [103]. Advances in the understanding of cancer biology have led to selection of improved targeting agents and a wider range of toxins; and a number of antibody conjugates are currently in various stages of clinical testing and preclinical development [99].

1.3.2. Fab

For some applications, the Fc-mediated effector functions and prolonged serum circulation times are not required or may even be detrimental to the desired activity of the therapeutic. Eliminating the Fc domain to create $F(ab')_2$ results in a molecule

TABLE 1.2. Recombinant Antibody Parenteral Products Approved

Nonproprietary Name[a]	Trade Name	IgG Type	Half-Life, days	Indications, Administration Route,[b] and Dose	Manufacturer and Approval Year
			Murine (... omab)		
Muromonab-CD3	Orthoclone OKT3®	2	0.75[c]	Reversal of kidney transplant rejection, IV bolus 5 mg/day for ≤ 2 weeks	Ortho Biotech; 1986 (US)
Ibritumomab, tiuxetan (^{90}Y)	Zevalin®	1	1.1[c]	Non-Hodgkin's lymphoma, IV 3.2 mg	IDEC (US) 2002; Schering AG (EU) 2004
Tositumomab (IgG, IgG-^{131}I)	Bexxar®	2	2.7[c]	Non-Hodgkin's lymphoma, IV 450 mg	Corixa/GSK; 2003 (US)
			Chimeric (... ximab)		
Rituximab	Rituxan®	1	9.4[c]	Non-Hodgkin's lymphoma, IV 375 mg/m^2; rheumatoid arthritis, 1 g, followed by 1 g in 2 weeks	Genentech/IDEC; 1997 (US)
Infliximab	Remicade®	1	9.5[c]	Anti-TNF for Crohn's disease, rheumatoid arthritis, psoriasis, IV 2–5 mg/kg	Centocor; 1998 (US), 1999 (EU)
Basiliximab	Simulect®	1	4.1[c]	Prophylaxis of organ transplant, IV bolus, or infusion 20 mg prior to transplantation, 20 mg after 4 days	Novartis; 1998 (US, EU)
Cetuximab	Erbitux®	1	4.8[c]	EGF receptor-expressing colorectal cancer; lung, head, and neck cancer; IV 400 mg/m^2 followed by weekly 250 mg/m^2	ImClone/BMS; 2004 (US)

			Humanized (. . . zumab)		
Daclizumab	Zenapax®	1	20[c]	Anti-IL-2 for prevention of kidney transplant rejection, IV 1 mg/kg	Roche/Protein Design Lab; 1997 (EU), 1999 (US)
Palivizumab	Synagis®	1	19–27[c]	Prophylaxis of pediatric respiratory syncytial virus, IM 15 mg/kg	MedImmune; 1998 (US) Abbott 1999 (EU)
Trastuzumab	Herceptin®	1	2.7–10[c]	HER2 overexpressed metastatic breast cancer, IV 4 mg/kg, followed by 2 mg/kg every week for 52 weeks	Genentech 1998 (US); Hoffmann-La Roche 2000 (EU)
Alemtuzumab	Campath®	1	12[c]	Chronic lymphocytic leukemia, escalation from IV 3 to 30 mg/day, 3 times per week	Berlex/ILEX/ Millenium 2001 (US) Millenium/ILEX 2001 (EU)
Gemtuzumab ozogamicin	Mylotarg®	4	1.9–2.5[c]	Acute myeloid leukemia, IV 9 mg/m^2, second dose in 14 days	Wyeth; 2000 (US)
Omalizumab	Xolair®	1	20[c]	Anti-IgE for asthma, SC 150–375 mg every 2 or 4 weeks	Genentech/ Novartis/Tanox; 2003 (US)
Efalizumab	Raptiva®	1	Dose-dependent	Binds LFA for plague psoriasis, SC single 0.7 mg/kg followed by weekly 1 mg/kg	Genentech/Xoma; 2003 (US)
Bevacizumab	Avastin®	1	20	Anti-VEGF colorectal cancer with chemotherapy, 5–10 mg/kg every 14 days; lung cancer, 15 mg/kg every 3 weeks; breast cancer, 10 mg/kg every 14 days IV	Genentech, 2004 (US)
Natalizumab	Tysabri® (Antegren®)	4	11	Multiple sclerosis and Crohn's disease, IV 300 mg every 4 weeks	Biogen/Elan; 2004 (US)

(continued)

TABLE 1.2. Recombinant Antibody Parenteral Products Approved (*Continued*)

Nonproprietary Name[a]	Trade Name	IgG Type	Half-Life, days	Indications, Administration Route,[b] and Dose	Manufacturer and Approval Year
Eculizumab	Soliris®	2/4	12	Paroxysmal nocturnal hemoglobinuria, IV 600 mg every 7 days, for 4 weeks, then 900 mg every 14 days	Alexion; 2007 (US)
			Human MAb (. . . umab)		
Adalimumab	Humira (US)®, Trudexa (EU)	1	14.7–19.3[c]	Anti-TNF for rheumatoid arthritis, SC 40 mg every 1–2 weeks; Crohn's disease 160 mg followed by 80 mg every 2 weeks	Cambridge Antibody/Abbott; 2002 (US), 2003(EU)
Panitumumab	Vectibix™	2	3.6–10.9	EGFR-expressing, metastatic colorectal carcinoma, IV 6 mg/kg every 14 days	Amgen/Abgenix; 2005 (US)
Golimumab	Simponi™	1	14	Anti-TNF for rheumatoid arthritis, active psoriatic arthritis and active ankylosing spondylitis, 50 mg SC once a month	Centocor; 2009 (US)
			Fab and Its Conjugates		
Abciximab (chimeric Fab)	ReoPro®	1	0.29[c]	Prevent blood clots, 0.25 mg/kg IV bolus followed by IV infusion of 0.125 $\mu g\ kg^{-1}\ min^{-1}$ for 12 h	Centocor (Lilly); 1994 (US)
Ranibizumab	Lucentis®	1		Age-related macular degeneration, 0.5 mg intravitreal once a month	Genentech; 2006 (US)
Certolizumab pegol	Cimzia®	1	14	Crohn's disease, 2× SC 200 mg each and on days 14 and 28	Celltech/UCB; 2008 (US)

[a]Nonproprietary name: U.S. adopted name.
[b]The IV route is indicated for intravenous infusion unless otherwise stated.
[c]From Table 1.1 in Reference [130]; all others from Reference [16].

that retains the bivalent character of the antibody structure in a smaller construct, whereas Fab's are monovalent (Fig. 1.6). Both constructs can enable bacterial production and may enhance tissue penetration. Ranibizumab (Lucentis, by Genentech in 2006), approved for treatment of wet acute macular degeneration, is a recombinant humanized IgG1 kappa isotype Fab that binds to and inhibits the biological activity of human VEGF-A [16]. Designed for intravitreal injection, the Fc domain functions were not needed; and elimination of complement-mediated or cell-dependent cytotoxity, as well as rapid systemic clearance were additional advantages [104]. In addition, point mutations were introduced into the CDR region of the parent anti-VEGF Fab to enhance binding potency [105]. Pharmacokinetic studies of primates showed that, after a single intravitreal administration, biologically effective retinal concentrations of ranibizumab were maintained for one month, while serum levels were 1000–2000-fold lower than the ocular levels [106].

Certolizumab pegol (Cimzia, by UCB in 2008) is a PEGylated version of an anti-TNFα monoclonal antibody fragment approved for treatment of Crohn's disease. The humanized antibody fragment is a $F(ab')_2$, (i.e., two Fab fragments linked by a disulfide bond), manufactured in *E. coli*, with a light chain of 214 amino acids and a heavy chain of 229 amino acids. In order to improve half-life without losing binding avidity, two PEG moieties (20 kDa each) are conjugated to one cysteine residue in the hinge region, distant from the antigen-binding region. This enhances the elimination half-life up to 311 hours. Given that the construct does not contain an Fc portion, it is incapable of inducing *in vitro* complement activation, ADCC or apoptosis [16,107].

1.3.3. Newer Antibody-Derived Structures

With increasing application in research, biotechnology, and medical therapy, it has become obvious that antibodies, with all their merits, also suffer inherent disadvantages under some circumstances. Immunoglobulin (Ig), molecules are rather large, thus limiting tissue penetration. For solid tumors, there are the additional problems imposed by penetration through the vasculature and dispersion against an interstitial pressure. In addition, IgGs are composed of two different polypeptides, the light and heavy chains, which requires complicated cloning steps for recombinant expression; they are also glycosylated, which requires eukaryotic cell culture, time-consuming optimization, and limited production capacities. The complex architecture of their antigen-binding site formed by six hypervariable loops, three from each chain, can be difficult to manipulate simultaneously if synthetic libraries are to be generated. The constant Fc region mediates immunological effector functions that are crucial for only a few biopharmaceutical applications and may lead to undesired interactions; in some cases the long serum half-life of a full-length MAb can be a disadvantage. Consequently, there is an emerging need for antibody alternatives with improved features. Beyond Fabs, even more radical approaches are being investigated, including dissecting the molecule into constituent domains and recombining them to generate monovalent, bivalent, or even multivalent fragments [19]. Some of these products are now in preclinical and clinical trials [108].

Single-chain variable fragments (scFv's) are an increasingly popular format in which the V_H and V_L domains are joined with a flexible polypeptide linker to prevent dissociation (Fig. 1.6). As with Fab's, scFv fragments, comprising both V_H and V_L domains, may retain the specific, monovalent, antigen-binding affinity of the parent IgG, while showing improved tissue penetration and rapid pharmacokinetics [108]. Despite early excitement concerning the functional activity of single V domains, these fragments were poorly soluble and prone to aggregation. More recent reports on camelid antibodies have revived interest in this class since, unlike mouse V_H domains, camelid VhH domains are in general soluble and can be produced as stable *in vitro* targeting reagents for diagnostic platforms and nanosensors [109]. For therapeutic applications, humanization may be required to reduce immunogenicity.

It may be desirable for some applications to design multivalent antibody fragments with greater functional avidity and slower dissociation rates for cell surface or multimeric antigens. In addition to increased avidity, multivalent engagement of cell surface receptors may result in increased activation of transmembrane signaling, which may improve the therapeutic efficiency of some antibodies [110,111]. Fab and scFv fragments have been engineered into dimeric, trimeric, or tetrameric conjugates using either chemical or genetic crosslinks [19]. One strategy to genetically encode multimeric scFv's has been the reduction of scFv linker length (to five or fewer residues), which directs self-assembly into these multivalent antibody forms [112]. The diabodies may prove to have stability advantages *in vivo*, although additional optimization of the linker sequences may be required to control protease degradation. Alternatively disulfide linkages can stabilize bivalent diabodies (Fig. 1.6). More recently, stable Fv tetramers have also been generated by noncovalent association [113].

For solid tumors, a number of approaches are being explored to increase vascular penetration and improve tissue distribution. One such approach is targeting fragments to surface molecules on endothelial *caveolae*, specialized plasmalemmal invaginations that can transcytose across the normally restrictive endothelial cell barriers [114]. This may allow greater penetration into and accumulation throughout solid tumors to enable more effective delivery of imaging agents or therapeutic payloads [108]. Another emerging format is produced by heterodimeric association of two domain-swapped scFv's of different specificity. These molecules are designed to bind two different epitopes on the same target, thereby increasing avidity and specificity, or to bind two different antigens for different applications such as recruitment of cytotoxic T and natural-killer cells. One class of bispecific scFvs and diabodies are designed to bind CD3 T-cell coreceptor and thereby recruit cytotoxic T cells to the tumor site. Diabody–Fc fusions can target two cell surface receptors such as IGFR and EGFR simultaneously, to elicit rapid IGFR internalization and recruit effective antibody-dependent cellular cytoxicity. Another application of multiple specificity constructs is pretargeting (e.g., multistep delivery of an antibody–strepavidin conjugate to a receptor followed by biotin labeled with a radioisotope) [108]. Additionally, scFv–cytokine fusion proteins are another format being explored as cancer therapeutics. In intial evaluations, poor protein production, adverse immune reactions, and vascular leak syndrome limited their preclinical evaluation. However, some scFv–cytokines have begun to show promise, such as scFv–IL12 fusion proteins that were found to be superior to IL-12 alone in a lung metastasis model [115].

1.4. THERAPEUTIC PROTEIN ALTERNATIVES TO ANTIBODY-DERIVED SCAFFOLDS

As crystallography and genome sequencing converge, it has become apparent that biological recognition is dependent on properties of shape and charge, rather than simply the primary sequences involved in protein binding [116]. This observation has triggered efforts to create alternative protein scaffolds with potential for high-affinity binding of a broad range of specific targets. Similar to the novel antibody-based constructs, it is hoped that these alternative formats may confer additional advantages or specific capabilities, such as bacterial production, enhanced stability, increased tissue penetration, multivalent structures, alternate routes of delivery, and/or other features. By exploring more diverse sequence and structural space, it may be possible to extend the performance range of protein therapeutics even further. The alternative scaffolds can be derived from natural proteins found in eukaryotic or bacterial organisms. Core structures may be modified by incorporating sequences for target binding and/or other functions, or novel binding surfaces may be derived from these sources. A few of these emerging scaffolds, intended to illustrate some of the basic approaches in this area, are described below.

Many of the emerging scaffolds attempt to exploit common families of human proteins involved in molecular recognition functions. In one example protein scaffolds have been engineered based on the fibronectin type III domain (FN3). Fibronectin is a large glycoprotein that is found in the extracellular matrix of tissues, and is involved in cell–substrate and cell–cell interactions. The FN3 domain is found in ~2% of all human proteins, including cell surface receptors, carbohydrate-binding domains, and cell adhesion proteins, with a configuration that has been characterized as a three-dimensional β-sandwich structure, similar to the antibody V_H domain with fewer β strands [117]. The molecular recognition function that is characteristic of these proteins resides in a binding site consisting of surface loops; and it is this feature that may be manipulated by introducing novel sequences or engineering the interfaces to modulate target specificity and binding affinity [68,118]. Another very common protein family that is being explored as a framework for novel therapeutics is the ankyrin repeat proteins (ARPs), which are among the most naturally abundant binding proteins encoded by the human genome [119] These proteins provide the backbone for DARPins (designed ankyrin repeat proteins), in which structural motifs are assembled onto a fairly rigid protein framework consisting of α helixes and β sheets [76,120–122]. Along with several other classes of emerging scaffolds (e.g. tetranectins and avimers, see below), DARPins can be engineered to produce multivalent molecules.

A number of other families of human proteins are being explored as potential scaffolds for therapeutic proteins. Lipocalins are composed of a central cylindrical β sheet (β barrel) with eight antiparallel strands carrying four loops that are variable in their number of amino acid residues, sequence, and conformation. In their natural form, lipocalins typically bind to low-molecular-weight targets, such as vitamins [123]. The β barrel provides a rigid backbone onto which specific binding sequences can be grafted to yield constructs, termed *anticalins*, which can be engineered for specificity and binding affinity to particular targets [124,125]. Affilin scaffolds are based on the rigid β-sheet folds of human ubiquitin and γ-crystallin. Randomized residues of eight amino acids are isolated from phage libraries against specific protein and peptide targets, to create a binding site that is grafted onto the γ-crystallin backbone [126]. The "knottin family" of proteins, known as *microbodies*, consist of an antiparallel β sheet and cystine knot motif with three disulfide bonds. These constructs are usually composed of 30 amino acids, with an introduced peptide ligand sequence for target presentation. Because of their relatively small size and stability, the cystine knot microproteins have been investigated for oral delivery [127,128].

Other constructs have been designed to provide modular structures consisting of several units for increased avidity or carrying multiple functions. Avimers are composed of binding domains based on the A domain of the human low-density lipoprotein receptor. Peptide sequences against specific targets are obtained via phage display, then "stitched" onto the ~35–amino acid A-domain backbone to form an a multiepitope binding scaffold with high avidity to its target [129]. Another construction is based on human tetranectin, which is found in tissues and plasma, and has a natural ability to bind to a range of proteins at the ligand-binding C-type lectin domain. Although initially monomeric, the protein trimerizes through coiled-coil helical trimerization at the *N*-terminal domain to present a multivalent interface. As a result of the multivalent binding domains, the modified tetranectins are able to bind to their targets with high avidity [130].

In efforts to exploit an even broader sequence and structural space, nonhuman proteins have gained some consideration as sources of human therapeutics. Phylomers, small, nonimmunoglobulin protein fragments encoded by bacterial DNA, are one example. Since bacteria contain the greatest known source of sequence divergence, highly diverse libraries of polypeptides can be generated through varying combinations of gene sequences. Sequence mutagenesis allows the selection of molecules with affinity on par with antibodies and other native proteins, in molecules that range between 15 and 80 amino acid residues [131].

These approaches highlight some of the efforts to produce novel classes of therapeutic proteins with the potential to address multiple targets, similar to antibodies, while also providing diverse characteristics that may enable new applications or confer other benefits. In addition to very challenging performance requirements, a common concern underlying all alternative protein scaffolds is the risk of immunogenicity. Nonetheless, if successful, these approaches have the potential to extend the range of protein therapeutics even beyond the very considerable achievements of today's native proteins, monoclonal antibodies, and antibody-related products in the therapeutic biologics space.

1.5. CONCLUSION

Proteins represent the most diverse range of functions in living organisms, with enormous potential to generate highly potent and specific therapeutics to treat a wide range of disease states. The advent of recombinant DNA technology combined with monoclonal antibody technology has greatly expanded the range of protein therapeutics, from a very limited number of factors that could be isolated from natural animal or human sources, to include close to the full range of human protein structures as potential therapeutics. Further these technologies have enabled production of this highly diverse range of protein structures and functions in formats that are generally acceptable in terms of their immunological and safety properties, allowing application of biological therapeutics for treatment of a widening range of indications and disease states. At the same time, the field continues to experiment with a variety of techniques to modify or engineer protein structures to expand further the range of activities and pharmaceutical properties of this highly diverse class. Although the requirement for injection still poses a barrier in some settings, adoption of more convenient injection devices, particularly for chronic home administration, has increased acceptance of biologics for a wider range of therapeutic applications. As a result, there are well over a hundred approved protein therapeutic products on the market today for treatment of a broad range of diseases and indications, and hundreds more in various stages of clinical and preclinical development, representing an impressive potential to alleviate currently unmet medical needs and benefit patients.

Nevertheless, there are still significant constraints and barriers to realizing the full potential of this therapeutic class. Many of these hurdles are related to the inherent size, flexibility, and structural complexity of this class of molecules; ironically, these are the same properties that also enable the broad range of activities and functions as well as the often exquisite specificity of action. Characterization of these complex and labile molecules remains a challenge, and the field continues to seek additional techniques to probe physical structure, association states, and degradation pathways. The requirement for injection still constitutes a hurdle in some disease and geographic settings, and more complex delivery devices pose their own challenges in terms of formulation and stability for many products. Further, administration by routes other than intravenous (IV) is limited in the amounts that can be delivered via a single injection; and the drive to higher doses introduces additional formulation and processing challenges, including solubility, viscosity, and aggregation [132]. While great strides have been made in the elimination of impurities and infectious agents during manufacturing, and the general safety of protein therapeutics has advanced tremendously, safety remains a significant consideration. Immunogenic reactions in particular can impose constraints on molecular design as well as strict standards for stability in production, storage, and handling. To maintain the integrity of these labile molecules, most protein therapeutics must be shipped and stored under refrigerated conditions for the majority of their shelf lives, posing additional challenges in manufacturing and distribution, as well as restricting the use of these products in areas where the cold chain is less developed or unavailable. Later chapters in this volume outline the great strides that have been made in the field in terms of the formulation of protein therapeutics, as well as the processes for manufacturing the final product, and describe some of the ongoing research to address these and other challenges.

The ultimate hurdles for protein therapeutics may lie in the realm of biology more so than chemistry. Because of their great specificity, the most problematic toxicities for biologics are often related to their mechanism of action. Incorporation of targeting strategies in the design and delivery of biological therapeutics may assist in overcoming some of these impediments. The limited capacity for tissue penetration and membrane transport further restricts the potential of this class of therapeutics. These constraints, which are again related to their size and structural complexity, significantly limit the targets that can be addressed with this class, despite their great range of activities and functions. For example, the inability of most proteins to penetrate the blood–brain barrier restricts their application in many neurological disorders, as well as other protected compartments. Moreover, many of the most desirable targets for therapeutic intervention are intracellular and hence beyond the range of the vast majority of current protein therapeutics. Hopefully the effort and ingenuity dedicated to the design new classes of protein therapeutics will eventually result in breakthroughs that will extend the application of these molecules, with their inherent advantages in terms of function, potency, and specificity, even to these most challenging frontiers.

REFERENCES

1. Hoffmann, F., (1900), U.S. Patent #644,677, "Acetylsalicylic acid" assigned to Farben-Fabriken of Elberfeld Co., New York.
2. Bliss, M. (1982), *The Discovery of Insulin*, Univ. Chicago Press.
3. Frasier, S. D. (1997), The not-so-good old days: Working with pituitary growth hormone in North America, 1956 to 1985, *J. Pediatr*. **131**: S1–S4.
4. Brown, P. et al. (2000), Iatrogenic Creutzfeldt-Jakob disease at the millennium, *Neurology* **55**: 1075–1081.
5. Cohn, E. J. (1945), Blood proteins and their therapeutic value, *Science* **101**: 51–56.
6. Berger, M. (1999), A history of immune globulin therapy, from the Harvard crash program to monoclonal antibodies, *Curr. Allergy Asthma Rep*. **2**: 368–378.
7. Resnik, S. (1999), *Blood Saga: Hemophilia, AIDS, and the Survival of a Community*, Univ. California Press, Berkeley.
8. James, R. (1968), *Wall Street J*. Hope for 'Bleeders', page 1.
9. Enck, R. E., Betts, R. F., Brown, M. R., and Miller, G. (1979), Viral serology (hepatitis B virus, cytomegalovirus, Epstein-Barr virus) and abnormal liver function tests in transfused patients with hereditary hemorrhagic diseases, *Transfusion* **19**: 32–38.
10. Curran, J. W. et al. (1984), Acquired immunodeficiency syndrome (AIDS) associated with transfusions, *New Engl. J. Med*. **310**: 69–75.
11. Bray, B. L. (2003), Large-scale manufacture of peptide therapeutics by chemical synthesis, *Nat. Rev*. **2**: 587–593.
12. Cohen, S. N., Chang, A. C. Y., Boyer, H. W., and Helling, R. B. (1973), Construction of biologically functional bacterial plasmids *in vitro*, *Proc. Natl. Acad. Sci. USA* **70**: 3240–3244.
13. Verma, R., Boleti, E., and George, A. J. (1998), Antibody engineering: Comparison of bacterial, yeast, insect and mammalian expression systems, *J Immunol. Meth*. **216**: 165–181.
14. Leader, B., Baca, Q. J., and Golan, D. E. (2008), Protein therapeutics: A summary and pharmacological classification, *Nat. Rev*. **7**: 21–39.
15. Kohler, G. and Milstein, C. (1975), Continuous cultures of fused cells secreting antibody of predefined specificity, *Nature* **256**: 495–497.

16. FDA Website (2009), Licensed product approvals for biological drugs. URL: www.accessdata.fda.gov/scripts/cder/drugsatfda/index.cfm
17. Jakobovits, A., Amado, R. G., Yang, X., Roskos, L., and Schwab, G. (2007), From XenoMouse technology to panitumumab, the first fully human antibody product from transgenic mice, *Nat. Biotechnol.* **25**: 1134–1143.
18. Reichert, J. M. and Valge-Archer, V. E. (2007), Development trends for monoclonal antibody cancer therapeutics, *Nat. Rev.* **6**: 349–356.
19. Carter, P. J. (2006), Potent antibody therapeutics by design, *Nat. Rev. Immunol.* **6**: 343–357.
20. Baker, M. (2005), Upping the ante on antibodies, *Nat. Biotechnol.* **23**: 1065–1072.
21. Itakura, K. et al. (1977), Expression in Escherichia coli of a chemically synthesized gene for the hormone somatostatin, *Science* **198**: 1056–1063.
22. Goeddel, D. V. et al. (1979), Expression in Escherichia coli of chemically synthesized genes for human insulin, *Proc. Natl. Acad. Sci. USA* **76**: 106–110.
23. Goeddel, D. V. et al. (1979), Direct expression in Escherichia coli of a DNA sequence coding for human growth hormone, *Nature* **281**: 544–548.
24. Walsh, G. (2003), *Biopharmaceuticals: Biochemistry and Biotechnology*, 2nd ed., Wiley, Hoboken, NJ.
25. Brange, J. and Langkjær, L. (1993), *Insulin Structure and Stability* in Stability and Characterization of Protein and Peptides Drugs, Y. John Wang and Rodney Pearlman ed., Springer Dordrecht p. 315.
26. Brange, J. and Volund, A. (1999), Insulin analogs with improved pharmacokinetic profiles, *Adv. Drug Deliv. Rev.* **35**: 307–335.
27. Mersebach, H. F. S., Sparre, T., and Jensen, L. B. (2008), in *Immunogenecity of Biopharmaceuticals*, van de Weert, M. and Moller, E. ed., Springer.
28. Frank, B., Pettee, J., Zimmerman, R., and Burck, P. (1981), *Peptides: Synthesis-Structure-Function*, Pierce Chemical Co.
29. Brange, J. et al. (1988), Monomeric insulins obtained by protein engineering and their medical implications, *Nature* **333**: 679–682.
30. Olsen, H. B. and Kaarsholm, N. C. (2000), Structural effects of protein lipidation as revealed by LysB29-myristoyl, des(B30) insulin, *Biochemistry* **39**: 11893–11900.
31. Yuan, L., Wang, J., and Shen, W. C. (2005), Reversible lipidization prolongs the pharmacological effect, plasma duration, and liver retention of octreotide, *Pharm. Res.* **22**: 220–227.
32. Moore, W. V. and Leppert, P. (1980), Role of aggregated human growth hormone (hGH) in development of antibodies to hGH, *J. Clin. Endocrinol. Metab.* **51**: 691–697.
33. Brown, P., Gajdusek, D. C., Gibbs, C. J., Jr., Asher, D. M. (1985), Potential epidemic of Creutzfeldt-Jakob disease from human growth hormone therapy, *New Engl. J. Med.* **313**: 728–731.
34. Haan, C., Kreis, S., Margue, C., and Behrmann, I. (2006), Jaks and cytokine receptors—an intimate relationship, *Biochem. Pharmacol.* **72**: 1538–1546.
35. Hill, C. P., Osslund, T. D., and Eisenberg, D. (1993), The structure of granulocyte-colony-stimulating factor and its relationship to other growth factors, *Proc. Nat. Acad. Sci.* **90**: 5167–5171.
36. Brandhuber, B. J., Boone, T., Kenney, W. C., and McKay, D. B. (1987), Three-dimensional structure of interleukin-2, *Science* **238**: 1707–1709.

37. Bazan, J. F. (1992), Unraveling the structure of IL-2, *Science* **257**: 410–413.
38. Geigert, J. and Ghrist, F. D. (1996), Development and shelf-life determination of recombinant human granulocyte—macrophage colony-stimulating factor (LEUKINE, GM-CSF), *Pharm. Biotechnol.* **9**: 329–342.
39. Kinstler, O., Molineux, G., Treuheit, M., Ladd, D., and Gegg, C. (2002), Mono-N-terminal poly(ethylene glycol)-protein conjugates, *Adv. Drug Deliv. Rev.* **54**: 477–485.
40. Kinstler, O. B. et al. (1996), Characterization and stability of N-terminally PEGylated rhG-CSF, *Pharm. Res.* **13**: 996–1002.
41. Elliott, S. et al. (2003), Enhancement of therapeutic protein *in vivo* activities through glycoengineering, *Nat. Biotechnol.* **21**: 414.
42. Macdougall, I. C. (2008), Novel erythropoiesis-stimulating agents: A new era in anemia management, *Clin. J. Am. Soc. Nephrol.* **3**: 200–207.
43. Seto, M. H. et al. (1995), Homology model of human interferon-{alpha}8 and its receptor complex, *Protein Sci.* **4**: 655–670.
44. Wang, Y. S. et al. (2002), Structural and biological characterization of pegylated recombinant interferon alpha-2b and its therapeutic implications, *Adv. Drug Deliv. Rev.* **54**: 547–570.
45. Rajender, R., Modi, M. W., and Pedder, S. (2002), Use of peginterferon alfa-2a (40KD) (Pegasys) for the treatment of hepatitis C, *Adv. Drug Deliv. Rev.* **54**: 571–586.
46. Karpusas, M. et al. (1997), The crystal structure of human interferon beta at 2.2-A resolution, *Proc. Natl. Acad. Sci.* **94**: 11813–11818.
47. Ealick, S. E. et al. (1991), Three-dimensional structure of recombinant human interferon-gamma, *Science* **252**: 698–702.
48. Jiang, X. et al. (2000), Structure of the active core of human stem cell factor and analysis of binding to its receptor kit, *Embo. J.* **19**: 3192–3203.
49. Osslund, T. D. et al. (1998), Correlation between the 1.6 A crystal structure and mutational analysis of keratinocyte growth factor, *Protein Sci.* **7**: 1681–1690.
50. NIH (2009), *Clinical Trials Monitored by the National Institutes of Health (NIH)*. URL: http://clinicaltrials.gov
51. Schreuder, H. et al. (1997), A new cytokine-receptor binding mode revealed by the crystal structure of the IL-1 receptor with an antagonist, *Nature* **386**: 194–200.
52. McDonald, N. Q. et al. (1991), New protein fold revealed by a 2.3-A resolution crystal structure of nerve growth factor, *Nature* **354**: 411–414.
53. Murray-Rust J. M. N., Blundell, T. L., Hosang, M., Oefner, C., Winkler, F., and Bradshaw, R. A. (1993), Topological similarities in TGF-beta 2, PDGF-BB and NGF define a superfamily of polypeptide growth factors, *Structure* **1**(2): 153–159.
54. FDA (2009), Website for device approvals. Center for Devices and Radiological Health URL: www.fda.gov/cdrh/mda
55. Walsh, G. (1998), *Biopharmaceuticals: Biochemistry and Biotechnology*, Wiley, New York.
56. Fay, P. J. (1993), Factor VIII structure and function, *Thromb. Haemost.* **70**: 63–67.
57. Vehar, G. A. et al. (1984), Structure of human factor VIII, *Nature* **312**: 337–342.
58. Pittman, D. D., Wang, J. H., and Kaufman, R. J. (1992), Identification and functional importance of tyrosine sulfate residues within recombinant factor VIII, *Biochemistry* **31**: 3315–3325.

59. McMullen, B. A., Fujikawa, K., Davie, E. W., Hedner, U., and Ezban, M. (1995), Locations of disulfide bonds and free cysteines in the heavy and light chains of recombinant human factor VIII (antihemophilic factor A), *Protein Sci*. **4**: 740–746.
60. Mikaelsson, M. E., Forsman, N., and Oswaldsson, U. M. (1983), Human factor VIII: A calcium-linked protein complex, *Blood* **62**: 1006–1015.
61. Bennett, W. F. et al. (1991), High resolution analysis of functional determinants on human tissue- type plasminogen activator, *J. Biol. Chem*. **266**: 5191–5201.
62. de Vos, A. M., Ultsch, M. H., Kelley, R. F., Padmanabhan, K., Tulinsky, A., Westbrook, M. L., and Kossiakoff, A. A. (1992), Crystal structure of the kringle 2 domain of tissue plasminogen activator at 2.4-A, *Biochemistry* **31**: 270–279.
63. Nguyen, T. H. and Ward, C. (1993), Stability characterization and formulation development of alteplase, a recombinant tissue plasminogen activator, *Pharm. Biotechnol*. **5**: 91–134.
64. Mattes, R. (2001), The production of improved tissue-type plasminogen activator in Escherichia coli, *Semin. Thromb. Hemost*. **27**: 325–336.
65. Shire, S. J. (1996), Stability characterization and formulation development of recombinant human deoxyribonuclease I [Pulmozyme (dornase alpha)], *Pharm. Biotechnol*. **9**: 393–426.
66. Edmunds, T. (1998), Cerezyme: A case study, *Devel. Biol. Stand*. **96**: 131–140.
67. AuBuchon, J. P., Birkmeyer, J. D., and Busch, M. P. (1997), Safety of the blood supply in the United States: Opportunities and controversies, *Ann. Intern. Med*. **127**: 904–909.
68. Kolls, J., Peppel, K., Silva, M., and Beutler, B. (1994), Prolonged and effective blockade of tumor necrosis factor activity through adenovirus-mediated gene transfer, *Proc. Natl. Acad. Sci. USA* **91**: 215–219.
69. Bussel, J. B. et al. (2006), AMG 531, a thrombopoiesis-stimulating protein, for chronic ITP, *New Engl. J. Med*. **355**: 1672–1681.
70. Mascelli, M. A. et al. (2007), Molecular, biologic, and pharmacokinetic properties of monoclonal antibodies: Impact of these parameters on early clinical development, *J. Clin. Pharmacol*. **47**: 553–565.
71. Presta, L. G. (2005), Selection, design, and engineering of therapeutic antibodies, *J. Allergy Clin. Immunol*. **116**: 731–736; quiz 737.
72. Harris, L., Skaletsky, E., and McPhorsen, A., (1998), Crystallographic structure of an intact IgG1 monoclonal antibody. *J. Mol. Biol*. **275**: 8610.
73. Honegger, A., et al. (2005), A mutation designed to alter crystal packing permits structural analysis of a tight-binding fluorescein-scFv complex. *Protein Sci*. **14**: 2537.
74. Carmichael, J., et al. (2003), The crystal structure of an anti-CEA scFv diabody assembled from T84.66 scFvs in V(L)-to-V(H) orientation: implications for diabody flexibility. *J. Mol. Biol*. **326**: 341.
75. Ewert, S., Huber, T., Honegger, A., and Pluckthun, A. (2003), Biophysical properties of human antibody variable domains, *J. Mol. Biol*. **325**: 531–553.
76. Stanfield, R. L., Zemla, A., Wilson, I. A., and Rupp, B. (2006), Antibody elbow angles are influenced by their light chain class, *J. Mol. Biol*. **357**: 1566–1574.
77. Caruthers, M. H. (1985), Gene synthesis machines: DNA chemistry and its uses, *Science* **230**: 281–285.

78. Roux, K. H., Strelets, L., Brekke, O. H., Sandlie, I., and Michaelsen, T. E. (1998), Comparisons of the ability of human IgG3 hinge mutants, IgM, IgE, and IgA2, to form small immune complexes: A role for flexibility and geometry, *J. Immunol*. **161**: 4083–4090.
79. Brekke, O. H., Michaelsen, T. E., and Sandlie, I. (1995), The structural requirements for complement activation by IgG: does it hinge on the hinge? *Immunol. Today* **16**: 85–90.
80. Wang, W., Singh, S., Zeng, D. L., King, K., and Nema, S. (2007), Antibody structure, instability, and formulation, *J. Pharm. Sci*. **96**: 1–26.
81. Salfeld, J. G. (2007), Isotype selection in antibody engineering, *Nat. Biotechnol*. **25**: 1369–1372.
82. Aalberse, R. C. and Schuurman, J. (2002), IgG4 breaking the rules, *Immunology* **105**: 9–19.
83. Lu, Y. et al. (2008), The effect of a point mutation on the stability of IgG4 as monitored by analytical ultracentrifugation, *J. Pharm. Sci*. **97**: 960–969.
84. Weiner, L. M. and Carter, P. (2005), Tunable antibodies, *Nat. Biotechnol*. **23**: 556–557.
85. Anand, B. (2008), in *Pharmaceutical Biotechnology*, 3rd ed., Crommelin R. S. D. and Meibohm, B., eds., Informa Healthcare, New York, p. 466.
86. Cohen-Solal, J. F., Cassard, L., Fridman, W. H., and Sautes-Fridman, C. (2004), Fc gamma receptors, *Immunol. Lett*. **92**: 199–205.
87. Sondermann, P., Kaiser, J., and Jacob, U. (2001), Molecular basis for immune complex recognition: A comparison of Fc-receptor structures, *J. Mol. Biol*. **309**: 737–749.
88. Lazar, G. A. et al. (2006), Engineered antibody Fc variants with enhanced effector function, *Proc. Natl. Acad. Sci. USA* **103**: 4005–4010.
89. Shields, R. L. et al. (2001). High resolution mapping of the binding site on human IgG1 for Fc gamma RI, Fc gamma RII, Fc gamma RIII, and FcRn and design of IgG1 variants with improved binding to the Fc gamma R, *J. Biol. Chem*. **276**: 6591–6604.
90. Roskos, L., Davis, C. G., and Schwab, G. M. (2004), The clinical pharmacology of therapeutic monoclonal antibodies, *Drug Devel. Res*. **61**: 108–120.
91. Amgen press release (2009), *Amgen Submits Biologics License Application for FDA Approval of Denosumab in Women with Postmenopausal Osteoporosis and in Patients Undergoing Hormone Ablation for Either Prostate or Breast Cancer*.
92. McClung, M. R. et al. (2006), Denosumab in postmenopausal women with low bone mineral density, *New Engl. J. Med*. **354**: 821–831.
93. Cohen, S. B. et al. (2008), Denosumab treatment effects on structural damage, bone mineral density, and bone turnover in rheumatoid arthritis: A twelve-month, multicenter, randomized, double-blind, placebo-controlled, phase II clinical trial, *Arthritis Rheum*. **58**: 1299–1309.
94. Reichert, J. M. and Wenger, J. B. (2008), Development trends for new cancer therapeutics and vaccines, *Drug Discov. Today* **13**: 30–37.
95. Lonberg, N. (2005), Human antibodies from transgenic animals, *Nat. Biotechnol*. **23**: 1117–1125.
96. Presta, L. G. (2006), Engineering of therapeutic antibodies to minimize immunogenicity and optimize function, *Adv. Drug Deliv. Rev*. **58**: 640–656.
97. Moolten, F. L. and Cooperband, S. R. (1970), Selective destruction of target cells by diphtheria toxin conjugated to antibody directed against antigens on the cells, *Science* **169**: 68–70.

98. Hurwitz, E., Levy, R., and Maron, R. (1975), Proceedings: Daunomycin and adriamycin antitumor cell antibody conjugates, *Isr. J. Med. Sci*. **11**: 1396.

99. Schrama, D., Reisfeld, R. A., and Becker, J. C. (2006), Antibody targeted drugs as cancer therapeutics, *Nat. Rev*. **5**: 147–159.

100. Wu, A. M. and Senter, P. D. (2005), Arming antibodies: Prospects and challenges for immunoconjugates, *Nat. Biotechnol*. **23**: 1137–1146.

101. Hulett, M. D. and Hogarth, P. M. (1994), Molecular basis of Fc receptor function, *Adv. Immunol*. **57**: 1–127.

102. Hinman, L. M. et al. (1993), Preparation and characterization of monoclonal antibody conjugates of the calicheamicins: A novel and potent family of antitumor antibiotics, *Cancer Res*. **53**: 3336–3342.

103. Carter, P. J. and Senter, P. D. (2008), Antibody-drug conjugates for cancer therapy, *Cancer J*. **14**: 154–169.

104. Ferrara, N., Damico, L., Shams, N., Lowman, H., and Kim, R. (2006), Development of ranibizumab, an anti-vascular endothelial growth factor antigen binding fragment, as therapy for neovascular age-related macular degeneration, *Retina* **26**: 859–870.

105. Chen, Y. et al. (1999), Selection and analysis of an optimized anti-VEGF antibody: Crystal structure of an affinity-matured Fab in complex with antigen, *J. Mol. Biol*. **293**: 865–881.

106. Gaudreault, J., Fei, D., Rusit, J., Suboc, P., and Shiu, V. (2005), Preclinical pharmacokinetics of Ranibizumab (rhuFabV2) after a single intravitreal administration, *Invest Ophthalmol. Vis. Sci*. **46**: 726–733.

107. Chapman, A. P. et al. (1999), Therapeutic antibody fragments with prolonged *in vivo* half-lives, *Nat. Biotechnol*. **17**: 780–783.

108. Holliger, P. and Hudson, P. J. (2005), Engineered antibody fragments and the rise of single domains, *Nat. Biotechnol*. **23**: 1126–1136.

109. Muyldermans, S. (2001), Single domain camel antibodies: Current status, *J. Biotechnol*. **74**: 277–302.

110. Linsley, P. S. (2005), New look at an old costimulator, *Nat. Immunol*. **6**: 231–232.

111. Teeling, J. L. et al. (2004), Characterization of new human CD20 monoclonal antibodies with potent cytolytic activity against non-Hodgkin lymphomas, *Blood* **104**: 1793–1800.

112. Holliger, P. et al. (1999), Carcinoembryonic antigen (CEA)-specific T-cell activation in colon carcinoma induced by anti-CD3 x anti-CEA bispecific diabodies and B7 x anti-CEA bispecific fusion proteins, *Cancer Res*. **59**: 2909–2916.

113. Kipriyanov, S. M. et al. (1996), Affinity enhancement of a recombinant antibody: Formation of complexes with multiple valency by a single-chain Fv fragment-core streptavidin fusion, *Protein Eng*. **9**: 203–211.

114. Oh, P. et al. (2004), Subtractive proteomic mapping of the endothelial surface in lung and solid tumours for tissue-specific therapy, *Nature* **429**: 629–635.

115. Halin, C. et al. (2002), Enhancement of the antitumor activity of interleukin-12 by targeted delivery to neovasculature, *Nat. Biotechnol*. **20**: 264–269.

116. Yang, A. S. and Honig, B. (2000), An integrated approach to the analysis and modeling of protein sequences and structures. III. A comparative study of sequence conservation in protein structural families using multiple structural alignments, *J. Mol. Biol*. **301**: 691–711.

117. Olson, C. A. and Roberts, R. W. (2007), Design, expression, and stability of a diverse protein library based on the human fibronectin type III domain, *Protein Sci*. **16**: 476–484.

118. Koide, A., Bailey, C. W., Huang, X., and Koide, S. (1998), The fibronectin type III domain as a scaffold for novel binding proteins, *J. Mol. Biol.* **284**: 1141–1151.
119. Lander, E. S. et al. (2001), Initial sequencing and analysis of the human genome, *Nature* **409**: 860–921.
120. Forrer, P., Binz, H. K., Stumpp, M. T., and Pluckthun, A. (2004), Consensus design of repeat proteins, *Chembiochem* **5**: 183–189.
121. Kohl, A. et al. (2003), Designed to be stable: Crystal structure of a consensus ankyrin repeat protein, *Proc. Natl. Acad. Sci. USA* **100**: 1700–1705.
122. Zahnd, C. et al. (2007), A designed ankyrin repeat protein evolved to picomolar affinity to Her2, *J. Mol. Biol.* **369**: 1015–1028.
123. Weiss, G. A. and Lowman, H. B. (2000), Anticalins versus antibodies: made-to-order binding proteins for small molecules, *Chem. Biol.* **7**: R177–R184.
124. Skerra, A. (2000), Lipocalins as a scaffold, *Biochim. Biophys. Acta* **1482**: 337–350.
125. Beste, G., Schmidt, F. S., Stibora, T., and Skerra, A. (1999), Small antibody-like proteins with prescribed ligand specificities derived from the lipocalin fold, *Proc. Natl. Acad. Sci. USA* **96**: 1898–1903.
126. Ebersbach, H. et al. (2007), Affilin-novel binding molecules based on human gamma-B-crystallin, an all beta-sheet protein, *J. Mol. Biol.* **372**: 172–185.
127. Werle, M. et al. (2006), The potential of cystine-knot microproteins as novel pharmacophoric scaffolds in oral peptide drug delivery, *J. Drug Target* **14**: 137–146.
128. Werle, M., Kafedjiiski, K., Kolmar, H., and Bernkop-Schnurch, A. (2007), Evaluation and improvement of the properties of the novel cystine-knot microprotein McoEeTI for oral administration, *Int. J. Pharm.* **332**: 72–79.
129. Silverman, J. et al. (2005), Multivalent avimer proteins evolved by exon shuffling of a family of human receptor domains, *Nat. Biotechnol.* **23**: 1556–1561.
130. Thogersen, C. and Holldack, J. (2006), A tetranectin-based platform for protein engineering, *Innov. Pharm. Technol.* 27–30.
131. Watt, P. M. (2006), Screening for peptide drugs from the natural repertoire of biodiverse protein folds, *Nat. Biotechnol.* **24**: 177–183.
132. Shire, S. J., Shahrokh, Z., and Liu, J. (2004), Challenges in the development of high protein concentration formulations, *J. Pharm. Sci.* **93**: 1390–1402.
133. Lobo, E. D., Hansen, R. J., and Balthasar, J. P. (2004), Antibody pharmacokinetics and pharmacodynamics, *J. Pharm. Sci.* **93**: 2645–2668.

2

CHEMICAL INSTABILITY IN PEPTIDE AND PROTEIN PHARMACEUTICALS

Elizabeth M. Topp, Lei Zhang, Hong Zhao, Robert W. Payne, Gabriel J. Evans, and Mark Cornell Manning

2.1. INTRODUCTION

Peptides and proteins contain multiple functional groups, many of which can undergo some type of chemical modification or damage [1]. To stabilize peptide and protein pharmaceuticals, one must have a detailed understanding of these degradation pathways. This review summarizes the primary routes of chemical instability and provides some insights into the effects of these chemical modifications on physical instability. General approaches toward diminishing hydrolysis and oxidation are also presented. Those interested in antibody stabilization and formulation are directed to several excellent reviews on the effects of chemical instability on the heterogeneity of monoclonal antibodies (MAbs) [2,3].

2.2. HYDROLYTIC REACTIONS

Hydrolysis is a general term for a class of chemical reactions in which bonds are broken by the addition of water. This section summarizes hydrolytic reactions that occur in peptides and proteins, along with some associated reactions: deamidation, aspartate (Asp) isomerization, proteolysis, and β elimination.

Formulation and Process Development Strategies for Manufacturing Biopharmaceuticals, edited by Jameel and Hershenson

2.2.1. Deamidation

2.2.1.1. Mechanism of Deamidation. Deamidation is one of the most common hydrolytic reactions of peptides and proteins, and has been discussed in detail in several excellent texts devoted to the subject [1,4–9]. More recent reviews address formulation issues for proteins susceptible to deamidation [9], biological implications of the reaction [8], and mass spectrometric approaches to detection and quantitation of deamidation products [10].

Deamidation occurs primarily at asparagine (Asn) residues. Although glutamine (Gln) may also be affected, the reaction is generally far slower and therefore less commonly observed at Gln. When deamidation occurs at Asn sites, it modifies the Asn side chain to produce aspartate (Asp) and/or isoaspartate (isoAsp) with loss of ammonia (Fig. 2.1), effectively mutating the amino acid sequence through a posttranslational chemical reaction. Since the Asp side chain is acidic while Asn is neutral, deamidation also changes the protein's net charge. At neutral to basic pH, a structural isomer of Asp, isoAsp, is also produced and is generally the major product (Fig. 2.1). The formation of isoAsp not only changes the amino acid sequence but also alters the peptide backbone by introducing an extra methylene group (i.e., a β-amino acid). The Asp- and isoAsp-containing products differ from the native protein in mass by +1 atomic mass unit (amu), a much smaller mass change than is typical of many oxidative degradation reactions.

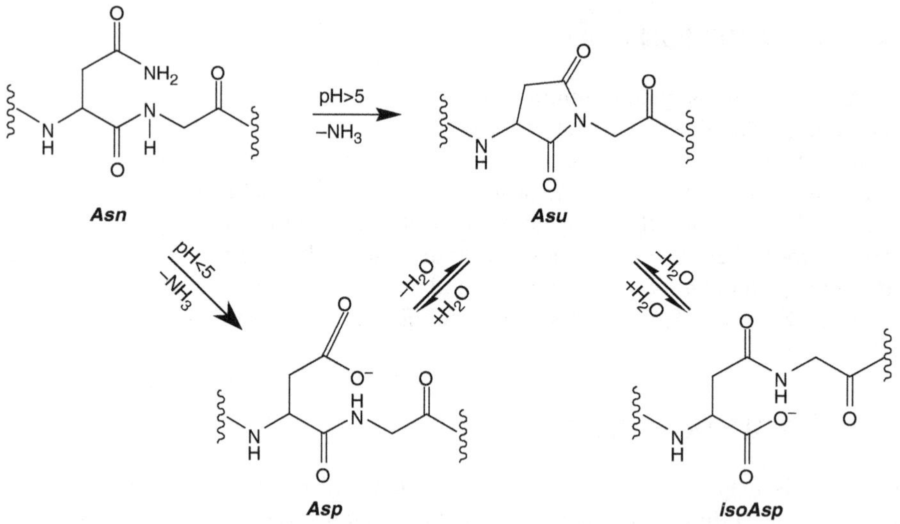

Figure 2.1. Asparagine (Asn) deamidation and aspartate (Asp) isomerization. At pH > 5, deamidation of Asn proceeds through a cyclic imide (succinimide, Asu) intermediate with loss of ammonia to produce the Asp- and isoasparate (isoAsp)-containing variants. At acidic pH, the Asn side chain undergoes direct hydrolysis, giving the Asp product exclusively. Isomerization of Asp to isoAsp also occurs through the Asu intermediate.

In acidic solution (pH<4), deamidation takes place by direct hydrolysis of the Asn side chain amide to form only Asp [1,4–6]. This process is subject to acid catalysis. In neutral to basic solution, the first step in deamidation is the nucleophilic attack of the $N+1$ nitrogen of the protein backbone on the carbonyl group of the Asn side chain (Fig. 2.1). This step is base-catalyzed, since abstraction or partial abstraction of the backbone amide proton makes the nitrogen more nucleophilic, accelerating the reaction. A cyclic imide (succinimide) intermediate is formed (Asu, Fig. 2.1) with loss of ammonia. Since ammonia is a gas and is typically not retained in solution, this step is effectively irreversible. While the Asu intermediate often can be detected, it is readily hydrolyzed in aqueous solution to form the Asp and isoAsp products (Fig. 2.1). The pH dependence of Asu formation and subsequent hydrolysis has been reported [11]. Hydrolysis of the Asu intermediate occurs by the attack of water on either of the carbonyl groups of the five-membered succinimide ring. The relative amounts of the Asp and isoAsp products formed depend on the ease with which hydrolysis occurs at these two carbonyl groups. In neutral solution, the isoAsp product is usually favored, with typical isoAsp:Asp molar ratios ranging from 3 : 1 to 5 : 1 observed in model peptides. While deamidation is classified as a hydrolytic reaction with isoAsp and Asp as the reaction products, the initial formation of Asu is not hydrolytic and is itself an irreversible modification of the protein. This somewhat semantic point may be particularly important for proteins in organic solvents or dry solids, in which low water content may preclude the hydrolytic step but not the formation of Asu. While the hydrolytic mechanism presented in Figure 2.1 is widely accepted, alternative mechanisms have also been proposed [12,13]. Quantum chemical modeling of the reaction has also been reported [14], along with a similar study on proteolysis relative to deamidation [15].

Rates of Asn deamidation are influenced by several factors, including protein primary sequence, solution pH and temperature. The reaction is strongly influenced by primary sequence [5], with the adjacent amino acid on the C-terminal side (i.e., the $N+1$ amino acid) having the greatest influence on the rate. In model peptides, faster rates have generally been observed with small, nonpolar amino acids at the $N+1$ site; the Asn–Gly sequence is particularly prone to deamidation and is considered a "hot spot" for the reaction. Bulky and/or hydrophobic amino acids at $N+1$ usually slow deamidation rates considerably. Higher-order structure also influences deamidation rates. For example, it has been shown that deamidation is more rapid within flexible, loop-type structures than within the core of the protein [16–19], but these effects have not been as studied as thoroughly and remain an active area of research. Even at the secondary structure level, deamidation rates are diminished within regular structures, such as α helices [20], β sheets [21], and β turns [22].

Deamidation is subject to both acid and base catalyses, with the slowest rates typically observed in the pH range of 4–5. Buffer catalysis of Asn deamidation has also been reported [4,5,23–25]. Phosphate, borate, and Tris have all been reported to increase deamidation rates, so using minimal buffer concentrations will slow deamidation as well. Deamidation rates increase with increasing temperature, and Arrhenius activation energies (E_a) in the range of 12–25 kcal/mol have been reported [26–30]. Since temperature changes may also influence deamidation rate by changing protein structure in the local vicinity of the labile Asn residue, reported E_a values should be regarded as composites of chemical and structural effects [5].

2.2.1.2. Effect of Deamidation on Physical Stability of Proteins. Deamidation may alter the local structure of the polypeptide chain [31]. Deamidation converts a neutral side chain (Asn) into an anionic side chain (pH > 4), which may alter the conformation surrounding the deamidation site. However, formation of the Asp product is not expected to perturb the local conformation as much as isoAsp, which introduces a methylene group into the peptide backbone. Deamidation can certainly promote aggregation, as was observed for a deamidated fragment of amylin [32] and β-B_1-crystallin [33]. In bovine growth hormone, there is a strong correlation between the level of aggregation and the extent of deamidation [34]. In γ-D-crystallin, deamidation at Gln residues lowers the conformational stability of the protein [35]. Similarly, deamidation of Gln residues within collagen reduces the stability of the triple helix as well [36]. The specific mechanisms by which protein structure is altered by deamidation are not fully understood, but the introduction of β-amino acids by isoAsp formation and the transformation of the neutral side chain (i.e., of Asn or Gln) to the anionic, acidic products (e.g., Asp, isoAsp) may be involved.

2.2.2. Asp Isomerization

In addition to Asn deamidation, isoAsp can also be produced in peptides and proteins by isomerization at Asp residues. The mechanisms of the two reactions are similar. Cyclization of endogenous Asp occurs through the attack of the $N+1$ backbone amide nitrogen on the carbonyl of the Asp side chain to produce the Asu intermediate with loss of water (Fig. 2.1). Hydrolysis of Asu produces either the Asp-containing starting material in an unproductive reverse reaction, or the isoAsp-containing variant (Fig. 2.1). In model peptides, rates of Asu formation from Asp are typically more than an order of magnitude slower than Asu formation from Asn [5]. As in Asn deamidation, Asp isomerization is sensitive to pH, temperature, and protein structure [5], although the literature on these influences on the isomerization reaction is less extensive than for deamidation. The reader is directed to the general reviews on Asn deamidation cited above for additional information on Asp isomerization.

The fact that both Asn deamidation and Asp isomerization proceed through the succinimide (Asu, Fig. 2.1) serves to highlight the reactivity of this intermediate. Succinimide formation has been implicated as a route to peptide bond hydrolysis, as described below, and as a route to racemization of L-Asn to D-Asn with subsequent formation of epimerized deamidation products (i.e., D-Asp, D-isoAsp) [38]. The attack of amine nucleophiles on the succinimide has also been suggested as a pathway for the formation of amide-linked covalent protein aggregates [38], an important type of degradant in protein drug products that is often poorly defined chemically.

2.2.3. Proteolysis

Hydrolysis of the amide bonds that connect amino acids to form the protein backbone is called *proteolysis*. A single proteolytic event splits the protein backbone in two, producing two shorter polypeptide chains (Fig. 2.2a); the total mass of these polypeptide products is equal to that of the initial protein chain plus the mass of water, as described in a number of review articles [39–41]. Proteolysis that occurs near the *N* or *C* terminus is often called "clipping." In vivo, the reaction is mediated by a variety of proteolytic enzymes, but the reaction can also occur in the absence of enzymes. In the simplest case, the direct attack of water on the peptide bond carbonyl leads to proteolysis (Fig. 2.2a). This attack is facilitated by protonation of the carbonyl, which serves to stabilize the tetrahedral intermediate that is formed, so that the reaction is acid-catalyzed. Intramolecular hydrogen bonding to the carbonyl group of the Asp side chain also accelerates the reaction, as evidenced by faster proteolytic rates when Ser or Tyr are in the $n+1$ position [42,43]. Peptide bonds located on the *C*-terminal side of Asp residues are particularly susceptible to proteolysis because of intramolecular catalysis by the carboxylate side chain [44]. At these sites, proteolysis occurs via the intramolecular attack of the Asp side chain on the *C*-terminal amide carbonyl to form a cyclic anhydride (Fig. 2.2b) with cleavage of the peptide bond. The subsequent attack of water at either of the anhydride carbonyl groups produces a *C*-terminal Asp residue (Fig. 2.2b).

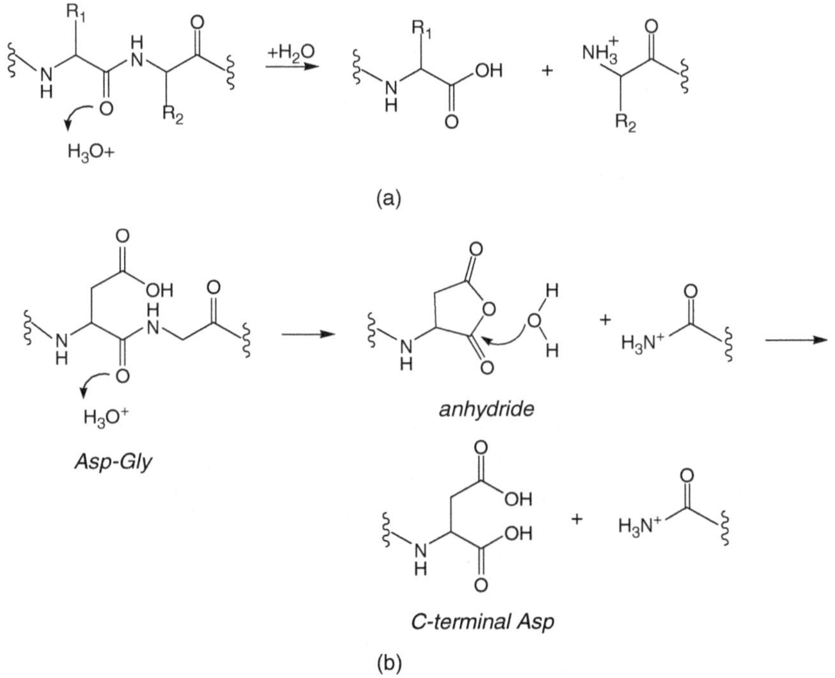

Figure 2.2. Proteolysis: (a) general mechanism of the acid-catalyzed reaction and (b) at Asp, catalyzed by the Asp side chain. In (a), attack of water on the protonated amide carbonyl cleaves the peptide bond, producing two shorter chains with new *C* and *N* termini. In (b), acid catalysis occurs intramolecularly at Asp, proceeding through the anhydride intermediate to produce amide bond cleavage.

2.2.3.1. Trp Hydrolysis. In addition to these better-known degradation processes, other functional groups are also sensitive to hydrolysis. For example, the side chain of Trp is known to undergo hydrolysis. The primary degradation product is called *kynurenine* [45–47], which fluoresces at a much longer wavelength (450 nm) than Trp itself.

2.3. *N*-TERMINAL CYCLIZATION REACTIONS

N-Terminal cyclization reactions are similar to the hydrolytic reactions just described in that both involve intramolecular nucleophilic attack. They differ in that, for the reactions described below, the nucleophile is located at the *N* terminus of the polypeptide chain. In general, two pathways have been described: diketopiperazine (DKP) formation and pyroglutamic acid (pGlu) formation.

2.3.1. Diketopiperazine (DKP) Formation

The *N*-terminal amino group can be a potent nucleophile, especially at $pH > 8$. If the amine attacks the second carbonyl group in the peptide backbone, a DKP ring is formed. Degradation of the *N* terminus by DKP formation has been commonly observed during long-term storage and peptide synthesis [48–50]. First observed in peptides [51], the DKP ring can rearrange with loss of the first two amino acids or reversal of their positions in the chain. Formation of DKP depends on the percentage of terminal amino groups in the free base form [38,52,53]. Under acidic conditions, the reaction is pH-independent. Kinetic analyses of DKP formation in peptides have examined the effects of pH, buffer type and concentration, and temperature [38,52–54]. The first-order rate constant generally increases with increasing buffer concentration, except for carbonate, which shows no concentration dependence [54]. Degradation caused by DKP formation was shown to be responsible for the *N*-terminal heterogeneity observed in hGH [48]. A detailed kinetic analysis of DKP in a model dipeptide has been presented [38].

To the extent that DKP formation leads to reduction in the length of the polypeptide chain, it can be considered a proteolytic reaction. Rearrangement of a DKP from the first two amino acids, via cleavage of the peptide bond *C*-terminal to the second amino acid, produces a clipped protein reduced in molecular weight by the mass of the two amino acids. In solution, DKP formation is common for proteins with the *N*-terminal sequence NH_2–Gly–Pro [41].

2.3.2. Pyroglutamic Acid (pGlu) Formation

Nonenzymatic formation of pyroGlu (pGlu) follows a mechanism similar to that for DKP formation [56] in that it involves nucleophilic attack of the *N*-terminal amine on the polypeptide chain of the *N*-terminal Glu with elimination of water (Fig. 2.3). This cyclized, *N*-terminal structure is often observed in monoclonal antibodies because of the frequency of Glu in the first position of the light chain and occasionally in the

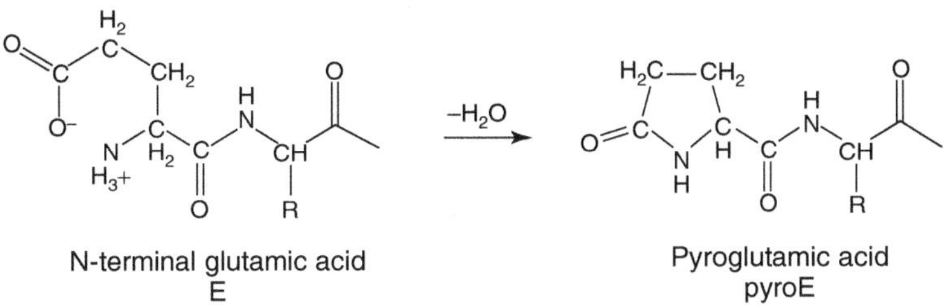

Figure 2.3. Conversion of *N*-terminal Glu (E) to pyroGlu (pyroE). Attack of the *N*-terminal amine on the Glu side chain produces the five-membered pyroglutamate ring with loss of water.

heavy chain [57–64]. The presence of pGlu is easily detected using mass spectrometric techniques, and a number of such methods have been reported [57–59,60,62–64].

As the reaction involves nucleophilic attack, rates of pGlu formation are typically pH-dependent. The pH dependence of the reaction has been reported [57,58], although the data are quite limited compared to the detailed pH profiles published for many hydrolytic reactions, and buffer effects on pGlu formation have been reported [57,65]. Phosphate appears to cause more rapid cyclization, at least in model peptides [65]. At lower pH, acetate appears to be the most favorable buffer species for slowing pGlu formation [57]. Finally, it has been reported that pGlu can be formed from *N*-terminal Gln residues as well as Glu, although the reaction appears to be slower with Gln [66,67].

2.4. RACEMIZATION AND β ELIMINATION

Racemization and β elimination are closely related. In both reactions, the initial step involves the abstraction of a proton from the α carbon of an amino acid, so that the reactions are favored under alkaline conditions. This proton can recombine with the resultant carbanion to reform the original amino acid, in either the D or L form (racemization) or can rearrange to eject a leaving group from the β carbon, forming a double bond between the α and β carbons [68]. Therefore, the reaction is favored if there is (1) a good leaving group (e.g., Ser, Cys) or (2) an aromatic ring attached to the β carbon (e.g., Phe, Tyr), as the double bond will be stabilized by conjugation with the ring system. Typically, these reactions require high pH and elevated temperature; otherwise, the published literature on conditions that accelerate these reactions is relatively limited. The thermal decomposition of cysteine residues via β elimination has been reported [69]. Similar reaction kinetics were observed for several proteins under the same conditions, suggesting that a common mechanism exists and that neither sequence nor local structure affect the reaction rate. At cystine groups, β elimination is initiated by the abstraction of a proton from the alpha carbon (C_α) of Cys by hydroxide or another strong base to produce a carbanion at C_α (Fig. 2.4). A vinyl bond

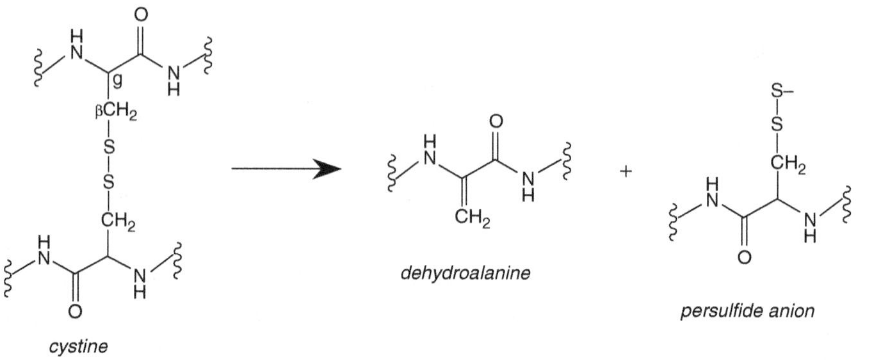

Figure 2.4. Beta-elimination at cystine. Cleavage of the C–S bond produces a vinyl bond between C_α and C_β, forming dehydroalanine with expulsion of a persulfide anion. Both products can undergo addition reactions that may contribute to intermolecular crosslinking and the formation of covalent protein aggregates.

is then formed between C_α and C_β to produce dehydroalanine at this site, with expulsion of the persulfide anion (i.e., $R{-}S{-}S^-$). Both dehydroalanine and the persulfide anion can undergo additional reactions, many of which have been implicated in the formation of intermolecular crosslinks and protein aggregates. For example, attack of the nucleophilic Lys side chain on dehydroalanine produces a covalent lysinoalanine crosslink, while persulfide anions have been implicated in thiol–disulfide scrambling and in the formation of reactive oxygen species (ROS) [70], which can lead to oxidative damage at various sites. β elimination at Ser occurs similarly, producing the reactive dehydroalanine with elimination of water. The reaction at Ser occurs more readily if the side chain has been modified by phosphorylation or glycosylation, since the modified side chains are better leaving groups than is the Ser hydroxyl. Since chirality is lost by the initial deprotonation at C_α, reprotonation of the carbanion may produce racemization even without subsequent β elimination [71].

There have been a few reports of β elimination in proteins of pharmaceutical relevance. For example, β elimination was observed for insulin in the solid state [72,73] and for IL-1ra in the solution [74]. Both show increased aggregation after chemical modification. Both β elimination and racemization have been observed in MAbs, with β elimination leading to proteolytic damage [75,76]. In these MAbs, the β-elimination pathway was observed at basic pH, consistent with abstraction of the C_α proton as the initial step.

Racemization of Asp residues in crystallins has been described in a number of studies [77–79]. It appears that the bulk of the preceding amino acid and the structure around the Asp residue affect racemization rates [77]. Other than that, little is known about the effect of intrinsic factors on racemization rates. Even the impact of racemization on subsequent physical stability is not well known. Formation of D-Asp^{23} reportedly increases the rate of fibril formation in Aβ peptides [80,81]. On the other hand, racemization of Asp has been reported not to be involved in amyloid plaque formation [82].

Finally, racemization can occur during deamidation at Asn residues, as noted above [83]. It was long believed that the racemization occurred via the cyclic imide intermediate as a result of increased acidity of the C_α–H bond in the five-membered ring [84]. However, a careful physical organic chemistry study has since shown that racemization occurs via a tetrahedral intermediate formed prior to cyclic imide ring formation [85].

2.5. OXIDATION REACTIONS

A number of amino acid side chains can be modified via oxidative processes (see Table 2.1). For the most part, the sensitive residues are those with aromatic or sulfur-containing side chains. Since oxidation proceeds through several different mechanisms, it important to distinguish the various oxidation processes and not just identify a list of oxidation-sensitive residues. As oxidation reactions can produce a wide range of degradation products, analytical methods become critically important in establishing which decomposition pathway is at work [86].

Oxidation of pharmaceutically relevant species has been reviewed [87–93]. However, a comprehensive summary of this field has not appeared for some time. There are three general types of oxidative degradation pathways for peptides and proteins: metal-catalyzed oxidation (MCO), photooxidation, and oxidation caused by a free-radical cascade. In MCO, oxidation occurs locally near the metal-binding site, usually modifying His or Cys, which are good ligands for metals. Photooxidation typically involves reactive oxygen species (ROS) that are less potent than peroxide and hydroxyl radical, such as singlet oxygen and superoxide. With a free-radical cascade, oxidation can be initiated by a variety of sources, including oxygen in the air, organic free-radical impurities in containers, trace levels of redox-active metals, or deliberate introduction of a ROS.

2.5.1. Metal-Catalyzed Oxidation (MCO)

Many proteins bind metals. In some cases, metal binding provides a degree of conformational stabilization, as in DNase [94] or factor VIII [95]. Some of these binding metals are not redox-active, as they cannot easily access more than one oxidation state. These include metal ions such as Zn^{2+}, Ca^{2+}, and Mg^{2+}. However, a number

TABLE 2.1. Common Products of Oxidation-Sensitive Amino Acid Residues

Amino Acid	Oxidation Product(s)
Methionine (Met)	Met-sulfoxide
Histidine (His)	2-Oxo-His
Tryptophan (Trp)	*N*-Formylkynurenine, kyurenine, hydroxy-Trp
Cysteine (Cys)	Cystine, sulfenic acid
Tyrosine (Tyr)	Dityrosine

of transition metals catalyze oxidative damage [88,90,96], including Fe, Cu, Mn, and Cr. These metals are often found in pharmaceutical preparations as impurities, either in the active pharmaceutical ingredient (API) itself or in the excipients (especially in sugars and polymers). Oxidatively active metals may also be leached from the containers used to store and process protein products, such as stainless-steel manufacturing equipment [93,97].

Once a redox-active metal binds to the protein, ROS can be generated. Although metal binding may be weak, with dissociation constants in the mM range, the interaction can generate ROS and lead to localized oxidative damage [98]. For example, human growth hormone (hGH) binds a metal in the center of the four-helix bundle that comprises the core of its globular structure. Zinc binding leads to some degree of conformational stabilization [99,100], although additional amounts of zinc can cause precipitation [99]. Redox-active metals such as iron can bind in the same binding site [101,102]; in hGH, two of the metal binding sites involve His residues [102–103]. On generation of a ROS at the metal-binding site, these highly reactive species do not have to diffuse very far before encountering a substrate, in this case the protein itself. As a result, MCO tends to cause oxidative damage very close to the metal-binding site. In hGH, the moieties that incur the most damage are the two nearby His residues, each forming 2-oxo-His (Fig. 2.1) [103]. Similar behavior has been observed in the MCO of prolastin [104], also resulting in oxidation of His residues. Similarly, MCO of Met residues has been reported [105], but this requires the Met side chain to be near the metal-binding center.

2.5.2. Free-Radical Oxidation

The initiation, propagation, and termination steps associated with a free-radical cascade have been reviewed widely, even for peptides and proteins [91,92,106]. Once free radicals are generated by initiation reactions, they can propagate easily until they result in some type of termination event, usually involving chemical modification of the peptide or protein. Some oxidation reactions require only the mildest of ROS, such as molecular oxygen or singlet oxygen. Such is the case for conversion of the thioether linkage in Met to the corresponding sulfoxide [90]. As a result, controlling oxygen content in the drug product is an effective strategy for minimizing Met oxidation. This approach is often taken with lyophilized products, where the vials can be sealed under vacuum or with a nitrogen blanket [107]. Maintaining low levels of oxygen and other ROS in liquid formulations is more challenging. Liquid formulations are likely to come into contact with the vial stopper, and stoppers are a common source of metals and organic free radicals [93]. In addition, since oxygen solubility increases with decreasing temperature [108], Met oxidation may occur more rapidly at 2°C–8°C than at room temperature despite the slowing due to Arrhenius behavior. Oxygen solubility in ice is extremely high [109], which may make development of frozen solutions for early-phase clinical trials even more problematic.

For pharmaceutical products, the site of oxidation is often Met residues [1], as they can easily be converted into the corresponding sulfoxide. The reaction occurs even in the presence of molecular oxygen, a relatively weak form of ROS. Not all

Met residues oxidize at the same rate; however, an observation similar to the finding that not all Asn residues deamidate at the same rate. Met oxidation rates appear to be correlated with the solvent accessible surface area [110–113], which may explain the differential reactivity of Met residues in hGH [114]. On the other hand, some studies indicate that Met oxidation is correlated with conformational stability of the protein [114,115]. For example, the reactivity of Met_{11} in GCSF correlates with C_m [117]. Interestingly, the reactivity of Met_{11} does not correlate with ΔG_u. While the molecular basis for this correlation is not readily apparent, it is another aspect of oxidation that deserves further study.

Oxidation of Trp side chains can produce a variety of products, including hydroxy-Trp, *N*-formylkynurenine (NFK), kynurenine, and 3-hydroxykynurenine. The distribution of products depends on the nature of the oxidizing species. While HOCl produces hydroxy-Trp in surfactant protein B, the use of Fenton's reagent (a reactive iron complex) leads to kynurenine-type products [116].

The most common oxidation reaction at Cys is the conversion of the free thiol to a disulfide bond (i.e., formation of cystine) [117] (see Fig. 2.5). When the reaction occurs intramolecularly, it can produce covalent dimers and higher-order aggregates [118,119], a mechanism of aggregation typically diagnosed by disappearance of the aggregates on addition of a reducing agent such as DTT or β-mercaptoethanol to regenerate the reduced form (i.e., free Cys residues) [117,119–121]. Controlling pH and redox potential can limit the formation of incorrect disulfide bonds. Since the thiolate anion is more reactive than Cys, the reaction is enhanced on deprotonation ($pK_a \sim 8.5$) [117] and can be effectively quenched at low pH. A second formulation strategy is to remove redox-active species, such as metals or redox-active organic compounds. Finally, since compounds with free Cys residues are more likely to undergo this type of reaction, they are frequently replaced by site-directed mutagenesis to generate more stable forms. A study on Cys oxidation illustrated the importance of sulfenic acid formation in oxidation of this residue [122], formed by the addition of an oxygen atom to the Cys sulfur atom.

2.5.3. Photostability and Photooxidation

Despite the requirement for photostability testing according to ICH guidelines [123], the literature on photostability of proteins is rather sparse, particularly if the work on heme proteins is excluded. Fortunately, an excellent review article has appeared on photostability of protein pharmaceuticals [124] in which the mechanisms for photolytic oxidation of various functional groups are described in detail. There are other, older reviews available as well [125,126].

The photooxidation of recombinant vascular epithelial growth factor (VEGF) has been reported [127]. Exposure to light results in oxidation of all six Met residues, with those residues having greater solvent exposure oxidizing to a greater extent. From a stability perspective, a solvent-exposed surface area of greater than 65 AA^2 is associated with rapid oxidation. If this is generally true, then this result could be used to identify particularly susceptible Met residues. The effect of oxidation on the biological activity was examined as well.

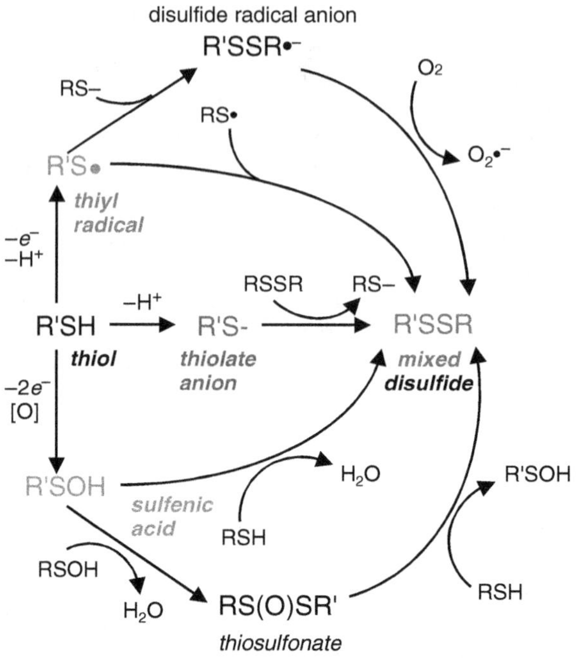

Figure 2.5. Some pathways for thiol–disulfide exchange. In solution, the thiolate anion (RS^-) is the reactive species, though alternative pathways involving the thiyl radical (R'S) or sulfenic acid (RSOH) are also possible. Intermolecular thiol–disulfide exchange has been implicated in the formation of covalent protein aggregates. (Adapted from Carballal et al. [188].)

Some photostability data have been reported for antibodies. The photostability of a Trp residue in a monoclonal antibody (MAb) was examined [128]. Replacement of the Trp residue with Phe improved the photostability dramatically. Photooxidation of two different MAbs was investigated, monitoring the damage at Trp105 and four different Met residues [128]. In both antibodies, Met^{255} degraded most quickly. A very detailed study on the oxidation of the monoclonal antibody, HER2 (herceptin) has also been published [129], which includes a particularly informative section on photostability. A singlet oxygen pathway was reported, with polysorbate 20 acting as a photosensitizer. Singlet oxygen is capable of converting Met to Met-sulfoxide [130]. Protection from light and the addition of free Met were both effective at minimizing photooxidation of Met residues in HER2.

Photostability of milk proteins has been reported. There is a concern over protein modification during prolonged storage under fluorescent lighting. Six different proteins were examined and the type and level of damage summarized [131]. They include dityrosine formation and oxidation of Trp side chains. One of the Trp degradation products, NFK was reported to be a photosensitizer [131], facilitating further oxidative damage.

Finally, it should be noted that photooxidation can proceed in the solid state as well as in aqueous solution. Photooxidation of bovine growth hormone (bGH) in the solid state was studied [132]. This protein undergoes photoionization of Trp and one-electron reduction of disulfides. This is followed by backtransfer of the electron, leading to the generation of two thiyl radicals [133]. These add molecular oxygen to produce a number of sulfur-based oxidation products. Not surprisingly, the extent of damage is proportional to the oxygen content in the system. These reactions, unlike the HER2 example described above, do not involve singlet oxygen formation. Similar degradation at Cys and Trp has been observed on irradiation of BSA [134].

2.5.4. Formulation Strategies for Controlling Oxidation

Unlike hydrolytic reactions, pH provides little control over oxidative pathways. In fact, the oxidation of Met has been shown to be almost invariant with changing pH, at least over the range usually used in commercial formulations [135]. Any pH sensitivity of Met oxidation is likely due to conformational changes in the protein, as reported for factor VIIa [136].

Possibly the most effective formulation approach is to use sacrificial additives or free-radical scavengers. Of these, the addition of the free amino acid, Met, has proven to be among the most effective [137,138]. It has been used to stabilize the MAb, HER2 (herceptin) [129], as has thiosulfate. In those studies, a fivefold excess of Met was effective at preventing oxidation-induced damage of HER2. By comparison, thiosulfate required a 25-fold excess to reduce the level of oxidative damage. Met has also been used to stabilize GCSF and PTH(13–34) [139] and CNTF and NGF [140]. The derivative *N*-Ac-Met also appears to be an effective sacrificial additive [141]. *N*-Acetyl tryptophan (*N*-Ac-Trp) may function in the same manner [142], and is often used to stabilize HSA.

While antioxidants are effective at slowing damage from free radical cascades [90–92], as occurs with radiation damage, they may actually accelerate damage via MCO [88,98,143]. Presumably, this is due to complexation of the metal by the antioxidant, in which case the resulting complex is more redox-active than the initial species. This serves to emphasize the point that some understanding of the mechanism of degradation is required to devise an effective formulation strategy for mitigating chemical and physical damage.

While formulation strategies typically focus on direct intervention to retard chemical degradation, there is some ability to modulate chemical reactivity by altering conformation. For example, excluded solutes, which are added to provide increased conformational stability, can be used to hinder oxidation. Excluded solutes, which include sugars, polyols, and most amino acids, tend to force the native state ensemble to its most compact forms. In doing so, residues that are partially solvent exposed become less so, thereby reducing the solvent exposed surface area and limiting access to these groups and slowing oxidation. This is especially true if the reactive Met is located in a highly mobile and flexible section of the protein.

While not always effective, there are some striking examples of using this approach to reduce the extent of Met oxidation. In subtilisin, there is a Met near the reactive site. Addition of increased amounts of sucrose compacts the structure, reducing the ability of ROS to reach the sensitive Met [144]. With the industrial enzyme alkaline protease, addition of polyols was observed to slow oxidation dramatically [145]. However, in some cases, addition of sucrose has been reported to increase oxidation rates, as observed for factor VII [146]. Unfortunately, no mechanistic explanation was provided for this effect. In other cases, no appreciable effect of excluded solutes was observed, as with GCSF [147]. Moreover, some polyols, such as mannitol, are purported to be radical scavengers. The ability of sugars and polyols to reduce oxidation, presumably via this mechanism, has been shown for relaxin [148].

The packaging or container system used can influence oxidative stability as well. For example, there have been reports of oxidation that appear to be due to leaching of metals from stainless steel, as in the case of the MAb, HER2 [129]. This protein oxidized more quickly in the presence of NaCl, suggesting that the salt may have played a role in leaching metals from stainless-steel containers, especially at low pH. In general, stoppers and other containers are known to be sources of leachable metals that can cause subsequent oxidation in protein pharmaceuticals [93]. There have been reports of tungsten oxide leaching into products packaged in prefilled syringes [149–151]. The tungsten is introduced during manufacturing of glass syringes when a hot filament is used to extrude the syringe port. Even plastic can be problematic. For example, plastic tubes have been implicated in the oxidation of IL-11 [152].

2.5.5. Effect of Oxidation on Physical Stability

The link between chemical degradation and aggregation is still poorly understood, although a growing literature is providing insight into this important relationship. Certainly, oxidation of the side chain of Met results in a significant decrease in hydrophobicity [153]. Oxidation can also decrease conformational stability [154–156]. Not surprisingly, Met oxidation can also have a significant impact on aggregation rates, possibly as a result of altered conformational stability. On one hand, it has been shown that Met oxidation in α-synuclein inhibits fibril formation significantly [153]. Only by decreasing the pH can oxidized α-synuclein be forced to form fibrils. Interestingly, another type of oxidation, formation of dityrosine, produces a dimer that is the prenucleus for fibril formation in this protein, causing increased aggregation rates [157]. Several groups have reported that oxidation of Met^{35} in Aβ(1–40) slows aggregation and increases solubility [158,159]. Oxidation of Met residues in a coiled-coil peptide abolished fibril formation altogether [160].

More often, however, oxidation increases the propensity to aggregate. For example, MCO of relaxin increases its propensity to aggregate [161,162]. Similarly, oxidation of calcitonin leads to greater fibril formation [163], while oxidation of Met residues in the F_c region of antibodies leads to greater aggregation during storage [155]. The immunoglobulin light chain, LEN, does not form amyloid deposits in vivo, but can be induced to form fibrils under mildly destabilizing conditions. LEN has been reported to be resistant to Met oxidation, but once oxidized, tends to form amorphous aggregates rather than fibrils [164].

Thus, it appears that Met oxidation may either reduce or increase aggregation propensity. In fact, in the case of HER2, photooxidation had no effect on aggregation [129]. The impact on aggregation may be due to modification of the surface properties, altering protein–protein interactions, or by altering the global conformation of the protein. When both the native and oxidized forms undergo oxidation, the mechanisms of aggregation may differ.

2.6. CONDENSATION REACTIONS

In a condensation reaction, two molecules combine to produce a covalently linked product with the elimination of a small molecule, typically water or a low-molecular-weight alcohol. Glycation is one of the most common condensation reactions in peptides and proteins. *Glycation* refers to the reaction between a reducing sugar and an amine, as found in the side chain of Lys residues, with elimination of water. Also called the *Maillard reaction*, the process begins with formation of a Schiff base, which can undergo subsequent rearrangement to form a number of products [165,166]. While the focus of this section will be on the Maillard reaction in aqueous solution, there are many examples of the reaction occurring in the solid state, especially in food systems [167–172], where the reaction is often referred to as *nonenzymatic browning*. In addition, Maillard degradation products are among those found during storage of pharmaceuticals in the solid state [7]. In those cases, the Maillard reaction may [116] or may not [173] be affected by the glass transition temperature T_g of the amorphous phase.

The Maillard reaction requires the presence of a reducing sugar, which as a result are usually contraindicated in formulation development for biomolecules [174], although a number of products include such sugars as stabilizers. Although they can and do function as structural stabilizers, the risk of chemical damage via glycation is significant. In almost all cases, nonreducing sugars or polyols can be selected as effective alternatives.

Glycation with glucose or fructose has been widely reported [165,175]. For example, the glycation of interferon (IFN), bovine serum albumin (BSA), and human serum albumin (HSA) in the presence of reducing sugars has been studied [176,177], as well as the glycation of bioactive peptides (enkephalins, GLP, etc.) by glucose and fructose [178,179]. Maillard products can undergo further rearrangement to form so-called advanced glycation end (AGE) products [175]. One of these AGE products is N^{ε}-(carboxymethyl)lysine [177]. Formation of AGE products in HSA has been shown to increase its propensity for thermally induced aggregation [180].

Glycation can occur in formulations that do not contain a reducing sugar. Hydrolysis of sucrose can result in formation of fructose and glucose, both known reducing sugars. This occurs during heat treatment of proteins intended to inactivate viruses [181,182]. In addition, glycation has been reported for sucrose formulations in acidic solutions [183], and glycation of an IgG2 was observed during storage in a sucrose formulation [184]. Maillard products were seen when samples were stored at 37°C, but not when stored at 2°C–8°C. The analytical method for detecting glycation in this IgG2 system was also described in a separate study [185]. Similarly, degradation of maltodextrins has been implicated in generation of reducing sugars, leading to the glycation of an industrial enzyme at elevated temperature [187].

2.7. SUMMARY

Common routes of chemical degradation of peptides and proteins include (1) hydrolytic reactions such as deamidation and proteolysis; (2) *N*-terminal cyclization reactions such as DKP and pGlu formation; (3) oxidation through metal-catalyzed, photoinduced, and free-radical cascade pathways; (4) and condensation reactions, particularly with sugars. Although a degree of mechanistic definition of these reactions has been achieved, the effects of higher-order structure are less well understood, and the selection of conditions and excipients to minimize these reactions in peptide and protein drug products is often largely empirical. The increasing use of high-resolution analytical methods and molecular simulation approaches is expected to improve both mechanistic understanding of the reactions at the whole-protein level and the selection of stabilizing excipients.

REFERENCES

1. Manning, M. C., Patel, K., and Borchardt, R. T. (1989), Stability of protein pharmaceuticals, *Pharm. Res.*, **6**: 903–918.
2. Wang, W., Singh, S., Zeng, D. L., King, K., and Nema, S. (2007), Antibody structure, instability, and formulation, *J. Pharm. Sci.*, **96**: 1–26.
3. Liu, H., Gaza-Bulseco, G., Faldu, D., Chumsae, C., and Sun, J. (2008), Heterogeneity of monoclonal antibodies, *J. Pharm. Sci.* **97**: 2426–2447.
4. Aswad, D. W., ed. (1995), *Deamidation and Isoaspartate Formation in Peptides and Proteins*, CRC Press, Boca Raton, FL.
5. Robinson, N. E. and Robinson, A. B. (2004), *Molecular Clocks: Deamidation of Asparaginyl and Glutaminyl Residues in Peptides and Proteins*, Althouse Press, Cave Junction, OR.
6. Wright, H. T. (1991), Nonenzymatic deamidation of asparaginyl and glutaminyl residues in proteins, *Crit. Rev. Biochem. Mol. Biol.* **26**: 1–52.
7. Lai, M. C. and Topp, E. M. (1999). Solid-state chemical stability of proteins and peptides, *J. Pharm. Sci.* **88**: 489–500.
8. Reissner, K. J. and Aswad, D. W. (2003), Deamidation and isoaspartate formation in proteins: Unwanted alterations or surreptitious signals? *Cell. Mol. Life Sci.* **60**: 1281–1295.
9. Wakanar, A. A. and Borchardt, R. T. (2006), Formulation considerations for proteins susceptible to asparagines deamidation and aspartate isomerization, *J. Pharm. Sci.* **95**: 2321–2336.
10. Barnes, C. A. S. and Lim, A. (2007), Applications of mass spectrometry for the structural characterization of recombinant protein pharmaceuticals, *Mass Spectrom. Rev.* **26**: 370–388.

11. Chu, G. L., Chelius, D., Xiao, G., Khor, H. K., Coulibaly, S., and Bondarenko, P. V. (2007), Accumulation of succinimide in a recombinant monoclonal antibody in mildly acidic buffers under elevated temperatures, *Pharm. Res.* **24**: 1145–1156.
12. Clarke, S., Stephenson, R. C., and Lowenson, J. D. (1992), Lability of asparagine and aspartic acid residues in proteins and peptides, in *Stability of Protein Pharmaceuticals, Part A: Chemical and Physical Pathways of Protein Degradation*, Ahern, T. J. and Manning, M. C., eds., Plenum Press, New York, pp. 1–29.
13. Catak, S., Monard, G., Aviyente, V., and Ruiz-Lopez, M. F. (2006), Reaction mechanism of deamidation of asparaginyl residues in peptides: Effect of solvent molecules, *J. Phys. Chem. A*, **110**: 8354–8365.
14. Heaton, A. L. and Armentrout, P. B. (2008), Thermodynamics and mechanism of the deamidation of sodium-bound asparagine, *J. Am. Chem. Soc.* **130**: 10227–10232.
15. Catak, S., Monard, G., Aviyente, V., and Ruiz-Lopez, M. F. (2008), Computational study on nonenzymatic peptide bond cleavage at asparagine and aspartic acid, *J. Phys. Chem. A* **112**: 8752–8761.
16. Kossiakoff, A. A. (1988), Tertiary structure is a principal determinant to protein deamidation. *Science* **240**: 191–194.
17. Xie, M., Shahrokh, Z., Kadkhhodayan, M., Henzel, W. J., Powell, M. F., Borchardt, R. T., and Schowen, R. L. (2003), Asparagine deamidation in recombinant human lymphotoxin: Hindrance by three-dimensional structures, *J. Pharm. Sci.* **92**: 869–880.
18. Capasso, S. and Salvadori, S. (1999), Effect of the three-dimensional structure on the deamidation reaction of ribonuclease A, *J. Pept. Res.* **54**: 377–382.
19. Capasso, S. and Di Cerbo, P. (2000), Kinetic and thermodynamic control of the relative yield of the deamidation of asparagine and isomerization of aspartic acid residues, *J. Pept. Res.* **56**: 382–387.
20. Stevenson, C. L., Donlan, M. E., Friedman, A. R., and Borchardt, R. T. (1993), Solution conformation of LEU27 HGRF (1–32) NH2 and its deamidation products by 2D NMR. *Int. J. Pept. Protein Res.* **42**: 24–32.
21. Rivers, J., McDonald, L., Edwards, I. J., and Beynon, R. J. (2008), Asparagine deamidation and the role of higher order protein structure, *J. Proteome Res.* **7**: 921–927.
22. Xie, M., Aube, J., Borchardt, R. T., Morton, M., Topp, E. M., Vander Velde, D., and Schowen, R. L. (2000), Reactivity toward deamidation of asparagine residues in beta-turn structures. *J. Pept. Res.* **56**: 165–171.
23. Patel, K. and Borchardt, R. T. (1990), Deamidation of asparaginyl residues in proteins: A potential pathway for chemical degradation of proteins in lyophilized dosage forms, *J. Parenter. Sci. Technol.* **44**: 300–310.
24. Tyler-Cross, R. and Schirch, V. (1991), Effects of amino acid sequence, buffers, and ionic strength on the rate and mechanism of deamidation of asparagine residues in small peptides, *J. Biol. Chem.* **226**: 22549–22556.
25. Zheng, J. Y. and Janis, L. J. (2006), Influence of pH, buffer species, and storage temperature on physiochemical stability of a humanized monoclonal antibody LA298, *Int. J. Pharm.* **308**: 46–51.
26. Brange, J. (1992), Chemical stability of insulin. 4. Mechanisms and kinetics of chemical transformations in pharmaceutical formulation, *Acta Pharm. Nordica* **4**: 209–222.
27. Fisher, B. V. and Porter, P. B. (1981), Stability of bovine insulin, *J. Pharm. Pharmacol.* **33**: 203–206.

28. Klotz, A. V. and Thomas, B. A. (1993), N5-Methylasparagine and asparagine as nucleophiles in peptides: Main-chain vs. side-chain amide cleavage, *J. Org. Chem.* **58**: 6985–6989.

29. Stratton, L. P., Kelly, R. M., Rowe, J., Shively, J. E., Smith, D. D., Carpenter, J. F., and Manning, M. C. (2001), Controlling deamidation rates in a model peptide: effects of temperature, peptide concentration, and additives, *J. Pharm. Sci.* **90**: 2141–2148.

30. Zhang, J., Lee, T. C., and Ho, C. T. (1993), Kinetics and mechanism of nonenzymic deamidation of soy protein, *J. Food Process. Preserv.*, **17**: 259–268.

31. Wearne, S. J. and Creighton, T. E. (1989), Effect of protein conformation on rate of deamidation- ribonuclease A, *Proteins: Struct. Funct. Genet.* **5**: 8–12.

32. Nilsson, M. R., Driscoll, M., and Raleigh, D. P. (2002), Low levels of asparagine deamidation can have a dramatic effect on aggregation of amyloidogenic peptides: implications for the study amyloid formation, *Protein Sci.* **11**: 342–349.

33. Harms, M. J., Wilmorth, P. A., Kapfer, D. M., Steel, E. A., David, L. L., Bachinger, H. P., and Lampi, K. J. (2004), Laser light-scattering evidence for an altered association of beta B1-crystallin deamidated in the connecting peptide, *Protein Sci.* **13**: 678–686.

34. Harn, N. R., Jeng, Y. N., Kostelc, J. G., and Middaugh, C. R. (2005), Spectroscopic analysis of highly concentrated suspensions of bovine somatotropin in sesame oil, *J. Pharm. Sci.* **94**: 2487–2495.

35. Flaugh, S. L., Mills, I. A., and King, J. (2006), Glutamine deamidation destabilizes human V D-crystallin and lowers the kinetic barrier to unfolding, *J. Biol. Chem.* **281**: 30782–30793.

36. Silva, T., Kirkpatrick, A., Brodsky, B., and Ramshaw, J. A. M. (2005), Effect of deamidation on stability for the collegen to gelatin transition, *J. Agric. Food Chem.* **53**: 7802–7806.

37. Li, B., Borchardt, R. T., Topp, E. M., Vander Velde, D., and Schowen, R. L. (2003), Racemization of an asparagine residue during peptide deamidation, *J. Am. Chem. Soc.* **125**: 11486–11487.

38. DeHart, M. P. and Anderson, B. D. (2007), The role of the cyclic imide in alternate degradation pathways for asparagine-containing peptides and proteins, *J. Pharm. Sci.* **96**: 2667–2685.

39. Inglis, A. S. (1983), Cleavage at aspartic acid, *Meth. Enzymol.* **91**: 324–332.

40. Goolcharran, C., Khossravi, M., and Borchardt, R. T. (2000), Chemical pathways of peptide and protein degradation, in *Pharmaceutical Formulation Development of Peptides and Proteins*, Frokjaer, S. and Hovgaard, L., eds., CRC Press, New York.

41. Frokjaer, S., and Otzen, D. E. (2005), Protein drug stability: A formulation challenge, *Nat. Rev. Drug Discov.* **4**: 298–306.

42. Joshi, A. B. and Kirsch, L. E. (2002), The relative rates of glutamine and asparagine deamidation in glucagon fragment 22–29 under acidic conditions, *J. Pharm. Sci.* **91**: 2331–2345.

43. Joshi, A. B. and Kirsch, L. E. (2004), The estimation of glutaminyl deamidation and aspartyl cleavage rates in glucagon, *Int. J. Pharm.* **273**: 213–219.

44. Schöneich, C., Hageman, M. J., and Borchardt, R. T. (1997), Stability of peptide and proteins, in *Controlled Drug Delivery. Challenges and Strategies*, Park, K., ed., ACS Books, Washington, DC.

45. Ledvina, M. and Labella, F. S. (1970), Fluorescent substances in protein hydrolyzates 1. Acid "Hydrolyzates" of individual amino acids, *Anal. Biochem*. **36**: 174–181.

46. Xing, D. K. L., Crane, D. T., Bolgiano, B., Corbel, M. J., Jones, C., and Sesardic, D. (1996), Physicochemical and immunological studies on the stability of free and microsphere-encapsulated tetanus toxoid in vitro, *Vaccine* **14**: 1205–1213.

47. Luykx, D. M. A. M., Casteleijn, M. G., Jiskoot, W., Westdijk, J., and Jongen, D. M. J. M. (2004), Physicochemical studies on the stability of influenza haemagglutin in vaccine bulk material, *Eur. J. Pharm. Sci*. **23**: 65–75.

48. Battersby, J. E., Hancock, W. S., Canovadavis, E., Oeswein, J., and O'Connor, B. (1994), Diketopiperazine formation and N-terminal degradation in recombinant human growth-hormone, *Int. J. Pept. Protein Res*. **44**: 215–222.

49. Fisher, P. (2003), Diketopiperazines in peptide and combinatorial chemistry, *J. Pept. Sci*. **9**: 9–35.

50. Marsden, B. J., Nguyen, T. M. D., and Schiller, P. W. (1993), Spontaneous degradation via diketopiperazine formation of peptides containing a tetrahydroisoquinoline-3-carboxylic acid residue in the 2-position of the peptide sequence, *Int. J. Pept. Protein Res*. **41**: 313–316.

51. Sepetov, N. F., Krymsky, M. A., Ovchinnikov, M. V., Bespalova, Z. D., Isakova, O. L., Soucek, M., and Lebl, M. (1991), Rearrangement, racemization and decomposition of peptides in aqueous solution, *Pept. Res*. **4**: 308–313.

52. Capasso, S., Vergara, A., and Mazzarella, L. (1998), Mechanism of 2,5-dioxopiperazine formation, *J. Am. Chem. Soc*. **120**: 1990–1995.

53. Capasso, S., Sica, F., Mazzarella, L., Balboni, G., Guerrini, R., and Salvadori, S. (1995), Acid catalysis in the formation of dioxopiperazines from peptides containing tetrahydroisoquinoline-3-carboxylic acid at position-2, *Int. J. Pept. Protein Res*. **45**: 567–573.

54. Capasso, S. and Mazzarella, L. (1999), Solvent effects on diketopiperazine formation from N-terminal peptide residues, *J. Chem. Soc. Perkin Trans. 2* **2**: 329–332.

55. Kertscher, U., Bienert, M., Krause, E., Sepetov, N. F., and Mehlis, B. (1993), Spontaneous chemical degradation of substance P in the solid-phase and in solution, *Int. J. Pept. Protein Res*. **41**: 207–211.

56. Blomback, B. (1967), Derivatives of glutamine in peptides, *Meth. Enzymol*. **11**: 398–411.

57. Yu, L., Vizel, A., Huff, M. B., Young, M., Remmele Jr., R. L., and He, B. (2006), Investigation of N-terminal glutamate cyclization of recombinant monoclonal antibody in formulation development, *J. Pharm. Biomed. Anal*. **42**: 455–463.

58. Liu, H., Gaza-Bulseco, G., and Sun, J. (2006), Characterization of the stability of a fully human monoclonal IgG after prolonged incubation at elevated temperature, *J. Chrom. B* **837**: 35–43.

59. Chelius, D., Jing, K., Lueras, A., Rehder, D. S., Dillion, T. M., Vizel, A., Rajan, R. S., Li, T., Treuheit, M. J., and Bondarenko, P. V. (2006), Formation of pyroglutamic acid from N-terminal glutamic acid in immunoglobulin gamma antibodies, *Anal. Chem*. **78**: 2370–2376.

60. Lewis, D. A., Guzzetta, A. W., Hancock, W. S., and Costello, M. (1994), Characterization of humanized anti-TAC, an antibody directed against the interleukin 2 receptor, using electrospray ionization mass spectrometry by direct infusion, LC/MS, and MS/MS, *Anal. Chem*. **66**: 585–595.

61. Werner, W. E., Wu, S., and Mulkerrin, M. (2005), The removal of pyroglutamic acid from monoclonal antibodies without denaturation of the protein chains, *Anal. Biochem.* **342**: 120–125.

62. Wang, L., Amphlett, G., Blatter, W. A., Lambert, J. M., and Zhang, W. (2005), Structural characterization of the maytansinoid-monoclonal antibody immunoconjugate, huN901-DM1, by mass spectrometry, *Protein Sci.* **14**: 2436–2446.

63. Wang, L., Amphlett, G., Lambert, J. M., Blatter, W., and Zhang, W. (2005), Structural characterization of a recombinant monoclonal antibody by electrospray time-of -flight mass spectrometry, *Pharm. Res.* **22**: 1338–1349.

64. Rehder, D. S., Dillion, T. M., Pipes, G. D., and Bondarenko, P. V. (2006), Reversed-phase liquid chromatography/mass spectrometry analysis of reduced monoclonal antibodies in pharmaceutics, *J. Chrom. A* **1102**: 164–175.

65. Dick, L. W., Kim, C., Qiu, D. F., and Cheng, K. C. (2007), Determination of the origin of the N-terminal pyro-glutamate variation in monoclonal antibodies using model peptides, *Biotechnol. Bioeng.* **97**: 544–553.

66. Busby Jr., W. H., Quackenbush, G. E., Humm, J., Youngblood, W. W., and Kizer, J. S. (1987), An enzyme(s) that converts glutaminyl-peptides into pyroglutamyl-peptides. Presence in pituitary, brain, adrenal medulla, and lymphocytes, *J. Biol. Chem.* **262**: 8532–8536.

67. Messer, M. (1963), Enzymatic cyclization of L-glutamine and L-glutaminyl peptides, *Nature* **197**: 1299–.

68. Bunnett, J. F. (1962), The mechanism of bimolecular beta-elimination reactions, *Angew. Chem.* **1**: 225–280.

69. Volkin, D. B. and Klibanov, A. M. (1987), Thermal destruction processes in proteins involving cystine residues, *J. Biol. Chem.* **262**: 2945–2950.

70. Chatterji, T., Keerthi, K., and Gates, K. S. (2005), Generation of reactive oxygen species by a persulfide (BnSSH), *Bioorg. Med. Chem. Lett.* **15**: 3921–3924.

71. Violand, B. N. V. and Siegel, N. R. (2000). Protein and peptide chemical and physical stability, in *Peptide and Protein Drug Analysis*, Reid, R. E. ed., Marcel Dekker, New York, pp. 257–284.

72. Costantino, H. R., Langer, R., and Klibanov, A. M. (1994), Solid-phase aggregation of proteins under pharmaceutically relevant conditions, *J. Pharm. Sci.* **83**: 1662–1669.

73. Costantino, H. R., Schwendeman, S. P., Langer, R., and Klibanov, A. M. (1998), Deterioration of lyophilized pharmaceutical proteins, *Biochem.* (Moscow) **63**: 357–363.

74. Chang, B. S., Kendrick, B. S., and Carpenter, J. F. (1996), Surface-induced denaturation of proteins during freezing and its inhibition by surfactants, *J. Pharm. Sci.* **85**: 1325–1330.

75. Cohen, S. L., Price, C., and Vlasak, J. (2007), Beta-elimination and peptide bond hydrolysis: Two distinct mechanisms of human IgG1 hinge fragmentation upon storage, *J. Am. Chem. Soc.* **129**: 6976–6977.

76. Gaza-Bulesco, G., and Liu, H. (2008), Fragmentation of a recombinant monoclonal antibody at various pH, *Pharm. Res.* **25**: 1881–1890.

77. Nakamura, T., Sakai, M., Sadakane, Y., Haga, T., Goto, Y., Kinouchi, T., Saito, T., and Fujii, N. (2008), Differential rate constants of racemization of aspartyl and asparaginyl residues in human alpha A-crystallin mutants, *Biochim. Biophys. Acta* **1784**: 1192–1199.

78. Nakamura, T., Sadakane, Y., and Fujii, N. (2006), Kinetic study of racemization of aspartyl residues in recombinant human alpha A-crystallin, *Biochim. Biophys. Acta* **1764**: 800–806.

79. Fujii, N., Ishibashi, Y., Satoh, K., Fujino, M., and Harada, K. (1994), Simultaneous racemization and isomerization at specific aspartic acid residues in alpha B-crystallin from the aged human lens, *Biochim. Biophys. Acta* **1204**: 157–163.

80. Sakai-Kato, K., Naito, M., and Utsunomiya-Tate, N. (2007), Racemization of the amyloidal beta Asp^1 residue blocks the acceleration of fibril formation caused by racemization of the Asp^{23} residue, *Biochem. Biophys. Res. Commun.* **364**: 464–469.

81. Tomiyama, T., Asano, S., Furiya, Y., Shirasawa, T., Endo, N., and Mori, H. (1994), Racemization of Asp^{23} residue affects the aggregation properties of Alzheimer amyloid beta protein analogues, *J. Biol. Chem.* **269**: 10205–10208.

82. Murakami, K., Uno, M., Masuda, Y., Shimizu, T., Shirasawa, T., and Irie, K. (2008), Isomerization and/or racemization at Asp^{23} of Abeta-42 do not increase its aggregative ability, neurotoxicity, and radical productivity in vitro, *Biochem. Biophys. Res. Commun.* **366**: 745–751.

83. Geiger, T. and Clarke, S. (1987), Deamidation, isomerization, and racemization at asparaginyl and aspartyl residues in peptides- succinimide-linked reactions that contribute to protein degradation, *J. Biol. Chem.* **262**: 785–794.

84. Radkiewicz, J. L., Zipse, H., Clarke, S., and Houk, K. N. (1996), Accelerated racemization of aspartic acid and asparagine residues via succinimide intermediates: An ab initio theoretical exploration of mechanism, *J. Am. Chem. Soc.* **118**: 9148–9155.

85. Li, B., Borchardt, R. T., Topp, E. M., Vander Velde, D., and Schowen, R. L. (2003), Racemization of an asparagine residue during peptide deamidation, *J. Am. Chem. Soc.* **125**: 11486–11487.

86. Schey, K. L. and Finley, E. L. (2000), Identification of peptide oxidation by tandem mass spectrometry. *Acc. Chem. Res.* **33**: 299–306.

87. Schöneich, C., Zhao, F., Yang, J., and Miller, B. L. (1997), Mechanisms of methionine oxidation peptides, *ACS Symp. Ser.* **675**: 79–89.

88. Hovorka, S. W. and Schöneich, C. (2001), Oxidative degradation of pharmaceuticals: Theory, mechanisms and inhibition, *J. Pharm. Sci.* **90**: 253–269.

89. Volkin, D. B., Mach, H., and Middaugh, C. R. (1997), Degradative covalent reactions important to protein stability, *Mol. Biotechnol.* **8**: 105–122.

90. Li, S., Schöneich, C., and Borchardt, R. T. (1995), Chemical instability of protein pharmaceuticals: Mechanisms of oxidation and strategies for stabilization, *Biotechnol. Bioeng.* **48**: 490–500.

91. Stadtman, E. R. (1990), Metal ion-catalyzed oxidation of proteins: Biochemical mechanism and biological consequences, *Free Radical Biol. Med.* **9**: 315–325.

92. Stadtman, E. R. (1993), Oxidation of free amino acids and amino acid residues in proteins by radiolysis and by metal-catalyzed reactions, *Annu. Rev. Biochem.* **62**: 797–821.

93. Nguyen, T. H. (1994), Oxidation degradation of protein pharmaceuticals, *ACS Symp. Ser.* **567**: 59–71.

94. Chen, B., Costantino, H. R., Liu, J., Hsu, C. C., and Shire, S. J. (1999), Influence of calcium ions on the structure and stability of recombinant human deoxyribonuclease 1 in the aqueous and lyophilized states, *J. Pharm. Sci.* **88**: 477–482.

95. Grillo, A. O., Edwards, K. T., Kashi, R. S., Shipley, K. M., Hu, L., Besman, M. J., and Middaugh, C. R. (2001), Conformational origin of the aggregation of recombinant human growth factor VIII, *Biochemistry* **40**: 586–595.
96. Schöneich, C. (2000), Mechanisms of metal-catalyzed oxidation of histidine to 2-oxo-histidine in peptides and proteins, *J. Pharm. Biomed. Anal.* **21**: 1093–1097.
97. Chen, B., Bautista, R., Yu, K., Zapata, G. A., Mulkerrin, M. G., and Shamow, S. M. (2003), Influence of histidine on the stability and physical properties of a fully human antibody in aqueous and solid forms, *Pharm. Res.* **20**: 1952–1960.
98. Buettner, G. R. and Jurkiewicz, B. A. (1996), Catalytic metals, ascorbate and free radicals: Combinations to avoid, *Rad. Res.* **145**: 532–541.
99. Yang, T.-H., Cleland, J. L., Lam, X., Meyer, J. D., Jones, L. S., Randolph, T. W., Manning, M. C., and Carpenter, J. F. (2000), Effect of zinc binding and precipitation on structures of recombinant human growth hormone and nerve growth factor, *J. Pharm. Sci.* **89**: 1480–1485.
100. Cunningham, B. C., Mulkerrin, M. G., and Wells, J. A. (1991), Dimerization of human growth hormone by zinc, *Science* **253**: 545–548.
101. Hovorka, S. W., Hong, J., Cleland, J. L., and Schöneich, C. (2001), Metal-catalyzed oxidation of human growth hormone: Modulation by solvent-induced changes of protein conformation, *J. Pharm. Sci.* **90**: 58–69.
102. Zhao, F., Ghezzo-Schöneich, E., Aced, G. I., Hong, J., Milby, T., and Schöneich, C. (1997), Metal-catalyzed oxidation of histidine in human growth hormone, *J. Biol. Chem.* **272**: 9019–9029.
103. Hovorka, S. W., Williams, T. D., and Schöneich, C. (2002), Characterization of the metal-binding site of bovine growth hormone through site-specific metal-catalyzed oxidation and high-performance liquid chromatography-tandem mass spectrometry, *Anal. Biochem.* **300**: 206–211.
104. Sadineni, V., Galeva, N. A., and Schöneich, C. (2006), Characterization of the metal-binding site of human prolactin by site-specific metal-catalyzed oxidation, *Anal. Biochem.* **358**: 208–215.
105. Hong, J. and Schöneich, C. (2001), The metal-catalyzed oxidation of methionine in peptides by Fenton systems involves two consecutive one-electron oxidation processes, *Free Radical Biol. Med.* **31**: 1432–1441.
106. Hawkins, C. L. and Davies, M. J. (2001), Generation and propagation of radical reactions on proteins, *Biochim. Biophys. Acta* **1504**: 196–219.
107. Kerwin, B. A., Akers, M. J., Apostol, I., Moore-Einsel, C., Etter, J. E., Hess, E., Lippincott, J., Levine, J., Mathews, A. J., Revilla-Sharp, P., Schubert, R., and Looker, D. L. (1999), Acute and long-term stability studies of deoxy hemoglobin and characterization of ascorbate-induced modifications, *J. Pharm. Sci.* **88**: 79–88.
108. Fransson, J., Florin-Robertsson, E., Axelsson, K., and Nyhlen, C. (1996), Oxidation of human insulin-like growth factor I in formulation studies: kinetics of methionine oxidation in aqueous solution and in solid state, *Pharm. Res.* **13**: 1252–1257.
109. Wisniewski, R. (1998), Developing large-scale cryopreservation systems for biopharmaceutical products, *BioPharm* (June) 50–60.
110. Griffiths, S. W., and Cooney, C. L. (2002), Relationship between protein structure and methionine oxidation in recombinant a 1-antitrypsin, *Biochemistry* **41**: 6245–6252.

111. Duenas, E. T., Keck, R., DeVos, A., Jones, A. J. S., and Cleland, J. L. (2001), Comparison between light-induced and chemically induced oxidation of rhVEGF, *Pharm. Res.* **18**: 1455–1460.

112. Chu, J. W., Yin, J., Wang, D. I. C., and Trout, B. L. (2004), Molecular dynamic situations and oxidation rates of methionine residues of granulocyte colony-stimulating factor at different pH values, *Biochemistry* **43**: 1019–1029.

113. Lu, H. S., Fausset, P. R., Nahri, L. O., Horan, T., Shinagawa, K., Shimamoto, G., and Boone, T. C. (1999), Chemical modification and site-directed mutagenesis of methionine residues in recombinant human granulocyte colony-stimulating factor: Effect on stability and biological activity, *Arch. Biochem. Biophys.* **362**: 1–11.

114. Teh, L. C., Murphy, L. J., Huq, N. L., Surus, A. S, Friesen, H. G., Lazarus, L., and Chapman, G. E. (1987), Methionine oxidation in human growth hormone and human chorionic somatomammotropin. Effects on receptor binding and biological activities, *J. Biol. Chem.* **262**: 6462–6477.

115. Thirumangalathu, R., Krishnan, S., Bondarenko, D., Speed-Ricci, M., Randolph, T. W., Carpenter, J. F., and Brems, D. N. (2007), Oxidation of methionine residues in recombinant human interleukin-1 receptor antagonist: implications of conformational stability on protein oxidation kinetics, *Biochemistry* **46**: 6213–6224.

116. Manzanares, D., Rodriguez-Capote, K., Liu, S., Haines, T., Ramos, Y., Zhao, L., Doherty-Kirby, A., Lajoie, G., and Possmayer, F. (2007), Modification of tryptophan and methionine residues is implicated in the oxidative inactivation of surfactant protein B, *Biochemistry* **46**: 5604–5615.

117. Gray, W. R. (1997), Disulfide bonds between cysteine residues, in *Protein Structure: A Practical Guide*, Creighton, T. E., ed., IRL Press, pp. 165–186.

118. Wedemeyer, W. J., Welker, E., Narayan, M., and Scheraga, H. A. (2000), Disulfide bonds and protein folding, *Biochemistry* **39**: 4207–4216.

119. David, C., Foley, S., Mavon, C., and Enescu, M. (2008), Reductive unfolding of serum albumins uncovered by Raman spectroscopy, *Biopolymers* **89**: 623–634.

120. Graham, L. and Gallop, P. M. (1994), Covalent protein crosslinks: General detection, quantitation, and characterization via modification with diphenylborinic acid, *Anal. Biochem.* **217**: 298–305.

121. Lorand, L., Hsu, L. K., and Siefring, G. E. (1981), Lens transglutaminase and cataract formation, *Proc. Natl. Acad. Sci. USA* **78**: 1356–1360.

122. Turell, L., Botti, H., Carballal, S., Ferrer-Sueta, G., Souza, J. M., Duran, R., Freeman, B. A., Radi, R., and Alvarez, B. (2008), Reactivity of sulfenic acid in human serum albumin, *Biochemistry* **47**: 358–367.

123. International Conference on Harmonization (1996), Guidance for industry: Q1B. Photostability testing of new drug substances and products; www.ich.org.

124. Kerwin, B. A. and Remmele, R. J., Jr. (2007), Protect from light: Photodegradation and protein biologics, *J. Pharm. Sci.* **96**: 1468–1479.

125. Davies, M. J. and Truscott, R. J. (2001), Photo-oxidation of proteins and its role in cataractogenesis, *J. Photochem. Photobiol. B* **63**: 114–125.

126. Grossweiner, L. I. (1984), Photochemistry of proteins: A review, *Curr. Eye Res.* **3**: 137–144.

127. Duenas, E. T., Keck, R., DeVos, A., Jones, A. J. S., and Cleland, J. L. (2001): Comparison between light induced and chemically induced oxidation of rhVEGF, *Pharm. Res.* **18**: 1455–1460.

128. Wei, Z., Feng, J., Lin, H.-Y., Mullapudi, S., Bishop, E., Tous, G. I., Casas-Finet, J., Hakki, F., Strouse, R., and Schenerman, M. A. (2007), Identification of a single tryptophan residue as critical for binding activity in a humanized monoclonal antibody against respiratory syncytial virus, *Anal. Chem.* **79**: 2797–2805.

129. Lam, X. M., Yang, J. Y., and Cleland, J. L. (1997), Antioxidants for prevention of methionine oxidation in recombinant monoclonal antibody HER2, *J. Pharm. Sci.* **86**: 1250–1255.

130. Sysak, P. K., Foote, C. S., and Ching, T.-Y. (1977), Chemistry of singlet oxygen-xxv. Photooxygenation of methionine, *Photochem. Photobiol.* **26**: 19–27.

131. Dalsgaard, T. K., Otzen, D., Nielsen, J. H., and Larsen, L. B. (2007), Changes in structures of milk proteins upon photo-oxidation, *J. Agric. Food Chem.* **55**: 10968–10976.

132. Miller, B. L., Hageman, M. J., Thamann, T. J., Barron, L. B., and Schöneich, C. (2003), Solid-state photodegradation of bovine somatotropin (bovine growth hormone): Evidence for tryptophan-mediated photoxidation of disulfide bonds, *J. Pharm. Sci.* **92**: 1698–1709.

133. Schöneich, C. (2008), Mechanisms of protein damage induced by cysteine thiyl radical formation, *Chem. Res. Toxicol.* **21**: 1175–1179.

134. Silvester, J. A., Timmins, G. S., and Davies, M. J. (1998), Photodynamically generated bovine serum albumin radicals: Evidence for damage transfer and oxidation at cysteine and tryptophan residues. *Free Radical Biol. Med.* **24**: 754–766.

135. Chu, J. W., Yin, J., Brooks, B. R., Wang, D. I. C., Ricci, M. S., Brems, D. N., and Trout, B. L. (2004), A comprehensive picture of non-site specific oxidation of methionine residues by peroxides in protein pharmaceuticals, *J. Pharm. Sci.* **93**: 3096–3102.

136. Kornfelt, T., Persson, E., and Palm, L. (1999), Oxidation of methionine residues in coagulation Factor VIIa, *Arch. Biochem. Biophys.* **363**: 43–54.

137. Li, T., Chang, B., and Sloey C. (2003), *L-Methionine as a Stabilizer for NESP/EPO in HSA-Free Formulations*, US Patent application 20,030,104,996.

138. Takruri, H. (1993), *Method for the Stabilization of Methionine-Containing Polypeptides*, US Patent 5,272,135, Dec. 21, 1993.

139. Yin, J., Chu, J., Ricci, M. S., Brems, D. N., Wang, D. I. C., and Trout, B. L. (2004), Effects of antioxidants on the hydrogen peroxide-mediated oxidation of methionine residues in granulocyte colony-stimulating factor and human parathyroid hormone fragment 13–34, *Pharm. Res.* **21**: 2377–2383.

140. Knepp, V. M., Whatley, J. L., Muchnik, A., and Calderwood, T. S. (1996), Identification of antioxidants for prevention of peroxide-mediated oxidation of recombinant human ciliary neurotrophic factor and recombinant human nerve growth factor, *PDA J. Pharm. Sci. Technol.* **50**: 163–171.

141. Anraku, M., Kouno, Y., Kai, T., Tsurusaki, Y., Yamasaki, K., and Otagiri, M. (2007). The role of N-acetyl-methioninate as a new stabilizer for albumin products, *Int. J. Pharm.* **329**: 19–24.

142. Anraku, M., Tsurusaki, Y., Watanabe, H., Maruyama, T., Kragh-Hansen, U., and Otagiri, M. (2004), Stabilizing mechanisms in commercial albumin preparations: Octanoate and N-acetyl-L-tryptophanate protect human serum albumin against heat and oxidative stress, *Biochim. Biophys. Acta* **1702**: 9–17.

143. Bridgewater, J. D. and Vachet, R. W. (2005), Metal-catalyzed oxidation reactions and mass spectrometry: The roles of ascorbate and different oxidizing agents in determining Cu-protein-binding sites, *Anal. Biochem.* **341**: 122–130.

144. DePaz, R. A., Barnett, C. C., Dale, D. A., Carpenter, J. F., Gaertner, A. L., and Randolph, T. W. (2000), The excluding effects of sucrose on a protein chemical degradation pathway: Methionine oxidation in subtilisin, *Arch. Biochem. Biophys*. **384**: 123–132.

145. Joo, H., Koo, Y., Choi, J., and Chang, C. (2005), Stabilization method of an alkaline protease from inactivation by heat, SDS and hydrogen peroxide, *Enzyme Microb. Technol*. **36**: 766–772.

146. Soenderkaer, S., Carpenter, J. F., Weert, M., Hansen, L. L., Flink, J., and Frokjaar, S. (2004), Effects of sucrose on rFVIIa aggregation and methionine oxidation, *Eur. J. Pharm. Sci*. **21**: 597–606.

147. Yin, J., Chu, J., Ricci, M. S., Brems, D. N., Wang, D. I. C., and Trout, B. L. (2005), Effects of excipients on the hydrogen peroxide-induced oxidation of methionine residues in granulocyte colony-stimulating factor, *Pharm. Res*. **22**: 141–147.

148. Li, S., Patapoff, T. W., Nguyen, T. H., and Borchardt, R. T. (1996), Inhibitory effect of sugars and polyols on the metal-catalyzed oxidation of human relaxin, *J. Pharm. Sci*. **85**: 868–872.

149. Wen, Z., Torraca, G., Yee, C., and Li, G. (2007), Investigation of contaminants in protein pharmaceuticals in pre-filled syringes by multiple micro-spectroscopies, *Am. Pharm. Rev*. **10**(5): 101–107.

150. Markovic, I. (2006), Challenges associated with extractable and/or leachable substances in therapeutic biologic protein products, *Am. Pharm. Rev*. **9**(6): 20–27.

151. Rosenberg, A. S., and Worobec, A. (2004), A risk-based approach to immunogenicity concerns of therapeutic protein products, Part 2: Considering host-specific and product-specific factors impacting immunogenicity, *BioPharm* (Dec.) 34–40.

152. Yokota, H., Saito, H., Masuoka, K., Kaniwa, H., and Shibanuma, T. (2000), Reversed phase HPLC of Met^{58} oxidized rhIL-11: Oxidation enhanced by plastic tubes, *J. Pharm. Biomed. Anal*. **24**: 317–324.

153. Uversky, V. N., Yamin, G., Souillac, P. O., Goers, J., Glaser, C. B., and Fink, A. L. (2002), Methionine oxidation inhibits fibrillation of human alpha-synuclein in vitro, *FEBS Lett*. **517**: 239–244.

154. Fisher, M. T. and Stadtman, E. R. (1992), Oxidative modification of *Escherichia coli* glutamine synthetase: Decreases in the thermodynamic stability of protein structure and specific changes in the active site conformation, *J. Biol. Chem*. **267**: 1872–1880.

155. Liu, D., Ren, D., Huang, H., Dankberg, J., Rosenfeld, R., Cocco, M. J., Li, L., Brems, D. N., and Remmele Jr., R. C. (2008), Structure and stability changes of human IgG1 Fc as a consequence of methionine oxidation, *Biochemistry* **47**: 5088–5100.

156. Gao, J., Yin, D. H., Yao, Y., Qin, H., Schöneich, C., Williams, T. D., and Squier, T. C. (1998), Loss of conformational stability of calmodulin upon methionine oxidation, *Biophys. J*. **74**: 1115–1134.

157. Krishnan, S., Chi, E. Y., Wood, S. J., Kendrick, B. S., Li., C., Garzon-Rodriguez, W., Wypych, J., Randolph, T. W., Narhi, L. O., Biere, A. L., Citron, M., and Carpenter, J. F. (2003), Oxidative dimer formation is the critical rate-limiting step for Parkinson's disease alpha-synuclein fibrillogenesis, *Biochemistry* **42**: 829–837.

158. Palmblad, M., Westlind-Danielsson, A., and Bergquist, J. (2002), Oxidation of methionine-35 attenuates formation of amyloid β-peptide 1–40 oligomers, *J. Biol. Chem*. **277**: 19506–19510.

159. Watson, A. A., Fairlie, D. P., and Craik, D. J. (1998), Solution structure of methionine-oxidized amyloid beta-peptide (1–40). Does oxidation affect conformational switching? *Biochemistry* **37**: 12700–12706.

160. Steinmetz, M. O., Garcia-Echeverria, C., and Kammerer, R. A. (2005), Design of a coiled-coil-based model peptide system to explore the fundamentals of amyloid formation. *Int. J. Pept. Res. Ther*. **11**: 43–52.

161. Khossravi, M., Shire, S. J., and Borchardt, R. T. (2000), Evidence for the involvement of histadine A(12) in the aggregation and precipitation of human relaxin induced by metal-catalyzed oxidation, *Biochemistry* **39**: 5876–5885.

162. Li, S., Nguyen, T. H., Schöneich, C., and Borchardt, R. T. (1995), Aggregation and precipitation of human relaxin induced by metal-catalyzed oxidation, *Biochemistry* **34**: 5762–5772.

163. Gaudiano, M. C., Colone, M., Bombelli, C., Chistolini, P., Valvo, L., and Diociaiuti, M. (2005): Early stages of salmon calcitonin aggregation: effect induced by aging and oxidation processes in water and in the presence of model membranes, *Biochim. Biophys. Acta* **1750**: 134–145.

164. Hu, D., Qin, Z., Xue, B., Fink, A. L., and Uversky, V. N. (2008), Effect of methionine oxidation on the structural properties, conformational stability, and aggregation of immunoglobulin light chain LEN, *Biochemistry* **47**: 8665–8677.

165. Horvat, S. and Jakas, A. (2004), Peptide and amino acid glycation: New insights into the Maillard reaction, *J. Pept. Sci*. **10**: 119–137.

166. Ledl, F. and Schleicher, E. (1990), New aspects of the Maillard reaction in foods and the human body, *Angew. Chem. Int. Ed Engl*. **29**: 565–594.

167. Bell, L. N., White, K. L., and Chen, Y. (1998), Maillard reaction in glassy low-moisture solids as affected by buffer type and concentration, *J. Food Sci*. **63**: 785–788.

168. Bell, L. N., Touma, D. E., White, K. L., and Chen, Y. (1998), Glycine loss and Maillard browning as related to the glass transition in a model food system, *J. Food Sci*. **63**: 625–628.

169. Ajandouz, E. H. and Puigserver, A. (1999), Nonenzymatic browning reaction of essential amino acids: Effect of pH on caramelization and Maillard reaction kinetics, *J. Agric. Food Chem*. **47**: 1786–1793.

170. Labuza, T. P., Tannenba, S. R., and Karel, M. (1970), Water content and stability of low-moisture and intermediate-moisture foods, *Food Technol*. **24**: 543–.

171. Schebor, C., del Pilar Buera, M., Karel, M., and Chirife, J. (2000), Color formation due to non-enzymatic browning in amorphous, glassy, anhydrous, model systems, *Food Chem*. **65**: 427–432.

172. Hill, S. A., MacNaughtan, W., Farhot, I. A., Noel, T. R., Parker, R., Ring, S. G., and Whitcombe, M. J. (2005), The effect of thermal history on the Maillard reaction in a glassy matrix, *J. Agric. Food Chem*. **53**: 10213–10218.

173. Kawai, K., Hagiwara, T., Takai, R., and Suzuki, T. (2005), The rate of non-enzymatic browning reaction in model freeze-dried food system in the glassy state, *Innov. Food Sci. Emerging Technol*. **6**: 346–350.

174. Carpenter, J. F., Pikal, M. J., Chang, B. S., and Randolph, T. W. (1997), Rational design of stable lyophilized protein formulations: some practical advice, *Pharm. Res*. **14**: 969–975.

175. Gerrard, J. A. (2002), Aspects of an AGEing chemistry—recent developments concerning the Maillard reaction, *Austral J. Chem*. **55**: 299–310.

176. Zheng, X., Wu, S., and Hancock, W. S. (2006), Glycation of interferon-beta-1b and human serum albumin in a lyophilized glucose formation: Part III: Application of proteomic analysis to the manufacture of biological drugs, *Int. J. Pharm.* **322**: 136–145.

177. Hinton, D. J. S. and Ames, J. M. (2006), Site specificity of glycation and carboxymethylation of bovine serum albumin by fructose, *Amino Acids* **30**: 425–435.

178. Jakas, A., Vinković, M., Smrečki, V., Šporec, M., and Horvat, Š. (2008), Fructose-induced N-terminal glycation of enkephalins and related peptides, *J. Pept. Sci.* **14**: 936–945.

179. Roščić, M. and Horvat, Š. (2006), Transformations of bioactive peptides in the presence of sugars—characterization and stability studies of the adducts generated via the Maillard reaction, *Bioorg. Med. Chem.* **14**: 4933–4943.

180. Bouma, B., Kroon-Batenberg, L. M. J., Wu, Y., Brunjes, B., Posthuma, G., Kranenburg, O., de Groot, P. G., Voest, E. E., and Gebbink, M. F. B. G. (2003), Glycation induces formation of amyloid cross-beta structure in albumin, *J. Biol. Chem.* **278**: 41810–41819.

181. Smales, C. M., Pepper, D. S., and James, D. C. (2001), Protein modifications during antiviral heat bioprocessing and subsequent storage, *Biotechnol. Progress* **17**: 974–978.

182. Smales, C. M., Pepper, D. S., and James, D. C. (2000), Mechanisms of protein modification during model anti-viral heat-treatment bioprocessing of beta-lactoglobulin variant A in the presence of sucrose, *Biotechnol. Appl. Biochem.* **32**: 109–119.

183. Hawe, A. and Friess, W. (2008), Development of HSA-free formulations for a hydrophobic cytokine with improved stability, *Eur. J. Pharm. Biopharm.* **68**: 169–182.

184. Gadgil, H. S., Bondarenko, P. V., Pipes, G., Rehder, D., McAuley, A., Perico, N., Dillon, T., Ricci, M., and Treuheit, M. (2007), The LC/MS analysis of glycation of IgG molecules in sucrose containing formulations, *J. Pharm. Sci.* **96**: 2607–2621.

185. Brady, L. J., Martinez, T., and Balland, A. (2007), Characterization of nonenzymatic glycation on a monoclonal antibody, *Anal. Chem.* **79**: 9403–9413.

186. Kennedy, D. M., Skillen, A. W., and Self, C. H. (1994), Glycation of monoclonal-antibodies impairs their ability to bind antigen, *Clin. Exp. Immunol.* **98**: 245–251.

187. Sutthirak, P., Dharmsthiti, S., and Lertsiri, S. (2005), Effect of glycation on stability and kinetic parameters of thermostable glucoamylase from *Aspergillus niger*, *Process Biochem.* **40**: 2821–2826.

188. Carballal, S., Alvarez, B., Turell, L., Botti, H., Freeman, B. A., and Radi, R. (2007), Sulfenic acid in human serum albumin, *Amino Acids* **32**: 543–551.

3

PHYSICAL STABILITY OF PROTEIN PHARMACEUTICALS

Byeong S. Chang and Bernice Yeung

3.1. INTRODUCTION

The unique three-dimensional structures of proteins are not only critical for specific interactions with related molecules for precise biochemical reactions but also important for other supportive functions such as feedback mechanism, transport, and solubility in physiological environments [1–3]. For these reasons, the native structures of proteins are designed to adapt to environmental changes, such as pH, ionic strength, hydrophobicity, surface, ligands, and other biological molecules, including proteins [4–6]. While the flexibility of structure is required for necessary biological function of proteins, it also presents unique challenges in the development of proteins for pharmaceutical applications by causing conformational changes, aggregation, and precipitation, which often are related to loss of biological activity as well as immunogenicity of the proteins [7–11]. In order to address this particular issue, various platforms for manufacturing process and formulation development have been introduced to the pharmaceutical industry. In practice, research focuses on comprehending the nature of protein structures, understanding the conditions that can cause the alteration from native structures and their contribution to the degradation of proteins, developing novel analytical tools that examine these processes and end results of

Formulation and Process Development Strategies for Manufacturing Biopharmaceuticals, edited by Jameel and Hershenson

degradation pathways, and designing approaches to utilize this information for maintaining the nativity of each protein for its safe pharmaceutical application (for related review articles, see Refs. 12–21).

Although fully understanding all the possible combinations of structural variants, especially under stressed conditions, would be virtually impossible, several decades' worth of research by both academic and pharmaceutical institutions has established general approaches that can be applied for the successful development of most biopharmaceutical products [17,19,22–24]. Among these, key accomplishments relevant for pharmaceutical development, including identification of key stress factors, understanding major degradation products, and developing analytical methods to determine the degradation products, are discussed in this chapter.

This chapter specifically focuses on the physical changes that occur in proteins and their contribution to the overall stability relevant to product development, so readers are referred to other chapters in this book for chemical degradations and the like.

3.2. PHYSICAL STABILITY OF PROTEIN PHARMACEUTICALS

3.2.1. Physical Forms of Proteins Found in Practice

Unlike conventional small-molecule pharmaceuticals, proteins present unique challenges related to the proper folding of their native states and their appropriate structural maintenance. The classic discussions of the primary, secondary, tertiary, and quaternary structure of proteins can be found in most basic biochemistry textbooks, so they are not repeated here [25,26]. It is generally agreed that proteins will spontaneously fold to assume their native structures to fit the environment in which they are naturally found, such as exposing hydrophilic residues and hiding hydrophobic residues in the core in an aqueous environment. When proteins are purified and kept in non-physiological environments and exposed to physicochemical stresses in particular, proteins undergo structural changes to adapt to their changing environments. The structural changes can result in combinations of structural variants during the handling and processing of these molecules [27,28]. The structural changes invariably introduce different physical properties to the molecule, such as solubility, a tendency to aggregate, and surface activity, ultimately resulting in the formation of undesirable degradations such as soluble aggregates, insoluble precipitates, and losses due to surface adsorption. Various physical forms of proteins that have been routinely observed are summarized in Scheme 3.1. The native structure of a protein (N) assumes a correctly folded form with the lowest free-energy state. In reality, their native conformations are in thermodynamic equilibrium with their environment. N* represents a rare state when the structure of protein becomes more compact on changes in the environment [29]. The probability of unfolding is smaller in this compact structural state of the molecule as evidenced by reduced rates of hydrogen-deuterium (H–D) exchange and/or exposure of varied residues. The native structure can also undergo partial structural changes, when either stress is induced or is found naturally in solution, and can convert into various forms of folding intermediate structures (I), which

$$\begin{array}{ccccccccc} N^* & \leftrightarrow & N & \leftrightarrow & I & \leftrightarrow & D & \rightarrow & D^{**} \\ & & & & \updownarrow & & & & \\ & & & & A & \rightarrow & P & & \\ & & & & \downarrow & & & & \\ & & & & A^{**} & & & & \end{array}$$

Scheme 3.1. Some physical forms of proteins (N = native structure, I = intermediate folding structure; A = aggregated form; D = denatured form; P = precipitated form).

have been routinely observed during the refolding processes of proteins [27,30]. Typically, subtle structural changes (e.g., changes in tertiary structure) are involved in this process even though the whole spectrum of folding intermediates can fit in the species. Among the intermediate species, certain species will have a greater tendency to interact with other protein molecules and form soluble or insoluble aggregates (A) or precipitates (P) [31]. These soluble aggregates may be reversible to a monomeric intermediate form, but they can turn into irreversible aggregates as the molecules undergo further structural changes for stable aggregate conformations (A**) [32]. From the intermediate forms, the proteins can further lose their crucial structural elements, such as secondary structures, resulting in a denatured form (D). This transition does not routinely occur in reality unless strong physical stresses, including heating above thermal unfolding temperature, or high concentrations of chaeotropes (e.g., guanidine hydrochloride), are introduced to the solution. Although the denaturation can be reversible, other physicochemical changes to the denatured form can render the process irreversible (D**). Stable proteins should be able to maintain the native form (N) despite the presence of some reversible forms such as N*, I, D, and A, although this may be inevitable in practice.

3.2.2. Measure of the Physical Stability of Proteins

3.2.2.1. Structural Stability. Proteins become soluble in aqueous solutions following the interaction of surface-exposed residues of their native structure with water, molecules, and other surfaces. The solubility of a protein may change as the structure undergoes changes in response to its exposure to different residues. Therefore, the physical stability of most proteins can be expressed as their resistance to unfolding forces because aggregation and/or precipitation can occur when the structural change results in a less soluble conformational state. This has also been expressed in thermodynamic terms, for example, thermal unfolding temperature or free energy of unfolding [33]. Essentially, proteins can undergo conformational changes from their native and biologically active forms to nonnative and/or inactive conformations. The resistance to unfolding, also known as *thermodynamic stability*, varies among different proteins and depends on a combination of various forces that contribute to the folding of proteins. These forces originate from covalent bonds, such as disulfide bonds, electrostatic interactions, hydrophobic interactions, hydrogen bonds, and van der Waals interactions [34]. The majority of information available from literature refers to the complete loss of structure, including loss of secondary structure. However, in practice, most degradation products observed in the development of protein therapeutics are derived from more subtle changes, such as quaternary or tertiary structural changes [35,36]. For example, thermal analysis of precipitated protein paste with microcalorimetry often shows a full endothermic unfolding curve, suggesting that proteins with intact secondary structure can still precipitate (unpublished data). This model of differential solubility of various structural ensembles is better understood from protein refolding research [34,37–39].

There have been enormous research efforts to understand the structural stability of proteins. These include structural analyses of thermophilic proteins [40–42]. Furthermore, site-specific mutation studies have been conducted to identify key stabilizing forces [43,44].

Generally speaking, the intrinsic physical stability of proteins at their native structure can be expressed as a resistance to the formation of transitional intermediate states that are prone to convert to irreversible conformations.

3.2.2.2. Colloidal Stability. Alternative measurements of a protein's structural stability can be derived from the protein's ability to remain as a monomeric native form. A protein's colloidal stability is influenced by interactions with other molecules (proteins, excipients, preservatives, metal ions, salts, ligands, leachables, nuclei, etc.) as well as other external factors such as surfaces and solvents [45–47].

In an ideal environment, proteins would always refold back to their native, active, and soluble forms. However, proteins possess the flexibility to assume structures other than their native forms, including soluble aggregates or insoluble forms, if lower free-energy states can be achieved in different environments, such as purified into simpler buffered systems [48]. These transitions to nonnative structures can be reversible, but some structural changes bear a significant activation energy barrier and for all practical purposes become irreversible under normal operating conditions [49]. Thus, it is important to understand the protein molecules as colloidal particles and their adaptations to the surrounding environment.

Various physical properties of proteins, including the collective interactive forces of exposed amino acid residues as well as the stability of the native structure, will contribute in keeping the native protein molecules from aggregating or unfolding [50]. These intermolecular interactions should include charge interactions, hydrophobic interactions, hydrogen bonds, and van der Waals interactions.

The colloidal stability of protein molecules contributes to important physical properties such as solubility, viscosity, surface interaction, foaming, thermal stability, and resistance to shear stress [45,51]. All possible transitional state conformations should be considered when the colloidal stability is concerned as certain unstable structures may become limiting factors.

Proteins colloidal stability relies on a number of factors:

- Intrinsic solubility of native forms
- Formation of transitional conformations under stress
- External factors such as solvents, solutes, and concentration

The colloidal stability of protein molecules has been expressed as a number of useful parameters such as preferential interaction parameters and excluded volume effects [52,53], as well as the second virial coefficient [45,54,55]. Preferential hydration has been useful in understanding the physical stability of proteins in solution in terms of cosolvent-induced changes. The second virial coefficient is another thermodynamic parameter that characterizes two-body interactions in dilute solutions by reflecting the magnitude and sign of interaction [54]. Using these parameters, the physical stability of various proteins has been successfully correlated with different formulation conditions [20,45].

3.2.2.3. Physical Degradation Products. Protein physical stability has been measured via a couple different methods: (1) analyzing the structural changes under stress: and (2) quantifying the changes resulting from the stress (see Section 3.3).

The former approach is useful in understanding both reversible and irreversible structural changes that contribute to the change in protein physical properties. Analytical methods useful for the former approach include various spectrometric analyses equipped with means to introduce various stresses to the samples as well as thermal analyses [56]. Monitoring the degree of unfolding by measuring the kinetics of association with external molecules, for example, probes that exert unique signals on interaction with proteins or specific residues, chemical reagents that bind to specific residues, hydrogen–deuterium exchange, is another approach for measuring the exposure of specific residues or surfaces [57]. This approach is useful for understanding all the relevant structural changes that need to occur before critical degradation takes place. Often, there are multiple structural changes involved before the final degradation products are formed. Any of them can be a rate-limiting process. While this approach is useful for various structural changes that occur under stress, it also provides information regarding irrelevant structural changes, such as reversible changes or the reactions that do not occur in real-life situations, which can potentially interfere with desirable signals (see Section 3.4.1).

The latter approach is routinely used during quality control of protein pharmaceuticals as it determines specific degradation products that result from specific stresses [58,59]. First, proteins are exposed to various stresses, such as forced degradation studies and/or accelerated stability studies, followed by analysis of the degradation products that are generated by irreversible reactions, or kinetic traps (see Section 3.4). Analytical methods useful for this approach include routine chromatographic analyses, particle analyses, electrophoresis, biological activity assays, mass spectrometry, and visual observation (see Section 3.3). Chromatographic analyses, primarily high- or ultra-high-performance liquid chromatography (HPLC or UPLC) methods, have been successfully utilized to determine changes in size, shape, surface charge, hydrophobicity, and other properties. Particle analyses have been also useful in charactering both soluble and insoluble particles with light-scattering methods and analytical ultracentrifuge. The limitation of this approach is a consequence of treating the whole degradation process as a black box; it lacks the ability to identify and understand the rate-limiting structural changes.

In theory, proteins should be able to recover their native structures on removal of stresses unless there is irreversible transition involved. Therefore, the following reactions that make the structural changes effectively irreversible will be considered as relevant degradation pathways. In general, the degradation pathways can be classified roughly into the following criteria:

- Formation of stable misfolded species
- Aggregation or precipitation of misfolded species
- Aggregation-induced structural changes
- Chemical modification of critical residues
- Interaction with other solutes (metal, preservatives)
- Surface-induced structural changes and/or aggregation
- Surface adsorption

3.2.2.3.1. Formation of Stable Misfolded Species. This is based on the assumption that proteins may assume structures other than the native structure during the course of refolding [37,38]. In theory, the native structure of each protein is the most thermodynamically stable state with the lowest free energy. However, different conformations have been observed during process development of refolding proteins from inclusion bodies [48,49]. In this metastable structural state, the misfolded proteins can remain reasonably stable until additional energy is introduced to overcome its activation energy barrier before refolding to the native structure.

3.2.2.3.2. Aggregation or Precipitation of Misfolded Species. Proteins aggregate and/or precipitate in their native structures and/or partially unfolded transitional ensembles, including both intermediate state ensembles, denatured state ensembles, and/or transitional state ensembles, when formation of aggregates is feasible [60,61]. Underlying principles for both soluble and insoluble aggregates are not significantly different from a practical perspective, so both forms of aggregates are treated interchangeably in this discussion. Aggregated forms of proteins can be classified into two important groups: reversible aggregates and irreversible aggregates.

3.2.2.3.2.1. Reversible Aggregates or Precipitates: The best example of reversible aggregates can be found in the aggregation, or precipitation, due to the equilibrium solubility of proteins [62–65]. If the concentrations of proteins increase above their critical concentrations or solubility limits, proteins will naturally form aggregates or insoluble precipitates. Unless further structural changes beyond the solubility-induced aggregation occur, the proteins will revert back to their native monomeric forms as soon as the concentrations fall below the solubility limit or the environment changes to permit greater solubility. This reversible aggregation process is considered as an equilibrium process similar to the solubility phenomena of other molecules [66].

3.2.2.3.2.2. Irreversible Aggregates or Precipitates: This is one of the most common as well as critical degradation products, as the smallest quantity of aggregated forms can routinely induce undesirable immune responses [8]. Moreover, it is one of major problems associated with the quality control of protein pharmaceuticals as small quantities of protein (e.g., 1–10 μg) of protein can form large numbers of visible particles. If aggregated forms of protein do not dissociate into native monomeric molecules when diluted into lower concentrations with the same formulation, then they are considered to be irreversible aggregates. Often, an aggregate is deemed irreversible if it fails to dissociate when the product is administered to its targeted delivery site, such as an injection site. Irreversible aggregates are generally formed during intermediate state ensembles [37] or molten globule states [67] through hydrophobic interactions.

Irreversibly aggregated forms of proteins can be dissociated into monomeric forms in many different ways depending on the nature and strength of their intermolecular forces. Some weakly associated noncovalent aggregates can be dissociated with a simple addition of surfactant(s) into the solution, adjustment of temperature, adjustment of pH to induce charge repulsion among molecules, and/or adjustment of ionic strength. Formulation strategies to stabilize proteins by introducing dissociation factors are discussed in better detail later in this chapter.

Stronger noncovalent aggregates can be dissociated by exposing the aggregates to harsher nonphysiological conditions such as extreme pH, strong detergents such as sodium dodecyl sulfate (SDS), and/or high concentrations of chaeotropes such as guanidine hydrochloride or urea. As most proteins adapt to the harsh environment by changing to their nonnative structures, these approaches are seldom recommended for pharmaceutical formulations.

Covalent aggregates, mostly with intermolecular disulfide bonds, currently can be dissociated into monomers only with the help of reducing agents such as mercaptoethanol, dithiothreitol, tris-[2-carboxyethyl]phosphine (TCEP), or glutathione. These covalent aggregates are usually formed directly from free cysteine residues found in native protein molecules but can also be formed from intermolecular disulfide scrambling of pre-existing intramolecular disulfide bonds [68,69].

3.2.2.3.2.3. Soluble Forms that Precipitate When Introduced to Physiological Conditions: Proteins can maintain integrity in their molten globule states or transitional states under certain solution conditions [59]. When these formulations are delivered into physiological conditions, the proteins may precipitate or aggregate during the course of refolding back to their native structures. For example, if some proteins have limited solubility or are prone to aggregation and/or precipitation during storage at neutral pH, the proteins may be formulated in an acidic environment. Alternatively, additional solubilizing agents could be added. If these formulations are introduced parenterally into a physiological environment, a fraction of the proteins can result in an aggregated or precipitated form at the delivery site.

3.2.2.3.3. AGGREGATION-INDUCED STRUCTURAL CHANGES. Proteins typically maintain their native structures in their natural environments in order to be functional. Furthermore, they are flexible so as to adapt to changing environments for their biological functions. For instance, most proteins undergo conformational changes when they interact with ligands or receptors for their activities. This implies that proteins may sacrifice the original structure and rearrange to different conformations when the environment changes, such as contact surface, temperature, concentration, and solution condition. A similar scenario occurs when protein molecules associate with each other while forming aggregates. Once the association begins, loosely linked native molecules can further squeeze into irreversibly aggregated forms through additional structural rearrangements such as the formation of β-sheets and domain swapping [70,71]. This is well illustrated by the formation of amyloid fibrils from soluble protein to prefibrillar species followed by protofilaments and ultimately branched fibrils [72].

3.2.2.3.4. Chemical Modifications of Critical Residues. During the formation of transitional state ensembles, some amino acid residues susceptible to chemical modifications can become exposed. Because of the change in physicochemical properties around the modified residues, the affected protein molecules may not be able to refold back into their native structures. The structure-induced exposure and modification of buried residues as well as the irreversible changes in the resulting structures of proteins are well illustrated by the oxidation of methionine residues [73,74]. Change in other residues by other chemical modifications can also affect the transitional state ensembles' ability to refold back to their native structures. For example, the rate of deamidation of asparagines residues can be influenced by the flexibility of the peptide backbone [75]. At the same time, the resulting changes in charge from amino residue of asparagines to the carboxylic residues of aspartic acid and/or isoaspartic acid may also alter the way in which the proteins will be refolded [76].

3.2.2.3.5. Interaction with Other Solutes (Metal, Preservatives). Proteins interact very closely with other solutes and can integrate into other stable structural states. The principle of preferential hydration for stabilizing solutes (e.g., sugars, salts, amino acid residues) has been widely discussed in the literature [29,53,77,78]. The stabilizing solutes have a tendency to be excluded from the hydration surface of proteins, thus leaving the protein molecules preferentially hydrated. At the preferentially hydrated state, the void that the protein molecules occupy is thermodynamically unfavorable and keeps the proteins with a minimal exposed surface area. As the denatured state of protein molecules would have a much larger exposed surface area, the larger difference in the free energy of unfolding in the presence of these cosolvents is the major driving force for stabilization. As the preferential hydration effect comes at the cost of a relatively high concentration of cosolvents, for example, 1 M or greater for sucrose, these cosolvents have not been widely used in pharmaceutical applications because of the hyperosmolality of resulting formulations.

Another group of stabilizing cosolvents works more specifically with the exposed chemical nature of proteins. The effectiveness of these cosolvents, such as charged polymers, is rather difficult to predict as the stabilizing effect comes from a very specific interaction with proteins [79,80].

Another group of excipients that showed direct physical interaction with proteins and contributed to the aggregation and/or precipitation of proteins is phenolic compounds. More relevant to pharmaceutical applications, these antimicrobial preservatives have shown greater impact on the physical stability of proteins. These parenterally approved preservatives include benzyl alcohol, cresol, phenol, parabenes, and benzalkonium chloride. The preservatives interact with proteins and keep the transitional state ensembles from folding back to their native structure [51,81,82].

Metal ions contribute to the intrinsic structural stability of proteins by increasing resistance to unfolding or directing their folding [83]. Removal of the metal ions from these proteins often results in a loss of the native structure and functionality. However, metal ions also foster intermolecular interactions and cause aggregation and/or precipitation. Zinc is typically added to promote association to hexamers and to increase chemical stability [84,85].

Additionally, metal impurities facilitate undesirable aggregation or precipitation. [86] showed metal induced conformational changes followed by the formation of fibrillar forms of α-synuclein. There have been presentations discussing the aggregation of therapeutic proteins in prefilled syringes caused by tungsten oxide leaching out of staked needles [8].

3.2.2.3.6. Surface-Induced Structural Changes and/or Aggregation. During product shipping and handling, proteins in aqueous solution formulations are subjected to agitation or shaking, which is one of the most common physical stresses that can cause protein aggregation [87,88]. Agitation-induced protein denaturation and/or aggregation is presumably the consequence of protein interactions with surfaces, such as the air–water and vial–water interfaces [87,89–92]. It is generally believed that protein unfolding at air–liquid interfaces may irreversibly expose the interior hydrophobic core, which may, in turn, lead to intermolecular associations of nonpolar residues. It has been observed that agitation-induced protein aggregation decreases at higher protein concentrations. Small amounts of surfactants are usually effective in stabilizing proteins against this type of stress [90,93].

A widely accepted argument seen in previous studies attributes protein aggregation during processing to its increased exposure to various surfaces, most commonly the air–water interfaces [94–96]. Proteins are known to form films of a nonnative conformation at such boundaries. It has been argued that turbulence caused by agitation can create additional surface area and facilitate an exchange between the native population in the bulk and the denatured species on the surface. Presumably, accumulation of these denatured species eventually leads to precipitation. By the same trend of thought, solid–liquid interfaces were ascribed for the aggregation seen during filtration [96,97] and pumping [98]. However, direct proof of these pathways is scarce, and little is known about the details of how such proteins form aggregates when exposed to these stresses.

These results challenge the "surface denaturation" hypothesis because the native forms of the most labile proteins do not adsorb to the surface fast enough to account for its rapid precipitation during agitation. It is based on the assumption that the adsorption of native protein to the surface is an instantaneous process. Instead, they suggest that the decrease in surface tension is a result of the proteins' need to be unfolded into a more hydrophobic configuration to adsorb to the air–water interface. This implication is consistent with an alternate hypothesis that claims shear stress is the primary source of protein denaturation during stirring, pumping, and filtration [99–103]. They discovered that the product of shear rate and time Gt correlates with the kinetics of enzyme inactivation. Residual protein activity with respect to Gt commonly indicates that a protein is approximately stable up to a certain Gt value but above this point the decrease in activity is directly proportional to $\ln(Gt)$.

The reversibility of surface adsorption has been debated in the literature. Earlier studies support the theory that the adsorption process, especially with additional denaturation at the interface, is irreversible [104,105]. This is further supported by the formation of insoluble proteins at the interface due to the changed conformation [106,107].

3.2.2.3.7. Surface Adsorption to Solid Surfaces. The presence of surfaces and interfaces can have a profound effect on aggregate development, and studies have shown that the growth rates and final aggregate morphologies are sensitive to the details of the surface chemistry of an interface, such as hydrophobicity and charge [92,108–110]. Among the more relevant therapeutic applications, extensive research has been conducted on the surface adsorption of insulin pump systems [111] and inactivation/surface adsorption of IL-2 to tubing materials [112–114].

The degree of structural and activity loss depends on the individual protein. While extensive loss is expected for labile proteins such as insulin and IL-2, a significantly smaller loss, about the quantity for monolayer adsorption on the order of a few micrograms per 3 mL vial/stopper configuration, has been observed for more stable proteins.

The reversibility of proteins adsorbed to solid surface has been well summarized by Norde [115] and MacRitchie [91]. The main points are summarized below:

- Only a small fraction of surface-adsorbed proteins desorb on dilution of the system with additional solvent.
- The degree of adsorption depends on multiple factors, including solution condition and temperature.
- Protein molecules may be exchanged between surface and bulk solutions.
- Adsorbed proteins may be displaced from the surface by molecules of other species.
- Protein molecules may not refold back to their original structures once they are unfolded at the interface.

It is generally agreed that surface-adsorbed protein molecules are difficult to desorb from the surface unless the solution condition is changed, for example, by additional surface active agents.

3.3. ANALYTICAL METHODS FOR PHYSICAL CHANGES IN PROTEINS

To characterize conformational changes in a protein, caused either by changes in its solution environment or by physical degradation in the protein itself, biophysical techniques are often employed to examine the secondary or tertiary structures of the molecule [116]. Results generated from these techniques not only lend insight into the understanding of protein structures but may also provide guidance in their formulation development. The common techniques used in formulation development include spectroscopic analyses (second-derivative ultraviolet spectroscopy, circular dichroism, Fourier transform infrared spectroscopy, Raman spectroscopy, and fluorescence and phosphorescence spectroscopy), thermal analysis (differential scanning calorimetry), and size-based analysis (analytical ultracentrifuge, light scattering). In addition, the three-dimensional structure of proteins may also be studied to gain a better understanding of their structure–function relationships as well as their interactions with various factors in their environment. Techniques for this characterization include hydrogen–deuterium (H/D) exchange mass spectrometry, nuclear magnetic resonance spectroscopy, X-ray crystallography, and electron paramagnetic spin resonance spectroscopy.

3.3.1. Spectroscopic Techniques

3.3.1.1. Second-Derivative Ultraviolet (UV) Spectroscopy. The environment of aromatic residues in a protein can be monitored using second-derivative UV spectroscopy. When the tertiary structure of a protein is perturbed as a function of pH, heat, or other factors, shifts in the UV peak positions can be observed in its second-derivative UV spectrum. An example of the application of second derivative UV spectroscopy includes the study of aggregation behavior of an IgG_2 antibody as a function of pH and temperature [117]. By monitoring the shifts in peak maxima of five selected peaks in their second derivative UV spectra, investigators determined stable pH and temperature ranges, gaining insight into the aggregation behavior of the antibody during the freeze–thaw process.

3.3.1.2. Circular Dichroism. Circular dichroism (CD) measures the difference in absorbance for the left-handed and right-handed circularly polarized light from a protein sample. This difference is contributed to the interaction of chromophores in a protein that are located in a chiral (asymmetric) environment. In CD analysis, both near-UV CD (310–255 nm) and far-UV CD (< 250 nm) regions of the spectrum can be measured. Aromatic amino acids (tryptophan, tyrosine, phenylalanine) and cystine contribute to the near-UV CD signals, which reflect the tertiary structure of the protein. The peptide bonds (amide) contribute to the far-UV CD signals, which represent the secondary structure of the protein (i.e. α-helix, β-sheet, β-turn structures).

Typical applications with CD analysis include (1) estimation of the content of protein secondary structures and (2) detection of conformational changes in a protein due to changes in its solution environment (pH, excipients, addition of denaturants) and temperature. In a more recent study, an *E. coli*-expressed IgG1 Fc was studied by CD before and after forced oxidation by H_2O_2, and the data indicate perturbation in the tertiary structure of the molecule, whereas its secondary structure was largely unchanged [118]. In another case, MAb samples at over 100 mg/mL concentration were directly analyzed with CD, in which two MAb molecules were compared by CD as a function of temperature. The results revealed a 7°C stabilization of one MAb over the other at the same protein concentration [119].

3.3.1.3. Fourier Transform Infrared Spectroscopy (FTIR). In infrared spectroscopy, a molecule absorbs infrared radiation as it undergoes a net change in dipole moment resulting from its vibrational or rotational motion. In terms of wavenumber, the infrared spectrum is divided into near- (from 12,800 to 4000 cm^{-1}), middle- (from 4000 to 200 cm^{-1}), and far-IR regions (from 200 to 10 cm^{-1}), with 1700–1600 cm^{-1} as the most commonly used region for protein analysis. This region, commonly referred to as the *amide I region*, corresponds to the carbonyl (C=O) stretching vibration of a peptide bond. Hydrogen bonding of the C=O group to the amide group yields information on the secondary structures of a protein with the following known maxima [120]: α helix (1654 cm^{-1}), β sheet (1633 and 1684 cm^{-1}), γ turn (1672 cm^{-1}), and random coil (1654 cm^{-1}). A comparison of the band intensities at these wavenumbers may indicate differences in secondary structures between protein samples.

Modern infrared spectrometers are usually Fourier transform infrared (FTIR) spectrometers, due to the fact that the detector signal is related by Fourier transformation into the measured spectrum. To avoid issues caused by short pathlength, in which interferences from water in the aqueous sample overlaps with the amide I band, the attenuated total reflectance (ATR) technique may be employed. In ATR, a thin film of the sample is placed on a crystal, and the infrared light is reflected several times through the crystal and the sample before measurements are made.

Fourier transform infrared has been widely applied to study changes in the secondary structures of proteins in formulation studies, because formulations that produce secondary structures closest to the native protein structure often lead to longer storage stability. More recent examples of FTIR applications have included the effects of mannitol content in spray-dried antibody formulations [121], effects of sucrose and sorbitol contents in lyophilized antibody formulations [122], and a comparison of various sucrose contents for an antibody in different presentations (freeze-dried, foam, and spray-dried) [123]. In an interesting application of FTIR, the structural integrity of rhBMP-2 microparticles prepared from controlled precipitation was compared to that of a reference formulation in solution. It was found that between 20°C and 70°C, the FTIR spectra of both the rhBMP-2 microparticles and the resolubilized microparticles showed largely the same pattern as that of the liquid formulation. Moreover, after 6 months of storage, the microparticle suspension maintained its secondary structure as those of the unstressed and stressed reference formulations in liquid [124].

3.3.1.4. Fourier Transform (FT) Raman Spectroscopy. The Raman effect arises when a beam of intense radiation, typically from a laser, passes through a sample where the molecules undergo a change in molecular polarizability as they vibrate. This change in polarizability results in information regarding the peptide backbones and secondary structures of protein samples. Information obtained by FT-Raman spectroscopy is often complementary to results from FTIR. One advantage of FT-Raman over FTIR is that water happens to be a weak Raman scatter. Consequently, proteins in the native (aqueous) state can be studied in addition to the solid state. Typical regions in FT-Raman spectra that are characteristic for proteins include the amide I region for α-helical structure (at ~1660 cm^{-1}) and the amide III region (1250–1350 cm^{-1}).

Three model proteins differing in size and structure were compared in terms of their native structures before and after lyophilization and spray drying. The results from their FT-Raman analysis showed good correlation to data from biological (enzyme activity) assays [125], which indicate that among the proteins studied, lysozyme and deoxyribonuclease I show better folding reversibility in the lyophilized form than in the spray-dried form.

3.3.1.5. Fluorescence Spectroscopy. In fluorescence measurements, the protein is excited at a wavelength corresponding to its excitation maximum, and light is emitted during the protein's transition from the excited state to the ground state. Aromatic amino acids, including tryptophan, tyrosine, phenylalanine, and histidine, provide intrinsic fluorescence signals that can be measured. The information obtained is related to the environment surrounding these residues, hence, the integrity of the tertiary structure of the molecule can be interpreted. For these reasons, fluorescence spectroscopy is a useful tool for studying the unfolding of a protein in different environments (pH, temperature, buffer excipients, etc.). A more recent publication demonstrates the application of fluorescence spectroscopy toward the direct analysis of lyophilized drug products in accelerated degradation. A close correlation was found between the diminishing fluorescence intensities of the lyophilized samples with the SE-HPLC results obtained from the reconstituted samples [126].

Besides intrinsic fluorescence, proteins may also be derivitized with fluorescent probes to enhance the sensitivity of the measurement at wavelengths that are characteristic of the probes used. Probes such as 8-amino-1-naphthalenesulfonate (ANS), Nile Red, and thioflavin T have been used for the detection of protein aggregates. The probes bind to the hydrophobic surface of proteins that are exposed during fibril formation. Once bound, their fluorescence maxima are shifted and their intensities can be measured. Aggregation in an antibody at high concentration (193 mg/mL) in different buffers has been characterized without dilution using Nile Red binding, and the results between fluorescence spectroscopy and fluorescence microscopy analyses correlated well [127].

3.3.1.6. Phosphorescence Spectroscopy. As in fluorescence spectroscopy, a molecule is excited from its lowest vibrational state to an excited state in phosphorescence spectroscopy. While in its excited state, a change in electron spin must occur in order for the molecule to phosphoresce. This means that a phosphorescence event is less probable than other spectroscopic events and the rate is also slow. Phosphorescence lifetimes are typically range from 10^{-4} to 10 s, in contrast to a fluorescence lifetime between 10^{-9} and 10^{-6} s.

Conventionally, phosphorescence measurements needed to be performed at liquid nitrogen temperature in order to prevent degradation of the phosphorescence signal because of collisional deactivation. Nowadays, room temperature phosphorescence may be performed in solution, as long as the sample is purged of all traces of oxygen.

A more recent example discusses the use of this technique to access differences in the microenvironment of two tryptophan residues in a glutamine binding protein. The difference in phosphorescence lifetimes of Trp^{220} and Trp^{32} is on the order of several hundred milliseconds, which is sufficient for determining Trp^{32} to be in a well-ordered environment [128]. The work illustrates the potential of using phosphorescence measurement to probe changes in the tertiary structure of proteins caused by ligand binding or other environmental changes.

3.3.2. Thermal Analysis

3.3.2.1. Differential Scanning Calorimetry (DSC). In DSC, the heat flow (or heat capacity) of a sample is measured as a function of temperature. The technique has been useful for studying the unfolding of secondary structures of protein molecules and for characterizing the conformational stability of proteins in different conditions, in both solid and liquid states. While protein unfolding produces endothermic peaks as measured by DSC, protein aggregation is detected as an exothermic event. With DSC, the melting or denaturation temperatures T_m of different protein samples can be compared, giving insight to the differences in their secondary structures. Another application of DSC has been its use in optimizing lyophilization parameters. Protein formulations may be analyzed by DSC to determine the optimal freezing and annealing temperatures during the lyophilization cycle, which would, in turn, lead to a more elegant cake appearance. A more recent example of applying DSC in formulation development shows the effects of sucrose on rHSA, which clearly shows sucrose adding protection to the protein by shifting the protein's melting point from 123°C to 147°C, whereas rHSA formulated with mannitol alone did not show the same benefits [129]. Other applications have included the comparison of two MAb molecules by DSC and the effects of various excipients on the thermal stability of recombinant adenoviruses [130].

3.3.3. Size-Based Analysis

3.3.3.1. Analytical Ultracentrifugation (AUC). For routine, high-throughput analysis of protein aggregates, size exclusion (SE-HPLC) is unquestionably the most sensitive and reproducible method available. However, SE-HPLC has a few drawbacks, due to its reliance on the use of a column, which can lead to poor mass recovery of aggregated species [131]. The nonspecific interaction between the proteins and the column matrix may necessitate the addition of salt and organic solvents that would, in turn, disrupt the aggregated species. On the contrary, AUC does not have the same issues as does SE-HPLC. Although the technique is not as sensitive as SE-HPLC and requires highly trained expertise for data analysis, it is a valuable technique for studying protein aggregation in solution and is often used to validate results obtained by SE-HPLC.

The basic AUC instrumentation includes a centrifuge, a rotor with sample compartments, and an optical system than can measure the protein concentration gradients under centrifugal force. Two basic modes of operation exist in AUC: sedimentation velocity and sedimentation equilibrium. Sedimentation velocity, which is conducted at high centrifugal speeds [40,000–60,000 rev/min (rpm)], yields information related to the sedimentation coefficient, diffusion coefficient, and molecular weight. Sedimentation equilibrium, which is performed at lower rotor speeds (10,000–30,000 rpm), yields information on molecular weight, stoichiometry, and binding affinity of self- or heteroassociating molecules.

3.3.3.2. Light Scattering. The principle of light scattering is based on the fact that the intensity of light scattered by a sample is proportional to the particle size and the concentration of that sample. For protein applications, two types of light-scattering techniques are commonly used: dynamic light scattering (DLS) and multiangle laser light scattering (MALLS), also referred to as *static light scattering*. With DLS, a hydrodynamic diameter of a sample in ranging from 1 nm to 1 μm can be measured and results reported in a volume-based distribution. A cytokine in liquid formulations at various pH was studied by DLS, and the results indicate an increase in volume percentage of aggregated species as a function of increasing pH [132]. Other examples of DLS applications include the study of effects of excipients on an immunoglobulin G in terms of aggregate prevention and thermodynamic stability [133] and an investigation into the opalescent appearance of an IgG1 formulation at high concentrations [134].

In MALLS, the weight-average molecular weight (M_w) and number-average molecular weight (M_n) are determined as a function of concentration. MALLS can be used as a standalone detector for batch analysis or it can be used as an online detector in conjunction with refractive index detection for SE-HPLC analysis. Most commonly, the 90° signal from a MALLS measurement is used for the determination of Mw and Mn. Provided that the *dn/dc* ratio of an analyte is known, MALLS measurements can provide the absolute determination of M_w and M_n that is independent of the hydrodynamic size of the analyte (i.e., its elution position on SE-HPLC relative to molecular weight standards). The stability samples of two antibody drug products are analyzed by SE-HPLC with MALLS detection and the MW of the various aggregated species can be determined [135].

3.3.4. Three-dimensional Structural Characterization of Proteins

3.3.4.1. Hydrogen–Deuterium Exchange Mass Spectrometry. Hydrogen–deuterium (H/D) exchange has been a powerful tool for studying the surface properties of proteins. Surface exposed hydrogens in OH, NH, and SH functional groups can be replaced with deuteriums on exposure to D_2O-containing buffers. The only exception is proline, which does not contain amide hydrogen. H/D exchange may take place either in properly folded proteins, in which only localized exchanges occur, or in partially or completely unfolded proteins. The H/D exchange reaction occurs readily, and the reaction can be followed by a variety of techniques including mass spectrometry, FTIR, electron paramagnetic resonance spectroscopy, and nuclear magnetic resonance spectroscopy. More recently, H/D mass spectrometry (MS) utilizing electrospray ionization has by far had the widest applications. Since mass information is obtained in MS, analysis of H/D exchange by MS yields information on the structure of the entire protein, rather than just a few specific residues as in the case with spectroscopy-based techniques.

To conduct MS analysis of an H/D exchange reaction in a protein, samples at fixed time intervals during the reaction can be quenched at low pH and low temperature prior to injection onto the MS system. Whole-mass measurements are performed, and the mass shift due to H/D exchange can be followed as a function of reaction time. To gain a better understanding of local H/D exchange rates, quenched samples may be digested using pepsin at low pH (2.5) and low temperature (0°C) for a short time, followed by peptide mapping using an HPLC coupled to the MS and tandem mass spectrometric analysis (MS/MS).

Li et al. [136] applied H/D MS analysis to the characterization of protein conformation in lyophilized solids. The model protein calmodulin was lyophilized in the presence of various carbohydrate excipients. The lyophilized cakes were then exposed to D_2O vapor under controlled humidity and temperature, and samples at specific time intervals were digested with pepsin and analyzed by LC-MS/MS. The results demonstrate the best protective effects by raffinose, dextran, and trehalose, while mannitol and sucrose show greater uptake of deuterium than do the other carbohydrates. Furthermore, the peptide map results show that different carbohydrates protect different regions of the protein.

3.3.4.2. X-Ray Crystallography. In X-ray crystallography, an X-ray beam strikes a single crystal and scatters into different angles. The intensities and angles at which the X ray scatters allow for the determination of electron density within the crystal, which, in turn, leads to the deduction of a three-dimensional structure of the crystal. A variety of molecules have been successfully analyzed by X-ray crystallography, including small-molecule drugs, proteins, and nucleic acids. In one example, the structure in insulin neutral protamine hagedorn (NPH) formulations was studied. Protamine is an arginine-rich small peptide that is cocrystallized with insulin in the presence of zinc in the formulations. Characterization of the insulin NPH crystals indicates that protamine adopts a disordered crystal structure, which suggests its primary function to be the balancing of overall charges of insulin in the formulations rather than any kind of specific binding to insulin [137].

3.3.4.3. Nuclear Magnetic Resonance Spectroscopy (NMR). In NMR, an analyte is studied in the presence of a constant magnetic field while an orthogonal magnetic field is applied. The response to the perturbing magnetic field forms the basis of NMR, in which all nuclei with an odd number of protons and/or neutrons give an intrinsic magnetic moment that can be measured. ^{1}H and ^{13}C NMR are the most commonly used nuclei in NMR. In general, large magnetic fields are preferred since they offer increased sensitivity. In more recent years, solid-state NMR has been useful in providing structural information on proteins that include membrane proteins and lyophilized proteins. One example demonstrates that the stability offered by arginine in freeze-dried antibody products is due to noncovalent interactions between the arginine side chain and the protein [138]. Results from this work suggest that charged amino acids may provide greater stabilizing effects than do the noncharged ones, perhaps because of the preferential binding of the stabilizer with the protein during freezing.

3.3.4.4. Electron Paramagnetic Resonance Spectroscopy. Electron paramagnetic resonance EPR, or electron spin resonance (ESR), spectroscopy is similar in principle to NMR, except that it is the magnetic moment of an unpaired electron in an analyte that is measured, rather than that of an atomic nucleus. Electron paramagnetic resonance is less commonly used than NMR because most stable molecules have their electrons paired. An example of application of EPR can be found in a report by Kempe et al. [139], in which chitosan was characterized as a potential biopolymer for drug delivery in the presence of glycerol-2-phosphate. Electron paramagnetic resonance has been utilized to monitor the dynamics of insulin, the drug candidate used in this study, inside the chitosan gel and during its release. From this study, the optimal percentage of glycerol-2-phosphate and pH were determined for gel formation at room temperature, and the release of insulin was found to be sustainable for a period of up to 2 weeks under optimal conditions.

3.3.5. Stability-Indicating Assays for Detection of Physical Degradation in Protein Therapeutics

Analytical assays used for the stability testing of protein therapeutics need to encompass all possible physical forms of degradation that result from normal stresses during a typical stability program. Common biochemical degradations include aggregation, deamidation, oxidation, and fragmentation (clipping). In the case of glycoproteins, additional degradation may include some degree of deglycosylation or desialylation. Cyclization of *N*-terminal glutamine has been reported in recombinant monoclonal antibodies after prolonged heat stress [140,141]. All of these degradations can usually be detected as product impurities by HPLC and electrophoresis-based assays, with separation principles based on size, charge, or hydrophobicity. The following are techniques typically used in the analysis of physical degradation in stability testing.

3.3.5.1. Size-Based Determination.

3.3.5.1.1. Size Exclusion Chromatography (SE-HPLC). In this technique, protein species are separated according to hydrodynamic size under native (aqueous) conditions. Although primarily intended for detection of soluble aggregates, SE-HPLC may also be applied for analysis of clipped species in a protein sample. Popular choices of columns used for protein analysis have included the Tosoh TSK gel and Shodex KW-800 series, both of which are silica-based with high recovery rates and resolution. Proteins in the range of ~10 kDa (cytokines) up to 150 kDa (monoclonal antibodies) have been successfully analyzed on these columns. For hydrophobic proteins that tend to have poor recovery, polymer-based columns may be utilized at the cost of resolution compared to their silica-based counterparts.

Typical mobile phase used for SE-HPLC includes phosphate-based buffers with 100–500 mM of sodium chloride at pH close to neutral (i.e., pH 6.5–7.5). As an alternative, sodium sulfate can be used as a substitute for sodium chloride at similar concentrations [142]. For proteins that tend to have excessive interaction with the silica columns, leading to tailing or poor recovery, arginine at 0.2–0.75 M may be used to minimize adsorption [143].

3.3.5.1.2. Sodium Dodecyl Sulfate–Polyacrylamide Gel Electrophoresis (SDS-PAGE) and Capillary Electrophoresis with Sodium Dodecyl Sulfate (CE-SDS). The SDS-PAGE method has been the conventional size-based technique for analysis of proteins. Unlike SE-HPLC, SDS-PAGE is performed in the presence of SDS and therefore offers complementary information on aggregates that are covalent in nature. Generally, SDS-PAGE offers better resolution than does SE-HPLC as both aggregated and clipped species can be separated within the same gel at high efficiency. For stability and lot release testing, Coomassie Blue staining has been the method of choice for visualization, and densitometry may be applied for a semiquantitative measurement of the separated bands. If greater sensitivity is required, silver or Sypro staining may be applied, although the results are typically not as quantitative as Coomassie staining and the techniques are not as quality control (QC)-friendly.

More recently, CE-SDS has been gaining acceptance as a substitute for SDS-PAGE in both the stability and lot release testing of protein therapeutics. Unlike the case in SDS-PAGE, in CE-SDS the analyte does not sieve through a solid polyacrylamide matrix; rather, a polymer liquid matrix is used for size based separation in a fused-silica capillary with a small internal diameter (50–100 μm). The polymer liquid gel includes polyethylene glycol/polyethylene oxide, dextran, and hydroxyalkyl cellulose, at various concentrations depending on the molecular weight range of interest. The exact makeup of commercially available CE-SDS gel is proprietary information; however, a gel kit optimized specifically for MAb analysis is available from Beckman Coulter.

The CE-SDS method may be performed using a conventional CE instrument where a physical capillary is used for separation (i.e., the P/ACE system, by Beckman Coulter, or the HPCE system, by Agilent) or a chip-based CE instrument where microfluidic channels are utilized for analysis (i.e., Bioanalyzer, by Agilent, or Labchip 90 by Caliper). In either case, the analyte is mixed with a SDS-containing sample buffer much like in SDS-PAGE, with the option of adding a reducing agent if reduced analysis is desired. The sample is then injected electrokinetically into the capillary for separation, under an applied voltage of 15–30 kV. Because of effective heat dissipation in a capillary or microfludic channel, much higher voltages can be applied than in SDS-PAGE with little current generated, leading to superior resolution in CE-SDS. In addition, CE-SDS data are quantitative and can be analyzed much like HPLC data, which is an added benefit over conventional SDS-PAGE analysis. Typical electropherograms are shown from the CE-SDS analysis of antibody molecules using a conventional CE instrument [144] and a chip-based CE system [145].

As CE-SDS analysis is usually performed using UV detection, it yields sensitivity equivalent to that of Coomassie Blue staining. For added sensitivity, the analyte may be derivatized with a suitable fluorescent tag during sample preparation and detected using laser-induced fluorescence (LIF). A CE-SDS assay utilizing LIF detection has been validated for use in a QC environment to support the stability monitoring of a

MAb molecule [146]. The sensitivity obtained by this approach is similar to that of silver staining.

3.3.5.2. Charge-Based Determination.

3.3.5.2.1. ION EXCHANGE CHROMATOGRAPHY. For lot release and stability testing of protein therapeutics, a charge-based purity method is typically expected from the regulatory agencies as part of the standard panel of test methods. Under conditions used in ion exchange chromatography (IEC), surface charges on a protein are exposed and interact with the ionic functional groups in the column. By choosing a mobile phase with an appropriate pH that is suitably above or below the pI of the protein, either anion or cation exchange can be performed. In either case, the separation mechanism in IEC is based on charge, which allows for the separation of deamidated, deglycosylated [147] or other charge-based variants [144,145,148,149] from the main product peak in stability monitoring.

3.3.5.2.2. CAPILLARY ISOELECTRIC FOCUSING (CIEF) AND CAPILLARY ZONE ELECTROPHORESIS (CZE). The CIEF method separates charge-based variants on the same principle as does conventional IEF, whereas the separation mechanism in CZE is based on both charge and mass. CIEF performed in a conventional CE instrument has been widely applied for the characterization of pI values for protein therapeutics, and is gaining acceptance for use in lot release and stability testing. More recently, CIEF performed with whole-column imaging detection ("imaged CIEF") is gaining attention for its advantage over conventional CIEF, in which the focused protein bands are imaged without the need for band mobilization [150], which can lead to peak broadening.

3.3.5.3. Hydrophobicity-Based Determination. Separation in reversed-phase (RP) and hydrophobic interaction chromatography (HIC) is based on hydrophobicity. These techniques are often used for the detection of product impurities that include oxidation (at either methionine or tryptophan), fragmentation (clipping), and sometimes deamidation. More recently, RP-HPLC has been applied to the detection of pyroglutamic acid formation at the *N*-terminus of the heavy chain in an IgG1 monoclonal antibody [140]. Although less frequently utilized than RP-HPLC, HIC-HPLC has been applied in the stability monitoring of protein therapeutics, and in one case, the isomerization of an aspartic acid residue was detected by HIC-HPLC along with RP-HPLC [152].

3.3.5.4. Other Techniques. Purity analysis by HPLC and electrophoresis-based methods is typically sufficient for the stability monitoring of protein therapeutics. By characterizing and identifying the minor forms detected in these purity methods, one can apply them toward direct stability monitoring without the need for more detailed structural characterization such as peptide mapping and oligosaccharide mapping. While the data obtained from these detailed analyses are valuable, peptide and oligosaccharide mappings are time-consuming and can be highly variable depending on the analyst. Therefore, these techniques are typically not included in stability monitoring programs and stability specifications; rather, they are reserved for characterization use only. For additional information about this field, readers are referred to the following articles [153,154].

3.4. STRESSES THAT CAUSE PHYSICAL DEGRADATION OF PROTEINS

Among the physical stresses, the effects of heat, agitation, surfaces, and pressure are relatively well characterized, and plenty of examples can be found in literature [155,156]. Other chemical stresses that are known to affect the structure of proteins include pH, metal ions, chaeotropic salts, detergents, preservatives, and organic solvents. While all of these can clearly induce structural changes of the proteins, one must seek the most relevant stresses to individual proteins, as some proteins are more susceptible to certain stresses than others. In this section, some common stresses and experimental approaches to stress induction are briefly discussed.

3.4.1. Heat

Like most reactions, the rates of most protein degradation pathways are accelerated by the increase in temperature. This is especially true for structural changes in proteins as the temperature dependence, or activation energy, of structural changes is on the order of 25–150 kcal/mol, much greater than the reactions of small molecules. At elevated temperatures, the solution conditions as well as the intrinsic flexibility of proteins alters so that the population of transitional state ensembles increases. This is the basis for incubating products at higher temperatures to accelerate degradation reactions so that expected degradation products and better formulations can be found in much smaller timescales than actual storage conditions. At the same time, relevant stability-indicating assays can be developed using the stressed samples. While the benefit of utilizing the accelerated reactions at elevated temperatures is large, one must always remember the importance of observations in real-life situations.

A common mistake found in the development of protein formulations is the use of thermal analyses for screening purposes. This can be achieved by differential scanning calorimetry and/or any other device that can determine the denaturation of proteins while equipped with a temperature control system. Information obtained from such instruments includes thermal unfolding temperatures, enthalpies of unfolding, and change of specific heat during thermal unfolding [157,158]. While these results are relatively quick to obtain and can very effectively differentiate testing samples, the information regarding the unfolding of proteins has little to do with most degradation products that are routinely observed during the refrigerated storage of protein products. While the activation energies for thermal unfolding of proteins are on the order of 100 kcal/mol [56], the activation energies for most pharmaceutically relevant degradation products are on the order of 30 kcal/mol (unpublished general observations). This is because most physical degradations of proteins result from subtle structural changes, specifically, transitional state ensembles, rather than a complete loss of secondary structure. Therefore, the information obtained from thermal analyses should be used for formulation development with discretion.

Similar consideration should be given when extrapolating the results obtained from higher temperatures to realistic storage conditions, such as refrigerated storage. Each protein degradation pathway may involve multiple species of structural ensembles as well as multiple processes of structural changes. Therefore, it is difficult to assume a simple two-state model. In addition, multiple degradations will have to be considered when optimizing formulation and/or process conditions. Each pathway will have its own unique reaction order and activation energy, so it is difficult to assume the same major degradation products will occur from storage at both 4°C and 37°C. In reality, empirical thermodynamic parameters obtained from actual stability studies may come from several different rate-limiting processes at different temperatures.

3.4.2. Surface and/or Shear

As discussed in earlier sections, proteins are sensitive to surface-induced denaturations. Most proteins form soluble aggregates and/or precipitates, and can unfold and/or adsorb to solid surfaces when sufficient agitation or shear forces are introduced [87,91,92,97,99]. This phenomenon is observed during routine pharmaceutical processing or handling such as pumping, mixing, ultrafiltration/diafiltration, stirring, membrane filtration, filling, and transportation. With the exception of very sensitive proteins, the problem can be easily resolved by including a small quantity of surface-active agents, as is discussed in Section 3.5.1.

Experimental results that unequivocally explain the underlying principles of the surface-induced denaturation of proteins are difficult to find in the literature. A couple of leading hypotheses are (1) *surface-induced denaturation*, which represents surface adsorption, denaturation/coagulation, recycling back to bulk solution, aggregation among the transitional state species, and/or adsorption to other solid surfaces [91,92] and (2) *shear-induced denaturation*, which represents the structural alteration of proteins in bulk solution state by shear force, aggregation among the transitional state species, and/or adsorption to solid surfaces [100,159–161]. For practical purposes, the process of these structural changes followed by irreversible aggregation or precipitation of protein molecules is considered identical. Because of the ambiguity, it has been difficult to design instrumentation to introduce quantitative stress. The most common approaches that introduce surface denaturation are vortexing and stirring. Vortexing with headspace air or additional beads has been successfully used to understand the susceptibility of surface denaturation as well as to identify necessary surface-active agents [90,162,163]. Stirring with a magnetic bar has also been used with equal effectiveness [164].

The simple agitation or stirring experiments for understanding each protein's susceptibility to surface-induced aggregation is equally important as are other incubation experiments, as incubation at elevated temperatures does not introduce an equivalent stress to mechanical stimulation. The agitation experiment takes only a few hours, and it is recommended that this study be performed prior to other accelerated stability studies. Once stable formulation is identified against the surface-induced denaturation, the protein should remain stable during various stressful processes such as pumping, filtration, filling, shipping, and shaking.

Relevant surfaces should include air–liquid interfaces, solid–liquid interfaces, and liquid–liquid interfaces. Proteins adsorb to air–liquid interfaces, concentrate at the interface, undergo structural changes, and then self-associate as well as aggregate [165,166]. Formation of insoluble films at interfaces has been observed for some labile proteins. The agitation of protein solutions with headspace air is also a routine source of aggregation. Likewise, proteins are known as good foaming agents because of their structural flexibility to adsorb to the air–liquid interface and rearrange their structure to form a strong cohesive viscoelastic film, which is required for stability of foam [167,168].

Proteins are also known to stabilize liquid–liquid interfaces via similar principles. Proteins stabilize these interfaces by rearranging their hydrophobic and hydrophilic residues to reduce free energy at these interfaces [169–171]. Good examples of protein degradations at the liquid–liquid interface include aggregation of proteins by silicon oil [172], oil-in water emulsion [173], and water–lipid bilayer [174].

Similar structural changes were also observed at solid–liquid interfaces. Protein aggregations resulting from interaction with container–closure systems, membranes, tubing, and pumps are well documented in the literature [26,114,175–177]. Most articles describe the effect of the solid–liquid interface in terms of process of adsorption, structural changes at the interface, irreversible adsorption, and/or possible recycling to bulk solution. Such interactions with solid surfaces may result in irreversible adsorption, and/or decrease in recovery, and aggregation, as well as loss of biological activity of the proteins.

The effect of smaller solid particles on nucleation of protein aggregation has been extensively studied in the area of amyloid aggregation [178–184]. The formation of small nuclei with the complementing surface is a rate-limiting process for the growth of aggregate. This phenomenon is well illustrated by the particle formation of an antibody product when it was pumped through a piston pump fostering stainless-steel particles [177].

3.4.3. Light

Exposure to light introduces various stresses to proteins [185,186]. Light-induced denaturation proteins can be largely divided into two areas: intrinsic and extrinsic. Intrinsic changes are induced as direct structural changes to proteins. Various amino acid residues, including tryptophan, tyrosine, phenylalanine, methionine, histidine, and cysteine/cystine, are oxidized during exposure to light. This may result in physical changes in proteins such as aggregation, denaturation, and inactivation.

Extrinsic changes result from light-induced changes in the surrounding solution environment, such as change in solvent and/or excipients. It is well known that light generates reactive oxygen species that further react with proteins and cause various types of oxidative damage [187–192]. The excipients that are prone to the oxidative damages and foster the reactive oxygen species or free radicals include metals, polyethylene glycols, polysorbates, ascorbic acid, lipids, and amino acids such as tryptophan, histidine, and tyrosine.

3.4.4. Dehydration

The native structures of proteins in aqueous solution are defined by their balancing forces primarily between the hydrophobic and hydrophilic interactions. Therefore, the presence of bulk water around proteins is required for maintaining their native structure [193]. When powder forms of proteins are preferred, the water molecules will have to be removed from the surface of proteins. It appears that the native structures of proteins are relatively well preserved until the water molecules from the hydration layer, or monolayer, are removed from the surface of proteins [194]. This phenomenon is better illustrated by the effective stabilization of native structure by carbohydrates during drying processes [195–197]. This is discussed further in Section 3.5.1, but the sugars protect the native structure of proteins during drying processes by substituting the molecular interactions from water molecules, such as hydrogen bonds.

3.4.5. Freezing

Proteins experience various physicochemical stresses during the freezing process [198–199]. In addition to the decrease in temperature and subsequent phase separation of water into ice crystals, the changes in the concentration of protein and other unfrozen solutes can become rather complicated. In the freeze-concentrated unfrozen fractions, solutes may undergo further phase separations into insoluble precipitates, crystals, and/or liquid–liquid phase separations. It is well known that the selective precipitation of buffer components may also result in pH changes, such as decrease in pH when disodium phosphate precipitates while monosodium phosphate remains amorphous [200,201].

Phase separation of stabilizing additives can also result in undesirable structural changes of proteins. Various excipients (sugars, amino acids, salts, polyols, etc.) have been used to effectively stabilize proteins during the freeze–thaw process [12,13,53,195]. Some of these stabilizers may lose efficacy as they precipitate or crystallize into a frozen state. A detailed discussion of the phase separation of excipient in frozen state is available in the literature [202,203].

The surface-induced structural changes are also experienced during the freezing and freeze-drying processes. Proteins undergo structural changes at the surface of ice crystals and/or other crystalline excipients, which may result in aggregation or precipitation [197]. A substantial body of literature is available demonstrating the positive correlation between the surface area of crystals and the degree of protein degradation [204–210].

These freeze-induced physical changes of proteins are directly relevant to the stability of freeze-dried products. Additional stresses induced by dehydration should apply for the freeze-dried formulation as discussed previously.

3.4.6. Cold Denaturation

The denaturation of proteins at lower temperature, specifically, cold denaturation, has also been reported [33,211–214]. As the solubility of hydrophobic residues increases, pK_a for acid–base changes, the strength for hydrogen bonds changes, and the hydrophobic interaction force decreases at lower temperature, some proteins may find lower free energy in other unfolded states or dissociated subunits [215,216]. In general, cold denaturation appears to be a reversible process, partly because of the enhanced solubility of hydrophobic species, and does not have a practical impact on the quality of protein products.

3.4.7. pH

One of the most critical variables in the physicochemical stability of proteins is pH. pH is important for maintaining the native structure of proteins by balancing interactions among amino acid residues in the molecule as well as interaction with external solution environment through ionization states of individual residues, electrostatic free-energy states, charge interactions, charge repulsions, and so on [217]. In addition, pH also affects the colloidal stability of various structural states by similar mechanisms. Dependent on the ionic strength as well as the nature of other ionic species in the solution, proteins can readily adapt to different structural states including aggregates and precipitates. In practice, the best way to find the most favorable pH condition of each protein is to expose that protein to a matrix of pH values and ionic strengths.

Proteins undergo significant conformational changes in the acidic pH range, which often results in severe aggregation, precipitation, and/or adsorption to container surfaces. The acid induced conformational change sometimes results in the molten globular state of the proteins [38,218–221]. In this molten globular state, proteins have native-like secondary structures [222] but with incomplete tertiary structures [223–225]. It appears that the colloidal stability of the molten globular-state molecules determines the rate of aggregation at acidic pH values. For some proteins, the reversible transition to and from the molten globular state at acidic pH values is a stable formulation condition for proteins that have limited stability in the neutral pH range [59].

Proteins generally have limited solubility around their isoelectric points because of balanced charges from ionic amino acid residues. Other pH-dependent chemical reactions, such as deamidation, formation of cyclic imide, disulfide formation, and scrambling, can also affect the structure of resulting molecules.

3.4.8. Pressure

Proteins also undergo structural changes at high pressures [226–228]. The structural changes and related population of unfolded species also resulted in aggregation and precipitation of proteins [173,229,230]. The pressure-induced aggregation has not been a major issue in the development of protein therapeutics as the pressure required to unfold proteins (e.g., >100 MPa) is beyond the range of routine exposure.

However, aggregated forms of proteins can be dissociated under pressure, and/or the solubility of misfolded forms of proteins is greater under pressure. This technique has been effectively used to increase the refolding efficiency at higher concentrations [231–233].

3.5. PRACTICAL APPROACHES TO MINIMIZING PHYSICAL DEGRADATIONS

In practice, proteins are exposed to various process conditions that are not always ideal for maintaining their physical structures. More often than not, we have little idea about the natural condition where the proteins perform their biological functions. These issues have been overcome by thorough scientific investigation and accumulated experience during the course of commercializing protein therapeutics. In general, proteins can be protected from undesirable physical degradations by following a few well-chosen approaches.

3.5.1. Optimization of Formulation for Native Structure

The native structure of a protein can be preserved by optimizing the solution conditions to minimize structural changes and to minimize intermolecular interactions. Identifying optimal pH and ionic strength is a priority as both intramolecular and intermolecular charge interactions of ionic residues are most critical for preserving native structure without aggregation [234,235].

Addition of sugars, such as sucrose, trehalose, sorbitol, and mannitol [77,78]; amino acids [236]; and/or salts [237], stabilizes the native structure of proteins by the principle of preferential hydration [53]. Even more compact native structure species, which have less flexibility to the transitional-state ensemble, are observed in the presence of high-concentration sugars, owing to the preferential hydration effect [29].

In a dried state, the native structures of proteins are preserved by including sugars, such as sucrose and trehalose, which can replace the hydrogen bonds at the proteins' exposed polar groups [195,196,238]. Preservation of the native structure in a dried state is also effective in stabilizing the proteins during subsequent storage [12,13,239]. For identifying the ideal sugar stabilizer for dry powder formulations, one of the most effective analytical tools is FTIR [195,196], because spectra from both liquid and dried states can be measured and directly compared.

3.5.2. Improving the Colloidal Stability

Attempting to keep proteins in their native structures is not practical for most proteins because of the routine manufacturing, handling, transportation, and storage processes that are required for commercialization. Alternatively, a more sensible approach is to identify the condition where the unstable species have better colloidal stability [240]. As long as the unstable transitional state species are kept intact during the course of stress, they will ultimately refold back to the native structure after the stresses subside. This approach can be universally applied to all proteins, from the ones with unstable intrinsic stability to others that form unstable species under stress.

The colloidal stability of proteins is characterized by the formation of irreversible physical degradation products. Among all ensembles of intermediate folding species, the ones that cannot refold back to their native structures will be identified as degradation products, which include inactive forms, aggregated forms, precipitated forms, and surface-adsorbed forms. Experimentally test, formulations are exposed to various relevant stresses followed by analysis specifically for the irreversibly modified species. Through this approach, relevant stresses, key degradation products, and stability-indicating assays are established. Using this methodology, formulations for improved physical stability are screened and the stability of proteins in the given formulations is demonstrated. Likewise, this is the basis of routine pharmaceutical stability studies, such as forced degradation studies, accelerated stability studies, and real-time stability studies.

Both pH and ionic strength are the most critical factors in colloidal stability of various structural variants of proteins. More often than not, a simple adjustment of pH and/or ionic strength is sufficient to provide a good colloidal stability of a critical species. Consequently, this makes the product more stable against relevant stresses. The aggregation of proteins induced by benzyl alcohol was reduced at lower pHs by stabilizing unstable species generated by the interaction of the protein with benzyl alcohol [82,241].

Additives can also enhance the colloidal stability of proteins. Surfactants, primarily nonionic surfactants such as polysorbates for formulation use, effectively stabilize proteins against agitation [162,163,242], freeze–thawing [12,13,90], and/or surface adsorption [243]. Surfactants not only prevent undesirable adsorption of proteins to surfaces by preoccupying them but also associate with unstable species and prevent their aggregation [244].

Ionic additives such as heparin are also effective in improving the colloidal stability of proteins [245] and dextran sulfate [246].

Specific amino acids have also been used to enhance the colloidal stability of proteins. Arginine is probably one the most widely used amino acids for inhibiting proteins from aggregation [247–251].

Naturally occurring chaparones are known to enhance the colloidal stability of various transitional state intermediates and mediate the correct folding of proteins [252–256].

Intrinsic colloidal stability of native forms as well as transitional state intermediates has been improved by covalent conjugation with soluble polymers or glycosylation. The decrease in the rate of protein aggregation by pegylation has been reported numerous times [257–259]. Glycosylation also has been effective in enhancing the colloidal stability of proteins and preventing aggregation [260–263].

Other stabilizers or formulation conditions that contribute to improved colloidal stability of proteins are often difficult to predict theoretically. However, some effective stabilizers have been discovered empirically by exposing proteins to relevant stresses followed by analyzing the presence of irreversible aggregates.

REFERENCES

1. Valente, A. P., Miyamoto, C. A., and Almeida, F. C. (2006), *Curr. Med. Chem.* **13**(30): 3697–703.
2. Luo, B. H., Carman, C. V., and Springer, T. A. (2007), *Annu. Rev. Immunol.* **25**: 619–647.

3. Goodey, N. M. and Benkovic, S. J. (2008), *Nat. Chem. Biol.* **4**(8): 474–482.
4. Shahrokh, Z., Sluzky, V., Cleland, J. L., Shire, S. J., Randolph, T. W. (1997), *Therapeutic Protein and Peptide Formulation and Delivery*, American Chemical Society, Washington, DC.
5. Carpenter, J. F. and Manning, M. C. (2002), *Rational Design of Stable Protein Formulations (Pharmaceutical Biotechnology*, Vol. 13): *Theory and Practice*, Springer-Verlag, New York.
6. McNally, E. (2000), *Protein Formulation and Delivery*, Marcel Dekker, New York.
7. Schellekens, H. (2008), *Biotechnol. Annu. Rev*. **14**: 191–202.
8. Rosenberg, A. S. (2006), *AAPS J*. **8**(3): E501–E507.
9. Bennett, C. L., Luminari, S., Nissenson, A. R., Tallman, M. S., Klinge, S. A., McWilliams, N., McKoy, J. M., Kim, B., Lyons, E. A., Trifilio, S. M., Raisch, D. W., Evens, A. M., Kuzel, T. M., Schumock, G. T., Belknap, S. M., Locatelli, F., Rossert, J., Casadevall, N. (2004), *New Engl. J. Med*. **351**(14): 1403–1408.
10. Hermeling, S., Schellekens, H., Maas, C., Gebbink, M. F., Crommelin, D. J., and Jiskoot, W. (2006), *J. Pharm. Sci*. **95**(5): 1084–1096.
11. Hermeling, S., Crommelin, D. J., Schellekens, H., and Jiskoot, W. (2004), *Pharm. Res*. **21**(6): 897–903.
12. Carpenter, J. F., Chang, B. S., Garzon-Rodriguez, W., and Randolph, T. W. (2002), Rationale design of stable lyophilized protein formulations: theory and practice, in *Rationale Design of Stable Protein Formulations–Theory and Practice*, Carpenter, J. F. and Manning, M. C., eds, Kluwer Academic/Plenum, New York, pp. 109–133.
13. Carpenter, J. F. and Chang, B. S. (1996), Lyophilization of biopharmaceutical products, in *Biotechnology Issues in Pharmaceutical Process Engineering*, Vol. 2, Avis, K. and Wu, V., eds., Intepharm Press, Buffalo Grove, IL; pp. 199–264.
14. Carpenter, J. F., Kendrick, B. S., Chang, B. S., Manning, M. C., and Randolph, T. W. (1999), *Meth. Enzymol*. **309**: 236–255.
15. Wang, W. (1999), *Int. J. Pharm*. **185**(2): 129–188.
16. Carpenter, J. F., Manning, M. C., and Randolph, T. W. (2002), *Curr. Protoc. Protein Sci*., Chapter 4, Unit 4.6.
17. Chang, B. S. and Hershenson, S. (2002), Practical approaches to protein formulation development, in *Rationale Design of Stable Protein Formulations—Theory and Practice*, Carpenter, J. F. and Manning, M. C., eds., Kluwer Academic/Plenum, New York, pp. 1–25.
18. Kendrick, B. S., Li, T., and Chang, B. S. (2002), *Pharm. Biotechnol*. **13**: 61–84.
19. Patro, S. Y., Freund, E., and Chang. B. S. (2002), *Biotech. Annu. Rev*. **8**: 55–84.
20. Chi, E. Y., Krishnan, S., Randolph, T. W., and Carpenter, J. F. (2003), *Pharm. Res*. **20**(9): 1325–1336.
21. Frokjaer, S. and Otzen, D. E. (2005), *Nat. Rev. Drug Discov*. **4**(4): 298–306.
22. Cleland, J. L., Powell, M. F., and Shire. S. J. (1993), *Crit. Rev. Ther. Drug. Carrier Sys.*, **10**: 307–377.
23. Nail, S. L. and Akers, M. J. (2002), *Development and Manufacture of Protein Pharmaceuticals (Pharmaceutical Biotechnology*, Vol. 14), Springer-Verlag, New York.
24. Cromwell, M. E. M., Hilario, E., and Jacobson, F. (2006), *AAPS J*. **8**(3): E572–E579.

25. Branden, C. and Tooze, J. (1999), *Introduction to Protein Structure*, Taylor & Francis, Oxford, UK.
26. Lesk, A. M. (2000), *The Structural Biology of Proteins*, Oxford Univ. Press, Oxford, UK.
27. Santucci, R., Sinibaldi, F., and Fiorucci, L. (2008), *Mini Rev. Med. Chem.* **8**(1): 57–62.
28. Ferreira, S. T., De Felice, F. G., and Chapeaurouge, A. (2006), *Cell. Biochem. Biophys.* **44**(3): 539–548.
29. Kendrick, B. S., Chang, B. S., Arakawa, T., Peterson, B., Randolph, T. W., Manning, M. C., and Carpenter, J. F. (1997), *Proc. Natl. Acad. Sci. USA* **94**: 11917–11922.
30. Tang, K. E. S. and Dill, K. A., (1998), *J. Biomol. Struct. Dyn.* **16**, 397–411.
31. Roberts, C. J. (2007), *Biotechnol. Bioeng.* **98**(5): 927–938.
32. Calamai, M., Canale, C., Relini, A., Stefani, M., Chiti, F., and Dobson, C. M. (2005), *J. Mol. Biol.* **346**(2): 603–616.
33. Pace, C. N. and Tanford, C. (1968), *Biochemistry* **7**: 198–208.
34. Dill, K. A. and Chan, H. S. (1997), *Nat. Struct. Biol.* **4**: 10–19.
35. Krishnan, S., Chi, E. Y., Webb, J. N., Chang, B. S., Shan, D., Goldenberg, M., Manning, M. C., Randolph, T. W., and Carpenter, J. F. (2002), *Biochemistry*, **41**(20): 6422–6431.
36. Raso, S. W., Abel, J., Barnes, J. M., Maloney, K. M., Pipes, G., Treuheit, M. J., King, J., and Brems, D. N. (2005), *Protein Sci.* **14**: 2246–2257.
37. King, J., Haasepettingell, C., Robinson, A. S., Speed, M., and Mitraki, A. (1996), *FASEB J.* **10**: 57–66.
38. Fink, A. L., Calciano, L. J., Goto, Y., Nishimura, A., and Swedberg, S. A. (1993), *Protein Sci.* **2**: 1155–1160.
39. Eaton, W. A., Muñoz, V., Thompson, P. A., Chan, C.-K., and Hofrichter, J. (1997), *Curr. Opin. Struct. Biol.* **7**: 10–14.
40. Kumar, S., Tsai, C. J., Ma, B., and Nussinov, R. (2000), *Proteins.* **38**(4): 368–383.
41. Trivedi, S., Gehlot, H. S., and Rao, S. R. (2006), *Genetics and Molecular Res.* **5**(4): 816–827.
42. Argos, P., Rossmann., M. G., Grau, U. M., Zuber., H., Frank., G., and Tratschin, J. D. (1979), *Biochemistry* **18**: 5698–5703.
43. van den Berg, J. S. P., Limburg, M., Kappelle, L. J., Pals, G., Arwert, F., and Westerveld, A. (1998), *Ann. Neurol.* **43**: 494–498.
44. Querol, E., Pervez-Pons, J. A., and Mozo-Villarias, A. (1996), *Protein Eng.* **9**: 265–271.
45. Valente, J. J., Payne, R. W., Manning, M. C., Wilson, W. W., and Henry, C. S. (2005), *Curr. Pharm. Biotechnol.* **6**: 427–436.
46. Ruckenstein, E. and Shulgin, I. L. (2006), *Adv. Colloid Interface Sci.* **123–126**: 97–103.
47. Winzor, D. J., Deszczynski, M., Harding, S. E., and Wills, P. R. (2007), *Biophys. Chem.* **128**(1): 46–55.
48. Leandro, P. and Gomes, C. M. (2008), *Mini Rev. Med. Chem.* **8**(9): 901–911.
49. Gianni, S., Ivarsson, Y., Jemth, P., Brunori, M., and Travaglini-Allocatelli, C. (2007), *Biophys. Chem.* **128**(2–3): 105–113.
50. Guo, J., Harn, N., Robbins, A., Dougherty, R., and Middaugh, C. R. (2006) *Biochemistry* **45**(28): 8686–8696.
51. Chi, E. Y., Kendrick, B. S., Carpenter, J. F., and Randolph, T. W. (2005), *J. Pharm. Sci.* **94**(12): 2735–2748.
52. Timasheff, S. N. (1993), *Annu. Rev. Biophys. Biomol. Struct.* **22**: 67–97.

53. Timasheff, S. N. (1998), *Adv. Protein Chem.* **51**: 355–432.
54. Neal, B. L., Asthagiri, D., Velev, O. D., Lenhoff, A. M., and Kaler, E. W. (1999), *J. Cryst. Growth.* **196**: 377–387.
55. Alford, J. R., Kendrick, B. S., Carpenter, J. F., and Randolph, T. W. (2008), *Anal. Biochem.* **377**(2): 128–133.
56. Privalov, P. L. (2009), *Meth. Mol. Biol.* **490**: 1–39.
57. Yan, X., Watson, J., Ho, P. S., and Deinzer, M. L. (2004), *Mol. Cell. Proteom.* **3**(1): 10–23.
58. Piedmonte, D. M. and Treuheit, M. J. (2008), *Adv. Drug Deliv. Rev.* **60**(1): 50–58.
59. Herman, A. C., Boone, T. C., and Lu, H. S. (1996), *Formulation, Characterization and Stability of Protein Drugs*, Pearlman, R. and Wang, Y. J., eds., Plenum Press, New York, pp. 303–328.
60. Tobler, S. A., and Fernandez, E. J. (2002), *Protein Sci.* **11**: 1340–1352.
61. Fawzi, N. L., Chubukov, V., Clark, L. A., Brown, S., and Head-Gordon, T. (2005), *Protein Sci.* **14**: 993–1003.
62. Shulgin, I. L. and Ruckenstein, E. (2006), *Biophys. Chem.* **120**(3): 188–198.
63. Keeler, C., Hodsdon, M. E., and Dannies, P. S. (2004), *J. Mol. Neurosci.* **22**: 43–49.
64. Dannies, P. (2003), *BioDrugs* **7**(5): 315–324.
65. Shih, Y. C., Prausnitz, J. M., and Blanch, H. W. (1992), *Biotechnol. Bioeng.* **40**(10): 1155–1164.
66. Liu, J., Nguyen, M. D., Andya, J. D., and Shire, S. J. (2005), *J. Pharm. Sci.* **94**(9): 1928–1940.
67. Safar, J., Roller, P. P., Gajdusek, D. C., and Gibbs Jr., C. J. (1994), *Biochemistry* **33**: 8375–8383.
68. Mamathambika, B. S. and Bardwell, J. C. (2008), *Annu. Rev. Cell. Devel. Biol.* **24**: 211–235.
69. Chang, J. Y., Lu, B. Y., and Li, L. (2005), *Anal. Biochem.* **342**(1): 78–85.
70. Janowski, R., Kozak, M., Jankowska, E., Grzonka, Z., Grubb, A., Abrahamson, M., and Jaskolski, M. (2001), *Nat. Struct. Biol.* **8**: 316–320.
71. Liu, Y. and Eisenberg, D. (2002), *Protein Sci.* **11**: 1285–1299.
72. Dobson, C. M. (2003), *Nature* **426**: 884–890.
73. Gao, J., Yin, D. H., Yao, Y., Sun, H., Qin, Z., Schoneich, C., Williams, T. D., and Squier, T. C. (1998), *Biophys. J.* **74**: 1115–1134.
74. Li, S., Schöneich, C., and Borchardt, R. T. (1995), *Pharm Res.* **12**(3): 348–355.
75. Robinson, N. E. and Robinson, A. B. (2001), *Proc. Natl. Acad. Sci. USA* **98**(22): 12409–12413.
76. Gupta, R. and Srivastava, O. P. (2004), *J. Biol. Chem.* **279**(43): 44258–44269.
77. Lee, J. C. and Timasheff, S. N. (1981), *J. Biol. Chem.* **256**(14): 7193–7201.
78. Arakawa, T. and Timasheff, S. N. (1982), *Biochemistry* **21**: 6536–6544.
79. Chen, B. L., Arakawa, T., Hsu, E., Narhi, L. O., Tressel, T. J., and Chien, S. L., (1994), *J. Pharm. Sci.* **83**: 1657–1661.
80. Volkin, D. B., Tsai, P. K., Dabora, J. M., et al., (1993), *Arch. Biochem. Biophys.* **300**: 30–41.

81. Thirumangalathu, R., Krishnan, S., Brems, D. N., Randolph, T. W., and Carpenter, J. F. (2006), *J. Pharm. Sci*. **95**(7): 1480–1497.
82. Zhang, Y., Roy, S., Jones, L. S., Krishnan, S., Kerwin, B. A., Chang, B. S., Manning, M. C., Randolph, T. W., and Carpenter, J. F. (2004), *J. Pharm. Sci*. **93**(12): 3076–3089.
83. DeGrado, W. F., Summa, C. M., Pavone, V., Nastri, F., and Lombardi, A. (1999), *Annu. Rev. Biochem*. **68**: 779–819.
84. Brange J. (1987), *Galenics of Insulin: The Physico-chemical and Pharmaceutical Aspects of Insulin and Insulin Preparations*, Springer-Verlag, New York, pp. 1–101.
85. Brange, J. (1992), *Acta Pharm. Nord*. **4**(4): 209–222.
86. Uversky, V. N., Li, J., and Fink, A. L. (2001), *J. Biol. Chem*. **276**: 44284–44296.
87. Henson, A. F., Mitchell, J. R., and Musselwhite, P. R. (1970), *J. Colloid Interface Sci*. **32**: 162–165.
88. Maa, Y. F. and Hsu, C. C. (1997), *Biotechnol. Bioeng*. **54**: 503–512.
89. Charman, S. A., Mason, K. L., and Charman, W. N. (1993), *Pharm. Res*. **10**: 954–961.
90. Kreilgaard, L., Jones, L. S., Randolph, T. W., Frokjaer, S., Flink, J. M., Manning, M. C., and Carpenter, J. F. (1998), *J. Pharm. Sci*. **87**(12): 1597–1603.
91. MacRitchie, F. (1989), *J. Colloid Interface Sci*. **41**: 25–34.
92. Sluzky, V., Tamada, J. A., Klibanov, A. M., and Langer, R. (1991), *Proc. Natl. Acad. Sci. USA* **88**: 9377–9381.
93. Chang, B. S., Kendrick, B. S., and Carpenter, J. F. (1996), *J. Pharm. Sci*. **85**: 1325–1330.
94. Cumper, C. W. N. and Alexander, A. E. (1950), *Australian J. Sci. Res., Ser A: Phys. Sci*. **5**: 189–197.
95. Thomas, C. R. and Dunhill, P. (1979), *Biotechnol. Bioeng*. **21**(12): 2279–2302.
96. Sluzky, V., Klibanov, A., and Langer, R., (1992), *Biotech. and Bioeng*. **40**: 895–903.
97. Truskey, G. A., Gabler, R., DiLeo, A., and Manter, T. (1987), *J. Parenter. Sci. Technol*. **41**: 180–193.
98. Narendranathan, T. J. and Dunnill, P. (1982), *Biotech. Bioeng*. **24**: 2103–2107.
99. Charm, S. E. and Wong, B. L. (1970), *Biotechnol. Bioeng*. **12**: 1103–1109.
100. Charm, S. E. and Wong, B. L. (1981), *Enzyme Microb. Technol*. **3**: 111–118.
101. Charm, S. E. and Lai, C. J. (1971), *Biotechnol. Bioeng*. **13**: 185–202.
102. Tirrell, M. and Middleman, S. (1979), *Biopolymers* **18**: 59–72.
103. Lencki, R. W., Tecante, A., and Choplin, L. (1993), *Biotechnol. Bioeng*. **42**(9): 1061–1067.
104. Langmuir, I. and Schaefer, V. J. (1930), *Chem. Rev*. **24**: 181.
105. Bull, H. B. (1947), *Adv. Protein Chem*. **3**: 95.
106. Bull, H. B. and Neurath, H. (1937), *J. Biol. Chem*. **118**: 163.
107. Kaplan, J. G. and Frazer, M. H. (1953), *Nature* **171**: 550.
108. Sharp, J. S., Forrest, J. A., and Jones, R. A. L. (2002), *Biochemistry* **41**: 15810–15819.
109. Adams, S., Higgins, A. M., and Jones, R. A. L. (2002), *Langmuir* **18**: 4854–4861.
110. Zhu, M., Souillac, P. O., Ionescu-Zanetti, C., Carter, S. A., and Fink, A. L. (2002), *J. Biol. Chem*. **277**: 50914–50922.
111. Sefton, M. V. (1982), in *Biomaterials: Interfacial Phenomena and Applications*, Cooper, S. L. and Peppas, N. A., eds., American Chemical Society, Washington, DC, pp. 511–522.

112. Vlasveld, L. T., Beijnen, J. H., Sein, J. J., Rankin, E. M., Melief, C. J. M., and Hekman, A. (1993), *Eur. J. Cancer* **29A**: 1979.

113. Miles, D. W., Bird, C. R., Wadhwa, M., Summerhayes, M., Balkwill, F. R., Thorpe, R., and Rubens, R. D. (1990), *Lancet* **335**: 1602.

114. Tzannis, S. T., Hrusheskyb, W. J. M., Wood, P. A., and Przybycien, T. M. (1997), *J. Colloid Interface Sci.* **189**(2): 216–228.

115. Norde, W., and Haynes, C.A., (1995), Reversability and the Mechanism of Protein absorption in *Protein at Interface II: Fundamentals and Applications*. ACS Symposium Series 602, ACS, Washington D.C.

116. Jiskoot, W. and Crommelin, D. (2005), *Methods for Structural Analysis of Protein Pharmaceuticals*, Springer, New York.

117. Kueltzo, L. A., Wang, W., Randolph, T. W., and Carpenter, J. F. (2008), *J. Pharm. Sci.* **97**: 1801–1812.

118. Liu, D., Ren, D., Huang, H., Dankberg, J., Rosenfeld, R., Cocco, M. J., Li, L., Brems, D. N., and Remmele, R. L. (2008), *Biochemistry* **47**: 5088–5100.

119. Harn, N., Allan, C., Oliver, C., and Middaugh, C. R. (2007), *J. Pharm. Sci.* **96**, 532–546.

120. Barth, A. (2007), *Biochim. Biophys. Acta* **1767**: 1073–1101.

121. Schule, S., Friess, W., Bechtold-Peters, K., and Garidel, P. (2007), *Eur. J. Pharm. Biopharm.* **65**: 1–9.

122. Chang, L., Sheperd, E., Sun, J., Tang, X., and Pikal, M. J. (2005), *J. Pharm. Sci.* **94**: 1445–1455.

123. Abdul-Fattah, A. M., Le, V. T., Yee, L., Nguyen, L., Kalonia, D. S., Cicerone, M. T., and Pikal, M. J. (2007), *J. Pharm. Sci.* **96**: 1983–2008.

124. Schwartz, D., Sofia, S., and Friess, W. (2006), *Eur. J. Pharm. Biopharm.* **63**(3): 241–248.

125. Elkordy, A. A., Forbes, R. T., and Barry. B. W. (2008), *Eur. J. Pharm. Sci.* **33**: 177–190.

126. Ramachander, R., Jiang, J., Li, C., Eris, T., Young, M., Dimitrova, M., and Narhi, L. (2008), *Anal. Biochem.* **376**: 173–182.

127. Demeule, B., Gurny, R., and Arvinte, T. (2007), *Int. J. Pharm.* **329**: 7–45.

128. D'Auria, S., Staiano, M., Varriale, A., Gonnelli, M., Marabotti, A., Rossi, M., and Strambini, G. B. (2008), *Proteins.* **71**: 743–750.

129. Han, J., Jin, B-S., Lee, S-B., Sohn, Y., Joung, J-W., and Lee, J-H. (2007), *Arch. Pharm. Res.* **30**: 1121–1131.

130. Ihnat, P. M., Vellekamp, G., Obenauer-Kutner, L. J., Duan, J., Han, M. A., Witchey-Lakshmanan, L. C., and Grace, M. J. (2005), *Biochim. Biophys. Acta* **1726**: 138–151.

131. Gabrielson, J. P., Brader, M. L., Pekar, A. H., Mathis, K. B. Winter, G., Carpenter, J. F., and Randooph, T. W. (2007), *J. Pharm. Sci.* **96**: 268–279.

132. Hawe, A. and Friess, W. (2008), *Eur. J. Pharm. Biopharm.* **68**: 169–182.

133. Ahrer, K., Buchacher, A., Iberer, G., and Jungbauer, A. (2006), *J. Biochem. Biophys. Meth.* **66**: 73–86.

134. Sukumar, M., Doyle, B. L., Combs, J. L., and Pekar, A. H. (2004), *Pharm. Res.* **21**: 1087–1093.

135. Ye, H. (2006), *Anal. Biochem.* **356**: 76–85.

136. Li, Y., Williams, T. D., and Topp, E. M. (2008), *Pharm. Res.* **25**: 259–267.

137. Norman, M., Hubalek, F., and Schluckebier, G. (2007), *Eur. J. Pharm. Sci.* **30**: 414–423.

138. Tien, F., Middaugh, R., Offerdahl, T., Munson, E., Sane, S., and Rytting, J. H. (2007), *Int. J. Pharm.* **335**: 20–31.
139. Kempe, S., Metz, H., Bastrop, M., Hvilsom, A., Contri, R. V., and Mader, K. (2008), *Eur. J. Pharm. Biopharm*, **68**: 26–33.
140. Liu, H., Gaza-Bulseco, G., and Sun, J., (2006), *J. Chrom.* **837**: 35–43.
141. Yu, L., Remmele, R. L., and He, B. (2006), *Rapid Commun. Mass Spectrom.* **20**: 3674–3680.
142. Qian, J., Tang, Q., Cronin, B., Markovich, R., and Rustum, A. (2008), *J. Chrom.* **1194**: 48–86.
143. Ejima, D., Yumioka, R., Arakawa, T., and Tsumoto, K. (2005), *J. Chrom.* **1094**: 49–55.
144. Han, M, Phan, D., Nightlinger, N., Taylor, L., Jankhah, S., Woodruff, B., Yates, Z., Freeman, S., Guo, A., Balland, A., and Pettit, D. (2006), *Chromatographia* **64**: 335–342.
145. Vasilyeva, E., Woodard, J., Taylor, F. R., Kretschmer, M., Fajardo, H., Lyubarskaya, Y., Kobayashi, K., Dingley, A., and Mhatre, R. (2004), *Electrophoresis* **25**: 3890–3896.
146. Salas-Solano, O., Tomlinson, B., Du, S., Parker, M., Strahan, A., and Ma, S. (2006), *Anal. Chem.* **78**: 6583–6594.
147. Gaza-Bulseco, G., Bulseco, A., Chumsae, C., and Liu, H. (2008), *J. Chrom.* **862**: 155–160.
148. Johnson, K. A., Raisley-Flango, K., Tangarone, B. S., Porter, T. J., and Rouse, J. C. (2007), *Anal. Biochem.* **360**: 75–83.
149. Lyubarskaya, Y., Houde, D., Woodard, J., Murphy, D., and Mhatre, R. (2006), *Anal. Biochem.* **348**: 24–39.
150. Bo, T. and Pawilszyn, J. (2006), *J. Separ. Sci.*, **29**: 1018–1025.
151. Han, Y., Jin, B.-S., Lee, S.-B, Sohn, Y., Joung, J.-W., and Lee, J.-H. (2007), *Arch. Pharm. Res.* **30**: 1121–1131.
152. Wakankar, A. A., Liu, J., Vandervelde, D., Wang, Y. J., Shire, S. J., and Borchardt, R. T. (2007), *J. Pharm. Sci.* **96**: 1708–1718.
153. Peter-Katalinić, J. (2005), *Meth. Enzymol.* **405**: 139–171.
154. Medzihradszky, K. F. (2005), *Meth. Enzymol.* **405**: 116–138.
155. Manning, M. C., Patel, K., and Borchardt, R. T., (1989), *Pharm. Res.* **6**: 903–917.
156. Carpenter, J. F., Pikal, M. J., Chang, B. S., and Randolph, T. W. (1997), *Pharm. Res.* **14**: 969–975.
157. Spink, C. H. (2008), *Meth. Cell. Biol.* **84**: 115–141.
158. Privalov, P. L. and Dragan, A. I. (2007), *Biophys. Chem.* **126**: 16–24.
159. Penarrubia, L. and Moreno, J. (1987), *Biochim. Biophys. Acta* **916**: 227–235.
160. Lee, Y. K. and Choo, C. L. (1989), *Biotech. Bioeng.* **33**: 183–190.
161. Elias, C. B. and Joshi, J. B. (1998), *Adv. Biochem. Eng.* **59**: 48–71.
162. Bam, N. B., Cleland, J. L., Yang, J., Manning, M. C., Carpenter, J. F., Kelley, R. F., and Randolph, T. W. (1998), *J. Pharm. Sci.* **87**(12): 1554–1559.
163. Chou, D. K., Krishnamurthy, R., Randolph, T. W., Carpenter, J. F., and Manning, M. C. (2005), *J. Pharm. Sci.* **94**(6): 1368–1381.
164. Kiese, S., Papppenberger, A., Friess, W., and Mahler, H. C. (2008), *J. Pharm. Sci.* **97**(10): 4347–4366.
165. Meinders, M. B. J. and de Jongh, H. H. J. (2002), *Biopolymers* **67**: 319–322.
166. Kudryashova, E. V., Visser, A. J. W. G., and De Jongh, H. H. J. (2005), *Protein Sci.* **14**: 483–493.

167. Horiuchi, T. and Fukushima, D. (1978), *Food Chem.* **3**: 35.

168. Townsend, A. A. and Nakai, S. (1983), *J. Food Sci.* **48**: 588.

169. Kato, A. and Nakai, S. (1980), *Biochem. Biophys. Acta* **624**: 13.

170. Kinsella, J. E. (1981), Relationship between structure and functional properties of food proteins, in *Food Proteins*, Fox, P. F. and Condon, J. J., eds., Applied Science Publishers, London and New York.

171. Halling, P. J. (1981), Protein stabilized foam and emulsions, *CRC Crit. Rev. Food Sci. Nutr.* **15**: 155.

172. Jones, L. S., Kaufmann, A., and Middaugh, C. R. (2005), *J. Pharm. Sci.* **94**: 918–927

173. Kim, H. J., Decker, E. A., and McClements, D. J. (2002), *J. Agric. Food Chem.* **50**(24): 7131–7137.

174. Gorbenko, G. P., Ioffe, V. M., and Kinnuneny, P. K. J. (2007), *Biophys. J.* **93**: 140–153.

175. Ruiz, L., Reyes, N., Aroche, K., Tolosa, V., Simanca, V., Rodríguez, T., and Hardy, E. (2005), *J. Pharm. Pharm. Sci.* **8**(2): 207–216.

176. Tzannis, S. T., Hrusheskyb, W. J. M., Wood, P. A., and Przybycien, T. M. (1996), *Proc. Natl. Acad. Sci. USA* **93**: 5460–5465.

177. Tyagi, A. K., Randolph, T. W., Dong, A., Maloney, K. M., Hitscherich Jr., C., and Carpenter, J. F. (2009), *J. Pharm. Sci.* **98**(1): 94–104.

178. Jarrett, J. T., and Lansbury Jr., P. T. (1993), *Cell* **73**: 1055–1058.

179. Lazol, N. D., Grant, M. A., Condron, M. C., Rigby, A. C., and Teplow, D. B. (2005), *Protein Sci.* **14**: 1581–1596.

180. Haass, C. and Steiner, H. (2001), *Nat. Neurosci.* **4**: 859–860.

181. Klein, W. L., Krafft, G. A., and Finch, C. E. (2001), *Trends Neurosci.* **24**: 219–224.

182. Klein, W. L., Stine Jr., W. B., and Teplow, D. B. (2004), *Neurobiol. Aging* **25**: 569–580.

183. Kirkitadze, M. D., Bitan, G., and Teplow, D. B. (2002), *J. Neurosci. Res.* **69**: 567–577.

184. Walsh, D. M., Hartley, D. M., Kusumoto, Y., Fezoui, Y., Condron, M. M., Lomakin, A., Benedek, G. B., Selkoe, D. J., and Teplow, D. B. (1999), *J. Biol. Chem.* **274**: 25945–25952.

185. Kerwin, B. A. and Remmele Jr., R. L. (2007), *J. Pharm. Sci.* **96**(6): 1468–1479.

186. Qi, P., Volkin, D. B., Zhao, H., Nedved, M. L., Hughes, R., Bass, R., Yi, S. C., Panek, M. E., Wang, D., DalMonte, P., and Bond, M. D. (2009), *J. Pharm. Sci.* **98**(9): 3117–3130.

187. Berlett, B. S. and Stadtman, E. R. (1997), *J. Biol. Chem.* **272**(33): 20313–20316.

188. Swallow, A. J. (1960), in *Radiation Chemistry of Organic Compounds*, Swallow, A. J., ed., Pergamon Press, New York, pp. 211–224.

189. Garrison, W. M. (1987), *Chem. Rev.* **87**: 381–398.

190. Garrison, W. M., Jayko, M. E., and Bennett, W. (1962), *Rad. Res.* **16**: 487–502.

191. Schuessler, H. and Schilling, K. (1984), *Int. J. Rad. Biol.* **45**: 267–281.

192. Stadtman, E. R. and Levine, R. L. (2000), Protein oxidation, *Ann. NY Acad. Sci.* **899**: 191–208.

193. Franks, F. (2002), *Biophys. Chem.* **96**: 117–127.

194. Hsu, C. C., Ward, C. A., Pearlman, R., Nguyen, H. M., Yeung, D. A., and Curley, J. G. (1992), *Devel. Biol. Stand.* **74**: 255–270.

195. Carpenter, J. F. and Crowe, J. H. (1988), *Cryobiology* **25**: 244.

196. Prestrelski, S. J., Tedischi, N., Arakawa, T., and Carpenter, J. F. (1993), *Biophys. J*. **65**: 661–671.
197. Chang, B. S., Beauvais, R. M., Dong, A., and Carpenter, J. F. (1996), *Arch. Biochem. Biophys*. **331**: 249–258.
198. Pikal-Cleland, K. A., Cleland, J. L., Anchordoquy, T. J., and Carpenter, J. F. (2002), *J. Pharm. Sci*. **91**: 1969–1679.
199. Cao, E., Chen, Y., Cui, Z., and Foster, P. R. (2003), *Biotechnol. Bioeng*. **82**: 684–690.
200. Van den Berg, L. (1959), *Arch. Biochem. Biophys*. **84**: 305–315.
201. Van den Berg, L. and Rose, D. (1959), *Arch. Biochem. Biophys*. **81**: 319.
202. Chatterjee, K., Shalaev, E. Y., and Suryanarayanan, R. (2005), *J. Pharm. Sci*. **94**(4): 798–808.
203. Chatterjee, K., Shalaev, E. Y., and Suryanarayanan, R. (2005), *Pharm. Res*. **22**: 303–309.
204. Jiang, S. and Nail, S. L. (1998), *Eur. J. Pharm. Biopharm*. **45**: 249–257.
205. Sarciaux, J. M., Mansour, S., Hageman, M. J., and Nail, S. L. (1999), *J. Pharm. Sci*. **88**: 1354–1361.
206. Eckhardt, B. M., Oeswein, J. Q., and Bewley, T. A. (1991), *Pharm. Res*. **8**: 1360–1364.
207. Heller, M. C., Carpenter, J. F., and Randolph, T. W. (1997), *Biotechnol. Progress* **13**: 590–596.
208. Heller, M. C., Carpenter, J. F., and Randolph, T. W. (1999), *Biotechnol. Bioeng*. **63**: 166–174.
209. Lueckel, B., Helk, B., Bodmer, D., and Leuenberger, H (1998), *Pharm. Devel. Technol*. **3**: 337–346.
210. Izutsu, K. and Kojima, S. (2002), *J. Pharm. Pharmacol*. **54**: 1033–1039.
211. Privalov, P. L. (1990), Cold denaturation of proteins, *Crit. Rev. Biochem. Mol. Biol*. **25**: 281–305.
212. Privalov, P. L. (2000), *Meth. Enzymol*. **323**: 31–62.
213. Franks, F. (1988), *Characterization of Proteins*, Humana Press, Clifton, NJ.
214. Tang, X. and Pikal, M. J. (2005), *Pharm. Res*. **22**(7): 1167–1175.
215. Franks, F. and Hatley, R. H. M. (1991), *Pure Appl. Chem*. **63**(10): 1367–1380.
216. Jaenicke, R. (1990), *Philos. Trans. Roy. Soc. Lond. B* **326**: 535–551.
217. Creighton, T. E. (1993), in *Proteins: Structures and Molecular Properties*, Creighton, T. E. ed., Freeman, New York, pp. 1–48.
218. Ikeguchi, M., Kuwajima, K., Mitani, M., and Sugai, S. (1986), *Biochemistry* **25**: 6965–6972.
219. Kuwajima, K. (1989), *Proteins Struct. Funct. Genet*. **6**: 87–103.
220. Goto, Y., Calciano, L. J., and Fink, A. L. (1990), *Proc. Natl. Acad. Sci. USA* **87**: 573–577.
221. Sugawara, T., Kuwajima, K., and Sugai, S. (1991), *Biochemistry* **30**: 2698–2706.
222. Dolgikh, D. A., Gilmanshin, R. I., Brazhnikov, E. V., Bychkova, V. E., Semisotnov, G. V., Venyaminov, S., and Ptitsyn, O. B. (1981), *FEBS Lett*. **136**: 311–315.
223. Marmorino, J. L., Lehti, M., and Pielak, G. J. (1998), *J. Mol. Biol*. **275**: 379–388.
224. Ptitsyn, O. (1992), The molten globule state, in *Protein Folding*, Creighton, T., ed., Freeman, New York, pp. 243–300.
225. Ohgushi, M. and Wada, A. (1983), *FEBS Lett*. **164**: 21–24.
226. Hawley, S. A. (1971), *Biochemistry* **10**: 2436–2442.

227. Jaenicke, R. (1981), *Annu. Rev. Biophys. Bioeng*. **10**: 1–67.
228. Heremans, K. (1995), High pressure effects on biomolecules, in *High Pressure Processing of Foods*, Ledward, D. A., Johnston, D. E., Earnshaw, R. G., and Hasting, A. P. M., eds., Nottingham Univ. Press, Leicestershire, UK.
229. Seefeldt, M. B., Kim, Y. S., Tolley, K. P., Seely, J., Carpenter, J. F., and Randolph. T. W. (2005), *Protein Sci*. **14**: 2258–2266.
230. Webb, J. N., Webb, S. D., Cleland, J. L., Carpenter, J. F., and Randolph, T. W. (2001), *Proc. Natl. Acad. Sci. USA* **98**: 7259–7264.
231. St. John, R. J., Carpenter, J. F., and Randolph, T. W. (1999), *Proc. Natl. Acad. Sci. USA* **96**(23): 13029–13033.
232. St. John, R. J., Carpenter, J. F., Balny, C., and Randolph, T. W. (2001), *J. Biol. Chem*. **276**(50): 46856–46863.
233. Lefebvre, B. G., Comolli, N. K., Gage, M. J., and Robinson, A. S. (2004), *Protein Sci*. **13**: 1538–1546.
234. Takahashi, T. (1997), *Adv. Biophys*. **34**: 41–54.
235. Nakamura, H. (1996), *Q. Rev. Biophys*. **29**(1): 1–90.
236. Arakawa, T. and Timasheff, S. N. (1983), *Arch. Biochem. Biophys*. **224**: 169–177.
237. Arakawa, T. and Timasheff, S. N. (1982), *Biochemistry* **21**: 6545–6552.
238. Allison, S. D., Chang, B. S., Randolph, T. W., and Carpenter, J. F. (1999), *Arch. Biochem. Biophys*. **365**: 289–298.
239. Andya, J. D., Hsu, C. C., and Shire, S. J. (2003), *AAPS Pharm. Sci*. **5**(2): 1–11.
240. Schlieker, C., Bukau, B., and Mogk, A. (2002), *J. Biotechnol*. **96**(1): 13–21.
241. Thirumangalathu, R., Krishnan, S., Brems, D. N., Randolph, T. W., and Carpenter, J. F. (2006), *J. Pharm. Sci*. **95**(7): 1480–1497.
242. Bam, N. B., Cleland, J. L., and Randolph, T. W. (1996), *Biotechnol. Progress* **12**(6): 801–809.
243. Joshi, O., Chu, L., McGuire, J., and Wang, D. Q. (2008), *J. Pharm. Sci*. **97**(11): 4741–4755.
244. Joshi, O., Chu, L., McGuire, J., and Wang, D. Q. (2009), *J. Pharm. Sci*. **98**(9): 3099–107.
245. Giger, K., Vanam, R. P., Seyrek, E., and Dubin, P. L. (2008), *Biomacromolecules* **9**(9): 2338–2344.
246. Chung, K., Kim, J., Cho, B. K., Ko, B. J., Hwang, B. Y., and Kim, B. G. (2008), *Biochim. Biophys. Acta*. **1774**(2): 249–257.
247. Taneja, S. and Ahmad, F. (1994), *Biochem. J*. **303**: 147–153.
248. Arakawa, T. and Tsumoto, K. (2003), *Biochem. Biophys. Res. Commun*. **304**: 148–152.
249. Shiraki, K., Kudou, M., Nishikori, S., Kitagawa, H., Imanaka, T., and Takagi, M. (2004), *Eur. J. Biochem*. **271**: 3242–3247.
250. Baynes, B. M., Wang, D. I. C., and Trout, B. L. (2005), *Biochemistry* **44**: 4919–4925.
251. Arakawa, T., Ejima, D., Tsumoto, K., Obeyama, N., Tanaka, Y., Kita, Y., and Timasheff, S. N. (2007), *Biophys. Chem*. **127**(1–2): 1–8.
252. Beissinger, M. and Buchner, J. (1998), *J. Biol. Chem*. **379**: 245–259.
253. Frydman, J. (2001), *Annu. Rev. Biochem*. **70**: 603–647.
254. Young, J. C., Barral, J. M., and Ulrich Hartl, F. (2003), *Trends Biochem. Sci*. **28**: 541–547.

255. Wegele, H., Muller, L., and Buchner, J. (2004), *Rev. Physiol. Biochem. Pharmacol.* **151**: 1–44.

256. Mogk, A. and Bukau, B. (2004), *Curr. Biol.* **14**: R78–R80.

257. Katre, N. V., Knauf, M. J., and Laird, W. J. (1987), *Proc. Natl. Acad. Sci. USA* **84**: 1487–1491.

258. Manish, D. and Park, T. G. (2001), *J. Control. Release* **73**: 233–244.

259. Rajan, R. S., Li, T., Aras, M., Sloey, C., Sutherland, W., Arai, H., Briddell, R., Kinstler, O., Lueras, A. M. K., Zhang, Y., Yeghnazar, H., Treuheit, M., and Brems, D. N. (2006), *Protein Sci.* **15**: 1063–1075.

260. Narhi, L. O., Arakawa, T., Aoki, K. H., Elmore, R., Rohde, M. F., Boone, T., and Strickland, T. W. (1991), *J. Biol. Chem.* **266**: 23022–23026.

261. Endo, Y., Nagai, H., Watanabe, Y., Ochi, K., and Takagi, T. (1992), *J. Biochem.* **112**: 700–706.

262. Ono, M. (1994), *Eur. J. Cancer* **30A**(Suppl. 3): S7–S11.

263. Sinclair, A. M. and Elliott, S. (2005), *J. Pharm. Sci.* **94**: 1626–1635.

4

IMMUNOGENICITY OF THERAPEUTIC PROTEINS

Steven J. Swanson

4.1. INTRODUCTION

The development of protein therapeutics is a complex process and presents unique challenges to manufacturers. Unlike small-molecule therapeutics that are not recognized by the recipient's immune system, protein therapeutics are of sufficient size and complexity that they can be recognized. For many of the protein therapeutics that are designed to approximate endogenous proteins manufactured by the recipient, the immune system has already established a level of tolerance toward that protein. *Tolerance* is the mechanism that the immune system uses to essentially ignore all of the proteins that are produced by an individual to prevent the occurence of an autoimmune reaction. An important purpose for the immune system is therefore to identify proteins that are not produced by the individual and to facilitate their removal. When foreign proteins, those not produced by the individual, are detected by the immune system, a cascade of events is triggered that results in the removal of those foreign proteins while leaving self-produced proteins intact.

Although a very close homology between a protein therapeutic and an endogenous protein helps protect tolerance, several aspects of the manufacturing process can help

Formulation and Process Development Strategies for Manufacturing Biopharmaceuticals,
edited by Jameel and Hershenson

the immune system recognize the therapeutic protein as a foreign protein and break through the tolerance. Once the immune system has been triggered to respond to the therapeutic protein, there are several levels of response that can be initiated, ranging from a very low level response that has very little impact on the ability of the therapeutic protein to perform its function to a robust immune response that can abrogate all of the biological effects of not only the therapeutic protein but also the endogenous counterpart protein. The ability of the immune system to block all of the effects of not only the protein therapeutic but also an endogenous counterpart is the reason why immunogenicity of therapeutic proteins is taken so seriously by manufacturers and is regarded as a significant safety factor by regulatory agencies throughout the world.

While much is known regarding initiation and progression of an immune response against therapeutic proteins, work continues to identify quantifiable risk factors to help explain why some protein therapeutics induce a robust immune response, while others do not. In this chapter we focus on how the immune system responds to a therapeutic protein, the risk factors that can lead to an immune response, different ways of identifying and characterizing an immune response, a strategy for monitoring patients for developing an immune response, and the consequences of an immune response to a therapeutic protein.

4.2. HOW THE IMMUNE SYSTEM RESPONDS TO PROTEIN THERAPEUTICS

The first stage in an immune system response to a foreign protein is recognition. In the case of therapeutic proteins, this can occur as soon as the therapeutic is injected into the patient. Depending on the route of administration, different cells in the immune system will be the first to respond. When the drug is administered subcutaneously, dendritic cells are typically first to encounter the protein. Once the dendritic cells encounter the protein, they can phagocytize it, break it down into small subunits, and then ship these subunits to the surface of the dendritic cell to present it to other cells in the immune system. Macrophages are also important for phagocytosis, inflammation, and subsequent antigen presentation. This antigen presentation is important for the next step in the immune response. Other cells that are involved include granulocytes, which can be further classified as eosinophils, basophils, or neutrophils. Eosinophils are very important in fighting parasitic infections and in allergic responses, while the role of basophils is not as well understood. Other cells involved in the early immune response are mast cells, which can release histamine as well as the other contents of their granules on stimulation by a foreign protein. This release results in a local inflammatory response that can trigger other response cells to be attracted to the area.

The inflammatory response is important for the next phase of an immune response because of the recruitment of other cells in the immune system to the local area. The other cells that are recruited include macrophages, neutrophils, and lymphocytes. When T cells encounter dendritic cells with protein fragments presented on their surfaces that are compatible with the T cell, the T cell binds to the target on the dendritic cell and becomes activated. This activation triggers proliferation of that T cell clone, which then can trigger binding to a specific B cell, leading to the stimulation and proliferation of the antibody-producing B-cell clone. The end result of the activation and proliferation cascade is production of B cells that are capable of producing antibodies that are released into circulation and also the generation of memory B cells that provide the person with the ability to respond more rapidly and robustly to subsequent encounters with that particular protein.

The initial antibodies that are produced and released into circulation are of the IgM class and are generally of low affinity. IgM antibodies are pentameric in nature with 10 identical binding sites available to bind the triggering protein. The low-affinity nature of the binding of these antibodies to the target protein means that these antibodies are generally not capable of binding tightly enough to neutralize the biological effect of the therapeutic protein. This early immune response that can occur after a single dose of a therapeutic protein generally does not have a significant clinical effect. However, subsequent exposure to the therapeutic protein and recognition by the immune system triggers a much more robust response.

While the initial response produces IgM antibodies at moderate concentrations (ranging from low picograms to low micrograms) that appear in the circulation within 2 weeks of exposure, the second exposure can result in a very robust response. This secondary response can cause antibodies to appear in the circulation within hours of exposure and the concentration ranges to hundreds of micrograms per milliliter of blood. In addition, the interaction of activated T cells induces a change in the immune response in that B cells switch from making IgM antibodies to making IgG antibodies. This class switching mediated by T cells is accompanied by other changes in the antibodies that are produced. A phenomenon known as *epitope spreading* occurs where the epitope, or region on the protein where the antibody binds, "spreads" across the protein such that different antibodies are generated that are capable of binding to different regions of the protein. This increases the chances that antibodies will be generated that are capable of binding and neutralizing the active region of the protein. Another change is that in addition to multiple regions of the protein now being recognized by antibodies, the affinity of the antibodies increases. This increase in affinity means that the antibodies are capable of binding more vigorously to the protein and there is a much greater chance that the antibody will have a clinical effect. Another attribute of this more robust immune response is that a higher concentration of antibodies is produced. The end result of this T-cell-induced immune maturation is that more antibodies of the IgG class are produced that bind tighter and recognize more regions of the protein. This cascade of events often leads to an immune response that is capable of neutralizing the biological effect of the drug. The impact for the patient is a lower potential efficacy of the protein therapeutic. As is described in more detail later in this chapter, the other potential consequence is that an endogenous counterpart to the therapeutic protein could also be the target of these antibodies, which could lead to complete neutralization of the endogenous protein.

The type of immune response that a patient is able to generate against a therapeutic protein can also depend on the general state of health of that person's immune system. For example, a patient who demonstrates a high level of autoantibodies (antibodies directed against "self-made" proteins) may have challenges in maintaining tolerance and could be more likely to induce an immune response. In contrast, patients undergoing chemotherapy that weakens the immune system are generally less likely to mount an immune response against a therapeutic protein. Perhaps the biggest challenge for manufacturers of therapeutic proteins is when proteins are produced for replacement therapy in patients who are genetically incapable of making a given protein. Because there is no equivalent self-made protein, any tolerance for the protein therapeutic is unlikely, and this situation can easily lead to an immune response. In these cases, it may be possible to suppress the immune system, similar to the suppression that is used to support organ transplants, to enable the patient to benefit from the protein therapeutic. Because of the general, rather than very specific, nature of immune suppression that current therapies provide, immunosuppression is warranted only in certain extreme situations, and for the potential for immunosuppression to provide relief against immunogenicity induced by therapeutic proteins in the majority of cases in the near future is very low.

4.3. RISK FACTORS FOR INDUCING IMMUNOGENICITY

The risk factors for inducing an immune response are those factors that can render the therapeutic protein more likely to be initially recognized by the immune system, those factors that are likely to increase general inflammation, which could lead to a progression of the immune response, or those factors that specifically promote maturation of the immune response.

While there is still much to learn regarding specifics of what induces a more robust immune response, there are several key factors to consider. Two of the most important risk factors are aggregation and sequence differences. There are many different forms of aggregates, and the most disconcerting would be insoluble high-molecular-weight aggregates. Aggregates of this type are more likely to remain deposited and scavenged by dendritic cells, especially with a subcutaneously administered protein. The risk can often be reduced by using an intravenous route of administration, as this path avoids contact with dendritic cells. The risk is not completely eliminated, as there are other cells such as monocytes that could still respond to an intravenously administered therapeutic. Small aggregates that contain loosely associated dimers of the therapeutic protein, in contrast, would present a lower risk factor. It would certainly be advantageous to ascribe a risk score to the presence of aggregates that would allow a risk assessment based on factors such as the amount of aggregates present in a given preparation of drug, the type of association holding the aggregate together (covalent or noncovalent), and the size of the aggregate. Unfortunately, there is presently insufficient information available to allow such a meaningful assessment. Therapeutic protein manufacturers are therefore left with the challenge of reducing aggregate levels as much as possible to minimize the risk of immunogenicity. As more data are collected that correlate protein aggregation data with clinical outcome and immunogenicity assessment, we may be able to better understand the risk of aggregation to immunogenicity in quantitative terms.

Significant differences in sequence homology with an endogenous counterpart protein can often lead to the immune system recognizing the protein therapeutic as a foreign entity. In cases where a therapeutic protein mimics an endogenous protein, there is a greater chance for an immune response to be generated when there is less sequence homology between the therapeutic and the endogenous protein, because tolerance for the protein therapeutic will be less likely. Areas where the homology differs that correspond to surface regions of the protein are more likely to induce an immune response than sequences corresponding to amino acid sequences that are not exposed on the surface. It is important to note, however, that changes in a buried amino acid sequence sometimes translate into conformational differences that can affect the positioning of surface residues. When considering chimeric antibodies as therapeutics, the regions of the protein that are nonhuman in origin are likely to be recognized as foreign and induce antibody formation. The strategy behind the development of humanized antibodies is to identify the regions of the therapeutic antibody that are nonhuman in origin and are not directly responsible for binding to the antibody target and to replace them with the amino acid sequence from a human antibody. This reduces the amount of protein with nonhuman amino acid regions and should reduce the opportunity for the therapeutic protein to be recognized as foreign, which should, in turn, reduce the immunogenicity of that protein. The next possible improvement beyond humanized antibodies is the ability to generate fully human antibodies. These are produced in such a way that the entire amino acid sequence is human in origin. This further reduces the potential for a therapeutic antibody to induce an immune response. While this evolution from murine to chimeric to humanized to fully human antibodies has reduced the theoretical risk of inducing an immune response, it is important to note that even fully human antibodies can be immunogenic to some degree. This underscores the complexity and the multiple variables that work together to determine the immunogenicity of a therapeutic protein.

Another area of concern when considering the amino acid sequence of a therapeutic protein candidate is to compare its amino acid sequence with a library containing sequences from microbial and other sources that are known to be immunogenic in humans. The presence of sequence homology between the therapeutic and known immunogens heightens the concern that the therapeutic protein could trigger a robust immune response due to cross-reactivity.

In addition to these factors, there is a growing literature that describes efforts to identify short amino acid sequences within therapeutic proteins that are likely to induce an immune response because they are predicted to be T-cell epitopes. Comparing the amino acid sequence of the protein with the sequence from known T-cell response regions can identify these T-cell epitopes. The T-cell response regions are small amino acid sequences that, when expressed on the surface of antigen-presenting cells such as macrophages or dendritic cells, trigger a robust immune response after interaction with the appropriate T-cell clone. The presence of suspected T-cell epitopes within a therapeutic protein can be compared with known T-cell epitopes contained on other therapeutic proteins where the immunogenicity is known, allowing manufacturers to modify potential immunogenic regions of a protein. Because the T-cell participation is required for the immune system to mount a mature response (including class switching from IgM to IgG, high concentrations of antibodies produced, and increased affinity of the antibodies against the target), the presence of T-cell epitopes is an important factor in determining whether the protein therapeutic will induce a clinically relevant immune response. There are ongoing efforts to predict whether a protein therapeutic will induce an important immune response based primarily on scanning the amino acid sequence for the presence of known T-cell epitopes. Immunogenicity prediction is still in the early stages of development, but it may provide an important tool in the future to help manufacturers develop therapeutic proteins with reduced immunogenicity.

A common feature of all of the risk factors for immunogenicity described here is that there are no data allowing us to quantify the risk imposed by these factors. However, even without such quantifiable data, the presence of these factors should be considered as potentially capable of increasing the immunogenicity of any therapeutic protein.

Additional risk factors include oxidation, presence of glycosylation patterns that differ from endogenous proteins, chemical instability resulting in a changed form of the therapeutic protein, and sequence homology with microbial agents that are known to elicit a robust immune response. Each of these presents a different challenge and mechanism for increasing immunogenicity.

Oxidation and deamidation can occur when there are less stable regions of the molecule that are exposed on the surface of the protein. When these chemical changes occur, they can alter the three-dimensional structure and thus create new and unique regions on the surface of the protein that could be recognized by the immune system as epitopes. It is important that the formulation of the protein be able to maintain stability to reduce the chance for critical chemical modifications.

The immunogenicity of a therapeutic protein is very complex and a result of interactions between many different factors. This complexity makes predictions regarding the immunogenicity of any given protein very challenging. Understanding the nature of a given protein therapeutic and the role of these various risk factors can help in the design of a protein therapeutic with minimal immunogenicity.

4.4. METHODS FOR IDENTIFYING AN IMMUNE RESPONSE

Many different documented methods have been used to identify whether a protein therapeutic has induced an immune response and to understand the nature and magnitude of that response. The types of assays that have been utilized can be categorized as screening immunoassays, confirmatory immunoassays, and neutralizing antibody assays. Each category of assay has a different purpose and set of attributes. What all immunogenicity assessment assays have in common is the need to understand the capabilities of the assay in order to correctly interpret any results from that assay. The process of assay validation provides an opportunity to learn about the limits of an analytical procedure, and understanding those limits helps prevent misinterpretation of a result. As an example, assume that an assay has a sensitivity of 200 ng/mL, meaning that it will identify a sample as positive for the presence of antibodies if there is at least a circulating level of that antibody of 200 ng/mL. If a sample is positive in the assay, this suggests that the subject has antidrug antibodies with circulating antibodies at a concentration of at least 200 ng/mL. If a sample tests negative, the interpretation is that the subject does not have antibodies against the protein therapeutic. However, when correctly interpreted in the context of the limits of the assay, the subject does not have circulating antibodies against the therapeutic protein that reach the concentration of 200 ng/mL. It is a subtle yet important difference in the interpretation of the result. Without understanding the sensitivity of an analytical procedure, it is not possible to interpret what a positive or negative result in that assay really means. Another criterion that is evaluated during assay validation is the specificity of an assay. Without understanding whether an assay scores a positive result only when there is antibody present that will only, or specifically, bind to the therapeutic protein, one cannot correctly interpret a result. Many times, an assay that lacks specificity falsely identifies a higher number of subjects as positive for antibodies against the therapeutic protein.

A screening immunoassay is the first analytical procedure used in identifying the presence of antitherapeutic protein antibodies. The screening assay should be one that can be run in high-throughput mode since many clinical trials generate a high number of samples to test. It is important that the screening assay have sufficient sensitivity to be able to identify all of the subjects who develop clinically meaningful antibodies in response to treatment with the therapeutic protein. The sensitivity that is required varies depending on the drug product, since some products, especially those used for protein replacement therapy, can have clinical effects with low levels of antibodies being produced. It is anticipated that screening assays will be able to detect the presence of at least 500 ng/mL, and most marketed products are supported by assays with considerably better sensitivity than that. Another important attribute of a screening assay is its level of robustness. Clinical development programs encompass years of trials, and it is very important that the assay used for detecting antibodies be consistent throughout that development cycle to ensure that a comparison of immunogenicity data can be made across all of the studies included in the drug's development.

Several different platforms have been utilized for screening assays. One of the earliest techniques used was the radioimmunoprecipitaton assay. This is a very straightforward assay where a subject's serum sample is tested by adding a measure of therapeutic protein that has been tagged with a radiolabel. The radiolabeled drug would bind to any antibodies against the therapeutic protein contained in the subject's sample. The resulting immune complex is then pulled out of solution using an agent such as protein A. The tube containing the sample is then centrifuged, and the radioactivity in the resulting pellet is measured. A high level of radioactivity in the pellet corresponds to a high level of antitherapeutic protein antibody in the sample. This platform of assays can be very sensitive. There are several drawbacks to this platform, including an inability to detect some low-affinity antibodies because the antibody does not remain bound tightly enough to the labeled drug to be precipitated. This results in a false-negative result. A second challenge is that many of the precipitating agents, including protein A, are not efficient at causing the precipitation of IgM antibodies. This leads to a poor ability to detect IgM antibodies in this assay. The IgM antibodies are typically the earliest antibodies produced in subjects, and the inability to detect them reduces the likelihood of detecting an early immune response to the therapeutic protein. It is important for manufacturers to identify the earliest indication of an immune response being generated so that the immunogenicity of the protein can be fully understood. While very popular at one time, other methods have largely replaced the radioimmune precipitation method as the screening method of choice. This is due primarily to the concern about using radioactivity and recognition that an early immune response is not always detected using this method.

One of the most widely used techniques currently used to screen for the presence of antibodies is *enzyme-linked immunosorbent assay* (ELISA), which immobilizes the therapeutic protein onto the surface of a microtiter plate. A serum sample to be tested is then added, and after repeated wash and incubation steps, any antibodies capable of binding to the therapeutic protein are detected. There are different forms of ELISA, including direct, indirect, and bridging. In direct ELISA, the therapeutic protein is directly coupled onto the surface of the microplate; with indirect ELISA, the protein is attached via a separate capture reagent. In both of these platforms, the captured antibody is detected with an enzyme-labeled secondary reagent that is capable of binding the antibody from the serum sample that has bound to the immobilized therapeutic protein. In the bridging platform, the detecting reagent is an enzyme-labeled form of the therapeutic protein. An advantage of this method is that it is highly specific and rarely demonstrates false-positive results. This is because it takes advantage of the bivalency of antibodies and requires that the antitherapeutic antibody bind two separate molecules of the protein therapeutic, one that is immobilized and the other that has the enzyme label. In all forms of ELISA, the substrate for the labeling enzyme is added and the colorimetric change observed is proportional to the amount of enzyme label that has been captured onto the microtiter plate. The amount of enzyme label is also proportional to the amount of antitherapeutic antibody that has been captured. While it is apparent that the bridging format provides significant advantages related to specificity, the format is not ideal for detecting an early immune response where the antibodies are of low affinity. In general, the early antibodies produced in response to a therapeutic protein tend to be of low affinity; that is, they do not bind as tightly as do high-affinity antibodies, which are produced by a robust and mature immune response. It is crucial for manufacturers to understand the number of subjects who develop this low-level immune response and also identify how many of those subjects progress to a robust and mature immune response. All of the ELISA methods described above can be optimized for automation and do not require expensive or sophisticated instrumentation to perform.

Another popular screening tool utilizes electrochemiluminescence (ECL). These assays are very similar to ELISAs but are typically more sensitive. The increase in sensitivity is due to the replacement of the colorimetric reaction with an ECL signal. Another advantage beyond the enhanced sensitivity is the wider dynamic range made possible by the ECL signal. The instruments required for this technology are more limited in distribution and are also more expensive. These ECL assays are also quite amenable to automation, which allows them to be run in a high-throughput mode. A challenge that is somewhat unique to ECL assays is a phenomenon referred to as a "hook effect." A hook effect occurs when the signal from the assay increases up to when a critical concentration of antitherapeutic antibody is reached, but if there is additional therapeutic antibody present, the signal no longer continues to go up, rather, it decreases. When examining the plot of antibody concentration compared with assay signal, one discerns the shape of a fish hook. In extreme cases, the signal can be depleted by high concentrations of antibody to a level where the assay result is negative! Care must be exercised to properly interpret results from these assays, and a simple technique for protecting against an unrecognized hook effect is to analyze the serum sample at two different dilutions. Under hook effect conditions, a more dilute serum sample will have a higher test result than will a more concentrated sample from the same subject despite the fact that there is less antibody present.

Another type of immunoassay that is gaining popularity is the Biacore™, an example of a biosensor instrument. This instrument detects the presence of antibodies in real time through monitoring increases in mass when antibodies bind to a therapeutic protein immobilized on the surface of a sensorchip. This real-time detection offers the advantage of antibody detection without the need for incubation with a labeled detection reagent. This translates into a better ability to detect low-affinity antibodies because the antibodies are detected before they have a chance to dissociate from the immobilized protein. These assays are also capable of being automated, although throughput with this instrument is less than can be accomplished with ELISA or ECL.

The second tier of immunoassay is the confirmatory assay. A 2008 published guidance document from the European Medicines Evaluation Agency (EMEA) (www.ema.europa.eu/pdfs/human/biosimilar/1432706enfin.pdf) describes the importance of a confirmatory assay to provide confidence that the screening assay positive result is a true positive. There are different ways to confirm the positivity of an antibody sample. One method is to use a completely different analytical platform to test the subject's sample. This approach follows the assumption that it would be unlikely for two separate and independent methods using different platforms to have the same nonspecific reactivity. Therefore, if a sample is positive using two different procedures, it is highly likely that the sample represents a true positive result. A different approach uses a sample treatment to eliminate any immunoglobulins and then reanalyzes the positive sample using the same assay. In this instance, a true positive sample would score negative in the confirmatory assay because the antibody recognizing the protein therapeutic would have been removed by the sample treatment. If the sample still scored positive in the assay after sample treatment, that reactivity indicates the sample contained a serum factor other than antibody, which caused the positive result in the screening assay, and hence the final result for the serum sample would be negative. If the Biacore™ instrument is utilized for the screening assay, it can be an easy additional assay step to confirm that the agent binding to the immobilized therapeutic protein is an antibody. This is accomplished in human serum samples by the addition of an antihuman immunoglobulin. This would then bind to the antibody captured onto the immobilized therapeutic protein and would result in a measurable signal. Another approach for confirming the presence of a positive antibody sample is to reanalyze the sample in the presence of added soluble drug. The soluble drug would compete with the immobilized drug as a surface for the antibody to bind. This would result in a reduction in the signal in the assay. If the binding observed in the screening assay is due to a nonspecific interaction, the additional soluble drug would be less likely to inhibit the signal. It is important to note that it is very difficult to inhibit the binding of IgM antibodies by adding soluble drug because of the lower affinity and the need to block 10 binding sites in order to inhibit the binding to the immobilized drug. While inhibition of >90% of the binding of IgG antibodies can be achieved by addition of excess soluble drug, IgM can rarely be inhibited by more than 50 %. This is an important consideration when interpreting the results from these inhibition tests. When samples are confirmed as positive for the presence of antibody, the next step involves an assay that can identify whether the antibody is capable of neutralizing the biological effect of the drug.

Neutralizing antibody assays are most often biological assays. The principle behind this type of assay is to take advantage of a measurable biological effect that a drug can have on a cell culture. The cells are then grown in the presence of the drug as well as a serum sample. If the biological effect is inhibited in the subject's serum sample, and the sample has been demonstrated by immunoassay to contain antibodies capable of binding to the protein therapeutic, the results of the bioassay indicate the presence of a neutralizing antibody. A *neutralizing antibody* is defined by its ability to neutralize a biological effect of the protein therapeutic. Subjects with neutralizing antibodies have the potential to have some or all of the biological effects of the protein therapeutic blocked. The concentration of the neutralizing antibodies is an important factor in determining the clinical significance of the immune response. In some instances the neutralizing antibodies can block not only the protein therapeutic but also an endogenous counterpart. These are the cases with the highest clinical impact. Biological assays are very different from immunoassays as they tend to provide much lower throughput and be more time-consuming. This extra time is necessary to allow cells to grow and see benefit from the protein therapeutic. Biological assays also are more labor-intensive and require that the analyst be skilled at cell culture techniques. Many different endpoints have been used to monitor a biological effect, including growth, cytokine release, and mRNA expression. These assays play a major role in understanding the likely clinical impact of a given subject's immune response. However, even if a subject has antibodies that bind but do not neutralize the protein therapeutic, there is still the potential for a clinical impact because the antibodies might act to accelerate clearance of the drug that could negatively impact any clinical benefit for the subject.

Another valuable application for the Biacore™ is in the characterization of an immune response. This can include determination of the concentration, specificity, isotype(s) of the antibodies, affinity of the antibody to the therapeutic protein, and the epitope or region of the protein where the antibody binds. The data obtained during characterization can be valuable toward understanding the nature and magnitude of the immune response. For example, if a subject initially has antibodies specific for the protein therapeutic that are only IgM but it is later found that there are IgG antibodies against the therapeutic, this indicates that the immune response has matured, which suggests a potentially increased likelihood for the antibodies to have clinical relevance. By observing that the concentration of antibodies specific for a protein therapeutic is decreasing, one may infer that the immune response is waning.

4.5. STRATEGIC APPROACH FOR MONITORING IMMUNOGENICITY

Monitoring clinical trials is the best way to understand the immunogenicity of a protein therapeutic. There are several important considerations when developing an immunogenicity monitoring strategy. The timing of when antibody samples should be taken requires careful consideration. Because circulating drug interferes with the ability of assays to detect antibodies, it is important to collect serum samples for antibody analysis at times when the level of circulating drug is as low as possible. This can become very challenging for therapeutics such as monoclonal antibodies, which tend to have long half-lives and remain in the circulation for long periods of time. The baseline sample, a sample taken prior to the first administration of the therapeutic protein, is critical for understanding whether a patient has developed an immune response. There are instances when subjects have preexisting antibodies that are capable of binding to certain therapeutics. The reason for some patients having baseline antibodies even prior to the first administration of the protein is not well understood. This might be due to prior exposure to a very similar therapeutic protein or could simply be the result of cross-reactivity. Whatever the cause, it is important to determine whether a subject has baseline antibodies. A subject who has baseline antibodies and subsequently develops a more robust immune response is less troubling than a subject who had no preexisting antibodies and then develops an immune response.

In addition to taking samples at baseline, it is important to take sufficient antibody samples throughout a clinical trial to understand the immunogenicity of the drug. If samples are taken only at the beginning and conclusion of a clinical trial, a manufacturer would miss any immune response that was transient in nature. This would give an immunogenicity rate that was biased downward. It is also important to consider the length of the clinical trial. Antibody development in a subject can occur at any time in a clinical trial. Most subjects who develop immune responses tend to initiate their response within the first year of treatment. When shorter trials are conducted, it is possible to underestimate the true immunogenicity rate.

The second component of an immunogenicity assessment strategy is selection of the most appropriate analytical procedures to use. A tiered approach to analytical procedures is most commonly employed. The screening assay is most commonly a high-throughput immunoassay designed such that up to 5% of the samples that test positive do not contain antibodies to the protein therapeutic. This provides confidence that all of the samples that do contain a concentration of antibodies higher than the sensitivity of the assay test positive in the assay. The next assay to be used is a confirmatory assay, which helps reduce the number of false-positive samples. The confirmatory assay has sensitivity comparable to that of the screening assay but a different specificity profile. Because the confirmatory assay is used only for samples that test positive in the screening assay, it is not so important that this assay be high-throughput, unless the scope of the clinical program is sufficiently high, or the number of positive samples in the screening assay is large. The final assay in the tier is a biological assay that is used to test samples that are confirmed as positive. If there is uncertainty as to the ability of the screening assay to detect all positive samples, the biological assay can be used with additional samples as verification for the performance of the screening assay. In cases where the biological assay has significantly better sensitivity than the screening immunoassay, it would be important to analyze all samples with the most sensitive method available, even if that assay is a lower-throughput biological assay.

There are many different assay types available, and care should be taken during assay development to pick the assay type and platform that is optimal for the therapeutic protein being developed. Prior to using the analytical procedure for a clinical trial, the assay should be validated to understand the performance limitations of the procedure to ensure that any results generated by these assays are properly interpreted.

4.6. CONSEQUENCES OF AN IMMUNE RESPONSE TO A THERAPEUTIC PROTEIN

When an immune response is generated in a subject treated with a therapeutic protein, the clinical consequences can range from having no observable clinical effect to having a dramatic impact that neutralizes both the protein therapeutic and the endogenous counterpart. While the likely consequences of an immune response in any given subject cannot be predicted, it is important in the setting of a clinical trial to fully test subjects so that their immunogenicity status is known and all appropriate measures are taken to minimize any significant clinical consequences for the subject. While most immune responses to current therapeutic proteins are minor, there are documented reports of the antibodies having an impact on the subject.

Depending on the therapeutic protein and the biological effect that it exerts, it can often be difficult to detect whether there is an impactful immune response in any given subject. In contrast, certain therapeutic proteins exert their effects in ways that are readily monitored. For example, erythropoietin induces red blood cell production, which can easily be monitored through simple screening tests. When subjects develop neutralizing antibodies, it is often recommended that the protein therapeutic be withdrawn. This can prevent the further exacerbation of the immune response and can also be warranted when the therapeutic is not providing benefit to the subject because of the antibodies.

When antibodies either neutralize (bind to the reactive region of the protein therapeutic and render the protein nonfunctional) or bind and rapidly clear the protein therapeutic, the result is that the protein therapeutic will no longer be efficacious for that patient. In addition, if those antibodies also react with an endogenous counterpart protein, the end result could be a deficiency syndrome because the antibodies block not only the protein therapeutic but also the endogenous counterpart. If the protein is a critical and nonredundant part of a pathway (e.g., erythropoietin is a critical and nonredundant component of the pathway for red blood cell production) when neutralized or rapidly cleared, the pathway can be completely blocked. As in the erythropoietin example, neutralizing antibodies against erythropoietin have resulted in pure red cell aplasia in patients.

If the antibodies generated against the protein therapeutic have the ability to sustain the protein in circulation by blocking clearance, any toxicities associated with that protein therapeutic could be exacerbated. Another potential consequence would be hypersensitivity reactions if IgE antibodies are generated against the protein therapeutic. These IgE antibodies could trigger histamine release that would lead to presentation of allergic symptoms.

4.7. NONCLINICAL IMMUNOGENICITY TESTING

Understanding immunogenicity in the nonclinical setting is different from looking at human trials. The majority of human therapeutic proteins would be foreign to the nonhuman immune system. For this reason, it is fully expected that human therapeutics would induce a robust immune response in the nonclinical setting. At this time, the presence of antibodies in nonclinical studies, including studies in monkeys, is not predictive of what would happen when the therapeutic is administered to humans. The purpose of testing for immunogenicity in nonclinical trials is to allow interpretation of the results from the study. In some cases, a very robust immune response is seen in nonclinical studies that completely neutralizes the biological effect of the drug in a matter of days. In these cases, it is important to determine whether the presence of these antibodies has invalidated the study. One valuable tool to help ascertain whether the animals were still exposed to the drug despite the presence of antibodies is the use of biologically active drug-level assays. These procedures allow an assessment to determine whether there was biologically active drug present in the animals. The absence of biologically active drug would suggest that the antibodies had effectively blocked the drug and eliminated its ability to exert any biological effect. When this happens, it is very difficult to assess the safety of the drug. Therefore, the value in determining whether animals in nonclinical trials develop antibodies is in helping interpret the results from toxicology studies.

4.8. SUMMARY

Immunogenicity of therapeutic proteins represents an important safety factor for protein therapeutics. It is critical to fully understand the immunogenic profile of therapeutic proteins as they progress through clinical development. While the mere presence of an immune response would not likely result in termination of a clinical development program, it is only through thoughtful analysis of the characteristics of that immune response that the safety profile of the protein can be determined. Immunogenicity of protein therapeutics remains an area of research as companies strive to better understand why some people mount an immune response against a given therapeutic while many other subjects do not. The industry continues to develop new procedures and tests to better understand immunogenicity.

5

PREFORMULATION RESEARCH: ASSESSING PROTEIN SOLUTION BEHAVIOR DURING EARLY DEVELOPMENT

Bernardo Perez-Ramirez, Nicholas Guziewicz, and Robert Simler

5.1. INTRODUCTION

5.1.1. Preformulation Research: A Component of the Development Strategy

Development of robust bioformulations is a dynamic process involving input from many stakeholders such as marketing, clinical research, pharmacy, and manufacturing, to name only a few. Preformulation research is the initial step during formulation development that evaluates the solution behavior of proteins a function of multiple variables such as pH, ionic strength, shear sensitivity, and temperature (Fig. 5.1). Initial studies should identify significant modalities of instability and support the development of strategies to minimize such instabilities. Beginning these investigations before protein solubility and stability challenges become critical factors during development is highly advisable. To achieve this goal, planning the development of suitable dosage forms should therefore start very early in the development process.

Ideally the formulator should become involved during the molecular biology research and initial purification efforts. For example, an early understanding of protein solubility as a function of pH and ionic strength (the most fundamental variables in protein formulation development) is of great value to teams engaged in optimizing host cell expression or protein recovery. This type of evaluation, usually with limited

Formulation and Process Development Strategies for Manufacturing Biopharmaceuticals, edited by Jameel and Hershenson

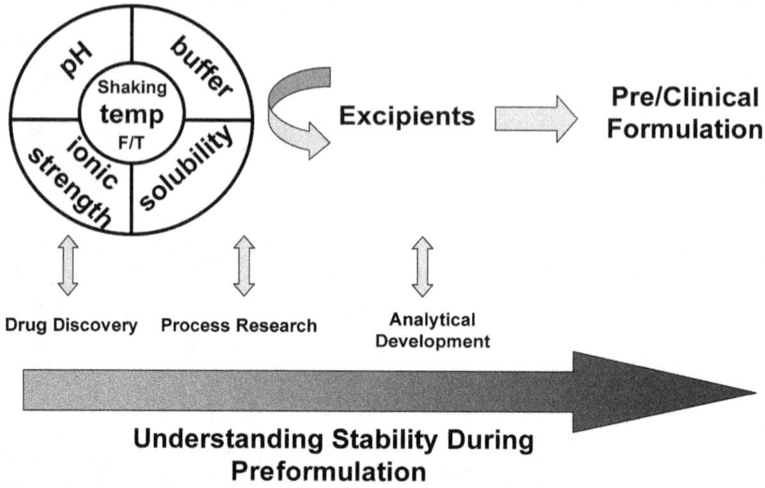

Figure 5.1. Preformulation research in the overall development strategy.

amounts of material, may be the main goal of preformulation research during the discovery phase of the product. If *in vitro* and initial animal model data support continuation of the product development process, then the preformulation research effort may expand the candidate physicochemical profile. Concurrent with the protein solution behavior assessment is the development of stability-indicating analytical assays.

Because of the complex nature of proteins, generic approaches to preformulation and formulation development are of limited value. For example, antibody molecules that share many structural similarities often have quite different solution behaviors. Thus, candidate-specific biophysical and thermodynamic analyses are helpful in designing initial protein stabilization strategies while keeping in mind the ultimate goal of developing robust formulations for commercialization. Also, the kinetic contribution to stability of biopharmaceuticals should not be overlooked. The subject matter of this chapter is to review the principles and general approaches used in preformulation research for proteins as therapeutic candidates.

5.2. FORCES IN PROTEIN STABILIZATION

There are many excellent reviews on protein conformation that the formulator may wish to become familiar with. Consultation of the classical literature such as the studies by Charles Tanford [1,2] and others [3–6] is highly advised. Nevertheless, some of the fundamental concepts useful to preformulation research are briefly summarized in this section. The native structure of proteins is the result of several opposing forces. Under physiological conditions, the native (folded) and the denatured[1] (unfolded) states of a

[1]The nonnative, denatured, and unfolded states of proteins are assumed to be equivalent for this analysis. This may not be always true.

protein are in equilibrium. Thus, from this equilibrium the standard Gibbs free energy of stabilization (ΔG°) can be obtained. The free-energy difference between the native and denatured states is referred to as the *conformational stability* of the protein. A major destabilizing effect on folded proteins is the greater conformational entropy of the unfolded state. The conformational stability of the native state is dictated basically by the amino acid sequence and solution variables such as pH, ionic strength, and interaction with particular ligands. The small free energy of stabilization of a folded protein arises from many interatomic forces, some stabilizing and others destabilizing. The free energy of stabilization of the native state typically accounts for no more than 5–20 kcal/mol [7]. This marginal conformational stability of proteins is perfectly suitable for the changing needs of living cells. Throughout evolution, proteins have gained only a marginal level of stabilization compared with the unfolded states. The reasons for this may reflect the many biological functions that proteins perform. Flexibility is required for optimal interaction of enzymes with their substrate or protein–protein interactions leading to optimal catalysis. In this regard, it is interesting to note the fine control of protein conformational flexibility observed in some elasmobranch fish species [8–10], where a precise accumulation of a stabilizer such as trimethylamine *N*-oxide (TMAO) and a classical denaturant such as urea coexist in a well-defined ratio to allow for "breathing modes" in the protein [11,12].

The disulfide bridge between cysteinyl residues is the only type of covalent bond that is formed during protein folding. Nevertheless, the native state of proteins is maintained by several weak forces that contribute to conformational stability, such as van der Waals electrostatic interactions, (London dispersion forces), hydrogen bonds, and the hydrophobic effect.

5.2.1. Weak Interacting Forces

The weak interacting forces that help in stabilizing the native protein state oscillate around 1–10 kcal/mol. These forces are quite weak compared with the energy contained in a covalent bond, namely, 50–60 kcal/mol. As an individual contribution, these weak interacting forces do not account for much stabilization, but the presence of several of these weak forces could be additive and therefore significant in their contribution to the native conformation.

5.2.2. Electrostatic Interactions

The most common type of weak interactions in proteins is electrostatics. This type of interaction is ubiquitous and often accounts for protein–ligand interactions and solute–solute contacts. The electrostatic effect can also give rise to large rate accelerations in dilute solutions, if both the substrate and catalyst are polyelectrolytes. The energy of the electrostatic interactions (salt bridges) depends on the dielectric constant of the medium and is entropically favored [13], thus releasing water involved in solvation. Most ion pairs (salt bridges) occur between amino acid residues that are distant in the sequence and in different segments of the protein secondary structure. The amide group in the protein backbone has a partial negative charge on the oxygen atom and a partial positive charge on the N–H atom, resulting in an effective amide dipole moment of about 1.2×10^{-29} coulomb-meter, pointing parallel to the C–O and N–H bonds [14–16]. In the α-helix structure, the dipoles are parallel to the helix axis pointing toward the *N* terminus [16]. The β-sheet structure, very common in IgGs, also has an effective dipole with a positive charge at the *N* terminus and negative charge at the *C* terminus. The consequence of this differential charge distribution along the protein is the potential positioning of negative ions toward the *N* termini of α helices, contributing to helix stabilization [16]. This might also explain why aspartic and glutamic acid residues in proteins are preferentially found toward the *N* termini of α helices [17].

5.2.3. London Dispersion Forces (Van der Waals Interactions)

Fritz London, around 1930, suggested that the motion of electrons within an atom or nonpolar molecule can result in a transient dipole moment. *London dispersion forces*, also known as *van der Waals interactions*, result from the generation of transient dipoles, namely, attractive forces between neutral molecules. These forces include dipole–dipole, charge–dipole, charge–induced dipole, dipole–induced dipole, and induced dipole–induced dipole interactions [18]. These interactions are relatively weak and short range. The characteristic features of dispersion interactions are their proportionality to the polarizability of the interacting molecules and their dependence on the sixth power of the intermolecular separation d, as described by Equation (5.1), where E is the energy of the dispersion interaction and α is the polarizability [18]:

$$E = -\frac{3\alpha_1\alpha_2}{2d^6} \tag{5.1}$$

As a result of electron repulsion, a temporary dipole on one atom can induce a similar dipole on a neighboring atom. As a consequence, neighboring atoms become attracted to one another. London dispersion forces are a contributing factor to protein conformational stabilization when atoms are close together. The torsion angles ϕ and ψ (on the two sides of the peptide bond) are favorable for most residues in the right-handed α helix, where the atoms pack together, making favorable van der Waals interactions.

5.2.4. Hydrogen Bonds

The hydrogen bond is a type of electrostatic interaction that energetically falls into a category between covalent bonding and ionic interactions. Values of the energy of typical hydrogen bonds lie in the range of 3–6 kcal/mol. Intramolecular hydrogen bonding in proteins is favored by the entropy gain from the release of bound water and is essential to the structure and stability of globular proteins [19,20]. Because of its unique structure, water can act as both a hydrogen donor and hydrogen acceptor; hence the formation of intermolecular hydrogen bonds between two solute molecules in water requires that the hydrogen bonds of each of these molecules be broken. The most significant hydrogen bonds in proteins are the ones between amide groups in the α helix and other structures [21]. The favorable formation of hydrogen bonds in water should produce a favorable decrease in enthalpy H. However, formation of hydrogen bonds requires fixation of the molecules, thus the entropy S tends to decrease on hydrogen bond formation. Nevertheless, these contributions to the standard free energy of hydrogen bond formation at constant temperature T and pressure P cancel out. As a consequence, large changes in enthalpy and entropy should produce little or no impact on the free energy of hydrogen bond formation.

5.2.5. Hydrophobic Effect

The hydrophobic interaction has long been considered the primary stabilizing force present in the folded protein [22–25], but not in the unfolded protein; it undoubtedly accounts for the heat capacity change on folding, and its favorable contribution to the free-energy change may be estimated from the accessible surface area buried in the folded state. Hydrophobicity[2] also plays a major role in self-association of protein molecules. Details of the three-dimensional structure of proteins have shown that in folded proteins, apolar side chains of isoleucine, leucine, valine, tryptophan, tyrosine, and phenylalanine tend to be buried away from the solvent, and charged and polar residues tend to be in contact with the solvent. This distribution of hydrophobic residues was predicted by Kauzmann [22] from studies of small molecules such as methane and benzene, and refined by Nozaki and Tanford [26], who measured the hydrophobicity for various amino acids as a model system.

The hydrophobic effect or "bond" can be understood as the consequence of two effects. The first effect is the increase in order or "structure" of the solvent molecules (eg., water at a concentration of ~55 M), which can lead to a negative change in entropy when a nonpolar molecule is placed in an aqueous solution [27,28]. A lower entropy means a greater free energy as can be deduced by examining the relationship between ΔG and ΔS as shown by Equation (5.2). Studies on the solubility of hydrocarbons also indicate an enthalpic contribution at high temperatures to the low solubility of apolar molecules in water [29]. The second origin of the hydrophobic effect can be accounted for by separation of the strongly interacting water molecules in order to create a "cavity" and insert a weakly interacting molecule into the solution. The large cohesive force of water molecules is manifested by the high surface tension. Thus, the free energy required to form a cavity in water and separate the strongly interacting water molecules is very large. The free energy required to form a cavity in a solvent for accommodating the solute is given by the product of the cavity surface area A and the surface tension of the solvent, $\sigma : \Delta G = \sigma A$. The formation of the "cavity" is not compensated by a favorable free energy of transfer of the apolar residues to water. For example, consider the impact on "cavity formation" of a small molecule such as propane (C_3H_8). Propane will create a cavity; some hydrogen bonds of the water network will be broken. Propane does not interact strongly

[2]Charles Tanford, in *Physical Chemistry of Macromolecules* (1961, Wiley, New York, p. 130), stated that "The word hydrophobic is really a misnomer. It implies that the dissolved substances dislike water, whereas, in fact, it is the water that dislikes the dissolved substance."

with water; in other words, it does not form hydrogen bonds. The water molecules around the propane molecule are oriented to rebuild the hydrogen bonds broken during the addition of the hydrocarbon molecule to the solution. The latter process is exothermic (i.e., energy is released), accounting for the negative values of ΔH in the transfer process. The net result is that water molecules around the hydrocarbon become ordered. Because there is only a small change in the number of hydrogen bonds, the enthalpy change is minimal. However, the ordering of water around the hydrocarbon molecules is associated with a decrease in entropy. Thus, a solvent or formulation in which the protein is dissolved exerts a profound effect on the structure of the protein, and solution variables may affect the secondary, tertiary, or quaternary protein structure.

5.3. PHYSICAL STABILITY

5.3.1. Essential Thermodynamics of Protein Stabilization

Under the term *physical stability*, which refers to changes in the protein secondary, tertiary, or quaternary structure that do not affect covalent bonds, we can group protein denaturation, adsorption, aggregation, self-association, and precipitation. The conformational stability of proteins is a critical factor that formulation development aims to control to maintain potency, safety, and purity among other considerations. The approach to determine the conformational stability of a given protein requires knowledge of the equilibrium constant and change in the standard Gibbs free energy (ΔG) for the reaction between the protein in the folded versus the unfolded states. The natural tendency of a system to move to lower energies, namely, decreased enthalpy H together with occupancy of multiple states, increased entropy S, as a result of thermal perturbations can be synthesized in the Gibbs free-energy relation:

$$\Delta G = \Delta H - T\,\Delta S \tag{5.2}$$

The equilibrium constant K_{eq} for any process is related to the standard Gibbs free-energy change by

$$\Delta G^\circ = -RT\ln(K_{\mathrm{eq}}) = \Delta H^\circ - T\,\Delta S^\circ \tag{5.3}$$

where R is the gas constant, 1.987 cal K^{-1} mol^{-1}).

The temperature dependence of enthalpy and entropy can be described in terms of the heat capacity (ΔC_p) at a given temperature and constant pressure:

$$\Delta H(T) = \Delta H(T_{\mathrm{ref}}) + \int_{T_{\mathrm{ref}}}^{T} \Delta C_p\,dT \tag{5.4}$$

$$\Delta S(T) = \Delta S(T_{\mathrm{ref}}) + \int_{T_{\mathrm{ref}}}^{T} \frac{\Delta C_p}{T}\,dt \tag{5.5}$$

Here, T_{ref} is a convenient reference temperature. If ΔC_p is assumed to be constant, as a reasonable approximation within a temperature range (not always true), then integration of Equations (5.4) and (5.5) give approximate values for the dependence of ΔH and ΔS on T_{ref}:

$$\Delta H(T) \approx \Delta H(T_{\text{ref}}) + \Delta C_p(T - T_{\text{ref}}) \tag{5.6}$$

$$\Delta S(T) \approx \Delta S(T_{\text{ref}}) + \Delta C_p \ln \frac{T}{T_{\text{ref}}} \tag{5.7}$$

It is usually convenient in protein-folding experiments, particularly when addressing the role of different excipients in protein stabilization during formulation, to choose as T_{ref}, the midpoint of the transition T_{m} as a convenient reference temperature. It follows that at T_{m} the ΔG for the folding $\leftrightarrow$ unfolding transition is zero and from Equations (5.6) and (5.7) that: $\Delta H(T_{\text{m}}) = T_{\text{m}} \Delta S(T_{\text{m}})$. By substituting Equations (5.6) and (5.7) into (5.2), we obtain the modified Gibbs–Helmholtz equation:

$$\Delta G = \Delta H^{\circ} - T\,\Delta S^{\circ} + \Delta C_p \left[(T - T_{\text{m}}) - T \ln \frac{T}{T_{\text{m}}} \right] \tag{5.8}$$

Equation (5.8) allows for a complete description of the Gibbs free energy of stabilization for a protein as function of temperature (Fig. 5.2). One important requirement for applying Equation (5.8) is the assumption of a two-state model of equilibrium. However, when dealing with large proteins that tend to stick, aggregate and precipitate, this is not always possible. Nevertheless, formulation development is not about absolute stability but rather having a proper shelf life. Hence, an *apparent* two-state model could be applied under certain circumstances to understand the relative stability of a given protein under different sets of conditions, as we will see later in this chapter. The equilibrium constant (K_{eq}) could be measured as function of temperature by different means such as calorimetry, UV–visible spectroscopy, or fluorescence. The direct way to obtain the enthalpy contribution of a reaction is by calorimetry, although spectroscopy approaches could be employed. Indirect methods to obtain enthalpy changes (ΔH) such as estimates from the Van't Hoff equation [Eq. (5.11)] from the dependence of the equilibrium constant with temperature for a two-state process could be

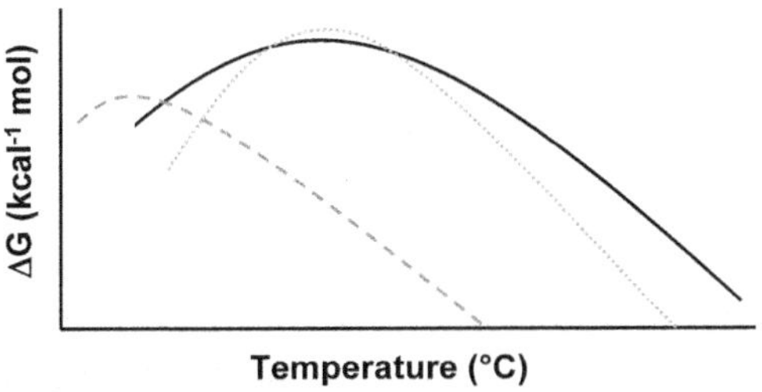

Figure 5.2. The free energy of unfolding for a protein is temperature-dependent.

very useful.

$$\Delta G = -RT \ln K_{eq} = \Delta H^\circ - T\,\Delta S^\circ \tag{5.9}$$

and

$$\ln K_{eq} = \frac{-\Delta H^\circ}{RT} + \frac{\Delta S^\circ}{R} \tag{5.10}$$

it follows that

$$\Delta H^\circ = -R\frac{d \ln K_{eq}}{d(1/T)} = RT^2\frac{d \ln K_{eq}}{dT} \tag{5.11}$$

A Van't Hoff analysis can be used to obtain enthalpy values of a reaction without the need for direct calorimetric measurements. As such, optical spectroscopy such as UV–visible or CD thermal melting studies could be very helpful in obtaining an initial enthalpy "fingerprint" of a protein during preformulation research.

5.3.2. Solvent–Protein Interactions

5.3.2.1. Preferential Exclusion. Proteins in solution are intrinsically not very stable. To achieve long-term storage stability of proteins, it is necessary to understand solution behavior under conditions that favor degradation. An understanding of how excipients interact with proteins to prevent instabilities is essential in designing useful stabilization strategies. It is well known through the pioneering studies of Timasheff and colleagues that the native conformation of proteins can be stabilized by sugars such as sucrose, glucose, lactose, and trehalose [30–32] at high concentrations (≥ 1 M), as well as by polyhydric alcohols (ethylene glycol, xylitol, sorbitol, mannitol, and inositol [33–35]), and by other compounds such as glycerol [36,37] or sodium glutamate [38]. Studies performed by Timasheff and co-workers at ambient temperature have shown that the mechanism of stabilization by these compounds is preferential exclusion from contact with the protein surface; specifically, in their presence the protein environment is enriched with water relative to the bulk solvent composition [39]. This means that addition of these compounds is thermodynamically unfavorable since it raises the chemical potential (the partial molar free energy) of both the protein and the additive. Multicomponent solution thermodynamic analyses of the interactions of glucose, sucrose, galactose, and trehalose [35] have led to the conclusion that these sugars stabilize the native structure of proteins by inducing a greater preferential exclusion of denaturated, unfolded proteins than of the native state [30,39]. Under those conditions the equilibrium will be displaced toward the more compact state with lower preferential exclusion of the excipient (reflected as negative binding of the excipient to protein) [30]. Preferential exclusion could also stabilize a given conformation of side chains in the active site of enzymes [40], if the extent of preferential exclusion differs between the active and inactive conformations [41].

Different excipients (or cosolvents) are preferentially excluded by different mechanisms [42]. The polyethylene glycols are excluded by steric interference [43,44], glycerol is excluded because of the solvophobic effect [36,45], and sucrose and sodium glutamate are excluded as a result of increased surface tension of water [30,38]. In general, Timasheff's studies have led to the classification of stabilizing excipients (cosolvents) into two groups. In the first group the preferential interaction (hydration, exclusion) is totally independent of the chemical nature of the protein surface and is driven mainly by surface exposure area such as the case with sugars, some amino acids, and salts. The second group of stabilizing excipients (cosolvents) includes those in which stabilization depends on the chemical nature of both the protein surface and the excipient. Alcohols and polyethylene glycols belong to this category.

5.3.2.2. Protons, Salts, and Buffer Effects in Preformulation. Buffer composition and pH usually have a large impact on the conformational stability of proteins. These are the first parameters that should be screened and optimized during preformulation research. Protein stability typically decreases sharply at the acidic and basic ranges of pH [46]. This behavior could be explained by the participation of protons in the unfolding process where the pK (the apparent ionization constant) of some ionizable residues differ in native and unfolded species. A group such as a side chain COO^- or $NH_3{}^+$ in a protein shows an observed pK_a that depends on the local charge environment. Nearby positively charged groups repel proton ions in solution and cause the pK_a to decrease [47–49], whereas neighboring negatively charged groups attract protons and cause the pK_a to increase [50]. Charge interactions usually cause the pK_a of a group to be lower in the native state than in the unfolded state at acidic pH (below the isoelectric point) but higher in the native state than the unfolded state at alkaline pH [46]. The reason is that the protein has a net negative charge above, but a net positive charge below the isoelectric pH, and a protein molecule expands to occupy a larger volume when it unfolds, therefore decreasing charge density [46]. When an ionizable group is part of a salt bridge, the sign of the pK_a shift for a COO^- group will be lowered in the proximity of opposite charges. Conversely, the pK for the NH_3^+ groups will be higher, providing stabilization to the salt bridges. Salt bridges and electrostatic interactions are important stabilizing factors at low salt concentrations and could be explained by the Debye–Huckel theory, (a detailed treatment of the Debye–Huckel theory to protein electrostatic interactions has been provided by Tanford [2]).

The fact that solubility of proteins increases at low salt concentrations, known as "salting in," may be explained by the effect of additional counterions in shielding multiple charges on the protein and therefore increasing solubility. Proteins are surrounded by the salt counterions, and this *screening* results in decreasing the electrostatic free energy of the protein and increasing the activity of the solvent, which leads to increased solubility. The direct relationship between the logarithm of solubility and the square root of the ionic strength of the solution applies. The situation becomes different in solutions where the concentration of salt is on the order of 1 M. This effect depends on the nature of the ions, as was observed by Hofmeister in 1888 during his studies on precipitation of serum globulins [52]. A series of the relative effect of salts (cations and ions) to induce protein precipitation is referred to as the *Hofmeister* or *lyotropic series*. The same series is responsible for the "salting out" effect. The range order of effectiveness in salting out is given by

$$\text{Cations}: \quad Mg^{2+} > Li^{+} > Na^{+} > K^{+} > NH_3{}^{+}$$

$$\text{Anions}: \quad S0_4{}^{2-} > HP0_4{}^{2-} > \text{acetate} > \text{citrate} > \text{tartrate} >$$

$$Cl^{-} > N0_3{}^{-} > Cl0_3{}^{-} > I^{-} > Cl0_4{}^{-} > SCN$$

At concentrations of ~1 M, the Hofmeister salts will increase hydrophobic interactions, and those interactions could be explained by the cavity model (see Section 5.2.5), which permits the partial interpretation of the effect of concentrated salts on protein solubility in terms of the increase in surface tension of water [53]. An alternative explanation for the salting out effect has been given in terms of preferential exclusion of salts from the surface of proteins [54] affecting hydrogen-bonding properties [55]. An additional factor to consider in concentrated solutions is excluded volume effects arising from the limited amount of water available to solvate the solutes. Therefore, the equilibrium would be shifted toward minimization of solute–water interactions. In general, the consensus is that salts that contain sulfate, phosphate, citrate, acetate, or fluoride tend to precipitate redundant and protect proteins against denaturation, whereas salts that contain thyocyanate, iodide, perchlorate, lithium, calcium, or barium tend to dissolve, denature, and dissociate proteins. The denaturant action of Hofmeister ions like SCN^{-} results from the fact that they salt in the peptide group and, as a consequence, interact much more strongly with the unfolded form of a protein than with its native form [56], driving the equilibrium to denaturation. Thus, the separation between protein denaturants and stabilizers ("null point") is the result of a balance between these two opposing classes of interactions: "salting out" nonpolar groups and "salting in" the peptide group as discussed by Baldwin [56]. However, the exact mechanism of salting out is still protein-specific. Nevertheless, at high salt concentration these effects become less important. The preferential exclusion mechanism described previously could minimize protein surface exposure [57] and therefore induce stabilization. The amount and type of salt in a given protein formulation could have an impact on chemical and physical stability. This reflects added salt and other excipients, residuals from purification, and specific protein–ion interactions, such as Gibbs–Donnan effects resulting from charged polyelectrolytes and ionized species in solution. Thus, the ionic strength of a protein solution constitutes the second cornerstone, after pH, in preformulation research. Since ionic strength is a critical parameter affecting protein conformational stability, we should define it before moving further. In 1921, Lewis and Randall defined the ionic strength of a solution I, as the semisum of the products of the molality of each one of the ions in the solution times the square of the charge of the ions [58], where m_i is the molality of the ith ion and Z_i is the charge of the ion:

$$I = \frac{1}{2} \sum_i m_i Z_i^2 \tag{5.12}$$

A buffer of any given composition has different ionic strengths at different pH values. Buffers of different compositions may have different ionic strengths at the same pH; this is an important consideration in studying the pH dependence of protein solubility, ligand binding, or conformational stability. An approach to separate the contribution of pH and ionic strength is to first estimate the buffer with the largest ionic strength and then adjust all the other buffers to a given ionic strength by adding an "inert" salt. Different buffer species at a given pH could be interacting with the protein of interest in such a way as to provide a significant difference in the physical and chemical properties of proteins. The use of polybuffers as an approach to provide constant buffering conditions and ionic strength throughout the desired pH range to be analyzed is helpful [59]. Some reactions will show a strong dependence on rate and equilibrium with ionic strength rather than concentration of a particular salt. Thus, during preformulation it is more informative to evaluate relevant parameters (solubility, aggregation, thermal stability, etc.) as functions of ionic strength rather than salt concentration. The electrostatic contribution to protein stability may be estimated from the change in summed electrostatic free-energy terms for all charge sites according to the modified Tanford–Kirkwood theory [60].

5.3.2.3. Ligand Binding and Stabilization. The three-dimensional structure of proteins in solution can be affected by ligands (stabilizers, destabilizers), pH, ionic strength, temperature, and pressure. In general, some of these variables show interrelations or linkages [61,62]. For instance, the binding of a ligand could be coupled to protons. The linkage is reflected in the change of the apparent binding constant of the ligand with pH, or as a change in the pK_a of a critical residue in the protein on ligand binding (i.e., proton affinity). If a reaction, let us say, $A + B \leftrightarrow C$, is affected by a ligand x, then, according to the fundamental Wyman linkage function [61,62]; we obtain

$$\left[\frac{\partial \ln K}{\partial \ln a_x}\right]_{T,P,a_{i\neq x}} = \Delta\nu = \Delta\nu_{\mathrm{P}} - \Delta\nu_{\mathrm{R}} \tag{5.13}$$

where K is the equilibrium binding constant for the reaction $A + B \leftrightarrow C$, a_x is the activity of ligand x, and $\Delta\nu$ is the number of ligand molecules participating in the reaction. A plot of ln K versus the activity of the ligand has as slope the number of ligand molecules $\Delta\nu$ participating in the reaction where T and P are temperature and pressure respectively.

The importance of the Wyman linkage functions [Eq. (5.13)] and its applications to formulation development is based on the following fact: *If ligand binding between two states is different, then the equilibrium between those states will be shifted to the state with higher affinity*. For a given equilibrium (e.g., folding/unfolding of a protein, oligomerization, enzymatic activity), if that equilibrium is shown to be linked to ligand binding, then one of the forms in solution could be stabilized by such ligand binding. Deviations from equilibrium will initiate a change in the system toward the equilibrium concentration of reactants and products. This is known as the *Le Châtelier's principle*.

Interaction of a ligand with the active or regulatory sites in enzymes often leads to increased stability. Figure 5.3a shows the apparent melting temperature of a model enzyme to be significantly increased by addition of a ligand. Thus, modulation of the concentration of the ligand could provide improved stability as reflected by higher apparent T_m values (Fig. 5.3b).

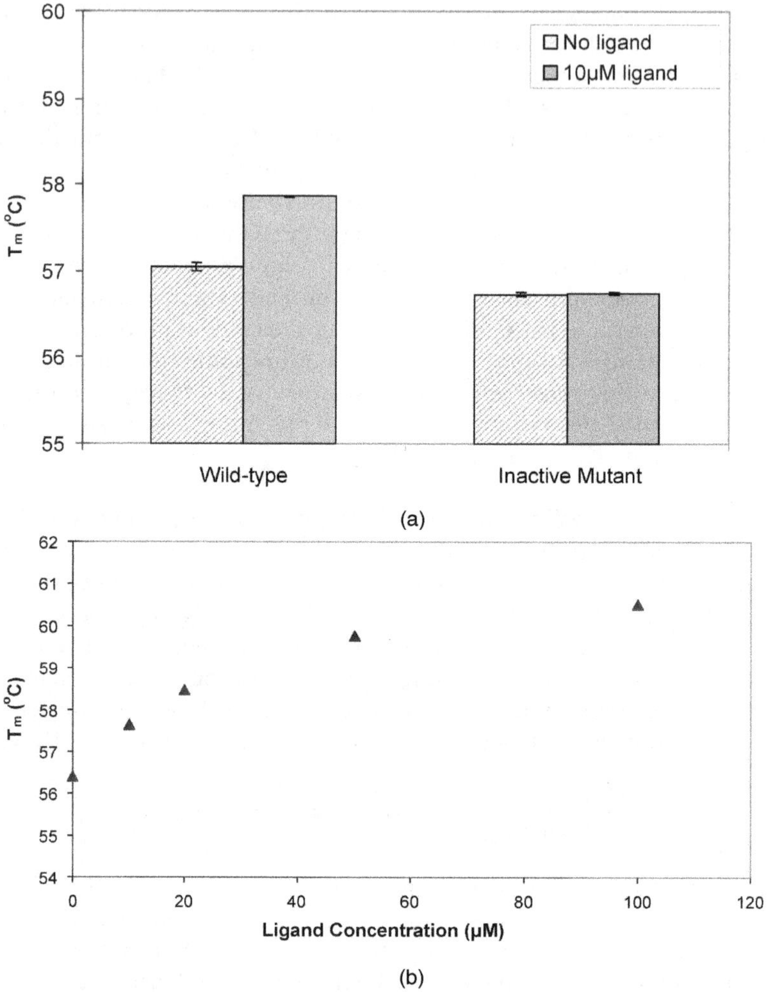

Figure 5.3. Stabilization through ligand binding for a model enzyme: (a) apparent T_m measured by differential scanning calorimetry; (b) relationship between concentration of ligand and increase in thermal stability for the model enzyme.

5.4. STRATEGIES TO ASSESS PHYSICAL STABILITY DURING PREFORMULATION RESEARCH

5.4.1. Biophysical Approaches

Preformulation research studies should be initiated as early as possible in the product development cycle. This is particularly important in programs with aggressive timelines. In addition, during the early stages, very small amounts of protein, perhaps only a few milligrams, are usually available.

TABLE 5.1. Essential Analytical Assays to Assess Physical Stability of Therapeutic Proteins

Parameter	Method	Technique
Conformation, structural changes, aggregation	Spectroscopy	CD, fluorescence, FTIR, Raman, NMR
Oligomerization, precipitation	Light scattering	Dynamic light scattering, Rayleigh (static) scattering
	Analytical ultracentrifugation	Sedimentation velocity, equilibrium centrifugation
	Chromatography	Size exclusion chromatography
	Elution	Field flow fractionation
	Electrophoresis	SDS-PAGE, native
Thermal stability	Calorimetry	DSC
	Spectroscopy	CD melts
Chemical denaturation[a]	Spectroscopy	Fluorescence, CD

[a] Urea or guanidynium hydrochloride unfolding curves.

During preformulation research, critical protein physical parameters such as conformation, stability, oligomerization state, and propensity for aggregation under a variety of conditions should be assessed using a complementary set of well-established analytical techniques. A summary of useful analytical techniques for evaluating physical stability is presented in Table 5.1. In particular, spectroscopy provides a facile means of obtaining information on the structure and stability of proteins, while using only small amounts of material [63] and may minimize the initial need for other assay development. For instance, far-UV circular dichroism analysis provides detailed information on the secondary structural elements in a protein and can require as little as 20 μg of protein [64,65]. Intrinsic fluorescence experiments can require even less protein (5–20 μg) and monitor the packing around tryptophan residues and thus provide a probe of tertiary structure in the molecule [66,67].

Other biophysical techniques, such as near-UV circular dichroism (CD), second-derivative absorbance spectroscopy, and analytical ultracentrifugation [68,69], can provide further insight into the tertiary and quaternary structure of proteins. For an even more immediate assessment of a protein's stability, differential scanning calorimetry (DSC) experiments can be performed to obtain a melting temperature T_{m} and other thermodynamic values such as ΔH and ΔC_p under a range of formulation conditions. These thermodynamic parameters may be good initial predictors as to which buffer conditions will provide the best candidates for long-term real-time stability studies [70]. Additionally, many spectroscopic techniques can monitor the unfolding of a protein, induced either thermally or chemically, and thus yield predictive thermodynamic parameters commonly obtained via calorimetric measurements. Spectroscopic experiments also offer the advantage of yielding more detailed structural information about the protein as it undergoes artificially induced denaturation.

The use of thermodynamics is very advantageous in comparing the stability of a molecule in different formulation conditions. First, the derivation of thermodynamic parameters from spectroscopic unfolding data, whether thermally or chemically induced, is well defined. These parameters provide for a robust assessment of protein stability [71,72]. Additionally, thermodynamic parameters describing the solution behavior of a protein are quantitative values that can be used to compare the stability of a protein as probed using different methods. This strategy is convenient because different spectroscopic methods monitor different features of protein structure, particularly as unfolding takes place. The disruptions of structure that each of these techniques monitors do not have to occur simultaneously. For instance, the classic signature for an intermediate present on unfolding is that the unfolding curve derived from secondary structure unfolding does not overlay that resulting from the denaturation of tertiary structure. As such, reconciling the results gained from different techniques is not a simple exercise. However, the use of thermodynamic principles allows for these problems to be identified and global analysis to be done yielding a single set of parameters that describe the entire unfolding of the protein. This is true even if more complex aspects of protein behavior, such as the presence of an intermediate or oligomerization, are operative.

5.4.2. Tools for Data Analysis

An equilibrium unfolding model is required in order to determine thermodynamic parameters from a set of protein denaturation data obtained via spectroscopic methods. Given the complexities in the solution behavior of a molecule, these models may feature intermediates and dissociation steps. Such steps are often difficult to detect and identify, making the determination of the appropriate equilibrium unfolding model a time-consuming process.

One tool which has been quite useful in assisting the determination of appropriate unfolding models is *singular value decomposition* (SVD) analysis [73], a model-independent mathematical algorithm that takes a given set of data and breaks it down into a minimum number of basis vectors to describe those data. For instance, with a thermal melt monitored by CD where complete spectra are obtained at every temperature unit, a set of vectors with their own spectrum and temperature dependence is determined. These vectors are used to create three matrices, one of which is based on wavelength [73,74]. By identifying the number of nonrandom vectors used to create this matrix, one can determine the number of distinct spectroscopic states that are populated during unfolding. This can reveal the presence of intermediates that need to be accounted for when constructing an unfolding model, but may be difficult to positively identify by merely looking at the response of a particular signal as a function of temperature.

Singular value decomposition analysis offers other advantages when fitting denaturation data obtained using spectroscopy. The standard method of analysis for fitting such data involves identifying the wavelength of maximum change and fitting its response to obtain thermodynamic parameters. However, the data collected at different wavelengths within a given experiment may be monitoring different processes, and thus, by utilizing only one wavelength with the standard analysis method, this additional information is lost. On the other hand, fitting the signal response as a function of temperature at every wavelength collected in a set of spectra in order to

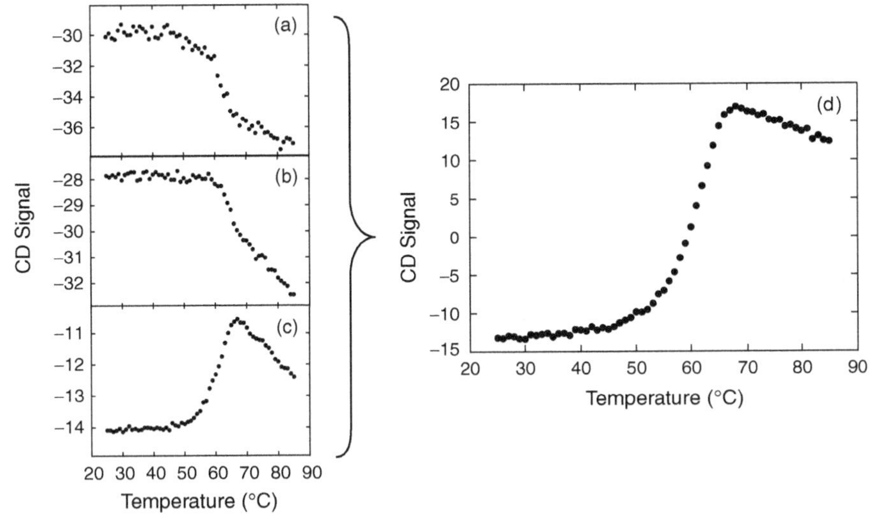

Figure 5.4. Schematic showing the incorporation of the thermal response of several far-UV CD wavelengths [(a) 210 nm, (b) 221.5 nm, (c) 230 nm] into one (d) composite SVD unfolding vector describing all the wavelengths for thermally induced unfolding of rhGAA.

ensure that these data are accounted for would be a daunting and time-consuming task. Using SVD analysis, one can solve this problem by constructing vectors that represent a composite signal response to temperature based on all the wavelengths collected, drastically reducing the analysis time to determine robust thermodynamic parameters (Fig. 5.4). This is particularly useful when a series of spectroscopic experiments is being employed and a single set of thermodynamic parameters is desired that describes the data collected with all the techniques. Singular value decomposition can provide unfolding vectors for each technique, thus significantly reducing the number of unfolding traces that need to be globally fit to obtain thermodynamic parameters while still making use of all the data collected during each of those experiments (Fig. 5.5). Additionally, because such vectors are based on the complete set of collected data at every wavelength, the signal-to-noise ratio (SNR) of the composite response is usually significantly better than the signal at a particular wavelength, thus leading to more robust fitting routines.

Table 5.2 shows the thermodynamic values determined by the thermal denaturation of recombinant human acid α-glucosidase (rhGAA) [75–77] as monitored by far-UV CD. In this case, SVD analysis was used to account for all the wavelengths in the far-UV CD spectra that were collected as a function of temperature to monitor the unfolding. SVD analysis also confirmed the presence of only two distinct spectroscopic species on unfolding, the native and "unfolded" state. Thus, a composite unfolding vector could be fit to a two-state[3] equilibrium unfolding model to

[3]If a single-step unfolding curve is observed, it does not necessarily prove that unfolding closely approaches a two-step mechanism.

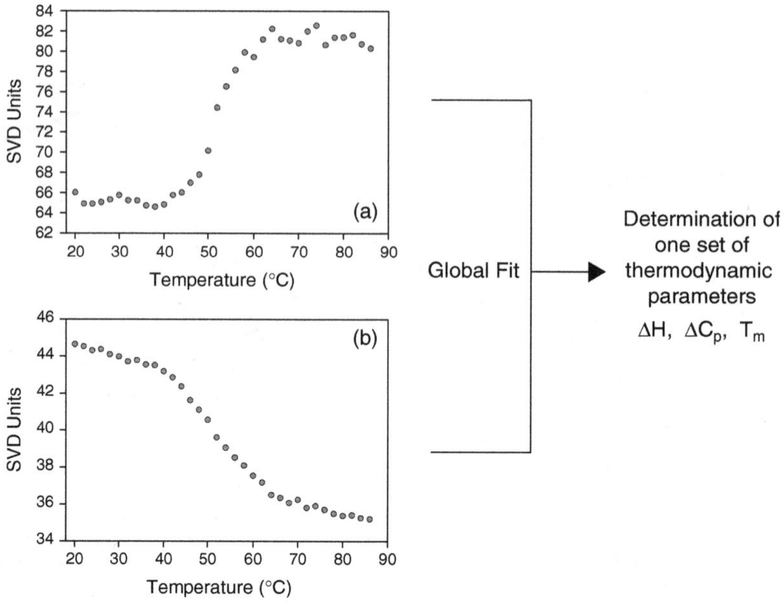

Figure 5.5. Schematic showing how a global fit can be done of SVD vectors from different techniques into one set of thermodynamic parameters. Panels (a) and (b) are the SVD unfolding vectors from far-UV CD and fluorescence data, respectively, for a model protein.

TABLE 5.2. Thermodynamic Parameters[a] for Thermal Unfolding of Recombinant Acid α-Glucosidase (50 μg/mL) in 25 mM Sodium Phosphate as Determined by Far-UV CD Spectroscopy and SVD Analysis

pH	$T^{\mathrm{app}}_{\mathrm{m\text{-}Den}}$ °C[a]	ΔH^{app}, kcal/mol^{-1}	ΔC_p, cal mol^{-1} K^{-1}
4	55.5 (0.4)	78.4 (5.4)	3.0 (0.4)
5	54.6 (0.3)	89.7 (6.2)	4.0 (0.5)
5.5	66.9 (0.4)	128.7 (7.3)	3.0 (0.5)
6	61.0 (0.5)	84.1 (5.8)	2.7 (0.4)
7	45.1 (0.4)	75.4 (5.3)	3.2 (0.4)
8	31.2 (0.4)	38.9 (4.3)	2.2 (0.3)

[a] Errors are shown in parentheses.
[b] All T_m values were determined at 5°C/h.

yield robust thermodynamic parameters as shown in Table 5.2. It should be noted that rhGAA does not unfold via a fully reversible process. This is often the case with biopharmaceutical proteins, which are often large, multidomain proteins that contain disulfide bonds, glycosylation sites, and other posttranslational modifications. Thus, the values listed in Table 5.2 have to be labeled as "apparent", not true, thermodynamic values. However, despite this caveat, the comparative quantitative values obtained via this analysis are still extremely valuable in guiding formulation development.

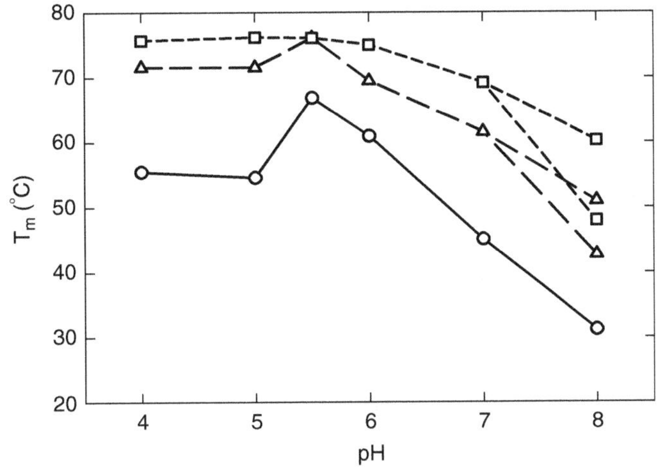

Figure 5.6. Midpoint temperature values T_m for rhGAA thermal denaturation experiments as monitored by far-UV CD at a scan rate of ~5°Chr^{-1} (circles, solid line) and DSC at scan rates of 30 (triangles, long dashed line) and 200°Chr^{-1} (squares, short dashed line). At pH 8.0 two transitions were observed by DSC.

Spectroscopy and DSC are both useful techniques for determining thermodynamic parameters that may be predictive for long-term stability. Spectroscopy offers the additional advantage of obtaining structural information about the molecule, whereas DSC allows much more rapid data collection. Whereas DSC experiments are often performed at scan rates of up to 200°Chr^{-1}, spectroscopy experiments are performed at scan rates as low as 5°Chr^{-1}. Therefore, the absolute values for the thermodynamic parameters obtained by each of the two techniques often differ as a result of this well-known scan rate dependence (Figs. 5.6 and 5.7). However, the trends in the data are often the same (Fig. 5.7b). The example in Figure 5.6 shows the T_m values obtained for the unfolding of rhGAA as a function of pH determined by spectroscopy (with a scan rate of ~5°Chr^{-1}) and DSC (performed at scan rates of both 30°Chr^{-1} and 200°Chr^{-1}). As can be seen, the T_m values at each respective pH vary between the three experiments. However, the trend, which identifies pH 5.5 as the pH of maximum stability, is common to all three collection methods. As such, the choice of method is largely dependent on other variables that must be accounted for when designing the study (time, material requirements, structural information, etc.).

5.5. POTENTIAL CHEMICAL INSTABILITIES IN PREFORMULATION

Chemical stability refers to those mechanisms in which chemical bonds are broken and/or new ones are formed. Thermodynamic approaches and assessment of the physical stability of a protein is important in designing formulation strategies

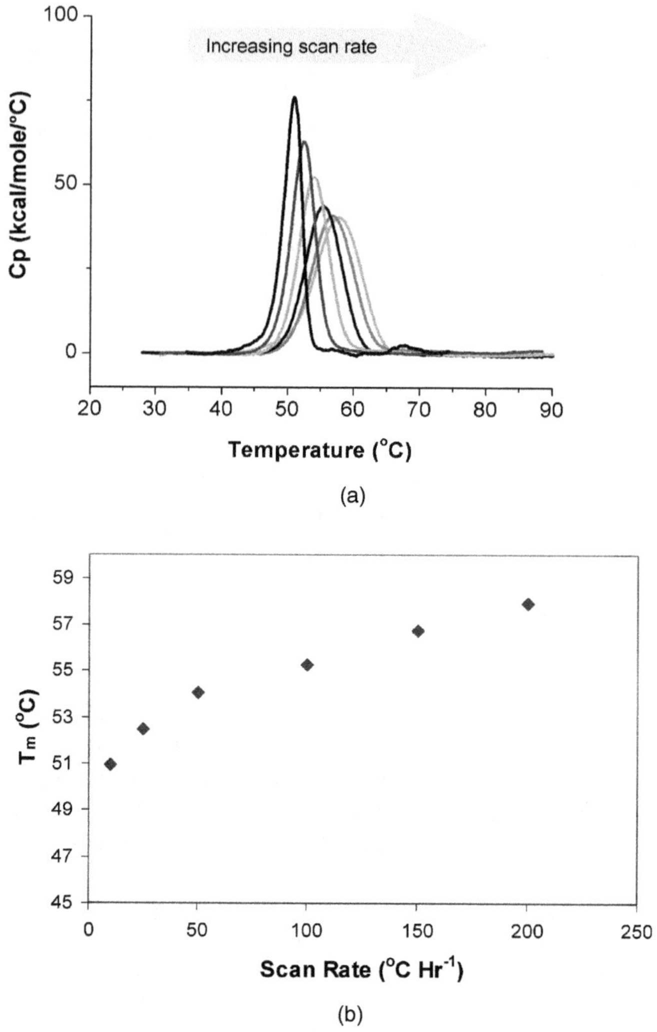

Figure 5.7. Scan rate dependence and effects. (a) DSC analysis of a model protein at different scan rates: 10°C, 25°C, 50°C, 100°C, 150°C, and 200°C per hour; (b) dependence of apparent T_m on scan rate.

when material is limited and other analytical assays are not yet developed. However, an assessment of the physical stability of a protein also has limitations that a formulator should note. Most noteworthy is the tendency of proteins to aggregate and not behave as ideal thermodynamic systems. Thus, thermodynamic parameters developed to assess stability are only *apparent* and could be valid within a limited set of experimental conditions. Thermodynamics cannot predict biochemical stability. Furthermore, a direct link to kinetics is not provided by physical stability alone.

Because of the need for proper shelf-life assignments in biotherapeutic products, kinetic stability becomes an important component during preformulation and formulation development of proteins. Some instabilities may be pathway-dependent; for example, a physical instability such as aggregation could be linked to oxidation of particular residues within the protein. Moreover, isomerization of critical residues in IgG-type molecules, not detected by physical techniques, could lead to changes in bioactivity or potency [78].

The most common chemical instabilities in proteins are hydrolysis, oxidation, deamidation, isomerization, β elimination, and disulfide exchange. One or a combination of the above mentioned mechanisms of chemical degradation can take place in a protein. Given the different types of chemical instabilities in proteins, it is necessary to analyze candidate proteins by a battery of biochemical assays to be able to identify likely degradation mechanisms. Relevant analytical methods to address chemical instabilities are presented in Table 5.3. Since several reviews have dealt with the subject in detail [79,80], only a brief synopsis of the main degradative mechanisms in proteins is presented below.

5.5.1. Hydrolysis

Hydrolysis is a common cause of degradation in proteins involving cleavage of the peptide bond in an acid catalized environment.

5.5.2. Deamidation

Under neutral-to-alkaline solution conditions, deamidation proceeds through formation of a cyclic imide intermediate [81]. Deamidation may be mitigated by decreasing the pH below 6.5. The rate of protein deamidation at acidic pH is often much slower. Deamidation can be detected by using molecular mass or charge differences. Isoelectric focusing may facilitate visualization of deamidated forms. Charge-variant quantitation may be obtained by ion exchange chromatography (IEC). This technique allows purification of the different species for further characterization by peptide mapping or mass spectrometry.

5.5.3. Isomerization

Isomerization of aspartic acid residues to isoaspartic acid is a common instability occurring at acidic pH [82,83]. Deamidation of asparaginyl residues occurs via a succinimide intermediate that subsequently hydrolyzes into a mixture of aspartyl- and isoaspartyl-residue-containing forms. Under acidic conditions, aspartic acid residues may also isomerize through a cyclic intermediate pathway. The isoaspartyl residue is generally difficult to detect because isoaspartic acid is isomeric with aspartic acid, showing no mass difference. More recent studies using electron-capture dissociation mass spectrometry have demonstrated an approach to differentiate these two isoforms in synthetic peptides [84,85].

TABLE 5.3. Analytical Methods to Characterize Chemical Instabilities During Preformulation Research

Method	Purpose
SDS-PAGE	Purity, hydrolysis, mass, disulfide-mediated bonds (aggregates)
Reversed-phase HPLC (RP-HPLC) peptide map	Identification, purity, charge modification (deamidation), oxidation
Ion exchange chromatography	Charge alteration, posttranslational modifications
Isoelectric focusing (IEF)	Charge heterogeneity, isoelectric point (pI)
Capillary electrophoresis	Charge heterogeneity in addition to structural microheterogeneity
Mass spectrometry methods	Mass changes, molecular weight, sequence confirmation

5.5.4. Oxidation

Oxidation is a common mechanism of degradation in proteins [86–88]. Methionine residues are typically most susceptible to becoming oxidized. The specific position within the molecule affects the propensity to oxidize. Cysteine oxidation is dependent on temperature, pH, and buffer. The reaction could also be affected by neighboring groups and active oxygen species. Other residues with propensity to photooxidize in visible light are histidine, tyrosine, methionine, cysteine, and tryptophan. Histidine is most sensitive to oxidation at neutral pH and also by metals. Tyrosine is oxidized at alkaline pH.

5.5.5. Beta Elimination

Beta elimination can occur at cysteine, serine, threonine, phenylalanine, and lysine residues [79]. These reactions may be accelerated by high temperature, alkaline pH, or metals.

5.5.6. Disulfide Exchange

Disulfide interchange of cysteine can occur at neutral or alkaline pH and is catalyzed by free-thiol groups [89].

5.6. STRESS STUDIES TO IDENTIFY POTENTIAL INSTABILITIES

Evaluating data obtained under stress conditions and accelerated stability studies is usually an important aspect of the initial protein stability evaluation [90,91]. For example, after short-term incubation at elevated temperatures or after freeze–thaw or shear by agitation, properties such as activity and purity may be assessed. The expectation is to gain an understanding of the modes of instability that could impact the drug product during storage, shipping, handling, and distribution; and to design formulation strategies that could mitigate protein instabilities or pathways of degradation. It is often difficult to predict the degradation profile for a protein. Even IgGs of the same subclass that share common structural features often behave quite differently in solution [92]. In order to evaluate likely instabilities of candidate proteins, the formulator should have available (or under development), analytical assays that indicate stability. Thus, the second goal of preformulation research is the development and/or qualification of stability-indicating assays. These methods could then be used in a formal stability program once a dosage form is identified.

The major analytical assays relevant to assess stability fall in the category of bioactivity assays (which permit evaluation of the impact of solution variables on potency) and methods that address biochemical properties. After evaluating the protein under stress conditions and identifying the likely degradation mechanism, one can then optimize the formulation in terms of pH and ionic strength and address the need for excipients to prevent aggregation, shear stress, oxidation, and other problems.

Experiments performed at multiple temperatures, ranging from 20°C to 40°C, can accelerate degradation and help evaluate meaningful instabilities that might occur at the recommended storage temperatures. Nevertheless, it is not always possible to extrapolate from accelerated temperatures what really might occur under refrigerated or recommended storage conditions. A common practice is to use Arrhenius kinetics to extrapolate reaction rates from one temperature to another by employing the following equation

$$k = A\,e^{-E_a/RT} \tag{5.14}$$

where E_a is activation energy, R is the gas constant, T is absolute temperature in degrees Kelvin, and A is the arrhenius constant. Equation (5.14) can also be written as follows:

$$\ln k = -\frac{E_a}{RT} + \ln A \tag{5.15}$$

Arrhenius kinetic extrapolation analysis should be employed with extreme care to avoid potential pitfalls. If degradation follows Arrhenius kinetics, then accelerated testing provides a convenient method for the "approximation" of shelf life. Nevertheless, with increasing temperature, other reactions could become rate-limiting that cannot be described by Arrhenius kinetics. Hence, the formulator needs to be aware that a specific degradation might occur much more rapidly under accelerated conditions but may be insignificant at lower temperatures. Thus, real-time stability testing remains the final proof of the solution behavior for the protein of interest. This permits the establishment of recommended storage conditions, retest periods, and shelf life. The issue of selecting proper accelerated storage temperatures to evaluate degradation likely to occur during the shelf life of the product is critical. Figure 5.8 shows an example where a combination of low pH and 37°C induces a rapid formation of high-molecular-weight species reaching 90% after only one week of storage. In this case, the use of DSC can allow for the identification of appropriate storage conditions

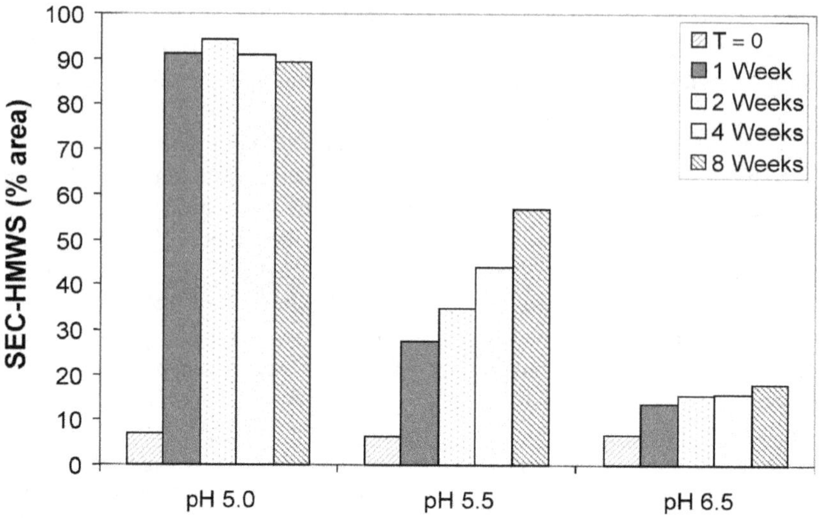

Figure 5.8. Choosing appropriate accelerated storage temperatures; pH screening of a model protein during storage at 37°C as monitored by size exclusion chromatography.

as shown in Figure 5.9a, where the rapid formation of high-molecular-weight species could be ascribed to a shift in the unfolding equilibrium due to low pH.

In contrast, a more stable protein will have a higher onset of melt as measured by DSC. This is the case in the example depicted in Figure 5.9b, which shows a particularly stable protein that manifests no changes after storage at 30°C. In this situation, increasing the temperature to 40°C may be appropriate to help in accelerating degradation and while maintaining the temperature well below the onset of melting at 60°C.

The general approach to carry out stress studies should include

- Timecourse analysis of samples under varying temperature, pH, ionic strength, and other gentle stress conditions.
- Identification of excipients that provide significant benefits against thermal, shear, light, and freeze–thaw stresses as well as preservation of bioactivity.
- Performing analysis using a battery of analytical assays, including physicochemical assessment.
- Demonstrating stability-indicating power of analytical assays. This information should be communicated by the preformulation team to the groups working on developing or improving the analytical assays.
- Isolating and identifying degradation products
- From the protein sequence, posttranslational modifications and identification of key degradation products, a map of the likely degradation profile, and pathways should emerge. This provides the basis for developing effective stabilizing strategies for robust dosage forms with long shelf life.

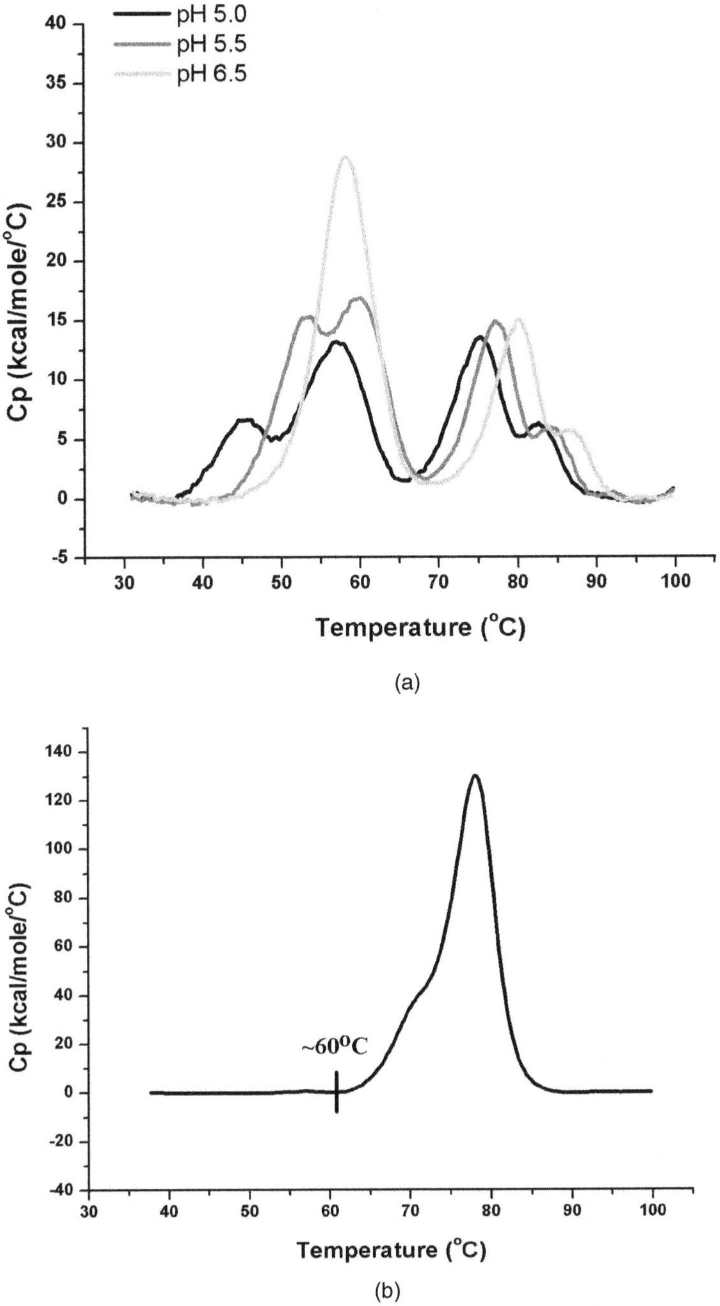

Figure 5.9. Setting up proper storage conditions by using DSC: (a) IgG with low onset of melting; (b) IgG with high onset of melting.

ACKNOWLEDGMENTS

The authors are grateful to Dr. Robert J. Mattaliano for critically reading the manuscript, Lisa Wood for formatting, and Trent Richardson for his help in preparing the figures. BPR is also indebted to many talented individuals who have worked in his laboratory throughout the years.

REFERENCES

1. Tanford, C. (1968), Protein denaturation, *Adv. Protein Chem.* **23**: 121–275.
2. Tanford, C. (1961), *Physical Chemistry of Macromolecules*, Wiley, New York, 1961, pp. 1–680.
3. Ramachandran, G. N. and Sasisekharan, V. (1968), Conformation of polypeptides and proteins, *Adv. Protein Chem.* **23**: 284–438.
4. Franks, F. (1993), *Protein Biotechnology, Isolation, Characterization and Stabilization*, Humana Press, Totowa, NJ, pp. 395–436.
5. Privalov, P. L. (1979), Stability of proteins, *Adv. Protein Chem.* **33**: 167–238.
6. Privalov, P. L. (1982), Stability of proteins. Proteins which do not present a single cooperative system, *Adv. Protein Chem.* **35**: 1–104.
7. Pace, C. N. (1975), The stability of globular proteins, *Crit. Rev. Biochem.* **3**: 1–43.
8. Yancey, P. H., Clark, M. E., Hand, S. C., Bowlus, R. D. and Somero, G. N. (1982), Living with water stress: Evolution of osmolyte systems, *Science* **217**: 1214–1222.
9. Somero, G. N. (1983), Environmental adaptation of proteins: Strategies for the conservation of critical functional and structural traits, *Comp. Biochem. Physiol. A* **76**: 621–633.
10. Yancey, P. H. and Somero, G. N. (1979), Counteraction of urea destabilization of protein structure by methyl amine osmoregulatory compounds of elasmobranch fishes, *Biochem. J.* **183**: 317–323.
11. Lin, T. Y. and Timasheff, S. N. (1994), Why do some organisms use a urea methylamine mixture as osmolyte? Thermodynamic compensation of urea and trimethylamine N-oxide interactions with proteins, *Biochemistry* **33**: 12695–12701.
12. Bennion, B. J. and Dagget, V. (2004), Counteraction of urea-induced protein denaturation by trimethylamine N-oxide: A chemical chaperone at atomic resolution, *Proc. Natl. Acad. Sci USA* **101**: 6433–6438.
13. Ross, P. D. and Subramanian, S. (1981), Thermodynamics of protein association reactions: Forces contributing to stability, *Biochemistry* **20**: 3096–3102.
14. Wada, A. (1976), The alpha-helix as an electric macro-dipole, *Adv. Biophys.* **9**: 1–63.
15. Sheridan, R. P., Levy, R. M., and Salemme, F. R. (1982), The alpha-helical dipole model and electrostatic stabilization of 4-alpha-helical proteins, *Proc. Natl. Acad. Sci. USA* **79**: 4545–4549.
16. Hol, W. G. J. (1985), The role of the α-helix dipole in protein function and structure, *Progress Biophys Mol. Biol.* **45**: 149–195.
17. Blagdon, D. E. and Goodman, M. (1975), Mechanisms of protein and polypeptide helix initiation, *Biopolymers* **14**: 241–245.
18. Jencks, W. (1987), *Catalysis in Chemistry and Enzymology*, Dover, New York, p. 412.

19. Myers, J. K. and Pace, N. (1996), Hydrogen bonding stabilizes globular proteins, *Biophys. J*. **71**: 2033–2039.
20. Wu, J., Bell, A. F., Luo, L., Stephens, A. W., Stankovich, M. T., and Tonge, P. J. (2003), Probing hydrogen-bonding interactions in the active site of medium-chain acyl-CoA dehydrogenase using Raman spectroscopy, *Biochemistry* **42**: 11846–11856.
21. Hagler, A. T. and Lifson, S. (1974), Energy functions for peptides and proteins II. The amide hydrogen bond and calculation of amide crystal properties, *J. Am. Chem. Soc*. **96**: 5327–5335.
22. Kauzmann, W. (1959), Some factors in the interpretation of protein denaturation, *Adv. Protein Chem*. **14**: 1–63.
23. Dill, K. A. (1990). Dominant forces in protein folding. *Biochemistry* **29**: 7133–7155.
24. Hodges, R. S., Zhou, N. E., Kay, C. M., and Semchuck, P. D. (1990), Synthetic model proteins: Contribution of hydrophobic residues and disulfide bonds to protein stability, *Pept. Res*. **3**: 123–137.
25. Rezaei-Ghaleh, N., Ramshini, H., Ebraim-Habibi, A., Moosavi-Movahedi, A. A., and Nemat-Gorgani, M. (2008), Thermal aggregation of alpha-chmymotrypsin: Role of hydrophobic and electrostatic interactions, *Biophys. Chem*. **132**: 23–32.
26. Nozaki, Y. and Tanford, C. (1971), The solubility of amino acids and two glycine peptides in aqueous ethanol and dioxane solutions; establishment of hydrophobicity scale, *J. Biol. Chem*. **246**: 2211–2217.
27. Pohorille, A. and Pratt, L. (1992), Theory of hydrophobicity: Transient cavities in molecular liquids, *Proc. Natl. Acad. Sci. USA* **89**: 2995–2999.
28. Nemethy, G. and Scheraga, H. A. (1962), Structure of water and hydrophobic bonding in proteins I. A model for the thermodynamic properties of liquid water, *J. Chem. Phys*. **36**: 3382–3400.
29. Baldwin, R. L. (1986), Temperature dependence of the hydrophobic interaction in protein folding, *Proc. Natl. Acad. Sci. USA* **83**: 8069–8072.
30. Lee, J. C. and Timasheff, S. N. (1981), The stabilization of proteins by sucrose, *J. Biol. Chem*. **256**: 7193–7201.
31. Arakawa, T. and Timasheff, S. N. (1982): Stabilization of protein structure by sugars, *Biochemistry* **21**: 6536–6544.
32. Lin, T.-Y. and Timasheff, S. N. (1996), On the role of surface tension in globular protein stabilization, *Protein Sci*. **5**: 372–381.
33. Gekko, K. and Morikawa, T. (1981), Preferential hydration of bovine serum albumin in polyhydric alcohol-water mixtures, *J. Biochem*. **90**: 39–50.
34. Xie, G. F. and Timasheff, S. N. (1997), Mechanism of the stabilization of ribonuclease A by sorbitol: Preferential hydration is greater for the denatured than for the native protein, *Protein Sci*. **6**: 211–221.
35. Xie, G-F. and Timasheff, S. N. (1997), The thermodynamic of protein stabilization by trehalose, *Biophys. Chem*. **64**: 25–43.
36. Gekko, K. and Timasheff, S. N. (1981), Mechanism of protein stabilization by glycerol: Preferential hydration in glycerol-water mixtures, *Biochemistry* **20**: 4667–4676.
37. Na, G. C. and Timasheff, S. N. (1981), Interaction of calf-brain tubulin with glycerol, *J. Mol. Biol*. **151**: 165–178.

38. Arakawa, T. and Timasheff, S. N. (1984), The mechanism of action of Na glutamate, lysine HCl and piperazine-N,N'-bis(2-ethanesulfonic acid) in the stabilization of tubulin and microtubule formation, *J. Biol. Chem.* **259**: 4979–4986.
39. Timasheff, S. N. (1993), The control of protein stability and association by weak interactions with water: How do solvents affect these processes? *Annu. Rev. Biophys. Biomol. Struct.* **22**: 67–97.
40. Perez-Ramirez, B., Shearwin, K. E., and Timasheff, S. N. (1994), The colchicine-induced GTPase activity of tubulin: State of the product, activation by microtubule-promoting cosolvents, *Biochemistry* **33**: 6253–6261.
41. Perez-Ramirez, B. and Timasheff, S. N. (1994), Cosolvent modulation of the tubulin-colchicine GTPase-activating conformational change: Strength of the enzymatic activity, *Biochemistry* **33**: 6262–6267.
42. Timasheff, S. N. (1995), Preferential interactions of water and cosolvents with proteins, chapter 11, in: *Protein Solvent Interactions*, Rober B., Gregory, ed., Marcel-Dekker, New York, pp. 445–482.
43. Arakawa, T., Timasheff, S. N. (1985). Mechanism of poly (ethylene glycol) interaction with proteins, *Biochemistry* **24**: 6756–6762.
44. Bhat, R. and Timasheff, S. N. (1992), Steric exclusion is the principal source of the preferential hydration of proteins in the presence of poly ethylene glycols, *Protein Sci.* **1**: 1133–1143.
45. Gekko, K. and Timasheff, S. N. (1981), Thermodynamic and kinetic examination of protein stabilization by glycerol, *Biochemistry* **20**: 4677–4686.
46. Baldwin, R. L. and Eisenberg, D. (1987), Protein stability, in *Protein Engineering*, Oxender, D. L. and Fox, C. F., eds., Alan Liss, New York, pp. 127–148.
47. Slebe, J. C. and Martinez-Carrion, M. (1976), Carbamylation of aspartate amino transferase, the pK value of the active site lysyl residue, *J. Biol. Chem.* **251**: 5663–5669.
48. Slebe, J. C. and Martinez-Carrion, M. (1978), Selective chemical modification and ^{19}F NMR in the assignment of a pK value to the active site lysyl residue in aspartate amino transferase, *J. Biol. Chem.* **253**: 2093–2097.
49. Czerwinski, R. M., Harris T. K., Massiah, M. A., Mildvan, A. S., and Whitman, C. P. (2001), The structural basis for the perturbed pKa of the catalytic base in 4-oxalocrotonate tautomerase: Kinetic and structural effects of mutations of phe50, *Biochemistry* **20**: 1984–1995.
50. Matthews, J. B. (1985), Electrostatic effects in proteins, *Annu. Rev. Biophys. Chem.* **14**: 387–417.
51. Fox, S. W. and Foster, J. F. (1957), *Introduction to Protein Chemistry*, J Wiley, New York, p. 242.
52. Hofmeister, F. (1888), Zur lehre von der wirkung der salze. Zweite mittheilung. *Arch. Phatol. Pharmakol.* **24**: 247–260.
53. Melander, W. and Howarth, C. (1977), Salt effect on hydrophobic interactions in precipitation and chromatography of proteins: An interpretation of the lyotropic series, *Arch. Biochem. Biophys.* **183**: 200–215.
54. Arakawa, T. and Timasheff, S. N. (1985), Theory of protein solubility, *Meth. Enzymol.* **114**: 49–77.
55. Collins, K. D. and Washabaugh, M. W. (1977), The Hofmeister effect and the behavior of water interfaces, *Q. Rev. Biophys.* **18**: 323–422.

56. Baldwin R. L. (1996), How Hofmeister ion interactions affect protein stability, *Biophys. J.* **71**: 2056–2063.

57. Arakawa, T. and Timasheff, S. N. (1982), Preferential interactions of proteins with salts in concentrated solutions, *Biochemistry* **21**: 6545–6552.

58. Lewis, G. N. and Randall, M. (1921), The activity coefficient of strong electrolytes, *J. Am. Chem. Soc.* **43**: 1112.

59. Ellis, K. J. and Morrison J. F. (1982), Buffers of constant ionic strength for studying pH-dependent processes, *Meth. Enzymol.* **87**: 405–426.

60. Havranek, J. J. and Harbury, P. (1999), Tanford-Kirkwood electrostatic for protein modeling, *Proc. Natl. Acad. Sci. USA* **96**: 11145–11150.

61. Wyman, J. (1964), Linked functions and reciprocal effects in hemoblobin: A second look, *Adv. Protein Chem.* **19**: 223–286.

62. Wyman, J. and Gill, S. J. (1990), *Binding and Linkage. Functional Chemistry of Biological Macromolecules*, Univ. Science Books, Mill Valley, CA, 1–303.

63. Meyer, J. D., Ho, B., and Manning, M. C. (2002), Effect of conformation on the chemical stability of pharmaceutically relevant polypeptides, *Pharm. Biotechnol.* **13**: 85–107.

64. Kelly, S. M., Jess, T. J. and Price, N. C. (2005), How to study proteins by circular dichroism, *Biochem. Biophys. Acta* **1751**: 119–139.

65. Kelly, S. M. and Price, N. C. (2000), The use of circular dichroism in the investigation of protein structure and function, *Curr. Protein Pept. Sci.* **1**: 349–384.

66. Jiskoot, W., Hlady, V., Naleway, J. J., and Herron, J. N. (1995), Application of fluorescence spectroscopy for determining the structure and function of proteins, *Pharm. Biotechnol.* **7**: 1–63.

67. Royer, C. A. (1995), Fluorescence spectroscopy, *Meth. Mol. Biol.* **40**: 65–89.

68. Laue, T. M., Stafford, W. F. 3rd (1999), Modern applications of analytical ultracentrifugation, *Annu. Rev. Biophys. Biomol. Struct.* **28**: 75–100.

69. Perez-Ramirez, B. and Steckert, J. J. (2005), Probing reversible self-association of therapeutic proteins by sedimentation velocity in the analytical ultracentrifuge, *Meth. Mol. Biol.* **308**: 301–318.

70. Remmele, R. L. Jr., Nightlinger, N. S., Srinivasan, S., and Gombotz, W. R. (1998), Interleukin-1 receptor (IL-1R) liquid formulation development using differential scanning calorimetry, *Pharm. Res.* **15**: 200–2008.

71. Baldwin, R. L. and Rose, G. D. (1999), Is protein folding hierarchic? I. Local structure and peptide folding, *Trends Biochem. Sci.* **24**: 26–33.

72. Chen, Y., Ding, F., Nie, H., Serohijos, A. W., Sharma, S., Wilcox, K. C., Yin, S., and Dokholyan, N. V. (2008), Protein folding: Then and now, *Arch. Biochem. Biophys.* **469**(1): 4–19.

73. Henry, E. R. and Hofrichter, J. (1992), Singular value decomposition: Application to analysis of experimental data, *Meth. Enzymol.* **210**: 129–192.

74. Ionescu, R. M., Smith, V. F., O'Neill, J. C. Jr., and Matthews, C. R. (2000), Multistate equilibrium unfolding of *Escherichia coli* dihydrofolate reductase: Thermodynamic and spectroscopic description of the native, intermediate, and unfolded ensembles, *Biochemistry* **39**: 9540–9550.

75. Hoefsloot, L. H., Hoogeveen-Weterveld, M., Reuser, A. J., and Oostra, B. A. (1990), Characterization of the human lysosomal α-glucosidase gene, *Biochem. J.* **272**: 493–497.

76. Hoefsloot, L. H., Willemsen, R., Kroos, M. A., Hoogeveen-Westerveld, M., Hermans, M. M., van der Ploeg, A. T., Oostra, B. A., and Reuser, A. J. (1990), Expression and routing

of human lysosomal α-glucosidase in transiently transfected mammalian cells, *Biochem. J.* **272**: 485–492.

77. Wisselaar, H. A., Kroos, M. A., Hermans, M. M. P., van Beeumen, J., and Reuser, A. J. J. (1993), Structural and functional changes of lysosomal acid-glucosidase during intracellular transport and maturation, *J. Biol Chem.* **268**: 2223–2231.
78. Wakankar, A. A., Borchardt, R. T., Eigenbrot, C., Shia, S. T., Wang, Y. J., Shire, S. J., and Liu, J. L. (2007), Aspartate isomerization in the complementary-determining regions of two closed related monoclonal antibodies, *Biochemistry* **46**: 1534–1544.
79. Manning, M. C., Patel, K., and Borchard, R. T. (1989), Stability of protein pharmaceuticals, *Pharm. Res.* **6**: 903–918.
80. Schoeneich, C., Hageman, M. J., and Borchardt R. T. (1997), Stability of peptides and proteins, in *Controlled Drug Delivery Challenges and Strategies*, Park, K., ed., American Chemical Society, Washington, DC, pp. 205–228.
81. Geiger, T. and Clark, S. (1987), Deamidation, isomerization and racemization at asparaginyl and aspartyl residues in peptides. Succinimided-linked reactions that contribute to protein degradation, *J. Biol. Chem.* **262**: 785–794.
82. Clarke, S. (1987), Propensity for spontaneous succinimide formation from aspartyl and asparaginyl residues in cellular proteins, *J. Pept. Protein Res.* **30**: 808–821.
83. Reissner, K. J. and Aswald, D. W. (2003), Deamidation and isoaspartic formation in proteins: Unwanted alterations or surreptitious signals? *Cell. Mol. Life Sci.* **60**: 1281–1295.
84. Cournoyer, J. J., Lin, C., and O'Connor, P. B. (2006), Detecting deamidation products in proteins by electron capture dissociation, *Anal Chem.* **78**: 1264–1271.
85. Cournoyer, J. J., Lin, C., Bowman, M. J., and O'Connor, P. B. (2007), Quantitating the relative abundance of isoaspartyl residues in deamidated proteins by electron capture dissociation, *J. Am. Mass Spectrom.* **18**: 48–56.
86. Franson, J., Florin-Robertsson, E., Axelson, K., and Nyhlen, C. (1996), Oxidation of human insulin-like growth factor I in formulation studies: Kinetics of methionine oxidation in aqueous solution and in solid state, *Pharm. Res.* **13**: 1252–1257.
87. Franson, J. and Hagman, A. (1996), Oxidation of human insulin-like growth factor I in formulation studies, II Effects of oxygen, visible light, and phosphate on methionine oxidation in aqueous solution and evaluation of possible mechanisms, *Pharm. Res.* **13**: 1476–1481.
88. Bobrowski, K., Hug, G. L., Pogocki, D., Marciniak, B., and Schöneich, C. H. (2007), Stabilization of sulfide radical cations through complexation with the peptide bond: Mechanisms relevant to oxidation of proteins containing multiple methionine residues, *J. Phys. Chem. B* **111**: 9608–9620.
89. Lowther, W. T., Brot, N., Weissback, H., Honek, J. F., and Matthews, W. (2000), Thiol-disulfide exchange is involved in the catalytic mechanism of peptide methionine sulfoxide reductase, *Proc. Natl. Acad. Sci. USA* **97**: 6463–6468.
90. Franks, F. (1994), Accelerated stability testing of bioproducts: Attractions and pitfalls, *Trends Biotechnol.* **12**: 114–117.
91. Kirkwood, T. B. L. (1984), Design and analysis of accelerated degradation tests for the stability of biological standards. III. Principles of design, *J. Biol. Stand.* **12**: 215–224.
92. Lu, Y., Harding, S. E., Michaelsen, T. E., Longman, E., Davis, K. G., Ortega, A., Gunter Grossmann, J., Sandlie, I., and Garcia de la Torre, J. (2007), Solution conformation of wild type and mutant IgG3 and IgG4 immunoglobulins using crystallohydrodynamics: Possible implications for complement activation, *Biohys. J.* **93**: 3733–3744.

6

FORMULATION DEVELOPMENT OF PHASE 1–2 BIOPHARMACEUTICALS: AN EFFICIENT AND TIMELY APPROACH

Nicholas W. Warne

6.1. INTRODUCTION

The field of protein formulations has, since the early to mid-1990s, seen a substantial increase in the level of sophistication of analytical tools, high-throughput approaches [1,2], and challenges in the development of commercially competitive dosage forms. One question that is raised, however, is whether the application of these intensive approaches is appropriate for early clinical phase protein products. Given the low probability of a candidate protein becoming a commercial product, recently cited as approximately 8% from initial investigational new drug application (IND) to successful license application [3], is it wise to invest resources to optimize a protein's formulation when the safe and effective dose and the route of administration are unknown and the target product profile is unconfirmed? The use of a broad array of formulation tools and formulation options requires a substantial investment of time and resources, which can be lacking during the early phases of product development. Rather, the more appropriate goal may be the efficient utilization of these limited resources to increase the throughput of new-drug candidates into phase 1 clinical trials. On this basis, the experienced formulation scientist should consider the objective of getting compounds to the clinic faster and more efficiently, with a high degree of success and an acceptable

Formulation and Process Development Strategies for Manufacturing Biopharmaceuticals, edited by Jameel and Hershenson

stability profile, so that an initial clinical assessment of safety and biological activity can be conducted.

The purpose of this chapter is to propose that, for an early-phase clinical project, a stable, simple formulation that permits flexibility in clinical testing is ideal and formulation optimization should be delayed until information on the initial safety profile, dose, route, and competitive environment are better defined. The goal of this approach is to rapidly reach a suitable formulation, in a standard container–closure system, that provides the clinician with flexibility to provide doses that span >100× ranges without undue manipulations within the clinical pharmacy and that are compatible with multiple modes of administration (intravenous, intramuscular, or subcutaneous). For example, a 50 mg/mL liquid dosage form can easily accommodate doses as low as 0.2 mg/kg for a small, 50-kg patient (10 mg dose or a 100-μL subcutaneous injection) and a ≤20 mg/kg dose for a larger, 100-kg patient (2000 mg dose or an infusion of 40 mL of drug product). The added utility of intravenous dosing and the use of syringe pumps provide further flexibility, especially at lower concentrations. Thus, the formulation scientist, when first presented with a new protein candidate, may be tempted to develop an optimized dosage form for late-stage clinical testing or commercial distribution. Rather, the formulator should choose to focus on designing a suitable dosage form that is stable for several years, provides broad dose range flexibility, and is easy to manufacture.

6.2. FORMULATION OPTIONS AND LESSONS LEARNED

If one were to fully assess the behavior of a protein during early formulation efforts, then formulation characteristics such as pH (from 4.0 to 8.0), ionic strength using sodium chloride, cryopreservatives such as sucrose, and the need for surfactants would all have to be examined. However, in the development of protein therapeutics, protein with which to perform early formulation studies can often be in short supply. This is particularly true in early lab-scale production leading up to the identification of the appropriate cell line, cell culture conditions, and purification schemes. The formulation scientist may be asked to identify a formulation that will be utilized to prepare the supplies to be used in the IND-enabling toxicology studies having had only 12–16 weeks and several hundred milligrams or grams of protein to work with. Further, there may be a strong desire to generate data that support a phase 1 formulation at increasingly elevated protein concentrations, often up to 100 mg/mL or more for an antibody. It is critical, during this early stage of product development, that all experiments be clearly defined in terms of materials needed and expected outcomes, and that proper execution of the studies result in identifying a suitable formulation. It may be critical to use one's experience in designing these studies to develop early clinical-stage formulations in a timely manner when protein supplies, analytical procedures, and time are in short supply.

The first step in designing a suitable dosage form and formulation is often to gather information on pH. The pH of the formulation, which is usually critical to the stability of the product during the manufacturing process and during extended storage, must be selected to minimize a wide variety of possible degradation mechanisms, including aggregation, poor solubility (partially based on the pI of the protein), and chemical degradation by oxidation, deamidation, and hydrolysis, to name only a few mechanisms [4]. Many of these chemical degradation mechanisms can be managed effectively by lyophilization [5]; however, lyophilization adds time and expense and, in any case, the formulator must also consider the effect of short-term hold times during the manufacturing process as well as immediately prior to administration. In order to assess a suitable pH while minimizing time, resources, and material requirement, one may wish to begin by first learning from the work of others.

Table 6.1 lists the formulations for more recently approved antibody and antibody-like products. The information contained in this table has been taken from the publicly available sources provided by the manufacturers [6–14]. Examination of Table 6.1 demonstrates some common themes in terms of pH and stabilizers. These commercial formulations show a mix of lyophilized and liquid presentations and a protein concentration range of 5–125 mg/mL. Close review of the Table 6.1 demonstrates several trends. First, the pH range of the nine formulations varies from 5.2 to 7.2 with an average of 6.4. Further, of the nine examples, three utilize histidine as the buffer, three utilize sodium phosphate, and three use sodium citrate. Also, while several liquid formulations utilize sodium chloride to add isotonic strength, none of the lyophilized formulations use sodium chloride, presumably because of its problematic characteristics during the freeze-drying process [15].

A review of Table 6.1 suggests that, if time and/or material are limited, it may be reasonable to focus initially on pH values between 5 and 7.5. If it is desirable to narrow these even further, one might note that several antibody products were suitably stable at pH∼6.0 in histidine, at concentrations of 5–40 mM as described in Table 6.1. One could argue that for the purpose of designing a suitable stable phase 1 dosage form for an antibody it may be adequate, or even optimal, to begin with pH 6.0 in 10–20 mM histidine.

Additionally, since seven of the nine examples presented in Table 6.1 contain either polysorbate 20 or polysorbate 80, there may be value in asking whether it is prudent, when time and materials are limited, to assess whether one should include polysorbate in the formulation. From the commercial experience of the formulation scientists who developed the products listed in Table 6.1, the simple recommendation would be to include some level of polysorbate in the formulation at a concentration between 0.005% and 0.2%. Thus, the decision is no longer whether to include a surfactant in the formulation, but rather to ask how much to include on the basis of its intended purpose, such as protection from mechanical agitation or possible assistance during the reconstitution of a lyophilized powder [16].

A similar review of the products listed in Table 6.1 can be applied to other parameters such as the need for NaCl, a cryoprotectant or other stabilizers, and/or other properties of the formulation. The use of the experience of others, combined with several simple confirmatory experiments, may provide an efficient means to develop a formulation very quickly that is suitable for early clinical development.

TABLE 6.1. Formulations of Several Commercial Products[a]

Product Name	Description	Formulation
Amevive (alefacept)	Dimeric fusion protein	Lyophilized; 15 mg/mL alefacept, 10 mg/mL glycine, 7.2 mg/mL sodium citrate dihydrate, 0.12 mg/mL citric acid pH 6.9
Enbrel (etanercept)	Dimeric fusion protein	Liquid; 50 mg/mL etanercept, 10 mg/mL sucrose, 5.8 mg/mL sodium chloride, 5.3 mg/mL L-arginine hydrochloride, 2.6 mg/mL sodium phosphate monobasic monohydrate, 0.9 mg/mL sodium phosphate dibasic anhydrous
Herceptin (trastuzumab)	Antibody	Lyophilized; 21 mg/mL trastuzumab, ~20 mg/mL α,α-trehalose dehydrate, 0.5 mg/mL L-histidine HCl, 0.32 mg/mL L-histidine, 0.09 mg/mL polysorbate 20, 1.1% benzyl alcohol, pH ~6
Humira (adalimumab)	Antibody	Liquid; 50 mg/mL adalimumab, 6.2 mg/mL sodium chloride, 0.86 mg/mL sodium citrate, 1.3 mg/mL citric acid monohydrate, 12 mg/mL mannitol, 1 mg/mL polysorbate 80, pH ~5.2
Raptiva (efalizumab)	Antibody	Lyophilized; 100 mg/mL Raptiva, ~82 mg/mL sucrose, 4.5 mg/mL L-histidine hydrochloride monohydrate, 2.9 mg/mL L-histidine, 2 mg/mL polysorbate 20, pH ~6.2.
Remicade (infliximab)	Antibody	Lyophilized; ~10 mg/mL infliximab, 50 mg/mL sucrose, 0.05 mg/mL polysorbate 80, 0.22 mg/mL monobasic sodium phosphate monohydrate, 0.61 mg/mL dibasic sodium phosphate dihydrate, pH ~7.2
Rituxan (Rituximab)	Antibody	Liquid; 10 mg/mL Rituximab, 9 mg/mL sodium chloride, 7.35 mg/mL sodium citrate dihydrate, 0.7 mg/mL polysorbate 80, pH ~6.5
Zenapax (daclizumab)	Antibody	Liquid; 5 mg/mL daclizumab, 3.6 mg/mL sodium phosphate monobasic monohydrate, 11 mg/mL sodium phosphate dibasic heptahydrate, 4.6 mg/mL sodium chloride, 0.2 mg/mL polysorbate 80, pH ~6.9
Xolair (omalizumab)	Antibody	Lyophilized; 125 mg/mL omalizumab, 90 mg/mL sucrose, 1.7 mg/mL L-histidine hydrochloride monohydrate, 1.1 mg/mL L-histidine, 0.3 mg/mL polysorbate 20

[a] All formulations are expressed as mg/mL after reconstitution. Because of fill volume variability, some concentrations are approximate.

These examples—the selection of an appropriate pH and the use of polysorbates in antibody products—demonstrate the concept of a "simple formulation." It is desirable for protein formulations to be inherently simple, and each excipient and condition should have a clearly defined purpose. More recent regulatory guidance [17] suggests that the selected excipients and their respective concentrations be justified on the basis of their impact on the product's stability, bioavailability, and manufacturability. While additional data are collected on these parameters during the late clinical stage or during commercial production (or line extension), much of this information will not exist during preclinical development prior to the preparation of early-stage clinical supplies. Limited information should encourage the formulation scientist to seek simple solutions to the design of the formulation space for the protein in question. With this in mind, the "simple formulation" approach may be broadly applicable across antibody products. If, for example, 10–20 mM histidine, some level of polysorbate, and pH 6.0 work for several antibodies, it may be a reasonable starting point to assume that these parameters may apply broadly. Rather than performing a broad pH screen, at pH 4–8, under accelerated storage conditions of 25°C and 40°C, the formulation scientist may simply wish to verify whether the selection of 10–20 mM histidine at pH 6.0 provides adequate stability for the early clinical dosage form. A similar approach could be taken in identifying the level of polysorbate. For example, the products listed in Table 6.1 encompass a range from 0.005% to 0.2% with either polysorbate 20 or polysorbate 80. A simple titration of polysorbate within the formulation, when suitably stressed, should provide a rationale for selection of an appropriate concentration. This is discussed later.

A buffer and a specific pH range are always required to ensure the stability of the product during manufacture and storage. While several buffer systems are available, the most widely utilized and accepted buffers are listed in Table 6.2. Again, for an early-stage clinical product, one may want to ask whether there is value in exploring obscure buffering agents that may require new raw-materials specifications, receipt tests, possible safety studies, and questions regarding the compendial nature of the chemical. Alternatively, one may simply wish to determine the optimal pH for the product and assess a well-known, globally accepted buffer at a reasonable concentration.

In utilizing previous experience with buffers, the first step is to use examples that are relevant for the type of compound (antibody, fusion protein, etc.) and the targeted manufacturing process (liquid or lyophilized). The nine examples in Table 6.1 utilize histidine, sodium phosphate, or citrate. These buffer systems are suitable for the pH range of the products in Table 6.1, but may not be suitable for nonantibody products

TABLE 6.2. Excipients Utilized for Buffering Protein Solutions

Buffer	pK_a (20°C)	Compendial Status[a]
Glutamic acid	4.25	Ph. Eur., JPC, FCC
Succinic acid	5.6	Ph. Eur., NF, JPE
Citrate	4.75	Ph. Eur., USP, JP
L-Histidine HCl	6.0	Ph. Eur., USP, JPC
Sodium phosphate	7.21	Ph. Eur., USP, JPE
Trometamol (Tris)	8.08	Ph. Eur., USP, JPC

[a] *Abbreviations*: Ph. Eur = *European Pharmacopoeia*; JPC = *Japanese Pharmacopoeia Codex*; NF = *National Formulary*; FCC = *Food Chemical Codex*; USP = *United States Pharmacopoeia*; JPE = *Japanese Pharmaceutical Excipients*; JP = *Japanese Pharmacopoeia*.

that would have different pH requirements. Further, several buffers have inherent weaknesses when used for protein formulations. For example, sodium phosphate will selectively crystallize when frozen, often leading to a substantial decrease in the pH of the solution, by as much as 2 pH units [18,19]. Other buffers such as acetate may be suitable for acidic liquid formulations, however they are not appropriate for a lyophilized dosage form because of their volatility. At pH <6, succinate has been utilized [20] and glutamate has been used as well [21]. For pH >7.0, tromethamine (Tris) may be considered [7,22]. For these reasons, the formulation scientist is encouraged to use standard excipients that have been utilized previously, have acceptable safety profiles, are widely available, and enjoy broad acceptability by the regulatory authorities.

In addition to a buffer and pH range, one typically requires a stabilizer to protect the protein from the rigors of ultrafiltration, lyophilization, and multiple freeze–thaws. While the rationale for excipient selection, for lyophilized products in particular, has been discussed elsewhere [5,23], the principles recommended for stabilizer selection are similar to those for selecting a suitable buffer, pH, and surfactant; that is, one may wish to learn from experience and avoid overinvesting time and effort unless the protein shows instability with commonly used excipients. If we review the antibody formulations presented in Table 6.1, it is apparent that the stabilizer of choice, for lyophilized products, is sucrose at a level of approximately 1%–9%. Trehalose is also used but is not as widely available. Drug substances are often produced in batches that must be stored for extended periods ranging from months to years. While some drug substances, such as antibodies, may be stored under refrigerated conditions, drug substances often must be frozen to preserve stability. A cryopreservative is often utilized to protect the protein of interest from the rigors of repeated freeze–thaw as well as freeze-drying. The use of sucrose as a cryoprotectant is well documented as is its use as a stabilizer for lyophilized dosage forms [23]. Preformulation experiments are key in determining the feasibility of a liquid formulation. The formulator must decide, for the early-stage clinical product, whether lyophilization should be an option. The use of sucrose, at neutral pH, provides the opportunity to either lyophilize the product or develop a liquid formulation that will be suitably stable against freeze–thaw-induced damage. One aspect to consider, however, is the use of sucrose in liquid dosage forms where the pH is slightly acidic; in this case hydrolysis will occur even under refrigerated conditions [24]. For a lyophilized dosage form, however, the use of sucrose, which is widely accepted, should be encouraged.

Whether one lyophilizes an early-stage clinical product is of considerable debate. Often, the choice of whether to lyophilize is driven by the availability of lyophilization facilities for early-stage clinical products. In addition, because of their commercial advantages in terms of convenience, liquid dosage forms are often utilized. In making this decision, it is critical to remember that, driven by product experience over time, the early-clinical-phase product is often likely to be distinct from the eventual late-clinical-stage or commercial product. It may be unjustified to assume that an early-clinical-phase product will be suitable for commercialization because the clinical product must serve a different purpose. The early clinical product must be designed to provide adequate flexibility to the clinician in terms of administered dose (often encompassing a 100 × range as well as intravenous, intramuscular, or subcutaneous administration), while the commercial dosage form is often tailored to a specific dose level, such as the case with a liquid prefilled syringe in which the full contents of the syringe should be administered. One should also consider the question of product comparability as a program progresses from an early-clinical-stage dosage form to a late-clinical-stage dosage form. While stability is a critical aspect of a suitable dosage form, manufacturing time is also important. The use of a lyophilized dosage form may enable the clinical manufacture of a single lot of vials that may be suitably stable for several years with less concern about degradation or failing product quality specifications. For a liquid dosage form, depending on the product, the possibility of degradation is greater, as is the risk of having to manufacture additional clinical supplies to resupply the clinic. For these reasons, it is worth considering the use of lyophilized products for early-phase clinical programs.

In summary, for an early-phase clinical program, one may want to avoid overinvesting precious time, effort, and limited material in optimizing a formulation when the dose, early clinical safety, biological activity profiles, route of administration, and competitive environment are unknown. There may be adequate knowledge and experience either in the lab or in the literature that will enable the formulation scientist to quickly define a suitable formulation for an early-clinical-phase product. This approach, of course, assumes that the experience can be directly applied to a known class of compounds such as antibodies or antibody-like proteins. If, in contrast, the product is unique, then a more traditional approach to pH and NaCl optimization may need to be taken.

6.3. PREFORMULATION OF UNUSUAL PROTEINS

Thus far, we have discussed the lessons that can be learned from the experiences of others with a well-characterized family of proteins, namely, antibodies. While exceptions exist broadly within the antibody family of proteins, it should be appreciated that if one is formulating a unique protein, the experience of others may not readily apply. That said, the same general criteria do apply, in that one needs to identify a suitable pH, buffer, surfactant, cryoprotectant, and, if required for subcutaneous administration, tonicity agent such as sodium chloride or mannitol. In these instances, it may be desirable to consider the use of automation and miniaturization as an approach to reduce the time and material requirements for early formulation development. The following example is such a case.

Recombinant human bone morphogenetic protein 2 (rhBMP-2) is a basic, disulfide-linked dimeric member of the TGFα superfamily of growth factors [25]. It induces de novo bone; and rhBMP-2, delivered on an absorbable collagen sponge, is approved for use in humans in several indications [26]. The commercial formulation is 1.5 mg/mL rhBMP-2 buffered with glutamic acid at pH 4.5 [21]. Studies have been performed in the author's lab, based on previous pH and solubility screens [27], to examine the use of a high-throughput approach to pH and ionic strength mapping of the solubility of rhBMP-2. This approach, using a 96-well microtiter dish, takes advantage of the behavior of rhBMP-2 to precipitate under certain conditions of salt and pH. This precipitation and compromised solubility is, in fact, the critical stability attribute of rhBMP-2. This latter point is important in this example because the formulation scientist will wish to focus on the important aspects of the protein of interest, as with the solubility of rhBMP-2, and be somewhat less concerned with protein characteristics that are seemingly well behaved.

In the author's lab, rhBMP-2 was diluted into a 96-well microtiter dish containing a matrix of pH (pH 2.5–8.0) and sodium chloride levels from 0 to 200 mM. In brief, the pH of the solutions was adjusted using a 100 mM MES at pH 3 and 100 mM HEPES pH 10 buffer system that, on titration with either dilute HCl or NaOH, provided a linear pH response (Fig. 6.1). Further, at each pH, varying levels of sodium chloride from 0 to 200 mM were introduced into the solution to vary ionic strength. The plates were prepared in a semiautomated manner using a programmable robotic titrator (Tecan Genesis RSP100). Once the plates were prepared, and the pH ranges and levels of ionic strength confirmed, 2 mg/mL rhBMP-2 was diluted into each well (200 μL per well) to a concentration of 0.2 mg/mL and allowed to incubate at room temperature for 30 min. Once incubation was complete, the plates were analyzed by UV–vis spectroscopy at 320 nm using a spectrophotometer. A parallel plate was prepared to provide an appropriate buffer blank with which to subtract the 320 nm measurements. The subtracted results are presented in Figure 6.2. This approach enabled us to characterize the pH/sodium chloride solubility profile for rhBMP-2 using very little protein (200 μL × 0.2 mg/mL × 96 wells = 3.8 mg) in a short time (one day).

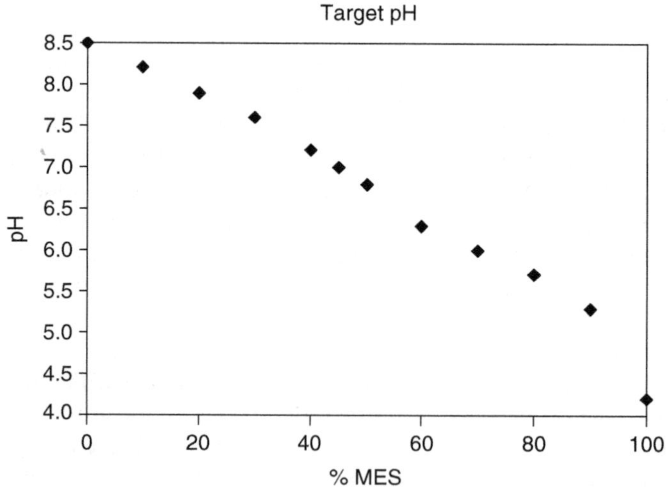

Figure 6.1. Effect of pH on MES titration in the presence of HEPES. As the level of MES buffer increases from 0 to 100 mM, the level of HEPES buffer decreases linearly from 100 to 0 mM. This titration yields a linear pH range of 5–9.

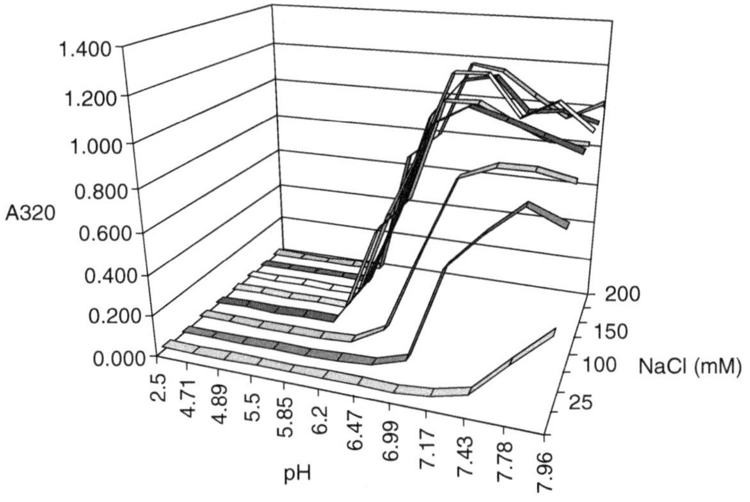

Figure 6.2. Effect of pH and NaCl on rhBMP-2 opalescence. Several rhBMP-2 solutions, at 0.2 mg/mL, were titrated to adjust pH and NaCl levels. The resultant solutions were examined by absorbance at 320 nm to assess opalescence.

In addition to assaying the solubility of a protein using a robotic titrator system, additional analyses can be performed within the same plate by using HPLC equipment with 96-well microtiter plates within the sample holder. Using this capability, a set of two to three plates can be utilized to assess the pH–ionic strength screen for a protein as it pertains to concentration, opalescence, size exclusion HPLC (SE-HPLC) for high-molecular-weight formation, cation exchange HPLC for isoform distribution modification, and other approaches such as differential scanning calorimetry (DSC) or capillary electrophoresis. With proper equipment and training, a substantial amount of information can be obtained using a relatively small quantity of protein (~10 mg) and time (several weeks). This approach has broad applicability and is particularly suited to the assessment of solubility and stability issues, presuming that this is a key factor to the stability of the dosage form. If poor solubility is a critical characteristic of the protein, then experiments can be performed with reduced amounts of protein. As the solubility of a target protein increases, however, the protein concentration at which these critical experiments are performed must also be increased, which will consume greater amounts of protein. A similar plate-based pH and sodium chloride titration procedure, performed at 20 mg/mL, for example, would require approximately 380 mg per plate. This may be prohibitive during early preformulation studies, but appropriate during later stages of formulation development. In summary, however, for proteins with which there is little experience, appropriate pH and ionic strength mapping can be performed quickly and with a minimal amount of protein and, if planned judiciously, can yield a tremendous amount of information pertaining to the solubility or thermal stability of the protein.

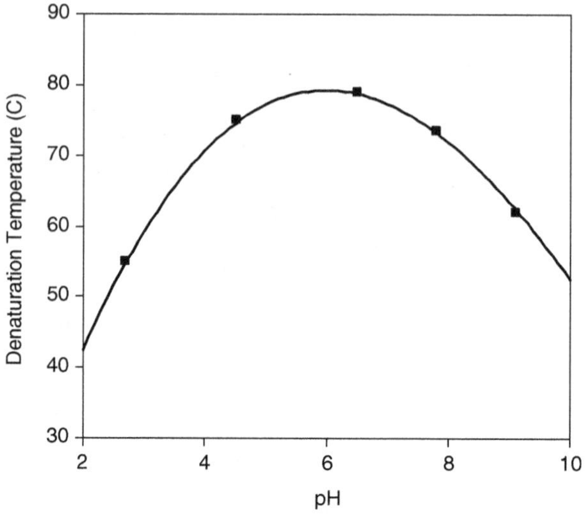

Figure 6.3. Effect of pH on rhBMP-2 denaturation temperature. Differential scanning calorimetry was performed on rhBMP-2 solutions to determine the thermal denaturation temperature.

Finally, one should be cautious in utilizing techniques such as DSC for short-term stability studies for the assessment of relative stability. While the DSC technique has seen a revolution in sample handling and sensitivity [28], one must consider the question that DSC is being used to answer. If the target protein has a lower than desired denaturation temperature, then characterizing the behavior of the protein at various pH levels and ionic strength adds substantial value to the formulator's ability to properly formulate the protein. If, however, thermal denaturation is not the critical stability parameter for the protein, then the use of DSC data as the basis for formulation decisions has decreased value. A pertinent example is presented in Figure 6.3, in which rhBMP-2, prepared at a variety of pH values, demonstrates that the optimal pH for the formulation would be approximately pH 6.0. This conclusion differs substantially from the previous study, which demonstrated that pH 4.5 is optimal for the solubility of rhBMP-2 [29]. Since solubility is a greater concern for the formulation of rhBMP-2, and because the optimization of the denaturation temperature from 75°C to 79°C provides marginal benefit, a pH of approximately 4.5 was selected for the rhBMP-2 formulation [21].

6.4. ASSESSMENT OF FORMULATION PERFORMANCE

No discussion of formulation development is complete without mention of the tools that one can use to assess whether a formulation is adequate to withstand the rigors of not only manufacturing but also storage, shipment, and use at the clinical site. While more recent review articles have focused on a wide variety of analytical tools that can be utilized to assess the inherent stability of a protein, many of these tools require larger amounts of protein, require advanced knowledge of how to collect and interpret data, and are designed to assess particular structural characteristics of a protein that may be a significant weakness of the protein. Very few examples exist in which UV–vis spectroscopy has been a critical stability-indicating assay for a protein; however, concentration is a critical piece of information for interpretation of other assay data (specific activity) and may also signal a propensity to adsorb to surfaces. In contrast, other spectroscopic techniques can add tremendous value when applied appropriately (e.g., the use of FTIR to monitor protein denaturation during lyophilization) [30]; however, these techniques add value only once a mature understanding of the proteins' primary instabilities has been achieved. Because the primary route of degradation of frozen drug substance is often high-molecular-weight formation, development of an antibody formulation may require the use of SE-HPLC as the primary screening tool. Further, once it is established that high-molecular-weight (HMW) formation is the primary route of degradation, an assessment of process steps that could lead to HMW formation, such as freeze–thaw, freeze drying, and agitation [18,19,23], should be the focus of the preformulation studies. While establishing the need for appropriate pH and ionic strength is critical to the solution stability of the target protein, assessing the lead candidate formulation's ability to withstand multiple freeze–thaw stress, as observed during the handling of frozen process intermediates, as well as mechanical agitation, as observed during thawing or mixing processes, is important [31]. Often, the use of surfactants such as polysorbate 20 and polysorbate 80 is also required. Of the protein formulations presented in Table 6.1, seven of the nine examples utilize a form of polysorbate in the formulation, presumably to better manage the protein's ability to withstand some form of mechanical stress or to aid reconstitution.

6.5. SUMMARY

The development of a suitable early-clinical-phase formulation can often be based, in part, on in-house experience or the experience of others. Utilization of all possible analytical tools and assessment of all possible stabilizers, buffers, and surfactants, tonicity agents may not be required. The application of experience and the assessment of a small number of formulation options in protecting the target protein from the rigors of stability and manufacturing may be more than adequate for the development of a suitable early clinical phase dosage form and may substantially reduce the time, staff, and material required at this stage. Additionally, it may be highly desirable to delay optimization of the dosage form for late-stage clinical or commercial applications until a safe and efficacious dose, route of administration, and success of the product are better defined.

REFERENCES

1. Capelle, M. A. H., Gurny, R., and Arvinte, T. (2007), High throughput screening of protein formulation stability: Practical considerations, *Eur. J. Pharm. Biopharm.* **65**: 131–148.

2. Das, D. (2007), Early stage protein formulation development and use of high throughput screening methods, *Proc. AAPS National Biotechnology Conf*., June 24–27, 2007, San Diego, CA.
3. FDA (2004), *Innovation or Stagnation: Challenge and Opportunity on the Critical Path to New Medical Products*, FDA report, March 2004.
4. Manning, M. C., Patel, K., and Borchardt, R. T. (1989), Stability of protein pharmaceuticals, *Pharm. Res*. **6**: 903–918.
5. Carpenter, J. F., Pikal, M. J., Chang, B. S., and Randolph, T. W. (1997), Rational design of stable lyophilized protein formulations: Some practical advice, *Pharm. Res*. **14**: 969–975.
6. See www.Amevive.com, Astellas Pharma US, Inc., Deerfield, IL 60015.
7. See www.enbrel.com, Amgen, Thousand Oaks, CA 91320 and Wyeth Pharmaceuticals Inc., Philadelphia, PA 19101.
8. See www.herceptin.com, Genentech Inc., South San Francisco, CA 94080–4990.
9. See www.humira.com, Abbott Laboratories, Abbott Park, IL.
10. See www.raptiva.com, Genentech Inc., South San Francisco, CA 94080–4990.
11. See www.remicade.com, Centocor, Inc., Horsham, PA 19044.
12. See www.rituxan.com, Genentech Inc., South San Francisco, CA 94080–4990.
13. See www.rocheusa.com/products/zenapax/, Hoffmann-La Roche Inc.
14. See www.xolair.com, Genentech Inc. and Novartis Pharmaceuticals.
15. Telang, C., Yu, L., and Suryanarayanan, R. (2003), Effective inhibition of mannitol crystallization in frozen solutions by sodium chloride, *Pharm. Res*. **20**: 660–667.
16. Chang, B. S., Kendrick, B. S., and Carpenter, J. F. (1996), Surface-induced denaturation of proteins during freezing and its inhibition by surfactants, *J. Pharm. Sci*. **85**: 1325–1330.
17. ICH (Int. Conf. Harmonization) (2005), *ICH Harmonized Tripartate Guideline on Pharmaceutical Development*, Q8, Step 4 version dated Nov. 10, 2005.
18. Sarciaux, J.-M., Mansour, S., Hageman, M. J., and Nail, S. L. (1999), Effects of uffer composition and processing conditions on aggregation of bovine IgG during freeze-drying, *J. Pharm. Sci*. **88**: 1354–1361.
19. Anchordoquy, T. J. and Carpenter, J. F. (1996), Polymers protect lactate dehydrogenase during freeze-drying by inhibiting dissociation in the frozen state, *Arch. Biochem. Biophys*. **332**: 231–238.
20. See www.actimmune.com, Intermune Inc., Brisbane, CA 94005.
21. Yim, K. W., Huberty, M. C., Northey, Jr., R. P., and Schrier, J. A. (January 31, 1995), *Formulations for Delivery of Osteogenic Proteins*, US Patent 5,385,887.
22. See www.leukine.com, Berlex, Seattle, WA 98101.
23. Allison, S. D., Randolph, T. W., Manning, M. C., Middleton, K., Davis, A., and Carpenter, J. F. (1998), Effects of drying methods and additives on structure and function of actin: Mechanisms of dehydration-induced damage and its inhibition, *Arch. Biochem. Biophys*. **358**: 171–181.
24. Farr, G. W. and Heitz, J. R. (1974), Imidazole catalysis of sucrose hydrolysis, *J. Dental Res*. **53**: 516–519.
25. Wang, E. A., Rosen, V., Cordes, P., Hewick, R. M., Kriz, M. J., Luxenberg, D. P., Sibley, B. S., and Wozney, J. M. (1988), Purification and characterization of other distinct bone-inducing factors, *Proc. Natl. Acad. Sci. USA* **85**: 9484–9488.

26. Termaat, M. F., Den Boer, F. C., Bakker, F. C., Patka, P., and Haarman, H. J. Th. M. (2005), Bone morphogenetic proteins. Development and clinical efficacy in the treatment of fractures and bone defects, *J. Bone Joint Surg*. **87**: 1367–1378.
27. Abatiello, S. E. and Porter, T. J. (1997), Anion-mediated precipitation of recombinant human bone morphogenetic protein (rhBMP-2) is dependent upon the heparin binding N-terminal region, *Proc. Protein Society Meeting*, Boston, July 13–16, 1997.
28. Remmele, R. L. Jr. and Gombotz, W. R. (2000), Differential scanning calorimetry: A practical tool for elucidating stability of liquid pharmaceuticals, *BioPharm* **13**: 36–46.
29. Warne, N. W., BMP-2/ACS: Influence of the delivery matrix on the behavior of rhBMP-2, and *Proc. Protein Stability Conference*, Breckenridge, July 2005.
30. Susi, H. and Byler, D. M. (1983), Protein structure by Fourier transform infrared spectroscopy: Second derivative spectra, *Biochem. Biophys. Res. Commun*. **115**: 391–397.
31. Vidanovic, D., Askrabic, J. M., Stankovic, M., and Poprzen, V. (2003), Effects of non-ionic surfactants on the physical stability of immunoglobulin G in aqueous solution during mechanical agitation, *Pharmazie* **58**: 399–404.

7

LATE-STAGE FORMULATION DEVELOPMENT AND CHARACTERIZATION OF BIOPHARMACEUTICALS

Adeola O. Grillo

7.1. INTRODUCTION

While early formulation development entails screening different solution conditions (pH, ionic strength, excipients, etc.) that confer stability to a product, late-stage formulation development involves optimizing and characterizing the formulation. Activities performed during characterization of the optimized product include identifying product degradants, identifying critical formulation attributes, and evaluating the robustness of the product. These activities are performed to identify aspects of the drug product that are critical to product quality and to justify control strategies [1].

Robustness is the quality of being able to withstand stresses, pressures, or changes in procedure. A robust product is stable not only from the date of manufacture to the end of its shelf life; but also to stresses and excursions that can occur during manufacturing, shipping, storage, and patient use. Robustness can be built into a product during formulation development. This is done by identifying how different stresses impact the drug substance and optimizing the formulation of the drug product so that the impact of stress is minimized. Information obtained from formulation

Formulation and Process Development Strategies for Manufacturing Biopharmaceuticals,
edited by Jameel and Hershenson

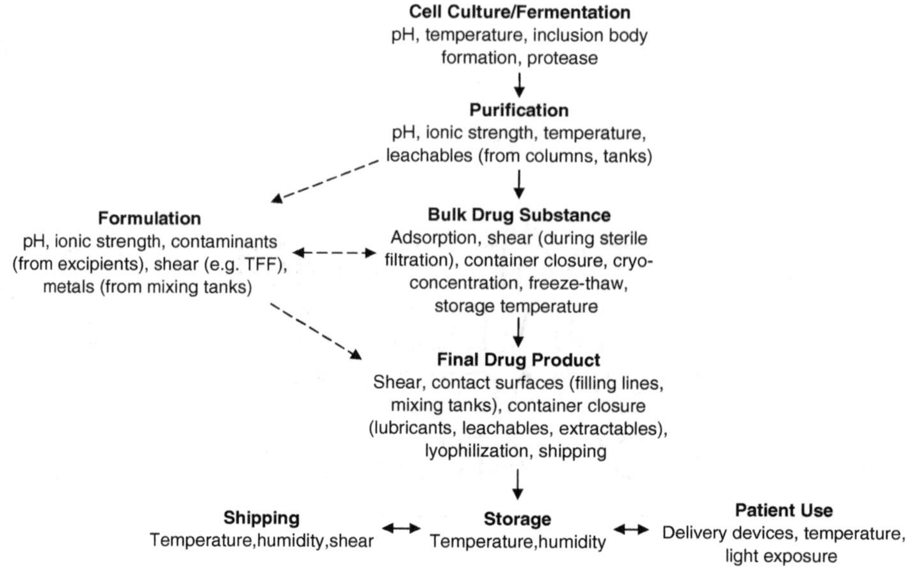

Figure 7.1. Typical stress conditions occurring during manufacture, shipping, storage, and patient use.

screening using biophysical and short-term accelerated stability studies are used to identify buffers and excipients that stabilize the product against physical and chemical degradation. Likewise, information obtained during formulation development, such as the effect of pH and ionic strength on solubility and stability, can be used to identify solution conditions that will result in an optimized formulation. Finally, the conditions and concentrations of the excipients are selected to ensure that the product is not formulated at an edge and small variations in formulation components do not have a major impact on product stability.

One of the tools that can be used to develop a robust formulation and to characterize the robustness of the formulated product is forced degradation studies. Forced degradation studies are performed to identify the impact of different stresses on the protein, to aid in the elucidation of degradation profiles, to identify the stability-indicating power of the analytical methods used to monitor product stability, and to aid with validation of the methods [2]. Figure 7.1 outlines some of the stress conditions that occur during manufacturing, shipping, storage, and patient use. The information obtained from evaluating the impact of stresses (such as heat, freeze–thaw, shear, and light-induced stress) on a product is used during early-stage formulation development to select a formulation that is robust to specific stresses. Forced degradation studies are also performed on the optimized formulation. The results from the forced degradation studies are then used to set acceptable limits of stress or excursions that the product can withstand without having an impact on product quality.

Another tool that can be used to build robustness in the formulated product, as well as evaluate robustness of the product, is design of experiments (DOE). Utilizing

DOE allows for the identification of critical factors and interaction between factors that impact the product. In contrast to evaluating factors one at a time, DOE makes it possible to experiment with different factors simultaneously to determine how they interact [3]. Thus DOE is a useful tool in defining and characterizing the formulation design space. It is widely used in the pharmaceutical industry for optimization and characterization of drug synthesis, cell culture–fermentation–purification processes, and analytical methods [4,5]. Its use in formulation development and characterization of biotechnology products has also been reported [6–9].

This chapter describes how forced degradation studies and DOE can be used to characterize and demonstrate robustness of protein formulations during late-stage formulation development. Results of characterization studies on monoclonal antibody (MAb) products from literature and studies performed in Human Genome Sciences (HGS) are used as illustrations, since MAbs represent the the most rapidly growing area of biopharmaceutical products [10]. The illustrations give an overview of how forced degradation studies and DOE can be used to identify attributes that are critical to product quality, justify control strategies, characterize the formulation design space, and, in essence, implement quality by design (QbD) into pharmaceutical development [1].

7.2. FORCED DEGRADATION STUDIES FOR CHARACTERIZING ROBUSTNESS TO STRESS

Temperature excursions, exposure to light, freeze–thaw/thermal cycling, and contact with different surfaces are some of the common stress conditions that the formulated product can be exposed to during manufacturing, shipping, storage, and patient use. By intentionally exposing the final drug product (FDP) to different stress conditions, the investigator can identify the impact of the stress and the allowable limits for stress exposure.

7.2.1. Heat Stress

Heat stress accelerates most degradation pathways of a protein and thus is a very useful stress condition for identifying potential degradation pathways. In addition, as a result of temperature excursions that occur during manufacturing, shipping, storage, and patient use, it is important to characterize the impact of heat stress on the drug substance and drug product. Heat stress studies are conducted by storing the drug substance or drug product at elevated storage temperatures and analyzing the stressed protein using a variety of analytical methods. Degradation products that have been identified for monoclonal antibodies (MAbs) include aggregation, deamidation, aspartic acid isomerization, fragmentation, disulfide scrambling, nonreducible covalent crosslinks, oxidation, pyroglutamic acid formation, and glycation [11]. These degradation pathways are identified using a variety of biophysical (e.g., spectroscopy, ultracentrifugation) and analytical methods (e.g., chromatography, electrophoresis, mass spectrometry). High-performance liquid chromatography (HPLC) and sodium dodecyl sulfate–polyacrylamide gel electrophoresis (SDS-PAGE) are standard industry methods used to routinely monitor degradation in MAbs. Fractions from HPLC and SDS-PAGE analysis can be isolated to further characterize the degradation products. In addition, separation methods such as HPLC or capillary electrophoresis (CE) can also be used in tandem with mass spectrometry to identify the degradation products.

Since many biotechnology products are susceptible to heat stress, the main control strategy for minimizing heat stress degradation is to control the storage, shipping, and patient use temperatures. Accelerated stability testing and temperature excursion studies are performed to identify limits for acceptable temperature excursions that would have minimal impact on product quality. In addition, formulation screening during early-stage formulation development can be used to identify excipients that can be incorporated into the formulation to stabilize the protein against heat stress. For products that are particularly labile in solution, freeze drying and spray freeze drying are formulation strategies that can be used to minimize degradation.

7.2.2. Photostability

It is well known that proteins are susceptible to UV rays. Amino acid residues susceptible to light-induced stress include the aromatic residues tryptophan, tyrosine, and phenylalanine, in addition to cystine and methionine. There are two options for performing photostability studies according to ICH Q1B [12]. Option 1 uses a light source designed to produce emission output similar to the D65/ID65 standard, which combines output simulating outdoor daylight (D65 standard) and indirect indoor daylight (ID65 standard). Option 2 uses separate light sources; a cool white fluorescent lamp designed to produce an output similar to that specified in ISO 10977 [13] and a near-UV fluorescent lamp with emission from 320 to 400 nm and maximum emission between 350 and 370 nm [12]. By performing the study under option 2, one can identify the effects of visible fluorescence light (indoor lighting) on the product. Photostability studies are performed by exposing samples side by side with control samples wrapped in foil to not less than (NLT) 1.2 million lux hours and integrated near-UV energy of NLT 200 $(W{\cdot}h)/m^2$ [12]. Guideline ICH Q1B also has a decision flowchart for performing photostability studies. Samples are directly exposed to light and if there is an unacceptable change in stability, samples are exposed in the immediate pack. If there is an unacceptable change, samples are then exposed in the marketing pack. The photostability study is ended at the step where there is an acceptable change in stability. If an unacceptable change is still observed in the marketing pack, then the formulation or packaging should be redesigned.

Forced degradation photostability studies are performed by exposing samples to levels in excess of the ICH guidelines. Forced degradation studies performed on two HGS MAbs showed similar degradation products, including aggregation, methionine oxidation, and nonreducible thioether crosslinks. In addition, tryptophan oxidation was observed in the Fab region of one of the antibodies [14]. The oxidation in the Fab also resulted in potency loss. Aggregation, methionine oxidation, nonreducible thioether cross-links, and tryptophan oxidation in the complementarity-determining region (CDR) region of a MAb leading to potency loss are degradation pathways that have been reported for other MAbs as a result of photodegradation [15–17]. An increase in free-thiol content, due to UV exposure, has also been reported [17].

A common strategy for minimizing the effect of light stress on biopharmaceuticals is to protect the product from light. Thus, many biotechnology products contain a "protect from light" statement included in the full prescribing information for the drugs [18]. Additional control strategies include the addition of stabilizing excipients in the formulation [18–20]. In light-induced degradation reactions mediated by oxygen, removal of oxygen by packaging under inert gases can also decrease the impact of the degradation [21].

7.2.3. Freeze–Thaw and Thermal Cycling Studies

Proteins undergo freeze–thaw on thawing of product stored in the frozen state, during shipping and potentially during patient use. The product can also undergo temperature excursions outside the recommended storage temperature during shipping and patient use. Thus it is important to evaluate the effect of these stresses to determine the impact on product quality and to identify acceptable temperature limits and excursion durations. There are no specific guidelines for performing thermal cycling studies for biopharmaceuticals; however, the same philosophy behind evaluating the impact of thermal cycling on nonbiologics [22] can also be used for biologic products. One approach to performing thermal cycling studies is to cycle the product to temperatures above and below the product recommended storage temperature, with samples incubated for 2 days at each cycled temperature for a total of three cycles [23]. To verify that cycling has no impact on product stability, cycled samples can also be stored alongside noncycled controls at recommended and accelerated temperatures and the stability monitored. In addition, differential scanning calorimetry (DSC) can be used to evaluate thermal transitions in the bulk drug substance (BDS) and FDP to aid in the design of thermal cycling studies or in data interpretation. For example, results from a thermal cycling study performed on an HGS MAb indicated that cycling the BDS from −80°C to −40°C or −20°C resulted in the formation of visible particulate matter and soluble aggregates while cycling from −80°C to 25°C had no major impact. Similarly, cycling the liquid FPD from 2°C–8°C to −20°C resulted in aggregation, while cycling from 2°C–8°C to −15°C, 25°C, or 40°C had no major impact. In addition, control samples stored at −40°C or −20°C over the duration of the cycling study also resulted in the formation of visible particulate matter and soluble aggregates. Low-temperature DSC characterization of the formulation revealed a eutectic transition at approximately −18°C and a glass transition temperature of the maximally frozen concentrate (T_g') at approximately −35°C. Factors that contribute to protein instability during freezing include cold denaturation, ice crystallization, and freeze concentration. Of these, evidence from literature suggests that ice crystallization and freeze concentration play a greater role in protein instability during freezing [24]. Protein stability is also negatively impacted by crystallization of excipients during storage. Thus prolonged exposure of the protein to temperatures close to the eutectic and glass transition temperatures of the product affected the stability of the protein.

Results for freeze–thaw and thermal cycling studies are used to set acceptable limits for exposure to stress. For example, if a product is known to be susceptible to freezing stress, a "do not freeze" statement can be included in the package insert. Additional control strategies for minimizing the effect of freeze–thaw stress include incorporating excipients that protect the protein from freeze–thaw stress in the formulation. Freeze–thaw studies performed as part of formulation screening during early formulation development aid in the selection of a formulation that is stable to freeze–thaw stress. Possible formulation strategies include the addition of a cryoprotectant and surfactant in addition to avoiding buffer components such as sodium phosphate that can increase the impact of freeze–thaw stress due to pH shifts during freezing.

7.2.4. Compatibility Studies

Proteins come into contact with different types of surfaces during the manufacturing process, storage in container–closure (container or closure) systems, and administration of the drug product. Contact surfaces during the manufacturing process include different types of plastic, glass, stainless steel, and material from tubing, purification columns, containers (for holding material during fermentation, purification, and fill), and filters. Containers for storing the BDS can be made from plastic, glass, or stainless steel. Similarly, FDP can come into contact with glass, plastic, rubber stoppers, metals, and silicone used to manufacture container–closure systems such as ampoules, vials, and prefilled syringes. During administration in the clinic, the product comes into contact with IV bags and tubing, as well as syringes and needles. Potential stresses during contact include adsorption of the protein to the surfaces as well as leachables that impact protein stability [25–27]. Thus studies need to be performed to verify that the different surfaces that the product comes into contact with will not adversely affect the product and ultimately impact patient safety. Compatibility studies are performed by storing the product in contact with the material of interest and evaluating the impact on protein concentration and stability. For contact surfaces that have the potential to leach contaminants, studies can be performed to evaluate the stability of the protein after exposure to the contact surfaces. This can be done by monitoring the stability of the protein at recommended and accelerated storage temperatures in an inert container–closure unit after storing the protein in contact with the material of interest. For example, to evaluate the impact of stainless-steel exposure, protein samples were stored in stainless-steel cups and glass vials over 2 weeks at 25°C and 40°C. The protein samples were then aliquoted into glass vials and monitored over 6 months at 2°–8°C and 25°C. Analysis with stability-indicating methods showed no major changes after storing in steel cups for 2 weeks or subsequent storage in glass vials at 2°C–8°C for 6 months. However, increased degradation (including aggregation and oxidation) was observed after storage in glass vials for 6 months at 25°C correlating with the initial length of exposure in the steel cups [28]. Increased aggregation in a MAb after undergoing freeze–thaw in a stainless-steel tank has also been reported [8].

Control strategies to minimize stress due to material contact include limiting contact with materials that negatively impact the stability of the product and adding excipients that minimize degradation. Examples include

1. Using glass tanks for holding or formulating the FDP during the fill–finish process if leachables from steel tanks have a negative impact on stability

2. Using stoppers with fluoroelastomer coatings to reduce leachables
3. Adding surfactants or stabilizers such as sucrose to reduce adsorption

7.3. DOE AS A FORMULATION CHARACTERIZATION TOOL

The forced degradation studies described above are typically done to evaluate the impact of stress one at a time. On the other hand, DOE is performed to evaluate the effect of multiple factors at the same time and to identify interactions between factors that affect a particular attribute. Examples of experimental designs include:

1. *Full Factorial Designs.* This design includes all possible combinations of factors being evaluated. For example, a two-level full factorial design evaluating four different factors would result in $2^4 = 16$ combinations. A full factorial design provides information on the impact of all possible interactions. But because it requires running all possible combinations, it is the most time-consuming and resource-intensive.
2. *Fractional Factorial Designs.* This design is used to reduce the number of runs in a factorial design and thus can be used to quickly screen different factors to identify which have the most significant effect. Examples of fractional factorial designs included Plackett–Burmann and Taguchi designs.
3. *Response Surface Design.* This design allows for evaluation of quadratic effects and thus is used to identify curvature in the response surface. Examples or response surface designs include central composite designs (CCDs) and Box–Behnken designs.

Thus, since DOE can be used to identify parameters that have the most significant effect on an attribute, it is a powerful tool for determining the aspects of the formulation that are critical to product quality [1]. It is also a useful tool for optimizing formulations, evaluating the robustness of the formulation, and defining the formulation design space. Examples illustrating the use of DOE for formulation optimization and robustness are described below.

7.3.1. DOE for Evaluation of the Impact of pH and Ionic Strength

The DOE method was used to evaluate the effect of pH and NaCl concentration on the aggregation of an HGS MAb stored at 2°C–8°C over 6 weeks. Since protein degradation as a function of pH results in a U-shaped profile, a response surface design was used. The experimental design and data analysis were performed using JMP software (Cary, NC). The study consisted of a central composite design evaluating pH 3–9 and 0–300 mM NaCl. The results predicted that greater stability would be observed between pH 6 and 8 and that at neutral pH, variations in NaCl concentration between 100 and 200 mM would not significantly impact MAb aggregation (Fig. 7.2). The interaction profiles shown in Figure 7.3 illustrate how changes in NaCl concentration would impact MAb aggregation at acidic and basic pH values. Below pH 7, increasing

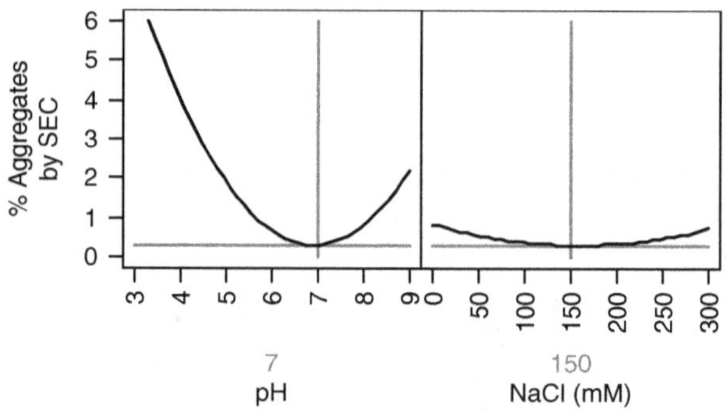

Figure 7.2. Prediction profile on the effect of pH and NaCl concentration on aggregation of an HGS mAb.

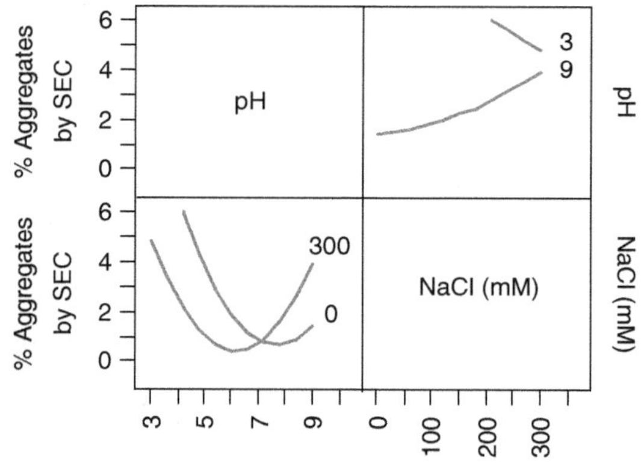

Figure 7.3. Interaction profiles on the effect of pH and NaCl concentration on aggregation of an HGS mAb.

NaCl concentrations would decrease aggregation, while above pH 7 increasing NaCl concentrations would increase aggregation. The results show that if the MAb were formulated below pH 7 to minimize degradation such as deamidation and oxidation that might increase at basic pH [29], the inclusion of NaCl in the formulation would, in addition to modulating the tonicity of the product, also stabilize the product against aggregation.

7.3.2. DOE in Characterization of Excipient Robustness

Figures 7.4 and 7.5 illustrate the prediction profile results of an experiment designed to verify the critical components of an optimized formulation and demonstrate formulation robustness. The study consisted of a screening design evaluating protein, buffer, tonicifying agent, cryoprotectant, surfactant, and pH levels. A $2^{(6-2)}$ fractional factorial design was used to determine significant factors and two-factor interactions

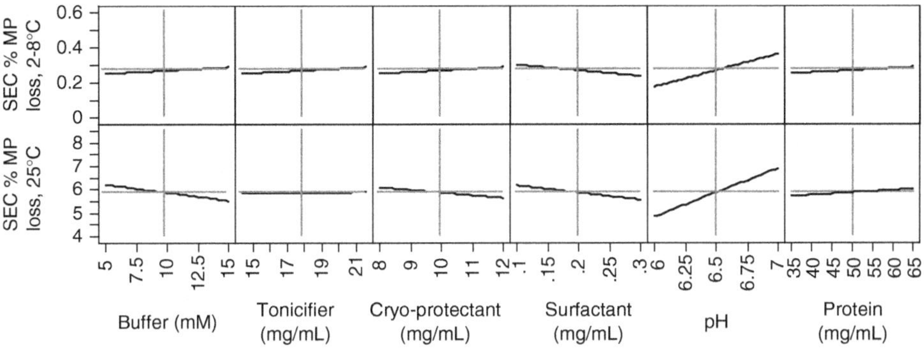

Figure 7.4. Effect of formulation components on mAb SE-HPLC.

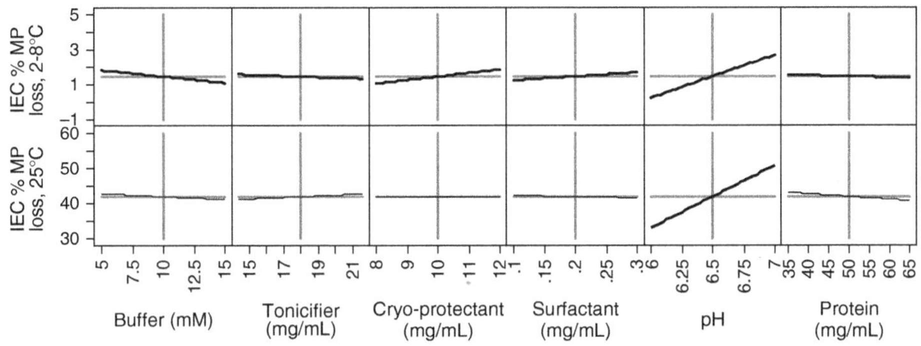

Figure 7.5. Effect of formulation components on mAb IE-HPLC.

in the formulation that impacted protein stability (monitored by size exclusion and ion exchange chromatography). Replicate center points were also included in the study to minimize the impact of noise in the data due to run-to-run or assay variation. The study demonstrated that the formulation was robust to a wide range of the formulation components, except for pH. Thus the pH of the formulation would need to be tightly controlled during the manufacturing process.

In addition to identifying formulation components that have a significant effect on product stability, the results of a DOE robustness study can be used to identify proven acceptable ranges and support deviations in the manufacturing process that result in excipient levels that are off target.

7.4. CONCLUSIONS

International Conference on Harmonization (ICH) Q9 gives guidelines on using a risk-based approach to drug development. A typical quality risk management includes risk assessment, risk control, and risk review [30]. Forced degradation studies and DOE are essential tools to evaluating potential risks to the formulated product. When these tools are used to characterize product critical quality attributes and the formulation design space, the risk of protein instability, and ultimately, the risk to patient safety can be minimized.

REFERENCES

1. ICH (Int. Conf. Harmonization) (2005), *Pharmaceutical Development*, Q8, Nov. 2005.
2. ICH (1995), *Quality of Biotechnological Products: Stability Testing of Biotechnological/Biological Products*, Q5C, Nov. 1995.
3. Box, G. E., Hunter, J. S., and Hunter, W. G. (2005), *Statistics for Experimenters: Design, Innovation, and Discovery,* 2nd ed., Wiley, Hoboken, NJ.
4. Godbert, S. (2000), Experiment design—identifying factors that affect responses, in *Design and Analysis in Chemical Research,* Vol. 3 of Sheffield Analytical Chemistry Series, Tranter, R. L., ed., Sheffield Academic Press, Sheffield, UK, pp. 188–236.
5. Prvan, T. and Street, D. J. (2000), An annotated bibliography of application papers using certain classes of fractional factorial and related designs, *J. Statist. Plan. Infer*. **106**: 245–269.
6. Gupta, S. and Kaisheva, E. (2003), Development of a multidose formulation for a humanized monoclonal antibody using experimental design techniques, *AAPS PharmSci*. **5**(2): E8.
7. Bedu-Addo, F., Moreadith, R., and Advant, S. J. (2002), Preformulation development of recombinant pegylated staphylokinase SY161 using statistical design, *AAPS PharmSci*. **4**(4): E19.
8. Chen, B, Bautista, R., Yu, K., Zapata, G. A., Mulkerrin, M. G., and Chamow, S. M. (2003), Influence of histidine on the stability and physical properties of a fully human antibody in aqueous and solid forms, *Pharm. Res*. **20**(12): 1952–1960.
9. Katayama, D. S., Kirchhoff, C. F., Elliott, C. M., Johnson, R. E., Borgmeyer, J., Thiele, B. R., Zeng, D. L., Qi, H., Ludwig, J. D., and Manning, M. C. (2004), Retrospective statistical analysis of lyophilized protein formulations of progenipoietin using PLS: Determination of the critical parameters for long-term storage stability, *J. Pharm. Sci*. **93**(10): 2609–2623.
10. Reichert, J. M., Rosensweig, C. J., Faden, L. B., and Dewitz, M. C. (2005), Monoclonal antibody successes in the clinic, *Nat. Biotechnol*. **23**: 1073–1078.
11. Liu, H., Gaza-Bulseco, G., Faldu, D., Chumsae, C., and Sun, J. (2008), Heterogeneity of monoclonal antibodies, *J. Pharm. Sci*. **97**(7): 2426–2447.
12. ICH (1996), *Photostability Testing of New Drug Substances and Products*, Q1B, Nov. 1996.
13. ISO (Int. Standardization Organization) (1993), *Photography—Processed Photographic Colour Films and Paper Prints—Methods for Measuring Image Stability*, ISO 10977.
14. Meeler, A., Penn, N., Bergerud, L., Grillo, A. O., and Perkins, M. (2007), *Investigation of a Light-Induced Color Change of an IGg1 Monoclonal Antibody*, poster presented at the Colorado Protein Stability Conf., Breckenridge, CO, July 19–21, 2007.
15. Wei, Z., Feng, J., Lin, H.-Y., Mullapudi, S., Bishop, E., Tous, G. I., Casas-Finet, J., Hakki, F., Strouse, R., and Schenerman, M. A. (2007), Identification of a single tryptophan residue as critical for binding activity in a humanized monoclonal antibody against respiratory syncytial virus, *Anal. Chem*. **79**(7): 2797–2805.

16. Lam, X. M., Yang, J. Y., and Cleland, J. L. (1997), Antioxidants for prevention of methionine oxidation in recombinant monoclonal antibody HER2, *J. Pharm. Sci*. **86**(11): 1250–1255.

17. Roy, S. (2006), Light induced aggregation of type I soluble tumor necrosis factor receptor, paper presented at the Workshop on Protein Aggregation, Breckenridge, CO, Sept. 26–27, 2006.

18. Kerwin, B. A., Remmele, R. L. Jr. (2007), Protect from light: Photodegradation and protein biologics, *J. Pharm. Sci*. **96**(6): 1468–1479.

19. Fransson, J. and Hagman, A. (1996), Oxidation of human insulin-like growth factor I in formulation studies, II. Effects of oxygen, visible light, and phosphate on methionine oxidation in aqueous solution and evaluation of possible mechanisms, *Pharm. Res*. **13**(10): 1476–1481.

20. Durchschlag, H. (2001), Strategies for the spectroscopic characterization of irradiated proteins and other biomolecules, *J. Mol. Struct*. **565–566**: 197–203.

21. Miller, B. L., Hageman, M. J., Thamann, T. J., Barròn, L. B., and Schöneich C. (2003), Solid-state photodegradation of bovine somatotropin (bovine growth hormone): Evidence for tryptophan-mediated photooxidation of disulfide bonds, *J. Pharm. Sci*. **92**(8): 1698–1709.

22. FDA (2003), *FDA Guidance for Industry; INDs for Phase 2 and Phase 3 Studies; Chemistry, Manufacturing, and Controls Information*, May 2003.

23. Lucas, T. L., Rafik, H. B., and Robers, H. S. (2004), A stability program for the distribution of drug products, *Pharm. Technol*. (July): 68–72.

24. Bhatnagar, B. S., Bogner, R. H., and Pikal, M. J. (2007), Protein stability during freezing: Separation of stresses and mechanisms of protein stabilization, *Pharm. Devel. Technol*. **12**(5): 505–523.

25. Tzannis, S. T., Hrushesky, W. J. M., Wood, P. A., and Przybycien, T. M. (1996), Irreversible inactivation of interleukin-2 in a pump-based delivery environment, *Proc. Natl. Acad. Sci. USA* **93**(11): 5460–5465.

26. McLeod, A. G., Walker, I. R., Zheng, S., and Hayward, C. P. (2000), Loss of factor VIII activity during storage in PVC containers due to adsorption, *Haemophilia* **6**(2): 89–92.

27. Fliszar, K. A., Walker, D., and Allain, L. (2006), Profiling of metal ions leached from pharmaceutical packaging materials, *PDA J. Pharm. Sci. Technol*. **60**(6): 337–342.

28. Penn, N., Palombo, J., Grillo, A., and Perkins, M. (2008), *Impact of Short-Term Stainless Steel Exposure on Stability Profile of a MAb,* poster presented at the IBC's Well-Characterized Biologicals Conf., Reston, VA, November 10–12, 2008.

29. Akers, M. J., Vasudevan, V., and Stickelmeyer M. (2002), Formulation development of protein dosage forms, in *Development and Manufacture of Protein Pharmaceuticals,* Vol. 14 of Pharmaceutical Biotechnology Series, Nail, S. L. and Akers, M. L., eds., Plenum, New York, pp. 47–127.

30. ICH (2005), *Quality Risk Management*, Q9, Nov. 2005.

8

AN EMPIRICAL PHASE DIAGRAM–HIGH-THROUGHPUT SCREENING APPROACH TO THE CHARACTERIZATION AND FORMULATION OF BIOPHARMACEUTICALS

Sangeeta B. Joshi, Akhilesh Bhambhani, Yuhong Zeng, and C. Russell Middaugh

8.1. INTRODUCTION

The stability of macromolecule-based drug products and vaccines, in contrast to that of traditional small-molecule drugs, is more often a significant problem for the pharmaceutical and biotechnology industry and often leads to major issues in development, manufacturability, storage, and shipment. The integrity of these biopharmaceuticals is often compromised by effects of temperature, pH, contaminants, and excipient incompatibility as well as the intrinsic instability of the molecules themselves. This can jeopardize not only therapeutic efficacy but also the ability to store and ship such agents in a pharmaceutically acceptable manner. Thus, regulatory agencies require that companies thoroughly understand their products' manufacturing control points prior to launch as well as have the ability to demonstrate long-term storage stability. It is therefore important to develop a product in which a comprehensive understanding of the ramifications of nonoptimal conditions during both manufacturing and storage is available.

While an empirical approach is often used to understand the stability and develop appropriate formulations of macromolecular systems, results of such studies often lead to less-than-optimal formulations. This can create a variety of difficult problems

Formulation and Process Development Strategies for Manufacturing Biopharmaceuticals, edited by Jameel and Hershenson

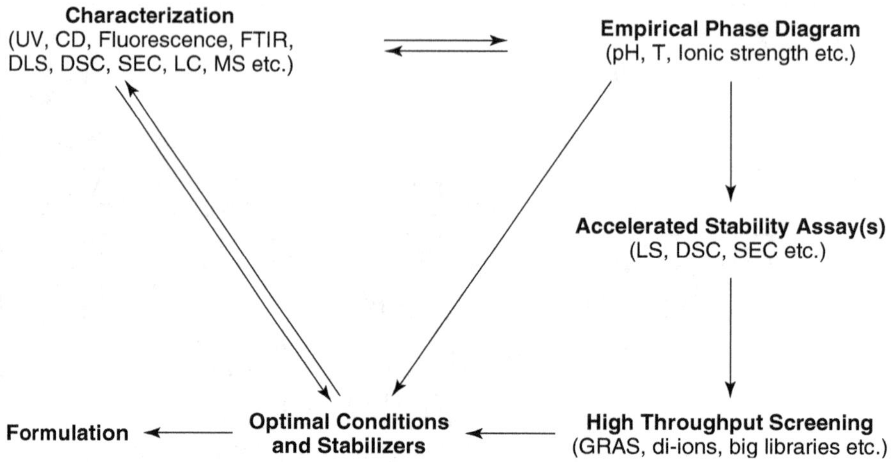

Figure 8.1. A general approach to the development of macromolecular formulations employing EPDs in conjunction with high-throughput screening to identify stabilizers.

in the drug development process. Furthermore, many smaller companies attempting to develop biopharmaceuticals have neither the resources nor the time to invest in prolonged investigations. Therefore, it is desirable to have in place a set of processes or systems that can produce a detailed understanding of stability while at the same time introducing both efficiency and speed to the preformulation and formulation processes. This review focuses on one such systematic approach developed in our lab for the rapid characterization and formulation of biopharmaceuticals.

A schematic representation of the "empirical phase diagram/high-throughput screening" approach described here to optimize macromolecule formulation is shown in Figure 8.1, and can be briefly described as follows. Extensive studies using a variety of techniques sensitive to different properties of the target macromolecules are performed to characterize the physical, chemical, and biological properties of the drug candidate under a range of solution conditions. The data generated in the process are combined to construct a more intuitive visual picture of the results; this is known as an *empirical phase diagram* (EPD). This not only permits a relatively straightforward interpretation of the complex datasets usually generated but also graphically depicts zones of defined structural behavior and regions of change of the molecule or complex of interest. This information is then used to design high-throughput screening assays to select potential stabilizers for the biopharmaceutical. The potential excipients thus obtained are then tested for their ability to stabilize the macromolecule against physicochemical degradation and are further optimized to obtain a desirable formulation. In this chapter, this approach to optimize macromolecule formulation is described in more detail and a few specific examples of studies using this method undertaken within the authors' laboratory are provided.

8.2. BIOPHYSICAL CHARACTERIZATION OF BIOPHARMACEUTICALS

Optimization of a formulation requires a detailed understanding of the macromolecule's properties under a variety of pharmaceutically relevant stresses. Despite the individual structural and functional differences, biopharmaceuticals ranging from simple peptides, proteins, and nucleic acids to more complex systems such as viruses, virus-like particles (VLPs), and bacteria are primarily comprised of proteins, nucleic acids, lipids, and polysaccharides as bioactive components. The functions of all such macromolecules are usually dictated by their structures. Thus, in most cases, finding the optimal conditions that maintain and/or enhance the structure and/or stability of macromolecules to be used as drugs or vaccines is the main objective of preformulation studies. In this regard, factors such as pH, temperature, ionic strength, buffer system, drug concentration, solute binding, shear forces, presence of contaminants, and adjuvants in the case of vaccines can all alter the intra- and intermolecular interactions that maintain the structures of macromolecules [1], resulting in a loss of their biological function. A wide variety of experimental techniques can be used to characterize these bioactive components, although in the review in this chapter we confine ourselves to the characterization of proteins and nucleic acids, since these are the two classes of macromolecules most commonly used in biopharmaceuticals.

Far-UV circular dichroism (CD) and Fourier transform infrared (FTIR) spectroscopy are the two techniques most commonly employed to monitor the secondary structures of proteins and DNA, although Raman spectroscopy is also used occasionally. Similarly, UV–visible (vis) absorption, fluorescence, and near-UV-CD are powerful spectroscopic techniques for detecting alterations in tertiary structure. Note that despite the superior resolving power of nuclear magnetic resonance (NMR) and X-ray crystallography, these techniques are not addressed here because they are still used primarily indirectly in formulation development. Techniques such as dynamic light scattering (DLS), optical density measurements, and size exclusion chromatography (SEC) are commonly employed to monitor the aggregation behavior of proteins and nucleic acids. These representative techniques, pertinent to the examples presented in this chapter, are briefly described below.

The circular dichroism of a macromolecule arises from the differences in absorption of left- and right-handed circularly polarized light in the presence of optically active chromophores. Both proteins and nucleic acids contain a chromophore (the peptide bond and the purine and pyrimidine bases, respectively), which is often arranged in a regular array to produce dramatic optical activity. The far-UV CD spectra (below 260 nm) of proteins are very sensitive to changes in protein secondary structure with alterations between 2% and 3% in helix or sheet content easily detected. Secondary structural elements of each protein exhibits characteristic far-UV CD signals, with the α helix exhibiting a strong double minimum at 222 and 208–210 nm and a stronger maximum at 191–193 nm. In contrast, β-sheet structures typically manifest a weaker single minimum between 215 and 217 nm and a stronger positive maximum between 195 and 200 nm [2]. Certain rarer forms of β structure as well as turns and irregular structure also manifest characteristic spectra. Changes in protein structure often cause loss of signal intensities or distortion of the peaks, inducing shifts in wavelength maxima of the CD spectra. For example, alteration from a typical spectrum of an α helix to that characteristic of a β sheet when the temperature is elevated, may suggest formation of intermolecular β sheets during protein association at high temperatures [3]. Circular dichroism spectra are also very sensitive to secondary structure changes in nucleic acids. Because of the differences in base-stacking interactions, the major secondary structures found in DNA, such as the A, B, C, D, T, and Z forms, can be identified by their unique CD spectra between 190 and 300 nm. Typically, the protein and plasmid DNA concentrations used for CD studies are in the range of 100–200 μg/mL and 50 μg/mL, respectively, using a 0.1-cm pathlength cell [4]. This technique has also proved useful in investigation of nonviral gene delivery complexes in which changes in DNA structure can be easily seen as various cationic delivery agents complex to nucleic acids (both DNA and RNA) [4,5].

Intrinsic and extrinsic fluorescence spectroscopy are often the methods of choice for studying alterations in the tertiary structures of proteins. The dominant natural fluorophore in proteins is usually the indole ring of tryptophan, which absorbs near 290 nm. In addition, the emission spectrum of tryptophan is highly sensitive to the polarity of its immediate environment. Thus, on unfolding of a protein and subsequent exposure of the tryptophan residues from the usual apolar environment of a protein's interior to the polar aqueous solvent, a redshift in the emission peak is typically observed. Also, specific environmental changes can unpredictably perturb the intensity and quantum yield of tryptophan fluorescence because of its unique interactions with water, oxygen, solutes, peptide bonds, and other amino acid sidechains [6]. One phenomenon commonly observed is a blueshift in emission maxima, due to protein association arising from an increased burial of indole sidechain. It is also possible to obtain light-scattering data during such experiments by monitoring the scattered light seen at the excitation wavelength. The preferred method for this is to use a second photomultiplier located at 180° to the fluorescence detector, but one can also simply scan through the entire emission spectrum to obtain such data. Thus, information can be obtained about association/dissociation phenomena simultaneous with fluorescence emission data.

Because of their sensitivities to the polarity of microenvironments, extrinsic probes have also become important tools with which to monitor protein structural changes. The fluorescent probes most commonly used to detect tertiary structural changes of proteins are 8-anilino-1-naphthalenesulfonate (ANS) and its dimeric analog (bis-ANS). These hydrophobic probes are essentially nonfluorescent in aqueous solution, but become strongly fluorescent in a less polar environment. Thus, ANS is, commonly used to identify partially folded intermediate forms of proteins such as molten globule states [7,8]. When ANS binds to apolar sites in a protein that become accessible as a result of structural perturbations, a blueshift of the fluorescence emission maxima and an increase in the fluorescence intensity are usually observed. One needs to be careful in interpreting ANS fluorescence data since ANS may be able to induce structural changes in the protein [9,10]. Also, potential interactions between the negatively charged ANS and oppositely charged residues on the protein can further complicate interpretation [11]. Fluorescent dyes can also be used to detect the presence of specific types of interactions between proteins. Some of the more common dyes used for this purpose are Congo Red [12], thioflavine S [13], and thioflavine T [14,15]. These dyes are selective for intermolecular β-sheet interactions that are present among proteins that form β-amyloid structures [16]. As such, these dyes are used as indicators of aggregation, as well as probes of intermolecular β-sheet contacts that occur between subunit–subunit interfaces of viral capsid proteins [17] and other multimeric protein complexes [18]. Usually, the protein concentrations used for fluorescence studies are about 100–200 μg/mL. In general, one wishes to keep the absorbance at the excitation wavelength below 0.1 to prevent inner filter (self-absorption) effects.

The intrinsic fluorescence of the DNA and RNA bases is too weak for practical applications. The existence of numerous nucleic acid probes, however, enables extrinsic fluorescence to be widely used in this context. Commonly used DNA fluorescent probes include ethidium bromide and other high-affinity dyes such as YOYO-1, acridine orange, propidium iodide, and Hoechst, et al. [4]. Such dyes can be used because of their direct interaction with nucleic acids (e.g., in the helix grooves or between the bases) or in displacement assays in which dyes are competitively removed by ligands of interest. Because of the high sensitivity, the DNA concentration used in extrinsic fluorescence study is often in the nanomolar range [4].

Ultraviolet absorption spectroscopy can also be used to probe changes in the tertiary structure of a protein. For this purpose, high-resolution derivative UV absorption spectra are resolved into five to seven peaks originating from Trp, Tyr, and Phe residues. This method can simultaneously monitor changes in the microenvironment of all three aromatic residues, thus providing a more global picture of the behavior of protein tertiary structure than the fluorescence-based techniques. When the native state of a protein changes, the microenvironment surrounding each individual aromatic residue is subject to alteration. This can produce shifts in the positions of the absorption peaks. In general, these absorbance peaks shift to lower wavelengths (blueshift) when the aromatic residues become more exposed to solvents and vice versa. For these studies, the protein concentration should be selected to keep the absorbance at 280 nm below 1.0. To obtain the high resolution needed, use of a diode array spectrometer is highly recommended with an interpolative technique such as splining to obtain the necessary effective resolution. In addition, this technique can also be used to detect significant protein aggregation induced by various stresses, such as pH, temperature and high concentration, by measuring changes in optical density (turbidity) in the near-UV region (e.g., 320–400 nm). Temperature- and/or pH-dependent changes in optical density (OD) are often used to provide profiles of protein stability that can easily be used in microtiter-plate-based high-throughput screening assays and subsequent formulation development, as discussed below. The UV absorbance spectra of nucleic acids can also be deconvoluted into contributions from individual bases, but this is seldom as useful as in proteins. Derivative spectra can, however, still be used to detect structural changes since alterations in the interaction between the bases produce large changes in absorbance [5]. Thus, absorbance-detected thermally induced melting of the double helix as well as the effects of other environmental perturbations are commonly used to monitor nucleic acid structural stability in both DNA- and RNA-based complexes as well as viruses.

Other techniques such as Fourier transform infrared spectroscopy (FTIR), dynamic light scattering (DLS) and differential scanning calorimetry (DSC) are also often employed in the construction of phase diagrams. A detailed description of these techniques can be found elsewhere [19,20].

8.3. EMPIRICAL PHASE DIAGRAMS

8.3.1. Construction of EPDs

A diverse collection of data is generated from the various biophysical, calorimetric, and hydrodynamic techniques employed to characterize different structural features of macromolecules and their complexes, as discussed above. An internally consistent interpretation of such complex datasets obtained is often quite difficult. To address this problem, all the datasets can be mathematically incorporated into a color map, known as the *empirical phase diagram* (EPD). The EPDs permit an objective analysis of such datasets, and one can predict the state of a macromolecule under various conditions of interest. This initially involves a somewhat different way of thinking about macromolecules or macromolecular complexes such as viruses. For example, the most common way to think of a protein is as a collection of atoms present in different locations in a three-dimensional space. This is the picture that typically arises from crystallography or NMR data. Here, we replace this conventional view with a more abstract (but quite useful) picture in which a macromolecule is represented by a vector in a highly dimensioned experimental space. This vector is modulated into different values as the environmental conditions such as pH, temperature, and ionic strength are varied. Thus, this approach provides a highly information-rich picture of a macromolecular system whose behavior can be used to analyze changes in stability and/or structure with corresponding changes in environmental conditions.

To reiterate, in this method large datasets obtained from a variety of experimental techniques are used to establish a color map (EPD) of the physical stability of a target molecule or complex over a wide range of conditions such as pH, temperature, ionic strength, concentration, freeze–thaw cycles, agitation, and redox potential. Formally, this is accomplished by constructing a vector of n dimensions at each combination of stress conditions examined (pH, temperature, ionic strength, concentration, etc.). The dimensions of this vector are defined by the accumulated data (fluorescence emission maxima and intensities, CD intensities, second-derivative UV absorbance peak positions, etc.). The experimental datasets are represented as n-dimensional vectors in a temperature–pH phase space (or other state conditions), where n refers to the number of variables (i.e., number of different types of data) included in the calculation (e.g., $n = 9$ for a dataset that includes six UV absorption peaks, a CD signal at 222 nm, intrinsic and ANS fluorescence emission peak positions). The data from each technique at individual values of pH and temperature serve as the basis for the individual vector's components. After normalization of the data to values between -1 and 1, an $n \times n$ density matrix combining all the individual vectors is then constructed and n sets of eigenvalues and eigenvectors of the density matrix are calculated. The complete dataset is subsequently truncated and reexpanded into three dimensions consisting of eigenvectors corresponding to the three largest eigenvalues (i.e., the experimental measurements that contribute the most to the final vector). The resultant three-dimensional vectors are then converted into a colored plot with each vector component corresponding to a color using an arbitrary red/green/blue (RGB) color scheme. All of the necessary mathematical calculations are easily performed with commercially available software packages such as Matlab (The Mathworks, Inc., Natick, MA) or Mathematica (Wolfram Research, Champaign, IL). More detailed discussions of the mathematical theory and calculation process for the construction of macromolecular phase diagrams can be found elsewhere [21,22].

Empirical phase diagrams can be generated using either a single technique, such as high-resolution derivative UV spectroscopy [21,23] which generates multiple sets of at least semiindependent data, or employing multiple independent techniques and resultant heterogenous datasets such as CD, fluorescence (intrinsic and extrinsic), differential scanning calorimetry (DSC), and/or dynamic light scattering (DLS) [23,24]. In fact, the phase diagram approach was originally used to display datasets obtained from a protein's second-derivative UV spectroscopy studies [21]. This was based on the assumption that the (usual) six peaks seen behave somewhat differently in response to environmental perturbation because of the widely dispersed locations of the three different types of aromatic sidechains found in proteins. For example, Phe residues are typically buried, Tyr are interfacial, and Trp indole sidechains are dispersed throughout the structure. The multiple-technique-based approach, however, has the potential to provide more definitive apparent phase boundaries than do the UV-absorption-based diagrams since the individual techniques provide more independent information about the different levels of protein structure. In addition, it provides information about the aggregation behavior of the macromolecules, although turbidity studies can be used to provide related information in the single-method UV absorbance approach. Moreover, a multiple-technique-based phase diagram may detect subtle conformational transitions that may be undetected by the data used to construct the UV-absorbance-based phase diagrams [23]. No matter which approach is used, the resulting colored maps define regions of color that correspond to different physical (not necessarily thermodynamic) states of the macromolecules under the stress conditions investigated. Changes in physical state are thus demarcated by changes in color producing pseudo–phase boundaries; that is, changes in color correspond to transitions between various physical states of the macromolecule. It needs to be emphasized that these are not necessarily equilibrium transitions (no reversibility is implied), and thus these are not equilibrium (i.e., thermodynamic) phase diagrams; hence the use of the word *empirical* to describe them. Furthermore, their use is primarily empirical and their employment in a more fundamental analysis of macromolecular systems needs to be very carefully considered. The pseudo–phase boundaries, however, are the crucial product of the phase diagram since they can be used to give us a fairly precise idea of the conditions under which various physical (and, indirectly, chemical) degradation events occur. Thus, this technique in no way describes absolute physical states of molecules or their complexes. The empirical phase diagram, rather, describes changes in molecular states that are simply calculated from the coherence/incoherence of the data accumulated. Such empirical maps can provide an initial view of the stability of a macromolecule that can, in turn, be used to design high-throughput screening assays for potential stabilizers [24,25] (see discussion below). By employing high-throughput instrumentation (CD, fluorometry, DLS, DSC, etc.), EPDs can be constructed in a few days to a few weeks for a wide variety of environmental variables.

The EPD approach has so far been successfully employed to examine a wide variety of vaccine candidates and recombinant proteins as well as other macromolecule pharmaceuticals (Table 8.1) [3,10,21,23–43]. These include peptide and protein pharmaceuticals [human fibroblast growth factor 1 (FGF-1), interferon-β-1a, monoclonal antibodies, botulinum A neurotoxin and its complex, etc.], recombinant-protein-based vaccines (anthrax rPA, ricin toxin A, malaria EBA-175, *Clostridium difficile* toxins, etc.), viruses, and virus-like particle (VLP)-based vaccines (Norwalk VLPs, adenovirus types 2 and 5, respiratory syncytial virus, rotavirus, measles virus, etc.), DNA, and bacterial vaccines (unpublished data) as well as gene delivery vectors. These studies are summarized in Table 8.1, where references to the individual studies can be found. Phase diagrams have also been constructed employing a wide variety of environmental variables such as pH and temperature, temperature and active pharmaceutical ingredient (API) concentration, temperature and excipient concentration and pH and ionic strength. A few representative examples of such phase diagrams and their use as formulation tools are discussed below.

8.3.2. EPDs Using a pH–Temperature Phase Space

8.3.2.1. Employing a Single Technique. The power of the phase diagram approach employing a single technique that produces multiple experimental parameters is well illustrated with the relatively unstable recombinant protective antigen (rPA) of anthrax [24], which is used in vaccines against this dangerous organism. Employing a UV–visible diode array spectrophotometer, UV absorption spectra up to 0.01 nm resolution can be obtained using interpolation techniques that permit very precise resolution of the spectra into distinct peaks by derivative analysis [21]. A protein's second-derivative UV absorbance spectrum typically displays six distinctive negative peaks from its three different types of aromatic amino acids (Fig. 8.2), all of which can be monitored as a function of temperature at various pH values (e.g., pH 3–8) to provide a comprehensive overview of the tertiary structure of a protein. Structural changes in the protein induced by changes in temperature and pH can be characterized by alterations in the positions of these peaks. Because the three different aromatic residues are usually extensively dispersed throughout a protein's matrix, they potentially provide a very sensitive monitor of the three-dimensional structure of a protein and possible conformational changes as it is stressed. Representative plots of changes in peak positions of the Phe, Tyr, and Trp residues of rPA as a function of temperature at six different pH values are shown in Figure 8.3. One difficulty with such an approach involves the large amount and complex behavior of the data generated. To better visualize such data and provide a more intuitive picture of the protein's behavior, the EPD approach becomes extremely helpful. A T/pH phase diagram of rPA developed from high-resolution second-derivative UV absorption spectroscopy studies is shown in Figure 8.4.

TABLE 8.1. Examples of EPD-Based Characterization of Biopharmaceuticals

Variable	Biopharmaceutical	Techniques Employed	Number of Apparent Phases Observed	Conditions Maintaining Native-Like Structure	Reference
pH versus temperature	Recombinant bovine granulocyte colony–stimulating factor (bGCSF)	Second-derivative UV absorbance	6	pH 5–7, 10°C–50°C	21
	Recombinant human interferon β-1a ($I = 0.1$)	Fourth-derivative UV absorbance, CD, intrinsic and ANS fluorescence	≥ 3	pH 4, 10°C–55°C; pH 5–6, 10°C–57°C; pH 7, 10°C–55°C; pH 8, 10°C–50°C	23
	Recombinant human interferon β-1a ($I = 1.0$)		≥ 3	pH 4, 10°C–50°C; pH 5–6, 10°C–60°C; pH 7, 10°C–55°C; pH 8, 10°C–52°C	23
	rPA of *Bacillus anthracis*	CD, intrinsic and ANS fluorescence	5	pH 4, 10°C–30°C; pH 5, 10°C–35°C; pH 6–8, 10°C–40°C	24
	Hsc70	Second-derivative UV absorbance, CD, and intrinsic fluorescence	6	pH 5–8, 10°C–40°C	10
	Gp96	Second-derivative UV absorbance, CD, and intrinsic fluorescence	5	pH 6, 10°C–40°C; pH 7–8, 10°C–55°C	10
	EBA-175 RII	CD, intrinsic and ANS fluorescence	4	pH 5 to 8, 10°C–50°C	31
	Fibroblast growth factor 20 (FGF 20)	Fourth-derivative UV absorbance	≥ 3	pH 5 and 5.5, 10°C–40°C; pH 6 to 8, 10°C–50°C	3

(continued)

TABLE 8.1. (*Continued*)

Variable	Biopharmaceutical	Techniques Employed	Number of Apparent Phases Observed	Conditions Maintaining Native-Like Structure	Reference
	Recombinant human gelatin (8.5 kDa)	CD, light scattering, and intrinsic fluorescence	≥ 3	Undefined	34
	Recombinant human gelatin (25 kDa)	CD, light scattering, and intrinsic fluorescence	≥ 3	Undefined	34
	Recombinant human gelatin (50 kDa)	CD, light scattering, and intrinsic fluorescence	≥ 3	Undefined	34
	Recombinant human gelatin (100 kDa)	CD, light scattering, and intrinsic fluorescence	≥ 3	Undefined	34
	Ricin toxin A chain	CD, intrinsic and ANS fluorescence	3	pH 4, 10°C–30°C; pH 5–7, 10°C–45°C; pH 8, 10°C–42°C	25
	Pramlintide (synthetic analog of human amylin)	Second-derivative UV absorbance, intrinsic fluorescence, and $OD_{350\,nm}$	≥ 4	pH 5–8, 10°C–50°C	35
	Clostridium botulinum type A neurotoxoin	CD, intrinsic and ANS fluorescence	5	pH 5–8, 10°C–40°C	27
	C. botulinum type A holotoxoin complex	CD, instrinsic and ANS fluorescence	5	pH 3–5, 10°C–40°C	27
	Human fibroblast growth factor 1 (RGF-1)	Second-derivative UV absorbance	≥ 4	pH 5–8, 10°C–40°C	26
	Clostridium difficile toxoid A	CD, intrinsic fluorescence, ANS fluorescence, and $OD_{350\,nm}$	4	pH 6–8, 10°C–50°C	36
	C. difficile toxoid B		5	pH 5.5–7.5, 10°C–50°C	

IgGκ MAb (5 mg/mL)	CD, intrinsic fluorescence, ANS fluorescence, and light scattering	4	pH 4–8, 20°C–55°C	52
	HR ultrasound spectroscopy, PPC, and lifetime fluorescence	5	pH 4–8, 20°C–52°C	52
IgG MAb1 (100 mg/mL)	CD, intrinsic fluorescence, ANS fluorescence, and light scattering	≥ 3	pH 6–8, 10°C–70°C	Unpublished data
IpaD from *Shigella flexneri*	Second-derivative UV absorbance, CD, intrinsic and ANS fluorescence	≥ 8	pH 7–8, 10°C–30°C	57
SipD from *Salmonella* spp.	Second-derivative UV absorbance, CD, intrinsic and ANS fluorescence	≥ 8	pH 7–8, 10°C–35°C	57
BipD from *Burkholderia pseudomallei*	Second-derivative UV absorbance, CD, intrinsic and ANS fluorescence	≥ 4	pH 4–8, 10°C–30°C	57
PcrV from *Pseudomonas aeruginosa*	Second-derivative UV absorbance, CD, intrinsic and ANS fluorescence	≥ 4	pH 5–8, 10°C–45°C	57
LcrV from *Yersinia* spp.	Second-derivative UV absorbance, CD, intrinsic and ANS fluorescence	≥ 5	pH 5–8, 10°C–50°C	57
MxiH from *Shigella flexneri*	Second-derivative UV absorbance, CD, and ANS fluorescence	≥ 4	pH 5–7, 10°C–40°C	38

(*continued*)

TABLE 8.1. *(Continued)*

Variable	Biopharmaceutical	Techniques Employed	Number of Apparent Phases Observed	Conditions Maintaining Native-Like Structure	Reference
	PrgI from *Salmonella typhimurium*	Second-derivative UV absorbance, CD, and ANS fluorescence	≥ 4	pH 5–7, 10°C–45°C	38
	BsaL from *Burkholderia pseudomallei*	Second-derivative UV absorbance, CD, and ANS fluorescence	≥ 4	pH 5–7, 10°C–45°C	38
	Adenovirus type 5 (low ionic strength)	Second-derivative UV absorbance, intrinsic fluorescence, PI fluorescence, $OD_{350\,nm}$, and DLS	≥ 4	pH 6–8, 10°C–40°C	37
	Adenovirus type 5 (high ionic strength)		≥ 3	pH 4–8, 10°C–40°C	37
	Adenovirus type 2	Second-derivative UV absorbance, CD, intrinsic fluorescence, PI fluorescence, $OD_{350\,nm}$, and DLS	4	pH 5–8, 10°C–37°C	32
	Adenovirus type 4	CD, intrinsic fluorescence, PI fluorescence, light scattering, $OD_{350\,nm}$, and DLS	5	pH 3, 10°C–30°C; pH 4, 10°C–40°C; pH 5, 10°C–45°C; pH 6–8, 10°C–40°C	39
	Measles virus	CD, intrinsic fluorescence, ANS fluorescence, laurdan fluorescence, DLS, and light scattering	6	pH 7, 10°C–20°C	40

	Respiratory syncytial virus	Second-derivative UV absorbance, CD, intrinsic fluorescence, ANS fluorescence, and $OD_{350\,nm}$	≥ 3	pH 6 and 8, 10°C–40°C	28
	Rotavirus G1 strain	CD, intrinsic fluorescence, light scattering, and DLS	4	pH 7 and 8, 10°C–50°C	41
	Rotavirus G3 strain	CD, intrinsic fluorescence, light scattering, and DLS	4	pH 7 and 8, 10°C–50°C	41
	Rotavirus G4 strain	CD, intrinsic fluorescence, light scattering, and DLS	3	pH 7 and 8, 10°C–50°C	41
	Recombinant vaults expressed in Sf9 insect cell line	CD, intrinsic fluorescence, ANS fluorescence, and light scattering	6	pH 4–8, 10°C–45°C	42
	Norwalk virus–like particles	Second-derivative UV absorbance, CD, intrinsic and ANS fluorescence	4	pH 3–7, 10°C–55°C	29
	Attenuated *Vibrio cholerae*	CD, extrinsic (*Bac*light kit) fluorescence and DSC	4	pH 7 and 8, 10°C–40°C	43
Ionic strength versus pH	Plasmid DNA pMB 290	CD, YOYO-1 fluorescence, and DLS	≥ 3	pH 5–8, 0–150 mM	33
	DNA/DOTAP lipoplexes (low charge ratio)	CD, YOYO-1 fluorescence, and DLS	≥ 3	pH 5–8, 50–150 mM	33
	DNA/DOTAP lipoplexes (high charge ratios: 4)	CD, YOYO-1 fluorescence and DLS	≥ 3	pH 5–8, 0–150 mM	33
	DNA/DOPE-DPTAP liposomes (low charge ratio)	CD, YOYO-1 fluorescence, and DLS	≥ 3	pH 4–6, 50–150 mM pH 7–8, 70–75 mM	33

(continued)

TABLE 8.1. (*Continued*)

Variable	Biopharmaceutical	Techniques Employed	Number of Apparent Phases Observed	Conditions Maintaining Native-Like Structure	Reference
	DNA/DOPE-DPTAP liposomes (high charge ratio)	CD, YOYO-1 fluorescence, and DLS	≥ 3	pH 5–8, 0–150 mM	33
	DNA/PEI polyplexes (low charge ratio)	CD, YOYO-1 fluorescence, and DLS	≥ 3	pH 5–8, 50–150 mM	33
	DNA/PEI polyplexes (high charge ratio)	CD, YOYO-1 fluorescence, and DLS	≥ 3	pH 5–7, 25–125 mM	33
Concentration versus temperature	IgG1κ MAb1	Second-derivative UV absorbance, CD, intrinsic fluorescence, and $OD_{350\,nm}$	≥ 2	Concentration 0.1–140 mg/mL; temperature 10°C–50°C	30
	IgG1κ MAb2	Second-derivative UV absorbance, CD, intrinsic fluorescence, and $OD_{350\,nm}$	≥ 2	0.1 mg/mL, 10°C–75°C 0.1–10 mg/mL, 10°C–57°C; $\geq$ 100 mg/mL, 10°C–50°C	30
	Pramlintide (synthetic analog of human amylin) at pH 4	Second-derivative UV absorbance, intrinsic fluorescence, and $OD_{350\,nm}$	≥ 6	1.8 mg/mL, 10°C–30°C; 3.5 mg/mL, 10°C–35°C; 5.3 mg/mL, 10°C–45°C; 7.0 and 8.8 mg/mL, 10°C–40°C	35

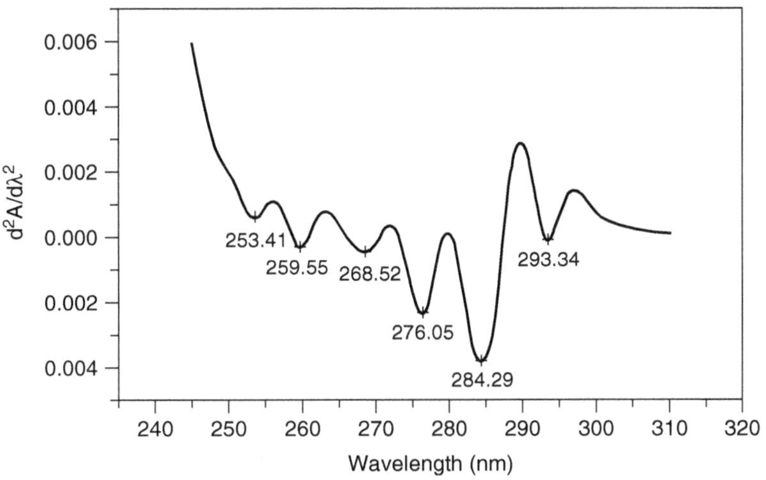

Figure 8.2. A second-derivative UV absorbance spectrum of a protein displaying six distinctive negative peaks from the three different types of aromatic amino acids.

To generate this phase diagram, second-derivative peak position data were obtained over the pH range of 3–8 and a temperature range of 10°C–82.5°C using a 20 mM citrate phosphate buffer ($I = 0.1$ adjusted with NaCl) (Fig. 8.3). This buffer was chosen to provide as flat a pH response as possible over this wide temperature range. At 10°C, the negative spectral peaks occurred at approximately 253 nm (peak 1: Phe), 259 nm (peak 2: Phe), 268.5 nm (peak 3: Phe), 276 nm (peak 4: Tyr), 284 nm (peak 5: Tyr/Trp), and 291.6 nm (peak 6: Trp). The six peak position data obtained as a function of temperature and pH from second-derivative UV absorption spectroscopy were used to construct an EPD of rPA using the multidimensional vector approach described above (Fig. 8.4). Thus, data at each unique value of pH and temperature were mapped to a vector consisting of the six derivative peak positions corresponding to that point. This particular phase diagram was constructed employing Matlab software. The phase diagram of rPA constructed in this manner suggests that the rPA exists in a similar physical state at lower temperatures (from 10°C to 40°C) over the entire pH range (3–8) studied. At temperatures greater than 45°C, a major structurally disruptive transition occurs at pH 6–8. The structural changes occur at much lower temperatures at the lower pH values (pH 4–5), indicating that the protein is more thermally stable in the pH range of 6–8. From these data, the region of maximum stability appears to be at pH 6–7 (up to 50°C). The transitions at various pH values in this phase diagram appeared to occur at slightly higher temperatures than those observed by other biophysical methods (Fig. 8.5). Thus, to further extend this approach to obtain a more global insight into the structural as well as the colloidal stability of the macromolecule, multiple techniques were also employed for construction of EPD of anthrax rPA, as discussed below. It should be mentioned that this approach can also be applied to other techniques, such as one-dimensional NMR

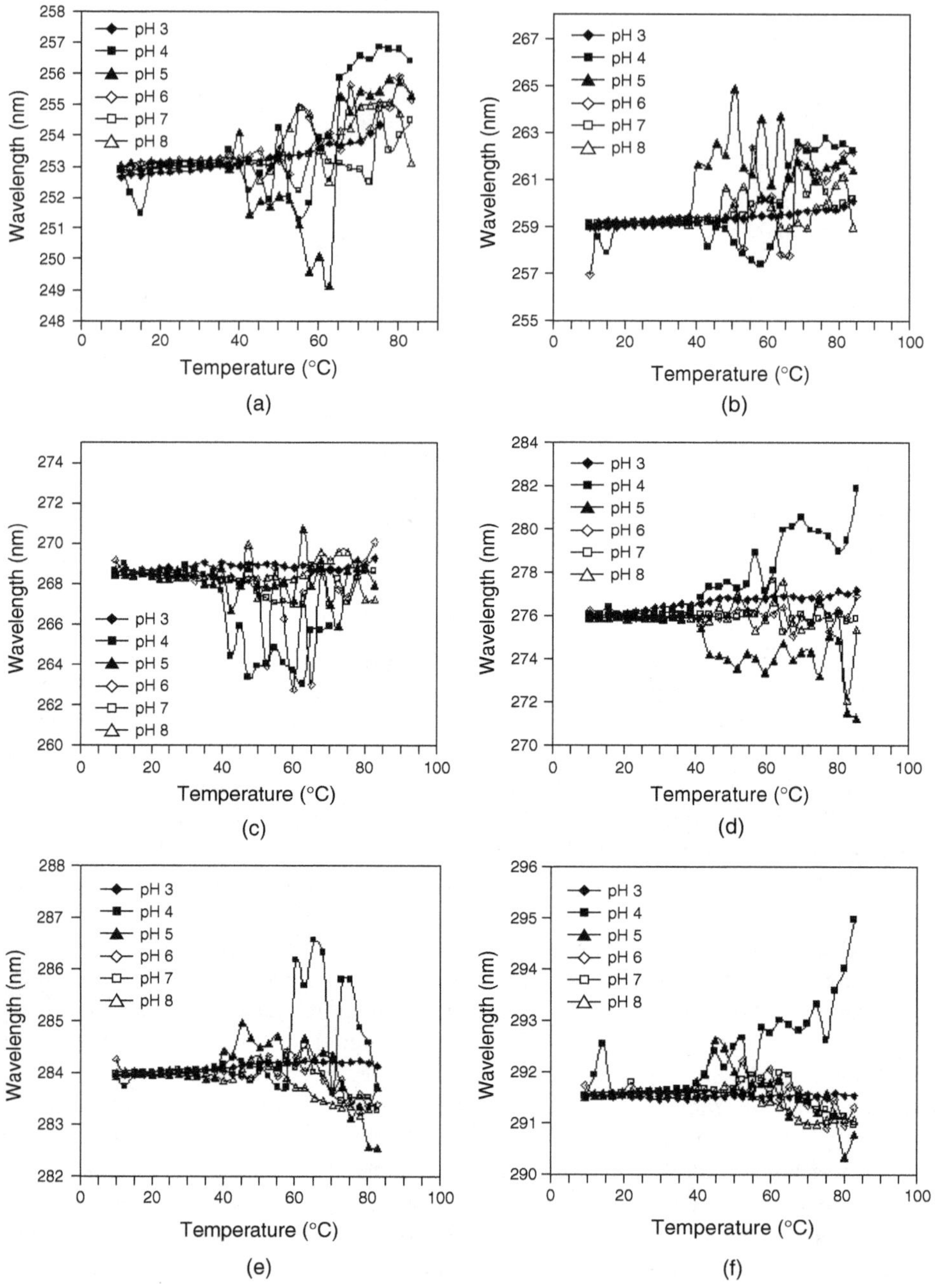

Figure 8.3. Second-derivative UV absorbance studies of rPA as a function of temperature and pH. The wavelength positions of six peaks are recorded: 1 (a), Phe; 2 (b) Phe; 3 (c), Phe, 4 (d), Tyr; 5 (e), Tyr/Trp; 6 (f), Trp. Error bars are not included to allow better visualization, but are generally of the order ± 0.04 nm.

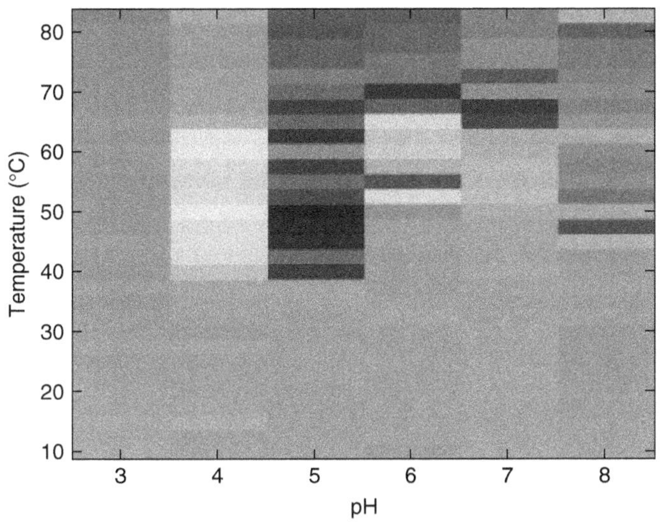

Figure 8.4. Empirical phase diagram of rPA on a temperature–pH axis constructed using high-resolution second-derivative UV spectroscopy data. The data employed to generate these images was obtained over the pH range of 3–8 (20 mM citrate phosphate buffer, $I = 0.1$ adjusted with NaCl) and temperatures of 12°C–82.5°C in 2.5°C increments. Blocks of continuous color represent homogenous phases, conditions under which the raw data-derived vectors behave similarly. See the insert for color representation of this figure.

or DSC in which multiple peaks (e.g., resonance, endotherms) can be monitored as a function of environmental perturbations.

8.3.2.2. Employing Multiple Techniques. The EPD approach is a general method that can use vectors of any finite dimension for the construction of diagrams [23]. Theoretically, there is no limit to the number of variables (datasets) that may be included in EPDs because of the truncation procedure (which could, of course, be more than the three terms currently employed for color display purposes). To detect more subtle changes and to incorporate additional structural information that may not be detected by the UV-absorption-based technique, data from multiple techniques have more commonly been used to construct EPDs [25,26]. When heterogeneous datasets from multiple techniques are used, great caution should be taken to keep experimental variables such as protein concentration, heating rates, and equilibration time at each temperature point as similar as possible. An example of a T/pH multiple-technique-based phase diagram for anthrax rPA with the corresponding phases is shown in Figure 8.6 [24]. This phase diagram was generated employing Mathematica software and utilizing far-UV CD, intrinsic, and ANS fluorescence data (Fig. 8.5a–c). In this case, each (T, pH) condition is mapped to a vector consisting of molar ellipticity at 222 nm obtained from CD, intrinsic tryptophan fluorescence spectral center-of-mass, and ANS fluorescence intensity data. The advantage of this approach is that

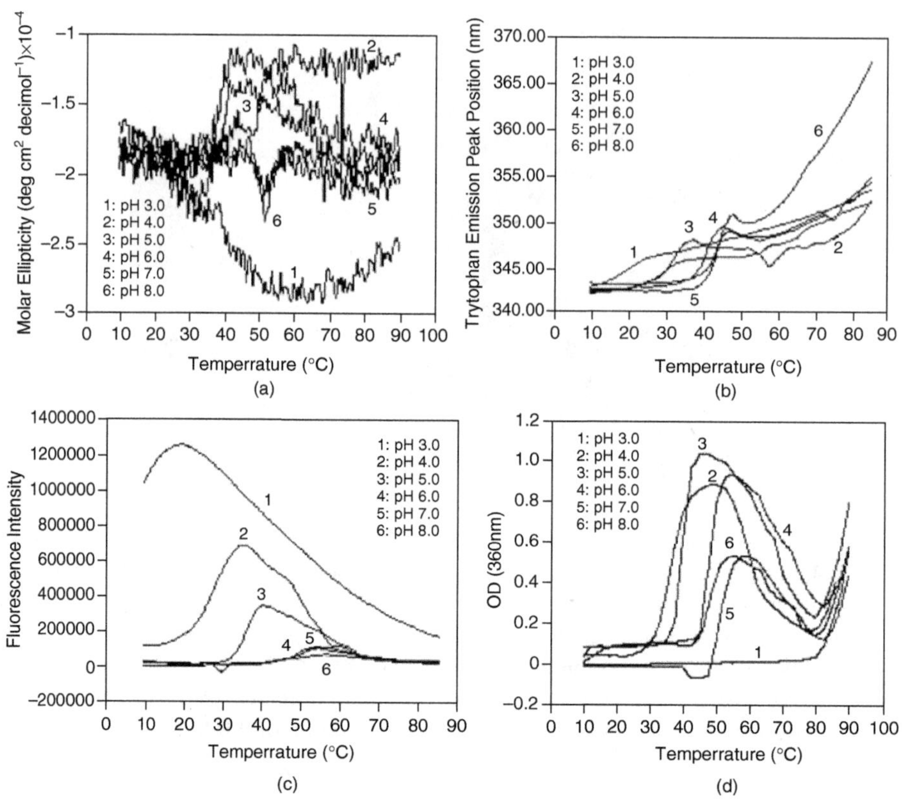

Figure 8.5. Physical stability of rPA as a function of pH and temperature in 20 mM citrate phosphate buffer ($I = 0.1$ adjusted with NaCl) as determined by CD thermal unfolding monitored at 222 nm (a); intrinsic Trp fluorescence peak positions employing a spectral center of mass procedure (b); ANS fluorescence intensity (c); turbidity measurements, OD at 360 nm (d). The thermal traces represent an average of two measurements in which each data point has a standard error of less than 0.5 [24].

the data employed reflect more global aspects of secondary and tertiary structure change as well as the exposure of the apolar binding sites of protein as a function of pH and temperature. This information can then be employed to assign the origin of the behavior of the protein with each apparent phase observed. The diagram for rPA obtained in this manner suggests that rPA adopts at least five distinct behaviors (Fig. 8.6). Again note that these do not correspond to true "thermodynamic phases" in the formal use of the concept but are simply empirical (nonthermodynamic) states defined by the experimental methods employed. The properties of the phases can be established by direct reference to the actual biophysical measurements themselves. Combining the multitechnique phase diagram with knowledge of the behavior of the protein on the basis of the individual measurements, it appears that the pinkish region in the lower right-hand corner of the diagram is the state of most continuous

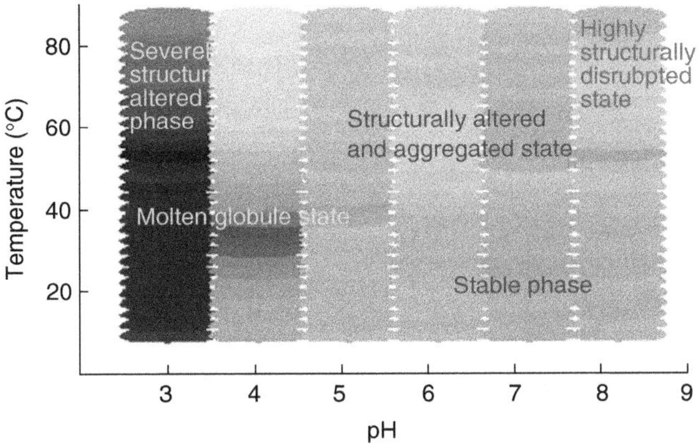

Figure 8.6. Empirical phase diagram of rPA on a temperature–pH plane on the basis of intrinsic and ANS fluorescence, light scattering ($OD_{360\,nm}$), and CD data. The data employed to generate these images was obtained over the pH range of 3–8 and at temperatures ranging from 12°C to 85°C in 2.5°C increments. Blocks of continuous color represent uniform phases, conditions under which the raw-data-derived vectors behave similarly [24]. See the insert for color representation of this figure.

stability (Fig. 8.6). This correlates well with CD, intrinsic, and ANS fluorescence, as well as the second-derivative UV absorbance data. The midpoints of the transitions observed by the biophysical measurements generally correspond to the color changes between the phases. A second phase (blue/purple) is apparent at pH 3 at temperatures below 45°C and also seems to encompass pH 4 at temperatures greater than 30°C and pH 5 at temperatures greater than 35°C. This state appears to have significant molten globule character since the tertiary structure changes appear to occur before the secondary structure alterations. As the temperature increases at pH 3, the protein rapidly enters another phase near 50°C. A third phase appears to be present at pH 5–8 in the higher-temperature region. This depicts a phase in which the protein is severely structurally altered and tends to aggregate and may also involve molten globule behavior. The aggregation behavior of the protein was further confirmed by light-scattering measurements (optical density at 360 nm), as shown in Figure 8.5d. At pH 8, a completely different phase appears at very high temperatures, which we assume corresponds to an even more structurally disrupted form. The information obtained from the phase diagram can be further used to select conditions for the development of a high-throughput assay to screen for potential stabilizers, as illustrated with rPA as an example (see Section 8.5).

8.3.3. EPDs Using Other Environmental Variables

8.3.3.1. Temperature Versus Concentration. Empirical phase diagrams can be constructed using any environmental variable that can be quantified. An example of an EPD of a monoclonal antibody (MAb) using temperature and protein

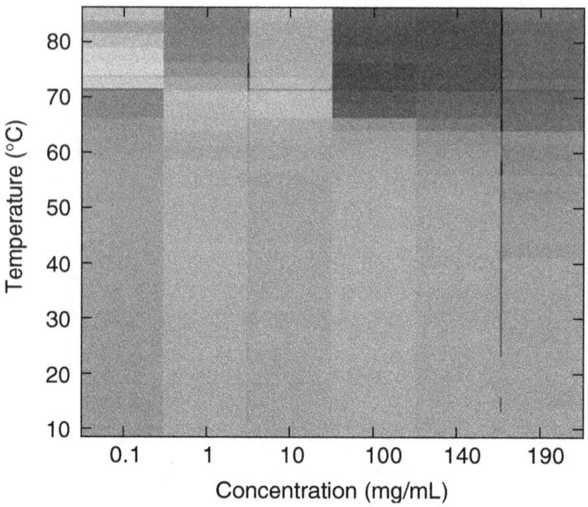

Figure 8.7. Empirical phase diagram of a monoclonal antibody on a temperature–protein concentration plane. Each temperature–concentration point on the diagram is constructed from data obtained from fluorescence emission peak position, $OD_{350\ nm}$, and second-derivative UV absorbance peak positions of Phe, Tyr, and Trp, and CD molar ellipticity. Note that the concentration axis is not linear [30]. See the insert for color representation of this figure.

concentration as independent variables is shown in Figure 8.7 [30]. The behavior of the immunoglobulins at ultrahigh concentrations has become of great interest because of a need to deliver low-volume/high-content doses of this class of proteins. This diagram is constructed of multidimensional vectors at each environmental condition (temperature and MAb concentration) from various normalized data comprising six second-derivative UV absorbance peaks, tryptophan emission peaks, turbidity (OD) at 350 nm, and CD molar ellipticity at 218 nm. Note that in this case the concentration data are representated on a nonlinear scale, making interpolation to other conditions somewhat more difficult.

One purpose of this study was to develop methods that can be employed to directly study the physical structure and thermal stability of proteins at high concentrations and then use a combination of these methods to construct a phase diagram to examine the effect of increasing protein concentration (from 0.1 to ~200 mg/mL) on thermal stability. Although highly concentrated protein formulations offer increased promise for more convenient treatment of a broad range of diseases, there remains a major obstacle to their development stemming from a lack of available methods to determine macromolecular structure, stability, and aggregation propensity within highly concentrated solutions [30,44]. Currently (as of 2009), most of the analytical methods employed in the development of highly concentrated protein formulations require dilution of the samples before acquiring data and subsequent extrapolation of stability-indicating parameters obtained from these studies may not be representative of the more highly concentrated solutions. Thus, biophysical characterization of a highly concentrated protein formulation requires the use of novel approaches. A detailed description of these methods (e.g., frontface fluorescence, short-pathlength CD, FTIR, DSC) is beyond the scope of this chapter and can be found elsewhere [30]. Thus, data obtained from the abovementioned techniques were used to generate the temperature–concentration phase diagram of the MAb (Fig. 8.7).

The information obtained from the individual experimental methods permits assignment of physical meaning to the colored phases observed in the diagram over the temperature and concentration range of the study (data not shown [30]). Briefly, the phase diagram of this monoclonal antibody (Fig. 8.7) suggests that at temperatures below 50°C, the antibody exists in a very similar state over the entire concentration range examined. Further inspection of the biophysical data collected in this study indicates that the protein is in a native or near-native state under these conditions. At temperatures above 65°C and at low concentrations, the immunoglobulin appears to be in a slightly unfolded and aggregated state. At higher concentrations, the phases at elevated temperatures correspond to different unfolded, gelled, and precipitated states. In addition, the apparent phase boundaries occur at somewhat decreased temperatures as the protein concentration is increased. Thus, for the first time a method was developed to directly detect the structural changes of a protein at high concentrations. Additionally, using this approach the thermal stability of the protein could be correlated to corresponding changes in its concentration. This study suggests that the decrease in structural stability observed at higher MAb concentrations is probably the result of aggregation or more limited self-association on heating in crowded solutions and not due to a decrease in the intrinsic structural stability of the MAb [30].

8.3.3.2. Ionic Strength Versus pH. The phase diagram approach has also been applied to determine changes in the physical stability of nonviral gene delivery complexes under different solution conditions [33]. In the past, the characterization of nonviral gene delivery systems was complicated by their size, complexity, and heterogeneity. This issue has been addressed with the creation of EPDs using several different physical techniques in combination. A representative ionic strength–pH EPD of a DNA/carrier complex is illustrated in Figure 8.8. This DNA/carrier complex was prepared by using a mixture of a cationic lipid [1,2-dioleyl-3-trimethyammonium propane (DOTAP)] and a helper lipid [1,2-dioleyl-*sn*-glycero-3-phosphoethanolamine (DOPE)]. Three experimental approaches (DLS, CD, and YOYO-1 dye fluorescence) were selected for this study to monitor different structural aspects of the particles related to size, DNA conformation, and the extent of DNA condensation, respectively [33]. The electrostatic nature of the interaction between the plasmid DNA and the cationic carriers prompted the selection of environmental variables pH (4–8, at one-unit intervals) and ionic strength (10–15 mM at 25 mM intervals). An EPD for (DOTAP/DOPE)/DNA complexes formulated at a high charge ratio indicates the presence of distinct structural forms of the complexes (Fig. 8.8). The regions of similar color represent at least somewhat homogenous phases, while structural changes of gene delivery complexes are marked by corresponding changes in color. In general, the EPD suggests the presence of at least three different physical states. At very

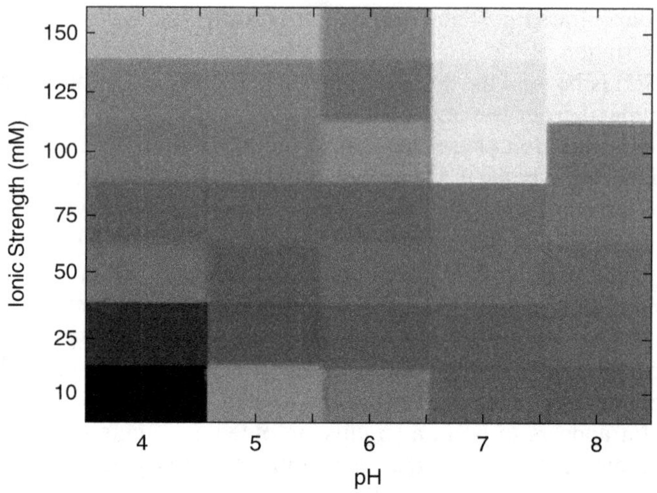

Figure 8.8. Ionic strength–pH EPD of a nonviral gene delivery complex. The EPD is generated from dynamic light scattering, CD, and fluorescence studies. Regions of similar color represent similar structural behavior, while the variation in color defines the conditions under which the structure of the gene delivery complex alters [33].

ionic strength (10 mM), a small and distinct phase is seen across pH 4–6, while a second phase is observed at ionic strength greater than 100 mM and pH 7–8. Presence of a third continuous phase is apparent in moving from pH 4–6 at 25–150 mM to pH 7–8 at ionic strength below 100 mM. The pH-dependent stability of these complexes may be attributed to varying ionization states of the carrier. The helper lipid DOPE, for example, is less ionized at high pH (7 and 8) and thus forms less stable complexes with DOTAP than at lower pH values. The light blue/green phase observed in the EPD corresponds to the presence of larger species as determined by DLS studies. Furthermore, protonation of the nucleic acid bases and phosphate groups at low pH values and condensation of the DNA plasmid mediated by salt could account for the observed structural alterations [45]. Thus, a comprehensive picture of the behavior of complexes with a corresponding identification of apparent phase boundaries in terms of relevant variables is obtained. Such an approach permits development of screening assays to identify potential stabilizers. Additionally, the correlation between the physical aspects of complexes with their ability to transfect cells can be explored. The EPD approach therefore provides a convenient and powerful tool to characterize the structure of the complexes over a wide range of experimental variables, especially for cases in which the use of higher-resolution techniques such as NMR or X-ray crystallography are difficult or impossible to apply. It should be noted that EPD analysis of gene delivery complexes reveals the presence of more gradual, subtle structural alterations than in the EPDs described above for other biopharmaceuticals (e.g., proteins, viral particles), in which well-defined changes in structure were reflected by abrupt color changes.

8.4. NEW APPROACHES TO EPDs

The EPD approach for the rapid preformulation of vaccines and macromolecule-based pharmaceuticals described above is based primarily on the static properties of the macromolecules involved. One defining characteristic of large molecules, however, is their dynamic nature. These internal motions range from small alterations of the sidechains and the peptide backbone to large-scale movement of structural domains relative to each other. The relationship between this property and the molecules' stability and formulation is somewhat controversial but is expected to be profound [46]. The phase diagram approach can be expanded to include dynamic data such as those obtained by isotope [hydrogen–deuterium (H/D)] exchange rates (k), red edge excitation spectra (REES) ($\Delta\lambda$), coefficients of thermal expansion (α), and compressibility (β) as well as fluorescence lifetime anisotropy measurements. Such data can be used to construct EPDs using the vector approach described above and should better reflect dynamic aspects of macromolecular structure. To elaborate, proteins in the presence of deuterium will undergo an exchange of amide hydrogen atoms for deuterium atoms at a rate dependent on the dynamic exposure of the peptide backbone. This rate of exchange is a function of both solvent penetration and local unfolding [47,48]. For this reason, extended measurements of the rate of H/D exchange using NMR, mass spectrometry, or FTIR, among other techniques, can be related to the internal amplitude motions within the protein that render interior regions accessible to the solvent. Red edge fluorescence is performed by exciting the target molecule at various excitation wavelengths and monitoring shifts in the wavelength emission maximum [49]. A redshift of the emission maximum reflects changes in the solvent relaxation time that occur with restricted motion of the fluorophore. These studies can thus be employed to probe the mobility of the fluorophore (intrinsic or extrinsic) as a function of stress variables such as pH, temperature, and ionic strength. A related technique can employ an extrinsic solute quencher such as acrylamide, iodide, or cesium ion to probe the accessibility of a fluorophore [50]. Another fluorescence-based approach sensitive to the dynamic nature of macromolecular behavior combines lifetime measurements and polarized light to obtain anisotropic information about the fluorophore that can be related to their mobility in terms of their rotational correlation times. A third technique, pressure perturbation calorimetry, can also be used to determine the thermal coefficient of expansion of a macromolecule [51]. When pressure is increased or decreased above a solution, heat is either absorbed or released, and this heat is proportional to the pressure change and the corresponding response of the macromolecules to these changes. This permits the calculation of volumetric properties of the protein. Since these measurements can be made over a wide range of temperatures, it is possible to calculate the coefficient of thermal expansion and volume change as a function of temperature of a macromolecule or particle. Another approach, ultrasonic spectroscopy, employs attenuation of sound measurements to determine the complementary parameter of compressibility [51]. In preliminary studies, we have found that "dynamic" EPDs provide information above and beyond that obtained by the static approach [52]. The utility of this approach in formulation activities has, however, yet to be established.

One can also construct EPDs using chemical stability data of proteins and vaccine antigens to further expand the utility of this approach. Typically, one- or two-dimensional reversed-phase HPLC coupled to mass spectrometry is used for this purpose. The conditions such as high pH and temperature (to induce deamidation) and the presence of oxidizing agents are used to accelerate chemical degradation processes and the resultant rate constants are used as input parameters. Lower-resolution methods such as isoelectric focusing or RP-HPLC alone can be used as well. Data from such analyses can either be combined into physically based phase diagrams or used separately as an independent EPD to provide a much more comprehensive view of the degradation of systems of interest. Comparisons of physically based EPDs with chemically based ones should be of special interest in establishing any relationship between the two phenomenon.

8.5. DEVELOPMENT AND OPTIMIZATION OF HIGH-THROUGHPUT STABILIZER SCREENING ASSAYS

The EPD approach can be used to comprehensively characterize the solution behavior of biopharmaceuticals under various stress conditions. Although not discussed here, it can also be used to characterize the solid-state behavior of dried formulations using alternate technologies. The identification of "apparent" phase boundaries is one of the most important applications of the approach. This locates conditions under which the macromolecules are marginally stable or undergo critical structural changes, as defined by the regions of abrupt or gradual color change. These conditions can be employed to accelerate degradation pathways to develop high-throughput screening assays for the identification of potential stabilizers for the rational formulation of biopharmaceuticals. The type of assay employed (static and dynamic light scattering for aggregation, intrinsic or extrinsic fluorescence for tertiary structural changes, CD for secondary structural alterations, etc.) is determined by the nature of the relevant physical changes and their potential adaptation to a high-throughput format to decrease analysis time. Techniques such as CD and DSC have also been automated, although they have not yet been adapted to microtiter-plate-based formats. Recent work by a variety of companies suggests that techniques such as FTIR and fluorescence lifetime anisotropy may soon be available in a high throughput format.

Once such assays are established, various libraries of compounds can be screened and potential stabilizers identified. Examples of compound libraries that can be screened include comprehensive selections of agents that are "generally regarded as safe" (GRAS), such as carbohydrates, amino acids, polymers, and detergents, as well as collections of osmolytes, polyanions, and low-molecular-weight di- and multiions [53]. Each compound should be examined over a wide range of concentration (e.g., 0.01%–20%w/v) and tested in at least triplicate over a range of positions on a microtiter plate. Data are typically recorded as a function of time, based on the initial optimization of the screening assay. Compounds that effectively stabilize the target system in terms of slowing or inhibiting the magnitude of the parameter change are selected for further study. If stabilizers are not identified in these small libraries, larger collections of hundreds of thousands of selected compounds used in drug-screening protocols can be employed although issues of safety become significant. Compounds identified in these libraries could also serve as starting points for combinatorial chemical attempts to generate unique stabilizers. Once potential excipients are identified, the concentration and their use in a combination that provide maximum stability can be initially determined by the screening assays. The optimal excipient(s) concentration is then used to verify, at admittedly low resolution, the mechanism of stabilization of the compound. For example, it is possible that a stabilizer may prevent aggregation but not provide conformational stabilization or vice versa; to obtain this information, individual techniques such as fluorescence and CD are used in the presence and absence of potential excipients to determine apparent thermal transition midpoint temperatures (T_ms values). An increase in T_m is taken as direct evidence of conformational stabilization. A typical goal is to obtain at least a 5°C–15°C increase in the thermal melting temperature in the presence of compounds tested. Similar studies of aggregation/association behavior using light scattering or some other size-sensitive technique are also performed. The information from the conformational and size-sensitive methods is then used to select the optimal stabilizers. Accelerated stability studies of this type are frequently but by no means always predictive of macromolecule stability behavior under more moderate (e.g., storage) conditions [54]. The major unknown here is whether the mechanism of physical degradation probed by the accelerated stability studies is similar to that seen under more moderate environmental conditions. Hence, long-term stability studies at lower temperatures (e.g., 0°C–4°C) need to be performed after the optimization of final formulations with independent considerations of chemical stability (see discussion above) and confirmation of stabilization by relevant biological assays.

Versions of the high-throughput screening approach for the identification of stabilizers described above has been successfully used in a wide variety of applications including formulation studies of peptides, proteins, viruses, and virus-like particles as well as other vaccine antigens (Table 8.2) [3,24,25,31,40,43,55,56,58]. For example, the phase diagram generated for anthrax rPA (Fig. 8.6) suggested that a high-throughput screening assay could be developed at pH 5 at 37°C using one or more techniques to generate signals reflecting physical degradation. Aggregation was one of the more apparent pathways of physical degradation of this protein and was therefore selected for stabilization analysis [24]. Therefore, a turbidity assay was developed for the rPA to produce conditions that could achieve significant aggregation by monitoring the turbidity of the solution at 360 nm in 96-well microtiter plates. Experiments were performed both with and without potential excipients and the aggregation seen with protein alone was used as the standard. The maximum OD observed in control samples at 70 min was used as a measure of the maximum extent of aggregation since the signal change was complete by this time. Inhibition by various agents was then characterized by their ability to lower the maximum OD obtained. A number of potent inhibitors were identified by this method, with trehalose, sorbitol, mannitol, dextrose, and sodium citrate found to be most effective. A representative example of the aggregation kinetics of the anthrax rPA at 37°C both alone and in the presence of selected compounds demonstrating a range of effectiveness, is shown in Figure 8.9.

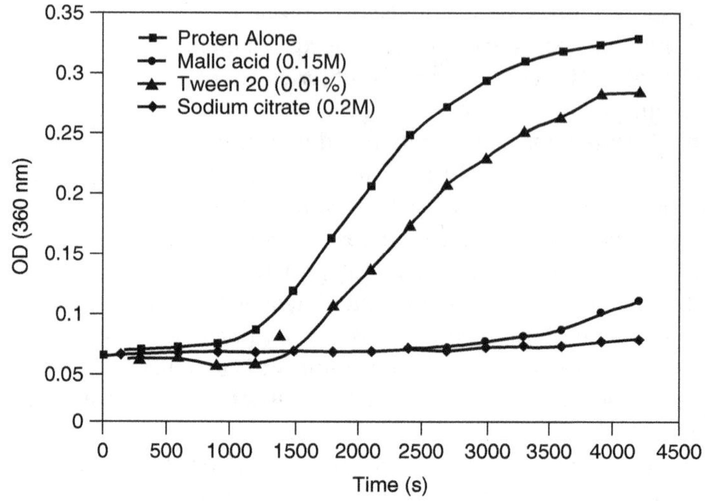

Figure 8.9. Aggregation of rPA in the presence and absence of excipients (0.15 M malic acid, 0.0 1% Tween 20, and 0.2% sodium citrate) [24].

This approach permitted efficient screening of a GRAS library and the rapid identification of potent stabilizers against aggregation. After optimization of the concentrations of selected stabilizers, studies were performed to ensure that the compounds also protected against conformational instability. To this end, thermal unfolding experiments in the presence of the potential stabilizers were performed using both intrinsic fluorescence and CD measurements. Representative data are shown in Figure 8.10, and the dramatic stabilizing effect of trehalose, one of the most effective inhibitors of aggregation, is clearly seen in terms of an increase in the thermal stability of rPA by almost 10°C over the unprotected protein. From the first measurement for construction of the EPDs to identification and optimization of stabilizers, the total time required for the general approach described here is typically less than 2 months and can often be accomplished in only a few weeks.

8.6. CONCLUSION

A critical step in the successful development of biopharmaceutical products is to find acceptable conditions for maintenance of stability. The less systematically based approaches that have typically been employed in the past for macromolecular stabilization are often both slow and produce less than optimal formulations. The "design of experiment" type of approach that is also sometimes employed suffers from significant holes in the experimental space examined but does offer the advantage that it can be very sparing in the use of sometimes scarce drug substance. The systematic approach to preformulation and formulation of biopharmaceuticals described here is both rapid and comprehensive and also provides a direct route into the screening of

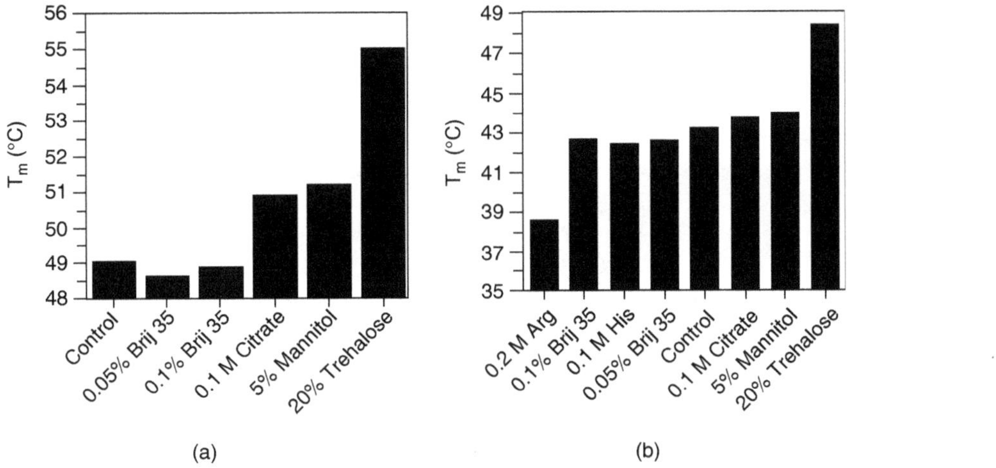

Figure 8.10. Increase in rPA thermal transition temperature in the presence of trehalose compared to several other excipients as determined by CD (a) and intrinsic Trp fluorescence (b). Note that the appearance of the fluorescence changes at much lower temperatures than do those detected by CD, which implies the existence of a molten globule-like state(s) between these two temperatures.

large libraries of potential stabilizers. It typically requires 3–20 mg of target macromolecules. Furthermore, this approach begins to move the formulation process from strategic experimentation to rational design. Although there is no limitation to the number of variables that may be included in the phase diagram because of the truncation procedure used, if more than three different types of measurements are used to describe a molecule's behavior, some information may be lost. Alternatively, other higher-dimensional display methods must be employed to portray the data. Another potential limitation to the EPD approach is the resolution between the colored phases. In some cases, the color differences between phases are not as distinctive as might be desired making the assignment of different phases difficult. Despite these limitations, the EPD approach appears to provide an intuitively attractive and powerful method to comprehensively characterize and display the solution behavior of biopharmaceuticals under various environmental conditions.

The EPD/high-throughput screening approach has been successfully used to characterize and formulate various biopharmaceuticals. To date this approach has been applied to a wide range of potential vaccine antigens and therapeutic agents, including peptides, proteins, and nucleic acids as well as macromolecular assemblies, including viruses, virus-like particles, bacteria, and toxins (Tables 8.1 and 8.2). With the availability of high-throughput instrumentation, the entire process of data acquisition, analysis, and interpretation described here can be accomplished in a relatively short period of time (weeks to a few months), potentially dramatically enhancing both the effectiveness and efficiency of the formulation development process.

TABLE 8.2. Optimal Stabilizers Identified for Selected Biopharmaceuticals Employing the EPD Approach[a]

Biopharmaceutical	Stabilizer(s)	Transition Temperature, °C				Reference
		Without Stabilizer(s)	With Stabilizer(s)	ΔT_m	Technique(s) Used to Measure T_m	
rPA of *Bacillus anthracis*	20% trehalose	49	55	6	CD	24
EBA-175 RII	20% sortibol	52	56	4	Intrinsic fluorescence	31
	20% sucrose	52	55	3		
Fibroblast growth factor 20 (FGF 20)	Heparin	55	67	12	DSC	3
Ricin toxin A chain	50% glycerol	50	66	16	Intrinsic fluorescence	25
Clostridium difficile toxoid A	20% sorbitol + 20% dextrose	60	67	7	Far-UV CD	56
C. difficile toxoid B	10% dextrose + 0.05% Tween 80	56	71	15	Far-UV CD	
IpaD from *Shigella flexneri*	10% trehalose + 5% dextrose	59	63	4	Intrinsic fluorescence	Unpublished data
SipD from *Salmonella* spp.	10% trehalose + 5% dextrose	57	63	6	Intrinsic fluorescence	Unpublished data
BipD from *Burkholderia pseudomallei*	10% trehalose + 5% dextrose	70	75	5	Intrinsic fluorescence	Unpublished data
PcrV from *Pseudomonas aeruginosa*	10% trehalose + 5% dextrose	60	65	5	Intrinsic fluorescence	Unpublished data

LcrV from *Yersinia* spp.	10% trehalose + 5% dextrose	75	78	3	Intrinsic fluorescence	Unpublished data
MxiH from *Shigella flexneri*	20% dextrose	41	47	6	Far UV-CD	Unpublished data
PrgI from *Salmonella typhimurium*	20% trehalose	41	45	4	Far UV-CD	Unpublished data
Respiratory syncytial virus	40% sucrose or trehalose	53	68	15	Far-UV CD	55
	40% sucrose or trehalose	55	65	10	Second-derivative UV	55
	40% trehalose	49	52	3	Laurdan fluorescence	55
	30% sucrose and 0.45 M arginine	50	70	20	Light scattering	55
Norwalk virus–like particles	20% sucrose	62(T_{m1}) 69(T_{m2})	64(T_{m1}) 71(T_{m2})	2(T_{m1}) 2(T_{m2})	DSC	58
Measles virus	10% mannitol	44	51	7	Laurdan fluorescence	40
Attenuated *V. cholerae*	0.15 M arginine or 0.075 M glutamic acid	32.5	55	2.5	extrinsic (*Bac*light kit) fluorescence	43

[a]Each measurement was conducted in triplicate and had ~ 1.0°C of uncertainty.

REFERENCES

1. Chang, B. S. and Hershenson, S. (2002), Practical approaches to protein formulation development, in *Rational Design of Stable Protein Formulations: Theory and Practice*, Carpenter, J. F. and Manning, M. C., eds., Kluwer Academic, New York, pp. 1–25.
2. Venyaminov, S. Y. and Yang, J. T. (1996), Determination of protein secondary structure, in *Circular Dichroism and the Conformational Analysis of Biomolecules*, Fasman, G. D., ed., Plenum Press, New York, pp. 69–107.
3. Fan, H., Vitharana, S. N., Chen, T., O'Keefe, D., and Middaugh, C. R. (2007), Effects of pH and polyanions on the thermal stability of fibroblast growth factor 20, *Mol. Pharm.* **4**: 232–240.
4. Ausar, S. A., Joshi, S. B., and Middaugh, C. R. (2008) Spectroscopic methods for the physical characterization and formulation of nonviral gene delivery systems, in *Methods in Molecular Biology: Gene Therapy Protocols*, (Le Doux, J. M. ed.) Humana Press, Totowa, NJ. pp. 55–80.
5. Braun, C. S., Kueltzo, L. A., and Middaugh, C. R. (2001), Ultraviolet absorption and circular dichroism spectroscopy of nonviral gene delivery complexes, in *Nonviral Vectors for Gene Therapy: Methods and Protocols*, Findeis, M. A., ed., Humana Press, Totowa, NJ, pp. 253–284.
6. Jiskoot, W., Visser, A. J. W. G., Herron, J. N., and Sutter, M. (2005), Fluorescence spectroscopy, in *Methods for Structural Analysis of Protein Pharmaceuticals*, Jiskoot, W. and Crommelin, D., eds., American Association of Pharmaceutical Scientists (AAPS), pp. 27–82.
7. Stryer, L. (1965), The interaction of a naphthalene dye with apomyoglobin and apohemoglobin. A fluorescent probe of non-polar binding sites, *J. Mol. Biol.* **13**: 482–495.
8. Cardamone, M. and Puri, N. K. (1992), Spectrofluorimetric assessment of the surface hydrophobicity of proteins, *Biochem. J.* **282**(Pt. 2): 589–593.
9. Ali, V., Prakash, K., Kulkarni, S., Ahmad, A., Madhusudan, K. P., and Bhakuni, V. (1999), 8-anilino-1-naphthalene sulfonic acid (ANS) induces folding of acid unfolded cytochrome c to molten globule state as a result of electrostatic interactions, *Biochemistry* **38**: 13635–13642.
10. Fan, H., Kashi, R. S., and Middaugh, C. R. (2006), Conformational lability of two molecular chaperones Hsc70 and GP96: Effects of pH and temperature, *Arch. Biochem. Biophys.* **447**: 34–45.
11. Matulis, D. and Lovrien, R. (1998), 1-Anilino-8-naphthalene sulfonate anion-protein binding depends primarily on ion pair formation, *Biophys. J.* **74**: 422–429.
12. Klunk, W. E., Pettegrew, J. W., and Abraham, D. J. (1989), Quantitative evaluation of congo red binding to amyloid-like proteins with a beta-pleated sheet conformation, *J. Histochem. Cytochem.* **37**: 1273–1281.
13. Guntern, R., Bouras, C., Hof, P. R., and Vallet, P. G. (1992), An improved thioflavine S method for staining neurofibrillary tangles and senile plaques in Alzheimer's disease, *Experientia* **48**: 8–10.
14. LeVine, H., 3rd. (1993), Thioflavine T interaction with synthetic Alzheimer's disease beta-amyloid peptides: Detection of amyloid aggregation in solution, *Protein Sci.* **2**: 404–410.
15. LeVine, H., 3rd. (1999), Quantification of beta-sheet amyloid fibril structures with thioflavin T, *Meth. Enzymol.* **309**: 274–284.

16. Vassar, P. S. and Culling, C. F. (1959), Fluorescent stains, with special reference to amyloid and connective tissues, *Arch. Pathol*. **68**: 487–498.
17. Alonso, L. G., Garcia-Alai, M. M., Smal, C., Centeno, J. M., Iacono, R., Castano, E., Gualfetti, P., and de Prat-Gay, G. (2004), The HPV16 E7 viral oncoprotein self-assembles into defined spherical oligomers, *Biochemistry* **43**: 3310–3317.
18. Hatters, D. M., MacPhee, C. E., Lawrence, L. J., Sawyer, W. H., and Howlett, G. J. (2000), Human apolipoprotein C-II forms twisted amyloid ribbons and closed loops, *Biochemistry* **39**: 8276–8283.
19. Wiethoff, C. M. and Middaugh, C. R. (2001), Light-scattering techniques for characterization of synthetic gene therapy vectors, in *Nonviral Vectors for Gene Therapy: Methods and Protocols*, Findeis, M. A., ed., Humana Press, Totowa, NJ, pp. 349–376.
20. Freire, E. (1995), Differential scanning calorimetry, *Meth. Mol. Biol*. **40**: 191–218.
21. Kueltzo, L. A., Ersoy, B., Darrington, T., Ralston, J. P., and Middaugh, C. R. (2003), Derivative absorbance spectroscopy and protein phase diagrams as tools for comprehensive protein characterization: A bGCSF case study, *J. Pharm. Sci*. **92**: 1805–1820.
22. Kueltzo, L. A. and Middaugh, C. R. (2003), Structural characterization of bovine granulocyte colony stimulating factor: Effect of termperature and pH, *J. Pharm. Sci*. **92**: 1793–1804.
23. Fan, H., Ralston, J., DiBase, M., Faulkner, E., and Middaugh, C. R. (2005), Solution behavior of IFN-ß-1a: An empirical phase diagram based approach, *J. Pharm. Sci*. **94**: 1893–1911.
24. Jiang, G., Joshi, S. B., Peek, L. J., Brandau, D. T., Huang, C. R., Ferriter, M. S., Woodley, W. D., Ford, B. M., Mar, K. D., Mikszta, J. A., Hwang, C. R., Ulrich, R., Harvey, N. G., Middaugh, C. R., and Sullivan, V. J. (2006), Anthrax vaccine powder formulations for nasal mucosal delivery, *J. Pharm. Sci*. **95**: 80–96.
25. Peek, L. J., Brey, R. N., and Middaugh, C. R. (2007). A rapid, three-step process for the preformulation of a recombinant ricin toxin A-chain vaccine, *J. Pharm. Sci*. **96**: 44–60.
26. Fan, H., Li, H., Zhang, M., and Middaugh, C. R. (2007), Effects of solutes on empirical phase diagrams of human fibroblast growth factor 1, *J. Pharm. Sci*. **96**: 1490–1503.
27. Brandau, D. T., Joshi, S. B., Smalter, A. M., Kim, S., Steadman, B., and Middaugh, C. R. (2007), Stability of the Clostridium botulinum type A neurotoxin complex: An empirical phase diagram based approach, *Mol. Pharm*. **4**: 571–582.
28. Ausar, S. F., Rexroad, J., Frolov, V. G., Look, J. L., Konar, N., and Middaugh, C. R. (2005), Analysis of the thermal and pH stability of human respiratory syncytial virus, *Mol. Pharm*. **2**: 491–499.
29. Ausar, S. F., Foubert, T. R., Hudson, M. H., Vedvick, T. S., and C.R., M. (2006), Conformational stability and disassembly of Norwalk virus like particles: Effect of pH and temperature, *J. Biol. Chem*. **281**: 19478–19488.
30. Harn, N., Allan, C., Oliver, C., and Middaugh, C. R. (2007), Highly concentrated monoclonal antibodies: Direct analysis of structure and stability, *J. Pharm. Sci*. **96**: 532–546.
31. Peek, L. J., Brandau, D. T., Jones, L. S., Joshi, S. B., and Middaugh, C. R. (2006), A systematic approach to stabilizing EBA-175 RII-NG for use as a malaria vaccine, *Vaccine* **24**: 5839–5851.
32. Rexroad, J., Martin, T. T., McNeilly, D., Godwin, S., and Middaugh, C. R. (2006), Thermal stability of adenovirus type 2 as a function of pH, *J. Pharm. Sci*. **95**: 1469–1479.

33. Ruponen, M., Braun, C. S., and C.R., M. (2006), Biophysical characterization of polymeric and liposomal gene delivery systems using empirical phase diagrams, *J. Pharm. Sci.* **95**: 2101–2114.

34. Thyagrajapuram, N., Olsen, D., and C.R.,M. (2007), The structure stability and complex behavior of recombinant human gelatins, *J. Pharm. Sci.* **96**: 3363–3378.

35. Nonoyama, A., Laurence, J. S., Garriques, L., Qi, H., Le, T., and Middaugh, C. R. (2008), A biophysical characterization of the peptide drug pramlintide (AC137) using empirical phase diagrams, *J. Pharm. Sci.*, **97**: 2552–2567.

36. Salnikova, M. S., Joshi, S. B., Rytting, J. H., Warny, M., and Middaugh, C. R. (2008), Physical characterization of Clostridium difficile toxins and toxoids: Effect of the formaldehyde crosslinking on thermal stability, *J. Pharm. Sci.* **97**: 3735–3752.

37. Rexroad, J., Evans, R. K., and Middaugh, C. R. (2006), Effect of pH and ionic strength on the physical stability of adenovirus type 5, *J. Pharm. Sci.* **95**: 237–247.

38. Barrett, B. S., Picking, W. L., Picking, W. D. and Middaugh, C. R. (2008), The response of type three secretion system needle proteins MxiHDelta5, BsaLDelta5, and PrgIDelta5 to temperature and pH, *Proteins* **73**(3): 632–643.

39. He, F., Joshi, S. B., Moore, D. S., Shinogle, H. E., Ohtake, S., Lechuga-Ballesteros, D., Martin, R. A., Truong-Le, V. L. and Middaugh, C. R. (2010), Using spectroscopic and microscopic methods to probe the structural stability of human adenovirus type 4, *Human Vaccines* **6**(2): 204–211.

40. Kissmann, J., Ausar, S. F., Rudolph, A., Braun, C., Cape, S. P., Sievers, R. E., Federspiel, M. J., Joshi, S. B., and Middaugh, C. R. (2008), Stabilization of measles virus for vaccine formulation, *Human Vaccines* **4**(5): 350–359.

41. Esfandiary, R., Yee, L., Ohtake, S., Martin, R. A., Truong-Le, V. L., Laurence, J. S., Lechuga-Ballesteros, D., Moore, D. S., Joshi, S. B., and Middaugh, C. R. (2010), Biophysical Characterization of Rotavirus Serotypes G1, G3, and G4, *Human Vaccines* **6**(5): (in press)

42. Esfandiary, R., Kickhoefer, V. A., Rome, L. H., Joshi, S. B., and Middaugh, C. R. (2009), Structural stability of vault particle, *J. Pharm. Sci.* **98**(4): 1376–1386.

43. Zeng, Y., Fan, H., Chiueh, G., Pham, B., Martin, R., Lechuga-Ballesteros, D., Truong, V. L., Joshi, S. B., and Middaugh, C. R. (2009), Towards development of stable formulations of a live attenuated bacterial vaccine: a preformulation study facilitated by a biophysical approach. *Human Vaccinces* **5**(5): 322–331.

44. Shire, S. J., Shahrokh, Z., and Liu, J. (2004), Challenges in the development of high protein concentration formulations, *J. Pharm. Sci.* **93**: 1390–1402.

45. Zimmer, C., Luck, G., and Venner, H. (1968), Studies on the conformation of protonated DNA, *Biopolymers* **6**: 563–564.

46. Kamerzell, T. J. and Middaugh, C. R. (2007), Two-dimensional correlation spectroscopy reveals coupled immunoglobulin regions of differential flexibility that influence stability, *Biochemistry* **46**: 9762–9773.

47. Miller, D. W. and Dill, K. A. (1995), A statistical mechanical model for hydrogen exchange in globular proteins, *Protein Sci.* **4**: 1860–1873.

48. Woodward, C. K. and Hilton, B. D. (1979), Hydrogen exchange kinetics and internal motions in proteins and nucleic acids, *Annu. Rev. Biophys. Bioeng.* **8**: 99–127.

49. Joshi, S. B., Kamerzell, T. J., McNown, C., and Middaugh, C. R. (2008), The interaction of heparin/polyanions with bovine, porcine, and human growth hormone, *J. Pharm. Sci.* **97**: 1368–1385.

50. Eftink, M. and Ghiron, C. (1975), Dynamics of a protein matrix revealed by fluorescence quenching, *Proc. Natl. Acad. Sci. USA* **72**: 3290–3294.

51. Kamerzell, T. J., Unruh, J. R., Johnson, C. K., and Middaugh, C. R. (2006), Conformational flexibility, hydration and state parameter fluctuations of fibroblast growth factor-10:effects of ligand binding, *Biochemistry* **45**: 15288–15300.

52. Ramsey, J. D., Gill, M. L., Kamerzell, T. J., Price, E. S., Joshi, S. B., Bishop, S. M., Oliver, C. N., and Middaugh, C. R. (2009), Using empirical phase diagrams to understand the role of intramolecular dynamics in immunoglobulin G stability, *J. Pharm. Sci.* **98**(7): 2432–2447.

53. Maclean, D. S., Qian, Q., and Middaugh, C. R. (2002), Stabilization of proteins by low molecular weight multi-ions, *J. Pharm. Sci*. **91**: 2220–2229.

54. Yoshioka, S. and Stella, V. J. (2000), *Stability of Drugs and Dosage Forms*, Kluwer Academic/Plenum Publishers, New York, pp. 187–199.

55. Ausar, S. F., Espina, M., Brock, J., Thyagarayapuran, N., Repetto, R., Khandke, L., and Middaugh, C. R. (2007), High-throughput screening of stabilizers for respiratory syncytial virus: Identification of stabilizers and their effects on the conformational thermostability of viral particles, *Human Vaccines* **3**: 94–103.

56. Salnikova, M. S., Joshi, S. B., Rytting, J. H., Warny, M., and Middaugh, C. R. (2008), Preformulation studies of *Clostridium difficile* toxoids A and B, *J. Pharm. Sci*. **97**: 4194–4207.

57. Markham, A., Birket, S., Picking, W. D., Picking, W. L., and Middaugh, C. R. (2008), pH sensitivity of type III secretion system tip proteins, in *Proteins: Structure, Function and Bioinformatics*, **71**: 1830–1842.

58. Kissmann, J., Ausar, S. F., Foubert, T. R., Brock, J., Switzer, M. H., Detzi, E. J., Vedvick, T. S., and Middaugh, C. R. (2008), Physical stabilization of Norwalk virus-like particles, *J. Pharm. Sci*. **97**: 4208–4218.

9

FLUORESCENCE AND PHOSPHORESCENCE METHODS TO PROBE PROTEIN STRUCTURE AND STABILITY IN ICE: THE CASE OF AZURIN

Giovanni B. Strambini

9.1. INTRODUCTION

Freezing of protein solutions may result in irreversible protein aggregation and severe loss of catalytic activity of enzymes, reasons for which many proteins cannot be stored in ice or lyophilized from it without partial inhibition of their function [1]. Notwithstanding the relevance of the phenomenon and the commercial importance for the growing polypeptide-based pharmaceutical industry, little is known on the structure and thermodynamic stability of proteins in ice. To date, the hypotheses advanced on the freeze damage mechanism are based on indirect evidence and are construed mainly around the presumed action of cryoprotectants, compounds that at relatively high concentrations prevent these alterations.

The paucity of structural information on proteins in the frozen state is due primarily to the insensitivity or the poor resolution of ordinary spectroscopic methods in a highly scattering and anisotropic medium as ice. For example, fourier transform infrared (FTIR) spectroscopy experiments, when conducted on freeze-labile proteins, have not detected significant changes in the secondary structure of the native fold [2]

Formulation and Process Development Strategies for Manufacturing Biopharmaceuticals, edited by Jameel and Hershenson

presumably because they are carried out at relatively large protein concentrations, a condition in which the perturbation is often attenuated as if the protein itself acted as a stabilizer. For the same reason, little is known on the effect of freezing on the thermodynamic stability of proteins. Indeed, formulations and protocols for the stabilization of pharmacoproteins during industrial freeze-drying processes are still largely worked out empirically, case by case, drawing mostly on long-time experience and guided by thermodynamic data pertaining to stabilizing additives in liquid solutions [3–5].

The more recent introduction of fluorescence and phosphorescence techniques capable of probing the protein tertiary structure and the thermodynamic stability of the native fold in ice holds promise for new insights on the nature of the ice perturbation as well as on the cryoprotection mechanisms of various additives. This chapter reviews recent (as of 2009) results obtained by these techniques on a model protein, azurin (and some appositely prepared mutants), on the effect of freezing on the integrity of the native fold structure and on its thermodynamic stability (ΔG°). Perturbations of the tertiary structure are probed directly both by alterations in the lifetime of tryptophan (Trp) phosphorescence emission and from freeze-induced binding of the fluorescence probe ANS (1-anilino-8-aminonaphthalene), whereas ΔG° is derived from guanidinium chloride denaturation of the protein, monitored by the change in Trp fluorescence spectrum and yield.

Azurin from *Pseudomonas aeruginosa* is a small (14-kDa), monomeric copper-binding protein well characterized in terms of its structure [6,7] and thermodynamic stability [8,9]. Other suitable features of azurin are:

1. The macromolecule has a single Trp residue (W48) located within the rigid inner core of the globular structure. Copper-free azurin exhibits the most blueshifted Trp fluorescence spectrum known in proteins [10] and intense longlived phosphorescence even at ambient temperature [11].
2. Residue W48 is wrapped inside a tight β-barrel motif, and its exposure to the solvent requires no less than global unfolding of the polypeptide as confirmed by the superposition of fluorescence and circular dichroism (CD) denaturation profiles [9].
3. The denaturation of Cu-free azurin by guanidinium chloride (Gdn) is a fully reversible process and the equilibrium is well represented by a two- state model [9,12].
4. Several mutants have been constructed of constant fold but varying plasticity or refolding kinetics useful in correlating molecular structure to thermodynamics.

9.2. TRYPTOPHAN PHOSPHORESCENCE AND ICE-INDUCED PARTIAL UNFOLDING

In aqueous solutions at ambient temperature the phosphorescence lifetime τ of tryptophan residues in globular proteins can span a wide range, from seconds to microseconds [13]. The main contributions to the shortening of τ are radiationless transitions from the excited triplet state, determined by the mobility of the protein–solvent microenvironment [14,15] plus the onset of intra-and intermolecular quenching reactions by specific side chains and trace impurities in the solvent. The outcome is that the phosphorescence lifetime is a sensitive probe of the protein structure surrounding the chromophore reporting on the flexibility of the site, the proximity of indole to quenching side chains or prosthetic groups, and on accessibility of the site to small solutes in the solvent. These features have proved to exhibit an unparallel sensitivity to the conformational state of the macromolecule, with the ability of τ to monitor even subtle alterations of the native fold often undetected by ordinary spectroscopic techniques. Examples are the relatively large variations in τ often induced by ligand-binding (metal ions [16–19], substrates, and cofactors [18,19]) subunit association [20,21], the addition of stabilizing [21] or denaturing [22] agents and cosolvents [23], and dehydration [24,25], besides variations in temperature and pressure [25]. Besides the unique sensitivity of τ to the dynamical structure, another feature of phosphorescence is that it is a delayed emission and as such is easily separated in time from excitation. The advantage of a delayed emission is that scattering of direct excitation, which is very large in ice and often precludes isolation of prompt fluorescence from the spurious background, does not interfere with phosphorescence measurements.

9.2.1. Effect of Ice Formation on Trp Phosphorescence Lifetime

Phosphorescence experiments in ice are typically carried out with 150-μL samples, at protein concentrations in the 1–10 μM range, within airtight cylindrical spectrosil cuvettes (4 mm ID) from which oxygen is completely removed prior to freezing [26]. Figure 9.1 shows that while the phosphorescence spectrum is practically indistinguishable between liquid and partly-frozen solutions, τ is markedly affected by the solidification of water. It is instructive first to consider how τ relates to temperature-induced variations of protein flexibility in solutions that do not undergo phase transitions. A typical example of a τ-versus-temperature profile is given in Figure 9.2, where the decay kinetics of azurin in a nonfreezing solvent (ethylene glycol buffer, 50/50, w/w) is compared with that in buffer. At 170 K the organic solvent mixture forms a rigid glass; protein motions are blocked by the high viscosity, and τ assumes its upper limiting value of 6 s characteristic of rigid matrices. When the solution is warmed above the glass transition temperature of the solvent, τ decreases progressively until, on approaching the temperature of thermal denaturation, it suddenly drops to below the 0.01 ms detection limit. While the gradual reduction of τ reflects a progressive gain in the macromolecule flexibility as a result of the decreased solvent viscosity and increased thermal energy, the steep high-temperature drop (not shown in the figure) signals the abrupt enhancement of segmental mobility that accompanies the unfolding of the globular structure. In contrast to the smooth τ-versus-temperature profile obtained in a nonfreezing organic solvent mixture in buffer τ exhibits an additional and sudden downward transition in correspondence of ice formation. At -12°C, as the supercooled solution solidifies, the average τ drops by about 30-fold, a clear indication of a dramatic gain in flexibility of the protein core. Besides a drastic reduction in τ, another salient feature of the phosphorescence emission in frozen samples is the emergence of pronounced lifetime heterogeneity, a sign that the protein conformation is no longer uniform in ice. The abovementioned phenomena appear to be quite general to proteins in ice, as it was observed at various degrees with just about all the proteins examined to date. Because the solidification of water is accompanied by an increase in viscosity of the unfrozen liquid (liquidus) and protein motions are coupled to solvent viscosity [23,27], a drastic increase in protein flexibility on freezing is totally unexpected. Thus, we deduce that the decrease in τ ice attests to a general and substantial loosening of the native fold, a partial unfolding of the polypeptide that involves even the rigid domain hosting the phosphorescence probe. From the multiphase phosphorescence decay we infer that in ice τ assumes a broad range of values. The distribution narrows down on approaching the melting temperature of ice, where an increasing fraction of the protein ensemble exhibits an unperturbed lifetime identical to that in supercooled liquid at the same temperature. In structural terms, this τ heterogeneity implies that the perturbation of protein structure induced by the solidification of water is not unique and that the two-phase system provides for a multitude of protein environments, some more stressful than others. Finally, it is observed that the strain on protein structure produced by freezing is, by and large, promptly relieved on thawing. For azurin, and other proteins, τ is readily reestablished to the value of the liquid phase and the freeze–thaw cycle can be repeated several times with only minor variations in spectroscopic properties. However, for enzymatic proteins such as lactic dehydrogenase that are subject to irreversible damage of enzymatic activity [28], the phosphorescence properties are not fully restored.

Among the hypotheses concerning the cause of freeze damage is the instability of the folded state at low temperature (subunit dissociation and cold denaturation), the low activity and peculiar properties of interstitial water (drying effect), changes in pH of the liquidus and state of ionization of amino acid side chains, and finally, protein–protein and protein–solute interactions elicited by the large freeze-induced concentration of solutes. Phosphorescence data argue that none of these factors seem to affect the integrity of azurin structure in ice. The long lifetime in supercooled solutions τ_0 proves that low temperature per se is not a destabilizing element. A drying effect is also readily discounted. The water of hydration crystallizes only in part and at much lower temperature ($T < 140$ K) [29]. Moreover, protein dehydration invariably yields enhanced structural rigidity, as evidenced by the large value of τ for fully desiccated protein powders [24]. Similarly, pH effects can also be ruled out. Between pH 4 and 10 the decrease in τ is largely independent of the starting solution's pH and of the buffering salt. Likewise, the interaction among protein molecules, which normally manifests itself in a concentration dependence of the phenomenon, cannot be responsible for the partial unfolding in ice as the lifetime varied little between 0.1 and 10 μM, whereas at even higher concentrations there is a distinct attenuation of the decrease in τ. The same conclusion is reached with respect to possible interactions with buffering salts or NaCl in the unfrozen liquid as τ_0 is only moderately sensitive to molar quantities of these solutes.

We are left to consider protein–ice interactions, which may take the form of an adsorption process, in the fashion of antifreeze proteins [30] or even of physical distortion of the protein globule by anisotropic compression in the intergranular space

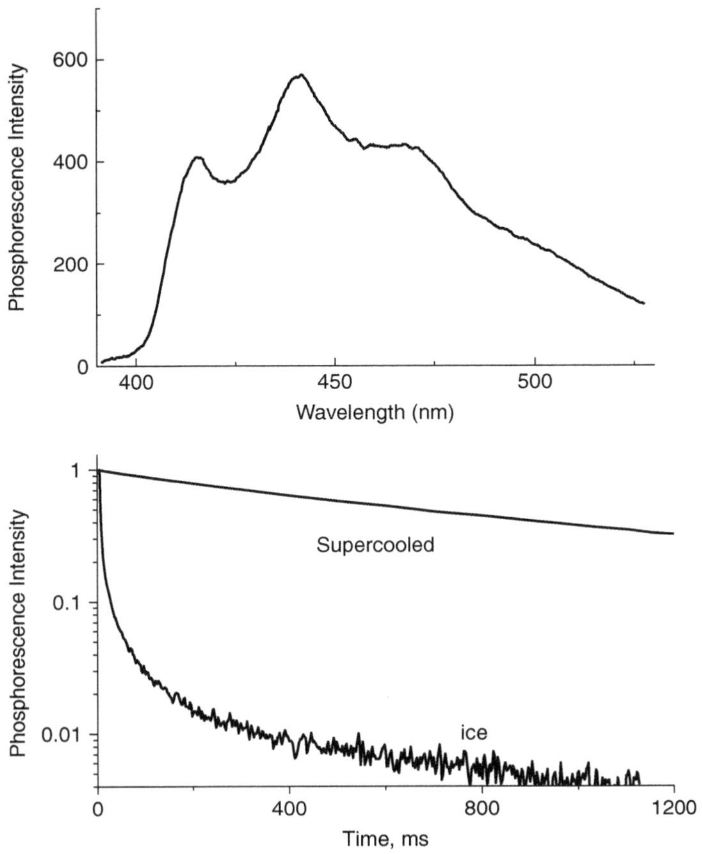

Figure 9.1. Effect of freezing on the phosphorescence spectrum and on the decay kinetics of azurin (5 μM, in 2 mM sodium phosphate buffer, pH 7) at −12°C. The spectrum in supercooled solution and in ice is superimposable.

of ice crystals [31]. In the first case the protein will distribute between the liquid and the interface to the solid phase and the extent of adsorption will be determined by the volume fraction of liquid water V_L and the surface area of ice. In the second case the perturbation is expected to be important only when the average size of intergranular spaces is comparable with or smaller than the protein molecular volume. Thus, whereas adsorption would manifest itself as a gradual process, purely mechanical effects of ice formation would exhibit a more-or-less sharp threshold at some specific V_L, a threshold governed by the molecular weight of the protein.

The importance of V_L on ice-induced perturbation was confirmed by the progressive drop in τ as the liquid fraction of water was reduced either by lowering the freezing temperature or decreasing the salt concentration, NaCl or KC1, in the starting liquid solution. The meaningful temperature range for phosphorescence measurements is between 0°C and −20°C; at colder temperatures the liquidus viscosity

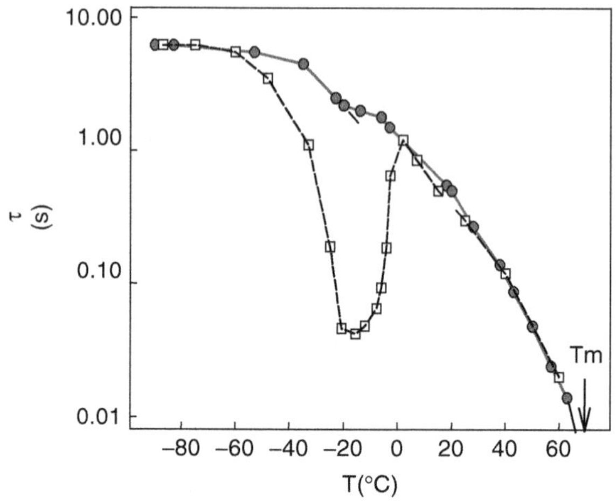

Figure 9.2. Dependence of the average phosphorescence lifetime τ of Trp48 of azurin on temperature and solvent composition: (•) 50/50 (w/w) ethylene glycol/potassium phosphate buffer (1 mM, pH 7) and (□) potassium phosphate buffer alone; T_m is the temperature of thermal denaturation. (Reprinted from Strambini and Gabellieri [26]).

increases rapidly and affects τ independently of how well structured the protein fold is (see Fig. 9.2). Figure 9.3a displays the average τ of azurin in ice as a function of the freezing temperature, for selected salt concentrations. Overall the results point out the dominant role of V_L on ice perturbation:

1. Independently of the salt concentration, τ decreases steeply on lowering the freezing temperature, and hence V_L. Even more significant is the reduction in τ on decreasing the salt concentration at constant temperature (data along vertical solid lines in Fig. 9.3a). Under these conditions what varies among the samples is only V_L as both the liquidus composition and temperature remain constant.
2. There is a net difference between the τ–T profiles of concentrated equimolar KC1 and NaCl solutions, whose eutectic temperatures are, respectively, −11°C and −23°C. Below −11°C KC1 crystallizes out and ice forms until V_L is reduced to the value of dilute solutions. The τ profile in KCl is analogous to that in NaCl down to −10°C, but in correspondence to the eutectic temperature (−11°C), τ drops abruptly to the value of dilute solutions.

When τ data are plotted against V_L (Fig. 9.3b), we see that irrespective of the temperature–salt concentration combination adopted, the variation of τ of azurin is a smooth function of V_L. There is no step-like threshold to support a perturbation mechanism on the basis of mechanical distortions of the protein globule. Moreover, other evidence based on the morphology of ice is consistent with protein adsorption to the solid phase. The extent of adsorption depends on the total surface area of ice in

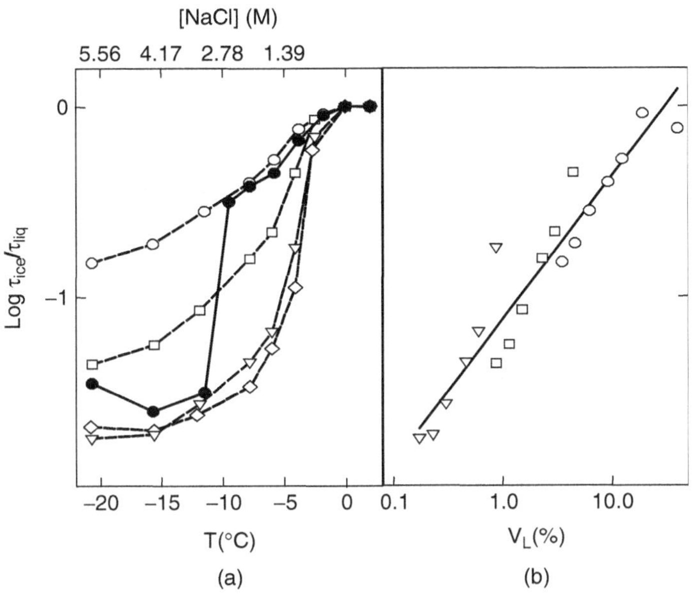

Figure 9.3. (a) Phosphorescence lifetime ($\tau/\tau_0 = \tau_{ice}/\tau_{supercooled\ solution}$)–temperature profiles of azurin in 1 mM sodium phosphate buffer, pH 7 (◇), and in the presence of NaCl or KCl: (▽)10 mM NaCl, (□) 50 mM NaCl, (○) 200 mM NaCl, and (•) 200 mM KCl. Indicated in the upper axis is the equilibrium concentration of NaCl in the liquidus calculated for an ideal solution. (b) Lifetime data from the three NaCl profiles of (a) plotted against V_L calculated assuming the solution to be ideal [$V_L = 3.6[\text{NaCl}]_{sol}/(273\text{-}T)$]. (Reprinted from Strambini and Gabellieri [26].)

contact with the liquidus, and for any V_L it depends on the size of ice crystals. Large crystals, and small ice surface area, are obtained by slow cooling and prolonged aging. Frozen protein solutions seeded with a small ice crystal at −2°C and then cooled at rates of −200°C or 1°C/min gave substantially different lifetimes, and larger τ values were associated with slower cooling rates. Likewise, rapidly cooled samples left to anneal for several hours showed a significant, up to twofold, recovery of the ratio τ/τ_0. In either test, rate of cooling or annealing, a smaller perturbation of protein structure is associated with the reduction in the ice surface area, again a correlation between τ and ice morphology that is consistent with a mechanism based on protein adsorption.

9.2.2. Effects of Cryoprotectants on Ice-Induced Perturbation

The findings discussed above suggest that the reasons for destabilization of the native fold in the frozen state must be sought in the direct interaction of the polypeptide with the solid phase. Sugars and polyols are substances that even in moderate concentrations afford a large degree of protection against freeze damage [28]. The influence of these cryoprotectants on the lifetime response to freezing suggests that there is a direct link

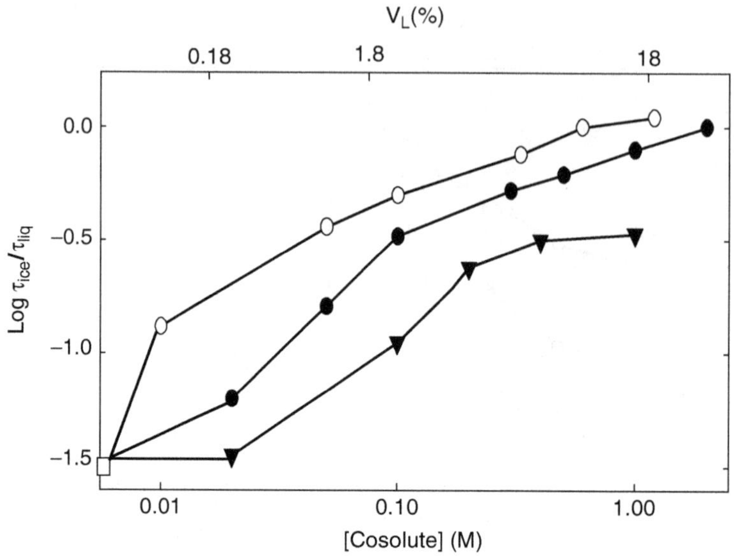

Figure 9.4. Effect of cosolutes on the lifetime ratio τ/τ_0 of azurin (1 mM sodium phosphate buffer, pH 7) at -10°C. The solutes are (○) sucrose, (•) glycerol, and (▼) NaCl. (Reprinted from Strambini and Gabellieri [26].)

between the loosening of the native structure revealed by τ and the onset of degenerative processes that lead to enzyme inactivation–denaturation. Figure 9.4 shows the influence of glycerol and sucrose, compounds commonly employed as cryoprotectants, on the lifetime ratio τ/τ_0 of azurin, at -10°C. The results are compared with those obtained with equimolar solutions of a "neutral" salt, NaCl, to account for the colligative effects (V_L) of the solutes. Although in liquid solutions molar concentrations of sucrose and glycerol barely affect the internal flexibility of azurin (τ_0), in ice even moderate quantities of them display a remarkable ability to attenuate the reduction in τ, particularly in the case of sucrose, which compared to glycerol is also a more potent cryoprotectant. This correlation between the attenuation of the ice-induced unfolding transition and the cryoprotective potential of the solute suggests a close connection between the structural alterations revealed by the phosphorescence probe and those implicated in freeze damage. In conclusion, the intrinsic phosphorescence emission of azurin demonstrates that the solidification of water alters the native fold deep down to the inner core of the macromolecule, the generalized increase in flexibility reflecting partial unfolding of the polypeptide. The results of various tests suggest that the dominant perturbation originates from a direct interaction of the macromolecule with ice. Structural studies of proteins adsorbed on various substrates such as metal oxides, silica, and charged polystyrene [32] have demonstrated an even more severe loss of secondary structure and a profound destabilization with respect to thermal unfolding. To date, the response of azurin to freezing was found to be quite general to the other proteins examined.

9.3. PERTURBATION OF AZURIN TERTIARY STRUCTURE IN FROZEN SOLUTIONS REVEALED BY ANS FLUORESCENCE

Tryptophan phosphorescence provides direct evidence that the solidification of water causes important alterations of the native globular fold, a change in structure that mimics the formation of molten globule states. However, structural dynamics affords only indirect information on the structure of the macromolecule and, further, is limited to specific sites of the protein interior, in that Trp phosphorescence probes exclusively the immediate environment of the chromophore. The aromatic chromophore 1-anilino-8-naphthalene sulfonate (ANS) is feebly fluorescent in water, but its spectrum is blueshifted and its intensity is dramatically increased in nonpolar solvents or when the chromophore binds to nonpolar sites of proteins [33,34]. Since the early 1990s, strong binding of ANS to molten globule states of proteins has been linked to the loss of tertiary structure [35], and since then the method has been applied to characterize transient states in protein denaturation [36–38]. Recently, studies conducted in our laboratory examined the possibility to unveil potential alterations of the protein tertiary structure in ice from binding of ANS. Here we summarize the results obtained with azurin [39].

9.3.1. Ice-Induced ANS Binding to Apoazurin

Figure 9.5 shows the change in ANS fluorescence intensity induced by freezing a solution containing 75 μM ANS and 2.5 μM azurin, at -13°C. Whereas ANS alone is weakly emitting in both liquid and frozen samples, in ice, the protein enhances the fluorescence intensity by sevenfold and shifts the spectrum 13 nm to the blue. This indicates that, in ice, ANS has accessibility to internal hydrophobic protein sites apparently not available when the protein is in solution. Before concluding that the creation of ANS binding sites is owed to ice-induced partial disruption of the protein tertiary structure other possibilities, peculiar to the frozen medium, must also be taken into account. One is the occupation of very weak binding sites reachable in ice, owing to the ~120-fold freeze-induced concentration of ANS ($[\text{ANS}]_{\text{ice}} = 9$ mM) in the liquidus; another is aspecific association of the probe to the polypeptide. Both these alternative explanations were ruled out. First, in liquid supercooled solutions (-13°C) no fluorescence enhancement was observed that could be attributed to the occupation of binding sites of mM affinity. Next, the fluorescence enhancement is totally abolished on azurin denaturation in the presence of 45 mM guanidine hydrochloride. We recall that the formation of ANS binding sites in a protein is generally associated with the loss of its tertiary structure and that these sites are destroyed on major denaturation of the globular fold. Further, freezing experiments conducted with Cd-azurin, which has the same fold of apoazurin but is more stable by > 3 kcal/mol, show that practically no ANS fluorescence enhancement is observed with the holoprotein. The net contrast in ANS binding capacity between apo and holo forms does suggest that the disruption of the tertiary structure in ice, inferred from ANS binding, is to be ascribed to the greater structural plasticity of the apo form. Certainly, the lack of fluorescence enhancement in the case of Cd-azurin rules out the possibility that the effects observed with apoazurin be ascribed to a superficial, aspecific association of ANS to the polypeptide.

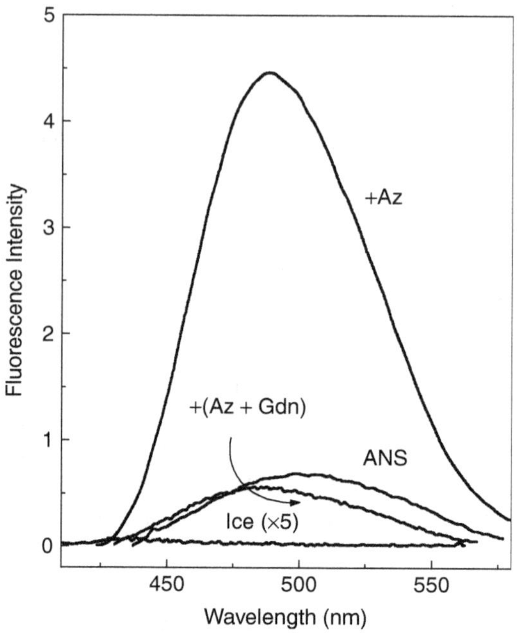

Figure 9.5. Enhancement of ANS (75 μM) fluorescence emission in ice (2 mM sodium phosphate, pH 7.5, plus 25 mM NaCl) at −13°C by the addition of azurin (2.5 μM). The lack of azurin enhancement when NaCl is replaced by 43 mM guanidinium hydrochloride (Gdn) and the low background luminescence from ice are also shown; $\lambda_{ex} = 350$ nm. (Reprinted from Gabellieri and Strambini [39].)

The increment in ANS fluorescence in ice caused by 2.5 μM Az, at −13°C, is shown in Figure 9.6 as a function of the starting ANS concentration in solution (10–200 μM). The intensity profile is reminiscent of a binding curve and shows that the enhancement tends to saturate above 75 μM, as at larger concentrations the intensity change is analogous to that of the protein free control. However, unlike a simple hyperbolic binding curve, the fluorescence increases more gradually in the low concentration range, conferring to the profile a slightly sigmoidal shape. The profile recalls cooperative binding reactions but, of course, multiple binding sites and a broad distribution of protein conformations in ice, inferred from the distribution of phosphorescence lifetimes, makes for ample variability in the affinity to ANS within the protein population. For this reason fluorescence profiles such as those of Figure 9.6 can at best provide a rough indication of the thermodynamic parameters governing ANS–protein complex. Their main utility is to compare relative changes in the ANS binding propensity brought about by varying experimental conditions such as the addition of solutes, freezing protocols, and temperature.

Independent support for the formation of protein–ANS complex is provided by the observation of fluorescence energy transfer from Trp (W48) to ANS. When the sample is excited at 280 nm, where both Trp and ANS absorb, the fluorescence intensity of

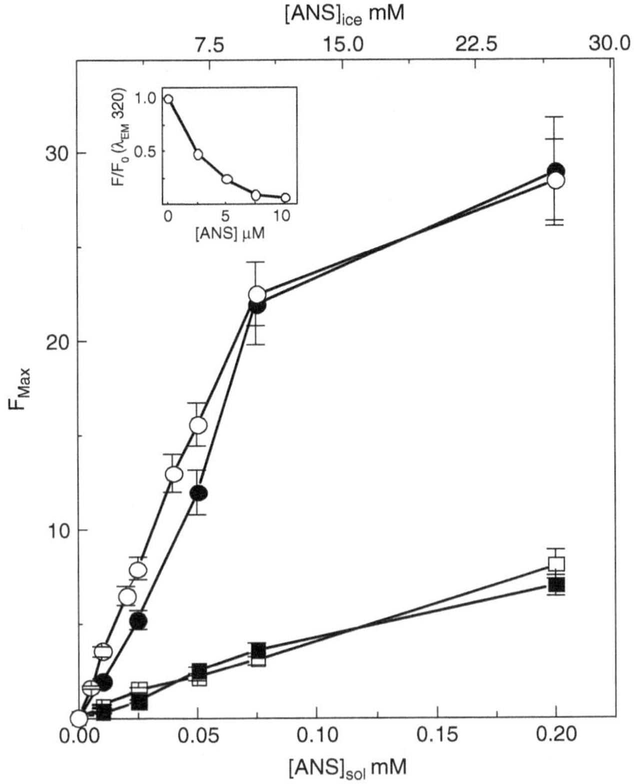

Figure 9.6. ANS fluorescence intensity (F_{max}) profiles in ice at −13°C as a function of its concentration. Direct excitation at 350 nm: (•) 2.5 μM azurin, (■) 2.5 μM cadmium azurin and (□) ANS control. Excitation at 280 nm: (○) 2.5 μM azurin. The latter profile was normalized for the change in absorbance of ANS between 280 and 350 nm. The inset shows the Trp fluorescence intensity of azurin ($\lambda_{ex} = 280$ nm) measured at 320 nm with a 305-nm longpass filter. Indicated on the top horizontal axis is the concentration of ANS in the water pools of ice, $[ANS]_{ice}$, estimated from a freeze concentration factor $f_c = 120$. (Reprinted from Gabellieri and Strambini [39].)

W48 decreases, sharply at increasing [ANS] (Fig. 9.6, inset), while the fluorescence of ANS significantly increases as compared to direct excitation of the probe at 350 nm. This finding demonstrates that the strongly fluorescent ANS molecules are within the range of energy transfer of W48, ($r \leq R_0 = 2.3$ nm) [34], as, given the size of the macromolecule, would be expected upon complex formation.

9.3.2. Effect of Temperature on ANS Binding to Azurin

To test whether the temperature-dependent structural alterations reported by the phosphorescence lifetime are also reflected on the ANS binding affinity/number of binding sites, the ANS fluorescence profiles were obtained at four different temperatures between −4°C and −18°C. The results are plotted in Figure 9.7 as a function of the $[ANS]_{ice}$ calculated from the respective freeze concentration factor. Two features are worth noting in the "binding" curves of Figure 9.7:

1. As the temperature is lowered, the fluorescence enhancement begins at progressively higher ANS concentrations (Fig. 9.7, inset) suggesting that the apparent affinity of the binding sites becomes smaller. The change in average binding constant is smooth and a factor of ~3. The trend need not necessarily point to a change in the nature of the azurin binding sites because, considering the drastic increase in solute concentration with freezing temperature, the apparent decrease in affinity is likely to reflect large deviations of activity coefficients form ideality and/or colder temperature itself.
2. The maximum fluorescence enhancement factor (the plateau value) increases at lower temperature, a sign that the number of binding sites is larger. The tendency is not a smooth one as most of the overall twofold change in intensity is found between −8°C and −13°C.

Presumably, nonequilibrium conditions may be responsible in part for the stepwise response and the measured enhancement may not be a strict indicator of the number of binding sites. For instance, below −13°C, ANS diffusion among increasingly smaller water pools may be severely blocked and no further binding may occur at lower temperature. At warmer temperatures, a sort of compensation between a decreased number of binding sites and a larger diffusion coefficient of ANS may conduct to similar plateau values for the curves at −8°C and −4°C. The main conclusion to be drawn from the temperature dependence of the ANS fluorescence enhancement is the net increase in the number of binding sites on lowering the freezing temperature a finding that is consistent with a greater perturbation of the protein tertiary structure.

9.3.3. Effect of Stabilizing and Destabilizing Agents

As opposed to glycerol, a classical stabilizing additive, the anion SCN^- is a mild destabilizer both in solution and in ice [2], and is used in preference to strong denaturants when loosening rather than complete unfolding the globular structure is required. Figure 9.8 compares ANS binding curves obtained in the presence of 20 mM glycerol and 5 mM NaSCN, at −13°C. The fraction of liquid water was maintained constant by adjusting the NaCl concentration to give the solution the same colligative properties. The results show that glycerol, even in the millimolar concentration range (inset of Fig. 9.8), completely abolishes ANS binding to azurin, whereas NaSCN more than doubles the fluorescence enhancement. The main effect of the destabilizer is not to increase the apparent affinity of ANS for azurin but to create additional binding sites with equal and also lower affinity. In the case of NaSCN a fluorescence plateau is not reached within the ANS concentration range explored. The opposite effect of the two additives on the ANS fluorescence enhancement is nicely correlated to their modulation of the protein stability to unfolding.

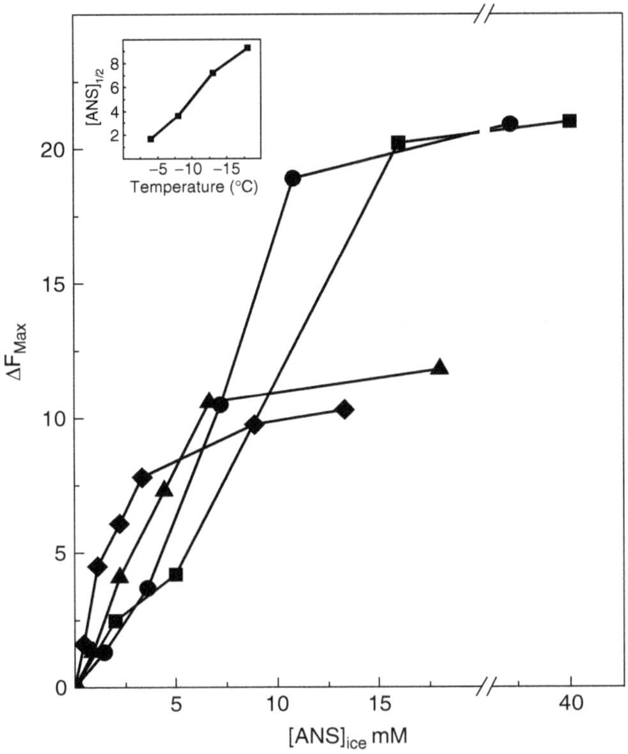

Figure 9.7. ANS fluorescence enhancement [ΔF_{max}=F_{max}(azurin) − F_{max}(control)] profiles as a function of the ice temperature: (■) −18°C, (•) −13°C, (▲) −8°C, and (♦) −4°C. The $[azurin]_{sol}$ is 2.5 μM. The $[ANS]_{ice}$ was calculated from a freeze concentration factor $f_c = 185$ (−18°C), 120 (−13°C), 82 (−8°C), and 41 (−4°C). The inset shows the dependence of the midpoint ANS concentration $[ANS]_{1/2}$ on the ice temperature. (Reprinted from Gabellieri and Strambini [39].)

9.3.4. Ice-Induced Enhancement of ANS Fluorescence in Mutant Apoazurins

Among the several azurin mutants available, I7S, F110S, and C3A/C26A maintain a native-like fold [40,41] and will therefore present the same surface topology and chemical composition to the ice–water interface. Both I7S and F110S are characterized by an internal cavity, caused by the substitution of bulky Ile and Phe with smaller Ser, whereas in C3A/C26A the disulfide bridge linking strands β1 and β2 has been severed. The stability of these mutants is about 3 kcal/mol lower relatively to the wild type (WT) [41–43] and the internal flexibility of the polypeptide is considerably larger, particularly with I7S and F110S [16]. Here, we compare the ANS binding capacity of the three azurin mutants to ascertain how these modified features modulate the ice-induced perturbation of tertiary structure [44]. The dependence of the ANS

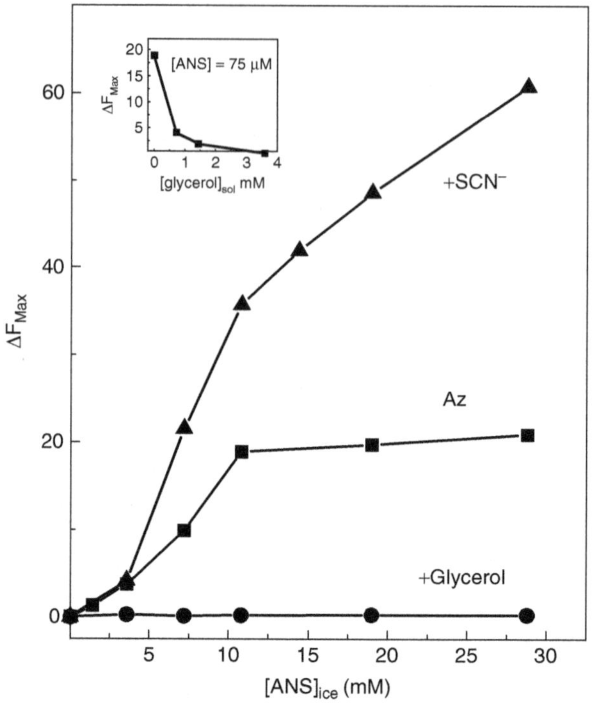

Figure 9.8. Fluorescence profiles for ANS in the presence of (•) glycerol (20 mM glycerol plus 15 mM NaCl) and of (▲) SCN^- (5 mM NaSCN plus 20 mM NaCl), in ice at −13°C. The data are compared to those obtained in the absence of additives (■). The $[azurin]_{sol}$ is 2.5 μM. The inset shows the dependence of ΔF_{max} on the glycerol concentration. (Reprinted from Gabellieri and Strambini [39].)

fluorescence intensity, in ice at −13°C, on the starting ANS concentration in solution (10–200 μM) is shown in Figure 9.9. The profiles of Figure 9.9 show that for the mutants, relative to the WT, onset of ANS binding occurs at lower concentrations and the intensity enhancement is significantly greater: 10.6-fold for C3A/C26A, 26-fold for I7S, and 30-fold for F110S. The larger ANS intensity indicates that the number of ANS binding sites and, presumably, the extent of the structural alteration are sensibly larger with the less stable mutant azurins. A rough estimate of binding affinities yields ANS dissociation constants in the range of 6–8 mM for wild type (WT), 2–4 mM for C3A/C26A, 2.5–5 mM for I7S, and 1.5–9 mM for F110S. It seems reasonable to associate the higher affinity and larger number of ANS binding sites of the mutant proteins to a less stable globular fold and conclude that their creation is favored by, or requires, structural plasticity. There are, however, significant differences among the mutants on the number of binding sites that is not explained in terms of absolute stability. The apparently much larger number of binding sites in I7S and F110S relative to C3A/C26A probably correlates with the greater internal hydration and structural flexibility of the former mutant azurins. In C3A/C26A the removal of the superficial

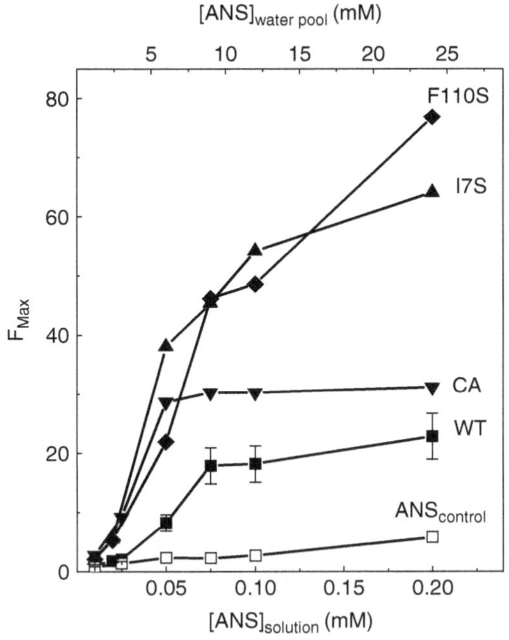

Figure 9.9. ANS fluorescence intensity (F_{max}) profiles of various azurin species (2.5 μM) in ice, at −13°C, as a function of the starting ANS concentration in solution: WT (!), C3A/C26A (B), I7S (7), and F110S (Λ). The control refers to a protein free sample. Indicated on the top horizontal axis is the concentration of ANS in the water pools of ice, $[ANS]_{(water\ pool)}$, estimated from a freeze concentration factor $f_c = 120$. (Reprinted from Gabellieri and Strambini [44].)

C3–C26 disulfide bridge has little effect on the compactness of the protein core and, judging from the phosphorescence properties of W48 embedded in t, internal motions remain restrained. On the contrary, the creation of an internal cavity of 40 cubic A° in I7S and 100 cubic A° in F110S, presumably filled with water [16,40], permits large amplitude structural fluctuations that increase remarkably the flexibility of the globular fold, as evidenced by the three to four orders-of-magnitude increase in permeability to acrylamide. A possible explanation for the correlation between internal hydration and ANS binding capacity is that the drastic drop in water activity on ice formation causes the cavities to dehydrate and collapse with consequent rearrangement of the internal peptide structure. This process may ensue in widespread deformations of the native fold with the creation of a large number of ANS binding pockets. Altogether, the considerations discussed above suggest that the ice-induced capacity to bind ANS, and presumably the magnitude of the underlying structural perturbation, will be greatest with deformable, flexible, and internally hydrated globular folds.

A direct correlation between the extent of ANS binding and the instability of azurin to the freeze–thaw process was also established from the degree of irreversible denaturation of WT and mutant species when subjected to repeated freeze–thaw cycles. The native protein fraction remaining after 15 freeze–thaw cycles decreases from WT to mutant proteins, and among the latter species it is significantly smaller with the cavity forming mutants I7S and F110S relative to compact C3A/C26A. Such linkage between instability to freeze–thaw and the propensity to bind ANS in ice supports the notion that irreversible damage to freeze-labile proteins, often resulting in enzyme inactivation and or protein aggregation, is likely to have originated from partial unfolding of the globular structure in ice. Tests extended to other proteins confirmed that a stressed condition of the globular structure in ice is widespread, indicating also that oligomers proteins are more susceptible to unfolding than are monomers presumably because the macromolecule can be further destabilized by subunit dissociation.

9.4. THERMODYNAMIC STABILITY OF AZURIN IN ICE

A quantitative description of protein stability in ice has been hampered by the difficulty in determining unfolding eqiulibria in a highly scattering and heterogeneous medium, characterized by spatially uneven concentration of solutes. As a result, the intensity of optical signals normally used to monitor the concentration of native and denatured species in equilibrium varies considerably from site-to-site, and in addition, they may be submerged by a conspicuous and unsteady background not readily accounted for. These limitations have been overcome by an experimental approach, based on the change of Trp fluorescence between native and denatured states of azurin, which permitted monitoring of unfolding eqiulibria and the thermodynamic stability (ΔG°) of the protein in frozen aqueous solutions [45]. The main innovations with respect to standard fluorimetry involved (1) drastic reduction in the spurious background from replacement of lamp with pulsed laser excitation and (2) simultaneous acquisition of the fluorescence spectrum by means of a charge-coupled device (CCD camera) combined with data analysis based on fluorescence intensities ratio, a procedure that avoids problems connected with site-to-site variation and sample to sample signal instability. The study was conducted with the azurin mutant C112S, for which, contrary to the WT, the unfolding equilibrium is promptly reversible in both liquid and frozen samples. For all details of the procedure and data analysis referral should be made to the original report [45].

9.4.1. Effect Freezing on the Unfolding Equilibrium at Constant Temperature and Solute Activity

The large redshift and quenching of W48 fluorescence on guanidinium hydrochloride (Gdn) induced azurin denaturation in ice is shown in Figure 9.10. The figure also shows a clearly identified isosbestic point in the spectrum together with the wavelengths used to calculate the fluorescence intensity ratio and, in turn, the fraction of native protein f_N in the denaturation equilibrium. Tests devised to check the spatial uniformity of f_N in ice consistently confirmed that despite an $\geq$2.5-fold variation of the fluorescence intensity across the sample, presumably reflecting distinct excitation/fluorescence collection efficiencies or spatially nonuniform protein concentration,

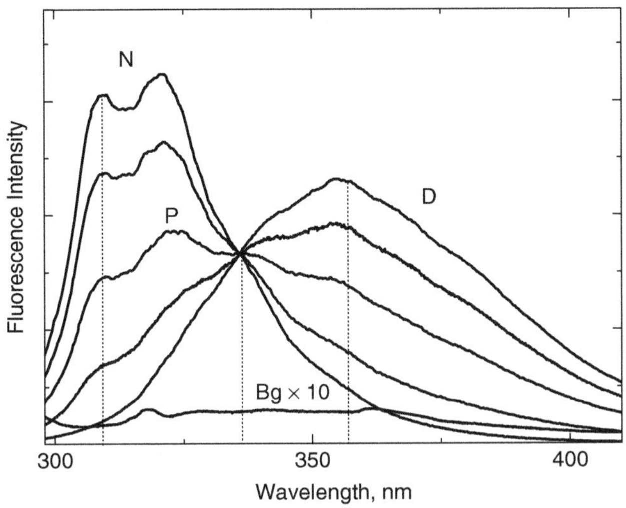

Figure 9.10. Fluorescence spectra of C112S azurin (6 μM) in ice, $V_L = 0.57\%$, at -14°C and various extents of denaturation. Native (N) and fully denatured (D) spectra refer to 0% and 100% Gdn content in the 19 mM Gdn-NaCl + 1 mM Tris (pH 7.5) salt mixture, respectively. A typical background signal (Bg) from ice is also included. Vertical dashed lines indicate the wavelengths used to calculate the fluorescence intensity ratio R and the isosbestic point. The excitation wavelength is 290 nm. All spectra are corrected for the instrumental response. (Reprinted from Strambini and Gonnelli [45].)

no significant differences could be detected in the degree of denaturation. Thus, it appears that at the lowest spatial resolution of 0.5 mm examined the composition and concentration of solutes in the liquidus is uniform and the stability of azurin in ice is constant throughout.

At constant freezing temperature, the fraction of liquid water in equilibrium with ice (V_L) can be reduced, and the amount ice in equilibrium with the liquidus increased, from 100% of liquid samples, by progressive dilution of the concentrated Gdn/NaCl salt mixtures used to denature the protein in the liquid state. Because samples prepared by increased dilution of the same Gdn/NaCl salt mixture have the same liquidus composition should ice not perturb the free energy of the protein in either native (N) or denatured (D) state the unfolding equilibrium would not be affected by freezing and f_N be independent of V_L. Figure 9.11 compares the value of f_N, at a fixed salt composition, between the liquid stock solution and frozen samples obtained from progressive dilution of the same stock from 3.51 M down to 10 mM total salt, corresponding to a decrease in V_L from 100% to 0.285%. The results show that, despite the constant activity of guanidinium in these samples, the fraction of native protein is a steep function of V_L, namely, that at some point the solid phase affects significantly the denaturation of azurin by Gdn. Unexpectedly, at small V_L the equilibrium shift toward the folded state, a change that in appearance would seem as a "stabilizing" influence of ice.

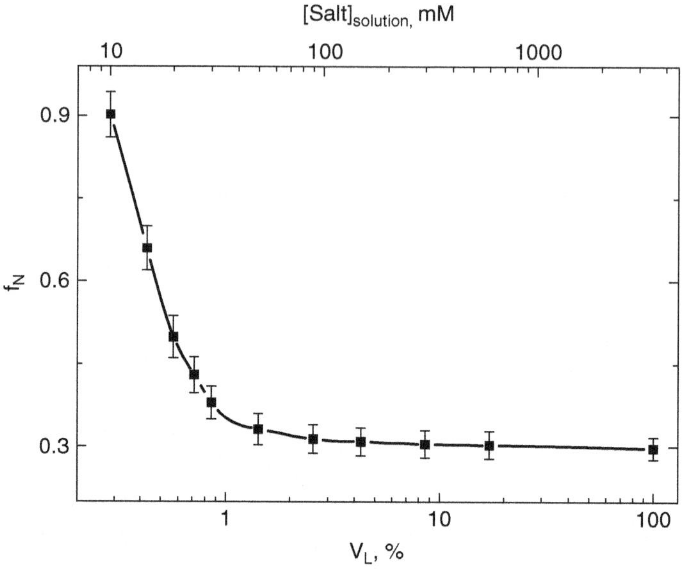

Figure 9.11. Fraction of native C112S azurin f_N in the denaturation equilibrium at $-14°C$, as a function of the fraction of liquid water V_L in equilibrium with ice. The total concentration of the salt mixture, at 20°C, is indicated on the top axis. The Gdn content in the mixture is 45.05%. The salt concentration in the liquidus is 3.51 M, and $V_L = 100\%$ refers to the liquid state. (Reprinted from Strambini and Gonnelli [45].)

9.4.2. Denaturation Profiles in Ice at Selected V_L

Equilibrium unfolding curves in liquid and in frozen samples at selected V_L, at $-14°C$, are compared in Figure 9.12a. At smaller V_L the profiles are less steep and shift to higher denaturant concentration. The thermodynamic parameters $\Delta G°$ and m value, obtained from the linear the free-energy plots [$\Delta G^{\circ}_{\text{Gdn}} = -RT\, ln(1/f_N - 1) = \Delta G° - m[\text{Gdn}]$ (Fig. 9.12b)], indicate that at large V_L the protein stability is practically identical between liquid and frozen samples, but that in ice $\Delta G°$ *decreases* progressively at smaller V_L, up to >3 kcal/mol (by roughly 35%) at $V_L = 0.285\%$. The data provide the first unequivocal proof that "dry" ice destabilizes the native fold of proteins. Because the main effect of reducing the size of the liquid water pool is to increase the surface area of ice in contact with the liquidus the mechanism underlying destabilization of the native(N) state in ice is reminiscent of an adsorption process in which the macromolecule is partitioned between the surface of ice and the surrounding liquidus. As V_L decreases, a greater fraction of the protein sample would adsorb to ice and change its stability. The decrease of $\Delta G°$ in ice implies that protein ice-interactions lower the free energy of the denatured (D) state to a greater extent than that of the N-state, suggesting that the interactions are stronger for the expanded protein fold having a larger solvent accessible surface area (ASA), which is reasonable for an adsorption process.

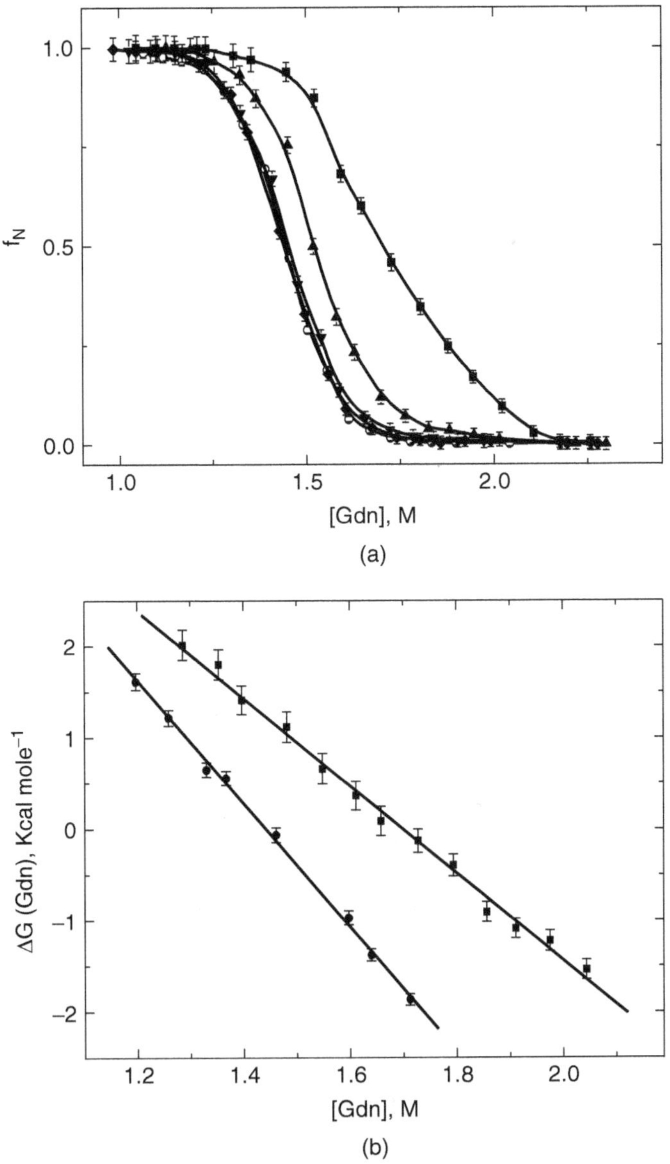

Figure 9.12. (a) Equilibrium denaturation curves of C112S azurin (6 μM), at −14°C, and at selected V_L : $V_L = 0.285\%$ (■), $V_L = 0.57$ % (▲), $V_L = 1.425\%$ (▼), and $V_L = 5.7\%$ (♦). The denaturation profile in the liquid state (○), 3.51 M total salt concentration, is included for comparison. (b) Linear free-energy plots [$\Delta G = -RT \ln(1/f_N^{-1}) = \Delta G^\circ - m$[Gdn]] for the denaturation of C112S azurin in ice, $V_L = 0.285\%$ (■), and in solution (•), at −14°C [data from part (a)]. (Reprinted from Strambini and Gonnelli [45].)

Interestingly, the lower protein stability in "dry" ice is correlated with the reduction in m value from 6.8 to about 4.0 kcal/mol. Several studies have demonstrated that the m value is directly correlated to the change in solvent-accessible surface area (ΔASA) of the protein during unfolding [46,47]. A smaller ΔASA in ice relative to the liquid state indicates that destabilization of the protein fold in ice is associated to either a larger ASA for the N state or a smaller ASA for the D state or both. Namely, the main effect of adsorption of the protein to ice is either partial unfolding of the N state, and consequent increase in ASA^N, or alternatively, partial compaction of the expanded D state, with consequent decrease of ASA^D. Practically no data are available on the structure of D state in ice and the fluorescence spectrum, which is sensitive to the microenvironment of the chromophore, does not report alterations at small V_L that could be interpreted as less than full exposure of W48 to the aqueous phase. On the contrary, there is direct spectroscopic evidence of partial unfolding of the N state at small V_L. Above we have seen that the phosphorescence lifetime of internal Trp residues in several proteins, azurin included, reports a drastic enhancement of the structural flexibility of the macromolecule in ice at small V_L that is consistent with substantial loosening of the compact protein core [26]. Likewise, ice-induced binding of ANS to azurin wildt type and mutants, confirmed that in frozen solutions the globular structure is generally perturbed tending to evolve into a molten globule-like state [39,44]. Both Trp phosphorescence and ANS binding probes report an altered, expanded N state that could largely account for decrease of ΔG° and m value observed here.

To test the effect of temperature on the ice perturbation, azurin denaturation studies were extended over the temperature interval from -8°C to -16°C. Overall, the stability of azurin is only slightly affected by temperature in both liquid and frozen solutions. In ice, ΔG° remains 30% to 35% smaller than in solution, and the decrease in protein stability is constantly associated with a corresponding reduction in m value. Such invariance to temperature attests to a predominant role of V_L (the extent of the ice surface in contact with the liquidus) on the ice perturbation.

The method presented here, based on azurin unfolding in ice as a model system, is presently employed to investigate the cryoprotective mechanisms of osmolytes (sugars, polyols, amino acids, and methylamines), salts of the Hofmeister series, and natural or synthetic polymers.

REFERENCES

1. Franks, F. (1985), *Biophysics and Biochemistry at Low Temperature*, Cambridge Univ. Press, London.
2. Allison, S. D., Dong, A., and Carpenter, J. F. (1996), Counteracting effects of thiocyanate and sucrose on chymotrypsinogen secondary structure and aggregation during freezing, drying, and rehydration, *Biophys. J*. **71**: 2022–2032.
3. Carpenter, J. F., Prestrelski, S. J., and Arakawa, T. (1993), Separation of freezing- and dryinginduced denaturation of lyophilized proteins using stress-specific stabilization. I. Enzyme activity and calorimetric studies, *Arch. Biochem. Biophys*. **303**: 456–464.
4. Chang, B. S., Kendrick, B. S., and Carpenter, J. F. (1996), Surface-induced denaturation of proteins during freezing and its inhibition by surfactants, *J. Pharm. Sci*. **85**, 1325–1330.

5. Prestrelski, S. J., Arakawa, T., and Carpenter, J. F. (1993), Separation of freezing- and drying-induced denaturation of lyophilized proteins using stress-specific stabilization. II. Structural studies using infrared spectroscopy, *Arch. Biochem. Biophys*. **303**: 465–473.
6. Nar, H., Messerschmidt, A., Huber, R., van de Kamp, M., and Canters, G. W. (1991), Crystal structure analysis of oxidized Pseudomonas aeruginosa azurin at pH 5.5 and pH 9.0. A pH-induced conformational transition involves a peptide bond flip, *J. Mol. Biol*. **221**: 765–772.
7. Nar, H., Messerschmidt, A., Huber, R., van de Kamp, M., and Canters, G. W. (1992), Crystal structure of Pseudomonas aeruginosa apo-azurin at 1.85 A resolution, *FEBS Lett*. **306**: 119–124.
8. Engeseth, H. R. and McMillin, D. R. (1986), Studies of thermally induced denaturation of azurin and azurin derivatives by differential scanning calorimetry: Evidence for copper selectivity, *Biochemistry* **25**: 2448–2455.
9. Sandberg, A., Leckner, J. and Karlsson, B. G. (2004), Apo-azurin folds via an intermediate that resembles the molten-globule, *Protein Sci*. **13**: 2628–2638.
10. Eftink, M. R. (1991), Fluorescence techniques for studying protein structure, *Meth. Biochem. Anal*. **35**: 127–205.
11. Strambini, G. B. and Gabellieri, E. (1991), Phosphorescence from Trp-48 in azurin—influence of Cu(II), Cu(I), Ag(I), and Cd(II) at the coordination site, *J. Phys. Chem*. **95**: 4352–4356.
12. Pozdnyakova, I. and Wittung-Stafshede, P. (2003), Approaching the speed limit for Greek key beta-barrel formation: transition-state movement tunes folding rate of zinc-substituted azurin, *Biochim. Biophys. Acta* **1651**: 1–4.
13. Gonnelli, M. and Strambini, G. B. (2005), Intramolecular quenching of tryptophan phosphorescence in short peptides and proteins, *Photochem. Photobiol*. **81**: 614–622.
14. Strambini, G. B. and Gonnelli, M. (1995), Tryptophan phosphorescence in fluid solution, *J. Am. Chem. Soc*. **117**: 7646–7651.
15. Gonnelli, M. and Strambini, G. B. (1995), Phosphorescence lifetime of tryptophan in proteins, *Biochemistry* **34**: 13847–13857.
16. Cioni, P., de Waal, E., Canters, G. W., and Strambini, G. B. (2004), Effects of cavity-forming mutations on the internal dynamics of azurin, *Biophys. J*. **86**: 1149–1159.
17. Horie, T. and Vanderkooi, J. M. (1982), Phosphorescence of tryptophan from parvalbumin and actin in liquid solution, *FEBS Lett*. **147**: 69–73.
18. Cioni, P. and Strambini, G. B. (1989), Dynamical structure of glutamate dehydrogenase as monitored by tryptophan phosphorescence. Signal transmission following binding of allosteric effectors, *J. Mol. Biol*. **207**: 237–247.
19. Strambini, G. B., Cioni, P., Peracchi, A., and Mozzarelli, A. (1992), Conformational-changes and subunit communication in tryptophan synthase—effect of substrates and substrate-analogs, *Biochemistry* **31**: 7535–7542.
20. Strambini, G. B., Cioni, P., and Puntoni, A. (1989), Relationship between the conformation of glutamate dehydrogenase, the state of association of its subunit, and catalytic function, *Biochemistry* **28**: 3808–3814.
21. Gabellieri, E. and Strambini, G. B. (2001), Structural perturbations of azurin deposited on solid matrices as revealed by Trp phosphorescence, *Biophys. J*. **80**: 2431–2438.

22. Strambini, G. B. and Gonnelli, M. (1986), Effects of urea and guanidine hydrochloride on the activity and dynamical structure of equine liver alcohol dehydrogenase, *Biochemistry* **25**: 2471–2476.
23. Gonnelli, M. and Strambini, G. B. (1993), Glycerol effects on protein flexibility—a tryptophan phosphorescence study, *Biophys. J*. **65**: 131–137.
24. Strambini, G. B. and Gabellieri, E. (1984), Intrinsic phosphorescence from proteins in the solid state, *Photochem. Photobiol*. **39**: 725–729.
25. Cioni, P. and Strambini, G. B. (2002), Tryptophan phosphorescence and pressure effects on protein structure, *Biochim. Biophys. Acta Protein Struct. Mol. Enzymol*. **1595**: 116–130.
26. Strambini, G. B. and Gabellieri, E. (1996), Proteins in frozen solutions: Evidence of ice-induced partial unfolding, *Biophys. J*. **70**: 971–976.
27. Gavish, B. (1980), Position-dependent viscosity effect on rate coefficient, *Phys. Rev. Lett*. **44**: 1160–1163.
28. Carpenter, J. F. and Crowe, J. H. (1988), The mechanism of cryoprotection of proteins by solutes, *Cryobiology* **25**: 244–255.
29. Sartor, G. and E. Mayer, (1994), Calorimetric study of crystal growth of ice in hydrated methemoglobin and of redistribution of the water clusters formed on melting the ice, *Biophys. J*. **67**: 1724–1732.
30. Chou, K. C. (1992), Energy-optimized structure of antifreeze protein and its binding mechanism, *J. Mol. Biol*. **223**: 509–517.
31. Haas, H., Brezesinski, G., and Mohwald, H. (1995), X-ray diffraction of a protein crystal anchored at the air/water interface, *Biophys. J*. **68**: 312–314.
32. Haynes, C. A. and Norde, W. (1995), Structures and stabilities of adsorbed proteins, *J. Colloid Interface Sci*. **169**: 313–328.
33. Stryer, L. (1965), The interaction of a naphthalene dye with apomyoglobin and apohemoglobin. A fluorescent probe of non-polar binding sites, *J. Mol. Biol*. **13**: 482–495.
34. Slavik, J. (1982), Anilinonaphthalene sulfonate as a probe of membrane composition and function, *Biochim. Biophys. Acta* **694**: 1–25.
35. Semisotnov, G. V., Rodionova, N. A., Razgulyaev, O. I., Uversky, V. N., Gripas, A. F., and Gilmanshin, R. I. (1991), Study of the molten globule intermediate state in protein folding by a hydrophobic fluorescent-probe, *Biopolymers* **31**: 119–128.
36. Das, B. K., Bhattacharyya, T., and Roy, S. (1995), Characterization of a urea induced molten globule intermediate state of glutaminyl-tRNA synthetase from Escherichia coli, *Biochemistry* **34**: 5242–5247.
37. Guha, S. and Bhattacharyya, B. (1995), A partially folded intermediate during tubulin unfolding: Its detection and spectroscopic characterization, *Biochemistry* **34**: 6925–6931.
38. Bismuto, E., Gratton, E., and Lamb, D. C. (2001), Dynamics of ANS binding to tuna apomyoglobin measured with fluorescence correlation spectroscopy, *Biophys. J*. **81**: 3510–3521.
39. Gabellieri, E., and Strambini, G. B. (2003), Perturbation of protein tertiary structure in frozen solutions revealed by 1-anilino-8-naphthalene sulfonate fluorescence, *Biophys. J*. **85**: 3214–3220.
40. Hammann, C., Messerschmidt, A., Huber, R., Nar, H., Gilardi, G., and Canters, G. W. (1996), X-ray crystal structure of the two site-specific mutants Ile7Ser and Phe110Ser of azurin from Pseudomonas aeruginosa, *J. Mol. Biol*. **255**: 362–366.

41. Bonander, N., Leckner, J., Guo, H., Karlsson, B. G., and Sjolin, L. (2000), Crystal structure of the disulfide bond-deficient azurin mutant C3A/C26A: How important is the S-S bond for folding and stability? *Eur. J. Biochem.* **267**: 4511–4519.

42. Guzzi, R., Sportelli, L., La Rosa, C., Milardi, D., Grasso, D., Verbeet, M. P., and Canters, G. W. (1999), A spectroscopic and calorimetric investigation on the thermal stability of the Cys3Ala/Cys26Ala azurin mutant, *Biophys. J.* **77**: 1052–1063.

43. Mei, G., Di Venere, A., Campeggi, F. M., Gilardi, G., Rosato, N., De Matteis, F., and Finazzi-Agro, A. (1999), The effect of pressure and guanidine hydrochloride on azurins mutated in the hydrophobic core, *Eur. J. Biochem.* **265**: 619–626.

44. Gabellieri, E. and Strambini, G. B. (2006), ANS fluorescence detects widespread perturbations of protein tertiary structure in ice, *Biophys. J.* **90**: 3239–3245.

45. Strambini, G. B. and Gonnelli, M. (2007), Protein stability in ice, *Biophys. J.* **92**: 2131–2138.

46. Courtenay, E. S., Capp, M. W., Saecker, R. M., and Record, Jr., M. T. (2000), Thermodynamic analysis of interactions between denaturants and protein surface exposed on unfolding: Interpretation of urea and guanidinium chloride m-values and their correlation with changes in accessible surface area (ASA) using preferential interaction coefficients and the local-bulk domain model, *Proteins Suppl.* **4**: 72–85.

47. Myers, J. K., Pace, C. N., and Scholtz, J. M. (1995), Denaturant m values and heat capacity changes: relation to changes in accessible surface areas of protein unfolding, *Protein Sci.* **4**: 2138–2148.

10

APPLICATIONS OF SEDIMENTATION VELOCITY ANALYTICAL ULTRACENTRIFUGATION

Tom Laue

10.1. INTRODUCTION

For over 75 years, analytical ultracentrifugation (AUC) has proved to be a powerful method for characterizing solutions of macromolecules and an indispensable tool for the quantitative analysis of macromolecular interactions [1–8]. It can be used to analyze the solution behavior of a variety of molecules in a wide range of solvents and over a wide range of solute concentrations. During analytical ultracentrifugation, samples are characterized in their native state under biologically relevant solution conditions, and in the absence of a matrix. Because AUC is nondestructive, samples may be recovered for further tests following the AUC procedure.

Two complementary views of solution behavior are available from AUC. Sedimentation velocity provides first-principle, hydrodynamic information about the size and shape of molecules [1,4,6], while sedimentation equilibrium provides first-principle, thermodynamic information about the solution molar masses, stoichiometries, association constants, and solution nonideality [3,6,8]. Different experimental protocols are used to conduct these two types of analysis. This chapter covers the fundamentals of sedimentation velocity.

Formulation and Process Development Strategies for Manufacturing Biopharmaceuticals, edited by Jameel and Hershenson

10.2. TYPES OF PROBLEMS THAT CAN BE ADDRESSED

Analytical ultracentrifugation provides useful information on the size and shape of macromolecules in solution with very few restrictions on the sample or the nature of the solvent. The fundamental requirements for the sample are that it (1) have an optical property that distinguishes it from other solution components, (2) sediments or floats (i.e., its density must differ from that of the solvent) at a reasonable rate at an experimentally achievable gravitational field, and (3) be chemically compatible with the sample holder. The fundamental solvent requirements are that it be chemically compatible with the sample cell and that it be compatible with the optical system. The range of molecular weights suitable for sedimentation velocity is vast, from a few hundred daltons (e.g., peptides, dyes, oligosaccharides), to several hundred million daltons (e.g., viruses, organelles).

Sample purity affects the sorts of questions that may be addressed by AUC. Detailed analyses are possible for highly purified samples with only a few discrete macromolecular components. Some of the thermodynamic parameters that can be measured by AUC include molecular weight, association state, and equilibrium constants for reversibly interacting systems. It also can provide hydrodynamic shape information. For samples containing many components, samples containing aggregates or lower-molecular-weight contaminants, or high-concentration samples, size distributions and average quantities may be determined. While these results may be more qualitative than those from more purified samples, the dependence of the distributions on macromolecular concentration, ligand binding, pH, and solvent composition can provide unique insights into macromolecular behavior.

10.3. BASIC THEORY

Mass will redistribute in a gravitational field until the gravitational potential energy exactly balances the chemical potential energy at each radial position. If the rate at which boundaries move during this redistribution is monitored, then one is conducting sedimentation velocity analysis. If the concentration distribution after equilibrium is reached is determined, then one is conducting sedimentation equilibrium analysis. This chapter focuses on sedimentation velocity analysis.

A sedimentation velocity experiment may be understood by considering the forces acting on a molecule during the experiment. The force on a particle due to the gravitational field is just $M_p\omega^2 r$, where M_p is the mass of the particle, ω is the rotor speed in radians per second [$\omega = (2\pi \cdot \text{rpm})/60$], and r is the distance from the center of the rotor. A counterforce is exerted on the particle by the mass of the solvent M, displaced as the particle sediments $M_s\omega^2 r$ leading to a net force of $(M_p - M_s)\omega^2 r$. The mass of solvent displaced is just the M_p times partial specific volume of the particle $\bar{v}$. (cm^3/g) times the density of the solvent ρ (g/cm^3). So the effective, or buoyant mass of the particle is $M_b = M_p(1 - \bar{v}.\rho)$. The last force to consider is the frictional force developed by the motion of the particle through the solvent, which given by fv, where f is the frictional coefficient and v is the velocity. Balancing these forces, we obtain the two definitions of the sedimentation coefficient:

$$s \equiv \frac{v}{\omega^2 r} = \frac{M_b}{f} \tag{10.1}$$

The first definition refers to the experimental measurements, the ratio of the velocity of the boundary to the gravitational field, and the second is the molecular definition, the ratio of the buoyant mass and the frictional coefficient.

Diffusion causes the sedimenting boundary to spread with time. Hence, by monitoring the motion and shape of a boundary it is possible to determine both the sedimentation coefficient and the translation diffusion coefficient D. From the Stokes–Einstein relationship, $D = RT/N_A f$, where R is the gas constant [erg/(mol · K)], T is the absolute temperature, and N_A is Avogadro's number.

The time evolution of the radial concentration distribution during sedimentation is given by the Lamm equation

$$\frac{dc}{dt} = D\left[\frac{d^2c}{dr^2} + \frac{1}{r}\frac{dc}{dr}\right] - s\omega^2\left[r\frac{dc}{dr} + 2c\right] \tag{10.2}$$

where c is the weight concentration of macromolecules and t is time. The optical systems on the analytical ultracentrifuge supply the radial concentration distribution at time intervals during the course of an experiment $c(r, t)$, and the instrument provides the rotor speed ω. The quantities sought in a velocity sedimentation experiment are s and D. While there are no exact solutions to the Lamm equation, both approximate [9,10] and numerical [11–13] solutions form the basis of many sedimentation velocity analysis programs used to extract s and D from AUC data. By taking the ratio s/D, we remove the frictional contribution to these parameters, and the result is proportional to the buoyant molar mass M_b through the Svedberg equation:

$$\frac{s}{D} = \frac{M_b}{RT} \tag{10.3}$$

The Lamm and Svedberg equations are the usual starting points for the equations that apply to real chemical systems [14–16], and the reader is referred to more thorough treatments of the density [17,42], the partial specific volume [18,19], and the frictional coefficient [20–24].

It is important to recognize that s and D are *system* properties and not properties of a molecule. A trivial example of this fact is that s and D will depend on pH, particularly if there is a pH-dependent conformational change. A less trivial example is the change in s and D at high concentrations. To a first approximation, changes in s and D for solutions containing one macromolecular component are approximated as $s = s_0/(1 + k_s c)$ and $D = D_0(1 - k_D c)$, where s_0 and D_0 are the sedimentation and diffusion coefficients extrapolated to zero concentration, c is the mass concentration, and k_s and k_D are experimentally determined coefficients. Values for k_s and k_D may be positive or negative depending on whether the concentration dependence is dominated by attractive or repulsive macromolecular interactions. For very high concentration solutions, or for solutions that contain more than one high-concentration macromolecular component, the approximations may not hold [14].

10.4. INSTRUMENTS AND OPTICAL SYSTEMS

The analytical ultracentrifuge is similar to a high-speed preparative centrifuge in that a spinning rotor provides a gravitational field large enough for the sedimentation of molecule-sized particles. What distinguishes the Beckman Coulter (Fullerton, CA) XLI analytical ultracentrifuge from a high-speed preparative centrifuge is the specialized rotors, sample holders, and optical systems that permit the observation of samples during sedimentation. The analytical rotor has through-holes that allow samples, contained in windowed sample cells, to be viewed while spinning. Each cell contains a centerpiece with chambers (called *channels*) to hold the liquid samples. The centerpiece, in turn, is sealed between windows to permit the passage of light through the channels. Centerpieces are made out of a variety of tough, inert materials, such as epoxy, anodized aluminum, or titanium. For biological materials, the epoxy-based centerpieces (either aluminum-filled or charcoal-filled for better thermal equilibration) typically are used. Depending on the type of experiment that will be performed, centerpieces are available that can hold several samples each. Rotors for the XLI are available that hold either four or eight cells; hence many samples may be analyzed at once.

10.4.1. Optical Systems

The fundamental measurements in AUC are radial concentration distributions. These concentration distributions, called "scans," are acquired at intervals ranging from minutes (for velocity sedimentation) to hours (for equilibrium sedimentation). As the rotor spins, each cell passes through the optical paths of detectors capable of measuring the concentration of molecules at closely spaced radial intervals in the cell. There are three commercially available optical detectors for the XLI to measure the concentration distributions: an absorbance spectrophotometer and Rayleigh interferometer from Beckman Coulter mentioned on 10-5, and the fluorescence detector from Aviv Biomedical (Lakewood, NJ) [25]. All subsequent analysis of sedimentation data relies on the quantity and quality of data available from these detectors. A detailed comparison of the absorbance and interference optical systems is available [26]. A summary of the properties of each optical system is presented below. In addition to these real-time optical systems, for size exclusion (SE) experiments tracer sedimentation methods have been described where the concentration gradients of labeled molecules are determined following centrifugation using a microfractionator [4,27].

10.4.1.1. Absorbance System. The absorbance optical system is the most frequently used type of detector for the analytical ultracentrifuge [26]. It is the easiest to use and operates as a standard double-beam spectrophotometer. Under conditions where the Beer–Lambert law holds, the absorbance signal is directly proportional to solute concentration: $A = \varepsilon cl$, where ε is the solute's weight extinction coefficient, c is the weight concentration, and l is the sample pathlength (1.2 cm for standard centerpieces). The rated precision of the absorbance system is ± 0.01 OD (where OD = optical density), although it is usually better than this. The noise is primarily stochastic. Hence, the noise appears as high-frequency "fuzz" around the signal. The scans may contain systematic noise that is either radially independent (e.g., the entire scan is shifted up or down) or time-independent (e.g., a feature, such as a scratch, that does not move from scan to scan). For the absorbance system, radially independent noise should be absent, and time-independent noise usually is limited to window imperfections (e.g., scratches, smudges). As described below, the other optical systems will have very different noise characteristics.

Although the absorbance optics function over a wavelength range from 190 to 800 nm, limited light intensity may restrict the useable range for two reasons:

1. Many standard biological solvent components absorb strongly at short wavelengths (e.g., disulfides, carbonyl oxygen, nitrogenous compounds, some detergents), so solvent components should be selected with care when data collection at short wavelengths is desired. A simple rule of thumb is that the solvent absorbance at the desired operating wavelength should be less than $\sim$0.5 OD, using water as the reference.
2. Output from the light source is blue-rich and "spiky," with the maximum output at 230 nm, and very low red light output. If there uncertainty about what wavelength to use, one can perform a wavelength scan using the XLI operating software. It is best to view these data as both intensities and absorbencies to ensure the data will have a good signal-to-noise ratio [26].

When preparing samples for the absorbance system, you should ensure that they have an absorbance between 0.2 and 1.0 OD at the chosen wavelengths. If you are interested in gathering data over a wide concentration range, you may want to scan different samples at different wavelengths. While this is permitted, the XLI wavelength selector is notoriously imprecise (± 3 nm) at setting the monochromator back to the same wavelength. Consequently, if your experimental protocol involves scanning samples at different wavelengths, you should make sure that the wavelengths used are in "flat" portions of the sample's absorbance spectrum, and not in regions where the absorbance is changing rapidly with wavelength. While some analysis programs (e.g., ULTRASCAN) have built in routines to adjust data for these variations, it is best to avoid the problem. Of the three optical systems, the absorbance system requires the longest to complete a scan. For sedimentation equilibrium, long scan times are not a problem. However, for sedimentation velocity experiments, long scan times may limit the amount of data acquired over the course of an experiment. Consequently, absorbance protocols for velocity experiments typically use a fairly coarse radial step size (0.003 cm) and no data averaging.

When used in a traditional double-beam mode (each sample having a corresponding reference solution), up to three (four-hole rotor) or seven (eight-hole rotor) samples may be analyzed. It is also possible to the use intensity data for sedimentation velocity analysis [28], thus doubling the number of samples per experiment. You should ensure that the material in the sample and reference channels have approximately the same absorbance reading, and that the absorbance is not too high (< 0.5OD). Otherwise, the automatic gain control logic of the XLI may produce unusable data.

10.4.1.2. Interference System. The signal from the Rayleigh interference optical system consists of equally spaced horizontal fringes whose vertical displacement ΔY is directly proportional to the optical path difference between light beams passing through the sample and reference solutions. Any refractive index difference Δn between the two solutions contributes to the optical pathlength so that $\Delta Y = \Delta n\, l/\lambda$, where l is the optical pathlength and λ is the wavelength of the light source [29,30]. For a nondialyzable solution component, the refractive index difference is proportional to the product of the weight concentration c and the refractive index increment dn/dc: $\Delta n = c(dn/dc)$. For proteins, dn/dc is relatively independent of composition with an average value of 0.186 mL/g [31]. For the XLI, $\lambda =\sim 670$ nm and the sample pathlength is 1.2 cm, so that a 1-mg/mL sample results in a fringe displacement of ~3.25 fringes [26].

Any material having a refractive index different from the reference will contribute to the signal. This is a useful characteristic but poses possible problems. Because the signal from the interference optical system does not rely on a chromophore, colorless compounds (e.g., polysaccharides, lipids) may be characterized by AUC using interference optics. However, it may be impossible to analyze a sample containing nondialyzable substances (e.g., detergents, lipid micelles). Thus, while the molecular weights and partial specific volumes of detergents may be characterized using interference optics [32], samples containing detergents are best studied using either absorbance or fluorescence optics.

Unlike the absorbance system, the interference signal has very little stochastic noise. However, since any pathlength difference between the sample and reference beams contributes to the fringe displacement, even tiny optical imperfections (e.g., dust, oil, dirt, scratches on the lenses and mirrors) are visible in the signal. Consequently, there is significant time-independent systematic noise. Furthermore, conversion of the interference image to fringe displacement measurements uses Fourier analysis to determine the fractional fringe displacement [33] for which the first radial position is arbitrarily assigned a zero fringe displacement. Since the fringes cannot be traced through certain image features (e.g., menisci), fringe displacement data also contain radially independent systematic noise. Both types of systematic noise must be removed prior to data analysis [14,34,35].

The precision and accuracy of the interference optical system places a premium on the optical components. Any variation in the window or centerpiece flatness $>0.01\ \lambda$ will cause a vertical shift in the image. A severe wedge ($>30\ \lambda$) will result in severe degradation or even loss of the image as the entire diffraction envelope can be displaced from the camera sensor. Stress on the optical components also may lead to refractive index changes. For this reason, sapphire windows *must* be used with the interference optical system. Also, in order to achieve the full accuracy of the interference optics, careful alignment and focusing are necessary [26,30]. It is not that the interference optics are particularly fussy with respect to focusing; rather, they offer precision and accuracy well beyond the other optical systems and hence require that more attention be paid to focus and alignment.

10.4.1.3. Fluorescence System. The fluorescence optical system is the most recent addition to the XLI. The AU-FDS (Aviv Biomedical, Lakewood, NJ) may be added to an existing XLI, and is based on previously described prototypes [37,38]. A laser light source must be used in order to achieve sufficient radial resolution (~20–50 μm). The AU-FDS laser provides excitation at 488 nm, and it must be used with suitable labels such as fluorescein, BODIPY, NBD, green fluorescent protein (GFP), and the many derivatives of these labels used for fluorescence microscopy. Information about specific labels and the chemistries available for attaching them to biomolecules may be found on the Web (see http://probes.invitrogen.com/handbook/). Alexa488 is an excellent choice because of its large extinction coefficient ($\sim 80,000$), insensitivity to pH, resistance to photobleaching, and the many coupling chemistries for covalently attaching the dye to specific functional groups on proteins and nucleic acids. The many variants of GFP may be used to generate transcriptionally labeled material for the AU-FDS. Because of the extraordinary sensitivity and selectivity of fluorescence detection, it is possible to characterize the sedimentation behavior of GFP-labeled proteins in cell lysates without further purification [25].

In AU-FDS optics, the emitted light passes through a pair of longpass (>505-nm) dichroic filters. This choice of filters captures the maximum amount of emitted light, providing good sensitivity, but offers no opportunity to select a label by its emission characteristics. Thus, there is currently no simple way to use multiple labels in the AU-FDS (e.g., for fluorescence resonance energy transfer).

Data from the fluorescence detector contain both high-frequency stochastic noise and low-frequency systematic noise. The stochastic noise level is about 1% of the signal. This observation holds over a wide range of sample concentrations and detector gain settings. Systematic noise tends to be time-independent and arises from two sources:

1. Fluorescent material may stick to the windows, particularly in places where there once was an air–liquid boundary. Hence, there can be regions where label stuck to the window while the cell was being handled (e.g., filled, placed in the rotor). The severity of this problem depends strongly on the nature of the sample, with some proteins exhibiting little sticking while other proteins and other materials (especially lipids) leaving an uneven coating over most of the window. While most analysis programs remove time-invariant noise, the resultant loss of materials to surfaces will affect the concentration of the labeled material (discussed below).
2. Background fluorescence from cell components (particularly epoxy centerpieces), another source of systematic noise, tends to be of lower magnitude and more uniform than that from adsorbed label, and also is removed during

data analysis. Sources of radially independent noise include variation in the source intensity and variation in detector sensitivity, both of which are small.

Conversion from fluorescence intensity to concentration is not trivial. As long as the signal is directly proportional to concentration, one can determine the sedimentation coefficient, diffusion coefficient, and molecular weight without needing to convert the data. Likewise, there are many qualitative observations (e.g., the sedimentation coefficient increases or decreases in response to some stimulus) that require only relative knowledge of the concentration. For these purposes data collected using AU-FDS may be handled in the same manner as absorbance or interference data. However, for concentration-dependent parameters (e.g., an association constant, or nonideality coefficient), fluorescence detection poses some difficulties.

The fluorescence intensity is proportional to the concentration, $F = I_0 Q \varepsilon c$, where ε is the extinction coefficient (either molar or weight, depending on the concentration units used for c), Q is the quantum yield (the fraction of photons absorbed that result in a fluorescence signal), and I_0 is the incident intensity of the excitation beam. While ε is relatively constant, Q is sensitive to the immediate surroundings of the dye (e.g., local polarizability and dipole moments) and to the specific solution conditions (e.g., how many and how uniform are the labels attached to the molecule of interest, the presence of quenchers). This means that it is more difficult to relate the fluorescence intensity to concentration than to absorbance or fringe displacement. Comparison of fluorescence intensities to standards may help, and special calibration centerpieces are available to hold several standards (Spin Analytical Inc., Durham, NH), though even using standards is not without problems [38].

Collisional quenching decreases Q, and hence decreases the fluorescence intensity. Removing quenchers uniformly (both from sample to sample and radially) is essential for good sensitivity and good reproducibility. While few common biological solvents contain quenching agents, some reagents (e.g., cesium ions, acetate ions, heavy metals, iodide, acrylamide) should be avoided. The most common quencher is molecular oxygen, which should be removed from samples by a nitrogen sparge or placing the samples under vacuum for a few minutes. It has been our experience that quenchers are not a problem in biological samples (e.g., serum, cell lysates).

One of the most common applications of AUC is the detection and characterization of molecular interactions. While it is straightforward to detect binding as changes in the sedimentation coefficient or changes in apparent molecular weight, determining an accurate association constant using fluorescence detection may be difficult. Specifically, if a label's environment changes on association (resulting from changes in polarizability, dipole moments, etc.), the quantum yield may be affected, and the fluorescence intensity will not be linear with concentration. At present, only one analysis program (SEDANAL, available free from http://rasmb.bbri.org/software) is equipped to handle changes in the quantum yield on molecular association. If one simply wants to get an estimate of the association constant, the wide dynamic range of the AU-FDS system typically allows a complete titration curve (e.g., S_W vs. c) to be obtained. The midpoint of the transition of the curve provides an estimate of the binding energy (as $\ln c$), and it may be possible to fit the titration curve to more sophisticated models [39–41].

While fluorescence optics may be used over a very broad concentration range, special care must be exercised when using samples containing at very low concentrations (< 10 nM) or high concentrations (>5 μM) of labeled material. For low concentrations, loss of material to surfaces can be a problem. Proteins, lipids, nucleic acids, and polysaccharides are "sticky" and tend to form layers on surfaces in contact with the solution. At low concentrations, the loss of material may be a significant fraction of the total material placed in the sample cell. The degree of stickiness varies from substance to substance. For the AUC sample holders, there are three surfaces to consider: the walls of the centerpiece, the cell windows, and the air–liquid meniscus. The simplest way to minimize these effects is to include some nonlabeled "carrier" protein in the sample buffer. Low concentration (0.1 mg/mL) ovalbumin, serum albumin, and κ-casein have all been used as carrier proteins. It is worthwhile to try more than one type of carrier protein to ensure that the carrier protein does not interact with the labeled material. The confocal design of the AU-FDS allows data acquisition at fairly high dye concentrations [38]. Nonetheless, the excitation beam intensity may be diminished by dye molecules not in the observation volume (inner filter effect) and lead to a nonlinear relationship between the concentration and fluorescence intensity. A similar problem will occur if the emitted light is absorbed by the fluorophore. The easiest fix for this is to reduce the concentration of the dye, either by diluting labeled material with unlabeled material or by decreasing the number of labels per molecule.

10.4.1.4. Optical Considerations for High-Concentration Formulations. High-concentration samples may be characterized by AUC with certain precautions. The signal for both the absorbance and interference optics is dependent on the optical pathlength. Decreasing the sample pathlength is the best way to extend their concentration range to high concentrations. Special 3-mm-thick centerpieces (and the adapters to use them with standard windows and cell housings) are available (Spin Analytical Inc., Durham, NH; Beckman Coulter, Inc., Indianapolis, IN) for this purpose. If accurate concentration-dependent parameters (equilibrium constants, nonideality coefficients) are sought, consideration must be given to the optical focus when using these centerpieces, particularly when high-concentration gradients are present [30].

The concentration gradients developed during sedimentation also are refractive index gradients, and these may affect any of the optical systems. The collimated light used in the absorbance and interference optical systems will be bent toward the base of the cell (for a sedimenting boundary, but toward the meniscus for a floating boundary). Ordinarily, the imaging optics will correct this distortion and bring the deviated light back to the correct radial position. However, if the gradient is steep enough and the optics improperly focused, the correction may not be entirely accurate [30]. If the gradient is steep enough, light even may be deviated entirely out of the optical path. A simple test for the absorbance system is to scan the cell at a nonabsorbing wavelength (e.g., 320 nm for a protein solution). This scan should appear as a flat line at 0 OD. If a an excessively steep gradient is present, there will be a "bump" in the scan centered at the boundary position. The height of the bump will diminish as the boundary spreads [26]. The only way to obtain accurate data is to reduce the steepness of the gradient. In some cases, this may be done by sedimenting at lower rotor speeds to let diffusion spread the boundary.

10.4.2. Sample Requirements

The sample requirements for AUC can vary greatly depending on the nature of the experiment, the optical detection system, and the extinction coefficient. The sample volumes required for AUC analysis are quite low. Sedimentation velocity experiments are generally performed using two sector cells that require 340–420 μL per sample, but for the fluorescence detection system special cells are available with volumes of 60 μL per sample are available (Spin Analytical Inc., Durham, NH).

The choice of sample concentrations can be challenging and involves balancing the desired concentration with the detector sensitivity and linearity. In many cases one simply wants to establish whether a sample is homogeneous, define the dominant association state, and possibly obtain some shape information. Here, the concentration range will be dictated by the optimal conditions for the AUC measurements. The low concentration limit for an AUC measurement is limited by the sensitivity of the detection system and the optical properties of the sample. The highest accessible concentrations are determined by the linearity of optical system, optical artifacts that occur at high concentration gradients, and thermodynamic and hydrodynamic nonideality, which becomes more pronounced at higher concentrations. Typical root-mean-square (RMS) noise levels for the absorption system are ~0.007 OD (SNR ~50 dB, calculated as 20*log S/N), and for the interference system the noise is ~0.01 fringes (SNR ~70–75 dB). Our experience is that a *minimum* usable SNR ratio is about 10 dB. Thus, reasonable (20 dB) SNR levels require a minimum sample concentration corresponding to ~0.06 OD or ~0.15 fringes. For a typical protein with a specific absorbance near 1 $(\text{mg/mL})^{-1}\text{cm}^{-1}$, 0.06 OD corresponds to a concentration of ~0.08 mg/mL (note that the usual centerpiece optical path is 1.2 cm). For the interference system, 0.15 fringe corresponds to about 0.05 mg/mL and the sensitivity interference optics is roughly comparable to that of the of the absorbance system operating at 280 nm. However, using the absorption optics, higher sensitivity measurements can be achieved at shorter wavelengths where the amide backbone absorption predominates. In the XLI, it is useful to work at 230 nm, where the light source has a strong output and the protein absorbance is approximately sevenfold higher than at 280 nm. Reasonably good SNR can be obtained at this wavelength with protein concentrations as low as 10–15 μg/mL. For the fluorescence system, the lowest usable concentration depends on the extinction coefficient and quantum yield. The typical SNR for this system is ~50 dB. For a fluorescein-like fluorescent dye ($\varepsilon \approx 65,000, Q \approx 0.95$), the lowest usable concentration is usually ~100 pM per dye molecule. The usable concentration decreases as more dye molecules are attached to the macromolecule.

The highest usable concentrations differ for the three optical systems. For the absorbance optical system, the highest usable signal is ~1.5 OD at 280 nm. This limit decreases to ~1.0 OD at 230 nm due to stray light considerations [26]. For the interference optical system, the upper concentration limit (using 12-mm centerpieces) is ~30 mg/mL. The upper limit for either the absorbance or interference optical systems can be increased about fourfold by using 3-mm centerpieces. However, regardless of the pathlength, the precautions listed in Section 10.4.1 regarding focusing and steep concentration gradients should be observed.

In experiments designed to measure the equilibrium constants for reversible associating systems, the concentration ranges must be chosen such that each species that participates in the equilibrium is present at an appreciable concentration. Thus, precise determination of K_d values for high-affinity reactions requires low sample concentrations that may lie below the detection limits discussed above. On the other end of the scale, weak interactions require high concentrations, where nonideality and optical artifacts can become problematic. The best way to choose sample concentrations and other experimental conditions, and to determine whether the equilibrium constants are even experimentally accessible for a given system, is by simulation. Synthetic data are generated using the appropriate molecular parameters, experimental conditions, and estimated equilibrium constants. Noise is added to the data to simulate the optical system being used. The data are then fit to determine whether the correct equilibrium constants can be recovered with reasonable confidence. Simulation routines are implemented in many AUC analysis software packages such as HETEROANALYSIS, SEDANAL, SEDFIT/SEDPHAT, and ULTRASCAN (discussed below).

10.4.2.1. Sample Preparation. In practice, for ease of data interpretation proteins should be at least 95% pure by SDS polyacrylamide gel electrophoresis (PAGE), and the mass spectrum should correspond to a single species consistent with the predicted molecular weight. Many proteins tend to form irreversible aggregates during purification or storage. While AUC is an excellent way to detect these, gel filtration is recommended to remove both aggregates and low-molecular-weight contaminants that may not be resolved on polyacrylamide gels. Samples that bind to the windows or centerpiece also may be encountered. The loss of soluble material may be assessed by measuring the signal (OD or fluorescence intensity) at low speed (3000 rpm) after loading the sample.

Samples should be equilibrated with the experimental buffer such that the compositions of the reference and sample solutions are identical. While solvent equilibration is good physical chemistry practice, it is an essential step when using the interference optics. Solvent equilibration can be accomplished by conventional gel filtration, using gel filtration spin columns or by dialysis. Most of the commonly used buffer components are compatible with AUC experiments. The major issues to keep in mind are ionic strength, absorbance (when using absorbance detection), viscosity, and the formation of density gradients during sedimentation. The ionic strength should be at least 10–50 mM to shield electrostatic interactions that contribute to thermodynamic nonideality. For absorbance measurements, the OD of the solvent at the detection wavelength should be minimized. For example, reductants such as mercaptoethanol and dithiothreitol absorb at 280 nm on oxidation; however, tris(2-carboxyethyl)phoshine (TCEP) is essentially transparent at this wavelength. At shorter wavelengths (e.g., 230 nm), many buffer constituents absorb and a buffer absorbance spectrum should be recorded using water as the reference. Highly viscous buffers slow sedimentation proportionately and should be avoided, or their presence should at least be taken into account to determine how long an experiment will take. Finally, density and viscosity gradients produced at high solute concentrations should be taken into account for sedimentation velocity (sedimentation velocity, SV) experiments [17].

Two critical parameters for interpretation of AUC experiments are ρ and $\overline{v}$.. Typically, ρ is calculated from the composition using SEDNTERP [42] or measured using a high-precision density meter. For proteins lacking prosthetic groups or posttranslational modification, $\overline{v}$ is commonly calculated from the amino acid composition [42]. However, these calculated values should be used with caution. Some buffer components are either excluded (e.g., glycerol) or concentrated (e.g., guanidine HCl) at the protein hydration layer, which affects $\overline{v}$ [43]. The effects of glycerol [44], salts and amino acids [45], guanidine hydrochloride [46,47], and urea [48] on $\overline{v}$ have been tabulated. Since $\overline{v}$ is a system (not molecular) parameter, any changes in the water density (e.g., through electrostriction) will affect it. In some cases the origin of an anomalous value of $\overline{v}$ may not be apparent from the protein structure [49], and it may be best to measure partial specific volumes experimentally. In the Edelstein–Schachman method, $\overline{v}$ is calculated from the linear change in the buoyant molecular weight in sedimentation equilibrium experiments performed in buffers where the density is increased by adding D_2O [5]. Alternatively, $\overline{v}$ may be obtained from the variation in solvent density with protein concentration using a high-precision density meter. Eisenberg [51] has shown that the buoyancy term commonly used in analytical ultracentrifugation experiments $M(1 - \overline{v}\rho)$ should be replaced when practical by the thermodynamically rigorous density increment $(\partial\rho/\partial c_2)_{p,\mu}$, where c_2 is the protein weight concentration and the subscript μ indicates a constant chemical potential of all other solution components.

10.5. DATA COLLECTION

Sedimentation velocity experiments are carried out in two channel cells with sector-shaped compartments in order to prevent convection. Care should be exercised in aligning the cells in the rotor. A typical experimental protocol is to run three sample concentrations spanning at least an order of magnitude, (e.g., 0.1, 0.3, 1.0 mg/mL).

For sedimentation velocity experiments using absorbance optics, the cells are assembled using standard double-sector centerpieces and quartz windows. The cells are filled with 430 μL of buffer in the reference sector and 420 μL of sample solution in the sample sector. Because of problems with wavelength reproducibility, a single wavelength is usually used with the absorbance optics, and sample concentrations chosen that will yield $OD_{280\ \mathrm{nm}}$ values of 1.2, 0.4, and 0.1 (the highest OD should be reduced to 1.0 for 230 nm) at the selected wavelength. The rotor, with properly counterbalanced cells, is loaded into the centrifuge and the vacuum system started. Once the rotor temperature has reached the set point, an additional hour is usually provided before starting the run in order to minimize temperature gradients.

The protocol for sedimentation velocity experiments using interference optics is somewhat different. In order to eliminate mismatches in refractive index due to solvent sedimentation, the solution and reference menisci heights should be matched using double-sector meniscus-matching centerpieces (Spin Analytical Inc., Durham, NH). The cells are assembled using sapphire windows. Test runs are conducted of the cells filled with water in order to preset the light source operating parameters for each cell, and to perform a radial calibration. This test run will also allow checking of the cells for leaks and proper liquid transfer, thus preventing possible sample loss. Once the test run is finished, the cells are removed from the rotor, the water is aspirated from the cells, and the assembled cells are dried in the vacuum chamber. One can also now replace the interference counterbalance with a fourth cell containing an additional sample dilution since the radial calibration has already been performed. For the actual run each meniscus-matching cell is loaded with 430 μL of buffer in the reference sector, and 420 μL of sample solution in the sample sector. The cells are placed in the rotor and the rotor is placed in the chamber, along with the monochromator–laser assembly. The rotor is accelerated to ~12,000 rpm, and the interference fringe pattern for each cell is checked to confirm that the excess buffer has transferred over to the sample sector from the reference side. At this point the rotor is stopped, removed from the centrifuge, and then gently inverted to thoroughly mix the contents of each cell. Now the rotor is placed back in the centrifuge and the temperature is equilibrated as previously described. A typical concentration series for four cells would be 1.5, 0.9, 0.3, and 0.1 mg/mL. Sample dilutions may be made immediately prior to the sedimentation velocity run unless it is suspected that slowly reversible reactions are taking place. In that case, dilutions are made, and then sufficient is time allowed for complete equilibration at the experimental temperature.

The instrument operating parameters include the temperature, the rotor speed, the time after speed is reached before the first scan is taken, the time interval between scans, and how many scans are to be acquired. For sedimentation velocity analysis, there should be no delay before data are acquired. Likewise, there is no reason to wait between scans, so there should be no interval between scans. These two parameters (scan delay and scan interval) should be set to zero in the method for either the Beckman Coulter ProteomeLab or Aviv-AOS software to maximize the number of datasets available for analysis.

Most experiments are conducted at 20°C, thus simplifying correction of the sedimentation and diffusion coefficients to standard conditions. The listed operating temperature range of the XLI is ranges from 0°C to 40°C. However, excessive oil vapor at operating temperatures above 35°C and difficulty maintaining temperatures below 4°C limit the useful temperature range. For a sedimentation velocity experiment, one wants to make sure that the samples have stabilized at the desired temperature prior to rotor acceleration.

Choosing the correct rotor speed for a sedimentation velocity experiment depends on what you want to know about your sample, what the expected component size distribution is, and which optical systems will be used. These considerations lead to competing needs. The resolution of solution components is proportional to ω^2, indicating that you should use the highest rotor speed possible, especially if you are trying to determine how many components there are in a solution. Thus, for samples with $s < 10$ S, it makes sense to use the highest rotor speeds (60,000 rpm). However, with modern global analysis software it is also beneficial to obtain a large number of scans. Thus, lower rotor speeds are required if components of interest have large sedimentation coefficients. Also, the Beckman absorbance optics require long scan times, and when scanning multiple samples and wavelengths, it may be useful to reduce rotor speeds. Although there is no simple formula for optimizing the rotor speed, we can use the definition of the sedimentation coefficient [Eq. (1.1)] to determine reasonable rotor speeds It should take a boundary at least 2 h to sediment the full length of the cell (1.5 cm maximum), to ensure that sufficient scans will be acquired. According to this criterion, the maximum recommended rotor speed decreases to 10,000 rpm if you wish to characterize 250-S material ($M \sim 20{,}000{,}000$) in the sample [41]. In addition, when using the absorbance system it is necessary to consider its longer scan times and adjust the rotor speed so that at least 30–40 scans are recorded during movement of the boundary across the cell.

10.6. DATA ANALYSIS

Methods for analyzing sedimentation velocity experiments have evolved rapidly, and many alternative approaches and software packages are available. Here, an approach is outlined that is useful in the initial stages of analyzing an unknown system. Note the detailed analysis of reversible interactions by sedimentation velocity is a complex topic [13,40,52–55]. The discussion here is limited to analysis of mixtures and how to determine whether reversible interactions are present. However, it is important to note that for interacting systems the boundaries do not correspond to discrete species; the apparent sedimentation coefficients and shape of boundaries are complex functions of the sedimentation coefficients of the species participating in the equilibrium, their concentrations, and the kinetic constants governing their interactions [53,56,57]. Software packages are available to fit sedimentation velocity data for such interacting systems (e.g., SEDANAL, SEDPHAT, ULTRASCAN).

In the initial stages of sedimentation velocity analysis is it useful to examine the data using methods that require the fewest assumptions about the nature of the system under investigation. The goal of these "model-free" approaches is to determine how many species are present and whether they interact. Information from these methods can be used to construct models and obtain starting parameters for more detailed analyses.

In the time derivative method, a group of scans taken over a short time interval are subtracted in pairs so that $\Delta c/\Delta t$ approximates dc/dt [35]. This subtraction step removes systematic noise from the data, which is particularly useful for interference data. The radial variable is then transformed to an apparent sedimentation coefficient (s^*) and $\Delta c/\Delta t$ transformed to the apparent sedimentation coefficient distribution function $g^*(s)$. These algorithms have been implemented in several software packages. One, DCDT+, is particularly easy to use without sacrificing rigor or accuracy. The $g^*(s)$ distributions resemble chromatographs and reveal whether a sample appears pure (one peak) or heterogeneous (multiple peaks or shoulders). It is important to inspect the distributions at multiple loading concentrations to check for reversible interactions. If the peak positions and relative peak areas are constant with concentration, the solution may be treated as containing a mixture of noninteracting components. For these samples the width of each peak is related to D, which allows the molecular mass of each component to be determined from the Svedberg equation.

However, systematic shifts in peak positions with concentration are diagnostic of mass–action equilibria. Likewise, $g^*(s)$ patterns in which the peak positions may remain constant or shift only slightly, but the relative area of the higher s^* features increase with loading concentration is diagnostic for slowly reversible interactions. More detailed analysis of these data will be required. It must be emphasized that the peak positions in $g^*(s)$ patterns for interacting systems cannot be interpreted as corresponding to particular solution components.

The main advantage of the time derivative method is simplicity. No models are assumed in the analysis. Its chief disadvantage is the limited number of scans that may be analyzed simultaneously, thus limiting the range of sedimentation coefficients available in an analysis. Also, the method does not work well with low-molecular-weight solutes (< 10 kDa) where diffusion dominates the sedimentation patterns.

In the $c(s)$ method implemented in the programs SEDFIT and SEDPHAT, the sedimentation coefficient distribution function is obtained from a direct fit of the data to the Lamm equation [60,61]. Complex and detailed models may be used with the $c(s)$ method. In its most basic implementation the program creates a grid of sedimentation coefficients covering the expected range of interest. By assuming that the same relationship exists between s and M for all species (using a constant f/f_0), a unique value of D is assigned for each point. The program then simulates the sedimentation boundaries for each point using a numerical solution of the Lamm equation. Finally, the data are fit as the linear least-squares sum of the Lamm equation solutions that define the concentration of each species in the grid. During this process, the systematic noise of the baseline (time-invariant noise) and the vertical displacements (jitter and integral fringe jumps) are removed by treating them as additional linear fitting parameters. The resulting $c(s)$ function is often quite "spiky," and a regularization procedure may be used to smooth the distribution function. Visual interpretation of $c(s)$ distributions determined at different loading concentration proceeds as described for $g^*(s)$ distributions.

For the $c(s)$ approach a poor fit to a particular model can be used to exclude a system's description. For example, a poor fit to the model of noninteracting discrete components indicates that concentration-dependent behavior (e.g., mass–action association) must be considered. Although the $c(s)$ distribution can be converted to a distribution of molar masses [$c(M)$ distribution], the derived masses will be accurate only if there is one dominant species present or if all the species have equal frictional ratios. More complex analysis procedures that do not assume a single f/f_0 value are also implemented in SEDFIT and SEDPHAT.

The main advantages of the $c(s)$ method are the excellent resolution and sensitivity. In contrast to the dc/dt method, there are no restrictions on the number of scans that can be included in the analysis, and the diffusional broadening is deconvoluted from the $c(s)$ distribution. The $c(s)$ method is thus very useful for characterizing homogeneity and quantifying impurities and aggregates. The main disadvantage of this approach is that it assumes a noninteracting mixture. Particular care must be exercised in the analysis of self- or heteroassociating systems since the resulting distributions are developed from an incorrect model. Nonetheless, for a system undergoing *rapid* association and dissociation, the distributions are reminiscent of those expected by limiting models [57], and useful semiquantitative information may be extracted [53,54]. For interacting systems undergoing reactions on the timescale of the sedimentation velocity experiment, peaks in the $c(s)$ distribution may not correspond to those found in true molecular species [54]. Provided that the $c(s)$ distribution is a good fit to the data, it is still feasible to extract thermodynamic parameters from the data by integration of the distribution and analyzing the dependence of weight-average sedimentation coefficients on the loading concentrations [40]. The only requirement for this analysis to be accurate is that all association reactions are at equilibrium prior to the start of sedimentation. This criterion may be met by incubating the sample dilutions a sufficient amount of time (e.g., overnight) at the sedimentation temperature prior to loading the cells.

The van Holde–Weischet approach [62] may also be used for the initial, qualitative analysis of SV experiments. Because sedimentation is proportional to the first power of time whereas diffusion is proportional to the square root of time, graphic extrapolation of the boundary to infinite time yields an integral sedimentation coefficient distribution, $G(s)$, in which the diffusional contribution has been removed. This method is implemented in ULTRASCAN, SEDFIT, and Beckman-Coulter software. More recent advances have extended this method for the analysis highly heterogeneous systems [63].

Although the information obtained from these model-free approaches may suffice to answer the relevant questions about the macromolecular system being studied, it is often useful to analyze the system using model-dependent procedures. For analysis of mixtures, the goal is usually to obtain the concentration, sedimentation coefficient, and mass of each species, as well as the concentration-dependent parameters describing mass–action associations, as well as thermodynamic and hydrodynamic nonideality. The available software uses either approximate (e.g., SVEDBERG, LAMM) or numerical (e.g., SEDANAL, SEDFIT/SEDPHAT, ULTRASCAN) solutions to the Lamm equation to fit data to various models. For noninteracting mixtures of components, the precision of the fitted parameters may be improved by analyzing several sets of data obtained at different loading concentrations. However, each of these programs may be used to characterize more complicated solutions than simple noninteracting mixtures. Both SEDPHAT and SEDANAL permit the analysis of interacting systems, allow the analysis of thermodynamic and hydrodynamic solution nonideality, and may simultaneously-analyze data acquired using any of the optical systems. SEDANAL provides a powerful "model editor" that allows users to create their own models for testing. It is possible to sequentially test various models (using algorithms to first try a monomer–dimer model, then try a monomer–dimer-trimer model, then try a monomer–tetramer model, etc.) using SEDPHAT, SEDANAL, or ULTRASCAN. However, ULTRASCAN provides a gateway to a suite of programs that run on a supercomputer cluster to apply unique ways to analyze data. In particular, the genetic algorithm suite in ULTRASCAN may be used to determine which model(s) best describe the data. For any of the methods described here, the absence of systematic deviations of the model from the data in a global fit is strong evidence that the model provides an accurate description of the solution. In some cases, more than one model may fit the data adequately, and further experiments must be conducted to resolve the ambiguity.

10.6.1. High-Concentration Solutions

Analysis of sedimentation velocity data acquired from high-concentration solutions is an active area of research. As pointed out in Section 10.6, the concentration dependence of s and D may be approximated using constant coefficients, k_s and k_D. All of the data analysis programs listed provide algorithms for fitting sedimentation velocity data to models that include k_s and k_D (usually referred to as *nonideal sedimentation*). Values of k_s or k_D are seldom interpreted since it is difficult to obtain accurate estimates, even though doing so seems an attractive way to detect weak interactions in a high-concentration solution. The fundamental problem is that when both k_s and k_D are positive (repulsive interactions dominate), they have opposite effects on a sedimentation boundary shape, with k_s leading to a sharpening of the boundary and k_D to boundary broadening with concentration. While estimates of k_s may be made on the extent of boundary slowing with increased concentration, estimates of k_D are rendered more difficult by the extent of boundary sharpening due to k_s. If there are simultaneous strong repulsive interactions and weak attractive interactions, the magnitudes of both k_s and k_D are decreased, but the difficulty in extracting accurate values remains. Finally, k_s and k_D are neither purely hydrodynamic nor purely thermodynamic parameters, making it interpret values in terms of a thermodynamic phenomenon (weak association).

10.6.2. Aggregate Detection

One common application of SV is to detect and quantify irreversible aggregates [4,5,41], and it is widely accepted that SV provides an excellent orthogonal method to SEC/MALS for this purpose. However, it must be recognized that SV analysis uses curve fitting to quantify aggregates, while there is a physical separation between species in SEC/MALS. Also, it has become apparent that variations in procedures (e.g., cell alignment) can impact $c(s)$ analysis, leading to considerable discussion about the accuracy of these analyses [66,67]. It appears from this recent work that for monoclonal antibodies it is feasible to detect $\sim$1% dimer with confidence, although the accuracy of the percent dimer is not nearly as well determined as it would be using SEC/MALS. On the other hand, since there is little sample dilution, SV offers advantages over SEC/MALS for proteins undergoing reversible aggregation. Also, SV can detect aggregates that are too large to fit through SEC guard columns and aggregates that bind to SEC media. Hence, SV provides an excellent orthogonal method for validating SEC/MALS analysis, and is excellent for characterizing very large or "sticky" aggregates, but is not considered useful for routine quality assurance/control (QA/QC) analysis.

10.7. SUMMARY

Sedimentation velocity analysis provides a first-principle means for characterizing the solution behavior of macromolecules. Both its broad molecular weight range and the variety of optical detectors available make sedimentation velocity applicable to nearly all biomolecules. All of the analysis methods and programs described here are evolving rapidly, and readers are advised to seek the most recent literature to determine which ones will best fit their needs [64,65].

REFERENCES

1. Cole, J. L. and Hansen, J. C. (1999), Analytical ultracentrifugation as a contemporary biomolecular research tool, *J. Biomol. Tech.* **10**: 163–174.
2. Hansen, J. C., Lebowitz, J., and Demeler, B. (1994), Analytical ultracentrifugation of complex macromolecular systems, *Biochemistry* **33**: 13155–13163.
3. Hensley, P. (1996), Defining the structure and stability of macromolecular assemblies in solution: The re-emergence of analytical ultracentrifugation as a practical tool, *Structure* **4**: 367–373.
4. Howlett, G. J., Minton, A. P., and Rivas, G. (2006), Analytical ultracentrifugation for the study of protein association and assembly, *Curr. Opin. Chem. Biol.* **10**: 430–436.
5. Scott, D. J. and Schuck, P. (2005), A brief introduction to the analytical ultracentrifugation of proteins for beginners, in *Analytical Ultracentrifugation*, Scott, D. J., Harding, S. E., and Rowe, A. J., eds., Royal Society of Chemistry, Cambridge, UK, pp. 1–25.
6. Laue, T. M. and Stafford, W. F. (1999), Modern applications of analytical ultracentrifugation, *Annu. Rev. Biophys. Biomol. Struct.* **28**: 75–100.
7. Lebowitz, J., Lewis, M. S., and Schuck, P. (2002), Modern analytical ultracentrifugation in protein science: A tutorial review, *Protein Sci.* **11**: 2067–2079.
8. Laue, T. M. (1995), Sedimentation equilibrium as thermodynamic tool, *Meth. Enzymol.* **259**: 427–452.
9. Behlke, J. and Ristau, O. (1997), Molecular mass determination by sedimentation velocity experiments and direct fitting of the concentration profiles, *Biophys. J.* **72**: 428–434.
10. Philo, J. S. (1994), Measuring sedimentation, diffusion, and molecular weights of small molecules by direct fitting of sedimentation velocity concentration profiles, in *Modern Analytical Ultracentrifugation*, Shuster, T. M. and Laue, T. M., eds., Birkhauser, Boston, pp. 156–170.
11. Demeler, B. and Saber, H. (1998), Determination of molecular parameters by fitting sedimentation data to finite-element solutions of the Lamm equation, *Biophys. J.* **74**: 444–454.
12. Schuck, P. (1998), Sedimentation analysis of noninteracting and self-associating solutes using numerical solutions to the Lamm equation, *Biophys. J.* **75**: 1503–1512.
13. Stafford, W. F. and Sherwood, P. J. (2004), Analysis of heterologous interacting systems by sedimentation velocity: Curve fitting algorithms for estimation of sedimentation coefficients, equilibrium and kinetic constants, *Biophys. Chem.* **108**: 231–243.
14. Fujita, H. (1975) *Foundations of ultracentrifugal analysis*, Wiley, New York.
15. Williams, J. W., van Holde, K. E., Baldwin, R. L., and Fujita, H. (1958), The theory of sedimentation analysis, *Chem. Rev.* **58**: 715–806.
16. Tanford, C. (1961), *Physical Chemistry of Macromolecules*, Wiley, New York.
17. Schuck, P. (2004), A model for sedimentation in inhomogeneous media. I. Dynamic density gradients from sedimenting co-solutes, *Biophys. Chem.* **108**: 187–200.
18. Perkins, S. J. (2001), X-ray and neutron scattering analyses of hydration shells: A molecular interpretation based on sequence predictions and modelling fits, *Biophys. Chem.* **93**: 129–139.

19. Cantor, C. R. and Schimmel, P. R. (1980), *Biophysical Chemistry*, Freeman, San Francisco, pp. 539–590.
20. Byron, O. (2000), Hydrodynamic bead modeling of biological macromolecules, *Meth. Enzymol.* **321**: 278–304.
21. Garcia De La Torre, J., Huertas, M. L., and Carrasco, B. (2000), Calculation of hydrodynamic properties of globular proteins from their atomic-level structure, *Biophys. J.* **78**: 719–730.
22. Rai, N., Nollmann, M., Spotorno, B., Tassara, G., Byron, O., and Rocco, M. (2005), SOMO (Solution Modeler) differences between X-Ray- and NMR-derived bead models suggest a role for side chain flexibility in protein hydrodynamics, *Structure* **13**: 723–734.
23. Richards, E. G. and Schachman, H.K. (1957), A differential ultracentrifuge technique for measuring small changes in sedimentation coefficients, *J. Am. Chem. Soc.* **79**: 5324–5325.
24. Hattan, S. J., Laue, T. M., and Chasteen, N. D. (2001), Purification and characterization of a novel calcium-binding protein from the extrapallial fluid of the mollusc, Mytilus edulis, *J. Biol. Chem.* **276**: 4461–4468.
25. Kroe, R. (2005), *Applications of Fluorescence Detected Sedimentation*, Ph.D. thesis, Univ. New Hampshire.
26. Laue, T. M. (1996), *Choosing Which Optical System of the Optima XL-I Analytical Centrifuge to Use*, Beckman Coulter Technical Report A-1821-A.
27. Rivas, G. and Minton, A. P. (2003), Tracer sedimentation equilibrium: A powerful tool for the quantitative characterization of macromolecular self- and hetero-associations in solution, *Biochem. Soc. Trans.* **31**: 1015–10159.
28. Kar, S. R., Kingsbury, J. S., Lewis, M. S., Laue, T. M., and Schuck, P. (2000), Analysis of transport experiments using pseudo-absorbance data, *Anal. Biochem.* **285**: 135–142.
29. Richards, E. G. and Schachman, H. K. (1959), Ultracentrifuge studies with Rayleigh interference optics. I. General applications, *J. Phys. Chem.* **63**: 1578–1591.
30. Yphantis, D. A. (1964), Equilibrium ultracentrifugation of dilute solutions, *Biochemistry* **3**: 297–317.
31. Huglin, M. B. (1972), Specific refractive index increments, in *Light Scattering from Polymer Solutions*, Huglin, M. B., ed., Academic Press, New York, pp. 165–332.
32. Reynolds, J. A. and McCaslin, D. R. (1985), Determination of protein molecular weight in complexes with detergent without knowledge of binding, *Meth. Enzymol.* **117**: 41–53.
33. DeRosier, D. J., Munk, P., and Cox, D. J. (1972), Automatic measurement of intereefernce photographs for the ultracentrifuge, *Anal. Biochem.* **50**: 139–153.
34. Schuck, P. and Demeler, B. (1999), Direct sedimentation analysis of interference optical data in analytical ultracentrifugation, *Biophys. J.* **76**: 2288–2296.
35. Stafford, W. F. (1992), Boundary analysis in sedimentation transport experiments: A procedure for obtaining sedimentation coefficient distributions using the time derivative of the concentration profile, *Anal. Biochem.* **203**: 295–301.
36. Richards, E. G., Teller, D. C., Hoagland, V. D. J., Haschemeyer, R. H., and Schachman, H. K. (1971), Alignment of Schlieren and Rayleigh optical systems in the ultracentrifuge. II. A general procedure, *Anal. Biochem.* **41**: 215–247.
37. Laue, T. M. Austin, J.B. and Rau, D.A. (2006), A light intensity measurement system for the analytical ultracentrifuge, in *Progress in Colloid and Polymer Science*, Coelfen, H., ed. Springer, Berlin, pp. 1–8.

38. MacGregor, I. K., Anderson, A. L., and Laue, T. M. (2004), Fluorescence detection for the XLI analytical ultracentrifuge, *Biophys. Chem.* **108**: 165–185.

39. Correia, J. J. (2000), Analysis of weight average sedimentation velocity data, *Meth. Enzymol.* **321**: 81–100.

40. Schuck, P. (2003), On the analysis of protein self-association by sedimentation velocity analytical ultracentrifugation, *Anal. Biochem.* **320**: 104–124.

41. Cole, J. L., Lary, J. W., Moody, T. P., and Laue, T. M. (2008), Analytical ultracentrifugation: Sedimentation velocity and sedimentation equilibrium, in *Methods in Cell Biology*, Vol. 84, Correia, J. J. Detrich, W., and Han, L., eds., Elsevier, Amsterdam, Chapter 6., pp. 143–179.

42. Laue, T. M., Shah, B. D., Ridgeway, T. M., and Pelletier, S. L. (1992), Computer-aided interpretation of analytical sedimentation data for proteins, in *Analytical Ultracentrifugation in Biochemistry and Polymer Science*, Harding, S., Rowe, A., and Horton, J., eds., Royal Society of Chemistry, Cambridge, UK, pp. 90–125.

43. Timasheff, S. N. (2002), Protein hydration, thermodynamic binding, and preferential hydration, *Biochemistry* **41**: 13473–13482.

44. Gekko, K. and Timasheff, S. N. (1981), Mechanism of protein stabilization by glycerol: Preferential hydration in glycerol-water mixtures, *Biochemistry* **20**: 4667–4676.

45. Arakawa, T. and Timasheff, S. N. (1985), Calculation of the partial specific volume of proteins in concentrated salt and amino acid solutions, *Meth. Enzymol.* **117**: 60–65.

46. Lee, J. C. and Timasheff, S. N. (1974), The calculation of partial specific volumes of proteins in guanidine hydrochloride, *Arch. Biochem. Biophys.* **165**: 268–273.

47. Lee, J. C. and Timasheff, S. N. (1974), Partial specific volumes and interactions with solvent components of proteins in guanidine hydrochloride, *Biochemistry* **13**: 257–265.

48. Prakash, V. and Timasheff, S. N. (1985), Calculation of partial specific volumes of proteins in 8M urea solution, *Meth. Enzymol.* **117**: 53–60.

49. Philo, J. S., Yang, T. H., and LaBarre, M. (2004), Re-examining the oligomerization state of macrophage migration inhibitory factor (MIF) in solution, *Biophys. Chem.* **108**: 77–87.

50. Edelstein, S. J. and Schachman, H. K. (1973), Measurement of partial specific volume by sedimentation equilibrium in H_2O-D_2O solutions, *Meth. Enzymol.* **27**: 82–98.

51. Eisenberg, H. (2000), Analytical ultracentrifugation in a Gibbsian perspective, *Biophys. Chem.* **88**: 1–9.

52. Rivas, G., Stafford, W., and Minton, A. P. (1999), Characterization of heterologous protein-protein interactions using analytical ultracentrifugation, *Methods* **19**: 194–212.

53. Dam, J. and Schuck, P. (2005), Sedimentation velocity analysis of heterogeneous protein-protein interactions: Sedimentation coefficient distributions c(s) and asymptotic boundary profiles from Gilbert-Jenkins theory, *Biophys. J.* **89**: 651–666.

54. Dam, J., Velikovsky, C. A., Mariuzza, R. A., Urbanke, C., and Schuck, P. (2005), Sedimentation velocity analysis of heterogeneous protein-protein interactions: Lamm equation modeling and sedimentation coefficient distributions c(s). *Biophys. J.* **89**: 619–634.

55. Stafford, W. F. (2000), Analysis of reversibly interacting macromolecular systems by time derivative sedimentation velocity, *Meth. Enzymol.* **323**: 302–325.

56. Cann, J. R. (1970), *Interacting Macromolecules*, Academic Press, New York.

57. Gilbert, G. A. and Jenkins, R. C. (1956), Boundary problems in the sedimentation and electrophoresis of complex systems in rapid reversible equilibrium, *Nature* **177**: 853–854.

58. Philo, J. S. (2000), A method for directly fitting the time derivative of sedimentation velocity data and an alternative algorithm for calculating sedimentation coefficient distribution functions, *Anal. Biochem.* **279**: 151–163.

59. Philo, J. S. (2006), Improved methods for fitting sedimentation coefficient distributions derived by time-derivative techniques, *Anal. Biochem.* **354**: 238–246.

60. Schuck, P. (2000), Size-distribution analysis of macromolecules by sedimentation velocity ultracentrifugation and Lamm equation modeling, *Biophys. J.* **78**: 1606–1619.

61. Dam, J. and Schuck, P. (2004), Calculating sedimentation coefficient distributions by direct modeling of sedimentation velocity concentration profiles, *Meth. Enzymol.* **384**: 185–212.

62. van Holde, K. E. and Weischet, W. O. (1978), Boundary analysis of sedimentation-velocity experiments with monodisperse and paucidisperse solutes, *Biopolymers* **17**: 1387–1403.

63. Demeler, B. and van Holde, K. E. (2004), Sedimentation velocity analysis of highly heterogeneous systems, *Anal. Biochem.* **335**: 279–288.

64. Vistica, J., Dam, J., Balbo, A., Yikilmaz, E., Mariuzza, R. A., Rouault, T. A., and Schuck, P. (2004), Sedimentation equilibrium analysis of protein interactions with global implicit mass conservation constraints and systematic noise decomposition, *Anal. Biochem.* **326**: 234–256.

65. Demeler, B. (2005), *UltraScan A Comprehensive Data Analysis Software Package for Analytical Ultracentrifugation Experiment*, version 9.0.

66. Gabrielson, J. P., Randolph, T. W., Kendrick, B. S., and Stoner, M. R. (2007), Sedimentation velocity analytical ultracentrifugation and SEDFIT/c(s): Limits of quantitation for a monoclonal antibody system, *Anal. Biochem.* **361**: 24–30.

67. Ejima, D., Tsumoto, K., Fukada, H., Yumioka, R., Nagase, K. Arakawa, T., and Philo, J. S. (2007), Effects of acid exposure on the conformation, stability, and aggregation of monoclonal antibodies, *Proteins Struct. Funct. Bioinform.* **66**: 954–962.

11

FIELD FLOW FRACTIONATION WITH MULTIANGLE LIGHT SCATTERING FOR MEASURING PARTICLE SIZE DISTRIBUTIONS OF VIRUS-LIKE PARTICLES

Joyce A. Sweeney and Christopher Hamm

11.1. INTRODUCTION

Field flow fractionation is a separation technology that is increasingly being used to discern populations in macromolecular assemblies such as viruses, virus-like particles, liposomal structures, polysaccharides, and monoclonal antibodies [1–8]. The application in this case study focuses on virus-like particles (VLPs), which are essentially macromolecular viral protein capsids devoid of DNA, and therefore are replication-incompetent. Virus-like particles generated from structural proteins of specific viruses can be used as vaccines that stimulate the humoral arm of the immune system to produce antibodies against specific viruses [9,10]. Other potential therapeutic applications include the use of VLPs as delivery vehicles in gene therapy [11,12]. The process of generating VLPs entails overexpression of a major viral structural protein in a host cell culture. The host cells can be bacterial, yeast, insect, or mammalian cells. The recombinant structural proteins self-assemble into substructures called *capsomeres*. Capsomeres subsequently form the virus-like particles or capsids when the physicochemical environment is conducive to self-assembly [13]. It has been demonstrated

Formulation and Process Development Strategies for Manufacturing Biopharmaceuticals, edited by Jameel and Hershenson

by cryo–transmission electron microscopy (TEM) that the VLPs form structures similar to that of the native virion [14]. Smaller virus structures with fewer capsomeres have also been known to form in many different types of analogous viral systems, including polyomavirus and simian virus 40 [15–19]. Thus, it is also expected that similar types of subpopulations of different size and morphology may form in the VLP assembly process. In fact, models derived by Stehle et al. [18] and Modis et al. [19] propose size distributions of populations. One such VLP model includes three discrete VLP populations with diameters of ∼20, 40, and 60 nm with symmetries of 12-ICOS (icosahedral with 12 capsomeres per VLP), 24-OCTA (octahedral with 24 capsomeres per VLP) and 72-ICOS (icosahedral with 72 capsomeres per VLP), respectively. Others have verified various size distributions for VLPs that undergo the self-assembly process [20].

The VLP assembly is known to be dependent on physicochemical conditions, such as pH, ionic strength, divalent cations, disulfide bonds, and presence and/or concentration of reducing agents, which can affect electrostatic or hydrophobic forces in the VLP [13,20–25]. Appropriate adjustment of these conditions can control the rate of the VLP assembly process, resulting in a more homogenous VLP population with a higher degree of stability.

Batch-mode dynamic light scattering is routinely used to monitor the Z-average diameters of VLPs during process development because of the ease of use and fast turnaround time. The Z-average diameter, however, represents a size-weighted average of all populations constituting the particle size distribution in the sample. Since the DLS measurement is based on scattering intensity, larger particles will have a tendency to bias the Z-average size value. The Rayleigh approximation, which shows that the scattering intensity from a particle is proportional to the sixth power of the diameter, provides a sense of the bias that can occur when even a very low concentration of large particles is present in a sample with a major population of much smaller size. Although the limited resolution of the batch-mode DLS method will not permit an accurate size measurement of the VLPs when multiple populations are present, but rather provides a size-weighted average of the entire distribution, batch-mode DLS can give information related to particle size and diffusion properties. For more accurate size distribution analysis it is necessary to apply a physical separation method such as field flow fractionation (FFF) interfaced to DLS or multiangle static light scattering (MALS). The addition of online light scattering analysis permits the size determination of monodisperse VLP populations as they are being separated, and thus eliminates any bias due to larger particles in the light-scattering size analysis. The use of online static light scattering and concentration analyses also permits the determination of VLP molecular weight. The molecular weight analysis requires an accurate extinction coefficient and/or refractive index increment (dn/dc) which are dependent on the VLP being analyzed. This analysis method is discussed in detail in Chapter 12.

11.2. FIELD FLOW FRACTIONATION: THEORY

The general concept of field flow fractionation (FFF) was first introduced and pioneered by J. Calvin Giddings in the mid-1960s [26]. Field flow fractionation is a separation technology based on differences in diffusion coefficients, and thus size of a particle or molecule [26–29]. The separation is achieved by first establishing a continuous flow of eluent through a very thin channel. The laminar channel flow will generate a parabolic velocity flow profile with flow velocity fastest in the center of the channel and continuously approaching zero velocity at the channel walls. Size selectivity is dependent on distributing the particles or molecules to be separated across the parabolic velocity profile. This distribution is achieved by a combination of particle diffusion in an upward direction within the channel and an opposing downward force externally applied across the channel. The externally applied force field is perpendicular to the laminar channel flow. The various types of FFF are distinguished by the form of the perpendicular force, such as flow, thermal, or electrical forces. The type of force used in these experiments is flow and is designated as flow FFF (FlFFF). In symmetric FlFFF, the upper and lower walls of the channel contain porous frits that permit the application of a perpendicular flow by means of a secondary pump [26–31]. The lower channel wall is covered by a filtration membrane with an appropriate molecular weight cutoff (MWCO). In these experiments, an alternative version of this method known as *asymmetric flow FFF* (A-FlFFF) was used. In A-FlFFF, crossflow is generated by means of a pressure differential that is established across the channel rather than using a secondary pump [32–35]. In addition to eliminating the need for a second pump, the A-FlFFF technique has the advantage of minimizing sample dilution by skimming eluent above the separation area of the channel prior to the channel flow entering the detector system.

When sample molecules or particles are injected into the FFF channel, the perpendicular flow creates a downward force that drives the particles toward the porous or *accumulation wall*, as it is often termed. At the same time, the molecules or particles will exhibit an opposing force in the opposite direction due to diffusion. Molecules or particles with the largest diffusion coefficients, and thus smallest size, will rise to the highest height in the lower half of the channel. Molecules or particles will distribute across the channel in the order of smallest to largest, with the largest residing at or near the accumulation wall. An equilibrium is established between the downward force of the perpendicular crossflow V_X and the diffusion forces of the molecule or particle. Figure 11.1 is a schematic diagram of the FFF principle.

Since it is optimal to establish the equilibrium at the head of the channel, a focusing step (as shown in Fig. 11.1a,b) is introduced prior to beginning the sample elution. The focusing step is achieved by reversing the channel outlet flow and controlling the magnitude of the inlet and outlet flow rates such that the two flows meet at a position at the head of the channel and effectively mimic a crossflow. Once the particle diffusion forces and the crossflow forces have reached equilibrium, with the particles distributed across the channel by size, the flow direction of the channel outlet is reestablished in the normal direction (i.e., flow to the detectors). This action allows the separation of the particles to begin as they travel down the channel (Fig. 11.1c) and experience the parabolic velocity flow.

The effect of critical experimental parameters on the retention time of the particles or molecules in the separation, and thus the impact on resolution, can be shown in the following expression [32–35]:

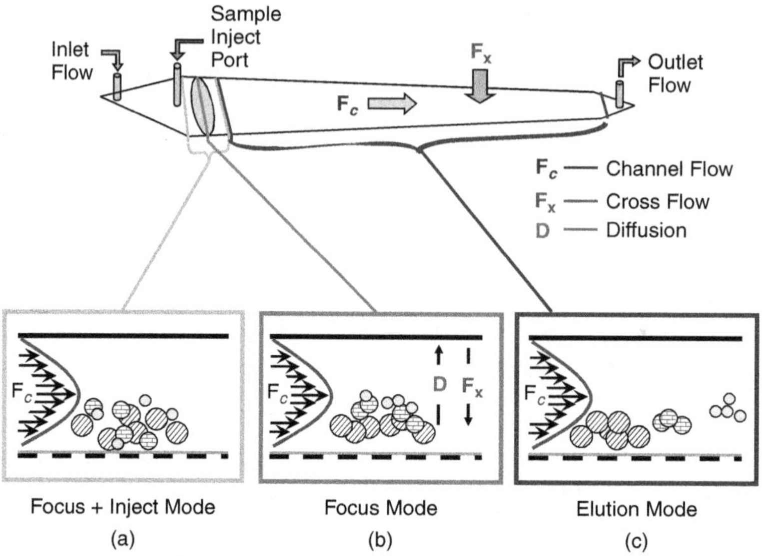

Figure 11.1. A schematic diagram showing the separation mechanism of FFF during the stages of sample injection (a), sample focusing (b), and elution (c).

$$RT \propto w^2 d \frac{V_x}{V_c} \tag{11.1}$$

where RT equals retention time, w^2 is the square of the channel height, d is particle or molecule diameter, V_x is cross flow velocity, and V_c is laminar channel flow velocity.

The channel height w has the largest impact on retention time since RT is proportional to the square of this parameter. Increasing the channel height, however, also has the effect of band broadening. The height of the channel can be changed using spacers of discrete size. Since the crossflow tends to keep particles or molecules in the channel by driving them toward the accumulation wall, it is directly proportional to retention time. The laminar channel flow tends to move particles down the channel and is therefore inversely proportional to channel flow. The appropriate combination of these two parameters is the most practical means of generating the desired resolution in the particle size separation. A more detailed explanation of the asymmetric flow dynamics is presented elsewhere [32,33].

The FFF analysis in this study was interfaced to a multiangle static light-scattering detection system (DAWN EOS, Wyatt Technology, Inc.) for measurement and analysis (ASTRA V software, Wyatt Technology, Inc., Santa Barbara, CA) of particle size and molecular weight. The online measurement at each elution point interval permitted the analysis of single size populations throughout the elution, eliminating any size bias as observed in batch-mode measurements. Particle size, expressed as the root-mean-square (RMS)

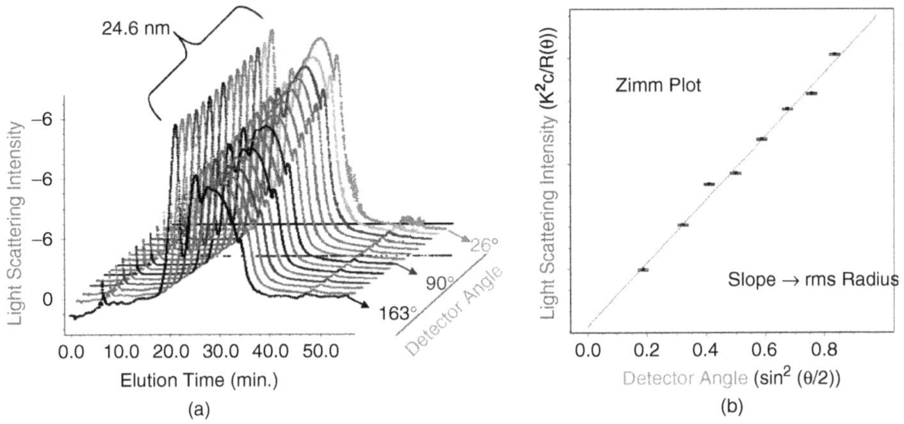

Figure 11.2. (a) An overlay of light-scattering elution profiles from 15 different detection angles ranging from 26° to 163° obtained simultaneously for an FFF-MALS/UV analysis of a VLP preparation; (b) a Zimm plot showing the light-scattering intensity dependence on detection angle for a sampling from a specific elution time. Such a plot is generated for each elution time interval throughout the elution. A RMS radius and molecular weight are calculated from the linear regression of each plot.

radius, and molecular weight were determined using the Rayleigh–Gans–Debye optical model [36]. This model does not assume any type of shape for the calculations. Figure 11.2a displays an overlay of the light-scattering elution profiles acquired using 15 of the 18 available MALS detection angles. The angular dependence of the light-scattering intensity on particle size is clearly demonstrated. At the earlier elution times when the smaller particles elute, the angular dependence is moderate. At later elution times, the larger particles show vastly different light-scattering intensities, with the highest intensity occurring at the lowest detection angles. When the intensity data (i.e., modified by instrumental and molecular parameters) for a specific elution point are plotted against detector angle [i.e., $\sin^2(\theta/2)$] using a Zimm format, as shown in Figure 11.2b, the RMS radius at each elution point is extracted from the slope of this plot and the molecular weight is determined from the y intercept. Plots and calculations were generated using Wyatt Technology ASTRA V software. The multiangle light-scattering theory and data analysis are discussed in detail in Chapter 12.

11.3. METHOD DEVELOPMENT FOR SIZE DISTRIBUTION ANALYSES OF VIRUS-LIKE PARTICLES

11.3.1. Materials and Methods

11.3.1.1. Virus-Like Particles. Virus-like particles (VLPs) were generated at Merck & Co., Inc. They were produced by standard recombinant technology and purification methods. The VLP preparations were maintained in a buffer system containing sodium chloride at pH ~ 6. Control of fermentation and purification conditions was optimized during development to control VLP aggregation and produced homogeneous size distributions. The VLP preparations analyzed and illustrated in this report were taken at various stages of process development. These samples were chosen to highlight the utility of the FFF application for process optimization.

11.3.1.2. FFF-MALS/UV Analysis. The particle size separation was conducted using a Wyatt Eclipse A-FlFFF system with a channel length of 24 cm interfaced to a Wyatt multiangle light-scattering (MALS) DAWN EOS detector and an Agilent UV diode array detector. The FFF channel was configured with a 350-μm spacer and a regenerated cellulose membrane with MWCO of 10,000. All sample analyses were conducted in the same buffer system as the sample matrix. Sample masses of approximately 10–20 μg were loaded into the channel in a 20 μL volume. Particle size separations were achieved using a crossflow ramp starting at 1.0 mL/min and decreasing to zero crossflow in 30 min, unless stated otherwise. The laminar channel flow was held constant at 1.0 mL/min throughout the analysis. The UV signal was monitored at 280 nm for concentration. Online particle size and molecular weight analyses at each elution interval in the separation were obtained using the light-scattering intensity signal at various angles and the Rayleigh–Gans–Debye optical model. The size and concentration data were ultimately used to calculate the particle size distributions for VLPs at various stages during development and final product.

11.3.2. FFF Method Optimization for VLPs

Asymmetric flow FFF-MALS/UV was used to analyze VLP preparations at various stages of development. The numbers of different types of populations within a sample, as well as the level of aggregation, are important sample parameters that need to be controlled as part of a process development. The preparations used for this report were obtained during developmental stages and thus display a wider size distribution to emphasize the utility of the FFF-MALS technique in optimizing process capabilities.

Figure 11.3 shows elution profiles for a VLP preparation as monitored by the 90° light-scattering detector (black trace) and the UV detector at 280 nm (red trace). Sample was injected at time zero and focused for six min. Elution was started with an initial crossflow of 1.5 mL/min ramping to zero crossflow by 36 min. In addition to the major population that elutes at approximately 23 min, additional populations are observed in this development lot. To ensure that the FFF experiment is not inducing artifactual aggregation in the sample as a result of focusing and/or crossflow strengths, a thorough assessment of optimal crossflow for these sample types was conducted.

The overlay displayed in Figure 11.4a of multiple elution profiles generated for a single sample shows the effect on elution time and resolution of various initial crossflow strengths ranging from 2.1 to 0.5 mL/min. In all cases the length of the gradient was held constant at 30 min, and therefore all gradients ramp to zero crossflow at an elution time of 36 minutes. Figure 11.4b shows the particle size distributions for this particular VLP preparation calculated from the datasets acquired at the various crossflow conditions. It is obvious in the elution profile and the corresponding calculated

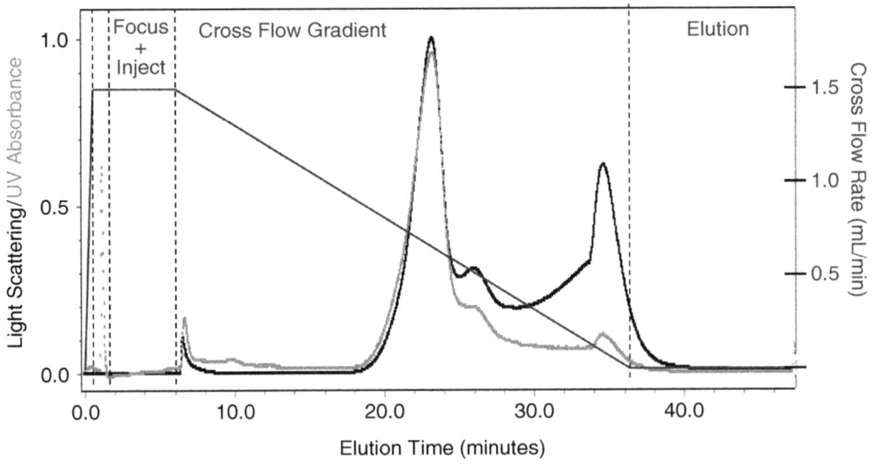

Figure 11.3. An overlay of the 90° light scattering elution profile (black trace) and the 280 nm absorbance elution profile (red trace) for a VLP development preparation. The blue schematic shows the strength of the crossflow throughout the analysis. See the insert for color representation of this figure.

size distribution profile that a crossflow of 0.5 mL/min is not strong enough to achieve optimal resolution. On the other hand, crossflows of 1.7, 1.9, and 2.1 mL/min, while achieving ideal resolution, also result in some high-level aggregation, as noted by the presence of populations in the region of >80 nm RMS radius. The crossflows of 1.0 and 1.5 mL/min seem to generate similar size distributions that demonstrate acceptable resolution and minimal artifactual aggregation. Since it is always desirable to use the lowest crossflow strength that will achieve the desired resolution, the crossflow of 1.0 mL/min was chosen as the optimal crossflow for the VLP samples.

Figure 11.5 shows the effect of mass load on the elution profiles. Panel (a) displays the elution profiles as monitored by the 90° light scattering. The RMS radius, calculated at each elution point, is overlaid with the corresponding elution profile according to color. The level of noise in the RMS radii throughout each profile can be used to determine the lowest mass load that can still provide reliable size data. The lower the mass load, the higher the attainable resolution. The sensitivity of the concentration detector, however, usually dictates the lower mass load limit, as observed in (b). In this case, a mass load of 4 μg results in RMS radii with a relatively low level of noise as observed in Figure 11.4a, but the sensitivity of the UV concentration detector is more reliable at a mass load of 10 μg (Fig. 11.5b), which appears to be the optimal mass load for the VLP samples.

11.4. RESULTS AND DISCUSSION

Figure 11.6 shows an overlay of analyses from six different preparations of VLP 1. Panel (a) displays the elution profiles as monitored by the 90° light-scattering detector.

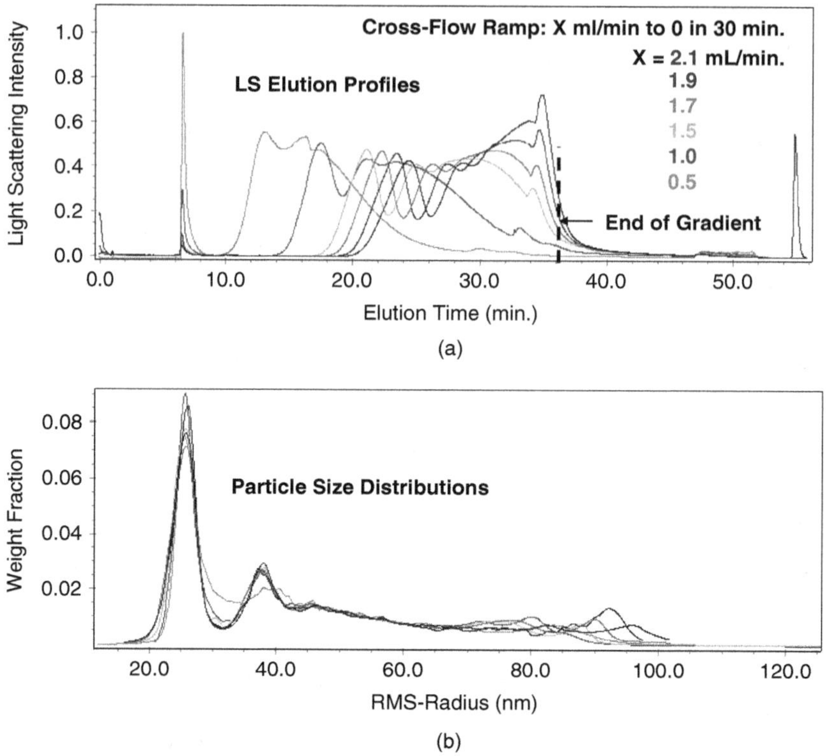

Figure 11.4. Effects of crossflow strength on sample resolution and aggregation during the FFF elution: (a) comparison of 90° light-scattering elution profiles generated for the same VLP preparation but using different starting crossflow strengths for each analysis; (b) calculated particle size distribution profiles generated from each elution profile of varying crossflow strength displayed in part (a). See the insert for color representation of this figure.

Panel (b) displays the particle size distribution analyses with particle size information derived from the angle-dependent light-scattering intensity and the mass amount at each size derived from the UV concentration detector. The preparations represented by the red traces were from early process development, and the blue traces show materials produced at a later stage. Regardless of development stage, all preparations show a highly monodisperse major population with an average RMS radius of approximately 25 nm. A second minor, but clearly distinct, population is centered at approximately 38 nm RMS radius. Any species in the size distribution profiles with a measured RMS radius of >60 nm is likely due to higher-order aggregation. This subpopulation is limited to very low levels, although the early preparations do show a somewhat higher level and an accompanying decrease in the amount of the major population centered at an RMS radius of approximately 25 nm. These observations demonstrate the utility of FFF-MALS/UV to monitor the effect of process improvements to changes in the particle size distribution.

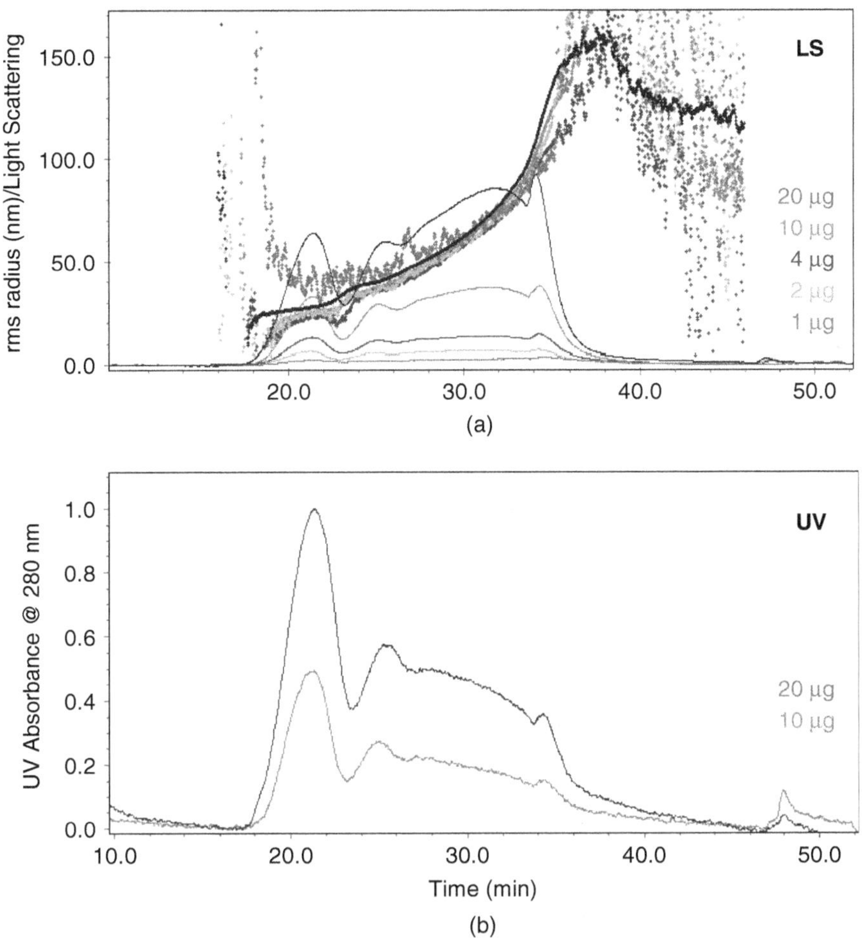

Figure 11.5. Effects of FFF mass load on signal-to-noise ratio in the calculated RMS data (a) and on sensitivity in UV absorbance detection (b).

The average RMS radii, converted to diameters, for the two distinct populations in each preparation are reported in Table 11.1 along with the Z-average hydrodynamic diameters acquired independently by batch-mode dynamic light scattering (DLS). The Z-average hydrodynamic diameters from the DLS measurements represent a size-weighted average of the two populations observed in each preparation by the FFF-MALS/UV technique. In fact, when the RMS radii from the FFF-MALS data are averaged over the two populations, the result is very close to the Z-average value obtained by DLS. The low polydispersity index of approximately 0.02 in the DLS results indicates that although larger size VLPs (i.e., RMS radii > 60 nm in the FFF-MALS data) are present in these preparations the amounts are actually quite low and do not significantly impact the DLS result.

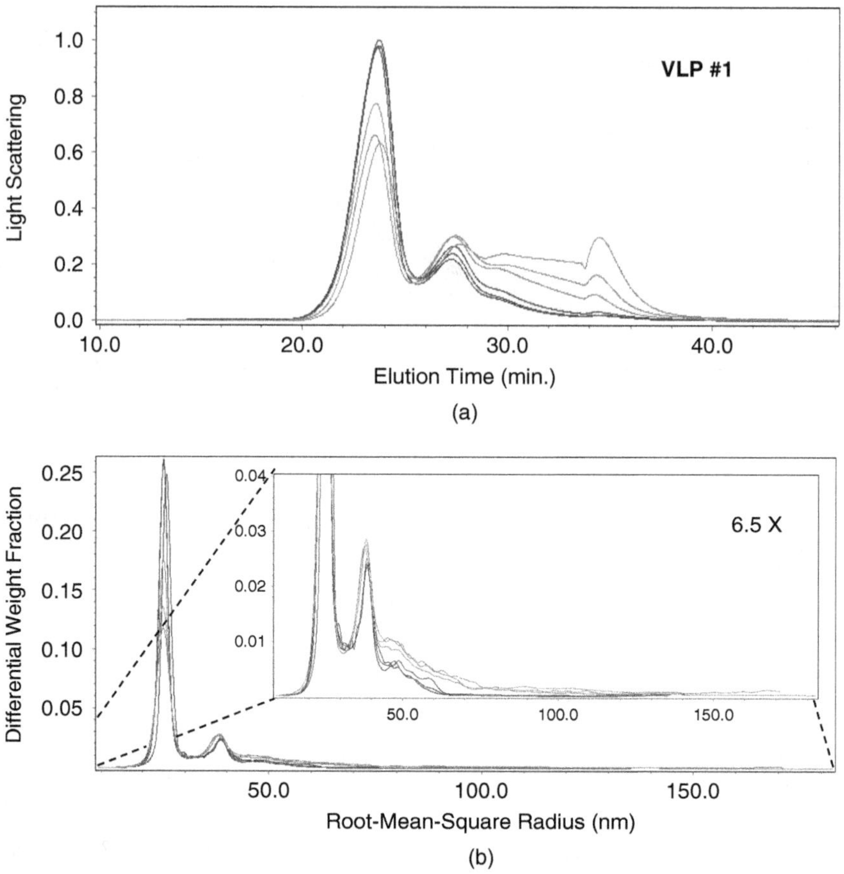

Figure 11.6. An overlay of (a) 90° light-scattering elution profiles for three early-stage-development lots (red traces) of VLP 1 and three later-stage-development preparations (blue traces); (b) calculated VLP size distribution profiles for each VLP preparation generated from the elution profiles in (a). Inset in (b) shows an expanded view of the *y* axis.

Figure 11.7 shows FFF-MALS data for another VLP, VLP 2. In this case, the major population in the size distribution profile is actually an overlap of two populations. This overlap is evident in the size distribution profile (see Fig. 11.7b), where the main peak has a peak width larger than that observed in the VLP 1 data, and reveals a slight shoulder at approximately 25 nm. There is also lower resolution between this overlapping population and the distinct peak centered at approximately 32 nm. Since the size range of the peaks is similar to that of VLP 1, the Z-average radius obtained by batch-mode DLS is also similar, as reported in Table 11.2. However, the higher polydispersity indices of approximately 0.08 for the VLP 2 data reflect the presence of the additional population in these preparations. An anomaly in the first VLP 2 preparation listed in Table 11.2 shows a polydispersity index of 0.18, more than twice the value

TABLE 11.1. Comparison of VLP 1 Diameters for Individual Populations in Particle Size Distributions Acquired by FFF-MALS/UV to *Z*-Average Hydrodynamic Diameters Determined from Batch-Mode DLS

	FFF-MALS		Dynamic Light Scattering
VLP 1 Preparation Number	Major Population Peak 1 RMS Radius × 2, nm	Peak 2 RMS Radius × 2, nm	*Z*-Average Diameter, nm (Polydispersity Index), 90° Detection Angle
1	50	75	64 (0.06)
2	49	74	62 (0.02)
3	50	76	63 (0.05)
4	51	76	60 (0.04)
5	51	76	61 (0.02)
6	50	76	61 (0.03)

TABLE 11.2. Comparison of VLP 2 Diameters for Individual Populations in Particle Size Distributions Acquired by FFF-MALS/UV to Z-Average Hydrodynamic Diameters Determined from Batch-Mode DLS

	FFF-MALS		Dynamic Light Scattering
VLP 2 Preparation Number	Major Population Peak 1 RMS Radius × 2, nm	Peak 2 RMS Radius × 2, nm	Z-Average Diameter, nm (Polydispersity Index), 90° Detection Angle
1	46	79	68 (0.18)
2	46	72	62 (0.09)
3	46	71	63 (0.08)
4	46	70	62 (0.08)
5	46	69	62 (0.08)

of the other preparations. This higher polydispersity index is attributed to a slightly higher level of large VLPs that is observed for this preparation in the FFF-MALS elution profile. This difference results in a shift of the average RMS radius for the second distinct peak to a higher value of 79 nm compared to approximately 70 nm for the other preparations. This observation also provides some insight into the sensitivity of the batch-mode DLS measurement for detecting true differences in preparations. An understanding of the correlation of the batch-mode DLS measurements and the FFF-MALS/UV data permits the use of the relatively fast DLS technique for routine quality control measurements. Any anomalies can be flagged and further investigated by FFF-MALS/UV to assess the significance of the differences measured by DLS.

The accuracy of the VLP size data is supported by independent measurements using cryo–transmission electron microscopy (cryo-TEM) [14]. Details of the cryo-TEM study are the subject of a future report. Figure 11.8 shows a comparison of the size distribution profiles for VLP 1 and VLP 2 as determined by FFF-MALS on the left and as determined by cryo-TEM on the right. The particle size in the cryo-TEM images was measured as the maximum feret diameter. An automated particle-counting

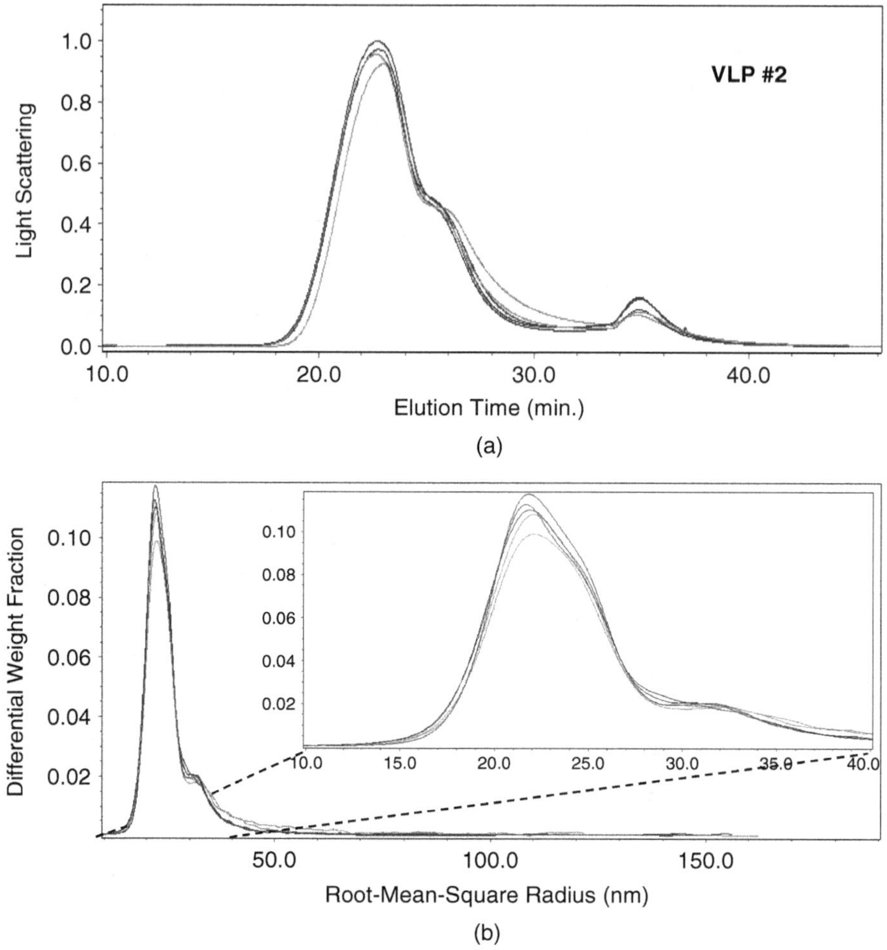

Figure 11.7. Overlay of (a) 90° light-scattering elution profiles for two very early-stage-development lots (red traces) of VLP 2 and three later-stage-development preparations (blue traces); (b) calculated VLP size distribution profiles for each VLP preparation generated from the elution profiles in (a). Inset in (b) shows an expanded view of the *x* axis to emphasize the overlap of two populations in the main peak.

algorithm was used in order to ensure a statistically relevant sampling of the particles and images. The particle selection criteria used in this early version of the automated algorithm registered only single particles with clearly discernable boundaries. Therefore, the populations that appear at approximately 76 and 70 nm diameters, for VLP 1 and VLP 2, respectively, do not appear in the cryo-TEM size distributions.

While it is also possible to measure the absolute molecular weight of the VLPs using the Rayleigh–Gans–Debye light-scattering model, the accuracy of the

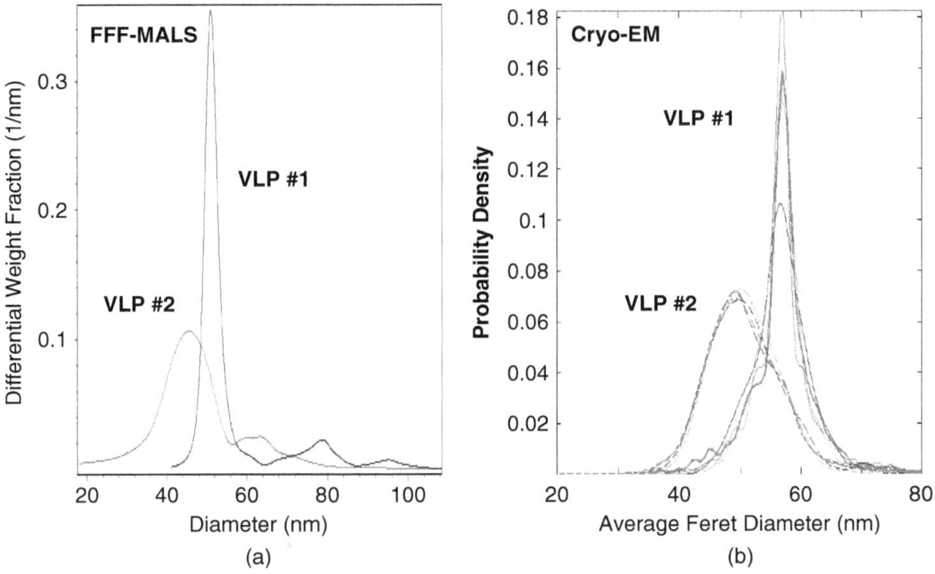

Figure 11.8. Comparison of size distribution profiles for VLPs 1 and 2 generated by FFF-MALS/UV (a) and Cryo-TEM [14] (b). Multiple traces in the cryo-TEM profiles for each VLP represent multiple analyses of the single preparation of each VLP. See the insert for color representation of this figure.

measurement will be impacted by the accuracy of the extinction coefficients determined for the VLPs. This information is the subject of a future report.

11.5. CONCLUSIONS

Accurate and reproducible size distribution profiles for virus-like particles can be acquired with relative ease using FFF interfaced to multiangle light scattering and 280 nm UV absorbance detection. Accuracy of the FFF-MALS/UV analysis is supported by independent measurements using the orthogonal technique of cryo-TEM. While size data generated by batch-mode DLS is more easily obtained in a somewhat shorter period of time than by FFF-MALS/UV analysis, batch-mode DLS does not provide a true measure of particle size when multiple populations are present. The technique of FFF-MALS/UV has the added advantage of providing molecular weight distributions as well. The ease of use and extremely fast turnaround times for batch-mode DLS analyses, however, cannot be overlooked. An extensive correlation between FFF-MALS/UV and batch-mode DLS databases, especially during process development, can provide insight to facilitate an understanding of anomalies that are sometimes observed during long-term production. Batch-mode DLS can serve as the quality control analysis, while FFF-MALS/UV can be used to establish true size distributions and provide insight into troubleshooting in future preparations.

ACKNOWLEDGMENTS

The authors would like to recognize the contributions of the following distinguished scientists to the efforts in this work: the Merck Fermentation and Cell Culture Department and the Merck Bioprocess Development Engineering Department within Merck Research Laboratories for generation of the VLP development preparations that were analyzed in this work; Drs. Phillip Wyatt and Michelle Chen of Wyatt Technology, Inc. and Dr. Dierk Roessner of Wyatt Technology Europe for many helpful discussions on MALS and FFF theory; Drs. Bridget Carragher and Clint Potter of The Scripps Research Institute, La Jolla, CA, and NanoImaging Services for the cryo-TEM analyses of the VLPs along with Drs. Paul Duncan and Michael Washabaugh at Merck, who were instrumental in establishing the collaboration with TSRI; Dr. Yang Wang and Ms. Assunta Ng, for batch-mode DLS analyses of the VLPs; and Drs. Paul Duncan, Juan Gimenez, John Hennessey, Michael Kosinski, Peter DePhillips, Robert Sitrin, Yang Wang, and Michael Washabaugh for many highly informative and useful discussions regarding the VLP vaccine bioprocess, development, and characterization.

REFERENCES

1. Hupfeld, S., Holsaeter, A. M., Skar, M., Frantzen, C. B., and Brandl, M. (2006), Liposome size analysis by dynamic/static light scattering upon size exclusion-/field flow-fractionation, *J. Nanosci. Nanotechnol*. **6**: 3025–3031.
2. Leeman, M., Wahlund, K.-G., and Wittgren, B. (2006), Programmed cross flow asymmetrical flow field-flow fractionation for the size separation of pullulans and hydroxypropyl cellulose, *J. Chrom. A* **1134**: 236–245.
3. Kowalkowski, T., Buszewski, B., Cantado, C., and Dondi, F. (2006), Field-flow fractionation: Theory, techniques, applications and the challenges, *Crit. Rev. Anal. Chem*. **36**: 129–135.
4. Korgel, B. A., van Zanten, J. H., and Monbouquette, H. G. (1998), Vesicle size distributions measured by flow field-flow fractionation coupled with multiangle light scattering, *Biophys. J*. **74**: 3264–3272.
5. Liu, J., Andya, J. D., and Shire, S. J. (2006), A critical review of analytical ultracentrifugation and field flow fractionation methods for measuring protein aggregation, *AAPS J*. **8**(3): E580–E589 (article 67).
6. Ratanathanawongs Williams, S. K., and Lee, D. (2006), Field-flow fractionation of proteins, polysaccharides, synthetic polymers, supramolecular assemblies, *J. Separ. Sci*. **29**: 1720–1732.
7. Yohannes, G., Sneck, M., Varjo, S. J. O., Jussila, M., Wiedmer, S. K., Kovanen, P. T., Oorni, K., and Riekkola, M.-L. (2006), Miniaturization of asymmetrical flow field-flow fractionation and application to studies on lipoprotein aggregation and fusion, *Anal. Biochem*. **354**: 255–265.
8. Litzen, A., Walter, J. K., Krischollek, H., and Wahlund, K.-G. (1993), Separation and quantitation of monoclonal antibody aggregates by asymmetrical flow field-flow fractionation and comparison to gel permeation chromatography, *Anal. Biochem*. **212**: 469–480.

9. McAleer, W. J., Buynak, E. B., Maigetter, R. Z., Wampler, E. E., Miller, W. J., and Hilleman, M. R. (1984), Human hepatitis B vaccine from recombinant yeast, *Nature* **307**: 178–180.

10. Netter, H. J., Woo, W.-P., Tindle, R., Macfarlan, R. I., and Gowans, E. J. (2003), Immunogenicity of recombinant HBsAg/HCV particles in mice pre-immunised with hepatitis B virus-specific vaccine, *Vaccine* **21**: 2692–2697.

11. Joyner, A., Keller, G., Phillips, R. A., and Bernstein, A. (1983), Retrovirus transfer of a bacterial gene into mouse haematopoietic progenitor cells, *Nature* **305**: 556–558.

12. Sorge, J., Wright, D., Erdman, V. D., and Cutting, A. E. (1984), Amphotropic retrovirus vector system for human cell gene transfer, *Mol. Cell. Biol*. **4**(9): 1730–1737.

13. Pattenden, L. K., Middelberg, A. P. J., Niebert, M., and Lipin, K. I. (2005), Towards the preparative and large-scale precision manufacture of virus-like particles, *Trends Biotechnol*. **23**(10): 523–529.

14. Potter, C. and Carragher, B. (2005), The Scripps Research Institute, Duncan, P. and Washabaugh, M., Merck & Co., Inc., personal communication, April 7, 2005.

15. Chen, X. S., Garcea, R. L., Goldberg, I., Casini, G., and Harrison, S. C. (2000), Structure of small virus-like particles assembled from the L1 protein of human papillomavirus 16, *Mol. Cell* **5**: 557–567.

16. Salunke, D. M., Caspar, D. L. D., and Garcea, R. L. (1989), Polymorphism in the assembly of polyomavirus capsid protein VP1, *Biophys. J*. **56**: 887–900.

17. Yan, Y., Stehle, T., Liddington, R. C., Zhao, H., and Harrison, S. C. (1996), Structure determination of simian virus 40 and murine polyomavirus by a combination of 30-fold and 5-fold electron-density averaging, *Structure* **4**: 157–164.

18. Stehle, T., Gamblin, S. J., Yan, Y., and Harrison, S. C. (1996), The structure of simian virus 40 refined at 3.1 A resolution, *Structure* **4**: 165–182.

19. Modis, Y., Trus, B. L., and Harrison, S. C. (2002), Atomic model of the papillomavirus capsid, *EMBO J*. **21**: 4754–4762.

20. McCarthy, M. P., White, W. I., Palmer-Hill, F., Koenig, S., and Suzich, J. A. (1998), Quantitative disassembly and reassembly of human papillomavirus type 11 virus-like particles in vitro, *J. Virol*. **72**(1): 32–41.

21. Montross, L., Watkins, S., Moreland, R. B., Mamon, H., Caspar, D. L. D., and Garcea, R. L. (1991), Nuclear assembly of polyomavirus capsids in insect cells expressing the major capsid protein VP1, *J. Virol*. **65**: 4991–4998.

22. Hanslip, S. J., Zaccai N. R., Middelberg, A. P. J., and Falconer, R. J. (2006), Assembly of human papillomavirus type-16 virus-like particles: Multifactorial study of assembly and competing aggregation, *Biotechnol. Progress* **22**: 554–560.

23. Chen, X. S. et al. (2001), Papillomavirus capsid protein expression in *Escherichia coli*: Purification and assembly of HPV11 and HPV16 L11, *J. Mol. Biol*. **307**: 173–182.

24. Zhang, W. et al. (1998), Expression of human papillomavirus type 16 L1 protein in *Escherichia coli*: Denaturation, renaturation, and self-assembly of virus-like particles in vitro, *Virology* **243**: 423–431.

25. Lai, W. B. and Middleberg, A. P. J. (2002), The production of human papillomavirus type 16 L1 vaccine product from *Escherichia coli* inclusion bodies, *Bioprocess Biosyst. Eng*. **25**: 121–128.

26. Giddings, J. C. (1966), A new separation concept based on coupling of concentration and flow nonuniformities, *Separ. Sci*. **1**: 123–125.

27. Giddings, J. C. (1973), The conceptual basis of field-flow fractionation, *J. Chem. Educ.* **50**: 667–669.
28. Beckett, R., Jue, Z., and Giddings, J. C. (1987), *Environ. Sci. Technol.* **21**: 289–295.
29. Giddings, J. C., Benincasa, M. A., Liu, M.-K., and Li, P. (1992), *J. Liq. Chrom.* **15**: 1729–1747.
30. Ratanathanawongs, S. K., Lee, I., and Giddings, J. C. (1991), Particle size distribution II: Assessment and characterization, in *ACS Symp. Ser.* **472**, Provder, T., ed., American Chemical Society, Washington, DC, pp. 229–246.
31. Ratanathanawongs, S. K. and Giddings, J. C. (1991), Chromatography of polymers: Characterization by SEC and FFF, in *ACS Symp. Ser.* **521**, Provder, T., ed., American Chemical Society, Washington, DC, pp. 13–29.
32. Wahlund, K.-G. and Giddings, J. C. (1987), Properties of an asymmetrical flow field-flow fractionation channel having one permeable wall, *Anal. Chem.* **59**: 1332–1339.
33. Litzen, A. and Wahlund, K.-G. (1991), Zone broadening and dilution in rectangular and trapezoidal asymmetrical flow field-flow fractionation channels, *Anal. Chem.* **63**: 1001–1007.
34. Caldwell, K. D. and Wahlund, K.-G. (2005), Field-flow fractionation, in *Methods for Structural Analysis of Protein Pharmaceuticals*, Jiskoot, W. and Crommelin, D. J. A., eds., AAPS Press, Arlington, VA, pp. 413–433 [reprinted in *Biotechnol. Pharm. Aspects* **3** (2005)].
35. Schimpf, M., Caldwell, K., Giddings, and J. C, eds. (2000), *Field-Flow Fractionation Handbook*, J. Wiley, New York.
36. Wyatt, P. J. (1993), Light scattering and the absolute characterization of macromolecules, *Anal. Chim. Acta* **272**: 1–40.

12

LIGHT-SCATTERING TECHNIQUES AND THEIR APPLICATION TO FORMULATION AND AGGREGATION CONCERNS

Michael Larkin and Philip Wyatt

12.1. INTRODUCTION

Light-scattering techniques have become essential tools for the study and measurement of many important properties of biological solutions. Their use and application cover a wide range of important areas, including formulation development, drug stability confirmation, confirmation of aggregate states, quality control, and various areas associated with manufacture and storage. This chapter is intended to provide an overview of two different classes of light scattering—multiangle light scattering (MALS) and dynamic light scattering (DLS)—as well as a discussion of the origins and interpretation of the signal measured in a DLS experiment. For each light-scattering method, we discuss detection limits and specific sample concentration and volume requirements, together with examples of experiments designed to measure targeted molecular or particle properties. We conclude the chapter with an exemplar system, utilizing a variety of light-scattering measurements to characterize the aggregation states of a protein.

12.2. OVERVIEW

We discuss two classes of light scattering that are in general use: static light scattering (SLS) and dynamic light scattering (DLS). Static light scattering includes its most

Formulation and Process Development Strategies for Manufacturing Biopharmaceuticals,
edited by Jameel and Hershenson

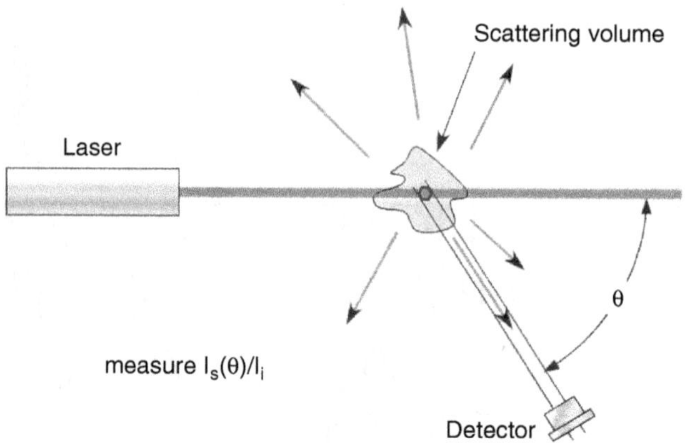

Figure 12.1. Detection of light scattered from a small volume of sample illuminated by a fine laser source beam.

general implementation, multiangle light scattering (MALS), wherein light scattered from a sample is collected over multiple scattering angles. Dynamic light scattering is also referred to as *quasi-elastic light scattering* (QELS) or *photon correlation spectroscopy* (PCS). For the light-scattering measurements discussed in this chapter, a fine beam of light, generally from a laser source, passes through a liquid sample. Some of this incident light, scattered from a small-volume element of suspending fluid (solvent) and the molecules or particles making up the sample, is detected by collimated photodetectors surrounding the sample. The general experimental setup, showing a single detector, is shown in Figure 12.1.

The particles (molecules, nanoparticles, viruses, etc.) within the solution undergo thermal translational diffusion (Brownian motion). This motion results in intensity fluctuations through time of the scattered light. The timescale of these fluctuations is related to the size of the particles, and is discussed in more detail later. These fluctuations typically span timescales of several microseconds up to several milliseconds, and if the photodetector collected intensities are averaged over 0.05 s or more, the fluctuations are unobservable in the recorded signals. Under these circumstances, the detector response is said to be proportional to the time-averaged intensity of scattered light. Measurement of this time-averaged intensity of scattered light corresponds to what we refer to as *static light scattering*. To first order, the time-averaged intensity of scattered light is proportional to the so-called weight-averaged molar mass of the molecules in the sample, times their concentration. For samples comprised of particles whose spatial extent is greater than about one-fiftieth of the wavelength of the incident light ($\sim$10 nm average radius for light of wavelength 660 nm in water), there is a measurable difference in the intensity of scattered light at different scattering angles: Light scattered into low angles (small value of θ in Fig. 12.1) has a greater intensity than does light scattered into larger angles. This asymmetry increases with particulate size. Thus, by measuring the angular variation of the scattered light intensity, one may determine an average size of the sample constituents. In a MALS measurement, multiple detectors collect light scattered by approximately the same volume element into different scattering angles θ. A MALS measurement, combined with a determination of sample concentration, permits us to deduce the sample's weight averaged molar mass and mean square radius [1–3].

Attractive or repulsive interactions between the solution constituents affect their spatial distribution, and therefore influence the amount of light scattered [4]. The influence of interactions on the distribution increases with sample concentration, and measuring the MALS intensities as a function of sample concentration, provides information regarding the strength of these interparticle interactions.

If a photodetector is able to measure more quickly than the time over which intensity fluctuations occur as a result of Brownian motion, then it is possible to characterize these intensity fluctuations. Smaller particles result in faster intensity fluctuations than do larger particles, and by measuring the timescale of intensity fluctuations, one can estimate the size or size distributions of particulates in solution. Using intensity fluctuations through time to estimate the size or size distributions of particulates in solution is called *dynamic light scattering*.

12.3. MULTIANGLE LIGHT SCATTERING

Many excellent reviews and detailed derivations of MALS theory may be found in the literature [1–3,5–7]. Here we present equations used in a typical MALS measurement to determine the weight-averaged molar mass, root-mean-square (RMS) radius, and interaction parameter A_2.

In a MALS experiment, the basic quantity measured is the excess Rayleigh ratio as a function of scattering angle and concentration: $R(\theta, c)$. This is proportional to the scattered intensity at the scattering θ angle and concentration c, in *excess* of the scattered intensity of the pure solvent in which the molecules of interest are to be measured. The relation between the Rayleigh ratio, the weight averaged molar mass M of the species present in the sample, the second virial coefficient A_2, and the concentration for vertically polarized incident light is given by Zimm [2] as

$$R(\theta, c) = K^*McP(\theta)[1 - 2A_2McP(\theta)] \tag{12.1}$$

where

$$K^* = \frac{4\pi^2 n_0^2}{N_A \lambda_0^4}\left(\frac{dn}{dc}\right)^2$$

where n_0 is the refractive index of the solvent, λ_0 is the vacuum wavelength of light, N_A is Avogadro's number, and dn/dc is the differential refractive index increment, that is, the change in the solution refractive index Δn for a concentration change Δc. The refractive index increment can have a significant wavelength dependence, and it is not unusual to find a 5%–10% change in the refractive index increment across the visible spectrum. It is a fairly straightforward process to measure the refractive index increment for a sample by measuring the refractive index as a function of concentration at the wavelength of light used for the light scattering measurement. Standard reference tables [8,9] with published values for the refractive index increment for many materials have been compiled. For unmodified proteins containing only amino acids (no glycosylation, pegylation, significant lipid content, etc.) in an aqueous solution, the value of the refractive index increment at 660–690 nm wavelengths is usually within 5% of 0.185 mL/g.

The second virial coefficient A_2 is a first-order correction to noninteracting behavior, and a measure of the average potential energy between particles in solution [2]. Depending on the light-scattering measurement being performed, it may be either an unknown or an assumed known. For particles that have no attractive or repulsive interaction in solution, A_2 has a value of zero and the light-scattering signal increases linearly with concentration. Particles that experience a net attractive interaction have a negative value for A_2, and the amount of light scattered by the sample increases over that of the noninteracting case. The converse is true for particles that experience a net repulsion. Coefficient A_2 may be determined by measuring the light-scattering signal as a function of concentration, with deviations from linearity being attributed to A_2. Attractive interactions between particles may alternatively be modeled using an equilibrium association coefficient k_a, as discussed further below. If the value of $2A_2Mc$ is much less than 1.0, then interactions do not significantly affect the intensity of scattered light. Methods which fractionate a sample, described below, often result in concentrations that are low enough for the value of this term to be quite small, and it is typical to assume an A_2 value of zero for measurements on fractionated samples.

The scattering function $P(\theta)$ is the angular dependence of light scattered from a particle, and is related to the size (spatial extent) of the particle. Fitting the angular dependence of $R(\theta, c)$ to the functional form of $P(\theta)$ allows one to determine the mean square radius of the particles. Function $P(\theta)$ has a maximum value of 1 for $\theta = 0$ and decreases from 1 with increasing angle θ (the angular dependence arises from the interference of light that scatters from different portions within the same scattering particle); $P(\theta)$ may be expressed as

$$P(\theta) = 1 - \mu_1 \sin^2 \frac{\theta}{2} \langle r_g^2 \rangle + \cdots, \qquad (12.2)$$

where $\mu_1 = [(16\pi^2 n_0^2)/3\lambda_0^2]$, $\langle r_g^2 \rangle = 1/M \int r^2 dM$ is the mean-square radius and r_g is the RMS radius, which is often termed the *radius of gyration*. The next term in this expansion is proportional to $\sin^4(\theta/2) \int r^4 dM$ and is significantly less than the first term, as long as r_g is small compared to the wavelength. Using this functional form for $P(\theta)$, one may determine the average mean square radius r_g without knowledge of the exact conformation of the particle. If the conformation of the sample is known a priori (e.g., solid sphere, hollow sphere, random coil, thin rod), then it is easy to calculate [1] the corresponding physical dimension for that conformation from r_g.

If the size of the particles being measured approaches or surpasses the wavelength of light being used for the measurement, then accurately fitting $R(\theta)$ data to Equation (12.2) requires including terms beyond $\sin^2(\theta/2)$. Within the limits discussed in Section 12.3.1, there are exact solutions [1,2] for $P(\theta)$ for a number of conformations (e.g., solid spheres, thin rods, random coils) that contain all higher-order terms in Equation (12.2), and allow the $R(\theta)$ data for large particles to be fit without increasing the number of fit parameters. Legitimately reducing the number of parameters when fitting data inevitably results in a more stable and higher-quality fit, and so if the sample being measured is known to have a conformation that has an exactly solved $P(\theta)$, it is generally preferable to use that functional form.

As r_g becomes significantly smaller than the wavelength of light, all models for $P(\theta)$ share approximately the same functional form and become indistinguishable from one another. All such models are valid, regardless of how small r_g becomes, but the change in the intensity of light as a function of the scattering angle is too weak to be reliably measured for a small r_g. The most sensitive MALS instruments available can resolve r_g down to slightly less than 10 nm. For samples with r_g less than 10 nm, a MALS measurement can determine the molar mass and A_2 value, but can set only an upper limit for the size of the particle. Proteins in a folded conformation with r_g greater than 10 nm typically have a molar mass of 1×10^6 g/mol or greater, and many proteins of interest are smaller than this. Proteins aggregating to form larger structures often are greater than 10 nm in size, and for such cases the size of the aggregates may be determined using MALS. Proteins conjugated with polyethylene glycol, saccharides, or surfactants also often have a value of r_g greater than 10 nm. Dynamic light scattering, discussed in detail in Section 12.4, can determine the size of particles with radii down to less than 1 nm, and so is often utilized to determine the size of proteins with radii less than 10 nm.

12.3.1. Limits of Applicability

Equation (12.1) is valid when the incident light is essentially unaffected by the scattering particle, and interactions between particles have only a small influence on the scattering signal. Two conditions must hold true for the incident light to be unchanged by the scattering particle: $|m - 1| \ll 1$ and $2ka|m - 1| \ll 1$, where $m = n/n_0$ is the ratio of the particle refractive index to that of the solvent, $2a$ is the approximate diameter of the particle, and wavenumber $k = 2\pi n_0/\lambda_0$. For interactions between particles to have only a small influence on the scattering signal, it must be true that $2A_2Mc \ll 1$. If these three conditions do not hold, then Equation (12.1) is not an accurate representation of the true state of affairs and the molar mass may not be reliably determined. However, if the single condition $2A_2Mc \ll 1$ is true, then the size may still be determined using a full solution of Maxwell's equations for the angular dependence of scattered light. Such exact solutions exist for only a few specific conformations, for example, solid spheres, coated spheres, and ellipsoids [5,10]. The full solution for arbitrary refractive index and size spherical particles is called the *Lorenz–Mie solution to Maxwell's equations*. Measurement of metallic spheres is an example for which Equation (12.1) may *not* be used, because of the difference in refractive index between the particles and the solvent, but for which the Lorenz–Mie solution is valid.

12.3.2. Alternate Expression for Light-Scattering Equation

Equation (12.1) may be inverted and rewritten as

$$\frac{K^*c}{R(\theta)} = \frac{1}{M}\left(1 + \frac{16\pi^2 n_0^2}{3\lambda_0^2}\sin^2\frac{\theta}{2}\langle r_g^2\rangle\right) + 2A_2 c \qquad (12.3)$$

under the assumption that $2A_2McP(\theta) \ll 1$ and $[(16\pi^2 n_0^2)/(3\lambda_0^2)]\sin^2(\theta/2)\langle r_g^2\rangle \ll 1$. Zimm [2] has shown that this alternate expression is valid to higher concentrations than is Equation (12.1). For systems satisfying the two conditions stated above, plotting $K^*c/R(\theta)$ against $\sin^2(\theta/2)$ yields a straight line with a positive slope proportional to $\langle r_g^2\rangle$. This expression is often used when fitting and/or displaying data.

12.3.3. Measurements on Fractionated Versus Unfractionated Samples

Fractionation is a term broadly used to describe a process by which multiple species within a sample can be separated from one another. Asymmetric flow field flow fractionation is an example of a method of fractionation discussed in detail in Chapter 11 of this book. Size exclusion chromatography (SEC) is yet another method of fractionation, whereby a fluid sample is pushed through a column containing a porous medium. As the sample traverses the column, the species in the sample moves out of the main flow of the fluid stream and into the pores of the medium. Larger species do not physically fit into as many such pores as do smaller species, and thus spend more time in the main flow of the fluid stream. This results in larger particles being pushed more quickly through the column than smaller particles, thus separating the species from one another. The fluid exits the column through narrow inner-diameter tubing, and the fractionated samples are then directed to flow "inline" through various instruments. Determination of the molar mass of a species postfractionation requires both a MALS detector and a concentration detector, such as a UV absorbance detector or a differential refractometer. The various species are made to pass sequentially through these detectors, and the signals from the detectors are combined in software in order to determine the species molar masses. Since a single species exiting (eluting from) a column passes sequentially through multiple detectors, in order to correctly interpret the resulting signals, the signals must be shifted in time to align with one another. While the goal for all fractionation methods is to completely separate a sample into its component species, any such method has a limited ability to resolve species that are similar to one another. It is often true that even postfractionation multiple species will be present at some level at the same time within a single instrument. In this case the measured value is an average of the species present in the measurement volume.

For MALS measurements made on unfractionated or incompletely fractionated samples, the molar mass measured is the weight-averaged molar mass, and the radius measured is weighted by the radii squared, molar mass, and concentration:

$$M_W = \frac{\sum_i M_i c_i}{\sum_i c_i}, < r_{\text{g-avg}}^2 > = \frac{\sum_i < r_g^2 >_i M_i c_i}{\sum_i M_i c_i} \tag{12.4}$$

Measurements on unfractionated samples are generally termed "batch" measurements.

For measurements postfractionation, the molar mass, concentration, and r_g are determined for each elution volume. Summing over each elution volume, one can calculate the averages in Equation (12.4) and compared them to unfractionated measurements of the same sample. This is generally good practice, ensuring that no sample is disproportionately lost during the fractionation process, or significantly altered by dilution.

12.3.4. Example Data

In Figure 12.2 we show MALS and concentration detector data for bovine serum albumin (BSA), fractionated via SEC. The data are acquired using a Wyatt DAWN® HELEOS ™ MALS detector, and a Wyatt Optilab® rEX™ differential refractometer.

For the data shown, 100 μL of 2 mg/mL BSA is injected into the fluid stream. The sample passes through the SEC column, where fractionation occurs, then through the light-scattering (LS) instrument, and finally through the differential refractometer, which serves as a concentration detector. For presentation in Figure 12.2, data are offset in time/elution volume to align the monomer peaks from the concentration

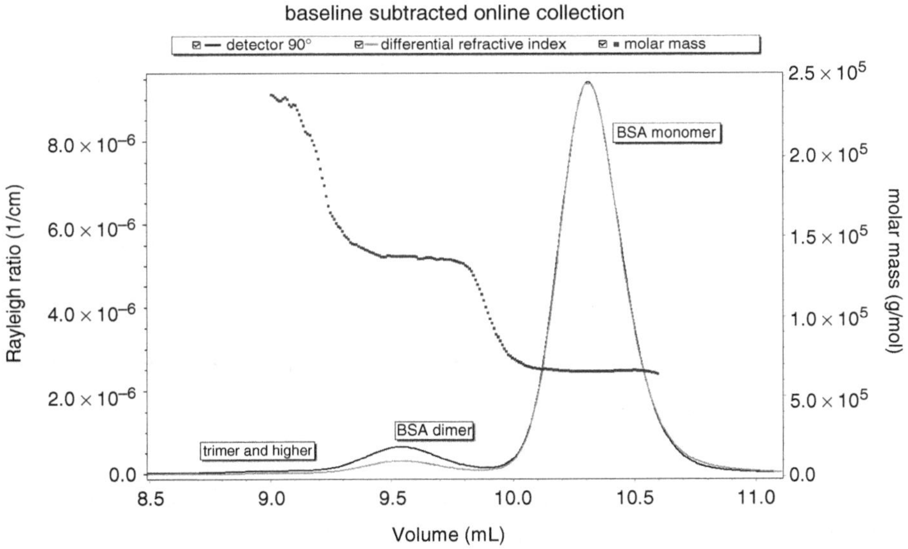

Figure 12.2. Static light scattering (black trace) at 90° scattering, differential refractive index concentration detector (red trace), and molar mass (blue points) data as a function of elution volume for BSA fractionated via SEC. See the insert for color representation of this figure.

detector and the LS detector, and the y-axis scales are selected such that the monomer peaks from the two detectors overlay well. The concentration detector in this case is a differential refractometer, which has a signal proportional to the mass concentration of the sample in solution. In contrast, the LS detector has a signal that is proportional to the concentration *times* the molar mass. With the monomer peaks made to overlay, the dimer peaks do not overlay, because the dimer molar mass is twice as large as the monomer molar mass. By simply overlaying the monomer peaks and examining the other peaks, we can observe that the dimer peak is approximately twice the molar mass of the monomer peak. Using Equation (12.1), we can determine the absolute molar masses for each peak, presented in blue in Figure 12.2.

The example data of BSA in Figure 12.2 do not illustrate the angular dependence of the intensity of scattered light, since r_g of BSA is less than 10 nm (it is approximately 3 nm), resulting in no measurable angular dependence. In such a case the molecular radius is indeterminable by MALS. However, averaging the signals from multiple detectors measuring the scattering at different angles improves the precision of the molar mass measurement (e.g., 16 detectors result in 4 times the precision of four similar detectors). The data shown in Figure 12.3 are from a sample that was fractionated also via SEC. The angular dependence of the MALS signal is seen clearly in Figure 12.3a. The data are plotted in the manner suggested by Equation (12.3), with $K^*c/R(\theta)$ on the y axis and $\sin^2(\theta/2)$ on the x axis. A Wyatt DAWN DSP MALS

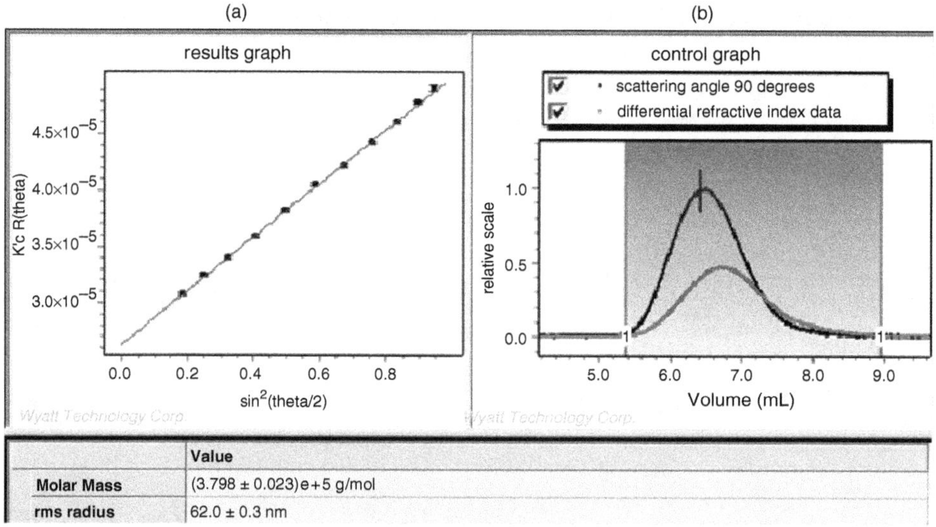

Figure 12.3. Multiangle light scattering measurements for a sample fractionated via SEC: (a) the MALS data showing angular dependence of intensity, with inverse of the intensity versus sin2(θ/2), for the elution volume indicated by the vertical line in (b). Black points in (a) are data, and the line is a fit to the data with $r_g = 62$ nm; (b) light-scattering signal (darker trace) at 90° scattering, and concentration detector signal (lighter trace) as a function of elution volume.

instrument and a Wyatt Optilab DSP differential refractometer were used to obtain the data shown in Figure 12.3.

In Figure 12.3b, we show the LS signal at 90° and the concentration detector signal, both as a function of elution volume. The LS signal and the concentration detector signal appear offset in elution volume, indicating a distribution of masses across the peak. Since we determine the absolute molar mass, radius, and concentration at each elution volume, it is possible to generate distributions of the absolute molar masses and radii (e.g., weight fraction as a function of molar mass). Such distributions are commonly used as QC specifications, for example.

12.3.5. Conformation Information

For each elution volume of a fractionated polydisperse (e.g., aggregating) sample, the molar mass and the RMS radius may be determined using MALS, and information regarding conformation of the sample may be obtained by comparing the molar masses to their respective radii. For an object growing as a solid sphere, the mass of the object increases in proportion to the radius cubed as $m = \rho\frac{4}{3}\pi r^3$, where ρ is the density. In general, the mass of an object is related to the RMS radius as $m = dr_g^a$, where d is a constant and a is the power by which the mass increases with radius ($a = 3$ for a solid sphere, 2 for an ideal random coil in a θ solvent or a hollow sphere, and 1 for a rod). This is usually rewritten as $r_g = (m/d)^{1/a}$, and finally $\log(r_g) = (1/a)\log(m) - (1/a)\log(d)$. Plotting the logarithm of the RMS radius against the logarithm of the molar mass yields a line with a slope $1/a$. Such a plot is termed a *conformation plot*. A conformation plot for several datasets is given in Figure 12.20 in Section 12.5. Those data are consistent with proteins initially aggregating as fibrillates, and eventually beginning to clump or coil.

12.3.6. Conjugate Analysis

Many samples of interest contain two distinct components: for example, a protein with polyethylene glycol or polysaccharides conjugated to the protein, or membrane proteins with surfactants associated with them. It is often important to characterize the manner and degree of conjugation or association. Using two different concentration detectors (e.g., a differential refractometer and a UV absorbance detector) in concert with an LS detector, it is possible to measure the molar mass of the protein, the molar mass of the conjugated material, and the total molar mass and size of the composite object [11].

In Figure 12.4 we show data acquired using a Wyatt DAWN EOS® MALS detector, a Wyatt Optilab DSP differential refractometer, and an Agilent 1100 UV absorbance detector operating at 280 nm. Utilizing the method described by Wen et al. [11], the Wyatt Astra® V software package combines the information from these three detectors to determine the molar masses of the two constituents of the complex, as well as the molar mass and radius of the full complex. Many examples exist in the literature demonstrating the use of this method [12,13].

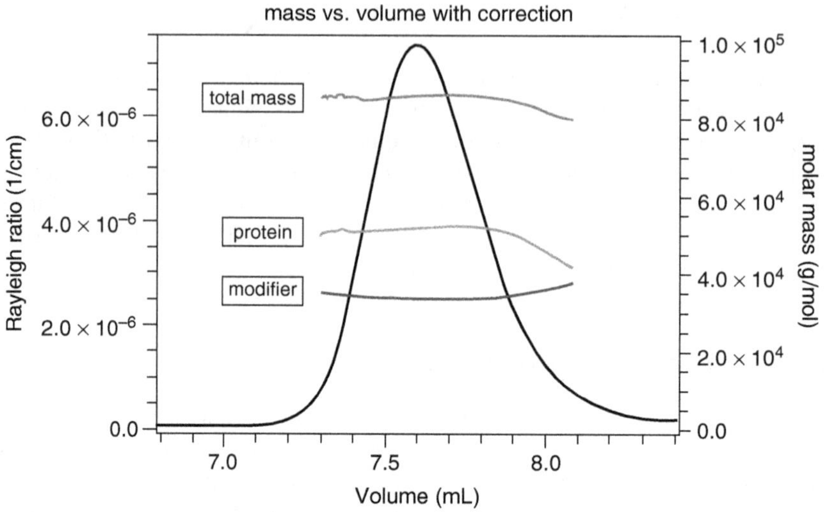

Figure 12.4. Multiangle conjugate analysis is used to determine the molar mass of protein, attached modifier, and the molar mass and radius of the complex.

12.3.7. MALS Measurements of Particle Number Densities

For particles for which Equation (12.1) is applicable and MALS and concentration detector signals are detectable, the number density of particles in solution may be determined in a straightforward way using the measured molar mass and concentration. For fairly monodisperse systems, MALS measurements also can be used to determine the number concentration for particles large enough to yield robust MALS signals, but dilute enough to yield no appreciable signal from a concentration detector (e.g., virus-like particles, liposomes). To accomplish this, the refractive index of the particles and the angular dependence of the MALS signal are used to determine the size, and calculate the absolute scattering, for a single particle in solution. Dividing the total MALS signal by the absolute scattering signal from a single particle yields the number density of the particles [14]. This technique has been used in conjunction with asymmetric flow field flow fractionation to determine the number density of virus-like particles in the concentration range of 10^8–10^{11}/mL with excellent accuracy [15]. The method works well, in theory, down to number densities as low as 10^6 /mL. This analysis method is patented, and is currently commercially available only in the Wyatt Astra software.

12.3.8. High-Concentration Measurements

More recent developments in antibody therapeutics, and the associated formulation challenges, make it increasingly desirable to obtain information regarding the association/aggregation state of proteins at extremely high concentrations, such as hundreds of grams per liter. It is usually quite easy to make high-quality MALS or DLS measurements at such high concentrations, but interpretation of the data remains a challenge. As discussed earlier, Equation (12.1) includes an A_2 term that accounts for interactions between particles in solution, but the equation is valid only at low (in theory vanishingly low) concentrations. Theories do exist that, with reasonably benign assumptions, can predict MALS and DLS results to very high concentrations for an arbitrary potential energy between molecules (e.g., charged spheres) [4]. However, these theories have mathematically complex solutions. Modeling data based on effective hard-sphere interactions is mathematically simple and has been relatively successful in fitting high-concentration results [16], and this model has since been extended to include scattering from associating species [17,18].

In Figure 12.5 we show LS data as a function of concentration, measured to very high concentrations. The vertical axis(ordinate) R/K^* in the figure is proportional to the intensity of scattered light. For concentrations below approximately 10 mg/mL, the data are interpretable using Equation (12.1). At higher concentrations, this model is no longer appropriate, but the data may be fit and interpreted using an effective hard-sphere model.

While it is not necessarily justified [20], it is not atypical to determine interaction parameters at low concentrations and assume that these parameters remain relatively unchanged at high concentrations. The model of stoichiometric associations, discussed in the next section, is often utilized in this way. Using this model, an association coefficient is measured at relatively low concentrations, and the percentage of the sample that is associating (e.g., as dimers) is predicted at high concentrations.

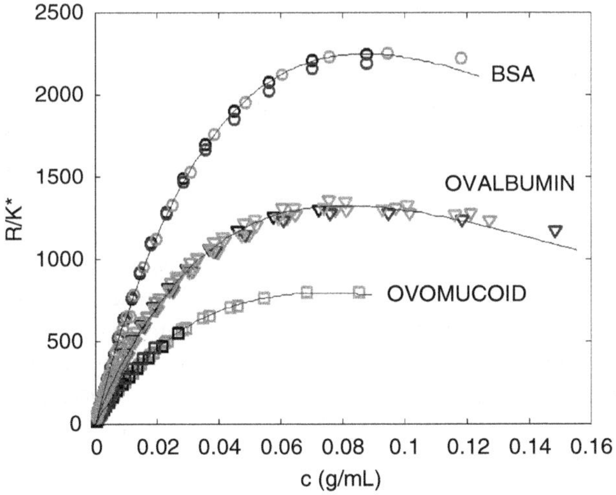

Figure 12.5. Multiangle LS data at 90° scattering taken as a function of concentration for BSA, ovalbumin, and ovomucoid. Points having different shades are data acquired during different experiments. The pronounced nonlinearity at higher concentrations arises from interactions. Solid lines are fits to a model of hard-sphere interactions. (Data from Fernandez and Minton [19].)

12.3.9. Free-Solution Measurements of Self-Association and Heteroassociation

The degree and stoichiometry of reversible self-associations for a single species in free solution, and the heteroassociation between multiple species in free solution, are of key interest in a number of fields for myriad reasons. For formulation development, such associations are typically investigated in order to predict later irreversible aggregation, which may develop for species having a greater degree of association. Irreversible aggregation is usually straightforward to detect and quantitate, since the aggregates remains present after dilution, such as after fractionation. Reversible associations are typically more difficult to measure, since the degree of association is concentration-dependent. Associations are measurable using a number of techniques, some of which require modifying (e.g., fluorescence tagging) or immobilizing the species on a surface, either of which may alter the association characteristics. Other techniques that are capable of measuring associations of unmodified proteins moving freely in solution often require a great deal of time to perform. In contrast, LS measurements that determine self- or heteroassociation properties may be performed within 1–2 h [21–23].

Particles that reversibly adhere to one another spend some proportion of their time bound, and the remainder of their time free in solution. The equilibrium association coefficient for such a system is a measure of the equilibrium condition for the number of particles in the bound state compared to the number of constituent particles free in solution

$$k_a = \frac{[AB]}{[A][B]} \tag{12.5}$$

where k_a is the equilibrium association coefficient with units of 1/molarity, [A] and [B] are the molar concentrations of free A and B in solution, and [AB] is the molar concentration of the associated species [24]. After a solution is prepared, the time required to essentially achieve equilibrium varies depending on the concentrations and association parameters specific to that system. For some systems equilibrium is essentially achieved in fractions of a second, while for others hours are required. Light-scattering models assuming associations are fundamentally different from virial expansion models incorporating A_2, in that association models assume that (1) some portion of the species are physically attached to one another, (2) there are no long-range interactions, and (3) for the duration of an association the associating species are not interacting in any way with any other particles in solution. Association models are accurate in the limit where such interactions dominate, and are not accurate at high concentrations.

12.3.9.1. Self-Association. For a single species having no self-interaction, either attractive or repulsive, the MALS signal increases linearly with sample concentration. If the MALS signal increase is less than linear with concentration, then there is a net repulsive interaction (e.g., unscreened charge), which may be accounted for using the A_2 term in Equation (12.1). If the MALS signal increase is greater than linear, then the signal may be accounted for using either an A_2 model with an attractive interaction as discussed above, or by assuming that the species is reversibly associating with itself. For a model of reversible associations, the association coefficient and the concentration will determine the equilibrium concentrations of monomer, dimer, trimer, and so on. For example, for a model of associations in which we assume that a species A associates with itself only as a dimer, the molar concentrations are given as

$$[A_{tot}] = 2[AA] + [A], \qquad [AA] = k_a[A][A] \tag{12.6}$$

where $[A_{tot}]$ is the total molar concentration of A in solution that is known from preparation, [AA] is the molar concentration of the dimer in solution, [A] is the molar concentration of free A in solution, and k_a is the association coefficient. Given $[A_{tot}]$ and k_a, these equations may be used to solve for the molar concentrations of [A] and [AA], and using Equation (12.4) and the molar mass of A, the corresponding MALS signal may be calculated. Fitting the data to this model, using k_a as a fit parameter, it is possible to thereby estimate k_a. The concentration dependence of a model that includes association as trimers is different from that of a model that incorporates only monomers and dimers, and therefore we can estimate the stoichiometry of self-association using MALS data measured as a function of concentration. In Figure 12.6 we show MALS data taken for a self-association experiment, where the self-association of chymotrypsin, pH 4.1, in 10 mM acetate, is measured at different NaCl concentration. The data are consistent with rising NaCl concentration resulting in an increased degree of self-association.

Sample dilution and data acquisition and analysis for the data shown in Figure 12.6 are automated using the Wyatt Calypso™ system, and acquired using a Wyatt DAWN HELEOS MALS detector and a Waters 2487 UV absorbance detector. Measurement of the five association constants displayed requires 6 h of automated run time, and 12 mL of 1.135 mg/mL chymotrypsin stock solution. Chymotrypsin has a molar mass of approximately 25 kDa. Similar measurements with samples having a larger molar mass require a smaller quantity of sample, while measurements for samples having a smaller molar mass require a larger quantity of sample.

12.3.9.2. Heteroassociation. For a mixture of two species, A and B, if the species are not associating, then the MALS signal is simply the weight average of A and B as in Equation (12.4). In this case, if MALS measurements are made with mixtures of solutions ranging from 100% stock solution A to 100% stock solution B, then the light-scattering signal changes linearly between these endpoints, as a function of the ratio of stock solutions. If a mixture of A and B results in a MALS signal in excess of this value, then the excess scattering may be understood in terms of A and B associating with one another, for example, forming an AB complex. For a model of associations in which we assume that species A associates with species B to form AB, with some association constant k_a, the molar concentrations of the various species are given as

$$[A_{tot}] = [AB] + [A], \qquad [B_{tot}] = [AB] + [B], \qquad [AB] = k_a[A][B] \tag{12.7}$$

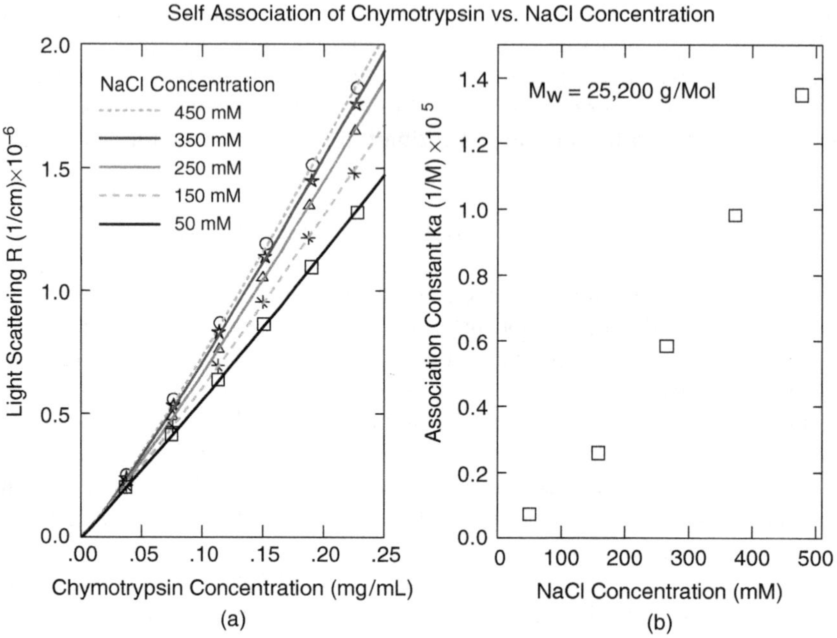

Figure 12.6. Multiangle LS free-solution self-association measurements of chymotrypsin: (a) MALS data as a function of chymotrypsin concentration for different NaCl concentrations solid lines are fits to the model of Eq. (12.6); (b) the resulting association constants k_a for the different NaCl concentrations. (Unpublished data of Hanlon and Some [25].)

where $[A_{tot}]$ and $[B_{tot}]$ are the total molar concentrations of A and B in solution that are known from preparation, [AB] is the molar concentration of the associated species, and [A] and [B] are the molar concentrations of free A and B in solution. Similar to the self-associating case, given $[A_{tot}]$, $[B_{tot}]$, the molar mass of A, and the molar mass of B, we can fit MALS data to estimate the association constant k_a. Moreover, the model of associations as AB may be checked, since for such associations the MALS signal will have a maximum value at a 1 : 1 molar ratio of A : B. If associations actually include AAB as an associating species, then the MALS signal will have a maximum signal between molar ratios of 2 : 1 and 1 : 1, and a model assuming only AB associations will simply not fit the data.

In the heteroassociation example above, it is possible that A or B may be self-associating in addition to associating with one another. This may be determined by measuring the concentration dependence of the MALS signal for A and B separately. With the self-association coefficients known from these experiments, the heteroassociation coefficient may then be determined from the concentration crossover experiment. These three measurements may be automated and performed within one experiment using three solutions: stock A, stock B, and buffer. In Figure 12.7 we show data taken in an automated self- and heteroassociation experiment.

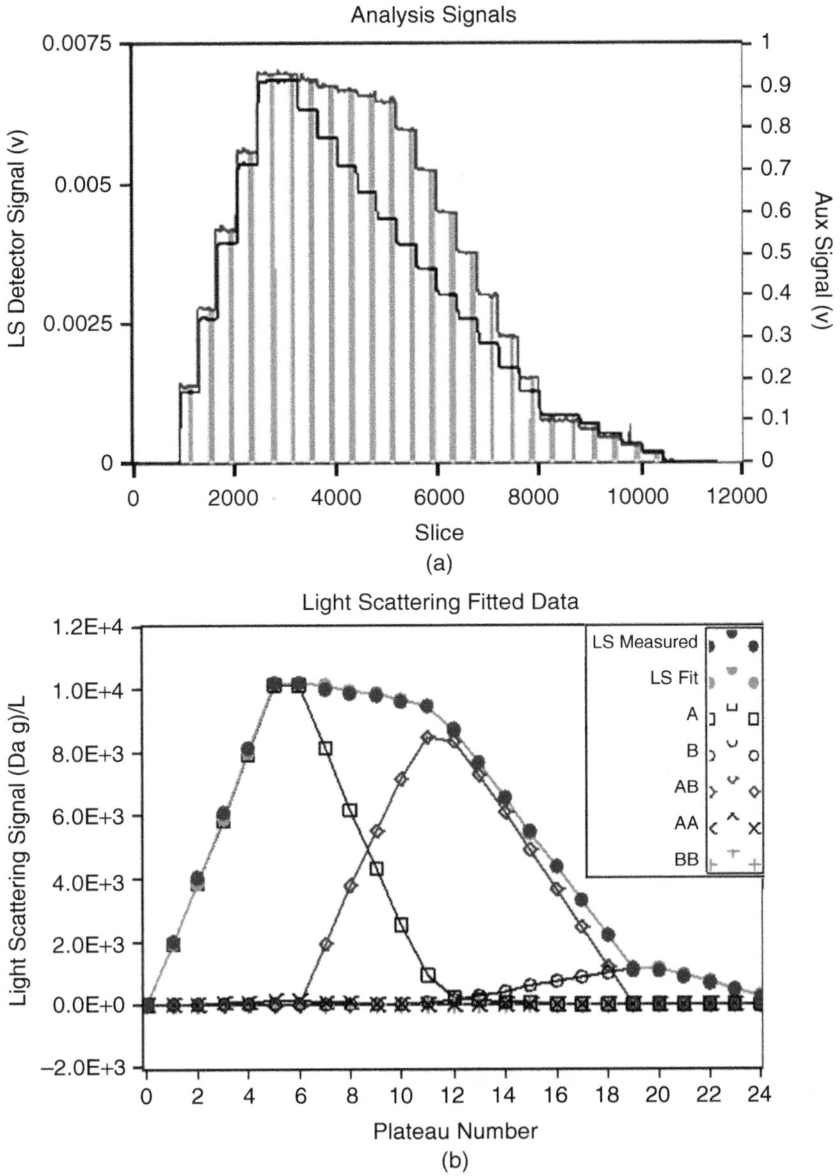

Figure 12.7. Multiangle LS free-solution self- and heteroassociation measurements of chymotrypsin and bovine pancreatic trypsin inhibitor (BPTI): (a) MALS data (upper trace) and concentration detector data the MALS trace is significantly above the concentration detector trace during the crossover from 100% chymotrypsin to 100% BPTI due to associations. In (b) MALS data (solid circles) derived from (a), along with the results of a fit (line through solid circles) to a model of self- and heteroassociation.

In this example, chymotrypsin and bovine pancreatic trypsin inhibitor (BPTI), in PBS at pH 7, form 1 : 1 complexes. Fitting the data to an AB association model yields an equilibrium association constant of 1.5×10^6 1/M. Data acquisition and analysis for data shown in Figure 12.7 are automated using the Wyatt Calypso system, and acquired using a Wyatt DAWN HELEOS MALS detector and a Waters 486 UV absorbance detector. Here the measurement of the three association constants (chymotrypsin and BPTI self-association coefficients, and the heteroassociation between chymotrypsin and BPTI) requires 1 h 40 min of automated run time, 22 mL of 0.45 mg/mL chymotrypsin stock solution, and 22 mL of 0.16 mg/mL BPTI stock solution. As with the self association measurements, larger molar mass molecules require less sample, while smaller molar mass molecules require more.

Previously we described MALS measurements of self- and hetero-associations between biomolecules. Recently, DLS has also been used for similar measurements [26] with the measured average hydrodynamic radius as the parameter indicating the presence of associations (as opposed to average molar mass for MALS measurements). MALS association measurements are more sensitive than similar DLS measurements; however, DLS measurements made in a multiwell plate require orders of magnitude less sample than similar MALS measurements.

12.3.10. Limits of Detection

The signal measured by a MALS instrument is proportional to the molar mass times the concentration. This means that for a given MALS instrument the minimum detectable concentration is lower for samples having a higher molar mass. Conversely, the molar mass of a very low molar mass molecule may be measured, provided that a very high concentration is used. For example, the molar mass of sucrose, at 342 g/mol, may be measured if approximately 0.5 mg/mL or greater concentrations are used. For proteins in aqueous solutions, state-of-the-art instruments are capable of measuring $M_W(\text{g/mol}) \times c(\text{g/L}) > 80[\text{g}^2/(\text{mol} \cdot \text{L})]$. Thus, a 100-kDa protein requires a minimum concentration of 0.8 μg/mL in the measurement volume. Fractionation techniques usually dilute samples, typically by a factor of 5–10, and this must be taken into account when estimating the minimum measurable concentration for a particular experiment.

12.4. DYNAMIC LIGHT SCATTERING

Dynamic light scattering is a standard technique used to determine the translational diffusion coefficients of particles in solution, from which the equivalent spherical hydrodynamic radii may be calculated. It is used to estimate the size, and size distributions, of particles in solution, and there are numerous texts and articles describing this technique [6,27,28]. Here we present a discussion of the origin of a DLS signal, following the description by Pecora [27].

With two particles in solution scattering light, there is usually a difference in the length of the path that light takes to travel from the light source, to the different particles, and then to the photodetector. This difference in the pathlength of the

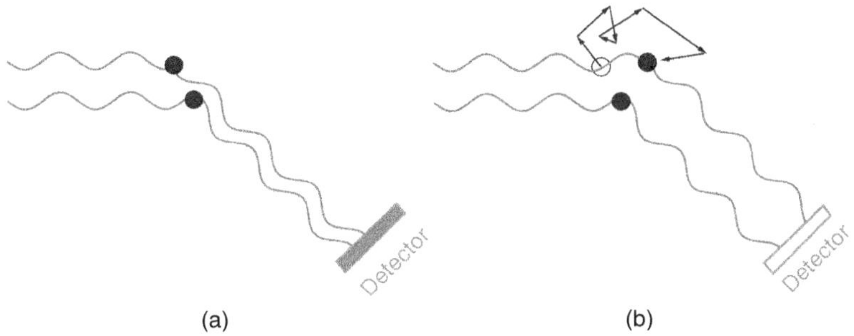

Figure 12.8. (a) Two particles have a spatial relation resulting in constructive interference of scattered light at the detector; (b) the particles have moved with respect to one another, causing the scattered light to interfere destructively at the detector. Thus, the intensity viewed by the detector varies through time.

light scattered from the two different particles results in a phase difference at the photodetector, and leads to interference of the light at the detector. This interference may span the range from constructive to destructive, and results in a corresponding intensity of light at the photodetector that is bright (up to 4 times the average intensity), dark, or somewhere in between. Thermal energy causes the particles in solution to move (diffuse), and so two particles in solution continually move with respect to one another. The resulting motion through time will change the pathlengths of light from light source, to the particles, and finally to the photodetector, thus altering the intensity of light on the photodetector. This is shown schematically in Figure 12.8. The same reasoning holds true for a macroscopic number of particles in the measurement volume, each thermally diffusing, whereby the light scattered from each particle interferes with the light scattered from every other particle.

Smaller particles diffuse more rapidly faster than larger particles do; therefore, the timescale of intensity fluctuations at the detector is shorter for smaller particles. In Figure 12.9 we show simulated data for intensity fluctuations for small versus large particles diffusing in water. The translation diffusion coefficient D for a *spherical* particle of radius a is given by the Stokes–Einstein relation

$$D = \frac{k_B T}{6\pi\eta a} \tag{12.8}$$

where k_B is the Boltzmann constant, T is the absolute temperature, and η is the solution viscosity [29]. The particle radius determined in a DLS measurement is an equivalent spherical radius, termed the *hydrodynamic radius* r_h, which is the radius of a hard sphere having the same diffusion coefficient as the particle being measured.

The timescale over which the intensity changes at the detector (i.e., the time it takes the intensity of light viewed by the detector to go from bright to dark) is approximately given by the time it takes a particle to move sufficiently to change the pathlength of light by half the wavelength of the light in the solution. A particle with translational diffusion coefficient D, diffusing for time t, will on average travel

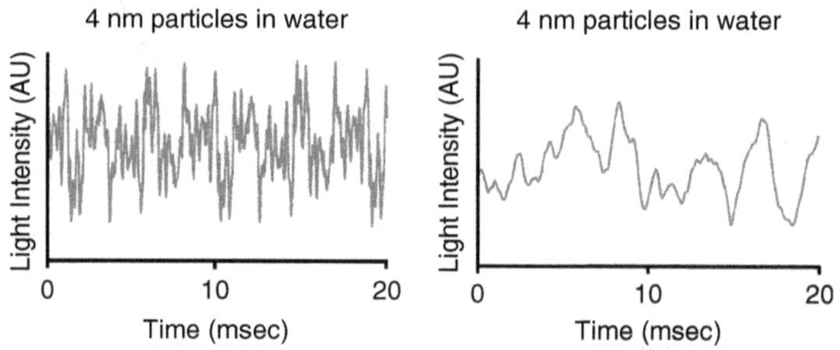

Figure 12.9. (a) Intensity of light through time expected for 4-nm-radius particles diffusing in water; (b) intensity through time expected for 40-nm radius particles.

a distance $\langle x \rangle = \sqrt{2Dt}$ from its starting point. The distance particles must travel to change the pathlength of light to the detector by half the wavelength of light depends on the scattering angle viewed by the detector. If the detector is exactly in the beam of light, at zero scattering angle, then all pathlengths of light for all particles in solution are identical. In such a case, the light scattered from all particles interferes constructively, and there is *never* any change in the pathlengths of light regardless of how far the particles move, and so there is no intensity fluctuation through time at the detector. If the detector observes light scattered into a low angle, near zero, then a difference in pathlength for different particles in solution can occur. However, the particles must move very far through space in order to change the pathlength of light by half the wavelength of light, thereby altering the intensity of light at the detector. If the detector is at a high angle, then the particles do not need to move very far to alter the intensity of light. The same particles in solution, diffusing at the same rate, will have very different timescales of intensity fluctuation, depending on the scattering angle of the light viewed by the detector. By changing the angle of measurement, it is possible to "speed up" or "slow down" intensity fluctuations that might otherwise be difficult to measure.

Quantifying the timescale(s) of intensity fluctuations is accomplished via the standard technique of *correlation analysis*, where two signals are compared in order to determine the degree of similarity between them. *Autocorrelation* is a correlation analysis where a function is compared to itself, in this case to determine how self-correlated the intensity is through time. The intensity autocorrelation function (ACF) is given by

$$g(\tau) = \frac{\langle I(t)I(t+\tau)\rangle}{\langle I \rangle^2} = \frac{1}{2T\langle I \rangle^2}\int_{-T}^{T} I(t)I(t+\tau)\,\mathrm{d}t \qquad (12.9)$$

where $\langle I \rangle$is the time-averaged intensity (i.e., the signal detected in a static LS measurement), $I(t)$ is the measured intensity as a function of time, and$I(t+\tau)$ is the same intensity through time but shifted in time by the delay time τ.

Figure 12.10 illustrates the mathematical process of determining the autocorrelation of a signal. In this figure we show a simulated intensity of light through time,

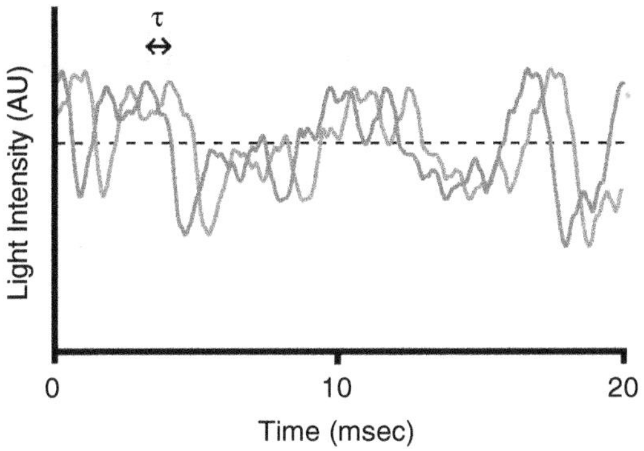

Figure 12.10. Simulated data of intensity through time, overlaid with the same intensity through time, offset in time by delay time τ. The intensity through time is correlated to itself at short delay times, but becomes uncorrelated to itself at long delay times.

overlaid with the same intensity through time, offset by delay time τ. If the delay time τ is zero, the two lines fall on top of one another. The intensity is always maximally correlated to itself for zero delay time τ, resulting in a maximal value for $g(\tau)$. As τ increases to the degree shown in Figure 12.10, the two traces remain correlated to one another, but not perfectly so since, throughout that interval, the particle positions with respect to each other have changed slightly. As τ increases further, the intensity becomes less and less correlated with itself. In the limit where τ is increased to the extent that no correlation exists, $\langle I(t)I(t+\tau)\rangle = \langle I\rangle^2$, and the autocorrelation function has a value of 1.0.

For a single species undergoing pure translational diffusion, the autocorrelation may be shown to be a pure exponential function, $g(\tau) = 1.0 + \beta\exp(-2\Gamma\tau)$, where β is the amplitude of the DLS signal, $\Gamma = q^2D$ is the correlation rate, $q = (4\pi n/\lambda_0)\sin(\theta/2)$ is the scattering wave vector set by the experimental conditions, n is the solvent refractive index, λ_0 is the vacuum wavelength of the light used in the measurement, and θ is the scattering angle.

If the ACF data fit well to a single exponential, then the resulting diffusion coefficient and Equation (12.9) may be used to calculate the hydrodynamic radius r_h. In Figure 12.11, we show autocorrelation function data and their fit to a simple exponential function for an immunoglobulin G (IgG). Data are acquired using a Wyatt DAWN EOS light-scattering instrument equipped with a Wyatt QELS dector. For this instrument, MALS and DLS(QELS) data may be obtained simultaneously, both in batch mode or in flow mode postfractionation. The data shown are acquired in flow mode after SEC separation. We display the data and fit, both with a linear and a logarithmic time axis. Correlation data are almost universally displayed with a logarithmic time axis, allowing a broad range of timescales to be viewed on the same graph.

If there is some distribution of species present, the species will in general have a distribution of diffusion coefficients D and corresponding correlation rates Γ. For

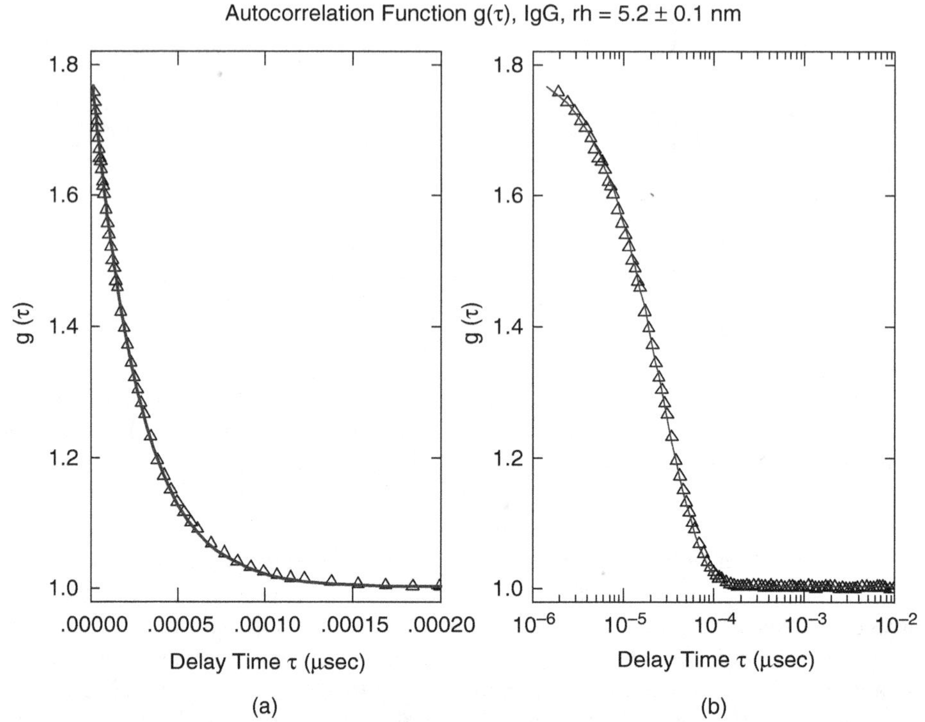

Figure 12.11. (a) Autocorrelation Function data (triangles) along with a fit of the data to a simple exponential function (line); (b) the same data and fit displayed with a log time axis. A log time axis is used in order to simultaneously view many time regions. Data are for an IgG sample and are taken online in flow mode, after SEC separation.

some distribution of correlation $G(\Gamma)$ rates normalized to 1, the ACF becomes

$$g(\tau) = B + \beta\left[\int_0^\infty e^{-\Gamma\tau}G(\Gamma)\,d\Gamma\right]^2 \tag{12.10}$$

The relative contributions to $G(\Gamma)$ by the different species depend on the relative intensities of light scattered by the various species and viewed by the detector. The relative contributions to $G(\Gamma)$ are therefore influenced by the distributions of molar mass, concentration, *dn/dc*, particle size r_g, and also the angle of scattered light viewed by the detector.

If the data do not fit well to a simple exponential function, then some distribution $G(\Gamma)$ exists, and we can analyze the data to estimate $G(\Gamma)$. In general, it is not possible to take a set of real data (by "real," we mean having finite resolution, extent, and noise), which consists of some distribution of exponential functions, and uniquely determine the true underlying distribution of exponentials. There are three possible ways to proceed: (1) the method of cumulants, in which we estimate general features of the distribution $G(\Gamma)$, such as the mean and width of the distribution; (2) the method of assuming some distribution for $G(\Gamma)$ and fitting the data to that assumed distribution; and (3) the method of regularization, whereby we determine a set of $G(\Gamma)$ that all equivalently fit the data, and then make some assumption to choose the "best" distribution between these equivalent fits (most typically the smoothest distribution is chosen). These three methods are discussed below.

12.4.1. Cumulants

The method of cumulants attempts to determine general features regarding the distribution $G(\Gamma)$, without assuming an underlying distribution [30]. In fact, since a functional form is used to fit the data, and the Laplace transform is unique [31], some underlying distribution is always assumed. In a cumulants analysis, ACF data are fit to the function

$$g(\tau) = B + \beta \exp\left(-2\kappa_1\tau + \kappa_2\tau^2 - \frac{\kappa_3}{3}\tau^3 + \cdots\right) \tag{12.11}$$

where κ_i are the cumulant terms. The first three cumulant terms are equal to the first three moments of the distribution of $G(\Gamma)$. Higher cumulant terms are related but not equivalent to higher moments of the distribution. Equation (12.11) is exactly true if all (infinitely many) cumulant terms are included in the expansion. Autocorrelation function data are seldom feature-rich to the point where more than the first two or three cumulant terms may be used for fitting. Therefore, when analyzing data, the cumulant series must be truncated, typically keeping just the first and second cumulant terms. In this case the functional form used for fitting becomes

$$g(\tau) = B + \beta \exp(-2\kappa_1\tau + \kappa_2\tau^2) \tag{12.12}$$

Fitting the data to this function works reasonably well, but since κ_1 and κ_2 must be positive, it is clear that the $\kappa_2\tau^2$ term in the exponential will always lead to a divergence in the fit function at a sufficiently large τ. Such a function cannot be physical, since any combination of decaying exponential functions must monotonically decrease. It may be shown that the distribution of $G(\Gamma)$ that gives rise to Equation (12.12) is a simple Gaussian distribution of correlation rates, including negative correlation rates, which, of course, in actuality do not exist. Nevertheless, fitting ACF data using Equation (12.12) usually work well to estimate κ_1 and κ_2, as long as the data at longer delay times are not included in the fit.

12.4.2. Assuming $G(\Gamma)$:

As discussed above, while a cumulant fit is in fact equivalent to assuming a distribution $G(\Gamma)$, the assumed distribution is rarely physically possible. One way to ensure a physically reasonable distribution $G(\Gamma)$ is to pick a distribution and use Equation (12.10) to calculate the expected $g(\tau)$. An obvious choice is the Gaussian distribution for $G(\Gamma)$ assumed by the standard cumulant model, provided we limit the distribution to positive correlation rates. Many other possibilities exist for physically reasonable distributions.

12.4.3. Regularization

The analysis methods for DLS data discussed thus far provide general information regarding the distribution $G(\Gamma)$ (e.g., the first and second moments of the distribution), but do not determine the actual distribution $G(\Gamma)$ from which the distribution of sizes can be derived. Given a purely mathematical function for $g(\tau)$, it *is* mathematically possible to use the inverse Laplace transform [31] to determine $G(\Gamma)$ uniquely. However, given a "real world" $g(\tau)$, which does not have infinite resolution and extent but does have finite noise, the distribution cannot be uniquely determined. Instead, we find an infinite number of $G(\Gamma)$ which all fit the data equally well, and choosing between these solutions will require some additional assumption unsupported by the data [6,32]. Most commonly the "smoothest" distribution—the one that minimizes the sum over the second derivative of $G(\Gamma)$—is chosen as the most reasonable solution. Finding a set of $G(\Gamma)$ that best fits the data, and then choosing among these $G(\Gamma)$ using some additional requirement, is generally called *regularization*. We illustrate the method of regularization analysis in Figure 12.12, where we present data drawn from three distinct samples.

Figure 12.12a,b presents data and results for fairly monodisperse samples, with measured hydrodynamic radii of 3.8 and 51 nm, respectively. Figure 12.12c shows data and results from a regularization analysis for a mixed sample containing the samples in both (a) and (b). For these disparate-sized species, regularization analysis clearly establishes the presence of two species. A rule of thumb is that for a regularization analysis to resolve two species as distinct, their hydrodynamic radii must differ by a factor of ~ 4. For the data in Figure 12.12, 5-μL samples are measured within a 1536-well plate using a Wyatt DynaPro™ DLS plate reader.

Regularization results are very easily and commonly overinterpreted. They are only an estimate of $G(\Gamma)$, and that estimate can easily be wrong. Regularization methods generally do produce quite good estimates of $G(\Gamma)$ for clean, simple systems such as bimodal distributions or broadly polydisperse systems. However, the technique is not able to resolve fine details of a distribution. For example, for a sample consisting of an equal mixture of monomer and dimer species, a regularization analysis can never resolve the presence of two distinct species. However, either regularization or a cumulant-type analysis can easily resolve the change in polydispersity (described below) and shift in average r_h associated with the presence of a significant percentage of dimer.

12.4.4. Polydispersity

The sample percent polydispersity is a measure of the breadth of the distribution of sizes in the sample with respect to the average size. It is calculated as $Pd = 100\mu_2/\mu_1^2$, where μ_1 and μ_2 are respectively the first moment (mean value) and second moment (width) of the distribution of correlation rates $G(\Gamma)$. The first and second moments of a distribution are equal to the first and second cumulant terms.

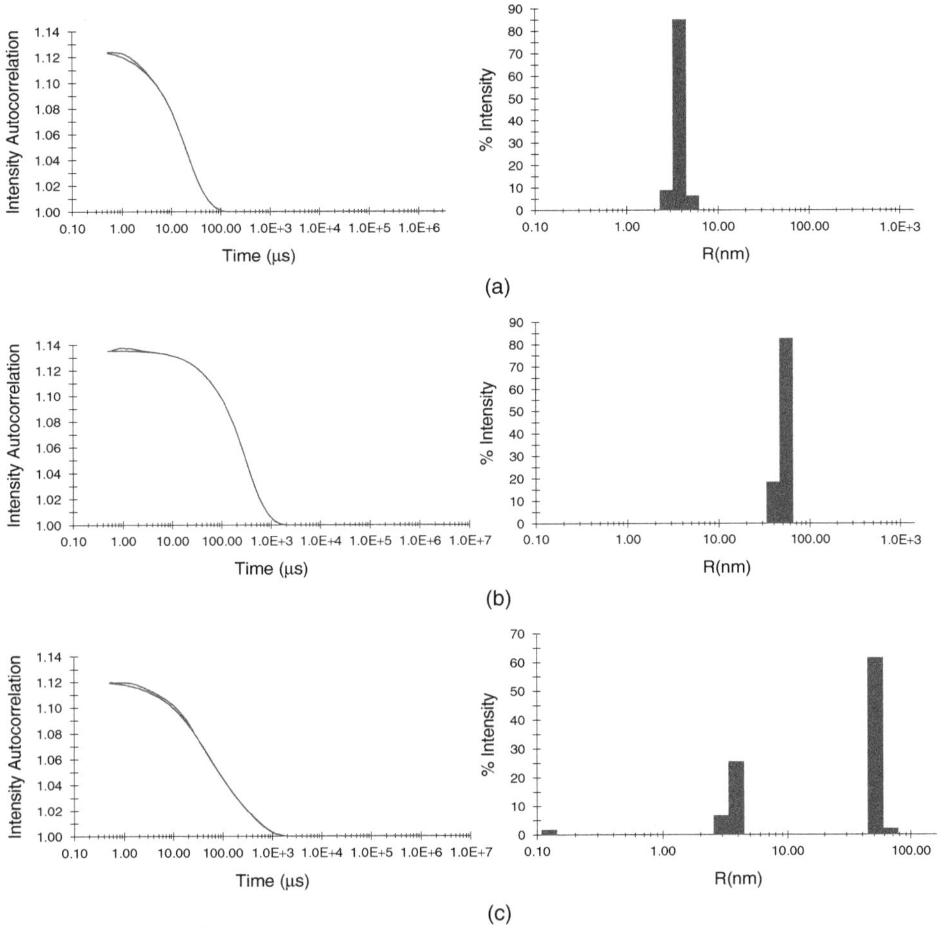

Figure 12.12. Data and regularization fits for three samples. Data for (a) and (b) are from two different monodisperse species, and data for (c) is from a mixture of the two.

12.4.5. Examples of Analysis Methods

The data displayed in Figure 12.13 do not fit well to a simple exponential function, but do fit well to a Gaussian distribution of correlation rates, indicating that there is some distribution of sizes present in the sample. The data shown in Figure 12.13b are from a sample containing a mixture of $r_h = 3.8$ nm and $r_h = 51$ nm species. For this bimodal distribution, a Gaussian distribution of correlation rates is clearly not an accurate description of the real underlying $G(\Gamma)$, and therefore results in a poor fit to the data. A regularization analysis yields a much closer correspondence between data and fit result.

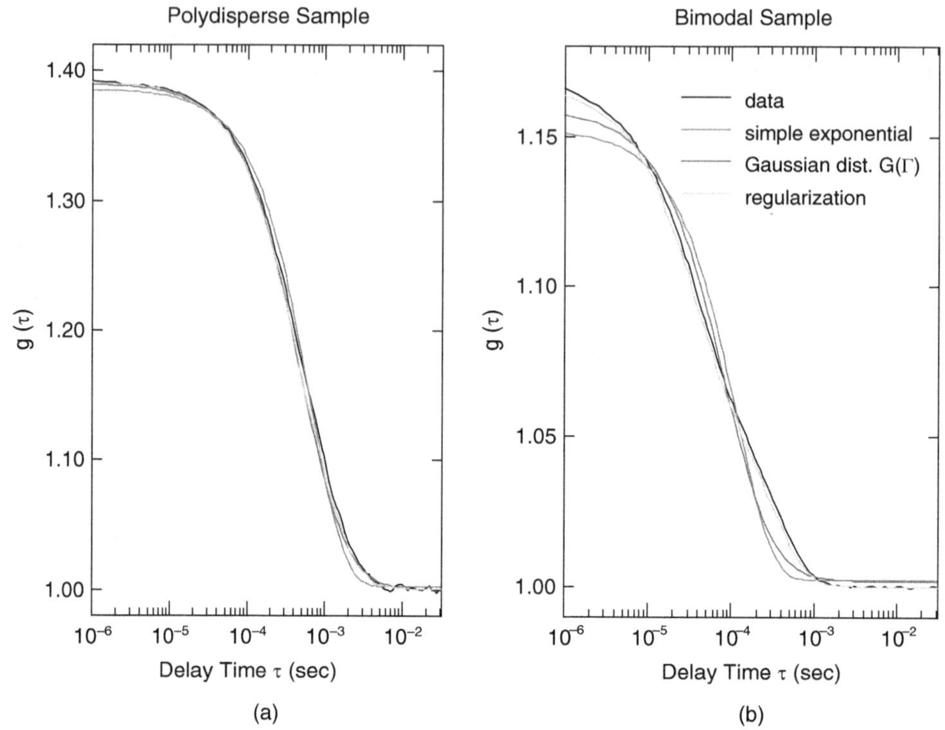

Figure 12.13. (a) Data and fits to a polydisperse sample; (b) data and fits to a sample prepared for a bimodal distribution (mix of $r_h = 3.8$ nm and $r_h = 50$ nm). Both 5-μL samples were measured within a 1536-well plate, using a Wyatt DynaPro DLS plate reader. See the insert for color representation of this figure.

12.4.6. Determination of Aggregate State

Dynamic light scattering is extremely sensitive to the presence of large aggregates, and may be used semiquantitatively to estimate the weight fraction of such aggregates in an unfractionated sample. It is also an excellent tool for assessing whether a sample is largely monodisperse, or has a significant degree of polydispersity. It does not have the resolution for measurements on unfractionated samples to quantitate the percentage of monomer, dimer, trimer, and so on present in a sample. However, if one assumes that the sample is a simple combination of monomer and dimer, with no higher oligomers, then it is possible from the shift in measured radius to estimate the degree of dimerization. For such a system, the measured hydrodynamic radius $r_{\text{h-avg}}$ may be seen to be

$$r_{\text{h-avg}} = \frac{1 + P_d}{(1 - P_d)/r_m + 2P_d/r_d}$$

where P_d is the percent dimer by weight, r_m is the hydrodynamic radius of the monomer, and r_d is the hydrodynamic radius of the dimer. For two connected hard spheres, the hydrodynamic radius is calculated to be 1.33 times that of a single sphere [33], and so it is reasonable to assume $r_{\mathrm{d}} = 1.33r_{\mathrm{m}}$. For a clean system it is possible to achieve a standard deviation of 1% on the measured radius in DLS measurements, corresponding to a detection limit of about 3% dimer, by weight, for batch measurements. Variations in the viscosity of the sample solution due to changes in formulation can easily have an influence on the reported radius of much greater than 1%. Achieving a 1% reproducibility on the reported radius requires either comparing only sample solutions that are known to have the same viscosity, or knowing the relative sample viscosities to an accuracy of much better than 1%. There is, of course, no way to know from such a measurement whether the shift in radius is due exclusively to the presence of dimers, but an upper limit on the percentage of dimers may be obtained in this way.

12.4.7. Number Fluctuations

In the discussion above we state that the autocorrelation is a simple exponential function for a single species undergoing pure translational diffusion. This is true as long as the intensity fluctuations of light through time are dominated by the processes described above, and are not significantly influenced by changes in the number of particles in the detected volume [34,35]. For a given volume containing some number of particles n, the expected fluctuation through time in the number of particles is $\sqrt{n}$, and so the fluctuations in the intensity of scattered light due to number fluctuations is given by $\sqrt{n}/n$. For n of 100, 10% fluctuations in intensity may be expected, and for n of 1000, 3% fluctuations in intensity may be expected from number fluctuations. For most DLS measurement systems, 3% fluctuations in intensity due to number fluctuations are difficult to observe. The DLS measurement volumes of commercial instruments tend to be 40–100 μm in diameter with approximate corresponding volumes of $(40\ \mu\mathrm{m})^3$–$(100\ \mu\mathrm{m})^3$. This yields the minimum number density of particles to be $1.5\times10^{10} - 1\times10^{9}\ /\mathrm{cm}^3$, for which the resulting correlation functions are exponential and interpretable using the models discussed previously. For number densities below this minimum, the measured intensity of light fluctuates through time with a timescale associated with the diffusion coefficient *and* the particle concentration. Such number fluctuations are only observable for large particles, since dilute solutions of smaller particles do not scatter sufficient light to be measurable. The ACF from a sample with number fluctuations has both the expected exponential decay, and also a much longer time-scale "foot" which is not exponential. It is possible, in theory, to use ACF data, which are strongly influenced by number fluctuations to determine the concentration of large particles in solution [36]. In practice, however, it is not easy to do this. Generally, if fits to data result in radii that are very large (close to 1 μm or more), we must take care to ensure that the timescale measured is the timescale of diffusion, rather than that of number fluctuations. This may be checked by measuring at different dilutions. If the signal is due to number fluctuations, lower concentrations will result in a larger amplitude of the "foot" in the ACF.

12.4.8. Multiple Scattering and Interactions

Multiple scattering (measurements of milky or turbid solutions) will result in a reduction in the apparent size measured by DLS. The degree to which the DLS signal is influenced by multiple scattering depends on instrumental details of light path length in solution, and other factors. Techniques exist that can largely reject multiply scattered light [37], but we will not discuss them in detail here. Measurements may be made as a function of sample concentration, to evaluate the effect of multiple scattering on DLS measurements. If no concentration dependence is seen, then multiple scattering is most likely *not* influencing the data, although there is always the small chance of competing effects canceling one another. If concentration dependence *is* seen, then it may be due to either multiple scattering or interparticle interactions.

If the mean distance between particles in solution $< x > = n^{-1/3}$, where n is the sample number density, is similar to the length scale of the potential energy between the particles, then interactions between particles may significantly influence the diffusion coefficient of the particles [27,38]. For example, the potential energy between charged molecules in solution is strongly influenced by the ionic strength of the solution, with the length scale of the interactions given by the Debye–Huckel screening length. For solutions with low ionic strength, the interaction length becomes long and interactions between molecules can cause the diffusion coefficient to increase, resulting in an apparent reduction in the hydrodynamic radius of the molecule. In this case, the apparent size decreases with decreasing ionic strength, or with increasing concentration at fixed ionic strength. This effect has been used to estimate the charge on proteins [39]. Once again, if DLS measurements indicate no concentration dependence, we can assume the effect to be negligible.

12.4.9. Limits of Detection

As with MALS measurements, the DLS signal is proportional to the sample molar mass and concentration. The quality of a DLS measurement may also be improved by a longer measurement time. In general, the data quality improves with the square root of the measurement time; thus, measuring 4 times as long results in an improvement in the signal-to-noise ratio (SNR) by a factor of 2. For proteins in aqueous solutions, state-of-the-art instruments are capable of measuring $M_W(\text{g/mol}) \times c(\text{g/L}) > 1400[\text{g}^2/(\text{mol} \cdot \text{L})]$ for 30-s measurements. For a 100-kDa protein, this corresponds to a minimum concentration of 14 μg/mL.

12.5. CASE STUDY: MALS AND DLS MEASUREMENTS OF AN AGGREGATING PROTEIN

An interesting example of the utility of LS techniques concerns a protein system where the protein aggregates in response to changes in an environmental condition, such as formulation pH, ionic strength, concentration, series of freeze–thaw cycles, and forced aging. We examined such a concept by heat-stressing BSA at 60°C for varying periods of time, from 0 to 175 mins, and performed a set of measurements on the heat-stressed samples. We determined the degree and manner of aggregation within the system both as a function of time exposed to 60°C, and as a function of time stored post exposure. Measurements conducted included (1) batch (unfractionated) DLS measurements immediately after heat stressing the sample and after 13 days storage poststress, (2) MALS measurements on two sample dilutions with each fractionated by SEC, and (3) batch MALS measurements. The time, material, and resources required, as well as the kind of information obtained, varies considerably between these measurements. In general, batch methods are faster and easier, while fractionation methods require more time and provide much more detailed information. Batch measurements allow measurement in the formulation "cocktail" at the concentration of interest, while fractionation methods often include dilution and solvent exchange during fractionation. In general, it is important to perform both batch and fractionated measurements, and to confirm that the fractionation process has not removed material from the sample. From measurements following fractionation, it is always possible to reconstruct the expected result of a batch measurement. If there is a discrepancy between these measurements, it must be explained. It is also important to measure samples prepared at different concentrations to examine any aggregations or associations found for possible reversibility.

In our experiment, BSA (Sigma-Aldrich) was prepared without dialysis at 2 mg/mL concentration in a phosphate buffered saline solution, pH 6.7, and filtered using a 0.020 μm Whatman Anotop™ filter. A vial containing the BSA sample was placed into a water bath held at 60°C, and aliquots were removed throughout a period of 175 min. Each aliquot was stored at 4°C–5°C. The data collected confirm the earlier studies of Holm [39] that elevated temperature can cause irreversible aggregation. The fraction of each aliquot in an aggregated state, the weight-averaged molar mass of the aggregate, and the average particle size all increased with time held at 60°C. The data were consistent with the protein aggregating as fibrillates; initial aggregates yielded a linear conformation while those produced by longer exposure assumed a more random structure (i.e., the fibrillates increasingly clump or coil).

12.5.1. DLS Measurements

For the DLS measurements, triplicate 30 μL samples were collected every 5 min, during the 175 min that the sample was exposed to the 60°C environment, and placed into a Corning low-volume 384-well plate (Corning part number 3540), resulting in a total of 108 aliquots collected for the 36 different exposures. All wells were topped with approximately 15 μL of silicon oil to prevent evaporation. Measurements were made using a Wyatt DynaPro DLS plate reader (Plate Reader Plus), with temperature set to 5°C. For this instrument the entire sample plate was held constant at the temperature of measurement, and measurements were made by directing a beam of light into the wells, without removing any sample from the wells and without physically introducing anything into the wells.

For each aliquot measured, DLS data were collected over 25 s. Including movement between wells and automatic adjustment of laser power and detector attenuation settings, the total measurement time for all wells was about 50 min. All error bars

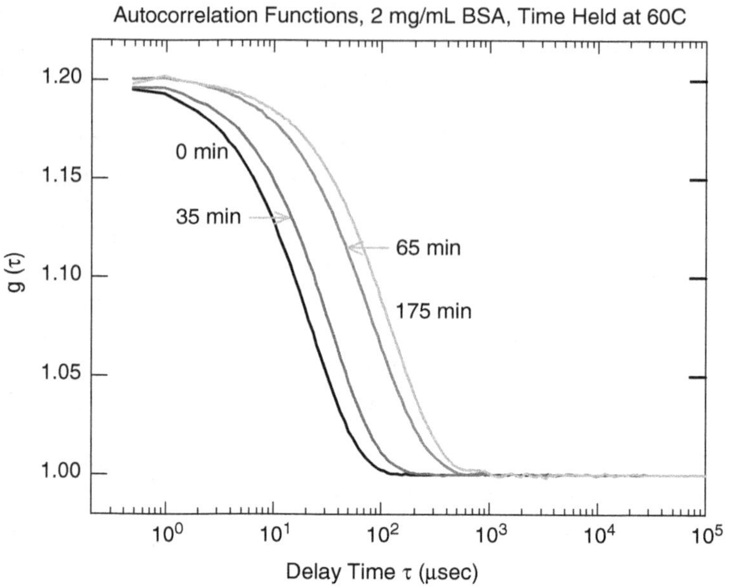

Figure 12.14. Autocorrelation functions for 2 mg/mL BSA aliquots exposed to 60°C for 0, 35, 65, and 175 min.

shown in the figures reflect the associated standard deviations of the measurements of the same sample in different wells. Data acquisition and data analysis were performed by the Wyatt Dynamics software v6.9.2. Samples were measured soon after being prepared, stored at 4°C–5°C for 13 days, and then measured again.

In Figure 12.14 we show the autocorrelation functions $g(\tau)$ for aliquots held at 60°C for 0, 35, 65, and 175 mins. The timescale of the autocorrelation function clearly increases with extended exposure at 60°C, confirming that the corresponding aliquots contain species diffusing more slowly than the initial aliquots and, therefore, have on larger average hydrodynamic radii.

The average hydrodynamic radius r_h for all the species present in an aliquot may be obtained by fitting the data to a cumulant model and selecting the mean of the distribution to determine the average r_h. In Figure 12.15a we show the average r_h obtained from a cumulant-type fit to the ACF data, as a function of time that the aliquots were exposed to 60°C. The data acquired on day 0 are quite similar to those acquired on day 13, indicating that the degree of sample aggregation did not significantly evolve over that time period. However, for aliquots exposed to 60°C for times 10–60 min, the average r_h, scattering intensities, and percent polydispersity are all somewhat smaller for data acquired on day 13 than for data taken on day 0. The well plate was not agitated prior to the day 13 measurement, and while the samples may have actually evolved through time, it is also possible that larger species fell out of solution or were otherwise removed from the region of measurement (~1 mm above the well bottom) during that time.

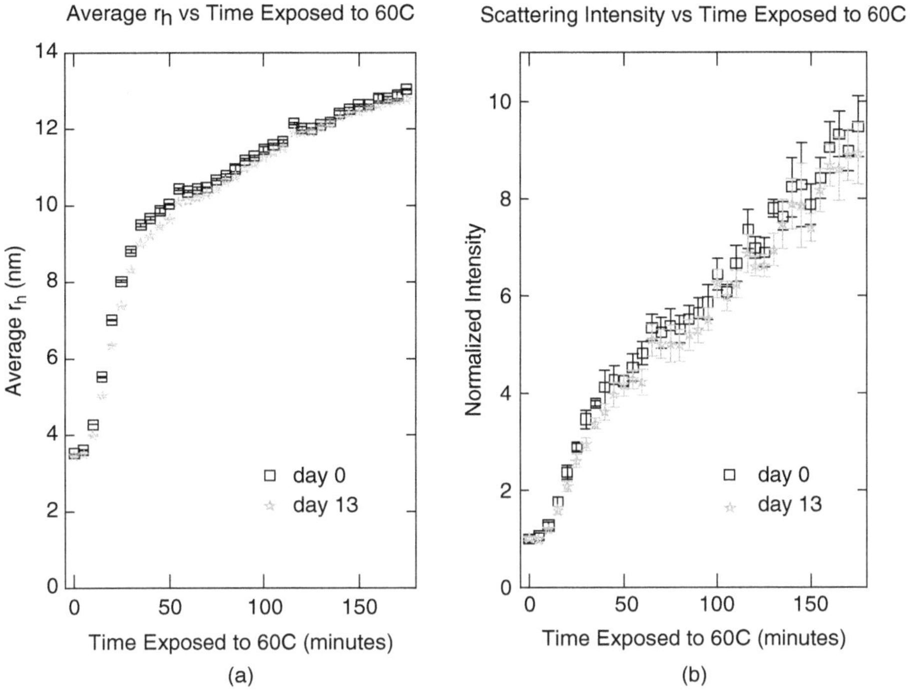

Figure 12.15. Measurements of 36 aliquots, all measurements made at 5°C, all samples stored at 4°C–5°C: (a) the average hydrodynamic radius as a function of time that the aliquot was exposed to 60°C; (b) the intensity of scattered light, which is proportional to the weight averaged molar mass, as a function of time that the sample was exposed to 60°C. The intensity data were normalized to the intensity of the time 0 sample.

In addition to measuring the autocorrelation function, a DLS instrument may be used to perform an approximate static LS measurement. The number of photons per second detected by the DLS single photon counting module detector is called the *count rate*. It is roughly proportional to the intensity of light scattered from the sample, after normalizing by the laser power and detector attenuator setting for each measurement. Single-photon counting module detectors are extremely sensitive, but they inevitably have a count rate that is not very linear with intensity. While the nonlinearity of the detectors may be determined and partially accounted for, absolute molar mass measurements cannot be made with precision using such detectors. However, as long as this is understood, it is still quite useful to view the relative intensity of light scattered from each sample. Such data are presented in Figure 12.15b. The intensity of scattered light is generally proportional to the weight-averaged molar mass of the sample, and is indicative of aggregation increasing as a function of time exposed to 60°C.

The aliquot polydispersity as a function of time exposed to 60°C is shown in Figure 12.16a. The polydispersity is seen to first increase, then reach a maximum

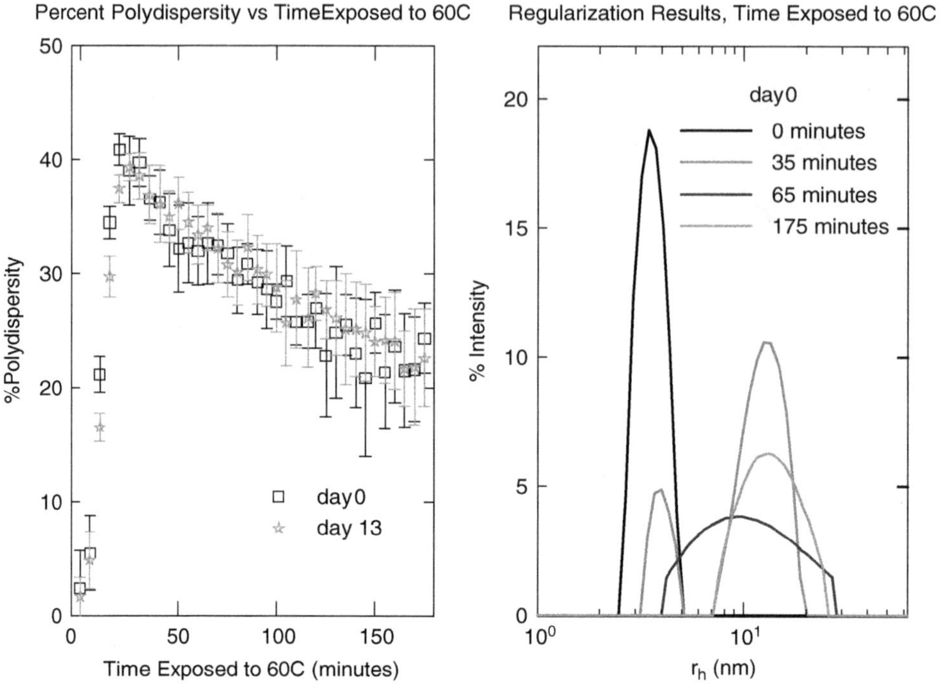

(a) (b)

Figure 12.16. All measurements made at 5°C, all samples stored at 4°C–5°C: (a) sample polydispersity as a function of time sample is exposed to 60°C, as measured on days 0 and 13; (b) results of regularization analysis, where the actual distributions of sizes present in the samples are estimated, for samples exposed to 60°C for 0, 35, 65, and 175 mins.

value at about 20 min of exposure, and then decrease. The reduction in polydispersity for times greater than 20 min seen in Figure 12.16a is consistent with the regularization results shown in Figure 12.16b. These results demonstrate a change from scattering from a bimodal distribution, consisting of small oligomers and high-molar-mass aggregates (HMMA), to a largely monomodal distribution where the light scattered is dominated by the HMMA. The aggregated protein is, of course, quite polydisperse, but the polydispersity with respect to the average size of the sample is reduced, since less and less of the sample consists of smaller unaggregated protein.

These unfractionated DLS measurements require minimal effort and sample volume, and provide a general idea of what is happening for this system. It is easy to observe trends and obtain insight regarding the evolution of the system as a function of environmental condition, such as time exposed to 60°C or time stored at 5°C. From these measurements we note that with exposure to 60°C there is early, rapid growth of a HMMA species, followed by a steady but reduced rate of growth. The system evolves somewhat with storage at 4°C–5°C for 13 days, and further investigation is needed to establish whether this change is meaningful, or an artifact resulting from sample handling.

12.5.2. MALS Measurements

For MALS measurements, we removed 3-mL aliquots at 60°C exposure times of 0, 35, 65, 100, 140, and 175 min. The MALS measurements were made (1) with SEC fractionation with duplicate injections of 50 μL of sample having concentrations of 2 and 0.33 mg/mL and (2) without fractionation, with samples diluted to 0.4 mg/mL and filtered with 0.2-μm Whatman Anotop filters. Measurements were made using Wyatt DAWN HELEOS multiangle light-scattering detectors, a Wyatt Optilab rEX differential refractometer (dRI) for concentration determination, and an Agilent 1200 series UV absorbance detector for additional concentration determination. Size exclusion chromatography separations were performed at a flow rate of 0.5 mL/min, using an Agilent 1200 series autosampler with a temperature controlled sample space set to 4°C, a guard column Shodex Protein KW-G, and a separation column Shodex Protein KW-803.

In Figure 12.17 we display SEC fractionated MALS data at 90° scattering and differential refractive index concentration data of BSA samples for several 60°C exposure times. A polydisperse HMMA is apparent for all data for which the sample was exposed for any time to 60°C. The MALS signal is proportional to the concentration times the molar mass, while the concentration detector signal is proportional only to the concentration. The significant difference between the MALS data and the concentration data in the relative peak amplitudes for the monomer and HMMA clearly highlights the difference in molar masses. In Figure 12.18 we show the MALS traces for the same datasets, overlayed with the molar masses measured for each elution time.

The molar mass of the monomer peak, of course, does not change with duration of time exposed to 60°C, and the mass population of the HMMA species evolves strongly as a function of time exposed to 60°C. Also, for elution times up to 21 min, the same elution time corresponds to somewhat different measured molar masses for the various times exposed to 60°C. Longer exposure times result in a larger molar mass for a particular elution time. Since SEC separations are based on hydrodynamic radius, a larger molar mass for the same elution time implies that the structure of the HMMA species becomes more compact with time exposed to 60°C, with the same hydrodynamic radius having a larger molar mass. Separation is clearly not complete for the HMMA species. This lack of separation could mimic the "compactness" effect discussed above, and additional effort to optimize separation for the HMMA is required to robustly make such a statement. However, conformation plots presented below are consistent with the HMMA species initially experiencing rod-like growth (e.g., fibrils), and becoming more compact with time exposed to 60°C.

The SEC data were analyzed by selecting peak regions for (1) the monomer peak; (2) the dimer, trimer, and higher peaks; and (3) the HMMA species. Peak regions are contiguous to one another, and cover all regions where MALS and concentration detector signals exist.

Figure 12.19a illustrates the results of molar mass measurements as a function of time exposed to 60°C. Data are shown from duplicate SEC runs at two concentrations, for four separate SEC runs for each exposure time. There is no discernable difference between the data taken at the two concentrations, indicating that if the aggregate is reversibly associated, then the timescale for disassociation is significantly longer than

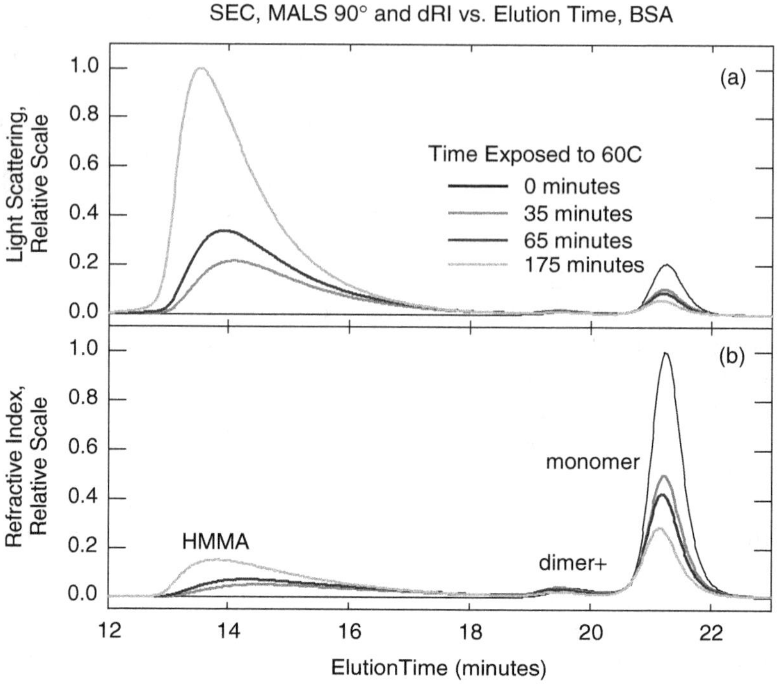

Figure 12.17. (a) Multiangle LS 90° light scattering; (b) differential refractive index concentration data for BSA exposed to 60°C for different periods of time, and fractionated via SEC. The contrast between the MALS and the concentration detector traces clarifies the molar mass difference between the different species present in the samples. See the insert for color representation of this figure.

the several hours that the samples were held at different concentrations before measurement. The monomer molar mass of course does not change with time exposed to 60°C. The mass for the "dimer" peak, which, as a result of incomplete separation, includes higher oligomers, increases somewhat as a function of time exposed to 60°C. The HMMA evolves to produce a larger weight-averaged molar mass as a function of time held at 60°C. There is always a possibility that some portion of the sample had been removed by the SEC column, and to check for this we measured the unfractionated sample in a "batch" measurement. For unfractionated measurements, the measured weight-averaged molar mass is given by Equation (12.4). Because the molar masses and concentrations are known for each elution volume of the measurements involving fractionation, we may use Equation (12.4) to calculate the expected results of batch measurements. A discrepancy between the SEC molar mass integrated over the complete run and the batch measurement is an indication that some portion of the sample has been lost or altered. For large species it is not unusual to find that the column has filtered some portion of the sample. In this case, the M_w from batch measurements are systematically *less than* the M_w obtained by averaging over the SEC

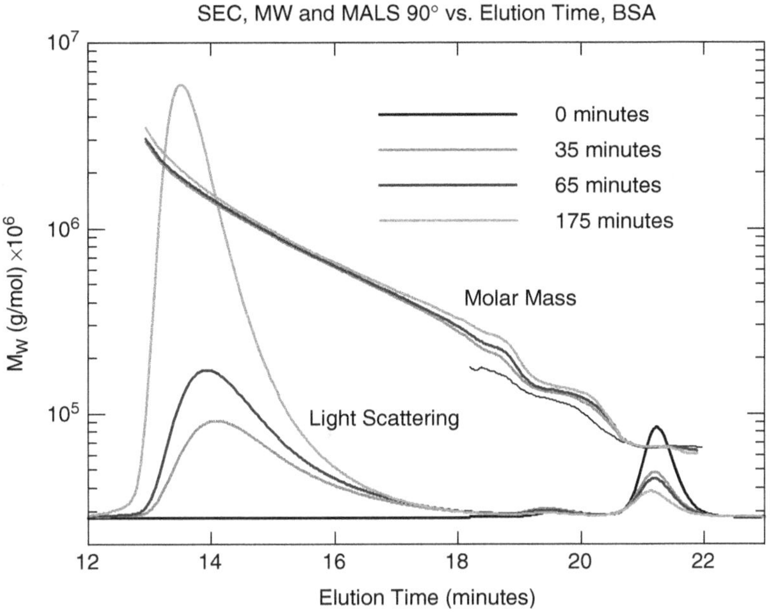

Figure 12.18. Multiangle LS 90° light scattering and molar mass results derived from both the MALS and the concentration detectors for BSA exposed to 60°C for different periods of time. See the insert for color representation of this figure.

data. If larger species had been filtered by the column, then the converse would be true. In our experiment, samples used for batch measurements were treated differently from samples used for SEC measurements, in that the batch samples were filtered using 200-nm-pore filters. It is possible that additional filtering would have removed some material from the batch measurement that had successfully traversed the SEC column. The discrepancy between the average SEC and batch results is sufficient to warrant further investigation; a possible first step might be to rerun the samples at one exposure time with both SEC and batch measurement, using both filtered and unfiltered samples. In Figure 12.19b we show data from the concentration detector for the weight fractions of the different peaks, as well as the total percent of recovered mass.

The HMMA species has a size that becomes large enough to allow measurement of the RMS radius r_g in addition to the molar mass. As discussed previously, for a polydisperse sample, plotting $\log(r_g)$ against $\log(M_w)$ and finding the slope a allows an estimate of the conformation of the sample. In Figure 12.20 we show the conformation plot for the HMMA species exposed to 60°C. The relation between mass and radius for samples exposed for ≤ 100 min is consistent with the growth of rod-like structures, such as fibrillates. For samples exposed to the same temperature for 140 and 175 min the slope becomes less than 1, consistent with a denser structure. Long structures clumping together or coiling would both result in a lower slope. Growth of fibrils in heat-stressed BSA has been reported previously [40].

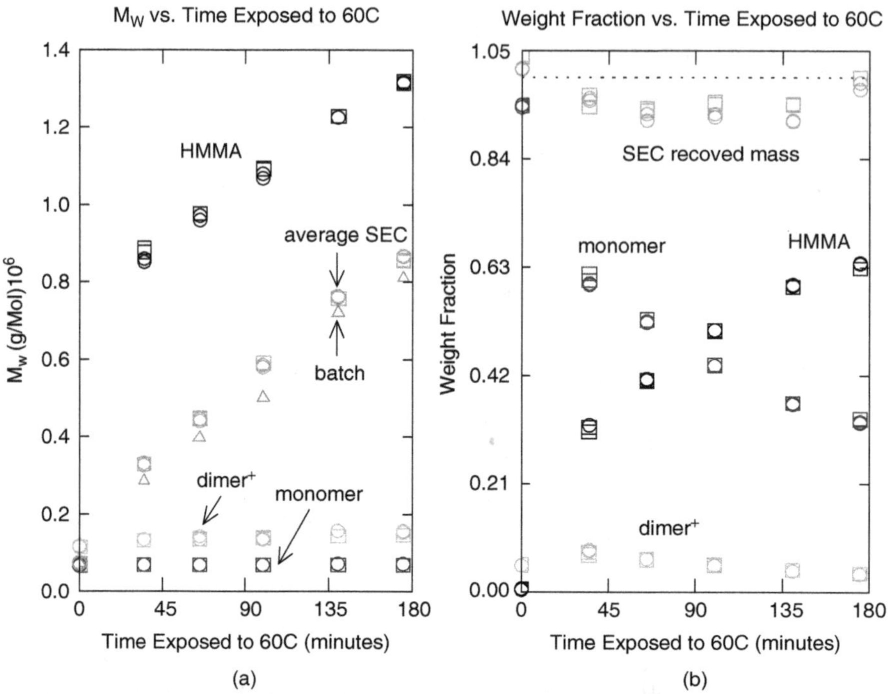

Figure 12.19. (a) Multiangle LS results for batch measurements and for duplicate SEC measurements at two concentrations. Squares indicate SEC data for 2 mg/mL concentration, circles indicate SEC data for 0.33 mg/mL concentration, and triangles indicate batch measurements at 0.4 mg/mL. Shown are the weight-averaged molar masses for (1) each peak in SEC as described in the text, (2) all SEC peaks in combination, and (3) batch measurements. Panel (b) shows concentration detector data for the weight fractions of each peak, and the percent SEC recovered mass. See the insert for color representation of this figure.

12.5.3. Aggregating Protein System Conclusions

We have demonstrated the use of MALS and DLS in determining the state of aggregation of the system under study. On exposure to 60°C, a HMMA species forms and evolves with time exposed to 60°C to have a larger average size and larger weight-averaged molar mass. The data are consistent with the sample initially aggregating as fibrils that eventually begin to clump or coil. The system does not evolve significantly during the 13 days held at 5°C postexposure to heat stress, although some change is seen, indicating the need for additional measurements. The system chosen for this example exhibits gross aggregation, which may not be typical of systems under study in a formulation environment. However, we hope that it is clear from the chapter as a whole that the measurements performed and the methods used can determine identically subtle degrees of aggregation and association, as well as detect the presence of small numbers of large aggregates.

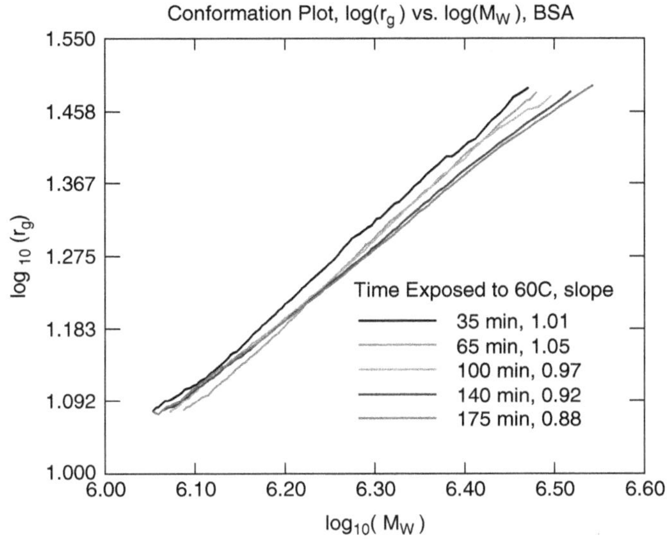

Figure 12.20. Conformation plot from HMMA species. The slope of the resulting line is equal to $1/a$, where a is the power of the radius by which the mass increases. For solid spheres $a = 0.33$, for ideal random coils under theta conditions $a = 0.5$, and for rods $a = 1.0$. See the insert for color representation of this figure.

12.6. CONCLUSION

The use of light-scattering techniques for the study, measurement, development, and control of most elements affecting biological products has been shown to be of fundamental importance and unequaled application. Indeed, the range of applications is so broad that these techniques have become among the most widely used and essential elements of the pharmaceutical industry's most important armamentarium.

ACKNOWLEDGMENTS

We wish to thank Dr. Michelle Chen for her many helpful comments and suggestions. We also thank Vivianna Lim and Ryan Reddick for their sample preparations and data acquisition.

REFERENCES

1. Wyatt, P. J. (1993), Light scattering and the absolute characterization of macromolecules, *Anal. Chem. Acta* **272**: 1–40.
2. Zimm, B. (1948), The scattering of light and the radial distribution function of high polymer solutions, *J. Chem. Phys*. **16**(12): 1093–1099.

3. Debye, P. (1947), Molecular-weight determination by light scattering, *J. Phys. Colloid. Chem.* **51**: 18–32.
4. Goodstein, D. L. (1985), *States of Matter*, Dover, Mineola, NY, pp. 231–236, 260–266, 283–301.
5. Bohren, C. F. and Huffman, D. R. (1983), *Absorption and Scattering of Light by Small Particles*, Wiley, New York.
6. Chu, B. (2007), *Laser Light Scattering Basic Principles and Practice*, 2nd ed., Dover, Mineola, NY, 2007.
7. Harding, S. E., Sattelle, D. B., and Bloomfield, V. A. eds. (1992), *Laser Light Scattering in Biochemistry*, The Royal Society of Chemistry, Cambridge, UK.
8. Fasman, G. D. (1976)., *CRC Handbook of Biochemistry and Molecular Biology*, 3rd ed., Vol. II, CRC Press, Boca Raton, FL, pp. 372–382.
9. Theisen, A., Johann, C., Deacon, M. P., and Harding, S. E. (2000), *Refractive Index Increment Data-Book for Polymer and Biomolecular Scientists*, Nottingham Univ. Press, Nottingham, UK.
10. Kerker, M., ed. (1988), *Selected Papers on Light Scattering*, SPIE Vol. 951, Parts I and II, The International Society for Optical Engineers, Bellingham, WA.
11. Wen, J., Arakawa, T., and Philo, J. S. (1996), Size exclusion chromatography with on-line light-scattering, absorbance, and refractive index detectors for studying proteins and their interactions, *Anal. Biochem.* **240**: 155–166.
12. Yernool, D., Boudker, O., Folta-Stogniew, E., and Eric Gouaux, E. (2003), Trimeric subunit stoichiometry of the glutamate transporters from Bacillus caldotenax and Bacillus stearothermophilus, *Biochemistry* **42**: 12981–12988.
13. Vasquez, V., Cortes, D. M., Furukawa, H., and Perozo, E. (2007), An optimized purification and reconstitution method for the MscS channel: Strategies for spectroscopical analysis, *Biochemistry* **46**: 6766–6773.
14. Wyatt, P. J. and Weida, M. J. (2004), *Method and Apparatus for Determining Absolute Number Densities of Particles in Suspension*, US Patent 6,774,994.
15. Wei, Z., Mcevoy, M, Razinkov, V., Polozova, A., Li, E., Casas-Finet, J., Tous, G. I., Balu, P., Pan, A. A., Mehta, H., and Schenerman, M. A. (2007), Biophysical characterization of influenza virus subpopulations using field flow fractionation and multiangle light scattering: Correlation of particle counts, size distribution and infectivity, *J. Virol. Meth.* **144**: 122–132.
16. Minton, A. P., and Edelhoch, H. (1982), Light scattering of bovine serum albumin solutions: Extension of the hard particle model to allow for electrostatic repulsion, *Biopolymers* **21**: 451–458.
17. Minton, A. P. (2007), Static light scattering from concentrated protein solutions, I: General theory for protein mixtures and application to self-associating proteins, *Biophys. J.* **93**: 1321–1328.
18. Fernandez, C. and Minton, A. P. (2009), Static light scattering from concentrated protein solutions II: Experimental test of theory for protein mixtures and weakly self-associating proteins, *Biophys. J.* **96**: 1992–1998.
19. Fernandez, C. and Minton, A. P. (2008), Automated measurement of the static light scattering of macromolecular solutions over a broad range of concentrations, *Anal. Biochem.* **381**: 254–257.

20. Zimmerman, S. B. and Minton, A. P. (1993), Macromolecular crowding: Biochemical, biophysical, and physiological consequences, *Annu. Rev. Biophys. Biomol. Struct*. **22**: 27–65.
21. Attri, A. K. and Minton, A. P. (2005), Composition gradient static light scattering (CG-SLS): A new technique for rapid detection and quantitative characterization of reversible macromolecular hetero-associations in solution, *Anal. Biochem*. **346**: 132–138.
22. Attri, A. K. and Minton, A. P. (2005), New methods for measuring macromolecular interactions in solution via static light scattering: basic methodology and application to nonassociating and self-associating proteins, *Anal. Biochem*. **337**: 103–110.
23. Kameyama, K. and Minton, A. P. (2006), Rapid quantitative characterization of protein interactions by composition gradient static light scattering, *Biophys. J*. **90**: 2164–2169.
24. Voet, D. and Voet, J. G. (1995), *Biochemistry*, 2nd ed., Wiley, New York.
25. Hanlon, A. and Some, D. (2008), Data presented at the 52nd Annual Meeting of the Biophysical Society, Long Beach, CA, Feb 2–6, 2008.
26. Hanlon, A., Larkin, M. I. and Reddick, R. M. (2010), Free-solution, label-free protein-protein interactions characterized by dynamic light ccattering, *Biophys. J*. **98**: 297–304.
27. Pecora, R. J., ed. (1985), *Dynamic Light Scattering—Applications of Photon Correlation Spectroscopy*, Plenum Press, New York.
28. Berne, B. J., and Pecora, R. (1976), *Dynamic Light Scattering*, Wiley-Interscience, New York.
29. Reif, F. (1965), *Fundamentals of Statistical and Thermal Physics*, McGraw-Hill, New York, Section 15.6.
30. Koppel, D. E. (1972), Analysis of macromolecular polydispersity in intensity correlation spectroscopy: The method of cumulants, *J. Chem. Phys*. **57**: 4814–4820.
31. Churchill, R. V. (1972), *Operational Mathematics*, Third Edition, McGraw-Hill, New York.
32. Provencher, S. W. (1979), Inverse problems in polymer characterization: Direct analysis of polydispersity with photon correlation spectroscopy, *Makromol. Chem*. **180**: 201–209.
33. de la Torre, J. G. and Bloomfield, V. A. (1977), Hydrodynamic properties of macromolecular complexes. I. Translation, *Biopolymers* **16**: 1747.
34. Schaefer, D. W. and Berne B. J. (1972), Light scattering from non-Gaussian concentration fluctuations, *Phys. Rev. Lett*. **28**: 475–478.
35. Elson, E. L. and Webb, W.W. (1975), Concentration correlation spectroscopy: A new biophysical probe based on occupation number fluctuations, *Annu. Rev. Biophys. Bioeng*. **4**: 311–334.
36. Nijman, E. J., Merkus, H. G., Marjinissen J. C. M., and Scarlett, B. (2001), Simulations and experiments on number fluctuations in photon-correlation spectroscopy at low particle concentrations, *Appl. Opt*. **40**(24): 4058–4063.
37. Meyer, W. V., Cannell, D. S., Tin, P., Cheung. H. M., Mann, Jr., J. A., Taylor, T. W., Lock, J. A., Zhu, J., and Smart. A. E. (1999), *Cross-Correlation Method and Apparatus for Suppressing the Effects of Multiple Scattering*, US Patent 5,956,139.
38. Ackerson, B. J. (1976), Correlations for interacting Brownian particles, *J. Chem. Phys*. **64**(1): 242–246.
39. Neal, D. G., Purich, D., and Cannell, D. S. (1984), Osmotic susceptibility and diffusion coefficient of charged bovine serum albumin, *J. Chem. Phys*. **80**(7): 3469–3477.
40. Holm, N. K, Jespersen, S. K., Thomassen, L. V., Wolff, T. Y., Sehgal, P., Thomsen, L. A., Christiansen, G., Andersen, C. B., Knudsen, A. D., and Otzen, D. E. (2007), Aggregation and fibrillation of bovine serum albumin, *Biochim. Biophys. Acta* **1774**: 1128–1138.

PART II

DEVELOPMENT OF A FORMULATION FOR LIQUID DOSAGE FORM

13

EFFECTIVE APPROACHES TO FORMULATION DEVELOPMENT OF BIOPHARMACEUTICALS

Rajiv Nayar and Mitra Mosharraf

13.1. INTRODUCTION

This chapter addresses a critical need in biopharmaceutical formulation development that has arisen from the shift in the drug research–development (R&D) paradigm in recent years. With the completion of the human genome sequence, advances in high-throughput instrumentation, high-throughput screening, and systems biology research, focus has shifted from drug discovery into drug development. There is now a critical need to perform effective formulation development of biopharmaceuticals in a more rational fashion rather than an empirical approach [1,2]. This, in turn, leads to development in shorter timespans and with limited material and resources. This chapter provides an algorithmic approach to formulation development of a biopharmaceutical that is not only high-throughput but also justifies the formulation components that stabilize the biomolecule. It is first important to grasp current global and industry perspectives on the current state of biotechnology and drug development challenges that formulation groups are facing.

Formulation and Process Development Strategies for Manufacturing Biopharmaceuticals, edited by Jameel and Hershenson

13.2. INDUSTRY PERSPECTIVE

Biopharmaceutical-based medicines (proteins, peptides, and genomic materials) now account for an increasing proportion of the world pharmaceutical markets. Over 125 biopharmaceuticals were approved by the FDA as of August 2008 [3]. Biotechnological sales in the United States grew to \$40.3 billion USD in 2006 (20% more than previous year) according to IMS Health. In 2007, \$58.8 billion was invested in R&D by the pharmaceutical research companies to support development of over 2000 new medicines [3]. Many of these medicines consist of novel peptides and protein based drug products.

Several new-drug discovery technological advances have created this immense impact. For example, unraveling of the human genome, systems biology, and pharmacogenomics have been able to provide new knowledge in unraveling the genetic factors responsible for numerous diseases. High-throughput screening (HTS) has contributed to reducing the time and cost of identifying potential therapeutic (bioactive) molecules as lead compounds.

13.3. THE MAJOR CHALLENGES

13.3.1. Drug Development Technical Challenges

Because of the increased "hit" rate for potential drugs from HTS, a biopharmaceutical drug can go into development with many potential candidates. The protein may be identified through genomic or proteomic activities or through more traditional medical research. It may initially be associated with a particular disease process or a certain metabolic event. In any case, its mechanism of action—as well as many of its structural characteristics and biochemical properties—may be unknown. This, in turn, creates a great sense of urgency to rapidly move the drug candidate through the development process into clinical trials. This creates significant expectations on analytic, formulation, and process development teams to rapidly advance the drug candidates into preclinical and clinical trials with pharmaceutically acceptable stable drug product formulations. Such drug products should be

- Safe and efficacious
- Stable to meet the desired shelf life
- Scalable to manufacturing lot sizes for clinical studies
- Economical to manufacture formulated with approved parenteral formulation excipients

13.3.2. Technical Challenges for a Pharmaceutically Acceptable Formulation

Hence, one of the more challenging aspects of developing protein pharmaceuticals is dealing with and overcoming the inherent physical and chemical instabilities of proteins [5–7]. This instability has the potential to alter the state of the protein from the desired (native) form to an undesirable form (on storage), compromising patient safety and drug efficacy [8]. The set of activities related to overcoming the inherent instability of the drug is referred to as *formulation development*. A successful formulation process has four stages:

1. Preformulation
2. Stabilization of the active substance in bulk form
3. Formulation in the designated dosage forms
4. Fill–finish aseptic manufacturing activities

Preformulation is an exploratory activity that begins early in biopharmaceutical development [9–11]. Preformulation studies are designed to characterize the molecule and develop a stability-indicating assay using stress conditions and physicochemical, and analytical investigation in support of promising experimental formulations. Data from preformulation studies provide the necessary groundwork for formulation attempts. Successful formulations take into account a drug's physico-chemical interactions with the other ingredients (and their interactions with each other) to produce a safe, stable, beneficial, and marketable product [12]. This is referred to as a *pharmaceutically acceptable formulation*.

Selection and development of an optimal drug product, including its inactive formulation ingredients, manufacturing details, and packaging system, are dependent on the properties of active drug, or its preformulation parameters (PPs). These parameters of the active pharmaceutical ingredient (API) need to be evaluated in initial stages of drug discovery and development, so that the most suitable delivery system can be selected. Thus, a well-coordinated, -planned, and -implemented preformulation program is necessary to minimize unforeseen failures in subsequent stages of the development. This has to be congruent with the product target profile which has been established by the product development teams including research, process development, manufacturing, preclinical, clinical, marketing representatives, and upper management.

13.3.3. Product Development Challenges

Unknown and unclear drug product profiles can present significant challenges in product development. Prior to any product formulation activities, a basic drug product profile is required to establish the mode of intented delivery (intravenous, subcutaneous, subdermal, or a special drug delivery pathway such as intranasal, buccal, or inhaled) and the intended dose or dosage form. Formulators also need to know early on whether the product would be sold as a single dose or in multidose vials. Single-dose formulations are easier to protect against contamination and can be formulated much more rapidly than multidose formulations. Multidose vials, which are handled several times over several weeks, can become contaminated, so preservatives must be chosen and tested for stability and compatibility with the protein. This can be addressed in preformulation studies. In addition, studies demonstrating the effectiveness of the preservatives against specific organisms may be required. International disagreement over preservatives in food and drugs may present a problem at this stage. The U.S.,

EU, and Japanese compendial standards differ regarding the timing of antimicrobial tests and which preservatives and excipients are allowed. Japan, for example, does not accept phenol, a preservative used commonly in the United States. At this early stage, companies must decide whether their products will be distributed outside the United States. The EU is known to have the toughest acceptance criteria for preservatives. The International Conference on Harmonization (ICH) is working toward a common standard in this regard. Nevertheless, initiating a multidose formulation program, if not essential, may not be the desired approach for rapid development of a drug candidate.

Some companies have a preformulation/formulation team specifically devoted to these tasks. Industry experts claim that preformulation teams are often understaffed, underresourced, and *not* well understood by company management. Formulation is sometimes considered a sort of "black art" in drug development. This is because in the protein formulation development has traditionally been an empirical science because of the unique characteristics of proteins being formulated. However, a better understanding of protein behavior and an increased awareness of the adverse effects of an improperly formulated protein product on clinical trials and on the marketed product have highlighted the need for a more rational approach to formulation development. With about 125 approved biologicals, there is now a reasonable database on protein and peptide formulations available for the formulation scientists to consider. A dedicated formulation development group that is adequately staffed and resourced has thus become necessary in order to successfully incorporate protein formulation development in the overall drug development strategy.

Therefore, in order to overcome product development challenges, it is imperative that the formulation team work closely with other development teams, such as, the device and manufacturing technology teams. This minimizes the risks as the drug enters into manufacturing and final stages of product launch. Also, as the development process progresses, it becomes more difficult to make significant changes to the formulation used in earlier clinical studies.

13.4. EFFECTIVE VERSUS EFFICIENT FORMULATION DEVELOPMENT

In order to rapidly formulate a drug, it is important to clearly understand the concepts of effectiveness versus efficiency [13]. Being *efficient* refers to the ability to do tasks "right" rather than the ability to get the "right things done." Being *effective* is to get the "right" tasks done. For rapid formulation development, the focus should be on using an effective approach rather than an efficient approach so that the "right things get done right the first time." The concept, although intuitive, is extremely important to understand, especially in view of the fact that the analytical tools that are now available to the formulation and analytical scientists are so powerful and efficient that it is easy to perform many analytics during formulation development, generating an immense amount of data requiring time-consuming analysis. This can complicate the process of obtaining relevant useful information and knowledge about the biopharmaceutical for formulation development. Therefore, an effective systematic approach to formulation development is essential as formulation groups are asked to formulate more drugs with limited resources, drug amounts, and short timelines.

13.5. EFFECTIVE ALGORITHM FOR FORMULATION DEVELOPMENT

An effective approach to formulating stable biopharmaceuticals should include the following set of tasks:

1. Define a specific product target profile (PTP) for the formulation.
2. Obtain a basic understanding of the biochemical and biophysical parameters of the protein on the basis of amino acid sequence analysis.
3. Develop an appropriate stability-indicating assay for the protein that is likely to reflect real-time instability.
4. Identify the appropriate analytics for the protein.
5. Use design of experiments (DOE) (multivariate experimental designs) to investigate multiple formulation variables simultaneously in a single experimental design.
6. Identify three to five potential lead formulations and perform acute stabilty studies.
7. Identify a potential formulation for preclinical and clinical trials.

13.5.1. Important First Step: Defining the Drug Product Target Profile (PTP)

Prior to any extensive formulation development, the drug PTP needs to be clearly defined. This profile defines the specific aims of the formulation group in developing a drug product that will address the preclinical and clinical requirements for the biopharmaceutical. The following questions might be included in the checklist:

1. What is the administration route of the drug product (e.g., intravenous, subcutaneous, or intradermal; or other noninvasive routes such as aerosol, buccal, intranasal, etc.)?
2. What is the preferred dosage form (frozen, liquid, lyophilized, special drug delivery system)?
3. What is the preferred drug concentration (for both preclinical and clinical dosing)?
4. Will the product treat a chronic or acute disorder (frequency of dosing)?
5. Will it need specific targeting for a broad or narrow therapeutic window?
6. Will it be administered at home or in the clinic or hospital as the final marketed product?
7. What is the competitive landscape? Are other drugs already treating the same indication? If so, should a novel formulation be developed?
8. What will give the project the advantage over existing treatments and those that may emerge? For example, marketing considerations arise early in product development for monoclonal antibodies (MAbs). Typically, MAbs are needed at high doses (hundreds of milligrams per dose) and are normally delivered

intravenously. The drive to reduce health care costs has created a need to administer MAb therapeutics more conveniently, at home, subcutaneously. Thus, MAbs must be available at high concentrations (~200 mg/mL) in the vial. At these high concentrations, MAb-containing solutions are viscous, making them difficult to administer conveniently. Hence, a preformulation activity that needs to be considered is a concentration study investigating solubility behavior, effect of concentration on viscosity, and increased potential for aggregation. These studies have the potential to strongly influence the target product profile as well as the design of the clinical trial.

13.5.2. Basic Understanding of the Biochemical and Biophysical Properties of the Protein

Before any experimental work on the protein candidate, a detailed profile of the biochemical properties of the protein needs to be developed by analyzing the amino acid sequence of the protein. Various programs and Websites are available to perform this task. For example, a useful Website is the EXPASY PROTPARAM site (http://www.expasy.ch/tools/protparam.html), a tool that allows the computation of various physical and chemical parameters for a given protein. The computed parameters include the molecular weight, theoretical pI, amino acid composition, atomic composition, extinction coefficient, estimated half-life, instability index, aliphatic index, and grand average of hydropathicity (GRAVY) of a protein. The sequence can be either inputed from the published sequence or entered manually. This information allows the formulator to develop an appropriate preformulation strategy for the molecule in addressing the potential instabilities that may be associated with the molecule. An example of this analysis is shown in Figure 13.1 for ovalbumin as an example of a model protein.

13.5.3. Appropriate Stability-Indicating Assay

Selecting an appropriate stability-indicating assay (SIA) for the protein is one of the most critical tasks for the formulator because it establishes a relevant high-throughput system for evaluating a large number of formulations using the appropriate analytics. This is essential if the formulation development is performed using multivariable statistical experimental designs (DOE). The SIA should be validated by

1. Appropriately stressing the protein so that the protein instability reflects the main instability pathway. This could be biochemical instability, such as, oxidation or deamidation or a physical instability, such as, protein aggregation.
2. Correlate the instability observed using the stress condition with decrease in the activity or potency of the molecule.
3. The stress condition for destabilizing the protein should be sufficient to cause a change in protein characteristics of around 20%–40% that can be measured accurately and reproducibly by the analytical assay.

OVAL_CHICK (P01012)

Ovalbumin (Egg albumin) (Plakalbumin) (Allergen Gal d II) (Allergen Gal d 2)
Gallus gallus (Chicken).
The computation has been carried out on a user selected segment from position 49 to position 384 in this sequence of 386 residues.

Considered sequence fragment:

```
         1          11         21         31         41         51
         |          |          |          |          |          |
    1                                                       ST RTQINKVVRF   60
   61 DKLPGFGDSI EAQCGTSVNV HSSLRDILNQ ITKPNDVYSF SLASRLYAEE RYPILPEYLQ  120
  121 CVKELYRGGL EPINFQTAAD QARELINSWV ESQTNGIIRN VLQPSSVDSQ TAMVLVNAIV  180
  181 FKGLWEKAFK DEDTQAMPFR VTEQESKPVQ MMYQIGLFRV ASMASEKMKI LELPFASGTM  240
  241 SMLVLLPDEV SGLEQLESII NFEKLTEWTS SNVMEERKIK VYLPRMKMEE KYNLTSVLMA  300
  301 MGITDVFSSS ANLSGISSAE SLKISQAVHA AHAEINEAGR EVVGSAEAGV DAASVSEEFR  360
  361 ADHPFLFCIK HIATNAVLFF GRCV
```

Number of amino acids: 336
Molecular weight: 37429.8
Theoretical pI: 5.17

CSV format

Amino acid composition:

Ala	(A)	28	8.3%
Arg	(R)	15	4.5%
Asn	(N)	15	4.5%
Asp	(D)	12	3.6%
Cys	(C)	4	1.2%
Gln	(Q)	15	4.5%
Glu	(E)	30	8.9%
Gly	(G)	16	4.8%
His	(H)	5	1.5%
Ile	(I)	21	6.2%
Leu	(L)	29	8.6%
Lys	(K)	17	5.1%
Met	(M)	13	3.9%
Phe	(F)	16	4.8%
Pro	(P)	12	3.6%
Ser	(S)	34	10.1%
Thr	(T)	15	4.5%
Trp	(W)	3	0.9%
Tyr	(Y)	8	2.4%
Val	(V)	28	8.3%
Pyl	(O)	0	0.0%
Sec	(U)	0	0.0%

Total number of negatively charged residues (Asp + Glu): 42
Total number of positively charged residues (Arg + Lys): 32

Atomic composition:

Carbon	C	1659
Hydrogen	H	2633
Nitrogen	N	441
Oxygen	O	508
Sulfur	S	17

Figure 13.1. ProtParam analysis of ovalbumin on the various physicochemical properties that can be deduced from the protein sequence. (http://www.expasy.ch/tools/protparam.html).

```
Formula: C1659H2633N441O508S17
Total number of atoms:  5258

Extinction coefficients:

Extinction coefficients are in units of M-1 cm-1, at 280 nm measured in
water.

Ext. coefficient     28670
Abs 0.1% (=1 g/l)    0.766, assuming ALL Cys residues appear as half
cystines

Ext. coefficient     28420
Abs 0.1% (=1 g/l)    0.759, assuming NO Cys residues appear as half
cystines
Estimated half-life:

The N-terminal of the sequence considered is S (Ser).

The estimated half-life is: 1.9 hours (mammalian reticulocytes, in
vitro).
          >20 hours (yeast, in vivo).
          >10 hours (Escherichia coli, in vivo).

Instability index:

The instability index (II) is computed to be 36.42
This classifies the protein as stable.

Aliphatic index:  90.54

Grand average of hydropathicity (GRAVY): -0.077
```

Figure 13.1. *(Continued)*

The most common stresses utilized for establishing a stability indicating assay are thermal stress, where the protein solution is incubated for a defined time period to initiate instability. An appropriate temperature is typically around five degrees below the onset melting temperature of the protein. Multiple freeze–thaw stresses or agitation under controlled conditions are also appropriate stresses for establishing a SIA. A typical multiple freeze–thaw stress may involve three freeze–thaw cycles of the protein formulations by freezing the formulations at -80°C and thawing the frozen solutions either at room temperature or rapidly at 30°C. Agitation under controlled conditions may involve shaking the protein formulations at defined agitation settings for fixed time periods [14]. Forced degradation stresses, such as, extreme pH levels, use of oxidizing agents, or extreme thermal conditions are not appropriate for establishing a SIA as these are not representative of actual storage conditions that the protein might experience during its shelf life and its degradation profile. These studies are useful in establishing the appropriate analytics for monitoring the chemical degradant of the protein.

The main objective for establishing a SIA should be to define a stress condition that causes the major degradation pathway of the protein. In preliminary studies it is usually prudent to expose the protein to multiple stress conditions in order to determine the appropriate stress condition. Establishing an appropriate SIA is, therefore, one of the most important tasks for the formulator, as selection of a suitable formulation will be based on testing a number of formulation candidates using this assay.

13.5.4. Identification of Proper Analytics for Formulation Development

With the current advances in technology, protein analysis has undergone a dramatic transformation. Readily available analytical equipment and technology in protein analytics has provided the analytical and formulation groups to rapidly amass large databases of analytical data on the protein. High-throughput instrumentation using robotic automations in biochemical and biophysical analytical techniques, such as HPLC, protein size analysis on a protein chip, circular dichroism (CD), differential scanning calorimetry (DSC), dynamic light scattering (DLS), mass spectroscopy [liquid chromatography–tandem mass spectrometry (LC-MS/MS), matrix-assisted laser desorption–ionization/time-of-flight MS (MALDI-TOF)], and UV–visible (UV–vis) spectroscopy allow analysis of multiple samples in short timeframes, small sample sizes, and less workforce without compromising the quality of the data. The advantages and disadvantages of these analyses have been discussed in numerous reviews [14–16].These techniques can provide quantitative data on the protein characteristics for analysis of different protein formulations.

Identification of the proper analytical tools for formulation development is key for identifying a stable formulation. Ideally, two or three orthogonal types of analytics that are stability-indicating for formulation development are sufficient to develop a suitable formulation prototype that can be refined during later stages of development. This approach allows the formulation group to establish correlations between the biochemical and biophysical properties of the protein to its activity or potency.

Another approach that is effective and utilizes less resource is to use a coupled analytical approach where one analysis gives multiple responses. An example of such an analysis is UV–vis spectroscopy using a diode array UV–vis spectrophotometer. Absorbance spectroscopy is a straightforward, nondestructive, and efficient method for the investigation of protein conformation and its aggregation propensity, and provides protein quantification without interference in most formulation matrices. A single UV measurement of a sample can provide the stability profile of a protein from these three important analytical responses:

1. Protein concentration using absorbance at 280 nm
2. Protein aggregation by measuring light scattering at 340 nm
3. Second-derivative analysis of the UV spectra between 250 and 350 nm that provides information on the tertiary conformational stability of the protein

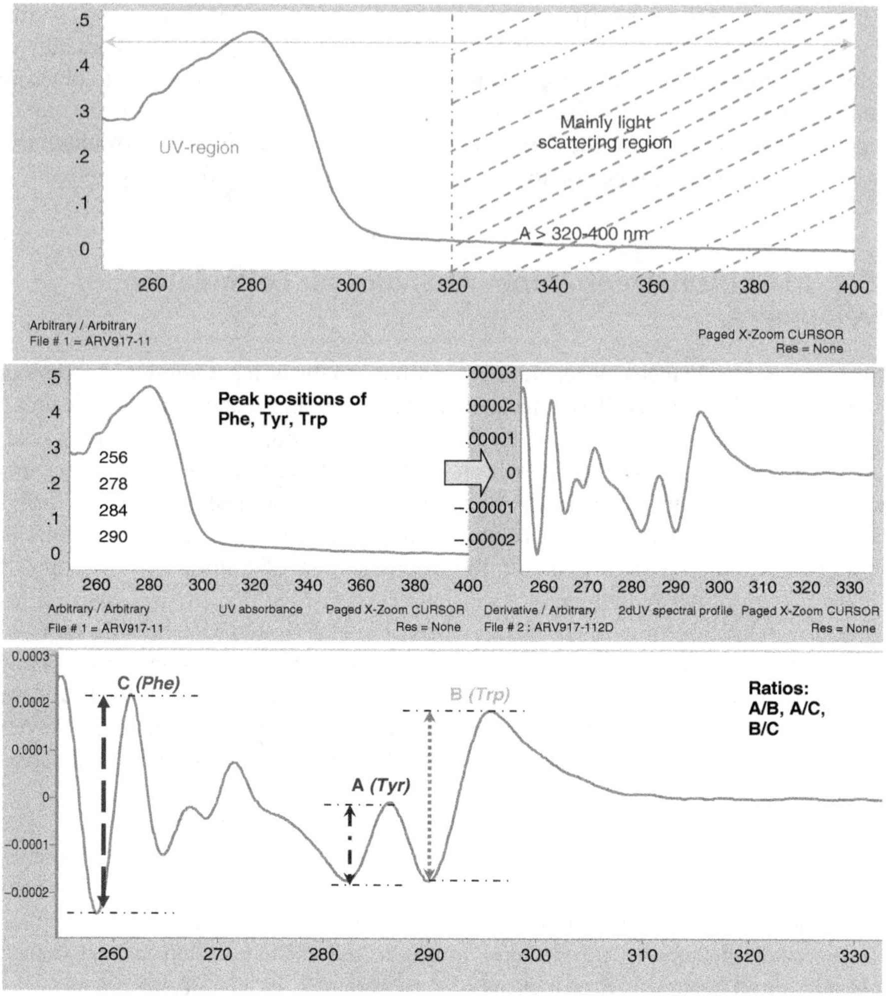

Figure 13.2. Use of UV–visible spectroscopy of a protein to obtain protein quantitation, protein aggregation, and tertiary conformational analyisis by second derivative.

This approach is illustrated in Figure 13.2. The speed and nondestructive nature of absorption spectroscopy coupled with the availability of multiple sample configurations and automated collection software in current instrumentation make it a desirable choice for high-thoroughput preformulation and formulation studies. The advantage of using second-derivative UV analysis (2dUV) is that it provides quantifiable parameters for tertiary conformational analysis, such as peak-to-valley ratios and peak positions for three different amino acids; phenylalanine, tyrosine, and tryptophan in the protein molecule. The average microenvironments of these three molecules affect the peak-to-valley ratios between the amino acids and their peak positions. Because the three aromatic residues are dispersed through the three-dimensional structure of most proteins, they potentially provide a series of signals that should be sensitive to a wide variety of both subtle and major structural transitions. The changes in these parameters are indicative of changes in the tertiary conformation of the protein and are likely preludes to biochemical and physical instabilities. By comparing the changes of these parameters to the native active form of the molecule, 2dUV analysis can provide information on any subtle changes in the tertiary structure of the protein after it has been stressed and become "unstable." In contrast, a stabilizing formulation would show little to minor changes in the tertiary conformation after the protein has been stressed. The inclusion of all three intrinsic chromophoric amino acids in the data provides a more global picture of protein tertiary structure in comparison to steady-state fluorescence methods, and a wealth of data are obtained from a single spectral reading of the sample. An example of differences in tertiary conformation of the native protein and a destablized protein is shown in Figure 13.3 for a protease inhibitor protein drug. The changes in conformational stability correlate strongly with loss in protease inhibitory activity.

With the use of such high-throughput techniques, the problem is not how to analyze a protein but what to do with all the data that have been generated. An effective approach that addresses this management of massive amounts of data to obtain useful information and subsequent knowledge about the protein is to use statistical methods, such as, principle component analysis (PCA) and principle latent analysis (PLA). These analysis can rapidly identify the critical components in the massive data sets that are responsible for the critical attributes of the protein. The using statistical analysis, such as, principle component analysis (PCA) and principle latent analysis (PLA). The use of design of experiments (DOE) in formulation development is an efficient use of resources that can minimize the experimental work and maximize the results. Proper

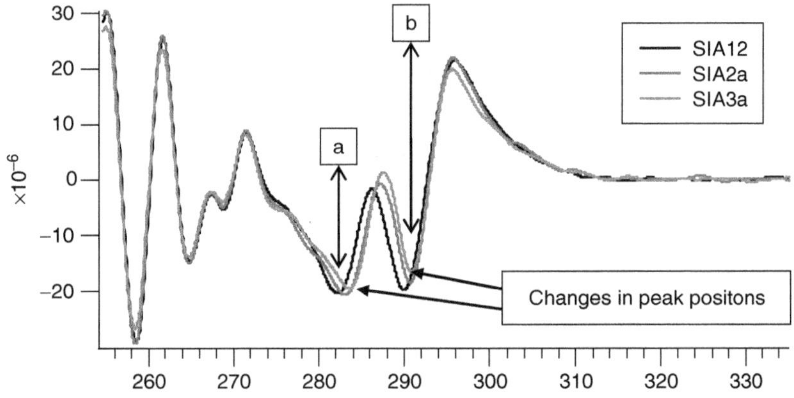

	a/b ratio	IU/mL
SIA 12 = 10 mg/ml untreated control	0.454	38.9
SIA 2a = 65 min/50°C	0.496	18.8
SIA 3a = 120 min/50°C	0.573	10.2

Figure 13.3. Correlations between tertiary conformational changes and protein activity. Significant changes are observed in both peak ratios (*a*/*b*), peak positions, and protein activity [in international units per milliliter (IU/mL)].

use of DOE is a powerful tool in formulation development, and improper use of DOE is a wasteful exercise that can lead to misleading interpretation of the results. The next section addresses the do's and don'ts of DOE in formulation development.

13.6. THE EXPERIMENTS BY DESIGN (E*B*D) APPROACH IN FORMULATION DEVELOPMENT

Proper use of the experiments by design approach is a powerful approach in formulation development. It allows for efficient experimentation during formulation development that can save time, materials, and resources. By definition, DOE is a set of statistically designed experimental trials that concurrently evaluates multiple variables in a formulation in a single set of experiments employing a minimum number of trials. These variables are typically pH, ionic strength, protein concentration, buffer concentration, and concentrations of one or two excipients. Use of DOE without knowledge of its advantages and disadvantages has been a hindrance in its widespread acceptance in biopharmaceutical formulation development. What makes DOE truly effective is in its use in a process defined as *experiments by design* (EbD).

The purpose of EbD is to use DOE congruently with the overall project objects (design process or the product target profile). This results in *effective* formulation development using *efficient* DOE to perform the right set of experiments. The EbD process is based on the concept of employing *quality by design* (QbD) in formulation development.

The four principles of EbD are the same as those outlined by W. Edwards Deming in the 1950s [17]. He proposed that business processes should be analyzed and measured to identify sources of variations that cause products to deviate from customer requirements. Deming created a diagram or "wheel of quality management" to illustrate this continuous process shown in Figure 13.4. The process is commonly known as the PDCA cycle:

1. *Plan:* Decide on the project goals and approach.
2. *Do:* Implement the plan and measure its performance (results).

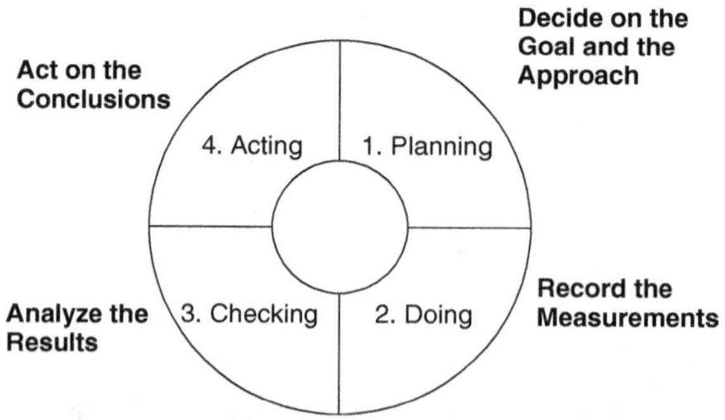

Figure 13.4. Deming's wheel of quality management.

3. *Check:* Assess the measurements and report the results to decisionmakers; that means to analyze and to validate the results.
4. *Act:* Decide on the changes needed to improve the process; that means to act on the conclusions.

13.6.1. Planning

Planning is the most important component of the EbD process and can be broken down into four components:

1. It involves clearly defining the project objectives or the product target profile (PTP).
2. The definition of success should be clearly defined without any ambiguity. For example, the definition of a suitable formulation can not be that it is stable, but rather, stability should be clearly defined in measurable terms. Is it 2 years at room temperature with retention of 80% of the protein activity, for example?
3. The feasibility of the project needs to be evaluated. A rough estimate of time, resources, and the amount of work needs to be determined. In cases where an appropriate SIA has not been established, such analysis may suggest that the DOE approach may not be suitable, because the experimental trials may be too time-consuming and resource-intensive for performing the DOE study. For example, a three-variable response surface experimental design requires 15 unique trials with five replicates for statistical significance resulting in 20 experimental trial formulations.
4. An estimate of the experimental design needs to be determined. It is important to clearly define the resolution of the responses being measured (i.e., least important difference) and the error in the measurements (i.e., standard deviation of the assay). This clearly defines the amount of time, resources, protein requirements, and most importantly, outlines the experimental design variables and responses for the whole development team's input.

13.6.2. Doing

The "doing" step involves performing the actual DOE experiment, collecting the necessary experimental responses (results), and analyzing the data. A critical component of doing is to perform the complete set of the DOE trials. Since only 30%–50% of the theoretical complete set of experimental trials are being done using DOE, it is imperative that data from all the trials be obtained and analyzed. Analysis of the DOE trials also offers numerous advantages in clarity of presentation through three-dimensional and contour plots, ranking of the formulation variables using Pareto plots, and statistical evaluation of the dataset. The statistical evaluation provides the inherent error in the experimental design (i.e., residual error or model error) and the analytical assay (i.e., replicate error or pure error). The model error illustrates the differences between the model predictions and the actual experimental observation. The replicate error is independent of the design model utilized in the DOE and is commonly also referred to the *replicate standard deviation*, which is the difference between replicate experimental data and the replicate average. These analyses give the resolution or "least important difference" or, in other words, the smallest detectable effect in the results. The DOE model is inadequate if the model error is greater than the pure error. This lack of fit can sometimes be remedied by data transformation or by using a more complex model (design surface) to fit the experimental results.

13.6.3. Checking

This step involves determining the optimal formulation and verifying the lead stable protein formulation. Since the data generated are empirical, the model generated from the DOE process is only as good as the quality of the data. With a response surface plot, it provides the ability to predict the behavior of the formulation at any point in the design window within statistical limits. An optimal formulation can be predicted and has to be verified experimental in a second round of experimental trials. This step is sometimes overlooked, and not validating the empirical DOE stability model can risk failure of the predicted formulation from the DOE design space.

13.6.4. Acting

The final step in EbD is to present the DOE data in a coherent fashion to the project team and use the verified DOE model to make accurate predictions on the properties of the lead formulation with respect to the PTP. If the objectives are met, the formulation progresses into the next stage of development. A validated model also allows accurate predictions about how changes in the formulations within the design space will affect the performance of the formulation without doing any additional work. Finally, the empirical formulation model can be used to rationalize and justify the formulation from a quality by design perspective.

13.7. A CASE STUDY FOR EFFECTIVE FORMULATION DEVELOPMENT OF A PROTEIN

To illustrate the principles of EbD and how effective formulation development can be done with limited amounts of protein, limited resources, and fast timelines, a case study for optimizing a stable formulation of ovalbumin as a model protein is presented. A similar approach has been applied to other proprietary protein drugs in biopharmaceutical formulation development by the authors. Ovalbumin was chosen as it is a well-characterized nonpropritory protein. Ovalbumin is highly susceptible to aggregation under thermal stress (50°C–60°C) that can be conveniently monitored using UV–vis spectroscopy [18].

13.7.1. Objective

The objective was to develop a pharmaceutically acceptable formulation that would significantly reduce the aggregation of ovalbumin using a convenient high-thoroughput assay for monitoring protein aggregation and using limited quantities of the protein.

The product target profile (PTP) was used to identify an ovalbumin formulation that did not aggregate in an isotonic (~320 milliosmol/kg) formulation.

13.7.2. Experimental Design

Ovalbumin has a denaturing melting temperature (T_m) of ~70°C, where it begins to precipitate [17]. On the basis of this property, a stability-indicating assay for the protein was to incubate it in various formulations at 60°C ($<T_m$ onset). Aggregation was monitored by UV–vis spectroscopy at 340 nm (light scattering of protein aggregates) after a 30 min incubation period in a 96-well plate reader. A quadratic response surface experimental design was chosen as only four variables were being evaluated and the objective was to obtain a well-defined quadratic surface of the formulation design space from the experimental trials. A total of 20 unique trials plus five replicates were required for the design and all the experiments were performed on a 96-well plate in a total volume of 300 μL per experimental trial using a total of 7.5 mg of protein. A predesigned 96-well plate (*i*Formulate™ formulation plate) of these four variables was obtained from HTD Biosystems (Hercules, CA).

13.7.3. Design Variables

Four key continuous variables that can affect protein stability were chosen as the design variables for a quadratic response surface experimental design: (1) pH in the range of 5–8, (2) ionic strength in the range of 0–200 mM NaCl, (3) buffer concentration in the range of 10–50 mM, and (4) sucrose, as a stabilizing excipient, in the concentration range of 0–10 wt%.

13.7.4. Experimental Trial

Ovalbumin was added to the *i*Formulate plate at final concentration of 1 mg/mL. The plate was incubated at 60°C for 30 min, and protein aggregation was measured by absorbance at 340 nm using a 96-well plate reader. The experiment utilized a total of 7.5 mg of ovalbumin in the experimental trials. The trail formulations and the absorbance values at 340 nm are shown in Table 13.1.

13.7.5. Results

The results of the DOE were analyzed by ECHIP DOE software, and the response surfaces shown in three-dimensional plots in Figure 13.5. The primary response (z axis) was absorbance at 340 nm that measured light scattering due to protein aggregation. The objective was to define an isotonic formulation with low aggregation following thermal stress. Figure 13.5a shows the response surface plots for pH and ionic strength. The steep curvature along the pH axis suggests the changes in pH significantly affect ovalbumin aggregation rather than NaCl concentration. Figure 13.5b shows the response-surface plot for pH and sucrose concentration. In this case also, pH is the more significant variable and sucrose appears to decrease

TABLE 13.1. Formulations Employed in Response Surface DOE Experimental Trial[a]

	Variables				Response
Trial Number	pH	NaCl	Buffer Concentration	Sucrose	Aggregation after 30 min at 60°C
11	**5.0**	**200**	**10.0**	**5**	0.658
15	6.5	200	30.0	10	0.000
17	6.5	200	10.0	0	0.001
16	6.5	0	10.0	10	−0.002
8	**5.0**	**200**	**50.0**	**10**	0.490
7	8.0	0	50.0	10	0.002
4	**5.0**	200	**50.0**	**0**	0.475
2	8.0	200	10.0	0	−0.001
3	8.0	0	50.0	0	−0.001
12	**5.0**	0	**50.0**	**5**	0.731
3	8.0	0	50.0	0	0.002
2	8.0	200	10.0	0	0.000
20	**5.0**	100	**10.0**	**10**	0.449
14	8.0	100	30.0	10	−0.001
18	8.0	0	30.0	0	0.000
5	**5.0**	0	**10.0**	**10**	0.380
13	6.5	100	50.0	10	0.000
5	**5.0**	0	**10.0**	**10**	0.428
4	**5.0**	200	**50.0**	**0**	0.363
19	**5.0**	100	**50.0**	**0**	0.859
6	8.0	200	10.0	10	0.000
1	5.0	0	**10.0**	**0**	0.472
9	8.0	200	50.0	5	0.000
10	8.0	0	10.0	5	−0.001
1	**5.0**	0	**10.0**	**0**	0.902

[a] Four formulation parameters were studied: pH, ionic strength, buffer concentration, and sucrose concentration. Boldfaced formulations indicate significant protein aggregation as measured by absorbance at 340 nm after 30 min incubation at 65°C.

protein aggregation at higher concentrations as shown by the downward trend in A340 values along the sucrose axis. Figure 13.5c is a more complex response surface plot showing interactions between sucrose and NaCl. A low sucrose concentrations, increasing concentrations of NaCl appear to reduce the amount of aggregation, whereas at high sucrose concentrations, the effect of NaCl is not as significant on ovalbumin aggregation. Analysis of these data plots suggests that the most critical formulation variable is pH, followed by sucrose concentration. Ionic strength or NaCl concentration is not a critical parameter affecting the propensity of ovalbumin to aggregate under thermal stress conditions. Buffer concentration was also not a critical variable (results not shown here). Figure 13.6 shows a contour response surface plot where, on the basis of the empirical response surface plots from the DOE trials, an optimized low aggregating formulation of ovalbumin is predicted as comprising of 10 wt% sucrose, 0 mM NaCl, and 25 mM buffer at pH 7.06. This formulation is also

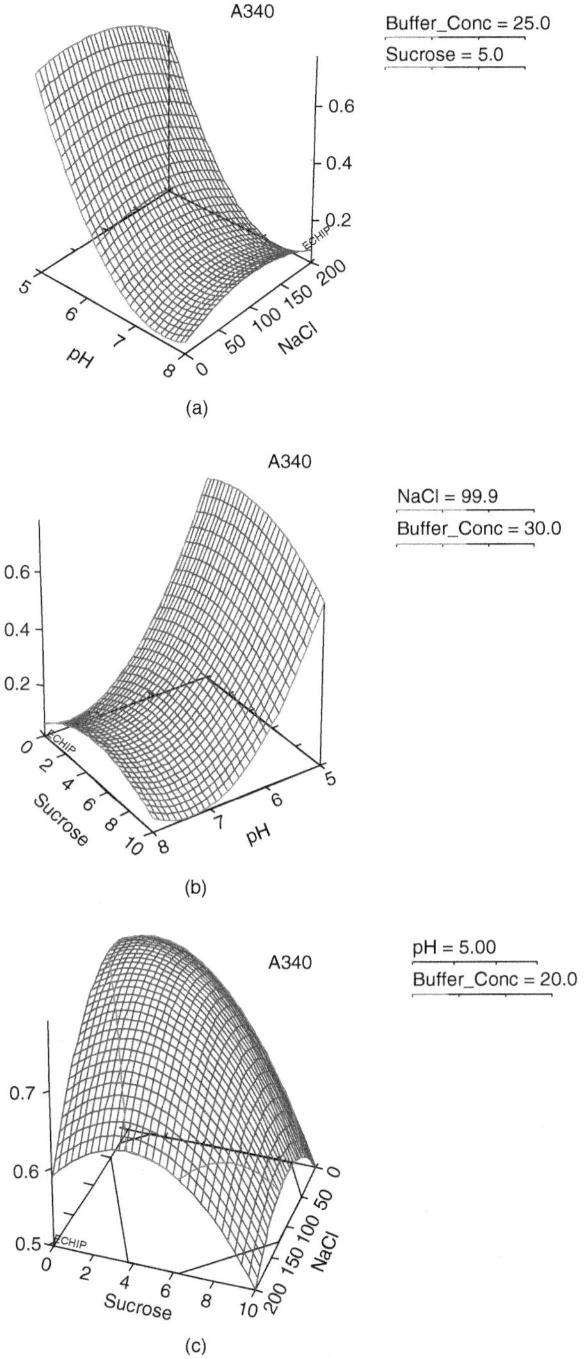

Figure 13.5. Response surface plots of A340 as a function of pH, sucrose, and NaCl.

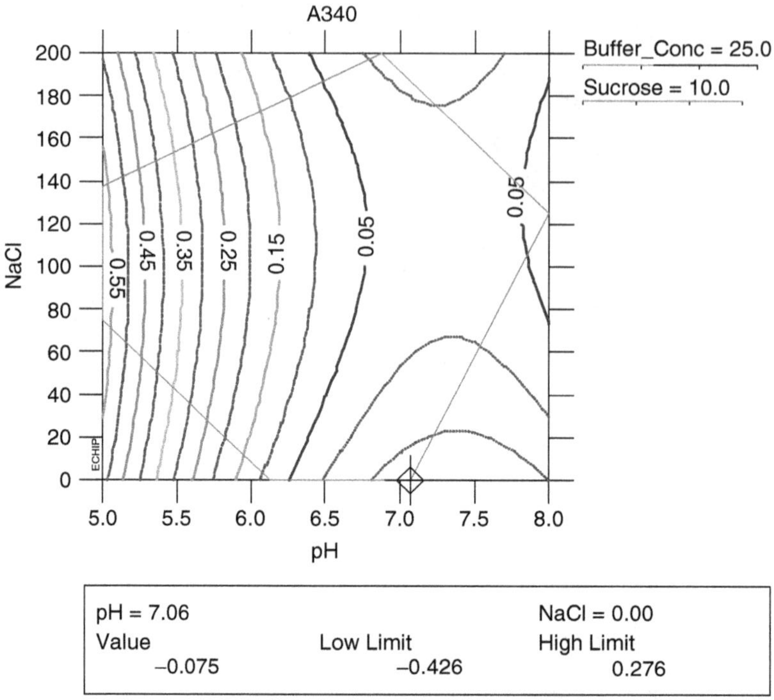

Figure 13.6. Optimal isotonic formulation of ovalbumin as predicted by the response surface formulation model.

an isotonic formulation as defined in the PTP of the ovalbumin formulation. This formulation was subsequently verified in additional formulation trials.

13.7.6. Summary

This case study demonstrates the principles of using EbD effectively in developing a non-aggregating formulation of ovalbumin that is isotonic and meets the formulation PTP objective. The formulation development utilized minimal quantities of protein, and used an appropriate stability-indicating assay and high-throughput *i*Formulate™ plates that evaluated four formulation variables in a DOE design space of 20 unique formulations with five replicates for statistical evaluation of the data.

13.8. OVERALL EXPERIMENT BY DESIGN STRATEGY

In summary, this chapter addresses the urgent need in the biopharmaceutical environment for high-throughput formulation development. Strategies for utilizing limited material, limited resources, and fast timelines are discussed and illustrated. It should be noted that rapid formulation development does not necessary suggest that formulation is done inefficiently and empirically with no rationalization of the formulation. Using the experiments by design (EbD) approach, a robust formulation can be developed that is in line with the QbD concepts that are now emphasized in pharmaceutical development. This is due in part to the availability of high-throughput analytical systems that are now available to the formulation and analytical groups that allows large number of experiments to be performed with limited material. The large datasets can be analyzed and presented effectively using DOE software so that rational decisions can be made for designing a robust formulation.

The EbD strategy can be summarized in six steps that provide an effective approach to generating a stable protein formulation using limited material, resources, and fast timescales:

1. Identify the relevant variables and responses.
2. Perform a preliminary screening DOE to reduce the number of formulation variables for a more complex response surface design.
3. Identify three or four key variables that affect the stability of the protein.
4. Run a response surface experimental design.
5. Get a predictive picture of the stability of the protein so that an optimal formulation can be defined.
6. Verify the optimal lead formulation experimentally and demonstrate correlation between experimental and predicted responses.

REFERENCES

1. Nayar, R. and Manning, M. C. (2002), High throughput formulation: Strategies for rapid development of stable protein products, *Pharm. Biotechnol*. **13**: 177–198.
2. Nayar, R. and Manning, M. C. (2002), High-throughput biopharmaceutical drug development: Meeting the coming challenge, *BioPharm* **15**(2): 20–28.
3. Landers, P. (2007), Cost of developing a drug increases to about $1.7 billion, in *PAREXEL's Pharmaceutical R&D Statistical Sourcebook 2006/2007*, Bain & Co. (report of a 2003 study), p. 107.
4. Accenture, Inc. (2007), *The Pursuit of High Performance through Research and Development*, PHRMa report by Accenture's Health and Life Sciences Group.
5. Ahern, T. J. and Manning, M. C., eds. (1992), *Stability of Protein Pharmaceuticals*, Vol. 3 of Pharmaceutical Biotechnology Series, Plenum Press, New York.
6. Wang, Y. J. and Pearlman, R., eds. (1993), *Stability and Characterization of Protein and Peptide Drugs: Case Histories*, Vol. 5 of Pharmaceutical Biotechnology Series, Plenum Press, New York.
7. Manning, M. C., Patel, K., and Borchardt, R. T. (1989), Stability of protein pharmaceuticals, *Pharm. Res*. **6**: 903–917.
8. Nail, S. L. and Akers, M. J., eds. (2002), *Development and Manufacture of Protein Pharmaceuticals*, Vol. 14 of Pharmaceutical Biotechnology Series, Plenum Press, New York.
9. Wang, Y.-C. J. and Hanson, M. A. (1988), Parenteral formulations of proteins and peptides. Stability and stabilizers, *J. Parenter. Sci. Technol*. **42**(Suppl.).

10. Wang, W. (1999), Instability, stabilization, and formulation of liquid protein pharmaceuticals, *Int. J. Pharm.* **185**(2): 129–188.
11. Wang, W. (2000), Lyophilization and development of solid protein pharmaceuticals, *Int. J. Pharm.* **203**(1–2): 1–60.
12. Patro, S. Y., Freund, E., and Chang, B. S. (2002), Protein formulation and fill-finish operations, *Biotechnol. Annu. Rev.* **8**: 55–84.
13. Drucker, P. F. (2006), *The Effective Executive: The Definitive Guide to Getting the Right Things Done*, Harperbusiness Essentials, Elsevier, Oxford, UK.
14. Volkin, D. B., Sanyal, G., Burke, C. J., and Middaugh, C. R. (2002), Preformulation studies as an essential guide to formulation development and manufacture of protein pharmaceuticals, in *Development and Manufacture of Protein Pharmaceuticals*, Nail, S. L. and Akers, M. J., eds., Vol. 14 of Pharmaceutical Biotechnology Series, Plenum Press, New York, pp. 1–39.
15. Middaugh, R. C. (1994), The analysis of protein pharmaceuticals: Near Future advances, *Cytotechnology* **15**: 187–194.
16. Liebler, D. C. (2002), *Introduction to Proteomics: Tools for the New Biology*, Humana Press, Totowa, NJ.
17. Deming, W. E. (1986), *Out of the Crisis*, MIT Center for Advanced Engineering study.
18. Krielgaard, L., Jones, L. S., Randolph, T. W., Frokjaer, S., Flink, J. M., Manning, M. C., and Carpenter, J. F. (1998), Effect of tween 20 on freeze-thawing- and agitation-induced aggregation of recombinant human factor XIII, *J. Pharm. Sci.* **87**(12): 1593–1603.

14

PREDICTION OF AGGREGATION PROPENSITY FROM PRIMARY SEQUENCE INFORMATION

Mark Cornell Manning, Gabriel J. Evans, Cody M. Van Pelt, and Robert W. Payne

14.1. INTRODUCTION

Aggregation of proteins remains the most common and problematic physical instability of proteins [1,2]. Certainly, when developing a protein pharmaceutical product it is the most vexing. In part, this is due to the increased immunogenicity exhibited by protein aggregates, placing the patent at greater risk [3–6]. If one could predict a priori whether a protein were prone to aggregation, selection of appropriate drug candidates would be improved significantly. This chapter discusses the current state of predicting aggregation behavior from the primary sequence.

Obviously, aggregation is highly dependent on solution conditions [1,7,8]. In particular, it is important to distinguish between amorphous aggregate formation (often called β *aggregation*) and fibrillation or formation of amyloid fibrils [9]. It appears that β aggregation is characterized by a greater degree of flexibility, while fibril formation is almost quasicrystalline [9]. This implies that sequence specificity may not be as great for β aggregation as it is for fibril formation. For solution conditions far from the isoelectric point (pI) of the peptide or protein, fibril formation can and does occur, suggesting that specific side chain interactions are important. However, as the pH moves

Formulation and Process Development Strategies for Manufacturing Biopharmaceuticals, edited by Jameel and Hershenson

closer to the pI, β aggregation becomes more prevalent or preferred [10]. In these cases, the sequence specificity may not be as important as an overall compositional description of the protein sequence in terms of physicochemical properties.

Most often, algorithms do not consider variation in solution conditions when predicting aggregation rates or propensities. Rather, they attempt to develop a rank ordering, even possibly predicting aggregation rates for a predetermined set of conditions. Therefore, they are not intended to give a complete description of the aggregation behavior of a protein, but, rather, focus on identifying sequences and motifs that appear to be involved in protein self-association. Still, some methods do try to account for some of the variations that may occur in pH, temperature, and other parameters, as discussed below.

The various approaches can be divided into composition-based and sequence-based methods [11–13]. The former focuses on the overall amino acid content, assessing characteristics such as hydrophobicity or overall charge. In general, these tend to be less reliable, as they do not account for the conformation or neighboring amino acids. The latter have become relatively well developed, thanks in part to the work by Dobson, Chiti, Serrano, Caflisch, and others. Together, we now have a variety of tools for predicting a priori the aggregating potential of a particular protein sequence.

Finally, this chapter discusses some of the emerging ideas regarding sequence and protein stability, such as the existence of hot spots, nucleation sites, conformational switches, and gatekeeper residues. The concept is that there are short sequences embedded within a protein that control, in large part, the conformational stability of the protein, including being the foci for aggregation or self-assembly. Some of these ideas were born out of protein folding studies [14], as researchers look at the initial stages of higher-order structure formation [15–17]. Not surprisingly, the same sequences that serve to initiate folding also appear to be trigger points for aggregate formation [15].

14.2. COMPOSITION-BASED APPROACHES

Some predictive algorithms focus on how the overall composition of the protein affects aggregation behavior. It is important to keep in mind that many of these algorithms are based on fibril formation in peptide systems, such as prions and amyloid peptides. In fact, there are only about 20 known peptides or proteins that undergo fibril formation under physiological conditions [8]. Still, fibrillogenesis represents an important aggregation pathway, but it may or may not be appropriate for predicting associative behavior of globular proteins.

Considering the composition of a protein and the effect on aggregation propensity, one might imagine that higher amounts of hydrophobic residues might lead to increased aggregation. Moreover, as proteins become more likely to associate near the isoelectric point (pI), the value of the pI is often considered in such predictive schemes. For example, Calamai et al. examined the relative hydrophobicity and net charge of two analogous proteins as being the basis for increased amyloidogenic proteins of one of them [18]. Using protein engineering, they were able to estimate that the differences in the overall charge accounted for 20%–25% of the difference in aggregation rates.

Charge effects have been investigated in systems where the overall charge could be manipulated. In some cases, the proteins can become very highly charged. Studies on green fluorescent protein (GFP) demonstrated that supercharging the proteins (+36 or −30) had only modest effects on thermodynamic stability, but greatly improved resistance to aggregation [19]. Such species fold correctly and appear to possess reasonable stability, arguing against the notion that highly charged species will necessarily unfold as a result of electrostatic repulsions. Another study on ovalbumin found that modulating the net charge did not discriminate between variants in terms of aggregation behavior [20]. However, these changes did affect the denaturation temperature. The authors concluded that any effects on aggregation propensity were due to alterations of conformational stability. Conversely, in 14 of the 16 mutations with increased aggregation propensity in acylphosphatase the overall charge on the protein was reduced, suggesting that net charge is an important factor to consider [21].

True composition-based algorithms, based on multiple protein sequences, have been reported. For example, the charge on the protein and the propensity for a sequence to form β-sheet structure has been found to correlate with aggregation rates [21–23]. Similarly, charge state and propensity to form β-sheet structure were found to be necessary to allow fibril formation to proceed in tetrapeptides [24]. In Aβ(1–40), substitutions at Asp^{23} or Glu^{22} affected amyloid formation significantly, suggesting that charge effects, likely via salt bridge formation, were important in fibril assembly [25]. Altogether, there appears to be both a general and specific role of electrostatics in controlling aggregation behavior.

Galzitskaya found that greater hydrophobicity and β-sheet propensity leads to increased likelihood of aggregate formation [26]. A study on mutants of α-synuclein demonstrated that fibril formation correlated with mean hydrophilicity, β-sheet propensity, and net charge on the protein [27]. Still other groups have found that, while overall composition is important in controlling aggregation, the composition of the N- and C-terminal residues are the most influential [28,29].

More recently, there has been some focus on aromatic residues and the role they play in affecting aggregation. The potentially significant role of hydrophobic interactions in aggregation has been proposed for some time [30]. However, the central question is whether the effect is general or whether specific interaction, as occur with aromatic side chains, is necessary. The relative abundance of aromatic residues in amyloid-forming peptides suggests that aromatic–aromatic interactions, that is, π stacking, may be responsible for initiation of fibril assembly [31]. The concept is similar to the possible role of π stacking in protein folding [32,33]. Other studies have also identified the importance of aromatics, while also emphasizing the role that dipole moment and solvent accessibility can play in governing aggregation rates [34]. Studies by Azriel and Gazit on a hexapeptide from islet amyloid polypeptide determined that aromatics, especially phenylalanine (Phe or F) residues, play an important role in fibril formation [35]. In fact, they describe a FF motif, where side-by-side Phe residues dramatically stabilize fibrils. Similarly, studies on a dodecapeptide also showed the ability of Phe residues to "zip together" adjacent strands of a β sheet [36]. The same study also found that salt bridge formation was important in stabilizing fibrils. On the other hand, replacement of Phe with Leu results in islet amyloid polypeptide mutants that exhibit similar aggregation propensity [37].

Conversely, a study on tetrapeptides found that the sequence, KVVE, formed fibrils as readily as did KFFE [24], suggesting that hydrophobic interactions, and not just π stacking, contribute to fibril stability. The same appear to be true for fibril formation in the Aβ(1–42) peptide, where the central hydrophobic region appear to be critical for fibrillation [38], but does not necessarily require aromatics to be present [38]. Moreover, a study on islet amyloid polypeptide also found that aromatic interactions were not required for fibril formation, although removing Phe residues does slow the rate of fibril formation [39].

Since many aggregates proceed from a partially unfolded state, it is likely that conformational stability is another global property that might control aggregation rates [1,30, 40–42]. A few studies have included this factor in their predictive schemes [43,44], as described below. Overall, composition-based methods have been successful, but not entirely accurate. One study demonstrated that shuffled sequences with the same amino acid composition display widely varying aggregation rates [45]. By clustering of residues with the greatest aggregation propensity, the aggregation rates can be increased by orders of magnitude. Therefore, sequence-based algorithms are more likely to be accurate. These are discussed below.

14.3. SEQUENCE-BASED ALGORITHMS

The homology between amyloid-forming peptides was identified some time ago [46]. This has produced a number of theoretical studies aimed at predicting aggregation propensity directly from primary sequence. As a result, a wide variety of approaches have been taken when performing sequence analyses. In some cases, the approaches are similar to pattern recognition. For example, hydrophobicity profiles were examined, looking for propensities to form local or nonlocal contacts [47]. These contacts, if intermolecular, would lead to aggregate or fibril formation. Indeed, prevention of hydrophobic interactions has been shown to slow aggregation [48]. In doing so, critical residues were identified for about a dozen proteins that could be the contact points for starting aggregate formation [47]. The importance of hydrophobic interactions, as well as hydrogen bonding in amyloid assembly is the basis for a prediction scheme described by Saiki et al. [49]. The basis parameters for these two properties are quite rough, yet the model has some predictive capacity.

By considering some of the compositional factors listed above, Chiti et al. developed an equation to predict aggregation rates of various mutants [23] [see Eq. (14.1)]. The factors A, B, and C were determined from the slopes of the plots of aggregation rate versus the individual properties. These are 0.633, 0.198, and −0.491 for A, B, and C, respectively:

$$\ln \frac{\nu_{\text{mut}}}{\nu_{\text{wt}}} = A\Delta(\text{hydrophobicity}) + B(\Delta\Delta G_{\text{coil-}\alpha} + \Delta\Delta G_{\beta\text{-coil}}) + C\Delta(\text{charge}) \quad (14.1)$$

The correlation coefficient for predicting the aggregation rates of mutations in 27 short peptides was 0.85. While other improvements have been made to this algorithm [12], it remains the most widely known prediction scheme for aggregation propensity. Although it is empirical [12], it is quite effective; however, its inability to predict aggregation rates accurately where there has been a substantial change in net charge has been criticized [12]. This is likely due to the fact that the hydrophobicity and charge parameters are not completely independent. Interestingly, the model has been used to redesign calcitonin, a 32-residue therapeutic peptide, so that it displays reduced aggregation propensity [50].

The same group developed an algorithm they call "Zyggregator" [42,51] (see also www.zyggregator.com to use the algorithm), whereby aggregation-prone sequences can be identified by comparing the aggregation propensity score of a given sequence to one for a random sequence of the same overall composition [51]. The first step is to calculate an intrinsic propensity, using the following equation

$$P_{\text{agg}} = \alpha_{\text{hydr}} I^{\text{hydr}} + \alpha_{\alpha} I^{\alpha} + \alpha_{\beta} I^{\beta} + \alpha_{\text{pat}} I^{\text{pat}} + \alpha_{\text{ch}} I^{\text{ch}} \tag{14.2}$$

where I^{hydr} represents the hydrophobicity of the sequence, I^{α} is the α-helical propensity, I^{β} is the β-sheet propensity, I^{pat} is the hydrophobic patterning, and I^{ch} is absolute net charge on the polypeptide. The α coefficients, the weighting factors, were determined by fitting to the reference dataset. *Patterning* refers to the alternating pattern of hydrophobic and hydrophilic residues often found in β-sheet structure common to fibril formation and β aggregation [40,41]. In fact, such patterns are underrepresented in protein sequences [52], probably because they do play a role in efficient assembly of intermolecular β sheets [53]. Patterning affects whether polypeptides can form parallel or antiparallel β-sheet structures. An algorithm called *PASTA* (prediction of amyloid structure aggregation) determines a pairwise energy function, leading to a propensity to form either parallel or antiparallel β-sheet structures [54]. This allows prediction of the registry of hydrogen bonds in an intermolecular β sheet, that is, within an amyloidogenic sequence. So, this approach is based on both intrinsic β-sheet propensity and patterning of hydrophobic and hydrophilic side chains. Applying this method to Aβ(1–40) produced predicted sequences of amyloid consistent with experimental results. Proline scanning mutagenesis indicates that residues 12–24 and 30–40 are involved in aggregate formation [55]. The PASTA algorithm found the best pairing was residues 12–20, and the second best was 31–40 [54].

Then, the aggregation score Z_{agg} can be obtained. It is calculated to be the difference in P_{agg} from the average P_{agg} over a set of random polypeptides (μ_{agg}), divided by the standard deviation (σ_{agg}) from the average [see Eq. (14.3)]. In doing so, one can assess the effect of a single mutation on aggregation propensity:

$$Z_{\text{agg}} = \frac{P_{\text{agg}} - \mu_{\text{agg}}}{\sigma_{\text{agg}}} \tag{14.3}$$

Using this methodology, they have calculated the Z_{agg} scores for a variety of mutations. For example, they calculated Z_{agg} for genetic mutations in Aβ(1–42) (see Table 14.1). In all four mutations, the correct trend in aggregation propensity was predicted. More

TABLE 14.1. Aggregation Propensity for Various Naturally Occurring Mutants of Aβ(1–42), Both Experimentally and Predicted by Various Algorithms

Mutation	Difference in Soluble Aβ(1–42) [23], fmol/mL	Difference in TANGO Score [43]	ΔZ_{agg} Score [51]	ΔAP [86]
Dutch (E22Q)	4.5	120	0.17	+0.44
Flemish (A21G)	−9.0	−325	−0.02	−1.22
Italian (E22K)	0.5	140	0.18	—
Arctic (E22G)	5.0	90	0.24	+2.14

recent studies on Aβ(1–40) using Zyggregator have shown a link between the thermodynamics of aggregation and the kinetics (as measured by the lag time and growth rate for fibrils) [42].

A statistical mechanism approach to aggregation prediction was taken by Fernandez-Escamilla et al., leading to an algorithm termed TANGO [43]. On the basis of β-sheet formation properties, they predicted the aggregation rates of 179 different peptides found in the literature. The algorithm was also able to correctly predict pathogenic mutations in a number of proteins implicated in fibril formation *in vivo*. For example, TANGO calculations indicate that the Dutch, Italian, and Arctic mutations in Aβ(1–42) are pathogenic, while the Flemish mutation is not. These results are consistent with the amount of soluble Aβ(1–42) recovered (Table 14.1), although TANGO does predict the Italian mutation to be cause more fibrillation than it actually does. Note that the predictions regarding these mutations are similar to those obtained from the Zyggregator method.

The same TANGO algorithm was used to compare the aggregation propensity of intrinsically disordered proteins (IDPs) with globular proteins, examining almost 300 proteins in each class [44]. The β-aggregation propensity was found to be similar among all structural classes or globular proteins, including membrane proteins. This suggests that β aggregation is not determined by hydrophobicity and β-sheet propensity alone. Interestingly, they also found that globular proteins contained about 3 times as many aggregation nucleating regions as the IDPs [44]. Whether this then translates into lower potential for amyloid formation from IDPs remains to be determined.

In studies on a hexapeptide [43,56], the TANGO algorithm was used to examine sequence space for β-aggregation and amyloidosis/fibril formation (see Table 14.2). Those residues that are unfavorable for either type of aggregation are not highlighted, while those mutations that lead to formation of both types of aggregate appear in boldface italic. Conversely, there are some mutations that lead to specific types of aggregate, with those in boldface only forming fibrils and those in italic only forming β aggregates. Note that Pro residues inhibit aggregate formation, consistent with their role in aggregation resistance (see Section 14.8, below).

Finally, the TANGO algorithm was used to examine 28 complete proteomes [57]. It appears that, although many proteins possess some number of aggregation-prone sequences, the frequency is rather low. Only a small portion is predicted to display high propensity to aggregate. The authors believe this to be due to a similarity between the nucleation sites for aggregation and those needed to initiate protein folding [53],

TABLE 14.2. Results from an Exhaustive Mutational Study on the Hexapeptide STVIIE[a]

Position 1	Position 2	Position 3	Position 4	Position 5	Position 6
L	***L***	***L***	***L***	*L*	***L***
V	***V***	***V***	*V*	*V*	***V***
I	***I***	*I*	***I***	***I***	***I***
F	***F***	***F***	***F***	***F***	***F***
A	***A***	*A*	*A*	*A*	***A***
W	*W*	***W***	***W***	*W*	***W***
Y	***Y***	*Y*	***Y***	***Y***	***Y***
T	***T***	*T*	***T***	*T*	***T***
M	***M***	*M*	*M*	*M*	***M***
C	**C**	**C**	C	C	**C**
S	**S**	**S**	S	S	**S**
G	**G**	G	G	G	**G**
H	H	H	H	H	H
N	**N**	**N**	**N**	N	**N**
Q	**Q**	**Q**	Q	Q	**Q**
D	**D**	D	D	D	**D**
E	**E**	E	E	E	**E**
K	K	K	K	K	K
R	R	R	R	R	R
P	P	P	P	P	P

Source: Data taken from [9].
[a]Residues with a preference for amyloid formation are shown in boldface, those with a preference for β aggregation are shown in italic, and those with both preferences are shown in boldface italic.

consistent with other views about aggregation and folding being finely balanced [22]. Interestingly, it appears that the packing density separates globular proteins from native unfolded proteins as well [15].

Another sequence-based method was developed by Caflisch and co-workers [34,58], who calculated the aggregation propensity π_{il} for a sequence of l residues, starting at residue i, from a calibration set of nine proteins [34]. Those authors [34,58] seek to distinguish their approach from the one developed by Chiti and co-workers [21–23] by attributing aggregation propensity based on parameter optimization rather than from first principles. The π_{il} value was calculated from the following equation

$$\pi_{il} = \phi_{il} + \Phi_{il} \tag{14.4}$$

where $\Phi_{il} = \exp(A_{il} + B_{il} + C_{il})$, where A_{il} refers to the aromaticity of the sequence, B_{il} to the β-sheet propensity, and C_{il} to the charge. This method allows one to calculate aggregation propensity, but like the others, to identify mutations that can increase or decrease aggregation rates. The authors found numerous five-residue sequences with large π_{il} scores in proteins that are known to aggregate or form fibrils [58].

Another algorithm has been developed to identify amyloidogenic sequences based on packing density from molecular dynamics simulations [15,26]. Amino acids were assigned packing densities based on analysis of protein crystal structures (Table 14.3).

TABLE 14.3. Amino Acid Scale for Packing Density

Residue	Packing Density (Number of Close Residues)
Gly	17.18 ± 0.03
Asp	17.39 ± 0.03
Glu	17.43 ± 0.03
Pro	17.53 ± 0.04
Lys	17.72 ± 0.03
Ser	18.35 ± 0.04
Asn	18.57 ± 0.04
Gln	19.19 ± 0.04
Thr	19.91 ± 0.04
Ala	19.97 ± 0.03
Arg	21.03 ± 0.04
His	21.64 ± 0.06
Cys	23.99 ± 0.07
Val	24.05 ± 0.03
Met	24.80 ± 0.07
Leu	25.53 ± 0.03
Ile	25.96 ± 0.04
Tyr	26.17 ± 0.05
Phe	27.42 ± 0.05
Trp	28.53 ± 0.09

Source: Date taken from [15].

These values were then used to identify short sequences that had unusually high packing density or efficiency, with the idea that these would form stable contact points between polypeptide chains, that is, nuclei for aggregate formation.

Another set of algorithms is based on the crystal structure of the hexapeptide, NNQQNY [59,60], similar to the approach described by Lopez de la Paz and Serrano and co-workers [11,56], who based identification on the behavior of the amyloidogenic peptide, STVIIE. By threading new structures on to this crystal structure, fibril-forming sequences can be identified [59]. However, the likelihood of false positives due to neglecting charge repulsions might be high [12]. Similarly, a three-dimensional (3D) profile was imposed on the structure, where the energy was then calculated and used to identify sequences of similar fibril-forming propensity [60]. Some of the driving force for fibrillogenesis in the NNQQNY system seems to be efficiency of packing of side chains, as discussed above. This concept that β propensity is hidden or coupled to tertiary structure interactions has also been described by Yoon and Walsh [61].

Other approaches involve sequence alignment or comparison of thermophilic proteins with their mesophilic counterparts. One such extensive comparison identified salt bridges and main-chain hydrogen bonds as being essential for increased conformational stability [62]. Presumably, this would also lead to slower aggregation rates [1]. On the other hand, the polar/nonpolar composition of the surface appears to be unimportant [62].

The effects of extrinsic factors, such as pH and temperature, are often ignored in aggregation prediction schemes, although not entirely. For example, aggregation propensities for the 20 naturally occurring amino acids were scaled based on pH, using values for pH 2, pH 7, and pH 13 [51]. In all cases, the most effective at promoting aggregation were the aromatic amino acids, consistent with the thoughts regarding π stacking discussed above. A much more detailed assessment of extrinsic factors was conducted for more than a dozen amyloid-forming peptides and proteins [63]. The aggregation rates varied over five orders of magnitude and covered a wide range of concentrations and pH values. A new equation was developed, including intrinsic factors and extrinsic ones:

$$\log(k) = \alpha_0 + \alpha_{\text{hydr}} I^{\text{hydr}} + \alpha_{\text{pat}} I^{\text{pat}} + \alpha_{\text{ch}} I^{\text{ch}} + \alpha_{\text{pH}} E^{\text{pH}} + \alpha_{\text{ionic}} E^{\text{ionic}} + \alpha_{\text{conc}} E^{\text{conc}} \tag{14.5}$$

Here, I terms refer to intrinsic properties, E terms refer to extrinsic properties, and k is the aggregation rate in reciprocal seconds (s^{-1}) [63]. Statistical analysis of the various factors found that, while all of the intrinsic factors (hydrophobicity, charge, and patterns) were significant, only ionic strength was significant among the extrinsic factors (pH and polypeptide concentration were not). Still, Equation (14.4) was able to reproduce the observed aggregation rates with a correlation coefficient of 0.92. One of the most significant factors was patterning, where five residues of alternating hydrophobic–hydrophilic residues was found to yield the most significant correlation. Of the extrinsic factors, it is likely that the ionic strength effects are due to alterations in the effective charge of the protein or peptide.

All of these studies point to an important question, which is whether a sequence-based approach is superior to one based solely on composition. A more recent article addresses this concept [10]. When a protein is in its native state, the sequence (and the associated physicochemical properties of the side chains) will govern the conformational stability and three-dimensional structure.

Under these conditions, a sequence-based methodology will be most appropriate. As the polypeptide undergoes some degree of unfolding, the influence of the side chains is no longer as specific. Rather, the average or sum of the side chain properties becomes more important and descriptive [10].

14.4. CHEMOMETRIC METHODS

Since the late 1980s, a number of reduced property scales (also called *principal properties*) have been reported for amino acids. These scales seek to capture the wide variation of chemical and physical properties in a relatively small number of numerical parameters [64–66]. The most widely used are referred to as *z scores* [64,65]. The initial set of z scores contained three terms that roughly correlated to hydrophilicity (z_1), steric bulk (z_2) and electronic factors (z_3). This initial scale was expanded to include two additional scores (now referred to as "zz_1 through zz_5"). However, the exact physical meaning of these new factors is less clear.

Our laboratory has been using these reduced property scales to quantify the physicochemical properties of amino acid side chains. Once a matrix of a series of sequences has been assembled, one can use projection to latent structures (PLS), a multivariate statistical approach to correlate sequence information to a quantifiable measure of aggregation propensity. For example, Hecht and co-workers have prepared series of mutations in positions 41 and 42 in Aβ(1–42) and linked them to green fluorescent protein (GFP) [38]. The resulting fluorescence intensity should correlate with reduced aggregation propensity. A quadratic PLS model was constructed and it was found that zz_1 and zz_3 in each position were statistically significant in their effects on aggregation propensity (unpublished results). Since both were positively correlated, this means that increasing hydrophilicity and charge/polarity reduced the likelihood of aggregation. Interestingly, steric bulk was not an important factor.

14.5. LATTICE THEORY AND COARSE-GRAINED MODELS

All-atom theoretical models of proteins folding are so computationally intensive that only the smallest systems can be studied. In the case of protein aggregation, where a number of peptide or protein molecules must be included, the restrictions are even greater. As a result, many studies have moved to "lattice" or "coarse-grained" models, where each amino acid is modeled as a bead on a necklace [67]. While the lattice concept has been most widely used for protein-folding studies, it has been applied to aggregation behavior as well. One early study found that the propensity to aggregation was extraordinarily sensitive to the sequence, with one amino acid making large changes [68], a finding echoed by later studies [69,70]. These studies also emphasized the importance of hydrophobic interactions in driving aggregation [67,68]. Finally, these studies illustrate the effect of the solvent, in terms of charge shielding and polarity, as being instrumental to altering aggregation rates [67].

Langevin dynamics simulations, another type of coarse-grained model, examining fibril assembly, indicate that decreasing β-aggregation propensity of a peptide leads to a rougher energy landscape and a more heterogeneous pathway to fibril formation [71]. Nguyen and Hall have published on discontinuous molecular dynamics of aggregation behavior, where a large number of peptides can be included [72–74]. In another theoretical approach, a so-called toy model has been proposed to explain aggregation rates for fully unfolded proteins, based on a collision encounter scheme [75]. These types of studies will likely provide greater understanding in the future of the molecular forces driving fibrillogenesis.

14.6. MUTAGENESIS STUDIES

Mutagenesis studies, while not true predictive schemes, can provide information pertinent to the construction and validation of these mathematical models [56]. Amyloidogenic peptides have been a particular target. For example, scanning proline mutagenesis of Aβ peptides identified residues 12–24 and 30–40 as being most critical for amyloid formation [55]. Using the same approach, a finer map of the critical residues in Aβ(1–40) was obtained [76]. Similarly, a mutagenesis study of position 18 in Aβ(1–40) provided insight into the role that position played in amyloid fibril formation [77]. Proline mutagenesis was used to examine amyloid formation in human amylin(20–29) as well [78].

A number of mutagenesis studies from Hecht and co-workers have been conducted on Aβ(1–42) using GFP as a reporter [38,79,80]. Chimeras of GFP and Aβ(1–42) mutants were constructed. Increased aggregation propensity produced lower recovery of the chimeric protein. Therefore, the intensity of the fluorescence signal could be correlated with aggregation propensity. This methodology provides another avenue for quantifying aggregation propensity, allowing chemometric analysis as described above.

14.7. ANTIBODY FRAMEWORKS

For quite some time, researchers have postulated whether different antibody technologies, producing slightly different constant regions, possessed different intrinsic stabilities. For a long time, this type of discussion was based on fragmentary and anecdotal information. More recently, some studies have appeared that indicate that different antibody frameworks indeed do display variation instability [81,82]. Combinatorial assembly of a large domain antibody was conducted using a small repertoire of aggregation-resistant domains [83]. Analysis of these data could lead to improved stability of full-length monoclonal antibodies. In addition, framework shuffling to reduce immunogenicity has been found to alter thermostability as well [84]. It is expected that as more of this sequence-stability information is available, there will be further theoretical studies into predicting antibody stability profiles.

14.8. HOT SPOTS AND GATEKEEPERS

As sequence-based methods have evolved, there have been indications that there are certain short stretches of amino acids (two to five residues in length) that appear to act as nucleation sites for subsequent aggregation (see discussion above). The same concept can be found in short sequences that have a dominant effect on the conformational stability of a globular protein. These are often referred to as "hot spots." Tiana et al. have described minimal model calculations that are capable of identifying hot spots [85]. The algorithm was tested experimentally by mutagenesis studies on four small proteins. Their conclusion was that the algorithm correctly identified 60%–80% of the hot spots in these proteins [85].

Another algorithm to predict aggregation hot spots was developed by Ventura and co-workers [86,87]. The program, called AGGRESCAN, allows one to detect these hot spots. The initial work examined the effect of single point mutations on the aggregation of Aβ *in vivo* [88]. These propensities are then used as the basis for

TABLE 14.4. Amino Acid Scale for Aggregation Propensity

Amino Acid	Aggregation Propensity
Ile	1.822
Phe	1.754
Val	1.594
Leu	1.380
Tyr	1.159
Trp	1.037
Met	0.910
Cys	0.604
Ala	−0.036
Thr	−0.159
Ser	−0.294
Pro	−0.334
Gly	−0.535
Lys	−0.931
His	−1.033
Gln	−1.231
Arg	−1.240
Asn	−1.302
Glu	−1.412
Asp	−1.836

Source: Data taken from [86].

the algorithm (see Table 14.4). This approach appears to predict accurately many of the fibril nucleation sites or hot spots that have been determined experimentally in β_2-microglobulin, transthyretin, and prion protein [86]. A larger number of sequences were analyzed in a subsequent study [87]. The same algorithm was used to predict the effects of the common genetic mutations in Aβ(1–42) (see Table 14.1). As with the other algorithms, the general trend in aggregation propensity is predicted correctly.

Similarly, there are residues that seem to act as gatekeepers, that is, prevent aggregation from propagating. In a study on fibrillation in S6 from *Thermus thermophilus*, there appears to be a contiguous region of residues that inhibit fibril formation [89]. These residues are mostly unstructured in the native state, suggesting that flexibility and solvent accessibility are important determinants of aggregate formation (see discussion above). These findings were consistent with previous studies on the same protein, reinforcing this view [90–92]. This study is also consistent with the concept that protein aggregation can proceed from a native-like state [93], often referred to as the *aggregation-competent state*, which is similar in terms of hydrodynamic radius and conformation to the native state [94,95].

In general, Gly and Pro residues appear to function as gatekeepers [89,96]. Likely, this is due to the fact that they are highly disfavored for forming β-sheet structure. Certainly, this is consistent with the studies on amyloid peptides summarized in Table 14.2. Another study demonstrated that glycine (Gly) residues were apparently able to inhibit aggregation, making them gatekeepers in human muscle acylphosphatase [96]. The only exception was the G15A mutation. Once again, this would infer that flexibility is an important aspect to gatekeeper regions. Other residues have been implicated in controlling sequence propensity for aggregation. Pawar et al. identified Lys and Arg residues as "sequence breakers" [51], reducing the aggregation propensity of a sequence, presumably by charge–charge repulsion.

14.9. CONFORMATIONAL SWITCHES

The propensity for β-sheet formation is an interesting aspect of aggregate or fibril formation. Indeed, what one is really seeking are sequences that are finely balanced between globular structure and intermolecular β-sheet formation. In other words, the emphasis appears to be on sequences that can easily switch form one conformation to another depending on environment or solution conditions. Kallberg et al. first described this behavior when they found α-helical segments that were predicted to have a greater β-sheet propensity by standard algorithms [97]. Subsequent studies used mutational analysis to design peptides that would easily undergo the transition from α helix to β sheet [98].

These types of studies led to the concept of a conformational switch, a molecular trigger that can initiate aggregate formation. The concept of a conformational switch has been described by a number of researchers [99–104]. In some cases, the existence of conformational switches was predicted from the sequence [99,104]. The most recent study, by Hamodrakas et al. [99], combined prediction of conformational switches with other amyloidogenic prediction methods (as found in Refs. 15,56, and 105). Examining 21 proteins with known ability to form amyloid fibrils, they found that the conformational switches often coincided with sequences that were amyloidogenic [99]. The same regions were often found to be ambivalent for preferences of α over β structure. Indeed, many of the algorithms identified the same sequences within a given protein. Furthermore, the study found new possible amyloidogenic stretches in some of the proteins. Finally, it appears that most conformational switches reside on the surface of the protein.

14.10. SUMMARY

The ability to predict aggregation propensity from primary sequence information has increased tremendously in recent years. In fact, many of these advances described herein have only been reported since 2007. While one can argue about the relative merits of a particular algorithm, all of them appear to have some predictive capacity and all provide some molecular insight into the bases for β-aggregate or amyloid formation.

REFERENCES

1. Chi, E. Y., Krishnan, S., Randolph, T. W., and Carpenter, J. F. (2003), Physical stability of proteins in aqueous solution: Mechanism and driving forces in nonnative protein aggregation, *Pharm. Res.* **20**: 1325–1336.

2. Idicula-Thomas, S. and Balaji, P. V. (2005), Protein aggregation: A perspective from amyloid and inclusion-body formation, *Curr. Sci.* **92**: 758–767.
3. Rosenberg, A. S. (2006), Effects of protein aggregates: An immunologic perspective, *AAPS J.* **8** (article 59).
4. Sharma, B. (2007), Immunogenicity of therapeutic proteins. Part 1: Impact of product handling, *Biotechnol. Adv.* **25**: 310–317.
5. Sharma, B. (2007), Immunogenicity of therapeutic proteins. Part 2: Impact of container closures, *Biotechnol. Adv.* **25**: 318–324.
6. Sharma, B. (2007), Immunogenicity of therapeutic proteins. Part 3: Impact of manufacturing changes, *Biotechnol. Adv.* **25**: 325–331.
7. Yon, J. M. (2002), Protein folding in the post-genomic era, *J. Cell. Mol. Med.* **6**: 307–327.
8. Dobson, C. M. (2004), Principles of protein folding, misfolding, and aggregation, *Semin. Cell Devel. Biol.* **15**: 3–16.
9. Rousseau, F., Schymkowitz, J., and Serrano, L. (2006), Protein aggregation and amyloidosis: Confusion of the kinds? *Curr. Opin. Struct. Biol.* **16**: 118–126.
10. Krebs, M. R. H., Devlin, G. L., and Donald, A. M. (2007), Protein particulates: Another generic form of protein aggregation? *Biophys. J.* **97**: 1336–1342.
11. Pastor, M. T., Esteras-Chopo, A., and Lopez de la Paz, M. (2005), Design of model systems for amyloid formation: Lessons for prediction and inhibition, *Curr. Opin. Struct. Biol.* **15**: 57–63.
12. Caflisch, A. (2006), Computational models for the prediction of polypeptide aggregation propensity, *Curr. Opin. Chem. Biol.* **10**: 437–444.
13. Bemporad, F., Calloni, G., Camponi, S., Plakoutsi, G., Taddei, N., and Chiti, F. (2006), Sequence and structural determinants of amyloid fibril formation, *Acc. Chem. Res.* **39**: 620–627.
14. Garbuzynskiy, O. V., Finkelstein, A. V., and Galzitskaya, O. V. (2005), On the prediction of folding nuclei in globular proteins, *Mol. Biol.* **39**: 906–914.
15. Galzitskaya, O. V., Garbuzinskiy, S. A., and Lobanov, M. Y. (2006), Prediction of amyloidogenic and disordered regions in protein chains, *PloS Comput. Biol.* **2**: 1639–1648.
16. Galzitskaya, O. V., Ivankov, D. N., and Finkelstein, A. V. (2001), Folding nuclei in proteins, *Mol. Biol.* **35**: 708–717.
17. Galzitskaya, O. V., Garbuzinskiy, S. A., and Lobanov, M. Y. (2006), Search for amyloidogenic regions of the protein chain, *Mol. Biol.* **40**: 910–918.
18. Calamai, M., Taddei, N., Stefani, M., Ramponi, G., and Chiti, F. (2003), Relative influence of hydrophobicity and net charge in the aggregation of two homologous proteins, *Biochemistry* **42**: 15078–15083.
19. Lawrence, M. S., Phillips, K. J., and Liu, D. R. (2007), Supercharging proteins can impart unusual resilience, *J. Am. Chem. Soc.* **129**: 10110–10112.
20. Broersen, K., Weijers, M., de Groot, J., Hamer, R. J., and de Jongh, H. H. J. (2007), Effect of protein charge on the generation of aggregation-prone conformers, *Biomacromolecules* **8**: 1648–1656.
21. Chiti, F., Calamai, M., Taddei, N., Stefani, M., Ramponi, G., and Dobson, C. M. (2002), Studies of the aggregation of mutant proteins in vitro provide insights into the genetics of amyloid diseases, *Proc. Natl. Acad. Sci. USA* **99**: 16419–16426.

22. Chiti, F., Taddei, N., Baroni, F., Capanni, C., Stefani, M., Ramponi, G., and Dobson, C. M. (2002), Kinetic partitioning of protein folding and aggregation, *Nat. Struct. Biol.* **9**: 137–143.

23. Chiti, F., Stefani, M., Taddei, N., Ramponi, G., and Dobson, C. (2003), Rationalization of the effects of mutations on peptide and protein aggregation rates, *Nature* **424**: 805–808.

24. Tjernberg, L., Hosia, W., Bark, N., Thyberg, J., and Johansson, J. (2002), Charge attraction and β propensity are necessary for amyloid fibril formation from tetrapeptides, *J. Biol. Chem.* **277**: 43243–43246.

25. Bitan, G., Vollers, S. S., and Teplow, D. B. (2003), Elucidation of primary structure elements controlling early amyloid beta-protein oliogomerization, *J. Biol. Chem.* **278**: 34882–34889.

26. Galzitskaya, O. V. (2006), Identification of β-aggregate sites in protein chains, *Mol. Biol.* **40**: 839–843.

27. Zibaee, S., Jakes, R., Fraser, G., Serpell, L. C., Crowther, R. A., and Goedert, M. (2007), Sequence determinants for amyloid fibrillogenesis of human α-synuclein, *J. Mol. Biol.* **374**: 454–464.

28. Goldsbury, C., Goldie, K., Pellaud, J., Seelig, J., Frey, P., Muller, S. A., Kistler, J., Cooper, G. J. S., and Aebi, U. (2000), Amyloid fibril formation from full-length and fragments of amylin, *J. Struct. Biol.* **130**: 352–362.

29. Jones, S., Manning, J., Kad, N. M., and Radford, S. E. (2003), Amyloid-forming peptides from β_2-microglobulin- insights into the mechanism of fibril formation in vitro, *J. Mol. Biol.* **325**: 249–257.

30. Fink, A. L. (1998), Protein aggregation: Folding aggregates, inclusion bodies and amyloid, *Fold. Design* **3**: R9–R23.

31. Gazit, E. (2002), A possible role for π-stacking in the self-assembly of amyloid fibrils, *FASEB J.* **16**: 77–83.

32. Sun, S. and Bernstein, E. R. (1996), Aromatic van der Waals clusters: Structure and nonrigidity, *J. Phys. Chem.* **100**: 13348–13366.

33. McGaughey, G. B., Gagné, M., and Rappé, A. K. (1998), π-stacking interactions. Alive and well in proteins, *J. Biol. Chem.* **273**: 15458–15463.

34. Tartaglia, G. G., Cavalli, A., Pellarin, R., and Caflisch, A. (2004), The role of aromaticity, exposed surface, and dipole moment in determining protein aggregation rates, *Protein Sci.* **13**: 1939–1941.

35. Azriel, R. and Gazit, E. (2001), Analysis of the minimal amyloid-forming fragment of the islet amyloid polypeptide, *J. Biol. Chem.* **276**: 34156–34161.

36. Makin, O. S., Atkins, E., Sikorski, P., Johanson, J., and Serpell, L. C. (2005), Molecular basis for amyloid fibril formation and stability, *Proc. Natl. Acad. Sci. USA* **102**: 315–320.

37. Tracz, S. M., Abedini, A., Driscoll, M., and Raleigh, D. P. (2004), Role of aromatic interactions in amyloid formation by peptides derived from human amylin, *Biochemistry* **43**: 15901–15908.

38. Kim, W. and Hecht, M. H. (2006), Generic hydrophobic residues are sufficient to promote aggregation of the Alzheimer's Aβ 42 peptide, *Proc. Natl. Acad. Sci. USA* **103**: 15824–15829.

39. Marek, P., Abedini, A., Song, B., Kanungo, M., Johnson, M. E., Gupta, R., Zaman, W., Wong, S. S., and Raleigh, D. P. (2007), Aromatic interactions are not required for amyloid

fibril formation by islet amyloid polypeptide but do influence the rate of fibril formation and fibril morphology, *Biochemistry* **46**: 3255–3261.

40. West, M. W., Wang, W. X., Patterson, J., Mancias, J. D., Beasley, J. R., and Hecht, M. H. (1999), *De novo* amyloid proteins from designed combinatorial libraries, *Proc. Natl. Acad. Sci. USA* **96**: 11211–11216.
41. Wang, W. X., and Hecht, M. H. (2002), Rationally designed mutations convert *de novo* amyloid-like fibrils into monomeric β-sheet proteins, *Proc. Natl. Acad. Sci. USA* **99**: 2760–2765.
42. Meinhardt, J., Tartaglia, G. G., Pawar, A., Christopeit, T., Hortschansky, P., Schroeckh, V., Dobson, C. M., Vendruscolo, M., and Fändrich, M. (2007), Similarities in the thermodynamics and kinetics of aggregation of disease-related Aβ(1–40) peptides, *Protein Sci.* **16**: 1214–1222.
43. Fernandez-Escamilla, A., Rousseau, F., Schymkowitz, J., and Serrano, L. (2004), Prediction of sequence-dependent and mutational effects on the aggregation of peptides and proteins, *Nat. Biotechnol.* **22**: 1302–1306.
44. Linding, R., Schymkowitz, J., Rousseau, F., Diella, F., and Serrano, L. (2004), A comparative study of the relationship between protein structure and β-aggregation in globular and intrinsically disordered proteins, *J. Mol. Biol.* **342**: 345–353.
45. Monsellier, E., Ramazzotti, M., de Laureto, P. P., Tartaglia, G. G., Taddei, N., Fontana, A., Vendruscolo, M., and Chiti, F. (2007), The distribution of residues in a polypeptide sequence is a determinant of aggregation optimized by evolution, *Biophys. J.* **93**: 4382–4391.
46. Turnell, W. G. and Finch, J. T. (1992), Binding of the dye Congo Red to the amyloid protein pig insulin reveals a novel homology amongst amyloid-forming peptide sequences, *J. Mol. Biol.* **227**: 1205–1223.
47. Grover, A., Dugar, D., and Kundu, B. (2005), Predicting alternate structure attainment and amyloidogenesis: A non-linear signal analysis approach, *Biochem. Biophys. Res. Commun.* **338**: 1410–1416.
48. Kundu, B. and Gluptasarma, P. (1999), Hydrophobic dye inhibits aggregation of molten carbonic anhydrase during thermal unfolding and refolding, *Proteins Struct. Funct. Genet.* **37**: 321–324.
49. Saiki, M., Konokahara, T., and Morii, H. (2006), Interaction-based evaluation of the propensity for amyloid formation with cross-β structure, *Biochem. Biophys. Res. Commun.* **343**: 1262–1271.
50. Fowler, S. B., Poon, S., Muff, R., Chiti, F., Dobson, C. M., and Zurdo, J. (2005), Rational design of aggregation-resistant bioactive peptides: Reengineering human calcitonin, *Proc. Natl. Acad. Sci. USA* **102**: 10105–10110.
51. Pawar, A. P., DuBay, K. F., Zurdo, J., Chiti, F., Vendruscolo, M., and Dobson, C. M. (2005), Prediction of "aggregation-prone" and "aggregation-susceptible" regions in proteins associated with neurodegenerative diseases, *J. Mol. Biol.* **350**: 379–392.
52. Broome, B. M. and Hecht, M. H. (2000), Nature disfavors sequences of alternating polar and non-polar amino acids: implications for amyloidogenesis, *J. Mol. Biol.* **296**: 961–968.
53. Gsponer, J. and Vendruscolo, M. (2006), Theoretical approaches to protein aggregation, *Protein Pept. Lett.* **13**: 287–293.
54. Trovato, A., Chiti, F., Maritan, A., and Seno, F. (2006), Insight into the structure of amyloid fibrils from the analysis of globular proteins, *PloS Comput. Biol.* **2**: 1608–1618.

55. Wood, S. J., Wetzel, R., Martin, J. D., and Hurle, M. R. (1995), Prolines and amyloidogenicity in fragments of the Alzheimer's peptide beta/A4, *Biochemistry* **34**: 724–730.

56. Lopez de la Paz, M. and Serrano, L. (2004), Sequence determinants of amyloid fibril formation, *Proc. Natl. Acad. Sci. USA* **101**: 87–92.

57. Rousseau, F, Serrano, L., and Schymkowitz, J. W. H. (2006), How evolutionary pressure against protein aggregation shaped chaperone specificity, *J. Mol. Biol.* **355**: 1037–1047.

58. Tartaglia, G. G., Cavalli, A., Pellarin, R., and Caflisch, A. (2005), Prediction of aggregation rate and aggregation-prone segments in polypeptide sequences, *Protein Sci.* **14**: 2723–2734.

59. Zhang, Z., Chen, H., and Lai, L. (2007), Identification of amyloid fibril-forming segments based on structure and residue-based statistical potential, *Bioinformatics* **23**: 2218–2225.

60. Thompson, M. J., Sievers, S. A., Karanicolas, J., Ivanova, M. I., Baker, D., and Eisenberg, D. (2006), The 3D profile method for identifying fibril-forming segments of proteins. *Proc. Natl. Acad. Sci. USA* **103**: 4074–4078.

61. Yoon, S. and Welsh, W. J. (2004), Detecting hidden sequences propensity for amyloid fibril formation, *Protein Sci.* **13**: 2149–2160.

62. Sadeghi, M., Naderi-Manesh, H., Zarrabi, M., and Ranjbar, B. (2006), Effective factors in thermostability of thermophilic proteins, *Biophys. Chem.* **119**: 256–270.

63. Dubay, K. F., Pawar, A. P., Chiti, F., Zurdo, J., Dobson, C. M., and Vendruscolo, M. (2004), Prediction of the absolute aggregation rates of amyloidogenic polypeptide chains, *J. Mol. Biol.* **341**: 1317–1326.

64. Sandberg, M., Eriksson, L., Jonsson, J., Sjostrom, M., and Wold, S. (1998), New chemical descriptors relevant for the design of biologically active peptides. A multivariate characterization of 87 amino acids, *J. Med. Chem.* **41**: 2481–2491.

65. Hellberg, S., Sjostrim, M., Skakerberg, B., and Wold, S. (1987), Peptide quantitative structure-activity relationships, a multivariate approach, *J. Med. Chem.* **30**: 1126–1135.

66. Cruciani, G., Baroni, M., Carosati, E., Clementi, M., Valigi, R., and Clementi, S. (2004), Peptide studies by means of principal properties of amino acids derived from MIF descriptors, *J. Chemometr.* **18**: 146–155.

67. Cellmer, T., Bratko, D., Prausnitz, J. M., and Blanch, H. W. (2007), Protein aggregation in silico, *Trends Biotechnol.* **25**: 254–261.

68. Fields, G. B., Alonso, D. O. V., Stigter, D., and Dill, K. A. (1992), Theory for the aggregation of proteins and copolymers, *J. Phys. Chem.* **96**: 3974–3981.

69. Bratko, D., Cellmer, T., Prausnitz, J. M., and Blanch, H. W. (2006), Effect of single-point alterations on the aggregation propensity of a model protein, *J. Am. Chem. Soc.* **128**: 1683–1691.

70. Clark, L. A. (2005), Protein aggregation determinants from a simplified model: Cooperative folders resist aggregation, *Protein Sci.* **14**: 653–662.

71. Pellarin, R., Guarnera, E., Caflisch, A. (2007), Pathways and intermediates of amyloid fibril formation, *J. Mol. Biol.* **364**: 917–924.

72. Nguyen, H. D. and Hall, C. K. (2004), Spontaneous fibril formation by polyalanines: discontinuous molecular dynamics simulations, *J. Am. Chem. Soc.* **128**: 1890–1901.

73. Nguyen, H. D. and Hall, C. K. (2004), Phase diagrams describing fibrillization by polyalanine peptides, *Biophys. J.* **87**: 4122–4134.

74. Nguyen, H. D. and Hall, C. K. (2005), Moelcular dynamics simulations of spontaneous fibril formation by random-coil peptides, *Proc. Natl. Acad. Sci. USA* **101**: 16180–16185.

75. Hall, D., Hirota, N., and Dobson, C. M. (2005), A toy model for predicting the rate of amyloid formation from unfolded protein, *J. Mol. Biol.* **351**: 195–205.

76. Williams, A. D., Portelius, E., Kheterpal, I., Guo, J., Cook, K. D., Xu, Y., and Wetzel, R. (2004), Mapping Aβ amyloid fibril secondary structure using scanning proline mutagenesis, *J. Mol. Biol.* **335**: 833–842.

77. Christopeit, T., Horschansky, P., Schroekh, V., Guhrs, K., Zandomeneghi, G., and Fandrich, M. (2005), Mutagenesis analysis of the nucleation propensity of oxidized Alzheimer's β-amyloid peptide, *Protein Sci.* **14**: 2125–2131.

78. Moriarty, D. F. and Raleigh, D. P. (1999), Effects of sequential proline substitutions on amyloid formation by human amylin$_{20-29}$, *Biochemistry* **38**: 1811–1818.

79. Wurth, C., Kim, W., and Hecht, M. H. (2006), Combinatorial approaches to probe the sequence determinants of protein aggregation and amyloidogenicity, *Protein Pept. Lett.* **13**: 279–286.

80. Wurth, C., Guimard, N. K., and Hecht, M. H. (2002), Mutations that reduce aggregation of the Alzheimer's Aβ42 peptide: an unbiased search for the sequence determinants of Ab amyloidogenesis, *J. Mol. Biol.* **319**: 1279–1290.

81. Garber, E. and Demarest, S. J. (2007), A broad range of Fab stabilities within a host of therapeutic IgGs, *Biochem. Biophys. Res. Commun.* **355**: 751–757.

82. Yasui, H., Ito, W., and Kurosawa, Y. (1994), Effects of substitutions of amino acids on the thermal stability of the Fv fragments of antibodies, *FEBS Lett.* **353**: 143–146.

83. Christ, D., Famm, K., and Winter, G. (2007), Repertoires of aggregation-resistant human antibody domains, *Protein Eng. Design* **20**: 413–416.

84. Damschroder, M. M., Widjaja, L., Gill, P. S., Krasnoperov, V., Jiang, W., Dall'Acqua, W. F., and Wu, H. (2007), Framework shuffling of antibodies to reduce immunogenicity and manipulate functional and biophysical properties, *Mol. Immunol.* **44**: 3049–3060.

85. Tiana, G., Simona, F., De Mori, G. M. S., Broglia, R. A., and Colombo, G. (2004), Understanding the determinants of stability and folding of small globular proteins from their energetics, *Protein Sci.* **13**: 113–124.

86. de Groot, N. S., Pallarés, I., Avilés, F. X., Vendrell, J., and Ventura, S. (2005), Prediction of "hot spots" of aggregation in disease-linked polypeptides, *BMC Struct. Biol.* **5**: 18.

87. Conchillo-Sole, O., de Groot, N. S., Avilés, F. X., Vendrell, J., Daura, X., and Ventura, S. (2007), AGGRESCAN: A server for the prediction and evaluation of "hot spots" of aggregation in polypeptides, *BMC Bioinform.* **8**: 65.

88. de Groot, N. S., Aviles, F. X., Vendrell, J., and Ventura, S. (2006), Mutagenesis of the central hydrophobic core in Abeta42 Alzheimer's peptide. Side chain properties correlate with aggregation propensity, *FEBS J.* **276**: 658–668.

89. Pedersen, J. S., Christensen, G., and Otzen, D. E. (2004), Modulation of S6 fibrillation by unfolding rates and gatekeeper residues, *J. Mol. Biol.* **341**: 575–588.

90. Otzen, D. E. and Oliveberg, M. (1999), Salt-induced detour through compact regions of the protein folding landscape, *Proc. Natl. Acad. Sci. USA* **96**: 11746–11751.

91. Morgensen, J. E., Ipsen, H., Holm, J., and Otzen, D. E. (2004), Elimination of a misfolded folding intermediate by a single point mutation, *Biochemistry* **43**: 3357–3367.

92. Otzen, D. E., Kristensen, O., and Oliveberg, M. (2000), Designed protein tetramer zipped together with a hydrophobic Alzheimer homology: A structural clue to amyloid assembly, *Proc. Natl. Acad. Sci. USA* **97**: 9907–9912.

93. Carpenter, J. F., Kendrick, B. S., Chang, B. S., Manning, M. C., and Randolph, T. W. (1999), Inhibition of stress-induced aggregation of protein therapeutics, *Meth. Enzymol.* **309**: 236–255.
94. Kendrick, B. S., Chang, B. S., Arakawa, T., Peterson, B., Randolph, T. W., Manning, M. C., and Carpenter, J. F. (1997), Preferential exclusion of sucrose from recombinant interleukin-1 receptor antagonist: Role in restricted conformational mobility and compaction of native state, *Proc. Natl. Acad. Sci. USA* **94**: 11917–11922.
95. Kendrick, B. S., Carpenter, J. F., Cleland, J. L., and Randolph, T. W. (1998), A transient expansion of the native state precedes aggregation of recombinant human interferon-gamma, *Proc. Natl. Acad. Sci. USA* **95**: 14142–14146.
96. Parrini, C., Taddei, N., Ramazzotti, M., Degl'Innocenti, D., Ramponi, G., Dobson, C. M., and Chiti, F. (2005), Glycine residues appear to be evolutionarily conserved for their ability to inhibit aggregation, *Structure* **13**: 1143–1151.
97. Kallberg, Y., Gustafsson, M., Persson, B., Thyberg, J., and Johansson, J. (2001), Prediction of amyloid fibril-forming proteins, *J. Biol. Chem.* **276**: 12945–12950.
98. Takahashi, Y., Ueno, A., and Mihara, H. (2000), Mutational analysis of designed peptides that undergo structural transition from α helix to β sheet and amyloid fibril formation, *Structure Fold. Design* **8**: 915–925.
99. Hamodrakas, S. J., Liappa, C., and Iconomidou, V. A. (2007), Consensus prediction of amyloidogenic determinants in amyloid fibril-forming proteins, *Int. J. Biol. Macromol.* **41**: 295–300.
100. Kelly, J. W. (1996), Alternative conformations of amyloidogenic proteins govern their behavior, *Curr. Opin. Struct. Biol.* **6**: 11–17.
101. Gross, M. (2000), Proteins that convert from α helix to β sheet implications for folding and disease, *Protein Pept. Sci.* **1**: 339–347.
102. Kallberg, Y., Gustafsson, M., Persson, B., Thyberg, J., and Johansson, J. (2001), Prediction of amyloid fibril-forming proteins, *J. Biol. Chem.* **276**: 12945–12950.
103. Pagel, K., Vagt, T., and Koksch, B. (2005), Directing the secondary structure of polypeptides at will: From helices to amyloids and back again? *Org. Biomol. Chem.* **3**: 3843–3850.
104. Young, M., Kirschenbaum, K., Dill, K. A., and Highsmith, S. (1999), Predicting conformational switches in proteins, *Protein Sci.* **8**: 1752–1764.
105. Combet, C., Blanchet, C., Geourjon, C., and Delange, G. (2000), NPS@: Network protein sequence analysis, *Trends Biochem. Sci.* **25**: 147–150.

15

HIGH-CONCENTRATION ANTIBODY FORMULATIONS

Steven J. Shire, Jun Liu, Wolfgang Friess, Susanne Jörg, and Hanns-Christian Mahler

15.1. INTRODUCTION

15.1.1. Antibody Therapeutics

As of 2007, twenty-six therapeutic monoclonal antibodies (including FAb and FAb'2 forms) are approved by the FDA Center for Biologics Evaluation and Research (CBER) [1]. In total, 132 monoclonal antibodies are reported to be in testing in clinical studies, and from one report in 2005, antibody sales were forecast to almost double from US$ 7.7 billion in 2004 to US$ 13.2 billion in 2008 [2]. To date, therapeutic antibody products focus on indications such as oncology, arthritis, inflammation and immune disorders besides cardiovascular, respiratory or infectious diseases [2].

15.1.2. Rationale and Need for High-Concentration Formulations—and Definition of "High"

Although the majority of antibody products on the market are administered intravenously (IV) [3], the alternate subcutaneous (SC) route of administration is highly preferred for therapeutic indications where home (self-) medication is desirable, for

Formulation and Process Development Strategies for Manufacturing Biopharmaceuticals, edited by Jameel and Hershenson

example, for chronic diseases such as asthma, and arthritic diseases. Since SC administration improves ease of use and avoids hospitalization for administration, the resulting increase in patient compliance as well as significant reduction in workload for clinical personnel help in reducing treatment costs.

However, (monoclonal) antibodies usually require relatively high clinical doses, typically in the range of 5–700 mg per patient—administered either as fixed dose or on the basis of body weight (kg) or body surface (m^2) of the patient, respectively (e.g., Campath, 3–30 mg; Herceptin, 2–4 mg/kg; Erbitux, 250–400 mg/m^2). The anticipated dose per single administration depends not only on the activity but also on the pharmacokinetic profile of the molecule and the desired dosing frequency for the intended route of administration. Typically, one to four IV injections or infusions per month are mandatory to maintain a pharmacodynamically effective concentration level owing to a circulatory half-life of approximately 12–48 h of murine and 3–10 days of chimeric and humanized monoclonal antibodies [3].

The subcutaneous route is currently limited by an injection volume of 1 up to perhaps 1.5 mL [4] as attributable to tissue backpressure and injection pain. The latter is, however, considered to depend also on the injected formulation, needle size, patient characteristics, and injection performance [4]. Therefore, one of the key challenges is to develop a sufficiently stable and robust formulation for the desired clinical dose per SC administration for a target volume of currently not more than 1–1.5 mL. For clinical SC doses in the range above 75 mg per patient, this results in the need for formulations significantly exceeding concentrations of 50 mg/mL.

15.1.3. The "High-Concentration Landscape"

In fact, only four approved antibody drug products, namely, adalimumab, efalizumab (recently removed from the market), palivizumab, and omalizumab, are currently marketed as "high-concentration formulations" and hence are administered extravascularily (per *Physician's Desk Reference*). Adalimumab (Humira), the first approved fully human phage display–derived monoclonal antibody for the second-line treatment of acute rheumatoid arthritis, is administered as a biweekly SC injection of 0.8 mL containing 40 mg adalimumab. Efalizumab (Raptiva) is a humanized antibody for the treatment of moderate to severe chronic plaque psoriasis with a dosage regime including an initial single dose of 0.7 mg/kg followed by weekly SC injections of 1.0 mg/kg efalizumab over a period of 12 weeks. Palivizumab (Synagis) is intended for the prevention of respiratory syncytial virus (RSV) disease in pediatric patients and administered IM (intramuscularly) with a dose of 15 mg/kg body weight. The anti-IgE-Fc antibody omalizumab (Xolair) is indicated for the treatment of moderate to severe persistent asthma, with 150–375 mg administered SC at biweekly or monthly intervals (per *Physician's Desk Reference*). To date, Xolair is the antibody product with the highest product concentration used for administration (125 mg/mL after reconstitution of the freeze-dried product) on the market. A significant number of antibodies are currently in development or clinical testing with concentrations as high as 100–150 mg/mL by many pharmaceutical companies, and a summary of the currently approved high-concentration antibody products is given in Table 15.1.

TABLE 15.1. Approved High-Concentration Monoclonal Antibodies as of January 2006[a]

Trade Name	INN	Company	Delivery Route	Dosage Form	Dosage Strength, mg/unit	Concentration, mg/mL
Humira™	Adalimumab	Abbott	SC	Liquid, prefilled syringe	40	50
Synagis®	Palivizumab	Abbott, MedImmune	IM	Lyophilisate, vial	100	100[b]
Raptiva™	Efalizumab	Serono, Genentech	SC	Lyophilisate, vial	150	100[b]
Xolair®	Omalizumab	Genentech, Novartis	SC	Lyophilisate, vial	150	125[b]

[a] Ref. [1].
[b] After reconstitution.

15.1.4. Upsides and Downsides of High-Concentration Formulations

As described above, high-concentration protein formulations can have multiple advantages for patients and the health-care system. From a marketing perspective, SC "ready to use" dosage forms also improve competitiveness or can be part of product life-cycle management. Besides that, high concentration formulations require less storage space and may reduce overall shipping costs due to the minimized storage volume of the concentrated bulk and drug products. The choice of a liquid dosage form would additionally offer the advantage of reducing manufacturing cost and increasing time savings [5].

However, as the concentration of an antibody formulation increases, an impact on the yield can be expected. For example, product (volume) losses or requirements during manufacturing due to dead volumes (e.g., TFF, tubings, filters), flushes, in-process controls, final product analytics (quality control) may contribute to undesirably high costs of goods. Viscous formulations also tend to adhere to various surfaces, and this can further contribute to the overall costs of goods due to unrecoverable product losses, such as in the TFF system, the product containers, and the filling equipment. Additionally, the overfill of the drug product in final vials can also increase significantly for viscous product (up to 30%) [5]. This should be taken into account to ensure the specified volume to be extracted, consistent with USP and European pharmaceutical requirements [6]. The use of narrower-based vials or prefilled syringes could minimize the viscosity related product losses of the final drug product. However, if using freeze drying as a concentration method, the vial size and format is often governed by the total amount of bulk material needed for the final dosage form [7]. Therefore, effort should be taken to minimize losses by modifying dead volumes in production equipment or to use formulations with low viscosity. Nevertheless, these problems should be counterbalanced partly by the cost savings due to the shortened fill and finishing procedures and the decreased storage and transport cost, and moreover by the greater benefits to the patients, health care system, and the pharmaceutical company, as discussed earlier.

Clinically, compared to IV, SC administrations may have the advantages of lymph targeting [8]. However, it should be considered that a drug absorption step is necessary, and that variability in bioavailabilites of 50%–100% after SC administration may be expected [9], potentially requiring an increased antibody dose.

A widespread concern of protein therapeutics is their potential to cause an immunogenic reaction. In the context of the administration route, biopharmaceuticals given SC are considered to be more immunogenic than the same protein given IM, which, in turn, is a more immunogenic route than IV administration [10]. Immunogenicity could lead to serious safety implications such as potentially altering pharmacokinetics of a drug, inhibition of the therapeutic effect of the recombinant protein, or—depending on the clinically used protein drug—interaction or depletion with physiological proteins due to antidrug antibodies. This has to be carefully evaluated in the treatment of chronic conditions where repeated dosing is required [10].

Furthermore, high-concentration formulations pose a significant technical challenge, for example, with regard to the following [6,7]:

- Manufacturing challenges (e.g., increased viscosity resulting in lower filtration flow rates)
- Stability challenges (e.g., increased aggregation at high protein concentration)
- Analytical challenges (e.g., analysis at high protein concentration avoiding dilution)
- Formulation and administration challenges

The mentioned aspects are to be discussed in more depth in the following sections.

15.2. MANUFACTURING ISSUES AND CHALLENGES

15.2.1. Pharmaceutical Development and Impact of Process

The development of a successful high-concentration commercial formulation of antibody for SC delivery often begins with the choice of a process to manufacture the formulation. Much of the preformulation and formulation work will be highly dependent on the type and scale of the process used to concentrate the monoclonal antibody (MAb). In particular, scale-up issues are important since development of a formulation that does not use scalable processes may be difficult and costly to introduce into a manufacturing plant, and thus no matter how good the formulation is in terms of stability, if it cannot be manufactured, then there is no product.

In general, any drying method that removes water can be used to produce a high-concentration protein solution. Several of these methods were summarized previously [6], but only a few of these methods are amenable to scale-up in a manufacturing plant. Although a high-concentration antibody product is highly desirable for administration by the SC route, the manufacturing of a high-concentration MAb at large scale remains a difficult and challenging process, and requires a careful consideration and balance of requirements for final product and the manufacturing process.

15.2.2. Common Concentration Processes

The common techniques currently available for manufacturing high-concentration antibody solution include tangential flow filtration (TFF) and drying techniques, such as lyophilization and spray drying [6]. Other techniques, such as crystallization [11] and microparticle-based technology [12], have also demonstrated the potential for producing an injectable high-concentration antibody formulation in a suspension. These techniques potentially can increase the protein concentration beyond its solubility as discussed by Shire et al. [6,13]. Because of the differences in basic principles, the design, optimization, and scale-up of these processes can be very different. A thorough understanding of process operations and process limitations is critical for the successful transfer of small-scale benchtop results to commercial manufacturing. The commonly used techniques for protein concentration are discussed in detail below.

15.2.2.1. Tangential Flow Filtration. Tangential flow filtration (TFF) is one of the most commonly used methods for formulating and concentrating therapeutic antibodies. The method is readily scalable and can provide high process rates with good efficiency without adversely affecting the product quality [14,15]. Tangential flow filtration is a pressure-driven separation technique that utilizes a selective ultrafiltration (UF) membrane or hollow fibers to separate molecules in solution on the basis of differences in size and charge. During the TFF process, as the antibody solution is circulated by pumping tangentially across the surface of an UF membrane, large antibody molecules will be retained and concentrated, while the small excipient molecules and water will cross the membrane. As an antibody concentrates, a thin layer of antibody also known as a *concentration polarization layer* may be generated that may lead to formation of a gel layer adjacent to the membrane surface. The antibody in this thin layer contains higher concentration antibody than in the bulk solution, and as a result can serve to limit the flow of fluid through the membrane. As an example, when a protein bulk solution is concentrated to 100 mg/mL, the concentration at the membrane surface may be as high as >125 mg/mL [6]. Depending on a protein's propensity to aggregate or precipitate, this could lead to decreased flux and eventually membrane clogging. The buildup of protein near the membrane surface can be largely overcome by increasing the tangential flow of liquid across the membrane surface [16].

Although TFF processes provide a rapid formulation and concentration method for most proteins, the development of a robust TFF process requires a careful consideration of several key factors. First, the rapid pumping and continuous circulation through narrow pathways may generate sufficient shear and cavitation stresses on a protein that results in protein unfolding and precipitation [17,18]. Additional care should be taken to minimize these stress conditions. Also, the viscosity of high-concentration product should be controlled below certain limits, since the viscous protein solution can significantly increase the process time and result in higher backpressure that may exceed the manufacturing limit of the system. In addition, the viscous sample may be difficult to recover and lead to significant loss of product. Some of these viscosity-related problems may be managed by improving equipment design and operation parameters. For antibodies that have adequate thermal stability, an increase of process temperature can significantly reduce the viscosity of a high-concentration antibody solution, and this approach has been successfully used for concentrating a viscous monoclonal antibody to a final concentration of 150 mg/mL [6]. Another effective approach to solve the viscosity issue is to use appropriate excipients, pH, and buffer conditions that can reduce the solution viscosity of final product [19]. These approaches have proven to be a very effective way to use TFF processes efficiently to manufacture high-concentration antibody formulations for SC delivery. In addition to the viscosity issue, the TFF process is also limited by the solubility of protein. As protein aggregates and precipitates, it can clog the membrane pores and significantly slow down the concentration process. As concentrations significantly higher than the target concentration are reached during TFF operations, as a result of concentration polarization, as well as the requirement for using a buffer flush to increase the product recovery, the maximum concentrate of the protein achievable during the TFF process should be well beyond the target concentration in the final formulation. Therefore, it is important that the maximum achievable concentration of the product be determined along with appropriate formulation screening studies to identify those conditions that can maximize the achievable concentration during processing.

15.2.2.2. Additional Concentration Techniques. In addition to the TFF process, high-concentration protein formulations can also be manufactured using a variety of drying techniques. These processes are particularly suitable for proteins with viscosity issues limiting TFF or products that are not stable in a liquid formulation. Possible drying techniques include spray drying [20,21], spray freeze drying [22], fluid-bed drying [23,24], vacuum drying [25,26], and lyophilization [27], with spray drying and lyophilization being the most applicable from an industry perspective. These methods, although more complicated, expensive, and inconvenient than the TFF process for manufacturing liquid products, potentially provide flexibility for the formulation scientist and benefits for manufacturing high-concentration product, while at the same time posing significant challenges to operate as a scalable and robust GMP (Good Manufacturing Practice) process in industry practice. Spray drying has been extensively used in the pharmaceutical industry for preparing dry powder for aerosol delivery. This method allows water to be separated from the formed particles in a continuous production process. The process involves the atomization of a liquid feedstock into a spray of droplets followed by exposure to hot air to remove the water. Although the heat can potentially damage the protein, this effect is generally limited, due to a short exposure time and an evaporative cooling effect during the critical drying period. The moisture level is usually higher than after lyophilization and has to be controlled to ensure the stability of the final product [28].

Lyophilization is another approach that can be used for concentrating protein. In this approach, a protein sample is first lyophilized in a relatively low concentration with a large load volume. Sufficient excipients and lyoprotectants are added to stabilize the protein from the freezing and drying stresses that occur during lyophilization. A high-concentration protein ready for SC administration is achieved by reconstituting the sample with a smaller volume of diluent [29]. The lyophilization process is usually performed by dehydrating final product in vials, although conceivably also as bulk processing in tray units. Lyophilizing in final vials results in a ready-to-use product, whereas bulk processing can be used to produce a high-concentration liquid formulation after reconstitution and prior to filling into vials. The bulk process has been used primarily to produce dry powders for aerosol delivery in the pharmaceutical industry, and has several limitations: (1) for parenteral protein products, bulk-drying processes require operation under aseptic conditions—this is not always as easy as a process whereby the lyophilization is done in the final vial configuration; and (2) the bulk processing often results in significant loss of materials during the process because of the large surface area of the trays and difficulty in quantitatively removing product. In comparison, lyophilized drug product in vials has proved to be a more cost-effective approach and has provided flexibility for generating high-concentration protein formulations.

Although most of the bulk solution in lyophilized vials can have a relatively lower concentration and lower viscosity, care needs to be taken to ensure that the

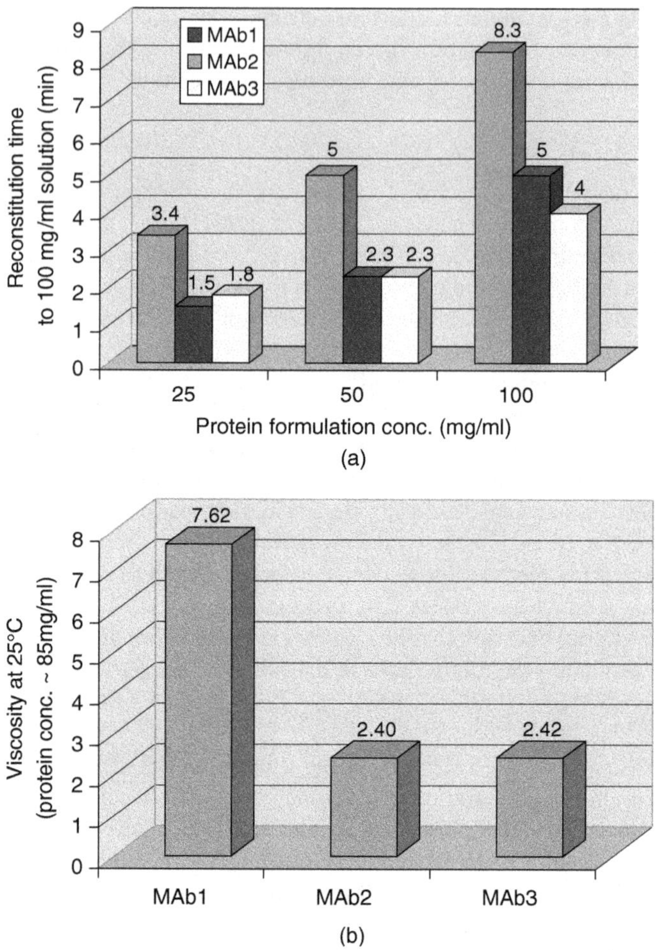

Figure 15.1. Three humanized IgG1 MAbs with the same IgG1 human Fc construct but different complementarity-determining regions (CDRs) formulated identically and lyophilized using the same lyophilization process: (a) time to reconstitute the lyophilized cake resulting in a final MAb concentration of 100 mg/mL; (b) viscosity at 25°C after reconstitution to a final MAb concentration of 85 mg/mL.

viscosity of final reconstituted products are minimized since a high viscosity can have a significant impact on the time of reconstitution, amount of overfill and ease of drug delivery. Figure 15.1a shows an example of lyophilized monoclonal antibodies targeted for a final delivery concentration of 100 mg/mL after reconstitution. As shown in Figure 15.1b, the increase in solution viscosity leads to a significant increase in the reconstitution time. In addition to the solution viscosity, the protein loading concentration can also affect the reconstitution time [6]. It was demonstrated that as the protein loading concentration decreased, reconstitution times also decreased.

Further investigation of the morphology of the lyophilized solid by scanning electron microscopy showed a gradual transition from a dense solid to a more loosely packed structure as protein-loading concentration was decreased. Thus, the ease of wetting of the cake and impact on reconstitution time is very likely related to the differences in this morphology.

For lyophilized formulations, a higher concentration of lyoprotectant can significantly improve the stability of the protein [30]. However, high lyoprotectant concentration can also result in a hypertonic solution that may not be desirable for SC delivery. In addition, the increase of lyoprotectant can also increase the time of the lyophilization process, solution viscosity, and reconstitution time. Therefore, the concentration of lyoprotectant in the final formulation needs to be optimized to balance the needs for product stability, pharmaceutical manufacturing, and delivery.

15.2.2.3. Protein Suspensions

15.2.2.3.1. Crystallization. Protein crystallization has been used for structure analysis, bioseparation, and controlled drug delivery [31]. It has been successfully used in the development of insulin formulations [32]. Protein crystallization has also been shown to improve product stability and half-life [33,34]. When protein forms the crystal, protein will be concentrated in the crystal. If the crystal is small enough, a crystal suspension will be formed. Depending on the size of the protein crystals, these crystals can be directly injected into patients. Crystallization of whole antibodies historically has been a very difficult task because of their large size, glycosylation, and flexibility at the hinge region. More recent progress in this area has resulted in a generation of a high-concentration, lower-viscosity crystalline preparations of therapeutic antibodies that are considered suitable for SC injection [31]. To produce high-volume crystalline suspensions, a batch process method was designed. Antibodies were first mixed with crystallization buffer, and then followed by tumbling or incubation with or without shaking. Crystal seeds were added into the mixture to facilitate the crystallization process. With the appropriate optimized process, the yield of crystallization process was described to be more than 90%. The size of the crystals can also be controlled by manipulating the temperature and mixing conditions, and therefore it is possible to deliver the crystals to the patient directly. Although the potencies of these antibodies in crystal form have been shown to be fully retained using *in vitro* assays, limited information is available about the long-term chemical and physical stability of these antibodies as a crystal suspension and about its behavior *in vivo*. Other publications in this area show, however, how challenging a task crystallization of an antibody to develop a protein suspension can be, especially considering yield and also the formation of amorphous precipitation occurring under crystallization conditions [35].

15.2.2.3.2. Protein Precipitation. In addition to protein crystals, proteins can also form amorphous precipitates in solution [36]. Protein precipitation has been widely used for protein recovery. Protein precipitation is usually induced by adding salts or organic solvents or by changing the pH of the solution. More recent progress has been made to improve the precipitation process and produce well-shaped particles that can be applied directly for SC injection [12]. Precipitated protein particles have been described to be prepared by formation of supersaturated protein and poly(ethylene glycol) (PEG) aqueous solutions [37], such as at elevated temperatures, followed by cooling of the solutions resulting in protein particle formation [12], as well as when using ammonium acetate or citrate at defined and constant temperature [37]. The choice of pH is critical for the formation of precipitation and particle formation, and the choice of a pH that is close to the protein isoelectric point has been recommended [12]. The formed particles are separated from the precipitating agent and other excipients using methods such as filtration or centrifugation. The particles then need to be washed with a buffer to remove the excessive excipient and to be formulated into the final formulation for long-term storage and delivery to patients. Although the suspension-based technologies have shown great potential, additional studies of physicochemical stability need to be conducted to ensure adequate processing conditions, scale-up, stability and molecular integrity, and adequate shelf life for long-term storage, which is considered challenging for a protein suspension.

15.3. PROCESS SCALE-UP

Although antibodies have been routinely concentrated at small scale using benchtop devices, the design of a robust manufacturing process that will allow for manufacture of concentrated dosage forms at a larger scale is critical for successful implementation of benchtop experimental findings. In the case of TFF, critical operating parameters for optimizing the TFF condition are transmembrane pressure, retentate flow rate, and permeate flux. High-shear and cavitation conditions should be avoided to minimize the potential damage to the protein. To be compatible with the TFF process, the antibodies should have adequate solubility and viscosity well beyond their target concentrations because of the membrane polarization and product recovery needs.

In the case of lyophilization, the critical operating parameters, such as shelf temperature range, chamber pressure ranges, condenser temperature, primary drying time, secondary drying time, and the load on shelves, require validation at commercial scale. Lyophilization cycles developed at small scale may need to be modified and optimized in order to suit large-scale production.

15.4. ANALYTICAL CONSIDERATIONS

Many of the analytical methods currently used for characterization of antibodies focus on the chemical structure of the antibody and the chemical degradation pathways that result from alteration of amino acid residues. Generally, the methods performed at low concentration can be easily adapted to analyze antibodies at high concentration [38,39]. They often involve a sample preparation step to allow for the dilution of products to a concentration range that is compatible with the analytical methods. However, the sample preparation step should be adequately assessed to ensure that the analytical result will not be altered. These methods include peptide mapping to determine chemical structure as well as identity of the antibody, ion exchange chromatography to determine charge variants, hydrophobic interaction chromatography to evaluate isomerized and oxidized species, reversed-phase chromatography to study oxidized species, capillary electrophoresis (CE-SDS or SDS-PAGE) to determine covalently linked aggregates and fragments, and size exclusion chromatography (SEC) to determine soluble aggregates and fragments. In addition, analytical methods used to determine solution osmolality and surfactant concentrations may also require dilution or extraction steps to minimize interference from high protein concentration in order to obtain accurate and reproducible results. All these methods need to be optimized and validated to ensure that the final results will not be compromised by high protein concentration.

Most of the analytical methods used for evaluating low-concentration protein in solid dosage forms are quite suitable for high-protein-concentration conditions in either liquid or solid condition. These techniques include Fourier transform infrared (FTIR) spectroscopy [40–42], Raman spectroscopy to study the protein's secondary structure, and fluorescence spectroscopy, more recently used to investigate possible tertiary structural alterations of protein during or after the freezing, drying, storage, and reconstitution process [43]. Differential scanning calorimetry (DSC) has also been used [44], particularly modulated DSC technology [45,46], to determine glass transition temperatures and thermal phase transitions of the solid-state matrix at high protein concentrations. There are also reports that DSC was not applicable to determine melting temperature (unfolding) in high-concentration antibodies as sample dilution was required [41].

Analytical methods used to monitor protein aggregation and particulate formation are most likely to be dependent on protein concentration and formulation conditions, particularly if the protein undergoes reversible self-association. This self-association is often thought to be less of a concern since on administration these types of aggregates are expected to dissociate on dilution back to the native monomeric protein. One has to keep in mind that these aggregates will be introduced into a crowded environment, whether in the blood or in the SC space. The impact of molecular crowding, as is discussed below, may alter the association. Reversible association can also lead to altered solution properties such as viscosity, which has potential implications for manufacturing, stability, and delivery of high-concentration protein drugs developed for SC administration. Analytical methods such as SDS-PAGE, and nongel-sieving SDS-CE with and without reducing agents are often used to determine the presence of covalently linked, SDS nondissociable and disulfide crosslinked aggregates. The size distribution of antibodies under nondenaturing conditions can be monitored using methods such as size exclusion chromatography (SEC) and field flow fractionation (FFF). The SEC method, which has been and still remains the main workhorse in the industry for analysis of protein size distribution, uses HPLC equipment that is found in virtually every pharmaceutical analytical laboratory. This method is fairly rapid and easy to use, but there are several limitations. The chromatographic separation by SEC is based on the hydrodynamic volume of the different species rather than the molecular mass. Therefore, proteins with highly asymmetric structures or extended residues, such as highly glycosylated protein, will be eluted early and behave more like a protein with greater apparent molecular mass. This can result in significant errors when retention time is used to determine molecular mass. Although this problem can be addressed by coupling light-scattering detection with the chromatography to obtain weight-average molecular weight across the elution profile, interactions with the column matrix may result in altering the number of aggregated species and change of retention time, which may result in poor resolution of different species. Some of these matrix–protein interactions can be decreased or eliminated by adding organic solvents or salts, but this then raises the possibility that such manipulations may impact the true size distribution. In addition, noncovalent reversible protein self-association, which is highly concentration-dependent, may be altered significantly by SEC because of the dilution that occurs during the chromatography.

Technologies such as static light scattering (SLS) [47], dynamic light scattering (DLS) [48–50], small-angle X-ray light scattering (SAX) [51], and analytical ultracentrifugation (AUC) [52] have been used to provide more information about protein self-association states at high protein concentrations as well as at low concentration. Moreover, these methods can be used to monitor reversible, irreversible, covalently linked, and noncovalently linked protein aggregates. Both sedimentation equilibrium and SLS analysis can provide information about molecular weight, binding constant, and activity coefficient for a self-association system. The analysis of these data often requires extensive mathematical fitting to models that include multiple interacting species [53]. Sedimentation velocity analysis can provide information about size, shape, and distribution of aggregates. This method has been greatly improved over the years and has been used routinely to monitor small amounts of protein aggregates and their sedimentation coefficients [54]. Despite this progress, the AUC method remains a difficult and low-throughput method and requires highly trained personnel for instrumentation and data analysis. The method is mostly used as an orthogonal approach to verify results from SEC [55]. Relatively high concentrations of protein solutions at about 20 mg/mL can be analyzed with interference optics in an analytical ultracentrifuge using centrifuge cells with the standard 1.2-cm optical path. Measurements using the UV absorption scanner are more limited because of the high absorptivity of proteins resulting in absorbance values well over 1.5, essentially beyond the detection limit of the photomultiplier tube. Radial scans at wavelengths different from that of maximum absorption may circumvent this problem, but optical measurements are still limited by the large refractive index gradients that form around the boundary during centrifugation of high-concentration protein solutions. These gradients result in an effect known as *Wiener skewing*, where the light beam is bent, resulting in incorrect concentration measurements or, worse, no signal at all, since the beam can be bent completely out of the optical path [56]. The method of decreasing the optical pathlength by using thin gaskets rather than standard centerpieces has been used to circumvent this problem [57], but loading of the cells can be complicated, and this technique can be difficult because of the flexibility of the gaskets. New cell centerpieces with pathlengths as narrow as 0.6 mm have been developed by Spin Analyticals and should help with some of these issues. An alternate technique that uses a preparative ultracentrifuge has been described [58] and used to analyze protein formulations at 100 mg/mL [19]. Essentially, samples are centrifuged to sedimentation equilibrium, and the centrifuge tubes are removed and the contents of each tube are collected in 10-μL fractions using a BRANDEL® automated microfractionator (minimum step size 5 μm). These collected fractions are placed into a 96-well UV plate and diluted with phosphate-buffered saline (PBS) for measuring protein concentration by UV absorption spectroscopy. This method works fairly well for high-concentration protein solutions because of the stability of the gradients, but still requires rapid handling. The biggest disadvantage of this method is the additional time required for sample collection, dilution, and subsequent measurement of the absorbance values. In addition the number of data points is about one-third the number typically obtained in an analytical ultracentrifuge. The recent development of a fluorescence detection system (FDS) offers an alternate way to analyze high-concentration protein solutions directly in the analytical ultracentrifuge. Figure 15.2 shows the result of a sedimentation

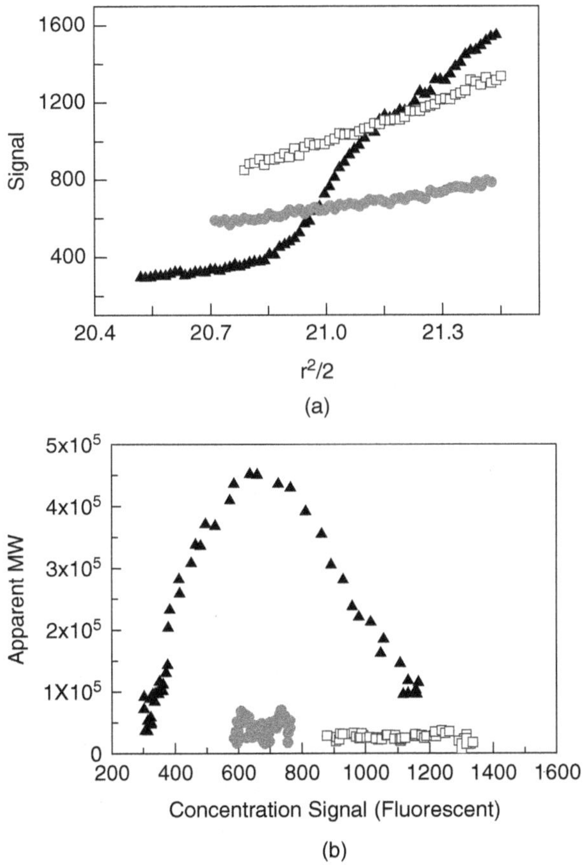

Figure 15.2. Sedimentation equilibrium of the same three MAbs as in Figure 15.1. The data were collected using the fluorescence detection system in a Beckman-AVIV XLF analytical ultracentrifuge. The MAbs were fluorescently labeled [59], characterized to ensure integrity, and added to unlabeled high-concentration solution to serve as a tracer for detection. (a) The raw fluorescence signal data as a function of radial position; (b) apparent molecular weight determined from the data in panel (a).

equilibrium study of two concentrated monoclonal antibodies. The data were collected using the FDS in a Beckman Aviv XLF analytical ultracentrifuge. The MAbs were fluorescently labeled [58], characterized to ensure integrity, and added to unlabeled high-concentration solution to serve as a tracer for detection. As the concentration of MAb1 increases, the apparent weight average molecular weight also increases. This indicates the presence of self-association for MAb1 at high protein concentration. In comparison, MAb2 undergoes very little association, and this result is consistent with previous observations with the preparative sedimentation equilibrium method [19].

Another biophysical method that has not been used a great deal to investigate protein self-association at high concentration is SLS. Some of the concern revolves around contributions from multiple scattering at the higher concentrations. It has been shown that it is possible to obtain good-quality data at high concentrations and the results are consistent with those obtained by AUC [60].

Technologies such as standard cone and plate rheometry [61] and ultrasonic rheometry [62] have also shown the potential usage for analyzing high-concentration protein formulations. The apparent increases of product viscosity and storage module G' for an IgG2 monoclonal antibody appear to correlate with increased protein self-association at high protein concentration [62]. Although the change in viscoelastic properties of an antibody is difficult to translate into a quantitative number for measuring levels of aggregates, these methods at least provide qualitative information about protein–protein interactions at high concentrations under formulation conditions.

15.4.1. Challenges in Analysis of Data at High Protein Concentration: Thermodynamic Nonideality

Although it is now possible to collect data at high protein concentrations, the analysis and interpretation of experimental data at nonideal condition with high-concentration protein remains to be a difficult and challenging task. Most data analyses yield only qualitative information about protein interaction at high-concentration conditions.

Idealized conditions, minimizing the effects of nonspecific interactions between the constituents of a solution, are not valid at high protein concentrations. The effects of thermodynamic nonideality, also referred to as *excluded volume* (macromolecular crowding), and the associated effects on rates and equilibria of macromolecular association reactions, have to be taken into account for high-concentration formulations [63,64]. On the basis of the mechanism of excluded volume, an increased volume fraction is occupied by the protein molecules at higher protein concentrations. As a result, the decrease in the effective volume available and, in turn, the higher apparent protein concentration pushes the reaction equilibrium of protein self-association toward the more associated state [63,64]. Thus, owing to a steady increase of the nonideal correction factor (Γ) of the equilibrium constant with an increased amount of fractional volume (Φ) occupied by the protein molecules, protein self-association, involving the formation of an n-mer from n protein monomers, is progressively fostered with increasing protein concentration (Fig. 15.3a). The parameter Γ is a composition-, temperature- and pressure-dependent measure of intermolecular interactions. The contribution of each protein monomer under the conditions of ideal thermodynamics to

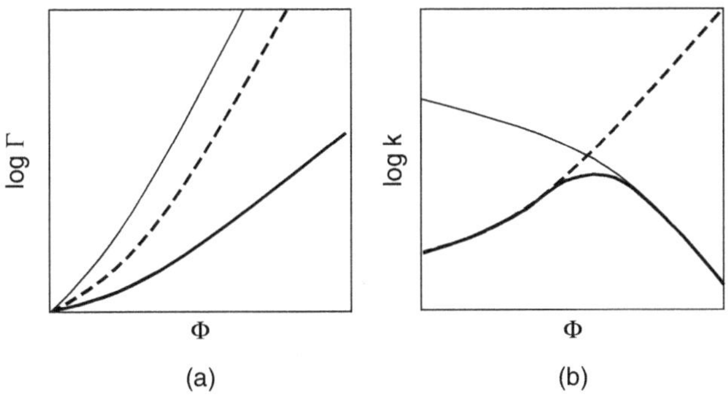

Figure 15.3. Dependence of the nonideality factor Γ (a) for various degrees of association ($n = 2$, solid line; $n = 3$, dashed line; $n = 4$, thin line) and of the reaction rate constant k (b) [solid line, being transition-state-limited (dashed line) at low volume occupancy and diffusion-limited (thin line) at high volume occupancy] from the fractional volume occupancy Φ. [Modified from Hall and Minton [63] and Zimmermann and Minton [64]).

the association process can be enhanced by the contribution of the volume exclusion mechanism. Moreover, in case of a constant fractional volume occupancy, the reaction of self-association is favored with increasing degree of self-association [63,64] (Fig. 15.3a).

The effect of high protein concentrations on the equilibrium of protein self-association has to be differentiated from the impact on the reaction rate. The reaction rate k of macromolecular association can be limited by either the conversion of the activated complex to a fully formed dimer or the diffusion-controlled formation of the activated complex. In the first transition state controlled case an enhanced reaction rate would result with increasing fractional volume occupancy. Thus, crowding would increase the reaction rate of self-association approximately to the same extent as the equilibrium constant is increased. In contrast, diffusion-limited reactions such as the formation of an activated complex result in a diminished reaction rate, owing to the fact that crowding considerably lowers the diffusional mobility of macromolecules—a hydrodynamic rather than thermodynamic consequence [63,64]. Since the overall bimolecular reaction rate is transition state limited at low fractional volume occupancy, the aggregation reaction rate increases initially with protein concentration, up to the point where protein concentration and molecular crowding become sufficiently greater than the contribution from protein concentration. Diffusional limitations set an upper limit to the reaction rate leading to a decrease in bimolecular reaction rate with further increasing fractional volume occupancy [64] (Fig. 15.3b).

The formation of native protein aggregates caused by the shift in the self-association equilibrium as a consequence of increased collision frequency could also be important in the formation of nonnative protein aggregates. The aggregates may serve as nuclei that can undergo or have already undergone conformational changes and, hence, can grow rapidly to form insoluble precipitates [65–67]. Accordingly, most aggregation reactions are reported to be of second or higher order and would be enormously accelerated in high-concentration protein formulations [66–68]. However, because of the larger and/or more extended structure of the protein in the denatured state, the equilibrium of the protein unfolding reaction can be driven toward the more compact native conformation by the volume exclusion effect as a consequence of increasing protein concentrations [63]. Therefore, an increase in protein concentration could also actually stabilize the protein against the formation of insoluble aggregates, either under conditions that favor protein self-association [69,70] or with respect to agitation-induced aggregation [71].

Finally, the relationship between protein concentration and nonnative aggregation tendency has to be evaluated on a case-by-case basis, whereas the formation of native protein aggregates is generally pronounced at higher protein concentrations.

15.5. FORMULATION DEVELOPMENT AND APPROACHES FOR HIGH-CONCENTRATION FORMULATIONS

15.5.1. General Formulation Considerations

The main challenges in designing stable formulations for high-concentration proteins are very similar to those for more standard low-concentration protein formulations. Many of the chemical degradation reactions such as deamidation, aspartate isomerization, oxidation, and peptide bond hydrolysis that occur in proteins are generally concentration-independent. A great deal is now known about the various hydrolytically driven chemical degradations that occur in proteins, and several reviews are available [72–74]. Protein aggregation, on the other hand, is expected to be one of the primary degradation pathways in high-concentration protein formulations since bi- or multimolecular collisions often dictate the formation of aggregates, and likely exacerbate as protein concentration increases. The actual relationship of concentration to aggregate formation will depend on the size of aggregates as well as the association mechanism [74,75]. Thus, as discussed in a review of irreversible noncovalent aggregation, this aggregation probably proceeds through interaction of exposed hydrophobic regions, in which case the protein unfolding step may be rate-limiting, resulting in pseudo-first-order behavior [66] rather than the expected higher-order concentration dependency. High concentration of protein may still have considerable impact on these types of mechanisms since the flexibility of a protein or alteration of volume required for a conformational change may very well be impacted by the neighboring interactions and excluded volume effects that can predominate at high concentrations. Moreover, even if there is a rate-limiting step that is concentration-independent at the lower concentration, the subsequent concentration-dependent steps may lead to overall more rapid kinetics at the higher concentration. Thus, in general, it is not surprising that protein aggregation may be a bigger challenge when developing higher-concentration protein formulations. However, the mechanism of aggregation may lead to unexpected results. As an example, aggregation and particulate formation due to protein interaction with air–water interfaces may have an inverse concentration dependence; that is, at higher protein concentrations the percentage of aggregate might decrease with an increase in protein concentration since the proteins are surface-active and there are fixed amounts of protein present at the air–water interface, essentially preventing damage to the remaining bulk protein [71].

As already mentioned, the chemical degradation pathways that dictate shelf life will occur irrespective of the protein concentration. The impact of formulating at higher concentration is more likely to influence physical properties such as viscosity as well as physical degradation pathways, especially protein aggregation. Formulation design to minimize protein degradation and enhance stability will require conditions that balance many of the different chemical degradations as well as the physical degradations. Creation of solid dosage forms using freeze drying or other drying methods that remove water from the final formulation will often help since many of the protein chemical degradations are hydrolytically driven and can be minimized over a large range of conditions. However, additional challenges may arise since lyoprotectants and/or cryoprotectants will need to be added to protect the protein from the additional stresses of the lyophilization process. Although solid dosage forms generally will have longer shelf life than will liquid formulations, the added convenience for dosing and greater ease of coupling with an injection device makes liquid protein formulations more desirable. However, design of aqueous liquid formulations can be much more challenging since in general the chemical degradations will occur at faster rates, thus requiring conditions that minimize physicochemical degradation kinetics. At higher protein concentrations, because of the potential for faster aggregation kinetics, this challenge can be even greater than at lower protein concentrations. Moreover, physical properties that may be related to protein self-association, such as viscosity (to be discussed in more detail later), can also increase the complexity and difficulty of designing a successful high-concentration liquid formulation. This difficulty is illustrated by a pH screen study for a humanized monoclonal IgG1 antibody formulated at 100 mg/mL shown in Figure 15.4. The chemical degradation routes in this antibody are dominated by an Asp isomerization and deamidation of Asn residues. The pH rate profile suggests that an optimum pH for control of both reactions is at pH ~6, whereas the viscosity of the formulation is at a maximum at that pH. Although the pH of a liquid formulation is a critical formulation parameter that can be used to minimize overall chemical degradation, this example shows that this strategy becomes more difficult because of the additional contributions of high viscosity at high protein concentration.

15.5.2. Primary Packaging Considerations

In general, similar considerations for the choice of primary packaging components exist as for lower concentration protein products. However, since high-concentration formulations are generally considered for use by SC administration, the use of prefilled syringes or cartridges as primary packaging is often desired. Such primary containers are actually devices and require sufficient lubrication to ensure that they function over the entire shelf life of the product. The most common lubricant used is silicone oil, and there have been several reports and studies [76] linking silicon oil contact to protein denaturation and aggregation. This is of crucial importance, since silicone oil represents a hydrophobic surface, possibly inducing direct protein–protein interactions [76]. Additionally, adverse effects of silicone oil on patients' safety, particularly

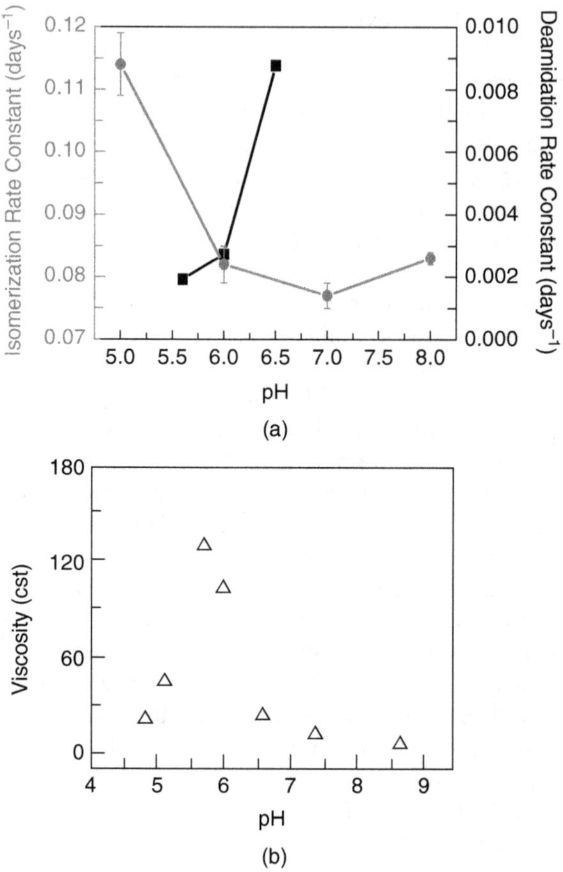

Figure 15.4. Humanized IgG1 MAb1 formulated at 100 mg/mL: (a) isomerization rate constant for an Asp residue and deamidation rate constant for an Asn residue as a function of pH; (b) kinematic viscosity as function of pH.

in terms of SC administration, have been discussed [77]. Particulate matter in a parenteral immunoglobulin in siliconized prefilled syringes could be attributed to the coalescence of freely permeable silicone oil as well as to antibody aggregates [78]. Additional concerns have been raised regarding levels of tungsten on the barrels of prefilled syringes with staked needles. The high tungsten levels in these syringes are due to the tungsten pins that are used to form the opening in the glass barrel for the staked needle. These high tungsten levels have been implicated in generating protein aggregates, especially at the higher protein concentrations used for SC delivery [79]. Several strategies are being pursued by the manufacturers of prefilled syringes, including use of proprietary glass coatings or baked-on siliconization processes as well as alternate material for the pins used in the manufacture of staked needle syringes [80].

In addition, more recent advances in prefilled syringe design where the contact with the needle and surrounding area is minimized may also solve many of these problems.

15.5.3. Viscosity and Administration Challenges

Viscosity is a measure of the resistance that a liquid system offers to flow when it is subjected to an applied shear stress. Newtonian fluids obey a linear relationship, whereas in non-Newtonian fluids the viscosity could be either enlarged or diminished with increasing shearing stress. Capillary viscometers are commonly utilized to assess viscosity of ideal solutions, whereas plate and rotational cylindrical rheometer allow analysis of suspensions, as well as high viscosity solutions and samples obeying non-Newtonian behavior [81,82].

Antibodies tend to form viscous solutions at high concentration because of their macromolecular nature and potential for intermolecular interactions [6,7]. Hence, viscosity η is a complex function of protein concentration c expressed as a power series relative to the viscosity of the pure solvent η_0 and including the terms k_1 for a single macromolecule, as well as k_2, k_3, also referred to as *two-body*, *three-body* virial coefficients, respectively, for describing the interaction of two or three or more protein molecules [81]:

$$\eta = \eta_0(1 + k_1 c + k_2 c^2 + \cdots) \tag{15.1}$$

A substantial increase in viscosity results when protein self-association prevails. Because macromolecular crowding also affects hydrodynamics properties [63,64], the hard quasi-spherical particle model, providing a realistic picture of the thermodynamic nonideality due to molecular crowding in high-concentration formulations, could also furnish a conceptual basis for the understanding of the hydrodynamic properties of a concentrated solution. Thus, viscosity η could be described by the following relationship

$$\eta = \eta_0\, e^{\nu\phi/(1-k\phi)} \tag{15.2}$$

where k is a crowding factor (usually in the range between 1.35 and 1.91 for hard spheres), ϕ is the volume fraction occupied by the macromolecule, and v is an universal molecular shape parameter (equal to 2.5 for spherical particles and exceeding 2.5 for nonspherical particles) that can be directly related to the shape of a particle independent of the volume [83].

The intrinsic viscosity η is an important structural parameter measured in infinite diluted protein solutions, as it is regarded as an intrinsic function of the individual dissolved or dispersed macromolecule. Accordingly, k_1 [see Eq. (1.1)] would be the quantity of interest, and the intrinsic viscosity could be summarized by the relation

$$[\eta] = \frac{\nu(V_\mathrm{h} N_0)}{M} = \lim_{\phi \to 0} \frac{\eta - \eta_0}{\eta_0} = \nu\phi = \lim_{c \to 0} \frac{\eta - \eta_0}{\eta_0} \tag{15.3}$$

where V_h is the hydrated volume of the protein molecule, N_0 is Avogadro's number, ν and ϕ are defined previously and M is the protein molecular weight [81,82]. Consequently, the intrinsic viscosity is affected by the shape and the hydration of the protein molecules, including chemically bound water via hydrogen bonds and physically entrained solvent. Additionally, the influence of the protein electrostatic charge and surface charge distribution on the viscosity has to be taken into consideration as well as the strong dependence of protein charge from pH and ionic strength. The contributions of electrostatic interactions to the intrinsic viscosity are based primarily on the diffuse double electrostatic layer surrounding the protein. Besides that, the repulsion between different protein molecules and an impact of the interparticle repulsion on protein shape could be involved [82].

Since a high viscosity affects the ability to manufacture drug product and bulk product, to withdraw a proper dose from a container into a syringe and to inject the dose into the patient, much effort is spent on the reduction of the viscosity of some viscous high-concentration formulations. Lowering of the viscosity of a concentrated antibody formulation was achieved by modifying the total ionic strength, by addition of different salt and buffer excipients or by altering the pH [19,84]. In case a protein suspension is envisioned, the viscosity depends, according to Einstein's equation of rheology for suspensions

$$\eta = \eta_0 + 2.5\Phi\eta_0 \tag{15.4}$$

mainly on the viscosity of the formulation vehicle and the volume fraction of the suspended particles [85]. Increased viscosity of the protein solution is a key factor in the ultrafiltration concentration, affecting the filtrate flux, processing times, membrane fouling and thus the reusability of the ultrafiltration membrane. Viscosity-related issues of the drug product fill–finishing procedure could arise from hindered flow through the filling equipment, impacting the reproducibility of fill volumes. Moreover, despite the increased viscosity, the passage of sterile filters has to be ensured in adequate time, especially if the product must be maintained at subambient temperatures [86]. These complications can be circumvented by using a drying technique as a concentration method, where the fill–finishing process takes place at lower concentrations.

15.5.4. Evaluation of Parameters Influencing Viscosity

Viscosity is an important factor in the development of high-concentration protein formulations, as it increases significantly with increasing protein concentration, and makes processing and application of liquid concentrated antibody preparations difficult. The concentration dependency of viscosity of macromolecules often can be accounted for by the influence of macromolecular crowding as described by the Mooney equation [Eq. (15.2)]. Substituting $c[\eta] = v\phi$ yields the modified Mooney equation relating viscosity to concentration:

$$\eta = \eta_0 \exp\left(\frac{[\eta]c}{1 - (k/v)[\eta]c}\right) \tag{15.5}$$

This equation theoretically should account for the concentration dependence of viscosity for any macromolecule as long as there are no intermolecular interactions; thus, the viscosity is explained purely by excluded volume effects. This equation as suggested by Minton and Ross can be used as a one-parameter fit model where k/v is the fitting parameter. Liu et al. evaluated the viscosity dependence of three closely related IgG1 monoclonal antibodies using a typical value of intrinsic viscosity at pH ~6–7 of 6.3 cm^3/g and a measured solvent viscosity of 0.0011 Pa · S [19]. As seen in Figure 15.5, two of the three monoclonal antibodies had reasonable fits to the Mooney equation, whereas one of them did not. A more recent study using another monoclonal antibody, MAb4, also shows higher viscosity at 100 mg/mL when compared to MAb2 and MAb3 (Fig. 15.5a). Preparative AUC as well as SLS determinations suggest that MAb2 and MAb3 do not appreciably self-associate under these conditions and thus have viscosity–concentration profiles that would be expected as a result of molecular crowding (excluded volume) for molecules of this size. Thus, the greater value

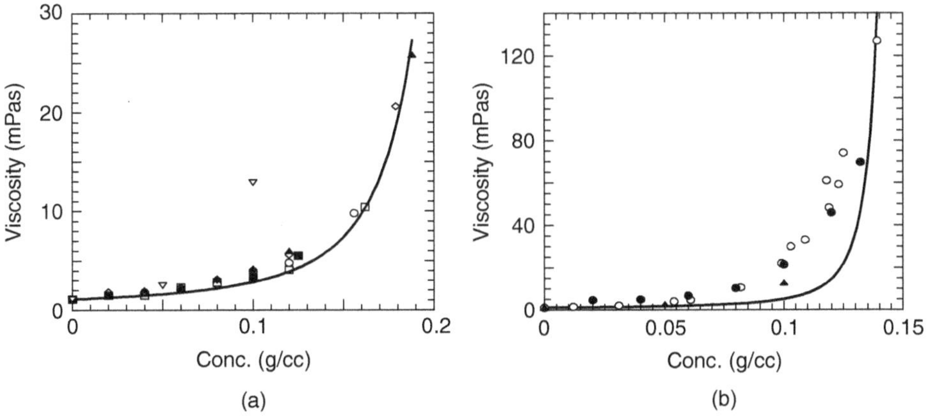

Figure 15.5. Viscosity of MAb as a function of protein concentration. (a) MAb2 with no added salt (□) and at 150 mM NaCl (O). MAb3 with no added salt (▲), and at 150 mM NaCl (◇). Nonlyophilized samples prepared by TFF have a composition of 266 mM sucrose and 16 mM histidine at pH 6.0 with and without 150 mM NaCl. Lyophilized and reconstituted MAb2 with no added salt (■). Lyophilized samples are at either 100 mg/mL MAb2 in 240 mM trehalose, 40 mM histidine 0.04% polysorbate 20, or 125 mg/mL in 300 mM trehalose, 50 mM histidine, and 0.05% polysorbate 20. MAb4 (▽) in 10 mM phosphate buffer at pH 5.5, no added salt. The solid curve is the result of a nonlinear regression of all the data to Equation (15.5) using a solvent viscosity of 1.1 mPas and intrinsic viscosity of 6.3 cm^3/g. The best fit with a $X^2 = 0.003$ resulted in a value for $k/\nu = 0.533$. (b) MAb1 lyophilized and reconstituted (O) and nonlyophilized (•) without any added NaCl. All samples contain 266 mM sucrose, 16 mM histidine, and 0.03% polysorbate 20 at pH 6. MAb4 (▲) in 10 mM phosphate buffer at pH 5.5, no added salt. The solid curve is the result of a nonlinear regression of all the data to Equation (15.5) using a solvent viscosity of 1.1 mPa and intrinsic viscosity of 6.3 cm^3/g. The best fit with a $\chi^2 = 1.16$ resulted in a value for $\chi/\nu = 0.936$.

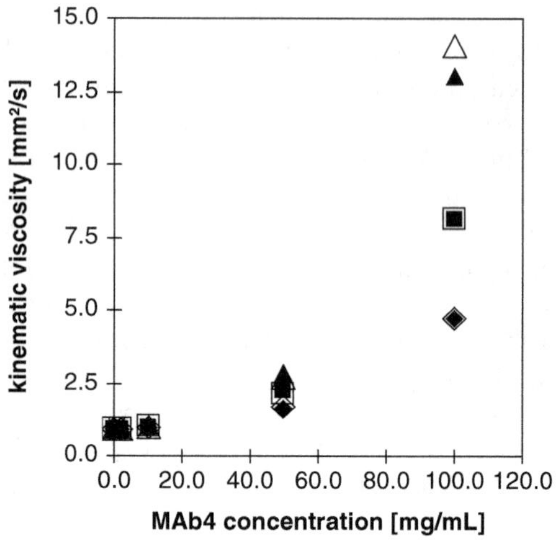

Figure 15.6. Kinematic viscosity of different Mab4 formulations with dependence on protein concentration. MAb4 solutions were formulated at different protein concentrations, pH values, and with or without addition of sodium chloride [citrate/pH 5.5 (△), phosphate/pH 5.5 (▲), citrate/pH 7.2 (□), phosphate/pH 7.2 (■), citrate/pH 7.2/145 mM sodium chloride (◊), and phosphate/pH 7.2/h, 145 mM sodium chloride (♦)].

determined for MAb4 suggests that there are additional interactions beyond excluded volume effects that account for the viscosity profiles of this antibody. In the case of MAb4, the kinematic viscosity was determined by capillary viscometry at various protein concentrations (2, 10, 50, 100 mg/mL of antibody). Besides that, the impact of two buffer systems (10 mM sodium citrate, 10 mM sodium phosphate), each formulated at a pH of 5.5 or 7.2 with the addition of 145 mM sodium chloride was investigated (Fig. 15.6) [87]. Lowering the viscosity of concentrated antibody formulations has been described in the literature by increasing the total ionic strength of the formulation using different salt and buffer excipients [19,84].

The viscosity of the tested antibody is highly dependent on the protein used, its protein concentration, the pH value, and formulation excipients. Kinematic viscosity was the same in both citrate and phosphate buffer. However, it was governed by the choice of pH value (pH 5.5 higher as compared to pH 7.2) and was reduced by addition of 145 mM sodium chloride (Fig. 15.6). A drastic increase in viscosity with increasing protein concentration was observed for all tested formulations (data not shown). The largest kinematic viscosity value of 14 mm^2/s measured for the 100 mg/mL formulation at pH 5.5 without addition of sodium chloride reflects an approximately 14-fold increase in viscosity as compared to common aqueous injection solutions.

As discussed earlier, the viscosity increase as a function of concentration could not be exclusively attributed to nonideality of the concentrated preparations because of excluded volume effects. Under conditions where the protein bears substantial net charge, the size of the effective hard particle significantly exceeds that of the actual model because of its mass [88]. Nevertheless, additional factors contributing to the unusually high viscosities have to be considered, such as electrostatic intermolecular interactions [19]. Since the viscosity is particularly increased at low pH values far away from the pI of the antibody, the electrostatic interactions responsible for the high viscosity at low pH might be repulsive ones. Since there are in-sufficient data on high pH, it is difficult to make additional conclusions, that is, whether the viscosity is minimized near the pI of this MAb. In contrast, an observed viscosity increase near the pI of another monoclonal antibody was attributed to attractive interactions between the IgG monomers, leading to a significant self-association as a function of protein concentration [19]. The differences in behavior of these antibodies are clearly demonstrated by the different dependence of viscosity on pH (Figs. 15.6 and 15.4b). In the case of MAb1 in Liu et al. [18]. a maximum viscosity is actually observed at low salt conditions at a pH value close to the pI value for this MAb. Sodium chloride was capable of decreasing the viscosity of antibody. It was suggested that the viscosity-lowering effect may be caused by a partial shielding of the repulsive intermolecular interactions by the chloride anion, as the antibody is positively charged in the pH range of interest. Moreover, the fit of the concentration dependence on the viscosity with a Mooney equation was slightly better in the presence of sodium chloride. Similarly, the presence of sodium chloride led to a reduction in protein self-association and, in turn, in the viscosity of the antibody example described above, where attractive interactions are hypothezised to be responsible for the unusual high viscosities [19]. In general it was observed that a series of salts (mainly chloride and sulfate) seemed to decrease viscosity as a function of ionic strength. However, it was noted that NH_4SCN and NaSCN salts were more effective and deviated from this simple ionic strength plot (Fig. 15.7). More recent published results [61] show that a series of cationic chloride salts do, indeed, reduce viscosity simply as a function of ionic strength, thus suggesting that the cations do not interact with the protein (Fig. 15.8). On the other hand, a series of anionic sodium salts appear to follow the Hofmeister series in the ability of the anions to reduce viscosity, suggesting that there are specific interactions of the anions that are responsible for the reduction of viscosity (Fig. 15.9). Liu et al. have discussed a particularly viscosity-lowering effect of the chaotropic thiocyanate anion, reducing protein-protein interactions [19].

15.5.5. Syringeability of High-Concentration Antibody Formulations

Syringeability and injectability need to be examined with appropriate *in vitro* studies during development of a high-concentration suspension or a liquid formulation. However, no standardized testing procedure has been established so far [89]. Syringeability testing may involve an evaluation of the ease of flow through a needle as well as the foaming tendency and dosing accuracy. The injectability of the high-concentration preparation is characterized mainly by the time, pressure, or force required for completing the injection at a constant flow. For high-concentration suspensions flow dynamics are also related to the particle characteristics of the suspension, and clogging of the needle on withdrawal or injection might be observed [90]. The ease of withdrawal could

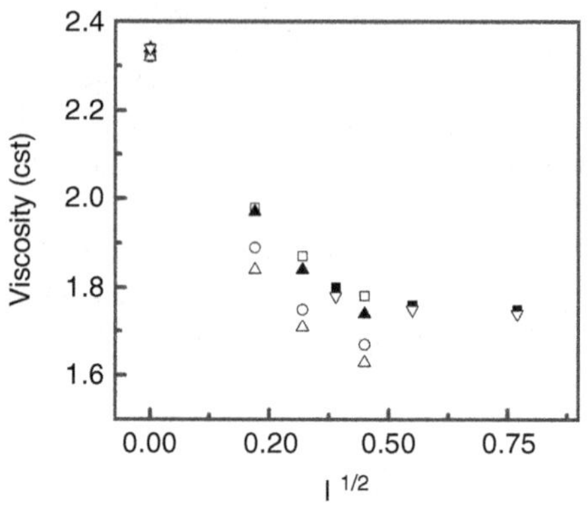

Figure 15.7. The effects of various salts as a function of ionic strength (plotted as $I^{1/2}$) on the kinematic viscosity of MAb1 solution containing 40 mg/mL of protein, 10 mM histidine, 250 mM sucrose, 0.01% polysorbate 20, and various amounts of salts at pH 6. The viscosity was measured using a Cannon–Fenske routine viscometer (Cannon Instruments, PA) at 25°C. The salts included NaCl (□), NaSCN (O), NH_4SCN (△), $(NH_4)_2SO_4$ (■), NH_4Cl (▲), and $CaCl_2$ (∇).

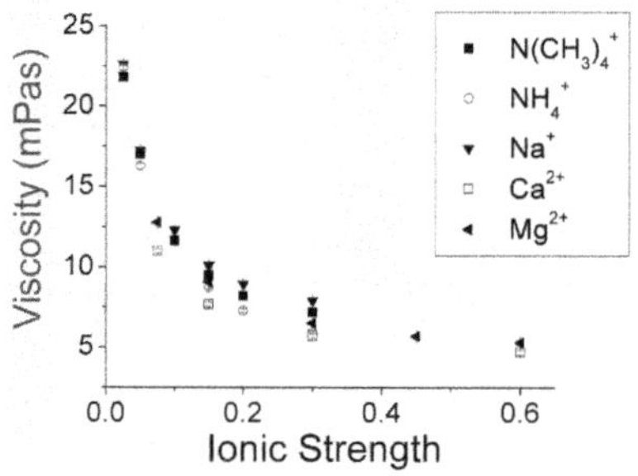

Figure 15.8. Solution viscosity of MAb1 in 30 mM histidine at pH 6 at 125 mg/mL with chloride salts as a function of ionic strength.

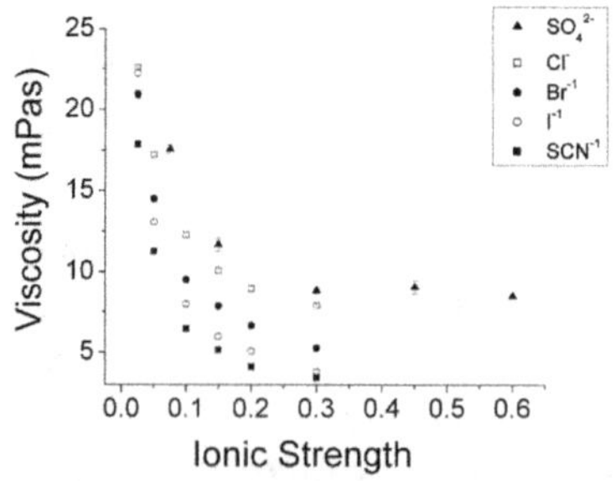

Figure 15.9. Solution viscosity of MAb1 in 30 mM huistidine at pH 6 at 125 mg/mL with sodium salts as a function of ionic strength.

be evaluated as time required to load a distinct volume of a high-concentration preparation into a syringe [6,7]. On the other hand, the injectability is typically assessed by measuring the injection force in terms of a constant injection speed, or vice versa using a specialized testing apparatus, such as a microcapillary rheometer or an automatic injector pistol [90,91]. Additionally, such studies may consider protein-stability-related issues on withdrawal and injection, as the use of force in manipulating the viscous formulation leads to shear stress and possibly excessive frothing, potentially resulting in denaturation and inactivation of the therapeutically active protein [6].

In addition to rheometric studies, the practical impact of viscosity on the ability to withdraw a proper dose of concentrated antibody solutions from a container into a syringe can be evaluated. Generally, it is assumed that the time period required for drawing a liquid into a syringe represents the resistance to flow (i.e., the viscosity η). Accordingly, the flow properties through the needles might be described by the Hagen–Poiseuille law for laminar flow in tubes

$$Q = \frac{\pi r^4 \Delta p}{8 \eta L} \tag{15.6}$$

where Q is the volumetric flow rate of the Newtonian model fluids, r is the needle radius, Δp is the pressure difference between the ends of the needle, and L is the length of the needle [92]. Syringeability testing, as the ability to flow easily through a needle, was performed using different needle sizes applicable for SC application (24, 26, 27 gauge). As shown previously [19] the correlation of syringeability with viscosity data can be shown for high-concentration antibody formulations. Matheus et al. [87] evaluated the syringeability of MAb4 depending on different diameters of the attached needles. Syringeability data yielded with various needle sizes [27 gauge with ultrathin wall, outer diameter (OD) of 0.40 mm corresponding to an inner diameter (ID) of 0.25 mm; 26 gauge with regular wall, OD 0.45 mm, ID 0.26 mm; 24 gauge with regular wall, OD 0.55 mm, ID 0.30 mm] are plotted against the kinematic viscosity (Fig. 15.10). As expected, the time required to draw 1 mL of the concentrated antibody preparation into a 1-mL syringe was increased with decreasing ID of the needles used and hence interpreted as the worst syringeability. Linear correlations between syringeability and viscosity could be found, with a steeper slope [5.48 (27 gauge), 3.01 (26 gauge), 0.974 (24 gauge)] and better linear fit occurring with decreasing ID of the attached needle [R^2 values: 0.91 (27 gauge), 0.75 (26 gauge), 0.64 (24 gauge)]. The time needed to withdraw 1 mL of antibody solution was acceptable for the 24-gauge needle in the high-viscosity range (<10 s). A longer time (≤ 50 s) was required when using the 26-gauge needle, and a rather inadequate syringeability resulted from the 27-gauge needle. However, the 27-gauge needle used is characterized by a short cannula length (20 vs. 25 mm for the 26- and 24-gauge needles) as well as by an ultrathin wand of the cannula, resulting in an increased inner diameter, which improves the flow of viscous solutions. Both the length and the inner diameter of a needle impose resistance to flow and hence flux through a syringe.

To answer the question of reasonable viscosities and times to draw the drug product into the syringe, syringeability of already marketed high-concentration antibody

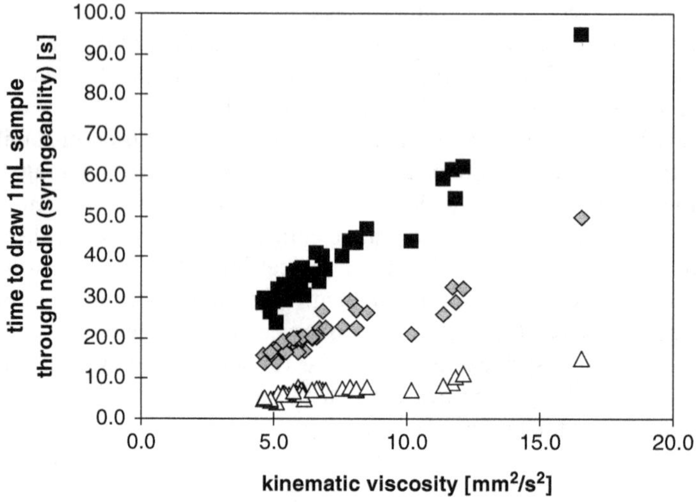

Figure 15.10. Syringeability versus kinematic viscosity of different antibody solutions tested. The time to draw a 1-mL antibody solution through a needle was measured with an attached 24- (△), 26- (◆), or 27 gauge (■) needle.

formulations was investigated. (Table 15.2). The antibody efalizumab formulation showed the highest viscosity of all antibody samples tested, though the recommended needle size was exceeded by approximately 25% in the 26-gauge needle. However, there are other published reports of antibody formulations with even higher viscosity and syringeability at a concentration of 100 mg/mL [7,19]. Administration of the preparation by a 24-gauge needle could be preferable over the other tested SC needle sizes of a smaller ID, due to a significantly lower timespan required for the viscous antibody solution to be drawn up. Since a linear viscosity–syringeability correlation was confirmed, the Hagen–Poiseuille law can be applied to calculate the syringeability of an antibody formulation of a given viscosity using a defined syringe needle system. Although the injectability correlated well with syringeability, other factors such as the forces driving the plunger and the resistance from local tissue should be taken into account for designing an adequate dosage form.

15.6. CONCLUSION AND SUMMARY

The SC route of administration is highly preferred for therapeutic indications where home (self-) medication is desirable, for example, for chronic diseases such as asthma and arthritic diseases. However, especially in the case of antibodies, it requires the delivery of high amounts of protein in a small volume. Apart from improving patient convenience and providing competiveness, high-concentration formulations require less storage space and may reduce overall shipping costs, owing to the minimized storage volume of the concentrated bulk and drug products.

TABLE 15.2. Syringeability[a] of Various Antibody Formulations Needle ($n = 1$)

Antibody	Formulation (Protein Concentration, Buffer, pH Value, and Addition of Salts)	Needle Used for SC Application	Syringeability[a]	
			26 gauge Needle	27 gauge Needle
Efalizumab	100 mg/ml in histidine, pH 6.2	25 gauge	14.5	28.0
Adalimumab	50 mg/ml in phosphate and citrate, pH 5.2+ 105 mM sodium chloride	27 gauge	7.5	13.5
Trastuzumab	21 mg/ml in histidine, pH 6.0	NA[b]	7.0	11.5

[a] *Syringeability* is a measurement of the time taken to draw one (or multiple) 1-mL sample(s) through one (or multiple) needle(s).
[b] Not applicable.

However, high-concentration antibody formulations also pose specific challenges to formulation and process scientists, for example, to overcome manufacturing challenges (e.g., increased viscosity-lowering filtration flow rates), stability challenges (e.g., increased aggregation at high protein concentration), analytical challenges (e.g., analysis at high protein concentration avoiding dilution), and formulation and administration challenges.

The common techniques currently available for manufacturing high-concentration antibody solution include tangential flow filtration (TFF) and drying techniques, such as lyophilization and spray drying. Although other techniques, such as crystallization, precipitation, or microparticle-based technology, have also demonstrated the potential for producing an injectable high-concentration antibody formulation in a suspension, there are several challenges yet to overcome in industrial development of those technologies.

Many of the analytical methods currently used for characterization of antibodies focus on the chemical structure of antibody and the chemical degradation pathways that result from alteration of amino acid residues. Generally, these methods performed at low concentration can be easily adapted to analyze antibodies at high concentration. However, since they often involve a sample preparation step to allow for the dilution of the products to a concentration range that is compatible with the analytical methods, those methods should be adequately assessed as to whether this impacts the analytical results. Analytical methods that can be used for the analysis of protein–protein interaction include SEC, light scattering, and AUC. Although it is now possible to collect data at high protein concentrations, the analysis and interpretation of experimental data at nonideal conditions with high-concentration protein remains a difficult and challenging task. Most data analyses yield only qualitative information about protein interaction at high-concentration conditions. With regard to stability, protein aggregation is expected to be one of the primary degradation pathways in high-concentration protein formulations since bi- or multimolecular collisions are likely to be increased as protein concentration increases. Another significant challenge for high-concentration antibody formulations is the tendency of antibodies to form viscous solutions at higher concentration because of their macromolecular nature and potential for intermolecular interactions. The viscosity, however, is not solely a function of concentration but is also influenced by formulation parameters such as pH and excipients. The increase in viscosity may influence manufacturability of such products as well as syringeability and administration. Often, subcutaneous formulations are intended for prefilled syringes or autoinjectors for self-administration, where the patient is then required to overcome the forces needed to deliver the viscous solution via needles of adequate size.

REFERENCES

1. Shire, S. J. and Gombotz, W. (2010), in *Current Trends in Monoclonal Antibody Development and Manufacturing*, Shire, S. J., Gombotz, W., Bechtold-Peters, K., and Andya, J., eds., Springer, New York, pp. 1–5.
2. Pavlou, A. K. and Belsey, M. J. (2005), The therapeutic antibodies market to 2008, *Eur. J. Pharm. Biopharm.* **59**(3): 389–396.

3. Roskos, L. K., Davis, C. G., and Schwab, G. M. (2004), The clinical pharmacology of therapeutic monoclonal antibodies, *Drug Devel. Res.* **61**(3): 108–120.
4. Gatlin, L. A. and Gatlin, C. A. B. (1999), Formulation and administration techniques to minimize injection pain and tissue damage associated with parenteral products, in *Injectable Drug Development: Techniques to Reduce Pain and Irritation*, Gapta, P. K. and Brazeau, G. A., eds., Interpharm Press, Denver, pp. 401–425.
5. Chang, B. S. and Hershenson, S. (2002), Practical approaches to protein formulation development, in *Rational Design of Stable Protein Formulations: Theory and Practice*, Carpenter, J. F. and Manning, M. C., eds., Kluwer Academic/Plenum, New York, pp. 1–25.
6. Shire, S. J., Shahrokh, Z., and Liu, J., (2004), Challenges in the development of high protein concentration formulations, *J. Pharm. Sci.* **93**(6): 1390–1402.
7. Harris, R. J., Shire, S. J., and Winter, C. (2004), Commercial manufacturing scale formulation and analytical characterization of therapeutic recombinant antibodies, *Drug Devel. Res.* **61**(3): 137–154.
8. Porter, C., Edwards, G., and Charman, S. (2001), Lymphatic transport of proteins after s.c. injection: Implications of animal model selection, *Adv. Drug Deliv. Rev.* **50**(1–2): 157–171.
9. Lobo, E. D., Hansen, R. J., and Balthasar, J. P. (2004), Antibody pharmacokinetics and pharmacodynamics, *J. Pharm. Sci.* **93**(11): 2645–2668.
10. Schellekens, H. (2002), Bioequivalence and the immunogenicity of biopharmaceuticals, *Nat. Rev. Drug Discov.* **1**(6): 457–462.
11. Klyushnichenko, V. (2003), Protein crystallization: from HTS to kilogram-scale, *Curr. Opin. Drug Discov. Devel.* **6**(6): 848–854.
12. Bromberg, L., Rashba-Step, J., and Scott, T. (2005), Insulin particle formation in supersaturated aqueous solutions of poly(ethylene glycol), *Biophys. J.* **89**(5): 3424–3433.
13. Schein, C. H. (1990), Solubility as a function of protein structure and solvent components, *Bio/Technology* **8**(4): 308–317.
14. Shiloach, J., Martin, N., and Moes, H. (1988), Tangential flow filtration, *Adv. Biotechnol. Processes* **8**: 97–125.
15. van Reis, R. and Zydney, A. (2001), Membrane separations in biotechnology, *Curr. Opin. Biotechnol.* **12**(2): 208–211.
16. Gabler, F. (1987), Principles of tangential flow filtration: Applications to biological processing, in *Filtration in the Pharmaceutical Industry*, Meltzer, T., ed., Marcel Dekker, New York, pp. 453–488.
17. Thomas, C. R., Nienow, A. W., and Dunnill, P. (1979), Action of shear on enzymes: Studies with alcohol dehydrogenase, *Biotechnol. Bioeng.* **21**(12): 2263–2278.
18. Watterson, J. G., Schaub, M. C., and Waser, P. G. (1974), Shear-induced protein-protein interaction at the air-water interface, *Biochim. Biophys. Acta* **356**(2): 133–143.
19. Liu, J., Nguyen, M. D., Andya, J. D., and Shire, S. J. (2005), Reversible self-association increases the viscosity of a concentrated monoclonal antibody in aqueous solution, *J. Pharm. Sci.* **94**(9): 1928–1940.
20. Masters, K. (1985), *Spray Drying Handbook*, 4th ed., Longman Scientific, New York.
21. Rouan, S. K. E. (1996), Biotechnology-based pharmaceuticals, in *Modern Pharmaceutics*, 3rd ed., Banker, G. S., and Rhodes, C. T., eds., Marcel Dekker, New York, pp. 843–873.
22. Sonner, C., Maa, Y. F., and Lee, G., (2002), Spray-freeze-drying for protein powder preparation: Particle characterization and a case study with trypsinogen stability, *J. Pharm. Sci.* **91**(10): 2122–2139.

23. Poupitch, G. (1994), Fluid bed drying in the laboratory, *Am. Biotechnol. Lab.* **12**(4): 30–34.
24. Wildfong, P. L., Samy, A. S., Corfa, J., Peck, G. E., and Morris, K. R. (2002), Accelerated fluid bed drying using NIR monitoring and phenomenological modelling: Method assessment and formulation suitability, *J. Pharm. Sci.* **91**(3): 631–639.
25. Mattern, M., Winter, G., Kohnert, U., and Lee, G. (1999), Formulation of proteins in vacuum-dried glasses. II. Process and storage stability in sugar-free amino acid systems, *Pharm. Devel. Technol.* **4**(2): 199–208.
26. Miller, D. P., Anderson, R. E., and de Pablo, J. J. (1998), Stabilization of lactate dehydrogenase following freeze thawing and vacuum-drying in the presence of trehalose and borate, *Pharm. Res.* **15**(8): 1215–1221.
27. Avis, K. E. (1990), Parenteral preparations, in *Remington's Pharmaceutical Sciences*, 18th ed., Gennaro, A. R., ed., Mack Publishing, Easton, PA, pp. 1545–1569.
28. Schule, S., Schulz-Fademrecht, T., Garidel, P., Bechtold-Peters, K., and Friess, W., (2008), Stabilization of IgG1 in spray-dried powders for inhalation, *Eur. J. Pharm. Biopharm.* **69**(3): 793–807.
29. Andya, J., Cleland, J., Hsu, C., Lam, X., Overcashier, D., Shire, S., Yang, J., and Wu, S.: *Protein formulation*. US Patent March 14,1996, US6267958 B1
30. Andya, J. D., Hsu, C. C., and Shire, S. J. (2003), Mechanisms of aggregate formation and carbohydrate excipient stabilization of lyophilized humanized monoclonal antibody formulations, *AAPS PharmSci.* **5**(2): E10.
31. Yang, M. X., Shenoy, B., Disttler, M., Patel, R., McGrath, M., Pechenov, S., and Margolin, A. L., (2003), Crystalline monoclonal antibodies for subcutaneous delivery, *Proc. Natl. Acad. Sci. USA* **100**(12): 6934–6939.
32. Brange, J. and Vølund, A. (1999), Insulin analogs with improved pharmacokinetic profiles, *Adv. Drug Deliv. Rev.* **35**(2–3): 307–335.
33. Harris, L. J., Skaletsky, E., and McPherson, A. (1995), Crystallization of intact monoclonal antibodies, *Proteins* **23**(2): 285–289.
34. Shenoy, B., Wang, Y., Shan, W., and Margolin, A. L. (2001), Stability of crystalline proteins, *Biotechnol. Bioeng.* **73**(5): 358–369.
35. Matheus, S., Friess, W., and Mahler, H. (2005), *High Concentration IgG1-Antibody Formulations by Precipitation and Crystallisation*, AAPS National Biotechnology Conf. poster, San Francisco.
36. Creighton, T. (1992), Proteins in solution and in membranes, in *Proteins: Structures and Molecular Properties*, 2nd ed., Freeman, New York, pp. 261–272.
37. Matheus, S., Mahler, H.-C. Solid forms of anti-EGFR antibodies. Patent US 2007/0122411 (A1), EP 1686961 (A1)
38. Jones, A. J. S. (1993), Analytical methods for the assessment of protein formulations and delivery systems, in *Formulation and Delivery of Peptides and Proteins*, Cleland, J. L. and Langer, R., eds., American Chemical Society, Washington, DC, pp. 22–45.
39. Pearlman, R. and Nguyen, T. H. (1991), Analysis of protein drugs, in *Peptide and Protein Drug Delivery*, Lee, V. H., ed., Marcel Dekker, New York, pp. 247–301.
40. Costantino, H. R., Chen, B., Griebenow, K., Hsu, C. C., and Shire, S. J. (1998), Fourier-transform infrared spectroscopic investigation of the secondary structure of aqueous and dried recombinant human deoxyribonuclease I, *Pharm. Pharmacol. Commun.* **4**: 391–395.
41. Matheus, S., Friess, W., and Mahler, H. C. (2006), FTIR and nDSC as analytical tools for high-concentration protein formulations, *Pharm. Res.* **23**(6): 1350–1363.

42. Prestrelski, S. J., Tedeschi, N., Arakawa, T., and Carpenter, J. F. (1993), Dehydration-induced conformational transitions in proteins and their inhibition by stabilizers, *Biophys. J.* **65**(2): 661–671.

43. Sharma, V. K., and Kalonia, D. S. (2003), Steady-state trytophan fluorescence spectroscopy study to probe tertiary structure of proteins in solid powders, *J. Pharm. Sci.* **92**(4): 890–899.

44. Hatley, R. H. M. (1990), *Proc. Int. Symp. Biological Product Freeze-Drying and Formulation*, Bethesda, MD, pp. 105–122.

45. Breen, E. D., Curley, J. G., Overcashier, D. E., Hsu, C. C., and Shire, S. J. (2001), Effect of moisture on the stability of a lyophilized humanized monoclonal antibody formulation, *Pharm. Res.* **18**(9): 1345–1353.

46. McPhillips, C. D. Q. M., Royall, P. G., and Hill, V. L. (1999), Characterization of the glass transition of HPMC using modulated differential scanning calorimetry, *Int. J. Pharm.* **180**: 83–90.

47. Georgalis, Y. and Saenger, W. (1999), Light scattering studies on supersaturated protein solutions, *Sci. Progress* **82**(Pt. 4): 271–294.

48. Bloomfield, V. A. (1981), Quasi-elastic light scattering applications in biochemistry and biology, *Annu. Rev. Biophys. Bioeng.* **10**: 421–450.

49. Pecora, R. (1972), Quasi-elastic light scattering from macromolecules, *Annu. Rev. Biophys. Bioeng.* **1**: 257–276.

50. Schurr, J. M. (1977), Dynamic light scattering of biopolymers and biocolloids, *CRC Crit. Rev. Biochem.* **4**(4): 371–431.

51. Wu, C. F. and Chen, S. H. (1988), Small angle neutron and x-ray scattering studies of concentrated protein solutions. II. Cytochrome C, *Biopolymers* **27**(7): 1065–1083.

52. Schuck, P. (1998), Sedimentation analysis of noninteracting and self-associating solutes using numerical solutions to the Lamm equation, *Biophys. J.* **75**(3): 1503–1512.

53. Johnson, M. L. and Straume, M. (1994), Comments on the analysis of sedimentation equilibrium experiments, in *Modern Analytical Ultracentrifugation*, Schuster, T. M. and Laue, T. M., eds., Birkhauser, Boston, pp. 37–63.

54. Schuck, P., Perugini, M. A., Gonzales, N. R., Howlett, G. J., and Schubert, D. (2002), Size-distribution analysis of proteins by analytical ultracentrifugation: strategies and application to model systems, *Biophys. J.* **82**(2): 1096–1111.

55. Liu, J., Andya, J. D., and Shire, S. J. (2006), A critical review of analytical ultracentrifugation and field flow fractionation methods for measuring protein aggregation, *AAPS J.* **8**(3): E580–E589.

56. Gonzalez, J. M., Rivas, G., and Minton, A. P. (2003), Effect of large refractive index gradients on the performance of absorption optics in the Beckman XL-A/I analytical ultracentrifuge: An experimental study, *Anal. Biochem.* **313**(1): 133–136.

57. Minton, A. P., and Lewis, M. S. (1981), Self-association in highly concentrated solutions of myoglobin: A novel analysis of sedimentation equilibrium of highly nonideal solutions, *Biophys. Chem.* **14**(4): 317–324.

58. Minton, A. P. (1989), Analytical centrifugation with preparative ultracentrifuges, *Anal. Biochem.* **176**: 209–216.

59. Moore, J., Kamerzell, T., Shen, Y., Kwong, Z., Zhu, Q., Laue, T., Liu, J., and Shire, S. (2007), High protein concentration analytical ultracentrifugation using fluorescence detection: Fluorescent labeling and characterization of monoclonal antibodies, *Proc. AAPS Natl. Biotechnology Conf.,* San Diego.

60. Scherer, T., Kanai, S., Liu, J., and Shire, S. (2007). Characterization of monoclonal antibodies at high concentrations by light scattering, *Proc. AAPS Natl. Biotechnology Conf.*, San Diego.
61. Kanai, S., Liu, J., Patapoff, T., and Shire, S. J. (2008), Reversible self-association of a concentrated monoclonal antibody solution mediated by Fab-Fab interaction that impacts solution viscosity, *J. Pharm. Sci.* **97**(3): 4219–4227.
62. Saluja, A., Badkar, A. V., Zeng, D. L., Nema, S., and Kalonia, D. S. (2006), Application of high-frequency rheology measurements for analyzing protein-protein interactions in high protein concentration solutions using a model monoclonal antibody (IgG2), *J. Pharm. Sci.* **95**(9): 1967–1983.
63. Hall, D. and Minton, A. P. (2003), Macromolecular crowding: Qualitative and semiquantitative successes, quantitative challenges, *Biochim. Biophys. Acta* **1649**(2): 127–139.
64. Zimmerman, S. B. and Minton, A. P. (1993), Macromolecular crowding: Biochemical, biophysical, and physiological consequences, *Annu. Rev. Biophys. Biomol. Struct.* **22**: 27–65.
65. Brange, J. (2000), Physical stability of proteins, in *Pharmaceutical Formulation Development of Peptides and Proteins*, Frokjaer, S. and Hovgaard, L., eds., Taylor & Francis, London, pp. 89–112.
66. Chi, E. Y., Krishnan, S., Randolph, T. W., and Carpenter, J. F. (2003), Physical stability of proteins in aqueous solution: Mechanism and driving forces in non-native protein aggregation, *Pharm. Res.* **20**(9): 1325–1336.
67. Wang, W. (2005), Protein aggregation and its inhibition in biopharmaceutics, *Int. J. Pharm.* **289**(1–2): 1–30.
68. Carpenter, J., Kendrick, B., Chang, B., Manning, M., and Randolph, T. (1999), Inhibition of stress induced aggregation of protein therapeutics, in *Amyloid, Prions, and Other Protein Aggregates*, Wetzel, R., ed., Academic Press, San Diego, pp. 236–255.
69. Dathe, M., Gast, K., Zirwer, D., Welfle, H., and Mehlis, B. (1990), Insulin aggregation in solution, *Int. J. Pept. Protein Res.* **36**(4): 344–349.
70. Kendrick, B., Li, T., and Chang, B. (2003), Physical stabilization of proteins in aqueous solution, in *Rational Design of Stable Protein Formualtions:Theory and Practice*, Carpenter, J. and Manning, M., eds., Kluwer Academic/Plenum Publishers, New York, pp. 61–84.
71. Treuheit, M. J., Kosky, A. A., and Brems, D. N. (2002), Inverse relationship of protein concentration and aggregation, *Pharm. Res.* **19**(4): 511–516.
72. Ahern, T. J. and Manning, M. C., eds. (1992), *Stability of Protein Pharmaceuticals. Part A: Chemical and Physical Pathways of Protein Degradation*, Plenum Press, New York, p. 434.
73. Cleland, J. L., Powell, M. F., and Shire, S. J. (1993), The development of stable protein formulations: A close look at protein aggregation, deamidation and oxidation, *Crit. Rev. Ther. Drug Carrier Syst.* **10**(4): 307–377.
74. Manning, M. C., Patel, K., and Borchardt, R. T. (1989), Stability of protein pharmaceuticals, *Pharm. Res.* **6**(11): 903–918.
75. Glatz, C. E. (1992), Modeling of aggregation-precipitation phenomena, in *Stability of Protein Pharmaceuticals*, Ahern, T. J. and Manning, M. C., eds., Plenum Press, New York, pp. 135–166.
76. Jones, L. S., Kaufmann, A., and Middaugh, C. R. (2005), Silicone oil induced aggregation of proteins, *J. Pharm. Sci.* **94**(4): 918–927.
77. Chantelau, E. (1989), Silicone oil contamination of insulin, *Diabet. Med.* **6**(3): 278.

78. Fraunhofer, W. and Winter, G. (2004), The use of asymmetrical flow field-flow fractionation in pharmaceutics and biopharmaceutics, *Eur. J. Pharm. Biopharm.* **58**(2): 369–383.

79. Rosenberg, A. (2006), Effects of protein aggregates: An immunologic perspective, *AAPS J.* **8**(3): E59

80. Harrison, B. and Rios, M. (2007), Big shot-developments in prefilled syringes, *Pharm. Technol.* **31**(3): 50–60.

81. Cantor, C. R. and Schimmel, P. R. (1980), Other hydrodynamic techniques, in *Biophysical Chemistry Part II: Techniques for the Study of Biological Structure and Function*, Cantor, C. and Schimmel, P., eds., Freeman, New York, pp. 643–685.

82. Harding, S. E. (1997), The intrinsic viscosity of biological macromolecules. Progress in measurement, interpretation and application to structure in dilute solution, *Progress Biophys. Mol. Biol.* **68**(2–3): 207–262.

83. Ross, P. D. and Minton, A. P. (1977), Hard quasispherical model for the viscosity of hemoglobin solutions, *Biochem. Biophys. Res. Commun.* **76**(4): 971–976.

84. Arvinte, T. and Fauquex, P. (2002), *Stable Liquid Formulations.* ed.

85. Basu, S. K., Govardhan, C. P., Jung, C. W., and Margolin, A. L. (2004), Protein crystals for the delivery of biopharmaceuticals, *Expert Opin. Biol. Ther.* **4**(3): 301–317.

86. Akers, M., Vasudevan, V., and Stickelmeyer, M. (2002), Formulation development of protein dosage forms, in *Development and Manufacture of Protein Pharmaceuticals*, Nail, S. and Akers, M., eds., Springer, pp. 47–114.

87. Matheus, S., Friess, W., and Mahler, H. (2005), Liquid high concentration IgG1-antibody formulations, *Proc. AAPS Natl. Biotechnology Conf.,* San Francisco.

88. Minton, A. P. and Edelhoch, H. (1982), Light scattering of bovine serum albumin solutions: Extension of the hard particle model to allow for electrostatic repulsion, *Biopolymers* **21**(2): 451–458.

89. Burgess, D. J., Hussain, A. S., Ingallinera, T. S., and Chen, M.-L. (2002), Assuring quality and performance of sustained and controlled release parenterals: AAPS workshop report, co-sponsored by FDA and USP, *Pharm. Res.* **19**(11): 1761–1768.

90. Allahham, A., Stewart, P., Marriott, J., and Mainwaring, D. E. (2004), Flow and injection characteristics of pharmaceutical parenteral formulations using a micro-capillary rheometer, *Int. J. Pharm.* **270**(1–2): 139–148.

91. Zingerman, J., Pope, D., Wilkinson, P., and Perfetto, L. (1987), Automatic injector apparatus for studying the injectability of parenteral formulations for animal health, *Int. J. Pharm.* **36**(2–3): 141–145.

92. Ritschel, W. and Suzuki, K. (1979), In vitro testing of injectability, *Pharm. Industr.* **41**: 468–475.

16

DEVELOPMENT OF FORMULATIONS FOR THERAPEUTIC MONOCLONAL ANTIBODIES AND Fc FUSION PROTEINS

Sampathkumar Krishnan, Monica M. Pallitto, and Margaret S. Ricci

16.1. INTRODUCTION

Monoclonal antibody-based therapies have become a huge area for biopharmaceutical development, with 18 monoclonal antibodies (Table 16.1) on the market and nearly 200 antibody molecules in clinical facilities [1–4]. Monoclonal antibodies for therapeutic and prophylactic indications over the years have moved from fully murine and humanized murine forms to completely human forms. There has also been a breakthroughs since the 1980s regarding purification [5], analytical methods including biological assays [6], and manufacturing aspects leading to the ability to prepare purer lots of monoclonal antibodies economically at large scales. The majority of monoclonal antibodies that are currently approved or in clinical development are focused on meeting therapeutic needs in the areas of oncology, autoimmune, and inflammatory diseases [1,4].

Antibodies and Fc fusion proteins are large macromolecules (typically >150 kDa), an order of magnitude larger than many other protein therapeutics such as cytokines, and are multidomain as well as typically glycosylated in nature, (if produced by mammalian cell culture [7]. Domains of antibodies have naturally evolved to associate with a variety of targets such as antigens and FcRn receptors with high

Formulation and Process Development Strategies for Manufacturing Biopharmaceuticals, edited by Jameel and Hershenson

TABLE 16.1. Detailed List of Antibody Products Approved and Marketed in the United States

Antibody Product	Manufacturing Company	Generic Name; Description	Delivery Route/ Dosage Form	Final Formulation Concentration,[a] mg/mL	Buffer Components; pH	Excipients
Avasatin®	Genentech	Bevacizumab; humanized IgG1, anti-VEGF	IV/liquid	25	Na phosphate; pH 6.4	Trehalose, polysorbate 20
Bexxar®	Corixam-GSK	Tositumomab-I131; murine IgG2a, anti-CD20, radiolabeled (^{131}I)	IV/liquid	14	Na phosphate; pH 7.2	NaCl, maltose
Campath®-1H	Millennium-ILEX	Alemtuzumab; humanized IgG1, anti-CD52	IV/liquid	30	Na,K-phosphate, pH 7	NaCl, KCl, NaEDTA, polysorbate 80
Erbitux®	Imclone Systems	Cetuximab; chimeric IgG1, anti-EGFR	IV/liquid	2	Na phosphate; pH 7.2	NaCl
Herceptin®	Genentech	Trastuzumab; humanized IgG1, anti-HER2	IV/lyophilized	21 (440 mg)	Histidine; pH 6	Trehalose, polysorbate 20, benzyl alcohol
Humira®	Abbott	Adalimumab; human IgG1, anti-TNFα	SC/liquid	50	Na phosphate, Na Citrate; pH 5.2	Mannitol, polysorbate 80
Mylotarg®	Wyeth	Gemtuzumab ozogamicin; humanized IgG4, anti-CD33, immunotoxin	IV/lyophilized	4 (5 mg)	Na phosphate	NaCl, sucrose, dextran 40
Othroclone OKT3®	J & J-Ortho Biotech	Muromonab-CD3; murine IgG2a, anti-CD3	IV/liquid	1	Na,K phosphate; pH 7	NaCl, polysorbate 80

Raptiva®	Genentech	Efalizumab; humanized IgG1, anti-CD11a	SC/lyophilized	100 (150 mg)	Histidine; pH 6.2	Sucrose, polysorbate 20
Remicade®	Centocorb	Infliximab; chimeric IgG1, anti-TNFα	IV/lyophilized	10 (100 mg)	Na phosphate; pH 7.2	Sucrose, polysorbate 80
ReoPro®	Centocor-Lilly	Abciximab; chimeric IgG1, anti-GPIIb/IIIa, Fab	IV/liquid	2	Na phosphate; pH 7.2	NaCl, polysorbate 80
Rituxan®	Idec-Genentech	Rituximab; chimeric IgG1, anti-CD20	IV/liquid	10	Na citrate; pH 6.5	NaCl, polysorbate 80
Simulect	Novartis	Basilixamab; chimeric IgG, anti-CD25	IV/lyophilized	4 (20 mg)	Na,K phosphate	NaCl, sucrose, mannitol, glycine; polysorbate 80
Synagis®	MedImmune	Palivizumab; humanized IgG1, anti-RSV	IM/liquid	100	Histidine	NaCl; glycine
Tysabri	Biogen Idec	Natalizumab; humanized IgG4, anti-4α-integrin	IV/liquid	20	Na phosphate; pH 6.1	NaCl; polysorbate 80
Vectibix	Amgen	Panitumumab; human IgG2, anti-EGFR	IV/liquid	20	Na acetate	NaCl
Xolair®	Genentech	Omalizumab; humanized IgG1, anti-IgE	SC/lyophilized	125 (202.5 mg)	Histidine	Sucrose, polysorbate 20
Zenapax®	Roche	Daclizumab; humanized IgG1, anti-CD25	IV/liquid	5	Na phosphate; pH 6.9	NaCl, polysorbate 80
Zevalin®	Biogen Idec	Ibritumomab tiuxetan; murine IgG1, anti-CD20; radiolabeled (^{90}Y or ^{111}In)	IV/liquid	1.6	Na acetate, Na,K-phosphate; pH 7	NaCl, human albumin

[a] From total lyophilized product if applicable.

affinity, which makes them useful protein therapeutics. In the body, antibodies typically have a half-life of 30 days. On the other hand, they are expected to be stable in storage for more than 2 years. The goal of a formulation program for therapeutic antibodies and Fc fusion proteins, as well as other protein therapeutics, is generally to develop a stable, robust formulation that minimizes physical and covalent degradation, ensures long-term storage stability, and prevents any adverse *in vivo* effects such as injection site, immunogenic or anaphylactic reactions. Additionally, instability of the antibody molecules can alter the pharmacology of the drug product as it affects both the pharmacokinetics in the serum and drug clearance from the body.

Antibodies and Fc fusion proteins, like other proteins, can be degraded under conditions where they are exposed to extremes of heat, freezing, light, pH, agitation, shear stress, metals, and substances such as silicone oil from prefilled syringes. Exposed surface residues of each antibody are unique and require specific formulation excipients to provide maximal stability against the aforementioned stresses. Assessment of the physicochemical and thermodynamic instability of antibodies using novel analytical technologies has led to the identification of several events that are more specific to the unique nature of this particular class of proteins such as variations in Fc glycosylation, Fc methionine oxidation, hinge region cleavage, and glycation of Lys residues [8]. An optimal formulation should minimize all such antibody degradation reactions in solution, or at minimum, mitigate those degradation reactions that impact critical quality attributes.

Biotechnology companies and contract research organizations are using improved analytical methodologies to monitor the degradation of the protein therapeutics during stability testing. Forced denaturation, agitation, and freeze–thaw studies are used to simulate the conventional stresses that a protein can undergo during production, shipping, storage, and administration. The effectiveness of forced degradation and stability studies done on a small scale to predict long-term, large-scale product stability depends on a number of factors: (1) the temperature dependence of the protein has to be understood; (2) accurate predictions of the shelf storage require that the protein system follow Arrhenius degradation kinetics over the temperature range that is used for the accelerated stability studies; and (3) the stability studies have to be conducted on multiple manufacturing lots that are representative of the commercial process.

One of the main challenges facing the manufacturers of biologicals in terms of formulation is demonstrating biocomparability in terms of product stability and clinical bioequivalence. The production of biological products is a complex process that undergoes continual development and refinement before commercialization and may continue post-launch. In most cases, any alterations in the process used to manufacture the antibody molecules can result in wide differences in the structural and functional properties of the molecules. These changes can alter the stability, clinical efficacy, and/or safety of the recombinant antibody therapeutic. Therefore, there is a need to perform formulation development studies on manufacturing lots that are representative of the commercial process for the approved and marketed drug product. Another issue is the complexity of *Escherichia coli* production processes for cytokines versus mammalian-cell-derived processes for antibody production, which results in heterogeneity in glycosylation patterns. Antibodies are often heterogeneous as a result of charge variants, glycosylation differences, and disulfide chemistry. Formulation screening must be initiated early in development even before knowing the commercial drug dose and before the commercial process is set. The purity of the excipients used in the formulation may present an additional challenge. Vendor or lot differences in the purity of the excipients can jeopardize the consistency of the drug product. With all of these considerations, formulation development can be a considerable challenge. In addition, the formulation screens must be efficient to accommodate limited amounts of protein available for early formulation studies.

Regulatory agencies require rigorous testing procedures to determine the stability of a pharmaceutical formulation over time. Regulatory perspectives of the characterization and stability testing procedures have changed with advances in analytical technologies, especially in the fields of mass spectrometry and chromatography, and there is an increased understanding of the biology of recombinant proteins, as well as preclinical and clinical experience with many approved products. Also, there is a regulatory requirement to demonstrate that material or process changes during antibody production generate bioequivalent drug product.

16.2. MECHANISMS OF DEGRADATION

16.2.1. Physical Instability

16.2.1.1. Aggregation and Particle Formation. Aggregation and related particle formation is a dominant degradation pathway of antibodies and can occur during all stages of protein therapeutic processing and storage [3,9,10]. Aggregation of light-chain antibody fragments and their deposition into amorphous precipitates or insoluble fibrils has also been linked to amyloid diseases such as systemic amyloidosis [11–13]. Knowledge of the mechanisms underlying the protein aggregation processes is essential to develop rational *in vitro* preventive strategies. The aggregation phenomena can be stipulated by protein structural changes or by colloidal effects affecting protein–protein interactions [14]. Such events for proteins in general could occur via a simple diffusion-limited mechanism [15] or involve nucleation as the primary stage for further growth and propagation of aggregates [16,17]. From earlier studies it has become evident that proteins with dominating β-sheet content are prone to aggregation [18,19] and can self-assemble into either amorphous precipitates [15] or well-defined fibrils [17]. The aggregation process is also sensitive to a wide range of factors such as protein concentration, hydrophobicity, and charge as well as solution pH, ionic strength, and temperature [3,9,10,20]. Particle formation due to aggregation is a major issue, and control of particle levels for parenteral administration is necessary to prevent potential adverse reactions, as well as potential clogging of intravenous lines and filters. When high therapeutic doses are required, the need for high volumes may be countered by increasing the concentration of the antibody (sometimes several orders of magnitude higher than conventional protein therapeutics). This, in turn, may result in increased problems relating to aggregation and particulation.

Antibody aggregation is complex and can proceed through covalent or noncovalent association that is highly dependent on the solution conditions, including pH, ionic strength, and excipients [21,22]. This association can be due to disulfide or nondisulfide covalent bonds, while the noncovalent associations can be due to hydrophobic or electrostatic interactions. Adding to the complexity, a given antibody can undergo multiple mechanisms of aggregation. Antibodies can undergo domain swapping, which can lead to altered structure and aggregation. Increasing the temperature and pH of the formulation often results in covalent crosslinking due to disulfide shuffling, while protein concentration, salt content, and other factors can promote non-covalent association. Antibodies have multiple intradomain as well as interdomain linkages through the disulfides [7], and these linkages have been found to undergo shuffling during processing leading to product heterogeneity and aggregation. Antibodies are also susceptible to photo-oxidation, which can lead to aggregation.

In most cases aggregation of protein molecules proceeds through a partially and reversibly unfolded conformational state that results from partial unfolding. This conformational state can be populated through the effects of solution conditions and the internal conformational stability of the molecule on the transition from native to unfolded states. A protein aggregate is formed between two or more molecules because of this higher-order structural disruption and exposure of hydrophobic regions leading to intermolecular interactions. This can eventually lead to aggregation and/or particulation [10,14].

After storage in solution under physiological conditions for a sufficiently long period, dimers may represent the main component of total aggregates [approximately 10%–30% (w/w) at a protein concentration of ~160 mg/mL] [23]. More recent studies have examined kinetic and thermodynamic aspects of the dimerization of IgG1s in solution [21,22]. Using a recombinant human monoclonal antibody that recognizes vascular endothelial growth factor (rhMAb-VEGF) as a model [21], it was found that aggregation rates were greater in slightly alkaline (pH 7.5–8.5) compared to slightly acidic (pH 6.5–7.5) conditions. A high-salt environment (1 M NaCl) also enhanced dimerization. The nature of the IgG1 dimers was found to be highly complex, resulting from different associations of the antibody domains [22]. In our laboratory, we have studied the structure, stability, and conformational dynamics of the Fab, Fab'_2, and Fc fragments of an IgG1 molecule [24]. Structural studies of the intact antibody and fragments showed that the structure of the Fc fragment is most susceptible to pH changes. Thermal, guanidine HCl–induced and urea unfolding studies at pH 7.4 and 5.0 showed differences in conformational stability of the various fragments at these two pH levels. Incubation studies performed with the intact protein and the fragments at 37°C and 50°C showed that the Fc fragment aggregated faster than did the Fab and the intact antibody. We proposed that Fc–Fc and possibly Fab–Fc are responsible for the aggregation and particle formation of the IgG1 antibody molecule as a result of temperature-induced stress. McAuley et al. [25] showed that disulfide bond formation in the human C_H3 domain plays an important role in antibody stability and dimerization. This domain contains a single buried, highly conserved disulfide bond. The authors showed that this disulfide bond significantly affects the stability and monomer–dimer equilibrium of the human C_H3 domain, which may have implications for the stability of the intact antibody.

Another study of empirical phase diagrams of monoclonal antibody solutions produced from spectroscopic data suggested (1) the existence of similar structural states at low temperatures independent of concentration and (2) a decrease in the temperature at which phase changes were observed with increasing concentration. The decrease in structural stability observed in these studies is probably the result of aggregation or self-association of the recombinant MAbs on heating in crowded solutions, and not a decrease in the intrinsic structural stability of the MAbs [26]. A related investigation [27] found that at a given concentration, the phase separation temperature for proteins in general strongly increases with the molecular weight of the oligomers. These findings imply that for phase separation, the detailed changes of the surface properties of the proteins are less important than the purely steric effects of oligomerization.

During manufacturing or shipment, proteins endure high mechanical or shear stress through filtration, mixing, and agitation and are exposed to various interfaces. Partially denatured molecules expose hydrophobic regions within the molecule, which can then result in interaction, protein aggregation, and particle formation [9,28]. Antibodies, like other proteins, can interact with air–water interfaces and surfaces such as metals and other hydrophobic components. Prior to delivery to the patient, protein pharmaceuticals often come in contact with a variety of surfaces (e.g., syringes and stoppers), which are treated to facilitate processing or to inhibit protein binding. One such coating, silicone oil, has previously been implicated in the induction of protein aggregation [29].

16.2.2. Factors Affecting Physical Instability

16.2.2.1. Solution Conditions. Vermeer and Norde have noted that pH has a strong influence on the antibody aggregation rate [30]. Table 16.2 outlines typical antibody-related degradation reactions and their mediation through use of appropriate solution conditions such as pH. Proteins in general are often stable against aggregation over narrow pH ranges, and may aggregate rapidly in solutions at pH values outside these ranges (Fig. 16.1). Examples include low-molecular-weight urokinase [31], relaxin [32], recombinant human granulocyte colony-stimulating factor (rhGCSF) [33], and insulin [34]. Both pH and salt can play a very important role for antibodies in solution, as they control physical properties such as conformational and colloidal stability as well as the chemical stability of the protein. In our studies, we found that different domains of the antibody have different pH sensitivities. It has also been shown that low pH (pH 4–6) and appropriate salt concentration reduce aggregation of antibodies in solution by affecting the noncovalent interactions between the antibody molecules. The Fab fragment is most sensitive to heat treatment, whereas the Fc fragment is most sensitive to decreasing pH. The structural transitions observed by DSC and CD studies in the whole IgG is the sum effect of those determined for the isolated Fab and Fc fragments [30,35].

The total charge on the protein is affected by the solution pH and electrostatic interactions within the antibody molecules and with the ions. Electrostatic interactions can affect protein stability in different ways. The amino acids in the antibody can be charged with increasing acidity or basicity of the solution. This can happen at a pH away from the isoelectric point (pI) of a protein [36]. The increasing charge repulsion between these charged groups of the antibody in such a solution can destabilize the folded or native state because of the high charge density. Thus, pH-induced

TABLE 16.2. Typical Antibody-Related Degradation Reactions in Therapeutic Formulations

Degradation	Causes	Possible Solutions
Noncovalent aggregation	Structural changes, colloidal stability, heat and other physical stress, sorbitol crystallization during freezing	pH, buffer salt, ionic additives, protein concentration, improving raw-material purity
Covalent aggregation	Disulfide rearrangement	pH, prevent association
Isomerization	pH ~ 5	pH, magnesium chloride
Deamidation	pH<5 and pH>6	pH
Clip formation	Proteases, metals, impurity	pH, chelation of metals, purity
Oxidation	Free radicals, reactive oxygen, metals, impurities, hydrogen peroxide	pH, free-radical and reactive oxygen scavengers, metal chelation
Surface denaturation	Low antibody concentration, binding to surfaces, hydrophobicity	Surfactants, protein concentration, pH

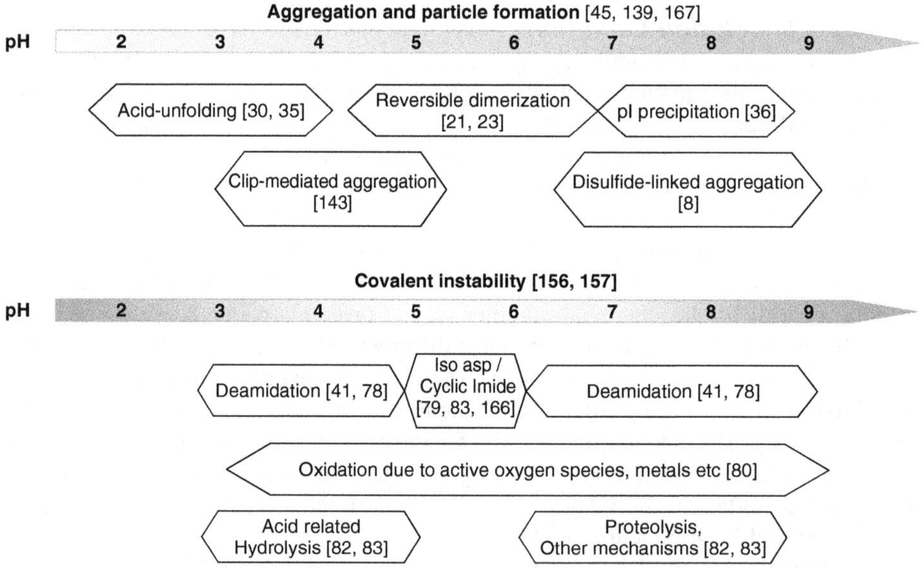

Figure 16.1. Different physical instabilities observed for antibodies at different pH levels along with some relevant references.

conformational unfolding can lead to a state of lower electrostatic free energy [14] Also, specific charge interactions, such as salt bridges, can affect antibody conformational stability. These salt bridges can stabilize the folded state and in some cases cause self-interaction [36,37]. When proteins possess both positively and negatively charged groups, the differential charge distribution on the surface of the antibody can cause protein–protein interactions, making assembly processes such as antibody aggregation energetically favorable [38].

16.2.2.2. Ligands and Cosolutes. Ligands and cosolutes are used in formulations to increase the physical stability of antibodies similar to other proteins. The Wyman linkage function applied by Timasheff [39] is commonly used to explain the effects of ligands and co-solutes in the formulation such as sucrose and salts. Through the Wyman linkage function, differential binding of ligand in two-state equilibrium will shift the equilibrium toward the state with the greatest binding. Binding of Zn^{2+} to human growth hormone or insulin is a very common example in which the free energy of unfolding for these proteins is increased and the native state of the protein is favored [40].

The Wyman linkage function can also be used to explain the effect of weakly interacting ligands (i.e., cosolutes), especially protein stabilizers such as sucrose and glycerol that are preferentially excluded from the surface of a protein molecule. In this case the degree of exclusion is proportional to the solvent-exposed surface area of the protein [39]. These cosolutes are excluded in the domain of the protein, and water takes its place in that domain, resulting in preferential hydration. Preferential exclusion can thus be interpreted as *negative* binding. During unfolding, protein surface area increases, leading to a greater degree of preferential exclusion. The net effect of greater negative binding to the unfolded state is to favor the native state.

Ligands and cosolutes that alter protein conformational stability also influence protein aggregate formation. For example, in the presence of polyanions, aggregation of acidic fibroblast growth factor [41] and native recombinant keratinocyte growth factor [42] are greatly inhibited. It has also been shown that the addition of weakly interacting, preferentially excluded solutes such as sucrose can inhibit aggregation of immunoglobulin light chains [17] and rhGCSF [15].

16.2.2.3. Salt Type and Concentration. Salts have complex effects on protein physical stability by modifying conformational stability and colloidal solubility, and may have different effects according to the surface charge of the protein or antibody. Salts bind to proteins, and destabilization of the protein can occur if the ions bind more strongly to the nonnative or unfolded state compared to the native state [43]. For example, the rate of aggregation of recombinant factor VIII SQ [44] was decreased in the presence of NaCl, while salt increased the aggregation rate for rhGCSF [33]. Moore et al. [21] found that salt increased dimer formation for IgG1 antibody.

Salts also modulate the strength of electrostatic interactions between the charged groups, at both intra and intermolecular levels. Thus, whereas intramolecular charge–charge interactions affect conformational stability, intermolecular electrostatic interactions affect degradation rates. The overall effect of salt on protein stability is a fine balance of multiple mechanisms by which salt interacts with protein molecules and affects protein–protein interactions. Because pH determines the type, total, and distribution of charges in a protein, salt binding effects may be strongly pH-dependent. These results suggest that protein stability can be increased by improving the coulombic interactions among charged groups on the protein surface [37].

16.2.2.4. Preservatives. Antimicrobial preservatives, such as benzyl alcohol, are often utilized in liquid protein and antibody formulations to prevent bacterial growth during storage. In particular, multidose formulations of proteins require effective preservatives to prevent microbial growth after opening and administration of the first dose. Preservatives are also required for certain drug delivery systems. However, preservatives can interact with proteins and often induce aggregation of protein in aqueous solution. For example, preservatives (e.g., phenol, *m*-cresol, benzyl alcohol) have been shown to induce aggregation of recombinant interleukin-1 receptor [45] and recombinant human interferon gamma (rhIFNγ) [46].

Preservatives can bind to the nonnative or unfolded states and make the molecule prone to aggregation. For example, it was observed that addition of benzyl alcohol perturbed the tertiary structure of rhINFγ without affecting its secondary structure, and the rate of rhINFγ aggregation increased as the molar ratio of benzyl alcohol to protein increased [46]. Also, preservatives reduced the apparent melting temperature of recombinant interleukin-1 receptor [45].

16.2.2.5. Surfactants. Nonionic surfactants are often utilized in protein and antibody formulations to prevent aggregation, surface denaturation, and adsorption during purification, filtration, transportation, freeze drying, spray-drying, and storage. *Surfactants* (surface-active agents) are amphiphilic molecules that tend to orient such that the exposure of the hydrophobic portion to the aqueous solution is minimized. For example, surfactants adsorb at air–water interfaces, forming a surface layer of surfactant molecules oriented so that only their hydrophilic ends are exposed to water. Such orientation and surface adsorption can also occur at solid–water interfaces such as those found in vials, syringes, tubing, and other containers. Protein molecules are also surface-active and adsorb at interfaces. Surface tension forces at interfaces can perturb protein structure, often resulting in aggregation. Surfactants inhibit interface-induced aggregation by limiting the extent of protein adsorption [47].

As in other cosolutes, differential binding of surfactants to native and unfolded states of protein influences the protein's conformational stability. In some cases surfactants still can kinetically inhibit protein aggregation at interfaces despite causing a reduction in the thermodynamic stability of the protein conformation. This often helps prevent adsorption of antibodies formulated at low concentrations to interfaces such as IV lines, bags, and storage containers. In addition, surfactants have been shown to act as chemical chaperones, increasing rates of protein refolding and thus reducing aggregation [48,49].

16.2.2.6. Freeze–Thaw-Related Damage. Freeze–thawing is a common stress to which a therapeutic protein can be exposed to during manufacturing, shipping, and storage. Therapeutic proteins are purposely frozen for storage of bulk drug substance or for storage of analytical samples. The final commercial product also may be frozen accidentally because of mishandling. This process may happen once or multiple times, with additional damage to the protein potentially occurring during each subsequent freeze–thaw cycle.

Protein aggregation during freeze–thawing has been attributed to partial unfolding of protein molecules caused by the perturbing conditions arising during the process [50]. Perturbation of the protein conformation can be caused by low temperature [51], freeze concentration of solutes [52], pH changes due to buffer crystallization [53], exposure of protein molecules to the ice–liquid interface, and/or adsorption to the container surface [54,55]. Additionally, freezing-induced increases in salt concentration can reduce intermolecular repulsion (i.e., colloidal stability) between protein molecules via charge shielding, resulting in more favorable intermolecular interactions that lead to aggregation [14].

During freezing, there is also an increase in the concentration of protein molecules [56] when ice crystallizes and phase separates from the remaining amorphous material. Additional excipients (e.g., salt, mannitol) may also crystallize. In aqueous solution, increased protein concentration typically corresponds to an increase in the rate of aggregation. Although freeze concentration of a protein would therefore be expected to promote aggregation, it has often been observed that increasing the initial concentration of a protein will actually reduce the percentage of aggregation occurring during freeze–thawing [50]. It has been suggested that increasing the initial protein concentration reduces the fraction of protein molecules that is exposed to the ice–liquid interface, resulting in reduced aggregation. Thus, the effect of changing the initial protein concentration on damage during freeze–thawing can be difficult to predict.

Numerous factors can affect the magnitude and nature of freezing-induced stresses, as well as the protein's responses to them. Among the most critical are the pH and ionic strength of the solution, because these factors, in general, modulate both the conformational and colloidal stability of protein [14] as well as a protein's response to physical stresses. In addition, the warming and cooling rates used during freeze–thaw can alter the degree of macroscopic freeze concentration, surface area of the ice–liquid interface, and duration of exposure of the protein to these potential stresses. The container material, geometry, and volume can also affect protein damage during freeze–thawing by modulating the effects of adsorption of protein molecules at the liquid–container interface, and by altering cooling and warming rates.

In some cases additional changes in physical state have been observed during frozen storage. Piedmonte et al. [57] observed protein aggregation during storage of sorbitol-containing formulations at −30°C. The aggregation correlated with DSC melts that are characteristic of crystalline substances and suggest that the sorbitol crystallizes over time in the formulation. During freezing, the excipient must remain in the same phase as the protein to provide protein stability. By crystallizing, the sorbitol is phase-separated from the protein, which leads to protein aggregation.

16.2.2.7. Lyophilization-Induced Stresses. Typically, it is desirable to formulate therapeutic proteins in liquid formulations for ease of administration and lower cost of production. However, proteins in liquid formulations are generally at a greater risk of and physicochemical degradation. Liquid formulations are also less robust with respect to stresses experienced during shipping and handling. If the desired shelf life cannot be achieved in a liquid formulation, lyophilization is often the alternative. Lyophilized proteins are typically less susceptible to physicochemical degradation because of the scarcity of water and the greatly reduced mobility of molecules in the dried state [9,58]. Spray drying is also a potential technology for producing fine protein powders for inhalation drug delivery [59,60].

Although biopharmaceuticals are generally more stable in the dried state, it is well known that lyophilization and spray-drying processes themselves can be greatly damaging to proteins. Lyophilization involves two major steps; freezing of a protein solution and drying of the frozen matrix under vacuum. The freezing step can potentially destabilize or denature proteins by a variety of mechanisms, including cold denaturation, concentration and pH effects, and ice–liquid interfacial effects. The drying step can potentially damage proteins by disruption and/or removal of the hydrogen-bonding network of water molecules. These dehydration-induced stresses are also present during spray drying. Moreover, gas–liquid interface and exposure to high temperatures used during spray drying can induce additional damage [61].

The molecular mobility of amorphous pharmaceutical materials is known to be a key factor in determining their stability, reactivity, and physicochemical properties [62]. Molecular motions in amorphous systems are usually characterized by measuring the time dependence of some bulk property such as enthalpy or volume, or by using spectroscopic techniques to monitor the motions of particular functional groups. Usually such molecular mobility is quantified using relaxation time constants. Chang et al. [63] found that stability correlated best with the preservation of native structure for sucrose-based formulations, but with a trehalose-based formulation, neither structural relaxation time nor extent of native structure was predictive of stability. However, it is possible that the β relaxations rather than the α relaxation are critical to the stability. Plasticizers like glycerol may decrease τ for "α motion" but increase τ for "β motion" and stabilize proteins [64].

16.2.2.8. Interfacial Stresses. Many packaging components for parenteral products (e.g., glass syringes, vial stoppers) require the use of some form of lubrication for their processability and functionality. Siliconization of packaging components (such as glass, elastomeric closures, plastic, and metal) places a thin lubricating film on the surface of the components. Silicone oils have very low surface tension (20–25 mJ/m^2) and good wetting properties that permit the oil to spread readily on most solid surfaces [29,65,66]. Silicone oil is applied to cartridges and barrels of plastic and glass syringes to facilitate smooth and easy movement of plungers within the barrels. Further, silicone oil is applied to exterior surfaces of hypodermic needles to reduce the frictional drag and pain when the needles pass through the tissue [66]. Given the wide use of silicone oil coating on containers and closures, the potential for silicone oil-induced aggregation of therapeutic proteins is a major issue for product development.

Contamination of therapeutic protein formulations with silicone oil was first reported in the 1980s following observation of elevated blood glucose levels in patients administered with "cloudy" insulin from plastic syringes [65]. Analysis of the insulin formulations revealed that the silicone oil was causing protein particle formation. Although this problem has been longstanding, it has been described infrequently in the literature. More recently, Jones et al. [29] showed that silicone oil at a concentration of 0.5% (w/v) induced aggregation of four model proteins (ribonuclease A, lysozyme, bovine serum albumin, and concanavalin A) during heating. However, there is a lack of published reports on silicone oil effects on aggregation of monoclonal antibodies, for which hundreds of new products are in various stages of development and marketing.

It has been suggested that problems associated with silicone oil–containing syringes may easily be overcome by replacement with silicone oil–free syringes or by substituting silicone oil with other substances. Unfortunately, the suggested replacements for silicone oil such as Teflon® may also cause problems. It is important to recognize that the hydrophobic surfaces of Teflon have also been implicated in causing protein aggregation due to adsorption of protein molecules at the Teflon–solvent interface [67]. Therefore, although other alternatives to silicone oil are under exploration, work to understand how silicone oil induces protein aggregation is critical. This will be of great benefit in suggesting rationale and practical strategies to develop formulations that prevent this route of product degradation.

16.2.2.9. Consequences of Physical Instability of Antibodies on Safety, Efficacy, and Immunogenicity. The capacity of protein aggregates to enhance immune responses to the native form of the protein has been documented since the 1950s. Protein antigens presented in a highly arrayed structure, such as might be found in large native-like aggregate species, are highly potent in inducing immune response even in the absence of humoral immunity or T-cell help. It has been shown that the potency to induce immune response may be related to the ability of multivalent protein species capable of crosslinking B-cell receptor, which activates other immune pathways such as activation of B cells and class II major histocompatability factor (MHC) [68].

The formation of soluble aggregates in a protein formulation can have a significant effect on the pharmacokinetics and immunogenicity of the protein [68–71]. Insoluble aggregates or particles are considered undesirable for administration to patients. There are studies in the literature that directly correlate aggregate formation to loss of activity [72,73]. Several studies also show that administration of aggregated protein (e.g., aggregates of human serum albumin, human growth hormone) results in increased immune response and production of antibodies in the body against the therapeutic protein, which may further result in autoimmunity [68,74,75], that is, immune responses against the body's native protein. For some proteins, improper formulation or incompatibility of the protein with certain physiological conditions can cause injection-site reactions as well as formation of soluble aggregates at the site of injection [76,77]. The latter issue can lead to slow dissolution of the drug from the site of injection and thereby altered pharmacokinetics.

Thus, it is considered critical to minimize aggregation and particulation in the drug product.

16.2.3. Covalent Degradation

The covalent stability of antibodies is a complex function of solution conditions. Antibodies are known to undergo deamidation, isomerization, oxidation, proteolysis, and other covalent modifications, as a function of pH, temperature, and excipients. Table 16.2 outlines typical antibody-related degradation and its mediation through use of appropriate solution conditions. Deamidation has been identified as a source of charge heterogeneity of antibodies [41,78]. Deamidation can occur at asparagine or glutamine residues, resulting in a charge variant (aspartate or glutamate), isoform (cyclic imide or succinate), or cleavage product. Isomerization and cyclization can occur after an asparagine or aspartate residue, particularly if the subsequent residue is sterically small (e.g., glycine or serine). Both the charged isoforms and isoaspartate have been identified and characterized in antibody degradation studies [79]. Oxidation of methionine residues has also been observed in antibodies exposed to elevated temperature or intense light [80]. In addition to methionine, other residues that are sensitive to oxidation include cysteine, histidine, tryptophan, and tyrosine. Photooxidation of tryptophan, tyrosine, phenylalanine, and cysteines can lead to free-radical formation, crosslinking, and yellowing effects [81]. Proteolysis is another critical instability of antibodies that occurs in liquid formulation conditions [82,83]. Peptide mapping and mass spectral analysis can be utilized to detect and monitor a variety of covalent modifications in antibodies [84,85] as a function of formulation solution conditions [86,87]. In addition, disulfide bond scrambling, cleavages, covalent dimer formation, and other chemical or covalent degradation reactions have been observed in antibodies as a function of storage time. In liquid formulations, both IgG1 and IgG2 antibodies are prone to clipping via nonenzymatic fragmentation [83,88]. Site-specific fragmentation of the peptide backbone and disulfide bond linkage utilizing capillary electrophoresis and matrix-assisted laser desorption/ionization (MALDI) mass spectrometry has identified site-specific fragmentation of the peptide backbone and disulfide bond linkage. The resulting species include one-armed complexes missing one light chain or one Fab arm, in addition to free Fab or free heavy and light chain [82]. The cleavage sites are located predominantly in the hinge region of the heavy chain. The nature of the site-specific fragmentation in the case of hinge cleavage may be driven by molecular kinetics and not by protease contaminants [83].

16.3. METHODS

16.3.1. Physical Characterization

Physical characterization of the small, soluble aggregates of monoclonal antibodies has commonly been accomplished by established techniques such as size exclusion chromatography (SEC) [89–91] and gel electrophoresis [91,92]. Further characterization of antibodies has also been accomplished by orthogonal methods, including laser light scattering [21,93], analytical ultracentrifugation [89–91], and field flow fractionation [91,94]. However, characterization of very large aggregates of monoclonal antibodies has been hampered by the lack of sufficient analytical methods that characterize small populations of subvisible and visible particulates well. Several techniques have been explored, including light obscuration [95], microscopy [96,97], electrical sensing zone instruments (Coulter counting) [98], dynamic light scattering [95], and turbidity/opalescence [95], although each has its own strengths and weaknesses. Aggregation propensity has been screened by examining thermal unfolding by differential scanning calorimetry [99–103] or Fourier transform infrared (FTIR) spectroscopy [101,102] and by determining second virial coefficients by techniques such as self-interaction chromatography [104] or light scattering [33].

Physical characterization of the formation of soluble aggregates such as dimers and small oligomers is commonly achieved by SEC. Gel electrophoresis is also used [92], but SEC has become the method of choice for high-throughput quantification of antibody aggregation. The molecular weight of early-eluting SEC peaks can be identified by online multiangle light scattering [21,93]. Light-scattering detection with SEC analysis also improves the ability to detect small amounts of large aggregates that may not have measurable UV signal, but because of their large size, scatter light [105]. Confirmation of the molecular weight by an orthogonal technique is important, as molecular mass standards can lead to erroneous results. For example, a dimer can be misidentified as a trimer [21], or there can be multiple peaks with the same molecular weight such as two-dimer peaks. There may even be an altered form or chemical modification to the monomer that elutes with a larger hydrodynamic radius [106].

However, it is also possible that sample preparation and SEC analysis itself may alter the distribution of monomeric and larger species. High-concentration antibodies are typically diluted prior to injection, which can allow dissociation of reversible aggregates. Analytical results would then vary depending on storage time and temperature between sample preparation and analysis. Methods where samples are injected without dilution can address these issues, but high-protein loads can lead to other problems. Injection into the mobile phase can also alter the aggregate content either by dissociation on dilution or by a change in solution conditions. Appropriate mobile-phase selection can also reduce column interactions resulting in very different elution profiles [107].

Sedimentation velocity analytical ultracentrifugation (SV-AUC) may be used to verify SEC analysis and characterize distributions in antibody formulations at 1–2 mg/mL up to possibly 10 mg/mL with minimal concerns over nonideality effects [89]. Measurements can be obtained at high concentrations (~40–50 mg/mL), but interpretation of the results is hindered by nonideal thermodynamic and hydrodynamic behavior [89]. The SV-AUC method can be used to aid in the selection of SEC mobile phases that give the most accurate aggregation results [89] and to quantitate aggregates present [108]. However, precision and accuracy of the aggregate levels determined by AUC is dependent on factors such as the quality of centerpieces used [108] and instrument condition [90].

The resolution of larger species by SEC is limited by column selection. For example, resolution on Tosoh Biosep G3000-type columns is limited to ~0.5 MDa (trimer to tetramer) or ~20 nm. The upper limit on G4000-type columns is 7 MDa (<~50mer), with an approximate hydrodynamic diameter of 50 nm. Flow field flow fractionation techniques with light-scattering detection potentially increase the detectable size range of antibody aggregates to ~50 μm [90,94]; however, care must be taken to investigate the transition between normal and steric mode [94]. Both flow field flow fractionation (F4; also abbreviated FFFF) [90] and asymmetric flow field flow fractionation (AF4) [91,94,109] have been used to characterize antibody aggregation. Litzen et al. [109] demonstrated improved resolution of aggregate species by AF4 as compared to SEC. Gabrielson et al. [91] compared the levels of aggregate detected by SEC, AF4, and SV-AUC. In this study, SEC significantly underestimated the percent of soluble aggregate detected. Sedimentation velocity AUC detected the most, and AF4 detected slightly less soluble aggregate than did SV-AUC for unstressed samples. Dimer was purified by SEC and analyzed. Reanalysis by SEC within 1 h of collection showed reduced dimer with increased monomer and trimer. Sedimentation velocity AUC also revealed some polydispersity, although much less dramatic than the SEC results [91]. Both SV-AUC and AF4 are powerful orthogonal techniques in comparison to SEC, but require specialized training and instrumentation such that they are unlikely to replace SEC as the workhorse.

As aggregates become extremely large, they may also be detected by light obscuration techniques as described in USP<788> [110]. The USP method specifies the characterization of 10- and 25-μm particulates. Depending on instrumentation, light obscuration may also detect particulates as small as 1 μm and up to 400 μm. The USP limits for particulates were initially focused on foreign particulate matter and not specifically for protein or antibody solutions that may contain low levels of product-related proteinaceous particulates. The increased viscosities of high-concentration antibody formulations may also cause sample analysis issues [110].

Methods to characterize and monitor visible proteinaceous particulates are limited; USP788 describes a filter test to manually count all particulate matter collected on a filter after failure of the light obscuration test or if light obscuration is not suitable. Results from the filter method can differ significantly from light obscuration results, possibly because of the fragile nature of proteinaceous particulates. The filter method can also be used when visible particulates are observed. After particulates are collected on a filter, analysis may be performed to identify observed particulates by infrared spectroscopy or energy-dispersive spectroscopy. We used digital quantitative microscopy to determine the size distribution of particulate matter collected on a filter. A sample result is shown for stressed samples of a monoclonal antibody formulation with and without polysorbate 20 (Fig. 16.2). The addition of polysorbate reduces the number of particulates detected. A further modification to the USP technique was also reported, in which proteinaceous particulates were identified by staining with protein-specific dyes [96]. A drawback of such methods is that they are highly labor-intensive. Fluorescence microscopy after staining antibody solutions with Nile Red has also been used to detect particles ~1–10 μm in solution without filtration or dilution [111]. Automated methods for characterization of visible particulates are emerging, such as flow microscopy with digital image capture [97]. These methods store digital images for all particles detected, allowing further characterization of morphology

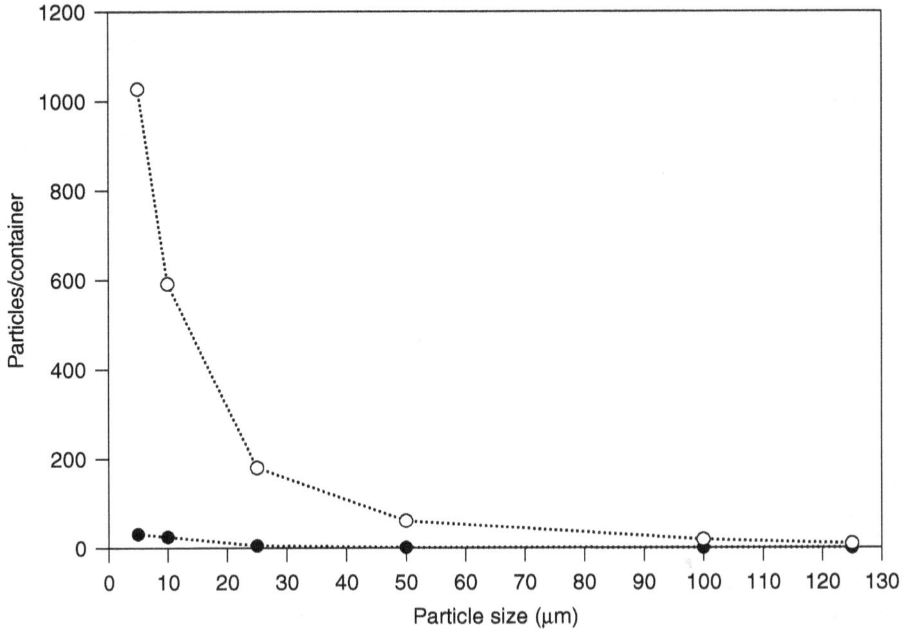

Figure 16.2. Particles counted as a function of size for a stressed monoclonal antibody formulation without polysorbate (○) and with polysorbate (•). Solutions were collected on a 0.2-μm white polycarbonate membrane filter. Particles with a dimension >10 μm were counted using a Clemex Vision PE system with a Nikon E600 microscope. Particle size is equivalent circular diameter. (Unpublished data.)

and size distributions, offering significant advantages over indirect detection by light obscuration [97]. This methodology is currently material- and labor-intensive, but is an excellent technique for orthogonal characterization of higher-throughput methods.

Multiple techniques are often necessary to characterize antibody aggregation. Demeule et al. [111] characterized antibody aggregates under two solution conditions (0.1% acetic acid with 50 mM magnesium chloride, and phosphate buffer) by light scattering, fluorescence microscopy, and AF4. Under stable conditions for the antibody no aggregates were observed by fluorescence microscopy with Nile Red, while AF4 showed a distribution of monomer (88%), low-molecular-weight aggregates (2%), and high-molecular-weight aggregate (10%) with a size $\sim 1.5 \times 10^6$ Da. Under unstable conditions, many spherical aggregates with a diameter ~3 μm were seen by fluorescence microscopy. These aggregates were fragile and easily disrupted by the crossflow during AF4 analysis. The large aggregates eluted in the steric mode under a mild separation method with diameter > 200 nm.

The presence of larger aggregates can also be monitored qualitatively by dynamic light scattering (DLS) [95,100,112] or opalescence/turbidity [87,95,112]. Kiese et al.

[113] monitored the timecourse of aggregation time by light obscuration, DLS, and turbidity of 2-mg/mL MAb samples subjected to mechanical stress (stirring or shaking). Increases in 10- and 25-μm subvisible particles were detected by light obscuration in samples subjected to mechanical stress. Dynamic light scattering also detected the presence of a second peak in some samples, although less reliably than light obscuration. Stressed samples became more turbid when monitored at 350 nm and 550 nm and by visual comparison with opalescent reference standards. Monitoring at 350 nm was found to be more sensitive than 550 nm, although both were qualitatively similar. Ahrer et al. [100] detected trace large aggregates by DLS and SEC with light scattering. Although DLS can be used to detect the presence of larger aggregates, care must be taken to avoid overinterpretation of results from heterogeneous samples.

Early formulation development requires assays that can quickly delineate more optimal solution conditions. Thermal stability is often used to try to predict physical stability [99,101,102,114] for long-term storage or physical stresses encountered during formulation and storage; DSC has historically been the method of choice [99]. More recently, FTIR has been compared to DSC [101,102] and evaluated as a way to predict stability at high concentrations [26,101].

Besides thermal stability, mechanical stress may also be used to evaluate aggregation propensity during formulation development. Mechanical stresses with air–liquid interfacial phenomena are frequently used [95,115], although the surface–liquid interface may also induce aggregation. Mahler et al. tested stirring in special cone-shaped vials and agitation by horizontal shaking [95], and found that stirring induced higher levels of smaller aggregates. Levine et al. [115] stressed samples by placing half-filled vials on an orbital shaker, maximizing the air–liquid interface. They also correlated lower-surface-tension measurements to increased susceptibility of monoclonal antibodies to precipitation.

Second virial coefficients (B22) can be used to determine solution conditions with increased protein solubility [116], and positive values have been correlated to decreased protein aggregation propensity [33]. Traditional methods to determine second virial coefficients, including static light scattering and membrane osmometry, are usually material- and labor-intensive [104,117]. Improved methods include self-interaction chromatography [104,118], dual-detector cells for light-scattering intensity and concentration by SEC [119] and composition-gradient light scattering [120]. The ultrasonic storage modulus was also found to correlate with second virial coefficients and can be used to monitor protein–protein interactions at high concentration [121]. However second virial coefficients do not always predict conditions of optimal formulation stability. In a study of IgG2 aggregation at 37°C, no correlation was found between aggregation and the second virial coefficient [119]. In fact, formulation conditions at pH 7.4 and 5.4 that did not show significant aggregation had more negative second virial coefficients than did those measured in pH 4.0 formulations, which did aggregate significantly. Interestingly, near-UV spectra did show differences as a function of pH that correlated with aggregation propensity.

16.3.2. Covalent Characterization

More recent advancements in analytical methodologies have advanced the covalent characterization of intact recombinant antibodies [122,123]. Typically, analytical methods such as cation exchange (CEX) chromatography, isoelectric focusing (IEF), capillary isoelectric focusing (CIEF), and capillary zone electrophoresis (CZE) are utilized to monitor charge variants of antibodies that arise from deamidation, glycosylation, and C-terminal lysine processing [79,124–131]. Likewise, hydrophobic interaction chromatography (HIC) is effective in separating oxidized or isomerized degradants [80,132]. Most of these analytical methods are incompatible with online mass spectral analysis because of the salts in the mobile phase. Under certain mobile-phase conditions, capillary electrophoresis can be utilized in tandem with mass spectrometry for identification of covalent heterogeneity and degradants [133].

Early attempts at reversed-phase analysis of intact antibodies had limited success in overcoming issues with poor recovery and column fouling [134]. Perfusion chromatography or HIC in reversed-phase mode have resolved light and heavy chains of antibodies, with insufficient separation capabilities for degradant species [131,134]. Analysis of antibody fragments and peptide maps by reversed-phase chromatography has been a reliable approach for full covalent characterization of antibodies, albeit time-consuming in sample preparation [131,134–138].

Dillon et al. have optimized reversed-phase chromatography conditions (Fig. 16.3) for intact recombinant antibody analysis with improved recovery, resolution, reproducibility, and column lifetime [122,123]. The key factors for improved chromatography are high column temperatures (70°C–80°C), mobile phases with high eluotropic strength solvents, and long alkyl chains of the stationary phase, such as the Zorbax stable bond [122] and Polaris ether column [123]. Reversed-phase analysis of intact antibodies is also capable of separating degradants such as cleavage products [122] and disulfide isoforms of the IgG2 subclass [123] (Fig. 16.3).

16.4. LIQUID FORMULATION STRATEGIES

16.4.1. Physical Stabilization

Several formulation papers report that antibodies in general are most stable at pH 4.5–5.5, and that aggregation increases dramatically near neutral pH or near the isoelectric point [21,103,131]. In certain cases, antibodies can form a reversible dimer under physiological solution conditions, and the self-association is driven by hydrophobic interactions [21]. In other circumstances, antibodies can self-associate to form primarily active covalent complexes [22]. The relative rates of clipping, irreversible aggregation, and reversible dimer formation are highly dependent on the solution pH and type of accelerated condition, such as elevated temperature or multiple freeze–thaw cycles [139]. However, in some cases antibody stability is not well correlated to protein concentration, buffer concentration, salt concentration, or agitation [139].

Physical stability of monoclonal antibodies can be affected by pH, and by the choice of buffer, stabilizers, and surfactants. To a lesser degree, selection of the buffer

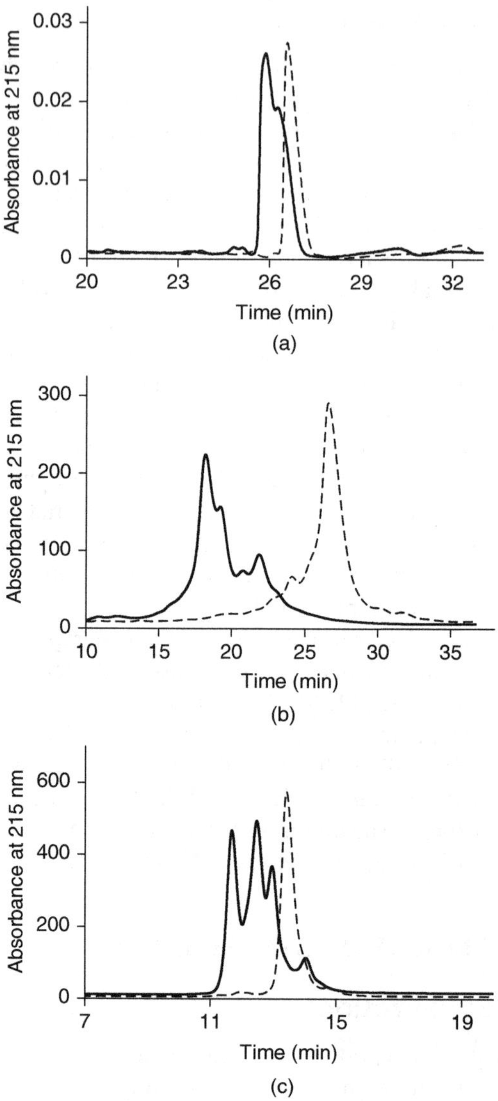

Figure 16.3. Analysis of an intact MAb expressed as an IgG1 (red) and IgG2 (blue) using (a) RP-HPLC, (b) CEX, and (c) CE-SDS. [More details are given in Wypych et al. [176] and Dillon et al. [123].

and buffer concentration at a given pH can also impact physical stability [140]. Stabilizing excipients include salts, polyols, and amino acids. Surfactants can minimize the formation of both soluble and insoluble aggregates due to surface interaction events, including agitation, and freeze–thaw. As concentrations of mAb formulations increase, drivers to identify formulations that optimize physical stability also increase because of concerns of potential immunogenicity of aggregates.

Criteria for the selection of pH include the pI of the MAb, structural stability, and sensitivity to other covalent reactions that could lead to aggregation. Formulating at a pH near the isoelectric point (pI) is undesirable as solubility may be reduced or opalescence/turbidity increased. If the pI of the antibody is near physiogical pH, formulation development activities may also need to include evaluation of physical behavior at physiological pH [141]. Although DSC can be used to explore the structural stability of monoclonal antibodies as a function of pH during preformulation studies, care must be taken in using the information as the primary criteria for pH selection, as thermodynamic stability does not always correlate with physical stability [114]. Covalent stability may also impact physical/structural stability. Covalent modifications could slightly alter the structural stability of the MAb and lead to aggregation [142]. Formation of clips due to acid hydrolysis can also lead to aggregation at elevated temperatures. Acid-induced clipping does not manifest at refrigerated temperatures; therefore, it is important to verify trends observed at high temperatures (>37°C) [143].

Early monoclonal antibody therapeutics were often formulated near neutral pH at relatively low concentrations. More recently, antibody therapeutics with liquid formulations has been trending toward pH ≤ 6, and is provided at increasing concentrations. Many antibodies have optimal physical stability near pH 5.5. The dimer dissociation kinetics for an antibody to VEGF were examined between pH 6.5 and 8.5, and dimer formation was found to be slowest at pH 6.5, where dissociation was fastest [21], while insignificant dimer formation occurred at pH 5.5 at 30 mg/mL. Incubation of a chimeric antibody at 60°C resulted in minimal amounts of higher-molecular-weight species at pH > 5.0, while freeze–thaw stability was optimal at pH < 5.5 or pH > 8.0 [139]. Significant precipitation was also observed in antibody solutions at pH < 4 [86]. Selection of pH can be the most critical parameter during formulation development. For example, while aggregation was not significantly influenced by buffer concentration, antibody concentration, or sodium chloride concentration, the dominant factor was pH [139] at all temperatures studied (from 30°C to 50°C).

Once the optimal pH range has been determined, stabilizing excipients can be used to further minimize aggregation. Salts have historically been used as the primary choice for tonicity modifiers in MAb formulations; however, addition of sodium chloride at pH 7.5 actually increased the rate of dimer formation for MAb VEGF [21]. Polyols and sugars and more recently amino acids have been recognized as desirable stabilizing agents. Addition of sugars (lactose, sucrose, and mannitol) or amino acids (L-arginine, L-lysine, and L-alanine) prevented freeze–thaw-induced aggregation [139]. Histidine was also found to have cryoprotectant properties comparable to sucrose when used at concentrations above 50 mM [140]. At accelerated temperatures, histidine combined with arginine outperformed sucrose in preventing aggregation [140]. Histidine also minimized aggregation as compared to citrate and succinate at pH 6 [140]. Additionally, increasing concentrations of histidine reduced the viscosity of a 150-mg/mL antibody formulation. However, there are concerns with compatibility of high concentrations of histidine HCl with stainless-steel containers, as leaching of iron, with potential coloration of protein product due to degradation, is possible [140,144,145]. Addition of sodium chloride and other salts to high-concentration antibody formulations reduced the solution viscosity [146]. Addition of low levels

(0.1% and 1%) polyvinylpyrrolidone significantly reduced heat-induced aggregation. However, levels greater than 1% led to increased aggregation and even precipitation [147]. A formulation that was stable when stored at −20°C or subjected to multiple freeze–thaws was observed to aggregate on long-term storage at −30°C, which was attributed to crystallization of sorbitol [57].

Addition of surfactants to antibody formulations can minimize the formation of both soluble and insoluble aggregates. Surfactants primarily provide protection against surface denaturation phenomena that occur during freeze–thaw or agitation stresses. Polysorbate 80 and pluronic F68 effectively prevented antibody precipitation due to agitation stresses designed to simulate shipping, and performed better than did Brij 700, a less active surfactant [115]. Addition of 0.01% polysorbate 80 to 2-mg/mL antibody formulations in phosphate buffered saline effectively prevented formation of subvisible particles and minimized turbidity due to stirring or shaking [95]. However, surfactants may not always be effective at preventing freeze–thaw-induced aggregation, especially if the aggregation is not driven by surface phenomenon as seen where addition of polysorbate 80 did not prevent freeze–thaw-induced aggregation [139].

Although counterintuitive, increasing the protein concentration can actually help stabilize the formulation to surface denaturation-induced aggregation, as the ratio of protein on the surface to protein in solution is greater at lower concentrations [148,149]. Also, as antibody concentrations increase, the solutions become more opalescent, which can be an indicator of self-association. For an IgG1, this opalescent appearance was found to be a consequence of Rayleigh scattering and not self-association or physical instability at room temperature [112]. However, at 5°C, the turbidity was nonlinear with concentration and indicated possible reversible noncovalent association [112]. Opalescence may be a natural characteristic of high-concentration antibody formulations, but further work to determine effects of formulation conditions is necessary, particularly at refrigeration temperatures.

16.4.2. Container–Closure Issues

The impact of primary packaging components on stability should be considered during formulation development activities. The protein or an excipient could leach materials from the primary packaging system. Examples of leachables observed in therapeutic protein products that led to unanticipated degradation include metal ions or vulcanizing agents from rubber stoppers, barium from glass vials, and tungsten oxide from prefilled syringes [150]. These issues are not limited to the primary packaging, as sometimes even short-term exposure to materials such as silicone oil during administration with disposable syringes can also lead to protein instability [150]. Silicone oil exposure can lead to aggregation [29] and is commonly used in both disposable and prefilled syringes. Stoppers are often siliconized for machinability, but crosslinked siliconization coating processes can be utilized to minimize exposure of the protein to silicone oil and other agents in the rubber stoppers. Formulation development can address such problems, for example, addition of EDTA to prevent metalloprotease activation by metal ions leached from stoppers [150]. Plastic vials have also been examined as potential alternatives to traditional glass vials for protein therapeutic agents [151].

16.5. HIGH CONCENTRATION LIQUID STRATEGIES

16.5.1. Formulation Issues

Development of formulations for high-concentration protein drugs can be quite challenging. Some proteins pose outright solubility issues at higher concentrations. Even when solubility is not limiting, high-concentration formulations place additional challenges on other areas, including manufacturing, stability, analytical, and delivery issues. Concentration and concentration-dependent degradation routes, such as aggregation, and particle formation may become much more critical at higher concentrations. Also, the high viscosity of these formulations complicates injectability of the formulation [146]. Additional analytical techniques may be required to allow for direct measurement in the formulation without substantial dilution of the protein [38]. Operationally, solubility of proteins could be described by the maximum amount of protein in the presence of cosolutes whereby the solution remains visibly clear (i.e., does not show protein precipitates, crystals, or gels), or does not sediment at 30,000*g* centrifugation for 30 min [152]. The dependence of protein solubility on ionic strength, salt form, pH, temperature, and certain excipients has been mechanistically explained by changes in bulk water surface tension and protein binding to water and ions versus self-association. Net charge of the antibody is a major determinant of solubility [153]. Binding of proteins to specific excipients or salts influences solubility through changes in protein conformation or masking of certain amino acids involved in self-interaction.

Common measures to minimize aggregation of antibodies in high-concentration formulation include ensuring raw-materials purity and product homogeneity; finding the optimal pH range for maximizing physical stability; adding excipients such as sugars, ionic additives, amino acids, and surfactants; and finding the right protein concentration range required for delivery of therapeutic doses in acceptable volumes, with minimal degradation. The relationship between concentration and aggregate formation depends on the mechanism of aggregation. As the protein concentration increases, the fraction of the total volume occupied by the protein increases. The resulting decrease in the effective volume available to the protein yields a higher apparent protein concentration that, in turn, favors self-association. This nonideal behavior of increasing the apparent theormodynamic association constant with increasing protein concentration may be shelf-life-limiting, especially when the resulting aggregates are irreversible.

In some cases it may be possible to take advantage of the requirement for a high protein concentration. For example, Gokarn et al. [154] explored a novel approach of controlling the formulation pH by harnessing the ability of MAbs to "self-buffer." Buffer capacities of four representative IgG2 molecules (designated "MAb1 through MAb4") were measured in the pH 4–6 range. The buffer capacity results indicated that the MAbs possessed a significant amount of buffer capacity, which increased linearly with concentration. The long-term stability of the self-buffered liquid formulation was comparable to the conventionally buffered formulations. No significant change in pH was observed in the self-buffered formulation after 12 months of storage at 37°C and 4°C. The 60-mg/mL self-buffered formulation was also observed to be stable to freeze–thaw cycling.

16.5.2. Viscosity

The majority of marketed MAb products are administered by intravenous infusion because of the requirement for high administered doses [155]. Although in certain indications this route of administration may be acceptable, in other cases it may present significant patient compliance, acceptability, and cost of treatment issues that render it undesirable. The limitations in the injection volumes required for subcutaneous (SC) administration (poor tolerance of volumes in excess of about 1.5 mL) and the large dose requirements of antibody treatments (in excess of 100 mg, typically 100–400 mg) necessitate their formulation as high-concentration solutions or suspensions. However, development of high-concentration protein formulations poses significant challenges, primarily because of their susceptibility toward concentration-dependent aggregation and their tendency to form viscous solutions, due to their high potential of intermolecular interactions and macromolecular crowding in solution. Macromolecular crowding (i.e., excluded volume effects) can impact protein physical properties such as viscosity, which can have a major impact on the ability to manufacture high-protein-concentration formulations as well as on the ability to administer the protein drug by injection. In general, the viscosity of a macromolecule in solution is dependent on interaction of two, three, or more protein molecules. In the case of significant protein self-association that results in formation of soluble aggregates, such interactions will lead to large increases in viscosity as a function of concentration.

These issues have motivated antibody innovators and drug delivery providers to evaluate strategies to enable SC delivery of MAbs. Some of these rely on improving solution stability and reducing the viscosity of the high-concentration protein formulations. Liu et al. [146] have proposed that manipulation of the solution conditions, such as ionic strength, buffer species, and pH, lead to a significant viscosity reduction of high-concentration solutions of recombinant anti-IgE MAb (rhuMAb E25 and E26).

16.5.3. Covalent Stabilization

Formulation approaches to covalent stabilization have been applied with moderate success. Often, the formulation pH is fixed according to solubility or aggregation constraints, and covalent stability may be compromised under such solution conditions [156]. If excipient strategies for covalent stabilization are insufficient, a lyophilized formulation may be a more effective approach to minimize covalent degradation under solution conditions constrained by physical stability requirements [157]. If no such constraints in physical stability or solubility exist, then the simplest approach to minimizing covalent degradation is to optimize the formulation pH. Oxidation, disulfide exchange, and deamidation can be minimized by lowering the pH. Conversely, increasing the pH can be an effective formulation strategy to control isomerization and hydrolysis, which are problematic under acidic solution conditions [9].

In addition to pH optimization, specific additives or formulation conditions have been found to be effective at controlling certain degradation reactions. General principles of formulation strategy for each specific mechanism of degradation are depicted in Figure 16.1. For oxidation issues, the addition of antioxidants, oxygen scavengers, or free-radical scavengers can inhibit degradation [80]. If the level of oxidizing agent is rate-limiting, free methionine in molar concentration exceeding that of the therapeutic antibody can serve as a scavenger to preferentially oxidize, instead of the antibody. Chelators, such as EDTA or citrate, may be effective at inhibiting metal-induced oxidation. A nitrogen overlay during drug product filling can prevent the oxidation that normally occurs by removing the oxygen in the air within the headspace that equilibrates with the drug product solution during the shelf life of the drug. Note that impurities in certain excipients, such as polysorbate [158] or benzyl alcohol [159,160], may contain free radicals from peroxides that can catalyze oxidation reactions, and proper sourcing and handling of these high-risk excipients can impact stability. Similarly, disulfide exchange reactions can be controlled by adding thiol scavengers and antioxidants or eliminating oxidizing excipients and trace-metal contaminants that can catalyze disulfide exchange [161,162]. Photooxidation is another related covalent degradation issue common to therapeutic antibodies [163]. Limiting UV exposure during purification and protecting the drug substance and drug product from light during storage, filling, distribution, and patient use help minimize photooxidation [81]. However, dark reactions continue to propagate after even brief exposure to intense light, such as that of automated visual inspection.

Deamidation is also a common covalent degradation mechanism of antibodies [41,78]. If pH reduction is ineffective or restricted because of other stability concerns, then buffer optimization may be an effective approach. Certain amine buffers, such as Tris, ammonium, or imidazole, can inhibit the deamidation reaction, and conversely, phosphate buffer can increase deamidation [164]. Excipients can also influence the rate of deamidation in both liquid and lyophilized states [165]. Under acidic formulation conditions, isomerization and hydrolysis at aspartic acid residues can occur [79,83], and the addition of magnesium has been found to inhibit the isomerization reaction [166]. Other forms of protease-mediated cleavage reactions can be inhibited by the addition of chelators to antagonize metalloproteases, protease inhibitors, pH optimization, and improved purification processes to remove the protease contaminants. Increased sucrose can also promote covalent stability by preferential hydration.

While liquid antibody formulations are less expensive, and generally easier to prepare for administration than are alternative formulation approaches, liquid antibody formulations are prone to oxidation [80], deamidation [41,78], aggregation [21,45,139,167], and fragmentation [82,83] as discussed above. In each of these events, water is the common culprit; water mediates electron transfer during oxidation and deamidation events and is also critical for fragmentation. The thermodynamic stresses that lead to protein aggregation result from exposure of hydrophobic protein surfaces to water and trying to find a lower-energy state by nonnative protein–protein interactions. Thus, removal of water can stabilize antibody-based drugs. Lyophilization or introduction into hydrophobic polymer systems can reduce the impact of water on antibody drug formulations.

16.6. LYOPHILIZED FORMULATION STRATEGIES

16.6.1. Stability and Reconstitution Issues of High-Concentration Lyophilized Formulations

Freeze drying or lyophilization is a widely used method for protein therapeutics to improve storage stability [9]. Lyophilization allows storage at high temperatures, and reduces conformational mobility due to slowed molecular motion resulting in decreased degradation of proteins. A lyophilized formulation may be the only practical choice for stabilizing many therapeutic proteins prone to degradation.

Antibodies may be damaged as a result of lyophilization as demonstrated by extensive formation of insoluble aggregates [54,149,168]. Improved native-like structure and a reduction in antibody aggregation can be obtained by incorporation of a carbohydrate excipient in sufficient quantities to fulfill the hydrogen-bonding requirements on the protein surface, suggesting a critical role for non-water molecules to act as replacements for water during drying and in the dry state [63,169–171]. Raman and FTIR spectroscopy have been used to monitor structural changes in the antibody in solid state [172].

The rate of rehydration during reconstitution of lyophilized formulations is a critical parameter. If the rate of rehydration is sufficiently slow to allow recovery of native conformation as water replaces nonwater excipient molecules, reconstitution will typically provide a satisfactory outcome for the antibody. If, however, the antibody does not have sufficient time or capacity to recover its native state during reconstitution, extensive aggregation can result. Excipients that have been reported to slow the reconstitution rate and thereby preventing such changes in the rehydrated antibody are glycerol [171] and other polymeric excipients.

Pharmaceutical scientists frequently use increased temperatures for accelerated formulation stability studies. Although such studies can be used to assess formulation parameters such as pH and ionic strength, antibodies lyophilized in the presence of a cryoprotective sugar excipient provide a more complicated picture. Lyophilized amorphous solids composed of a monoclonal antibody and a sugar cryoprotectant material such as sucrose or trehalose form glassy structures, liquids that are too viscous to flow. Antibody–excipient complexes exhibit a glass transition temperature (T_g), and materials have different properties above and below this value. To predict shelf life for accelerated degradation studies, it is important to use testing temperatures that do not exceed the T_g threshold of the complex or the collapse temperature. The apparent glass transition temperature of formulations in frozen condition is denoted as T_g'. Both T_g' and collapse temperature are close together for most proteins at low protein concentrations [58]; however, at higher protein concentrations, the difference between collapse temperature and T_g' are larger. Primary drying can be shortened by using conditions in which the product temperature substantially exceeds T_g' without any apparent detrimental effect to the product [172]. Annealing of rapidly cooled solutions results in significantly less aggregation in reconstituted freeze-dried solids than in nonannealed controls, with a corresponding decrease in specific surface area of the freeze-dried, annealed system [54]. Increased concentration of the IgG significantly improved the stability of the IgG against freeze-drying-induced aggregation, which may be explained by a smaller percentage of the protein residing at the ice/freeze–concentrate interface as the IgG concentration was increased. Added salts such as NaCl or KCl contribute markedly to insoluble aggregate formation [54].

Sugars are commonly used as stabilizers in lyophilized protein and antibody formulations. Various molar ratios of sugar to protein have been tested, and the stability of the resulting lyophilized formulations has been determined by measuring aggregation, deamidation, and oxidation of the reconstituted protein and IR spectroscopy (secondary structure) of the dried protein. In one case a 360:1 molar ratio of lyoprotectant to protein was required for storage stability of the protein, resulting in a sugar concentration three- to four-fold below the isoosmotic concentration. Formulations with combinations of sucrose (20 mM) or trehalose (20 mM) and mannitol (40 mM) had stability comparable to those with sucrose or trehalose alone at 60 mM concentration [63,170,173]. Figure 16.4 shows an example in which the level of aggregation is reduced with increasing concentration of sucrose in an etanercept formulation at 50 mg/mL (unpublished data). However sucrose had no effect on low-molecular-weight (LMW) clips.

16.6.2. Reconstitution

The mechanism of stabilization of proteins by lyophilization has been relatively well studied, but not much is known about reconstitution of lyophilized cakes. There has been increased interest in high dose formulations of therapeutic proteins, including antibodies as an option to reduce dosing frequency. This comes with different challenges with respect to reconstitution of the lyophilized cake. Typical issues are long

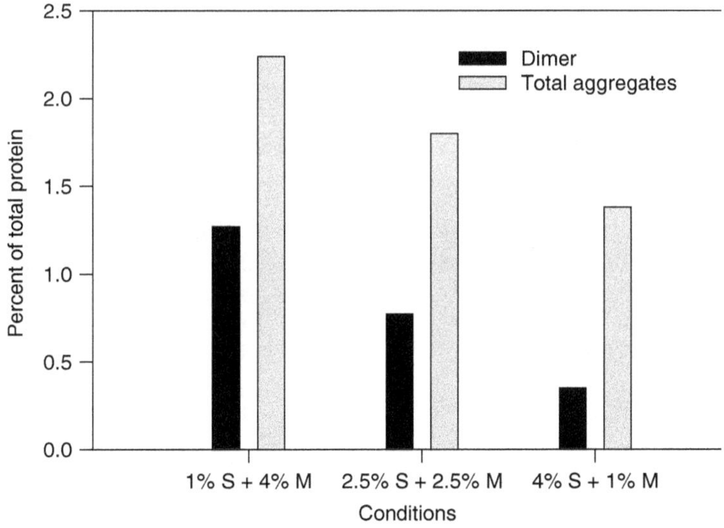

Figure 16.4. Size exclusion HPLC analysis of the Etanercept 50 mg/mL lyophilized samples on reconstitution containing different sucrose (S) and mannitol (M) concentrations. (Unpublished data.)

reconstitution times, foam layer, undissolved particles, protein denaturation, aggregation leading to turbidity, and high viscosity. The reconstitution step may potentially affect the protein stability and delivery to patients by several mechanisms. Rapid reconstitution may not allow a dried protein to rehydrate as slowly as the dehydration step [9].

We have studied the effects of different lyophilization cycle parameters, stabilizer concentration, and diluents on the reconstitution, structure, and stability of etanercept at 50 mg/ml. Annealing at −12°C and utilizing a secondary drying temperature at 25°C were important to improve the lyophilization characteristics of etanercept at high concentration (unpublished data), without compromising on the structure and stability of the protein (Table 16.3). Reducing air bubbles in the formulation prior to lyophilization and the diluent by degassing those solutions, coupled with vacuum during the freezing step of the cycle, was effective in improving the reconstitution properties of the etanercept at the high dose. Finally, diluents such as pluronic F68 decreased effervescence and foam formation during reconstitution of the cake.

16.6.3. High-concentration Strategies for Lyophilized Formulations

Noncovalent aggregation is a common limitation to high-concentration protein formulations.. To circumvent this problem, a formulation strategy was developed for proteins undergoing a slow self-association reaction. To minimize aggregation, the protein is lyophilized at low concentration in buffer containing half the standard excipient levels;

TABLE 16.3. Reconstitution Properties of Etanercept 50 mg/mL Lyophilized at Different Annealing and Secondary Drying Temperatures

Samples	Reconstitution Time, (s)	Height of Foam, mm	Comments
−20°C annealing, 45°C secondary drying	~90	6.4 ± 0.8	Effervescence observed on reconstitution; some solid material observed on the side of the vial but dissolve
−20°C annealing, 25°C secondary drying	41.6 ± 5.8	5.0 ± 1.4	Effervescence observed on reconstitution; some solid material observed initially but dissolve
−15°C annealing, 25°C secondary drying	42.5 ± 4.5	4.8 ± 1.4	Effervescence observed on reconstitution
−10°C annealing, 25°C secondary drying	68.3 ± 7.6	3.4 ± 2.2	Reduced effervescence on reconstitution compared to the other two annealing temperatures; slow dissolution of the cake
−5°C annealing, 25°C secondary drying	94.5 ± 5.0	2.9 ± 0.7	Reduced effervescence on reconstitution compared to the other two annealing temperatures; slow dissolution of the cake

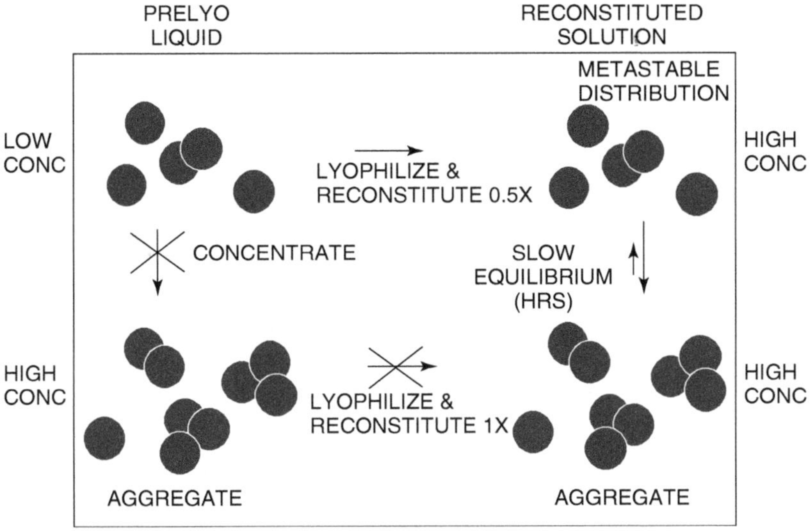

Figure 16.5. High-concentration lyophilized formulation strategy for proteins susceptible to aggregation. This strategy involves processing the protein at low concentration at 0.5× excipient levels, and then reconstituting the lyophilized cakes to half the original fill volume. (Unpublished data.)

then the lyophilized cake is reconstituted with "water for injection" to a final volume half that of the original fill volume to deliver a high concentration dosage form at full-strength excipient levels (Fig. 16.5) On reconstitution, the solution resembles the low-concentration material in terms of its aggregate content, and over the course of hours, the reconstituted solution undergoes a slow equilibrium to obtain an aggregate content characteristic of a high-concentration formulation. Assuming that the aggregation rate is sufficiently slow, the material will be injected before the aggregates grow. This strategy locks in the aggregate profile associated with the low concentration and takes advantage of the slow equilibrium between monomers and noncovalent aggregates. The method avoids processing and filling the purified bulk solution at high concentrations, where room temperature exposure may induce significant precipitation, membrane clogging, yield losses, and aggregation in the formulation. From a clinical standpoint, this strategy allows patients to have increased room temperature handling of the formulation without adverse impact on product quality attributes.

An example of the utility of the high concentration lyophilized formulation method was shown (Fig. 16.6) for r-met-Hu-Fc-Leptin (unpublished data). The protein was formulated and lyophilized at 50 mg/mL with 0.5 × excipient levels (5 mM histidine, 1% mannitol, 1% arginine HCl, 0.1% polysorbate 20, pH 5.0) and reconstituted with 0.5 × volume of water to result in a final concentration of 100 mg/mL with full-strength excipient levels (10 mM histidine, 2% mannitol, 2% arginine HCl, 0.2% polysorbate 20, pH 5.0). Aggregation kinetics were measured using size exclusion chromatography (SEC) to quantify the loss of main peak over time at room temperature

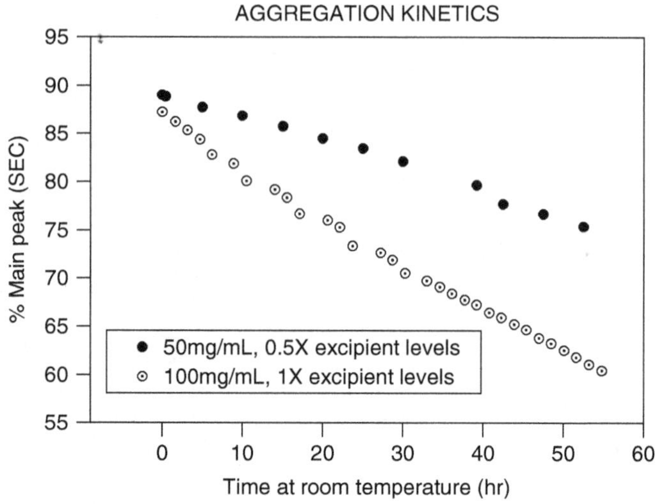

Figure 16.6. Aggregation kinetics using size exclusion chromatography to quantify the loss of main peak over time at room temperature for r-met-Hu-Fc-Leptin formulated at 50 mg/mL with 0.5 × excipient levels (5 mM histidine, 1% mannitol, 1% arginine HCl, 0.1% polysorbate 20, pH 5.0) and at 100 mg/mL with full-strength excipient levels (10 mM histidine, 2% mannitol, 2% arginine HCl, 0.2% polysorbate 20, pH 5.0). (Unpublished data.)

(Toso Haas G3000 SWxl column; mobile phase of 100 mM NaCl, 0.5 M NaCl, pH 6.9; flow rate 0.5 mL/min). The data show that the bulk drug substance at 50 mg/mL (0.5 × excipient levels) was more stable than at 100 mg/mL (1 × excipient levels), thus demonstrating the utility of this formulation strategy (Fig. 16.6) (unpublished data). This formulation strategy may be applied to other therapeutic proteins that undergo slow aggregation kinetics at high protein concentrations. Other variations on this method that employ different concentration factors other than a twofold increase in deliverable dosage may also be utilized.

16.7. ISSUES AFFECTING FORMULATION

16.7.1. Product Heterogeneity

Physical and covalent heterogeneity can confound the detection and interpretation of antibody stability properties. Known types of heterogeneity found in antibodies include glycosylation [127,128,174], *N*-terminal pyroglutamate formation [88,129], *C*-terminal processing [124–126], charge heterogeneity [78,79], disulfide connectivity [175], and conformational heterogeneity [88,122]. Reversed-phase HPLC analysis of intact antibodies revealed that the IgG2 subclass displayed significant heterogeneity, not seen in the IgG1 subclass [88,122]. Nonreduced peptide mapping and mass spectral characterization identified disulfide connectivity as the source of heterogeneity for the IgG2 subclass [176]. Disulfide connectivity has been shown to affect the overall conformation and flexibility of the antibody [123]. The heterogeneity in the IgG2 subclass was exhibited not only for recombinant monoclonal antibodies but also in endogenous human samples. Therefore, the existence of the human IgG2 subclass as an ensemble of structural isoforms is a natural phenomenon with structural and functional implications.

Redox (reduction–oxidation) treatment of IgG2 antibodies was implemented to populate isoforms with different disulfide connectivity of the light to heavy chains [123]. Redox treatment in the absence and presence of denaturant (approximately 1 M guanidine HCl) populated the two main structural isoforms (Fig. 16.7). As confirmed by non-reduced peptide mapping, redox treatment in the presence of denaturant populates the isoform termed "IgG2-A," which contains the expected disulfide connectivity with the light chain linked to the CH1 loop of the heavy chain. In contrast, redox treatment in the absence of denaturant populates "IgG2-B," a disulfide isoform Fab arm linked to the hinge region [123]. The IgG1 subtype and each of the IgG2 isoforms have distinctly different stability properties, presumably due to the conformational and/or flexibility effects of disulfide connectivity and hinge length differences.

In addition to disulfide heterogeneity of the IgG2 subclass, redox treatment was utilized to remove cysteinylation. An IgG1 antibody was refolded to remove cysteinylation in the variable region, and the refolded form with a free sulfhydryl was more stable than the cysteinylated form by DSC, equilibrium denaturation, and stability studies [177].

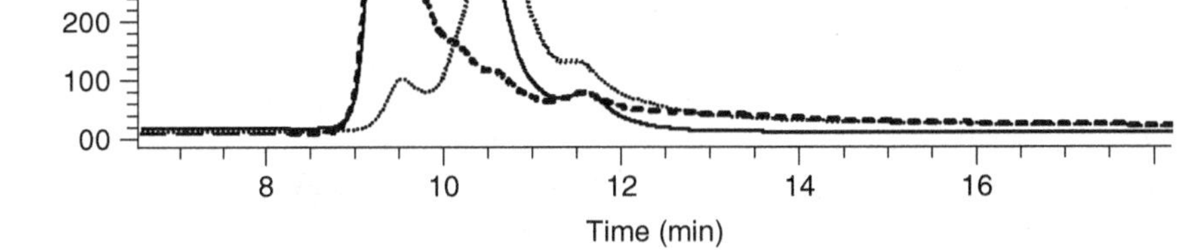

Figure 16.7. Reversed-phase chromatograms of the IgG2 control (solid line), native refold (dashed line), and guanidine HCl refold (dotted line) samples. The control represents a heterogeneous mixture of disulfide isoforms. Oxidative refolding in the absence and presence of denaturant can populate different isoforms (IgG2-B and IgG2-A, respectively). [More details can be found in Wypych et al. [176] and Dillon et al. [123].

16.7.2. Formulation Process: Filterability, Filling, and Other Operations

Antibody formulations may be impacted by processing conditions. For example, one must consider the impact of the formulation to contact surfaces such as stainless steel, nonideality effects during ultrafiltration/diafiltration operations [178], and the filling equipment. Contact with stainless steel can lead to methionine oxidation in high salt formulations [80] or in histidine solutions. Storage of high salt solutions in stainless-steel vessels can lead to pitting of the vessel and possibly leaching of metals.

As antibodies are formulated at higher concentrations, nonideality in the formulation process becomes more frequent. Excipients may coconcentrate or be excluded during ultrafiltration/diafiltration (UF/DF) operations as described by Stoner et al. [178]. At high protein concentrations, differences can sometimes be explained by use of molal concentrations instead of molar concentrations to account for volume exclusion effects. Antibodies are typically formulated away from their isoelectric point, and are charged. The Donnan effect accounts for partitioning of charged solutes across a membrane and is more pronounced in formulations with low concentrations of charged excipients. Non-specific interactions between proteins and solutes may also account for solute concentration differences across a membrane. If a solute has a positive affinity for the protein, the solute concentration in the formulated solution will be higher than diafiltration buffer. However the opposite effect may also be observed where the solute concentration is reduced by preferential exclusion [178]. These phenomena should be evaluated during formulation development, especially if much of the formulation optimization and testing are performed at lower antibody concentrations, where the effects may not be as pronounced or if multiple concentrations are desirable for the final antibody formulation.

The presence of certain formulation excipients during UF/DF can reduce turbidity or filter clogging. However, there may be other factors that preclude use of certain excipients during UF/DF. For example, during clinical development of an antibody in one study, an expensive excipient was removed from the UF/DF buffer, since 90% of the excipient went to waste when 10 diavolumes were used [179]. Experiments showed that this excipient reduced turbidity of in-process samples, but had no impact on the final product quality or long-term stability [179], justifying removal of the excipient from the process. The presence of surfactants such as polysorbates during UF/DF operations may minimize aggregation and filter clogging, but can present a control problem to consistently reach the target surfactant concentration.

The fill process itself can also lead to aggregation. Unlike the UF/DF operation, the fill process entails no further filtration; therefore, minimizing aggregation during filling is critical. Changes in fill equipment sometimes impact aggregation. For example, an antibody was initially filled using a rolling diaphragm pump, a change to a rotary piston pump in which the antibody solution, itself, is used to lubricate the piston was determined to be unacceptable [179]. Evaluation of different filler types during formulation development would allow the most flexibility; however, predicting all desired process changes is seldom feasible and needs to be dealt with on a case-by-case basis.

16.8. OVERVIEW

16.8.1. Holistic Cross-Functional Approach to Antibody Formulation Stability

Antibody formulation stability trends need to be examined holistically and appropriately balanced. Multiple mechanisms of degradation with different pH and temperature dependences can lead to complex stability trends and counterintuitive conclusions. Accelerated stability at elevated temperatures is a critical tool for formulation development [180]. An inherent assumption in the utility of accelerated data is that the stability properties at high temperatures are predictive of long-term shelf life from 2°C to 8°C. Presumably in the absence of conformational perturbations, these accelerated conditions would promote degradation reactions representative of actual storage temperatures, and the degradation reactions would follow Arrhenius kinetics [9]. However, multiple mechanisms of degradation can make the overall formulation trends complex and lead to non-Arrhenius kinetics. Degradation reactions, such as clipping and aggregation, have different temperature and pH dependences. The degradation reactions that predominate at elevated temperatures may not be the rate-limiting degradation reaction under refrigerated storage conditions. In addition, degradation reactions can be interdependent, such as the clip-mediated aggregation reaction observed in high-concentration antibody formulations at elevated temperatures.

The complex pH dependence of aggregation as a function of temperature renders formulation decisions based on accelerated data questionable. The most stable formulation at elevated temperatures may not be the optimal formulation under normal storage conditions. In addition, dramatically different aggregation properties at elevated temperatures have storage and handling implications for antibodies. This is particularly a concern for take-home products that may experience temperature excursions in transport, storage, and administration of the product by the patient. In addition, varying levels of structural isoforms between lots can result in differences in stability with respect to higher-order aggregates, clipping, temperature dependence, and so on. Differences in steric hindrance or flexibility of the hinge region may be responsible for the aggregation and clipping propensity.

Throughout the commercialization of a therapeutic, multiple processes may be used to produce the drug substance. Process changes provide benefits such as increased process yield, removal of less desirable raw materials, and reduced manufacturing time. Individual changes are typically evaluated for impact on the quality of the bulk drug substance. Evaluation of the process for impact to the formulation stability profile of the drug substance and drug product is frequently assumed to be out of scope. However, seemingly innocuous process changes can have significant impact on formulation stability for products at any stage of development. The commercial formulation recommendation for antibody therapeutics in development is based on the stability properties of early process material, but commercial process material can display unexpected increases in aggregation, particle propensity, or other degradation. These issues may manifest in the product quality of the drug substance, on freeze–thaw, or within the drug product formulation. Process changes in cell culture, viral inactivation pH, diafiltration buffer conditions, filter changes, or residual process impurities can contribute to stability differences. The physicochemical mechanism for these stability differences can be due to glycosylation, irreversible conformational perturbation, binding effects, or product heterogeneity. Therefore, it is important to assess the impact on drug product formulation stability as part of the comparability assessment. Process and formulation scientists must communicate regarding process changes throughout the stages of commercial process development, recognize potential formulation consequences of process changes, and demonstrate that there is no impact on formulation stability.

Another challenge to antibody formulation strategy is the analytical methods available for physical and covalent characterization. In some cases, assessment of the formulation stability is limited by the quality of the analytical methods available. With more recent analytical advancements, formulation scientists are becoming aware of new and interesting chemistries and conformational heterogeneity of antibodies. In many ways, antibody stability is not as straightforward as originally assumed, and may become more challenging to demonstrate as advanced tools used to characterize heterogeneity become available.

REFERENCES

1. Baker, M. (2005), Upping the ante on antibodies, *Nat. Biotechnol.* **23**: 1065–1072.
2. Reichert, J. M., Rosensweig, C. J., Faden, L. B., and Dewitz, M. C. (2005), Monoclonal antibody successes in the clinic, *Nat. Biotechnol.* **23**: 1073–1078.
3. Daugherty, A. L. and Mrsny, R. J. (2006), Formulation and delivery issues for monoclonal antibody therapeutics, *Adv. Drug Deliv. Rev.* **58**: 686–706.
4. Kozlowski, P. S. (2006), Current and future issues in the manufacturing and development of monoclonal antibodies, *Adv. Drug Deliv. Rev.* **58**: 707–722.
5. Low, D., O'Leary, R., and Pujar N. S. (2007), Future of antibody purification, *J. Chrom. B Anal. Technol. Biomed. Life Sci.* **848**(1): 48–63.
6. Mire-Sluis, A. R. (2001), Progress in the use of biological assays during the development of biotechnology products, *Pharm. Res.* **18**: 1239–1246.
7. Padlan, E. A. (1994), Anatomy of the antibody molecule, *Mol. Immunol.* **31**(3): 169–217.
8. Harris, R. J. (2005), Heterogeneity of recombinant antibodies: Linking structure to function, *Devel. Biol.* (Basel) **122**: 117–127.
9. Cleland, J. L., Powell, M. F., and Shire, S. J. (1993), The development of stable protein formulations: A close look at protein aggregation, deamidation, and oxidation, *Crit. Rev. Ther. Drug Carrier Syst.* **10**(4): 307–377.
10. Carpenter, J. F., Kendrick, B. S., Chang, B. S., Manning, M. C., and Randolph, T. W. (1999), Inhibition of stress-induced aggregation of protein therapeutics, *Meth. Enzymol.* **309**: 236–255.
11. Kyle, R. A. (1994), Monoclonal proteins and renal disease, *Annu. Rev. Med.* **45**: 71–77.
12. Koo, E. H., Lansbury, P. T., and Kelly, J. W. (1999), Amyloid diseases: Abnormal protein aggregation in neurodegeneration, *Proc. Natl. Acad. Sci. USA* **96**: 9989–9990.
13. Hardy, J., and Selkoe, D. J. (2002), The amyloid hypothesis of Alzheimer's disease: Progress and problems on the road to therapeutics, *Science* **297**: 353–356.

14. Chi, E. Y., Krishnan, S., Randolph, T. W., and Carpenter, J. F. (2003), Physical stability of proteins in aqueous solution: Mechanism and driving forces in nonnative protein aggregation, *Pharm. Res.* **20**: 1325–1336.
15. Krishnan, S., Chi, E. Y., Webb, J. N., Chang, B. S., Shan, D., Goldenberg, M., Manning, M. C., Randolph, T. W., and Carpenter, J. F. (2002), Aggregation of granulocyte colony stimulating factor under physiological conditions: Characterization and thermodynamic inhibition, *Biochemistry* **41**: 6422–6431.
16. Ferrone, F. (1999), Analysis of protein aggregation kinetics, *Meth. Enzymol.* **309**: 256–274.
17. Kim, Y. S., Cape, S. P., Chi, E., Raffen, R., Wilkins-Stevens, P., Stevens, F. J., Manning, M. C., Randolph, T. W., Solomon, A., and Carpenter, J. F. (2001), Counteracting effects of renal solutes on amyloid fibril formation by immunoglobulin light chains, *J. Biol. Chem* **276**: 1626–1633.
18. Kendrick, B. S., Chang, B. S., Arakawa, T., Peterson, B., Randolph, T. W., Manning, M. C., and Carpenter, J. F. (1997), Preferential exclusion of sucrose from recombinant interleukin-1 receptor antagonist: Role in restricted conformational mobility and compaction of native state, *Proc. Natl. Acad. Sci. USA* **94**: 11917–11922.
19. Webb, J. N., Webb, S. D., Cleland, J. L., Carpenter, J. F., and Randolph, T. W. (2001), Partial molar volume, surface area, and hydration changes for equilibrium unfolding and formation of aggregation transition state: High-pressure and co-solute studies on recombinant human IFN-gamma, *Proc. Natl. Acad. Sci. USA* **98**: 7259–7264.
20. Fink, A. L. (1998), Protein aggregation: Folding aggregates, inclusion bodies and amyloid, *Fold. Design* **3**, R9–R23.
21. Moore, J. M., Patapoff, T. W., and Cromwell, M. E. (1999), Kinetics and thermodynamics of dimer formation and dissociation for a recombinant humanized monoclonal antibody to vascular endothelial growth factor, *Biochemistry* **38**(42): 13960–13967.
22. Remmele, R. L., Jr., Callahan, W. J., Krishnan, S., Zhou, L., Bondarenko, P. V., Nichols, A. C., Kleemann, G. R., Pipes, G. D., Park, S., Fodor, S., Kras, E., and Brems, D. N. (2006), Active dimer of Epratuzumab provides insight into the complex nature of an antibody aggregate, *J. Pharm. Sci.* **95**(1): 126–145.
23. Gronski, P., Bauer, R., Bodenbender, L., Kanzy, E. J., Schmidt, K. H., Zilg, H., and Seiler, F. R. (1988), On the nature of IgG dimers: 1. Dimers in human polycolonal IgG preparations: Kinetic studies, *Behring. Inst. Mitt.* **82**: 127–143.
24. Krishnan, S., Callahan, W., Dong, A., Gandhi, S., Brych, S., Paterson, S., Kolvenbach, C., Dillon, T. M., Remmele, R. L., Jr., and Brems, D. N. Effect of solution conditions on structure, conformational stability and H/D exchange dynamics of an IgG1 monoclonal antibody and its fragments (manuscript in preparation for J Mol. Biol.).
25. McAuley, A., Jacob, J., Kolvenbach, C. G., Westland, K., Lee, H. J., Brych, S. R., Rehder, D., Kleemann, G. R., Brems, D. N., and Matsumura, M. (2008), Contributions of a disulfide bond to the structure, stability, and dimerization of human IgG1 antibody CH3 domain, *Protein Sci.* **17**: 95–106.
26. Harn, N., Allan, C., Oliver, C., and Middaugh, C. R. (2007), Highly concentrated monoclonal antibody solutions: Direct analysis of physical structure and thermal stability, *J. Pharm. Sci.* **96**(3): 532–546.
27. Asherie, N., Pande, J., Lomakin, A., Ogun, O., Hanson, S. R., Smith, J. B., and Benedek, G. B. (1998), Oligomerization and phase separation in globular protein solutions, *Biophys. Chem.* **75**(3): 213–227.

28. Kendrick, B. S., Li, T., and Chang, B. S. (2002), Physical stabilization of proteins in aqueous solution, *Pharm. Biotechnol.* **13**: 61–84.
29. Jones, L. S., Kaufmann, A., and Middaugh, C. R. (2005), Silicone oil induced aggregation of proteins, *J. Pharm. Sci.* **94**(4): 918–927.
30. Vermeer, A. W. P., and Norde, W. (2000), The thermal stability of immunoglobulin: Unfolding and aggregation of a multi-domain protein, *Biophys. J.* **76**: 394–404.
31. Vrkljan, M., Foster, T. M., Powers, M. E., Henkin, J., Porter, W. R., Staack, H., Carpenter, J. F., and Manning, M. C. (1994), Thermal-stability of low-molecular-weight urokinase during heat-treatment. 2. Effect of polymeric additives, *Pharm. Res.* **11**(7): 1004–1008.
32. Nguyen, T. H. and Shire, S. J. (1996), Stability and characterization of recombinant human relaxin, in *Formulation, Characterization, and Stability of Protein Drugs*, Pearlman, R. and Wang, Y. J., eds., Plenum Press, New York, pp. 247–211.
33. Chi, E. Y., Krishnan, S., Kendrick, B. S., Chang B. S., Carpenter, J. F., and Randolph, T. W. (2003), Roles of conformational stability and colloidal stability in the aggregation of recombinant human granulocyte colony-stimulating factor, *Protein Sci.* **12**(5): 903.
34. Nielsen, L., Khurana, R., Coats, A., Frokjaer, S., Brange, J., Vyas, S., Uversky, V. N., and Fink, A. L. (2001), Effect of environmental factors on the kinetics of insulin fibril formation: Elucidation of the molecular mechanism, *Biochemistry* **40**(20): 6036–6046.
35. Vermeer, A. W. P., Norde, W., and Amerongen, A. V. (2000), The unfolding/denaturation of immunogammaglobulin of isotype 2b and its Fab and Fc fragments, *Biophys. J.* **79**: 2150–2154.
36. Dill, K. A. (1990), Dominant forces in protein folding, *Biochemistry* **29**(31): 7133–7155.
37. Grimsley, G. R., Shaw, K. L., Fee, L. R., Alston, R. W., Huyghues-Despointes, B. M., Thurlkill, R. L., Scholtz, J. M., and Pace, C. N. (1999), Increasing protein stability by altering long-range coulombic interactions, *Protein Sci.* **8**(9): 1843–1849.
38. Shire, S. J., Shahrokh, Z., and Liu, J. (2004), Challenges in the development of high concentration formulations, *J. Pharm. Sci.* **93**(6): 1390–1402.
39. Timasheff, S. N. (1998), Control of protein stability and reactions by weakly interacting cosolvents: The simplicity of the complicated, *Adv. Protein Chem.* **51**: 355–432.
40. Cunningham, B. C., Mulkerrin, M. G., and Wells, J. A. (1991), Dimerization of human growth-hormone by zinc, *Science* **253**(5019): 545–548.
41. Tsai, P. K., Bruner, M. W., Irwin, J. I., Ip, C. C., Oliver, C. N., Nelson, R. W., Volkin, D. B., and Middaugh, C. R. (1993), Origin of the isoelectric heterogeneity of monoclonal immunoglobulin h1B4, *Pharm. Res.* **10**(11): 1580–1586.
42. Chen, B. L. and Arakawa, T. (1996), Stabilization of recombinant human keratinocyte growth factor by osmolytes and salts, *J. Pharm. Sci.* **85**(4): 419–422.
43. MacLean, D. S., Qian, Q. S., and Middaugh, C. R. (2002), Stabilization of proteins by low molecular weight multi-ions, *J. Pharm. Sci.* **91**(10): 2220–2229.
44. Fatouros, A., Osterberg, T., and Mikaelsson, M. (1997), Recombinant factor VIII SQ—influence of oxygen, metal ions, pH and ionic strength on its stability in aqueous solution, *Int. J. Pharm.* **155**(1): 121–131.
45. Remmele, R. L., Jr., Nightlinger, N. S., Srinivasan, S., and Gombotz, W. R. (1998), Interleukin-1 receptor (IL-1R) liquid formulation development using differential scanning calorimetry, *Pharm. Res.* **15**(2): 200–208.
46. Lam, X. M., Patapoff, T. W., and Nguyen, T. H. (1997), The effect of benzyl alcohol on recombinant human interferon-gamma, *Pharm. Res.* **14**(6): 725–729.

47. Jones, L. S., Bam, N. B., and Randolph, T. W. (1997), Surfactant-stabilized protein formulations: A review of protein-surfactants interactions and novel analytical methodologies, in *Therapeutic Protein and Peptide Formulation and Delivery*, Shahrokh, Z., ed., American Chemical Society, Washington, DC, *ACS Symp. Ser.* **675**: 206–222.

48. Bam, N. B., Cleland, J. L., and Randolph, T. W. (1996), Molten globule intermediate of recombinant human growth hormone: Stabilization with surfactants, *Biotechnol. Progress* **12**(6): 801–809.

49. Bam, N. B., Cleland, J. L., Yang, J., Manning, M. C., Carpenter, J. F., Kelley, R. F., and Randolph, T. W. (1998), Tween protects recombinant human growth hormone against agitation-induced damage via hydrophobic interactions, *J. Pharm. Sci.* **87**: 1554–1559.

50. Carpenter, J. F., Randolph, T. W., and Chang, B. S. (2004), Physical damage to proteins during freezing, drying and rehydration, in *Lyophilization of Biopharmaceuticals*, Pikal, M. J., and Constantino, H. R., eds., Springer, New York, pp. 423–442.

51. Privalov, P. L. (1990), Cold denaturation of proteins, *Crit. Rev. Biochem. Mol. Biol.* **25**: 281–305.

52. Arakawa, T., and Timasheff, S. N. (1982), Stabilization of protein-structure by sugars, *Biochemistry* **21**(25): 6536–6544.

53. Gomez, G., Pikal, M. J., and Rodriguez-Hornedo, N. (2001), Effect of initial buffer composition on pH changes during far-from-equilibrium freezing of sodium phosphate buffer solutions, *Pharm. Res.* **18**: 90–97.

54. Sarciaux, J. M., Mansour, S., Hageman, M. J., and Nail, S. L. (1999), Effects of buffer composition and processing conditions on aggregation of bovine IgG during freeze-drying, *J. Pharm. Sci.* **88**(12): 1354–1361.

55. Kreilgaard, L., Jones, L. S., Randolph, T. W., Frokjaer, S., Flink, J. M., Manning, M. C., and Carpenter, J. F. (1998), Effect of Tween 20 on freeze-thawing- and agitation-induced aggregation of recombinant human factor XIII, *J. Pharm. Sci.* **87**: 1597–1603.

56. Heller, M. C., Carpenter, J. F., and Randolph, T. W. (1999), Protein formulation and lyophilization cycle design: Prevention of damage due to freeze-concentration induced phase separation, *Biotechnol. Bioeng.* **63**: 166–174.

57. Piedmonte, D. M., Summers, C., McAuley, A., Karamujic, L., and Ratnaswamy, G. (2007), Sorbitol crystallization can lead to protein aggregation in frozen protein formulations, *Pharm. Res.* **24**(1): 136–146.

58. Pikal, M. J. (1999), Mechanisms of protein stabilization during freeze-drying and storage: The relative importance of thermodynamic stabilization and glassy state relaxation dynamics, in *Freeze-Drying/Lyophilization of Pharmaceutical and Biological Products*, Vol. **96**, Rey, L. and May, J. C., eds., Marcel Dekker, New York, pp. 161–198.

59. Mumenthaler, M., Hsu, C. C., and Pearlman, R. (1994), Feasibility study on spray-drying protein pharmaceuticals: Recombinant human growth hormone and tissue-type plasminogen activator, *Pharm. Res.* **11**: 12–20.

60. Broadhead, J., Edmond-Rouan, S. K., and Rhodes, C. T. (1996), The spray drying of pharmaceuticals, *Drug Devel. Industr. Pharm.* **22**: 813–822.

61. Maa, Y. F., Nguyen, P.-A. T., Andya, J. D., Dasovich, N., Shire, S. J., and Hsu, C. C. (1998), Effect of spray drying and subsequent processing on residual moisture content and physical/biochemical stability of protein inhalation powders, *Pharm. Res.* **15**: 768–775.

62. Shamblin, S. L., Hancock, B. C., Dupuis, Y., and Pikal, M. J. (2000), Interpretation of relaxation time constants for amorphous pharmaceutical systems, *J. Pharm. Sci.* **89**(3): 417–427.

63. Chang, L., Shepherd, D., Sun, J., Ouellette, D., Grant, K. L., Tang, X. C., and Pikal, M. J. (2005), Mechanism of protein stabilization by sugars during freeze-drying and storage: Native structure preservation, specific interaction, and/or immobilization in a glassy matrix? *J. Pharm. Sci.* **94**(7): 1427–1444.

64. Cicerone, M. T., Tellington, A., Trost, L., and Sokolov, A. (2003), Substantially improved stability of biological agents in dried form, *BioProcess Int.* **1**: 36–47.

65. Baldwin, R. N. (1988), Contamination of insulin by silicone oil: A potential hazard of plastic insulin syringes, *Diabet. Med.* **5**(8): 789–790.

66. Smith, E. J. (1988), Siliconization of parenteral packaging components, *J Parenter. Sci. Technol.* **42**: S1–S13.

67. Sluzky, V., Tamada, J. A., Klibanov, A. M., and Langer, R. (1991), Kinetics of insulin aggregation in aqueous solutions upon agitation in the presence of hydrophobic surfaces, *Proc. Natl. Acad. Sci. USA* **88**(21): 9377–9381.

68. Rosenberg, A. S. (2005), Effects of protein aggregates: An immunologic perspective, *AAPS J.* **8**(3): E501–E507.

69. Baert, F., Noman, M., Vermeire, S., et al. (2003), Influence of immunogenicity on the long term efficacy of infliximab in Crohn's disease, *New Engl. J. Med.* **348**: 601–608.

70. Ring, J., Stephan, W., and Brendel, W. (1979), Anaphylactoid reactions to infusions of plasma protein and human serum albumin, *Clin. Allergy* **9**: 89–97.

71. Ellis, E. and Henney, C. (1969), Adverse reactions following administration of human gamma globulin, *J. Allergy* **43**: 45–54.

72. Braun, A., Kwee, L., Labow, M. A., and Alsenz, J. (1997), Protein aggregates seem to play a key role among the parameters influencing the antigenicity of interferon alpha (IFN-alpha) in normal and transgenic mice, *Pharm. Res.* **14**: 1472–1478.

73. Seefeldt, M. B., Ouyang, J., Froland, W. A., Carpenter, J. F., and Randolph, T. W. (2004), High-pressure refolding of bikunin: Efficacy and thermodynamics, *Protein Sci.* **13**: 2639–2650.

74. Moore, W., and Leppert, P. (1980), Role of aggregated human growth hormone (hGH) in development of antibodies to hGH, *J. Clin. Endocrinol. Metab.* **51**: 691–697.

75. Barandun, S., Kistler, P., Jeunet, F., and Isliker, H. (1962), Intravenous administration of human γ-globulin, *Vox Sang* **7**: 157–174.

76. Casadevall, N., Nataf, J., Viron, B., et al. (2002), Pure red-cell aplasia and antierythropoietin antibodies in patients treated with recombinant erythropoietin, *New engl. J. Med.* **346**: 469–475.

77. Schellekens, H. (2002), Bioequivalence and immunogenecity of biopharmaceuticals, *Nat. Rev. Drug Discov.* **1**: 457–462.

78. Harris, R. J., Kabakoff, B., Macchi, F. D., Shen, F. J., Kwong, M., Andya, J. D., Shire, S. J., Bjork, N., Totpal, K., and Chen, A. B. (2001), Identification of multiple sources of charge heterogeneity in a recombinant antibody, *J. Chrom. B Biomed. Sci. Appl.* **752**(2): 233–245.

79. Zhang, W. and Czupryn, M. J. (2003), Analysis of isoaspartate in a recombinant monoclonal antibody and its charge isoforms, *J. Pharm. Biomed. Anal.* **30**(5): 1479–1490.

80. Lam, X. M., Yang, J. Y., and Cleland, J. L. (1997), Antioxidants for prevention of methionine oxidation in recombinant monoclonal antibody HER2, *J. Pharm. Sci.* **86**(11): 1250–1255.

81. Kerwin, B. A. and Remmele, R. L., Jr. (2007), Protect from light: photodegradation and protein biologics, *J. Pharm. Sci.* **96**(6): 1468–1479.

82. Alexander, A. J. and Hughes, D. E. (1995), Monitoring of IgG antibody thermal stability by micellar electrokinetic capillary chromatography and matrix-assisted laser desorption/ionization mass spectrometry, *Anal. Chem.* **67**(20): 3626–3632.

83. Cordoba, A. J., Shyong, B. J., Breen, D., and Harris, R. J. (2005), Non-enzymatic hinge region fragmentation of antibodies in solution, *J. Chrom. B Anal. Technol. Biomed. Life Sci.* **818**(2): 115–121.

84. Wang, L. T., Amphlett, G., Lambert, J. M., Blattler, W., and Zhang, W. (2005), Structural characterization of a recombinant monoclonal antibody by electrospray time-of-flight mass spectrometry, *Pharm. Res.* **22**(8): 1338–1349.

85. Kroon, D. J., Baldwin-Ferro, A., and Lalan, P. (1992), Identification of sites of degradation in a therapeutic monoclonal antibody by peptide mapping, *Pharm. Res.* **9**(11): 1386–1393.

86. Jiskoot, W., Beaver, E. C., de Koning, A. A., Herron, J. N., and Crommelin, D. J. (1990), Analytical approaches to the study of monoclonal antibody stability, *Pharm. Res.* **7**(12): 1234–1241.

87. Usami, A., Ohtsu, A., Takahama, S., and Fujii, T. (1996), The effect of pH, hydrogen peroxide and temperature on the stability of human monoclonal antibody, *J. Pharm. Biomed. Anal.* **14**(8–10): 1133–1140.

88. Dillon, T. M., Bondarenko, P. V., Rehder, D. S., Pipes, G. D., Kleemann, G. R., and Ricci, M. S. (2006), Optimization of a reversed-phase high-performance liquid chromatography/mass spectrometry method for characterizing recombinant antibody heterogeneity and stability, *J. Chrom. A* **1120**(1–2): 112–120.

89. Berkowitz, S. A. (2006), Role of analytical ultracentrifugation in assessing the aggregation of protein biopharmaceuticals, *AAPS J.* **8**(3): E590–E605.

90. Liu, J., Andya, J. D., and Shire, S. J. (2006), A critical review of analytical ultracentrifugation and field flow fractionation methods for measuring protein aggregation, *AAPS J.* **8**(3): E580–E589.

91. Gabrielson, J. P., Brader, M. L., Pekar, A. H., Mathis, K. B., Winter, G., Carpenter, J. F., and Randolph, T. W. (2007), Quantitation of aggregate levels in a recombinant humanized monoclonal antibody formulation by size-exclusion chromatography, asymmetrical flow field flow fractionation, and sedimentation velocity, *J. Pharm. Sci.* **96**(2): 268–279.

92. Schuurman, J., Perdok, G. J., Gorter, A. D., and Aalberse, R. C. (2001), The inter-heavy chain disulfide bonds of IgG4 are in equilibrium with intra-chain disulfide bonds, *Mol. Immunol.* **38**(1): 1–8.

93. Wen, J., Arakawa, T., and Philo, J. S. (1996), Size-exclusion chromatography with on-line light-scattering, absorbance, and refractive index detectors for studying proteins and their interactions, *Anal. Biochem.* **240**(2): 155–166.

94. Fraunhofer, W., and Winter, G. (2004), The use of asymmetrical flow field-flow fractionation in pharmaceutics and biopharmaceutics, *Eur. J. Pharm. Biopharm.* **58**(2): 369–383.

95. Mahler, H. C., Muller, R., Friess, W., Delille, A., and Matheus, S. (2005), Induction and analysis of aggregates in a liquid IgG1-antibody formulation, *Eur. J. Pharm. Biopharm.* **59**(3): 407–417.

96. Li, B., Flores, J., and Corvari, V. (2007), A simple method for the detection of insoluble aggregates in protein formulations, *J. Pharm. Sci.* **96**(7): 1840–1843.

97. Sharma, D. K., King, D., Moore, P., Oma, P., and Thomas, D. (2007), Flow microscopy for particulate analysis in parenteral and pharmaceutical fluids, *Euro. J. Parenter. Pharm. Sci.* **12**(4): 97–101.

98. Petenate, A. M. and Glatz, C. E. (1983), Isoelectric precipitation of soy protein. II. Kinetics of protein aggregate growth and breakage, *Biotechnol. Bioeng.* **25**(12): 3059–3078.

99. Vidanovic, D., Milic, A. J., Stankovic, M., and Poprzen, V. (2003), Effects of nonionic surfactants on the physical stability of immunoglobulin G in aqueous solution during mechanical agitation, *Pharmazie* **58**(6): 399–404.

100. Ahrer, K., Buchacher, A., Iberer, G., Josic, D., and Jungbauer, A. (2003), Analysis of aggregates of human immunoglobulin G using size-exclusion chromatography, static and dynamic light scattering, *J. Chrom. A* **1009**(1–2): 89–96.

101. Matheus, S., Friess, W., and Mahler, H. C. (2006), FTIR and nDSC as analytical tools for high-concentration protein formulations, *Pharm. Res.* **23**(6): 1350–1363.

102. Matheus, S., Mahler, H. C., and Friess, W. (2006), A critical evaluation of Tm (FTIR) measurements of high-concentration IgG1 antibody formulations as a formulation development tool, *Pharm. Res.* **23**(7): 1617–1627.

103. Szenczi, A., Kardos, J., Medgyesi, G., and Zavodszky, N. (2006), The effect of solvent environment on the conformation and stability of human polyclonal IgG in solution, *Biologicals* **34**(1): 5–14.

104. Tessier, P. M., Lenhoff, A. M., and Sandler, S. L. (2002), Rapid measurement of protein osmotic second virial coefficients by self-interaction chromatography, *Biophys. J.* **82**(3): 1620–1631.

105. Ahrer, K., Buchacher, A., Iberer, G., and Jungbauer, A. (2004), Detection of aggregate formation during production of human immunoglobulin G by means of light scattering, *J. Chrom. A* **1043**(1): 41–46.

106. Yang, J., Wang, S., Liu, J., and Raghani, A. (2007), Determination of tryptophan oxidation of monoclonal antibody by reversed phase high performance liquid chromatography, *J. Chrom. A* **1156**(1–2): 174–182.

107. Higley, T. J., Eksteen, R., and Lauder, S. (2002), Mobile phase optimization for the analysis of an antibody-based fusion protein and aggregates using a TSKgel Super SW3000 size-exclusion column, *Liq. Chrom. Gas Chrom. N. Am.* (Suppl. S): 12.

108. Pekar, A. and Sukumar, M. (2007), Quantitation of aggregates in therapeutic proteins using sedimentation velocity analytical ultracentrifugation: practical considerations that affect precision and accuracy, *Anal. Biochem.* **367**(2): 225–237.

109. Litzen, A., Walter, J. K., Krischollek, H., and Wahlund, K. G. (1993), Separation and quantitation of monoclonal antibody aggregates by asymmetrical flow field-flow fractionation and comparison to gel permeation chromatography, *Anal. Biochem.* **212**(2): 469–480.

110. *US, Pharmacopoeia* (2006), < 788 > *Particulate Matter in Injections, US Pharmacopoeia National Formulary*, USP 29-NF, 24th ed.

111. Demeule, B., Gurny, R., and Arvinte, T. (2007), Detection and characterization of protein aggregates by fluorescence microscopy, *Int. J. Pharm.* **329**(1–2): 37–45.

112. Sukumar, M., Doyle, B. L., Combs, J. L., and Pekar, A. H. (2004), Opalescent appearance of an IgG1 antibody at high concentrations and its relationship to noncovalent association, *Pharm. Res.* **21**(7): 1087–1093.

113. Kiese, S., Papppenberger, A., Friess, W., and Mahler, H. C. (2008), Shaken, not stirred: Mechanical stress testing of an IgG1 antibody, *J. Pharm. Sci.* **97**(10): 4347–4366.

114. Ahrer, K., Buchacher, A., Iberer, G., and Jungbauer, A. (2006), Thermodynamic stability and formation of aggregates of human immunoglobulin G characterized by differential scanning calorimetry and dynamic light scattering, *J. Biochem. Biophys. Meth.* **66**(1–3): 73–86.

115. Levine, H. L., Ransohoff, T. C., Kawahata, R. T., and McGregor, W. C. (1991), The use of surface tension measurements in the design of antibody-based product formulations, *J. Parenter. Sci. Technol.* **45**(3): 160–165.

116. Ruppert, S., Sandler, S. I., and Lenhoff, A. M. (2001), Correlation between the osmotic second virial coefficient and the solubility of proteins, *Biotechnol. Progress* **17**(1): 182–187.

117. Tessier, P. M., Verruto, V. J., Sandler, S. I., and Lenhoff, A. M. (2004), Correlation of diafiltration sieving behavior of lysozyme-BSA mixtures with osmotic second virial cross-coefficients, *Biotechnol. Bioeng.* **87**(3): 303–310.

118. Valente, J. J., Fryksdale, B. G., Dale, D. A., Gaertner, A. L., and Henry, C. S. (2006), Screening for physical stability of a Pseudomonas amylase using self-interaction chromatography, *Anal. Biochem.* **357**(1): 35–42.

119. Bajaj, H., Sharma, V. K., and Kalonia, D. S. (2004), Determination of second virial coefficient of proteins using a dual-detector cell for simultaneous measurement of scattered light intensity and concentration in SEC-HPLC, *Biophys. J.* **87**(6): 4048–4054.

120. Attri, A. K., and Minton, A. P. (2005), Composition gradient static light scattering: A new technique for rapid detection and quantitative characterization of reversible macromolecular hetero-associations in solution, *Anal. Biochem.* **346**(1): 132–138.

121. Saluja, A., Badkar, A. V., Zeng, D. L., Nema, S., and Kalonia, D. S. (2006), Application of high-frequency rheology measurements for analyzing protein-protein interactions in high protein concentration solutions using a model monoclonal antibody (IgG2), *J. Pharm. Sci.* **95**(9): 1967–1983.

122. Dillon, T. M., Bondarenko, P. V., and Ricci, M. S. (2004), Development of an analytical reversed-phase high-performance liquid chromatography-electrospray ionization mass spectrometry method for characterization of recombinant antibodies, *J. Chrom. A* **1053**(1–2): 299–305.

123. Dillon, T. M., Ricci, M. S., Vezina, C., Flynn, G. C., Liu, Y. D., Rehder, D. S., Plant, M., Henkle, B., Li, Y., Deechongkit, S., Varnum, B., Wypych, J., Balland, A., and Bondarenko, P. V. (2008), Structural and functional characterization of disulfide isoforms of the human IgG2 subclass, *J. Biol. Chem.* **283**(23): 16206–16215.

124. Perkins, M., Theiler, R., Lunte, S., and Jeschke, M. (2000), Determination of the origin of charge heterogeneity in a murine monoclonal antibody, *Pharm. Res.* **17**(9): 1110–1117.

125. Weitzhandler, M., Farnan, D., Horvath, J., Rohrer, J. S., Slingsby, R. W., Avdalovic, N., and Pohl, C. (1998), Protein variant separations by cation-exchange chromatography on tentacle-type polymeric stationary phases, *J. Chrom. A* **828**(1–2): 365–372.

126. Weitzhandler, M., Farnan, D., Rohrer, J. S., and Avdalovic, N. (2001), Protein variant separations using cation exchange chromatography on grafted, polymeric stationary phases, *Proteomics* **1**(2): 179–185.

127. Ma, S., and Nashabeh, W. (1999), Carbohydrate analysis of a chimeric recombinant monoclonal antibody by capillary electrophoresis with laser-induced fluorescence detection, *Anal. Chem.* **71**(22): 5185–5192.

128. Ma, S., and Nashabeh, W. (2001), Analysis of protein therapeutics by capillary electrophoresis, *Chromatographia* **53**(Suppl. S): S75–S89.

129. Moorhouse, K. G., Nashabeh, W., Deveney, J., Bjork, N. S., Mulkerrin, M. G., and Ryskamp, T. (1997), Validation of an HPLC method for the analysis of the charge heterogeneity of the recombinant monoclonal antibody Idec-C2b8 after papain digestion, *J. Pharm. Biomed. Anal.* **16**(4): 593–603.

130. Hunt, G., and Nashabeh, W. (1999), Capillary electrophoresis sodium dodecyl sulfate nongel sieving analysis of a therapeutic recombinant monoclonal antibody: A biotechnology perspective, *Anal. Chem.* **71**(13): 2390–2397.

131. Wan, H. Z., Kaneshiro, S., Frenz, J., and Cacia, J. (2001), Rapid method for monitoring galactosylation levels during recombinant antibody production by electrospray mass spectrometry with selective-ion monitoring, *J. Chrom. A* **913**(1–2): 437–446.

132. Cacia, J., Keck, R., Presta, L. G., and Frenz, J. (1996), Isomerization of an aspartic acid residue in the complementarity-determining regions of a recombinant antibody to human IgE—identification and effect on binding affinity, *Biochemistry* **35**(6): 1897–1903.

133. Eriksson, J. H., Mol, R., Somsen, G. W., Hinrichs, W. L., Frijlink, H. W., and de Jong, G. J. (2004), Feasibility of nonvolatile buffers in capillary electrophoresis-electrospray ionization-mass spectrometry of proteins, *Electrophoresis* **25**(1): 43–49.

134. Battersby, J. E., Snedecor, B., Chen, C., Champion, K. M., Riddle, L., and Vanderlaan, M. (2001), Affinity-reversed-phase liquid chromatography assay to quantitate recombinant antibodies and antibody fragments in fermentation broth, *J. Chrom. A* **927**(1–2): 61.

135. Lewis, D. A., Guzzetta, A. W., Hancock, W. S., and Costello, M. (1994), Characterization of humanized Anti-Tac, an antibody directed against the interleukin 2 receptor, using electrospray ionization mass spectrometry by direct infusion, Lc/Ms, and Ms/Ms, *Anal. Chem.* **66**(5): 585–595.

136. Kannan, K., Mulkerrin, M. G., Zhang, M., Gray, R., Steinharter, T., Sewerin, K., Baffi, R., Harris, R., and Karunatilake, C. (1997), Rapid analytical tryptic mapping of a recombinant chimeric monoclonal antibody and method validation challenges, *J. Pharm. Biomed. Anal.* **16**(4): 631–640.

137. Bongers, J., Cummings, J. J., Ebert, M. B., Federici, M. M., Gledhill, L., Gulati, D., Hilliard, G. M., Jones, B. H., Lee, K. R., Mozdzanowski, J., Naimoli, M., and Burman, S. (2000), Validation of a peptide mapping method for a therapeutic monoclonal antibody: What could we possibly learn about a method we have run 100 times? *J. Pharm. Biomed. Anal.* **21**(6): 1099–1128.

138. Chelius, D., Huff-Wimer, M. E., and Bondarenko, P. V. (2006), Reversed-phase liquid chromatography in-line with negative ionization electrospray mass spectrometry for the characterization of the disulfide-linkages of an immunoglobulin gamma antibody, *J. Am. Soc. Mass Spectrom.* **17**(11): 1590–1598.

139. Paborji, M., Pochopin, N. L., Coppola, W. P., and Bogardus, J. B. (1994), Chemical and physical stability of chimeric L6, a mouse-human monoclonal antibody, *Pharm. Res.* **11**(5): 764–771.

140. Chen, B., Bautista, R., Yu, K., Zapata, G. A., Mulkerrin, M. G., and Chamow, S. M. (2003), Influence of histidine on the stability and physical properties of a fully human antibody in aqueous and solid forms, *Pharm. Res.* **20**(12): 1952–1960.

141. Heymsfield, S. B., Greenberg, A. S., Fujioka, K., Dixon, R. M., Kushner, R., Hunt, T., Lubina, J. A., Patane, J., Self, B., Hunt, P., and McCamish, M. (1999), Recombinant leptin for weight loss in obese and lean adults: A randomized, controlled, dose-escalation trial, *JAMA* **282**(16): 1568–1575.

142. Liu, D., Ren, D., Huang, H., Dankberg, J., Rosenfeld, R., Cocco, M. J., Li, L., Brems, D. N., and Remmele, R. L., Jr. (2008), Structure and stability changes of human IgG1 Fc as a consequence of methionine oxidation, *Biochemistry* **47**(18): 5088–5100.

143. Perico, N., Purtell, J., Dillon, T. M., and Ricci, M. S. (2009), Conformational implications of an inversed pH-dependent antibody aggregation, *J. Pharm. Sci.* **98**(8): 3031–42.

144. Chang, S. H., Teshima, G. M., Milby, T., Gillece-Castro, B., and Canova-Davis, E. (1997), Metal-catalyzed photooxidation of histidine in human growth hormone, *Anal. Biochem* **244**: 221–227.

145. Schoneich, C. (2000), Mechanisms of metal-catalyzed oxidation of histidine to 2-oxo-histidine in peptides and proteins, *J. Pharm. Biomed. Anal.* **21**: 1093–1097.

146. Liu, J., Nguyen, M. D. H., Andya, J. D., and Shire, S. J. (2005), Reversible self-association increases the viscosity of a concentrated monoclonal antibody in aqueous solution, *J. Pharm. Sci.* **94**(9): 1928–1940 [erratum appears in *J. Pharm. Sci.* **95**(1): 234–235 (Jan. 2006)].

147. Gombotz, W. R., Pankey, S. C., Phan, D., Drager, R., Donaldson, K., Antonsen, K. P., Hoffman, A. S., and Raff, H. V. (1994), The stabilization of a human IgM monoclonal antibody with poly vinylpyrrolidone, *Pharm. Res.* **11**(5): 624–632.

148. Treuheit, M. J., Kosky, A. A., and Brems, D. N. (2002), Inverse relationship of protein concentration and aggregation, *Pharm. Res.* **19**(4): 511–516.

149. Chang, B. S., Kendrick, B. S., and Carpenter, J. F. (1996), Surface-induced denaturation of proteins during freezing and its inhibition by surfactants, *J. Pharm. Sci.* **85**(12): 1325–1330.

150. Marcovic, I. (2006), Challenges associated with extractable and/or leachable substances in therapeutic biologic protein products, *Am. Pharm. Rev.* (Sept./Oct.): 20–27.

151. Qadry, S. S., Roshdy, T. H., Char, H., Del Terzo, S., Tarantino, R., and Moschera, J. (2003), Evaluation of CZ-resin vials for packaging protein-based parenteral formulations, *Int. J. Pharm.* **252**(1–2): 207–212.

152. Schein, C. H. (1990), Solubility as a function of protein structure and solvent components, *BioTechnol* **8**(4): 308–317.

153. Collins, K. D. (2004), Ions from the Hofmeister series and osmolytes: effects on proteins in solution and in the crystallization process, *Methods* **34**(3): 300–311.

154. Gokarn, Y. R., Kras, E., Nodgaard, C., Dharmavaram, V., Fesinmeyer, R. M., Hultgen, H., Brych, S., Remmele, R. L. Jr., Brems, D. N., and Hershenson, S. (2008), Self-buffering antibody formulations, *J. Pharm Sci.* **97**(8): 3051–3066.

155. Dani, B., Platz, R., and Tzannis, S. T. (2007), High concentration formulation feasibility of human immunoglubulin G for subcutaneous administration, *J. Pharm. Sci.* **96**: 1504–1517.

156. Wang, W., Singh, S., Zeng, D. L., King, K., and Nema, S. (2007), Antibody structure, instability, and formulation, *J. Pharm. Sci.* **96**(1): 1–26.

157. Harris, R. J., Shire, S. J., and Winter, C. (2004), Commercial manufacturing scale formulation and analytical characterization of therapeutic recombinant antibodies, *Drug Devel. Res.* **61**: 137–154.

158. Ha, E., Wang, W., and Wang, Y. J. (2002), Peroxide formation in polysorbate 80 and protein stability, *J. Pharm. Sci.* **91**(10): 2252–2264.

159. Abend, A. M., Chung, L., Bibart, R. T., Brooks, M., and McCollum, D. G. (2004), Concerning the stability of benzyl alcohol: Formation of benzaldehyde dibenzyl acetal under aerobic conditions, *J. Pharm. Biomed. Anal.* **34**(5): 957–962.

160. Thirumangalathu, R., Krishnan, S., Brems, D. N., Randolph, T. W., and Carpenter, J. F. (2006), Effects of pH, temperature, and sucrose on benzyl alcohol-induced aggregation of recombinant human granulocyte colony stimulating factor, *J. Pharm. Sci.* **95**(7): 1480–1497.

161. Qi, H. and Heller, D. L. (1995), Stability and stabilization of insulinotropin in a dextran formulation, *PDA J. Pharm. Sci. Technol.* **49**(6): 289–293.

162. Tsai, P. K., Volkin, D. B., Dabora, J. M., Thompson, K. C., Bruner, M. W., Gress, J. O., Matuszewska, B., Keogan, M., Bondi, J. V., and Middaugh, C. R. (1993), Formulation design of acidic fibroblast growth-factor, *Pharm. Res.* **10**: 649–659.

163. Duenas, E. T., Keck, R., De Vos, A., Jones, A. J., and Cleland, J. L. (2001), Comparison between light induced and chemically induced oxidation of rhVEGF, *Pharm. Res.* **18**(10): 1455–1460.

164. Zheng, J. Y. and Janis, L. J. (2006), Influence of pH, buffer species, and storage temperature on physicochemical stability of a humanized monoclonal antibody LA298, *Int. J. Pharm.* **308**(1–2): 46–51.

165. Li, B., O'Meara, M. H., Lubach, J. W., Schowen, R. L., Topp, E. M., Munson, E. J., and Borchardt, R. T. (2005), Effects of sucrose and mannitol on asparagine deamidation rates of model peptides in solution and in the solid state, *J. Pharm. Sci.* **94**(8): 1723–1735.

166. Wakankar, A. A., Liu, J., Vandervelde, D., Wang, Y. J., Shire, S. J., and Borchardt, R. T. (2007), The effect of cosolutes on the isomerization of aspartic acid residues and conformational stability in a monoclonal antibody, *J. Pharm. Sci.* **96**(7): 1708–1718.

167. Riggin, A., Clodfelter, D., Maloney, A., Rickard, E., and Massey, E. (1991), Solution stability of the monoclonal antibody-vinca alkaloid conjugate, KS1/4-DAVLB, *Pharm. Res.* **8**(10): 1264–1269.

168. Carpenter, J. F., Prestrelski, S. J., and Arakawa, T. (1993), Separation of freezing- and drying-induced denaturation of lyophilized proteins using stress-specific stabilization. I. Enzyme activity and calorimetric studies, *Arch. Biochem. Biophys.* **303**: 456–464.

169. Carpenter, J. F., Pikal, M. J., Chang, B. S., and Randolph, T. W. (1997), Rational design of stable lyophilized formulations: Some practical advice, *Pharm. Res.* **14**: 969–975.

170. Cleland, J. L., Lam, X., Kendrick, B., Yang, J., Yang, T. H., Overcashier, D., Brooks, D., Hsu, C., and Carpenter, J. F. (2001), A specific molar ratio of stabilizer to protein is required for storage stability of a lyophilized monoclonal antibody, *J. Pharm. Sci.* **90**(3): 310–321.

171. Chang, L., Shepherd, D., Sun, J., Tang, X. C., and Pikal, M. J. (2005), Effect of sorbitol and residual moisture on the stability of lyophilized antibodies: Implications for the mechanism of protein stabilization in the solid state, *J. Pharm. Sci.* **94**(7): 1445–1455.

172. Colandene, J. D., Maldonado, L. M., Creagh, A. T., Vrettos, J. S., Goad, K. G., and Spitznagel, T. M. (2007), Lyophilization cycle development for a high-concentration monoclonal antibody formulation lacking a crystalline bulking agent, *J. Pharm. Sci.* **96**(6): 1598–1608.

173. Sane, S. U., Wong, R., and Hsu, C. C. (2004), Raman spectroscopic characterization of drying-induced structural changes in a therapeutic antibody: Correlating structural changes with long-term stability, *J. Pharm. Sci.* **93**(4): 1005–1018.

174. Masuda, K., Yamaguchi, Y., Kato, K., Takahashi, N., Shimada, I., and Arata, Y. (2000), Pairing of oligosaccharides in the Fc region of immunoglobulin G, *FEBS Lett.* **473**(3): 349–357.

175. Schauenstein, E., Dachs, F., Reiter, M., Gombotz, H., and List, W. (1986), Labile disulfide bonds and free thiol groups in human IgG. I. Assignment to IgG1 and IgG2 subclasses, *Int. Arch. Allergy Appl. Immunol.* **80**(2): 174–179.

176. Wypych, J., Li, M., Guo, A., Zhang, Z., Martinez, T., Allen, M., Fodor, S., Kelner, D., Flynn, G. C., Liu, Y. D., Bondarenko, P. V., Ricci, M. S., Dillon, T. M., and Balland, A. (2008), Human IgG2 antibodies display disulfide mediated structural isoforms, *J. Biol. Chem.* **283**(23): 16194–16205.

177. Gadgil, H. S., Bondarenko, P. V., Pipes, G. D., Dillon, T. M., Banks, D., Abel, J., Kleemann, G. R., and Treuheit, M. J. (2006), Identification of cysteinylation of a free cysteine in the Fab region of a recombinant monoclonal IgG1 antibody using Lys-C limited proteolysis coupled with LC/MS analysis, *Anal. Biochem.* **355**(2): 165–174.

178. Stoner, M. R., Fischer, N., Nixon, L., Buckel, S., Benke, M., Austin, F., Randolph, T. W., and Kendrick, B. S. (2004), Protein-solute interactions affect the outcome of ultrafiltration/diafiltration operations, *J. Pharm. Sci.* **93**(9): 2332–2342.

179. Cromwell, M. E., Hilario, E., and Jacobson, F. (2006), Protein aggregation and bioprocessing, *AAPS J.* **8**(3): E572–E579.

180. Chang, B. S. and Hershenson, S. (2002), Practical approaches to protein formulation development, *Pharm. Biotechnol.* **13**: 1–25.

17

REVERSIBLE SELF-ASSOCIATION OF PHARMACEUTICAL PROTEINS: CHARACTERIZATION AND CASE STUDIES

Vikas K. Sharma, Harminder Bajaj, and Devendra S. Kalonia

17.1. INTRODUCTION

Proteins have a tendency to undergo oligomerization, the nature and extent of which are defined by a protein's three-dimensional structure, surface distribution of charged and hydrophobic residues, conformational flexibility, concentration, and the solution environment (i.e., pH, temperature, ionic strength, presence of cosolutes). Under certain conditions, the net influence of these factors may lead to the formation of irreversible or reversible aggregates. *Irreversible aggregates* can be defined as those that do not participate in an equilibrium reaction and, once formed, will typically retain their higher-order state on dilution. These aggregates can be further categorized into two types: one that involves nonnative (altered tertiary/secondary conformation), noncovalently linked, higher-order species, and the other, which involves covalently linked higher-order species (e.g., disulfide linked), wherein each individual monomer unit is not significantly altered in its conformation.

On the other hand, *reversible aggregates* are those whose formation is dependent on protein concentration and in which equilibrium exists between the native state and the higher-order species. These include *soluble* reversible aggregates, such as dimer, trimer, and tetramer, in equilibrium with a monomer or another higher-order species,

Formulation and Process Development Strategies for Manufacturing Biopharmaceuticals, edited by Jameel and Hershenson

or *insoluble* reversible aggregates such as amorphous precipitate or protein crystals, which exist in equilibrium with the soluble form. While irreversible aggregates as well as the reversible insoluble precipitates are relatively easy to observe and characterize analytically, it is the soluble reversible aggregates that present a challenge in terms of analytical characterization, due primarily to the strong concentration dependence of their formation.

The formation of reversible soluble aggregates of a given protein species, which exist in equilibrium with each other and/or with the native state, is commonly referred to as *self-association*. The phenomenon of protein self-association is ubiquitous in nature and is essential for several physiological and biochemical processes at the cellular and molecular levels [1–4]. Understanding the phenomenon of protein self-association is crucial as often the monomeric and the oligomeric states have different biological/physicochemical properties and consequences. Furthermore, the interactions involved in the phenomenon of self-association can be easily perturbed by solution properties, namely, pH [5], ionic strength, temperature [6,7], and cosolutes [8,9]. Abundant literature is available on reporting the self-association of a wide variety of proteins, along with the effect of this phenomenon on various biological processes. Some of the notable examples relevant to human physiology include hemoglobin (a tetramer) [10], insulin (a hexamer in pancreas) [11], IgM (a pentamer) [12], and collagen (triplex fibril arrangement of long fibrous proteins) [13]. In addition to understanding the phenomenon of self-association at low concentrations, there has been an increasing level of interest on studying protein behavior, especially that of self-association, at relatively high concentration such as in the presence of crowding agents (sugars and nonionic polymers), to understand the physiological relevance of molecular crowding on protein structure and function [8,14,15].

A comprehensive presentation of this widespread phenomenon is beyond the length and scope of this chapter. Therefore, considering the overall theme of this book, we specifically discuss the relevance of protein self-association in terms of its implications on pharmaceutically relevant therapeutic proteins. Reversible self-association of proteins could have an impact on protein formulation design, stability, and solution properties such as viscosity. Protein association will also warrant development of appropriate analytical tools to characterize this behavior during the early development phase. Reversible self-association may further affect *in vivo* activity, pharmacokinetics, and safety. Often, the presence of higher-order species for a prolonged period of time may also increase the tendency of a protein to form irreversible aggregates. Therefore, it is critical that the tendency of a protein to undergo self-association be assessed thoroughly during its early development phase.

In this chapter, we first present a brief background on common mathematical expressions used to describe the phenomenon of self-association along with a general understanding of the mechanisms involved. In Section 17.3, analytical tools that are commonly utilized during protein pharmaceutical development to characterize protein association are reviewed. Later, in Section 17.9, a few case studies, exemplifying self-association behavior of pharmaceutically relevant proteins including characterization of self-association in highly concentrated antibody solutions, are presented.

17.2. BACKGROUND

In a simplified formula, self-association can be represented as

$$n\mathrm{P} \underset{k_\mathrm{d}}{\overset{k_\mathrm{a}}{\rightleftharpoons}} \mathrm{P}_n \tag{17.1}$$

where n is the number of molecules involved in the self-association of a protein molecule P, k_a is the rate constant of association, and k_d is the rate constant of dissociation. For example, a monomer dimer system could be represented as follows:

$$2\mathrm{P} \underset{k_\mathrm{d}}{\overset{k_\mathrm{a}}{\rightleftharpoons}} \mathrm{P}_2 \tag{17.2}$$

The self-association reaction is often defined by an association constant K_a or by a dissociation constant K_d. By definition, $K_\mathrm{a} = 1/K_\mathrm{d}$. For the abovementioned monomer–dimer equilibrium, the association constant can be defined as

$$K_\mathrm{a} = \frac{\mathrm{P}_2}{[\mathrm{P}]^2} \tag{17.3}$$

or alternatively

$$K_\mathrm{a} = \frac{k_\mathrm{a}}{k_\mathrm{d}} \tag{17.4}$$

The equilibrium reaction scheme can be defined for a variety of reversibly associating proteins involving various oligomeric states such as a monomer–trimer, dimer–tetramer, monomer–dimer–tetramer, and dimer–tetramer–octamer. In certain cases, the self-association is not defined or is restricted to a certain oligomer and may involve continuous growth of the oligomer units to large species. Examples of indefinite self-association include FtsZ association [16,17], tubulin association [18], or growth of amyloid fibrils [19]. A simple case of indefinite self-association can be described as follows [17]:

$$2\mathrm{P} \rightleftharpoons \mathrm{P}_2 + \mathrm{P} \rightleftharpoons \mathrm{P}_3 + \mathrm{P} \rightleftharpoons \mathrm{P}_4 \cdots \mathrm{P}_{i-1} + \mathrm{P} \rightleftharpoons \mathrm{P}_i \tag{17.5}$$

The association constant for a given oligomerization step is defined as

$$K_i = \frac{c_i}{c_1 c_{i-1}} \tag{17.6}$$

where c_i is the concentration of the i-mer, $K_1 \equiv 1$, and $i \geq 2$. For conservation of mass, we obtain

$$c_\mathrm{tot} = \frac{c_\mathrm{w}}{M_1} = \sum i c_i = \sum i Q_i c_1^i \tag{17.7}$$

and

$$Q_i = \prod_{j=1}^{i} K_i \tag{17.8}$$

where c_w is the total molar concentration of protein defined in terms of the monomer and M_1 is the mass of the monomer.

A detailed investigation into the mechanism of self-association may often involve determination of the rate of association and/or the dissociation reaction. The second-order rate constants of protein–protein association have been reported to vary from 10^3 to 10^9 M/s, and typically the rate constants of protein dimerization vary in the range of $0.5 - 5 \times 10^6\ M^{-1}\ s^{-1}$ [15,20]. For example, the association between hemoglobin dimers and tetramers has an association rate constant of $0.4 - 0.6 \times 10^6\ M^{-1}\ s^{-1}$. On the other hand, a higher association rate constant of $10^8\ M^{-1}\ s^{-1}$ has been reported for insulin dimerization. The typical rate constants observed for most protein complexes are on the order of $10^6\ M^{-1}\ s^{-1}$ and are much slower (~1000-fold) than the diffusion-limited association rate constant of $7 \times 10^9\ M^{-1}\ s^{-1}$ as predicted by Smoluchowski's equation ($k_a = 4\pi DR$, where D is the relative translational diffusion coefficient and R is the contact distance between two sphere centers). Yet, these protein association rate constants are considered to be large enough for oligomerization, rather than those predicted for very specific docking of two interacting sites. The observed rate constants of association (between the predicted diffusion-limited association and the highly specific docking association) are attributed mainly to two phenomena: (1) alignment of specific sites during multiple collisions before association occurs, even within the same pair of molecules; and (2) anisotropic distribution of reactivity over the protein surface that is overall steered by long-range attractive electrostatic and hydrophobic interactions. Overall, measurement of the association rates (or the dissociation rates) can provide further insights into the mechanism of protein self-association.

Mechanistically, the two types of interactions commonly used to describe the process of self-association include entropically driven hydrophobic interaction and/or enthalpically driven ionic and nonionic interactions. In order to understand the mechanism of self-association, thermodynamic parameters have been determined experimentally either by van't Hoff enthalpies from the temperature dependence of the equilibrium constants [21–24] or by tracking enthalpy changes calorimetrically [25].

Entropically driven protein self-association proceeds through the coalescence of nonpolar groups, initially accessible to solvent in isolated subunits, which then become buried on complex formation, leading to the release of the structured water. The basic criteria for hydrophobic interaction, as illustrated by Kauzmann, is overall a negative free-energy change with a positive entropic change, negligible change in enthalpy, and a large increase in the heat capacity [26]. Positive contributions of van't Hoff entropy have been observed for proteins such as the human platelet factor, human growth hormone, human insulin growth factor 1, human hemoglobin, and chicken erythrocyte histone, thus suggesting a role of hydrophobic interaction in the association process.

For certain other examples of protein association such as lysozyme dimerization [27], glucagon trimerization [28,29], α-chymotrypsin dimerization [30,31], and human myeloma [24], self-association has been observed along with net negative enthalpic and negative entropic changes. Such self-association behavior could not be explained on the basis on hydrophobic or electrostatic interactions and has been attributed to other nonionic interactions such as van der Waals interactions and hydrogen bonding [25]. Besides hydrophobic and other nonionic interactions, electrostatic attractive interactions can contribute toward the self-association of proteins. Examples that involve ionic interactions include β-lactoglobulin, α-chymotrypsin, and insulin. Often, however, self-association would be an outcome of a net balance between various interactions, including hydrophobic, van der Waals, hydrogen bonding, and electrostatic repulsive and attractive interactions.

17.3. ANALYTICAL TOOLS FOR CHARACTERIZATION OF SELF-ASSOCIATION

Characterization of the self-association behavior of a protein essentially involves characterization of the species involved in the equilibrium such as monomer–dimer or dimer–tetramer, and determination of the association constant K_a (or the dissociation constant K_d), defining the specific equilibrium under given solution conditions. A low value of K_a typically indicates a weaker complexation, whereas a higher value indicates a stronger complexation. Several techniques have been used for characterization of the species involved in reversible self-association equilibrium and for determination of the K_a value. These include osmometry [32], light-scattering-based techniques (i.e., static light scattering) [33,34], dynamic light scattering [35], small-angle neutron scattering [36], analytical centrifugation [37–41], chromatography-based techniques [42,43], isothermal titration calorimetry [44], and spectroscopy-based techniques [45]. In this chapter we briefly review the commonly used techniques of sedimentation equilibrium using analytical ultracentrifugation, static light scattering, and chromatography for the characterization of self-association equilibrium.

17.4. SEDIMENTATION EQUILIBRIUM ANALYTICAL ULTRACENTRIFUGATION

An analytical centrifuge, based on the centrifugation speeds, can be utilized to perform two different types of experiments: sedimentation velocity and sedimentation equilibrium. While sedimentation velocity studies are carried out at higher speeds and provide information about the sedimentation coefficient (s), at times the molecular weight, and the diffusion coefficient (D), sedimentation equilibrium studies provide information about the molecular weight and the thermodynamic parameters for an associating species, including the general nonideality of a protein solution. The technique of sedimentation equilibrium has been reviewed and utilized extensively and by far remains the method of choice for characterization of a self-associating species [46–49].

In a typical sedimentation equilibrium study, employing relatively lower centrifugation speeds, the distribution of the concentration of protein molecules along the radial distance of a centrifugal cell is measured after equilibrium has been established between the process of sedimentation (due to centrifugal force) and diffusion (due to concentration gradient). For an ideal, noninteracting, single-component system, the concentration gradient as a function of radial distance can be represented as [38,50]

$$\frac{d\ln c}{d(r^2)} = \frac{\omega^2(1-\overline{v}\rho)}{2RT}M_{\mathrm{wr}} \tag{17.9}$$

or on integration as

$$c(r) = c(a)\exp[AM_{\mathrm{wr}}(r^2 - a^2)] \tag{17.10}$$

and

$$A = \frac{(1-\overline{v}\rho)\omega^2}{2RT} \tag{17.11}$$

where $c(r)$ and c represent the concentration of the protein at a distance r, $c(a)$ is the concentration at a reference position, a is the radial distance of the reference, ω is the angular velocity of the rotor (in radians per second), R is the gas constant, T is the temperature, ρ is the solution density, $\overline{v}$ is the partial specific volume of the solute, and M_{wr} is the weight-average molecular weight of the species. Equations (17.9)–(17.11) describe the concentration gradient at equilibrium over a distance r in a centrifugation cell. For a single, nonassociating solute, assuming that partial molar volume of the species does not vary, the molecular weight can be determined from the slope of $\ln c$–r^2 plot.

If the sample is ideal and monodisperse, a plot of $\ln c$ versus r^2 will be linear (or M_{app} vs. c or r^2 will be constant); however, these plots become nonlinear if the system is more complex, that is, exhibits nonideality and/or self-association. For an ideal associating system such as a monomer–n-mer association, the total concentration at any radius, $c_{r,i}$, of the ith component can be expressed in terms of the monomer concentration $c_{r,1}$ and an association constant K_n by [51]

$$c_{r,i} = c_{r,1} + K_n(c_{r,1})^n \tag{17.12}$$

Similarly, for a nonideal system, the concentration gradient equation for an ith component can be written as [51]

$$\frac{d\ln c_i}{d(r^2)} = \frac{M_iA}{1 + (\partial\ln\gamma_i)/(\partial\ln c)} = \frac{M_iA}{1 + 2Bc + 3Cc^2 + \cdots} \tag{17.13}$$

where A is as defined by Equation (17.11), γ_i is the activity coefficient of the component i, B represents the second virial coefficient, C represents the third virial coefficient, and so forth. For an associating species, a series of association constants can be defined as

$$K_i = \frac{c_i}{(c_1)^{L(i)}} \tag{17.14}$$

where K_i is the association constant between the monomer and the ith aggregate, $L(i)$ is the degree of association for the ith aggregate, and $K_1 = 1$. Assuming dilute solutions such that the higher virial coefficients can be ignored, the total concentration at a radial position r can be written as [38]

$$c_r = \sum_{i=1}^{n+1} c_{i0} e^{L(i)[AM_1(r^2 - r_0^2) - BM_1(c_r - c_{r0})]} \tag{17.15}$$

where r_0 is the radial reference position, c_{r0} is the protein concentration at reference position, and B is the second virial coefficient.

At the reference radial position r_0, the total concentration can be written as

$$c_{r0} = \sum_{i=1}^{n+1} e^{\ln K_i + L(i) \ln c_{10}} \tag{17.16}$$

where c_{10} is the concentration of the monomer at the reference radial position. Equations 17.14–17.16 represent generalized forms of the sedimentation equilibrium equation to analyze a wide variety of discrete self-association systems and can be modified for each specific system.

For a protein molecule undergoing indefinite self-association, where the free-energy change is same is the same for the addition of monomer to an oligomer of any size, an intrinsic association constant, K, can be defined using the relationship [38]

$$c_i = iK^{(i-1)}(c_1)^i \tag{17.17}$$

For $Kc_1 < 1$, the total concentration c is defined as

$$c = \frac{c_1}{(1 - Kc_1)^2} \tag{17.18}$$

where

$$c_1 = c_{10} e^{[AM_1(r^2 - r_0^2) - BM_1(c_r - c_{r0})]} \tag{17.19}$$

Similar to the case for discrete self-association, for indefinite association the total concentration at a reference radial position, c_0, is given as

$$C_0 = \frac{c_{10}}{(1 - Kc_{10})^2} \tag{17.20}$$

Analysis of the data obtained in the form of total concentration as a function of radial distance and at a reference radial position, following a sedimentation equilibrium run, requires a global simultaneous fitting of a few equations and iterative calculations to estimate a number of parameters and produce the calculated c_r as a function of radial distance. On the basis of Equation (17.15), for a system of n–oligomers, the fitting parameters would involve n different k_n values, the monomer concentration c_{10}, and B. For Equation (17.19), the fitting parameters would involve monomer concentration at reference radial position, intrinsic association constant K, and B. The fit of a model is evaluated typically from the distribution of residuals between the calculated curve employing the fitted parameters and the experimental data.

17.4.1. Nonideality and Self-Association

In Equations (17.15) and (17.19), B, the second virial coefficient, represents the general nonideality of a protein solution arising from a two-body interaction. In the absence of the self-association term in the mathematical analysis, B can attain positive or negative values, representing net repulsive or net attractive interactions, respectively, in a dilute solution. In the case, however, where the self-association term is included to analyze a given data, interpretation of B is not straightforward and this parameter must be evaluated with caution. It should be noted that, in the case of self-association, even though multiple species are present in solution, each with their own specific and cross-coefficients, for the sake of simplicity, a single value of B is commonly used to fit the data and to represent the overall nonideality of the system [47,52,53].

In general, for a self-associating species, the value of B is constrained to be positive to include nonideality arising only from repulsive interactions, as negative B values are embedded in the data representing self-association. The concentration dependence of the molecular weight could serve as a means to distinguish between nonideal systems with nonspecific attractive interactions as represented through B and specific self-association behavior. An apparent M_w–concentration plot would be linear with either a negative or positive slope in the absence of self-association; however, a curvature should be expected, if self-associating species are present in solution. This formalism is not only restricted to analytical ultracentrifugation studies but also extended to data obtained through light-scattering and osmometry studies. For further analysis and interpretation of B values in analytical centrifugation in the presence of self-associating species, the reader should refer to articles by Wills and Winzor [54–57].

A number of analytical as well as numerical approaches have been developed to analyze the sedimentation equilibrium data for fitting of the equilibrium equations. Technological advances have improved instrumentation toward precision in data (e.g., concentration), and the availability of computational software has improved the handling of the complex data generated in sedimentation equilibrium experiments [41,58–60]. Various equations to model specific self-associating species are routinely incorporated in the sedimentation equilibrium analysis software and can be utilized to fit any given set of data. However, the results generated are often complex and could be easily misinterpreted if not assessed with caution and without incorporating proper controls. Sedimentation equilibrium studies require thorough knowledge and expertise in the subject, experimental setup, and data analysis, which could be a challenge to relatively inexperienced researchers.

17.5. STATIC LIGHT SCATTERING

One of the earlier techniques utilized for characterization of self-association of proteins involved static light scattering. Scattered light intensity increases with an increase in both molecular weight and concentration of macromolecules. Therefore, changes observed in the intensity of the scattered light due to increased molecular weight resulting from self-association, driven by protein concentration can, in principle, be utilized to extract the equilibrium association constant.

Static light scattering (SLS) measures the excess time-averaged intensity of the scattered light and relates it to the osmotic compressibility and second virial coefficient B of the interacting particles. The SLS data are processed by the classical method of Zimm [61], using the equation

$$\frac{Kc}{R_\theta} = \left(\frac{1}{M} + 2Bc\right)\left[1 + \frac{16\pi^2 r_g^2 \sin^2(\theta/2)}{3\lambda^2}\right] \tag{17.21}$$

where R_θ is the measured excess Rayleigh ratio of protein solution and is measured as a function of concentration c at multiple angles θ, M is the weight-average molecular weight, r_g is the radius of gyration, and λ is the wavelength of the incident light.

The constant K is defined as

$$K = \frac{4\pi^2(n\,dn\,/\,dc)^2}{\lambda^4 N_A} \tag{17.22}$$

where n is the refractive index of the solvent, dn/dc is the refractive index increment of the solute in a given solution, and N_A is the Avogadro's number. A Zimm plot is obtained by plotting Kc/R_θ against $\sin^2(\theta/2) + kc$, where k is an arbitrary constant. The molecular weight is obtained from the intercept following extrapolation to zero concentration, and B is determined from the slope of the concentration data, when extrapolated to zero angles.

Assuming isotropic scattering, Equation (17.21) can be simplified for proteins, since proteins do not typically exhibit the dependence on the angle of the incident light in the wavelength region commonly employed for light-scattering measurements (400–800 nm) and hence, measurements at a single angle are sufficient. This simplified equation, commonly known as the *Debye equation*, is represented as follows:

$$\frac{Kc}{R_\theta} = \frac{1}{M_w} + Bc \tag{17.23}$$

For generating a Debye plot (Kc/R_θ vs. c), the measurements are usually carried out at 90° to minimize contribution from dust particles. The inverse of the intercept provides the molecular weight, and the slope is related to the second virial coefficient B. A linear plot of Kc/R_θ versus c is often utilized to extract the B values without relating it to the phenomenon of self-association.

However, if the Debye plot yields an upward-curving plot with a net negative slope, it is indicative of self-association for a solution containing a single protein species. The curvature is attributed to a nonlinear increase in the weight-average molecular weight of the protein because of oligomerization on increase in protein concentration [62]. This is illustrated in Figure 17.1 [63,64]. The data can be fitted to obtain the association constant using a modified Debye equation to incorporate equations that define the association equilibrium. The formalism used to develop such an equation for simple monomer–dimer equilibrium has been described in detail elsewhere, and is briefly reviewed below [64].

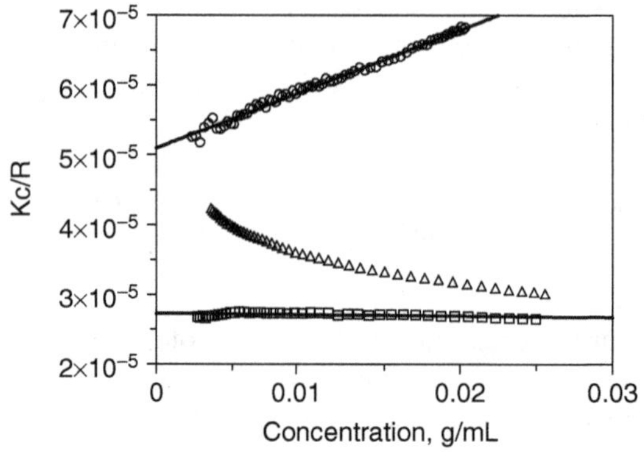

Figure 17.1. Debye plots of β-lactoglobulin A at pH 2.3, 0 M NaCl (○), pH 3.0, 0.1 M NaCl (△) and at pH 3.0, 1.0 M NaCl (□), generated using simultaneous light scattering and concentration detector using the assembly as described previously [63]. The linear plots at pH 2.3 (0 M NaCl) and at pH 3.0 (1.0 M NaCl) represent primarily the monomeric and dimeric proteins, respectively. The upward-curving Debye plot with a net negative slope at pH 3.0 (0.1 M NaCl) indicates presence of reversible self-associating species in solution [64].

For an associating system, the Debye equation [Eq. (17.23)] is modified to include the weight-average molecular weight of all the species present in the solution, and the term B is replaced with the term B_{avg} to represent general two-body interactions between all macromolecular species (e.g., monomer–monomer, dimer–dimer, and monomer–dimer in the case of a monomer–dimer association equilibrium). The modified Debye equation is written as

$$\frac{Kc_{tot}}{R_\theta} = \left(\frac{1}{M_{avg}} + B_{avg}c_t\right) \tag{17.24}$$

where M_{avg} is the weight average molecular weight of all the species present in the solution. For an associating system, the change in the chemical potential of the solvent with solute concentration is written as [65]

$$\frac{\partial \mu_1}{\partial c_{tot}} = \frac{\partial c_m}{\partial c_{tot} \bullet M_m} + \frac{\partial c_d}{\partial C_{tot} \bullet 2M_m} + \frac{\partial (B_{avg}c_{tot}^2)}{\partial c_{tot}} \tag{17.25}$$

The association constant for a simple monomer–dimer equilibrium can be written as

$$K_a = \frac{[c_d]}{[c_m]^2} \tag{17.26}$$

Where $[c_d]$ is the molar concentration of the dimer and $[c_m]$ is the molar concentration of the monomer. The total molar concentration $[c_{tot}]$ of the protein can be written in terms of the monomer concentration as

$$[c_{\text{tot}}] = [c_{\text{m}}] + 2[c_{\text{d}}] \tag{17.27}$$

Combining Equations (17.26) and (17.27) and converting molar concentration to grams per milliliter (g/mL), we can express the monomer and the dimer concentrations as follows:

$$c_{\text{m}} = \frac{-1 + (1 + 8000K_{\text{a}}c_{\text{tot}}/M_{\text{m}})^{1/2}}{4000K_{\text{a}}/M_{\text{m}}} \tag{17.28}$$

$$c_{\text{d}} = \frac{1 + 4000K_{\text{a}}c_{\text{tot}}/M_{\text{m}} - (1 + 8000K_{\text{a}}c_{\text{tot}}/M_{\text{m}})^{1/2}}{4000K_{\text{a}}/M_{\text{m}}} \tag{17.29}$$

The terms 4000 and 8000 in Equations (17.28) and (17.29) are used to maintain consistency between the volume units in K_{a} (liters per mole) and concentration (g/mL). Substituting for c_{m} and c_{d} from Equations (17.28) and (17.29) into Equation (17.25), taking partial derivatives, and using the result in the derivation of atoms Rayleigh's light-scattering equation, we obtain the following Debye equation:

$$\frac{Kc_{\text{tot}}}{R_{\theta}} = \frac{(1 + 8K_{\text{a}}c_{\text{tot}}/M_{\text{m}})^{1/2} + 1}{2(1 + 8K_{\text{a}}c_{\text{tot}}/M_{\text{m}})^{1/2}M_{\text{m}}} + Bc_{\text{tot}} \tag{17.30}$$

Equation (17.30) can be used to fit the Debye plot to obtain the association constant and the second virial coefficient.

Static light scattering was one of the first techniques used to establish the self-association properties of insulin [66]. While the data obtained from static light scattering can be easily analyzed for systems involving $2\text{P} = \text{P}_2$ type of equilibrium (monomer–dimer, dimer–tetramer), equations involving more complex equilibria have not been widely developed or utilized. Furthermore, the advancements in analytical centrifugation have lessened the use of light scattering techniques as a major technique for characterization of self-association systems. Although the technique of light scattering involves direct measurement of scattered light intensity and is noninvasive, the use of this technique in aqueous solutions is often challenging and requires tedious cleaning procedures to collect reproducible data. Because of enhanced variation and limited reproducibility, experiments are both labor- and time-intensive in order to minimize contribution from dust particles from the samples and samples tubes. While static light scattering in batch mode may be challenging for low-molecular-weight proteins and at low concentrations, the technique will provide more reproducible data for high-molecular-weight proteins and at high concentrations.

Beside batch-mode applications, static light scattering has also been utilized in a flow mode (involving measurement of scattered light intensity of a solution under flow) to characterize protein self-association. In such techniques, the protein solution is introduced in the detector through either a syringe pump or an HPLC system employing a SEC column. The former technique has been developed by Attri and Minton and has been used to characterize homo- as well as heteroassociation of proteins [67]. The latter technique has been reported by us for the characterization of homoassociation using β-lactoglobulin as the model associating protein [63,64]. In the Attri–Minton technique, a dual-syringe pump is utilized to generate different concentrations of protein by varying the flow rate of the two syringes followed by inflow mixing of the two solutions. The solution is allowed to equilibrate and then analyzed through a concentration and a light-scattering detector. The concentration and scattering data thus generated are used to characterize the self-association behavior of a given protein. This method, if used appropriately, can be applied to characterize both the association and dissociation behaviors of a self-associating species and can also to study association of two different protein species. However, caution must be exerted for association/dissociation kinetics that are slow enough to accommodate the duration of experiment and mixing allowed between each dilution. Furthermore, the sample must be pure and free from macromolecular impurities to prevent interference by irreversible and other nonproteinaceous material.

We have reported the design and use of a dual-source dual-detector system combining the concentration detector and the light-scattering detector in a single cell that can be used in conjunction with SEC-HPLC to characterize protein–protein interactions [63]. A concentration range required to generate the Debye plot can be obtained from a single injection and through utilization of the concentration distribution in the eluting peak. Each data point on the eluting peak can be converted to concentration and light-scattering intensity using signals from the appropriate detectors, and these data can be utilized to investigate the self-association behavior as well as nonideality of a given protein in a solution. Figure 17.2 [64] illustrates the usefulness of this technique to quantitatively assess the self-association of β-lactoglobulin.

The advantages of this technique include removal of macromolecular impurities and separation of irreversible aggregates by the SEC column, short duration of

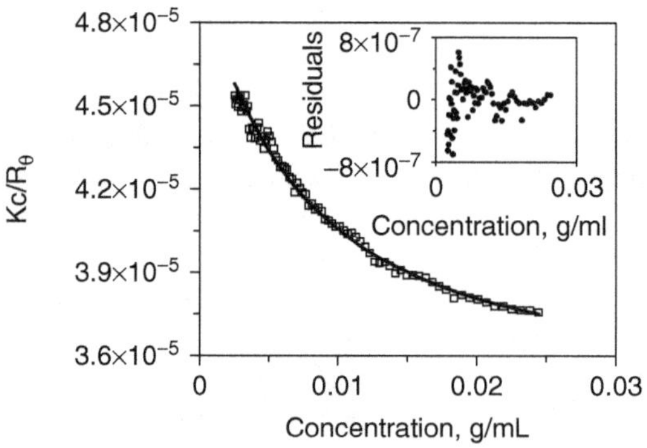

Figure 17.2. Analysis of the nonlinear Debye plots of β-lactoglobulin A by nonlinear least-squares regression. The data represented by markers were obtained using the dual-source dual-detector cell assembly in conjunction with SEC for β-lactoglobulin at pH 3.0 and 0.05 M NaCl concentrations. The lines represent the fit to the experimental data (single injection) using Equation (17.30), and the inset shows the distribution of residuals between the experimental data and the theoretical fit [64].

analysis, and availability of data from a single injection. However, the technique may not provide reproducible data at low concentrations, such as below 1 mg/mL (at the detector), and furthermore the chromatograms must be interpreted judiciously to distinguish between rapidly equilibrating versus slowly equilibrating species. While this technique is well suited for rapidly equilibrating species, for a slowly equilibrating species, the monomer and oligomer peaks could appear as separate peaks. A dependence of the chromatographic profile and the distribution of different species as a function of flow rate and concentration will help distinguish between slow and fast association equilibria.

17.6. CHROMATOGRAPHY-BASED TECHNIQUES

A few chromatography-based techniques have been reported for characterization of protein self-association. One of these techniques includes frontal gel chromatography [43,68]. In frontal gel chromatography, the solute is introduced into the column, previously saturated with the protein molecules, and an elution profile is generated, which can be differentiated into advancing and migrating boundaries. Protein association behavior is characterized based on the differences in the leading- and the trailing-edge profiles [69]. In the advancing boundary, the leading edge exhibits the migration characteristics of the monomer, because of the low protein concentration, whereas the migrating boundary is generated at high protein concentration. If the protein undergoes self-association, a sharply advancing boundary (primarily monomer) would be observed, and the trailing boundary would exhibit spreading compared to that obtained in a nonassociating system [70,71]. The existence of monomer/oligomer mixture in a rapidly associating system can be readily identified by failure of the advancing and trailing elution profiles to be mirror images of each other.

The equilibrium constants can be determined from knowledge of the weight-average elution volume $\overline{V}$ of the respective species in the system, which can be expressed as

$$\overline{V} = \frac{[C_1^\alpha V_1 + (C^\alpha - C_1^\alpha)V_n]}{C^\alpha} \tag{17.31}$$

where V_1 and V_n are the elution volumes of the monomer and the oligomer, respectively; C_1^α is the weight concentration of monomer in the applied protein solution with total concentration of C^α; and V_1 and V_n are obtained from the intercept of the plot of weight-average elution volume versus total concentration, and void volume of the column through a suitable selection of gel phase. To determine the weight concentration of monomer in a given profile, the elution volume of monomer and the oligomer needs to be determined. The association equilibrium constant can then be determined from the weight concentration of the monomer and oligomer in a typical chromatogram of advancing and migrating boundary using the following expression

$$K_n = \frac{C_n^\alpha}{(C_1^\alpha)^n} = \frac{C^\alpha - C_1^\alpha}{(C_1^\alpha)^n} \tag{17.32}$$

where K_n is the association constant, C_n^α and C_1^α are weight concentrations of the oligomer and monomer, respectively, at a given timepoint in the chromatogram.

Although this method does provide a distinctive and definitive method of determining equilibrium constants, it suffers from a few drawbacks. First, excessive amount of protein is required to saturate the column. Furthermore, the determination of equilibrium constants in frontal gel chromatography is based on indirect measurement of molecular weights, which depends on the weight concentration of monomer and oligomer species in a mixture. Weight concentration of species could de determined by the magnitude of elution volume in a chromatogram. Since elution volume is also dependent on the solvent conditions, one has to perform calibration of molecular weight *versus* elution volume by an absolute method [68].

In principle, size exclusion chromatography (SEC) by itself can be utilized to characterize protein self-association. In SEC, while irreversible aggregates can be resolved by choosing appropriate column and solution conditions, the behavior of reversibly associating species would depend on the equilibrium and the interaction of these species with column, thus affecting the profile of the eluting peaks [70,72]. Three cases could exist: (1) slow interconversion between the two species, (2) intermediate interconversion, and/or (3) rapid interconversion. In case of slow interconversion, the column separates the associated species present in solution on the timescale of the experiment, and these species elute as distinctly different peaks. In this scenario, the ratio of the monomer-to-dimer peak area changes as a function of concentration because of the different proportions of the two species present in the injection solution as defined by K_a. In the case of rapidly equilibrating associating species, the associating species rapidly reequilibrates following elution from column. The dimer and the monomer are not resolved, the elution time will vary with concentration, and the peak will contain a polydisperse molecular weight distribution. For the intermediate case (case 2, above), one would expect some separation as well as asymmetry in the eluting peak depending on K_a and initial concentration.

The three cases could be distinguished by injecting solutions of different initial protein concentration and at various rates. For example, for rapid interconversion, flow rate does not affect the distribution of the associating species, whereas for slow interconversion, the flow rate will have a distinct effect.

In one example, short-column SEC has been utilized to characterize the association behavior of recombinant human growth hormone [42]. The method separates the monomer and dimer of preequilibrated solutions of different protein concentrations at high flow rates through a short column and allows estimation of the dissociation constant through consideration of the dilution of protein and subsequent dissociation on elution. Through computer simulation of the moving boundaries of eluting peaks, the dissociation constant, the rate of dissociation, and the rate of association were determined. The method offers advantages of using low concentrations of protein and rapid characterization of association. However, the analysis could be complicated by the presence of higher-molecular-weight impurities and incomplete separation of the irreversible aggregates.

17.7. SPECTROSCOPY-BASED TECHNIQUES

Concentration-dependent changes in the circular dichroic spectra of such proteins as insulin [11,22] and glucagons [22] have been used to determine the association constant of their dimerization and trimerization, respectively. A prerequisite for these experiments was that each species of protein molecule (monomer and oligomer) exhibit distinct circular dichroic spectra, such that observed mean residue ellipticity at a given wavelength could be expressed in terms of contribution from fraction of monomer and dimer. The advantage of this methodology is that self-association can be studied at low concentrations; however, not all proteins meet the requirement of distinct for the pure monomer and the oligomer.

17.8. CASE STUDIES

17.8.1. Insulin

Insulin is a relatively small protein (6000 Da) composed of two chains, A chain (21 residues) and B chain (30 residues), linked through two disulfide bonds. Biologically, while the monomeric form of insulin is the active form and interacts with its receptor, insulin is, in fact, stored as a hexamer in complex with Zn^{2+} in the pancreas [45]. The physiological presence of these two forms has generated interest in studying the self-association behavior of this molecule for several decades. Siogren and Svedberg were the first to examine the self association of insulin, followed by Gutfreund [73], demonstrating an increase in the molecular weight of insulin as a function of concentration using osmometry. A nonlinear change in the ratio of osmotic pressure to concentration as a function of concentration suggested the possibility of insulin self-association. In the earlier days of investigation into insulin oligomerization, because of the lack of precise molecular weight measurements, the monomer molecular weight of insulin was determined to be 12,000 Da, and not 6000 Da, indicating that insulin dimer formed a stable species in solution. The multimeric nature of insulin has long been known.

Over the years, through several studies, the self-association behavior of insulin has been understood to a large extent. It is now well accepted that the oligomerization of insulin follows the monomer–dimer–tetramer–hexamer or the monomer–dimer–hexamer pathway and that oligomerization does not proceed through addition of one monomer at a time. Several factors have been shown to affect this equilibrium. These include zinc ion concentration, pH, and ionic strength [37,66, 74–76]. (Zn^{2+}) plays a critical role in stabilizing the higher-order species, such as the trimer, tetramer and the hexamer. In the absence of Zn^{2+}, dimer is the principal oligomeric species formed in solution. Insulin oligomerization also exhibits a strong pH and ionic strength dependence. At pH 2.2 and 0.1 M ionic strength, insulin monomer associates strongly to a form dimer at low concentrations (2 mg/mL) [37]. Increasing the ionic strength and pH to 4.0 favor formation of tetramer and hexamers [74].

Insulin has a tendency to form dimers in solutions or higher oligomers, primarily as a result of hydrogen bonding between the *C* termini of two B chains. However, in the presence of zinc ions, three dimers form a hexamer through coordination of two zinc ions with the histidine side chains of B10 residues. In addition, zinc further prevents the association of insulin to form higher-order aggregates [75,77–79]. The oligomerization tendency of insulin has been attributed to only a few key amino acids. Thus Pro at B28 and Lys at B29 were demonstrated to be critical for dimerization of insulin [80]. Alteration of insulin sequence at these two residues dramatically reduced the tendency of insulin to dimerize and also affected the formation of hexamer.

The oligomerization of insulin in the presence of zinc and its inhibition through alteration of select amino acids in the sequence have been utilized to produce different preparations of insulins for therapeutic use. Thus, a long-acting preparation of insulin, also known as *ultralente insulin*, is primarily the crystalline form of hexameric insulin in complexation with Zn^{2+} [81]. Since the active form of insulin is its monomeric form, the pharmacokinetics and duration of action of the ultralente insulin form largely depends on the dissociation of insulin zinc-hexamer. On the other hand, fast-acting insulin analogs have been prepared through substitutions of the critical amino acids, for example, of B28 Pro and B29 Lys, thus retaining the monomeric form and making insulin available immediately following administration [80]. One of the approaches to retain the monomeric characteristic of insulin utilizes the reversal of amino acids at positions 28 and 29. This insulin is commercially available as insulin lispro and has Lys at B28 and Pro at B29 [82]. Another approach utilizes the substitution of B29 Lys to Glu and B3 Asp to Lys to maintain the monomeric attribute of insulin. This fast-acting insulin is available as insulin glulisine [83].

Clearly, the detailed understanding of the self-association behavior of insulin and the mechanism involved has affected the drug development and the type of insulin available for therapeutic use.

17.8.2. Recombinant Human Relaxin

Relaxin is a peptide hormone belonging to the insulin family, and similar to insulin, is composed of two chains linked by two disulfide bonds. Relaxin is produced primarily by the corpus luteum in both pregnant and nonpregnant females and biologically appears to modulate the restructuring of tissues in organs in order to generate changes during pregnancy and parturition. In males, relaxin has been found to occur in heart artria. Clinically, relaxin has been investigated for its potential use in cervical ripening agent prior to labor induction and in scleroderma. Most recently, a clinical trial has been initiated for its potential use in acute heart failure and other acute care illnesses [84].

The self-association behavior of recombinant human relaxin has been studied, and this peptide hormone has been shown to exhibit a monomer–dimer equilibrium using sedimentation equilibrium analytical ultracentrifugation studies [38]. The data from equilibrium studies were fitted to several models, including monomer–dimer, monomer–dimer–trimer, monomer–dimer–tetramer, and an indefinite association model. Although the data could be fitted reasonably well for a monomer–dimer–trimer as well as a monomer–dimer–tetramer model, these possibilities were ruled out because the higher aggregate association constant (dimer–trimer or dimer–tetramer) was about 100-fold lower than the monomer–dimer association constant; the latter was similar for all models. Furthermore, on the basis of the multimer association model, the concentration of trimers or tetramers would be less than 0.1% under given solution conditions. Hence, the reversible monomer–dimer equilibrium was considered to be sufficient to describe the self-association behavior of human relaxin. The equilibrium association constant for the monomer–dimer system of human relaxin was reported to be on the order of $6 \times 10^5\ M^{-1}$, indicating strongly associating species, in a solution of 10 mM citrate buffer at pH 5.0 and at an ionic strength of 0.154 M, adjusted using NaCl.

Near-UV circular dichroism studies on relaxin as a function of its concentration indicated changes in the microenvironment of tyrosine on dissociation, indicating the role of this amino acid toward self-association of relaxin. Tyrosine is replaced by arginine in porcine relaxin, which did not exhibit the self-association behavior under similar solution conditions. In the case of relaxin, analytical ultracentrifugation provided a better evaluation of the association behavior than did SEC. As mentioned in another review [85], SEC could not provide information on the association behavior of relaxin, since the concentrations employed for SEC combined with the dilution effect on elution resulted in concentrations low enough to allow this protein to elute primarily as a monomer

17.8.3. RhuMAB VEGF

The self-association of a monoclonal antibody (IgG1), rhuMAb vascular endothelial growth factor (VEGF), was reported in a study by Moore et al. [86]. In general, the association of immunoglobulins, especially of the isotype IgG1, has not been commonly observed, except for in certain diseased states or at high concentrations. On the other hand, certain specific cryoglobulins, such as of the isotype IgG3, have been shown to undergo self-association and even more so in diseased states, for example, in rheumatoid arthritis and idiopathic cryoglobulinemia [87].

In the specific case of rhuMAb VEGF, the association/dissociation was primarily characterized using SEC and was observed during the early phase of product development. The evidence of rhuMAb VEGF self-association came primarily from the observation that a decrease in the aggregate amount was observed over a period of time when a 40-mg/mL solution, previously incubated at 40°C, was diluted to 1 mg/mL and subsequently stored at the same temperature (Fig. 17.3). In the detailed study of self-association of this particular antibody, it was observed that the rates of association and dissociation were strongly dependent on solution conditions such as pH, ionic strength, and temperature. In all cases the primary oligomeric species was a dimer, and the antibody primarily exhibited a monomer–dimer equilibrium under different solution conditions. From the relative rate of association versus the rate of dissociation at given solution condition, it was observed that the dimer formation was maximally favored in the pH range of 7.5–8.5 and no dimer was observed at pH 5.5. Although the association constant was found to be independent of temperature, the

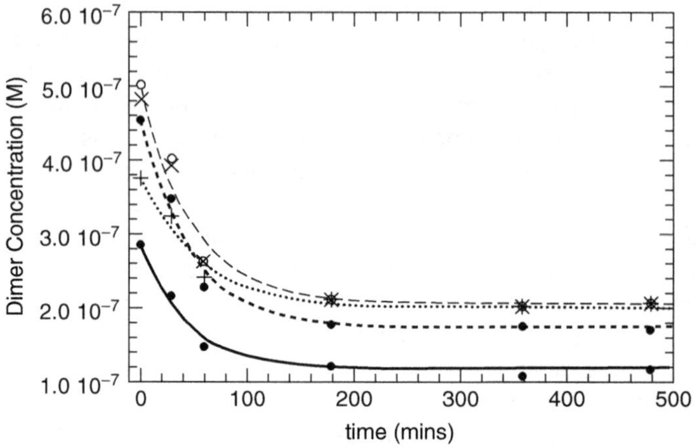

Figure 17.3. The pH dependence of rhuMAb VEGF dimer dissociation. The figure displays the observed kinetics of dimer dissociation at pH 6.0 (filled circles), pH 7.0 (filled diamonds), pH 7.5 (open circles), pH 8.0 (×), and pH 8.5 (+) at 40°C. The curves through individual points represent the best fit of the data with a single exponential using Equation (17.2) in Reference [86]. The loss of the dimer was followed by monitoring the area under the curve of the dimer peak in an SEC chromatogram [86].

rate of association and dissociation increased on increasing temperature. The dimer formation was further favored by an increase in NaCl concentration up to 1 M at pH 7.5. Overall, the measured association constants for dimerization of rhuMAb VEGF were on the order of 23–112 M^{-1} min^{-1}, indicating a relatively weak self-association of this antibody.

Thermodynamic analysis of the dissociation and association kinetics as function of temperature indicated primarily involvement of hydrophobic interactions in mediating the self-association of rhuMAb VEGF. This data were further supported by a decrease in the dissociation constant in D_2O compared to that in H_2O, as well as an enhanced rate of association under conditions of high salt concentration.

From a product development perspective, the characterization of the dimer was reported to be challenging because of the reversibility of monomer–dimer equilibrium and its dependence on solution conditions, and due to the slow kinetics of this process. Furthermore, the reversible dimer coeluted with the irreversible dimer (present in almost all antibodies) and hence necessitated the development of a method to distinguish the two forms of dimers. The irreversible dimer fraction typically increases on storage and is not affected by dilution. In addition, since the implications of reversible dimer on the *in vivo* efficacy and safety were unknown, it was essential that the amount of dimer present be characterized and controlled critically. This resulted in a more rigorous SEC method for the characterization of rhuMAb VEGF size heterogeneity during its product development phase.

17.9. HIGH-CONCENTRATION ANTIBODY SOLUTIONS

There has been tremendous interest in the characterization of self-association behavior of proteins at relatively high concentrations (typically higher than 100 mg/mL). This interest is primarily focused on two distinct areas: (1) understanding the behavior of proteins in crowded environments such as that in the interior of cells and in certain vital organs, for example, crystallines in the eye [8,9,15,88] and (2) in the area of pharmaceutical biotechnology for development of high-concentration antibody solutions for the chronic treatment of disorders such as autoimmune diseases [89]. Such indications often demand higher of doses the order of more than 1 mg/kg or 100 mg/dose. Since a subcutaneous delivery is desirable for these chronic administrations for outpatient or at-home treatments, a high concentration formulation of antibodies is preferable.

Formulation of antibodies at high concentrations of typically higher than 100 mg/mL, however, brings in additional challenges, notably that of enhanced viscosity. The viscosity of macromolecules including antibodies can be expressed in the form of a virial expansion as

$$\eta = \eta_0(1 + k_1 c_p + k_2 c_p^2 + k_3 c_p^3 + \cdots) \tag{17.33}$$

where viscosity η is related to solvent viscosity η_0 and concentration of the protein c_p (in g/mL) by a power series. In this equation, k_1 represents intrinsic viscosity and is related to the contribution from single molecules, whereas k_2, k_3, k_4, and so on are related to contributions from intermolecular interactions. At high concentrations, the higher-order interaction terms could become important and could lead to substantial increase in viscosity. In these circumstances, the phenomenon of self-association would represent the higher-order (two-body or higher) attractive interactions. The phenomenon of self-association is also favored thermodynamically in highly concentrated or crowded solutions as a result of "excluded volume" effects. It has been proposed that for macromolecules that may have an intrinsic tendency to self-associate, in crowded or concentrated solutions, this tendency will be further enhanced to result in a reduction in the excluded volume and hence lowering of the free energy of the system [90–92].

Typically, proteins, which have a tendency to weakly associate, may not exhibit experimentally measurable self-association behavior at low concentrations. Such a tendency, however, can lead to formation of oligomeric species at sufficiently high concentrations. Various techniques as described above have been developed and utilized mainly for assessing self-association behavior of proteins at dilute concentrations, and the theory developed behind such techniques may not be applicable to high concentrations.

Minton and co-workers have made significant progress in developing a theoretical background to assess the self-association behavior as well as nonideality of highly concentrated protein solutions [88,90,93,94]. The theoretical background has been adapted to the techniques of sedimentation equilibrium, osmotic pressure, and light scattering for characterization of self-association at high concentration. The self-associating behavior of proteins such as bovine serum albumin, aldolase, ovalbumin and most recently, of IgG, has been reported through the application of these conventional techniques through using the appropriate theoretical models and slight modifications of the existing methods [95,96]. These proteins, which typically exist as monomer at concentration of 10 mg/mL or lower, demonstrated association to higher oligomers such as dimers, trimers, and tetramers at concentrations of 30 mg/mL or above.

From a pharmaceutical development perspective, enhanced viscosity as a result of self-association can lead to several challenges in the development and delivery of antibodies. For example, concentration-dependent enhanced viscosity could impact the tangential flow filtration step commonly used to formulate and concentrate antibody solutions. In terms of delivery, a high-viscosity solution could be challenging to inject through a small-bore needle for a subcutaneous delivery.

This phenomenon of self-association has been reported for at least one therapeutic monoclonal antibody, when formulated at high concentrations [97]. In a comparison of three antibodies at concentrations exceeding 100 mg/mL, only one of the monoclonal antibody exhibited such behavior, thus demonstrating that this increase was presumably due to specific protein–protein interactions involved for this particular antibody and not simply to an increase in protein concentration. In a detailed investigation by Liu et al., the method of sedimentation equilibrium studies using preparative ultracentrifugation was used to assess whether self-association contributes to the enhanced viscosity of this particular antibody. In this method, previously described and developed by Minton [40,95], a high concentration solution of antibody was layered on top of a fluorocarbon in thick-walled polycarbonate tubes. The samples were then centrifuged at 15,000 rpm in a swinging bucket rotor for a period of time, until equilibrium is established. Following centrifugation, the contents of each tube were collected using a microfractionator, and the concentration of each fraction was determined in a 96-well UV plate following dilution. The apparent weight average molecular weight $M_{w,app}$ at each radial position was determined using the following equation

$$c(r) = c_0 \exp\left[\frac{M_{w,app}\omega^2(1-\overline{v}\rho)}{2RT}\xi\right] \qquad (17.34)$$

where $c(r)$ is the radial position at each position r, c_0 is the initial protein loading concentration, $\overline{v}$ is the partial specific volume, ρ is the buffer density, ω is the angular velocity [$\xi = (r^2 - r^2{}_0)/2$], and r_0 is the reference radial position.

The authors in this particular study note the complications that may arise in analysis and interpretation of such data, due to contribution of nonideality behavior at high concentrations. To correct for the nonideality contribution and the excluded volume contribution toward increase in viscosity simply due to the size and concentration of an antibody, the authors utilized a semiempirical approach. In this approach, the sedimentation equilibrium data from those antibodies, which do not show an anomalous viscosity behavior as a function of concentration and whose concentration-dependent viscosity increase could be simply described using a hard-sphere model, were utilized to correct for the increase in the apparent molecular weight beyond the nonideality and excluded volume effects.

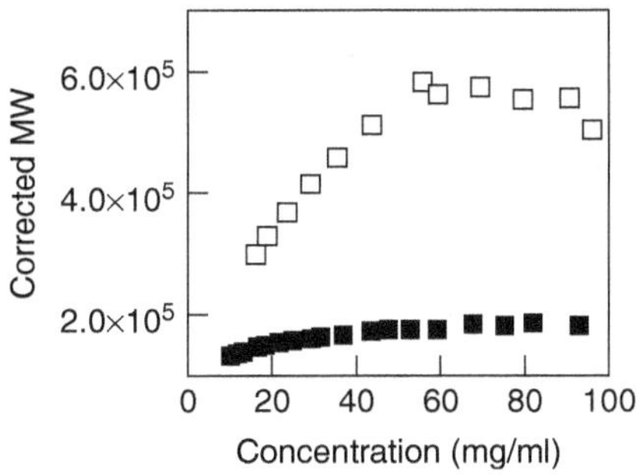

Figure 17.4. Corrected average molecular weight of an IgG1 monoclonal antibody obtained using sedimentation equilibrium studies through preparative ultracentrifugation in the presence (solid squares) and in the absence (hollow squares) of NaCl. The self-association behavior of this antibody is evident in the absence of NaCl [97].

On the basis of this analysis, it was shown that an increase in the viscosity of an antibody, beyond the excluded volume effect from hard-sphere contributions, was indeed due to the phenomenon of self-association at higher concentration and that this association was mediated primarily through electrostatic interactions. The viscosity and hence the self-association behavior were affected by pH and by the presence of salt (Fig. 17.4). The viscosity was observed to be highest in low-ionic-strength solutions prepared near the isoelectric point (pI) of the antibody. In a more recent report by Kanai et al. [98], this self-association behavior was specifically attributed to the Fab–Fab intermolecular interactions. The overall self-association behavior was attributed to relatively weak interactions that were not measurable at low concentrations using ultracentrifugation.

17.10. SUMMARY

The presence of oligomers as a result of reversible association can impact the pharmaceutical development of proteins, including the long-term stability of protein formulations, product characteristics such as viscosity for high-concentration solutions, and *in vivo* profile, including the activity, safety, and pharmacokinetics. Therefore, during the pharmaceutical development of a protein therapeutic, the tendency of a protein to self-associate must be assessed carefully and completely utilizing the appropriate analytical tools. The dissociation/association kinetics of protein self-association can vary from seconds to several days, and hence a variety of analytical tools should be utilized simultaneously to detect and characterize this behavior. Solution parameters such as pH, ionic strength, temperature, and presence of cosolutes can affect the kinetics of association and hence the equilibrium. Therefore, the protein may exist as a monomer under certain conditions and yet exhibit self-association under completely different conditions. As described in this chapter, various techniques have been utilized to characterize the self-association behavior of proteins, of which analytical centrifugation, light scattering, and size exclusion chromatography remain the methods of choice for investigating this behavior. While sedimentation equilibrium using analytical centrifugation can provide an accurate and quantitative assessment of reversible self-association, light scattering and chromatography-based methods can serve as quick screening analytical tools, especially in support of the early development of a protein therapeutic.

REFERENCES

1. Weiss, A. and Schlessinger, J. (1998), Switching signals on or off by receptor dimerization, *Cell* **94**(3): 277–280.
2. McNally, F. (2000), Capturing a ring of samurai, *Nat. Cell Biol*. **2**(1): E4–E7.
3. Milligan, G. (2001), Oligomerisation of G-protein-coupled receptors, *J. Cell Sci*. **114**(7): 1265–1271.
4. Marianayagam, N. J., Sunde, M., et al. (2004), The power of two: Protein dimerization in biology, *Trends Biochem. Sci*. **29**(11): 618–625.
5. Parkinson, E. J., Morris, M. B., et al. (2000), Acid denaturation of recombinant porcine growth hormone: Formation and self-association of folding intermediates, *Biochemistry* **39**(40): 12345–12354.
6. Timasheff, S. N. and Aune, K. C. (1971), Dimerization of a-chymotrypsin. I. pH dependence in the acid region, *Biochemistry* **10**(9): 1609–1617.
7. Timasheff, S. N., Aune, K. C., et al. (1971), Dimerization of a-chymotrypsin. II. Ionic strength and temperature dependence, *Biochemistry* **10**(9): 1617–1622.
8. Patel, C. N., Noble, S. M., et al. (2002), Effects of molecular crowding by saccharides on alpha-chymotrypsin dimerization, *Protein Sci*. **11**(5): 997–1003.
9. Snoussi, K. and Halle, B. (2005), Protein self-association induced by macromolecular crowding: A quantitative analysis by magnetic relaxation dispersion, *Biophys. J*. **88**(4): 2855–2866.
10. Brnjaskraljevic, J. and Maricic, S. (1978), Hydration and selfassociation of hemoglobin in solution, *Biochem. Biophys. Res. Commun*. **83**(3): 1048–1054.
11. Pocker, Y. and Biswas, S. B. (1981), Self-association of insulin and the role of hydrophobic bonding—a thermodynamic model of insulin dimerization, *Biochemistry* **20**(15): 4354–4361.
12. Baker, M. D., Wu, G. E., et al. (1986), A region of the immunoglobulin-mu heavy-chain necessary for forming pentameric Igm, *J. Immunol*. **137**(5): 1724–1728.
13. Tsilibary, E. C. and Charonis, A. S. (1986), The role of the main noncollagenous domain (Nc1) in type-Iv collagen self-assembly, *J. Cell Biol*. **103**(6): 2467–2473.
14. Nichol, L. W., Ogston, A. G., et al. (1981), Effect of inert polymers on protein self-association, *FEBS Lett*. **126**(1): 18–20.
15. Kozer, N. and Schreiber, G. (2004), Effect of crowding on protein-protein association rates: Fundamental differences between low and high mass crowding agents, *J. Mol. Biol*. **336**(3): 763–774.
16. Romberg, L., Simon, M., et al. (2001), Polymerization of FtsZ, a bacterial homolog of tubulin—is assembly cooperative? *J. Biol. Chem*. **276**(15): 11743–11753.

17. Kameyama, K. and Minton, A. P. (2006), Rapid quantitative characterization of protein interactions by composition gradient static light scattering, *Biophys. J*. **90**(6): 2164–2169.
18. Melki, R. and Carlier, M. F. (1993), Thermodynamics of tubulin polymerization into zinc sheets—assembly is not regulated by Gtp hydrolysis, *Biochemistry* **32**(13): 3405–3413.
19. Smith, A. M., Jahn, T. R., et al. (2006), Direct observation of oligomeric species formed in the early stages of amyloid fibril formation using electrospray ionisation mass spectrometry, *J. Mol. Biol*. **364**(1): 9–19.
20. Northrup, S. H. and Erickson, H. P. (1992), Kinetics of protein association explained by brownian dynamics computer-simulation, *Proc. Natl. Acad. Sci. USA* **89**(8): 3338–3342.
21. Diggle, J. H., McVittie, J. D., et al. (1975), The self-association of chicken-erythrocyte histones, *Eur. J. Biochem./FEBS* **56**(1): 173–182.
22. Formisano, S., Johnson, M. L., et al. (1977), Thermodynamics of the self-association of glucagon, *Proc. Natl. Acad. Sci. USA* **74**(8): 3340–3344.
23. Valdes, R., Jr. and Ackers, G. K. (1977), Thermodynamic studies on subunit assembly in human hemoglobin. Self-association of oxygenated chains (alphaSH and betaSH): Determination of stoichiometries and equilibrium constants as a function of temperature, *J. Biol. Chem*. **252**(1): 74–81.
24. Hall, C. G. and Abraham, G. N. (1984), Reversible self-association of a human myeloma protein. Thermodynamics and relevance to viscosity effects and solubility, *Biochemistry* **23**(22): 5123–5129.
25. Ross, P. D. and Subramanian, S. (1981), Thermodynamics of protein association reactions: Forces contributing to stability, *Biochemistry* **20**(11): 3096–30102.
26. Kauzmann, W. (1959), Some factors in the interpretation of protein denaturation, *Adv. Protein Chem*. **14**: 1–63.
27. Banerjee, S. K., Pogolotti, A., Jr., et al. (1975), Self-association of lysozyme. Thermochemical measurements: Effect of chemical modification of Trp-62, Trp-108, and Glu-35, *J. Biol. Chem*. **250**(20): 8260–8266.
28. Sasaki, K., Dockerill, S., et al. (1975), X-ray analysis of glucagon and its relationship to receptor binding, *Nature* **257**(5529): 751–757.
29. Johnson, R. E., Hruby, V. J., et al. (1979), Thermodynamics of glucagon aggregation, *Biochemistry* **18**(7): 1176–1179.
30. Shiao, D. D. F. and Sturtevant, J. M. (1969), Calorimetric investigations of the binding of inhibitors to a-chymotrypsin. I. Enthalpy of dilution of a-chymotrypsin and of proflavine, and the enthalpy of binding of indole, N-acetyl-D-tryptophan, and proflavine to a-chymotrypsin, *Biochemistry* **8**(12): 4910–4917.
31. Vandlen, R. L. and Tulinsky, A. (1973), Changes in the tertiary structure of alpha-chymotrypsin with change in pH: p4 4.2–6.7, *Biochemistry* **12**(21): 4193–4200.
32. Stein, A. and Page, D. (1980), Core histone associations in solutions of high salt—an osmotic-pressure study, *J. Biol. Chem*. **255**(8): 3629–3637.
33. Herskovits, T. T., Russell, M. W., et al. (1984), Light-scattering investigation of the subunit structure and sequential dissociation of Homarus-Americanus hemocyanin, *Biochemistry* **23**(8): 1875–1881.
34. Inoue, T., Fukushima, K., et al. (1984), Light-scattering study on subunit association-dissociation equilibria of bovine liver glutamate-dehydrogenase, *Biochim. Biophys. Acta* **786**(3): 144–150.

35. Sharma, M., Haque, Z. U., et al. (1996), Association tendency of beta-lactoglobulin AB purified by gel permeation chromatography as determined by dynamic light scattering under quiescent conditions, *Food Hydrocoll.* **10**(3): 323–328.

36. Verheul, M., Pedersen, J. S., et al. (1999), Association behavior of native beta-lactoglobulin, *Biopolymers* **49**(1): 11–20.

37. Jeffrey, P. D. and Coates, J. H. (1966), An equilibrium ultracentrifuge study of the self-association of bovine insulin, *Biochemistry* (Moscow) **5**(2): 489–498.

38. Shire, S. J., Holladay, L. A., et al. (1991), Self-association of human and porcine relaxin as assessed by analytical ultracentrifugation and circular-dichroism, *Biochemistry* **30**(31): 7703–7711.

39. Liu, J. and Shire, S. J. (1999), Analytical ultracentrifugation in the pharmaceutical industry, *J. Pharm. Sci.* **88**(12): 1237–1241.

40. Minton, A. P. (2000), Quantitative characterization of reversible macromolecular associations via sedimentation equilibrium: An introduction, *Exp. Mol. Med.* **32**: 1–5.

41. Vistica, J., Dam, J., et al. (2004), Sedimentation equilibrium analysis of protein interactions with global implicit mass conservation constraints and systematic noise decomposition, *Anal. Biochem.* **326**(2): 234–256.

42. Patapoff, T. W., Mrsny, R. J., et al. (1993), The application of size exclusion chromatography and computer simulation to study the thermodynamic and kinetic parameters for short-lived dissociable protein aggregates, *Anal Biochem.* **212**: 71–78.

43. Beeckmans, S. (1999), Chromatographic methods to study protein-protein interactions, *Methods* **19**(2): 278–305.

44. Bello, M., Perez-Hernandez, G., et al. (2008), Energetics of protein homodimerization: Effects of water sequestering on the formation of beta-lactoglobulin dimer, *Proteins Struct. Funct. Bioinform.* **70**(4): 1475–1487.

45. Uversky, V. N., Garriques, L. N., et al. (2003), Prediction of the association state of insulin using spectral parameters, *J. Pharm. Sci.* **92**(4): 847–858.

46. Williams, R. C., Jr. and Yphantis, D. A. (1971), Ultracentrifugation, *Encycl. Polym. Sci. Technol.* **14**: 97–116.

47. Cole, J. L. (1996), Characterization of human cytomegalovirus protease dimerization by analytical centrifugation, *Biochemistry* **35**(48): 15601–15610.

48. Hensley, P. (1996), Defining the structure and stability of macromolecular assemblies in solution: The re-emergence of analytical ultracentrifugation as a practical tool, *Structure* (Lond.) **4**(4): 367–373.

49. Minton, A. P. (1997), Alternative strategies for the characterization of associations in multicomponent solutions via measurement of sedimentation equilibrium, *Progress Colloid Polym. Sci.* **107**: 11–19.

50. Hansen, J. C., Lebowitz, J., et al. (1994), Analytical ultracentrifugation of complex macromolecular systems, *Biochemistry* **33**(45): 13155–13163.

51. Johnson, M. L., Correia, J. J., et al. (1981), Analysis of data from the analytical ultracentrifuge by non-linear least-squares techniques, *Biophys. J.* **36**(3): 575–588.

52. Kim, H., Deonier, R. C., et al. (1977), The investigations of self-association reactions by equilibrium ultracentrifugation, *Chem. Rev.* **77**: 659–690.

53. Godfrey, J. E., Eickbush, T. H., et al. (1980), Reversible association of calf thymus histones to form the symmetrical octamer (H2AH2BH3H4)2: A case of mixed associating system, *Biochemistry* **19**: 1339–1346.

54. Wills, P. R., Hall, D. R., et al. (2000), Interpretation of thermodynamic non-ideality in sedimentation equilibrium experiments on proteins, *Biophys. Chem.* **84**(3): 217–225.
55. Wills, P. R., Jacobsen, M. P., et al. (2000), Analysis of sedimentation equilibrium distributions reflecting nonideal macromolecular associations, *Biophys. J.* **79**(4): 2178–2187.
56. Wills, P. R. and Winzor, D. J. (2001), Studies of solute self-association by sedimentation equilibrium: Allowance for effects of thermodynamic non-ideality beyond the consequences of nearest-neighbor interactions, *Biophys. Chem.* **91**(3): 253–262.
57. Winzor, D. J., Deszczynski, M., et al. (2007), Nonequivalence of second virial coefficients from sedimentation equilibrium and static light scattering studies of protein solutions, *Biophys. Chem.* **128**(1): 46–55.
58. Ralston, G. B. and Morris, M. B. (1992), The use of the omega function for sedimentation equilibrium analysis, *Anal. Ultracentrifug. Biochem. Polym. Sci*. 253–274.
59. Winzor, D. J. and Harding, S. E. (2001), Sedimentation equilibrium in the analytical ultracentrifuge, *Protein-Ligand Interact. Hydrodyn. Calorim*. 105–135.
60. Lebowitz, J., Lewis Marc, S., et al. (2002), Modern analytical ultracentrifugation in protein science: A tutorial review, *Protein Sci*. **11**(9): 2067–2079.
61. Zimm, B. H. (1946), Application of the methods of molecular distribution to solutions of large molecules, *J. Chem. Phys*. **14**: 164–179.
62. Eisenberg, H. K., ed. (1976), *Biological Macromolecules and Polyelectrolytes in Solution*, Monographs on Physical Biochemistry: Biological Macromolecules and Polyelectrolytes in Solution series, Oxford Univ. Press, London, pp. 134–140.
63. Bajaj, H., Sharma, V. K., et al. (2004), Determination of second virial coefficient of proteins using a dual-detector cell for simultaneous measurement of scattered light intensity and concentration in SEC-HPLC, *Biophys. J.* **87**: 4048–4055.
64. Bajaj, H., Sharma, V. K., et al. (2007), A high-throughput method for detection of protein self-association and second virial coefficient using size-exclusion chromatography through simultaneous measurement of concentration and scattered light intensity, *Pharm. Res*. **24**: 2071–2083.
65. Tanford, C., ed. (1961), *Thermodynamics*, Physical Chemistry of Macromolecules series, Wiley, New York, pp. 202–204.
66. Doty, P., Gellert, M., et al. (1952), The association of insulin. I. Preliminary investigations, *J. Am. Chem. Soc*. **74**: 2065–2069.
67. Attri, A. K. and Minton, A. P. (2005), Composition gradient static light scattering: A new technique for rapid detection and quantitative characterization of reversible macromolecular hetero-associations in solution, *Anal. Biochem*. **346**: 132–138.
68. Winzor, D. J. and Scheraga, H. A. (1964), Chemically reacting systems on Sephadex. II. Molecular weights of monomers in rapid association equilibrium, *J. Phys. Chem*. **68**(1): 338–343.
69. Winzor, D. J. (2003), The development of chromatography for the characterization of protein interactions: a personal perspective, *Biochem. Soc. Trans*. **31**(5): 1010–1014.
70. Ackers, G. K. and Thompson, T. E. (1965), Determination of stoichiometry and equilibrium constants for reversibly associating systems by molecular sieve chromatography, *Proc. Natl. Acad. Sci. USA* **53**: 342–249.
71. Chun, P. W., Kim, S. J., et al. (1969), Determination of the equilibrium constants of associating protein systems. III. Evaluation of the weight fraction of monomer from the

weight-average partition coefficient (application to bovine liver glutamate dehydrogenase), *Biochemistry* **8**(4): 1625–1632.

72. Prochazka, K., Mandak, T., et al. (1990), Behavior of reversibly associating systems in size exclusion chromatography. Interpretation of experimental data based on theoretical model, *J Liq. Chrom.* **13**: 1765–1783.
73. Gutfreund, H. (1948), Molecular weight of insulin and its dependence on pH, concentration, and temperature, *Biochem. J.* **42**: 544–548.
74. Lord, R. S., Gubensek, F., et al. (1973), Insulin self-association. Spectrum changes and thermodynamics, *Biochemistry* **12**(22): 4385–4391.
75. Jeffrey, P. D. (1986), Self-association of des-(B26-B30)-insulin. The effect of calcium and some other divalent cations, *Biol. Chem. Hoppe-Seyler* **367**(5): 363–369.
76. Helmerhorst, E. and Stokes, G. B. (1987), Self-association of insulin. Its pH dependence and effect of plasma, *Diabetes* **36**(3): 261–264.
77. Milthorpe, B. K., Nichol, L. W., et al. (1977), The polymerization pattern of zinc(II)-insulin at pH 7.0, *Biochim. Biophys. Acta* **495**(2): 195–202.
78. Mark, A. E. and Jeffrey, P. D. (1990), The self-association of zinc-free bovine insulin: Four model patterns and their significance, *Biol. Chem. Hoppe-Seyler* **371**(12): 1165–1174.
79. Hansen, J. F. (1991), The self-association of zinc-free human insulin and insulin analog B13-glutamine, *Biophys. Chem.* **39**(1): 107–110.
80. Brems, D. N., Alter, L. A., et al. (1992), Altering the association properties of insulin by amino acid replacement, *Protein Eng.* **5**(6): 527–533.
81. Yip, C. M., DeFelippis, M. R., et al. (1998), Structural and morphological characterization of ultralente insulin crystals by atomic force microscopy: Evidence of hydrophobically driven assembly, *Biophys. J.* **75**(3): 1172–1179.
82. Howey, D. C., Bowsher, R. R., et al. (1994), [Lys(B28), Pro(B29)]-human insulin—a rapidly absorbed analog of human insulin, *Diabetes* **43**(3): 396–402.
83. Becker, R., Wollmer, A., et al. (2007), Self-association of insulin glulisine in solution—results of a comparative circular dichroism study, *Diabetes* **56**: A124–A125.
84. Jeyabalan, A., Shroff, S. G., et al. (2007), *The Vascular Actions of Relaxin*.
85. Cromwell, M. E. M., Felten, C., et al. (2007), Self-association of therapeutic proteins, in *Misbehaving Proteins*, Murphy, R. M. and Tsai, A. M., eds., Springer, New York, pp. 313–330.
86. Moore, J. M. R., Patapoff, T. W., et al. (1999), Kinetics and thermodynamics of dimer formation and dissociation for a recombinant humanized monoclonal antibody to vascular endothelial growth factor, *Biochemistry* **38**(42): 13960–13967.
87. Panka, D. J. (1997), Glycosylation is influential in murine IgG3 self-association, *Mol. Immunol.* **34**(8–9): 593–598.
88. Minton, A. P. (2007), Static light scattering from concentrated protein solutions, I: General theory for protein mixtures and application to self-associating proteins, *Biophys. J.* **93**(4): 1321–1328.
89. Shire, S. J., Shahrokh, Z., et al. (2004), Challenges in the development of high protein concentration formulations, *J. Pharm. Sci.* **93**(6): 1390–1402.
90. Minton, A. P. (2000), Implications of macromolecular crowding for protein assembly, *Curr. Opin. Struct. Biol.* **10**(1): 34–39.
91. Ellis, R. J. (2001), Macromolecular crowding: Obvious but underappreciated, *Trends Biochem. Sci.* **26**(10): 597–604.

92. Minton, A. P. (2001), The influence of macromolecular crowding and macromolecular confinement on biochemical reactions in physiological media, *J. Biol. Chem.* **276**(14): 10577–10580.

93. Attri, A. K. and Minton, A. P. (2005), New methods for measuring macromolecular interactions in solution via static light scattering: Basic methodology and application to nonassociating and self-associating proteins, *Anal. Biochem.* **337**(1): 103–110.

94. Minton, A. P. (2007), The effective hard particle model provides a simple, robust, and broadly applicable description of nonideal behavior in concentrated solutions of bovine serum albumin and other nonassociating proteins, *J. Pharm. Sci.* **96**(12): 3466–3469.

95. Muramatsu, N. and Minton, A. P. (1989), Hidden self-association of proteins, *J. Mol. Recogn.* **1**: 166–171.

96. Jimenez, M., Rivas, G., et al. (2007), Quantitative characterization of weak self-association in concentrated solutions of immunoglobulin G via the measurement of sedimentation equilibrium and osmotic pressure, *Biochemistry* **46**(28): 8373–8378.

97. Liu, J., Nguyen, M. D. H., et al. (2005), Reversible self-association increases the viscosity of a concentrated monoclonal antibody in aqueous solution, *J. Pharm. Sci.* **94**(9): 1928–1940.

98. Kanai, S., Liu, J., et al. (2008), Reversible self-association of a concentrated monoclonal antibody solution mediated by Fab-Fab interaction that impacts solution viscosity, *J. Pharm. Sci.*

PART III

DEVELOPMENT OF A FORMULATION FOR LYOPHILIZED DOSAGE FORM

18

DESIGN OF A FORMULATION FOR FREEZE DRYING

Feroz Jameel and Mike J. Pikal

18.1. INTRODUCTION

Most biomolecules are quite stable in aqueous solutions for short periods of time. However, a pharmaceutical product is expected to have adequate stability over a shelf life of typically 18 months or more. Not all proteins possess this desirable long-term stability in the aqueous state because of their fragile complex higher-order structures and their sensitivity to the microenvironment. When a product does not exhibit sufficient stability in aqueous solution, it must be produced in a stable, solid form. It is well known that the stability of most (small) molecules normally increases in the order of solution < glassy solid < crystalline solid, due to the increasingly restricted mobility of the reacting species in these phases [1–3]. There are several drying techniques such as lyophilization, spray drying, spray freeze drying, vacuum drying, and supercritical fluid technology that provide a solid-state product; however, all these techniques have pros and cons associated with them. Spray drying, which is increasingly becoming the most important method for dehydration in the food industry, is seeing success in producing biopharmaceutical powders where the size and morphology are central

Formulation and Process Development Strategies for Manufacturing Biopharmaceuticals, edited by Jameel and Hershenson

to the delivery and performance of the protein powder. However, control of residual moisture content and stresses imposed by atomization and interfaces continue to be challenging factors in meeting stability requirements and process yields.

Lyophilization, has been the method of choice for products intended for parenteral administration because of the following advantages: (1) it is a low-temperature process, and hence is expected to cause less thermal degradation compared to a "high temperature" process, such as spray drying; (2) it does not involve a terminal sterilization step and maintains sterility and "particle free" characteristics of the product much more easily than do other processes; (3) it offers a method of controlling residual moisture content and headspace gas composition in the vial for those products whose storage stability is influenced by residual moisture content and vial headspace gas composition such as oxygen; and (4) its scale-up is easy and it provides reasonable process yields.

Although lyophilization is recognized in the biopharmaceutical industry as a technique of choice for stabilization of labile protein molecules, it comes with several challenges and requirements that the formulation scientist must be aware of before embarking on formulation design.

18.2. REQUIREMENTS AND EXPECTATIONS OF A LYOPHILIZED PRODUCT

First, lyophilization itself can cause reversible or irreversible denaturation of the protein during processing. A lyophilized product is expected to have adequate stability, quality, and commercial manufacturing viability. The formulation and process should be designed in such a way that the protein will be able to withstand stresses imposed during the various steps or phases of freeze drying, namely, in-process stability, as well as long-term storage stability and sufficient stability postreconstitution. The lyophilized product should appear pharmaceutically elegant without collapse or meltback; should have low residual moisture content, rapid/easy reconstitution, and ease of administration; and should be compatible with the device (reconstitution kit). The process should be efficient, low-cost, and allow manufacturing without "major equipment restrictions" and be robust enough to be implemented on any typical production freeze dryers at any site. The design of the formulation directly or indirectly influences all the attributes mentioned above.

Typically, prior to design of the formulation, a thorough preformulation study is performed to understand the physicochemical characteristics of the protein molecule, the instabilities relating to solution conditions (pH, ionic strength etc.), and stability on exposure to various unit operations of manufacturing such as the freeze–thaw process of the bulk drug substance (BDS), the mixing, filtration, shaking (agitation), filling, and interactions with the containers of the final presentations and devices. This information is critical and aids in the design of a robust formulation. However, the effects of processing conditions on the integrity of the molecule are studied at the lab scale, and these studies can potentially be misleading as they are not necessarily representative of or mimic the large-scale process. It is suggested here that scaled-down models be built that would be representative of large-scale unit operations and that these models be used to screen formulation candidates to identify the best ones and backup units. The instabilities that are observed in the solution state continue to persist in the solid state. The advantage of the lyophilization process is that it converts the solution into solid dry powder by the removal of water and slows the rates of the reactions that involve water as reactant, thereby enhancing the pharmaceutical stability and achieving the desired shelflife. But drying comes with a price; it adds a new set of instabilities and challenges. Thus, during the design of the "formulation for freeze drying," the formulation scientist needs to take into account the instabilities observed in the solution state and the additional instabilities imparted by the lyophilization process, which will be mostly physical instabilities: both conformational and interfacial.

18.3. IN-PROCESS INSTABILITY

18.3.1. Freezing Phase

Freezing, which is the first step in the lyophilization process, is actually a dehydration process where the main objective is the removal of water as a solvent and essentially complete conversion of the water into ice. Depending on the freezing protocol, the product can potentially undergo freezing-induced denaturation by two common mechanisms: cryoconcentration [4–6] and ice–liquid surface denaturation [7]. In addition to these, some proteins tend to undergo spontaneous unfolding at low temperatures; this is termed *cold denaturation* [8].

18.3.1.1. Freeze Concentration (Cryoconcentration). As the temperature is lowered after the initial ice nucleation and crystallization, the product cools with continuous conversion of water into ice, thereby decreasing the amount of water in the remaining liquid phase and increasing the concentration of all solutes in the remaining solution. The process of solute concentration continues until the excipients either crystallize or until the system increases in viscosity sufficiently to transform into a solid amorphous system, or glass, which is deposited in the spaces between the ice dendrites. This process continues and slows down since the freezing point increasingly lowers as the solutes concentrate. Complete conversion of water into ice becomes practically impossible at the freezing temperatures that are typically employed. This result in a glassy system with almost 20% water remaining unfrozen and dispersed within the concentrated solutes, which reach concentrations as high as 20–50 times their initial concentrations [4–6]. This is the freeze concentrate state, where adverse reactions may potentially occur. Freeze concentration causes an increase in viscosity, decrease in water activity, and changes in surface and interfacial tension. It also results in an increase in concentration of protein and will dramatically increase the probability of colloidal instability, which can potentially lead to denaturation and aggregation of a protein [9,13]. For example, if we have a system that contains 1% solute and is susceptible to a second-order degradation reaction with the typical activation energy of 25 kcal/mol, if the temperature is decreased from 5°C to −25°C, the rate constant would decrease by a factor of 200. However, an increase in concentration from 1% to 76% would increase the concentration factor in the rate equation by a factor of 5800, resulting in a factor of 29 net increases in reaction rate (M. J. Pikal, unpublished data). Other excipients such as ionic salts and buffer species, if present in the formulation, will also concentrate during the freezing process. For example, if a formulation contains 0.15 M NaCl, the concentration of NaCl will increase to 6 M before NaCl forms a eutectic with ice. An exposure of protein to such high ionic strengths could contribute to the instability of the native conformation [10–12]. Although buffers are included in the formulations with the objective of maintaining the pH constant, an increase in concentration and decrease in temperature may result in selective crystallization of the less soluble buffer component, causing dramatic shifts in pH. The classic example is the sodium phosphate buffer system. It can show a dramatic decrease in pH of about four units following crystallization of the basic components of the buffer system. On the other hand, the potassium phosphate system shows an increase in pH with freezing [14,26]. Additionally, the amount of oxygen that becames trapped and concentrated during freezing can increase to 1150 times that at ambient conditions [13]. The formation of antioxidants resulting from interaction of amino acids with aldehydes is believed to occur maximally at −24°C, and the dielectric constant is believed to be more favorable in promoting nucleophilic associations in ice than in water [13].

18.3.1.2. Ice–Liquid Surface Denaturation. Proteins are known for their tendency to denature at interfaces, and the ice–aqueous interface offers a potential site for denaturation. Strambini and Gabellieri, using phosphorescence techniques to monitor the decay lifetime of tryphtophan residues, demonstrated that freezing of aqueous solutions of proteins causes proteins to denature at the ice–liquid interfaces. This denaturation often results in the perturbation or loosening of the native conformation, leading to loss of tertiary and secondary structure [7], which in some cases is reversible and in others is irreversible, thereby producing significant losses. The ice-specific surface area depends on the degree of underccoling, which, in turn, depends on the cooling rate. Thus, the higher the degree of undercooling, the smaller will be the size of the ice and the larger will be the ice-specific surface area. If the active drug, the protein, is susceptible to ice–liquid interface denaturation, a large ice surface area means more interfacial denaturation, perhaps leading to increased aggregation during the freezing process [16–18]. The nucleation temperature is stochastic and a colligative property; it varies with number and nature of solute. Additionally, anything that provides a nucleation site, such as the container surfaces, particles present in the liquid solution, and protein aggregates, will also affect the degree of undercooling [18].

18.3.1.3. Cold Denaturation. As proteins tend to undergo denaturation or structural transitions at elevated temperatures, which is often denoted T_m, some cold-labile proteins may similarly undergo unfolding on exposure to subambient temperatures. Cold denaturation is a specific interaction between the nonpolar groups of protein and water and is dependent on temperature. Hydration of these nonpolar groups is believed to be thermodynamically favored; in other words, Gibbs free energy of hydration is negative and increases with the decrease in temperature, leading to unfolding and resulting in buried nonpolar groups becoming exposed to water [8]. Cold denaturation is commonly observed with multimeric proteins such as LDH, asparaginase, phosphofructokinase, and even smaller proteins like lysozyme, which are termed *cold-labile*.

18.3.2. Drying Phase

During the drying phase the protein can potentially undergo additional stresses arising from the removal of "bound" water and the elevated temperatures that are used to drive water off. Hageman concluded that the protein surface contains up to 7% of water bound to the charged residues on the surface and at least this amount of water is critical for a protein to stay in the compact native conformation [21]. Hageman also estimated that antibodies, in general, have over 500 water-binding sites on their surfaces and require approximately 7% water to interact with these sites.

After the end of primary drying phase, depending on the composition of formulation, there will be 10%–20% residual water present. Secondary drying, which is employed to reduce the moisture levels to $\leq$1%–2% in an attempt to improve the storage stability, often unbalances the hydrogen bonding, perturbing the native conformation, which results in reversible or irreversible unfolding [19,20]. Replacing this monolayer of water with hydrogen-bonding excipients is commonly believed to be critical to the stabilization of the structure of the molecule in the solid state [21].

18.3.3. Storage Phase

No new stresses are developed during the storage that tend to destabilize the protein and compromise product stability. However, in addition to chemical degradation reactions, the stresses that were developed and identified during processing will continue to remain and pose challenges to the storage stability of the product. Additionally, the subtle stresses that developed during the processing and failed to be identified or picked up at the timescale of processing because of lack of sensitivity of the characterization techniques will manifest and may present surprises during storage. Higher levels of higher-order aggregates and/or low-molecular-weight (LMW) species and/or chemical degradants arising from oxidation or deamidation does not imply more faster reaction in the solid sate but is because of the timescale of storage, which is much longer than the timescale of processing [19,20]. It is believed that the conformational alterations due to stresses that occurred during processing continue to occur during storage and that a fraction of these conformationally altered species aggregate spontaneously on reconstitution of the lyophile, the degree of which depends on the nature of the protein system, procedure of reconstitution, and the reconstitution medium [18]. The remaining fraction tends to refold to the native conformation.

18.4. EXCIPIENTS

18.4.1. Bulking Agents

Formulations that contains small amounts of drug, such that the drug concentration amounts to much less than 1%, require a bulking agent that acts as "filler" to increase the density of the product; otherwise, one may observe a product "blowout." Inclusion of bulking agents in the formulations that contain concentrations >3% seldom require a bulking agent, and use of a bulking agent under these circumstances requires a strong rationale, as such use can potentially complicate the freeze-drying process. Besides their use as fillers, bulking agents are included in the formulations for various other uses, such as to enhance the pharmaceutical elegance of the cake and increase the maximum allowable product temperature during primary drying. Depending on the demands of the final product, one can select from a pool of freeze-drying excipients that give either a crystalline or an amorphous cake. Disaccharides such as sucrose, trehalose, and lactose; Trisaccharides such as raffinose; and other sugars such as hydroxethyl starch (HES), dextran, and Polyvinylpyrrolidine (PVP), can be used as amorphous bulking agents. They tend to remain amorphous during processing and on storage under dry conditions. However, most of them except HES and dextran have low collapse temperatures and will require drying at very low temperatures, making the process time much longer, an attribute not desirable in pharmaceutical manufacturing. The use of lactose needs to be questioned for protein pharmaceuticals as it is a reducing sugar and can easily form adducts with the amine groups on proteins. Dean and co-workers used dextran in combination with dissacharides in the lyophilization of actin and reported formation of a strong elegant cake structure [63]. Glycine and mannitol are the most commonly used crystalline bulking agents and are often preferred over amorphous bulking agents because of their relative ease of freeze drying. Mannitol and glycine both are easily crystallizable, are easy to reconstitute, and possess high eutectic temperatures ranging from approximately -1°C to -3°C, an attribute very useful in carrying out the primary drying at a high product temperature without collapse and loss of elegance. Although mannitol is used more commonly than glycine, mannitol-based formulations are susceptible to vial breakage during lyophilization if (1) proper freezing protocols are not designed, (2) high fill volume is used, and (3) mannitol is present at high concentration. Both bulking agents, mannitol and glycine, form various polymorphs on lyophilization. However, variation in crystal form is normally of little concern from a freeze drying perspective; however, the hydrate form of mannitol, if present, could prove to be deleterious to storage stability of the product [22–25]. If mannitol hydrate is not desolvated during the freeze-drying process, desolvation can occur during storage, thereby releasing water of hydration, depressing the glass transition temperature T_g of the product, and compromising the stability. In our experience, secondary drying at temperatures in excess of 50°C appears necessary to efficiently desolvate the mannitol hydrates [84]. The formulation scientist and the process engineer should not only ensure that the intended crystallization of excipients are complete during the freezing phase but also appreciate the formulation factors that favor complete crystallization. Presence of other solutes in the formulation, regardless of whether they remain amorphous or crystallize, will tend to inhibit or impede the crystallization of the bulking agent; hence, as a general formulation rule, the weight ratio of crystalline bulking agent to all other solutes should be a factor of $\geq$3.

18.5. BUFFERS

The attributes expected from a buffer for formulations intended for lyophilization are no different from those with liquid dosage forms:

1. The buffer is expected to be nonvolatile so that its components are not lost during mixing, filtration, fill, and finish, especially during lyophilization. If acetic acid or hydrochloric acid is used to adjust the pH, the lyophilization process can cause significant shifts in pH.
2. The buffer is expected to have high T_g'–collapse temperature, so that the overall formulation T_g'–collapse temperature values are not significantly depressed, enabling frozen storage of bulk drug substance and freeze drying at normal temperatures. Alternatively, only a very small amount of buffer is used so that depression of Tg′ is minimal.
3. The buffer must not crystallize on freezing, causing shifts in pH, and should not have a pK_a that is highly temperature-sensitive.

Tris-HCl buffer is quite sensitive to temperatures; with a decrease in temperature from 25°C to 0°C, an increase in pH by 0.8 unit is normally observed [26]. Buffers are included in the formulations with the intent of controlling and keeping the pH constant; however, shifts in pH may occur during the freezing stage as a result of selective crystallization of one of the buffer components. Such shifts in pH can lead to degradation, compromising the quality of the product as both the solubility and stability of the molecule are dependent on pH [27,28]. As stated above, the classic example is a phosphate buffer system, where sodium phosphate buffer may show a dramatic decrease in pH by four units due to crystallization of dibasic component, $Na_2HPO_4 \cdot 2H_2O$ [29], while the potassium phosphate system exhibits only a modest increase in pH of about 0.8 pH unit on freezing, even under equilibrium conditions. However, as one would expect, under nonequilibrium conditions and lower buffer concentrations, the pH shifts are moderated since the degree of crystallization decreases as the concentration decreases [29]. Gomez et al. have demonstrated that crystallization behavior of phosphate buffer is dependent on the initial concentration, cooling rates, and sample size [30]. Data collected on large-volume phosphate systems using freezing rates representative of freezing in vials are summarized in Table 18.1. Data indicate that the higher the concentration of buffer, the more prone is the buffer to larger pH shifts. Potassium phosphate buffers also may undergo large shifts in pH during freezing. As indicated in Table 18.1, if the initial pH is 5.5, the 100 mM potassium phosphate buffer increases in pH by 3.1 units during freezing. Larsen [27,28] has shown with a low-temperature electrode that citrate buffers in general are resistant to pH shifts on freezing; here, only a slight increase in pH (from 6.0 to 6.4) was observed. Crystallization was also observed with succinate buffer on warming of frozen solution [26].

TABLE 18.1. Shifts in pH During Nonequilibrium Freezing of Phosphate Buffers

Concentration, mM	Initial pH	Frozen pH	ΔpH
	Sodium Phosphate		
100	7.5	4.1	−3.4
8	7.5	5.1	−2.4
	Potassium Phosphate		
100	7.0	8.7	+1.7
100	5.5	8.6	+3.1
10	5.5	6.6	+1.1

Sources: Data taken from Gomez et al. [30].

18.6. STABILIZERS

Proteins in the solid state are susceptible to conformational changes, in addition to chemical degradation such as oxidation and deamidation. A stabilizer is an excipient that is added to the formulation to maintain the physico chemical stability of the drug active during processing and on storage.

18.6.1. Minimizing Oxidation

Protein oxidation is the covalent modification of a protein induced either directly by reactive oxygen species or indirectly by reaction with secondary byproducts of oxidative stress. Several residues on the proteins and peptides such as methionine, cysteine, histidine, tyrosine, and tryptophan are susceptible to oxidation under some conditions. The mechanism by which they undergo oxidation is quite complex and is potentially different for each amino acid, but they all have one thing in common. They all require the conversion of molecular oxygen into reactive oxygen by transition metals [iron(III) or copper(II)] or an appropriate electron donor. Methionine undergoes metal-induced oxidation to methionine sulfoxide or sulfones and other products, histidine residues are easily oxidized in metal-catalyzed autooxidation pathways to 2-oxohistidine, tryptophan and tyrosine may be degraded by light-catalyzed oxidations [41–44], and cysteine oxidizes to form either "nonnative" intra- or intermolecular disulfide bonds.

Chemical degradation by oxidation involves partial unfolding of the protein structure unless the oxidative residues are on the surface of the protein. In cases where prior structural change is required for the oxidation processes, stabilization can be facilitated by stabilizing the protein's native state through the modulation of solution conditions such as pH, temperature, and the presence of cosolutes [64]. Alternately, one may remove the sensitive residue by site-specific mutagenesis, replacing the labile residues with stable residues [65]. Despite modulating solution conditions to stabilize the structure, and even though the freeze drying is carried out under vacuum and the vials are backfilled with a blanket of N_2, some level of oxidation occurs in the solid state during processing and on storage. Oxygen trapped in the amorphous phase during freezing that was not removed during drying will be available for reaction, and not much oxygen, on a weight basis, is required for significant oxidation of a large molecule such as a protein. Specific chemical components may be included in the formulation to retard the oxidation rate, but one needs to be aware of their limitations. Ethylenediaminetetraacetic acid (EDTA) is commonly used as a chelating agent to complex with trace metals, and it works effectively in many instances. However, in some cases the complex can bind to the metal binding site on the protein and generate reactive oxygen exactly where it will do the damage. Similarly, ascorbates function as antioxidants but also may act as a prooxidant. Scavengers (i.e. thiourea) for reactive oxygen species such as hyroxyl radical or singlet oxygen may be included to retard the oxidation process. Li et al. found mannitol to be an effective scavenger for the dithiothreitol prooxidant system [44]. Care needs to be taken to eliminate or minimize the sources of transition metal and reducing agent contamination. Obviously, oxygen content in the solution being processed and leaching of metals from the equipment needs to be minimized. Poly(ethylene glycols) and nonionic surfactants are commonly included in the formulations to stabilize, but they are known to contain peroxides. Hence, both bulk drug substance and the excipients should be tested for purity, and tighter purity specifications for transition metal ions, prooxidants, and peroxides need to be established.

18.7. DESIGN OF FORMULATION

The first step toward the design of a robust formulation is the identification of instability(ies) during the various phases of freeze drying. This information is critical to the efficient selection of appropriate excipients and screening of formulation candidates; otherwise, much time will be wasted in following a trial-and-error approach. In a single well-designed experiment one can identify the problem. Perform a freeze drying run on a protein solution without any excipients, take out a few vials after the end of the freezing phase, and complete the freeze drying with the rest of the vials. Examine freeze–thaw and freeze-dried stability to isolate the problem.

If the freeze–thaw studies determine that the protein molecule is sensitive to freezing-induced denaturation and variations in freezing process parameters does not completely resolve the problem, addition of a stabilizer is required. Sometimes a combination of stabilizer and optimization of freezing protocol will be more effective than addition of stabilizer alone. As indicated in Section 18.3.1, several adverse reactions can occur during freezing, and a number of distinct "stresses" may develop during freezing. Thus, the proper choice of stabilizer is not always a straightforward task. Certainly, instability could be due to a shift in pH in the frozen state caused by selective crystallization of one of the buffer components if a protein's structural integrity is sensitive to pH shifts.

18.8. SELECTION CRITERIA FOR VARIOUS MATERIALS

18.8.1. Buffers

One should avoid the use of buffers that tend to crystallize on freezing or at least use small amounts such that the weight ratio of buffer to other solutes is so low that

TABLE 18.2. Effect of Sucrose on pH Shift in Sodium phosphate Buffer Systems During Nonequilibrium Freezing[a]

Weight Ratio, Sucrose : Buffer	pH Shift at −10°C
0.0	−3.4
0.36	−3.3
1.0	−2.9
2.0	−2.5
6.0	0.0[b]

Source: Data taken from Gomez et al. [30].

[a]The buffer concentration is 100 mM with an initial pH of 7.5; pH shift is final pH minus initial pH.

[b]No crystallization of buffer observed in freeze-dried sample (X-ray powder diffraction).

TABLE 18.3. Glass Transition Temperatures T_g of Selected Excipients Measured by DSC[a]

Compound	T_g,°C	Reference
Citric acid	11	54
Glycine	(~30)	13
Lactose	114	55
Maltose	100	55
Mannitol	13	21,56
Raffinose	114	55
Sorbitol	−1.6	56
Sucrose	75	55
Trehalose	118	55
Maltodextrin 860	169	55
PVP k90	176	55

[a]Consult the references listed at the end of this chapter for details on the techniques. Value in parentheses is extrapolated fom mixtures using the Fox equation and is highly approximate.

crystallization does not occur. The influence of the presence of other solutes on buffer crystallization and resulting pH shift is indicated by the data in Table 18.2 [30]. The fact that the buffer capacity increases with the concentration may motivate one to add high levels of buffer salts, but such additions must be carefully reviewed. High buffer concentrations increase the possibility of pH shifts during freezing; even if crystallization does not occur, high levels of amorphous buffer salts are problematic since salts normally depress the glass transition temperature of the freeze concentrate [33,34] of the formulation, thereby lowering the collapse temperature. The common buffer HEPES is a good example of this effect. In these situations, the Fox equation [Eq. (18.1)] may be used to estimate the magnitude of the reduction in T_g' and collapse temperature

$$\frac{1}{T_g} = \frac{W_1}{T_{g1}} + \frac{W_2}{T_{g2}} \tag{18.1}$$

where W_1 is the weight fraction of component 1 and T_{g1} is the glass transition temperature of pure component 1. For example, if the T_g' of a buffer salt is on the order of 20°C lower than that of the remainder of the formulation, a buffer level of ~25% of the total solids would lower the collapse temperature by about 5°C, which, in turn, would increase the primary drying time by a factor of ~2. Clearly, the formulator should employ low levels of buffer, as this practice would minimize both the impact of the buffer on collapse temperature and prevent crystallization-induced pH shifts. Additionally, the formulator needs to know that proteins have ionizable groups whose pK_a values range from 1 to 12 and for high-concentration protein formulations, with the protein itself acting as a universal buffer, and one perhaps does not need to add a buffer. Acetate buffer has a very low T_g'–collapse temperature (ranging from −80°C to −64°C) and is somewhat volatile in nature (i.e., acetic acid); phosphate buffers crystallize and have the potential for significant pH shifts, and thus are not the best choices for freeze drying. Buffers that do not crystallize during freezing and have ninimal pH change include citrate, histidine, and Tris. Biological buffers like HEPES and MOPS are also good choices. Citrate buffer is perhaps the best choice, as long as the pH range is adequate, as citrate does not crystallize and has a high collapse temperature [26].

18.8.2. Surfactants

If aggregates or particulates are observed after the freeze–thaw cycle, the instability could be due to partial unfolding caused by adsorption of protein at one of the several interfaces that are created during the process of freeze drying and/or spray–freeze drying and during reconstitution of the powder. Unfolding can occur at the ice–liquid interface that forms during freezing. Reconstitution of the dried powder releases the air that was trapped during freezing and forms an air–liquid interface. Spray–freeze drying creates a large air–liquid interface during atomization in addition to the interfaces created during normal freezing, drying, and reconstitution. Of these potential sites for adsorption, the ice–liquid interface is generally the most problematic in conventional freeze drying. Several techniques have been used to determine and confirm whether protein is involved in ice–liquid interface denaturation.

Webb et al. [35] studied the amount of protein that would maximally adsorb at the air–liquid interface during atomization of the solution in spray drying applications using electron spectroscopy and rhINFγ solution. They also investigated whether any correlation exists between amount adsorbed and amount denatured as aggregates. They reported that approximately 0.54 mg of rhIFNγ/mL would adsorb at the air–liquid interface during atomization, and this number closely corresponded to the actual level of aggregates, suggesting that all the protein adsorbed at the interface subsequently results in aggregation. Using several proteins and Teflon beads as hydrophobic surfaces, Chang et al. [18] demonstrated that the particulate levels after freeze–thaw were well correlated with particulate levels after shaking protein solutions with the Teflon beads. Using phosphorescence lifetime measurements, Strambini and Gabellieri [7] demonstrated that conformational damage may, indeed, result from the interaction of proteins with the ice surface.

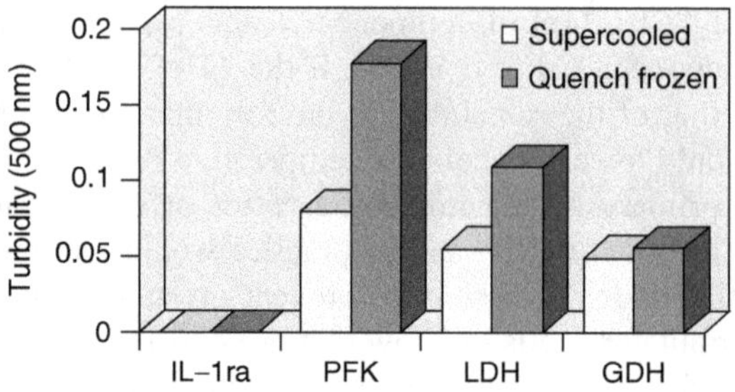

Figure 18.1. Effect of freezing rate on protein precipitation during freezing. (Courtesy Byeong Chang.)

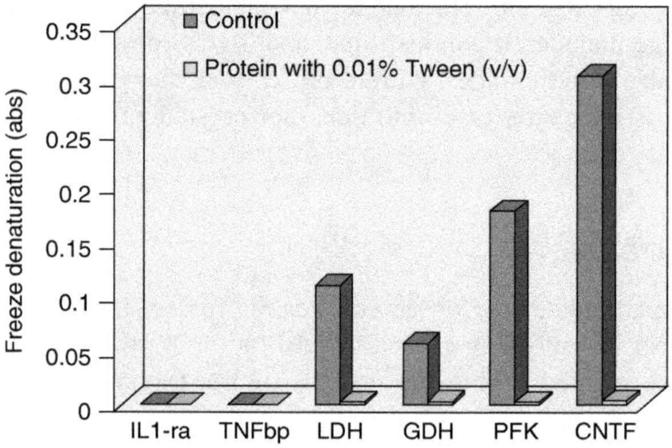

Figure 18.2. Effect of polysorbate on protein stability during freezing. (Courtesy Byeong Chang.)

Surfactants are widely used in the protein formulations for a variety of purposes: (1) as a wetting agent in the reconstitution medium to aid in reconstitution of powder; (2) to prevent loss of native protein due to adsorption at the surfaces of filters, equipment, and containers, especially in formulations with low protein concentrations; and (3) as a chaperone to aid in refolding of structurally altered forms and thus prevent aggregation. In all these applications, it is a nonionic surfactant, typically either polysorbate 20 or 80 that is used, usually in the range from 0.01% to 0.1% w/v.

There are several examples in the literature that clearly demonstrate that addition of surfactants protects the protein from interface-induced denaturation and therefore prevents the formation of particles or aggregates. Mumenthaler et al. [36] studied the effect of interfaces on the integrity of protein with spray drying and freeze drying of human growth hormone (hGH) and found that most of the aggregates, soluble and insoluble, were formed during atomization, and addition of 0.1% polysorbate 20 to the formulation greatly reduced formation of both soluble and insoluble aggregates due to interfacial denaturation.

Chang et al. [18] studied the effect of freezing rate on protein precipitation of various proteins (LDH, PFK, GDH, and IL-1ra) and found a linear correlation between the turbidity and freezing rate. They also found that the precipitation of protein can be eliminated by the addition of 0.01% polysorbate 20 (see Figs. 18.1 and 18.2). Jameel et al. studied the effect of freeze–thaw on the emission spectra and elution profiles of rhFVIII (recombinant human factor VIII) molecule in the presence and absence of polysorbate and found that the decreased fluorescence intensity, which is reflective of conformational changes, and formation of aggregates due to freeze–thaw can be eliminated by the addition of polysorbate 20 (see Fig. 18.3). Kerwin et al. reported that aggregation of hemoglobin during freeze-thaw can be essentially eliminated by addition of polysorbate 80 at concentrations from 0.0125% to 0.1% [39]. In the absence of polysorbate 80, particle formation increased with the increase in the number of freeze–thaw cycles, and after five of these cycles, particle concentration increased by more than an order of magnitude when frozen to

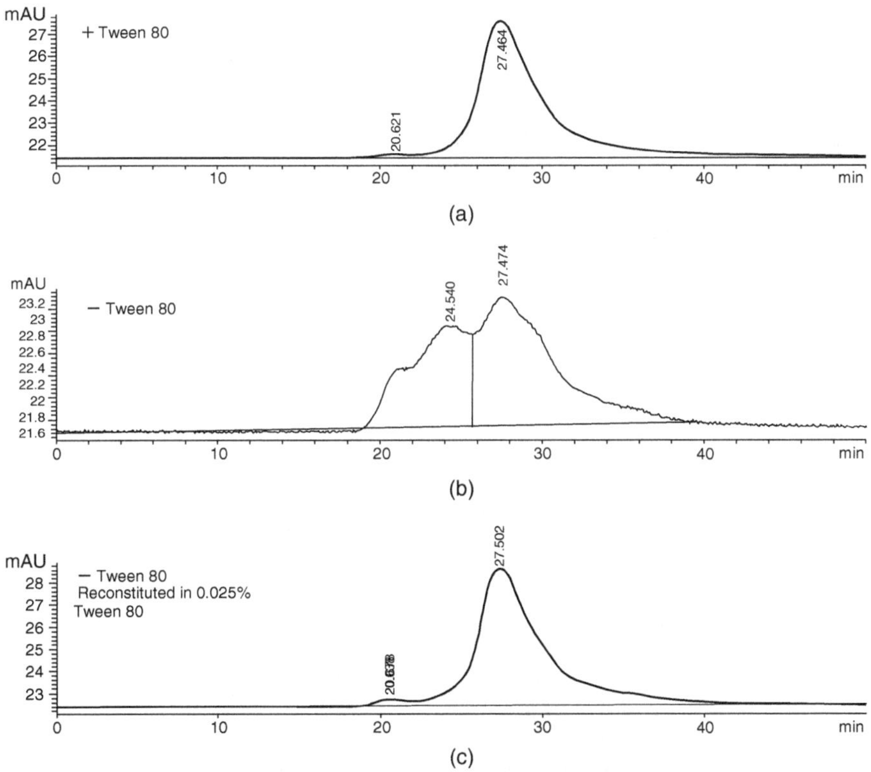

Figure 18.3. Elution profiles of rhFVIII post-freeze–thaw with and without polysorbate 80 post-freeze–thaw: (a) pure rhFVIII without exposure to freeze thaw; (b) after exposure to freeze–thaw in the absence of polysorbate 80; (c) after exposure to freeze–thaw in the presence of polysorbate 80.

−20°C and increased by a factor of ~5 when freezing to −80°C. Investigation into the mechanism of stabilization of haemoglobin by polysorbate 80 suggested that the stabilization mechanism does not involve binding of surfactant to the protein. However, it was speculated that a competitive displacement mechanism may be operative. However, in all cases, addition of low levels (~0.01%) of surfactants, either polysorbate 20 or 80, greatly minimized particulate formation. Thus, it is certainly prudent to test the effect of surfactants in situations where protein aggregation is a problem during freeze–thaw or during solution handling. If the addition of surfactant and practical variations in the freezing process does not eliminate the formation of aggregates and particulates or loss of activity, then the protein may be susceptible to be cold lability and addition of cryoprotectant may be required.

18.8.3. Cryoprotectants

The protein structure or conformational stability of a protein is normally a delicate balance between various interactions. These interactions are sensitive to extreme temperatures (hot and cold) on either side of an optimal temperature, which is termed the *stable temperature* (see Fig. 18.4). When proteins are exposed to elevated temperatures, the free energy for unfolding decreases and the native conformation is ultimately lost on thermal denaturation. Similarly, some proteins termed *cold-labile* tend to undergo spontaneous unfolding at cold temperatures, and the process is called *cold denaturation* [8,45]. Solutes such as sugar, polyols, certain amino acids, methylamines, and "salting out" salts are added in molar concentrations into the formulations to protect proteins against denaturation. When these solutes are used to prevent conformational changes and degradation during freezing, they are termed *cryoprotectants* and work same way as in nonfrozen aqueous solutions against thermal or chaotrope

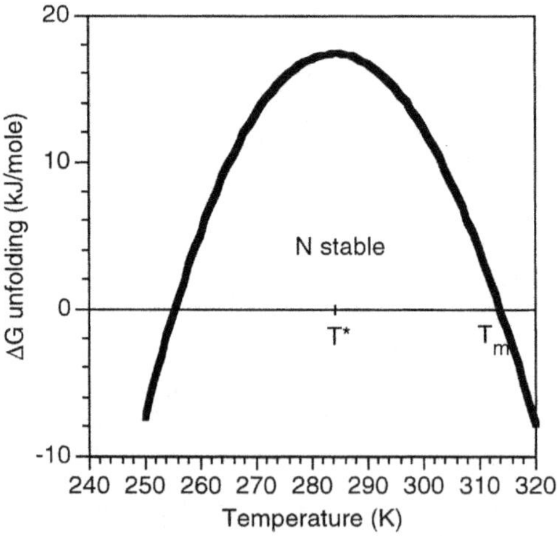

Figure 18.4. Schematic representation of the protein stability curve (Becktel and Schellman, 1987) illustrating the temperature dependence of the free energy of unfolding, ΔG.

unfolding. They work effectively in the concentration range of 300–600 mM by a mechanism called *preferential exclusion of solutes* proposed by Arakawa and Timasheff [46] and reviewed and demonstrated by Carpenter and Crowe [47,48]. According to Timasheff and co-workers, the solutes are excluded from the vicinity of protein, and their presence increases the chemical potential of the protein. This increase in chemical potential is a function of surface area; the larger the surface area, the higher will be the increase in chemical potential. Since the denatured form of protein will have a larger surface area, the chemical potential of the denatured form will be higher than that of the native form; hence the native form is thermodynamically favored. This hypothesis was tested for various protein systems by different research groups. Chang and co-workers found that sucrose was able to protect recombinant interleukin-1 receptor antagonist against the formation of aggregates from species whose tertiary structures were slightly perturbed but whose secondary structures were largely intact, similar to the molten globule state [83]. Anchordoguy et al studied a multimeric protein, LDH, and bovine serum albumin (BSA) as model proteins with polyvinylpyrrolidone (PVP) as a cryoprotectants, and found that both BSA and PVP were able to protect the quaternary structure of LDH against the dissociation into subunits during the freezing phase of lyophilization [15].

Since the increase in chemical potential of a protein due to the presence of a cryoprotectant does not necessarily correlate quantitatively with the effectiveness of that compound as a cryoprotectant, it is always advisable to screen several compounds [e.g., glycine, mannitol, disaccharides, poly(ethylene glycol)s (PEGs), PVP, certain amino acids, methylamines, and "salting out" salts) as potential cryoprotectants. High-molecular-weight PEGs are more effective than LMW ones; the classic example is lactate dehydrogenease [49]. In cases where the protein undergoes chemical degradation through oxidation, PEGs would not be a good choice as they may contain peroxides. Note that for cryoprotection, it is the concentration of the cryoprotectant in solution that is the key concentration variable, regardless of the concentration of protein in the formulation.

Depending on the protein molecule and its particular instabilities, one might need to include more than one excipient to stabilize. For example, PEG stabilizes the proteins during freezing due to preferential exclusion of PEG from the surface of the protein, while sugars such as trehalose protect the proteins during drying perhaps by hydrogen bonding to the dried protein. However, one has to be aware of the fact that freeze-concentration-induced phase separation between the excipients can occur, and such phase separation may result in loss of protein native structure [66]. An excellent example is provided by hemoglobin. Heller et al. have demonstrated that when PEG and dextran were used in the hemoglobulin formulation, freezing-induced phase separation occurred between PEG and dextran, which, in turn, caused changes in protein structure. They also demonstrated that addition of sugars that remained amorphous, such as sucrose and trehalose, did not prevent the loss of native structure. Surprisingly, mannitol in quantities that crystallize did protect the protein's structure. It was suggested that the mechanism involved division of the interstitial concentrated solution volume into small and segregated regions by crystalline mannitol, thereby inhibiting the nucleation and/or growth of the PEG–dextran phase separation [61,62].

Such phase separation problems can be moderated by increasing cooling rate, allowing less time for the system to phase-separate.

18.8.4. Lyoprotectants: Stabilization During Drying and Storage

In addition to possible factors unique to the protein of interest, the long-term storage stability of a lyophilized product depends on two critical factors that need to be addressed during the drying phase: (1) the physical state, that is, creating a glass with high T_g and low residual moisture content, and (2) retention of native structure in the dried state.

We freeze-dry to increase storage stability relative to the solution state, assuming that the restricted molecular mobility of the solid state will lead to stabilization. In general, stability of an amorphous solid (i.e., glass) is several orders of magnitude more stable than the corresponding aqueous solution, and at least with small molecules, crystalline drugs are much more stable than glasses [52]. In an attempt to remove water to create a glassy solid with high T_g during secondary drying, the hydrogen bonding between protein and water is broken. Thus, it is often argued that the hydration shell responsible for keeping the native structure intact is lost, causing a significant thermodynamic destabilization of the protein molecule that may result in aggregation. The use of lyoprotectants as "water substitutes" allows a hydrogen-bonding interaction with protein similar to that between water and the protein to occur, so that protein is thermodynamically stabilized. The stabilization mechanism can be interpreted both thermodynamically and kinetically. The thermodynamic interpretation of stabilization by water substitutes is that the stabilizer by hydrogen bonding with protein similar to water maintains or restores the free energy of unfolding such that thermodynamically native conformation is preferred over the unfolded form. The kinetic interpretation of stabilization via the vitreous hypothesis argues that since protein is molecularly dispersed in the amorphous component of the stabilizer, which will be very viscous and behave as a glass, the mobility will be restricted. Because of this viscous microenvironment around the protein, the mobility is restricted and the rate of conformational changes leading to unfolding of protein will be retarded or arrested and the protein will be stabilized.

Nonreducing di- and trisaccharides, such as sucrose, trehalose, or raffinose, are normally good lyoprotectants [19,20] as they would be expected to function as good water substitutes and good glass formers. Polymers are generally much less effective, possibly because of their steric hindrance in forming hydrogen bonds to the protein. One requirement of a stabilizer is that it be inert in the sense that it should be incapable of chemical reaction with the protein; hence, reducing sugars such as lactose, maltose, and glucose should not be used in protein formulations. It was reported that although the use of lactose provides high T_g and stabilizes hGH against the formation of aggregates lactose reacts with the hGH forming a Schiff base adduct [17]. Similarly, Shihong et al. studied the effects of reducing sugars on the chemical stability of recombinant human relaxin in the lyophilized state and reported covalent adducts of glucose with amino groups on the side chains of the protein (Lys and Arg) formed via Maillard reaction [67]. Under conditions of acid pH, particularly below pH 5, trehalose is a much more effective stabilizer than sucrose or raffinose, which hydrolyze to their reducing sugar components much more rapidly in the solid state than does trehalose, thereby leading to degradations initiated by reducing sugars [50]. This issue is discussed in more detail later. For stabilization during drying or storage, the key concentration variable is the weight ratio (or mole ratio) of saccharide to protein, with the weight ratio of stabilizer to protein normally ranging between 1 : 1 and 10 : 1.

The residual water content in the cake significantly influences the storage stability of the product. In addition to its role as a reactant promoting the chemical reactions, water also acts as a plasticizer affecting the physical state of the product. The effect of residual moisture content on the T_g of the commonly used amorphous stabilizers is illustrated in Figure 18.5. It can be seen that with the increase in water levels, there is a roughly linear decrease in the T_g values, making the system "less glass-like" at a given temperature, and eventually causing the glass to transform into a fluid at the storage temperature, compromising the stability requirement that protein be dispersed in the glassy phase [56]. In situations where it is not practically feasible to achieve low levels of residual moisture content in the product, use of trehalose will be beneficial as the T_g of trehalose (118°C) is much higher than that of sucrose (74°C), and any depression in T_g values will yield T_g values that are still high enough to meet the stability specification that the product glass transition temperature be well above all anticipated storage conditions. However, as long as the residual water content is maintained low and the product is not plasticized by other LMW formulation components, a sucrose formulation should be able to provide the required T_g to achieve stability at normal storage conditions. Because of a very low T_g, use of sorbitol as a major formulation component should be avoided. One should also be aware of the

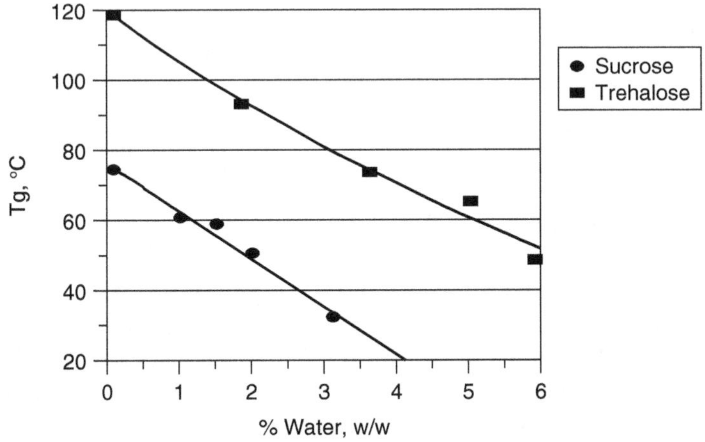

Figure 18.5. Effect of residual water on the glass transition temperature of sucrose and trehalose. (Data taken from Saleki-Gerhardt [56].)

fact that the incomplete crystallization of bulking agents such as glycine or mannitol may also depress product glass transition temperatures.

In order to determine the optimum amount of the stabilizer to meet the two important requirements mentioned above, a wide range of sugar: protein weight ratio with different sugars must be screened as this optimum weight ratio depends on both the nature of the protein and the presence of other excipients. In addition to physical properties, the screening process should involve evaluation of native structure of the protein in the solid state to get an early read of the effectiveness of both the formulation and the process; Raman and FTIR spectroscopy are commonly employed to evaluate the secondary structure of the protein in solid state. It has been shown by various groups [51–55] that stability during long-term storage in the dried solid is correlated with retention of native protein during freeze drying. Prestrelski et al. reported FTIR data on rIL-2 that exhibited a good correlation between storage stability and structure, with a more native conformation associated with greater storage stability [53]. Even with samples stored well below the T_g of the dried product, damage rose rapidly when the protein was unfolded in the dried solid. However, one will not always observe a good correlation between what is assessed with FTIR and the aggregation level observed postreconstitution. In many cases FTIR spectra may suggest changes in secondary structure, but on reconstitution one may not find a loss of activity or aggregate formation. Thus, the structural perturbations noted in the solid are usually reversed by dissolution; that is, the protein refolds to the native structure. On the other hand, FTIR spectra may suggest no change in native structure and complete protection by the stabilizer, but one may still observe degradation postreconstitution. In those situations, one might circumvent the problem by including surfactant in the diluent; the surfactant may act as chaperone and prevent any intermolecular interactions and/or function as a wetting agent that expedites the dissolution process [18].

If oxidation is a major degradation pathway, stabilization requires attention to the chemical composition of the system, including both the effects of impurities and addition of “stabilizers” as described earlier. David et al. characterized the solid-state reaction product from a mannitol-based lyophilized formulation and determined that degradation arises from oxidation. They concluded that reducing sugar impurities in mannitol were responsible for the oxidative degradation of the product via a mechanism that involves Schiff base intermediates [68].

18.9. IMPACT OF FORMULATION ON PROCESS

It is clear that one of the goals of formulation design should be to provide the highest collapse temperature that is practical within the constraints of maintaining protein stability. Higher concentration of protein in the formulation is desirable for two reasons: (1) Increasing the protein concentration in the formulation tends to increase the collapse temperature as the T'_g value of a protein is typically $-10°C$, and (2) a high concentration of protein tends to protect the protein against freezing induced denaturation. Thus, from a process throughput perspective, it is desirable to have a high weight ratio of protein to stabilizer; however, one of the stability requirements is that the stabilizer : protein weight ratio be high enough that the protein is molecularly dispersed in the amorphous component of the stabilizer with significant dilution. It is also true that exceeding the total solid mass by >10% will have significant impact on the drying process performance and that, at times, it becomes difficult to dry. Striking a fine balance between these competing requirements is one of the challenges that a formulation scientist and a process engineer need to juggle during the design of formulation and process. Details of formulation optimization are discussed in the next section.

Drying with retention of structure requires the product to be dried below the collapse temperature, which depends on the composition of the formulation. Selection of excipients should not be based exclusively on their ability to stabilize the protein but must also recognize their ability to meet the manufacturing requirements. Commercial lyophilization processes require the selection of those excipients that possess relatively high collapse temperatures and are sufficiently robust in physical state, such that slight variations in process conditions do not cause problems. The lower the T_g', the longer and more expensive will be the process. Moreover, drying of formulations with extremely low collapse temperatures may not be feasible in commercial lyophilization operations. The rate of sublimation is directly proportional to the difference in vapor pressure of ice at the product temperature and chamber pressure, and inversely proportional to product resistance:

$$\frac{dm}{dt} = \frac{(P_{\text{ice}} - P_{\text{c}})}{R_{\text{p}} + R_{\text{s}}} \qquad (18.2)$$

Equation (18.2) means that an efficient process is the one that runs at high product temperature, and a robust formulation is the one that gives high maximum allowable product temperature, specifically, collapse temperature, and low product resistance.

18.10. PRACTICAL CONSIDERATIONS IN THE DESIGN OF FORMULATIONS AND STABILITY TESTING

18.10.1. Formulation Design

18.10.1.1. General Principles. One should generally strive to minimize the complexity of the formulation, which normally means restricting the number of components to only those whose functions are important to product quality. Commonly, buffers are employed, and bulking agents and/or saccharide stabilizers may be included. Surfactants are often added to avoid product loss due to adsorption and/or degradation, particularly when the drug is a protein at very low concentration, and thus perform a stability function even though surfactants are seldom classified as stabilizers. Other types of materials may be added in special cases of instability, such as antioxidants and chelating agents, to sequester ions that may accelerate oxidation (i.e., iron). With proteins, one may add components to stability issues peculiar to a given compound, such as Ca^{2+} to stabilize conformation. Highly specific scenarios are beyond the scope of this chapter; here, we consider only stabilization principles of general applicability.

Although buffers are intended to control pH, as noted earlier, massive pH changes can occur during freezing when one component of the buffer system crystallizes. This problem is minimized by use of only a small amount of buffer relative to the sum total of other solutes that remain amorphous. Under such conditions, the buffer rarely undergoes crystallization, and instead remains amorphous, thereby avoiding the large pH shift that could otherwise occur. Stabilizers of various types are often necessary, but some data are necessary to allow the formulator to rationally proceed with formulation development. Thus, assessment of potential instability during process and storage is critical during the early phase of a project. One typically characterizes the purity (i.e., type and level of degradation products) in the starting solution and then carries out a freeze-drying process where samples are taken at various stages of the process for assessment of degradation. Samples are taken after freezing, thawed, and assayed. Samples are also taken after freeze drying is over, sometimes at various times during secondary drying, and these samples are assayed to determine the degradation during the freeze-drying process. Next, the freeze-dried samples are subjected to a brief accelerated stability test at 40°C or 50°C, but at least 10°C below the glass transition temperature, for 2 or perhaps as long as 4 weeks. The purpose of this test is to evaluate the scope of the stability problem, if any, and not to do formulation screening. Thus, one studies the pure drug, or if the fill concentration is much below 1%, a combination of the drug and 3% mannitol as a bulking agent. The stability during freeze–thaw and during freeze drying and reconstitution is examined. The preliminary assumption is that if the degradation after freeze drying and reconstitution is greater than that during freeze–thaw, the difference is attributable to degradation during the drying process, likely during secondary drying. The hidden assumption is that degradation during thawing is equivalent to that during reconstitution, a plausible but hardly exact approximation. If the drug concentration is low and significant degradation is noted during freeze–thaw, addition of a nonionic surfactant (i.e., polysorbate) might improve the stability. In addition, one might add compounds that behave as "excluded solutes" (i.e., as noted earlier, amino acids, nonreducing disaccharides, PEGs) in an attempt to thermodynamically stabilize the native conformation and thereby minimize degradation. If degradation appears to occur during drying, addition of a nonreducing saccharide stabilizer is indicated. Finally, if there is additional degradation during storage, and moisture content has been well controlled, the formulation requires addition of a nonreducing saccharide stabilizer in an amount of the same magnitude as the amount of drug. Historically, stabilizers have traditionally been nonreducing saccharides, which tend to resist crystallization, are chemically inert, and also hydrogen-bond effectively, thereby tending to maintain miscibility with the drug, particularly proteins [69]. Polysaccharides have not been found particularly effective [69]. The relevant properties are summarized in Table 18.4. Obviously, trehalose has some potential advantages over both sucrose and raffinose in that the glass transition temperatures are higher than for sucrose, and trehalose is not susceptible to acid hydrolysis. Although none of the saccharides listed in Table 18.4 are reducing sugars, acid hydrolysis leads to the formation of the constituent "reducing"

TABLE 18.4. Properties of Common Stabilizers and Bulking Agents

Stabilizer or Bulking Agent	T_g'°C	T_g°C	History of Commercial Use	Comments
			Stabilizers	
Sucrose	−35	75	Extensive	Not stable to acid hydrolysis
Trehalose	−28	118	Some	Stable to acid hydrolysis
Raffinose	−27	114	None	Not stable to acid hydrolysis
			Bulking Agents	
Mannitol	−30	13	Extensive	Good cake, vial breakage
Glycine	(−62)	31	Moderate	Fair cake, no vial breakage

Source: Data taken from Pikal [69].

monosaccharides, which can then lead to the sequence of chemical reactions with amine functionality commonly denoted as *Maillard reactions*, with subsequent degradation of the drug [69,70]. The data (Fig. 18.6) demonstrate that sucrose and raffinose hydrolyze readily at low pH, even in the solid state, but trehalose is relatively stable. Thus, for low-pH applications, trehalose is vastly superior to sucrose, and because of the relatively high T_g, trehalose is also superior to sucrose in applications where moisture control is difficult. However, in our experience [71], and in some literature

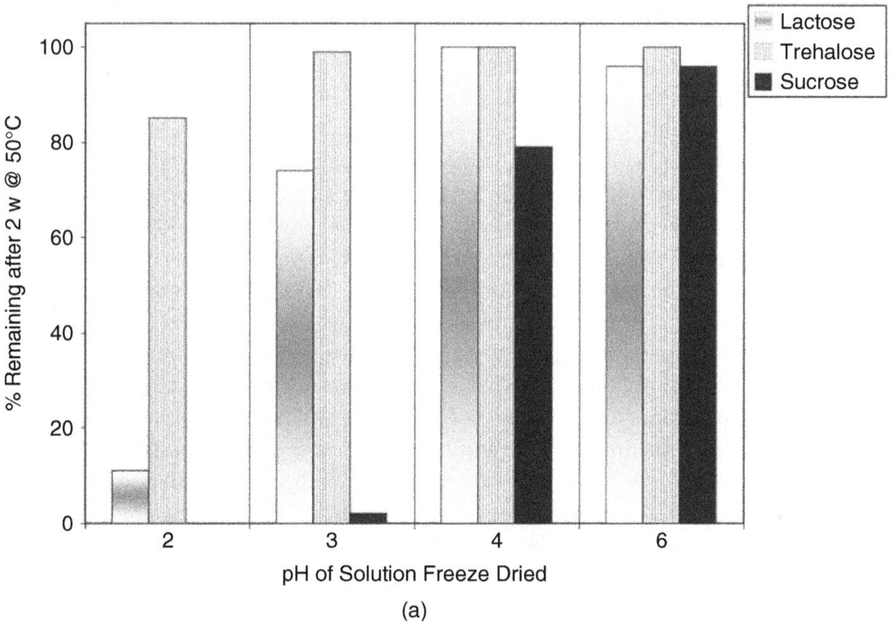

Figure 18.6. Hydrolysis of freeze-dried amorphous saccharides at acid pH: storage at 50°C (a) and 60°C (b) for 2 weeks. Assay was via HPLC with Dionex Carbopak PA1 column, and detection was performed using a pulsed ampereometric detector. (Data from Pikal et al. [50].)

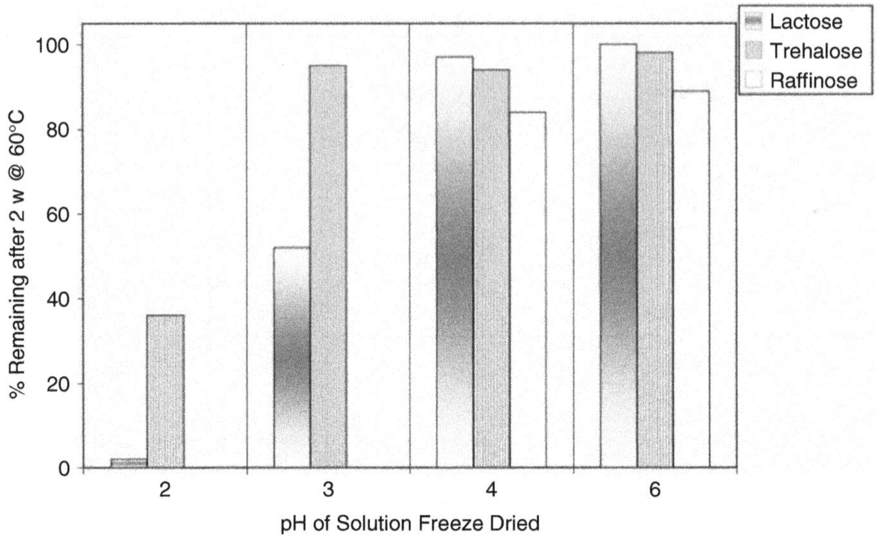

(b)

Figure 18.6. *(Continued)*

studies [72], sucrose is found to provide superior stabilization, at least for proteins. Furthermore, as long as the formulation is above pH 5 and one has the ability to control residual moisture, the problems of hydrolysis and reduction of T_g by water do not manifest themselves, and therefore, in most cases sucrose seems to be the best choice for a protein stabilizer.

Historically, mannitol has been the favorite choice for a bulking agent, although glycine also is quite satisfactory. Both have eutectic temperatures only slightly lower than 0°C, and both crystallize quite well even in the presence of other components, as long as they are the major component. However, it is critical that crystallization of the bulking agent be essentially complete as the failure to completely crystallize could produce a system with a very low glass transition temperature T_g' and therefore very low collapse temperature, particularly for glycine, and with a low glass transition temperature for the amorphous phase in the final freeze-dried product, particularly for mannitol. These problems arise because of low values of T_g' for the pure components, as well as low glass transition temperatures in dried amorphous solids for both mannitol and glycine [69]. While mannitol is unquestionably the most popular bulking agent, it does have one disadvantage; it tends to break vials when the fill depth is high [69,73]. This problem can be minimized or entirely eliminated if the system is not subjected to temperatures much below the T_g' of mannitol until after most of the mannitol has crystallized. Specifically, the initial stage of freezing should involve forming ice above −30°C and then increasing the temperature to around −20°C to allow complete crystallization of mannitol from a "slush" rather from a glassy system that can transmit mechanical stress from the expansion that occurs when mannitol and ice both crystallize.

18.10.1.2. Optimum Levels of Bulking Agents and Stabilizers. In general, it is advantageous to freeze-dry a system whose solids content is between about 3% w/w and 10%. Less than 3% may provide less elegance and may risk "blowout," but more than 10% may dry slowly because of high product resistance to vapor transport. Also, experience has shown that stabilization is normally better with higher volume (weight) ratios of stabilizer to protein (or presumably, other drug type). Thus, if possible, we attempt to use stabilizer : drug weight ratios ranging from 1 : 1 to 10 : 1. Ratios much higher than 10 : 1 tend to be impractical if the drug is present at anything except trace amounts. Note that while greater stabilization will result from higher levels of stabilizer, one does not always need the maximum stabilization, and at least with monoclonal antibodies, sufficient stabilization is sometimes achieved with weight ratios of only 0.5 : 1 [74]. With a bulking agent, at least with mannitol or glycine, crystallization of essentially all the bulking agent is critical, and to achieve complete crystallization, the crystallizable bulking agent needs to be present in significant excess; a weight ratio of 3 : 1 of bulking agent to all other amorphous solute components is recommended. Even at 3 : 1, annealing during the freezing process might be necessary to achieve the needed degree of crystallinity. Suggested guidelines for levels of bulking agents and saccharide stabilizers, consistent with the abovementioned principles, are given in Figures 18.7 and 18.8. Note that for high-dose drugs, to maintain total solute concentration below about 10%, it may be necessary to use levels of stabilizer lower than what might be desired for maximum stability. The significant question, however, is whether the level selected provides the stability needed. Finally, it should be emphasized that these are general guidelines, not hard rules, and on some

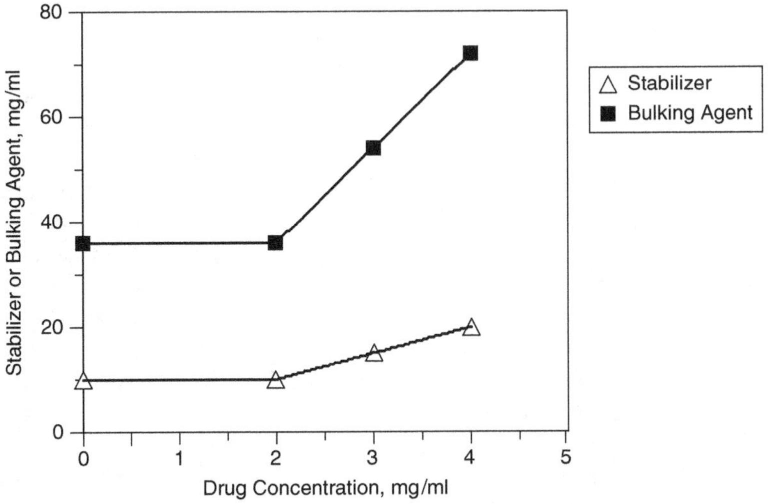

Figure 18.7. Suggested (approximate) levels of stabilizer and bulking agent for low-dose drugs.

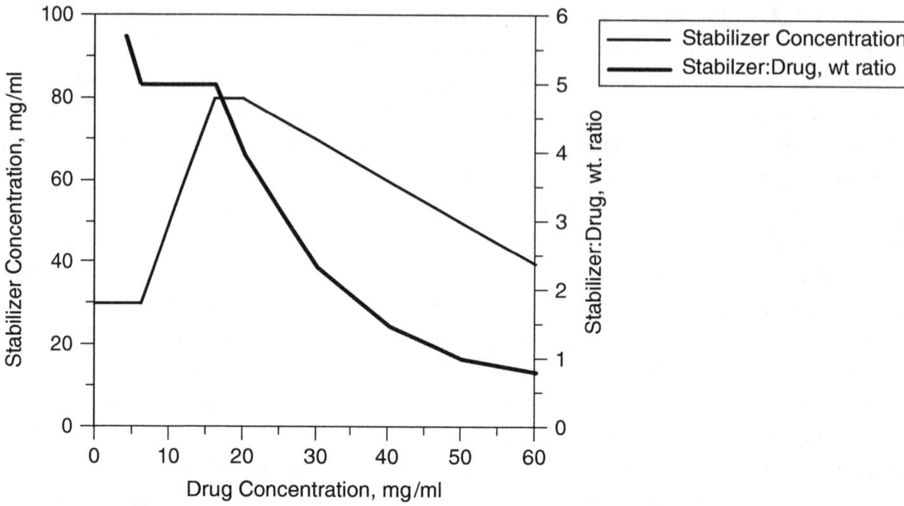

Figure 18.8. Suggested (approximate) levels of stabilizer concentration and resulting stabilizer : drug weight ratio for high-dose drugs.

occasions it may be necessary to formulate well outside these guidelines. For example, we have successfully freeze-dried from systems of 30% w/w solids.

While the discussion above has focused on selection of a single stabilizer, typically either sucrose or trehalose, we note that more recent studies suggest that enhanced stability may be produced by including a few percent of a "small molecule," such as glycerol or sorbitol, in the disaccharide stabilizer system [78,79]. Although such addition plasticizes the disaccharide : protein system ratio and lowers the T_g, it appears that "fast dynamics" is antiplasticized [80], and the stability clearly improves [78,79]. Thus, whenever a disaccharide stabilizer is selected, it would be useful to also evaluate a disaccharide stabilizer system to which has been added glycerol or sorbitol at roughly 2%–5% of the disaccharide content.

Combinations of stabilizer system and bulking agents can also be effectively utilized whenever a drug is present in a very small amount. Thus, with low levels of drug, a high stabilizer : drug weight ratio can be utilized with the total amorphous solute still being very low in concentration. Thus, a bulking agent, such as mannitol, can also be added, even at the recommended 3 : 1 weight ratio of bulking agent to amorphous phase, without driving the total solids content beyond the recommended 10%. Thus, one might develop a formulation for a drug at 1 mg/mL by using a 10-mg/mL stabilizer system (i.e., disaccharide-based) and a 50-mg/mL bulking agent (i.e., mannitol). The total concentration is about 6%, well below the "limit" of 10%, and is consistent with the recommended relative amounts of both stabilizer and bulking agent. Yet, such a system could be dried well above the collapse temperature of the amorphous phase without giving the appearance of macroscopic collapse. If so, elegance would be preserved, the sample would maintain good pore structure and would therefore undergo secondary drying well, and it would reconstitute well. As long as drying above the amorphous phase T_g' would not pose a stability problem, the net result would be a process that significantly reduces the process time yet maintains all valid measures of product quality.

Patient compliance and convenience requirements may warrant flexibility in presentations, such as multiple protein concentrations, and multiple doses from the same configuration. These will pose additional challenges on formulation design. For instance, the presence of a bacteriostatic agent necessary for a multidose presentation may have stability implications for the active drug. Isotonicity is critical for subcutaneous administrations but not critical for IV administrations, where isotonicity is usually controlled by the large-volume parenteral used in administration. Adjustment of isotonicity with sodium chloride needs to be questioned as its presence in the formulation significantly alters the freeze-drying properties of the formulation. In such situations an effective strategy would be to keep the bacteriostatic agent and any tonicity modifiers such as sodium chloride out of the formulation and include them only in the diluent [69]. Similarly, changes in diluent volume and composition can be used to deliver multiple protein concentrations from a single lyophilized concentration.

18.10.2. Stability Testing

18.10.2.1. Practical Considerations. One needs to recognize that many pharmaceutical systems, including the excipients, are not absolutely pure. Impurities may include trace transition metal ions that may catalyze oxidations, synthetic intermediates, and degradation products, as well as dissolved oxygen—even in freeze-dried solids. The phase chemistry is often complex with multiple crystalline phases and most important, we may have components that are nominally crystalline but remain partially amorphous during processing and recrystallize slowly on storage. Stability is normally characterized in systems with modest overall levels of degradation, simply because degradation in excess of about 10% is not normally consistent with regulatory constraints. Thus, the system has typically not undergone sufficient degradation to define reaction order. Further, the very concept of reaction order loses validity with either crystalline or amorphous solids [71]. Most reactions are complex, making it difficult to even define a rate constant, and in many cases the empirically defined rate constant may, in reality, be a combination of rate constants for fundamental reactions and/or combinations of rate constants and equilibrium constants. Thus, extrapolation of data from high temperatures using the Arrhenius equation is far from rigorous. Given the limited accuracy in assays, even via HPLC, it is normally better to measure levels of degradation products than simply characterize loss of parent. Assay for degradation product provides much higher sensitivity and allows degradation to be quantified in situations where the level of degradation is slight. However, one needs to ensure that all significant degradation products are being measured by a comparison of amount of loss of parent to the sum total of all degradation products, thereby performing a "mass balance." Good temperature control during stability studies is also important if quantitative comparisons are to be made. The usual temperature control of about $\pm 1°C$ is marginal. Simple calculation shows that a $\pm 1^\circ C$ temperature variation will typically result in a variation in rate constant of 10%–25%! Finally, unless one has some preliminary information on stability in the types of formulations being evaluated, a preliminary stability study is needed to properly choose the study temperatures and assay times. Guessing about conditions may well lead to poor or useless data and a large waste of time and material.

18.10.2.2. Kinetics of Degradation in the Amorphous Solid State. While classically one thinks of kinetics in terms of reaction order, the concept of order has meaning only in liquid or gaseous solutions [71]; with solids, the *elucidation of kinetics* refers to the evaluation of the time dependence of the reaction rate, or alternately, the time dependence of the loss of parent or appearance of degradation products. With degradation in the amorphous solid state, or glassy state, the kinetics is often "stretched time" kinetics, meaning that the relationship between parent (or purity, P) and time is given by an equation such as,

$$P(t) = P(t = 0) \cdot \exp(-k \cdot t^{\beta}) \tag{18.3}$$

where k is the rate constant on the stretched timescale t^{β} and β is a constant between zero and unity. Stretched time kinetics is consistent with the fact that the amorphous solid is composed of a distribution of microstates, each with a different degradation rate, and these microstates are not in structural equilibrium [71,75–77]. Normally, we find empirically, that $\beta \approx \frac{1}{2}$, and the kinetics is "square-root of time kinetics."

18.10.2.3. Control of Moisture and Oxygen. Nominally, when one seals the containers in the controlled environment of the freeze dryier, the level of water attained at the end of the process will be maintained, and oxygen will not contaminate the headspace of the container. While moisture permeation from ambient does not seem to be an issue, the moisture contained in the stoppers, usually as a result of steam sterilization, will slowly diffuse out of the stopper and into the hygroscopic amorphous product, thereby elevating the moisture content during storage. The available data [81] suggest that an equilibrium level of moisture is reached after several weeks at 40°C, about 6 months at 25°C, but longer than 24 months at 5°C. The equilibrium level of water is dependent on the stopper composition and stopper treatment history and increases as the mass of amorphous product in the vial decreases. This problem may be minimized by selection of low-moisture release stoppers and rigorous drying after steam sterilization.

Oxygen permeation can be an issue, even with butyl rubber stoppers, and one may observe several percent oxygen in the vial headspace after 2 years of storage (Fig. 18.9). Even several percent oxygen in the headspace can be problematic, particularly with protein drugs, where the mass of oxygen needed to stoichiometrically oxidize most of the dose in the vial can be very small. At least if oxidation is a suspected degradation route, oxygen should be excluded from the vial headspace, and the level of headspace oxygen needs to be monitored during stability testing to at least ensure that variability in oxygen levels between different studies is not responsible for significant variations in stability.

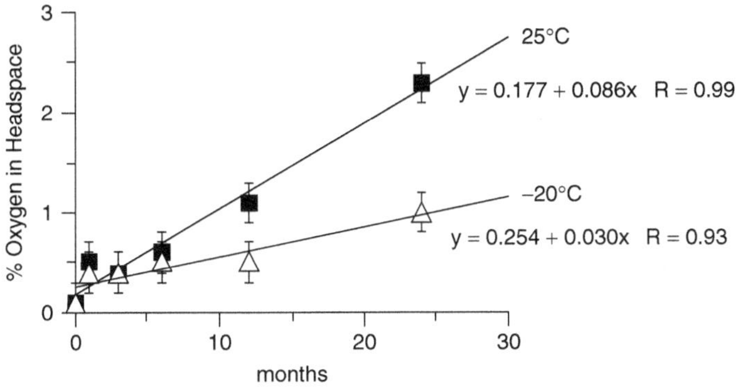

Figure 18.9. Oxygen permeability through stoppers at −20°C and 25°C, using 10-cm^3 vials with 20-mm finish gray butyl stoppers from West Co.

18.10.2.4. Selection of Appropriate Accelerated Testing Conditions, Temperature, and Time. The design of an accelerated stability test can prove difficult, and if care is not exercised, the test results may be useless or misleading. The test temperature needs to be sufficiently high such that the level of degradation characteristic of that produced during the shelf life of the product is produced in a relatively short period of time, but the difference between the test temperature and the anticipated storage temperature should not be so great that nonrepresentative results are obtained. For example, if one has a system where two (or more) degradation pathways are significant, but these pathways are characterized by very different activation energies, one may find that the dominant reaction at the accelerated test condition is not even important at the actual temperature of interest. Figure 18.10 illustrates such a shift in dominant reaction for an example system that undergoes both a hypothetical oxidation, with an activation energy of 15 kcal/mol, and a hypothetical deamidation reaction, with an activation energy of 25 kcal/mol. At high temperature, particularly above 50°C, the reaction is predominantly a higher-activation-energy reaction, but in the range of refrigerated storage, deamidation is insignificant, and oxidation is the dominant reaction. Thus, if a formulation is being optimized using high-temperature stability data, optimization is being carried out for a reaction that does not occur to an appreciable extent at the intended storage temperature of 5°C, and the test results may well be meaningless. Clearly, one needs to verify that the degradation products that are produced at the accelerated test conditions are essentially those, in roughly the same proportions, as produced at the intended storage temperature. Early in a project, this is difficult at best, and thus there is significant risk in accepting predictions based on accelerated test conditions run at very high temperatures.

The time required for an accelerated test to produce the same level of a particular degradation product as would be found at the end of the shelf life at the intended product storage conditions may be estimated from the Arrhenius equation and assumed time dependence of the degradation process. Such estimates are provided in Figure 18.11, where calculations were performed for both reactions with a low activation energy (15 kcal/mol) and a high activation energy (25 kcal/mol). Of course, the

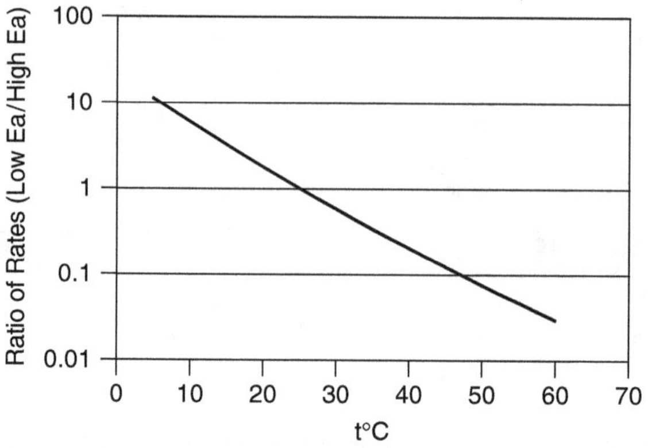

Figure 18.10. Effect of temperature on nature of the dominant reaction in a system of multiple reaction pathways, showing ratio of oxidation rate ($E_a = 15$ kcal/mol) to deamidation rate ($E_a = 25$ kcal/mol).

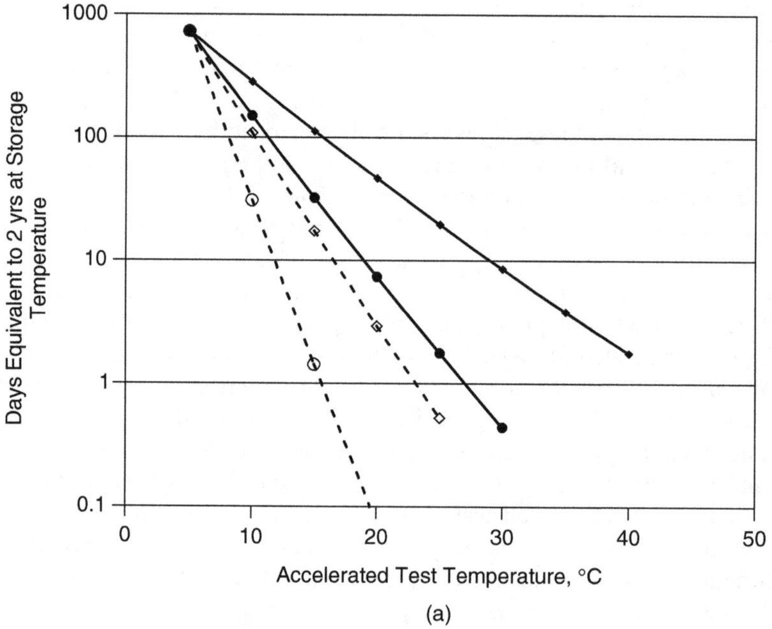

Figure 18.11a. Equivalent times for accelerated stability testing for refrigerated (5°C) storage. (a) and room-temperature (25°C). *Symbols key*: Diamonds = 15 kcal/mol; circles = 25 kcal/mol; filled symbols = exponential kinetics (First-order); open symbols = square root of time kinetics.

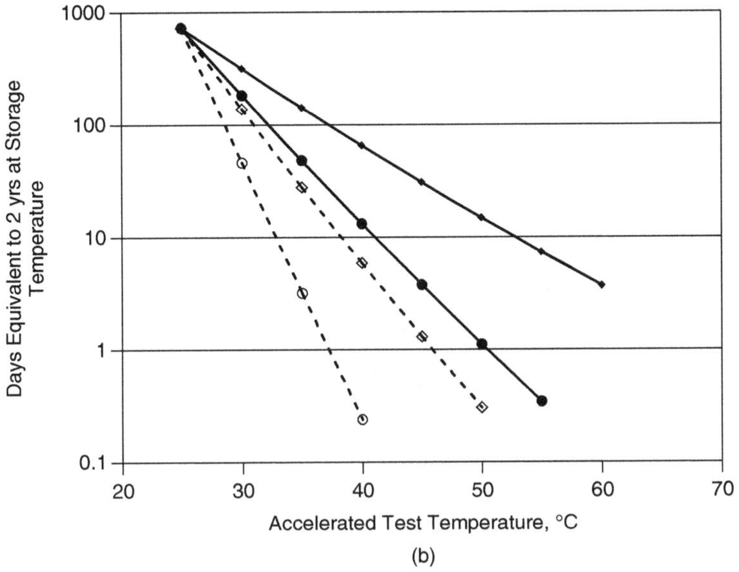

Figure 18.11b. Equivalent times for accelerated stability testing for storage at room temperature (25°C). (b) storage. *Symbols key*: Diamonds = 15 kcal/mol; circles = 25 kcal/mol; filled symbols = exponential kinetics (First-order); open symbols = square root of time kinetics.

times are shorter with the higher activation energy, and are also much shorter with the square root of time kinetics than for exponential (effective First-order) degradation kinetics. Note that the equivalent times are relatively short, even when the activation energy is low and the difference between accelerated test temperature and intended storage temperature is only 10°C–15°C, particularly for the square root of time kinetics. This is a significant point since degradation in glassy systems commonly follows the square root of time kinetics. The implication of this observation is that one does not necessarily need to employ accelerated test temperatures enormously higher than the intended storage temperature to ensure rapid turnaround of data. However, it must also be recognized that if the objective is to produce sufficient degradation to allow evaluation of a precise rate constant that, in turn, allows a quantitative assessment of differences in stability of trial formulations, one may need to produce significantly more degradation during the accelerated test than would result during the shelf life of a product. In such cases, the times would obviously need to be longer than indicated by Figure 18.11.

REFERENCES

1. Pikal, M. J. Lukes, A. L., and Lang, J. E. (1977), Thermal decomposition of amorphous β-lactam antibacterials, *J. Pharm Sci*. **66**: 1312.

2. Pikal, M. J. Lukes, A. L. Lang, J. E. and Gaines, K. (1978), Quantitative crystallinity determinations for beta-lactam antibiotics by solution calorimetry: Correlations with stability, *J. Pharm Sci*. **67**: 767.
3. Pikal, M. J. and Dellerman, M. (1989), International stability testing of pharmaceuticals by high-sensitivity isothermal calorimetry at 25°C: Cephalosporins in the solid and aqueous solution states, *J. Pharm*. **50**: 233–252.
4. Franks, F., ed. (1982), *The Properties of Aqueous Solutions at Subzero Temperatures*, Plenum Press, New York, pp. 215–338.
5. Franks, F. (1992), Freeze-drying: A combination of physics, chemistry, engineering and economics, *Jpn. J. Freeze Drying* **38**: 5–16.
6. Hatley, R. H. M. and Mant, A. (1993), Determination of the unfrozen water content of maximally freeze concentrated carbohydrate solutions, *Int. J. Biol. Macromol*. **15**: 227–232.
7. Strambini, G. B. and Gabellieri, E. (1996), Proteins in frozen solutions: Evidence of ice-induced partial unfolding, *Biophys. J*. **70**(2): 971–976.
8. Privalov, P. L. (1990), Cold denaturation of proteins, *Crit. Rev. Biochem. Mol. Biol*. **25**: 281–305.
9. Poulsen, K. Porsdal, F. L. R. V. (1981), *Acceleration of Chemical Reactions Due to Freezing Water Activity: Influences on Food Quality*, Academic Press, pp. 651–677.
10. Ahmad, F. and Bigelow, C. (1986), Thermodynamic stability of proteins in salt solutions: A comparison of the effectiveness of protein stabilizers, *J. Protein Chem*. **5**: 355–367.
11. Koseki, T. et al. (1990), Freezing denaturation of ovalbumin at acid pH, *J. Biochem*. **107**: 389–394.
12. Murase, N. and Franks, F. (1989), Salt precipitation during the freeze-concentration of phosphate buffer solutions, *Biophys. Chem*. **34**: 293–300.
13. Hyclone Labs, Inc. (1992), Freezing and thawing serum and other biological materials: Optimal procedures minimize damage and maximize shelf-life, in *Art to Science in Tissue Culture*, Vol. 2, #2, Hyclone Labs., Inc. 11 (spring 1992).
14. van den Berg, L. and Rose, D. (1959), Effect of freezing on the pH and composition of sodium and potassium phosphate solutions: The reciprocal system: $KH_2PO_4-Na_2HPO_4-H_2O$, *Arch. Biochem. Biophys*. **81**: 319–329.
15. Anchordoquy, T. J. and Carpenter, J. F. (1996), Polymers protect lactate dehydrogenase during freeze-drying by inhibiting dissociation in the frozen state, *Arch. Biochem. Biophys*. **332**: 231–238.
16. Eckhardt, B. M. Oeswein, J. Q., and Bewley, T. A. (1991), Effect of freezing on aggregation of human growth hormone, *Pharm. Res*. **8**(11): 1360–1364.
17. Pikal, M. J., Dellerman, K. M., Roy, M. L., and Riggin, R. M. (1991), The effects of formulation variables on the stability of freeze-dried human growth hormone, *Pharm. Res*. **8**: 427–436.
18. Chang, B. S., Kendrick, B. S., and Carpenter, J. F. (1996), Surface-induced denaturation of proteins during freezing and its inhibition by surfactants, *J. Pharm. Sci*., **85**: 1325–1330.
19. Carpenter, J. F., Pikal, M. J. Chang, B. S., and Randolph, T. W. (1997), Rational design of stable lyophilized protein formulations: Some practical advice, *Pharm. Res*. **14**: 969–975.
20. Pikal, M. J. (1999), Mechanisms of protein stabilization during freeze-drying and storage: The relative importance of thermodynamic stabilization and glassy state relaxation dynamics, in *Freeze-Drying/Lyophilization of Pharmaceutical and Biological Products*, Rey, L. and May, J., eds., Marcel Dekker, New York, Chapter 21. chap6.

21. Hageman, M J. (1988), The role of moisture in protein stability, *Drug Devel. Industr. Pharm.* **14**: 2047–2070 [see also Hsu, C. C., Ward, C. A., Pearlman, R., Ngyuyen, H. M., Yeung, D. A., and Curley, J. G. (1991), Determining the optimum residual moisture in lyophilized protein pharmaceuticals, *Drug Devel. Biol. Stand.* **74**: 235–271].
22. Akers, M. J., Milton, N., Byrn, S. R., and Nail, S. (1995), Glycine crystallization during freezing: The effects of salt form, pH, and ionic strength, *Pharm. Res.* **12**: 1457–1461.
23. Yu, L. Milton, N. Groleau, E, Mishra, D. and Vansickle, R. (1999), Existence of a mannitol hydrate during freeze-drying and practical implications, *J. Pharm. Sci.* **88**: 196–198.
24. Kim, A. J. Akers, M. J. and Nail, S. L. (1998), The physical state of mannitol after freeze-drying: Effects of mannitol concentration, freezing rate, and a noncrystallizing cosolute, *J. Pharm. Sci.* **87**: 931–935.
25. Cavatur, R. K. and Suryanarayanan, R. (1998), Characterization of phase transitions during freeze-drying by in situ X-ray powder diffractometry, *Pharm. Devel. Technol.* **3**: 579–586.
26. Shalaev, E., Johnson-Elton, T., Chang, L., and Pikal, M. J. (2002), Thermophysical properties of pharmaceutically compatible buffers at sub-zero temperatures: Implications for freeze drying, *Pharm. Res.* **19**: 195–201.
27. Larsen, S. S. (1971), Studies on stability of drugs in frozen systems. IV. The stability of benzylpenicillin sodium in frozen aqueous solutions, *Dansk Tidsskr. Farm.* **45**: 307–316.
28. Larsen. S. S. (1973), Studies on stability of drugs in frozen systems. VI. The effect of freezing on pH for buffered aqueous solutions, *Arch. Pharm. Chem. Sci. Ed.* **1**: 41–53.
29. Murase, N. and Franks, F. (1989), Salt precipitation during the freeze-concentration of phosphate buffer solutions, *Biophys. Chem.* **34**: 293–300.
30. Gomez, G. Pikal, M. J., and Rodriguez-Horned, N. (2001), Effect of initial buffer composition on pH changes during far-from equilibrium freezing of sodium phosphate buffer solutions, *Pharm. Res.* **18**: 90–97.
31. Rodriguez-Hornedo, N., Szkudlarek, B., and Pikal, M. J. (1994), Midwestern AAPS meeting, Chicago, Poster 24.
32. Eriksson, J. H. C., Hinrichs, W. L. J., de Jong, G. J., Sonsen, G. W., and Frijlink, H. W. (2003), Investigations into the stabilization of drugs by sugar glasses: III. The influence of several high-pH buffers, *Pharm. Res.* **20**: 1437–1443.
33. Chang, S. and Randall, C. S. (1992), Use of subambient thermal analysis to optimize protein lyophilization, *Cryobiology* **29**: 632–656.
34. Her, L. M. and Nail, S. L. (1994), Measurement of glass transitions of the maximally freeze-concentrated solutes by differential scanning calorimetry, *Pharm. Res.*, **11**: 54–59.
35. Webb, S. D., Golledge, S. L., Cleland, J. L., and Carpenter, J. F. (2002), Surface adsorption of recombinant human interferon-γ in lyophilized and spray-lyophilized formulations, *J. Pharm. Sci.* **91**: 1474–1487.
36. Mumenthaler, H., Hsu, C., and Pearlman, R. (1994), Feasibility study on spray-drying protein pharmaceuticals; Recombinant growth harmone and tissue-type plasminogen activator. *Pharm. Res.* **11**: 12–20.
37. Barn, N. B. Cleland, J. L., Yang, J., Manning, M. C. Carpenter, J. F. Kelley, R. F. and Randolph, T. W. (1998), Tween protects recombinant human growth hormone against agitation-induced damage via hydrophobic interactions. *J. Pharm Sci.* **87**: 1554–1559.
38. Barn, N. B., Randolph, T. W., and Cleland, J. L. (1995), Stability of protein formulations: investigation of surfactant effects by a novel EPR spectroscopic technique. *Pharm. Res.* **12**: 2–11.

39. Kerwin, B. A., Heller, M. C., Levin, S. H., and Randolph, T. W. (1998), Effects of Tween 80 and sucrose on acute short-term stability and long-term storage at-20 degrees of a recombinant haemoglobin, *J. Pharm. Sci*. **87**: 1062–1068.

40. Pikal, M. J., Dellerman, K., and Roy, M. L. (1991), Formulation and stability of freeze-dried proteins: Effects of moisture and oxygen on the stability of freeze-dried formulations of human growth hormone, Devel. Biol. Stand. **74**: 323–340.

41. Li, S., Schoneich, C., and Borchardt, R. (1995), Protein characterisation and quality control chemical instability of protein pharmaceuticals: Mechanisms of oxidation and strategies for stabilization, *Biotechnol. Bioeng*. **48**: 490–500.

42. Li, S., Schoneich, C., Wilson, G., and Borchardt, R. (1993), Chemical pathways of peptide degradation. V. Ascorbic acid promotes rather than inhibits the oxidation of methionine to methionine sulfoxide in small model peptides, *Pharm. Res*. **10**: 1572–1579.

43. Stadman, R. (1990), Oxidation of free amino acids and amino acid residues in proteins by radiolysis and by metal-catalyzed reactions, *Free Radical Biol. Med*. **9**: 315–325.

44. Li, S., Schoneich, C., and Borchardt, R. (1995), Chemical pathways of peptide degradation. VIII. Oxidation of methionine in small model peptides by prooxidant/transition metal ion systems: Influence of selective scavengers for reactive oxygen intermediates, *Pharm. Res*. **12**: 348–355.

45. Franks, F. (1985), *Biophysics and Biochemistry at Low Temperatures*, Cambridge Univ. Press, London.

46. Arakawa, T. and Timasheff, S. N. (1985), The stabilization of proteins by osmolytes, *Biophys. J*. **47**: 411–414.

47. Arakawa, T., Carpenter, J. F., Kita, Y., and Crowe, J. (1990), The basis for toxicity of certain cryoprotectants: A hypothesis, *Cryobiology* **27**: 401–415.

48. Carpenter, J. F. and Crowe, J. H. (1988), The mechanism of cryoprotection of proteins by solutes, *Cryobiology* **25**: 244–255.

49. Mi, Y. L., Wood, C. G., Thomas L., and Rashed, S. (2002), Effects of polyethylene glycol molecular weight and concentration on lactate dehydrogenase activity in solution and after freeze-thawing, *PDA J. Pharm. Sci. Technol*. **56**: 115–123.

50. Pikal, M. J. Busse, J., and Kovach, P. (no date), Eli Lilly & Co., unpublished results.

51. Carpenter, J. F., Kreilgaardd, L., Allison, S. D., and Randolph T. W. (2000), Roles of protein conformation and glassy state in the storage stability of dried protein formulations, in *Pharmaceutical Formulation Development of Peptides and Protein*, Frokjaer, S. and Hovgaard, L., eds., Taylor & Francis, London, pp. 178–188.

52. Pikal, M. J. (1999), Impact of polymorphism on the quality of lyophilized products, in *Polymorphism in Pharmaceutical Solids*, Brittain, H. G., ed., Marcel Dekker, New York, Chapter 54. chap10.

53. Prestrelski, S. J., Pikal, K. A., and Arakawa, T. (1995), Optimization of lyophilization conditions for recombinant human interleukin-2 by dried-state conformational analysis using FT-IR, *Pharm. Res*. **12**: 1250–1259.

54. Carpenter, J. F., Prestrelski, S. J., and Dong, A. (1998), Application of infrared spectroscopy to development of stable lyophilized protein formulations, *Eur. J. Pharm. Biopharm*. **45**: 231–238.

55. Sane, S. U., Wong, R., and Hsu, C. C. (2004), Raman spectroscopic characterization of drying-induced structural changes in a therapeutic antibody: Correlating structural changes with long-term stability, *J. Pharm. Sci*. **93**(4).

56. Saleki-Gerhardt, A. (1993), Role of water in the solid-state properties of crystalline and amorphous sugars, Ph.D. thesis, Univ. Wisconsin—Madison.

57. Shalaev, Y. and Kanev, A. N. (1994), *Cryobiology* **31**: 374–382.

58. Lueckel, A., Bodmer, D. Helk, B. and Leuenberger, H. (1998), *Pharm. Devel. Technol*. **3**: 325–336.

59. Kassraian, K, Spitznagel, T. Juneau, J. and Yim, K. (1998), *Pharm. Devel. Technol*. **3**: 233–239.

60. Hancock, B. C. and Zografi, G. (1994), *Pharm. Res*. **11**: 471–477.

61. Heller, M. C., Carpenter, J. F., and Randolph, T. W. (1997), Protein phase separation during freeze drying: Implications for pharmaceutical proteins, *Biotechnol. Progress* **13**: 590–596.

62. Heller, M. C., Carpenter, J. F., and Randolph, T. W. (1997), Manipulation of lyophilization induced phase separation, *Biotechnol. Progress* **13**: 590–596.

63. Dean S. A., Manning, M. C., Randolph, T. W., Middleton, K., Davis, A., and Carpenter, J. F. (2000), Optimization of storage stability of lyophilized actin using combinations of disaccharides and dextran, *J. Pharm. Sci*. **89**(2): 199–214.

64. Depaz, R. A., Barnett, C. C., Dale, D. A., Gaether, A. L., Carpenter, J. F., and Randolph, T. W. (2001), The excluding effects of sucrose on a protein degradation pathways: Methionine oxidation in subtilisin, *Arch. Biochem. Biophys*. **384**: 123–132.

65. Kim, Y. H., Berry, A. H., Spencer, D. S., and Stites, W. F. (2001), Comparing the effect on protein stability of methionine oxidation versus mutagenesis; steps towards engineering oxidative resistance in proteins, *Protein Eng*. **14**: 343–347.

66. Randolph, T. W. (1997), Phase separation of excipients during lyophilization: Effects on protein stability, *J. Pharm. Sci*. **86**(11): 1198–1203.

67. Li, S., Patapoff, T. W., Overcashier, D., Hsu, C., Nguyen, T. H., and Borchardt, R. T. (1996), Effects of reducing sugars on the chemical stability of human relaxin in the lyophilized state, *J. Pharm. Sci*. **85**(8): 873–877.

68. Dubost, D. C., Kaufman, M. J., Zimmerman, J. A., Bogusky, M. J., Coddington, A. B., and Pitzenberger, S. M. (1996), *Pharm. Res*. **13**(12): 1811–1814.

69. Pikal, M. (2002), Lyophilization, in *Encyclopedia of Pharmaceutical Technology*, Vol. 6, Swarbrick, J. and Boylan, J. C. eds., Marcel Dekker, New York, pp. 1299–1326.

70. Townsend, M. W. and S., DeLuca, P. P. (1988), Use of lyoprotectants in the freeze-drying of a model protein, ribonuclease A, *J. Parenter. Sci. Technol*. **42**: 190–199.

71. Pikal, M. J., Rigsbee, D., Roy, M. L., Galreath, D., Kovach, K. J., Wang, W., Carpenter, J. F., and Cicerone, M. T. (200), Solid state chemistry of proteins: II. The correlation of storage stability of freeze-dried human growth hormone (hGH) with structure and dynamics in the glassy solid, *J. Pharm. Sci*. **97**(12): 5106–5121 in press.

72. Duddu, S. P. and Dal Monte, P. R. (1997), Effect of glass transition temperature on the stability of lyophilized formulations containing a chimeric therapeutic monoclonal antibody, *Pharm. Res*. **14**: 591–595.

73. Williams, N. A. and Dean, T. (1991), Vial breakage by frozen mannitol solutions: Correlation with thermal characteristics and effect of sterioisomerism, additives, and vial configuration, *J. Parenter. Sci. Technol*. **45**: 94–100.

74. Chang, L., Shepherd, D., Sun, J., Ouellette, D., Grant, K., Tang, X., and Pikal, M. J. (2005), Mechanism of protein stabilization by sugars during freeze drying and storage:

Native structure preservation, specific interaction and/or immobilization in a glassy matrix? *J. Pharm. Sci.* **94**(7): 1427–1444.

75. Pikal, M. J. and Rigsbee, D. R. (1997), The stability of insulin in crystalline and amorphous solids: Observation of greater stability for the amorphous form, *Pharm. Res.* **14**, 1379–1387.
76. Yoshioka, S., Aso, Y., and Kojima, S. (2001), Usefulness of the Kohlrausch-Williams-Watts stretched exponential function to describe protein aggregation in lyophilized formulations and the temperature dependence near the glass transition temperature, *Pharm. Res.* **18**: 256–260.
77. Abdul-Fattah, A. M., Dellerman, K., Bogner, R. H., and Pikal, M. J. (2007), The effect of annealing on the stability of amorphous solids: Chemical stability of freeze-dried moxalactam, *J. Pharm. Sci.* **96**: 1237–1250.
78. Cicerone, M. T., Soles, C. L., Chowdhuri, Z., Pikal, M. J., and Chang, L. L. (2005), Fast dynamics as a diagnostic for excipients in preservation of dried proteins, *Am. Pharm. Rev.*, **8**(6): 24–27.
79. Chang, L., Shepherd, D., Sun, J., Tang, X., and Pikal, M. J. (2005), Effect of sorbitol and residual moisture on the stability of lyophilized antibodies: Implications for the mechanism of protein stabilization in the solid state, *J. Pharm. Sci.* **94**(7): 1445–1455.
80. Cicerone, M. T. and Soles, C. L. (2004), Fast dynamics and stabilization of proteins: Binary glasses of trehalose and glycerol, *Biophys. J.* **86**: 3836–3845.
81. Pikal, M. J. and Shah, S. (1991), Moisture transfer from stopper to product and resulting stability implications, *Devel. Biol. Stand.*, **74**: 165–179.
82. Pikal, M. J. and Shah, S. (1988), Eli Lilly & Co., unpublished observations.
83. Chang, S. Beauvais, R. M., Arakawa, T., Narhi, L. O., Dong, A., Aprisio., D. L., and Carpenter, J. F. (1996), Formation of an active dimmer during storage of interleukin-l receptor antagonist in aqueous solution, *Biophys. J.* **71**: 3399.
84. Jameel, F., Bjornson, E., Hu, L., Kabingue, K., Besman, M., and Pikal, M. J. (2000), Effects of formulation and process variations on the stability of lyophilized recombinant human factor VIII, *Haemophilia.* **6**(4) (June).

19

PROTEIN CONFORMATION AND REACTIVITY IN AMORPHOUS SOLIDS

Lei Zhang, Sandipan Sinha, and Elizabeth M. Topp

19.1. INTRODUCTION

Protein drugs are subject to a variety of chemical and physical degradation processes that reduce potency, limit shelf life, and increase the potential for immunogenic side effects. Chemical degradation produces covalent changes in particular amino acids through reactions such as deamidation and oxidation. Physical degradation changes protein conformation or phase behavior through noncovalent processes such as unfolding and precipitation. Protein drugs are often formulated as lyophilized solids in an attempt to control or prevent both chemical and physical degradation. Lyophilized proteins are generally amorphous rather than crystalline, and are lower in thermodynamic energy and therefore less susceptible to many types of degradation than corresponding aqueous solutions. Nevertheless, physicochemical degradation can still occur at rates sufficient to preclude a stable lyophilized formulation. An understanding of protein physicochemical degradation in lyophilized solids is therefore central to the efficient development of therapeutically active proteins into marketable lyophilized drug products.

This chapter reviews common chemical degradation reactions of proteins in lyophilized solids and the dependence of these reactions on protein structure in the

Formulation and Process Development Strategies for Manufacturing Biopharmaceuticals, edited by Jameel and Hershenson

solid state. Section 19.2 describes the mechanisms of reactions such as deamidation and methionine oxidation and summarizes the literature on these reactions in lyophilized solids. In solution, the rates of many protein chemical degradation reactions can be highly dependent on higher-order structure [1–5], and a similar dependence has been proposed for proteins in the solid state [6]. Section 19.3 reviews and critiques the evidence for a relationship between higher-order structure and chemical reactivity in amorphous solids. The chapter concludes with a brief summary.

19.2. CHEMICAL REACTIONS OF PROTEINS IN AMORPHOUS SOLIDS

19.2.1. Hydrolytic Reactions

Hydrolytic reactions break covalent bonds by reaction with water. Three major types of hydrolytic reactions are common in solid formulations of proteins: deamidation, peptide bond hydrolysis, and diketopiperazine formation.

19.2.1.1. Deamidation. Nonenzymatic deamidation of asparagine (Asn) or glutamine (Gln) is a major degradation pathway of peptides or proteins in solution. In deamidation, the amide group in the side chain of an Asn or Gln residue is hydrolyzed to form a free carboxylic acid. The reaction occurs much more rapidly at Asn than at Gln. Numerous studies on the reaction mechanism in solution have been published [7–12]. Protein deamidation in the solid state has also been observed, and the data support a similar mechanism.

Degradation of the model Asn–hexapeptide (VYPNGA) was studied in solid formulations lyophilized from acidic solutions [11]. In samples lyophilized from pH 3 solutions, deamidation took place via direct hydrolysis of the Asn side chain to produce the Asp–hexapeptide (Fig. 19.1). As the pH of the solution prior to lyophilization was increased from 3 to 5, intramolecular attack of the carbonyl carbon of the Asn side chain by the amide nitrogen anion of the succeeding amino acid to form a succinimidyl intermediate became more prominent. This intramolecular attack formed a succinimidyl group (cyclic imide) detectable as an intermediate in the product distribution. Because of the low level of water available for hydrolysis in the solid state, the formation of iso-Asp–hexapeptide was not always observed, though iso-Asp–hexapeptide is often the dominant degradation product in solution at pH 5 [7]. In solids with higher moisture content, however, iso-Asp and Asp degradation products are frequently observed.

Li et al. studied the deamidation kinetics of seven model peptides (VYPNGA, VYGNGA, VFGNGA, VIGNGA, VGGNGA, VGPNGA, and VGYNGA) in solution and in lyophilized solid formulations containing polyvinylpyrrolidone (PVP) [13]. The degradation of the model peptides from solution and solid-state formulations followed apparent first-order kinetics. The rate of deamidation was significantly influenced by the residues immediately N-terminal to asparagine (N-1, N-2) in the solid state but less so in solution. Increases in the volume and hydrophobicity of the N-1 and N-2 residues decreased the rate of deamidation in the solid state. An empirical model showed that

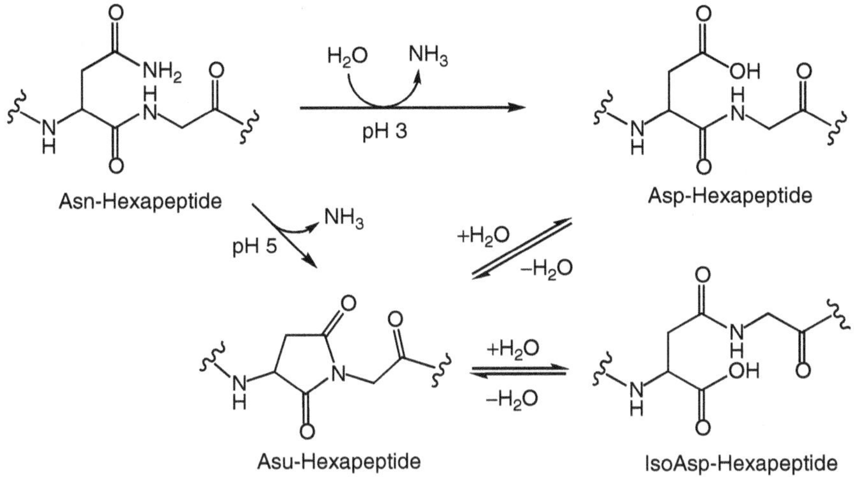

Figure 19.1. Deamidation of asparaginyl peptides.

the influences of volume and hydrophobicity are approximately equal for the N-1 and N-2 residues. In a similar deamindation study on four other model peptides (AcGQNGG, AcGQNDG, AcGQNEG and AcGQNQG; AC=acetyl), the group found that AcGQNGG and AcGQNQG had up to 60 times slower rates of deamidation in the solid state than in solution [14]. In contrast, the deamidation rates of AcGQNEG and AcGQNDG in the solid state were similar to those in solution. The results suggest that an $N + 1$ Glu or Asp residue may enhance local hydration, so that deamidation of Asn in the solid formulations actually proceeds in a solution-like environment.

Human insulin has also been observed to undergo deamidation in the solid state. Strickley and Anderson verified that insulin deamidation proceeds via formation of a cyclic anhydride by using aniline trapping of the intermediate [15,16]. While the succinimidyl intermediates formed during Asn–hexapeptide deamidation in the solid state tended to accumulate [17], the cyclic anhydride intermediate did not. The cyclic anhydride intermediate may react with water to form [desamidoA21] insulin, and may also react with another molecule of insulin to form covalent dimers [15,16]. Consistent with these findings, Pikal and Rigsbee observed that the deamidation of human insulin occurs predominantly at AsnA21, except at high relative humidity, when deamidation at AsnB3 is more prevalent [18].

19.2.1.2. Peptide Bond Hydrolysis. In the solid state, direct peptide bond hydrolysis is not favored because of the low water content. However, peptide bonds of Asp residues (especially Asp–Pro) are cleaved under acidic conditions at a rate at least 100 times faster than other peptide bonds. Intramolecular catalysis by a carboxyl group of the Asp residue is thought to be involved in the hydrolysis process. Either the N-terminal and/or C-terminal peptide bonds adjacent to the Asp residue can be hydrolyzed. Inglis [19] has described the mechanism for such hydrolysis as shown in

Figure 19.2. Cleavage of the *N*-terminal peptide bond proceeds via a six-membered ring intermediate, while the *C*-terminal peptide bond is hydrolyzed via an intermediate containing a five-membered ring.

The presence of nucleophilic group (e.g., –OH) in the side chain also leads to peptide bond cleavage via formation of a reactive intermediate with excipients. Lyophilized human relaxin (Rlx) formulated with glucose can undergo hydrolytic cleavage of the *C*-terminal serine (Ser) residue on the B chain (Trp28–Ser29–COOH) on storage at 40°C [20]. The resulting degradation product, des-Ser Rlx, was not observed when mannitol or trehalose was used as the excipient. Hydrolysis of the Trp28–Ser29 peptide bond was shown to be much more significant in the solid state than in solution. This result suggests that a high content of glucose is necessary for the reaction to occur. This reaction was thought to involve an initial reaction of the Ser hydroxyl group with glucose, followed by subsequent hydrolysis of the Trp–Ser bond via a cyclic intermediate.

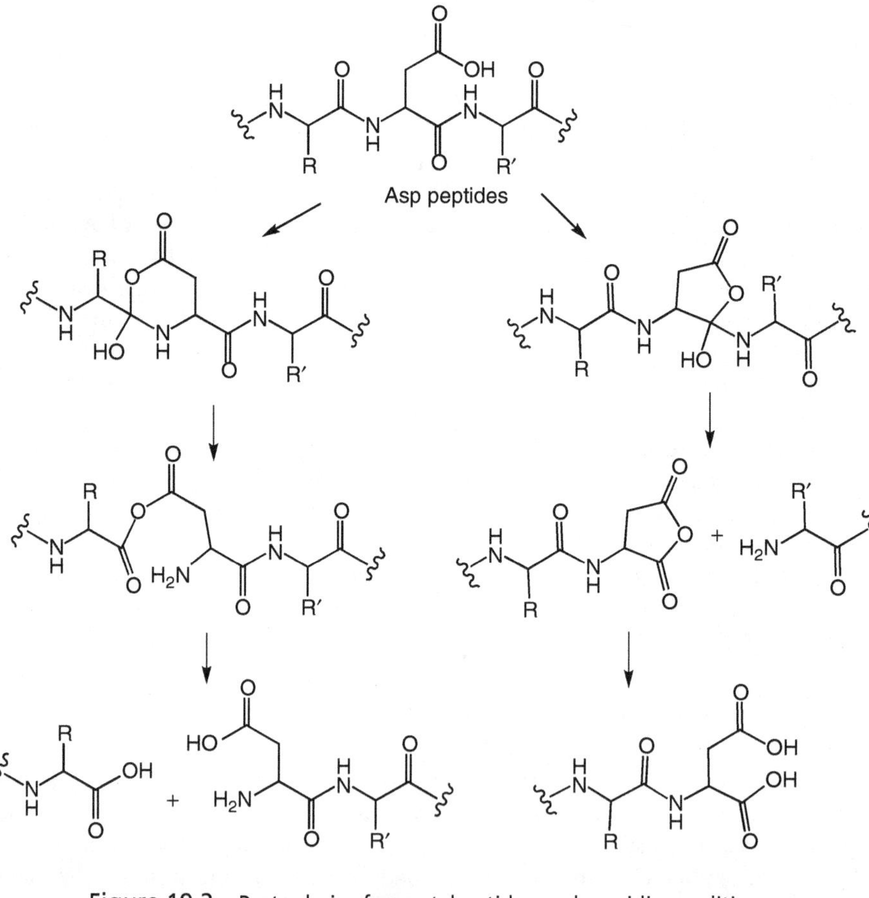

Figure 19.2. Proteolysis of aspartyl petides under acidic conditions.

19.2.1.3. Diketopiperazine Formation. Diketopiperazine (DKP) formation is another important mechanism for peptide bond cleavage. When a free *N*-terminal amino group attacks the carbonyl of the penultimate peptide bond and cleaves that bond, a DKP is formed from the two *N*-terminal amino acids, resulting in a "clipped" peptide or protein that lacks these two residues. Proteins with the *N*-terminal sequence H_2N–Gly–Pro are particularly prone to DKP formation. The degradation of the undecapeptide substance P (SP) in the solid state proceeds through DKP formation. The main pathway of decomposition consists of the sequential release of *N*-terminal dipeptides via their diketopiperazines, cyclo(Arg–Pro) and cyclo(Lys–Pro) [21]. Another example of DKP formation is degradation of aspartame [*R*-aspartylphenylalanine methyl ester (APM)] in the solid state. The major pathway is intramolecular cyclization to form exclusively DKP with the elimination of methanol [22]. Since water was essentially absent in the solid state, no further hydrolysis products were observed. The formation of DKP was confirmed by HPLC analysis of heated aspartylphenylalanine (AP) samples.

19.2.2. Oxidation

The side chains of His, Met, Cys, Trp, and Tyr residues in proteins are potential sites for oxidation in the presence of oxidizing agents. For example, hydrogen peroxide can modify imidazole (His), thioether (Met), sulfhydryl (Cys), indole (Trp), and phenol (Tyr) groups at neutral or slightly alkaline conditions, while under acidic conditions, oxidation of Met to Met sulfoxide is the primary reaction. Depending on the reaction conditions, Cys (RSH) can be successively oxidized to sulfenic acid (RSOH), disulfide (RSSH), sulfinic acid (RSO_2H), and sulfonic acid (RSO_3H). Among the factors that influence the rate of oxidation, the spatial positioning of the thiol group in the protein is particularly important. When one thiol group does not have contact with another intramolecular disulfide bonds are not formed. These spatial constraints may be particularly important in the relatively rigid environment of lyophilized solid formulations.

Of all oxidative reactions, Met oxidation has probably been documented most extensively. A major chemical decomposition pathway for human growth hormone (hGH) in the solid state is oxidation at Met14 to form the sulfoxide [23]. Even with minimal oxygen (0.05%) in the vial headspace, decomposition via oxidation is comparable to or greater than that due to deamidation [24]. As in solution, atmospheric oxygen can easily oxidize Met residues in the solid state, leading to chemical instability and loss of biological activity.

Human insulin-like growth factor I (hIGF-I), lyophilized from phosphate buffer, also undergoes oxidation at its Met residue [25]. The reaction was found to be second-order with respect to the amount of protein and dissolved oxygen. Interestingly, the reaction rate was similar in solution and in lyophilized formulations; however, Met oxidation in the solid state constitutes a greater fraction of the total protein modification than in solution. The reaction was also found to be accelerated on light exposure, which increased the oxidation rate by a factor of 30. This increase suggests that photooxidation may be involved in the generation of radicals. However, no further experiments were conducted to determine the nature of the radicals involved or the mechanism of Met oxidation in solids.

An unusual example of solid-state oxidation is the oxidative deamidation of a cyclic heptapeptide, acetylcysteine–asparagine-(5,5-dimethyl-4-thiazolidinecarbonyl)-(4-aminomethyl-phenylalanine)-glycine–aspartic acid–cysteine cyclic disulfide, in a lyophilized mannitol formulation [26]. The oxidation reaction occurs at the aminomethyl phenylalanine moiety to form a benzaldehyde derivative. Since this oxidative degradant was not detected in the neat solid drug stored under atmospheric oxygen, oxygen was not likely to be the reactive species. Instead, the decomposition of the heptapeptide may be due to the presence of reducing sugar impurities in the mannitol excipient. The proposed mechanism involves (1) formation of a Schiff base from the peptide primary amine reacting with the aldehyde group on the reducing sugar, (2) tautomerization of the Schiff base to a more stable configuration, and (3) hydrolysis of the new Schiff base to generate the aldehyde.

19.2.3. Condensation

In a condensation reaction, a small molecule such as water is eliminated. The condensation reaction between amino groups of amino acids (Lys and Arg) and carbonyl groups of excipients (e.g., sugars, ploymers) or carboxyl groups in the side chain of Asp or Glu is a particular problem in the solid state as the formation of Schiff bases or amide bonds is accelerated at low water activity.

The nonenzymatic Maillard reaction involves the condensation of reducing sugars with amino groups of amino acids (Lys and Arg) in proteins to form glycosylamino acids, which may undergo subsequent rearrangement (Fig. 19.3). Maillard reactions have been recognized as one important factor in determining long-term storage stability of lyophilized protein formulations containing sugars as excipients. Lyophilized human relaxin (Rlx) formulated with glucose was observed to degrade via the Maillard reaction to form adducts with glucose [20]. The resulting adducts were shown by liquid chromatography/mass spectroscopy (LC/MS) to contain up to four glucose molecules covalently attached to Rlx.

During stability studies of an asparagine-containing hexapeptide (VYPNGA) in solid polyvinylpyrrolidone (PVP) matrixes at high temperature (70°C) and low relative humidity (~0%), D'Souza et al. found that total recovery of the peptide and its known deamidation products was only 30% of the peptide load [27]. Size exclusion chromatography with fluorescence detection indicated the formation of a PVP–peptide adduct that was stable in the presence of 6 M guanidine hydrochloride. Control stability studies using *N*-acetyl phenylalanine, phenylalanine ethyl ester, and *N*-acetyl tyrosine ethyl ester demonstrated that the reaction involves the peptide *N* terminus. Disruption of the adduct in the presence of carboxypeptidase A further confirmed the formation of an amide bond between the peptide and PVP. Solid-state NMR spectroscopy using ^{15}N-labeled valine as a model of the peptide *N* terminus showed different populations of ^{15}N, suggesting that noncovalent peptide–polymer interactions precede amide bond formation.

Recombinant bovine somatotropin (rbSt) was observed to form nonreducible dimers in the solid state [28]. When lyophilized rbSt was stored at 30°C and 96%

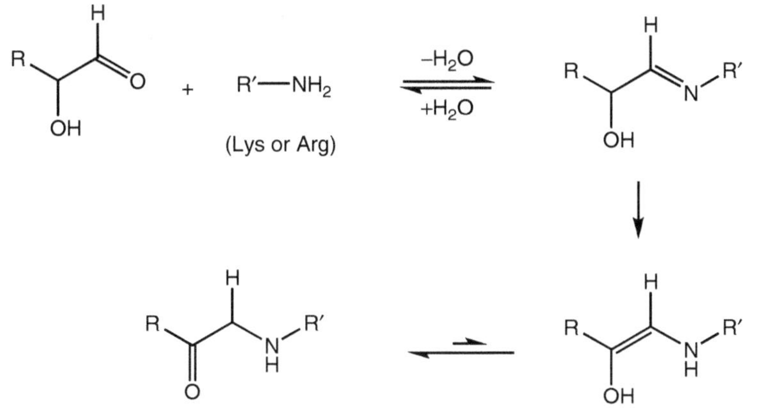

Figure 19.3. Maillard reaction.

relative humidity (RH), the rate of loss of monomeric rbSt was greater than or equal to the rate of loss in solution. Hageman et al. [28] hypothesized that such covalent interaction could be due to the condensation of a lysine ϵ-amine and a carboxyl side chain of asparagine or glutamine. A similar mechanism was proposed for the aggregation of lyophilized ribonuclease A during storage at 45°C [29]. Townsend et al. postulated that these covalent linkages did not result from an oxidation reaction but from the lysine residues reacting with Asp or Glu. The loss of free lysine residues in the insoluble aggregates, as determined by amino acid analysis, is consistent with this mechanism.

Condensation reactions between amino groups and carboxyl groups were also employed to form zero-length covalent crosslinks between protein molecules in the lyophilized state without the use of chemical reagents [30]. By simply sealing lyophilized RNase A under vacuum in a glass vessel and heating at 85°C, up to one-third of the total protein present became crosslinked, and dimer was the major product.

19.2.4. β Elimination

Elimination reactions proceed via loss of elements (leaving groups) from the starting material to form a new π bond in the product. In β elimination, elements are lost from two adjacent atoms: the α carbon bonded to the leaving group and the β carbon adjacent to it. Cys, Ser, and Thr residues degrade via β elimination at alkaline conditions. Lyophilized insulin has been observed to degrade rapidly with increased water content to produce both covalent and noncovalent aggregates [31]. A fivefold increase in the level of free thiols was measured in lyophilized insulin after incubation for 24 h at 50°C and 96% relative humidity. Costantino et al. postulated that free thiols produced by β elimination catalyzed the reshuffling of intact protein disulfides and led to intermolecular disulfide crosslinking. The proposed mechanism for β elimination involves hydroxide-ion-catalyzed cleavage of a carbon–sulfur bond (cysteine), resulting in two new residues, dehydroalanine and thiocysteine. Dehydroalanine then reacts with lysine to form a lysinoalanine crosslink, while thiocysteine undergoes further decomposition to yield thiol-containing products, such as cysteine and hydrosulfide ion (free thiols).

19.2.5. Covalent Aggregation

Aggregation is not the result of a single chemical reaction. In fact, many different types of chemical changes can induce the formation of aggregates. Aggregation via covalent crosslinking can be categorized into two types: reducible (via disulfide exchange) and nonreducible.

19.2.5.1. Reducible Aggregation. Disulfide exchange usually leads to reducible aggregation, which is detected via aggregate disappearance on treatment with a reducing agent such as dithiothreitol (DTT). Lyophilized bovine serum albumin, ovalbumin, glucose oxidase, β-lactoglobulin, and humanized monoclonal antibody (rhuMAb) have been observed to form covalent intermolecular disulfide linkages via a thiol–disulfide exchange reaction [32–34]. Intermolecular thiol–disulfide exchange is proposed to result from the ionized thiol on one albumin molecule attacking the disulfide linkage of another

$$P_1{-}S^- + P_2{-}S{-}S{-}P_2 \longrightarrow P_1{-}S{-}S{-}P_2 + P_2{-}S^-$$

where P_1 and P_2 are the first and second protein molecules [32]. Thiolate ions rather than thiols are thought to be the reactive species, since a decrease in the initial solution pH results in a slower aggregation rate. Free thiols are both necessary and sufficient for the moisture-induced aggregation to occur, since *S*-alkylated BSA (no free sulfhydryl groups) does not undergo aggregation. In solution, thiol–disulfide exchange is reversible with poor regioselectivity, and equilibrium is reached rapidly [35–38]. However, in a study of thiol–disulfide exchange between tocinoic acid (TA) and glutathione (GSH, reduced) in lyophilized solids, we observed that the exchange reaction is irreversible and highly regioselective [39]. This suggests that limited conformational flexibility in solids effectively prevents the backward reaction, and that the thiolate ion preferentially attacks the less hindered site of the disulfide bond in the forward reaction.

Lyophilized recombinant human keratinocyte growth factor (rhKGF) is prone to aggregation at elevated temperatures [40]. Aggregation is proposed to proceed initially with unfolding of the protein, leading to formation of large soluble aggregates, which can form disulfide bonds. These precipitates are converted to scrambled disulfides and/or nondisulfide crosslinked oligomers. Recombinant human albumin (rHA) also undergoes intermolecular thiol–disulfide exchange [41]. The unfolding or loss of tertiary structure in lyophilized rHA is suspected to have initiated the covalent thiol–disulfide exchange.

19.2.5.2. Nonreducible Aggregation. Lyophilized human insulin can undergo covalent dimerization at the AsnA21 site to form [AspA21–PheB1] and

[AspA21–GlyA1] insulin dimers [15]. Dimerization occurs via attack of the free *N*-terminal amine of an insulin molecule on the reactive cyclic anhydride intermediate formed by deamidation of the *C*-terminal AsnA21. Lyophilized tumor necrosis factor (TNF) is also susceptible to the formation of nonreducible aggregates consisting of dimers, trimers, and higher oligomers, which stayed intact on treatment with β-mercaptoethanol [42].

Schwendeman et al. observed a loss of lysine and histidine residues in studies of the moisture-induced aggregation of tetanus and diphtheria toxoids [43]. In this case, because the aggregates were not soluble under either reducing (10 mM DTT, 1 mM EDTA) or denaturing (6 M urea) conditions, and because the toxoids had been treated with formaldehyde ("formalinized") during production, aggregation was thought to be caused by formaldehyde-induced crosslinking of the proteins.

19.3. RELATING PROTEIN STRUCTURE AND REACTIVITY IN LYOPHILIZED SOLIDS

Although chemical and physical instabilities of proteins in lyophilized solids are often treated as independent processes, it is reasonable to expect that they are closely related in many cases. For example, the production of isoAsp by deamidation at Asn residues inserts a nonnative β-amino acid in the protein backbone, which may drive changes in conformation. Conversely, the "exposure" of normally buried groups by protein unfolding during lyophilization may promote their degradation during storage in the solid phase. While there is abundant literature relating protein structure to various chemical degradation reactions in solution [1–5], the parallel literature in the solid state is much more limited, and is reviewed in this section. Establishing such relationships in the solid state is complicated by the lack of high-resolution analytical methods to determine protein structure in amorphous solids. The available analytical methods are reviewed in Chapter 22 in this volume. In addition to the lack of high-resolution structural methods, there is some ambiguity in the literature with regard to the use of the term "stability" for solid protein samples. The term has been used to indicate (1) the preservation of secondary structure (i.e., structural stability), (2) the lack of detectable degradation products (e.g., aggregates, deamidation products) on reconstitution (i.e., pharmaceutical stability), (3) the retention of biological activity (i.e., pharmacological stability), (4) the preservation of primary sequence (i.e., chemical stability), or (5) some combination of these. Here, we specifically examine the literature relating protein structure in the solid state (1) to particular chemical degradation reactions (4).

Covalent protein aggregation in lyophilized solids has been particularly well studied, and it is generally accepted that aggregation is related to protein conformation in the solid state [44], although the experimental evidence is somewhat mixed. Andya et al. examined aggregation in lyophilized formulations of a recombinant human monoclonal antibody (rhuMAb) [34]. A more native-like structure was maintained in the solid state when the protein was lyophilized with sucrose or trehalose, as monitored by FTIR. During storage, the protein formed aggregates linked by intermolecular disulfide bonds. Circular dichroism (CD) spectroscopy showed no evidence of changes in protein structure in the soluble aggregates, however. The authors proposed that loss of protein structure facilitates the formation of intermolecular disulfide bonds and covalent aggregation. They further proposed that the inclusion of sugars fulfills protein hydrogen-bonding requirements, preventing noncovalent association of protein molecules that may precede covalent interactions. In studies of lyophilized formulations of a monoclonal antibody (rhuMAb HER2), Cleland et al. observed that protein degradation by aggregation and deamidation was reduced in formulations that minimized protein unfolding during lyophilization [45]. The inclusion of excipients such as sucrose and trehalose both inhibited deamidation and showed preservation of native structure by FTIR. Chen et al. observed increases in protein deamidation and aggregation in samples of human deoxyribonuclease (rhDNase) lyophilized without calcium. However, calcium did not produce detectable changes in rhDNase structure as measured by FTIR or CD spectroscopy [46].

Deamidation and other hydrolytic reactions have also been shown to have some dependence on protein structure in lyophilized solids. In addition to the studies described above in which both deamidation and aggregation occurred, several additional reports examine the relationship between structure and deamidation in lyophilized solids in greater detail. Krogmeier et al. studied deamidation in model cyclic peptides containing the deamidation-prone Asn–Gly sequence in a β turn [47]. In lyophilized solids, deamidation was 1.2- to 8-fold slower in the cyclic peptides than in their linear analogs, suggesting that secondary structure stabilizes the Asn residue against deamidation in solids. Results were similar to previous studies of cyclic peptides in solution [48]. The effect of peptide structure was less than the effect of moisture, however; deamidation rates in the solid samples varied by several orders of magnitude with changes in relative humidity (RH). As discussed above, Strickley and Anderson conducted detailed studies of the mechanisms of degradation of human insulin in solution and in lyophilized solids [15,16]. Both the solution and solid reactions proceeded through a cyclic anhydride intermediate formed at the *C*-terminal Asn residue of the A chain (AsnA21), which degraded further to produce desamido insulin and covalently linked insulin dimers. The authors inferred that the protein had sufficient conformational flexibility to permit the cyclization reaction, even in glassy solids. The distribution of products (i.e., desamido vs. dimerized insulin) depended on solid water content, effective "pH," and excipient selection. While an effect of protein conformation on the reactions was inferred, no measures of protein structure in the lyophilized samples were reported.

There are very few studies exploring a possible relationship between protein structure and oxidative reactions in lyophilized solids. Miller et al. studied the photoxidation of recombinant bovine somatotropin (rbST) in solid powders exposed to near-UV light (305–410 nm) [49]. Photooxidation exclusively targeted rbST disulfide bonds by mechanisms thought to involve photoionization of Trp followed by the one-electron reduction of the disulfide. The authors reported some evidence for conformational changes in the solid as a result of these reactions. However, SDS-PAGE under nonreducing conditions suggested the formation of a less compact structure following photooxidation. Fluorescence spectroscopy showed a loss of quenching of Trp fluorescence after photooxidation, indicating a change in the Trp environment (i.e., local structure) and/or chemical modification of disulfide bonds. However, similarities in the amide I envelopes of Raman spectra and a lack of frequency shifts in fluorescence emission suggested a lack of conformational change. DePaz et al. investigated oxidation in lyophilized formulations of subtilisin [50]. In solution, the enzyme is susceptible to oxidation by H_2O_2 at a partially buried methionine (Met222), resulting in loss of catalytic activity. Inclusion of disaccharides inhibited lyophilization-induced unfolding of the enzyme, as measured by second-derivative FTIR spectroscopy. These additives did not inhibit oxidation of Met222 during storage at 37°C and 20% or 75% relative humidity, however.

Taken together, these reports provide some indication of a relationship between chemical degradation reactions of proteins in lyophilized solids and protein structure. The evidence is somewhat stronger for covalent aggregation and deamidation than for oxidative reactions. The development of higher-resolution methods for assessing protein structure in the solid state, such as H/D exchange techniques, is expected to improve our ability to address this question.

19.4. SUMMARY

Proteins in lyophilized solids are subject to a variety of chemical degradation processes, including hydrolytic, oxidative, condensation, and β-elimination reactions. These processes produce covalent modifications in individual amino acids and form covalent aggregates when they occur intermolecularly. While it is generally accepted that the preservation of native protein structure in lyophilized solids will minimize both chemical and physical degradation reactions, the experimental evidence for this claim is mixed. There is a need for higher-resolution methods for monitoring both structural and chemical changes in proteins in the solid state. The development of such methods is expected to improve our ability to understand and control protein degradation reactions in lyophilized solids.

REFERENCES

1. Xie, M. and Schowen, R. L. (1999), Secondary structure and protein deamidation, *J. Pharm. Sci*. **88**(1): 8–13.
2. Wakankar, A. A., Liu, J., Vandervelde, D., Wang, Y. J., Shire, S. J., and Borchardt, R. T. (2007), The effect of cosolutes on the isomerization of aspartic acid residues and conformational stability in a monoclonal antibody, *J. Pharm. Sci*. **96**(7): 1708–1718.
3. Huus, K., Havelund, S., Olsen, H. B., van de Weert, M., and Frokjaer, S. (2006), Chemical and thermal stability of insulin: Effects of zinc and ligand binding to the insulin zinc-hexamer, *Pharm. Res*. **23**(11): 2611–2620.
4. Kerwin, B. A. and Remmele, R. L., Jr. (2007), Protect from light: Photodegradation and protein biologics, *J. Pharm. Sci*. **96**(6): 1468–1479.
5. Pan, B., Abel, J., Ricci, M. S., Brems, D. N., Wang, D. I., and Trout, B. L. (2006), Comparative oxidation studies of methionine residues reflect a structural effect on chemical kinetics in rhG-CSF, *Biochemistry* **45**(51): 15430–15443.

6. Carpenter, J. F., Prestrelski, S. J., and Dong, A. (1998), Application of infrared spectroscopy to development of stable lyophilized protein formulations, *Eur. J. Pharm. Biopharm.* **45**(3): 231–238.
7. Bhatt, N. P., Patel, K., and Borchardt, R. T. (1990), Chemical pathways of peptide degradation. I. Deamidation of adrenocorticotropic hormone, *Pharm. Res.* **7**(6): 593–599.
8. Patel, K. and Borchardt, R. T. (1990), Chemical pathways of peptide degradation. II. Kinetics of deamidation of an asparaginyl residue in a model hexapeptide, *Pharm. Res.* **7**(7): 703–711.
9. Geiger, T. and Clarke, S. (1987), Deamidation, isomerization, and racemization at asparaginyl and aspartyl residues in peptides. Succinimide-linked reactions that contribute to protein degradation, *J. Biol. Chem.* **262**(2): 785–794.
10. Patel, K. and Borchardt, R. T. (1990), Chemical pathways of peptide degradation. III. Effect of primary sequence on the pathways of deamidation of asparaginyl residues in hexapeptides, *Pharm. Res.* **7**(8): 787–793.
11. Oliyai, C., Patel, J. P., Carr, L., and Borchardt, R. T. (1994), Solid state chemical instability of an asparaginyl residue in a model hexapeptide, *J. Pharm. Sci. Technol.* **48**(3): 167–123.
12. Hekman, C., DeMond, W., Dixit, T., Mauch, S., Nuechterlein, M., Stepanenko, A., Williams, J. D., and Ye, M. (1998), Isolation and identification of peptide degradation products of heat stressed pramlintide injection drug product, *Pharm. Res.* **15**(4): 650–659.
13. Li, B., Schowen, R. L., Topp, E. M., and Borchardt, R. T. (2006), Effect of N-1 and N-2 residues on peptide deamidation rate in solution and solid state, *AAPS J.* **8**(1): E166–E173.
14. Li, B., Gorman, E. M., Moore, K. D., Williams, T., Schowen, R. L., Topp, E. M., and Borchardt, R. T. (2005), Effects of acidic N + 1 residues on asparagine deamidation rates in solution and in the solid state, *J. Pharm. Sci.* **94**(3): 666–675.
15. Strickley, R. G. and Anderson, B. D. (1996), Solid-state stability of human insulin. I. Mechanism and the effect of water on the kinetics of degradation in lyophiles from pH 2–5 solutions, *Pharm. Res.* **13**(8): 1142–1153.
16. Strickley, R. G. and Anderson, B. D. (1997), Solid-state stability of human insulin. II. Effect of water on reactive intermediate partitioning in lyophiles from pH 2–5 solutions: Stabilization against covalent dimer formation, *J. Pharm. Sci.* **86**(6): 645–653.
17. Oliyai, C., Patel, J. P., Carr, L., and Borchardt, R. T. (1994), Chemical pathways of peptide degradation. VII. Solid state chemical instability of an aspartyl residue in a model hexapeptide, *Pharm. Res.* **11**(6): 901–908.
18. Pikal, M. J. and Rigsbee, D. R. (1997), The stability of insulin in crystalline and amorphous solids: Observation of greater stability for the amorphous form, *Pharm. Res.* **14**(10): 1379–1387.
19. Inglis, A. S. (1983), Cleavage at aspartic acid, *Meth. Enzymol.* **91**: 324–332.
20. Li, S., Patapoff, T. W., Overcashier, D., Hsu, C., Nguyen, T. H., and Borchardt, R. T. (1996), Effects of reducing sugars on the chemical stability of human relaxin in the lyophilized state, *J. Pharm. Sci.* **85**(8): 873–877.
21. Kertscher, U., Bienert, M., Krause, E., Sepetov, N. F., and Mehlis, B. (1993), Spontaneous chemical degradation of substance P in the solid phase and in solution, *Int. J. Pept. Protein Res.* **41**(3): 207–211.
22. Leung, S. S. and Grant, D. J. (1997), Solid state stability studies of model dipeptides: Aspartame and aspartylphenylalanine, *J. Pharm. Sci.* **86**(1): 64–71.

23. Becker, G. W., Tackitt, P. M., Bromer, W. W., Lefeber, D. S., and Riggin, R. M. (1988), Isolation and characterization of a sulfoxide and a desamido derivative of biosynthetic human growth hormone, *Biotechnol. Appl. Biochem.* **10**(4): 326–337.
24. Pikal, M. J., Dellerman, K. M., Roy, M. L., and Riggin, R. M. (1991), The effects of formulation variables on the stability of freeze-dried human growth hormone, *Pharm. Res.* **8**(4): 427–436.
25. Fransson, J., Florin-Robertsson, E., Axelsson, K., and Nyhlen, C. (1996), Oxidation of human insulin-like growth factor I in formulation studies: Kinetics of methionine oxidation in aqueous solution and in solid state, *Pharm. Res.* **13**(8): 1252–1257.
26. Dubost, D. C., Kaufman, M. J., Zimmerman, J. A., Bogusky, M. J., Coddington, A. B., and Pitzenberger, S. M. (1996), Characterization of a solid state reaction product from a lyophilized formulation of a cyclic heptapeptide. A novel example of an excipient-induced oxidation, *Pharm. Res.* **13**(12): 1811–1814.
27. D'Souza, A. J., Schowen, R. L., Borchardt, R. T., Salsbury, J. S., Munson, E. J., and Topp, E. M. (2003), Reaction of a peptide with polyvinylpyrrolidone in the solid state, *J. Pharm. Sci.* **92**(3): 585–593.
28. Hageman, M. J., Bauer, J. M., Possert, P. L., and Darrington, R. T. (1992), Preformulation studies oriented toward sustained delivery of recombinant somatotropins, *J. Agric. Food Chem.* **40**: 348–355.
29. Townsend, M. W. and DeLuca, P. P. (1991), Nature of aggregates formed during storage of freeze-dried ribonuclease A, *J. Pharm. Sci.* **80**(1): 63–66.
30. Simons, B. L., King, M. C., Cyr, T., Hefford, M. A., and Kaplan, H. (2002), Covalent cross-linking of proteins without chemical reagents, *Protein Sci.* **11**(6): 1558–1564.
31. Costantino, H. R., Langer, R., and Klibanov, A. M. (1994), Moisture-induced aggregation of lyophilized insulin, *Pharm. Res.* **11**(1): 21–29.
32. Liu, W. R., Langer, R., and Klibanov, A. M. (1991), Moisture-induced aggregation of lyophilized proteins in the solid state, *Biotechnol. Bioeng.* **37**: 177–184.
33. Jordan, G. M., Yoshioka, S., and Terao, T. (1994), The aggregation of bovine serum albumin in solution and in the solid state, *J. Pharm. Pharmacol.* **46**(3): 182–185.
34. Andya, J. D., Hsu, C. C., and Shire, S. J. (2003), Mechanisms of aggregate formation and carbohydrate excipient stabilization of lyophilized humanized monoclonal antibody formulations, *AAPS PharmSci.* **5**(2): E10.
35. Yeo, P. L. and Rabenstein, D. L. (1993), Characterization of the thiol/disulfide chemistry of neurohypophyseal peptide hormones by high-performance liquid chromatography, *Anal. Chem.* **65**(21): 3061–3066.
36. Rabenstein, D. L. and Yeo, P. L. (1994), Kinetics and equilibria of the formation and reduction of the disulfide bonds in arginine-vasopressin and oxytocin by thiol/disulfide interchange with glutathione and cysteine, *J. Org. Chem.* **59**(15): 4223–4229.
37. Gilbert, H. F. (1995), Thiol/disulfide exchange equilibria and disulfide bond stability, *Meth. Enzymol.* **251**: 8–28.
38. Fernandes, P. A. and Ramos, M. J. (2004), Theoretical insights into the mechanism for thiol/disulfide exchange, *Chemistry* (Germany) **10**(1): 257–266.
39. Zhang, L., Williams, T. D., and Topp, E. M. (2009), Reversibility and regioselectivity in thiol/disulfide interchange of tocinoic acid with glutathione in lyophilized solids. *J Pharm Sci.* **98**(8): 3312–3318.

40. Chen, B. L., Arakawa, T., Morris, C. F., Kenney, W. C., Wells, C. M., and Pitt, C. G. (1994), Aggregation pathway of recombinant human keratinocyte growth factor and its stabilization, *Pharm. Res.* **11**(11): 1581–1587.
41. Costantino, H. R., Langer, R., and Klibanov, A. M. (1995), Aggregation of a lyophilized pharmaceutical protein, recombinant human albumin: Effect of moisture and stabilization by excipients, *Bio/Technology* **13**(5): 493–496.
42. Hora, M. S., Rana, R. K., and Smith, F. W. (1992), Lyophilized formulations of recombinant tumor necrosis factor, *Pharm. Res.* **9**(1): 33–36.
43. Schwendeman, S. P., Costantino, H. R., Gupta, R. K., Siber, G. R., Klibanov, A. M., and Langer, R. (1995), Stabilization of tetanus and diphtheria toxoids against moisture-induced aggregation, *Proc. Natl. Acad. Sci. USA* **92**: 11234–11238.
44. Costantino, H. R., Schwendeman, S. P., Langer, R., and Klibanov, A. M. (1998), Deterioration of lyophilized pharmaceutical proteins, *Biochemistry* (Moscow) **63**(3): 357–363.
45. Cleland, J. L., Lam, X., Kendrick, B., Yang, J., Yang, T. H., Overcashier, D., Brooks, D., Hsu, C., and Carpenter, J. F. (2001), A specific molar ratio of stabilizer to protein is required for storage stability of a lyophilized monoclonal antibody, *J. Pharm. Sci.* **90**(3): 310–321.
46. Chen, B., Costantino, H. R., Liu, J., Hsu, C. C., and Shire, S. J. (1999), Influence of calcium ions on the structure and stability of recombinant human deoxyribonuclease I in the aqueous and lyophilized states, *J. Pharm. Sci.* **88**(4): 477–482.
47. Krogmeier, S. L., Reddy, D. S., Vander Velde, D., Lushington, G. H., Siahaan, T. J., Middaugh, C. R., Borchardt, R. T., and Topp, E. M. (2005), Deamidation of model beta-turn cyclic peptides in the solid state, *J. Pharm. Sci.* **94**(12): 2616–2631.
48. Xie, M., Aube, J., Borchardt, R. T., Morton, M., Topp, E. M., Vander Velde, D., and Schowen, R. L. (2000), Reactivity toward deamidation of asparagine residues in beta-turn structures, *J. Pept. Res.* **56**(3): 165–171.
49. Miller, B. L., Hageman, M. J., Thamann, T. J., Barron, L. B., and Schoneich, C. (2003), Solid-state photodegradation of bovine somatotropin (bovine growth hormone): Evidence for tryptophan-mediated photooxidation of disulfide bonds, *J. Pharm. Sci.* **92**(8): 1698–1709.
50. DePaz, R. A., Dale, D. A., Barnett, C. C., Carpenter, J. F., Gaertner, A. L., and Randolph, T. W. (2002), Effects of drying methods and additives on the structure, function, and storage stability of subtilisin: Role of protein conformation and molecular mobility, *Enzyme Microb. Technol.* **31**(6): 765–774.

20

THE IMPACT OF BUFFER ON SOLID-STATE PROPERTIES AND STABILITY OF FREEZE-DRIED DOSAGE FORMS

Evgenyi Y. Shalaev and Larry A. Gatlin

20.1. INTRODUCTION

Buffers are used in many freeze-dried formulations to control pH, in order to improve stability during processing and shelf life as well as to facilitate dissolution on reconstitution. Buffering capacity and the potential for buffer catalysis are two main factors that should be considered during formulation development of freeze-dried and other injectable dosage forms. In addition to these general properties, freeze drying imposes special requirements on buffer selection, including buffer crystallization behavior during freezing, buffer impact on both the collapse temperature of the freeze-concentrated solution and the T_g of the dried formulation, and the volatility of buffer components under vacuum [1]. Buffer selection is an important step in formulation development, as appropriate choice of buffer would help a formulator optimize both processability, stability, and performance of a freeze-dried formulation. Buffer behavior during freeze drying has being reviewed elsewhere, with an emphasis on crystallization and pH changes [1]; this chapter deals with the impact of buffers on the stability of freeze-dried formulations.

Factors that influence stability of freeze-dried materials are extensively discussed in the literature and relatively well understood, albeit on a qualitative basis. Molecular

Formulation and Process Development Strategies for Manufacturing Biopharmaceuticals,
edited by Jameel and Hershenson

mobility is usually considered to play a major role in destabilization of freeze-dried systems [2–4]. It has been recognized, however, that molecular mobility is not the only factor in stability of freeze-dried forms. Other stability-related properties to consider include structure of the active component (e.g., crystalline long-range order for small molecules, retention of secondary structure for proteins), and—more recently—media properties, such as polarity and solid-state acid–base relationships [5,6]. It has been observed that freeze-dried formulations prepared with different buffers have different solid-state stability in many cases. Buffer type, buffer concentration, and solution pH before lyophilization are important formulation variables during formulation development of both clinical and commercial dosage forms. In this chapter, we consider the impact of buffers on stability-related physical chemical properties of freeze-dried formulations, including the glass transition temperature, acid–base relationships, the structure of an active compound, and their correlations with stability.

20.2. MOLECULAR MOBILITY: THE GLASS TRANSITION TEMPERATURE

The glass transition temperature T_g is an important reference point for describing temperature dependence of both molecular mobility and stability, and determination of the T_g is crucial for lyophilized formulations and other amorphous materials. In some cases, freeze-dried formulations with higher T_g had better stability, although multiple exceptions do exist (e.g., see Ref. [7]). Other measures of molecular mobility were reported to provide better correlation with stability than the T_g, including the rate of enthalpy relaxation and NMR spin–lattice relaxation time [8,9], although studies on the impact of buffers on the alternative measures of the molecular mobility are lacking. There are excellent general reviews on properties and significance of the T_g and other measurements of the molecular mobility (e.g., relaxation time) in pharmaceuticals [3,4,10].

The glass transition temperature of a complex mixture such as a freeze-dried formulation depends on both the glass transition temperatures of the individual components and their weight fraction. The relationships between composition of the mixture and the T_g can be described by a number of different expressions; the Fox and Gordon–Taylor equations are the most common. The lower the T_g of a buffer and the higher its fraction in a formulations, the lower the T_g of the formulation. The T_g behavior of lyophilized buffers as well as examples of correlations between buffer-induced lowering T_g and destabilization of freeze-dried formulations are considered below.

Acidic forms of carboxylic acid buffers have relatively low T_g values. For example, the T_g values of citric acid, tartaric acid, malic acid, and lactic acid were reported to be 11°C [11], 16°C [12], −20°C, and −60°C [13], respectively. Similarly, the T_g of succinic acid is probably below 30°C, as can be estimated from data for the T_g of binary PVP/succinic acid blends [14]. Some amino acid–based buffers have somewhat higher T_g values than do carboxylic acids. For example, the T_g of L-histidine, L-lysine hydrochloride, and glycine are approximately 37°C [15], 45°C [16], and 7°C [17], respectively. The T_g of the salt forms of the buffers are expected to be higher; for example, the T_g values for sodium citrate salts were reported to be significantly higher than that of citric acid ($T_g = 11°C$), at 69°C, 115°C [11], and 158°C [18] for monosodium citrate, disodium citrate, and trisodium citrate, respectively. A similar trend was observed for another carboxylic acid, indomethacin, where the T_g of the sodium salt (121°C) was much higher than that for the free acid (45°C) [19]. In addition, the T_g of a buffer would depend on the nature of a counterion; for example, a significant difference in the T_g of potassium and sodium salts was reported, with a lower T_g for potassium indomethacin (109°C vs. 121°C) [19].

With such significant difference in the T_g of a free acid and salt form of a buffer, one could expect that prelyophilization solution pH would have an impact on the T_g of freeze-dried buffer. Indeed, for sodium citrate buffer, for example, the T_g of the freeze-dried samples increased with an increase in prelyophilization solution pH (i.e., higher sodium ion content), with the T_g values increasing from 29°C to 66°C as the prelyophilization solution pH increased from 3.0 to 4.3 [11]. The increase in T_g with solution pH is related to the higher salt : free-acid ratio because a salt has a higher T_g. Note, however, that the magnitude of the pH impact on the T_g of freeze-dried form may depend on a buffer type; for example, no difference in the T_g values of the freeze-dried formulations, which were prepared from solutions at pH 6.0 and 9.8, was reported for four other buffers, namely, ammediol, Tris, tricine, and Ches [20]. An interesting behavior was reported for borax buffer, when a significant increase in the T_g was observed for borax buffer–sugar formulations lyophilized from solutions at near-neutral and alkaline pH range 6–10. This was attributed to a chemical reaction between borax ion and hydroxyl sugar groups with a formation of a new chemical entity or a molecular complex having an elevated T_g [20,21].

Because of relatively low T_g of many common buffers, one would expect a buffer to decrease the T_g of a freeze-dried formulation, especially if used at a relatively high concentration. Indeed, plasticization of the freeze-dried formulations by buffers has been reported [20,22]. In both cases, the decrease in the T_g correlated with increased solid-state instability. For example, the impact of four different buffers on the T_g and stability of lyophilized alkaline phosphatase in trehalose matrix has been studied [20]. These formulations were prepared from the same prelyophilization solution pH of 9.8. All four buffers studied decreased the T_g of freeze-dried formulations, although to different extents; ammediol and Tris had a more significant plasticizing effect on the T_g, followed by tricine and CHES. For example, the T_g was depressed by approximately 50°C in trehalose : ammediol 1 : 0.17 (w/w) mixture, whereas a much smaller T_g depression of approximately 10°C was observed for trehalose–Ches mixture at the same weight ratio. The trend in stability correlated with the T_g of the freeze-dried formulations. Tricine and CHES buffers provided a better stability, whereas formulations prepared with ammediol and Tris buffers were less stable. In addition, an increase in ammediol and Tris buffers concentration resulted in both depression of the T_g and increase in solid-state instability.

Citrate buffer destabilized lyophilized amorphous quinapril, when an increase in the degradation rate was observed with an increase in citrate concentration when the initial solution pH was above 2.75 [22]. The reaction rate increased by a factor of ~2 as the quinapril : citrate molar ratio decreased from 6 : 1 to 1 : 1. The decreased chemical stability was related to the depressed T_g (and corresponding increase in the molecular mobility) at a higher citrate concentration. However, in formulations that were prepared from solutions at pH 2.49, the reaction rate was the same and independent of the quinapril : citrate ratio. In order to explain such a difference in the effect of buffer concentration on stability, it was suggested that the reaction mechanism varied at different pH values, with the mobility-dominated trans–cis transition to be the rate-limiting step at pH 2.75–3.05, that is, in formulations where stability correlations with the T_g were observed. This study highlights the need to treat any extrapolation of the solid-state stability trend to a different pH range with caution.

20.3. MEDIA EFFECT: ACID–BASE RELATIONSHIPS

The pH of solution before freeze drying is an important formulation variable as it has a significant impact on the rate of solid-state chemical reactivity in the freeze-dried cake [23–26]. To describe the apparent parallelism in the pH–stability profiles in solution and dried states, the term "pH memory" has been used [27,28]. However, stability in the dehydrated state does not always parallel the pH dependence in solution [29,30], suggesting that knowledge of pH–stability relationships in solution may not be sufficient to predict the solution pH that would yield a freeze-dried formulation with maximal solid-state stability. Indeed, the pH of the solution that yields a lyophile of maximum stability could differ from the pH of maximum solution stability. Such shifts in the "pH stability profile" from solution to lyophilized material can be attributed to changes in either acid–base relationships during freezing and/or drying or the degradation mechanism. Two possible mechanisms for change in the apparent acidity from a solution to a freeze-dried material, namely, crystallization of buffer components and a more speculative "true" medium effect as a result of either proton activity or/and pK_a change from solution to lyophile, are considered below.

Crystallization of buffer components is one of the mechanisms possibly responsible for changes in solid-state acidity and compromised solid-state stability. The ability of buffer to maintain a target pH is defined by both buffer concentration and a ratio of acid–base buffer species; therefore, any change in this ratio during processing (e.g., by crystallizing out a buffer component during cooling) would cause a significant change in proton activity. Buffers that have a greater tendency to crystallize are usually expected to have a detrimental impact on stability of lyophilized formulations. In particular, phosphate and succinate buffers tend to crystallize during freezing, therefore producing significant pH changes and destabilization of an active ingredient, whereas citrate, glycolate, and malate are more resistant to crystallization [1].

An example of the destabilizing impact of phosphate buffer is reported by Almarsson et al. [31], who studied lyophilized formulations of a novel carbapenem compound that were prepared with either phosphate, citrate, bicarbonate, or without buffer. The rate of degradation of the formulation with phosphate buffer was significantly higher, by a factor of 2, than with bicarbonate buffer (prelyophilization solution pH 6.5). The stability of the formulation without buffer was slightly better than with bicarbonate buffer. Also, the stability of lyophilized formulation with citrate buffer prepared at pH 6.0 was better than that of either phosphate, carbonate, or no buffer formulations prepared at pH 6.5. The destabilization effect of phosphate buffer was attributed to crystallization of disodium phosphate and a corresponding decrease in pH, although direct tests that would allow confirmation of phosphate crystallization in the formulation studied were not performed.

Another case of a detrimental impact of a buffer with a high crystallization potential, namely, succinate, was reported in another study [32], where a significant improvement in the stability of a lyophilized protein was achieved by replacing succinate buffer with glycolate. It was suggested that succinate buffer crystallized during lyophilization, resulting in a pH shift to acidic values, which, in turn, caused significant protein unfolding (loss of α helix). Glycolate buffer, which did not crystallize during lyophilization, prevented the pH shift and provided better stability.

An increase in buffer concentration is expected to facilitate buffer crystallization, thereby increasing instability of the formulation. An example of the detrimental impact of a higher phosphate buffer concentration on stability was reported in another study [33], where an increase in the buffer concentration resulted in an accelerated degradation of aspartame in lyophilized formulations that also contained agar and microcrystalline cellulose. The degradation rate was higher in the formulation that was prepared at a higher phosphate concentration (0.1 M) as compared to 0.01 M buffer. In this study, phosphate concentration was the only variable; other formulation variables, namely, concentration of the agar–microcrystalline cellulose matrix before lyophilization (approximately 7%) and solution pH before lyophilization (pH 5), were identical. Both buffer catalysis and the possibility of phosphate crystallization were discussed to explain the destabilizing effect of the higher phosphate concentration, although no definite conclusion on the mechanism was achieved. It should be noted, however, that an increase in phosphate buffer concentration does not necessary result in destabilization. For example, an apparent lack of destabilization of a protein with increasing sodium phosphate buffer concentration has been reported [25] for human growth hormone (hGH), when the buffer concentration varied from 0.114 to 0.454 mg buffer/mg hGH. Although the stability of the formulations studied was significantly different, with up to 50% difference in the reaction rate, no clear trend in stability with buffer concentration was observed.

The nature of metal counterion may impact both buffer crystallization potential and stability of a freeze-dried cake. In particular, potassium phosphate buffer has a lower crystallization tendency than does sodium phosphate, which might have an impact on stability of freeze-dried formulations. Indeed, one such example was described by Sarciaux et al. [34], who reported the aggregation of IgG after freeze drying to be lower with potassium phosphate than with sodium phosphate buffer, possibly because potassium phosphate did not crystallize, thus minimizing the pH shift. Similar results were reported [35] for β-galactosidase, where the denaturation extent after freeze drying was significantly higher with sodium phosphate than that with potassium phosphate. Also, an increase in a potassium phosphate buffer concentration, ranging from 0.02 to 0.15 M, did not have any noticeable impact on asparagine deamidation of a lyophilized hexapeptide in PVP matrix (solution pH 6.6) [36]. A possible reason for the lack of concentration effect is that potassium salt was used in this study, which is more resistant to crystallization than is sodium salt.

In addition to the crystallization effects, there is another possible mechanism that could impact solid-state acid–base relationships and thereby stability of lyophilized formulations. We suggest, after an earlier study [5], that media effects, in a traditional physical organic chemistry sense, might contribute to differentiation in the stabilities of lyophilized formulations prepared with different buffers. Indeed, during freeze drying, composition of the formulation changes from a dilute aqueous solution, specifically, a polar media with a high dielectric constant, to an organic matrix, a media with a lower polarity. In solution, such changes in the dielectric constant of a system cause significant changes in processes that involve polar or charged species, including acid–base relationships. In particular, decrease in water content of a mixed solvent may result in significant pH changes; the pH of acetate buffer in water–ethanol solvent, for example, increased from approximately 4.5 to 7 as water content decreased from 100% to 0% [37]. Furthermore, the pK_a of ionizable groups of buffers and other species are media-sensitive, and can shift by several units with a change in a composition of the solution. For example, the apparent pK_a of acetic acid increases significantly with decreased polarity (e.g., 4.76 in water vs. 10.32 in ethanol) [37]. It should be noted that the magnitude of such changes depends on the chemical nature of a buffer; for instance, the pK_a of an amino group (–NH_3^+/–NH_2 equilibria) exhibited relatively minor changes with the dielectric constant [37]. Hence, a significant difference in the magnitude of the apparent pK_a changes on drying might be a source of a "buffer effect" on chemical stability of lyophilized formulations. One way to evaluate this contribution is to measure acidity in the lyophilized state by, for example, using probe molecules and the Hammett acidity function as discussed briefly below.

The Hammett acidity function, proposed in the 1920s [38], is based on determination of ionization of pH indicators. The Hammett acidity function for a case of a neutral acidic indicator molecules, HA $\leftrightarrow$ H^+ + A^-, is

$$H_- = \log\left[\frac{c_{A-}}{c_{HA}}\right] + pK_a = -\log a_{H+}\left[\frac{f_{A-}}{f_{HA}}\right] \tag{20.1}$$

where c, f, and a represent concentrations, activity coefficients, and activities, respectively, for the protonated (HA) and deprotonated (A^-) species of the indicator and K_a is the dissociation constant of the acid HA. The Hammett acidity function and pH indicators have been used to study acid–base relationships in strong acids and nonaqueous systems [39–41], on solid surfaces [42–46], in frozen aqueous solutions [20,47,48], and in the freeze-dried materials [49]. Despite the widespread use of the Hammett acidity function in physical organic chemistry and related fields, its exploration in freeze-dried pharmaceutical materials has been initiated more recently, and there is a lack of information on the Hammett acidity function of formulations freeze-dried with different buffers. One available example is considered here using results reported in a study [50]; in which the Hammett acidity function was used to compare solid-state acidity of lyophilized trehalose-based formulations with either citrate or phosphate buffers prepared from solutions with pH values ranging from 5.5 to 7.5.

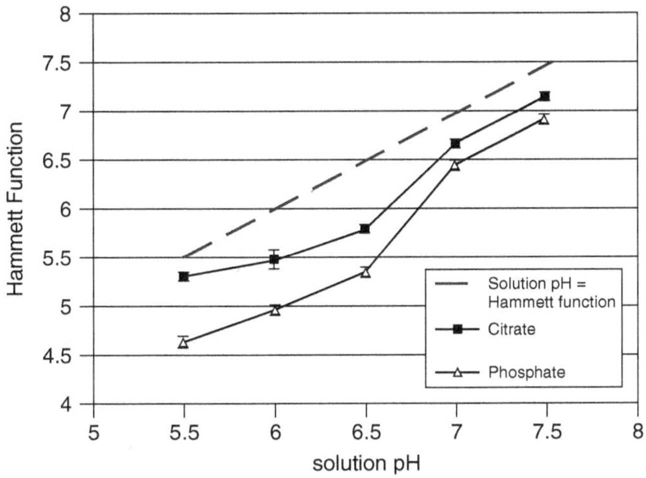

Figure 20.1. Hammett acidity functions of lyophilized trehalose/buffer formulations as a function of prelyophilization solution pH of citrate (■) and phosphate (△) buffered systems. The bold dashed line indicates a hypothetical situation in which the Hammett functions a numerically equal to the prelyophylization solution pH. The other two lines are shown as visual aid. (The figure was prepared using data reported by Chatterjee [50].)

The results are presented in Figure 20.1, where it can be seen that phosphate lyophiles had lower Hammett function values (higher acidity) than citrate lyophiles. The reason for a higher solid-state acidity of the phosphate buffer formulation is not clear; note that crystallization of phosphate buffer was not detected in the systems studied. From the practical perspective these results imply that, in order to achieve the same solid-state acidity with citrate and phosphate buffer, a formulation with phosphate buffer needs to be prepared at a higher solution pH. These preliminary results as well as studies on the correlations between the solid acidity and chemical reactivity of lyophilized formulations [51] suggest that the Hammett acidity function might be a useful tool in formulation development, including buffer screening for lyophilized dosage forms.

Finally, an interesting example of a buffering action in the solid state was reported by Sugimoto et al. [52], who studied the degradation of lyophilized sodium prasterone sulfate and observed that the solid-state degradation of an unbuffered formulation was inherently autocatalytic. They attributed this finding to the fact that (1) degradation produced acidic degradants and (2) prasterone sulfate is acid-sensitive. Therefore, they suggested that the apparent autocatalytic effect observed was a result of acidification caused by degradation during storage. The degradation was much slower in a formulation with phosphate citrate buffer that supposedly neutralized the acidification of the formulation at the initial stage of the degradation; a high drug : buffer ratio of 1 : 1 was used in this study.

20.4. STRUCTURE OF AN ACTIVE INGREDIENT

In this chapter, the term *structure* is used in two different contexts depending on the nature of an active pharmaceutical ingredient (API). For low-molecular-weight (LMW) API, *structure* reflects long-range order (or a lack of thereof), dividing materials between highly ordered crystalline and disordered amorphous states, with intermediate crystalline mesophase structures (e.g., liquid crystals). For proteins, which exist in freeze-dried formulations in an amorphous state, *structure* means secondary and ternary structure, such as α helices and β sheets. For LMW APIs, a crystalline material is usually much more stable than the same molecule in an amorphous state [53]. Therefore, a lyophilized formulation with an active ingredient in the crystalline state would manifest superior chemical stability. In addition, there are different noncrystalline states with variable degrees of order ranging between highly ordered crystalline and very disordered amorphous state, such as liquid crystals, plastic crystals, and conformationally disordered crystals [54]. The chemical stability of such states of the intermediate order (i.e., crystalline mesophases) is expected to be intermediate between crystalline (highest stability) and amorphous (lowest stability) states [54]. For proteins, a retention of a native-like secondary structure in a freeze-dried state is usually considered as a prerequisite for long-term stability. Therefore, testing the structural state of an active ingredient in lyophilized formulations, such as by small- and wide-angle X-ray diffraction for small molecules or FTIR for proteins, is a must in formulation development of freeze-dried dosage forms.

Although there are reasons to expect that solution pH as well as buffer type and concentration can modify crystallization behavior of the active compound during freeze drying, there is a lack of systematic studies on the impact of buffer on structure of an active ingredient in freeze-dried materials; a few examples are considered below. Both promotion and inhibition of crystallization of an active compound during lyophilization by excipients have been reported (see Ref. [55] for review). For example, glycine and other amino acids facilitated crystallization of three different compounds: atropine sulfate, cefoxitin sodium, and doxycycline hyclate [56]. On the other hand, phosphate buffer inhibited crystallization of a nonionizable solute, mannitol [57]. For protein formulations, buffer effect on protein secondary structure was reported in one study [35], where a better retention of native protein structure during freezing was observed with potassium phosphate as compared with sodium phosphate buffer, presumably because a more significant pH shift was expected with the sodium buffer.

20.5. OTHER CONSIDERATIONS

It would be naive to expect that physical chemical properties considered above to cover all mechanisms of buffer impact on stability of freeze-dried formulations. In some cases, other mechanisms were invoked to explain the buffer effect, in addition to molecular mobility, media effect, and structure considerations. A specific chemical mechanism was suggested by Osterberg and Wadsten [58], who suggested that an ability of L-histidine buffer to scavenge metal ions and therefore minimize oxidation of a protein could explain the stabilization effect of the buffer. On the other hand, degradation of a buffer itself might produce highly reactive species and have a destabilizing effect; for example, degradation of Tris buffer during accelerated storage of a freeze-dried material generated formaldehyde [59]. An interesting result of a significant "buffer effect" of an unknown nature was reported in another study [60], where a slight increase in histidine concentration from 4 to 6 mM (concentration in prelyophilization solution) resulted in substantial increase in stability. Finally, it would be worthwhile to mention a subset of acid–base relationships, namely, general buffer catalysis, which is commonly considered for solution stability but rarely for freeze-dried states. The possibility of a general buffer catalysis in freeze-dried material was also considered [36]. A significant buffer catalysis was observed in solution using carbonate, whereas no buffer catalysis was observed in the lyophilized formulations, where the stability of the peptide lyophilized with carbonate buffer (pH 9.5 and 10.4) did not change as the carbonate buffer concentration increased from 0.02 to 0.15 M.

20.6. CONCLUSION

From the practical lyophilization perspective, citrate buffer appears to be a good choice for formulations that are prepared at acidic or near neutral pH. It has a reasonably high T_g, especially when prepared from solutions at higher pH values, has low crystallization potential, and was reported to provide a better long-term stability than other buffers. Injection-site tolerance, however, might be a potential issue with citrate buffer [61]. Glycine or histidine might be a reasonable choice for alkaline region, although the crystallization, T_g, and collapse behavior of these buffers have not been studied in the same level of detail as phosphate or citrate buffers.

The most common buffer for injectable formulations, phosphate, was reported to have a negative impact on the stability of lyophilized formulations. The detrimental impact of phosphate buffer is commonly explained to be a consequence of the buffer crystallization and significant pH changes during lyophilization. It should be mentioned, however, that direct evidence of phosphate crystallization during freeze drying in real pharmaceutical formulations is scarce, in part because buffer is usually a minor part of a formulation on a weight basis, and the sensitivity provided by X-ray diffraction might not be sufficient to detect buffer crystallization.

Buffer type and concentration, as well as solution pH before lyophilization, are important formulation variables that determine post-freeze-drying recovery, long-term stability, and shelf life of lyophilized pharmaceuticals. In many cases, the impact of a buffer on the solid-state stability of amorphous lyophilized products can be understood through either molecular mobility (e.g., glass transition temperature) or acidity expressed as solution pH before lyophilization. In addition, our ability to predict and optimize stability of lyophilized products would benefit from studies in new areas, such as buffer effect on the structure of active pharmaceutical ingredients and determination of acid–base relationships directly in the lyophilized state using, for example, molecular probes and utilizing the Hammett acidity function.

ACKNOWLEDGMENTS

The authors thank Dr. B. Hancock, Professor R. Suryanarayanan, Drs. K. Chatterjee and R. Govindarajan, and Professor G. Zografi for helpful discussions and use of unpublished data.

REFERENCES

1. Shalaev, E. Y. (2005), The impact of buffer on processing and stability of freeze-dried dosage forms. Part 1. Solution freezing behavior, *Am. Pharm. Rev*. **8**(5): 80–87.
2. Hancock, B. C. and Zografi, G. (1997), Characteristics and significance of the amorphous state in pharmaceutical systems, *J. Pharm. Sci*. **86**: 1–12.
3. Shamblin, S. L., Tang, X., Chang, L., Hancock, B. C., and Pikal, M. J. (1999), Characterization of the time scales of molecular motion in pharmaceutically important glasses, *J. Phys. Chem. B* **103**: 4113–4121.
4. Hancock, B. C., Shamblin, S. L., and Zografi, G. (1995), Molecular mobility of amorphous pharmaceutical solids below their glass transition temperatures, *Pharm. Res*. **12**: 799–806.
5. Shalaev, E. Y. and Zografi, G. (1996), How does residual water affect the solid-state degradation of drugs in the amorphous state? *J. Pharm. Sci*. **85**: 1137–1141.
6. Li, R., D'Souza, A. J., Laird, B. B., Schowen, R. L., Borchardt, R. T., and Topp, E. M. (2000), Effects of solution polarity and viscosity on peptide deamidation, *J. Pept. Res*. **56**: 326–334.
7. Allison, S. D., Chang, B., Randolph, T. W., and Carpenter, J. F. (1999), Hydrogen bonding between sugar and protein is responsible for inhibition of dehydration-induced protein unfolding, *Arch. Biochem. Biophys*. **365**: 289–298.
8. Liu, J., Rigsbee, D. R., Stotz, C., and Pikal, M. J. (2002), Dynamics of pharmaceutical amorphous solids: The study of enthalpy relaxation by isothermal microcalorimetry, *J. Pharm. Sci*. **91**: 1852–1862.
9. Yoshioka, S. (2004), Molecular mobility of freeze-dried formulations as determined by NMR relaxation, and its effect on storage stability, *Freeze-drying/lyophilization of pharmaceutical and biological products,* L. Rey, J. C. May (Eds) **137**: 187–212.
10. Yu, L. (2001), Amorphous pharmaceutical solids: Preparation, characterization and stabilization, *Adv. Drug Deliv. Rev*. **48**: 27–42.
11. Li, J., Chatterjee, K., Medek, A., Shalaev, E., and Zografi, G. (2004), Acid-base characteristics of bromophenol blue-citrate buffer systems in the amorphous state, *J. Pharm. Sci*. **93**: 697–712.
12. Fukuoka, E., Makita, M., and Yamamura, S. (1989), Glassy state of pharmaceuticals. III. Thermal properties and stability of glassy pharmaceuticals and their binary glass mixtures, *Chem. Pharm. Bull*. **37**(4): 1047–1050.
13. Maltini, E., Anese, M., and Shtylla, I. (1997), State diagrams of some organic acid-water systems of interest in food, *Cryo-Letters* **18**: 263–268.
14. Li, X. D. and Goh, S. H. (2002), Specific interactions and miscibility of ternary blends of poly(2-vinylpyridine), poly(N-vinyl-2-pyrrolidone) and aliphatic dicarboxylic acid: Effect of spacer length of acid, *Polymer* **43**: 6853–6861.

15. Mattern, M., Winter, G., Kohnert, U., and Lee, G. (1999), Formulation of proteins in vacuum-dried glasses. II. Process and storage stability in sugar-free amino acid systems, *Pharm. Develop. Technol.* **4**(2): 199–208.

16. Buera, M. del P., Levi, G., and Karel, M. (1992), Glass transition in poly(vinylpyrrolidone): Effect of molecular weight and diluents, *Biotechnol. Progress* **8**: 144–148.

17. Suzukit, T. and Franks, F. (1993), Solid-liquid phase transitions and amorphous states in ternary sucrose-glycine-water systems, *J. Chem. Soc. Faraday Trans.* **89**: 3283–3288.

18. Zografi, G. (2004), personal communication.

19. Tong, P. and Zografi, G. (1999), Solid-state characteristics of amorphous sodium indomethacin relative to its free acid, *Pharm. Res.* **16**: 1186–1192.

20. Eriksson, J. H. C., Hinrichs, W. L. J., de Jong, G. J., Somsen, G. W., and Frijlink, H. W. (2003), Investigations into the stabilization of drugs by sugar glasses: III. The influence of several high-pH buffers, *Pharm. Res.* **20**: 1437–1443.

21. Miller, D. P., Anderson, R. E., and De Pablo, J. J. (1998), Stabilization of lactate dehydrogenase following freeze thawing and vacuum drying in the presence of trehalose and borate, *Pharm. Res.* **15**: 1215–1221.

22. Li, J., Guo, Y., and Zografi, G. (2002), Effects of a citrate buffer system on the solid-state chemical stability of lyophilized quinapril preparations, *Pharm. Res.* **19**: 20–26.

23. Liu, W. R., Langer, R., and Klibanov, M. A. (1991), Moisture-induced aggregation of lyophilized proteins in the solid state, *Biotechnol. Bioeng.* **37**: 177–184.

24. Lai, M. C. and Topp, E. M. (1999), Solid-state chemical stability of proteins and peptides, *J. Pharm. Sci.* **88**: 489–500.

25. Pikal, M. J., Dellerman, K. M., Roy, M. I., and Riggin, R. M. (1991). The effects of formulation variables on the stability of freeze-dried human growth hormone, *Pharm. Res.* **8**: 427–436.

26. Shalaev, E. Y., Lu, Q., Shalaeva, M., and Zografi, G. (2000), Acid-catalyzed inversion of sucrose in the amorphous state at very low levels of residual water, *Pharm. Res.* **17**: 366–370.

27. Yang, Z., Zacherl, D., and Russell, A. J. (1993), pH dependence of subtilisin dispersed in organic solvents, *J. Am. Chem. Soc.* **115**: 12251–12257.

28. Costantino, H. R., Griebenow, K., Langer, R., and Klibanov, A. M. (1997), On the pH memory of lyophilized compounds containing protein functional groups, *Biotechnol. Bioeng.* **53**: 345–348.

29. Strickley, R. G., Visor, G. C., Lin, L.-H., and Gu, L. (1989). An unexpected pH effect on the stability of moexipril lyophilized powder, *Pharm. Res.* **6**: 971–975.

30. Bell, L. N. and Labuza, T. P. (1991), Potential pH implications in the freeze-dried state, *Cryo-Letters* **12**: 235–244.

31. Almarsson, Ö., Seburg, R. A., Godshall, D., Tsai, E. W., and Kaufman, M. J. (2000), Solid-state chemistry of a novel carbapenem with a releasable sidechain, *Tetrahedron* **56**: 6877–6885.

32. Lam, X. M., Costantino, H. R., Overcashier, D. E., Nguyen, T. H., and Hsu, C. C. (1996), Replacing succinate with glycolate buffer improves the stability of lyophilized interferon-gamma, *Int. J. Pharm.* **142**: 85–95.

33. Bell, L. N. and Labuza, T. P. (1991), Aspartame degradation kinetics as affected by pH in intermediate and low moisture food systems, *J. Food Sci.* **56**: 17–20.

34. Sarciaux, J.-M., Mansour, S., Hageman, M. J., and Nail, S. L. (1999), Effects of buffer composition and processing conditions on aggregation of bovine IgG during freeze-drying, *J. Pharm. Sci*. **88**: 1354–1361.

35. Pikal-Cleland, K. A. and Carpenter, J. F. (2001), Lyophilization-induced protein denaturation in phosphate buffer systems: Monomeric and tetrameric b-galactosidase, *J. Pharm. Sci*. **90**: 1255–1268.

36. Song, Y., Schowen, R., Borchardt, R. T., and Topp, E. M. (2001), Effect of "pH" on the rate of asparagine deamidation in polymeric formulations: "pH"-rate profile, *J. Pharm. Sci*. **90**: 141–156.

37. Bates, R. G. (1973), *Determination of pH. Theory and Practice*, 2nd ed., Wiley, New York, pp. 156, 165, 239.

38. Hammett, L. P. (1928), The theory of acidity, *J. Am. Chem. Soc*. **50**: 2660–2673.

39. Boyer, J. P. H., Corriu, R. J. P., Perz, R. J. M., and Reye, C. G. (1975), Acidity in water-organic solvent mixtures. I. Relative importance of differing coefficients occurring in the acidity function, *Tetrahedron* **31**: 377–379.

40. Hammett, L. P. and Deyrup, A. J. (1932), A series of simple basic indicators. I. The acidity functions of mixtures of sulfuric and perchloric acids with water, *J. Am. Chem. Soc*. **54**: 2721–2739.

41. Steiner, E. C. and Gilbert, J. M. (1965), The acidities of weak acids in dimethyl sulfoxide, II. The Hammett acidity function, *J. Am. Chem. Soc*. **87**: 382–384.

42. Moffat, J. B. (1994), The acidity and basicity of solids: Intrinsic properties of the surface, its structure and composition, in *Acidity and Basicity of Solids. Theory, Assessment and Utility*, Fraissard, J. and Petrakis, L., eds., NATO ASI series, Series C: Mathematical and Physical Sciences, Vol. **444**, pp. 237–254.

43. Noda, L. K., de Almeida, R. M., Goncalves, N. S., Probst, L. F. D., and Sala, O. (2003), TiO_2 with a high sulfate content-thermogravimetric analysis, determination of acid sites by infrared spectroscopy and catalytic activity, *Catal. Today*. **85**: 69–74.

44. Haw, J. F. (2002), Zeolite acid strength and reaction mechanisms in catalysis, *Phys. Chem. Chem. Phys*. **4**: 5431–5441.

45. Glombitza, B. W., Oelkrug, D., and Schmidt, P. C. (1994) Surface acidity of solid pharmaceutical excipients. Part 1. Determination of the surface acidity, *Eur. J. Pharm. Biopharm*. **40**: 289–293.

46. Scheef, C. A. and Schmidt, P. C. (1998), Influence of surface acidity of excipients on the solid state stability of pirenzepine, *STP Pharma Sci*. **8**: 91–97.

47. Orii, Y. and Morita, M. (1977), Measurement of the pH of frozen buffer solutions by using pH indicators, *J. Biochem*. **81**: 163–168.

48. Croyle, M., Roessler, B. J., Davidson, B. L., Hilfinger, J. M., and Amidon, G. L. (1998), Factors that influence stability of recombinant adenoviral preparations for human gene therapy, *Pharm. Devel. Technol*. **3**: 373–383.

49. Govindarajan, R., Chatterjee, K., Gatlin, L., Suryanarayanan, R., and Shalaev, E. Y. (2006), Impact of freeze-drying on ionization of sulfonephthalein probe molecules in trehalose-citrate systems, *J. Pharm. Sci*. **95**(7): 1498–1510.

50. Chatterjee, K. (2004), Chemical Reactivity and Physical Stability of Lyophilized Solids Ph.D. thesis, Univ. Minnesota.

51. Chatterjee, K., Shalaev, E. Y., Suryanarayanan, R., and Govindarajan, R. (2008), Correlation between chemical reactivity and the Hammett acidity function in amorphous solids using inversion of sucrose as a model reaction, *J. Pharm. Sci*. **97**: 274–286.

52. Sugimoto, I., Ishihara, T., Habata, H., and Nakagawa, H. (1981), Stability of lyophilized sodium prasterone sulfate, *J. Parenter. Sci. Technol*. **35**: 88–92.

53. Pikal, M. J., Lukes, A. L., Lang, J. E., and Gaines, K. (1978), Quantitative crystallinity determination for beta-lactam antibiotics by solution calorimetry: Correlations with stability, *J. Pharm. Sci*. **67**: 767–773.

54. Shalaev, E. and Zografi, G. (2002), The concept of "structure" in amorphous solids from the perspectives of the pharmaceutical sciences, in *Progress in Amorphous Food and Pharmaceutical Systems*, Levine, H. ed., The Royal Society of Chemistry, pp. 11–30.

55. Shalaev, E. and Franks, F. (2002), Solid-liquid state diagrams in pharmaceutical lyophilisation: Crystallisation of solutes, in *Progress in Amorphous Food and Pharmaceutical Systems*, Levine, H., ed., The Royal Society of Chemistry, pp. 200–215.

56. Korey, D. J. and Schwartz, J. B. (1989), Effects of excipients on the crystallization of pharmaceutical compounds during lyophilization, *J. Parenter. Sci. Technol*. **43**: 80–83.

57. Izutsu, K.-I., Ocheda, S. O., Aoyagi, N., and Kojima, S. (2004), Effects of sodium tetraborate and boric acid on nonisothermal mannitol crystallization in frozen solutions and freeze-dried solids, *Int. J. Pharm*. **273**: 85–93.

58. Osterberg, T. and Wadsten, T. (1999), Physical state of L-histidine after freeze-drying and long-term storage, *Eur. J. Pharm. Sci*. **8**: 301–308.

59. Song, Y., Schowen, R. L., Borchardt, R. T., and Topp, E. M. (2001), Formaldehyde production by Tris buffer in peptide formulations at elevated temperature, *J. Pharm. Sci*. **90**: 1198–1203.

60. Chen, B., Bautista, R., Yu, K., Zapata, G. A., Mulkerrin, M. G., and Chamow, S. M. (2003), Influence of histidine on the stability and physical properties of a fully human antibody in aqueous and solid forms, *Pharm. Res*. **20**: 1952–1960.

61. Laursen, T., Hansen, B., and Fisker, S. (2006), Pain perception after subcutaneous injections of media containing different buffers, *Basic Clin. Pharmacol. Toxicol*. **98**(2): 218–221.

21

STABILIZATION OF LYOPHILIZED PHARMACEUTICALS BY CONTROL OF MOLECULAR MOBILITY: IMPACT OF THERMAL HISTORY

Suman Luthra and Michael J. Pikal

21.1. INTRODUCTION

This chapter summarizes the current understanding of molecular dynamics in pharmaceutically relevant amorphous systems, particularly those prepared by drying operations such as freeze [1,2] and spray [3,4] drying. A key application for such systems is the vitrification of biotechnology products in a single-phase amorphous solid containing a stabilizer (e.g., disaccharide). This process has proved to be the method of choice for long-term stabilization of macromolecules such as protein drug products. Amorphous systems are chemically and physically unstable relative to the crystalline state, which presents a significant challenge in the development of amorphous pharmaceutical dosage forms. A critical issue is the need to better understand the solid-state dynamics of the drug molecule and the relationship between these dynamics and degradation processes. In this chapter, the motions in the glassy state below T_g are classified and discussed, with an emphasis on comparison of dynamic processes that occur on different timescales. For each class of mobility, extrinsic factors (e.g., temperature, residual water) and intrinsic factors (e.g., formulation composition) that may influence dynamics are first discussed, followed by an analysis of the correlation of chemical and physical stability with mobility. Finally, the impact of annealing in reducing molecular mobility and improving chemical stability is discussed. Since most amorphous

Formulation and Process Development Strategies for Manufacturing Biopharmaceuticals, edited by Jameel and Hershenson

solids for pharmaceutical use are produced by lyophilization, our discussion focuses on lyophilized systems.

21.2. MOLECULAR MOTIONS IN THE AMORPHOUS STATE

Amorphous materials generally exhibit greater molecular mobility as compared to their crystalline counterparts. Figure 21.1 is a schematic of the temperature dependence of enthalpy as a supercooled liquid is cooled through the glass transition temperature T_g. For $T > T_g$, the molecules continuously undergo configurational changes (i.e., vibrational and translational motions) and are free to access all possible states of the system. Here, the average of an arbitrary property of the molecules over all the states (i.e., the statistical ensemble) is equal to the time-averaged property of a given molecule. Such a system is known as *ergodic* in statistical mechanics [5]. As the liquid supercools, the viscosity increases progressively, leading to reduced molecular mobility. This reduction in mobility increases the time required for the molecules to sample all the possible states. Eventually, the system dynamically becomes a solid, but one that is not in thermodynamic equilibrium. The ensemble average of the measured

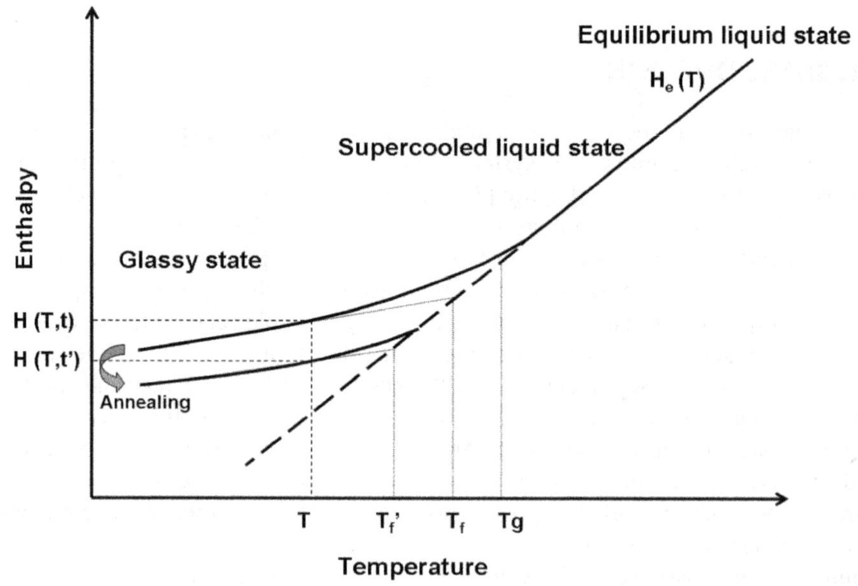

Figure 21.1. Schematic representation of enthalpy (H) as a function of temperature (T) for a glass-forming solid. On annealing at temperature T, from time t to t', enthalpy decreases and fictive temperature changes from T_f to T_f'. (Figure adapted from Luthra et al. [88]. Reproduced with permission from the *Journal of Pharmaceutical Sciences*, copyright © 2008, John Wiley & Sons.)

properties (e.g., specific volume or heat capacity) is no longer equivalent to the time average on the restricted timescale, and the system becomes nonergodic [6]. In the glass transition region, the structure becomes arrested because the displacement of an atom or group of atoms requires highly cooperative motions of the neighboring atoms, which are not possible with in a reasonable timeframe with the thermal energy provided. Such changes in molecular configurations that occur as a glass proceeds toward the lower-energy state are known collectively as *structural relaxation*. The characteristic relaxation time of this relaxation, known as the *structural relaxation time*, becomes greater than the experimental timescale. As a result, the system falls out of equilibrium and a glass is formed. In a glass, the timescale for molecular motion typically varies between 10 and 10^4 s at T_g [6], orders of magnitude longer than the picosecond–nanosecond rotation times characteristic of liquids.

In many respects, amorphous materials are like crystalline solids. While the long-range order characteristic of the crystalline state is absent, short-range order exists. The glassy state differs from the liquid state in that configurations in liquids change essentially immediately in response to change in an external variable, such as temperature. In glasses, however, the configurations are "frozen in" at T_g during cooling. Unlike the scenario in the stable crystalline phase, configurational mobility does exist in the glass, even though the timescale can be extremely long. Through these configurational rearrangements, i.e. structural relaxation, the system slowly transitions toward the supercooled liquid state [7], [8]. The glass transition temperature (T_g) has traditionally been used to identify the temperature region in which the change from liquid dynamics to solid dynamics occurs. It is important to stress that mobility, particularly mobility relevant to pharmaceutical stability, is not zero below T_g and that the attainment of solid-like dynamics is not equivalent to zero mobility [9].

Two broad classes of dynamics in glasses have been identified: (1) primary or global relaxation, also called "α relaxation" and (2) secondary relaxation, or "β relaxation." Alpha relaxations are directly coupled to viscosity and involve highly cooperative reorientation via rotational and/or translational motions. Local high-frequency dynamics of the glass, corresponding to collective small-length-scale motions occurring with timescales of a nanosecond or less, are known as the "fast dynamics of glasses" [10]. There is a lack of literature in the pharmaceutical arena that addresses mobility in glasses over a broad range of timescales, and only limited experimental support is available for the concept that dynamics other than the α-relaxation process will also have a bearing on stability of biological agents prepared as a glass [11]. However, extensive theoretical and experimental work has been carried out on dynamics in the glassy and supercooled liquid states. The many different modes of molecular motion (e.g., rotational or translational), changes in the length scale and type of motions with temperature, and cooperativity or coupling of different classes of molecular motions all complicate the analysis of molecular dynamics in glasses [12]. We shall attempt to address these issues in this chapter.

21.2.1. Alpha Relaxation or Global Dynamics: General Features

The characteristic timescale and length scale of α motions are long compared to other classes of motion, the timescale varying from minutes near the glass transition temperature to many years well below T_g. The activation energies of "α" molecular motion around T_g typically vary from 200 to 500 kJ/mol. Viscosity in supercooled liquids is frequently used as a measure of molecular mobility in glasses and liquids, with diffusion coefficient D and viscosity η related by the Stokes–Einstein equation:

$$D = \frac{kT}{6\pi r\eta} = \frac{l^2}{2\tau} \tag{21.1}$$

Here, r is the hydrodynamic radius of the molecule, k is the Boltzmann constant, T is the absolute temperature, l is the jump distance, and τ is the diffusional relaxation time. The pathway in a given reaction can involve a large number of diffusional jumps to produce the change in configuration required for a physical or chemical change [9].

Traditionally, glasses are defined in terms of viscosity, specifically, when viscosity reaches above 10^{14} poise. However, through the Stokes–Einstein equation, a glass can be defined using D as well, which in this equation is inversely proportional to τ. *Molecular mobility* may be defined as the reciprocal of this relaxation time. Note that the Stokes–Einstein equation is not a rigorous result, and large deviations from this relationship between diffusion coefficient and viscosity may occur. The temperature dependence of molecular motions above T_g is frequently described using the empirical Vogel–Tammann–Fulcher (VTF) equation

$$\tau(T) = \tau_0 \exp\left(\frac{D'T_0}{T - T_0}\right) \tag{21.2}$$

where τ is the mean molecular relaxation time, τ_0 is the relaxation time at high temperature in liquid state, T_0 is the temperature at which the configurational entropy vanishes, and here D' is related to the activation energy for molecular motion. Amorphous materials with large D' values show a weak dependence of relaxation time on temperature and are known as "strong" glasses. Materials with small D' values show a strong dependence of relaxation time on temperature, and are known as "fragile" glasses. This classification of glasses, based on the temperature dependence of relaxation time, implies that not all glasses are equivalent in their deviation from Arrhenius behavior; in other words, while strong glasses such as proteins and many polymers deviate little from the Arrhenius relationship, the deviation is large for fragile materials such as disaccharides. Figure 21.2 shows the relaxation times for glasses formed by rapidly cooling from above T_g, as predicted from heat capacity data. The characteristic dependence of relaxation time on temperature is shown for liquids above T_g (VTF) and Arrhenius behavior below T_g (i.e., extreme strong glass behavior). Typical fragile glasses (e.g., sucrose, Fig. 21.2) show a temperature dependence between these two extremes. The preceding discussion emphasizes that the α-relaxation timescale of a glass is a function of fragility as well as storage temperature and T_g [13]. The change in T_g with heating and/or cooling rate is routinely used to determine the activation energy for the glass transition, which can then be used to characterize the fragility of an amorphous system [14]. Most organic molecules, including drugs and polymers, are fragile, whereas systems with strongly bonded networks (e.g., SiO_2) are strong

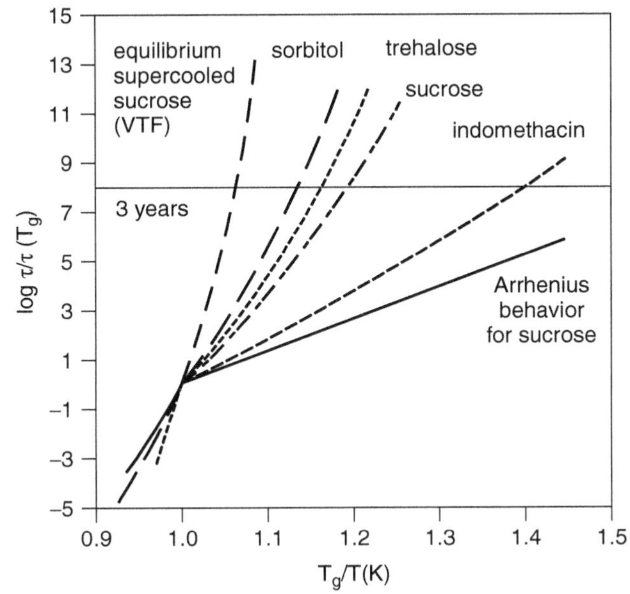

Figure 21.2. Mean relaxation times for real glasses formed by rapidly cooling from above T_g predicted from heat capacity data. (Figure adapted from Shamblin et al. [13]. Reproduced with permission from the *Journal of Physical Chemistry*, copyright © 1999, American Chemical Society.)

glasses [7]. However, as noted earlier, proteins exhibit many features of strong glass behavior, although they do not possess strong intermolecular bonded systems [15].

It is assumed that dynamics of glass-forming liquids is governed by the collective motions of molecules over a length scale characteristic of cooperative motion (ξ). The value of ξ increases as temperature decreases because of the decrease in configurational entropy (or free volume) [16,17]. Hempel et al. concluded that the characteristic length near T_g was about 1.0 and 3.5 nm, on the basis of complex heat capacity or dynamic calorimetry measurements in about 30 glass formers [18]. Erwin et al. reviewed the temperature dependence of relaxation times and length scales of cooperative motions and reported average ξ values in organic small molecules to range between 1.6 and 6.4 nm near T_g, from dielectric spectroscopy and viscosity measurements [19].

The timescale of molecular motions in a glass varies from around 100 s (i.e., around T_g) to years at room temperature for many pharmaceutical systems. In almost all cases, the molecular relaxation processes that occur in glasses are nonexponential, an observation normally attributed to the spatial heterogeneity in independently relaxing substates. These substates are not in equilibrium and therefore relax independently and at different rates, resulting in a distribution of relaxation times. Relaxation following a temperature change is described using an empirical non-exponential decay function [7] called the Kohlraush–Williams–Watt (KWW) or "stretched" exponential function [20]:

$$\phi(t) = \exp\left[-\left(\frac{t}{\tau}\right)^{\beta}\right] \tag{21.3}$$

where $\phi(t)$ is the normalized relaxation function, τ is the characteristic structural relaxation time, and β is a measure of nonexponentiality and characterizes the distribution of relaxation times. A value of β close to zero indicates a wide distribution of relaxing microstates, while a value close to 1 indicates a homogenous system with populations relaxing at similar relaxation rates [21]. Note that the KWW equation is an *empirical* representation of a multiexponential decay process.

Enthalpy relaxation measured using differential scanning calorimetery (DSC) and isothermal microcalorimetery (IMC) are the most common methods of assessing global α motions in pharmaceutical glasses. The timescales of motions measured using calorimetric techniques vary between 10^3 and 10^{10} s. The measurement of structural relaxation in pharmaceutically relevant glasses is a rather recent activity derived from similar studies in polymer science. The motions measured in these techniques are thought to represent those involved in viscous flow, typically whole-molecule translational and rotational motion. The molecular mobility is measured by (1) using the area under the enthalpy recovery peak after systematic aging for selected times below T_g in a DSC and (2) directly measuring the rate of enthalpy relaxation using isothermal microcalorimetry [22]. When an aged sample is heated in DSC, the enthalpy lost during the aging process is recovered and appears as an "overshoot" superimposed on T_g, with the magnitude of the overshoot dependent on the aging time.. The kinetics of the relationship between the enthalpy recovered and storage time are analyzed by a fit of the KWW equation [Eq. (21.3)] to calculate the molecular mobility parameters τ and β. The experimental values of τ and β may easily be obtained as a function of the aging temperature. The kinetic model (i.e., the KWW equation) assumes that the relaxation time is a constant during the experiment, that is, independent of time. However, during the calorimetric experiment, aging does occur, and τ increases. A theoretical analysis has shown that aging during the experiment translates into fitted values of τ and β as too large and too small, respectively, with values of τ^{β} as more accurate [22,23]. Thus, τ^{β} is a more robust parameter for characterizing structural relaxation in calorimetric experiments, or, indeed, in any experiment where aging occurs in the experiment itself. A review by Kawakami et al. provides a detailed account of the theoretical and experimental analysis of structural relaxation parameters using various thermal techniques including DSC and IMC [23].

Dielectric relaxation spectroscopy (DRS) is a noninvasive and rapid technique used to characterize molecular motions over a broad range of frequency (from 10^{-11} to 10^5 s) from above T_g to below T_g, at least in ideal situations. The wide accessible frequency range of dielectric techniques contributes to their popularity in the study of molecular dynamics. The wide frequency range is obtained through a combination of different analyzers: the reflectometry frequency analyzer (RFA) for the high-frequency range (10^6–10^9 Hz), the impedance analyzer for the low and medium frequency range (10^2–10^6 Hz), and thermally stimulated current (TSC) for the ultralow frequency range

($10^{-5}-10^{2}$ Hz) [24]; TSC is discussed in the next section. These methods essentially involve measuring the rotational motion of dipole-bearing groups. The dielectric techniques probe molecular motions by stimulating the different electrical dipole groups. This allows a clear differentiation between localized relaxing entities involved in secondary relaxation processes and the larger species involved in the longer-length-scale and longer-timescale primary relaxation. Dielectric relaxation studies involve the application of an alternating signal to a sample and the measurement of the response in terms of the in-phase and out-of-phase components. The response is described in terms of the complex permittivity, which is expressed in terms of the real and imaginary components. The inverse of the frequency corresponding to the maximum in the imaginary permittivity is taken as the relaxation time. It is possible to estimate the temperature dependence of the relaxation times by measuring the frequency-dependent dielectric behavior over a range of temperatures. There are good introductory reviews [25] and books [26] available on the measurement of dielectric relaxation properties and the application of the techniques to pharmaceutical systems. Bhugra et al. correlated the relaxation times obtained using TSDC with the relaxation times data measured using IMC and DSC in indomethacin and ketoconazole [27]. The authors observed that all measures of molecular mobility below the glass transition temperature, although divergent at lower temperatures, extrapolated to essentially the same relaxation time at the glass transition. The data suggested that the relaxation times measured below T_g are dependent on the technique used to measure the molecular mobility, and the relaxation times followed the trend TSDC < TAM < MDSC.

Dynamic mechanical analysis (DMA) ($10^{-2}-10^{2}$s) is used to characterize the viscoelastic nature of materials by measuring elastic and loss modulus. An oscillating mechanical force is applied to the sample, and the resulting displacement of the sample is measured. The viscosity data are analyzed using the VTF equation [Eq. (21.2)], and relaxation times of the glassy state are estimated [28] assuming the temperature dependence of the relaxation time and viscosity to be identical. Because of the limited frequency scale, this technique cannot directly measure relaxation times of glasses.

21.2.1.1. Impact of Environmental Factors on α Mobility: Temperature and Residual Water. As the temperature increases, viscosity decreases and translational motions become faster; that is, τ decreases [29,30]. The temperature dependence of τ is very strong (large E_a) as compared to the β relaxation (low E_a) (Fig. 21.3). The structural relaxation time (τ), which is the time required by a molecule to undergo large-scale, or "whole molecule," configurational motion, decreases at higher temperatures, and the glass undergoes aging sooner. The concept of aging is discussed further in Section 21.4. It has been suggested that at temperatures more than 50°C below T_g, the mobility relevant to pharmaceutical degradation becomes insignificant [7,31]. This suggestion is based on the fact that the relaxation time calculated from the VTF equation becomes infinite at $T = T_0$ and T_0 commonly (but not always!) is $\sim 50°$ below T_g. However, it must be recognized that the VTF equation does not apply in the nonequilibrium glassy state. The relaxation kinetics of glasses are determined not only by the thermodynamic temperature but also by the instantaneous structure of the glass. Tool suggested the use of a fictive temperature (T_f) as a means of characterizing this instantaneous structure, which depends on the thermal treatment, a parameter that is a function of the entire thermal history of the sample [32]. Hodge suggested the nonlinear functional form of the Adams–Gibbs model for the glassy state

$$\tau(T, T_f) = A' \exp\left\{\frac{B}{T(t)[1 - T_K/T_f(t)]}\right\} \tag{21.4}$$

where B is the normalized activation energy for cooperative relaxation [33,34]. This model has been shown to be appropriate for estimating τ in glasses [13]. It is important to note that τ evaluated from calorimetric data using the KWW equation [Eq. (21.3)] gives the most probable relaxation time [21]. The distribution of relaxation times should always be recognized when working with amorphous pharmaceutical materials, especially when attempting to use relaxation time constants for predicting their physical or chemical stability [21]. Although the most probable relaxation time at a particular temperature may be on the order of years, a small population may have the τ value on the timescale of days, and this faster relaxing substate may be critical to the stability problem of interest.

As indicated earlier, τ^{β} is a robust measure of molecular mobility. However, calorimetry measures heat regardless of its origin, so that processes other than structural relaxation (i.e., chemical degradation) may also be measured by sensitive calorimeters, rendering a fit by the KWW equation meaningless [35]. Liu et al. found that the time derivative of the KWW equation represented the relaxation kinetics of quenched trehalose in an isothermal calorimeter and provided physically reasonable results for τ^{β} [22]. In freeze-dried sucrose at higher temperatures, the authors found that the KWW expression could not fit the data well and resulted in "nonphysical" τ and β values. The authors made the empirical observation that when the change in power (i.e., rate of heat production) with time is very rapid, the two-parameter KWW derivative expression is not a particularly good description of the data, and the modified stretched exponential (MSE) is a better model [22]. The temperature dependence of τ^{β} in sucrose, trehalose, and raffinose glasses showed that sucrose is more fragile than trehalose and raffinose [22,36]. Considering the difference in fragility, the authors suggested that stability differences between formulations may be highly temperature dependent, even well below T_g. Because of the dependence of τ above T_g, measured using DRS, Duddu et al. extrapolated τ to temperatures below T_g using the VTF equation [36]. The VTF equation has traditionally been used to describe the temperature dependence of τ in the supercooled liquid state ($T > T_g$) [12], but, as discussed earlier, its use below T_g is not strictly correct. Similarly, Andronis et al. determined viscosity and the frequency-dependent shear modulus of indomethacin to be near and above T_g [28]. The viscosity data were fitted to the VTF equation, and relaxation times of indomethacin in the glassy state were estimated [28].

For reactions occurring in amorphous solids, reactivity generally increases with an increase in water content [37]. It is well known that water acts as a plasticizer, hence increasing molecular mobility in the solid. Water can affect the stability of a compound in the glassy state in other ways, by acting (e.g.) as a nucleophilic reactant, as a catalyst for proton transfer, or by modifying the polarity of the solid [38,39].

Only the effect of water on mobility is discussed here. The effect of water on T_g can be roughly approximated using the Fox equation [40]

$$\frac{1}{T_g} = \frac{w_1}{T_{g1}} + \frac{w_2}{T_{g2}} \tag{21.5}$$

where T_g is the observed glass transition temperature, w_1 is the weight fraction of the compound, T_{g1} is the T_g of the compound, w_2 is the weight fraction of water, and T_{g2} is the T_g of water (−108°C [41]). Roy et al. studied the effect of moisture and reported that the logarithm of the rate constant for degradation in freeze-dried monoclonal antibody–vinca conjugate systems was proportional to $T - T_g$, with water acting as a plasticizer and significantly impacting T_g [38]. Liu et al. reported the effect of water on α mobility measured using IMC. In quenched trehalose, as the moisture content increased from nearly zero to 2.7%, τ^{β} (at 40°C) decreased approximately from 31.2 to 17.6 (τ in hours), while T_g changed from 110°C to 75.7°C. Oksanen et al. measured diffusion coefficients of water using pulse field gradient spin echo NMR [42]. In the region of overlap where diffusion coefficients could be measured using both NMR and gravimetric methods, the results were consistent and showed that as water content increased, the D_{H_2O} also increased. The rate of decomposition (dimerization and deamidation) of lyophilized recombinant bovine somatotropin increased as residual water content increased [43]. The authors described the effect of water in terms of the impact of water on the flexibility of the protein, as well as the ability of the water to solubilize/mobilize reactants. Taken together, the current evidence indicates that, consistent with "intuition," global mobility increases with increases in temperature and residual water content.

21.2.1.2. Impact of Formulation Variables on α-Relaxations. As the weight fraction of a higher T_g compound increases, the T_g of the whole system increases. Proteins have relatively high T_g, but when formulated with low-T_g sugars like sucrose, the T_g of the mixture varies with the composition [44]. This generalization implies that if a homogenous system forms on the addition of a higher T_g additive, the α mobility of the system generally decreases. Shamblin et al. measured enthalpy relaxation rates using DSC for sucrose alone and sucrose colyophilized with PVP (polyvinylpyrrolidone), PVP/VA (polyvinylpyrrolidone–vinyl acetate), and trehalose at 10% concentration at various temperatures [35]. The molecular mobility measurements revealed that relative to sucrose alone, the system mobility is reduced by the presence of additives having a T_g greater than that of sucrose. Thus, while water and low-molecular-weight (LMW) additives such as sorbitol [45] plasticize the glass, high-molecular-weight (HMW) excipients such as PVP and dextran "antiplasticize" the amorphous matrix, that is, lower the overall α mobility. Prestrelski et al. reported that storage stability of recombinant human interleukin−2 increased when lyophilized with excipients that have high glass transition temperatures, although dehydration induced unfolding was still not prevented [46]. There is, however, at least one exception to the rule that structural relaxation time decreases with a reduction of T_g. In protein(IgG1) : sucrose systems, the addition of sucrose ($T_g \approx 74^\circ$C) to the pure protein ($T_g \approx 130^\circ$C) caused an *increase* in structural relaxation time at low to moderate levels of sucrose, but at weight ratios of sucrose higher than 1 : 1, structural relaxation time decreased with further addition of sucrose [44].

The discussion and the results reported above suggest that marked changes in the various extrinsic and intrinsic factors can result in changes in molecular mobility of glasses, with the effect governed by the nature and extent of molecular interactions. The following section discusses the nature of these interactions and their effects on chemical stability. Literature reports noting significant correlation of mobility with stability are also discussed.

21.2.1.3. Correlation of Physicochemical Stability with α Mobility. Every degradation reaction requires some degree of molecular mobility to occur at a significant rate in an amorphous matrix below T_g [9]. The validity of this assertion has been demonstrated for a number of compounds for which the lower mobility crystalline state was more stable than the amorphous form [47]. Traditionally, T_g has been used as an indicator of molecular mobility in amorphous systems, and an attempt has been made to correlate chemical stability with the difference between the storage temperature and T_g, i.e., $T_g - \mathrm{T}$ [36,38,48]. For example, a correlation between residual water content, T_g, and rate of deamidation of an asparagine-containing hexapeptide (VYPNGA; Asn–hexapeptide) in lyophilized poly(vinyl alcohol) (PVA, $T_g = 73^\circ$C) and polyvinylpyrrolidone (PVP, $T_g = 148^\circ$C) at 50°C was reported [49]. It was found that between the glassy state and near T_g, the deamidation rate increased by only three orders of magnitude in PVA and two orders of magnitude in PVP, but the decrease in viscosity was five orders of magnitude from glassy to rubbery states. The authors suggested that this difference in degradation rate and viscosity changes at T_g was due to the incomplete coupling of deamidation rate to matrix mobility. In general, it is not wise to consider T_g alone as a measure of molecular mobility, since this neglects the contributions of fragility, thermal history, and other contributions [6,50].

As discussed above, τ^β is a more reliable measure of molecular mobility, at least as compared to T_g. If it is assumed that the molecular motions responsible for instability and structural relaxation are coupled to viscosity, a coupling between instability and structural relaxation is expected

$$\mathrm{k} \propto (\mathrm{mobility})^c \tag{21.6}$$

where k is the degradation rate constant and c is the coupling coefficient between instability and relaxation. For diffusion controlled reactions in which the Stokes–Einstein equation is valid (i.e., that the diffusion coefficient is inversely proportional to viscosity), one expects, $c = 1$ [51]. Note that if degradation and relaxation both occur on a "stretched" timescale, as is normally observed, the value of k is defined in terms of stretched time and mobility would be defined in terms of τ^β, consistent with Equation (21.6).

Pikal et al. reported that an amorphous freeze-dried *cefamandole* sodium system had a lower chemical degradation rate and a higher relaxation time than did a similar formulation of *cephalothin* sodium prepared under the same freeze-drying conditions [50]. Shamblin et al. reported a coupling coefficient of 0.43 between dimerization and structural relaxation parameters in ethacrynate sodium formulations colyophilized with sucrose, trehalose, and PVP [51]. Duddu et al. found that aggregation in IgG1 formulations colyophilized with sucrose and trehalose was correlated with reduced correlation time (t/τ) [36]. The authors reported the interesting observation that at $t/\tau < 100$, aggregation in sucrose formulations was less than in the trehalose system, even if the molecular mobility was higher in the former. This finding, as with many others in the literature where chemical degradation does not correlate well with α mobility, suggests there are factors other than dynamics of α motions that are critical for stability behavior [44,51].

Yoshioka et al. have assessed the relative importance of molecular mobility and activation energy in determining the reactivity in lyophilized insulin systems [52,53]. The authors used a formalism that combines the rate in the solid state, in which diffusion is limited, and the rate in solution (high diffusivity) as proposed by Karel et al. [54]

$$k' = k\left(\frac{\alpha D_r}{k + \alpha D_r}\right); \qquad k = A\exp\left(\frac{-\Delta H}{RT}\right) \tag{21.7}$$

where k' is the apparent rate constant for degradation in the amorphous matrix, D_r is the diffusion coefficient, and α is a coefficient relating k' and D_r; k is the rate constant in solution state, in which it is assumed that diffusion is not a rate-limiting factor. As in conventional activated state theory, k is related to activation energy (ΔH) and the frequency factor A. A ratio of diffusion coefficients between the two states is given in terms of the corresponding ratio of relaxation times by

$$\frac{D_{r2}}{D_{r1}} \approx \left(\frac{T_2}{T_1}\right)\left(\frac{\tau_1}{\tau_2}\right)^{\xi} \tag{21.8}$$

where ξ is the coupling coefficient indicating the degree of decoupling between D_r and structural relaxation time, and is analogous to the constant c in Equation (21.5). The Vogel–Tammann–Fulcher (VTF) equation and the Adam–Gibbs–Vogel (AGV) theory correlate D_r with τ above and below T_g, respectively. These relationships in Equation (21.7) yield the following expression for the reaction rate constant, which can then be used to estimate t_{90}, the time to degrade to 90% of initial:

$$k' = \frac{A\exp(-\Delta H/RT)\alpha T(1/\tau)^{\xi}}{A\exp(-\Delta H/RT) + \alpha T(1/\tau)^{\xi}} = -\frac{-\ln(0.9)}{t_{90}} \tag{21.9}$$

Using nonlinear regression and assuming reasonable values of ΔH, the dependence of k'/k at T_g can be estimated using Equation (21.9). At lower moisture levels, degradation in insulin formulations lyophilized with trehalose showed a smaller k'/k ratio at T_g, suggesting that the degradation rate was governed by molecular mobility [52]. However, at higher moisture levels, larger values of k'/k were taken as proof that "activation energy" was the rate-limiting factor. Another study in lyophilized aspirin–sulfasalazine formulations containing PVP, dextran, and isomalto oligomers of different molecular weights was performed by the same group [55]. The t_{90} values calculated from the bimolecular reaction rates and the temperature dependence of τ estimated using the AGV equation were compared as a function of T_g/T. It was suggested that at $T > T_g$, the reaction rates were limited by the "activation energy" barrier. However, at $T < T_g$, prediction of the degradation rates could be based on the temperature dependence of τ, assuming that degradation is diffusion-limited.

Generally, it has been assumed that the temperature that marks the transition from a solid-like state to a liquid-like state is T_g. The *critical mobility temperature* T_{mc}, is defined as the temperature above which NMR relaxation behavior resembles that of a liquid, and is also sometimes used to assess mobility; T_{mc} is related to NMR mobility as T_g is to viscosity. Protein aggregation in lyophilized solids has been shown to correlate better with T_{mc} than with T_g. In this system, the values differed by 20°C–30°C, with $T_g > T_{mc}$ [56]. The result suggests that the formulations contained highly mobile protons even at temperatures below T_g. Aggregation rates of lyophilized BSA [57] and β-galactosidase [58] lyophilized with poly(vinyl alcohol) (PVA) changed abruptly around T_{mc}; T_{mc} is more closely related to protein stability than is the calorimetric T_g in lyophilized serum γ-globulin (BGG) formulations containing PVA and dextran. The T_{mc} of formulations containing higher-molecular-weight excipients is generally higher. In other studies, T_{mc} has also been used to evaluate the effects of different polymeric excipients [e.g., PVP, dextran, carboxymethylcellulose, and methyl cellulose (MC)] on the molecular mobility of polymer and water protons in lyophilized solids [56]. The authors concluded that polymers with higher hydrophilicity bind water molecules more strongly, decreasing the overall plasticizing effect of water on the lyophilized formulation.

21.2.2. Local Secondary Dynamics: General Features

In glasses, the "slow" molecular mobility includes the very slow main relaxation (α) and the faster secondary relaxations (β) or Johari–Goldstein (JG) relaxations corresponding to intramolecular reorientations. There is no universal classification of secondary relaxations in terms of details of atomic and molecular motion. In this section, we discuss secondary dynamics data measured using three techniques: (1) dipolar techniques, including DRS (from 10^{-11} to 10^5 s) and TSC (from 10^3 to 105 s), (2) NMR spin–lattice relaxation (from 10^{-6} to 10^{-3} s), and (3) inelastic neutron scattering (from 10^{-10} to 10^{-7} s).

The Johari–Goldstein process is a form of β relaxation often observed in glass-forming materials at lower temperatures [59]. It has been postulated that β relaxations occurring at high frequencies originate in intramolecular motion, such as rotational motion of a methyl group or a polymer side chain, but such processes have also been observed in supercooled simple liquids. In 1970, Johari and Goldstein demonstrated secondary relaxations in rigid molecules having no conformational flexibility, and concluded that at least some types of secondary relaxations must involve rearrangements in the orientation and position of at least one or more molecules in a local region [59]. Such relaxations were later known as "JG β relaxation" after these authors. Although

β relaxation has been observed for many years, references to it in the pharmaceutical literature have been sparse. Johari later attributed β relaxations occurring in the glassy state to "islands of mobility," that is, the noncooperative hindered motion of some molecules in large cages of other molecules, in other words, motion in low-density regions [60]. Consistent with this view, Williams and Watts assumed that the molecule as a whole takes part in the β-relaxation process [61]. Even small rigid molecules like disaccharides, which constitute most of the matrix in the lyophilized formulations, can have β mobility in addition to α mobility. In principle, such β mobility might be important to the reaction pathway involved in instability, and therefore, might be well correlated with stability. Beta relaxations are believed to be local, noncooperative processes in the sense that motion of the molecule in question does not require simultaneous motion of a large number of neighboring molecules constituting the matrix. In comparison to the α motions, the secondary relaxations may have a very small activation energy (10–30 kJ/mol) [62]. Johari et al. reported the activation energy of β relaxations for small rigid molecules to be small, in the range of 40–70 kJ/mol [59].

Most of the initial studies on secondary relaxations were carried out using DRS and thermally stimulated current depolarization (TSDC), which allowed processes of very small relaxational strength (e.g., rotational mobility) to be explored. Thermally stimulated current spectroscopy can measure the dynamics of molecular motions arising from the homogeneous substates that form the complete distribution of states [63]. Figure 21.3 illustrates the thermal sampling or windowing technique of TSDC in salicylic acid for obtaining the relaxation times for the different mobility substates [62,64]. Each elementary relaxation peak was analyzed to measure the corresponding temperature-dependent relaxation time of the relevant molecular motion for that substate as a function of temperature [24,62]. Bhugra et al. found TSDC to be a useful tool for characterizing molecular mobility in the sub-T_g region for indomethacin and ketoconazole [27].

Nuclear magnetic resonance relaxation times, such as, spin–spin relaxation time (T_2) and spin–lattice relaxation times (T_1 and $T_{1\rho}$), have been used as measures of molecular mobility in glassy solids [57,65,66]. The spin relaxation times are a measure of molecular dynamics because thermal motions lead to stochastic variations in the spin Hamiltonian, driving nuclear spin transitions. The average time that a molecule spends in a given position before it diffuses to another location or configuration is called the *correlation time* (τ_c). For heteronuclear spin systems, that is, formulations in which different spins are probed (e.g., ^{13}C and ^{1}H), the following relationship holds between relaxation time and the correlation time [67]:

$$\frac{1}{T_1} = \frac{3\gamma_I^4 h^2}{10 r_{II}^6 (2\pi)^2}\left[\frac{\tau_c}{1+\omega_I^2\tau_c^2} + \frac{4\tau_c}{1+4\omega_I^2\tau_c^2}\right] \qquad (21.10)$$

Here, γ_I is the gyromagnetic ratio of ^{13}C, h is Planck's constant, r_{II} is the internuclear distance between two ^{13}C nuclei, and ω_I is the Larmor frequency of ^{13}C. The correlation time of motions probed by T_1, $T_{1\rho}$, and T_2 relaxation times are on frequency scales on the order of 100–500 MHz, 20–100 kHz, and less than 10 kHz (ms to s), respectively [68]. In a typical plot of T_1 (or $T_{1\rho}$) versus correlation time of

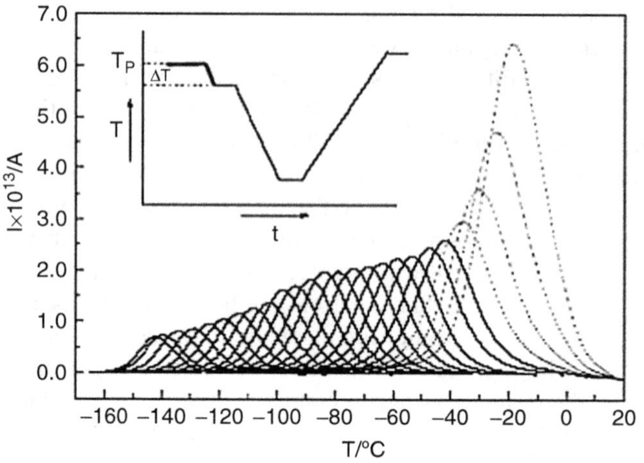

Figure 21.3. Thermal sampling or windowing technique of TSDC: partial polarization (PP) components of the β relaxation of salicylic acid. The polarization temperatures (T_p) used were varied between −143°C and −48°C with 5°C intervals. The polarization time of 5 min and polarization window of 2°C with 4°C/min heating rate were employed. The higher-temperature dotted peaks, with increasing intensity, were obtained respectively with $T_p = -43°C$, −38°C, −33°C, and −28°C, corresponding to the superposition of the α and β relaxations. The inset schematically depicts the experimental procedure used in the TS technique. (Figure adapted from Ramos et al. [62]. Reprinted from *Journal of Non-Crystalline Solids*, copyright © 2006, with permission from Elsevier.)

motion, T_1 undergoes a minimum [67]. As the temperature increases from deep in the solid state, the correlation time decreases, and T_1 decreases; that is, mobility increases. This is known as the *slow motional regime*. On the other side of the minimum in the fast motional or "liquid" regime, T_1 increases with increasing temperature. The position of the minimum depends on the frequency of observation or Larmor frequency. Preliminary measurements of T_1 as a function of temperature help determine whether the temperature range of interest for a given system is in the fast or slow regime of motion. Within that system, the impact of different formation variables on mobility can be analyzed with respect to the trends in spin relaxation times. When comparing different chemical systems, however, it is necessary to measure correlation times to unambiguously judge differences in mobility between systems.

Cicerone et al. have introduced the concept that dynamics on the nanosecond timescale may be important for protein stabilization in glasses [10,11,69]. Dynamic neutron scattering provides information about fast local dynamics by probing atomic fluctuations [70]. Cicerone et al. have used the high-flux backscattering spectrophotometer (HFBS), which enables very high-energy resolution studies of the high-frequency dynamics of many systems. A comprehensive review of the technique has been reported [10,71]. In the backscattering spectrometer, the sample is bombarded with a monoenergetic beam of neutrons, which interact with the protons of the sample.

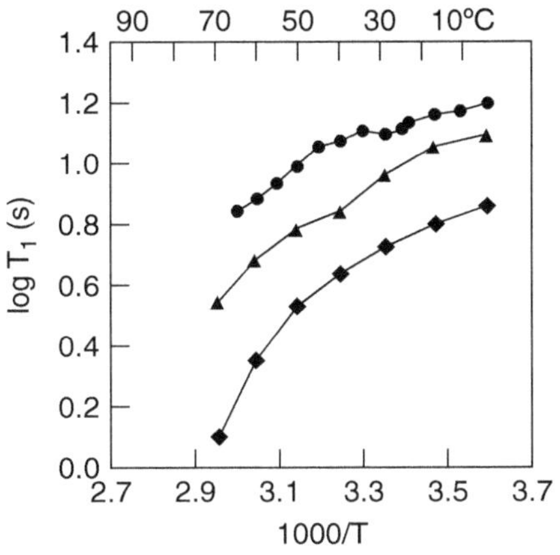

Figure 21.4. Spin–lattice relaxation time T_1, with methine carbon of dextran in lyophilized BGG formulation as a function of temperature. *Key*: (■) 12% RH; (•) 60% RH; (▲) 75% RH. (Figure adapted from Yoshioka et al. [72]. Reprinted from the *Journal of Pharmaceutical Sciences*, copyright © 2002, with permission from Wiley.)

Only the neutrons with a particular final energy deflected by the protons are detected. The motions in the sample are probed by varying the energy of the incident neutrons and measuring the gain or loss in energy that they undergo in their interaction with the sample [71]. The motions are analyzed in terms of the Debye–Waller factor (a harmonic oscillator model), and the hydrogen-weighted mean-square atomic displacement $< u^2 >$. The higher the neutron scattering parameter $< u^2 >$, the greater the amplitude of the local fast motions.

21.2.2.1. Impact of Environmental Factors on β Relaxation. Both T_1 and $T_{1\rho}$ measurements as a function of humidity show that as water content in a glass increases, T_1 decreases and molecular mobility increases (Fig. 21.4). This suggests that motions faster than α motions are also affected by the water content of the solid sample [72]. Decreases in ^{13}C $T_{1\rho}$ as water content increased have also been reported for insulin alone and for insulin : trehalose systems. In studies of lyophilized bovine serum γ-globulin, $T_{1\rho}$ decreased through a minimum as temperature increased for samples stored at higher RH (75% and 86% RH). Duddu et al. measured the activation energy of β relaxation as a function of water content in a PVP–water system [73]. As the water content increased $\leqq$ up to 31%, the activation energy increased from 36 to 46 kJ/mol, suggesting that thermally activated rotational diffusion of water played a major role in the β relaxation of PVP. The carbon T_1 relaxation times in lyophilized lysozyme decreased with increasing relative humidity, that is, residual water content, indicating increased local mobility. The correlation of decrease in T_1 relaxation times with increased rate of aggregation (loss of activity) for this system is shown in Figure 21.5.

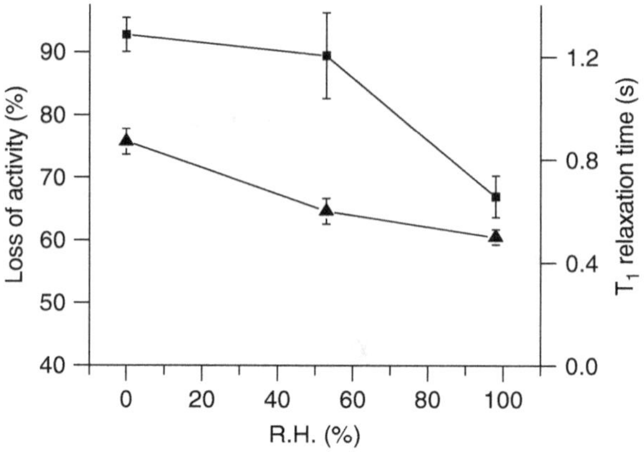

Figure 21.5. The effect of relative humidity on ^{13}C T_1 relaxation times and loss of activity in lyophilized lysozyme. *Key*: (■) T_1 relaxation time; (▲) Loss of activity (Figure adapted from Separovic et al. [75]. Reproduced from *Pharmaceutical Research*, copyright © 1998 with kind permission of Springer Science and Business Media.)

21.2.2.2. Impact of Formulation Variables on β Relaxation. It has been proposed in the literature that inclusion of a lower-molecular-weight (LMW) additive such as glycerol in an amorphous solid produces antiplasticization of fast dynamics [74]. These additives are thought to fill some of the free volume of the glass, hence decreasing mobility. What is not clear, however, is the mechanism by which these additives serve as an *antiplasticizer* for β motions, decreasing β mobility, while at the same time acting as a *plasticizer* for α motions. In keeping with this concept, Cicerone et al. observed that adding glycerol to glassy trehalose glass matrices at 5% w/w suppresses the short-time local dynamics [10]. Using incoherent neutron scattering, the authors measured the amplitude of motions that occur on a nanosecond timescale. The amplitude $< u^2 >$ was found to be smallest for matrix with 5% glycerol at all temperatures (Figure 21.6). The "spring constant," which signifies stiffness of a glass assuming the motions to be harmonic vibrations, was also estimated to be largest for the glass containing 5% glycerol. As the glycerol concentration increased, the T_g decreased monotonically, presumably reflecting an increase in α mobility. As discussed above, water increased the β mobility. Interestingly, glycerol, which is also a small molecule, decreased the β mobility. The results suggest that the size of the small molecule and/or the nature of its interactions with the matrix are/is critical in determining the behavior observed.

21.2.2.3. Correlation of Chemical Stability with β Relaxation. In the past, solid-state NMR has been used successfully to study hydration and stability of proteins [75]. Some support for the assertion that faster relaxation in the glassy state adversely affects chemical and physical stability has been provided using various NMR parameters, for example, T_1, T_2, and $T_{1\rho}$ of protons and carbons. In a 2006 report,

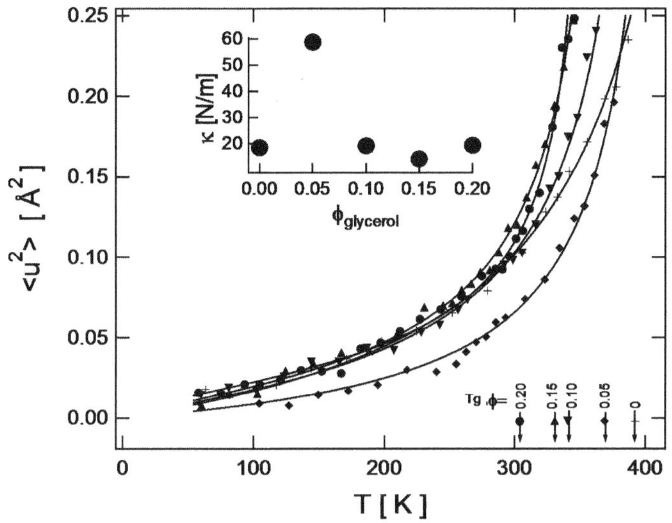

Figure 21.6. Debye–Waller factors from trehalose glasses with glycerol; 0 (+), 0.05 (♦), 0.10 (▼), 0.15(▲), and 0.20 (•). Inset shows spring constants of glasses. The *x* axis also shows the variation in T_g of the matrix as a function of glycerol fraction [10]. Reprinted from *Biophysical Journal*, copyright © 2004, with permission from Elsevier.

Yoshioka et al. studied the correlation between storage stability and ^{13}C $T_{1\rho}$ (rotating frame spin–lattice relaxation time) of freeze-dried insulin formulations, using solid-state nuclear magnetic resonance (ssNMR) [76]. The effects of trehalose and dextran on $T_{1\rho}$ were compared at various humidity levels. The authors demonstrated that the degradation rate of insulin in formulations containing dextran was greater than in those containing trehalose at low relative humidity (12% RH), while no difference in stability was observed at 44% RH. Similar ^{13}C $T_{1\rho}$ measurements for insulin alone and for insulin : dextran, formulations also showed no difference in stability. The insulin : trehalose system showed biexponential relaxation behavior; the authors proposed that the longer $T_{1\rho}$ was due to the carbonyl backbone, which was affected by trehalose; in other words, the mobility of these insulin carbonyl carbons was restricted in the presence of trehalose. The authors attributed the improvement in stability with trehalose to the inhibition of β relaxations, as evidenced by higher $T_{1\rho}$ values.

Cicerone et al. studied the stability of two enzymes, horseradish peroxide and yeast alcohol dehydrogenase, in glycerol–trehalose glasses with various compositions by measuring the deactivation times of the enzymes at various temperatures [11]. The deactivation times showed a maximum at 5% glycerol concentration in the trehalose matrix. Later, Cicerone and Soles reported a neutron scattering study in which they observed the suppression of the short-time local dynamics at 5% glycerol, suggesting that there is a correlation between the preservation of enzymes and the suppression of short-time local dynamics [10]. Improvements in stability for both enzymes were consistent with a decrease in mobility in the glassy matrices associated with fast

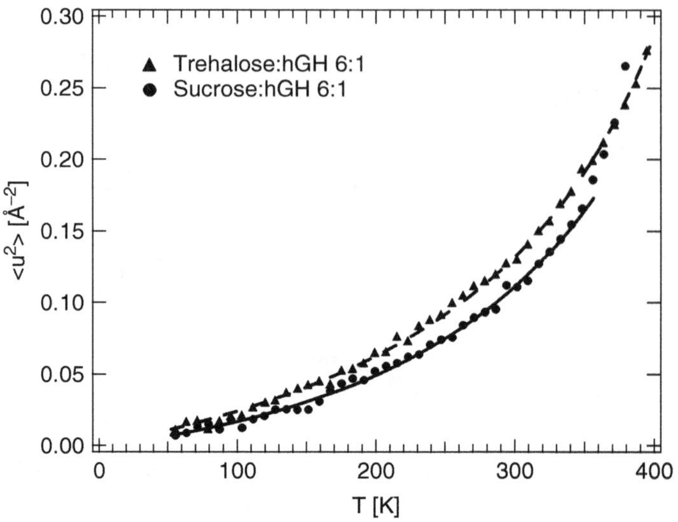

Figure 21.7. Amplitude of fast dynamics as measured by neutron scattering for 1 : 6 hGH : sucrose and hGH : trehalose formulations. The mean-square amplitude of hydrogen motion is plotted as a function of temperature. (Figure adapted from Pikal et al. [77]. Reprinted from the *Journal of Pharmaceutical Sciences*, copyright © 2008, John Wiley and Sons, Inc.)

dynamics as measured by neutron scattering studies. Pikal et al. reported that an hGH : sucrose formulation was much more stable than an hGH : trehalose formulation, but showed that the sucrose formulation had greater molecular mobility as measured using structural relaxation rates [77]. However, the lower-amplitude fast dynamics measured using neutron scattering was found to be slower in the sucrose system than in the trehalose system (Fig. 21.7). These measurements, reflecting fast dynamics, correlated well with stability; that is, well below T_g, fast dynamics controlled stability rather than global mobility. Dirama et al. presented a molecular dynamics simulation over short times (ns) in trehalose : glycerol glasses and reported that the (95 : 5) trehalose : glycerol matrix, which best stabilized the enzyme guest, had the strongest H-bonding network and the greatest suppression of fast dynamics [78]. Abdul-Fattah et al. measured the impact of drying method and formulation in systems containing sucrose and IgG1 [79]. As the protein fraction increased in the IgG1 formulation, $< u^2 >$ increased significantly, while the structural relaxation time τ^β increased only slightly, indicating a slight decrease in global mobility. The authors concluded that the fast dynamics and global dynamics were partially decoupled for these samples. In stabilizer-rich formulations, stability differences were best correlated to differences in "fast dynamics" molecular mobility. Small amounts of sorbitol in IgG1 : sucrose and IgG1 : trehalose formulations improved the storage stability even though global mobility increased on addition of the plasticizer [45], suggesting that mobility other than α relaxation is involved.

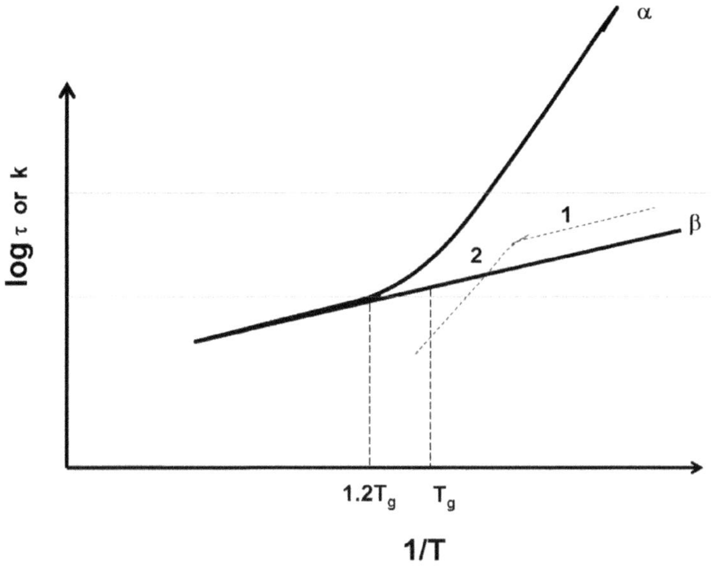

Figure 21.8. Schematic of α and β motions as a function of temperature. (The dashed-line curve) shows the degradation reaction rate; segment 1 represents the temperature window when reactivity is correlated with β mobility; segment 2 shows the temperature window when reactivity is correlated with α mobility [90].

The idea that fast dynamics affects chemical stability is physically appealing. However, the question remains as to why chemical and physical degradation processes in solids, which typically occur on a timescale of months or years, show better correlation with fast motion occurring on a micro- or nanosecond timescale than with global mobility, which has a timescale more similar to those of degradation processes. Fast dynamics are present at all temperatures, yet only seem to only control stability in the glassy state well below T_g. The schematic shown in Figure 21.8 is an attempt to explain this apparent dilemma. It is well appreciated that α motions have a very strong temperature dependence, so as temperature decreases to much below T_g, the global motions occur on timescales of years and should be insignificant for pharmaceutical stability. However, β mobility is still quite fast at temperatures below T_g. We suggest that there are two "reaction pathways" between the initial and final states, one involving large-length-scale, highly cooperative "α motions" and the other involving many small length scale "fast dynamics" motions. At high temperature, the α-motion pathway is fastest and dominates reaction behavior. However, as temperature decreases, the α-motion process slows greatly until the fast dynamics pathway is fastest and dominates reaction behavior much below T_g. This pathway reflects small-length-scale motion on a nanosecond timescale having very small cooperativity and low activation energy [77]. Depending on the system and the degradation mechanism, the crossover temperature can vary.

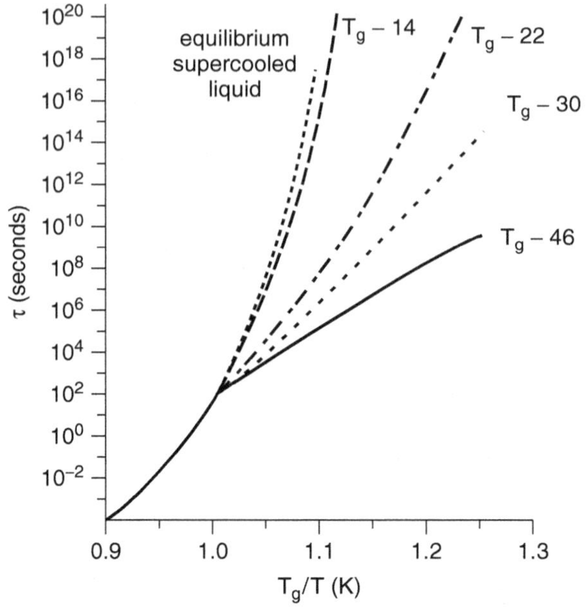

Figure 21.9. Variation in mean relaxation time with temperature for quenched sucrose glasses aged for 16 h at different temperatures. Note that the relaxation time increases as the aging temperature approaches T_g, reflecting the more rapid access to supercooled liquid state at a higher aging temperature. (Figure adapted from Shamblin et al. [13]. Reproduced with permission from the *Journal of Physical Chemistry*, copyright © 1999, American Chemical Society.)

21.3. IMPACT OF ANNEALING ON MOBILITY AND STABILITY IN GLASSY STATE

On aging, glassy systems experience an increase in relaxation time and a decrease in overall molecular mobility. The spontaneous process by which different properties of a glass stored below its T_g continue to change over time, in an attempt to attain equilibrium, is called *physical aging* or *structural relaxation* [8]. In contrast, the term *annealing* refers to heating an amorphous material at a temperature below T_g. A glass will approach the *equilibrium glassy state* asymptotically during an annealing process. The structural relaxation time (τ), which is the time required for a molecule to undergo large scale configurational motion, increases and mobility decreases on aging at temperatures below T_g. Figure 21.9 shows the effect of aging temperature on relaxation time in sucrose. As the aging temperature increases, the approach of the glass toward the equilibrium liquid is faster. Diffusivities of small molecules in polymer matrices have been known to decrease fully investigated with aging. Mardaleishvili et al. demonstrated this decrease in their study on poly(methyl methacrylate) (PMMA) films with an initiator for formation of free radicals via a photochemical reaction by annealing at two different temperatures [80]. This was followed by sample irradiation at a wavelength of 365 nm to initiate free-radical formation, which was measured as a function of time. The sample annealed at the higher temperature for a longer time showed the smallest rate of accumulation of free radicals. Hill et al. compared the extent of the Maillard reaction between lysine and glucose in annealed and untreated glasses [81]. The authors suggested that the matrix became densified on aging, which slowed the rate of reaction, although the effect was not significant.

The first report of stabilization by high-temperature treatment in pharmaceuticals was presented by Pikal et al. [82] and briefly discussed again in a review [83]. The authors noted that the stability of moxalactam disodium was superior when secondary drying was conducted at 60°C rather than 40°C, although the moisture contents at the two temperatures were essentially identical. As a follow-up to this observation, a more extensive study was performed in a system of moxalactam disodium formulated with 12% w/w mannitol, and the relationship between physical aging and chemical stability was systematically investigated [84]. The authors reported that as annealing temperature and time increased, the global mobility measured using τ^{β} increased. The chemical stability assay results showed a 1.7-fold decrease in decarboxylation rate in moxalactam : mannitol systems on annealing at 70°C (T_g = 122°C) for 8 h. A similar observation was made in a related study, which showed a small increase in the chemical stability of a lyophilized cefovecin formulation on annealing [85].

In another study, Smits et al. measured $T_{1\rho}$ relaxation times in starch as a function of aging time at different humidity levels [86]. At 30% RH, $T_g > T$, so as the material was annealed, molecular mobility decreased, and $T_{1\rho}$ increased. At 60% and 90% RH, the T_g decreased to below T and annealing did not occur, presumably because of high water sorption under these conditions. This study demonstrated the potential utility of NMR in the evaluation of physical aging.

Luthra et al. investigated the effect of controlled annealing on molecular mobility in lyophilized aspartame: disaccharide glasses [29,87]. The broad objectives of this research were (1) to study the correlation between the state of aging and global and/or local dynamics in the two component lyophilized systems and (2) to determine whether the annealing conditions required for achieving a minimum in fictive temperature T_f (TNM modeling, DSC) [29,88], a maximum in τ^{β} (IMC) [29], and a minimum in local mobility measured using NMR [87] also produced the maximum for chemical stability. The optimum aging conditions determined from both DSC and the best-fit parameters of the TNM model showed that annealing at a temperature about 15°C–25°C below T_g resulted in maximum structural relaxation [88]. Figure 21.10 shows that global mobility decreased in aspartame formulations as annealing proceeded. Accelerated stability tests on annealed formulations showed that thermal annealing of samples under protocols that were predicted using mathematical modeling and shown calorimetrically to produce optimum annealing for "mobility" were also close to those conditions that produced optimum chemical stability [89]. It is significant to note that improvement in sample purity was achieved via annealing after a short storage time in these studies (Fig. 21.11). In other words, although some degradation did occur during annealing, the lower degradation rate produced by annealing resulted in better purity at the end of the stability study for the annealed sample.

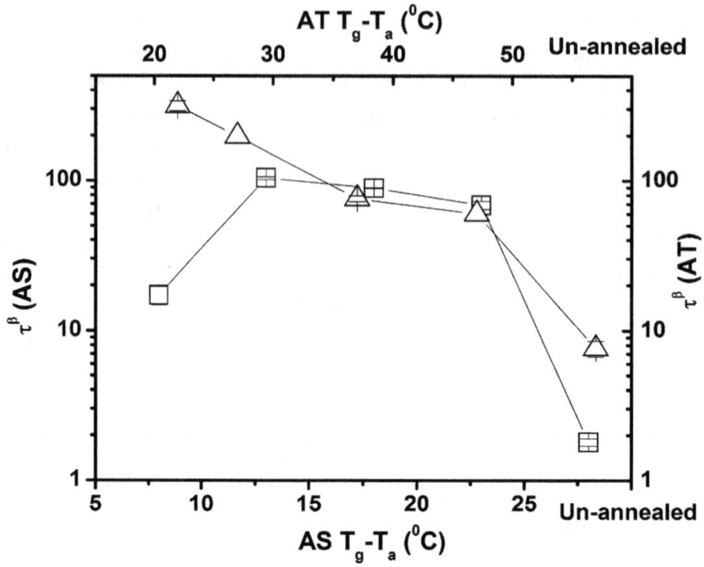

Figure 21.10. Structural relaxation parameters (τ^{β}, in hours) of annealed aspartame : sucrose (annealed for 20 h, □) and aspartame : trehalose (annealed for 10 hs, Δ) formulations, measured using isothermal microcalorimetry at 50°C [29].

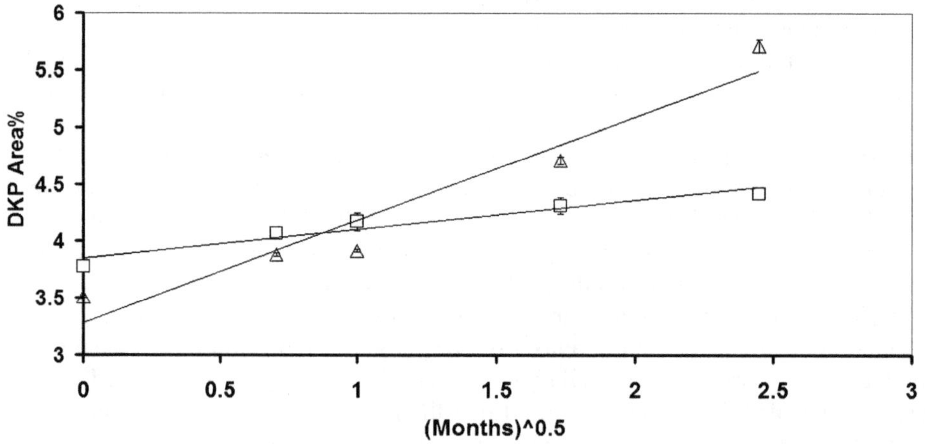

Figure 21.11. Appearance of 5-benzyl-3,6-dioxo-2-piperazineacetic acid (DKP) as a function of time at 50°C storage temperature. Aspartame : sucrose formulation subjected to annealing postlyophilization at 60°C for 20 hs. *Key*: (Δ) not annealed, $k = 0.90 \pm 0.05$; (□) annealed, $k = 0.25 \pm 0.03$. (Figure adapted from Luthra et al. [89]. Reprinted from the *Journal of Pharmaceutical Sciences*, copyright © 2008, John Wiley & Sons, Inc.)

21.4. SUMMARY AND CONCLUSIONS

The current understanding of the relationship between structural relaxation dynamics in amorphous solids and pharmaceutical stability in these systems is incomplete. In this chapter, we have attempted to develop a perspective on the effect of motions present in solid state on the stability of pharmaceuticals. Since molecular mobility, which is a key parameter affecting chemical stability in the amorphous state, is affected by thermal history, variations in processing conditions that influence thermal history also play an important role. Global and local dynamics are present in all lyophilized pharmaceutically relevant systems. There are documented cases for which the α motion seems to control stability, while in other cases "fast dynamics" correlates better with stability. While progress has been made in understanding this complex area and new concepts in stabilization have emerged, additional work is needed to define the motions most relevant to stability and to develop methods to quench them effectively.

REFERENCES

1. Carpenter, J. F., Pikal, M. J., Chang, B. S., and Randolph, T. W. (1997), Rational design of stable lyophilized protein formulations: Some practical advice. *Pharm. Res.* **14**(8): 969–975.
2. Costantio, H. R. and Pikal, M. J. (2004), *Lyophilization of Biopharmaceuticals*, AAPS Press, Arlington, VA, p. 686.
3. Maa, Y.-F. and Prestrelski, S. J. (2000), Biopharmaceutical powders: Particle formation and formulation considerations, *Curr. Trends Pharm. Biotechnol.* **1**(3): 283–302.
4. Lee, G. (2002), Spray-drying of proteins, *Pharm. Biotechnol.* (special issue on rational design of stable protein formulations) **13**: 135–158.
5. Beck, V. A. and Shaqfeh, E. S. G. (2006), Ergodicity breaking and conformational hysteresis in the dynamics of a polymer tethered at a surface stagnation point, *J. Chem. Phys.* **124**(9): 094902/1–094902/16.
6. Ediger, M. D., Angell, C. A., and Nagel, S. R. (1996), Supercooled liquids and glasses, *J. Phys. Chem.* **100**(31): 13200–13212.
7. Hancock, B. C, Shamblin, S. L., and Zografi, G. (1995), Molecular mobility of amorphous pharmaceutical solids below their glass-transition temperatures, *Pharm. Res.* **12**(6): 799–806.
8. Struik, L. C. E. (1978), *Physical Aging in Amorphous Polymers and Other Materials*, Elsevier, p. 242.
9. Pikal, M. J. (2004), Mechanisms of protein stabilization during freeze-drying and storage: The relative importance of thermodynamic stabilization and glassy state relaxation dynamics, *Drugs Pharm. Sci.* (special issue on freeze-drying/lyophilization of pharmaceutical and biological products) **137**: 63–107.
10. Cicerone, M. T. and Soles, C. L. (2004), Fast dynamics and stabilization of proteins: Binary glasses of trehalose and glycerol, *Biophys. J.* **86**(6): 3836–3845.
11. Cicerone, M. T., Tellington, A., Trost, L., and Sokolov, A. (2003), Substantially improved stability of biological agents in dried form: The role of glassy dynamics in preservation of biopharmaceuticals, *BioProcess Int.* **1**(1): 1–9.

12. Hancock, B. C. and Zografi, G. (1997), Characteristics and significance of the amorphous state in pharmaceutical systems, *J. Pharm. Sci*. **86**(1): 1–12.
13. Shamblin, S. L., Tang, X. L., Chang, L. Q., Hancock, B. C., and Pikal, M. J. (1999), Characterization of the time scales of molecular motion in pharmaceutically important glasses, *J. Phys. Chem. B* **103**(20): 4113–4121.
14. Angell, C. A. (1995), Formation of glasses from liquids and biopolymers, *Science* **267**(5206): 1924–1935.
15. Pikal, M. J. and Roy, M. L. (2007), Solid state chemistry of proteins: I. Glass transition behavior in freeze dried disaccharide formulations of human growth hormone (hGH), *J. Pharm. Sci*. **96**(10): 2765–2776.
16. Adam, G. and Gibbs, J. H. (1965), The temperature dependence of cooperative relaxation properties in glass-forming liquids, *J. Chem. Phys*. **43**(1): 139–146.
17. Cicerone, M. T. and Ediger, M. D. (1996), Enhanced translation of probe molecules in supercooled o-terphenyl: Signature of spatially heterogeneous dynamics? *J. Chem. Phys*. **104**(18): 7210–7218.
18. Hempel, E., Hempel, G., Hensel, A., Schick, C., and Donth, E. (2000), Characteristic length of dynamic glass transition near Tg for a wide assortment of glass-forming substances, *J. Phys. Chem. B* **104**(11): 2460–2466.
19. Erwin, B. M. and Colby, R. H. (2002), Temperature dependence of relaxation times and the length scale of cooperative motion for glass-forming liquids, *J. Non-Cryst. Solids* 307–310: 225–231.
20. Williams, G., and Watts, D. C. (1970), Non-symmetrical dielectric relaxation behaviour arising from a simple empirical decay function, *Trans. Faraday Soc*. **66**: 80.
21. Shamblin, S. L., Hancock, B. C., Dupuis, Y., and Pikal, M. J. (2000), Interpretation of relaxation time constants for amorphous pharmaceutical systems, *J. Pharm. Sci*. **89**(3): 417–427.
22. Liu, J., Rigsbee, D. R., Stotz, C., and Pikal, M. J. (2002), Dynamics of pharmaceutical amorphous solids: The study of enthalpy relaxation by isothermal microcalorimetry, *J. Pharm. Sci*. **91**(8): 1853–1862.
23. Kawakami, K., and Pikal, M. J. (2005), Calorimetric investigation of the structural relaxation of amorphous materials: Evaluating validity of the methodologies, *J. Pharm. Sci*. **94**(5): 948–965.
24. Boutonnet-Fagegaltier, N., Menegotto, J., Lamure, A., Duplaa, H., Caron, A., Lacabanne, C., and Bauer, M. (2002), Molecular mobility study of amorphous and crystalline phases of a pharmaceutical product by thermally stimulated current spectrometry, *J. Pharm. Sci*. **91**(6): 1548–1560.
25. Smith, G., Duffy, A. P., Shen, J., and Olliff, C. J. (1995), Dielectric relaxation spectroscopy and some applications in the pharmaceutical sciences, *J. Pharm. Sci*. **84**(9): 1029–1044.
26. Raju, G. G. (2003), *Dielectrics in Electric Fields*, Marcel Dekker, New York, pp. 1–578.
27. Bhugra, C., Shmeis, R., Krill, S. L., Pikal and Pikal, M. J. (2008) Different measures of molecular mobility: Comparison between calorimetric and thermally stimulated current relaxation times below Tg and correlation with dielectric relaxation times above Tg, *J. Pharm. Sci*. **97**(10): 4498–4515.
28. Andronis, V. and Zografi, G. (1997), Molecular mobility of supercooled amorphous indomethacin, determined by dynamic mechanical analysis, *Pharm. Res*. **14**(4): 410–414.

29. Luthra, S., Hodge, I. M., and Pikal, M. J. (2008), Investigation of the impact of annealing on global molecular mobility in glasses: Optimization for stabilization of amorphous pharmaceuticals, *J. Pharm. Sci*. **97**(9): 3865–3882.
30. Bhugra, C., Shmeis, R., Krill, S. L., and Pikal, M. J. (2006), Predictions of onset of crystallization from experimental relaxation times I—Correlation of molecular mobility from temperatures above the glass transition to temperatures below the glass transition, *Pharm. Res*. **23**(10): 2277–2290.
31. Van den Mooter, G., Augustijns, P., and Kinget, R. (1999), Stability prediction of amorphous benzodiazepines by calculation of the mean relaxation time constant using the Williams-Watts decay function, *Eur. J. Pharm. Biopharm*. **48**(1): 43–48.
32. Tool, A. Q. (1946), Relation between inelastic deformability and thermal expansion of glass in its annealing range, *J. Am. Ceramic Soc*. **29**(9): 240–253.
33. Hodge, I. M. (1987), Effects of annealing and prior history on enthalpy relaxation in glassy polymers. 6. Adam-Gibbs formulation of nonlinearity, *Macromolecules* **20**(11): 2897–2908.
34. Hodge, I. M. (1997), Adam-Gibbs formulation of enthalpy relaxation near the glass transition, *J. Res. Natl. Inst. Stand. Technol*. (*NIST*) **102**(2): 195–205.
35. Shamblin, S. L. and Zografi, G. (1998), Enthalpy relaxation in binary amorphous mixtures containing sucrose, *Pharm. Res*. **15**(12): 1828–1834.
36. Duddu, S. P., Zhang, G., and Dal Monte, P. R. (1997), The relationship between protein aggregation and molecular mobility below the glass transition temperature of lyophilized formulations containing a monclonal antibody, *Pharm. Res*. **14**(5): 596–600.
37. Shalaev, E. Y. and Zografi, G. (1996), How does residual water affect the solid-state degradation of drugs in the amorphous state? *J. Pharm. Sci*. **85**(11): 1137–1141.
38. Roy, M. L., Pikal, M. J., Rickard, E. C., and Maloney, A. M. (1992), The effects of formulation and moisture on the stability of a freeze-dried monoclonal antibody-vinca conjugate: A test of the WLF glass transition theory, *Devel. Biol. Stand*. (special issue on freeze-drying formulation of biological products) **74**: 323–340.
39. Xiang, T.-X. and Anderson, B. D. (2005), Distribution and effect of water content on molecular mobility in poly(vinylpyrrolidone) glasses: A molecular dynamics simulation, *Pharm. Res*. **22**(8): 1205–1214.
40. Fox, T. G. (1956), Influence of diluent and of copolymer composition on the glass temperature of a polymer system, *Bull. Am. Phys. Soc*. **1**: 123.
41. Velikov, V., Borick, S., and Angell, C. A. (2001), The glass transition of water, based on hyperquenching experiments, *Science* **294**(5550): 2335–2338.
42. Oksanen, C. A. and Zografi, G. (1993), Molecular mobility in mixtures of absorbed water and solid poly(vinylpyrrolidone), *Pharm. Res*. **10**(6): 791–799.
43. Hageman, M. J., Possert, P. L., and Bauer, J. M. (1992), Prediction and characterization of the water sorption isotherm for bovine somatotropin, *J. Agric. Food Chem*. **40**(2): 342–347.
44. Chang, L., Shepherd, D., Sun, J., Ouellette, D., Grant, K. L., Tang, X., and Pikal, M. J. (2005), Mechanism of protein stabilization by sugars during freeze-drying and storage: Native structure preservation, specific interaction, and/or immobilization in a glassy matrix? *J. Pharm. Sci*. **94**(7): 1427–1444.
45. Chang, L., Shepherd, D., Sun, J., Tang, X., and Pikal, M. J. (2005), Effect of sorbitol and residual moisture on the stability of lyophilized antibodies: Implications for the mechanism of protein stabilization in the solid state, *J. Pharm. Sci*. **94**(7): 1445–1455.

46. Prestrelski, S. J., Pikal, K. A., and Arakawa, T. (1995), Optimization of lyophilization conditions for recombinant human interleukin-2 by dried-state conformational analysis using Fourier-transform infrared spectroscopy, *Pharm. Res.* **12**(9): 1250–1259.
47. Pikal, M. J. and Dellerman, K. M. (1989), Stability testing of pharmaceuticals by high-sensitivity isothermal calorimetry at 25°C: Cephalosporins in the solid and aqueous solution states, *Int. J. Pharm.* **50**(3): 233–252.
48. Streefland, L., Auffret, A. D., and Franks, F. (1998), Bond cleavage reactions in solid aqueous carbohydrate solutions, *Pharm. Res.* **15**(6): 843–849.
49. Lai, M. C., Hageman, M. J., Schowen, R. L., Borchardt, R. T., and Topp, E. M. (1999), Chemical stability of peptides in polymers. 1. Effect of water on peptide deamidation in poly(vinyl alcohol) and poly(vinyl pyrrolidone) matrixes, *J. Pharm. Sci.* **88**(10): 1073–1080.
50. Pikal, M. J. (2002), Chemistry in solid amorphous matrices: Implication for biostabilization, in *Amorphous Food and Pharmaceutical Systems*, Harry, L., ed., The Royal Society of Chemistry, Cambridge, UK, pp. 257–277.
51. Shamblin, S. L., Hancock, B. C., and Pikal, B. C. (2006), Coupling between chemical reactivity and structural relaxation in pharmaceutical glasses, *Pharm. Res.* **23**(10): 2254–2268.
52. Yoshioka, S. and Aso, Y. (2005), A quantitative assessment of the significance of molecular mobility as a determinant for the stability of lyophilized insulin formulations, *Pharm. Res.* **22**(8): 1358–1364.
53. Yoshioka, S., Miyazaki, T., and Aso, Y. (2006), Degradation rate of lyophilized insulin, exhibiting an apparent Arrhenius behavior around glass transition temperature regardless of significant contribution of molecular mobility, *J. Pharm. Sci.* **95**(12): 2684–2691.
54. Karel, M. and Saguy, I. (1991), Effects of water on diffusion in food systems, *Adv. Exp. Med. Biol.* **302**: 157–173.
55. Yoshioka, S., Aso, Y., and Kojima, S. (2004), Temperature- and glass transition temperature-dependence of bimolecular reaction rates in lyophilized formulations described by the Adam-Gibbs-Vogel equation, *J. Pharm. Sci.* **93**(4): 1062–1069.
56. Yoshioka, S., Aso, Y., and Kojima, S. (1999), The effect of excipients on the molecular mobility of lyophilized formulations, as measured by glass transition temperature and NMR relaxation-based critical mobility temperature, *Pharm. Res.* **16**(1): 135–140.
57. Yoshioka, S., Aso, Y., and Kojima, S. (1996), Determination of molecular mobility of lyophilized bovine serum albumin and g-globulin by solid-state 1H NMR and relation to aggregation-susceptibility, *Pharm. Res.* **13**(6): 926–930.
58. Yoshioka, S., Tajima, S., Aso, Y., and Kojima, S. (2003), Inactivation and aggregation of b-galactosidase in lyophilized formulation described by Kohlrausch-Williams-Watts stretched exponential function, *Pharm. Res.* **20**(10): 1655–1660.
59. Johari, G. P., and Goldstein, M. (1970), Viscous liquids and the glass transition. II. Secondary relaxations in glasses of rigid molecules, *J. Chem. Phys.* **53**(6): 2372–2388.
60. Johari, G. P. (1973), Intrinsic mobility of molecular glasses, *J. Chem. Phys.* **58**(4): 1766–1770.
61. Williams, G. and Watts, D. C. (1970), Non-symmetrical dielectric relaxation behavior arising from a simple empirical decay function, *Trans. Faraday Soc.* **66**(1): 80–85.
62. Ramos, J. J. M., Correia, N. T., and Diogo, H. P. (2006), TSDC as a tool to study slow molecular mobility in condensed complex systems, *J. Non-Cryst. Solids* **352**(42–49): 4753–4757.

63. Lacabanne, C., Lamure, A., Teyssedre, G., Bernes, A., and Mourgues, M. (1994), Study of cooperative relaxation modes in complex systems by thermally stimulated current spectroscopy, *J. Non-Cryst. Solids* **172–174**(Pt. 2): 884–890.

64. Correia, N. T., Alvarez, C., Ramos, J. J. M., and Descamps, M. (2001), The b-a branching in D-sorbitol as studied by thermally stimulated depolarization currents (TSDC), *J. Phys. Chem. B* **105**(24): 5663–5669.

65. Aso, Y., Yoshioka, S., and Kojima, S. (2001), Explanation of the crystallization rate of amorphous nifedipine and phenobarbital from their molecular mobility as measured by 13C nuclear magnetic resonance relaxation time and the relaxation time obtained from the heating rate dependence of the glass transition temperature, *J. Pharm. Sci*. **90**(6): 798–806.

66. Masuda, K., Tabata, S., Sakata, Y., Hayase, T., Yonemochi, E., and Terada, K. (2005), Comparison of molecular mobility in the glassy state between amorphous indomethacin and salicin based on spin-lattice relaxation times, *Pharm. Res*. **22**(5): 797–805.

67. Becker, E. D. (2000), *High Resolution NMR: Theory and Chemical Applications*, 3rd ed., Academic Press.

68. Harris, R. K. (1993), State of the art for solids, *Chem. Britain* **29**(7): 601–604.

69. Cicerone, M. T., Tellington, A., Trost, L., and Sokolov, A. (2003), *Plasticized Hydrophilic Glasses for Improved Stabilization of Biological Agents*, PCT Int. Appl., 45, CODEN: PIXXD2 WO 2003035827 A2 20030501.

70. Hill, J. J., Shalaev, E. Y., and Zografi, G. (2005), Thermodynamic and dynamic factors involved in the stability of native protein structure in amorphous solids in relation to levels of hydration, *J. Pharm. Sci*. **94**(8): 1636–1667.

71. Meyer, A., Dimeo, R. M., Gehring, P. M., and Neumann, D. A. (2003), The high-flux backscattering spectrometer at the NIST Center for Neutron Research. *Rev. Sci. Instrum*. **74**(5): 2759–2777.

72. Yoshioka, S., Aso, Y., and Kojima, S. (2002), Different molecular motions in lyophilized protein formulations as determined by laboratory and rotating frame spin-lattice relaxation times, *J. Pharm. Sci*. **91**(10): 2203–2210.

73. Duddu, S. P. and Sokoloski, T. D. (1995), Dielectric analysis in the characterization of amorphous pharmaceutical solids. 1. Molecular mobility in poly(vinylpyrrolidone)-water systems in the glassy state, *J. Pharm. Sci*. **84**(6): 773–776.

74. Lourdin, D., Ring, S. G., and Colonna, P. (1998), Study of plasticizer-oligomer and plasticizer-polymer interactions by dielectric analysis: Maltose-glycerol and amylose-glycerol-water systems, *Carbohydr. Res*. **306**(4): 551–558.

75. Separovic, F., Lam, Y. H., Ke, X., and Chan, H. K. (1998), A solid-state NMR study of protein hydration and stability, *Pharm. Res*. **15**(12): 1816–1821.

76. Yoshioka, S., Miyazaki, T., and Aso, Y. (2006), b-Relaxation of insulin molecule in lyophilized formulations containing trehalose or dextran as a determinant of chemical reactivity, *Pharm. Res*. **23**(5): 961–966.

77. Pikal, M. J., Rigsbee, D. R., Roy, M. L., Galreath, D., Kovach, K. J., Wang, B. S., and Cicerone, M. T. (2008), Solid state chemistry of proteins: II. The correlation of storage stability of freeze-dried human growth hormone (hGH) with structure and dynamics in the glassy solid, *J. Pharm. Sci*. **97**(12): 5106–5121.

78. Dirama, T. E., Carri, G. A., and Sokolov, A. P. (2005), Role of hydrogen bonds in the fast dynamics of binary glasses of trehalose and glycerol: A molecular dynamics simulation study, *J. Chem. Phys*. **122**(11): 114505/1–114505/8.

79. Abdul-Fattah, A. M., Vu, Truong-Le, Yee, L. Nguyen, L., Kalonia, D. S., Cicerone, M. T., and Pikal, M. J. (2007), Drying-induced variations in physicochemical properties of amorphous pharmaceuticals and their impact on stability (I): Stability of a monoclonal antibody, *J. Pharm. Sci*. **96**(8): 1983–2008.
80. Mardaleishvili, I. R. and Anisimov, V. M. (1987), b-Transition and relaxation of impurities in molecules in their reactions in glass polymers, *Izvest. Akad. Nauk SSSR Ser. Khim*. (6): 1431–1432.
81. Hill, S. A., MacNaughtan, W., Farhat, I. A., Noel, T. R., Parker, R., Ring, S. G., and Whitcombe, M. J. (2005), The effect of thermal history on the Maillard reaction in a glassy matrix, *J. Agric. Food Chem*. **53**(26): 10213–10218.
82. Pikal, M. J. (1996), *Thermometric Seminars on Calorimetry in Materials Sciences*, Stockholm.
83. Pikal, M. J. (1999), Mechanisms of protein stabilization during freeze-drying and storage: The relative importance of thermodynamic stabilization and glassy state relaxation dynamics, in *Freeze-Drying/Lyophilization of Pharmaceutical and Biological Products*, Louis Rey and Joan May, Editors, Marcel Dekker pp. 161–198.
84. Abdul-Fattah, A. M., Bogner, R. H., and Pikal, M. J. (2007), The effect of annealing on the stability of amorphous solids: Chemical stability of freeze-dried moxalactam, *J. Pharm. Sci*. **96**(5): 1237–1250.
85. Pikal, M. J., Reddy, R. D., Shalaev, E. Y., and Ziegler, C. B. (2005), *Method of Stabilizing Disordered Cefovecin Sodium Salt Using Thermal Annealing*, Pfizer Products, Inc., p. 19.
86. Smits, A. L. M., Ruhnau, F. C., Vliegenthart, J. F. G., and Van Soest, J. J. G. (1998), Aging of starch based systems as observed with FT-IR and solid state NMR spectroscopy, *Starch/Staerke* **50**(11–12): 478–483.
87. Luthra, S., Utz, M., and Pikal, M. J. (2008), Solid state 13C NMR investigation of impact of annealing in lyophilized glasses, *J. Pharm. Sci*. **97**(10): 4336–4346.
88. Luthra, S., Hodge, I. M., and Pikal, M. J. (2008), Effects of annealing on enthalpy relaxation in lyophilized disaccharide formulations: Mathematical modeling of DSC curves, *J. Pharm. Sci*. **97**(8): 3084–3099.
89. Luthra, S., Hodge, I. M., Utz, M., and Pikal, M. J. (2008), Correlation of annealing with chemical stability in lyophilized pharmaceutical glasses, *J. Pharm. Sci*. **97**(12): 5240–5251.
90. Cicerone, M. T. (2007), Correlation of stability with fast dynamics, personal communication with Luthra, S. and Wang, B.

22

STRUCTURAL ANALYSIS OF PROTEINS IN DRIED MATRICES

Andrea Hawe, Sandipan Sinha, Wolfgang Friess, and Wim Jiskoot

22.1. INTRODUCTION

The development of stable formulations is a prerequisite for a safe application of proteins in pharmacotherapy. Proteins are complex molecules, susceptible to various degradation reactions, including chemical modifications, aggregation, fragmentation, or conformational changes during production or storage [1,2]. These instability reactions are to be avoided, as they can result in a loss of activity, severe side effects, and immunogenicity [3–5]. A drying step is frequently required for stabilization, with the goal to preserve the native protein structure. In this context, lyophilization is still the prevailing technology, but also spray-drying or other drying technologies, which produce free-flowable powders, such as for the inhalative application route, are on the upswing [6–8]. However, proteins do undergo degradation in the solid state as well, as reviewed in Chapter 19 in this volume.

As compared to analysis after reconstitution, direct structural analysis of proteins within solid matrices can be beneficial to gain deeper insight into the effect of the manufacturing process or the impact of added stabilizers on the proteins in their storage environments [9,10], with the goal to find optimum stabilizing conditions. When

Formulation and Process Development Strategies for Manufacturing Biopharmaceuticals, edited by Jameel and Hershenson

analyzing the protein only after reconstitution, reversible structural changes, which can negatively affect storage stability, could be overseen. A loss of the native structure on drying often correlates with loss of structure and biological activity after reconstitution [11,12] or long-term storage. Furthermore, changes in the secondary structure are sometimes associated with chemical instability (see Chapter 19). For several proteins it could be demonstrated that preservation of the native structure during lyophilization can be correlated with an improved storage stability of the formulation (Table 22.1).

Only few methods are suitable for this approach, because the high optical density of the solid matrices and the related scattering effects render the use of most

TABLE 22.1. Examples from the Literature Showing Correlation Between Maintained Protein Structure After Lyophilization and Improved Storage Stability

Protein	Excipients	Method for Solid-State Characterization	Storage Stability Assessed by	Reference
Interleukin-2 (IL-2)	Glucose, sucrose, trehalose, raffinose, stachyose, β-CD, dextran 10, 40	FTIR	SDS-PAGE, HP-SEC	67
Interleukin-1 receptor antagonist		FTIR	—	16
Antibody	Sucrose, trehalose, histidine	FT-Raman	HP-SEC	12
IgG1, recombinant human serum albumin (rHSA)	Sorbitol	FTIR, KBr	HP-SEC, IEC	68
Actin	Glucose sucrose trehalose dextran	FTIR, KBr	CD, activity testing	10
Recombinant humanized monoclonal antibody	Trehalose, sucrose	FTIR, KBr	HP-SEC, SDS-PAGE, CD	31
Recombinant human interleukin-11	Sucrose, trehalose, hydroxy-ethyl starch	FTIR, KBr	HP-SEC, RP-HPLC	69
Recombinant humanized IgG	Lactose	FTIR, KBr	HP-SEC	27
Monoclonal antibody	Sucrose	FTIR, KBr	HP-SEC	70

Abbreviations: HP-SEC = high-pressure size exclusion chromatography, RP-HPLC = reversed-phase HPLC; IEC = ion exchange chromatography; CD = circular dichroism.

spectroscopic techniques such as UV–visible absorption and CD spectroscopy difficult or impossible. The most frequently used technique for solid-state protein analysis is Fourier transform infrared (FTIR) spectroscopy. Other methods are near-infrared (NIR) spectroscopy, Raman spectroscopy, solid-state nuclear magnetic resonance (ssNMR) spectroscopy, hydrogen–deuterium (H/D) exchange in the solid state, and fluorescence spectroscopy. The basic principles of these methods are explained briefly with focus on their use for solid-state protein characterization, and their use for solid-state structural protein analysis is illustrated by selected applications.

22.2. FOURIER TRANSFORM INFRARED (FTIR) SPECTROSCOPY

Fourier transform infrared spectroscopy measures the light energy absorbed by protein molecules in the infrared region due to vibrational motions. Therefore, FTIR spectra are characteristic of bond types and highly sensitive to changes in the protein backbone, specifically, secondary structure. Only three of possibly nine absorption bands are typically used for analysis: amide I (1600–1700 cm^{-1}), amide II (1500–1600 cm^{-1}), and amide III (1220–1330 cm^{-1}) [13]. The amide I band, which arises as a result of C=O stretching vibrations with minor contributions from C–N stretching, is directly correlated to the protein backbone conformation. Secondary structure analysis of proteins in both solutions and solids is generally performed after taking the second derivative or deconvolution of the amide I band. The amide II band, arising from N–H bending and with C–N stretching vibrations, has been used to characterize protein secondary structure by employing H/D exchange experiments. The exchange of the imide hydrogen with deuterium leads to a band shift from approximately 1545 cm^{-1} to 1450 cm^{-1}, enabling quantification of H/D exchange kinetics. The amide III band reflects a number of coordinated vibrations, and although it has been used for quantitative secondary structure analysis, it has failed to gain importance because of its weak signal. For further in-depth description of FTIR spectroscopy, the interested reader is referred to the literature [13,14].

Fourier transform infrared spectroscopy enables measurement of protein structure in samples of almost any physical state and morphology, including dried formulations. It can measure samples that scatter light in the UV region, as long-wavelength IR radiation is less prone to light scattering and stray light is not modulated at the appropriate frequency by the interferometer and therefore is excluded by the Fourier transformation [15]. Suitable sampling techniques for solid-state analysis of proteins are potassium bromide (KBr) pellets, diffuse reflectance infrared Fourier transform (DRIFT), attenuated total reflectance (ATR), and photoacoustic FTIR (PA-FTIR).

Dried protein samples are predominantly analyzed in potassium bromide (KBr) pellets, where sample preparation plays an important role for the quality of the FTIR analysis. Potassium bromide pellets are prepared by compressing a mixture of about 1–5 mg of dried sample (at least 0.2–0.5 mg protein) in 100–300 mg KBr, with the exact ratio depending on the particular formulation. For sample preparation it is essential to keep the residual moisture of the KBr pellets as low as possible. This can be achieved by storing KBr and the pressed pellets in a desiccator, as well as by attaching the KBr pellet compactor to a vacuum pump. The mechanical stress imposed on the protein during grinding and compressing may affect the protein, but in general does not induce major structural changes [16]. A slight loss in the band intensity in the FTIR spectra of dried lysozyme and α-chromotrypsinogen after KBr processing was observed, mainly when employing high pressure during compression and exposing the protein formulation to moisture [17].

For diffuse reflectance FTIR (DRIFT), the protein sample is mixed with KBr or another IR-transparent salt. Compression is not necessary, which reduces the mechanical stress for the protein. The IR light scattered by the salt–protein mixture is absorbed by the protein and collimated by the detector [13].

Attenuated total reflectance (ATR) FTIR is another setup for measuring solid protein samples, without further sample preparation [13]. The penetration depth in ATR measurements is limited (1–2 μm), and therefore mainly protein in close proximity with the ATR crystal, typically ZnSe, Ge, or diamond, is measured.

Another approach requiring only minimal sample preparation is photoacoustic FTIR spectroscopy (PA-FTIR), which is used mainly in polymer science. On illumination with modulated radiation, the sample produces heat through nonradiative decay. The resulting thermal gradient within the sample heats gas molecules in contact with the sample and causes acoustic waves, which are detected by a sensitive microphone. The acoustic waves can be mathematically transformed to absorbance spectra [18]. The measurements are rather insensitive to scattered light, and it is possible to determine semiquantitatively the depth from which the signal originates [18]. By using PA-FTIR van de Weert et al. [19] demonstrated that lysozyme was homogenously distributed through PLGA microspheres and that only minor amounts of lysozyme were associated with the surface of the particles [19]. Correct background subtraction, which is essential for the quality of the spectra, is more difficult in the solid state as compared to the liquid state. For instance, the background spectrum of the excipients can be impacted by their morphology, such as the presence of different polymorphic forms or amorphous material. Generally, it is difficult to produce morphologically equal placebo samples for this purpose, especially, when the excipient spectrum is overlapping with the amide I region of the protein. This holds true for amino acids (e.g., glycine or histidine) and also for polymers such as hydroxyethyl starch or poly(lactic-*co*-glycolic acid) (PLGA) [20].

Further care is needed when comparing spectra collected by different techniques, as the sampling technique can impact the peak position and intensity (Fig. 22.1). The differences between ATR-FTIR spectra and transmission spectra collected either from KBr pellets or by the mull method were attributed to artifacts of the ATR method. Even after correction for penetration depth and refractive index dispersion, the secondary structure was determined less accurately by ATR-FTIR [21].

Alternatively to the sampling methods described above, solid samples and also partially hydrated protein formulations can be analyzed after suspending them in fluorinated hydrocarbon (mull method) [22,21] or directly without further sample preparation between two calcium fluoride windows [22].

An illustrative example of the application of FTIR spectroscopy to monitor structural changes in dried matrices is given by Prestrelski et al. [24], who compared the

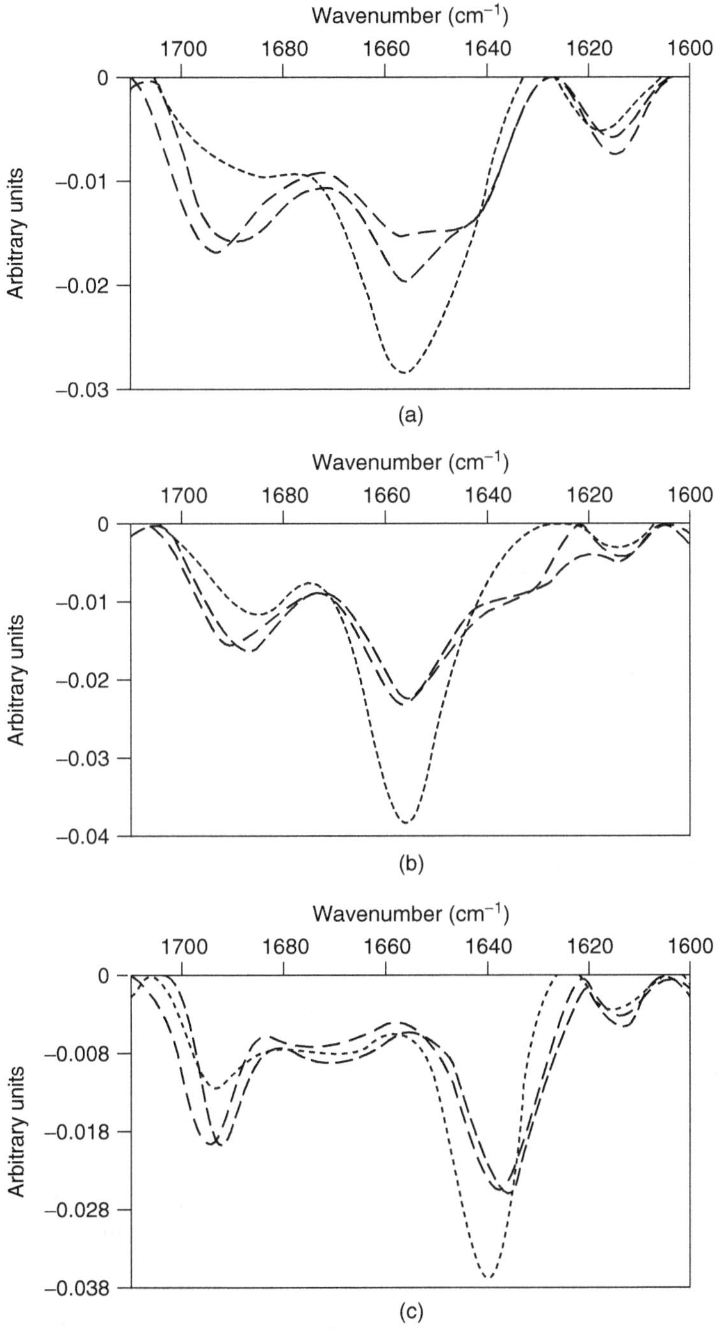

Figure 22.1. Second-derivative FTIR spectra of lyophilized lysozyme (a), BSA (b), and IgG (c) showing the KBr pellet transmission (dashed line), mull transmission (dashed–dotted line), and corrected attenuated total reflectance (dotted line) [21].

structure of several proteins in the solid state and after rehydration. Fourier transform IR revealed that GCSF was conformationally stable during both drying and rehydration, α-lactalbumin reversibly unfolded during the drying process, but was native-like after reconstitution, whereas γ-interferon and LDH denatured irreversibly during lyophilization, leading to aggregation in the rehydrated solution [24,25]. Furthermore, FTIR spectroscopy indicated that the initial secondary structures are correlated with drying-induced structural changes of proteins. In predominantly α-helical proteins (e.g., rhGH) or proteins with both α-helical and β-sheet structures (e.g., rhDNase), an increase in β-sheet in favor of α-helices can be generally observed. For β-sheet-rich proteins such as rhuIgG, the overall structure is less affected, as β-sheets are more stable on drying [22,26,27].

One of the major applications of FTIR spectroscopy is the evaluation of the stabilizing properties of excipients on the preservation of the secondary structure of a protein in the dried matrix, including studies on protein–excipient interactions [28–30]. When the retention of the native structure of a protein can be correlated with improved storage stability, the stabilizing properties of excipients can be estimated and the formulation optimized for long-term storage. Lyophilization of a recombinant human antibody (rhuMAb) without excipients resulted in reversible solid-state structural alterations detected by FTIR, and resulted in elevated aggregate formation during 12-month storage at 30°C. The structural alterations on lyophilization and correspondingly the aggregate formation on storage could be prevented by adding sucrose or trehalose [31]. Examples from the literature, where the retention of the native secondary structure in the solid state can be correlated with improved storage stability, are listed in Table 22.1. Nevertheless, the retention of the native structure in the solid state is not always a general predictor for stability, as was illustrated, for example, by moisture-induced aggregation of lyophilized tetanus toxoid on storage. The stability of this toxoid does not correlate with the initial solid-state structure determined by FTIR, as the aggregation mechanism is independent of the initial protein conformation [32].

22.3. NEAR-INFRARED SPECTROSCOPY

Near-infrared (NIR) spectroscopy has gained immense popularity as a simple, fast, noninvasive, and nondestructive technique, which allows measurement of both physical and chemical properties of samples, including residual moisture content, crystallinity, molecular interactions in solids, and protein structure [33,34]. The region between 13,300 and 4000 cm^{-1} is generally referred to as the *NIR region*. With respect to protein structural analysis in the solid state, NIR spectroscopy measures overtone (5000–13,000 cm^{-1}) and combination bands (4000–5000 cm^{-1}) of absorption bands involving C–H, N–H, and O–H bonds. These bands are the outcome of forbidden transitions from ground vibrational energy levels and are therefore several-fold less intense than their corresponding mid-IR absorption bands. The low molar absorptivity of NIR light permits generally rapid measurements in the reflectance mode, such as through a glass vial.

The combination bands (4000–5000 cm^{-1}) and first overtone bands (5600–6600 cm^{-1}) of proteins measured by NIR are often overlapping, with the result that they are broad and unresolved. To extract information from NIR spectra, data analysis techniques such as multiple linear regression (MLR) analysis, principal-components analysis (PCA), and partial least-squares regression (PLS) analysis are helpful tools. For an in-depth description of NIR spectroscopy and data analysis, the interested reader is referred to the literature [33,34].

The assignment of bands to certain structural elements is more complex than in FTIR spectroscopy because of the overlap. However, it was feasible to assign several NIR bands to α-helical (4369 and 4604 cm^{-1}) and β-sheet structures (4323, 4417, 4525–4535 cm^{-1}) [35]. So far, freeze-drying-induced conformational changes of several proteins and the stabilizing impact of excipients on different proteins were monitored by NIR spectroscopy [35–37]. Izutsu et al. [35] detected a slight reduction in α-helical structures with a concomitant rise in β-sheet structures on lyophilization and at elevated temperature, by comparing NIR spectra of seven proteins in solution and in the solid state. These studies have shown the potential of using NIR as non-destructive, fast technique for monitoring structural analysis of proteins. Limitations of NIR compared to FTIR are the relatively weak and overlapping signals in NIR. However, it is possible to simultaneously monitor changes in secondary structure of proteins and other physicochemical parameters of lyophilized formulations, such as crystallinity. For routine application, such as in process control by direct measurement of lyophilized products within vials, further studies are required.

22.4. RAMAN SPECTROSCOPY

Owing to technological advances, Raman spectroscopy has emerged as an important tool to monitor secondary structure of proteins. Like FTIR, Raman spectroscopy has the flexibility to study protein structure in different physical forms, including the solid state. A major advantage of Raman spectroscopy is that solid samples can be directly measured independent of their morphology and structure. An excellent review by Wen enlists the more recent advances of Raman spectroscopy [38]. Raman spectroscopy detects vibrational transitions within molecules and can be regarded as a complementary technique to IR spectroscopy. While FTIR is based on absorption, Raman spectroscopy is a two-photon process; after absorption the molecules transit to virtual states and emit photons with changed energies as Stokes Raman scattering (weaker energy), which is typically used in conventional Raman spectroscopy, and anti-Stokes Raman (higher energy) [38]. The Raman effect is independent of the wavelength of the incident light. So, light of any single wavelength from deep UV to near infrared can be used, thus offering great selectivity. However, Raman signals are weaker for higher wavelengths, as the intensity of the scattered light is proportional to λ^{-4}. A prerequisite for Raman scattering is a change in the electronic polarizability of the molecules within the vibrational mode [39,40]. Raman-active bands are found within proteins –CH, $-CH_2$, disulfide bridges, and aromatic amino acids, which have a high polarizability and symmetric vibrational modes [38]. Thus, changes in the conformation of disulfide bridges and hydrogen bonding of tyrosine can easily be monitored, which is a unique advantage of Raman spectroscopy [39].

For Raman spectroscopy of proteins, mainly light in the NIR region (mostly 1064 nm lasers in FT-Raman) is employed to overcome disturbance from fluorescent impurities [38]. Raman spectroscopy enjoys a few advantages over FTIR. Raman spectroscopy provides information on highly polarizable group (having symmetric vibrational modes), which may or may not be IR-active. This attribute of Raman spectroscopy helps to identify the disposition of the side chains of amino acid residues, such as hydrogen bonding, and freedom of S–H bonds.

One disadvantage is that Raman scatter intensity is intrinsically low. As a result, a large fraction of protein in the dried formulation is required to obtain good-quality spectra. Care also needs to be taken when adjusting the intensity of the laser, as the samples can be locally destroyed above a certain threshold.

Forbes et al. [41] analyzed the impact of thermal stress on secondary structure, as well as the activity of trypsin in solution, spray-dried, and crystallized solid samples. Spray drying resulted in a small drift in the amide I band at 1688 cm^{-1}, as well as the occurrence of a new band at 1785 cm^{-1} in FT-Raman spectra, which was also observed for heat-stressed samples but not on crystallization of trypsin. The changes in secondary structure detected by FT-Raman were associated with loss of biological activity of trypsin [41]. By analyzing the amide I band in Raman spectra Sane et al. [12] quantified drying-induced structural changes in dried antibody formulations, mainly an increase in turn and unordered structures at the expense of ß-sheet structures. The extent of these structural changes in the presence of excipients correlated with the long-term stability of the antibody formulations, assessed by HP-SEC [12].

22.5. MICROSCOPIC TECHNIQUES (IR, RAMAN)

Infrared spectroscopy and Raman spectroscopy can be combined with microscopy. Herewith the information on protein structure can be correlated to a certain location in the sample. Furthermore, when linked to a freeze-drying microscope, IR and Raman spectroscopy allow *in situ* monitoring of freezing or freeze-drying induced structural changes in the protein.

Schwegman et al. [42] employed IR microscopy to monitor changes in the secondary structure of lysozyme *in situ* during freezing, freeze–thawing, and freeze drying, and on reconstitution. Without excipients a reduction of spectral similarity (calculated from the area of overlap) on freezing and freeze drying was measured, whereas the similarity was almost fully recovered after reconstitution of the lyophilized samples (Table 22.2). This indicates that the structural changes in the freeze-dried state were at least partially reversible. In presence of mannitol, the secondary structure in the freeze-dried state was better preserved, but not after reconstitution, whereas a combination of mannitol with sucrose also preserved the secondary structure after reconstitution [42]. *In situ* measurements have the advantage that the impact of formulations on protein structure can be measured for the different process steps and thus can be a valuable tool for the screening of formulation excipients.

TABLE 22.2. Spectral Similarity of Lysozyme FTIR Spectra During Different Stages of Lyophilization Calculated from Area of Overlap for Different Formulations[a]

Excipients	Frozen Solution	Freeze-Dried	Reconstituted
None	91.3	73.4	96.0
20 mg/mL mannitol	90.1	88.7	90.1
20 mg/mL mannitol	—	87.2	96.4
10 mg/mL sucrose	—	—	—

[a]Values (expressed as percentage) taken from Schwegman et al. [42].

22.6. SOLID-STATE NUCLEAR MAGNETIC RESONANCE SPECTROSCOPY

Previously, solid-state NMR (ssNMR) spectroscopy was used almost exclusively for characterizing small molecules, but it is rapidly emerging as a technique for characterizing macromolecular formulations. Solid-state NMR measures the absorbance of radiofrequency radiation by certain atomic nuclei (^{1}H, ^{13}C, ^{15}N) within a strong magnetic field. It can provide atom-level detail about protein structure and degradation in liquid and in lyophilized formulations [43]. In solution NMR, each structure is reported as an ensemble of many structures, all of which are consistent with experimentally derived restraints. An energy-minimized-averaged structure is then derived by aligning the structures and then finding the average position in Cartesian space. In ssNMR the molecules are fixed in space and therefore result in broad peaks. This can be attributed to the strong coupling interactions and chemical shift anisotropy (CSA). These problems are overcome by development of suitable probes (e.g., ^{13}C- or ^{15}N-labeled proteins) and spinning the sample at an angle of 54°44′ between the spin axis and the direction of the magnetic field. This angle is also known as the "magic angle" and the technique as "magic-angle spinning" (MAS). Also, by cross-polarization of magnetization from protons to carbons (^{13}C), the sensitivity to the carbon nucleus can be enhanced leading to the expansion of capabilities of ssNMR as a viable method for looking at solids [44,45].

For biological systems, ssNMR application has been limited mainly to elucidation of structure of proteins with low water solubility, including membrane proteins and α-amyloid fibrils [46–50]. In lyophilized powders the molecular orientation is more random, making it more difficult to obtain detailed information on protein structure as compared to membrane proteins or amyloid fibrils. Still, ssNMR is very useful with respect to protein structural information in solid formulations. Although limited in its applications for lyophilized formulations it has been used for assessing secondary structure, chemical reactivity, and molecular mobility of both protein and excipient in amorphous solids [51,52]. Examples of the use of solid-state NMR are analysis of site-specific interaction between excipients and proteins [53], as well as molecular mobility studies of excipients and protein within dried matrices [54,55]. The main application of ssNMR focuses on the detection of chemical changes, which are discussed further in Chapter 19.

22.7. HYDROGEN–DEUTERIUM (H/D) EXCHANGE

Hydrogen–deuterium exchange has long been used to analyze protein structure, protein folding/unfolding mechanisms, and protein–ligand interactions in solution [56]. Traditionally H/D exchange was used in conjunction with NMR, but more recently H/D exchange in combination with FTIR [57] and with mass spectrometry [58,59] has emerged as powerful technique capable of investigating protein structure and dynamics in solids. When comparing mass spectrometry with NMR in H/D exchange experiments, mass spectrometry (ESI and MALDI) has superior sensitivity, requiring minute quantities of protein for analysis, and can handle larger proteins and their complexes, On the other hand, NMR requires high protein concentrations but can provide information at the amino acid level, that is, determining exchange rates at specified amide linkages.

Hydrogen–deuterium exchange is based on the rate at which the protein backbone amide hydrogens undergo exchange with deuterons when incubated in presence of D_2O vapor (e.g., solid samples) or solution (liquid samples). The amide hydrogens of the protein backbone are involved in numerous hydrogen bonds and therefore stabilize the secondary and tertiary structures of the protein. For each individual amide hydrogen, the exchange rate will depend on its solvent accessibility and its involvement in intramolecular hydrogen bonding to any other part of the protein [60]. Typically, a H/D exchange experiment involves exposing the protein sample to a D_2O environment (vapor or solution) for predetermined time intervals followed by withdrawal of samples and quenching by a low-pH buffer (typically pH 2.4) at low temperature (typically $0^\circ C$) to arrest backexchange with nondeuterated solvent [61]. After H/D exchange the proteins can be analyzed (e.g., by LC-ESI MS) as intact proteins or subjected to pepsin digestion after H/D exchange, which enables the assessment of deuterium uptake by different regions of the protein.

French et al. [57] used the isotopic shifts for the amide II/II' band in FTIR spectra to characterize the accessibility of human granulocyte colony-stimulating factor (rhGCSF), as well as recombinant consensus interferon (rConIFN) and trehalose in spray-dried powders to H/D exchange. At 33% RH the H/D exchange was found to decrease with increasing trehalose content, suggesting increased protection of the protein amide bonds due to hydrogen bonding with trehalose. At 75% RH the H/D exchange increased for both proteins, and it could be shown that trehalose has slight a stabilizing effect on α-helical structure. The H/D exchange data correlated well with results obtained at the two RH values with conventional FTIR [57].

Li et al. [62,63] used LC ESI-MS to perform H/D exchange studies with calmodulin as model protein lyophilized with calcium chloride, trehalose, sucrose, dextran, and mannitol. By pepsin digestion it could be shown that calcium chloride primarily stabilized the calcium binding loops, whereas trehalose stabilized mainly the α helices at 33% RH. In comparison, FTIR did not show appreciable perturbation of the secondary structure in any of the formulations [63]. A comparison of excipients furthermore revealed that trehalose and sucrose showed greatest protection against H/D exchange, with the effect exerted primarily on the α-helical portions. Dextran also showed slight protection, whereas mannitol lost its protective effect because of its recrystallization on storage [62].

22.8. SOLID-STATE FLUORESCENCE SPECTROSCOPY

Only a few studies have been reported on the employment of fluorescence spectroscopy for analyzing protein structure in the frozen state [64] or in lyophilized protein formulations [65,66]. Trypthophan fluorescence is affected by its microenvironment; therefore, it is possible to measure changes in tertiary structure of a protein. Note that drying itself creates a change in microenvironment (i.e., removal of water), so altered fluorescence signals do not necessarily point to protein conformational changes. The major obstacle is the high optical density of the dried formulations and the larger interference by scattered light. These issues can be overcome by measuring dried samples in the frontface mode.

22.9. SUMMARY AND CONCLUSIONS

Several methods are available for analyzing protein structure in the solid state. Although FTIR is still the key method for measuring secondary structure and aggregation in the solid state, technological advances and new insights in NIR spectroscopy, Raman spectroscopy, ssNMR, and H/D exchange have cleared the way for their utilization in solid-state protein analysis. In particular, noninvasive methods such as NIR spectroscopy could be further developed with respect to routine application in quality control of dried formulations within vials. Both IR and Raman microscopy allow *in situ* monitoring of protein structure during freezing or freeze drying. More specific information can be gained by H/D exchange studies, which can assess the stabilizing impact of stabilizers on defined positions within the protein, and by ssNMR, which can provide insight into site-specific interactions of excipients with proteins. Most importantly, solid-state analysis can give insight into structural changes of proteins in the dried state, which could be overlooked when focusing only on the reconstituted liquid formulations. Therefore, it can help optimize dried formulations with respect to improved stability.

REFERENCES

1. Manning, M. C., Patel, K., and Borchardt, R. T. (1989), Stability of protein pharmaceuticals, *Pharm. Res.* **6**: 903–918.
2. Chi, E. Y., Krishnan, S., Randolph, T. W., and Carpenter, J. F. (2003), Physical stability of proteins in aqueous solutions: Mechanisms and driving forces in nonnative protein aggregation, *Pharm. Res.* **20**: 1325–1336.
3. Demeule, B., Gurny, R., and Arvinte, T. (2006), Where disease pathogenesis meets protein formulation: Renal deposition of immunoglobulin aggregates, *Eur. J. Pharm. Biopharm.* **62**: 121–130.

4. Hermeling, S., Crommelin, D. J., Schellekens, H., and Jiskoot, W. (2004), Structure-immunogenicity relationships of therapeutic proteins, *Pharm. Res.* **21**: 897–903.
5. Schellekens, H. (2002), Bioequivalence and the immunogenicity of biopharmaceuticals, *Nat. Rev. Drug Discov.* **1**: 457–462.
6. Abdul-Fattah, A. M., Kalonia, D. S., and Pikal, M. (2007), The challenge of drying method selection for protein pharmaceuticals: Product quality implications. *J. Pharm. Sci.* **96**: 1886–1916.
7. Maa, Y. F. and Prestrelski, S. J. (2000), Biopharmaceutical powders: particle formation and formulation considerations, *Curr. Opin. Pharm. Biotechnol.* **1**: 283–302.
8. Jovanovic, N., Bouchard, A., Hofland, G. W., Witkamp, G. J., Crommelin, D. J. A., and Jiskoot, W. (2004), Stabilization of proteins in dry powder formulations using supercritical fluid technology, *Pharm. Res.* **21**: 1955–1969.
9. Dong, A., Prestrelski, S. J., Allison, S. D., and Carpenter, J. F. (1995), Infrared spectroscopic studies of lyophilization- and temperature-induced protein aggregation, *J. Pharm. Sci.* **84**: 415–424.
10. Allison, S. D., Manning, M. C., Randolph, T. W., Middleton, K., Davis, A., and Carpenter, J. F. (2000), Optimization of storage stability of lyophilized actin using combination of disaccharides and dextran, *J. Pharm. Sci.* **89**: 199–214.
11. Klibanov, A. M. and Schefiliti, J. A. (2004), On the relationship between confirmation and stability in solid pharmaceutical protein formulations, *Biotechnol. Lett.* **26**: 1103–1106.
12. Sane, S. U., Wong, R., and Hsu, C. C. (2004), Raman spectroscopic characterization of drying-induced structural changes in a therapeutic antibody: Correlating structural changes with long-term stability, *J. Pharm. Sci.* **93**: 1005–1018.
13. van de Weert, M., Hering, J. A., and Haris, P. I. (2005), Fourier transform infrared spectroscopy, in *Methods for Structural Analysis of Protein Pharmaceuticals*, Jiskoot, W. and Crommelin, D. J. A., eds., AAPS Press, Arlington, VA, pp. 131–166.
14. Chalmers, J. M. and Griffiths, P. R. *Handbook of Vibrational Spectroscopy*, Wiley, Chichester, UK, 2001.
15. Braiman, M. S. and Rothschild, K. J. (1988), Fourier transform infrared techniques for probing membrane protein structure, *Annu. Rev. Biophys. Chem.* **17**: 541–570.
16. Chang, B. S., Beauvais, R. M., Dong, A., and Carpenter, J. F. (1996), Physical factors affecting the storage stability of freeze-dried interleukin-1 receptor antagonist: Glass transition and protein confirmation, *Arch. Biochem. Biophys.* **331**: 249–258.
17. Meyer, J. D., Manning, M. C., and Carpenter, J. F. (2004), Effects of potassium bromide disk formulation on the infrared spectra of dried model proteins, *J. Pharm. Sci.* **93**: 496–506.
18. Dittmar, R. M. and Palmer, R. A. (1994), Fourier transform photoacoustic spectroscopy of polymers, *Appl. Spectrosc. Rev.* **29**: 171–231.
19. van de Weert, M., van't Hof, R., van der Weerd, J., Heeren, R. M. A., Posthuma, G., Hennink, W. E., and Crommelin, D. J. A. (2000), Lysozyme distribution and conformation in a biodegradable polymer matrix as determined by FTIR techniques, *J. Control. Release* **68**: 31–40.
20. Meyer, J. D., Bai, S. J., Rani, M., Suryanarayanan, R., Nayar, R., Carpenter, J. F., and Manning, M. C. (2004), Infrared spectroscopic studies of protein formulations containing glycine, *J. Pharm. Sci.* **93**: 1359–1366.

21. Luthra, S., Kalonia, D. S., and Pikal, M. J. (2007), Effect of hydration on the secondary structure of lyophilized proteins as measured by fourier transform infrared (FTIR) spectroscopy, *J. Pharm. Sci.* **96**: 2910–2921.

22. Costantino, H. R., Chen, B., Griebenow, K., Hsu, C. C., and Shire, S. J. (1998), Fourier-transform infrared spectroscopic investigations of the secondary structure of aqueous and dried recombinant human deoxyribonuclease I, *Pharm. Pharmacol. Commun.* **4**: 391–395.

23. Sharma, V. K. and Kalonia, D. S. (2004), Polyethylene glycol-induced precipitation of interferon alpha-2a followed by vacuum drying: Development of a novel process for obtaining a dry, stable powder, *AAPS PharmSci.* **6**: 1–14.

24. Prestrelski, S. J., Tedeschi, N., Arakawa, T., and Carpenter, J. F. (1993), Dehydration-induced conformational changes in proteins and their inhibition by stabilizers, *Biophys. J.* **65**: 661–671.

25. Prestrelski, S. J., Arakawa, T., and Carpenter, J. F. (1993), Separation of freezing- and drying-induced denaturation of lyophilized proteins by stress-specific stabilization: II. Structural studies using infrared spectroscopy, *Arch. Biochem. Biophys.* **303**: 465–473.

26. Costantino, H. R., Nguyen, T. H., and Hsu, C. C. (1996), Fourier-transform infrared spectroscopy demonstrates that lyophilization alters the secondary structure of recombinant human growth hormone, *Pharm. Sci.* **2**: 229–232.

27. Costantino, H. R., Andya, J. D, Shire, S. J., and Hsu, C. C. (1997), Fourier-transform infrared spectroscopic analysis of the secondary structure of recombinant humanized immunoglobulin G, *Pharm. Sci.* **3**: 121–128.

28. Carpenter, J. F. and Crowe, J. H. (1989), An infrared spectroscopic study of the interactions of carbohydrates with dried proteins, *Biochemistry* **28**: 3916–3922.

29. Izutsu, K., Aoyagi, N., and Kojima, S. (2004), Protection of protein secondary structure by saccharides of different molecular weights during freeze-drying, *Chem. Pharm. Bull.* **52**: 199–203.

30. Souillac, P. O., Middaugh, C. R., and Rytting, J. H. (2002), Investigation of protein/carbohydrate interactions in the dried state. 2. Diffuse reflectance FTIR studies, *Int. J. Pharm.* **235**: 207–218.

31. Andya, J. D., Hsu, C. C., and Shire, S. J. (2003), Mechanisms of aggregate formation and carbohyrate excipient stabilization of lyophilized humanized monoclonal antibody formulations, *AAPS PharmSci.* **5**: 94–104.

32. Costantino, H. R., Schwendman, S. P., Griebenow, K., Klibanov, A. M., and Langer, R. (1996), The secondary structure and aggregation of lyophilized tetanus toxoid, *J. Pharm. Sci.* **85**: 1290–1293.

33. Reich, G. (2005), Near-infrared spectroscopy and imaging: Basic principles and pharmaceutical applications, *Adv. Drug Deliv. Rev.* **57**: 1109–1143.

34. Roggo, Y., Chalus, P., Maurer, L., Lema-Martinez, C., Edmond, A., and Jent, N. (2007), A review of near infrared spectroscopy and chemometrics in pharmaceutical technology, *J. Pharm. Biomed. Anal.* **44**: 683–700.

35. Izutsu, K., Fujimaki, Y., Kuwabara, A., Hiyama, Y., Yomota, C., and Aoyagi, N. (2006), Near-infrared analysis of protein secondary structure in aqueous solutions and freeze-dried solids, *J. Pharm. Sci.* **95**: 781–789.

36. Bai, S., Nayar, R., Carpenter, J. F., and Manning, M. C. (2005), Noninvasive determination of protein conformation in the solid sated using near infrared (NIR) spectroscopy, *J. Pharm. Sci.* **94**: 2030–2038.

37. Robert, P., Devaux, M. F., Mouhous, N., and Dufour, E. (1998), Monitoring the secondary structure of proteins by near-infrared spectroscopy, *Appl. Spectrosc.* **53**: 226–232.

38. Wen, Z. Q. (2007), Raman spectroscopy of protein pharmaceuticals, *J. Pharm. Sci.* **96**: 2861–2878.

39. Nielsen, O. F. (2005), Raman spectrosocpy, in *Methods for Structural Analysis of Protein Pharmaceuticals*, Jiskoot, W. and Crommelin, D. J. A., eds., AAPS Press, Arlington, VA, pp. 167–198.

40. Strachan, C. J., Rades, T., Gordon, K. C., and Rantanen, J. (2007), Raman spectroscopy for quantitative analysis of pharmaceutical solids, *J. Pharm. Pharmacol.* **59**: 179–192.

41. Forbes, R. T., Barry, B. W., and Elkordy, A. A. (2007), Preparation and characterization of spray-dried and crystallized trypsin: FT-Raman study to detect protein denaturation after thermal stress, *Eur. J. Pharm. Sci.* **30**: 315–323.

42. Schwegman, J. J., Carpenter, J. F., and Nail, S. L. (2007), Infrared microscopy for in situ measurement of protein secondary structure during freezing and freeze-drying, *J. Pharm. Sci.* **96**: 179–195.

43. Wishart, D. (2005), Nuclear magnetic resonance spectrosocpy, in *Methods for Structural Analysis of Protein Pharmaceuticals*, Jiskoot, W. and Crommelin, D. J. A., eds., AAPS Press, Arlington, VA, pp. 199–244.

44. Bugay, D. E. (1993), Solid-state nuclear magnetic resonance spectroscopy: Theory and pharmaceutical applications, *Pharm. Res.* **10**: 317–327.

45. Yannoni, C. S. (1982), High-resolution NMR in solids: The CPMAS experiment, *Acc. Chem. Res.* **15**: 201–208.

46. Hong, M. and Wi, S. (2006), Torsion angle determination in biological solids by solid-state nuclear magnetic resonance, *NMR Spectrosc. Biol. Solids* **34**: 87–122.

47. Gong, X. M., Franzin, C. M., Thai, K., Yu, J., and Marassi, F. M. (2007), Nuclear magnetic resonance structural studies of membrane proteins in micelles and bilayers, *Meth. Mol. Biol.* **400**: 515–529.

48. Naito, A. and Kawamura, I. (2007), Solid-state NMR as a method to reveal structure and membrane-interaction of amyloidogenic proteins and peptides, *Biochim. Biophys. Acta* **1768**: 1900–1912.

49. Hong, M. (2007), Structure, topology, and dynamics of membrane peptides and proteins from solid-state NMR spectroscopy, *J. Phys. Chem.* **111**: 10340–10351.

50. Tycko, R. (2006), Characterization of amyloid structures at the molecular level by solid state nuclear magnetic resonance spectroscopy, *Meth. Enzymol.* **413**: 103–122

51. McDermott, A. E. (2004), Structural and dynamic studies of proteins by solid-state NMR spectroscopy: Rapid movement forward, *Curr. Opin. Struct. Biol.* **14**: 554–561.

52. Tishmack, P. A., Bugay, D. E., and Byrn, S. R. (2003), Solid-state nuclear magnetic resonance spectroscopy–pharmaceutical applications, *J. Pharm. Sci.* **92**(3): 441–474.

53. Tian, F., Middaugh, C. R., Offerdahl, T., Munson, E., Sane, S., and Rytting, J. H. (2007), Spectroscopic evaluation of the stabilization of humanized monoclonal antibodies in amino acid formulations, *Int. J. Pharm.* **335**: 20–31.

54. Lam, Y. H., Bustami, R., Phan, T., Chan, H. K., and Separovic, F. (2002), A solid-state NMR study of protein mobility in lyophilized protein-sugar powders, *J. Pharm. Sci.* **91**: 943–951.

55. Suihko, E. J., Forbes, R. T., and Apperley, D. C. (2005), A solid-state NMR study of molecular mobility and phase separation in co-spray-dried protein-sugar particles, *Eur. J. Pharm. Sci.* **25**: 105–112.

56. Yan, X., Watson, J., Ho, P. S., and Deinzer, M. L. (2004), Mass spectrometric approaches using electrospray ionization charge states and hydrogen-deuterium exchange for determining protein structures and their conformational changes, *Mol. Cell. Proteom.* **3**(1): 10–23

57. French, D. L., Arakawa, T., and Li, T. (2004), Fourier transformed infrared spectroscopic investigation of protein conformation in spray-dried protein/trehalose powders, *Biopolymers* **73**: 524–531.

58. Kaltashov, I. A. and Eyles, S. J. (2002), Studies of biomolecular conformations and conformational dynamics by mass spectrometry, *Mass Spectrom. Rev.* **21**: 37–71.

59. Wales, T. E. and Engen, J. R. (2006), Hydrogen exchange mass spectrometry for the analysis of protein dynamics, *Mass Spectrom. Rev.* **25**: 158–170.

60. Englander, S. W. and Kallenbach, N. R. (1983), Hydrogen exchange and structural dynamics of proteins and nucleic acids, *Q. Rev. Biophys.* **16**: 521–655.

61. Englander, S. W. and Poulsen, A. (1969), Hydrogen-tritium exchange of the random chain polypeptide *Biopolymers* **7**: 379–393.

62. Li, Y., Williams, T. D., and Topp, E. M. (2008), Effects of excipients on protein conformation in lyophilized solids by hydrogen/deuterium exchange mass spectrometry, *Pharm. Res.* **25**: 259–267.

63. Li, Y., Williams, T. D., Schowen, R. L., and Topp, E. M. (2007), Characterizing protein structure in amorphous solids using hydrogen/deuterium exchange with mass spectrometry, *Anal. Biochem.* **1**: 18–20.

64. Gabellieri, E. and Strambini, G. B. (2003), Perturbation of protein tertiary structure in frozen solutions revealed by 1-anilino-8-naphthalene sulfonate fluorescence, *Biophys. J.* **85**: 3214–3220.

65. Ramachander, R., Dimitrova, M., Young, M., Narhi, L., and Jiang, Y. (2006), Solid-state fluorescence to analyze lyophilized protein formulations, *Abstracts 232nd ACS National Meeting*, San Francisco, Sept. 10–14.

66. Sharma, V. K. and Kalonia, D. S., (2003), Steady-state tryptophan fluorescence spectroscopy study to probe tertiary structure of proteins in solid powders, *J. Pharm. Sci.* **92**: 890–899.

67. Prestrelski, S. J., Pikal, K. A., and Arakawa, T. (1995), Optimization of lyophilization conditions for recombinant human interleukin-2 by dried-state conformational analysis using Fourier-transform infrared spectroscopy, *Pharm. Res.* **12**: 1250–1259.

68. Chang, L., Shepherd, D., Sun, J., Ouellette, D., Grant, K. L., Tang, X., and Pikal, M. J. (2005), Mechanism of protein stabilization by sugars during freeze-drying and storage: Native structure preservation, specific interaction and/or immobilization in a glassy matrix? *J. Pharm. Sci.* **94**: 1427–1444.

69. Garzon-Rodriguez, W., Koval, R. L., Chongprasert, S., Krishnan, S., Randolph, T. W., Warne, N. W., and Carpenter, J. F. (2004), Optimizing storage stability of lyophilized recombinant human interleukin-11 with disacchariede/hydroxyethyl starch mixtures, *J. Pharm. Sci.* **93**: 684–696.

70. Abdul-Fattah, A. M., Truong-Le, V., Yee, L., Nguyen, L., Kalonia, D. S., Cicerone, M. T., and Pikal, M. J. (2007), Drying-Induced variations in physicochemical properties of amorphous pharmaceuticals and their impact on stability (I): Stability of a monoclonal antibody, *J. Pharm. Sci.* **96**: 1983–2008.

23

THE IMPACT OF FORMULATION AND DRYING PROCESSES ON THE CHARACTERISTICS AND PERFORMANCE OF BIOPHARMACEUTICAL POWDERS

Vu L. Truong and Ahmad M. Abdul-Fattah

23.1. INTRODUCTION

In simple terms, the solid-state properties of pharmaceutical powders can be divided into delivery and/or stability-impacting properties, as highlighted in Figure 23.1 [1–4]. Additionally, there are solid-state properties that can be indirectly impacted by storage stability or affect delivery such as powder flow properties, powder appearance (e.g., color, cake appearance in a lyophilized cake), reconstitution time, and change in pH after reconstitution. Solid-state properties are impacted by both formulation and processing conditions used to dry the drug (or the "thermal history" to which the drug is subjected through during drying). Chapter 28 reviewed the various drying methods that have been applied to biopharmaceuticals, the kinds of stresses associated with each drying methods that are applied to the drug, and the different solid-state properties that arise from each processing methods. In the present chapter, we examine how formulation and drying-induced variations in some important solid-state properties such as molecular mobility, protein structure, and surface properties can affect the storage stability of biopharmaceuticals in the glassy state.

Formulation and Process Development Strategies for Manufacturing Biopharmaceuticals, edited by Jameel and Hershenson

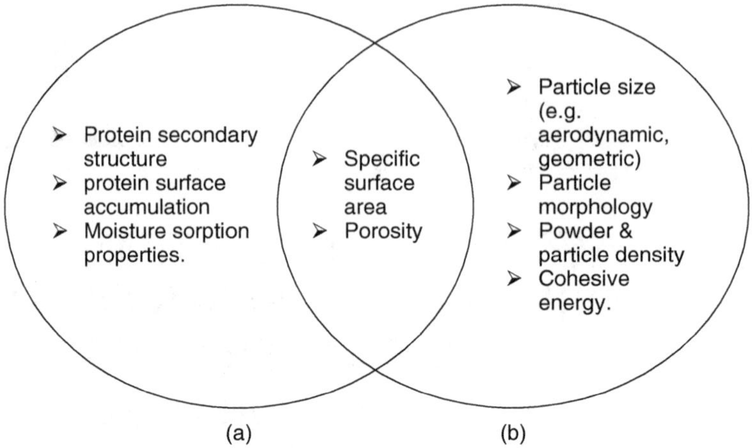

Figure 23.1. Examples of stability- (a) and delivery-impacting (b) solid-state properties.

23.2. THE IMPACT OF FORMULATION ON SOLID-STATE PROPERTIES AND STORAGE STABILITY OF BIOPHARMACEUTICALS

The most important parameter to be considered during formulation is choosing the right type of stabilizer. A suitable stabilizer needs to be compatible with the active biopharmaceutical ingredient (ABI). Pharmaceutical stabilizers that are inert and do not covalently react nor tightly bind to the ABI are typically preferred. Inappropriate selection of stabilizer could lead to unwanted reactions, leading to loss of activity. For example, using lactose as a stabilizer for human growth hormone (hGH) lead to Maillard reactions and activity loss [5]. As a general requirement, the stabilizer needs to be in the same phase as the ABI to achieve long-term storage stability [6,7]. This is also supported by the observations that mannitol + glycine [6–8] or polyethylene glycol (PEG) + dextran [9], when used as sole stabilizers tended to crystallize and phase-separate from the ABT in the formulation, which resulted in poor stabilization. Once the appropriate stabilizer has been selected, other aspects that need to be considered include, but are not limited to, the ratio of stabilizer to active ingredient, the pH and type of buffer system, and the presence or absence of surfactants [7,10–12].

"Stability-impacting" solid-state properties such as moisture adsoprtion/desorption behavior protein structure (secondary and/or tertiary), molecular mobility (global and/or local motions), and surface properties (total surface accumulation of active biopharmaceutical) may change if the excipients and/or excipients:protein ratio are changed. One or more of the stability-impacting solid-state properties usually controls the storage stability of the ABI. Hence, a change in one or more of these properties may alter the storage stability profile of the ABI.

23.2.1. Structure, Moisture Sorption Behavior, and Storage Stability

There is sufficient evidence in literature to support a correlation between preservation of protein secondary strutures (alpha helices, beta sheet s, beta turns, etc.) in the solid state and storage stability [13–16]. According to the water substitution hypothesis, a good stabilizer can form hydrogen bonds at specific sites on the surface of proteins and substitute for the thermodynamic stabilization function of water that is lost during drying. Protein unfolding would therefore be inhibited during the drying process by the addition of sufficient amount of stabilizer, which effectively increases the free energy of denaturation; that is, as water is removed, greater maintenance of protein secondary structure is achieved with the addition of the stabilizer. With greater preservation in secondary structure, better storage stability of the ABI is expected. Secondary structure in the solid state is commonly assessed by Fourier transform infrared spectroscopy (FTIR), and more recently by Raman spectroscopy. With FTIR studies, the spectral correlation coefficient value (r), a semiquantitative measure of the degree of structural preservation, can be determined. The closer r is to 1, the more "native" the secondary structure is [7,10,17]. Pikal et al showed an inverse relationship between initial r value and physical degradation of hGH freeze-dried with different stabilizers, as well as with increasing levels of trehalose [18,19]. The storage stability (physical stability) of Met-hGH freeze-dried with different stabilizers at both 50°C and 60°C also improved as initial r values (i.e., post-freeze drying) approached 1 [20]. Different stabilizers produced a similar outcome with the storage stability of recombinant human interleukin-11 in lyophilized form [21]. Formulations of interleukin-2 (IL-2) freeze-dried from different starting pH values also showed a similar relationship between storage stability and r [13]. There is currently growing speculation on a similar relationship between protein tertiary structure and storage stability in the solid state [22]. The interaction between protein and stabilizer increases up to a certain point, after which saturation of the binding sites on the protein surface is reached. No further thermodynamic stabilization is expected beyond this point [14,18,23]. For example, studies by Cleland et al. [24] show an improvement in the storage stability (chemical and physical) of a recombinant humanized monoclonal antibody (rhuMAb HER2) as a function of stabilizer ratio in the lyophilized formulations. Improvement in storage stability was consistent with an improvement in secondary structure with more added stabilizer, but beyond a certain rhuMAb:stabilizer molar ratio, no further improvement in secondary structure and storage stability of HER2 in the solid state was observed. Similar findings were made by Chang et al. [14].

Among the impacts of stabilizer replacing water-binding sites on the protein surface during dehydration is alteration of the moisture sorption property of the powder [16,25,26]. Studies on powder moisture sorption properties typically report monolayer hydration values (M_0) [27,28]; M_0 is the amount of water necessary to saturate the protein surface with a monolayer by binding to specific polar sites. The presence of a water monolayer inhibits the undesirable interactions between polar groups on adjacent protein molecules, thereby preserving their biological activity. Hence, M_0 is believed by some researchers to be the point for maximum shelf life in the dry state, above which the rate of chemical and/or physical reactions begin to increase

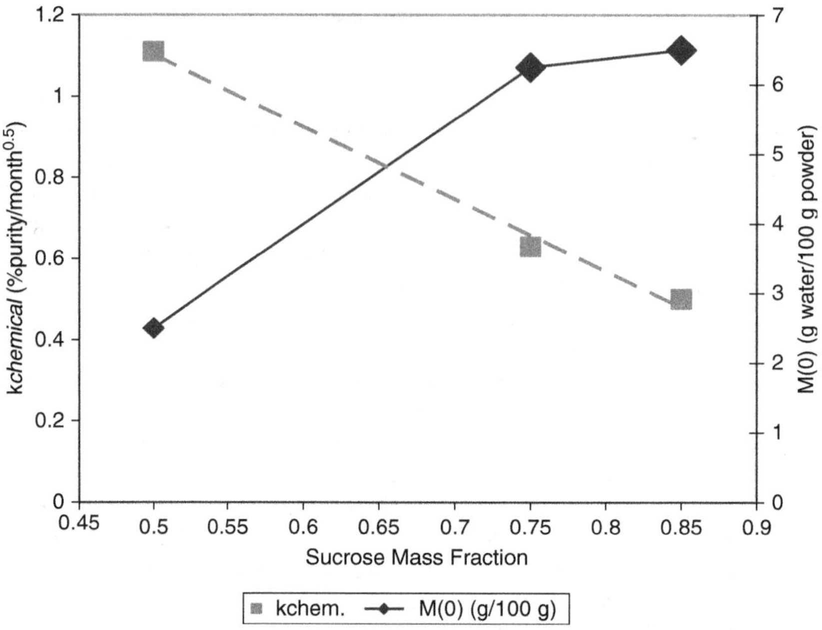

Figure 23.2. Correlation between chemical stability of hGH and M_0 in freeze-dried formulations with different fractions of sucrose. (Figure constructed from data in Constantino et al. [26] and Pikal et al. [29].)

significantly because of greater solubility and mobility of reactants [27,28]. In studies on hGH systems with sucrose and trehalose, M_0 was observed to decrease moderately up to a certain point (50:50), after which an increase was observed [26]. Figure 23.2 shows an improvement in the chemical stability of hGH in different sucrose-based formulations with an increase in M_0, brought about by an increase in the fraction of sucrose in the freeze-dried formulations [26,29]. Moisture levels were less than 1% w/w in these formulations. A similar relationship was observed with physical stability (aggregation rate constants) of hGH in the same formulations, and with both physical and chemical stabilities of hGH in trehalose-based formulations [26,29].

The previous observations with proteins such as hGH indicated that storage stability was a function of both protein secondary structure and M_0. To our knowledge, however, there is no published study showing a clear-cut relationship between all three parameters simultaneously.

23.2.2. Molecular Mobility and Storage Stability

Multiple studies show a strong correlation between molecular mobility in the solid state and storage stability [14,16,18,30–32]. Formulation, moisture content, and thermal treatment all influence molecular mobility in the solid state. Molecular mobility is covered in depth elsewhere in this book. Briefly, molecular mobility has been divided into global motions and local motions (fast dynamics) [18,31–34]. *Global motions* (also known as α *relaxations*) refer to motions that take place over a timescale of seconds or more and involve diffusion and/or rotation of entire molecules or large segments of molecules (i.e., motions related to viscosity of the medium). A good measure of α relaxation is relaxation time (τ) which proportional to the reciprocal of molecular mobility by global motions [18]. *Local motions* (also known as β *relaxations*), on the other hand, refer to motions of side chains and smaller groups within the molecule, and occur over faster timescales (e.g., on the order of nanoseconds). A good measure of β relaxation is the mean-square amplitude of atomic motions $<\mu^2>$, a value measured by incoherent elastic neutron scattering [33–35]. Other reported measures of β relaxation are rotating frame spin–lattice relaxation time ($T_{1\rho}$), a value measured by ^{13}C solid-state nuclear magnetic resonance spectroscopy [36].

It is well known that the glass transition temperature (T_g) is a function of both moisture content and formulation composition [16,37]. The higher the moisture content, and the closer T_g is to the storage temperature (T), the greater the global motions are (Fig. 23.3), thereby enhancing physical and/or chemical degradation during storage (16, 37, 38). Consistent with the glass dynamics theory, several studies have shown that a good correlation exists between moisture content and/or molecular mobility with storage stability. The physical stability of a recombinant humanized monoclonal antibody (rhuMAb E25) spray-dried with different stabilizers was a function of both type

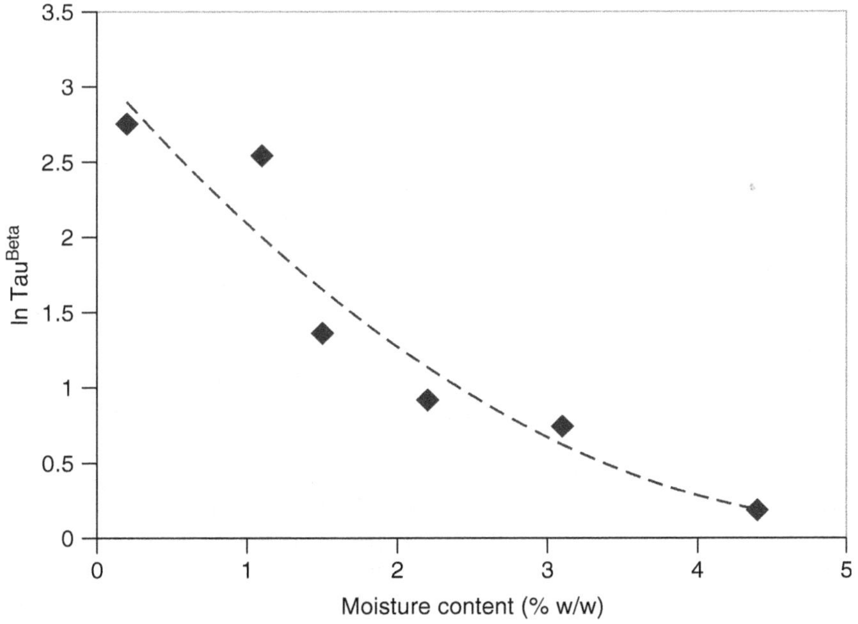

Figure 23.3. Relationship between molecular mobility and moisture content in a representative IgG_1 formulations. (Figure constructed from data in Chang et al. [16].)

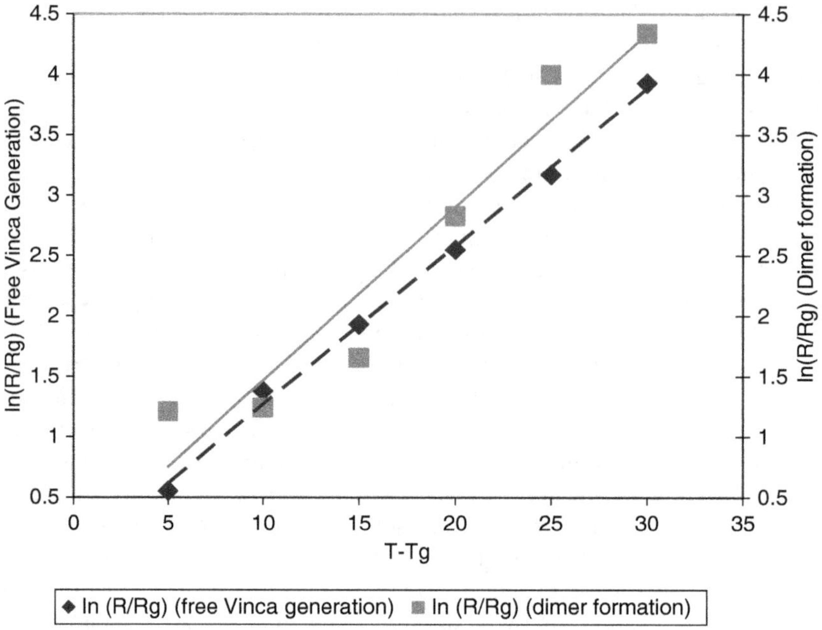

Figure 23.4. Relationship between chemical degradation and $T - T_g$ in a Vinca alkaloid antibody. (Figure constructed from data in Roy et al. [40].)

of stabilizer and moisture content [39]. Roy et al. showed that at various temperatures and moisture contents, the degradation of lyophilized vinca alkaloid-antibody congugate, leading to free vinca or dimer formation, was a function of $T - T_g$ (Fig. 23.4) [40]. Miao et al. found a similar relationship between the rate of nonenzymatic browning reaction (Maillard reaction) and $T - T_g$ at different moisture contents in both freeze-dried and spray-dried systems of lactose/trehalose L-lysine/D-xylose [41]. In other studies, direct measurements of global molecular mobility showed an improvement in the storage stability with an increase in relaxation time (i.e., lower global molecular mobility) [18,19,30,37].

In addition to the dependence of storage stability on global molecular mobility, a few case studies have emerged demonstrating the importance of fast dynamics in storage stability of biopharmaceuticals. Yoshioka et al. (36) showed that the chemical stability of insulin was better in trehalose-based formulations than in dextran-based formulations, owing to faster local dynamics with dextran. It has been suggested that plasticizing glasses by the addition of an appropriate amount of small-molecule plasticizer (such as glycerol or sorbitol) can lead to an improvement in the storage stability of the ABI (while maintaining T below T_g of the formulation). The stabilization mechanism associated with plasticizers is thought to be the coupling of these excipients to the bulk glass surrounding and the dynamic suppression of molecular vibrations [42]. For example, Cicerone et al. [35] demonstrated that the physicochemical stability of

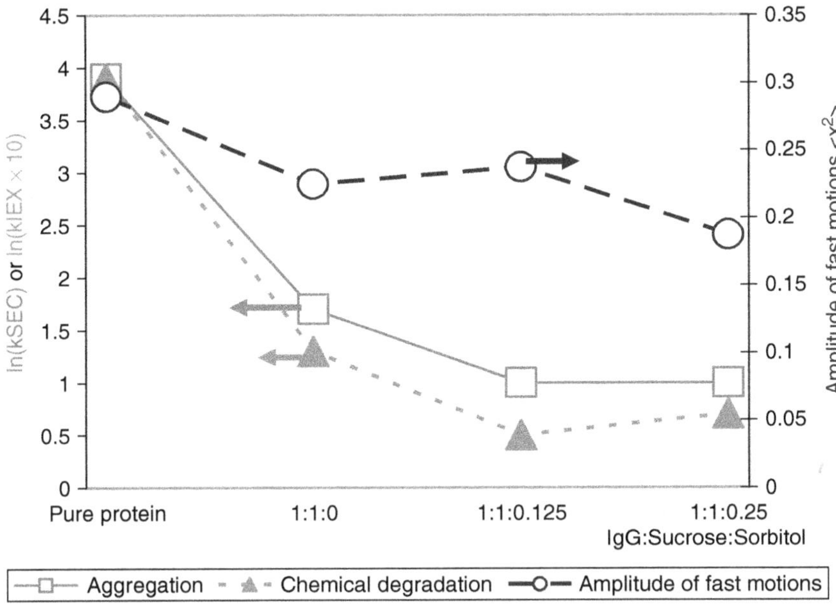

Figure 23.5. Relationship between fast dynamics and storage stability (chemical and physical) of different freeze-dried formulations of IgG_1. (Figure constructed from data in Cicerone et al. [35].)

an IgG in different formulations improved with the addition of small quantities of sorbitol to freeze-dried sucrose-based systems. The addition of appropriate amounts of sorbitol caused a decrease in the mean-square amplitude of atomic motions, that is, a decrease in molecular mobility by fast dynamics (Fig. 23.5). The dynamic suppression of local motions and the correlation to improved storage stability, could suggest that the plasticizer (conceptually) acts as a molecular shock absorber, dampening the vibrations of the sucrose-based excipient matrix, and reducing the physical stresses brought about by fast dynamics.

23.2.3. Correlation of Storage Stability to Other Solid-State Properties

Surface properties of dried formulations also play a role in the storage stability of biopharmaceuticals. A measure of specific surface area (SSA), density, and surface composition of dried powders can provide insight into the accumulation of the active biopharmaceutical ingredient at the powder surface [43]. The most commonly used technique in the pharmaceutical industry for measuring surface composition is *X-ray photoelectron spectroscopy* (XPS) [also known as *electron spectroscopy for chemical analysis* (ESCA)] [44–46]. Other tools have also been used such as auger electron spectroscopy, secondary-ion mass spectroscopy, and fluorescence techniques [47,48].

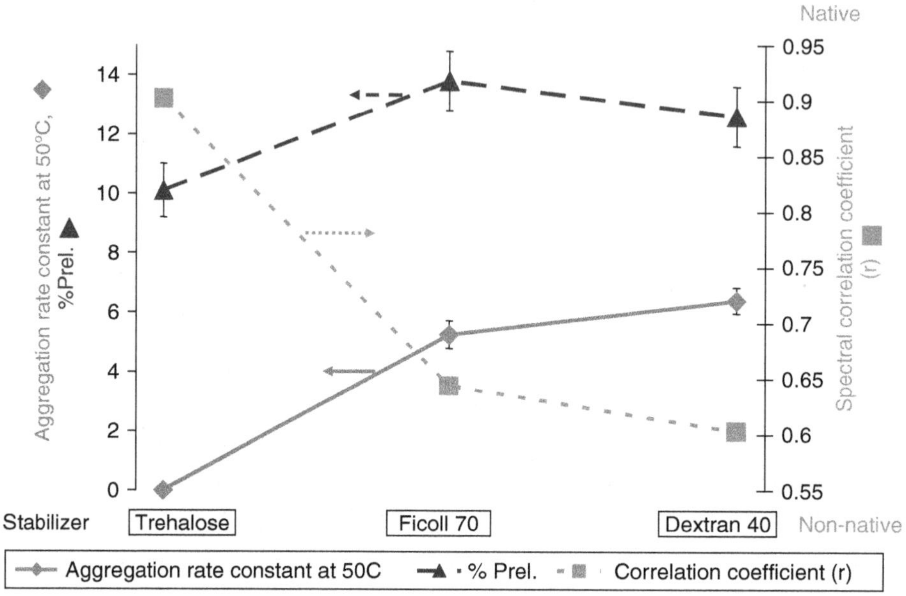

Figure 23.6. Relationship between protein secondary structure, protein surface accumulation, and physical stability of Met-hGH in different freeze-dried formulations. (Figure constructed from data in Abdul-Fatah et al. [20].)

The physical stability of Met-hGH in different formulations was found to improve with lower accumulation of Met-hGH at the surface of the dried powders (Fig. 23.6). The relationship between stability-impacting solid-state properties and storage stability in different amorphous formulations can become more complex because more than one factor may be involved in determining the stability profile of the active biopharmaceutical ingredient. For example in the met-hGH study, where surface enrichment was observed, stability could be improved with more "native" protein secondary structure content. This suggests that neither water substitution theory nor glass dynamics hypothesis alone will suffice for the interpretation of storage stability outcome. A study by Chang et al. [14] showed that the physicochemical stability of an IgG_1 freeze-dried with sucrose improved as the weight fraction of sucrose increased, consistent with a decrease in molecular mobility and a more native protein structure in the solid state (Fig. 23.7). Moreover, the same study done with recombinant human serum albumin (rhSA) showed similar trends as with IgG.

23.3. IMPACT OF DRYING METHOD ON SOLID-STATE PROPERTIES AND STORAGE STABILITY OF BIOPHARMACEUTICALS

It has been shown that not only does the formulation components strongly influence the solid state properties, including stability-impacting properties, but also the drying

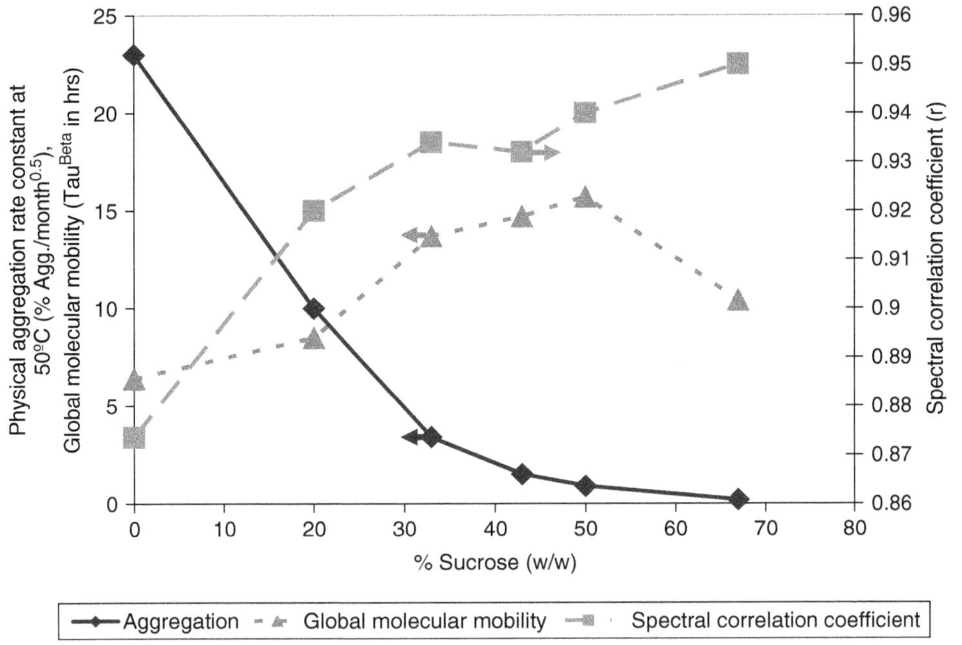

Figure 23.7. Relationship between protein secondary structure, global molecular mobility, and physical stability of an IgG_1 in different freeze-dried formulations. (Figure constructed from data in Chang et al. [14].)

process method that is used to arrive at the dry state [1–4]. However, there are few cases in the literature demonstrating that changes in solid-state properties brought about by changes in thermal history (especially drying method) impact the solid-state stability of the ABI. Causal relationships between thermal history and storage stability have not been fully appreciated because there has been a lack of storage stability data for identical systems dried by different methods as well as a lack of thorough characterization of the systems after drying. One study reports marginally better storage stability of interferon alpha-2a produced by precipitation followed by vacuum drying, as opposed to the same formulation freeze-dried [49]. However, there were insufficient comparative solid-state characterization data such as moisture content or protein secondary structures to support causal relationships [49]. Another study reports superior storage stability of a recombinant humanized monoclonal antibody (rhuMAb) in freeze-dried formulations compared to their spray-dried counterpart preparations [50]. No differences were observed between the secondary structure of rhuMAb formulations prepared by spray drying or freeze drying, as studied by Raman spectroscopy. However, no further attempt was made at exploring other factors that may lead to such differences. It can be argued that the storage stability profile of the ABI in a formulation dried by different methods can differ depending on a number of factors such as drying stresses encountered during the drying process, thermal history during the drying process (e.g., a high-temperature annealing step after drying), polyamorphism, and phase separation. To understand the underlying factor(s) involved, we shall further review the stability-impacting solid-state properties that are shaped by the drying methods used to manufacture the dry powder.

Several studies have shown that both spray drying and vacuum drying produce powders with a higher moisture content than does freeze drying [51,52]. Therefore unless an additional (secondary) drying step is employed to further reduce the final moisture content, the differences in storage stability profiles may be driven mainly by differences in moisture content. Whether differences exist in moisture sorption behavior due to drying methods remains a topic of debate, as is the implications of these differences, if any, on storage stability. One study shows little difference between moisture sorption isotherms of a freeze-dried system of hGH with trehalose and surfactant and its equivalent spray-dried preparation. However, the same study shows greater differences in moisture sortion behavior due to formulation (e.g., use of different stabilizers, presence or absence of surfactant) (Fig. 23.8) [53]. A study by Miao et al. shows that spray-dried preparations had higher M_0 values (as modeled by both GAB and BET isotherms) than did their equivalent freeze-dried preparations in different systems (lactose/reactant, lactose/trehalose/reactant, and trehalose/reactant)

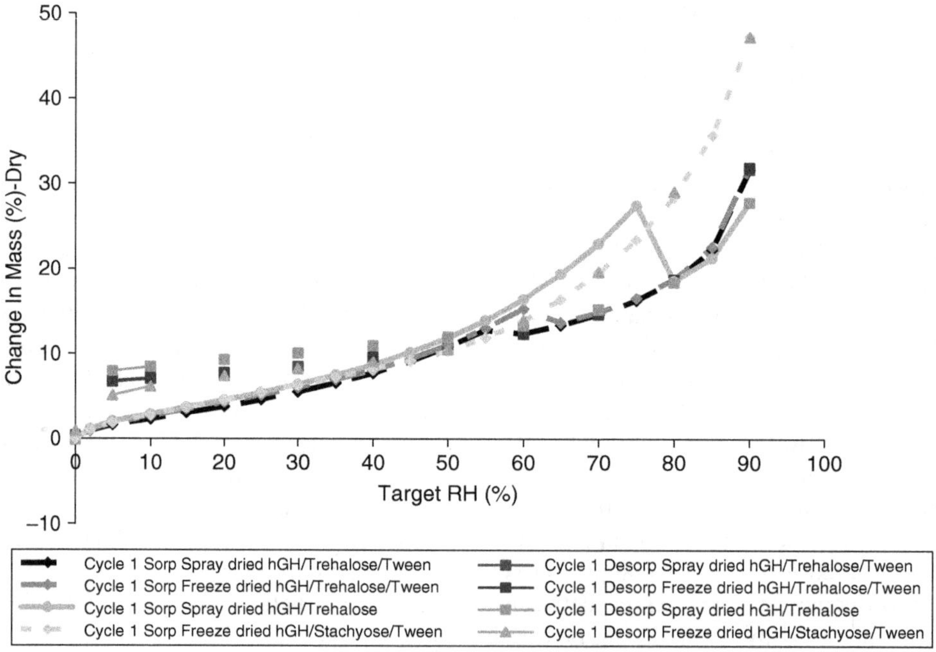

Figure 23.8. Differences between moisture sorption isotherms of spray-dried and freeze-dried formulations of hGH in trehalose or stachyose matrices with or without surfactant (Tween) [53], Abdul-Fattah et al., unpublished data.

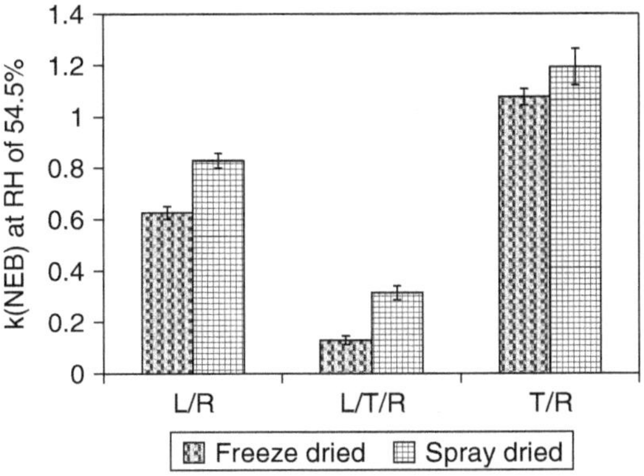

Figure 23.9. Rate of nonenzymatic browning (NEB) reaction in freeze-dried and spray-dried systems of lactose/reactant (L/R), lactose/trehalose/reactant (L/T/R), and trehalose/reactant (T/R) stored at a RH of 54.5%. (Figure constructed from data in Miao and Ross [54].)

[54]. Furthermore, the rate of nonenzymatic browning in the three systems (irrespective of spray-dried or freeze-dried) equilibrated at different relative humidity (RH) levels was lower in freeze-dried powders at all RH values (33.2%, 44.1%, 54.5%, and 65.6%), and independent of T_g (Fig. 23.9).

Different drying methods can result in powders with significantly different specific surface areas (SSAs) [44,45,55,56]. Also, different drying methods can result in powders with significantly different surface composition [44,45,56]. It is expected that powders with very high SSA and high composition heterogeneity will be inferior in storage stability, and the theoretical basis behind such relationship is explained in detail elsewhere [20,43]. In a study by Scherer et al. [57], an optimized formulation developed for a vaccine against influenza was dried by freeze drying, spray drying, and foam drying. Storage stability of the vaccine monitored at 37°C showed that the vaccine was most stable in foam dried powder and least stable in the spray-dried powder. Although extensive characterization studies were not performed, SSA measurements showed that the spray-dried powder had the highest SSA among all powders, while the foam-dried powder had the lowest SSA (Fig. 23.10). Webb et al. determined a rough correlation between the total surface accumulation of recombinant human interferon-γ (rhIFNγ) in a trehalose matrix and its storage stability (physical), as a function of both formulation and drying method [spray–freeze drying (SFD) and freeze drying] [58]. Freeze-dried preparations were more stable than their equivalent SFD preparations, consistent with a lower SSA and total rhIFNγ surface accumulation.

In a recent study to address the drying method-induced variation in surface heterogeneity and how it impacts storage stability, we dried a model vaccine with and without surfactant using spray drying, foam drying, and freeze drying with & without annealing [59]. BET and ESCA were used to measure the specific surface area (SSA) and the distribution of the vaccine on the solids' surface (as indicated by the nitrogen atom ['N'] content), respectively Figure 23.11. In the absence of surfactant,

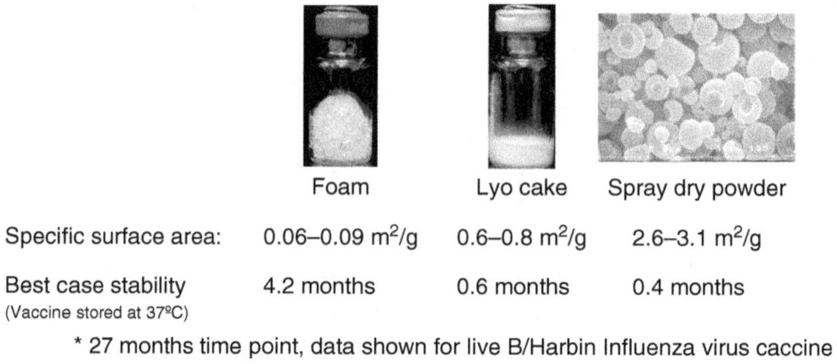

Figure 23.10. Different drying methods applied to the same formulation resulted in different solid-state structures and properties [57] Truong et al., unpublished data.

the spray-dried powder showed highest SSA, while the foam-dried powder showed extremely low SSA, with intermediate SSA values among freeze-dried powders. A high SSA combined with high percent surface N on the powder surface, indicates higher enrichment of vaccine on the surface of the spray-dried powder. On addition of surfactant, there was a significant drop in both SSA and percent surface N of most preparations (except freeze-dried powder with an annealing step after the freezing step). Foam-dried powder with surfactant remained lowest in terms of SSA. Overall, the foam-dried powder with surfactant had the lowest vaccine surface coverage (nearly none) as evidenced by the absence of N surface signal by ESCA. Storage stability studies with surfactant-free powders at both 25°C and 37°C showed that foam-dried powder had stability superior to that of the other powders. On addition of surfactant, storage stability of the vaccine improved significantly with both foam- and spray-dried powders. The same trend, however, was not observed with freeze drying. Perhaps this result may be due to some sort of structural damage caused by ice to the vaccine at the ice–water interface, which a surfactant failed to protect against. Overall, foam dried powder with surfactant was the most stable powder in all preparations studied. Thus this study shows a good correlation between storage stability of the vaccine and surface heterogeneity triggered by different drying methods and formulation, except when an ice–water interface is involved (Fig. 23.12).

A more extensive study [60] was performed in which sucrose-based formulations of an IgG_1 were dried by the same drying methods used with the vaccine study above. Formulations were made with different protein-to-sucrose ratios in the presence and absence of surfactant. For the sake of simplicity, we will discuss important findings for two surfactant-free systems that we will designate as stabilizer-rich (protein:sucrose ratio 1:4) and protein-rich (protein:sucrose ratio 4:1) systems. The dried formulations were examined for protein secondary structure, surface properties, and molecular mobility by both global dynamics and fast dynamics. Additionally, accelerated stability studies were performed to measure physical aggregation rate by size exclusion chromatography. The findings for both systems are summarized as follows:

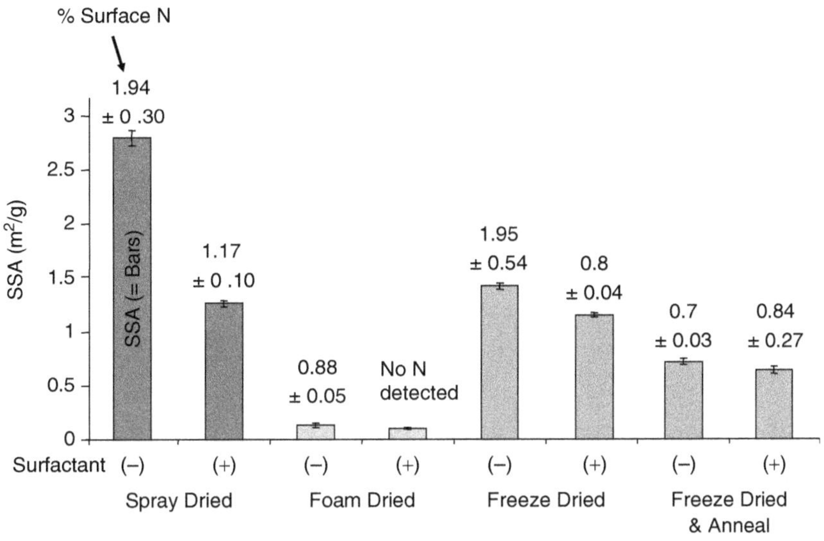

Figure 23.11. Surface heterogeneity of two formulations for a vaccine of PIV3 virus dried by different methods (colored bars represent specific surface area, while values above colored bars represent percentage of surface N (Nitrogen) as measured by ESCA). (Figure constructed from data in Abdul-Fatah et al. [59].)

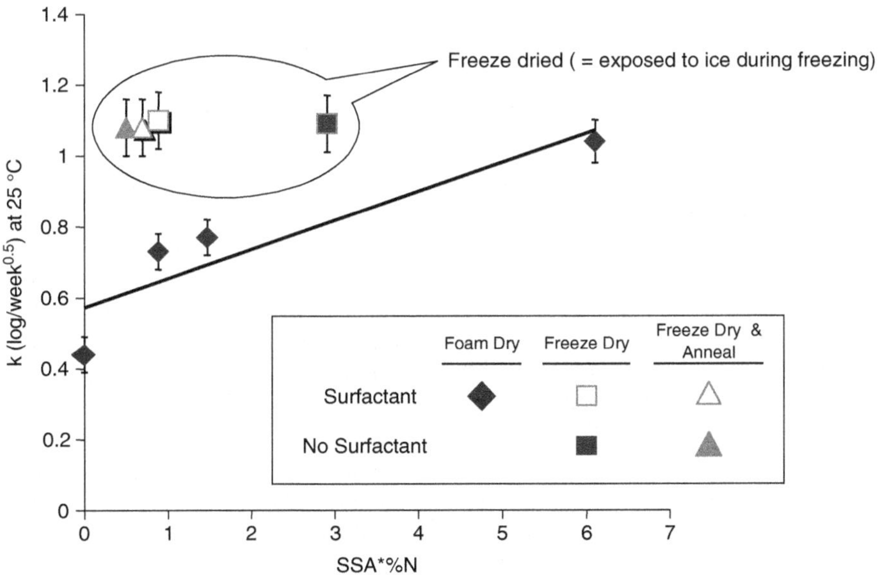

Figure 23.12. Relevance of specific surface area and surface composition to PIV3 vaccine stability at 25°C. A similar trend was observed at 37°C. (Figure constructed from data in Abdul-Fatah et al. [59].)

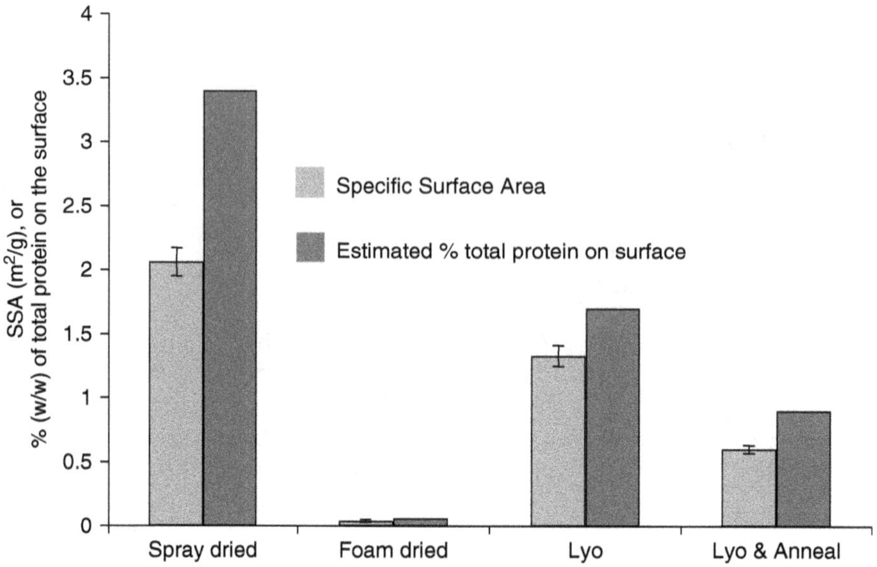

Figure 23.13. Specific surface area (SSA) and surface composition in a 1:4 system of IgG1:sucrose. (Figure constructed from data in Abdul-Fatah et al. [60].)

1. The FTIR data showed that in the stabilizer-rich system, the foam-dried powder experienced the greatest perturbation in secondary structure, followed by freeze-dried, followed by spray-dried powders. Therefore, the FTIR data would project a superior storage stability for spray-dried powder and poor storage stability for the foam. As was observed earlier with the vaccine study both SSA and total protein surface accumulation were highest with spray drying and extremely low with the foam (Fig. 23.13). No significant differences in global molecular mobility, as well as amplitude of rapid motions, were observed between spray-dried and freeze-dried powders. However, global motions dramatically slowed down and the amplitude of fast motions decreased with the foam powder (Fig. 23.14), thus projecting superior storage stability of the foam to all other powders in the 1 : 4 formulation. Such an outcome contradicts the expected storage stability outcome from solid-state FTIR studies. Storage stability studies at 50°C showed superior stability of foam compared to all other powders, consistent with a lower global and molecular mobility as determined by neutron scattering and TAM, respectively (Fig. 23.15). Moreover, in this system there was a rough correlation of storage stability to total protein surface accumulation (Fig. 23.16). Overall, foam drying resulted in lowest SSA, heterogeneity, global and local dynamics, and the best stability.
2. In the protein-rich system, both freeze-dried formulations experienced a greater perturbation in secondary structure, a result that projects poor storage stability with freeze drying. Total protein surface accumulation, SSA, and global

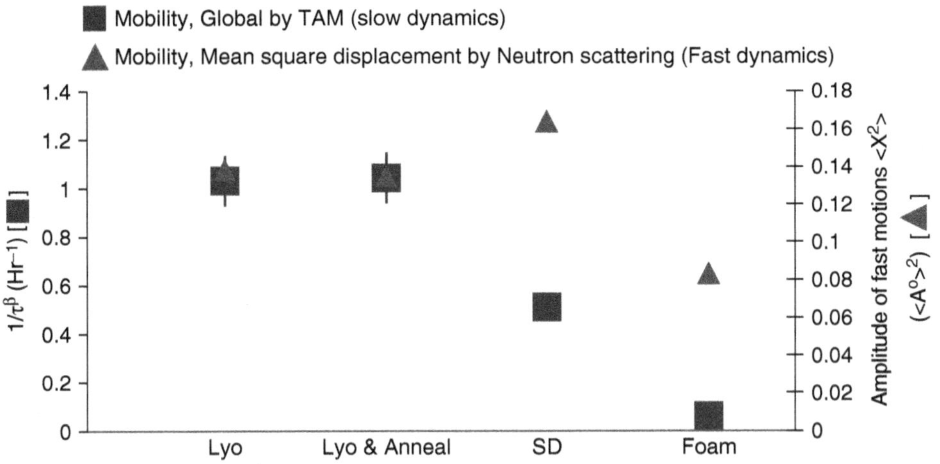

Figure 23.14. Molecular mobility at 50°C in a 1:4 system of IgG1:sucrose. (Figure constructed from data in Abdul-Fatah et al. [60].)

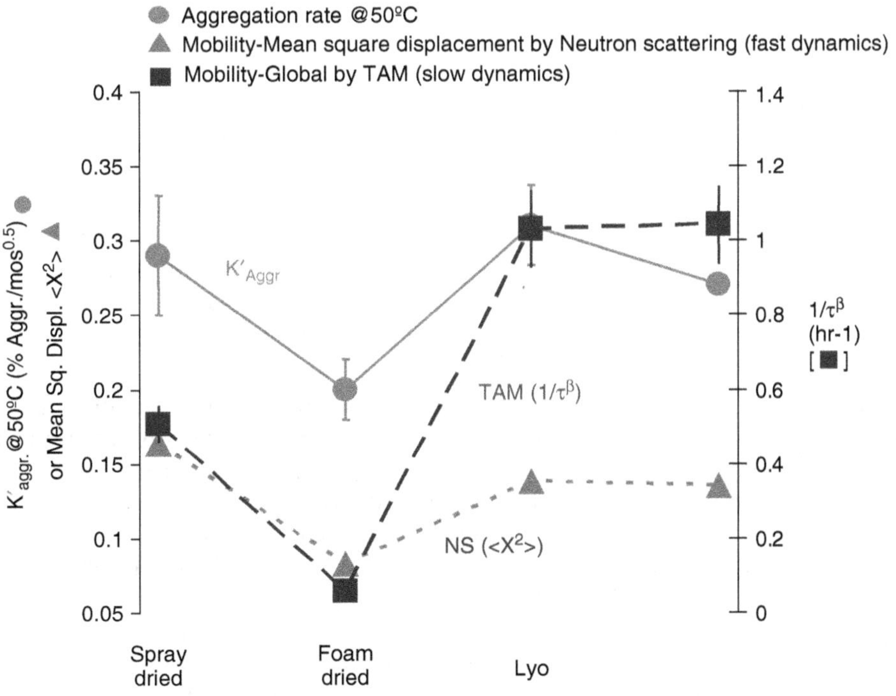

Figure 23.15. Correlation of storage stability at 50°C to molecular mobility in a 1:4 system of IgG1:sucrose. (Figure constructed from data in Abdul-Fatah et al. [60].)

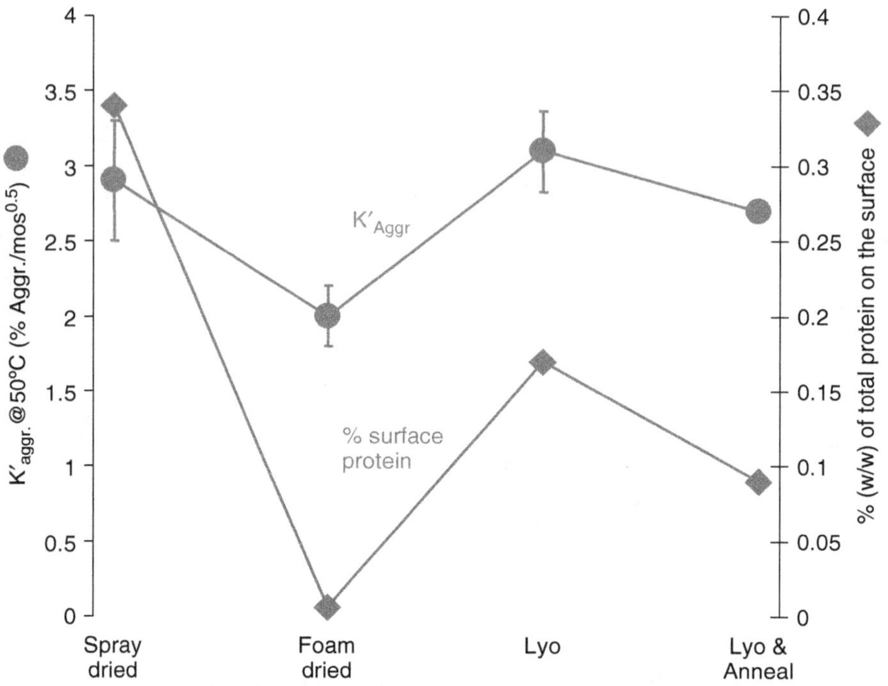

Figure 23.16. Correlation of storage stability at 50°C to surface composition in a 1:4 system of IgG1:sucrose. (Figure constructed from data in Abdul-Fatah et al. [60].)

molecular mobility all did not significantly differ by drying method in the protein-rich system. Storage stability studies at 50°C showed superior stability of foam and spray-dried powders compared to freeze dried powders, consistent with a more native-like protein structure in the solid state (Fig. 23.17).

In summary, for the stabilizer-rich system, chemical stability in the solid state did *not* correlate well with protein secondary structure but correlated reasonably well with molecular mobility (especially fast dynamics) and protein surface accumulation. Conversely, for the protein-rich system, only the solid-state secondary structure correlated well with physical stability. More findings from this study showed a good correlation between chemical stability and total protein surface accumulation in a stabilizer-rich system with surfactant prepared by different drying methods. The observation that preferential enrichment of protein on the surface of the spray dried IgG1 and the vaccine powder correlated with poorer stability further illustrates the impact of process method on solid-state properties and storage stability. Collectively, the data showed that given the same formulation that is dried by different process methods, the differences in solid-state properties can be significant, and can profoundly impact storage stability.

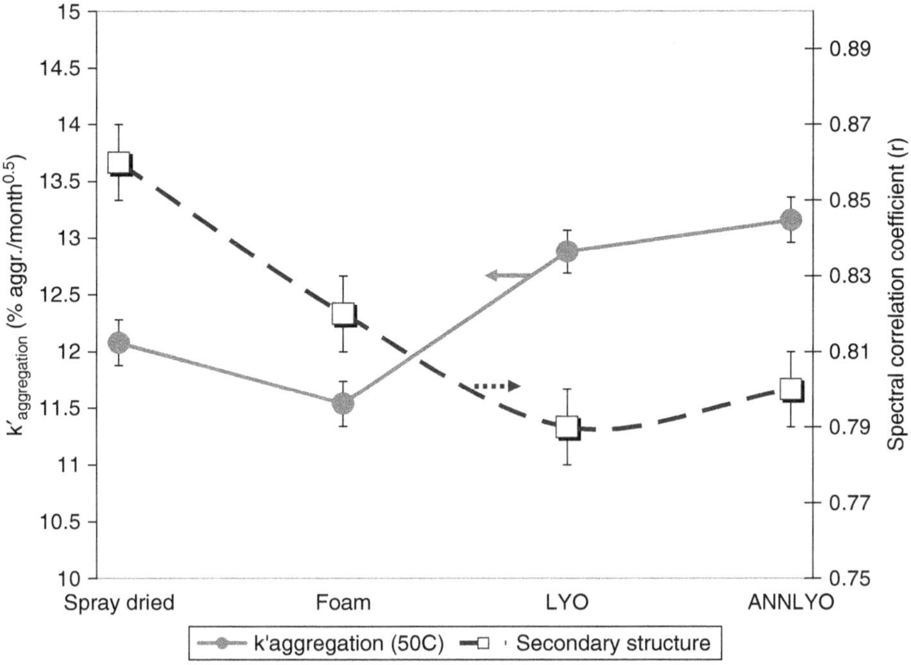

Figure 23.17. Correlation of storage stability at 50°C to protein secondary structure in a 4:1 system of IgG1:sucrose. (Figure constructed from data in Abdul-Fatah et al. [60].)

23.4. THE FUTURE—OPTIMIZING DRYING METHOD AND FORMULATION TO OPTIMIZE SOLID-STATE PROPERTIES

The data suggest that the drying method does impact storage stability, according to the type of formulation. In protein-rich systems protein structure may be the more influential determinant of storage stability, whereas molecular mobility and/or composition heterogeneity may have a greater effect on storage stability in stabilizer-rich systems and in systems with surfactants. However, it appears that formulation composition has a more profound effect on storage stability than does thermal history by drying. More research needs to be conducted in this arena to strengthen this conclusion with a broader range of biopharmaceuticals and drying methods. Overall, storage stability of biopharmaceuticals can be maximized by optimizing conditions of formulation as well as thermal history.

REFERENCES

1. Hickey, A. J., Mansour, H., Telko, M. J., Xu, Z., Smyth, H. D., Mulder, T., McLean, R., Langridge, J., and Papadopoulos, D. (2007), Physical characterization of component particles included in dry powder inhalers. I. Strategy review and static characteristics, *J. Pharm. Sci.* **96**: 1282–1301.

2. Hickey, A. J., Mansour, H., Telko, M. J., Xu, Z., Smyth, H. D., Mulder, T., McLean, R., Langridge, J., and Papadopoulos, D. (2007), Physical characterization of component particles included in dry powder inhalers. II. Dynamic characteristics, *J. Pharm. Sci*. **96**: 1302–1319.
3. Okamoto, H., Todo, H., Iida, K., and Danjo, K. (2002), Dry powders for pulmonary delivery of peptides and proteins, *Kona* **20**: 71–83.
4. Liu, J. (2006), Physical characterization of pharmaceutical formulations in frozen and freeze-dried solid states: Techniques and applications in freeze-drying development, *Pharm. Devel. Technol*. **11**: 3–28.
5. Pikal, M. J., Dellerman, K., and Roy, M. L. (1991), Formulation and stability of freeze-dried proteins: Effects of moisture and oxygen on the stability of freeze-dried formulations of human growth hormone, *Devel. Biol. Stand*. **74**: 21–38.
6. Akers, M., Milton, N., Byrn, S. R., and Nail, S. L. (1995), Glycine crystallization during freezing: The effects of salt form, pH, and ionic strength, *Pharm. Res*. **12**: 1457–1461.
7. Carpenter, J. F., Pikal, M. J., and Randolph, T. W. (1997), Rational design of stable lyophilized protein formulations: Some practical advice, *Pharm. Res*. **14**: 969–975.
8. Lu, X. and Pikal, M. J. (2004), Freeze-drying of mannitol-trehalose-sodium chloride-based formulations: The impact of annealing on dry layer resistance to mass transfer and cake structure, *Pharm. Devel. Technol*. **9**: 85–95.
9. Randolph, T. W. (1997), Phase separation of excipients during lyophilization: Effects on protein stability, *J. Pharm. Sci*. **86**: 1198–1203.
10. Carpenter, J., Chang, B., Rodriguez, W., and Randolph, T., (2002), Rational design of stable lyophilized protein formulation: Theory and practice, in Carpenter, J. F. and Manning, M. C., eds. *Pharmaceutical Biotechnology, Volume 13*, Kluwer/Plenum, New York, pp. 109–127.
11. Shalaev, E. Y., Johnson-Elton, T. D., Chang, L., and Pikal, M. J. (2002), Thermophysical properties of pharmaceutically compatible buffers at sub-zero temperatures: Implications for freeze-drying, *Pharm. Res*. **19**: 195–201.
12. Szkudlarek, B. (1996), pH changes of phosphate buffer solutions during freezing and their influence on the stability of a model protein, lactate dehydrogenase, *Book of Abstracts, 211th ACS Natl. Meeting*, New Orleans, March 24–28, 1996, p. BIOT-138.
13. Prestrelski, S. J., Pikal, K. A., and Arakawa, T. (1995), Optimization of lyophilization conditions for recombinant human interleukin-2 by dried-state conformational analysis using Fourier-transform infrared spectroscopy, *Pharm. Res*. **12**: 1250–1259.
14. Chang, L., Shepherd, D., Sun, J., Tang, X., and Pikal, M. J. (2005), Mechanism of protein stabilization by sugars during freeze-drying and storage: Native structure preservation, specific interaction, and/or immobilization in a glassy matrix? *J. Pharm. Sci*. **94**: 1427–1444.
15. Klibanov, A. and Schefiliti, J. (2004), On the relationship between conformation and stability in solid pharmaceutical protein formulations, *Biotechnol. Lett*. **26**: 1103–1106.
16. Chang, L., Shepherd, D., Sun, J., Tang, X., and Pikal, M. J. (2005), Effect of sorbitol and residual moisture on the stability of lyophilized antibodies: Implications for the mechanism of protein stabilization in the solid state, *J. Pharm. Sci*. **94**: 1445–1455.
17. Kendrick, B. S., Dong, A., Allison, S. D., Manning, M. C., and Carpenter, J. F. (1996), Quantitation of the area of overlap between second-derivative amide I infrared spectra to determine the structural similarity of a protein in different states, *J. Pharm. Sci*. **85**: 155–158.

18. Pikal, M. J. (2007), Mechanisms of protein stabilization during freeze-drying and storage: The relative importance of thermodynamic stabilization and glassy state relaxation dynamics, in *Drugs and the Pharmaceutical Sciences: Freeze-Drying/Lyophilization of Pharmaceutical and Biological Products*, Vol. 137, Koval, R. L. and Pikal, M. J., eds., Informa Healthcare, New York, pp. 63–107.

19. Pikal, M. J. (2002), Chemistry in solid amorphous matrices: Implication for biostabilization, in *Amorphous Food and Pharmaceutical Systems*, Levine, H., ed., The Royal Society of Chemistry, Cambridge, UK, pp. 257–272.

20. Abdul-Fattah, A. M., Lechuga-Ballesteros, D., Kalonia, D., and Pikal, M. J. (2008), The impact of drying method and formulation on the physical properties and stability of methionyl human growth hormone in the amorphous solid state, *J. Pharm. Sci*. **97**: 163–184.

21. Garzon-Rodriguez, W., Koval, R. L., Chongprasert, S., Krishnan, S., Randolph, T. W., Warne, N. W., and Carpenter, J. F. (2004), Optimizing storage stability of lyophilized recombinant human interleukin-11 with disaccharide/hydroxyethyl starch mixtures, *J. Pharm. Sci*. **93**: 684–696.

22. Kalonia, D. and Sharma, V., (2003), Steady-state tryptophan fluorescence spectroscopy study to probe tertiary structure of proteins in solid powders, *J. Pharm. Sci*. **92**: 890–899.

23. Franks, F., Hatley, R., and Mathias, S. (1991), Materials science and the production of shelf-stable biologicals, *BioPharm* **4**(9): 38–55 and the article itself is on pages 38, 40–42, 55.

24. Cleland, J. L., Lam, X., Kendrick, B., Yang, J., Yang, T. H., Overcashier, D., Brooks, D., Hsu, C., and Carpenter, J. F. (2001), A specific molar ratio of stabilizer to protein is required for storage stability of a lyophilized monoclonal antibody, *J. Pharm. Sci*. **90**: 310–321.

25. Newman, A., Reutzel-Edens, S. and Zografi, G. (2007), Characterization of the hygroscopic properties of active pharmaceutical ingredients, *J. Pharm. Sci*. **97**: 1047–1059.

26. Costantino, H. R., Curley, J. G., Wu, S., and Hsu, C. C. (1998), Water sorption behavior of lyophilized protein-sugar systems and implications for solid-state interactions, *Int. J. Pharm*. **166**: 211–221.

27. Belland, L. and Labuza, T. (2000), *Moisture Sorption: Practical Aspects of Isotherm Measurement and Use*, 2nd ed., C.H.I.P.S., Weimar, TX.

28. Sun, W. Q. (2002), *Methods for the Study of Water Relations Under Desiccation Stress*, CABI Publishing, Cambridge, MA.

29. Pikal, M. J., Rigsbee, D., Roy, M., Galbreath, D., Kovach, K., Wang, B., Carpenter, J., and Cicerone, M. T. (2008), Solid state chemistry of proteins: II. The correlation of storage stability of freeze-dried human growth hormone (hGH) with structure and dynamics in the glassy solid, *J. Pharm. Sci*. **97**: 5106–5121.

30. Yoshioka, S. and Aso, Y. (2007), Correlations between molecular mobility and chemical stability during storage of amorphous pharmaceuticals, *J. Pharm. Sci*. **96**: 960–981.

31. Pikal, M. J., Reddy, R. D., Shalaev, E. Y., and Ziegler, C. B. (2005), *Method of Stabilizing Disordered Cefovecin Sodium Salt Using Thermal Annealing*, Pfizer Products Inc., Patent application WO, 2005.

32. Shalaev, E. and Zografi, G. (2002), The concept of "structure" in amorphous solids from the perspective of the pharmaceutical sciences, *J. Roy. Soc. Chem*. (special publication) **281**: 11–30.

33. Cicerone, M. T. and Soles, C. L. (2004), Fast dynamics and stabilization of proteins: Binary glasses of trehalose and glycerol, *Biophys. J*. **86**: 3836–3845.
34. Cicerone, M. T., Tellington, A., Trost, L., and Sokolov, A. (2003), Substantially improved stability of biological agents in dried form: The role of glassy dynamics in preservation of biopharmaceuticals, *BioProcess Int*. **1**(1): 36–47 and the article itself is on pages 36–38, 40, 42, 44, 46–47.
35. Cicerone, M. T., Soles, C. L., Chowdhuri, Z., Pikal, M. J., and Chang, L. (2005), Fast dynamics as a diagnostic for excipients in preservation of dried proteins, *Am. Pharm. Rev*. **8**(6): 22–27 and the article itself is on pages 22, 24–27.
36. Yoshioka, S., Miyazaki, T., and Aso, Y. (2006), b-Relaxation of insulin molecule in lyophilized formulations containing trehalose or dextran as a determinant of chemical reactivity, *Pharm. Res*. **23**: 961–966.
37. Yoshioka, S., Aso, Y., and Kojima, S. (1996), Determination of molecular mobility of lyophilized bovine serum albumin and gamma-globulin by solid-state 1H NMR and relation to aggregation susceptibility, *Pharm. Res*. **13**: 926–930.
38. Sarciaux, J. M. and Hageman, M. J. (1997), Effects of bovine somatotropin (rbSt) concentration at different moisture levels on the physical stability of sucrose in freeze-dried rbSt/sucrose mixtures, *J. Pharm. Sci*. **86**(3): 365–371.
39. Andya, J., Maa, Y. F., Costantino, H. R., Nguyen, P., Dasovich, N., Sweeney, T., Hsu, C., and Shire, S. (1999), The effect of formulation excipients on protein stability and aerosol performance of spray-dried powders of a recombinant humanized anti-IgE monoclonal antibody, *Pharm. Res*. **16**: 350–358.
40. Roy, M. L., Pikal, M. J., Rickard, E. C., and Maloney, A. M. (1992), The effects of formulation and moisture on the stability of a freeze-dried monoclonal antibody-vinca conjugate: A test of the WLF glass transition theory, *Devel. Biol. Stand*. **74**: 323–339; discussion on p. 340.
41. Miao, S. and Roos, Y. H. (2004), Comparison of nonenzymatic browning kinetics in spray-dried and freeze-dried carbohydrate-based food model systems, *J. Food Sci*. **69**: e322–e331.
42. Cicerone, M. T., Tellington, A., Trost, L., and Sokolov, A. (2003), *Plasticized Hydrophilic Glasses for Improved Stabilization of Biological Agents*, Patent no. for the application is WO2003035827 USA, 2002.
43. Abdul-Fattah, A. M., Kalonia, D. S., and Pikal, M. J. (2007), The challenge of drying method selection for protein pharmaceuticals: Product quality implications. *J. Pharm. Sci*. **96**: 1886–1916.
44. Millqvist-Fureby, A., Burns, N., Landstrom, K., Faldt, P., and Bergenstahl, B. (1999), Surface activity at the air-water interface in relation to surface composition of spray-dried milk protein-stabilized emulsions, *J. Roy. Soc. Chem*. (special publication) **227**: 236–245.
45. Millqvist-Fureby, A., Elofsson, U., and Bergenstahl, B. (2001), Surface composition of spray-dried milk protein-stabilised emulsions in relation to pre-heat treatment of proteins, *Colloids Surf. B Biointerfaces* **21**: 47–58.
46. Webb, S. D., Cleland, J. L., Carpenter, J. F., Randolph, T. W. (2003), Effects of annealing lyophilized and spray-lyophilized formulations of recombinant human interferon-gamma, *J. Pharm. Sci*. **92**: 715–729.
47. Landström, K., Alsins, J., and Bergenstahl, B. (1999), A fluorescence method for quantitative measurements of specific protein at powder surfaces, *Colloids Surf. B Biointerfaces* **12**: 429–440.

48. Landström, K., Alsins, J., and Bergenstahl, B. (2000), Competitive protein adsorption between bovine serum albumin and -lactoglobulin during spray-drying, *Food Hydrocoll.* **14**: 75–82.

49. Sharma, V. and Kalonia, D. S. (2004), Polyethylene glycol-induced precipitation of interferon alpha-2a followed by vacuum drying: Development of a novel process for obtaining a dry, stable powder, *AAPS PharmSci*. **6**(1): 31–44.

50. Sane, S. U., Wong, R., and Hsu, C. C. (2004), Raman spectroscopic characterization of drying-induced structural changes in a therapeutic antibody: Correlating structural changes with long-term stability, *J. Pharm. Sci*. **93**: 1005–1018.

51. Pisal, S., Wawde, G., Salvankar, S., Lade, S., and Kadam, S. (2006), Vacuum foam drying for preservation of LaSota virus: Effect of additives, *AAPS PharmSciTech*. **7** (article 60).

52. Maa, Y. F., Nguyen, P., Andya, J., Dasovich, N., Sweeney, T., Shire, S. J., and Hsu, C. (1998), Effect of spray drying and subsequent processing conditions on residual moisture content and physical/biochemical stability of protein inhalation powders, *Pharm. Res*. **15**: 768–775.

53. Abdul-Fattah, A. M., Lechuga-Ballesteros, D., and Miller, D. P. (2003), Unpublished data, Nektar Therapeutics, Inc.

54. Miao, S. and Roos, Y. H. (2006), Isothermal study of nonenzymatic browning kinetics in spray-dried and freeze-dried systems at different relative vapor pressure environments, *Innov. Food Sci. & Emerging Technol*. **7**: 182–194.

55. Maa, Y., Nguyen, P., Sweeney, T., Shire, S. J., and Hsu, C. C. (1999), Protein inhalation powders: Spray drying vs spray freeze drying, *Pharm. Res*. **16**: 249–254.

56. Yu, Z., Johnston, K. P., and Williams, R. O. (2006), Spray freezing into liquid versus spray-freeze drying: Influence of atomization on protein aggregation and biological activity, *Eur. J. Pharm. Sci*. **27**: 9–18.

57. Scherer, T., Yee, L., Pan, E., Pham, B., and Truong-Le, Vu. (2002), Unpublished data, MedImmune Vaccines.

58. Webb, S. D., Golledge, S. L., Cleland, J. L., Carpenter, J. F., and Randolph, T. W. (2002), Surface adsorption of recombinant human interferon-g in lyophilized and spray-lyophilized formulations, *J. Pharm. Sci*. **91**: 1474–1487.

59. Abdul-Fattah, A. M., Truong-Le, Vu., Yee, L., Pan, E., Aso, Y., Kalonia, D. S., and Pikal, M. J. (2007), Drying-induced variations in physico-chemical properties of amorphous pharmaceuticals and their impact on stability II: Stability of a vaccine, *Pharm. Res*. **24**: 715–727.

60. Abdul-Fattah, A. M., Truong-Le, Vu., Yee, L., Nguyen, L., Kalonia, D. S., Cicerone, M. T., and Pikal, M. J. (2007), Drying-induced variations in physico-chemical properties of amorphous pharmaceuticals and their impact on stability (I): Stability of a monoclonal antibody, *J. Pharm. Sci*. **96**: 1983–2008.

PART IV

MANUFACTURING SCIENCES

24

MANUFACTURING FUNDAMENTALS FOR BIOPHARMACEUTICALS

Maninder Hora

24.1. INTRODUCTION

Biotechnology derived protein drugs and vaccines, or biopharmaceuticals, have come of age [1]. At the onset of the recombinant protein era in the 1980s, biotechnology companies were so overwhelmed by issues relating to production, testing, and the associated regulatory considerations that often the pharmaceutical development issues failed to receive the scientific attention they deserved. That changed rapidly as companies discovered that proteins behaved differently from small molecules in how they needed to be formulated, tested, handled, and processed to the final drug product. Specifically, biopharmaceuticals were found to be much more sensitive to degradation when subjected to perturbations considered normal for small molecules, causing unavoidable failures in drug product processing. Such failures were monumentally expensive as one considers the great difficulties encountered during production of protein drug substances in those early days, even notwithstanding the time, cost, and effort involved. Over the years, as appreciation for complexities associated with biopharmaceutical drug product manufacturing has increased, more emphasis on this aspect

Formulation and Process Development Strategies for Manufacturing Biopharmaceuticals,
edited by Jameel and Hershenson

has been placed. Tremendous gains in knowledge and understanding of drug product pharmaceutics, formulation characteristics, and associated processes have resulted in design and execution of robust and consistent manufacturing processes. The following chapters address various issues germane to manufacturing of biopharmaceutical drug products in detail. Several case studies illustrating potential problems and solutions are also presented. In this chapter, an overview of the underlying concepts and considerations covering biopharmaceutical manufacturing processes is provided.

24.2. MANUFACTURING PROCESSES FOR BIOLOGICAL DRUG SUBSTANCES

Biopharmaceutical drug substances are manufactured via highly specialized processes. A production process for a recombinant protein or antibody generally consists of a series of steps as shown in Figure 24.1. The protein is expressed in an appropriate host system, either bacterial such as *Escherichia coli* or yeast, or mammalian cells. A vial from the production cell bank is used to build an inoculum starting from small flasks or bioreactors and then transferred to the intended manufacturing-scale bioreactor containing the growth production medium. The cell-culture process operates in a batch, fed-batch, or continuous-perfusion mode, runs for 2–30 days, and produces the protein or antibody of interest in the cells or culture medium.

After the cell culture or fermentation steps in which sufficient growth of the target antibody or protein is achieved, a concentration step is employed to collect the protein of interest. In the case of a secreted protein, the cell supernatant is collected and concentrated using techniques such as microfiltration or centrifugation. When the

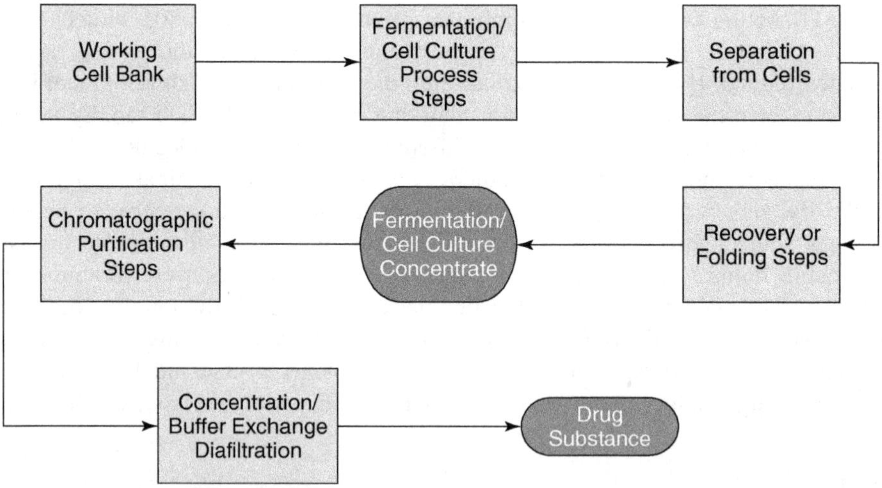

Figure 24.1. A flowchart of a biopharmaceutical drug substance manufacturing process.

protein is expressed within the host cells (such as in *E. coli*), the cells are collected, are disrupted by homogenization, and the protein is extracted using a chaotrope.

At this point, a crude concentrate is generated for further purification. Often, proteins derived from bacterial systems are in partially misfolded form requiring additional refolding steps. Monoclonal antibodies and other soluble proteins from cell culture, on the other hand, do not require refolding as these molecules are expressed and excreted in their original glycosylated, soluble state. Purification steps usually consist of several chromatographic steps (e.g., affinity, reversed phase, size exclusion or ion exchange) coupled with other steps such as precipitation and diafiltration. For cell culture- derived products, such as antibodies and subunit protein vaccine antigens, additional steps are inserted to inactivate any adventitious viruses potentially carried from the cells. Typically, a combination of low pH treatment and nanofiltration are employed during the purification process for this purpose. The final purification step generally consists of diafiltration, which, in one step concentrates the protein to a desired concentration while removing unwanted small-molecule solutes carried over from the purification process, and exchanging it into the preferred buffer solution for formulation.

At the conclusion of the entire set of cell culture, recovery, and purification steps, a bulk protein or antibody solution is obtained, which is designated as the *drug substance*.

In general, a biopharmaceutical drug substance manufacturing process is an involved, complex procedure requiring a variety of skills, facilities, equipment, and prolonged timeframes (several weeks to months) to accomplish. Therefore, tight process monitoring and controls are included at each stage to ensure the successful completion of various steps in the process train. The involved and thorough nature of a biopharmaceutical process explains the relatively high cost of producing a finished biopharmaceutical.

Fermentation or cell culture and recovery operation can have significant influence on molecular properties of the target protein. Misfolding, unintended incorporation of undesired amino acids, inappropriate disulfide bond formation, altered glycosylation, and proteolysis could occur in addition to the more usual changes such as aggregation, deamidation, and oxidation.

During purification, the target protein is subjected to wide variations in solvent exposure, pH, ionic strength, and concentration leading to accentuated degradative stresses similar to those described above. Some of these changes carry over into the eventual biophysical and biochemical characteristics of the biopharmaceutical. An excellent review has discussed protein aggregation during bioprocessing in detail [2]. Although the authors have quoted examples for aggregation changes only, similar effects are seen in proteins with respect to other degradation pathways as well.

24.3. ANALYTICAL CHARACTERISTICS OF BIOMOLECULES

Biomolecules are complex structures with three-dimensional or higher levels of configuration required for full biological activity. In contrast to small-molecule drugs, one needs to analytically evaluate a biomolecule from many angles to ensure that the set of characteristics that confer the desired biological activity to the molecule are preserved. Therefore, when analyzing of a biomolecule, we think of defining its analytical profile to define it completely.

Impurities in biological molecules originate from two major sources: the manufacturing process and degradation of the active molecule. From the previous section, the reader can appreciate that biomolecules are generated from living cells. Although we manage the expression and generation of the desired molecule in the cell by using our extensive understanding of such systems and application of process controls, we cannot *beat* the inherent variability of a biological system. This results in the generation of many variants (or *isoforms*) of the molecule that differ from the parent molecule in very small increments, such as the deletion or alteration of an amino acid residue in the molecule or the degree and type of its glycosylation. Such diversity is similar to what is seen for native proteins (interferons, interleukins, or growth hormones) present in the body. Additional changes in the molecule occur during its passage through the various processing steps, resulting in generation of new isoforms or increases in those already present. Impurities formed after isolation of the drug substance due to degradation could similarly change the isoform mix in the drug product. Therefore, a biomolecule drug product generally displays multiple variants within its analytical profile.

A comprehensive definition of analytical profile requires consideration of the following areas of characterization:

1. Physiochemical and biochemical
2. Biological and immunological

While these topics are adequately addressed in detail elsewhere in this book, we briefly discuss their potential application to the manufacturing aspects of drug products here.

Physical characterization of biomolecules by methods such as visual appearance, pH, moisture level (for lyophilized products), ultraviolet spectrophotometry, and particulate analyses is critical for assessing gross changes in drug products. These parameters also are most observable operationally, both at the manufacturing floor and during use by the health care provider or patient. The next level of information is obtained from protein purity analyses. The list comprises methods that measure soluble, insoluble, and dissociable aggregates (SDS-PAGE and size exclusion chromatography), charge heterogeneity (ion exchange chromatography, isoelectric focusing and capillary electrophoresis), and protein integrity (reversed-phase HPLC, peptide mapping, etc.).

In addition to the routine methods mentioned above, much conformational information can be obtained by biophysical techniques such as circular dischroism, FTIR, and differential scanning calorimetry. Structural information is supplemented by the various mass spectrometric techniques.

Biological activity techniques, usually indicative of potency, are most comprehensive gauges of conformational of the desired biomolecule. These are usually cell-based assays relying on some property of the molecule to alter the cell's growth, inhibition, or lytic potential. Because of the inherent complexity and variability of such assays, the results are relatively imprecise. Therefore, the use of these methodologies is restricted to monitoring large molecular changes as smaller effects are convoluted or masked by their variability. Nevertheless, biological assays carry immense value for evaluating the overall viability of the molecule.

Immunological assays are additional characterization tools for evaluating the functionality of biomolecules. A set of enzyme-linked immunosorbent assays (ELISAs) are used to detect interaction with various immunological epitopes. For MAb drugs and recombinant vaccine antigens, immunological activities are considered important for their functionalities. For therapeutic proteins, any decreases in their immunological profile may signal loss of activity, or increases in the same may signal partial unfolding of the molecule, potentially rendering it more immunogenic and less functionally active.

24.4. PROCESS COMPARABILITY

The reader must appreciate by now the highly complex and involved nature of biopharmaceutical processes and the associated analytical assays. In the early years of biotechnology, the available analytical methodologies were unlikely to detect minor changes in drug product caused by changes introduced in the manufacturing process of drug substance or product. This led to the frequent use of the phrase that "process is the product" to characterize biopharmaceutical manufacturing processes. This became an enormous impediment to sponsors considering routine manufacturing changes such as scale-up or move of a manufacturing process from one facility to another. Over the years since then, as understanding of biotechnology processes have increased and new analytical tools become available, more rational views have emerged leading to the "comparability" concept. The ICH quality guideline on comparability states that "demonstration of comparability does not necessarily mean that the quality attributes of the pre-change and post-change product are identical, but that they are highly similar and that the existing knowledge is sufficiently predictive that any changes in quality attributes have no adverse impact on the safety or efficacy of the drug product" [3]. This change in thinking has allowed both the regulatory authorities and industrial sponsors to define and follow an appropriate path when process changes, either in drug substance or product, are introduced. It also puts increased emphasis on creating an in-depth scientific understanding of manufacturing processes for biopharmaceuticals.

24.5. BIOPHARMACEUTICALS DOSAGE FORM TYPES

Most biotechnology drug products on the market are administered to the patient by injection. Based on the author's evaluation of the FDA Website [4] and other drug databases, a total of 83 approved products are included in this survey (Table 24.1). Please note that this survey is not exhaustive; it includes most of the recombinant drugs and vaccines, and the author has combined some products (e.g., mixed insulin

TABLE 24.1. Dosage Forms for Marketed Biopharmaceutical Drug Products

Generic Name	Proprietary Name	Manufacturer	Physical State(s)	Route
Abatacept	Orencia	Bristol-Myers Squibb	Lyophilized	IV
Abciximab	ReoPro	Centecor (J & J)	Solution	IV
Adalimumab	Humira	Abbott	Solution	SC
Agalsidase beta	Fabrazyme	Genzyme	Lyophilized	IV
Aldesleukin or Interleukin-2	Proleukin	Novartis V & D (Chiron)	Lyophilized	IV
Alefacept	Amevive	Astellas (Biogen)	Lyophilized	IV, IM
Alglucosidase alfa	Myozyme	Genzyme	Lyophilized	IV
Alemtuzumab	Campath	Genzyme (Ilex)	Solution	IV
Alteplase	Activase	Genentech	Lyophilized	IV
Anakinra	Kineret	Amgen	Solution	SC
Antihemophilic factor	Kogenate	Bayer	Lyophilized	IV
Arcitumomab	CEA-Scan	Immunomedics	Lyophilized	IV
Asparaginase	Elspar	Merck	Lyophilized	IV, IM
Basiliximab	Simulect	Novartis	Lyophilized	IV
Becaplermin	Regranex	OMJ Pharma/Chiron	Liquid gel	Topical
Bevacizumab	Avastin	Genentech	Solution	IV
Capromab Pendetide	ProstaScint	Cytogen	Solution	IV
Cetuximab	Erbitux	ImClone/BMS	Solution	IV
Coagulation factor V11a	NovoSeven	Novo Nordisk	Lyophilized	IV
Coagulation factor IX	BeneFix	Wyeth (Genetic Institute)	Lyophilized	IV
Daclizumab	Zenepax	Roche	Solution	IV
Darbepoeitin alfa	Aranesp	Amgen	Solution	IV, SC
Denileukin diftitox	Ontak	Eisai (Seragen)	Solution	IV
Dornase alfa	Pulmozyme	Genentech	Inhalation solution	Inhalation
Drotrecogin alfa (activated)	Xigris	Lilly	Lyophilized	IV
Eculizumab	Soliris	Alexion	Solution	IV
Efalizumab	Raptiva	Genentech	Lyophilized	SC
Epoetin alfa	Epogen, Procrit	Amgen, J & J	Solution	IV, SC
Epoetin alpha	Eprex	Ortho Biologics	Solution	IV, SC
Etanercept	Embrel	Amgen (Immunex)	Solution or lyophilized	SC
Filgrastim	Neupogen	Amgen	Solution	IV, SC
Galsulfase	Naglazyme	BioMarin	Solution	IV
Glucagon	Glucagon	Lilly	Lyophilized	IV, IM, SC
Hepatitis B vaccine (recombinant)	Recombivax HB	Merck	Liquid suspension	IM

TABLE 24.1. (*Continued*)

Generic Name	Proprietary Name	Manufacturer	Physical State(s)	Route
Hepatitis B vaccine (recombinant)	Engerix-B	Glaxo SmithKline	Liquid suspension	IM
Human papillomavirus recombinant vaccine	Gardasil	Merck	Liquid suspension	IM
Ibritumomab tiuexetan	Zevalin	Biogen-Idec	Solution	IV
Idursulfase	Elaprase	Shire	Solution	IV
Imiglucerase	Cerezyme	Genzyme	Lyophilized	IV
Infliximab	Remicade	Centecor (J & J)	Lyophilized	IV
Insulin aspart (recombinant)	NovoLog	Novo Nordisk	Solution	SC
Insulin detemir	Levemir	Novo Nordisk	Solution	SC
Insulin, human (recombinant)	Humulin R	Lilly	Solution	SC
Insulin, human (recombinant) isophane suspension	Humulin N	Lilly	Suspension	SC
Insulin, human (recombinant) isophane suspension and insulin solution	Humulin 70/30 and Humulin 50/50	Lilly	Suspension	SC
Insulin, human (recombinant), inhalation	Exubera	Pfizer/Nektar	Spray-dried powder	Inhalation
Insulin, lispro	Humalog	Lilly	Solution	SC
Insulin lispro suspension and Interforon alfa-2a insulin lispro solution	Humalog Mix 75/25 and Humalog Mix 50/50	Lilly	Suspension	SC
Interferon alfa-2a	Roferon A	Roche	Solution	SC
Interferon alfa-2b	Intron A	Schering Plough	Solution or lyophilized	IV, IM, SC or intraleisonal
Interferon alfacon-1	Infergen	Intermune (Amgen)	Solution	SC
Interferon beta-1a	Avonex	Biogen-Idec	Lyophilized	IM
Interferon beta-1a	Rebif	Serono	Solution	SC
Interferon beta-1b	Betaseron	Novartis V & D (Chiron)	Lyophilized	SC

(continued)

TABLE 24.1. Dosage Forms for Marketed Biopharmaceutical Drug Products (*continued*)

Generic Name	Proprietary Name	Manufacturer	Physical State(s)	Route
Interferon gamma-1b	Actimmune	Intermune (Genentech)	Solution	SC
Laronidase	Aldurazyme	BioMarin	Solution	IV
Lyme disease vaccine (recombinant OspA)	LYMErix	Glaxo SmithKline	Liquid suspension	IM
Muromonab-CD3	Orthoclone OKT3	Ortho Biotech	Solution	IV
Natalizumab	Tysabri	Biogen-Idec/Elan	Solution	IV
Nofetumomab	Ventuma	Boehringer-Ingelheim	Solution	IV
Omalizumab	Xolair	Genentech	Lyophilized	SC
Oprelvekin	Neumega	Wyeth (Genetic Institute)	Lyophilized	SC
Palifermin	Kepivance	Amgen	Lyophilized	IV
Palvizumab	Synagis	MedImmune	Solution	IM
Panitumumab	Vectibix	Amgen (Abgenix)	Solution	IV
Pegaspargase	Oncaspar	Enzon	Solution	IV, IM
Pegfilgrastim	Neulasta	Amgen	Solution	SC
Peginterferon alfa-2a	Pegasys	Roche	Solution	SC
Peginterferon alfa-2b	PEG-Intron	Schering Plough	Lyophilized	SC
Pegvisomant	Somavert	Pfizer (Pharmacia)	Lyophilized	SC
Ranibizumab	Lucentis	Genentech	Ophthalmic solution	Intravitreal
Rasburicase	Elitek	Sanofi-Synthelabo	Lyophilized	IV
Reteplase	Retavase	PDL (Boehringer Mannheim)	Lyophilized	IV
Rituximab	Rituxan	Genetech/Biogen-Idec	Solution	IV
Sargramostim	Leukine	Bayer (Berlex, Immunex)	Solution or lyophilized	IV
Somatropin	Genotropin	Pfizer (Pharmacia)	Lyophilized	SC
Somatropin	Humatrope	Lilly	Lyophilized	SC, IM
Somatropin	Nutropin	Genentech	Lyophilized	SC
Somatropin	Nutropin AQ	Genentech	Solution	SC
Somatropin	Norditropin	Novo Nordisk	Solution	SC
Tenecteplase	TNKase	Genentech	Lyophilized	IV
Tositumomab	Bexaar	GSK (Corixa)	Solution	IV
Trastuzumab	Herceptin	Genentech	Lyophilized	IV

dosage forms) for the purposes of analyzing trends. In total, 95% of these are drug products for parenteral administration; 37 are given intravenously, 26 subcutaneously, 6 intramuscularly, and 10 by multiple parenteral routes. The remaining products are administered as follows: two by inhalation (Pulmozyme and Exubera), one by intravitreal (Lucentis), and one by topical (Regranex) routes. Another examination of this list of drug products by the physical state of their dosage form(s) reveals that 38 are solutions, 31 are lyophilized, 7 are liquid suspensions, 1 is a solution for inhalation, 1 is an ophthalmic solution, 1 one is a spray-dried powder, and 1 one is a semisolid gel. Three products are available in both solution and lyophilized dosage forms. Not too surprisingly, dosage forms for the early biopharmaceuticals were lyophilized; the recent trend is toward solutions. In part, this is due to a greater understanding of protein stabilization in the liquid state, but another reason is that the newer drug products are predominantly monoclonal antibodies, which, in general, are more stable than the early therapeutic proteins. Of course, the overwhelming reason for preferring solutions over lyophilized dosage forms is their convenience by the end user as well as the ease of manufacturing.

Given the data presented above, biopharmaceutical scientists have overwhelmingly focused their attention toward development of parenteral dosage forms.

Figure 24.2 presents a schematic for a drug product manufacturing process. Biopharmaceutical drug substances are normally stored frozen at −20°C or −70°C, or as a solution at 5°C. The drug substance may be stored in bottles or in specially outfitted steel containers that are able to freeze and thaw solutions in a gentle and controlled manner. Thawing of frozen proteins from bottles could be performed in several ways: (1) by placing the bottle at 5°C for 24–48 h, (b) by thawing the bottle at room temperature for 4–12 h, or (3) by placing the bottle containing frozen solution in a 30°C–40°C water bath for short durations (~1 h). The thawing protocol needs to be worked out for each protein as different proteins tend to respond to these treatments differently. Aggregation or deamidation of proteins are normally observed under unfavorable thawing conditions [2].

Thawed or liquid drug substance solutions must be mixed using gentle mixing techniques as proteins tend to concentrate and "settle" at the bottom of the storage container even as solutions. Mixing ensures that drug product batches are produced at desired final concentrations.

Next, thawed and remixed drug substance is formulated to the desired final composition by a combination of diafiltration, dilution, or addition steps. At this stage, the product is known as the *final formulated bulk*. For products that cannot be sterilized by final filtration, the final formulated bulk is prepared by combining prefiltered components aseptically. Otherwise the final formulated bulk is subjected to a final 0.2-μm filtration. The sterile liquid is dispensed into the desired final containers (vials, syringes, etc.), and lyophilized if required for stability. The finished drug product is closed, sealed, and labeled. Sealed product is then subjected to a thorough inspection, usually by allowing it to pass through a visual observation station fitted with black-and-white backgrounds. Finally, the drug product is labeled and packaged into cartons. For nonparenteral products, the fill–finish procedures are modified as appropriate.

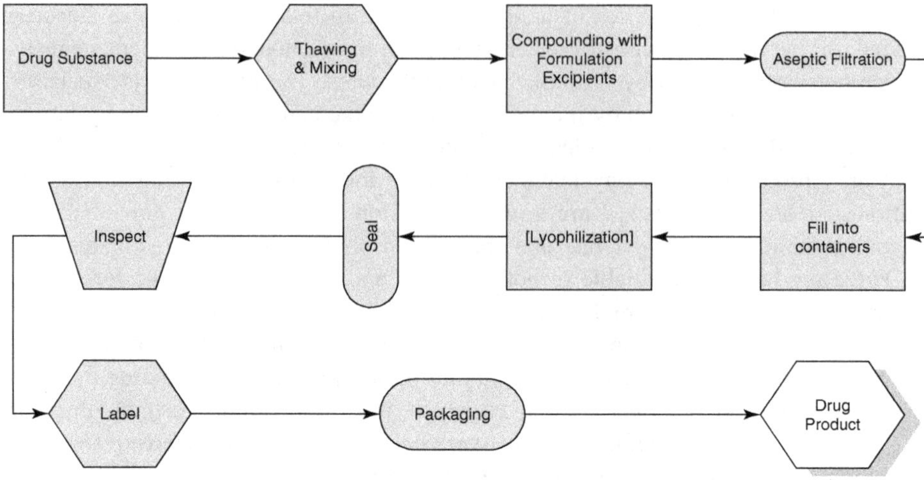

Figure 24.2. A flowchart of a biopharmaceutical drug product manufacturing process.

24.6. STABILITY CONSIDERATIONS

The unit steps for drug product processing indicated above seem quite innocuous. However, they could have a marked detrimental effect on the drug product if impact of the various steps is not understood at limit conditions. For example, if an unusually large number of containers fail the visual observation criteria, a reinspection is triggered. Thus the drug product is subjected to room temperature and light exposure conditions [5] for extended periods, causing further degradation. Labeling and packaging are operations that are often overlooked by the pharmaceutical scientist. However, since these are conducted at room temperature, product stability issues need to be considered. Thus, in-process stability of drug product is an issue that needs to be carefully thought through (see Table 24.2).

To account for exposures of the drug substance and drug product to manufacturing conditions, it is a good idea to think through the various possibilities and design a main study capturing most of the aspects. In addition, substudies addressing some issues may be necessitated. Eventually, a cumulative stability exposure limit needs to be assigned that becomes an important tool for resolving subsequent deviations that are encountered during routine manufacturing operations.

Consider an example of a cumulative exposure of the protein molecule during a drug product manufacturing process. The various possibilities for a drug product lot are shown in Figure 24.2. The cumulative time duration for these operations could be in days. Therefore, the pharmaceutical scientist should perform a stability test for at least a week in the case shown to cover for possible contingencies.

TABLE 24.2. In-Process Stability Evaluation

Process Stream	Exposure	Typical Time Durations,[a] h	Stresses	Comments
Drug substance (DS) after final purification step	Hold between final step and transfer to cold chamber or freezer	2–4	Temperature, light	—
Cold or frozen DS thawed to RT for drug product (DP) manufacturing	Time duration after thaw, during mixing, compounding, and filtration prior to initiation of fill	3–6	Heat, light, mechanical agitation	—
Duration of the fill–finish process (up to sealing)	Liquids: filling into containers, capping, and seal	6–20	Heat, light, mechanical agitation	For lyophilized products, consider liquid and lyophilized stabilities separately
	Lyophilized: filling into containers, partial capping, transfer to the lyophilizer, seal postlyophilization			
Inspection	Time duration after sealing, inspection, then storage	12–15	Heat, light, mechanical agitation	Single inspection
Label	—	12–15	Heat	—
Package	—	12–15	Heat	—
Total	—	*47–75*	—	—

[a] Assuming a lot size of 50,000 units.

24.7. COMPONENT PREPARATION

Component preparation for all drug products is of utmost importance to ensure patient safety. For biopharmaceuticals, this becomes even more important as exposure to residual amounts of substances in components could decrease purity or potency of the molecule because of their sensitive nature. Unbuffered solutions are subject to pH changes due to acid leaching from glass vials. Loss of proteins to the glass from protein solutions could also occur. It is well known that stoppers could cause increases in moisture content of lyophilized products [6]. Appropriate preparation procedures must be used to prevent components from being a source for destabilization of drug products.

24.8. SCALE-UP AND TECHNOLOGY TRANSFER CHALLENGES WITH DRUG PRODUCT MANUFACTURING

Scale-up of drug substance manufacturing normally gets all the limelight in a biopharmaceutical company. There is no doubt that biotechnology-derived drug substance processes are multistep and highly complex as discussed before, deserving extensive focus and attention. However, many companies have paid a heavy price for neglecting scale-up and technology transfer issues for drug products. From a business perspective, economic consequences of losing expensive drug substance and valuable time during large-scale drug product operation are extremely high. Therefore, companies are spending time and money to ensure successful scale-up and technology transfer of drug products.

During scale-up of drug products, some key areas for attention are:

1. *Thawing Procedure and Logistics*. As described earlier, unless fully integrated and controlled freeze–thaw units are used, one must consider how the thawing process should be conducted. Uncontrolled thawing (by placing the drug-containing container on the bench) are time-consuming and subject to time fluctuations. Also, thawing of small-scale containers is seldom predictive of large-scale thawing. A thick frozen layer at the outside peripheries of frozen solution slows down and prevents uniform thawing. Trial runs with frozen placebo solutions at the large scale help to better define thawing conditions.
2. *Mixing*. Typical lab-scale experiments are conducted using magnetic stirrers. At large scale, mechanical mixers are generally employed. Mixer type, speed, and duration are all important parameters that need to be optimized for maintaining integrity of the protein while achieving uniformity.
3. *Compounding*. Addition of large amounts of liquids (excpient or buffer solutions) is performed using pumps, mixing tanks, and other equipment while such operations are accomplished manually at the bench scale. Because of the large volumes involved, rate and order of addition of solutions become important in order to avoid local concentration buildup of the protein or excipients, or transient pH and ionic strength shifts; these operations could lead to denaturation, precipitation, or degradation by other mechanisms.
4. *Scale-Up of Filtration*. This step is sometimes not as straightforward as simply increasing the surface area proportionately. Many types of filter configuration are available for large-scale operation, and careful experiments with scale-down models should be done prior to selecting the right component. Some products generate proteinaceous matter during compounding or other earlier operations that could clog up the filter. Filter clogging not only reduces flow, thus increasing duration of the operation, but could also compromise filter or protein integrity. Use of a prefilter of a larger pore size ($\sim$0.45 μm) normally resolves this problem.
5. *Risk Assessment*. An issue of prime importance to consider is the material(s) used for product contact parts throughout the drug product process.

A risk assessment spreadsheet should be created to evaluate this and necessary compatibility or leachate/extractible studies performed to ensure successful scale-up.

6. *Lyophilization Scale-Up*. This is a topic by itself, and the reader is referred to the many excellent articles in the literature [7]. Suffice is to state that both the process and equipment for lyophilization are linked closely and transfer normally takes much longer, with additional time, and greater effort than does that needed for a solution fill–finish process.
7. *Raw Materials and Components*. Identifying raw materials and components for large-scale production well in advance is critical for implementing scaled-up processes. As proteins are sensitive to trace impurities present in excipients (such as peroxides present in polysorbate surfactants), pharmaceutical scientists sometimes come up with elegant but impractical solutions to stabilize them during formulation development. Sourcing of such materials for large-scale production (from nonanimal sources), qualifying them for production (technical and quality risk assessment or auditing of the vendor), and getting them into the manufacturing "system" is a multimonth process involving many departments of the company. Similarly, any new components (such as filters and stoppers) require similar treatment before use in production.
8. *Engineering or Test Runs*. An important milestone during scale-up of a process is running the process in test or engineering runs before formal production runs are performed. Engineering runs provide assurance that the consolidated process runs as expected. They also provide one final opportunity to the development scientist and his/her manufacturing counterpart to make any last-minute adjustments in the process. Since large-scale runs are expensive, requiring production capacity, plant staff, and systems, management usually prefers to minimize them while the scientists' tendency is to perform as many runs as possible to obtain maximal data and ensure a successful process scale-up and transfer. Obviously, both sides have to settle for a middle ground to ensure the company's success.

In planning a new-product technology transfer, it is immensely useful to gather a cross-functional team consisting of development, engineering, manufacturing, quality control, quality assurance, regulatory, purchasing, project management, and other relevant groups to discuss various real and potential issues regarding the incoming product and its impact on the existing product streams.

Most companies follow a formal process commonly referred to as *new-product introduction*. Key issues are checked off on this form with signed approval from various stakeholders identified above. Additionally, pharmaceutical development scientists generate detailed technology transfer reports covering wide-ranging studies providing manufacturing personnel with supplementary knowledge about the product and process. Table 24.3 presents a hypothetical checklist of items covered by technology transfer reports and new-product introduction forms.

TABLE 24.3. Drug Product Technology Transfer Checklist

1. Product name:
2. Dosage strength:
3. Vial size and type:
4. Stopper size and type:
5. Stopper treatment (siliconization, drying):
6. Fill volume:
7. Number of vials to be filled:
8. Formulation composition:
9. Formulation process:
10. Special filtration requirements:
11. Stability of formulated bulk:
12. Stability of product to multiple freeze–thaws:
13. Density:
14. Product handling guidelines:
15. Material compatibility issues:
16. Raw materials
17. Validation requirements:
18. Documentation (batch records, standard operating procedures):
19. Other unusual product issues:
20. Need for engineering runs:

24.9. PROCESS VALIDATION AND REGISTRATION

Process validation is a regulatory requirement before a product can be launched in the marketplace. Following successful conduct of the drug substance process validation, the process for manufacturing drug product must be validated. From a business perspective also, process validation ensures that the drug product can be manufactured in a consistent, reproducible, and controlled manner, providing safe and efficacious drug product in a reliable manner during routine production.

Before initiating process validation, the development scientist must perform experiments to define the "design space" and "operating space" for the process. On the basis of a design of experiments (DOE) approach and selected worst-case studies, critical parameters are identified. It is important to document development and other design space activities to create the knowledge base for the product. Formal protocols must be written before process validation is performed. A master validation protocol outlines the validation approach, sampling plan, assays, and deliverables and also describes the substudies that form part of the entire validation campaign. As three consecutive successful runs are required for validating the process, usually extra runs are scheduled to cover for possible contingencies. Substudies include hold time, mixing, and filter validation studies. Filter studies are often performed by the filter vendor in collaboration with the drug product manufacturer. Finally, a master validation report is written, on the basis of the results from all studies, also including any justification needed to resolve deviations observed from the validation protocol.

A strong process validation package is an important piece of documentation for the future inspections as well as a reference document during day-to-day manufacturing of the product after approval.

Preparation of documents for registration of a biopharmaceutical drug is normally a massive effort requiring writing, formatting, and collation of hundreds to thousands of documents. Manufacturing details are a component of the "Quality" section (formerly known as the "Chemistry Manufacturing and Control," or the "CMC" section). Section "3.2.P.3 Manufacture" encompasses the manufacturing aspects of a drug product to be reported in the *Common Technical Document—Quality*.

24.10. ROUTINE MANUFACTURING

Experience level with a given process is usually limited by the total number of drug product lots required for the conduct of the entire set of clinical studies to determine the drug's safety and efficacy prior to approval. For vaccines, this number is quite small (<10) as vaccines are normally given in very small doses. For antibodies, this number could be high as they are normally administered in relatively high doses. In any case, the number of lots during development is raely greater than 20 or 30. When the product is manufactured routinely after launch, 4 or 5 times this number of lots may be produced in one year alone for some high-volume products. During the first 1–2 years on the market, the knowledge base increases significantly. New problems often arise that were not seen during development. In such cases, one can go back to the design space studies to determine whether the data fall within those confines. If key or critical parameters fall outside the design space, then investigations are needed to evaluate the impact of this deviation. Certain other types of deviations are also quite common. For example, extended hold times during intermediate stages may be encountered because of equipment failure or for some other reason. These instances need to be evaluated on a case-by-case basis, and additional studies may be required to justify further processing or use of the product.

A technical troubleshooting group often provides technical support during routine manufacturing of a product. Close cooperation between technical support, process development, manufacturing, quality, and regulatory groups is required to ensure that the troubleshooting solutions are within the available knowledge base for the product.

24.11. CONCLUSION

Manufacturing of biopharmaceuticals is a highly involved and complex exercise. Scale-up and technology transfer of a process developed at the small scale is often a quite challenging exercise because of the sensitive nature of these molecules. In this chapter, the author has identified the various aspects of biopharmaceutical drugs that could possibly be impacted during processing, scale-up, technology transfer, and routine production. Preventive measures based on knowledge base of the molecule are suggested to mitigate or resolve such occurrences. The greater the knowledge base for the product in question, the less the chances for appearance of unexpected complications during manufacturing of the biopharmaceutical drug product.

ACKNOWLEDGMENTS

The author would like to thank his many colleagues at Chiron Corporation (now Novartis) and PDL BioPharma (now Facet Biotech) whose collaborative contributions have made this entire experience possible.

REFERENCES

1. Kelley, B. D. (2001), Bioprocessing of therapeutic proteins, *Curr. Opin. Biotechnol.* **12**: 173–174.
2. Cromwell, M. E. M., Hilario, E., and Jacobson, F. (2006), Protein aggregation and bioprocessing, *AAPS J.* **8**: E572–E579.
3. ICH (2004), *Comparability of Biotechnological/Biological Products Subject to Changes in Their Manufacturing Process*, ICH Guideline QSE (available online at http://www.ich.org, Nov. 2004).
4. FDA Website, http://www.fda.gov/.
5. Kerwyn, B. A. and Remmelle, R. L., Jr. (2007), Protect from light: Photodegradation and protein biologics, *J. Pharm. Sci.* **96**: 1468–1479.
6. Hora, M. S., and Wolfe, S. N. (2004), Critical steps in the preparation of elastomeric closures for biopharmaceutical freeze-dried products, in Rey, L. and May, J. C., eds., *Freeze-Drying/Lyophilization of Pharmaceutical and Biological Products*, Marcel Dekker, New York, pp. 309–323.
7. Tang, X. and Pikal, M. J. (2004), Design of freeze-drying processes for pharmaceuticals: Practical advice, *Pharm. Res.* **21**: 191–200.

25

PROTEIN STABILITY DURING BIOPROCESSING

Mark Cornell Manning, Gabriel J. Evans, and Robert W. Payne

25.1. INTRODUCTION

During processing, proteins undergo various steps or unit operations. Each of these can generate stresses severe enough to damage the protein, either physically or chemically. Yet, most of the attention of the pharmaceutical literature has been on the stability of proteins during storage in the final container. This chapter reviews what is known about the stability of proteins during various processing steps. Although there has been a review as recently as 2006 on aggregation during processing [1], a more comprehensive discussion of this topic is needed. The focus of this chapter is on unit operations performed after isolation of crude material, so any issues involving protein instability in cells or in the presence of cell debris are not addressed.

25.2. PURIFICATION

The literature on protein degradation during purification is rather modest, despite considerable knowledge about protein purification. This is likely due to the fact that, as purification processes are optimized, the bases for low yields or poor recovery

Formulation and Process Development Strategies for Manufacturing Biopharmaceuticals,
edited by Jameel and Hershenson

remain unexplored and unknown. Certainly, each biopharmaceutical company has a great deal of experiential information on this topic, even if little has found its way into the open literature. Occasionally, there are studies that focus on solubility or suppression of aggregation that indirectly address some of these stability issues [2].

25.2.1. Protein A Affinity Chromatography

One of the most powerful purification methods for monoclonal antibodies (MAbs) is the use of affinity chromatography using protein A. Protein A binds antibodies with very high specificity and avidity [3]. The release of bound antibody requires low pH conditions, which could lead to aggregation of some antibodies. This has been reported for Campath, a currently marketed MAb. According to the report, Campath aggregates more than 25% on release from the protein A column [4]. Clearly, these issues have been resolved as this is a marketed product. However, the report serves to illustrate the extent to which a protein can exhibit physical instability.

25.2.2. Ion Exchange Chromatography

Given the ionic nature of the stationary phase and mobile phases used in ion exchange chromatography (IEC), one can envision both conformational and colloidal instability as being potentially problematic. In the case of IEC separation of equine immunoglobulin preparations, Lewis and Nail report aggregation, especially for the IgGb (i.e., IgG2) fraction [5]. The aggregation occurs at relatively low pH, suggesting that equine IgG2 possesses lower conformational stability at acidic pH than do other IgG fractions. A general overview of salt effects on protein behavior during chromatography has just appeared. As high ionic strengths are often found in IEC, this chapter provides some useful insight into the impact of various salts on binding and protein solubility.

25.2.3. Hydrophobic Interaction Chromatography (HIC)

Fernandez and co-workers have investigated protein behavior during HIC, as the conditions are known to be denaturing [6,7]. Moreover, the recoveries from HIC can be quite variable [8]. For α-lactalbumin, there is initial adsorption to the stationary phase, leading to conformational changes. The extent of adsorption is increased by low-salt conditions.

25.3. FREEZE–THAW CYCLING

Nearly any protein product will undergo freezing and thawing at some point. It may occur during an intermediate holding step, until processing can begin again. A protein may need to be shipped to another site for further processing, being sent as a frozen bulk solution. Almost certainly, a protein may be frozen as a method for sample preservation until a particular analytical test can be performed. In many cases, the freezing and thawing of a protein is innocuous. In others, the freezing and/or thawing can be quite detrimental. This section explores some of the mechanisms of freeze–thaw-induced damage to proteins. A review by Bhatnagar et al. also provides some overview of the pathways by which a protein can undergo degradation during freezing [9].

During freezing and thawing, a number of stresses can occur that may lead to protein degradation. This fact is often overlooked. In fact, there is a common misconception that developing a frozen formulation for early-phase clinical testing is easy. Yet, there are a number of difficulties that can and do arise when an aqueous solution is lowered below the freezing point of water. It is worthwhile to consider some of these potential degradation pathways.

For half a century, it has been known that freezing sodium phosphate solutions leads to marked acidification as the dibasic form of sodium phosphate selectively crystallizes out of solution [10]. As a result, the acidity can increase with a drop in pH of three units or more. When this occurs, some proteins undergo dissociation or conformational changes that lead to subsequent aggregation [11–14]. Because of this phenomenon, the use of sodium phosphate [and phosphate-buffered saline (PBS)] is usually considered to be contraindicated when developing freeze-dried formulations [15]. However, one must also realize that the presence of other solutes will inhibit crystallization, so this problem can be minimal in actual formulations [16–18].

Another possible mechanism for protein damage during freezing is freeze concentration. As the volume of a solution increases, the rate at which the outer portion freezes relative to the inner becomes more distinct. As the bulk water is removed from solution by freezing, there is an effective concentration of the remaining solute(s), in the same way that removing solvent through evaporation would. The extent of freeze concentration varies, but most consider a factor of ≥ 10 to be easily achievable. In the case of normal saline, Franks reports a freeze concentration effect of 24-fold [19]. Freeze concentration of sugars is thought to be able to reach a final concentration of approximately 80% (w/v) [20]. Similar effects could be achieved for buffers and other salts. So, in some cases, protein degradation could be due to exposure to high-ionic-strength environments. Conversely, freeze-concentration-induced aggregation of the protein could also result from markedly higher protein concentrations.

Most likely, the predominant mechanism for freeze-induced damage is unfolding at the air–ice interface [21]. The first reports were from Chang et al. for IL-1ra (interleukin-1 receptor antagonist) [22] and from Strambini and Gabellieri [23]. Since then, a number of other reports on damage occurring at the ice–water interface can be found. Interfacial damage leads to increased aggregation for hemagglutinin (HA) variants [24]. If a protein is sensitive to this type of degradation pathway, the effect should be seen in freeze–thaw cycling, an important aspect of preformulation characterization.

The same mechanism may also be a primary mechanism for damage during freeze-drying [15,21] as well as during handling and storage. In any case, the most effective intervention may be addition of a surfactant to the formulation [15,22,25]. However, interfacial damage will still occur if there is insufficient surfactant present in the formulation to prevent damage at a surface. It has been reported that the critical micelle concentration (cmc) changes as a function of temperature, at least for Tween 80 [26].

Any damage caused by freezing and thawing is usually cumulative; that is, the extent of damage increases with the number of freeze–thaw cycles. This has been demonstrated for hemoglobin [27] and factor XIII [28]. The rate of freezing and thawing can also impact protein stability [11,29,30]. Keep in mind that freezing rate and cooling rate are not the same. In fact, they are often inversely correlated. For example, cooling slowly can lead to rapid freezing due to supercooling [31]. Dependence of stability on freezing rate has been reported for a number of systems [11,30,32]. A less obvious aspect of freeze–thaw instability is aging of the samples. The exact origin of the instability is unclear, but reports of stability diminishing over time in frozen solutions have been published [33]. If this does occur, one must consider the actual mobility of the solutes in the frozen state at the actual storage temperature.

This issue of mobility in the frozen state is an important one to discuss for frozen formulations (including bulk drug substances). In lyophilized solids, the stability of a protein is usually greater below the glass transition temperature [34–37]. There is an analogous transition temperature for the maximally freeze-concentrated state denoted as T_g'. Clearly, storage at −80°C will produce a truly glassy state (well below T_g') for most formulations, commensurate with greatly reduced mobility. However, maintenance of a −80°C cold chain is difficult and expensive. Therefore, most frozen products use −20°C as the preferred storage condition. Unfortunately, for most pharmaceutical formulations, T_g' is well below −20°C. In other words, most frozen formulations are not really frozen; that is, the translational mobility is similar to that in fluid solution. As a result, the benefits of lowered mobility in the frozen state are not realized unless specific formulation strategies are taken. Furthermore, one must be aware of the actual temperature of a storage device listed as a −20°C freezer. In some cases, this may mean that the temperature never exceeds −20°C (as required by some regulatory bodies). In others, it means that the average temperature is −20°C (usually ±5°C).

Keep in mind that the T_g' of the amorphous frozen state is a weight average of all the components. If the mixture is dominated by additives with T_g' values well below −20°C, then the final formulation will not be entirely frozen when stored at −20°C. In order to overcome this limitation, one must use of polymeric additives (e.g., PEGs, starches), which tend to have T_g' values above −20°C (Table 25.1).

If the protein is not immobilized in a rigid matrix as the temperature is lowered, cold denaturation becomes possible [38]. All proteins undergo cold denaturation, that

TABLE 25.1. Glass Transition Temperature Values of Common Additives

Compound	T_g', °C	Reference
Sucrose	−34	139
Sucrose	−31.3	140
Trehalose	−28.6	140
Dextran	−13.5	139
HES	−11	139
HES	−9.4	140

is, unfolding of the globular structure at a high temperature (often denoted T_m) and also a low temperature [38]. For most proteins, this so-called cold denaturation occurs well below the freezing point of water. However, if the protein is in a solution that has a low T_g', then sufficient mobility may occur to allow cold denaturation. Furthermore, a study on the protein pharmaceutical, IL-1ra, found the cold denaturation temperature for this pharmaceutical protein to be just below freezing [39].

A less obvious issue is that of oxygen solubility in water. Unlike the case with solids, the solubility of gases in water increases as the temperature is lowered. Fransson et al. have reported a nearly threefold increase in oxygen solubility in water on going from room temperature to 4°C [40]. Even more worrisome is the report that the solubility of oxygen in ice can be more than 1000-fold greater than that of water [41]. In either case, oxidation-sensitive proteins are more at risk when the formulation is frozen.

The last point to be made regarding frozen formulations is that the analytical challenges associated with frozen solutions are significant. It is much more difficult to characterize protein structure by spectroscopic methods for frozen samples than for liquid ones. Transmission types of spectroscopies, such as circular dichroism and UV absorbance, are no longer viable analytical methods for monitoring protein structure. Phosphorescence spectroscopy could be used [42], although exclusion of oxygen is required. In addition, infrared spectroscopy also can be used for frozen systems [43–45].

25.4. MEMBRANE FOULING

Filtration of protein solutions occurs in two contexts: sterile filtration (also known as *aseptic processing*) and ultrafiltration/diafiltration (UF/DF), where proteins undergo repetitive cycling through a pump–membrane system to exchange a buffer (DF) or reduce the volume, thereby leading to increased concentration (UF). In all cases, the protein can adsorb irreversibly to the membrane in a phenomenon termed membrane fouling.

An often overlooked aspect of membrane fouling is protein–protein interaction. Once the initial adsorption occurs to the membrane, the remainder of the process is protein adsorbing to a layer of protein [46]. As a result, a number of studies have indicated that modulating protein–protein interactions reduces membrane fouling [47–52]. Even in studies where protein–protein interactions were not explicitly considered, one can find evidence of colloidal properties affecting fouling behavior. For example, in a study on membrane fouling with human growth hormone (hGH), a highly surface-active protein, the focus was on interfacial phenomena [53]. Yet, the pH dependence of fouling is consistent with modulating intermolecular protein interactions as one approach to maintaining flux through the membrane. Similar findings were reported with bovine serum albumin [54].

The nature of the aqueous solution controls the degree of interaction between protein molecules. As higher and higher protein concentrations are used during processing, a new issue has arisen with UF/DF steps. As the protein concentration rises, the amount of low-molecular-weight species associated with the protein increases as well. The result is that there can be a significant disparity between the concentration of a certain solute in the retentate and the concentration in the filtration buffer. In the case of 200-mg/mL IL-1ra, the chloride concentration varied by as much as 30% on either side of the membrane [55].

High concentrations can also occur at the membrane surface during UF/DF [56], and this may contribute to protein destabilization. While this is difficult to prove, some studies seem to support the idea that this is a contributing factor to aggregation during UF processing [57]. In this study, protein concentrations exceed 400 mg/mL at the surface.

Aggregation of proteins during and after UF/DF processing has also been widely reported. For example, Maruyama et al. reported aggregation of BSA during ultrafiltration [58]. These aggregates were primarily covalent in nature, as capping the free Cys residue of BSA reduced aggregation significantly. The exact cause of the degradation was not specified, although stirring alone did not damage the protein. More recently, FGF20, a member of the acidic fibroblast growth factor superfamily, was reported to aggregate extensively during dialysis [59]. Increase in pH and addition of stabilizers reduced the aggregation significantly.

25.5. PROTEIN INSTABILITY DURING FILLING

As filling lines continue to operate under faster and faster conditions, there is great concern over protein stability. Moreover, as higher concentration formulations are being developed, issues with solution viscosity also become increasingly important. Unfortunately, there is little open literature on this particular unit operation regarding protein stability. Even the number of articles on the viscosity of liquid formulations of therapeutic protein is limited. In addition, concerns regarding shear damage have risen as protein concentrations increase and filling systems move more quickly. The issues regarding shear are discussed separately below.

25.5.1. Pumping Issues

There are reports that persistent exposure to pumps leads to aggregation of proteins. Recirculation of a monoclonal antibody through a piston pump leads to progressively larger amounts of particles, as measured by light obscuration, as described in USP monograph <788> [1]. The turbidity of the solution increases as well, both indicating that protein aggregation is increasing as filling proceeds. The most likely cause of protein damage during pumping is cavitation within the pump. The generation of microbubbles can lead to unfolding at the air–water interface. Collapse of the microbubbles can lead to free-radical generation and transient high temperatures. Certainly some earlier reports of protein degradation are consistent with pump-induced damage [60–62]. Reports of different degrees of enzyme inactivation with different types of pumps are consistent with such behavior [63]. Another report about BSA damage during diafiltration focused on the nature of the pumping system, with a screw pump causing more extensive degradation than a gear pump [54]. Furthermore, the damage was greater at 22°C than when the entire system was cooled to 8°C. The exact mechanism of the damage was not delineated, but it is clearly a combination of temperature and interface-induced damage.

25.5.2. Viscosity

Long ignored, viscosity is now becoming an essential physical property in the development of liquid formulations of proteins. High viscosity will affect filling of final containers, and it will also limit the ability of the drug to be given by injection if the viscosity is too high. Unfortunately, there are few reports about the viscosity of proteins, with even fewer about how to reduce viscosity through proper formulation. Viscosity effects, in the context of high-concentration formulations of MAbs, have been discussed by Liu et al. [64]. These authors associate reversible self-association with increased viscosity. In some cases, the addition of salt reduces the viscosity of a MAb significantly, although the effect is far from general [64,65]. We do know that IgGs from different species display different viscosities [66].

From a formulation perspective, it appears that pH can impact viscosity [67]. As seen above, salt may be beneficial. One other report claims that maltose can reduce viscosity in an antibody preparation [68].

25.6. CONTAINER INCOMPATABILITIES

There are two primary mechanisms of protein degradation as a result of contact with container closure systems: (1) adsorption to the container surface can lead to conformational changes that produce more aggregation-competent species; even if there is no structural change, adsorption to the container will lower the dose available to the patient, and (2) low-molecular-weight materials can leach from the container and/or stopper, causing either chemical or physical damage. This can occur even in vials and in prefilled syringes. A brief summary of these types of issues has been presented by the FDA [69] and industrial researchers [70].

25.6.1. Protein Adsorption

While the topic of protein adsorption has been recognized and reviewed for many years [71,72], the literature on little about protein adsorption during processing is relatively sparse. There are reports of adsorption to glass vials [73,74] and IV bags [75], not to mention antibody binding to plastic [76], rubber, and metal [77]. Unfortunately, there are only a few reports on adsorption to stainless-steel or production vessels. Most of the literature in this area has focused on leachables (see Section 25.6.2). Adsorption studies have tended to focus on stainless steel that has been implanted in the body. As a result, the adsorption studies have primarily used plasma proteins. For example, Desroches et al. used reflectance spectroscopy to examine the adsorption of human albumin to stainless steel [78]. Similar studies have been reported for BSA [79]. Others have focused on methods for diminishing adsorption to stainless steel through surface modifications [80]. More general studies have looked at adsorption of lysozyme, β-lactoglobulin, HSA, and ovalbumin [81] and whey proteins [82] to stainless-steel surfaces.

25.6.2. Protein Instability from Leachables and Extractables

Proteins encounter a wide variety of materials during processing. Apart from adsorption, where conformational changes can occur, there is also the potential for proteins to be exposed to low-molecular-weight (LMW) species that are leached or extracted from containers. Generally, in regulatory terminology, *leachable* refers to materials that enter the drug product under normal storage conditions, whereas an *extractable* is one that enters under stress or accelerated conditions [83].

In the case of stainless-steel vessels, redox-active metals such as iron (Fe) and copper (Cu) can cause oxidation via metal-catalyzed mechanisms [84–86]. In addition, there are reports of Cu-mediated cleavage of the hinge region of antibodies [87]. It has been reported that histidine, commonly used as a buffer for protein formulations, can exacerbate leaching of metals from stainless steel, leading to increased degradation rates [88]. At the same time, small amounts of His were tolerated and even stabilizing.

In addition to LMW species, it is possible for foreign matter to be desorbed from surfaces during processing, forming potential nuclei for subsequent aggregation. The potential for this has been seen for delamination of glass (i.e., glass shedding), causing nucleation-dependent aggregation of PAFase [89]. The desorption could be exacerbated by shear forces (see Section 25.7).

With more products being packaged in prefilled syringes, the potential for leachables causing protein degradation has increased dramatically [70]. In this particular report, tungsten particles were found, originating from the wires used to extrude the syringe tips. In addition, organic compounds from the plunger were also identified. Metals, such as barium, have been found as leachates [69]. Some of these have been shown to activate copurified proteases that result in subsequent degradation of the active protein [69,90].

Finally, there have been great concerns over silicone oil. Potentially, the droplets of silicones could be released from the syringe surface and could lead to interfacial damage of proteins. One study on the susceptibility of proteins to silicones showed that the amount needed to cause aggregation was relatively high [91]. Still, the concern over the ability of silicones in prefilled syringes to cause damage persists. This may be due to long storage times or the contributions of other degradation mechanisms. Certainly, more information is needed to understand fully the role that silicones play in protein stability. As more products move to prefilled syringes, approaches to deal with silicones, or devices that can perform without them, will be increasingly necessary.

25.7. SHEAR FORCES

The literature describing shear-induced damage to proteins goes back to the early 1980s. There is a wide variety of reports regarding shear-induced damage to proteins. Therefore, it is necessary to review the literature and summarize what really is known.

Early reports described shear-induced denaturation of proteins, especially enzymes [92–95]. The idea was that the force applied by shear was sufficient to unfold the protein, ultimately leading to aggregation. However, an increasing number of papers have appeared contradicting this view [62,96,97]. It is now widely believed that proteins, even larger proteins like antibodies, are too small to allow sufficient force to be applied to unfold them. In fact, one study estimated that the shear rate required to unfold a small protein would be in excess of $10^7\ s^{-1}$ [96]. Part of the difficulty in interpreting results from shear studies is that, appropriate controls are often lacking. Moreover, may articles purporting to be shear studies are, in fact, agitation studies [60,61,98–101]. In addition, even if the air–water interface is rigorously excluded, the ability of the protein to interact with solid interfaces (glass, plastic, metal, etc.) is rarely considered and may be a source of damage in of itself.

Still, there are cases where protein aggregation is clearly attributed to shear forces [102–104]. If protein denaturation is not the mechanism, how does shear promote aggregation? Four potential mechanisms can be envisioned:

1. *Cavitation-Producing Microbubbles*. As the microbubbles collapse, they generate high transient temperature that leads to free-radical formation, resulting in S–S crosslinked aggregates [102,105].
2. *Turbulent Flow*. Turbulent flow that occurs during high shear could dislodge unfolded protein from solid surfaces [106].
3. *Increased Mixing*. This is caused by high shear, leading to higher growth rates for aggregation [107,108].
4. *Shear*. Shear forces can dislodge foreign material from surfaces, which are capable of acting as prenuclei for nucleation-controlled aggregation [89].

While the appearance of foreign matter in protein products has not yet been attributed directly to shear, it is a likely result. When foreign matter does appear, it could serve as the prenucleus for subsequent aggregation. The foreign matter would serve as a seed to which the protein could adsorb. Certainly, having partially native protein adsorbed to large particles could arise serious immunogenicity concerns [90,109–112]. What is becoming apparent is that shear and interfacial stress are interrelated, as shown by a study on a monoclonal antibody [113]. While interfacial stress causes the initial damage, it appears that shear forces can exacerbate these instabilities. These factors will likely make shear forces on proteins an active area of research for some time.

25.8. PATHOGEN INACTIVATION

Over the years, there has been a proliferation of methods for inactivation of pathogens. This includes prolonged heating, acidification, solvent detergent treatments, and nanofiltration. In addition, there is interest in developing true terminal sterilization methods using high-energy radiation, as used in the food industry. Often, more than one of these approaches is used for a given product to ensure maximal safety. Each of these methods is valuable and has been evaluated for its ability to reduce pathogen loads. However, all of them have the potential for causing significant damage to protein pharmaceuticals as well. In this section, we focus on two of these techniques, heat treatment and γ-irradiation.

25.8.1. Heat Treatment

For proteins in solution, the standard protocol has been to heat at 60°C for 10 h. Not surprisingly, many proteins do not survive such extended heating. In the case of human serum albumin (HSA), one of the first proteins to be treated using this method, the denaturation temperature of the protein, is very near to 60°C if stabilizers are not present. As a result, researchers in the plasma fractionation industry discovered excipients that stabilized HSA specifically and significantly. Two of them are used in nearly all HSA products: sodium caprylate and *N*-acetyltryptophan (*N*-AcTrp). In the presence of these two, the T_m value for HSA rises to well above 70°C [114,115]. In addition, these two stabilizers may help stabilize HSA with respect to solvent–detergent treatment as well. In the case of HSA, it appears that the impurity protein, haptoglobin, is involved in the thermally induced aggregation of HSA during heat treatment (Lin, Meyer, Carpenter, and Manning, unpublished results). Haptoglobin is much less stable than HSA. On heating, it unfolds and begins to aggregate. Because the HSA is present at much higher concentration (up to 1000-fold higher), the growing haptoglobin aggregate incorporates some HSA as well. Therefore, it is important to note that minor protein components may lead to loss of the primary active ingredient. In addition, HSA aggregated on freeze drying after the initial plasma fractionation step [116]. Subsequent purification could not remove the small amount of aggregate. Then, heat treatment led to a higher level of aggregate, presumably where the initially damaged material served as a nucleus.

A screening study was performed for urokinase, which was derived from circulatory organs. During the 60°C heat treatment, a significant amount of the protein precipitated. Therefore, a search for conditions or excipients that would reduce the loss was initiated. The result was identification of some compounds, most notably gelatin, that reduced the extent of damage significantly [117,118]. Of note is the fact that urokinase itself is thermally stable enough to survive the 60°C heat treatment. However, there was an impurity, some *N*-terminally clipped material, in each batch. This protein was much less stable and unfolded during the heating. As it aggregated, it captured some of the intact urokinase in the aggregate [117,118]. This is an important lesson for formulation scientists. If there are other protein species present, their stability profile may be such that aggregation of the less stable protein might result in loss of the active ingredient.

Another type of degradation process has also been reported to occur during heat treatment. Smales and co-workers report that prolonged heating of sucrose-containing formulations results in thermal breakdown of the sucrose and subsequent chemical modification of the protein in a Maillard-type reaction [119–121]. Similarly, glycation can occur at elevated temperature in the presence of maltodextrins [122]. More recently, glycation of an IgG2 monoclonal antibody was reported in sucrose formulations [123]. Although no glycation was seen for samples stored at 4°C, even after

18 months, there was significant glycation for samples stored at room temperature or above. As many as 10 different lysine residues were identified as glycation sites in this study. The reactivity of the each lysine appears to correlate with the degree of solvent exposure. A similar study has been reported for interferon-β and HSA [124].

Glycated proteins appear to be more likely to aggregate [125]. There are few reports on how to formulate against glycation-induced damage. One study revealed that trehalose and 6-aminohexanoic acid were effective stabilizers for the enzyme glucsose-6-phosphate dehydrogenase [126].

25.8.2. Gamma Irradiation

The potential for high-energy irradiation to inactivate pathogens has been known for some time. The difficulty is preventing damage to a biological, such as a protein, while inactivating bacteria and viruses. This has become even more challenging as smaller and smaller viruses are being found, not to mention attempts to inactivate prions. The use of high-energy radiation for sterilization of biopharmaceuticals has been reviewed (i.e., γ rays, electron beams) in some detail [127,128]. Specific issues related to protein instability are the focus of this section.

High-energy irradiation, whether from γ rays, which have been used in attempts to sterilize pharmaceutical products directly, or from electron beams, used to sterilize equipment, causes damage by one of two primary mechanisms. The first is direct interaction of the radiation with the protein, leading to free-radical generation and/or chain scission. The second is irradiation of the solvent (usually water), producing a host of free radicals, especially hydroxyl radical. This second pathway is the most common, as the protein usually occupies a very small volume fraction of the sample. Generation of these highly reactive oxygen species produces a variety of oxidation products.

Two approaches can reduce the extent of damage: (1) one can move to lyophilized formulations, although the chance for the generated free radicals to react with the protein will be great on reconstitution [129]; or (2) one can use free-radical scavengers (also known as *radioprotectants* or *antioxidants*) to reduce the damage that occurs to the active ingredient. A number of radioprotectants have been identified, or at least proposed, in various studies [130–136]. A radioprotectant is a compound that acts a free-radical scavenger and reduces or limits the extent of degradation to an irradiated protein. Some improved schemes for evaluating and rank ordering of antioxidants/radical scavengers have been reported [137,138]. While radioprotectants are valuable excipients, the extent of damage to proteins, even though it can be reduced or limited, is probably still too high to allow it to be an effective general measure that can be taken to reduce viral loads. Certainly, additional work is needed to assess whether γ-irradiation can be an effective sterilization tool for biopharmaceutical products.

25.9. SUMMARY

Fortunately, there is increasing awareness of stability issues that can arise with protein pharmaceuticals during processing. On the other hand, there are few systematic studies.

This is due in part to the time and financial pressures of bringing a new product to market. Certainly, there is a hesitancy to admit that there are issues with the manufacturing process for obvious reasons. As a result, our understanding of the potential for protein degradation during bioprocessing is limited, apart from anecdotal examples. This chapter captures what is known in this area, one that will continue to be of prime importance as the biopharmaceutical industry continues to evolve and mature.

REFERENCES

1. Cromwell, M. E. M., Hilario, E., and Jacobson, F. (2006), Protein aggregation and bioprocessing, *AAPS J.* **8**: E572–E579.
2. Bondos, S. E., and Bicknell, A. (2003), Detection and prevention of protein aggregation before, during, and after purification, *Anal. Biochem.* **316**: 223–231.
3. Hober, S., Nord, K., and Linhult, M. (2007), Protein A chromatography for antibody purification, *J. Chrom. B* **848**: 40–47.
4. Philips, J., Drumm, A., Harrision, P., Bird, P., Bhamra, K., Berrie, E., and Hale, J. (2001), Manufacture and quality control of CAMPATH-1 antibodies for clinical trials, *Cytotherapy* **3**: 233–242.
5. Lewis, J. D. and Nail, S. L. (1997), The influence of ion exchange chromatography conditions on aggregation of equine IgG, *Process. Biochem.* **32**: 279–283.
6. Xiao, Y., Freed, A. S., Jones, T. T., Makrodimitris, K., O'Connell, J. P., and Fernandez, E. J. (2006), Protein instability during HIC: Describing the effects of mobile phase conditions on instability and chromatographic retention, *Biotechnol. Bioeng.* **93**: 1177–1189.
7. Jones, T. T., and Fernandez, E. J. (2003), α-Lactalbumin tertiary structure changes on hydrophobic interaction chromatography surfaces, *J. Colloid Interface Sci.* **259**: 27–35.
8. Fausnaugh, J. L., Kennedy, L. A., and Regnier, F. E. (1984), Comparison of hydrophobic interaction and reverse phase chromatography of proteins, *J. Chrom.* **317**: 141–155.
9. Bhatnagar, B. S., Bogner, R. H., and Pikal, M. J. (2007), Protein stability during freezing: Separation of stresses and mechanisms of protein stabilization, *Pharm. Devel. Technol.* **12**: 505–523.
10. van den Berg, L. (1959), The effect of addition of sodium and potassium chloride to the reciprocal system: KH_2PO_4-Na_2HPO_4-H2O on pH and composition during freezing, *Arch. Biochem. Biophys.* **84**: 305–315.
11. Sarciaux, J. M., Mansour, S., Hageman, M. J., and Nail, S. L. (1999), Effects of buffer composition and processing conditions on aggregation of bovine IgG during freeze-drying, *J. Pharm. Sci.* **88**: 1354–1361.
12. Pikal-Cleland, K. A. and Carpenter, J. F. (2001), Lyophilization-induced protein denaturation in phosphate buffer systems: Monomeric and tetrameric beta-galactosidase, *J. Pharm. Sci.* **90**: 1255–1268.
13. Anchordoquy, T. J. and Carpenter, J. F. (1996), Polymers protect lactate dehydrogenase during freeze-drying by inhibiting dissociation in the frozen state, *Arch. Biochem. Biophys.* **332**: 231–238.
14. Follmer, C., Pereira, F. V., Silveira, N. P., and Carlini, C. R. (2004), Jack bean urease (EC 3.5.1.5) aggregation monitored by dynamic and static light scattering, *Biophys. Chem.* **111**: 79–87.

15. Carpenter, J. F., Pikal, M. J., Chung, B. S., and Randolph, T. W. (1997), Rational design of stable lyophilized protein formulations: some practical advice, *Pharm. Res.* **14**: 969–975.
16. Shalaev, E. Y. and Kanev, A. N. (1994), Study of the solid-liquid state diagram of the water-glycine-sucrose system, *Cryobiology* **31**: 374–382.
17. Fitzpatrick, S. and Saklatvala, R. (2003), Understanding the physical stability of freeze dried dosage forms from the glass transition temperature of amorphous components, *J. Pharm. Sci.* **92**: 2504–2510.
18. Pyne, A., Chatterjee, K., and Suryanarayanan, R. (2003), Crystalline to amorphous transition of disodium hydrogen phosphate during primary drying, *Pharm. Res.* **20**: 802–803.
19. Franks, F. (1990), Freeze-drying- from empiricism to predictability, *Cryo-Letters* **11**: 93–110.
20. Roos, Y. H. (1993), Melting and glass transitions of low molecular weight carbohydrates, *Carbohydr. Res.* **238**: 39–48.
21. Martin, S. W. H. and Mo, J. (2007), Stability considerations for lyophilized biologics, *Am. Pharm. Rev.* **10**(4): 31.
22. Chang, B. S., Kendrick, B. S., and Carpenter, J. F. (1996), Surface-induced denaturation of proteins during freezing and its inhibition by surfactants, *J. Pharm. Sci.* **85**: 1325–1330.
23. Strambini, G. B. and Gabellieri, E. (1996), Proteins in frozen solutions: Evidence of ice-induced partial unfolding, *Biophys. J.* **70**: 971–976.
24. Luykx, D. M. A. M., Casteleijn, M. G., Jiskoot, W., Westdijk, J., and Jongen, P. M. J. M. (2004), Physicochemical studies on the stability of influenza haemagglutinin in vaccine bulk material, *Eur. J. Pharm. Sci.* **23**: 65–75.
25. Randolph, T. W. and Jones, L. S. (2002), Surfactant-protein interactions, in *Rational Design of Stable Protein Formulations: Theory and Practice*. Carpenter, J. F. and Manning, M. C., eds., Kluwer Academic/Plenum Publishers, New York, pp. 159–175.
26. Hillgren, A., Lindgren, J., and Alden, M. (2002), Protection mechanism of Tween 80 during freeze-thawing of a model protein, LDH, *Int. J. Pharm.* **237**: 57–69.
27. Kerwin, B. A., Heller, M. C., Levin, S. H., and Randolph, T. W. (1998), Effects of Tween 80 and sucrose on acute short-term stability and long-term storage at −20°C of a recombinant hemoglobin, *J. Pharm. Sci.* **87**: 1062–1068.
28. Krielgaard, L., Jones, J. S., Randolph, T. W., Frokjaer, S., Flink, J. M., Manning, M. C., and Carpenter, J. F. (1998), Effect of Tween 20 on freeze-thawing and agitation-induced aggregation of recombinant human factor XIII, *J. Pharm. Sci.* **87**: 1597–1603.
29. Cao, E., Chen, Y., Cui, Z., and Foster, P. R. (2003), Effect of freezing and thawing rates on denaturation of proteins in aqueous solutions, *Biotechnol. Bioeng.* **82**: 684–690.
30. Jiang, S. and Nail, S. L. (1998), Effect of process conditions on recovery of protein activity after freezing and freeze-drying, *Eur. J. Pharm. Biopharm.* **45**: 249–257.
31. Hatley, R. H. M. and Franks, F. (1989), The cold-induced denaturation of lactate dehydrogenase at sub-zero temperatures in the absence of perturbants, *FEBS Lett.* **257**: 171–173.
32. Hsu, C. C., Nguyen, H. M., Yeung, D. A., Brooks, D. A., Koe, G. S., Bewley, T. A., and Pearlman, R. (1995), Surface denaturation at solid-void interface—a possible pathway by which opalescent participates form during the storage of lyophilized tissue-type plasminogen activator at high temperatures, *Pharm. Res.* **12**: 69–77.
33. Bhatnagar, B. S., Nehm, S. J., Pikal, M. J., and Bogner, R. H. (2005), Post-thaw aging affects activity of lactate dehydrogenase, *J. Pharm. Sci.* **94**: 1382–1388.

34. Prestrelski, S. J., Pikal, K. A., and Arakawa, T. (1995), Optimization of lyophilization conditions for recombinant interleukin-2 by dried-state conformational-analysis using Fourier-transform infrared-spectroscopy, *Pharm. Res.* **12**: 1250–1259.
35. Schebor, C., del Pilar Buera, M., and Chirife, J. (1996), Glassy state in relation to the thermal inactivation of the enzyme invertase in amorphous dried matrices of trehalose, maltodextrin and PVP, *J. Food Eng.* **30**: 269–282.
36. Chang, B. S., Beauvais, R. M., Dong, A., and Carpenter, J. F. (1996), Physical factors affecting the storage stability of freeze-dried interleukin-1 receptor antagonist: glass transition and protein conformation, *Arch. Biochem. Biophys.* **331**: 249–258.
37. Zhou, P. and Labuza, T. P. (2007), Effect of water content on glass transition and protein aggregation of whey protein powders during short-term storage, *Food Biophys.* **2**: 108–116.
38. Privalov, P. L. (1990), Cold denaturation of proteins, *Crit. Rev. Biochem. Mol. Biol.* **25**: 281–305.
39. Thirumangalathu, R., Krishnan, S., Bondarenko, P., Speed-Ricci, M., Randolph, T. W., Carpenter, J. F., and Brems, D. N. (2007), Oxidation of methionine residues in recombinant human interleukin-1 receptor antagonist: Implications of conformational stability on protein oxidation kinetics, *Biochemistry* **46**: 6213–6224.
40. Fransson, J., Florin-Robertsson, E., Axelsson, K., and Nyhlen, C. (1996), Oxidation of human insulin-like growth factor I in formulation studies: Kinetics of methionine oxidation in aqueous solution and in solid state, *Pharm. Res.* **13**: 1252–1257.
41. Wisniewski, R. (1998), Developing large-scale cryopreservation systems for biopharmaceutical products, *BioPharm* **11**: 50–60.
42. Kerwin, B. A., Chang, B. S., Gegg, C. V., Gonnelli, M., Li, T. S., and Strambini, G. B. (2002), Interactions between PEG and type I soluble tumor necrosis factor receptor: Modulation by pH and by PEGylation at the N terminus, *Protein Sci.* **11**: 1825–1833.
43. Carpenter, J. F., Prestrelski, S. J., and Dong, A. (1998), Application of infrared spectroscopy to development of stable lyophilized protein formulations, *Eur. J. Pharm. Biopharm.* **45**: 231–238.
44. Remmele, R. L., Stushnoff, C., and Carpenter, J. F. (1997), Real-time in situ monitoring of lysozyme during lyophilization using infrared spectroscopy: Dehydration stress in the presence of sucrose, *Pharm. Res.* **14**: 1548–1555.
45. Schwegman, J. J., Carpenter, J. F., and Nail, S. L. (2007), Infrared microscopy for in situ measurement of protein secondary structure during freezing and freeze-drying, *J. Pharm. Sci.* **96**: 179–195.
46. Huisman, I. H., Prádanos, P., and Hernández, A. (2000), The effect of protein-protein interactions on membrane fouling in ultrafiltration, *J. Membrane Sci.* **179**: 79–90.
47. Saksena, S. and Zydney, A. L. (1997), Influence of protein-protein interactions on bulk mass transport during ultrafiltration, *J. Membrane Sci.* **125**: 93–108.
48. Koehler, J. A., Ulbricht, M., and Belfort, G. (1997), Intermolecular forces between proteins and polymer films with relevance to filtration, *Langmuir* **13**: 4162–4171.
49. Palecek, S. P. and Zydney, A. L. (1994), Intermolecular electrostatic interactions and their effect on flux and protein deposition during protein filtration, *Biotechnol. Progress* **10**: 207–213.
50. Huisman, I. H., Tragardh, G., and Tragardh, C. (1999), Particle transport in crossflow microfiltration—II. effects of particle-particle interactions, *Chem. Eng. Sci.* **54**: 281–289.

51. Maa, Y.-F. and Hsu, C. C. (1996), Membrane fouling in sterile filtration of recombinant human growth hormone, *Biotechnol. Bioeng.* **50**: 319–328.

52. Belfort, G. and Lee, C. S. (1991), Attractive and repulsive interactions between and within adsorbed ribonuclease A layers, *Proc. Natl. Acad. Sci. USA* **88**: 9146–9150.

53. Salgin, S. (2007), Effects of ionic environments on bovine serum albumin fouling in a cross-flow ultrafiltration system, *Chem. Eng. Technol.* **30**: 255–260.

54. Meireles, M., Aimar, P., and Sanchez, V. (1991), Albumin denaturation during ultrafiltration—effects of operating- conditions and consequences on membrane fouling, *Biotechnol. Bioeng.* **38**: 528–534.

55. Stoner, M. R., Fischer, N., Buckel, S., Benke, M., Austin, F., Randolph, T. W., and Kendrick, B. S. (2004), Protein-solute interactions affect the outcome of ultrafiltration/diafiltration operations, *J. Pharm. Sci.* **93**: 2332–2342.

56. Zydney, A. L. (1992), Concentration effects on membrane sieving: Development of a stagnant film model incorporating the effects of solute-solute interactions, *J. Membrane Sci.* **68**: 183–190.

57. Kim, K. J., Chen, V., and Fane, A. G. (1993), Some factors determining protein aggregation during ultrafiltration, *Biotechnol. Bioeng.* **42**: 260–265.

58. Maruyama, T., Katoh, S., Nakajima, M., and Nabetani, H. (2001), Mechanism of bovine serum albumin aggregation during aggregation, *Biotechnol. Bioeng.* **75**: 233–238.

59. Fan, H., Vitharana, S. N., Chen, T., O'Keefe, D., and Middaugh, C. R. (2007), Effects of pH and polyanions on the thermal stability of fibroblast growth factor 20, *Mol. Pharm.* **4**: 232–240.

60. Thomas, C. R., Nienow, A. W., and Dunnill, P. (1979), Action of shear on enzymes—studies with alcohol-dehydrogenase, *Biotechnol. Bioeng.* **21**: 2263–2278.

61. Watterson, J. G., Schaub, M. C., and Waser, P. G. (1974), Shear-induced protein-protein interaction at air-water-interface, *Biochem. Biophys. Acta* **356**: 133–143.

62. Maa, Y. F. and Hsu, C. C. (1997), Protein denaturation by combined of shear and air-liquid interface, *Biotechnol. Bioeng.* **54**: 503–512.

63. Bommarius, A. S. and Karau, A. (2005), Deactivation of formate dehydrogenase (FDH) in solution and at gas-liquid interfaces, *Biotechnol. Progress* **21**: 1663–1672.

64. Liu, J., Nguyen, M. D. H., Andya, J. D., and Shire, S. J. (2005), Reversible self-association increases the viscosity of a concentrated monoclonal antibody in aqueous solution, *J. Pharm. Sci.* **94**: 1928–1940.

65. Liu, J. and Shire, S. J. (2005), *Reduced-Viscosity Concentrated Protein Formulations*, US Patent 6,875,432 (April 5, 2005).

66. Monkos, K. and Turczynski, B. (1999), A comparative study on viscosity of human, bovine and pig IgG immunoglobulins in aqueous solutions, *Int. J. Biol. Macromol.* **26**: 155–159.

67. Geirsdottir, M., Hlynsdottir, H., Thorkelsson, G., and Sigurgisladottir, S. (2007), Solubility and viscosity of herring (*Clupea Harengus*) proteins as affected by freezing and frozen storage, *J. Food Sci.* **72**: 376–380.

68. Waris, B. N., Hasan, U., and Srivastava, N. (2001), Stabilisation of ovalbumin by maltose, *Thermochim. Acta* **375**: 1–7.

69. Markovic, I. (2006), Challenges associated with extractable and/or leachable substances in therapeutic biologic protein products, *Am. Pharm. Rev.* **9**: 20–27.

70. Wen, Z., Torraca, G., Yee, C., and Li, G. (2007), Investigation of contaminants in protein pharmaceuticals in pre-filled syringes by multiple micro-spectroscopies, *Am. Pharm. Rev.* **10**: 101–106.

71. Wahlgren, M. and Arnebrant, T. (1991), Protein adsorption to solid surfaces, *Trends Biotechnol.* **9**: 201–207.

72. Gray, J. J. (2004), The interaction of proteins with solid surfaces. *Curr. Opin. Struct. Biol.* **14**: 110–115.

73. Wang, P. L., Udeani, G. O., and Johnston, T. P. (1995), Inhibition of granulocyte colony stimulating factor (G-CSF) adsorption to polyvinyl chloride using a nonionic surfactant, *Int. J. Pharm.* **114**: 177–184.

74. Reyes, N., Ruiz, L., Aroche, K., Geronimo, K., Brito, O., and Hardy, E. (2005), Stability of Ala125 recombinant human interleukin-2 in solution, *J. Pharm. Pharmacol.* **57**: 31–37.

75. Twardowski, Z. J., Nolph, K. D., McGary, T. J., and Moore, H. L. (1983), Influence of temperature and time on insulin adsorption to plastic bags, *Am. J. Hosp. Pharm.* **40**: 583–586.

76. Doran, P. M. (2006), Loss of secreted antibody from transgenic plant tissue cultures due to surface adsorption, *J. Biotechnol.* **122**: 39–54.

77. Young, B. R., Pitt, W. G., and Cooper, S. L. (1988), Protein adsorption on polymeric biomaterials 1. Adsorption isotherms, *J. Colloid Interface Sci.* **124**: 28–43.

78. Desroches, M. J., Chaudhary, N., and Omanovic, S. (2007), PM-IRRAS investigation of the interaction of serum albumin and fibrinogen with a biomedical-grade stainless steel 316LVM surface, *Biomacromolecules* **8**: 2836–2844.

79. Mutlu, S., Cokeliler, D., and Mutlu, M. (2007), Modification of food contacting surfaces by plasma polymerization technique. Part II: Static and dynamic adsorption behavior of a model protein "bovine serum albumin" on stainless steel surface, *J. Food Eng.* **78**: 494–499.

80. Muller, R., Abke, J., Schnell, E., Macionczyk, F., Gbureck, U., Mehrl, R., Ruszczak, Z., Kujat, R., Englert, C., Nerlich, M., and Angele, P. (2005), Surface engineering of stainless steel materials by covalent collagen immobilization to improve implant biocompatibility, *Biomaterials* **26**: 6962–6972.

81. Matsumura, H. and Saburi, M. (2006), Protein adsorption and wetting of the protein adsorbed surfaces studied by a new type of laser reflectometer, *Colloid. Surf. Biointerfaces* **47**: 146–152.

82. Santos, O., Nylander, T., Schillen, K., Paulsson, M., and Tragardh, C. (2006), Effect of surface and bulk solution properties on the adsorption of whey protein onto steel surfaces at high temperature, *J. Food Eng.* **73**: 174–189.

83. Osterberg, R. E. (2005), Potential toxicity of extractables and leachables in drug products, *Am. Pharm. Rev.* **8**: 64.

84. Hovorka, S. W., Williams, T. D., and Schoneich, C. (2002), Characterization of the metal-binding site of bovine growth hormone through site-specific metal-catalyzed oxidation and high-performance liquid chromatography- tandem mass spectrometry, *Anal. Biochem.* **300**: 206–211.

85. Hovorka, S. W., Hong, J., Cleland, J. L., and Schöneich, C. (2001), Metal-catalyzed oxidation of human growth hormone: modulation by solvent-induced changes of protein conformation, *J. Pharm. Sci.* **90**: 58–69.

86. Khossravi, M., Shire, S. J., and Borchardt, R. T. (2000), Evidence for the involvement of histidine A(12) in the aggregation and precipitation of human relaxin induced by metal-catalyzed oxidation, *Biochemistry* **39**: 5876–5885.
87. Smith, M. A., Easton, M., Everett, P., Lewis, G., Payne, M., Riveros Moreno, V., and Allen, G. (1996), Specific cleavage of immunoglobulin G by copper ions, *Int. J. Pept. Protein Res.* **48**: 48–55.
88. Chen, B., Bautista, R., Yu, K., Zapata, G. A., Mulkerrin, M. G., and Chamow, S. M. (2003), Influence of histidine on the stability and physical properties of a fully human antibody in aqueous and solid forms, *Pharm. Res.* **20**: 1952–1960.
89. Chi, E. Y., Weickmann, J., Carpenter, J. F., Manning, M. C., and Randolph, T. W. (2005), Heterogeneous nucleation-controlled particulate formation of recombinant human platelet-activating factor acetylhydrolase in pharmaceutical formulation, *J. Pharm. Sci.* **94**: 256–274.
90. Rosenberg, A. S. and Worobec, A. S. (2005), A risk-based approach to immunogenicity concerns for therapeutic protein products, *BioPharm* **18**: 32.
91. Jones. L. S., Kaufman, A., and Middaugh, C. R. (2005), Silicone oil induced aggregation of proteins, *J. Pharm. Sci.* **94**: 918–927.
92. Charm, S. E. and Wong, B. L. (1981), Shear effects on enzymes, *Enzyme Microb. Technol.* **3**: 111–118.
93. Reese, E. T. and Ryu, D. Y. (1980), Shear inactivation of cellulase of *Trichoderma reesei, Enzyme Microb. Technol.* **2**: 239–240.
94. Charm, S. E. and Wong, B. L. (1970), Enzyme inactivation with shearing, *Biotechnol. Bioeng.* **12**: 1103–1109.
95. Charm, S. E. and Wong, B. L. (1970), Shear degradation of fibrinogen in circulation, *Science* **170**: 466–468.
96. Jaspe, J. and Hagen, S. J. (2006), Do protein molecules unfold in a simple shear flow? *Biophys. J.* **91**: 3415–3424.
97. Speigel, T. (1999), Whey protein aggregation under shear conditions—effects of lactose and heating temperature on aggregate size and structure, *Int. J. Food Sci. Technol.* **34**: 523–531.
98. Oliva, A., Santove na, A., Fari na, J., and Llabr es, M. (2003), Effect of high shear rate on stability of proteins: kinetic study, *J. Pharm. Biomed. Anal.* **33**: 145–155.
99. Byrne, E. P. and Fitzpatrick, J. J. (2002), Investigation of how agitation during precipitation, and subsequent processing affects the particle size distribution and separation of α-lactalbumin enriched whey protein precipitates, *Biochem. Eng. J.* **10**: 17–25.
100. Colombie, S., Gaunand, A., Rinaudo, M., and Lindet, B. (2000), Irreversible lysozyme inactivation and aggregation induced by stirring: kinetic study and aggregates characterization, *Biotechnol. Lett.* **22**: 277–283.
101. Yu, Z., Johnston, K. P., and Williams III, R. O. (2006), Spray freezing into liquid versus spray-freeze drying: Influence of atomization on protein aggregation and biological activity, *Eur. J. Pharm. Sci.* **27**: 9–18.
102. Morel, M.-H., Redl, A., and Guilbert, S. (2002), Mechanism of heat and shear aggregation of wheat gluten protein upon mixing, *Biomacromolecules* **3**: 488–497.
103. Akkermans, C., Venema, P., Rogers, S. S., van der Goot, A., Boom, R. M., and van der Linden, E. (2006), Shear pulses nucleate fibril formation, *Food Biophys.* **1**: 144–150.

104. Hill, E. K., Krebs, B., Goodall, D. G., Howlett, G. J., and Dunstan, D. E. (2006), Shear flow induced amyloid fibril formation, *Biomacromolecules* **7**: 10–13.

105. Maa, Y-F. and Hsu, C. C. (1996), Effect of high shear on proteins, *Biotechnol. Bioeng.* **51**: 458–465.

106. Santos, O., Nylander, T., Paulsson, M., and TrägÅrdh, C. (2006), Whey protein adsorption onto steel surfaces- effects of temperature, flow rate, residence time and aggregation, *J. Food Eng.* **74**: 468–483.

107. Belmar-Beiny, M. T., Gotham, S. M., Paterson, W. R., and Fryer, P. J. (1993), The effect of Reynolds number and fluid temperature in whey protein fouling, *J. Food Eng.* **19**: 119–139.

108. Simmons, M. J. H., Jayaraman, P., and Fryer, P. J. (2006), The effect of temperature and shear rate upon the aggregation of whey protein and its implications for milk fouling, *J. Food Eng.* **79**: 517–528.

109. Rosenberg, A. S. (2006), Effects of protein aggregates: An immunologic perspective, *AAPS J.* **8**: (article 59).

110. Sharma, B. (2007), Immunogenicity of therapeutic proteins. Part 1: Impact of product handling, *Biotechnol. Adv.* **25**: 310–317.

111. Sharma, B. (2007), Immunogenicity of therapeutic proteins. Part 2: Impact of container closures, *Biotechnol. Adv.* **25**: 318–324.

112. Sharma, B. (2007), Immunogenicity of therapeutic proteins. Part 3: Impact of manufacturing changes, *Biotechnol. Adv.* **25**: 325–331.

113. Biddlecombe, J. G., Craig, A. V., Zhang, H., Uddin, S., Mulot, S., Fish, B. C., and Bracewell, P. G. (2007), Determining antibody stability: Creation of solid-liquid interfacial effects within a high shear environment, *Biotechnol. Progress* **23**: 1218–1222.

114. Ross, P. D., Finlayson, J. S., and Shrake, A. (1984), Thermal-stability of human-albumin measured by differential scanning calorimetry. 2. Effects of isomers of n-acetyltryptophanate and tryptophanate, pH, reheating, and dimerization, *Vox Sang.* **47**: 19–27.

115. Shrake, A., Finlayson, J. S., and Ross, P. D. (1984), Thermal-stability of human-albumin measured by differential scanning calorimetry. 1. Effects of caprylate and N-acetyltryptophanate, *Vox Sang.* **47**: 7–18.

116. Lin, J.-J., Meyer, J. D., Carpenter, J. F., and Manning, M. C. (2000), Stability of human serum albumin during bioprocessing: Denaturation and aggregation during processing of albumin paste, *Pharm. Res.* **17**: 391–396.

117. Porter, W. R., Staack, H., Brandt, K., Carpenter, J. F., and Manning, M. C. (1993), Thermal stability of low molecular weight urokinase during heat treatment. I. Effects of protein concentration, pH and ionic strength, *Thromb. Res.* **71**: 265–279.

118. Vrkljan, M., Powers, M. E., Foster, T. M., Henkin, J., Porter, W. R., Carpenter, J. F., and Manning, M. C. (1994), Thermal stability of low molecular weight urokinase during heat treatment. II. Effects of polymeric additives, *Pharm. Res.* **11**: 1004–1008.

119. Smales, C. M., Pepper, D. S., and James, D. C. (2001), Protein modification during antiviral heat bioprocessing and subsequent storage, *Biotechnol. Progress*, **17**: 974–978.

120. Smales, C. M., Pepper, D. S., and James, D. C. (2000), Protein modification during antiviral heat bioprocessing, *Biotechnol. Bioeng.* **67**: 177–188.

121. Smales, C. M., Pepper, D. S., and James, D. C. (2000), Mechanisms of protein modification during model anti-viral heat-treatment bioprocessing of β-lactoglobulin variant A in the presence of sucrose, *Biotechnol. Appl. Biochem.* **32**: 109–119.

122. Sutthirak, P., Dharmsthiti, S., and Lertsiri, S. (2005), Effect of glycation on stability and kinetic parameters of thermostable glucoamylase from *Aspergillus niger, Process Biochem.* **40**: 2821–2826.

123. Gadgil, H. S., Bondarenko, P. V., Pipes, G., Rehder, D., Mcauley, A., Perico, N., Dillon, T., Ricci, M., and Treuheit, M. (2007), The LC/MS analysis of glycation of igG molecules in sucrose containing formulations, *J. Pharm. Sci.* **96**: 2607–2621.

124. Zheng, X., Wu, S. L., and Hancock, W. S. (2006), Glycation of interferon-beta-1b and human serum albumin in a lyophilized glucose formulation: Part III: Application of proteomic analysis to the manufacture of biological drugs, *Int. J. Pharm.* **322**: 136–145.

125. Bouma, B., Kroon-Batenburg, L. M. J., Wu, Y. P., Brunjes, B., Posthuma, G., Kranenburg, O., de Groot, P. G., Voest, E. E., and Gebbink, M. F. B. G. (2003), Glycation induces formation of amyloid cross-β structure in albumin, *J. Biol. Chem.* **278**: 41810–41819.

126. Ganea, E. and Harding, J. J. (2005), Trehalose and 6-aminohexanioc acid stabilize and renature glucose-6-phosphate dehydrogenase inactivated by glycation and by guanidinium hydrochloride, *Biol. Chem.* **386**: 269–275.

127. Jacobs, G. P. (2007), Radiation sterilization of parenterals, *Pharm. Technol. Suppl.* s32.

128. Yamamoto, O. (1992), Effect of radiation on protein stability, in *Stability of Protein Pharmaceuticals*, Ahern, T. J. and Manning, M. C., eds., Plenum Press, New York, pp. 361–421.

129. Molina, M. D., Armstrong, T. K., Zhang, Y., Patel, M. M., Lentz, Y. K., and Anchordoquy, T. J. (2004), The stability of lyophilized lipid/DNA complexes during prolonged storage, *J. Pharm. Sci.* **93**: 2259–2273.

130. Grieb, T., Forng, R-. Y., Brown, R., Owolabi, T., Maddox, E., McBain, A., Drohan, W. N., Mann, D. M., and Burgess, W. H. (2002), Effective use of gamma irradiation for pathogen inactivation of monoclonal antibody preparations, *Biologicals* **30**: 207–216.

131. Yu, H., Sabato, S. F., D'Aprano, G., and Lacroix, M. (2004), Effect of the addition of CMC on the aggregation behaviour or proteins, *Rad. Phys. Chem.* **71**: 129–133.

132. Ness, G. C., Pendleton, L. C., and McCreery, M. J. (2005), Target size analysis by radiation inactivation: the use of free radical scavengers, *Exp. Biol. Med.* **230**: 455–463.

133. Durschlag, H., Hefferle, T., and Zipper, P. (2003), Comparative investigations of the effects of X- and UV-irradiation on lysozyme in the absence or presence of additives, *Rad. Phys. Chem.* **67**: 479–486.

134. Krumhar, K. C., and Berry, J. W. (1990), Effect of antioxidant and conditions on solubility of irradiated food proteins in aqueous solutions, *J. Food Sci.* **55**: 1127–1132.

135. Moon, S. and Song, K. B. (2001), Effect of γ-irradiation on the molecular properties of ovalbumin and ovomucoid and protection by ascorbic acid, *Food Chem.* **74**: 479–483.

136. Cho, Y., Yang, J. S., and Song, K. B. (1999), Effect of ascorbic acid and protein concentration on the molecular weight profile of bovine serum albumin and β-lactoglobulin by γ-irradiation, *Food Res. Int.* **32**: 515–519.

137. Bertolini, F., Novaroli, L., Carrupt, P.-A., and Reist, M. (2007), Novel screening assay for antioxidant protection against peroxyl radical-induced loss of protein function, *J. Pharm. Sci.* **96**: 2931–2944.

138. Tomer, D. P., McLeman, L. D., Ohmine, S., Scherer, P. M., Murray, B. K., and O'Neill, K. L. (2007), Comparison of the total oxyradical scavenging capacity and oxygen radical absorbance capacity antioxidant assays, *J. Med. Food* **10**: 337–344.
139. Rambhatla, S., Ramot, R., Bhugra, C., and Pikal, M. J. (2004), Heat and mass transfer scale-up issues during freeze drying: II. Control and characterization of the degree of super cooling, *AAPS PharmSciTech.* **5**: (article 58).
140. Allison, S. D., Molina, M. D. C., and Anchordoquy, T. J. (2000), Stabilization of lipid/DNA complexes during the freezing step of the lyophilization process: the particle isolation hypothesis, *Biochim. Biophys. Acta* **1468**: 127–138.

26

FREEZING AND THAWING OF PROTEIN SOLUTIONS

Satish K. Singh and Sandeep Nema

26.1. INTRODUCTION

Freezing is an increasingly common unit operation employed in the production of biologics. Protein drug substance is often frozen to allow the bulk to be maintained for a longer period prior to its conversion to drug product. Interactions with water and with the solid phase of water, ice, play a critical role in the stabilization of biologics. This chapter discusses the behavior of solution as it is being frozen, the impact of the freezing (and thawing) process on protein structure (stability), and critical factors that should be considered in designing and developing a successful process for freezing protein solutions. Significant work on freezing has been done in the areas of food, tissue, cells, and organ preservation. Although much of this work has direct application in freezing of proteins, cell viability and complex cellular matrix effects are not involved, making the protein solution system somewhat simpler!

Formulation and Process Development Strategies for Manufacturing Biopharmaceuticals,
edited by Jameel and Hershenson

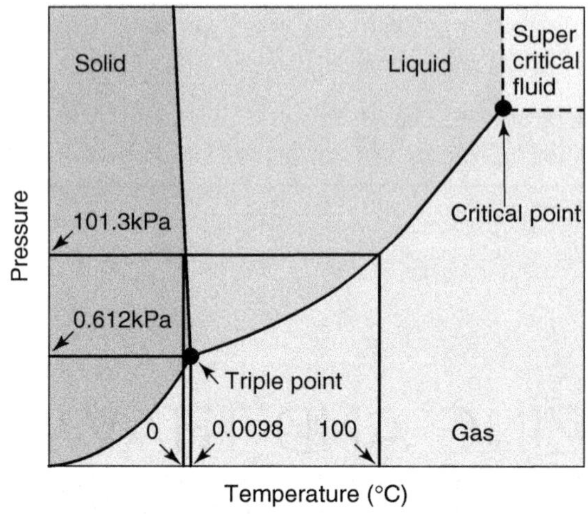

Figure 26.1. Equilibrium phase diagram of water.

26.2. THE BASICS OF FREEZING

26.2.1. Freezing of Water (Ice–Water Phase Diagram)

Equilibrium phase diagrams of water provide a starting point toward understanding the behavior of frozen systems. Water is a universal solvent used for biologics. Its phase behavior in the normally accessible and relevant temperature ranges can be represented by the equilibrium phase diagram shown in Figure 26.1. Pure liquid water under equilibrium conditions will turn into ice (I_h, hexagonal form) at about 0°C at 1 atm pressure. An important feature of this phase diagram is a slight negative slope of the water–ice freezing curve, indicating the expansion in molar volume of water when it freezes. The freezing point of water, under pressure, is lowered slightly, but it takes significant pressure, for instance, almost 600 bar for a reduction to −4°C. At triple point (273 K or 0°C) the vapor pressures for ice and water are the same: 4.59 mm Hg (Torr). At this temperature, all three phases (ice, water, and vapor) coexist, and in accordance with the Gibbs phase rule, there are no degrees of freedom. At temperatures above 647 K (374°C), water cannot exist in a liquid state. The fluid then shares the properties of gas, and no vapor pressure beyond this temperature is measurable. The point is called the *critical point*.

26.2.2. Equilibrium Freezing of Aqueous Solutions

Addition of solutes (excipients) to water results in the well-known colligative lowering of the freezing point in accordance with Raoult's law. The extent of lowering for aqueous solutions amounts to about 1.8°C per molal unit of dissolved molecules or ions. More important to the processing of biologics is the depression of freezing point due to supercooling, which is considered later in this section.

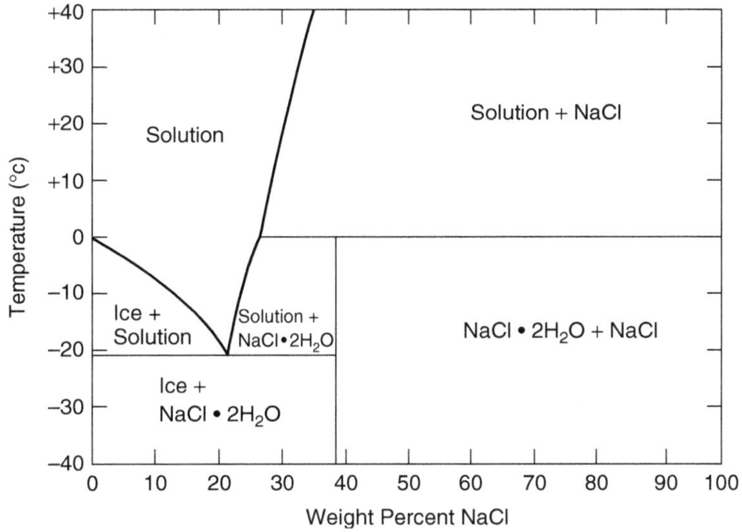

Figure 26.2. Equilibrium freezing of sodium chloride–water solution. (Adapted with permission from Cocks and Brower [1].)

Two (or more) component equilibrium phase diagrams describe the order of crystallization and phase compositions at any temperature. The common binary system sodium chloride (NaCl)–water diagram is illustrated in Figure 26.2 [1]. The important feature to note is the presence of the eutectic at −21.2°C with a composition of 23.3 wt% NaCl. The freezing of normal saline solution, containing about 150 mM (0.9 wt%) NaCl, all the way to the eutectic point will result in portions of the bulk (and the protein therein) experiencing an approximately 26-fold increase in salt concentration. It also shows that the common "frozen" storage temperature of −20°C may not be a truly frozen state. Other commonly useful binary phase diagrams are available: sucrose–water [2–4], trehalose–water [5], sorbitol–water, and maltitol–water [6].

Addition of a third common formulation ingredient (e.g., sucrose) to the NaCl–water system results in a complex ternary phase diagram [3,4] containing anhydrous and hydrated crystalline phases, and a hydrated stoichiometric compound of sucrose and NaCl. Again, as with the binary NaCl–water system, dramatic concentration changes occur as pure water crystallizes out in the form of the ice phase on cooling. This cryoconcentration effect depends on the initial composition as shown in Figure 26.3 [3], and varies with the components constituting the ternary system.

Detailed or partial phase diagrams are available for only a limited number of other systems: water–glycerol–NaCl [7,8], water–DMSO–NaCl [1,7], water–hydroxyethyl starch–NaCl [9], water–sucrose–glycine [10,11], water–glycine–raffinose, and water–glycine–trehalose [12]. Since most formulations for biologics use a buffer instead of simply water as the solvent, the equilibrium behavior will be altered compared to that indicated in the phase diagrams.

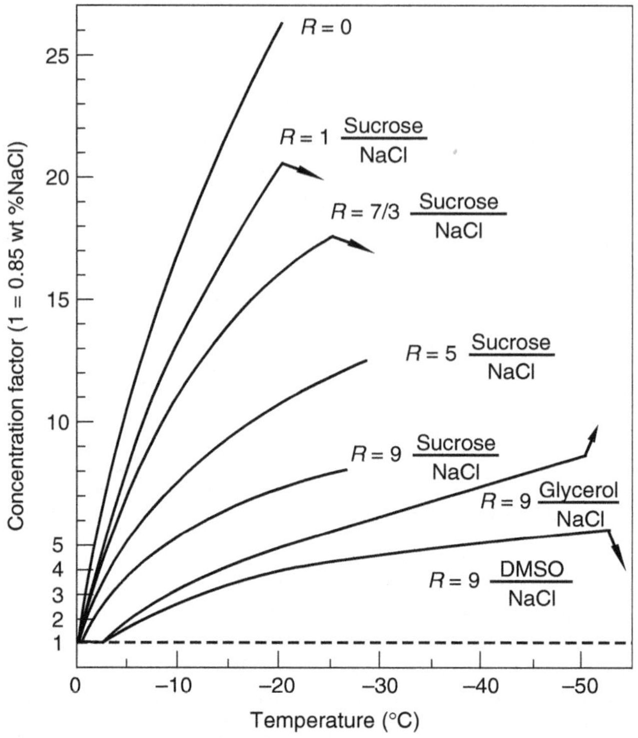

Figure 26.3. Changes in concentration of sodium chloride as a function of temperature and sucrose ratio during equilibrium freezing of sucrose–NaCl–water ternary system; R is the wt% ratio of sucrose (or cosolute) to NaCl. Curves for DMSO and glycerol as the cosolute instead of sucrose are also provided. A normal saline solution ($R = 0$) undergoes cryoconcentration by a factor of almost 26 (0.85% to ~22.5%) under equilibrium freezing. However, this is strongly attenuated by the presence of the co-solute. By nature of the solute concentration phenomenon, the curves either terminate when a glass is formed, or change slope discontinuously when a second phase precipitates. The direction of the arrow indicates the change in concentration of NaCl in the secondary phase. (Adapted with permission from Gayle et al. [3].)

26.3. NONEQUILIBRIUM FREEZING AND STATE DIAGRAMS

While equilibrium phase diagrams provide good guidance on the phases possible for a formulation undergoing freezing, practical systems and processes seldom lead to equilibrium behavior.

26.3.1. The Freezing Process

Freezing of an aqueous solution in practice seldom proceeds as an equilibrium process, and the events that occur cannot be predicted by equilibrium phase diagrams,

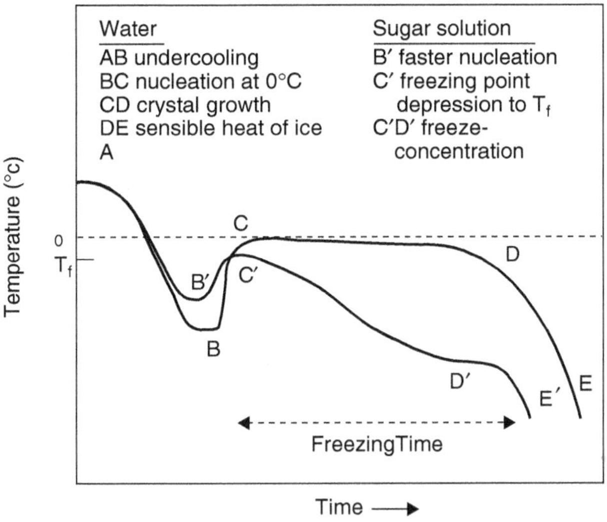

Figure 26.4. Time–temperature relationship during freezing of water and of an aqueous solution, illustrating the different stages of the process. (Adapted with permission from Akyurt et al. [13] and Goff [14].)

even where such data exist. The physicochemical changes that occur during a practical freezing process can be described by the time–temperature relationship diagram shown in Figure 26.4 for freezing of pure water ($ABCDE$) and an aqueous solution ($AB'C'D'E'$) [13,14]. As pure water is cooled from A, the first deviation from equilibrium behavior is in the form of undercooling (or supercooling) below the freezing point ($\sim$0°C) to B. Water in this chilled state is thermodynamically unstable, and the state is terminated by ice nucleation at B. The undercooling enables the activation energy for ice nucleation to be lowered enough to form stable nuclei and propagate the bulk freezing process. In aqueous solutions, presence of solute will generally promote heterogeneous nucleation, thereby accelerating the nucleation process, resulting in lesser undercooling (B'). (Homogeneous and heterogeneous ice nucleation phenomena are considered in greater detail later in this chapter.)

Once the critical mass of nuclei is reached at point B or B', most of the water starts to convert to ice and the latent heat that is released raises the temperature of the system almost instantaneously to the freezing/melting temperature (0°C or T_f) to point C or C'. The enthalpy of fusion releases -333.6 kJ/kg of energy, which is significant in relation to the heat capacity of water at $\sim$0°C (4.2 kJ kg^{-1} K^{-1}) and that of hexagonal ice I_h at 0°C (2.1 kJ kg^{-1} K^{-1}). The presence of solute results in a depression of freezing point as discussed earlier, and therefore $C'(= T_f)$ is lower than C (= 0°C). Presence of solute lowers the degree of undercooling for the solution compared to that for pure water, due to depression of the freezing point itself, and by enabling faster nucleation. In very concentrated solutions, it becomes progressively difficult to induce any degree of undercooling.

The ice crystals continue to grow as long as there is "freezable" water available. Temperature of the partially frozen mixture also does not drop during this period. In pure water, this occurs from C to D, after which point continued cooling lowers the temperature of ice along DE. *Freezing time* is defined as the time between onset of nucleation (C) and completion of crystal growth phase (D). The nature of the ice structure formed along CD depends on the freezing rate, with faster freezing rates promoting many small crystals compared to few large crystals at slower rates.

As the aqueous solution freezes, conversion of the water fraction into ice results in a progressive freeze concentration (cryoconcentration), of the unfrozen mixture. Solutes seldom partition into the ice phase. The increasing concentration of the non-ice-phase solute results in a continuously decreasing freezing point of the remaining solution along $C'D'$. The viscosity of the unfrozen mixture also increases as a consequence of the low temperature and high solute concentration, reducing diffusion coefficients and hindering the ability of water molecules to rearrange and form ice crystals. At D', one of the solutes may become supersaturated and crystallize (e.g., NaCl), release its enthalpy of crystallization, and cause a slight jump in temperature. This would indicate the presence of a eutectic. Further cooling would result in continued crystallization of water and possible solute until the system is completely solidified. If the solution does not contain a crystallizable solute (e.g., sucrose), the increasing viscosity would eventually result in the entire unfrozen fraction converting to glass ($C'D'E'$). A freezing time is difficult to assign, and therefore *process time* is defined as the time taken to reach a predetermined temperature below the initial freezing point [13].

26.3.2. State Diagram and the Glass Transition

The process described above can be advantageously shown on a state diagram; an augmented version is presented schematically in Figure 26.5 [15,16]. The reason the diagram is termed *state diagram* as opposed to the *phase diagram* becomes clear later since it incorporates information about the nonequilibrium states in which the system can exist. These nonequilibrium states are metastable from a thermodynamic perspective but kinetically stable in the timescales of practical interest.

In Figure 26.5, the freezing curve (ABC) and solubility curve (BD) are shown in relation to the glass transition curve (EF). The glass transition curve is not a phase boundary, unlike (ABC) and (BD), and instead represents an isoviscosity curve of a homogenous solid solution, the freeze concentrate. According to the state diagram, a dilute solution (protein and solute) on cooling will become progressively more concentrated as water component freezes out as pure ice. At low temperatures, ice formation will be faster than crystallization/precipitation of other solutes, and the composition of the non–ice phase will continue past point B. At sufficiently high solute concentrations, the remaining solution will become sufficiently viscous to be called a *glass* (viscosity $>10^{10}$ Pa s). In an ideal situation, the concentrated solution will reach point Q, which is the intersection between the extended freezing curve and the glass transition curve, defined as T_g' and W_g'. Thus, dilute solutions will usually exhibit a glass transition at T_g' on freezing. On the other hand, concentrated solutions will pass through

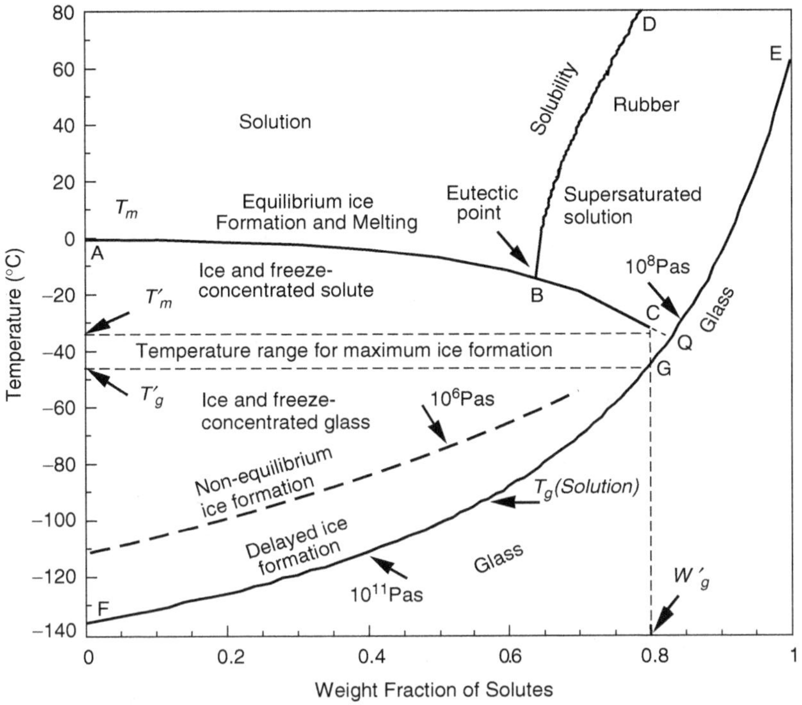

Figure 26.5. State diagram depicting different regions encountered during nonequilibrium freezing of a noncrystallizing solute in water. Curve *ABC* represents the equilibrium freezing boundary, while *BD* is the solubility curve; *EF* is an isoviscosity curve of homogeneous solid solution. During the freezing of a dilute solution, composition of the nonice phase continues past *B*, eventually producing a highly viscous glass at point *Q* with a characteristic glass transition temperature T'_g and composition of maximum freeze concentration W'_g. Maximum ice formation occurs in the region above T'_g but below the equilibrium ice melting temperature T'_m of ice in contact with the maximally freeze-concentrated matrix. Ice formation is, however, time-dependent because of the high viscosity of the unfrozen matrix, and rapid cooling through this temperature range can lead to partial freeze concentration and nonequilibrium ice formation. Therefore, devitrification of the freeze-concentrated glass is possible in case of nonequilibrium or delayed ice formation while stored below T'_g, leading to further ice formation and freeze concentration. (Adapted with permission from Roos and Karel [15,16].)

the solubility curve *BD* on cooling and eventually reach the glass transition curve *EF*. While passing through curve *BD*, precipitation of the solute should occur, but this is often hindered by the high viscosity of such systems. The extent (or lack) of precipitation determines the final glass composition. An individual concentration-dependent T_g is observed. Thus, T_g is concentration-dependent for concentrated systems, while T'_g is concentration-independent for dilute systems. [It should be noted that there is a small difference in the way T'_g and W'_g are depicted on the state diagram. They are variously ascribed to point *G* or *Q* (see, e.g., Refs. 2 and 16–18); point *Q* is more commonly used.] Different zones can also be marked in the state diagram showing different characteristics of the states produced [19].

The glass transition temperature depends on the nature of the glass former, its molecular weight and shape, and the concentrations of any other components (e.g., buffer salts) in the glass [20]. Subtle changes in chemical structure and water bonding impact the glass transition curve, exemplified by the T_g curves for trehalose and sucrose; the curve for trehalose is left-shifted by about 12% for the same T_g [2]. A comprehensive compilation of T_g' and W_g' is available [18,21], although some values are disputed [22]. The T_g' values generally increase with molecular weight, but W_g' is accepted to be approximately 80% for most carbohydrates [16,23]. The T_g' for pure water is generally accepted to be around 135–140 K (∼−135°C) and is the lower end of the glass transition curve, although Angell [2] suggests that the extrapolation by which this value is obtained may be misleading.

Proteins in solution also do not crystallize on cooling, and, because they are polymeric, exhibit a glass transition temperature. The T_g' for proteins tends to range around −10°C, irrespective of size and structure, for example, −11°C for ovalbumin, −13°C for lysozyme, −15°C for myoglobin, −11°C for BSA, and −9°C for lactic dehydrogenase [24].

26.3.3. The Glassy Matrix

Sugars, polyhydric alcohols, certain amino acids, and higher oligosaccharides are commonly used as stabilizers for biologics in liquid, lyophilized, and frozen states [25]. The utility of these excipients in the frozen state is a result of their inability to crystallize and the lack of eutectic phase separation. Most are subject to vitrification or glass formation as the freezing proceeds. A notable exception to this behavior of carbohydrates is mannitol, which readily crystallizes out [26,27]. Another important characteristic of the stabilizers used in protein formulations is that they should not phase-separate during freezing in order to maintain their protective effect [28–30]. Common parenteral formulation excipients and their likely behavior at the concentrations normally used in drug products are summarized below. Note that in the case of polymeric excipients, the behavior is dependent on the molecular weight, while the actual behavior of a solute in a solution depends on its concentration and interference by other solutes:

- Excipients and components that are likely to crystallize (and therefore phase-separate) are mannitol, raffinose, inositol, glycine, dibasic sodium phosphate, sodium chloride, polyethylene glycol, sodium lacate, and pegylated proteins.
- Excipients and components that are likely to remain amorphous are sucrose, trehalose, lactose, sorbitol,[1] glycerol, histidine, tromethamine, glycolate, succinate, acetate, citrate, polyvinylpyrrolidone (PVP), dextran, and nonionic surfactant.

[1]Sorbitol has been found to crystallize on storage at −30° C (above its T_g' of −44° C), although not at −20° C [31].

- Amorphous excipients and components that can phase-separate are trehalose, PVP, and dextran.

The relevance of the glass formation phenomenon and storage of a frozen biologic arises from the extremely high viscosity of the glassy matrix. The high viscosity hinders diffusion and therefore slows down any degradation reactions that require mobility of reactants (oxygen, water, side groups, peptide bonds). However, the high concentration of solutes in the glassy matrix results in reaction rate enhancements and more than counteracts the effect of low mobility, as has been shown in a number of studies [32–35]. In fact, an undercooled solution may provide greater stability than the frozen state [36]. Interestingly, under frozen conditions, endoproteinases (α-chymotrypsin, papain, ficin, etc.) are capable of coupling free amino acids, functioning as reverse carboxypeptidases. Competitive reactions are suppressed, and peptide yields are increased. The effect is ascribed to cryoconcentration, but other factors such as favorable conformation changes, increased proton mobility in ice, changes in dielectric behavior, and reduced water activity are also considered to contribute to the reversal of action [37–40].

Arrhenius kinetics applies for homogeneous supercooled solutions but fail as the solution freezes and the solute concentration increases. As a consequence, kinetics of many processes in the glassy state switch over to Williams–Landel–Ferry (WLF) temperature dependence and diffusion-controlled kinetics [18,20,21, 41–43], given by

$$\ln\left(\frac{k}{k_0}\right) = \frac{-C_1(T - T_0)}{C_2 + (T - T_0)} \qquad (26.1)$$

where T_0 is a reference temperature, generally taken to be T_g. The exact choice of T_0 is debated when applying the WLF equation, originally developed to describe viscosity changes, to other rate parameters [41,44–46]. The applicability of WLF equation implies that molecular relaxation times and related viscosity and molecular mobility phenomena acquire a temperature-dependent activation energy, while the Arrhenius model supposes constant activation energy. In the WLF regime, reaction rates are impacted by diffusional limitations of the reactants in the viscous matrix. The diffusion coefficient is inversely proportional to viscosity, and the viscosity follows a WLF dependence, resulting in WLF reaction kinetics. If applicable, the WLF equation shows that the reaction rate constant is more sensitive to temperature near T_g than would be predicted by the Arrhenius equation. A 10° change in temperature will change the rate by a factor of 2 in the Arrhenius model and by 10^4–10^5 in the WLF model [47].

A theoretical analysis of diffusion-controlled reactions in frozen sucrose solutions by Parker and Ring [48] suggests the existence of two reaction regimes. In the higher temperature range, between 0°C and −16.5°C, the effect of freeze concentration dominates, especially at low sucrose concentrations, and reaction rate enhancement is predicted. For low sucrose concentrations, a small reduction in temperature results in a large freeze concentration effect and a correspondingly large reaction rate enhancement. Below this temperature range, the diffusivity of reactant becomes rate-limiting and the reaction rate drops progressively as glass transition is approached. As sucrose concentration is increased, the effect of cryoconcentration is attenuated and the diffusive effect dominates. Therefore, in practical situations in which the effects of cryoconcentration are deleterious, the impact can be reduced by the use of concentrated solutions [48,49].

The cryoconcentration effect is not limited to solute; it also applies to dissolved gases. Solubility of oxygen in water increases significantly with a decrease in temperature from 8.11 mg/L at 25°C, 1 atm to 14.16 mg/L at 0°C, 1 atm [50]. However, solubility of air in ice is lower than in water, leading to cryoconcentration of the gas in the amorphous region between ice crystals and possible bubble formation in the final glassy matrix. Progression of cooling and freezing therefore proceeds with an increase in oxygen concentration in the nonfrozen liquid phase, with the possibility of promoting oxidation reaction in susceptible molecules. Equilibrium solubility of gases will also be higher in the amorphous interfacial region between ice crystals.

26.4. FREEZING

26.4.1. Ice Nucleation and Growth

Thermodynamic equilibrium between different phases of water (pure or in solution) exists along well-defined phase boundary lines. Changes of phase (liquid $\leftrightarrow$ vapor, vapor $\leftrightarrow$ solid, solid $\leftrightarrow$ liquid) take place at conditions either slightly or significantly away from these lines. These deviations are usually minor for changes from the more ordered to the less ordered states (liquid $\rightarrow$ vapor, solid $\rightarrow$ liquid, solid $\rightarrow$ vapor), but can be very large for the reverse processes, even though these changes are toward states of lower free energy. Phase changes toward lower energy states occur via metastable states [51]. A phase map for a water solution is represented schematically in Figure 26.6 [16], and illustrates the relationship between the various stable and metastable states and the pathways for transformation between them.

Supercooled liquid water is one such metastable state. The process of conversion from the supercooled state to the stable phase is initiated by nucleation from within the liquid phase. The process of nucleation and subsequent crystal growth is important in determining the actual distribution of phases or states in the frozen bulk.

26.4.1.1. Nucleation. The molecules in the liquid (water) phase are subject to small, transient energy and density fluctuations as a result of Brownian diffusion, described by a Boltzmann distribution. Occasionally, these fluctuations will lead to spontaneous formation of clusters in which groups of molecules adopt configurations that resemble the solid (ice) phase. These clusters are shortlived since the free energy of the cluster is higher than the sum of energies of the molecules making up the cluster. However, as the liquid is supercooled, the energetics of the process changes and the bulk free energy per molecule of the molecules in the cluster becomes lower than outside it. The lower bulk energy is opposed by the surface energy involved in creating the interface between the cluster and the dispersed phase, but the balance changes in favor of the cluster as the cluster grows larger. Beyond a certain size, further increase actually leads to a lowering of total free energy, and growth becomes more favorable than diminution. The size at which this occurs is called a *critical size* and the cluster (variously called *germ* or *embryo* also) at that point becomes a nucleus for further growth of the solid phase. The event is called *nucleation* [51].

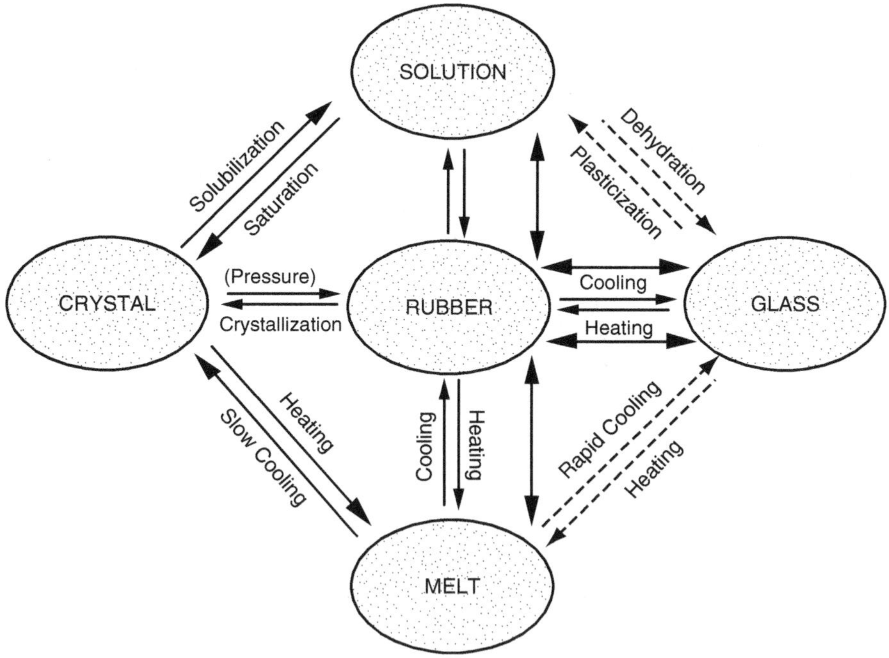

Figure 26.6. Phase map for water-soluble materials showing equilibrium state (crystal, melt, solution) and nonequilibrium metastable states (glass, rubber) with time-dependent properties, and the pathways for transition. Changes between equilibrium and glassy states occurs through the rubbery state. (Adapted with permission from Roos and Karel [16].)

26.4.1.2. Homogeneous and Heterogeneous Nucleation. In the case of pure water or a very clean solution, a stable nucleus has to form solely through internal fluctuations leading to the process of homogenous nucleation. In the case of a supercooled phase in contact with some foreign material or surface, the cluster can attach to the substrate and increase its stability. The substrate thus serves to lower the energy barrier for formation of the nucleus and leads to heterogeneous nucleation. Under otherwise identical conditions, heterogeneous nucleation always occurs at a temperature higher than that of homogeneous nucleation. Hence homogeneous nucleation serves as the lower temperature limit to which a liquid can be supercooled without freezing. The actual ice nuclei are, however, identical in both cases [51,52].

The temperature at which heterogeneous nucleation occurs depends strongly on the presence and suitability of the substrate. Suitability is determined by the match between the structure of the nucleating site and the ice crystal lattice [53]. In contrast, the homogeneous nucleation temperature (T_{hom}) is an intrinsic property of the metastable liquid itself. Since the requirement for homogenous nucleation is quite stringent (no foreign matter in the sample of liquid) the requirement is seldom met by anything except the smallest sample volumes. The homogenous ice nucleation temperature of pure water has been measured in microliter-sized samples and is in the region of 233 K (−40°C) [17,51,52]. Heterogeneous nucleation can occur anywhere between the melting temperature and T_{hom}, thereby determining the degree of supercooling.

Presence of solutes in water depresses T_{hom}, that is, inhibits homogeneous ice nucleation. An interesting relationship is observed between the extent of reduction in homogeneous nucleation temperature and the depression of freezing point T_{m}. A number of pharmaceutically relevant solutes (sucrose, glucose, NaCl, urea, glycerol, ethylene glycol) appear to fall within a narrow band of slope 2 ± 0.2 (i.e., $-\Delta T_{\text{hom}} \approx -2\Delta T_m$), with polymeric excipients, PEG and PVP, yielding plots of higher slopes [54]. Koop [52] explains this behavior by the "weak" temperature dependence of water activity in the solutions. However, Wilson et al. [55], in a reanalysis of the data from MacKenzie [54], state that the ratio varies between 1.3 and 2.5 and suggest that if sufficiently accurate measurements were made taking into account the stochastic nature of the nucleation phenomena, the ratio would be closer to unity. The relationship is independent of nature of solute for nonionic solutes.

Similar inhibition by added solute has been seen on heterogeneous nucleation temperature in a study of the impact of sucrose on nucleation due to freeze-dried *Psuedomonas syringae* bacteria [56,57]. However, ice nucleation can also be inhibited by mechanisms other than freezing-point depression, possibly through disruption of the embryonic ice crystals [58]. Antifreeze proteins and glycoproteins [also referred to as *ice structuring proteins* or (ISPs) found in some fish, insects, and plants prevent growth of ice crystals at very low concentrations (on the order of ≤1%). The mechanism is purported to be through adsorption onto ice surfaces, thereby pinning the surface by distinctly placed protein molecules. This causes the ice front to grow with local positive curvature, which is less thermodynamically favorable. The result is a nonequilibrium and noncolligative depression of the nucleation temperature below the melting point [58–61]. Other additives that can inhibit growth of nuclei include poly(vinyl alcohol) and certain nonionic and anionic surfactants [58,59].

The *nucleation rate* is defined as the probability per unit time and unit volume that a critical cluster (germ, embryo) size will be obtained. Homogenous nucleation rate is determined by an internal energy term, ΔG_t^*, incorporating terms for interfacial energy and critical radius. It can be written as

$$J(T) = A \exp \frac{-\Delta G^*}{kT} \tag{26.2}$$

with A dependent on the characteristics of the medium; $J(T)$ is the temperature-dependent volume nucleation rate coefficient of ice in water [62]. From a large number of measurements, a consensus empirical parameterization of $J(T)$ has been obtained, and shows a strong temperature dependence of J with temperature. Rate of nucleation increases by a factor of nearly 50 for each 1°C lowering of temperature. Critical cluster sizes have been estimated for homogenous freezing. At −40°C, −20°C, and −5°C, the critical radius are 0.8, 1.8, and 7.0 nm, respectively, corresponding roughly to 70, 650, and 45,000 molecules of water in the cluster [17,51]. This model also explains why −40°C is considered to be a limit for supercooling of pure water, although under extreme cooling rates ($\sim 10^{6}$°C/s), water is deemed to have been turned into glass even after supercooling down to −196°C [63,64]. Critical cluster size increases rapidly above −40°C. Hence the probability of homogenous nucleation becomes negligibly small just a few degrees above −40°C. However, practical systems usually undergo heterogeneous nucleation, and are nearly impossible to undercool down to −40°C [51,55].

A corresponding theoretical treatment for heterogeneous nucleation rate that relates to observation is difficult to formulate, although attempts have been made using surface energies and contact angles [51,52,65]. The rate is often based on the homogenous rate modified by the activity of the foreign surface toward nucleation (incorporated into the exponential term). The activity term is a function of the contact angle between the crystal and substrate. This treatment suggests that the critical free energy is a fraction of that required for homogenous nucleation, and is a function of the contact angle. The smaller the contact angle (i.e., the greater the wetting), the lower is the energy barrier. The size of the critical cluster is also smaller. For example, the critical cluster size for heterogeneous (contact angle 30°) nucleation at −5°C is about 600 molecules compared to about 45,000 for homogeneous nucleation [51,53,65].

In pharmaceutical terms, even sterile filtered solutions filled into clean tubular glass vials in a nearly particle-free environment cannot fulfill the requirements for homogeneous nucleation. Presence of microscopic particulate(s) or surface (roughness) in the container–closure system almost inevitably leads to heterogeneous nucleation. Commonly seen supercooling levels in the authors' laboratories as well as reported in the literature lie in the range from 5°C to 15°C [66].

26.4.1.3. Ice Growth. Once a nucleus is formed, ice grows at a rate that is proportional to the degree of supercooling [67]. However, a self-limiting factor is the capacity for removal of the enthalpy of fusion from the system. Ice grows much more slowly when free in solution than on a surface, where the heat can be conducted away more readily. The cooling system geometry, design, and the volume being processed are important factors that determine the overall process time. It was found that the linear ice propagation velocity in a solution at first increases with increased supercooling [56,57]. However, above a critical undercooling level the propagation velocity decreases as the undercooling increases.

Most studies in the literature on the impact of freezing stresses on proteins are done on small samples on the order of microliters L to a few milliliters L. These small samples freeze rapidly when compared to large-scale systems, despite efforts to distinguish between "slow" and "fast" freezing. Information generated from such studies should be therefore be analyzed in the context of the rate "achieved" as opposed to the "intended" rate and will be considered later in the chapter. Jennings [66] estimates that a glass vial containing 1 mL of water, when undercooled by 10°C, will fill up with ice in about 0.20 s once nucleation has been triggered.

26.4.1.4. The Mushy Zone: Planar Versus Dendritic Ice. When supercooling is present, the phase transition region may have a distinct thickness. This region is referred to as the "mushy zone" [13,66]. The microstructure in this region can appear to be dendritic, columnar, or amorphous as shown in Figure 26.7. Among these, the

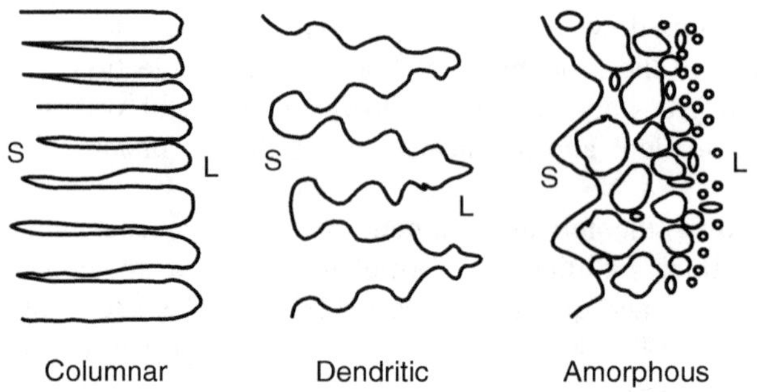

Figure 26.7. Possible interfacial morphologies during growth of ice at the phase transition zone. The presence of such interfacial zones can attenuate the degree of resultant cryoconcentration. (Adapted with permission from Akyurt et al. [13].)

dendritic structure is of particular interest, as the phenomenon has been stated as a key factor in the design and function of large-scale freeze–thaw units designed for storage of biologics [68].

As liquid water turns to ice, the latent heat of fusion is liberated at the interface between ice and liquid. If the ice crystal is in contact with a heat sink, then heat flows through the crystal as a result of its higher thermal conductivity. This situation leads to a planar ice front, because any instability that advances in front of the interface (into the liquid) will be at a disadvantage for shedding the latent heat, due to the greater distance between the surface and the heat sink and to the greater surface area at which more heat is generated. Conversely, if nucleation occurs in bulk solution (i.e., the crystal is not in contact with a heat sink), then the latent heat must be shed into the liquid. This situation leads to an unstable planar interface, because the instability that moves into the liquid, in front of the planar interface, will shed its heat to a greater volume of liquid [69].

A growing ice crystal will also exclude any solutes present in the solution. These will concentrate at the interface and lower the freezing point of the solution just in front of the interface. If the interface is planar, then the high solute concentration at the interface will stop ice growth at a temperature below the freezing point of the bulk solution, thus leaving the solution further away from the ice front in a supercooled state. This phenomenon, known as *constitutional supercooling*, is probably a significant consideration for the cryopreservation of tissues and organs. When it occurs, the ice front is even more unstable, as any protuberance will not only shed latent heat better but also grow into a supercooled solution, once it passes through the initial region of concentrated solute [69,70].

Ice growth in a normal physiological solution creates such a situation, and instabilities grow into the supercooled compartment via *dendritic growth*. With dendritic growth, the ice crystal extends throughout the solution and encapsulates solute in unfrozen channels. There is no region of the solution that is shielded from the ice crystal; thus supercooling does not occur to any significant degree. Furthermore, ice reaches all regions of the solution, as the initial planar ice front is broken down and dendritic crystals extend throughout the container. The effect is a lower macrocryoconcentration, although in the amorphous regions between the ice crystals, the trapped solute will still be subject to micro- or localized cryoconcentration to the maximal freeze concentrated composition, W_g' (Figure 26.5).

In summary, the effect of added solutes in an aqueous solution is to depress the nucleation temperature (homogeneous or heterogeneous), increase nucleation rate, and inhibit crystal growth [53]. This has important consequences for the stability of the frozen biologic.

26.4.1.5. Partitioning of Solute Between Ice and Water. During the freezing process, the rate at which ice crystallizes out of solution is determined largely by the rate of removal of the enthalpy of crystallization. If this rate is higher than the diffusion of solutes away from the growing ice front, the ice may entrap areas of concentrated solution between ice crytals. Although less common, if the solute interacts with or adsorbs to the ice front, it may partition itself between the solution and the ice interface [71]. Chen et al. [72] found that the solute inclusion in ice increases with increasing bulk concentration of solute, increasing ice growth rate, and decreasing stirring at the ice front. This phenomenon is used in progressive freeze concentration for either concentrating a solute or purifying a solvent [73–75].

When freezing a biologic, the opposite effect is desired, namely, avoidance of freeze concentration and solute polarization. Clearly, this effect can be avoided by increasing ice growth rate and avoiding stirring or mixing in the container being frozen. Furthermore, promotion of dendritic ice growth will lead to local solute entrapment and thus a lowering of the macroscopic cryoconcentration effect [68].

Gaseous solutes are also rejected by an advancing ice front and therefore become more concentrated in the liquid phase. If the solubility in the liquid phase is exceeded, gas bubbles may nucleate. A distribution coefficient of 0.05 has been estimated for oxygen in water ahead of a planar ice front [70,76]. Gases, however, have a much larger solubility in ice than do electrolytes such as NaCl, which are commonly used with biologics [70].

26.4.2. Rate of Cooling or Freezing

Nucleation and growth are kinetic processes. The final structure of ice and the distribution of solutes is therefore determined by the rates at which these processes occur, which are governed by the design of the system (geometry, size, heat distribution patterns, cooling rate, etc.). We restrict our discussion here to the applied cooling rate. Note that the cooling rate (= rate of heat removal) and freezing rate are related but the latter will change over time as freezing proceeds.

Cooling rate determines the degree of supercooling that can be achieved but does not directly control it. In the absence of nucleation, higher cooling rates will produce greater supercooling. Nucleation, when it does occur in such a situation, will be rapid and lead to a greater amount of nuclei being formed since the critical cluster size is smaller at lower temperatures. At completion of freezing, the average ice crystal ice size will therefore be smaller compared to a slowly cooled system with a lower degree

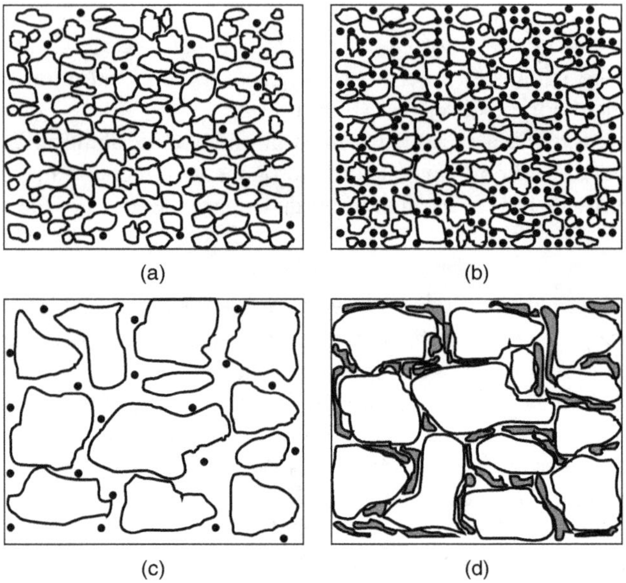

Figure 26.8. Schematic depiction of ice and solute morphologies in (a) dilute solution with high supercooling and small ice crystals, (b) concentrated solution with high supercooling and small ice crystals, (c) dilute solution with low supercooling and large ice crystals, and (d) concentrated solution with low supercooling and large ice crystals. (Adapted with permission from Engstrom et al. [78].)

of supercooling. Freezing rate and the resulting ice structure are therefore dependent on the degree of supercooling achieved prior to nucleation [77].

This general picture of process-dependent ice morphology has implications for the distribution of the solutes, illustrated in Figure 26.8 [78]. The fine and dense ice formed as a result of high degree of supercooling (i.e., large number of nuclei) implies narrow channels and small domains of glassy matrix containing the solute. If the solution is dilute, the solute domains may remain finely divided and may not become contiguous over the timecourse of the freezing process, and will finally be held in place by the glass (Fig. 26.8a). If the solution is concentrated, there is a greater likelihood for the solute domains to coalesce (Fig. 26.8b). If the degree of supercooling is low (i.e., if there are fewer nuclei), the ice will have more time to grow, leading to fewer and larger ice domains, with broader domains for the glassy matrix also. Again, in dilute solutions, the solutes in the glass domains will not encounter each other to coalesce (Fig. 26.8c). However, in concentrated solutions, the slow timescale of the freezing process provides time for the coalescence to occur (Fig. 26.8d). Such maximally freeze concentrated solutes may allow incompatibilities between solutes to manifest themselves at high concentrations [79].

In practice, the rate of cooling is determined by the engineering aspects of the system, including coolant temperature, coolant flow rate, heat transfer dimensions, and size of the bulk being cooled. Most industrial lyophilizers achieve cooling rates of 1—10 K/min, which is not particularly rapid. However, pharmaceutical solutions filled into clean vials in a class 100 environment could possibly be supercooled down to -15°C, leading to relatively high nucleation rates and a dense ice structure [80]. In freezing large volumes of bulk protein solutions, cooling rates are slower and tend to decrease as freezing proceeds. Undercooling down to a range between approximately -5°C and -10°C has been seen in the authors' laboratories, leading to a nucleation cascade that starts at the cooling surfaces and rolls across the bulk solution.

Since the degree of supercooling at nucleation determines the number of nuclei and the subsequent size distribution of the ice crystals, systems have been proposed to actively promote nucleation at the desired degree of supercooling. In pharmaceutical applications, these technologies have been aimed at the freezing step in lyophilization. Rambhatla et al. [81] have tested the ice-fogging technology, where a stream of cold nitrogen is streamed into the chamber once the desired degree of supercooling has been achieved. The nitrogen (cooled by circulating though a coil submerged in liquid nitrogen) causes moisture in the chamber to freeze, fall down, and seed the vials. Another system designed by Nakagawa et al. [82] uses ultrasound transmitted through the shelf to trigger nucleation in the vials. However, such active nucleation technologies have to date not been put into practice in the pharmaceutical industry.

26.4.3. Annealing

When a dense array of small ice crystals is undesirable, such as in freezing for lyophilization or in situations where the biologic is prone to aggregation when stored with high ice interface area, annealing can be used to grow the ice crystals after the freezing process is complete. Annealing is performed by maintaining the sample for a period of time at a subfreezing temperature but above T_g' or T_g, before reducing the temperature back again to the desired storage or processing temperature. The premise is that no or little change in ice crystal size occurs below T_g' in the timeframe of normal interest. If annealed above T_g', ice will melt and dilute the solute regions prior to resolidification [22]. During the annealing process, the smaller ice crystals, because of their higher surface energy, melt and preferentially freeze on the larger crystals, causing them to grow [83]. However, if the T_g of the amorphous phase is exceeded, the water component of that phase will also begin to crystallize, leading to further cryoconcentration [22].

The rate of this Ostwald ripening process is a function of the temperature above T_g' since this determines the mobility required for the molecules to diffuse and rearrange. The cubic mean size of ice crystals was shown to increase linearly with time, and the temperature dependence was shown to follow the WLF equation [84], implying a diffusion-controlled process.

26.4.4. Vial Breakage During Freezing and Frozen Storage

The inability of freezing systems to follow an equilibrium profile during practical processing occasionally leads to the observation of vial breakage during frozen storage or lyophilization. This has been most frequently reported with mannitol-based solutions

[85]. It was found to occur during warming of a previously frozen solution, and not during the freezing process and associated expansion in volume of ice as might be expected. Breakage is ascribed to the crystallization of mannitol during the warming of a previously frozen solution. However, it is not the crystallization (of mannitol) in itself but the associated crystallization of water associated with the amorphous phase that results in the volumetric expansion (water to ice I_h) leading to strain and vial breakage. In principle, the phenomenon may occur with other systems where a metastable state is formed on freezing, which subsequently crystallizes with release of water on heating [86,87]. The effect is exacerbated by faster cooling, increased concentration, and higher fill volume. Careful cooling to allow equilibrium crystallization of the water in solution was found to be able to prevent breakage [85]. Presence of small amounts of additives such as NaCl, lactose, glycine, and potassium chloride also prevented breakage by inhibiting the crystallization of hexagonal ice [88,89]. Hirakura et al. [86] found that by allowing time for the buffer salt to crystallize while holding it below the eutectic temperature during the freeze process, ampoule breakage during the thaw process was prevented. In this case, the protein (rhIL-11) was not sensitive to acidification due to sodium dibasic phosphate salt crystallization. Jiang et al. [90] also confirmed that large fill volumes correlated with increase in vial cracking. Other factors included concentration of mannitol and rate of freeze–thaw, which were positively correlated with frequency of vial breakage. They found that an intermediate hold at −30°C prior to further decrease in temperature reduced breakage by allowing mannitol to crystallize out. In case of noncrystallizing formulations (such as those containing proteins), breakage was found to increase with protein concentration and fill volume and during cooling to low temperatures such as −70°C. The cause of cracking in these solutions was attributed to the shrinkage of the frozen protein solution as it froze below −30°C [91].

26.4.5. Liquid–Liquid Phase Separation

Solutions containing polymeric excipients are known to show liquid–liquid phase separation. In a solution of two polymers, the free energy of mixing is a sum of the energies of interaction between the various polymeric segments and solvent, and their configurational entropy. If in the solvent mixture, the interaction energy between unlike polymer segments is even modestly unfavorable, the total free energy can be lowered by minimizing contacts by forming two phases, each one rich in one of the polymers even at the cost of an entropy decrease [92]. Such a change can be triggered by change in solvent conditions (induced, e.g., by temperature or addition of electrolytes). A number of phase separating systems have been reported, including protein–polymer(s) and protein–polyelectrolyte–nonionic polymer, and these systems are further modulated by additives such as sugars or electrolytes [79,93–100]. Some polymeric systems that are subject to this behavior include PEG:dextran, PVP:dextran, PVP:ficoll, PVP:sodium phosphate, and DEAE–dextran:dextran. If a protein is placed in such systems, the protein may also partition into the two phases. Carpenter and co-workers explored the impact of liquid–liquid-phase separation on recombinant hemoglobin in a PEG:dextran mixture. When the mixture was frozen, the conformation of hemoglobin was altered but each phase provided similar levels of structural protection. Replacement of NaCl with KCl, or addition of crystallizing excipient such as mannitol in the buffer, eliminated the phase separation. An appropriate choice of formulation composition as well as process control was suggested to eliminate the phase separation [79,98–100].

Presence of two glass transitions (either T_g' or T_g) in a DSC thermogram has been proposed as evidence for phase separation. Systems where this has been observed include PVP–sodium phosphate, BSA–PVP 10 K, lysozyme–PVP–Na_2SO_4, ovalbumin–PVP 10 K–NaCl, and similar systems have been found to show this behavior on freezing [93,95]. While polymeric systems are not commonly used in formulations of proteins, this is a phenomenon about which the formulator should be aware.

26.5. FREEZING OF BIOLOGICALS

Discussion about low-temperature behavior of proteins must distinguish between the impact of the low temperature or cold per se and the impact of the freezing process. Franks [101,102] distinguishes the former as the impact of "chill" and involves biological changes produced by changes in the property of liquid water with temperature (e.g., ionization constant, dielectric constant, hydrogen bond energies, hydrophobic interactions). The changes involved in protein structure as a consequence of low temperatures are reversible and are known as *cold denaturation*. In contrast, the freezing process, as discussed in earlier sections of this chapter, subjects proteins to other stresses as a consequence of the removal of water as ice. The resulting cryoconcentration and desiccation of protein can be classified as osmotic stresses. Other freezing-process-induced stresses include ice interface formation, pH changes, and phase separation. The changes in protein structure as a consequence of these stresses have a greater probability of being irreversible, and are classified as *freezing denaturation* [101,102].

26.5.1. Process-Induced Stress

26.5.1.1. Interaction Between Proteins and Ice. In their seminal paper on the behavior of protein in frozen solutions, Strambini and Gabellieri [103] showed that in a frozen matrix, proteins interact with the surface of ice. This is accompanied by loosening of the native structure (weakening of hydrophobic forces), visualized by an increase in tryptophan phosphorescence lifetimes normally buried in the rigid core of globular proteins. This results in perturbation of secondary and tertiary structure. The phenomenon is largely reversible on thawing, although a small fraction of the protein is irreversibly damaged. They also showed that "slow" cooling (at 1°C/min after seeding nucleation at −2°C) yields a smaller perturbation of the structure. Annealing of "fast" (rapidly) cooled samples (at 200°C/min after seeding nucleation at −2°C) shows that the structure recovers. Correlation between phosphorescence lifetimes and ice morphology was consistent with extent of protein adsorption. The perturbation is also related to the extent (fraction) of the liquid water pool in equilibrium with the solid phase formed on freezing. For azurin (mutant C112S), they established a threshold of

1.5%, below which the free energy of unfolding decreases sharply relative to the liquid state. As the fraction of liquid water decreases, a greater fraction of protein adsorbs to the ice. The decrease of free energy in ice implies that protein–ice interaction lowers the free energy of the denatured state to a greater extent than that of the native state. Interactions are therefore stronger for an expanded protein with a larger solvent accessible surface area, which aids in the adsorption [104]. 8-anilino-1-napthalenesulfonate (ANS) binding studies also show that in frozen solutions, the globular structure evolves into a molten globule-like state [105]. A strong positive correlation has also been shown between tendency of proteins to freeze-denature and to surface-denature (induced by shaking with Teflon beads representing a hydrophobic surface) [106].

26.5.1.2. Impact of Processing Rate. An obvious parameter to control during freeze–thaw processing is cooling or heating rate. A number of studies have reported on the effect of processing rate on recovery of structure of proteins, although a majority of these are on enzymes assessed solely by activity. Loss of activity, while generally easy to measure, can represent a range of underlying mechanisms, such as subunit dissociation, loss of secondary or tertiary structure, or even aggregation. It may be an insensitive measure if small structural changes occur that are distant from binding or catalytic regions. Nardid et al. [107] show that although the conformation of cytochrome-*c*-oxidase is perturbed by freeze–thaw treatment, the activity is not impacted. On the other hand, Edwards et al. [108] found that large changes in secondary structure of β-galactosidase begin to occur only after activity has been lost. Appropriate criteria must therefore be defined for each molecule on a case-by-case basis to properly assess the impact of freeze–thaw processing. Possible criteria include degree of aggregation, bioactivity, and loss of native secondary or tertiary structure.

Pikal-Cleland et al. [109] found that β-galactosidase in potassium phosphate buffer lost a certain amount of activity irrespective of processing protocol. This loss was attributed primarily to exposure to concentrated buffer salts or cold denaturation. However, the same enzyme in sodium phosphate buffer lost a greater fraction of its activity as a result of pH changes during freezing, and therefore fast freezing and fast thawing were recommended in this buffer to minimize exposure. Cao et al. [110] found slow freezing and fast thawing to be better for recovery of LDH activity, with recovery poorest in the sodium phosphate buffer. Nema and Avis [111] recommend "fast" freezing for LDH activity recovery. Anchordoquy and Carpenter [112] found that fast frozen LDH can still lose activity if thawed slowly because of acidic pH shifts caused by crystallization of solutes under increasing mobility. Jiang and Nail also reported the greatest loss of activity (LDH and β-galactosidase) with fast cooling, intermediate loss with slow cooling, and best recovery with intermediate cooling rates. Among studies looking directly at structural changes, Eckhardt et al. [113] found that fast freezing led to greater insoluble aggregates formation in hGH. Sarciaux et al. [114] did not observe any aggregate formation for an IgG subject to freeze thaw with either fast or slow freeze rates. They proposed that any denaturation that occurs during the freeze process is reversible on thawing, since aggregation was observed when the frozen material was subject to lyophilization. In a study correlating activity loss of L-asparaginase during freeze–thaw cycling, Jameel et al. [115] found that tetramers of the active enzyme dissociate into monomers and subsequently cleave into shorter fragments. The fragments then associate and precipitate. The greatest loss was found after the first cycle. The enzyme lost a large fraction of activity when frozen slowly at −40°C compared to complete deactivation on fast quenching in liquid nitrogen [116]. Carrasquillo et al. [117] concluded that protein structural alterations (slight decrease in α-helix content, significant increase in β sheet) observed after lyophilization (of α- and γ-chymotrypsin) were due to dehydration and not due to the freezing step, which was carried out by fast cooling in liquid nitrogen at a protein concentration of 10 mg/mL. A similar lack of structural changes during freezing was observed by Allison et al. [118] for α-chymotrypsinogen at 50 mg/mL. Pikal and co-workers found that the recovery of activity of LDH after freeze–thaw was dependent on the rate of thawing and also seemed to diminish with time after thawing. The hypothesis proposed is the formation of the active tetrameric state during freeze concentration, which dissociates on thawing to the inactive dimeric state [119]. The observation of "age"-dependent recovery of activity for multimeric enzymes implies that the results from freeze–thaw studies on enzymes that report activity as a readily measured parameter of quality must be examined more closely.

While some of the observations are undoubtedly protein-specific, a difficulty with interpretation of the literature is in the use of terms "fast" or "slow" freezing–thawing. Rates reported in the literature vary widely. For example, "fast" freezing in Anchordoquy and Carpenter [112] and Pikal-Cleland et al. [109] is immersion into liquid nitrogen, for Cao et al. [110] it is rates >20°C/min, while Eckhardt et al. [113] used 50°C/min. Slow freeze also varies from 0.1°C to 0.2°C/min [120], 0.4°C/min [109], and 1°C/min [110]. Thaw rates are less well defined, with air or refrigerator thawing considered slow while active thawing in a water bath defined as fast. Pikal and co-workers have classified cooling rates as slow (< 1°C/min), intermediate (1°C–10°C/min), and rapid (10°C–900 C°/min), and thaw rates as slow (1°C–5°C/min) and intermediate (>5°C/min) [121]. For practical purposes, descriptors such as "fast" and "slow" should be considered with caution. Almost all of the studies published in the literature use small (on the order of a few milliliters) to minute (hundreds of microliters) amounts of solution volumes and therefore cannot truly represent full-scale processing irrespective of how "slowly" the freeze or thaw is performed.

Some general considerations can, however, be outlined. Since proteins interact with the ice interface and are prone to surface denaturation, a smaller interfacial area with larger ice crystals, or a slower freezing rate would be preferable. However, a slower rate can lead to a greater degree of macrocryoconcentration, as well as longer exposure to the cryoconcentrated and/or pH-altered medium (discussed in the next section), leading to destabilization. During thawing, slower rates can lead to recrystallization, especially in frozen matrices created by rapid cooling, leading to additional perturbation at the ice–liquid interface. Again, slow thawing will also lead to longer exposure to the cryoconcentrated or pH-altered medium. The basic rule that emerges is that an optimum freezing rate needs to be defined for each protein depending on the predominant mechanism of destabilization during freezing. It is quite likely that proteins with multimeric or multidomain structures are more sensitive to freeze–thaw-induced stress than monomeric proteins.

26.5.1.3. pH Changes. Precipitation or crystallization of buffer salts during the freezing process can also change the pH in the frozen glass, resulting in a less-than-optimal pH environment for the biologic. Among the common buffers used for biologics, the sodium phosphate buffer mixture is particularly susceptible with pH changing from 7 to about 3.7 on cooling from room temperature to below -10°C, resulting from precipitation of the disodium salt [122–124]. Change in pH as a function of temperature for some common buffers is shown in Figure 26.9 [125,126]. As in all situations involving equilibrium versus practical processes, the actual crystallization of buffer or other solution components will depend on sample size, cooling rate, nucleation rates, and presence of other solutes and their concentrations. Under freezing conditions, salt crystal nucleation and growth are slow compared to ice crystal growth. As ice growth proceeds rapidly, the cryoconcentrated residual unfrozen solution phase tends to be compartmentalized into microdomains between advancing ice crystals. Salt crystallization in these domains seldom goes to completion, and it is common for frozen salt solutions to become partly or completely supersaturated [127]. The impact of process was also illustrated by Larsen [125], who showed that seeding a buffer while in the process of freezing can lead to a different pH–temperature profile compared to unseeded. Precipitation-induced pH shifts are also dependent on the concentration of the precipitating ions, and are determined by both initial pH and buffer salt concentrations [128]. The impact of supersaturation can be avoided by cooling to a low temperature and then heating back to the desired frozen storage temperature [125].

The T_g' values of several buffers in the frozen state (sodium tartrate, sodium malate, potassium citrate, and sodium citrate), have been reported by Shalaev et al. [129]. Trends in T_g' were seen with these buffers. Succinate and tartrate buffers were observed to crystallize, suggesting that the use of these buffers in freezing applications should be carefully monitored. Lyophilization in succinate buffer was found to reduce the stability of IFNγ due to this effect [130]. Glycolate buffer used as an alternative did not show the lowering of pH.

Proteins that are sensitive to pH-induced changes in conformation or stability are particularly susceptible to freezing in buffers that are temperature-sensitive. Carpenter and co-workers [109,131] showed significant perturbation of the monomeric and tetrameric structure of β-galactosidase due to freezing (as well as subsequent freeze drying) in sodium phosphate buffer due to pH shift from 7.0 to as low as 3.8. The same changes were not seen with potassium phosphate buffer, whose pH increases slightly on freezing. Optimal recovery of activity in the sodium phosphate buffer therefore required rapid cooling and rapid thawing to minimize exposure to the destabilizing conditions. Anchordoquy and Carpenter [112] showed that pH change could be attenuated by added polymers, thereby protecting LDH from freezing-induced pH shifts. The protective mechanism was inhibition of eutectic crystallization of salts. Pikal-Cleland et al. [132] showed that addition of glycine to sodium phosphate buffer had a mixed impact on stability of LDH and rhIFNγ. Low concentrations (<50 mM) suppressed pH decrease, while higher concentrations (>100 mM) enabled complete buffer salt

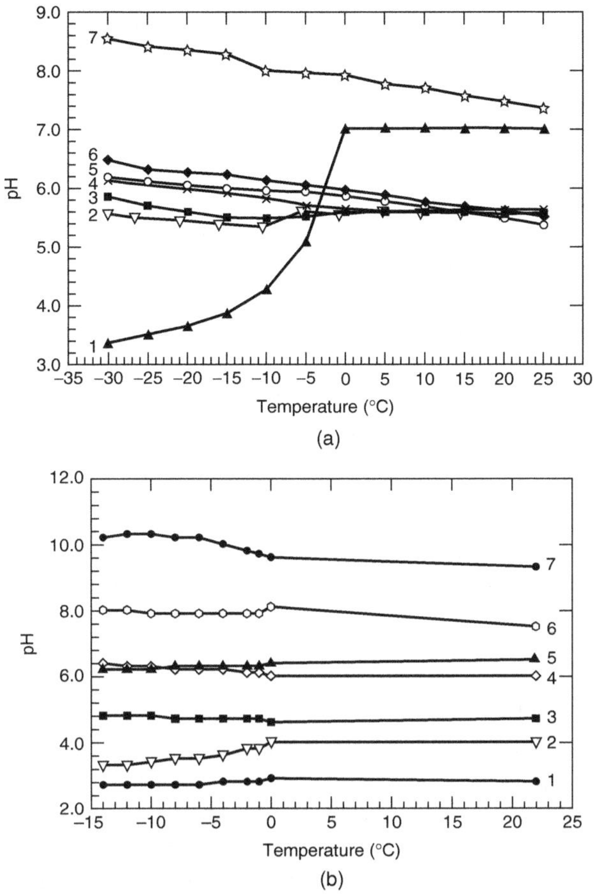

Figure 26.9. Impact of temperature change on pH of some common buffers: (a) curve 1-sodium phosphate, pH 7.0; curve 2-sodium citrate, pH 6.2; curve 3-sodium succinate, pH 5.5; curve 4-sodium acetate, pH 5.6; curve 5-histidine HCl, pH 5.4; curve 6-histidine acetate, pH 5.5; curve 7-tris HCl, pH 7.4 (Data from Kolhe et al. [126].) (b) curve 1—glycine–HCl, pH 2.8; curve 2—disodium citrate–HCl, pH 4.0; curve 3—acetic acid–sodium acetate, pH 4.7; curve 4—potassium hydrogen phthalate–NaOH, pH 6.0; curve 5—disodium citrate–NaOH, pH 6.5; curve 6—Tris (hydroxymethylaminomethane) HCl, pH 7.5; curve 7—glycine–NaOH, pH 9.3. (Data from Larsen [125].)

crystallization. While the added glycine may have a protective effect through the preferential exclusion mechanism, the results show that the impact of additives in general can be unpredictable.

A simple strategy to avoid pH-shift-induced damage is to avoid such buffers where possible. Common buffers such as citrate and acetate show minimal change on pH upon freezing [126] (Fig. 26.9). Alternatively, a balance has to be reached between damage induced by rapid freezing versus damage induced by longer exposure to destabilizing pH.

Apart from pH changes induced by precipitation, temperature changes also induce changes in the dissociation constants of the constituent species. The pK_w of water changes from 14.0 at 25°C to 17.3 at −25°C, a decrease by two orders of magnitude in degree of ionization. This is expected to impact equilibrium and kinetic processes that involve H^+ and OH^- ions. In an aqueous environment, the solvent acts as a conjugate acid or base, and any change in K_w will produce changes in the respective dissociation constants of acids and bases that define the ionization equilibria in solution. Since most proteins are polyelectrolytes, it is expected that even small changes in amino acid pK_a will affect their conformational stabilities and biological activities [17]. The impact of these changes is, however, difficult to assess. For example, the pK_a of ionizable groups differ in the folded and unfolded states of the protein because of differences in their local environments. The change in pK_a as a function of temperature for the same residue in a protein therefore varies with position (e.g., the 11 histidine imidazole groups range from pK_a of 5.5 to 7.8 at 299 K in myoglobin). The current prediction tools are inadequate to a priori understand the impact of changes in the ionization state of residue(s) and correlate them to changes in the structure of a protein since such changes are also accompanied by changes in the protein hydration [133]. The impact on structure through the change in ionization constants is, however, an intrinsic effect of low temperatures and is one of the factors in the phenomena of cold denaturation, discussed below.

26.5.2. Desiccation

Anchordoquy and Carpenter [112] showed that pH change could be attenuated by added polymers, thereby protecting LDH from freezing-induced pH shifts. However, part of the activity loss was also attributed to desiccation. As the freezing proceeds, water is removed as ice, leading to desiccation of the protein in the amorphous phase. Each protein molecule consists of a hydration shell consisting of water molecules directly attached to the protein. Water molecules aid in the proper folding and by lubricating the movement of the amino acid backbone and side groups by rapid formation and exchange of hydrogen bonds [134,135]. Hydration is therefore important for maintaining the three-dimensional structure of the protein. Structural changes have been recorded for lysozyme (amide I band narrowing implying protein–protein interaction) as water is removed during freezing in the absence of cryoprotectants such as sucrose [136]. The steepest changes were recorded at low hydration levels of under approximately 15%. This structural change was prevented in the presence of 10% sucrose.

26.5.3. Cold Denaturation

As temperature drops, properties of the aqueous solvent medium change, including the dielectric constant, acid–base ionization, diffusion rates and mobility, solubility of hydrophobic residues, and hydrogen bond energies. For example, dielectric permittivity of the water on cooling changes from 79 at 25°C to 109 at −25°C, and this will have an impact on ionic interactions. Hydrophobic interactions also weaken with lowering of temperature. These changes in themselves, without the complicating effects of freezing and phase changes, lead to reversible changes in the protein structure and destabilization called *cold denaturation*. Franks refers to this phenomenon as the impact of low temperature per se (of chill), as distinct from the impact on protein structure resulting from the actual freezing and its consequent effects (e.g., cryoconcentration, phase changes, ice surface denaturation) discussed above [17,101,137]. Cold-induced unfolding or cold denaturation is a physical consequence of the temperature sensitivity of the noncovalent electrostatic and hydrophobic interactions that determine protein stability, becoming weaker at lower temperature [101,138,139]. It is a thermodynamic consequence of the large and positive ΔC_p of unfolding of proteins (and that, within experimental error, can be taken as a constant for a given protein) [140]. Under the simplest mechanism of reversible two-state folding $\leftrightarrow$ unfolding, the temperature-dependent variation of thermodynamic stability of a protein is given by the Gibbs–Helmholtz equation

$$\Delta G_D(T) = \Delta H_D\left(1 - \frac{T}{T_D}\right) + \Delta C_p\left[(T - T_D) - T\ln\left(\frac{T}{T_D}\right)\right] \qquad (26.3)$$

where ΔG_D (T) is the Gibbs free-energy change between the denatured (D) and native (N) states at a given temperature T, ΔH_D is the enthalpy change between the two states at the melting temperature (T_D), while ΔC_p is the heat capacity change between the two states. The use of experimentally accessible temperature ranges and values determined for $\Delta H_D, \Delta C_p$, and T_D yields a skewed parabola-shaped protein stability curve as shown in Figure 26.10.

While the high-temperature denaturation (T_D) is experimentally accessible, the corresponding cold denaturation temperature (denoted as T'_D) has never been experimentally observed. In almost all cases, the extrapolations suggest that the T'_D lies below 0°C, a region where separation of the "chill"-induced effect is difficult to separate from the impact of freezing [141]. Experimental studies have, however, confirmed the parabolic nature of the free-energy curve, although a change in solvent conditions such as pH, addition of chaotropic agents, and other perturbants has to be used to make the T'_D accessible (see, e.g., Refs. 142–145). The implication of the impact of solvent conditions is that from a pharmaceutical perspective, commonly added sugars and polyols move the T'_D toward lower temperatures, thus stabilizing the molecule [146]. This aspect of protein (de)stabilization under freeze processing has not received much attention. The timescale of denaturation is much slower than timescales of freeze drying in the presence of sucrose. This is a consequence of the high solution viscosity in such systems at low temperatures [121,147].

In practical terms, the reversible nature of unfolding due to chill-induced cold denturation is in itself not detrimental to the storage stability of the protein. If the stress were strictly limited to cold denaturation, stability in the frozen state would probably not be an issue. Chill-induced unfolding likely renders the molecule more susceptible to the other "freeze"-induced stresses, leading to the observations of aggregation and loss of activity.

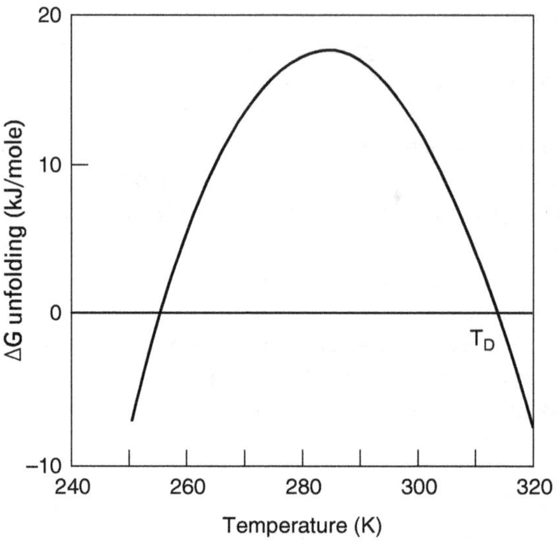

Figure 26.10. Hypothetical protein stability curve depicting the difference in free energy between denatured and native states as a function of temperature, in the absence of ice-induced stresses.

26.5.4. Mechanisms of Cryoprotection

Considering the stresses outlined above, it is clear that except for pH changes, and to some extent phase separation, elimination of the other stress is not an option. Cold denaturation, osmotic stress (cryoconcentration, desiccation), and ice interfacial stress are inevitable consequences of the freezing process. Formulations therefore have to be designed to be cryoprotective against all of these stresses. The pharmaceutically acceptable additives that have been empirically found to be useful include sugars, polyhydric alcohols, higher oligosaccharides, amino acids, and surfactants. Other additives include methylamines and lyotropic salts [30,111,148]. A number of these are proposed to function through the preferential exclusion mechanism (against cold denaturation by lowering the cold denaturation temperature and against osmotic stresses by stabilizing the native state), while surfactants interfere with the interaction at the ice interface. Since all stresses are present concurrently, more than one additive is usually required for complete protection.

26.5.4.1. Thermodynamic and Kinetic Stabilization. *Preferential exclusion* means that there is a deficiency in the stabilizing agent in the immediate vicinity of the protein (relative to the bulk phase), and the protein is therefore preferentially hydrated. This creates a thermodynamically unfavorable situation since the chemical potential of both protein and additive is increased. Consequently, the native structure of the protein is stabilized since denaturation would lead to an even greater thermodynamically unfavorable situation, due to the increase in contact surface area between protein and solvent. The invocation of the preferential exclusion mechanism for cryoprotection relies on the notion that the basic thermodynamic principles governing protein stability in the frozen state should not differ from those in the aqueous state. The effect is maintenance of the hydration shell of the protein under conditions of lowered water activity, irrespective of the temperature [30,49]. An assumption that dilute solution state concepts also apply in the freeze concentrated state is required [121].

Another mechanism that may enable protective solutes to function is kinetic stabilization by the glass state–vitrification mechanism [121]. Kinetic stabilization implies that the cryoprotectant simply slows down the rate of interconversion between native and denatured states without impacting the equilibrium. Formation of the glass at the end of the freezing process results in an increase in viscosity that immobilizes the protein, thus reducing the rate of denaturation.

The two mechanisms described above are not mutually exclusive, and both are likely to be involved in cryoprotection. As mentioned earlier, sucrose has been shown to decrease cold denaturation temperature (thermodynamic stabilization through preferential exclusion) as well as decrease unfolding time (kinetic stabilization) [147]. However, the implications of the presence of these two mechanisms is important. During formulation development using freeze–thaw cycling studies, inhibition of denaturation by addition of sugars or polyols implies that preferential exclusion (thermodynamic stabilization) against cold-induced denaturation is at work. In this situation, material that is subject to freeze–thaw cycling is found to be unchanged when examined after thaw. If this mechanism were the sole destabilization pathway in the frozen state, the formulation could then be stored for a length of time at any frozen-state temperature. If, on the other hand, frozen state storage results in denaturation or aggregation, the implication is that despite the thermodynamic stabilization, the kinetics of conversion to the denatured state has not been slowed down. The latter scenario would require that the dynamics of the glassy state be further reduced, normally by reducing the storage temperature further, preferably to below the T_g'.

26.5.4.2. Interfacial Stabilization. Finally, for protection against ice-interface-induced denaturation, stabilization can be achieved by preventing (competing for) adsorption to the ice surface. Chang et al. [106] showed that the presence of even small amounts of surfactants (e.g., >0.01% polysorbate 80; CMC ≅ 0.0016% at 20°C with 50 mM Na^+) was able to protect LDH and GDH during freeze–thaw. The protective effect was not specific to a chemical structure but was afforded by a variety of nonionic surfactants as well as ionic surfactants (only sodium lauryl sulfate was tested). Solutions with high protein concentrations are also more stable to freeze–thaw. One explanation for this is that a smaller fraction of the protein population is impacted by the ice interface in a concentrated solution. Another view is that the protein itself functions as a surfactant, adsorbing to the surface of ice. The protein can be self (see, e.g., Refs. 110,112, and 149) or foreign (see, e.g., Ref. 120). Anchordoquy and Carpenter [112] also show that addition of BSA protects LDH from freeze–thaw-induced denaturation by preventing freezing-induced pH shifts. However, it is entirely likely that the protective effect is to some extent a result of the BSA competing for ice surface interaction. The amount of surfactant required for complete protection will depend on the ice interface area and therefore on the time–temperature (processing) history of the material. Low interfacial area will require less surfactant. However, at the critical micelle concentration of the surfactant, the surfactant molecule would preferably form micelles and could therefore potentially allow some protein adsorption to occur [150]. The quantity of surfactant sufficient to provide appropriate protection under varying process conditions can be determined only empirically.

Presence of PEG in formulations can also reduce denaturation of proteins at the ice–water interface [25,148]. Although PEG itself is not surface-active and has little tendency to bind to ice surface, PEG hydrates crystallize slowly during freezing. The hydrated PEG chain can hydrogen-bond with ice. This behavior is seen with both low- and high-molecular-weight PEGs, but the low-molecular-weight PEGs cannot form as many hydrogen bonds. Such bonding between the PEG hydrate and the ice prevents protein from reaching the ice surface and denaturing [151].

26.6. FREEZE–THAW PROCESSING OF PROTEIN SOLUTIONS AT LARGE SCALE

A practical application of the freeze–thaw principles discussed earlier arises in the need to preserve and store large volumes of biologics or intermediates over several weeks to years. Production of biologics is an expensive process, and in order to optimize capacity utilization, the bulk protein solution is often produced in campaigns. The resulting drug substance has to be stored and converted into drug product according to market needs. In order to decouple the production of the bulk drug substance from the drug product, the bulk protein solution is often stored frozen. Storage volume requirements for biological bulk solution can be fairly large, especially with monoclonal antibodies. Furthermore, transport between facilities may be required, which is often easier with frozen material than with liquid since agitation during shipment is eliminated and it is simpler to maintain a temperature of less than −20°C than 2°C–8°C; the latter is the preferred storage temperature of most biologics drug product (liquid or lyophilized). Bulk storage information for a number of marketed biologics (including monoclonal antibodies) is summarized in Tables 26.1 and 26.2. Most of the data are taken directly from the EMA [European Public Assessment Report (EPAR) and FDA [Summary Basis of Approval, (SBA)] websites (accessed at the time of preparation of this chapter).

The large-scale processing of biologics solutions involves factors that are ignored in the numerous studies that have been published on freezing and thawing of biologics, most of which are carried out at microliter–milliliter scale. A factor which was alluded to in the discussion earlier was the "rate" of freezing. Larger dimensions and larger volumes lead to rates of heat removal that are generally lower than those utilized in laboratory studies. Furthermore, the rate of heat transfer changes with time as the ice phase grows on the surfaces from which heat is being removed. Considerable temperature gradients have to be maintained between the outermost frozen layers and the freezing front. Convective flows are generated as a result of density gradients

TABLE 26.1. Bulk Storage Information on Some Marketed Monoclonal Antibodies

Proprietary Name (INN Name)	Drug Substance Manufacturer	Bulk Storage State	Bulk Storage Temperature, °C	Bulk Storage Time or Retest Period, months	Bulk Storage Volume, L	Bulk Storage Container	Bulk Transport Involved
Avastin (bevacizumab)	Genentech, US	Frozen	−20	24 (45 days at 2–8°C); ≤5 freeze–thaw cycles	120–300	120-L and 300-L freeze–thaw stainless-steel (SS) tanks	Yes
(Mab)Campath (alemtuzumab)	Genzyme, Belgium	Liquid	2–8	6 (14 days at 30°C)	Batch size of bulk: 385 L	100-L HyClone bags	Yes
Erbitux (cetuximab)	ImClone, US BI, Germany	Liquid	2–8	12	—		Yes
Herceptin (trastuzumab)	Genentech, US	Frozen	<−20	—	120	120-L freeze–thaw SS tanks	Yes
Humira (adalimumab)	Abbott, US	Frozen	<−70	18	—	1.5 L in 2-L PETG bottles	Yes
LeukoScan (sulesomab)	Immunomedics, US	Liquid	2–8	—	—		No
Lucentis (ranibizumab)	Genentech, US	Frozen	≤−20	36 (90 days in-process hold time at 2°C–8°C); ≤3 freeze–thaw cycles	—	Vessel (not specified)	Yes
Raptiva (efalizumab)	Genentech, US	—	—	Long term	—	Tanks	No
Remicade (infliximab)	Centocor, Netherlands	Frozen	−70	—	—	—	Yes

(continued)

TABLE 26.1. *(Continued)*

Proprietary Name (INN Name)	Drug Substance Manufacturer	Bulk Storage State	Bulk Storage Temperature, °C	Bulk Storage Time or Retest Period, months	Bulk Storage Volume, L	Bulk Storage Container	Bulk Transport Involved
Rituxanor Mabthera (rituximab)	IDEC, US	Frozen	−20	18	—	—	Yes
Simulect (basiliximab)	Novartis, Switzerland	Frozen	<−60	36	—	—	Yes
Soliris (eculizumab)	Lonza, US	Liquid	2–8	18	—	—	Yes
Synagis (palivizumab)	BI, Germany	Liquid	2–8	6	—	—	No
Tysabri (natalizumab)	Biogen, US	Liquid	2–8	24	—	Polypropylene containers	No
Xolair (omalizumab)	Genentech, US	Frozen & Liquid	−20 or 2–8	Long term	—	High-capacity SS tanks	Yes
Zenapax (daclizumab)	Hoffman-La Roche, US	Liquid	2–8	—	20–200	Vessel (not specified)	No
Zevalin (ibritumomab tiuxetan)	Biogen, US	Liquid	2–8	—	—	Container (not specified)	Yes

−: Information not available.
Source: http://www.ema.europa.eu/.

TABLE 26.2. Bulk Storage Information on Some Commercial Biologics Other than Monoclonal Antibodies

Proprietary Name (INN Name)	Bulk Storage State	Bulk storage temperature (°C)	Bulk storage time or retest period (months)	Bulk Storage Container	Bulk Transport Involved
Advate (octocog alfa)	Frozen	−80	18	Bottle/screwcap/seal	Unknown
Angiox (bivalirudin)	Frozen	From −20 to −10	12	Multilayer polyethylene bags, aluminum foil	Unknown
Avonex (interferon beta-1a)	—	—	36	—	Unknown
Beromun (TNF alfa-1a)	Frozen	−70	24 (3-month in-process hold time at 2°C–8°C)	—	Unknown
Neorecormon (epoetin beta)	Frozen	−70 or −20	60	—	Unknown
Neulasta (pegfilgrastim)	Liquid	2–8	—	Polypropylene containers	Yes
Replagal (aglasidase alfa)	Frozen	—	—	—	Unknown
Revasc (desidurin)	Liquid	2–8	18	—	Unknown
Viraferon (interferon alfa-2b)	Frozen	−80	24	—	Unknown
Xigris (drotrecogin alfa)	Frozen	—	24	—	Unknown

Source: http://www.ema.europa.eu/.

created by temperature gradients. This leads to solutes and proteins concentrating near the bottom of the container while the upper regions are depleted. The same concerns about heat transfer rates are relevant for the thawing of large volumes as well. An adequate amount of heat has to be transferred rapidly without raising the temperature of the solution (and protein) in direct contact with the heat transfer surface to a point that will cause thermal degradation. Convective flow actually becomes desirable during the thaw process to ensure uniform heat transfer throughout the solution, as well as rapidly distributing any solutes subject to cryoconcentration during the freeze process [152].

The cooling rate applied to the system (equivalent to the rate of heat removal) and the rate of freezing are related, but the latter decreases with time as the thickness of the frozen layer increases. Thus, no single parameter can be taken as properly representative of the freezing process, especially in large-scale systems. It is possible to monitor the last point to freeze (if accessible) and use various time differences as parameters for characterizing the process, at least during development, qualification, and validation, if not during regular processing. Interestingly, Hartmann et al. [153] found that for an extended linear freezing geometry, the geometric center of the sample is the least representative for the entire volume. In the linear system considered by them, the optimum location was one-thirds away from the center and two-thirds away from the inside surface of the sample container. The rate of temperature decrease at this location represented at least 80% of the entire sample volume over a 50% range of variation. The main message is that the no single point of measurement can represent the system at large scale. The protein in such cases will be subject to a range of rates, and the formulation and process development efforts must therefore cover these ranges.

26.6.1. Large-Scale Freezing Systems

Large-scale freezing of biologics is carried out in various ways from improvised to purpose-designed systems.

The simplest storage concept involves filling the bulk solution into bottles or carboys of appropriate size and storing them in freezers. The advantage is simplicity. Disadvantages include lack of active control and potential variability between containers. Loading a number of filled containers at room temperature into a freezer can easily overwhelm the cooling capacity of a freezer, leading to long and variable freezing times between the containers. Thawing is generally performed by placing the container in a refrigerator or at room temperature. In the absence of an active thawing mechanism, the thaw times can be quite long (possibly days) depending on the size of the container. During this period significant concentration and temperature gradients can exist in the container if it is not actively shaken or agitated. When processing a number of these containers at a time, the spatial placement in the processing "equipment" (freezer, thaw area) can contribute to variability, since the containers will interact with each other during phase transitions and the relative position of the containers in the freezer will determine how rapidly they cool or warm. Size and handling of the containers are also factors to consider, with 20-L carboys likely to be a practical limit. However, the system is simple, and if the protein formulation is stable under a wide range of freeze–thaw conditions and can withstand cryoconcentration, the bottle/carboy system is the preferred mode of operation for many companies.

In order to overcome the limitations of passive freezing described above, variations include blast cooling via (vaporized) liquid nitrogen or dry ice–ethanol baths; although these systems have been designed, the authors are not aware of any published studies on the freezing of bulk-scale proteins with a system of the latter type.

A system from KVS-GmbH, Marburg, Germany utilizes liquid nitrogen coolant to freeze material in rotating stainless-steel cans. Maximum fill volume is 15 L in a container of nominal capacity of 29 L. The cans rotate on their cylindrical axis while lying at an angle of about 6° to the horizontal. Heat removal occurs through an air circulation system. With the use of liquid nitrogen, practical freezing temperatures down to −150°C can be achieved. The cooling rate is controlled by setting the temperature of the circulating air to the desired temperature profile. Thawing is accomplished by heating the circulating air over an electric heater.

A large-scale bulk solution freezing in stainless-steel vessels (CryoVessels®) is available from Sartorius-Stedim (http://www.sartorius-stedim.com) (see also the tabular list in Section 26.6.4.4, below). These vessels consist of jacketed stainless-steel tanks and an internal radial finned heat exchanger, which effectively divides the tank into several longitudinal sections. This has the effect of reducing the heat transfer distance and improved heat transfer across the entire volume (Fig. 26.11). The technology called CryoFin™ is said to promote dendritic ice formation, allowing ice and solute glass to be distributed somewhat uniformly in the tank [68,154,155]. The vessels are cooled and heated by an external refrigeration system that circulates heat transfer fluid through the jacket and core of the fin system. The temperature profile of the heat transfer fluid is programmable and results in reproducible temperature profiles in the vessel. The vessel is kept stationary throughout the freezing process below 0°C, but is gently agitated by rocking during the thawing process. The lack of agitation during freezing prevents movement of solute and promotes formation of dendritic ice. Agitation during thawing, however, promotes rapid mixing of the thawed material, thereby removing concentration hot spots and maintaining uniform temperature in the solution with rapid thawing. The lowest working temperature for the equipment is −60°C, and vessel volumes range from 20 to 300 L.

A scaled-down system of the CryoVessels is available called the *CryoWedge*™ which represents of one small wedge from the tank as shown in Figure 26.11. The critical heat transfer dimensions of the full tank are mimicked in this system, allowing process and formulation development to be studied at small scale. A typical temperature profile in a CryoWedge is shown in Figure 26.12. The stepwise reduction of the heat transfer fluid temperature (process) in Figure 26.12 is designed to create a temperature profile in the wedge similar to one that would be obtained in a full-scale tank.

A variation on the bulk freezing technology is the FreezeContainer® from Zeta Holdings (http://www.zeta.com/bio-und-verfahrenstechnik.html). Jacketed vessels (currently limited to 300 L capacity) are cooled or heated through an internal circulation system (mounted in the lid). Heat exchange is accomplished by an

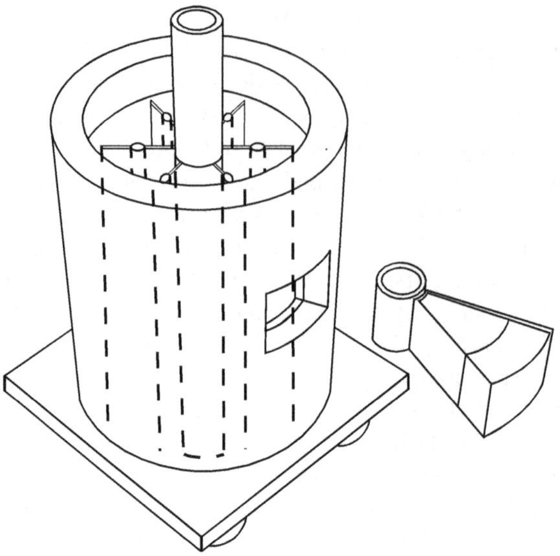

Figure 26.11. Schematic drawing of a CryoVessel with a cutout to illustrate the CryoWedge as a representative model of the vessel in terms of similarity of heat and mass transfer dimensions (see www.sartorius-stedim.com).

external refrigeration system via a circulating heat transfer fluid (Fig. 26.13). The temperature profile is programmable. The entire container is agitated during thawing.

Sartorius-Stedim has also developed a large-scale bag freezing system called Celsius™ (http://www.sartorius-stedim.com). The system uses upright bags made of Stedim71 film (ethylene vinyl acetate product contact material) that are filled with the solution to be frozen and held with slight compression between two plates that serve as heat exchange surfaces. These plates are cooled or heated by circulating heat transfer fluid from an external programmable refrigeration unit. The slight compression provides improved contact and heat transfer resulting in a frozen bag in the shape of a pillow. The bag is kept in frames so as not to stress the material during handling and transport. Nominal bag sizes are 16.6 and 8.3 L, with fill volumes ranging between 4.2 and 16 L and 2.1 and 8.3 L, respectively. Six bags can be simultaneously processed in the cryo unit.

A scaled-down system for the Celsius is available called the S3™, which utilizes small bags with the linear heat transfer dimension the same as that for the bags in Celsius. The process in the full-scale system can therefore be simulated in small bags between 30 and 100 mL.

Shamlou et al. [156] designed a bulk freezing container (up to 2 L capacity) using rectangular geometry so that the heat transfer path remained the same between a small-scale device for process development (30 mL volume) and the larger container. The container was designed to be blast frozen using liquid nitrogen blown into a chest freezer. Thaw was carried out by controlled electrical heaters without agitation.

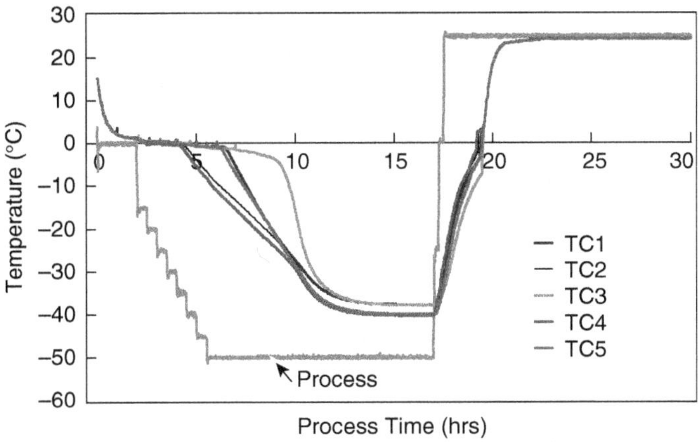

Figure 26.12. Illustrative freeze–thaw profile in a cryowedge. Thermocouples are placed at different positions in the wedge. The "process" probe represents the temperature of the heat transfer fluid.

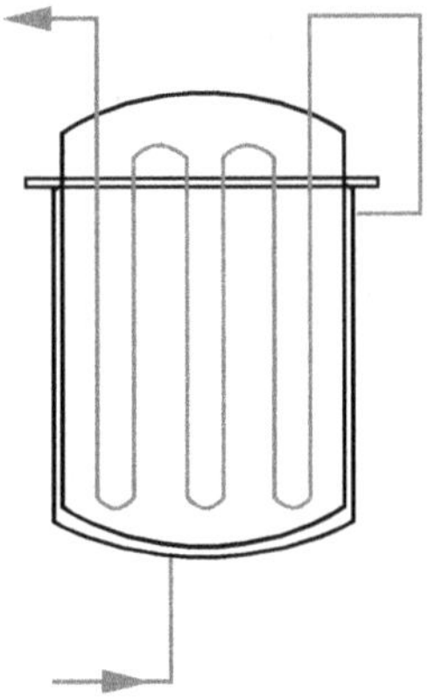

Figure 26.13. Schematic drawing of a FreezeContainer (Zeta Holdings) (www.zeta.com/bio-und-verfahrenstechnik.html).

26.6.2. Scale-Down Model Studies

Chen et al. [157] studied cryoconcentration in a specially designed radial cooling cell (15 mm diameter) with sodium permanganate to facilitate measurement of solute distribution. Depending on the cooling regimen, different degrees of cryoconcentration were seen for the sodium permanganate, with the greatest increase caused by slow cooling. However, when LDH was tested in an NaCl solution, the effect on NaCl was not very significant, while the cryoconcentration effect of the protein was not seen at all. On the other hand, in a comparative analysis of the freezing process for a monoclonal antibody in carboys (20 L capacity) and small bags in the S3 system, cryoconcentration by a factor of 10 was found in the carboys compared to a factor of ~2.5 in the small bags [158]. They concluded that high-freeze-front velocities (small dimensions, high rates) reduce cryoconcentration effects. Comparison of cryoconcentration effects in a 1-L bottle with that in a CryoWedge system showed similar results [159]. Their study also showed substantial time differences between bottles in a 3 × 3 array placed either in a −20°C or a −80°C freezer. Larger carboys required substantial time to both freeze and to thaw. However, neither of these studies showed increase in aggregation as a result of the cryoconcentration effect. Shamlou et al. [156] studied the freeze–thaw impact on rhGH in their scalable-design freeze–thaw unit and found that solutions of up to 30 mg/mL in phosphate buffer could undergo up to five cycles without any significant impact on aggregation levels. Zippelius [160] has conducted an extensive study on the bulk freeze–thaw behavior of (pegylated) interferon α-2a, trastuzumab, and daclizumab in the KVS GmbH rotational freezing system as well as the Sartorius-Stedim CryoWedge. The results show that the KVS system resulted in a lot of foam formation, and also about 20% of the bulk solution was impacted by cryoconcentration effects as opposed to about 0.5% in the CryoWedge.

26.6.3. Thawing of Biologicals at Scale

While freezing is the principal factor that determined the stability of a biologic when stored under frozen conditions, the process of thawing deserves some consideration. The major energy requirement for melting is to provide the latent heat of melting. Although simple in principle, the process must be controlled properly so as to ensure that the wall temperatures at the heat transfer surfaces do not exceed allowable limits for the product. In order to ensure that the thawed material does not overheat while a remainder is still in the frozen state, the mass should be agitated during processing, thereby ensuring efficient heat transfer as well as preventing hot spots [152].

Apart from the macroscale phenomenon described above, the thawing process can lead to recrystallization and annealing behavior in the ice crystals. Small ice crystals can melt and refreeze as discussed earlier, leading to Ostwald ripening effects. In most situations, this should not, however, be of any practical concern if the process is carried out reasonably rapidly.

26.6.4. Developing Frozen Storage Option for Proteins—Practical Advice

Formulation development for proteins has traditionally focused on the product and the final presentation. Bulk storage has been incorporated as an afterthought. In order to ensure a proactive approach to this step, some practical guidelines are provided in this section.

26.6.4.1. Excipients for Freeze–Thaw Stability. Sugars, polyhydric alcohols, certain amino acids and salts, and poly(ethylene glycol) (PEG) have cryoprotective properties [25,148]. In addition, a surfactant may be required to protect against ice interface induced denaturation, although PEG is proposed to protect by the same mechanism. This list provides a starting point for the identification of a formulation with the appropriate protective properties for the studies suggested below.

26.6.4.2. Impact of Freeze–Thaw Cycle (Assessing Thermodynamic and Interfacial Stability). Freeze–thaw (F/T) studies should be designed to cover a range of cooling and heating rates. Depending on the volumes, freezing with a significant degree of supercooling and rapid nucleation, and slow rates of thaw without agitation provide the greatest stress. Freeze–thaw cycles should cover a variety of temperatures, but thawing or holding between −20°C and 5°C will likely be worst-case scenario. Once the worst-case F/T process conditions have been identified for the molecule, formulation development can be conducted with these conditions. Freeze–thaw studies at low concentrations of the biologic are more likely to cause changes that can be detected. However, impact of F/T cycling at high concentrations should also be evaluated since cryoconcentration will likely occur in practice. A concentration of up to 5 × the actual concentration intended to be frozen may be worth evaluating. These studies can be used to identify the composition of excipients that function best as stabilizers, and will likely include excipients that function through preferential exclusion and through prevention of unfolding at the ice interface.

Experimental designs would incorporate the following:

- *Process parameters*—rate of cooling, rate of thawing, final freeze temperature, thaw temperature, presence/absence of agitation during thaw, concentration of biologic, and number of cycles
- *Composition parameters*—Surfactant concentration, protein concentration, and cyoprotectant concentrations
- *Quality variables*—Visual appearance, turbidity, subvisible particulates, concentration, aggregation, and biophysical characteristics

26.6.4.3. Impact of Long-Term Frozen Storage (Assessing Kinetic Stability). While F/T cycles provide a picture of the susceptibility of the protein to such stresses, the ability to withstand prolonged storage under frozen conditions is a stress that cannot be mimicked by the cycling test. It is possible that a fraction of the molecules exist in a (partially) unfolded state either as a consequence of freezing or through interaction with the ice interface. If aggregation is regarded as a nucleation and growth phenomenon, these unfolded molecules can serve as the nuclei. The storage temperature in relation to the T_g' determines the degree of mobility in the system. For proteins, this may result in slow aggregation as a consequence of diffusion and rearrangement while stored at or above T_g'. It should be noted that the temperature of zero mobility is often several degrees below T_g'. Any possible damage will therefore become evident only after a period of time. Long-term storage will also enable evaluation of other potential degradation mechanisms in relation to storage temperature.

Experimental designs would incorporate the following:

- *Process parameters*—T_g', storage temperature (range of 10°C above and below T_g' if possible), storage time, and rate of cooling applied
- *Composition parameters*—surfactant concentration, protein concentration, and cyoprotectant concentrations

- *Quality variables*—visual appearance, turbidity, subvisible particulates, concentration, aggregation, oxidation, and biophysical characteristics

26.6.4.4. Technology: Active Versus Passive Processes. For the purpose of this discussion, we are calling systems incorporating active cooling and heating, such as the Sartorius-Stedim CryoVessels, *active* and passive systems such as bottles and carboys in the freezer, *passive*. At the outset, it must be recognized that the commercially available controlled freeze–thaw systems can only control the rate of heat removal or heat application. A reproducible and predefined temperature ramp can be applied by these systems. During the freeze process, this enables the reproducible application of a predetermined optimum rate of heat removal. During the thaw process, the ability to actively heat the frozen mass while providing agitation is an advantage, since it enables practical reduction of time of thaw, and consequently reduced time of exposure to potentially deleterious conditions. However, these systems do not control the degree of supercooling, the rate of nucleation, or the actual rate of freezing experienced by any batch that is processed, nor do these systems control the ice nucleation and the final ice interfacial area achieved, which are ultimately important for the viability of the process. They simply increase the probability that these parameters would be similar from batch to batch.

The factors that could be considered in selection and utilization of the technology for bulk processing are tabulated below:

Materials of construction	*Passive* systems imply availability of a variety of materials of contact, polyolefins (LDPE, PP), and PETG; stainless steel are common. Extractables and leachables must be evaluated, especially for plastics. Variability in the material of construction from the supplier may require control, especially if leachables are critical. In addition, T_g of the plastic is important as material can become brittle at storage temperature. *Active* systems are currently limited to stainless steel, hastalloy, and ethylene vinyl acetate (EVA). Extractables and leachables must be evaluated, especially for plastics. Variability in the material of construction from the supplier may require control, especially if leachables are critical. Stainless steel can also leach metal ions (Fe, Ni) into solution. Surface finish specifications and maintenance are important to control this aspect. If sensitivity to metals is critical, tanks are available in hastalloy, although the increase in cost is significant.
Process development	*Passive* systems are generally modeled by small-scale containers, but their ability to represent the full-scale situation is very poor. Final viability must be tested at scale, or by use of cleverly designed alternatives (e.g., bottle in bottle).

	Active systems offer a scale-down alternative that simplifies process development.
Capacity	*Passive* systems are limited by storage and handling considerations. Current knowledge suggests 20-L maximum/carboy.
	Active systems vary from 2.1 to 300 L (nominal).
Capital	*Passive* systems require appropriate freezer dimensioning. It is important to determine whether the containers are disposable or reusable.
	Active systems require specialized equipment with significant capital investment.
Capability (storage temperature)	*Passive* systems can be processed down to −70°C or −80°C and stored. Volume limitations may apply at these temperatures. For all plastics-based containers, handling is important since fragility becomes a concern below the T_g of the plastic.
	Active systems can process down to −50°C in general, and operational storage for stainless steel systems is probably also limited to this temperature. Bags can be stored at −70°C or −80°C. For all plastics-based containers, handling is important since fragility becomes a concern below the T_g of the plastic.
	Storage at −20°C is quite common and widely available. Any fluctuations around −20°C that occur in the facility must be monitored.
Performance	*Passive* systems will generate a significantly greater degree of variability between containers of one batch and between batches. The product must be robust to this degree of variation in process and potentially exacerbated by subsequent storage time. Appropriate testing procedures, specifications, and pooling protocols must be defined.
	Active systems allow tighter control on process. Variability on storage can occur. Appropriate testing procedures, specifications, and pooling protocols must be defined.
Process Validation	*Passive* systems must have well-defined protocols for handling, fill volumes, preconditioning, spatial placement, freezer loads, and so on. In-process monitoring (positions) must be identified and documented as to relevance. The thaw process must be similarly defined. Again, a robust product will allow wider ranges of operating parameters. If using actively cooled systems based on liquid nitrogen or dry ice–ethanol bath, some of the considerations listed above still apply.
	Active systems are generally easier to validate since performance parameter reproducibility and control is available.

Stability models	Selection of a technology for bulk protein storage impacts the design and execution of ICH stability studies also. For instance, ICH Q5C states that "drug substance entered into a stability program should be stored in containers that properly represent the actual holding containers used during manufacture." While traditionally the "representation" has been limited to the material of construction, consideration from a process perspective may also be required. *Passive* systems can be easier to simulate using simple scaled-down containers with the same materials of construction. *Active* systems offer the possibility of using their scaled-down models, although these introduce a certain degree of complication in the execution of the stability program. Uncontrolled processing for a stability model may be justified by arguing that it represents a worst-case situation, although such an approach could have pitfalls.
Other	*Passive* systems such as bottles may require multiple containers for storage of one batch of solution. Multiple container–closures will be involved, implying risk for contamination. If using disposable containers, cost and disposal have to be addressed. If multiple-use containers are selected, maintenance (including special consideration of closure), cleaning (validation) program, lifetime, and related issues must be evaluated. If transport is to be considered, multiple containers increase the contamination risk, unless proper handling and testing protocols are in place. *Active* systems using stainless-steel vessels also require significant maintenance and cleaning (validation) programs. The disposable bag system (Celsius) is advantageous in this respect, although disposal has to be considered. Options are available for transportation.

26.6.5. Alternate Freezing Technologies

Although the current technologies available can result in reproducible freezing and thawing rates of protein solution, they lack the ability to control the ice nucleation process and resulting ice interface and structure. Several technologies have been developed to control nucleation in other areas of application, and thus have the potential to address this gap for processing bulk protein solutions:

- *Pressure Shift freezing* (PSF). In this process, the solution is subjected to high pressure (100–250 MPa); temperature is then lowered to approximately −20°C, and finally pressure is released either slowly (several minutes) or rapidly (seconds). This promotes metastable conditions and high ice nucleation velocities, resulting in small ice crystals of granular shape. The PSF method has been applied for food preservation since the formation of smaller ice crystals results in maintenance of texture and microstructure. It is important to consider the impact of high pressure on protein structure and function. The material can be thawed at high pressure and results in a reduction in thawing times when compared to atmospheric thawing at identical temperatures. This is due to two phenomenon: (1) as pressure is increased from 1 atm to 210 MPa, there is a linear depression in freezing point to −21.9°C. Above that pressure, there is a increase in freezing point since it involves different ice polymorphs); and (2) a reduction in latent heat values with increasing pressure. After thawing under pressure, temperature of the solution should be increased to prevent ice recrystallization during pressure drop (adiabatic expansion).

- *Isochoric Freezing*. Current freezing technologies involve freezing of solution at constant pressure (atmospheric pressure, known as *isobaric freezing*). In isochoric freezing, the volume of solution is kept constant, and as solution freezes, pressure within the system increases. The major advantage of isochoric freezing is to reduce the cryoconcentration of solutes at each temperature by almost a factor of 10 relative to the concentration during isobaric freezing (at 1 atm) [161].
- *Ultrasound-Induced Nucleation*. The technology creates bubble cavitation in cold solution that initiates the ice nucleation process. Use of ultrasound triggers nucleation at a lower degree of undercooling (e.g., a 10% mannitol solution in vials can nucleate at ∼−3°C, compared to ∼9°C in unaided conditions). This results in large and unidirectional ice crystals (dendritic type), while small and numerous heterogenous ice crystals form in samples with a high degree of supercooling [82].
- *Spray Freezing into Liquid Nitrogen*. This technology eliminates large gas–liquid interfaces, as is the case for spray freezing using cryogenic gas, by immersing the spray nozzle in cryogenic liquid. The process also results in rapid freezing (cooling rate ∼10^3 K/s), due to small volume and very low temperature of liquid nitrogen [78,162]. The spray-freezing technology was employed during the manufacturing of prolease–hGH microsphere [163].

Several other techniques have been used to control the rate of cooling and nucleation in solution, including electrofreezing, which involves high voltage applied to metal electrodes, liquid nitrogen fogging, thin layer freezing, and adding ice-nucleating substances such as *Pseudomonas syringae* bacterium or silver iodide [78,162,164,165]. Most of these techniques are experimental and not pharmaceutically relevant, but they do provide tools for understanding freezing behavior.

26.7. CONCLUDING REMARKS

Freeze–thaw processing of biologics is of increasing importance as several biopharmaceutical companies utilize this as a long-term storage option for bulk drug substance solution at large scale. A good understanding of the basics of freeze–thaw behavior and the impact on proteins is required for success in this area [166,167]. Apart from formulating for robustness toward freeze–thaw stresses, the formulation scientist needs to understand the impact of large-scale processing (i.e., cooling rate, probability of nucleation, cryoconcentration). The primary literature in the area of freezing of therapeutic biologics has been aimed at lyophilization and is limited to small volumes, which does not truly represent the complications of large-scale processing. This is an area of research that would benefit from a rigorous application of chemical engineering principles, cryobiology, biophysics, and related fields.

REFERENCES

1. Cocks, F. H. and Brower, W. E. (1974), Phase diagram relationships in cryobiology, *Cryobiology* **11**: 340–358.
2. Angell, C. A. (2002), Liquid fragility and the glass transition in water and aqueous solutions, *Chem. Rev.* **102**: 2627–2650.
3. Gayle, F. W., Cocks, F. H., and Shepard, M. L. (1977), The H_2O-NaCl-sucrose phase diagram and applications in cryobiology, *J. Appl. Chem. Biotechnol.* **27**: 599–607.
4. Shalaev, E. Y., and Franks, F. (1995), Equilibrium phase diagram of the water-sucrose-NaCl system, *Thermochim. Acta* **255**: 49–61.
5. Green, J. L. and Angell, C. A. (1989), Phase relationships and vitrification in saccharide-water solutions and the trehalose anomaly, *J. Phys. Chem.* **93**: 2880–2882.
6. Siniti, M., Jabrane, S., and Letoffe, J. M. (1999), Study of the respective binary phase diagrams of sorbitol with mannitol, maltitol and water, *Thermochim. Acta* **325**: 171–180.
7. Fahy, G. M. (1980), Analysis of solution effects injury. Equations for calculating phase diagram information for the ternary systems NaCl-dimethylsulfoxide-water and NaCl-glycerol-water, *Biophys. J.* **32**: 837–850.
8. Shepard, M. L., Goldston, G. S., and Cocks, F. H. (1976), The H_2O-NaCl-glycerol phase diagram and its application in cryobiology, *Cryobiology* **13**: 9–23.
9. Jochem, M. and Korber, C. (1987), Extended phase diagrams for the ternary solutions H_2O-NaCl-glycerol and H_2O-NaCl-hydroxyethylstarch (HES) determined by DSC, *Cryobiology* **24**: 513–536.
10. Shalaev, E. Y. and Kanev, A. N. (1994), Study of the solid-liquid state diagram of the water-glycine-sucrose system, *Cryobiology* **31**: 374–382.
11. Suzuki, T. and Franks, F. (1993), Solid-liquid phase transitions and amorphous states in ternary sucrose-glycine-water systems, *J. Chem. Soc. Faraday Trans.* **89**: 3283–3288.
12. Chatterjee, K., Shalaev, E. Y., and Suryanarayanan, R. (2005), Partially crystalline systems in lyophilization: I. Use of ternary state diagrams to determine extent of crystallization of bulking agent, *J. Pharm. Sci.* **94**: 798–808.
13. Akyurt, M., Zaki, G., and Habeebullah, B. (2002), Freezing phenomena in ice-water systems, *Energy Conserv. Manage.* **43**: 1773–1789.
14. Goff, H. D. (2008), *Time-Temperature Relationship for Freezing Pure Water and Aqueous Solutions*, in Dairy Science and Technology Education series (http://www.foodsci.uoguelph.ca/dairyedu/home.html, accessed April 2008).

15. Roos, Y. and Karel, M. (1991), Amorphous state and delayed ice formation in sucrose solutions, *Int. J. Food Sci. Technol.* **26**: 553–566.
16. Roos, Y. and Karel, M. (1991), Applying state diagrams to food processing and development, *Food Technol.* **45**: 66–71, 107.
17. Franks, F., Mathias, S. F., Hatley, R. H. M., Baust, J. G., Hvidt, A., Chapman, D., and Jaenicke, R. (1990), Water, temperature and life, *Phil. Trans. Roy. Soc. Lond. B Biol. Sci.* **326**: 517–533.
18. Levine, H. and Slade, L. (1988), Thermomechanical properties of small carbohydrate-water glasses and rubbers. Kinetically metastable systems at sub-zero temperatures, *J. Chem. Soc. Faraday Trans. I* **84**: 2619–2633.
19. Rahman, M. S. (2006), State diagram of foods; its potential use in food processing and product stability, *Trends Food Sci. Technol.* **17**: 129–141.
20. Franks, F. (1991), Freeze-drying: From empiricism to predictability. The significance of glass transitions, *Develop. Biol. Stand.* **74**: 9–19.
21. Levine, H. and Slade, L. (1988), Principles of "cryostabilization" technology from structure/property relationships of carbohydrtae/water systems—a review, *Cryo-Letters* **9**: 21–63.
22. Ablett, S., Izzard, M. J., and Lillford, P. J. (1992), Differential scanning calorimetric study of frozen sucrose and glycerol solutions, *J. Chem. Soc. Faraday Trans.* **1992**: 789–794.
23. Matveev, Y. I. and Ablett, S. (2002), Calculation of the Cg' and Tg' interaction point in the state diagram of frozen solutions, *Food Hydrocoll.* **16**: 419–422.
24. Chang, B. S. and Randall, C. S. (1992), Use of subambient thermal analysis to optimize protein lyophilization, *Cryobiology* **29**: 632–656.
25. Carpenter, J. F., Pikal, M. J., Chang, B. S., and Randolph, T. W. (1997), Rational design of stable lyophilized protein formulations: Some practical advice, *Pharm. Res.* **14**: 969–975.
26. Aldous, B. J., Auffret, A. D., and Franks, F. (1995), The crystallization of hydrates from amorphous carbohydrates, *Cryo-Letters* **16**: 181–186.
27. Pyne, A., Surana, R., and Suryanarayanan, R. (2002), Crystallization of mannitol below T_g' during freeze-drying in binary and ternary aqueous systems, *Pharm. Res.* **19**: 901–908.
28. Chang, L. L., Shepard, D., Sun, J., Ouelette, D., Grant, K. L., Tang, X. C., and Pikal, M. J. (2005), Mechanisms of protein stabilization by sugar during freeze-drying and storage. Native structure preservation, specific interaction, and/or immobilization in a glassy matrix, *J. Pharm. Sci.* **94**: 1427–1444.
29. Izutsu, K., Yoshioka, S., and Terao, T. (1993), Decreased protein stabilizing effects of cryoprotectants due to crystallization, *Pharm. Res.* **10**: 1232–1237.
30. Carpenter, J. F. and Crowe, J. H. (1988), The mechanism of cryoprotection of proteins by solutes, *Cryobiology* **25**: 244–255.
31. Piedmonte, D. M., Summers, C., McAuley, A., Karamujic, L., and Ratnaswamy, G. (2007), Sorbitol crystallization can lead to protein aggregation in frozen protein formulations, *Pharm. Res.* **24**: 136–146.
32. Hatley, R. H. M., Franks, F., and Day, H. (1986), Subzero-temperature preservation of reactive fluids in the undercooled state. II. The effect on the oxidation of ascorbic acid of freeze concentration and undercooling, *Biophys. Chem.* **24**: 187–192.
33. Hatley, R. H. M., Franks, F., Day, H., and Byth, B. (1986), Subzero temperature preservation of reactive fluids in the undercooled state. I. The reduction of potassium ferricyanide by potassium cyanide, *Biophys. Chem.* **24**: 41–46.

34. Pincock, R. E. and Kiovsky, T. E. (1966), Reactions in frozen solutions. VI. The reaction of ethylene chlorohydrin with hydroxyl ion in ice, *J. Am. Chem. Soc.* **88**: 4455–4459.

35. Takenaka, N., Ueda, A., Daimon, T., Bandow, H., Dohmaru, T., and Maeda, Y. (1996), Acceleration mechanism of chemical reaction by freezing: The reaction of nitrous acid with dissolved oxygen, *J. Phys. Chem.* **100**: 13874–13884.

36. Hatley, R. H. M., Franks, F., and Mathias, S. F. (1987), The stabilization of labile biochemicals by undercooling, *Process Biochem.* **22**: 169–172.

37. Haensler, M. and Jakubke, H.-D. (1996), Nonconventional protease catalysis in frozen aqueous solutions, *J. Pept. Sci.* **2**: 279–289.

38. Haensler, M. and Jakubke, H.-D. (1996), Reverse action of hydrolases in frozen aqueous systems, *Amino Acids* **11**: 379–395.

39. Haensler, M., Schuerer, H., Hahn, U., and Jakubke, H.-D. (1998), Reverse action of ribonuclease T1 variants in ice, *Nucleos. Nucleot.* **17**: 1267–1274.

40. Schuster, M., Aaviksaar, A., Haga, M., Ullman, U. and Jakubke, H.-D. (1991), Protease-catalyzed peptide synthesis in frozen aqueous systems: The freeze-concentration model, *Biomed. Biochem. Acta* **50**: S84–S89.

41. Buera, M. D.-P. and Karel, M. (1993), Application of the WLF equation to describe the combined effects of moisture and temperature on nonenzymatic browning rates in food systems, *J. Food Process. Preserv.* **17**: 31–45.

42. Reid, D. S., Kerr, W., and Hsu, J. (1994), The glass transition in the freezing process, *J. Food Eng.* **22**: 483–494.

43. Karel, M. and Saguy, I. (1991), Effects of water on diffusion in food systems, in *Water Relationships in Food*, Levine, H. and Slade, L., eds., Plenum Press, New York, pp. 157–173.

44. Peleg, M., Engel, R., Gonzalez-Martinez, C., and Corradini, M. G. (2002), Non-Arrhenius and non-WLF kinetics in food systems, *J. Sci. Food Agric.* **82**: 1346–1355.

45. Bhandari, B. R. and Howes, T. (1999), Implication of glass transition for the drying and stability of dried foods, *J. Food Eng.* **40**: 71–79.

46. Peleg, M. (1992), On the use of the WLF model in polymers and foods, *Crit. Rev. Food Sci. Nutr.* **32**: 59–66.

47. Schenz, T. W. (1995), Glass transitions and product stability—an overview, *Food Hydrocoll.* **9**: 307–315.

48. Parker, R. and Ring, S. G. (1995), A theoretical analysis of diffusion-controlled reactions in frozen solutions, *Cryo-Letters* **16**: 197–208.

49. Franks, F. (1985), *Biophysics and Biochemistry at Low Temperatures*, Cambridge Univ. Press.

50. Battino, R., Rettich, T. R., and Tominaga, T. (1983), The solubility of oxygen and ozone in liquids, *J. Phys. Chem. Ref. Data* **12**: 163–178.

51. Vali, G. (1995), Principles of ice nucleation, in *Biological Ice Nucleation and Its Applications*, Lee, J., R. E., Warren, G. J., and Gusta, L. V., eds., APS, St. Paul, MN, pp. 1–28.

52. Koop, T. (2004), Homogeneous ice nucleation in water and aqueous solutions, *Z. Phys. Chem.* **218**: 1231–1258.

53. Leloux, M. S. and Rikilt, D. (1999), The infuence of macromolecules on the freezing of water, *J. Mass Spectrom. Rev. Macromol. Chem. Phys.* **C39**: 1–16.

54. MacKenzie, A. P., Derbyshire, W., and Reid, D. S. (1977), Non-equilibrium freezing behavior of aqueous systems, *Phil. Trans. Roy. Soc. Lond. B Biol. Sci.* **278**: 167–189.

55. Wilson, P. W., Heneghan, A. F., and Haymet, A. D. J. (2003), Ice nucleation in nature: Supercooling point (SCP) measurements and the role of heterogeneous nucleation, *Cryobiology* **46**: 88–98.

56. Charoenrein, S. and Reid, D. S. (1989), The use of DSC to study the kinetics of heterogeneous and homogeneous nucleation of ice in aqueous systems, *Thermochim. Acta* **156**: 373–381.

57. Charoenrein, S., Goddard, M., and Reid, D. S. (1991), Effect of solute on the nucleation and propagation of ice, in *Water Relationships in Food*, Levine, H. and Slade, L., eds., Plenum Press, New York, pp. 191–198.

58. Holt, C. B. (2003), Substances which inhibit ice nucleation: A review, *Cryo-Letters* **24**: 269–274.

59. Inaba, H., Inada, T., Horibe, A., Suzuki, H., and Usui, H. (2005), Preventing agglomeration and growth of ice particles in water with suitable additives, *Int. J. Refrig.* **28**: 20–26.

60. Scotter, A. J., Marshall, C. B., Graham, L. A., Gilbert, J. A, Garnham, C. P., and Davies, P. L. (2006), The basis for hyperactivity of antifreeze proteins, *Cryobiology* **53**: 229–239.

61. Knight, C. A. (2000), Adding to the antifreeze agenda, *Nature* **406**: 249–250.

62. Clausse, D., Lanoiselle, J.-L., and Toumi, S. (2006), Ice nucleation in bulk and dispersed water: Application to freezing of foods, in *Water Properties of Food, Pharmaceutical and Biological Materials*, Buera, M. D.-P., Welti-Chanes, J., Lillford, P. J., and Corti, H. R., eds., CRC Press, Boca Raton, FL, pp. 133–150.

63. Speedy, R. J. (1996), Two waters and no ice please, *Nature* **380**: 289–290.

64. Tanaka, H. (1996), *Nature* **380**: 328–330.

65. Vali, G. (1999), Ice nucleation—theory, a tutorial; paper presented at the *NCAR/ASP 1999 Summer Colloquium* (available online at http://www-das.uwyo.edu/~vali, accessed April 2008).

66. Jennings, T. A. (1999), *Lyophilization. Introduction and Basic Principles*, Interpharm Press, Englewood, CO.

67. Lindenmeyer, C. S., Orrok, G. T., Jackson, K. A., and Chalmers, B. (1957), Rate of growth of ice crystals in supercooled water, *J. Chem. Phys.* **27**: 822.

68. Wisniewski, R. (1998), Developing large-scale cryopreservation systems for biopharmaceutical products, *BioPharm* **11**: 50–56.

69. Muldrew, K. (1999), *Mechanical Damage from Freezing. Planar vs Dendritic Ice Growth in Cryobiology—a Short Course* (available online at http://www.ucalgary.ca/~kmuldrew/cryo_course/cryo_chap9_1.html#04, accessed April 2008).

70. Korber, C. (1988), Phenomena at the advancing ice-liquid interface: Solutes, particles and biological cells, *Q. Rev. Biophys.* **21**: 229–298.

71. Telford, J. and Lillford, P. J. (2006), Incorporation of solute in the ice phase during freezing, in *Water Properties of Food, Pharmaceutical and Biological Materials*, Buera, M. D.-P., Welti-Chanes, J., Lillford, P. J., and Corti, H. R., eds., CRC Press, Boca Raton, FL, pp. 677–682.

72. Chen, P., Chen, X. D., and Free, K. W. (1998), Solute inclusion in ice formed from sucrose solutions on a sub-cooled surface—an experimental study, *J. Food Eng.* **38**: 1–13.

73. Janson, J.-C., Ersson, B., and Porath, J. (1974), The concentration of protein solutions by normal freezing, *Biotechnol. Bioeng.* **16**: 21–29.

74. Miyawaki, O., Liu, L., and Makamura, K. (1998), Effective partition coefficient of solute between ice and liquid phases in progressive freeze-concentration, *J. Food Sci.* **63**: 756–758.

75. Pradistsuwana, C., Theprugsa, P., and Miyakawi, O. (2003), Measurement of limiting partition coefficeint in progressive freeze-concentration, *Food Sci. Technol. Res.* **9**: 190–192.

76. Lipp, G., Korber, C., Englich, S., Hartmann, U., and Rau, G. (1987), Investigation of the behavior of dissolved gases during freezing, *Cryobiology* **24**: 489–503.

77. Pikal, M. J. (1994), Freeze-drying of proteins. Process, formulation, and stability, in *Formulation and Delivery of Proteins*, Cleland, J. F. and Langer, R. S., eds., American Chemical Society, Washington, DC., pp. 120–133.

78. Engstrom, J. D., Simpson, D. T., Cloonan, C., Lai, E. S., Williams, R. O., Kitto, G. B., and Johnston, K. P. (2007), Stable high surface area lactate dehydrogenase particles produced by spray freezing into liquid nitrogen, *Eur. J. Pharm. Biopharm.* **65**: 163–174.

79. Heller, M. C., Carpenter, J. F., and Randolph, T. W. (1999), Protein formulation and lyophilization cycle design: Prevention of damage due to freeze-concentration induced phase separation, *Biotechnol. Bioeng.* **63**: 166–174.

80. Pikal, M. J. (2002), The impact of freezing stage in lyophilization: Effects of the ice nucelation temperature on process design and product quality, *Am. Pharm. Rev.* **5**: 48–53.

81. Rambhatla, S., Ramot, R., Bhugra, C., and Pikal, M. J. (2004), Heat and mass transfer scale-up issues during freeze-drying: II. Control and characterization of the degree of supercooling, *AAPS PharmSciTech.* **5**: E58.

82. Nakagawa, K., Hottot, A., Vessot, S., and Andrieu, J. (2006), Influence of controlled nucleation by ultrasounds on ice morphology of frozen formulations for pharmaceutical proteins freeze-drying, *Chem. Eng. Process.* **45**: 783–791.

83. Searles, J. A., Carpenter, J. H., and Randolph, T. W. (2001), Annealing to optimize the primary drying rate, reduce freezing-induced drying rate heterogeneity, and determine T_g' in pharmaceutical lyophilization, *J. Pharm. Sci.* **90**: 872–887.

84. Hagiwara, T. and Hartel, R. W. (1996), Effect of sweetener, stabilizer, and storage temperature on ice recrystallization in ice cream, *J. Diary Sci.* **79**: 735–744.

85. Williams, N. A., Lee, Y., Polli, G. P., and Jennings, T. A. (1986), The effects of cooling rate on solid phase transitions and associated vial breakage occuring in frozen mannitol solutions, *PDA J. Parenter. Sci. Technol.* **40**: 135–141.

86. Hirakura, Y., Kojima, S., Okada, A., Yokohama, S., and Yokota, S. (2004), The improved dissolution and prevention of ampoule breakage attained by the introduction of pretreatment into the production process of the lyophilized formulation of recombinant human interlukin-11 (rhIL-11), *Int. J. Pharm.* **286**: 53–67.

87. Milton, N., Gopalrathnam, G., Craig, G. D., Mishra, D. S., Roy, M. L., and Yu, L. (2007), Vial breakage during freeze-drying: Crystallization of sodium chloride in sodium chloride-sucrose frozen aqueous solutions, *J. Pharm. Sci.* **96**: 1848–1853.

88. Ito, T., Yoshioka, Y., Sonoda, T., Yoshihashi, Y., Yonemouchi, E., and Terada, K. (2006), Mechanism of glass ampoule breakage prevention during the freeze-drying process of sodium thiopental lyophilization products on addition of sodium chloride, *J. Therm. Anal. Cal.* **85**: 731–739.

89. Williams, N. A. and Dean, T. (1991), Vial breakage by frozen mannitol solutions: Correlation with thermal charcteristics and effect of steroisomerism, additives, and vial configuration, *PDA J. Parenter. Sci. Technol.* **45**: 94–100.

90. Jiang, G., Akers, M., Jain, M., Guo, J., Distler, A., Swift, R., Wadhwa, M.-V. S., Jameel, F., Patro, S., and Freund, E. (2007), Mechanistic studies of glass vial breakage for frozen formulations. I. Vial breakage caused by crystallizable excipient mannitol, *PDA J. Parenter. Sci. Technol.* **61**: 441–451.

91. Jiang, G., Akers, M., Jain, M., Guo, J., Distler, A., Swift, R., Wadhwa, M.-V. S., Jameel, F., Patro, S., and Freund, E. (2007), Mechanistic studies of glass vial breakage for frozen formulations. II. Vial breakage caused by amorphous protein formulations, *PDA J. Parenter. Sci. Technol.* **61**: 452–460.

92. Walter, H., Johansson, G., and Brooks, D. E. (1991), Partitioning in aqueous two-phase systems, *Anal. Biochem.* **197**: 1–18.

93. Her, L.-M., Deras, M., and Nail, S. L. (1995), Electrolyte-induced changes in glass transition temperatures of freeze-concentrated solutes, *Pharm. Res.* **12**: 768–772.

94. Izutsu, K.-I., Yoshioka, S., Kojima, S., Randolph, T. W., and Carpenter, J. F. (1996), Effects of sugars and polymers on crystallization of poly(ethylene glycol) in frozen solutions: Phase separation between incompatible polymers, *Pharm. Res.* **13**: 1393–1400.

95. Izutsu, K.-I. and Kojima, S. (1999), *Phase Separation of Protein and Polymer Excipients in Frozen Solutions*, AAPS Annual Meeting, AAPS Poster 3728, New Orleans.

96. Izutsu, K.-I. and Kojima, S. (2000), Freeze-concentration separates proteins and polymer excipients into different amorphous phases, *Pharm. Res.* **17**: 1316–1322.

97. Randolph, T. W. (1997), Phase separation of excipients during lyophilization: Effects on protein stability, *J. Pharm. Sci.* **86**: 1198–1203.

98. Heller, M. C., Carpenter, J. F., and Randolph, T. W. (1996), Effect of phase separating systems on lyophilized hemoglobin, *J. Pharm. Sci.* **85**: 1358–1362.

99. Heller, M. C., Carpenter, J. F., and Randolph, T. W. (1997), Manipulation of lyophilization-induced phase separation: Implications for pharmaceutical proteins, *Biotechnol. Progress* **13**: 590–596.

100. Heller, M. C., Carpenter, J. F., and Randolph, T. W. (1999), Application of a thermodynamic model to the prediction of phase separations in freeze-concentrated formulations for protein lyophilization, *Arch. Biochem. Biophys.* **363**: 191–201.

101. Franks, F. (1995), Protein destabilization at low temperatures, *Adv. Protein Chem.* **46**: 105–139.

102. Franks, F. (2003), Nucleation of ice and its management in ecosystems, *Phil. Trans. Roy. Soc. Lond. A* **361**: 557–574.

103. Strambini, G. B. and Gabellieri, E. (1996), Proteins in frozen solutions: Evidence of ice-induced partial unfolding, *Biophys. J.* **70**: 971–976.

104. Strambini, G. B. and Gonnelli, M. (2007), Protein stability in ice, *Biophys. J.* **92**: 2131–2138.

105. Gabellieri, E. and Strambini, G. B. (2006), ANS fluorescence detects widespread perturbations of protein tertiary structure in ice, *Biophys. J.* **90**: 3239–3245.

106. Chang, B. S., Kendrick, B. S., and Carpenter, J. F. (1996), Surface-induced denaturation of proteins during freezing and its inhibition by surfactants, *J. Pharm. Sci.* **85**: 1325–1330.

107. Nardid, O., Dyubko, T. S., Rozanova, E. D., Naumenko, E. I., and Morozova, T. F. (1999), A study of structural and functional state modification of certain proteins when exposed to low temperatures, *Cryo-Letters* **20**: 283–286.

108. Edwards, R. A., Jacobson, A. L., and Huber, R. E. (1990), Thermal denaturation of β-galactosidase and of two site-specific mutants, *Biochemistry* **29**: 11001–11008.

109. Pikal-Cleland, K. A., Rodriguez-Hornedo, N., Amidon, G. L., and Carpenter, J. F. (2000), Protein denaturation during freezing and thawing in phosphate buffer systems: Monomeric and tetrameric β-galactosidase, *Arch. Biochem. Biophys.* **384**: 398–406.

110. Cao, E., Chen, Y., Cui, Z., and Foster, P. R. (2003), Effect of freezing and thawing rates on denaturation of proteins in aqueous solutions, *Biotechnol. Bioeng.* **82**: 684–690.

111. Nema, S. and Avis, K. E. (1993), Freeze-thaw studies of a model protein, lactate dehydrogenase, in the presence of cryoprotectants, *J. Parenter. Sci. Technol.* **47**: 76–83.

112. Anchordoquy, T. J. and Carpenter, J. F. (1996), Polymers protect lactate dehydrogenase during freeze-drying by inhibiting dissociation in the frozen state, *Arch. Biochem. Biophys.* **332**: 231–238.

113. Eckhardt, B. M., Oeswein, J. Q., and Bewley, T. A. (1991), Effect of freezing on aggregation of human growth hormone, *Pharm. Res.* **8**: 1360–1364.

114. Sarciaux, J.-M., Mansour, S., Hageman, M. J., and Nail, S. L. (1999), Effects of buffer composition and processing conditions on aggregation of bovine IgG during freeze-drying, *J. Pharm. Sci.* **88**: 1354–1361.

115. Jameel, F., Bogner, R., Mauri, F., and Kaolnia, D. (1997), Investigation of physicochemical changes to L-asparaginase during freeze-thaw cycling, *J. Pharm. Pharmacol.* **49**: 472–477.

116. Jameel, F., Kalonia, D., and Bogner, R. (1995), The effect of hetastarch on the stability of L-asparaginase during freeze-thaw cycling, *PDA J. Pharm. Sci. Technol.* **49**: 127–131.

117. Carrasquillo, K. G., Sanchez, C., and Briebenow, K. (2000), Relationship between conformational stability and lyophilization-induced structural changes in chymotrypsin, *Biotechnol. Appl. Biochem.* **31**: 41–53.

118. Allison, S. D., Dong, A., and Carpenter, J. F. (1996), Counteracting effects of thiocyanate and sucrose on chymotrypsinogen secondary structure and aggregation during freezing, drying, and rehydration, *Biophys. J.* **71**: 2022–2032.

119. Bhatnagar, B. S., Nehm, S. J., Pikal, M. J., and Bogner, R. H. (2005), Post-thaw aging affects activity of lactate dehydrogenase, *J. Pharm. Sci.* **94**: 1382–1388.

120. Tamiya, T., Okahashi, N., Sakuma, R., Aoyama, T., Akahane, T., and Matsumoto, J. J. (1985), Freeze denaturation of enzymes and its prevention with additives, *Cryobiology* **22**: 446–456.

121. Bhatnagar, B., Bogner, R. H., and Pikal, M. J. (2007), Protein stability during freezing: Separation of stresses and mechanisms of protein stabilization, *Pharm. Devel. Technol.* **12**: 505–523.

122. Murase, N. and Franks, F. (1989), Salt precipitation during the freeze concentration of phosphate buffer solutions, *Biophys. Chem.* **34**: 293–300.

123. van Den Berg, L. and Rose, D. (1959), Effect of freezing on the pH and composition of sodium and potassium phosphate solutions; the reciprocal system KH_2PO_4-Na_2HPO_4-H_2O, *Arch. Biochem. Biophys.* **81**: 319–329.

124. Van Den Berg, L. (1966), pH changes in buffers and foods during freezing and subsequent storage, *Cryobiology* **3**: 236–242.

125. Larsen, S. S. (1973), Studies on stability of drugs in frozen systems. VI. The effect of freezing upon pH for buffered aqueous solutions, *Arch. Pharm. Chem. Sci. Ed.* **1**: 41–53.

126. Kolhe, P., Amend, E., and Singh, S. K. (2010), Impact of freezing upon pH of buffered solutions and consequences for protein aggregation, *Biotech. Progr.* DOI: 10.1002/btpr.377.

127. Franks, F. and Murase, N. (1992), Nucleation and crystallization in aqueous systems during drying: Theory and practice, *Pure Appl. Chem.* **64**: 1667–1672.

128. Gomez, G., Pikal, M. J., and Nair, R.-H. (2001), Effect of initial buffer composition on pH changes during far-from equilibrium freezing of sodium phosphate buffer solutions, *Pharm. Res.* **18**: 90–97.

129. Shalaev, E. Y., Johnson-Elton, T. D., Chang, L., and Pikal, M. J. (2002), Thermophysical properties of pharmaceutically compatible buffers at sub-zero temperatures: Implications for freeze-drying, *Pharm. Res.* **19**: 195–201.

130. Lam, X. M., Costantino, H. R., Overcashier, D. E., Nguyen, T. H., and Hsu, C. C. (1996), Replacing succinate with glycolate buffer improves the stability of lyophilized interferon-γ, *Int. J. Pharm.* **142**: 85–95.

131. Pikal-Cleland, K. A. and Carpenter, J. F. (2001), Lyophilization-induced protein denaturation in phosphate buffer systems: Monomeric and Tetrameric β-Galactosidase, *J. Pharm. Sci.* **90**: 1255–1268.

132. Pikal-Cleland, K. A., Cleland, J. L., Anchordoquy, T. J., and Carpenter, J. F. (2002), Effect of glycine on pH changes and protein stability during freeze-thawing in phosphate buffer systems, *J. Pharm. Sci.* **91**: 1969–1979.

133. Bhattacharya, S. and Lecomte, J. T. J. (1997), Temperature dependence of histidine ionization constants in myoglobin, *Biophys. J.* **73**: 3241–3256.

134. Chaplin, M. (2008), Protein hydration in water structure and science (available online at http://www.lsbu.ac.uk/water/protein.html, accessed April 2008).

135. Zhang, L., Wang, L., Kao, Y.-T., Qiu, W., Yang, Y., Okobiah, O., and Zhong, D. (2007), Mapping hydration dynamics around a protein surface, *Proc. Natl. Acad. Sci. USA* **104**: 18461–18466.

136. Remmele, R. L., Stushnoff, C., and Carpenter, J. F. (1997), Real time in situ monitoring of lysozyme during lyophilization using infrared spectroscopy: Dehydration stress in the presence of sucrose, *Pharm. Res.* **14**: 1548–1555.

137. Franks, F. and Hatley, R. H. M. (1991), Stability of proteins at subzero temperatures: Thermodynamics and some ecological consequences, *Pure Appl. Chem.* **63**: 1367–1380.

138. Jaenicke, R. (1990), Protein structure and function at low temperature, *Phil. Trans. Roy. Soc. Lond. B Biol. Sci.* **326**: 535–553.

139. Privalov, P. L. (1990), Cold denaturation of proteins, *Crit. Rev. Biochem. Mol. Biol.* **25**: 281–305.

140. Becktel, W. J. and Schellman, J. A. (1987), Protein stability curves, *Biopolymers* **26**: 1859–1877.

141. Hatley, R. H. M. and Franks, F. (1989), The cold-induced denaturation of lactate dehydrogenase at sub-zero temperatures in the absence of perturbants, *FEBS Lett.* **257**: 171–173.

142. D'Amico, S., Gerday, C., and Fellert, G. (2001), Structural determinants of cold adaptation and stability in a large protein, *J. Biol. Chem.* **276**: 25791–25796.

143. Griko, Y. V., Privalov, P. L., Sturtevant, J. M., and Venyaminov, S. Y. (1988), Cold denaturation of staphylococcal nuclease, *Proc. Natl. Acad. Sci. USA* **85**: 3343–3347.

144. Ibarra-Molero, B., Makhadatze, G. I., and Sanchez-Ruiz, J. M. (1999), Cold denaturation of ubiquitin, *Biochim. Biophys. Acta* **1429**: 384–390.

145. Naik, M. T. and Huang, T.-H. (2004), Conformational stability and thermodynamic characterization of the lipoic acid bearing domain of the human mitochondrial branched chain α-ketoacid dehydrogenase, *Protein Sci.* **13**: 2483–2492.

146. Tang, X. and Pikal, M. J. (2005), The effect of stabilizers and denaturants on the cold denaturation temperatures of proteins and implications for freeze drying, *Pharm. Res.* **22**: 1167–1175.

147. Tang, X. and Pikal, M. J. (2005), Measurement of the kinetics of protein unfolding in viscous systems and implications for protein stability in freeze-drying, *Pharm. Res.* **22**: 1176–1185.

148. Arakawa, T., Kita, Y., and Carpenter, J. F. (1991), Protein-solvent interactions in pharmaceutical formulations, *Pharm. Res.* **8**: 285–291.

149. Jiang, S. and Nail, S. L. (1998), Effect of process conditions on recovery of protein activity after freezing and freeze-drying, *Eur. J. Pharm. Biopharm.* **45**: 249–257.

150. Hillgren, A., Lindgren, J., and Alden, M. (2002), Protection mechanism of Tween 80 during freeze-thawing of a model protein, LDH, *Int. J. Pharm.* **237**: 57–69.

151. Hillgren, A. and Alden, M. (2004), Differential scanning calorimetry iinvestigation of formation of poly(ethylene glycol) hydrate with controlled freeze-thawing of aqueous protein solution, *J. Appl. Polym. Sci.* **91**: 1626–1634.

152. Wisniewski, R. and Wu, V. (1996), Large-scale freezing and thawing of biopharmaceutical products, in *Biotechnology and Biopharmaceutical Manufacturing, Processing, and Preservation*, Avis, K. E. and Wu, V. L., eds., Interpharm Press, Buffalo Grove, IL, pp. 7–59.

153. Hartmann, U., Nunner, B., Korber, C., and Rau, G. (1991), Where should the cooling rate be determined in an extended freezing sample, *Cryobiology* **28**: 115–130.

154. Wilkins, J., Sesin, D. S., and Wiskiewski, R. (2001), Large-scale cryopreservation of biotherapeutic products, *Innov. Pharm. Technol.* **1**: 174–180.

155. Wisniewski, R. (1998), Large-scale cryopreservation of cells, cell components, and biological solutions, *BioPharm* **11**: 42–50.

156. Shamlou, P. A., Breen, L. H., Bell, W. V., Pollo, M., and Thomas, B. A. (2007), A new scaleable freeze-thaw technology for bulk protein solutions, *Biotechnol. Appl. Biochem.* **46**: 13–26.

157. Chen, Y. H., Cao, E., and Cui, Z. F. (2001), An experimental study of freeze concentration in biological media, *Trans. Inst. Chem. Eng.* **79**(Pt. C): 35–40.

158. Lashmar, U. T., Vanderburgh, M., and Little, S. J. (2007), Bulk freeze-thawing of macromolecules. Effect of cryoconcentration on their formulaiton and stability, *BioProcess Int.* **5**(6): 44–54.

159. Webb, S. D., Webb, J. N., Hughes, T. G., Sesin, D. F., and Kincaid, A. C. (2002), Freezing biopharmaceuticals using common techniques—and the magnitude of bulk-scale freeze-concentration, *BioPharm* **15**: 22–34.

160. Zippelius, R. (2000), *Investigation of freeze and thaw process on human pharmaceutical protein solutions at large scale*, Ph.D. dissertation, Ludwig-Maximilian Univ. Munich.

161. Rubinsky, B., Perez, P. A. and Carlson, M. E. (2005), The thermodynamic principles of isochoric cryopreservation, *Cryobiology* **50**: 121–138.

162. Engstrom, J. D., Simpson, D. T., Lai, E. S., Williams, R. O., and Johnston, K. P. (2007), Morphology of protein particles produced by spray freezing of concentrated solutions, *Eur. J. Pharm. Biopharm.* **65**: 149–162.

163. Tracy, M. A. (1998), Development and scale-up of a microsphere protein delivery system, *Biotechnol. Progress* **14**: 108–115.

164. Hew, C. L. and Yang, D. S. C. (1992), Protein interaction with ice, *Eur. J. Biochem.* **203**: 33–42.

165. Petersen, A., Schnieder, H., Rau, G., and Glasmacher, B. (2006), A new approach for freezing of aqueous solutions under active control of the nucelation temperature, *Cryobiology* **53**: 248–257.

166. Singh, S. K., Kolhe, P., Wang, W., and Nema, S. (2009), Large-Scale Freezing of Biologics – A Practitioner's Review. Part I: Fundamental aspects, *BioProcess Int.* **7**(9): 32–44.

167. Singh, S. K., Kolhe, P., Wang, W., and Nema, S. (2009), Large-Scale Freezing of Biologics – A Practitioner's Review. Part II: Practical advice, *BioProcess Int.* **7**(10): 34–42.

27

STRATEGIES FOR BULK STORAGE AND SHIPMENT OF PROTEINS

Feroz Jameel, Chakradhar Padala, and Theodore W. Randolph

27.1. INTRODUCTION

Continual increase in the yields of purified proteins due to increased process efficiencies is steadily increasing the need for improvements in safe and efficient storage methods, and the need for flexibility in pooling and shipping bulk drug substance. The bulk drug substance can be stored and shipped in the liquid state, as a dry powder, or in a frozen form, with each method having its own merits and limitations. For instance, protein stored in the liquid state can be dispensed conveniently as needed, but requires more diligence in order to avoid instability, microbial growth, and proteolytic degradation. In contrast, dry powders enable long-term storage of protein with very little threat of degradation, but their production involves a lengthy process of drying and reconstitution. Once frozen, proteins are relatively stable, but the freezing and thawing processes may damage proteins, and maintaining temperature control for frozen products is more cumbersome than for liquids. Hybrid processes can be envisioned as well; a relatively new technique called *cryogranulation* that creates free-flowing, frozen granules by exposing solution or slurries to a cryogenic material such as liquid nitrogen is being explored as an alternative means of storage for bulk drug substance [1].

Formulation and Process Development Strategies for Manufacturing Biopharmaceuticals,
edited by Jameel and Hershenson

However, regardless of the method of storage, the drug substance is expected to have a minimum of 2 years of storage stability prior to fill and finish, and an additional minimum of 2 years of stability in the drug product state. Hence, long-term storage should be treated as a separate unit operation deserving careful consideration. It is important to ensure that the protein in the drug substance storage phase retains its native state with minimal to no presence of nonnative protein aggregates, as even small levels of these soluble aggregates can serve as nuclei for the formation of visible and/or subvisible aggregates when exposed to fill and finish processing stresses [2].

Since there is substantial interdependence between formulation and processes, for successful implementation of the bulk storage technology, it is imperative to select the storage method early on so that the formulation and processes can be designed and developed to be compatible with each other. The technology selection process should involve logistics and technical feasibility in addition to cost analysis based on market projections. Once the decision on the choice of bulk storage and shipping technology is made, the development of the process should go hand in hand with formulation development. The focus of this chapter is to discuss in detail the challenges and considerations during the development, characterization, and scale-up of each storage technology, with emphasis on cryopreservation, which is the most commonly used method.

27.2. CONTAINERS FOR LIQUID AND FROZEN SOLUTIONS

27.2.1. Carboys

For frozen storage and shipment, polycarbonate carboys are preferred over polypropylene because of their lower glass transition temperatures and brittleness temperatures. Polycarbonate material remains rubbery and does not become brittle until −135°C. Their use for temperatures lower than −135°C should generally be avoided to prevent breakage [3]. Carboys are manufactured in various sizes; 10 and 20 L are the most commonly used in commercial manufacturing. In general, 20-L carboys are preferred over 10-L carboys because they reduce the number of carboys required, reduce the number of samples for ID and quality testing, minimize space required for storage and shipment, and turn out to be economical. A drawback to the use of 20-L carboys is their significant filled weight, which poses challenges for operators to lift and to mix manually to ensure solution homogeneity before processing. If 20-L carboys are used, dynamic thawing process using shakers or roller racks may need to be employed.

27.2.2. Stainless-Steel Containers

Type 316 stainless steel is an alloy commonly used in the pharmaceutical industry for containers conforming to the 3A sanitary code that applies for storage of both liquid and frozen bulk drug substances. However, long-term storage of bulk formulations that are high in chlorides or other electrolytes can contribute to corrosion of wetted metal surfaces, causing damage to both equipment and protein. It is well documented that interactions between stainless steel and halides can result in metal-catalyzed oxidation of protein formulation components [4]. Bei Chen et al. [5] observed formation of high levels of aggregates when ABX-IL8 solution containing high concentrations of histidine (60 mM) was subjected to freeze–thaw operations in stainless-steel tanks and subsequently stored at 40°C, compared to solutions without exposure to stainless-steel tanks. They attributed formation of aggregates to leaching of high levels of iron from the tanks [5]. Similar results were reported by Cleland et al. [6] on rhuMAb HER2 and by Roberts et al. in a literature review [7].

The effects of common formulation components on 316L stainless steel can be studied using various techniques such as electrochemical pitting corrosion testing [8]. Such techniques help in ranking various formulations by their probability of "corrosiveness" and identifying formulation components that have the greatest impact on corrosion.

27.2.3. Single-Use Containers

Formulation vessels represent a substantial capital cost. In addition, there are increasing concerns surrounding reliability and validation of cleaning operations, and the significant impact that cleaning processes have on production capacity. Thus disposable containers, such as BioProcess Containers, offer an attractive alternative to 316 stainless-steel vessels [9]. These disposable vessels are presterilized, are easy to handle and transport, and do not require clean-in-place or steam-in-place technology. Hence, they eliminate potential protein product contamination by cleaning products and occupy less space compared to metal tanks. These containers are designed to meet various requirements of unit operations; Labtainers® (Thermo Scientific) and Stedim Flexel® (Sartorius-Stedim) 2D bags can be used for transportation and storage of bulk intermediate product, and can handle stresses arising from freezing–thawing, heat inactivation, and irradiation of products. Mixtainers™ (Thermo Scientific) are designed to handle mixing of formulation buffers and other solutions, whereas the Smartainer II stainless-steel bin is designed to be a support vessel for single-use disposable liquid-handling systems (e.g., BioProcess Container systems) such as tank liners that are designed to contain formulation buffers, intermediates and final product. The outer layer of these bags is made up of EVOH as a barrier layer whereas the product-contacting inner layer is either made up of either ultra-low-density polyethylene or EVA.

27.2.4. Leachable/Extractable Criteria for Containers

With any container material, the potential interaction of the drug substance or drug product with the primary container is a primary safety concern, and a *Code of Federal Regulations* (CFR) relating to it has been issued, clearly stating that the drug product containers and closures shall not be reactive, additive, or absorptive so as to alter the safety, identity, strength, quality, or purity of the drug beyond the official or established requirements [10]. Regulatory agencies recommend performing the qualification of the primary container early on during commercial process development. This process involves the identification of extractables and leachables in the container, developing the testing criteria and the testing processes, and risk analysis to justify the proposed limits either by structure–activity analysis or by additional toxicological evaluation. If plastic containers such as carboys are used as primary containers, the extractables may include additives in the plastic (e.g., stabilizers) such as butylated hydroxytoluene (BHT), lubricants such as calcium or zinc stearates, and colorants. It is common for these additives to include both organic and inorganic components. There may also be some extractable monomer of the plastic in the final molded product. Although these extractables occur in very low concentrations [typically parts per million or billion (ppm or ppb) range], their possible effects on products need to be established through proper toxicological studies. Storage of protein solutions in metal containers such as stainless steel can result in leaching out of metals, and if the protein contains exposed methionine, cysteine, or sulfhydryl groups, these groups may then be susceptible to metal-catalyzed oxidation.

27.3. STORAGE AND SHIPPING METHODS

27.3.1. Liquid Storage and Shipment

If the final dosage form is going to be liquid, then the storage and shipment of bulk drug substance in the liquid state should be less challenging, although the additional stresses encountered during transportation must be taken into account. Storage and shipment of the bulk drug substance in the liquid state is the most convenient, cost-effective, and simplest of all the methods from both operational and logistics perspectives. The reasons for this are obvious—except for filtration followed by filling into the container, there are no major processes such as freezing, thawing, drying, and mixing needed; no dry ice is required during transportation; there is no risk of CO_2 ingress; and there are no significant scale-up issues. Depending on the desired storage time and the protein stability, liquid bulk drug substances allow for flexibility in container selection and storage conditions. They can be stored either at ambient or refrigerated temperatures (depending on the robustness of the formulation) in various types of containers such as stainless-steel vessels, disposable plastic bags, or carboys. However, the technical challenges that arise in liquid formulations due to the inherent susceptibility or sensitivity of the protein to stresses encountered during storage and transportation make this method less attractive than other methods. Proteins in general are relatively more susceptible to chemical and physical instabilities in the liquid state, and exhibit only marginal stability in solution [11,12]. However, some proteins (including most monoclonal antibodies) are quite rugged in nature, and when appropriately formulated, they are quite stable under a wide variety of adverse conditions and maintain native structure in the liquid state for extended period of time [13]. Additionally, antibodies frequently are administered in high doses that require high-concentration formulations, which, in turn, may increase solution stability and minimize the fraction of protein that may become damaged by adsorption to container surfaces.

To enhance and achieve the required stability of the bulk drug substance in the liquid state, each type of degradation species should be well characterized and the pathway(s) leading to their formation should be well understood such that they are properly inhibited in the final formulation. Additionally, some characterization work specific to bulk storage and shipment should be included in the characterization studies during the early development of product. The additional characterization work should constitute evaluation of impact of photoexposure, agitation, foaming, protein concentration, and container compatibility (leachable/extractable).

27.3.1.1. Stresses Encountered During Transportation of Liquid Formulations. Some proteins and antibodies are sensitive to light exposure, which can trigger degradation reactions that can continue even in the dark. Even relatively low levels of light, such as those associated with fluorescent bulbs under normal light intensities, can provoke photoinduced degradation. Product degradation during transportation can be minimized through use of light blockers [14].

Stresses that can potentially harm the protein also can come from agitation of the solution during handling and transportation. This agitation can increase the exposure of protein to damaging interfaces such as the air–water interface, and cause the fragile native conformation to partially or completely unfold, leading to aggregation [15–18]. Because proteins are inherently surface-active, protein solutions by themselves can create foam on agitation even in the absence of any surface-active agents in the formulation. Foam is well known to be a potential site for interfacial denaturation of proteins. Agitation and foaming during transportation can be minimized by minimizing underfill; however, there will be always some headspace. In addition to agitation and foaming due to headspace, one must be concerned about oxidation, which can be mitigated by nitrogen overlay.

The effect of protein concentration on storage stability should be evaluated as well since some proteins are more stable when stored at higher concentrations, and this advantage can be utilized by storing drug substance at higher concentrations, saving costs on number of containers required, storage space, and sampling for quality control (e.g., ID and bioburden) [19]

In addition to characterization, some precautions are necessary early on during downstream processing (especially the viral inactivation and purification steps) as these processing steps can cause stress manifested as the growth of small number of aggregates that can serve as nuclei for further growth of proteinaceous particulates or higher-order aggregates during long-term storage of the bulk drug substance. Efficient removal of proteases and bioburden from the bulk drug substance, coupled with secured container–closure integrity (CCI), is also critical to the subsequent long-term storage stability of the drug substance in the liquid state. Efficient removal of proteases can be verified through presence/absence of protein fragments on the SDS-PAGE. If fragments are observed, then either the protein solution needs to be purified further or protease inhibitors need to be added (see Section 27.3.3.5.1 for details).

27.3.1.2. Transport Validation. In either the liquid or the frozen state, a number of critical process parameters must be validated during transportation. In frozen formulations, the transportation packaging configuration needs to be validated for temperature adherence and container closure integrity during the transportation process. The shipment of the container via air and ground transportation should not be shipment- or destination-specific but rather time-dependent and should be validated for shipping time. The transportation validation runs can be performed with buffer-only placebo formulations, since it has been noted that the use of buffer will not affect the integrity of the data collected. Validation runs need to be performed with worst-case scenarios (e.g., with fill volumes at extreme end of the fill volume specifications).

For liquid formulations, transportation can be validated for miles or time against the impact on product integrity or potency. During the air and ground transport process, product is potentially exposed to a number of environmental and process-related variables that could adversely affect product quality. Shock and vibration occur as a result of handling and are inherent in the air and ground transport processes. For frozen shipments, the integrity of shippers used for transportation and their ability to maintain the desired temperature for the entire transit time must be validated.

27.3.2. Storage as Salted-Out Precipitates

Some proteins, especially enzymes, are stored as salted-out precipitates. The salting-out process does not result in nonnative aggregation, but rather a reduction in the protein's solubility, which causes proteins to precipitate in their native conformational state. The mechanism of solubility reduction is described by Arakawa and Timasheff for protein stabilization [20]. Preferential exclusion of certain added solutes increases the chemical potential of both the native and unfolded states of the protein, but because the unfolded state has a higher surface area and thus excludes more solute, its chemical potential is increased even more than that of the native state, resulting in an increased stability of the native state relative to that of the unfolded state. If sufficient excluded solute is added, the chemical potential of the native state eventually exceeds that of protein in the solid state, and the protein precipitates, but the precipitation occurs under conditions where the native state is highly stabilized with respect to unfolding, which enhances storage stability. Since the protein is both conformationally stabilized and in a state with greatly reduced mobility, aggregation processes are inhibited.

The most common excluded solute used to induce precipitation is ammonium sulfate at concentration as high as 70%. Ammonium sulfate is commonly chosen as the salt since high concentrations can be easily achieved and it seldom denatures proteins. When ammonium sulfate is added to the protein solution, a large number of water molecules bind to the sulfate ion, which reduces the water activity and thus water binding to the protein. If a protein is not hydrated by binding to water molecules, it will precipitate. Different proteins precipitate at different concentrations of ammonium sulfate. Furthermore, the presence of "excluded solutes" increases the free energy of unfolding, forcing the protein to stay in the folded state. These two positive effects can be further strengthened and storage stability can be further enhanced by storing these salted-out precipitates at subzero temperatures as these solutes also act as cryoprotectants. However, their large-scale use is limited as ammonium sulfate is corrosive, forms dense solutions (presenting problems to the collection of the precipitate by centrifugation), and may release gaseous ammonia, particularly at alkaline pH.

27.3.3. Storage as Frozen Solution (Cryopreservation)

Cryopreservation is considered when the bulk drug substance cannot be stored in the liquid state because of marginal storage stability. When proteins are stored in the frozen state, the rates of many of the common reactions leading to physical and/or chemical degradation are retarded [21]. In addition to enhancement of product stability, frozen storage of the bioproduct offers several other advantages, including minimization of the risk of microbial growth, elimination of agitation and foaming during transport, and flexibility for the manufacturing processes, such as allowing a hold step for pooling batches of intermediates, and storing final bulk drug substance before fill–finish operations. The performance of the freezing process is characterized by several critical process parameters, including freezing rate, freezing temperature, and time.

Freezing may also pose risks to proteins. During freezing, pure water is removed from the solution as ice, resulting in cryoconcentration of the remaining solution. Typically, solutes (e.g., buffer salts, stabilizing excipients, tonicity modifiers) as well as the drug substance itself become 10–15 times more concentrated in the non-ice phase of the formulation. Clearly, proteins may not be stable in solutions with such high concentrations. Freeze concentration of buffer salts (e.g., sodium phosphate) can result in differential precipitation of buffer salts and dramatic swings in pH in the freeze-concentrated solution. For example, the pH in a solution containing sodium phosphate originally formulated at pH 7.0 may drop below pH 4.0 in the freeze concentrate. Furthermore, low temperatures themselves may thermodynamically destabilize the protein at temperatures below the cold denaturation temperature. Finally, the presence of large amounts of ice–water interfacial area, as well as possible solid–liquid interfaces due to freeze-concentration-induced solute crystallization, may denature the protein. The extent of protein damage arising from these effects depends on several factors, including the intrinsic stability of the protein itself, the freezing method, the formulation composition, the container type or design, and the environment [25–27]. Therefore, in addition to effective and efficient design of the process, minimization of these adverse effects is central to the development and optimization of bulk freezing and thawing processes.

27.3.3.1. Development and Characterization of "Uncontrolled Rate" Freeze–Thaw Process

27.3.3.1.1. Process Development. Freezing and thawing parameters required to characterize freezing and thawing times, rates, and temperatures generally can be determined using a formulation buffer and do not necessarily require drug substance as dilute solutions represent a worst-case scenario because of the higher heat capacity associated with water. The minimum and maximum working fill volumes for various container sizes need to be established with respect to the relative integrity of the container. For freezing studies, it is advisable to use the same type of freezer (i.e., −20°C, −30°C, or −80°C) and mimic practices that would be typically used in the commercial manufacturing setting. Use of frost-free and upright freezers must be avoided as their freezing and thawing rates may not be representative because of the cycling of temperature to enable the frozen condensate to defrost. So the product may experience several freeze–thaw effects depending on the melting temperature of the product; exceptions and variations to this tendency also need to be studied and captured through characterization studies. For dynamic thawing processes, the optimal mixing speed (rpm), needs to be established in conjunction with the stability of the product.

During freezing, since the freezing fronts move from all sides of the container, the last point to freeze (LPTF) will be usually the middle center of the container (carboy) and the last point to thaw (LPT) would be the top center, a few centimeters below the surface of the liquid as the ice pieces detach and float to the surface. Thus, the temperature profiles at these points typically constitute the worst-case scenarios. The freezing process time, which is normally characterized by effective freeze time, that is, the time for the temperature measured at the LPTF to go from +10°C to −30°C (or the target freezing temperature), and the thaw process, characterized by the time required to raise the temperature at the LPT to +5°C (or the target thawing temperature), can be determined by placing temperature probes in the bulk material at LPTF and LPT. The temperature of the bulk material can be monitored using thermocouples and/or a data logger. The temperature and time profiles of both freezing and thawing processes determined through the placement of thermocouples is usually biased and not representative of the whole batch. Thermocouples act as nucleating sites and tend to cause the solution to begin to freeze earlier. Hence, it is always advisable to allocate an additional soak period of 15%–20% of the total freezing time.

The thawing process is relatively longer compared to the freezing process. The reasons for this are twofold: (1) ice has a lower thermal conductivity than does liquid water; and (2) to avoid the risk of (local) overheating during thawing, the temperature differential that provides the driving force for heat transfer is limited. It is also important to minimize the residence time that the protein spends in the freeze concentrate during thawing by expediting the process. Hence, dynamic thawing at warmer temperature should be considered over static thawing at 2°C–8°C, wherever feasible. Concentration gradients created by freeze concentration typically remain after freezing and thawing, especially under static thawing conditions (see Figs. 27.1 and 27.2). Therefore, adequate mixing of thawed solution is required to achieve homogeneity in the bulk container. Agitation can be accomplished through gentle rocking, swirling, or rolling of the container. The thawed liquid may also be recirculated from the top to the bottom of the container if the container has the proper outlets designed for this purpose. Bubbles created during thawing from release of air trapped in the frozen matrix, or foam created by mixing or recirculation, can be deleterious to protein products as air–liquid interfaces may serve as a site for denaturation [28]. Therefore, agitation should be approached cautiously. The acceptable thawing temperature and intensity of mixing and shaking depends on the formulation and needs to be established after careful evaluation of the effect of agitation on foam formation and the effect of thawing temperature on the stability of protein.

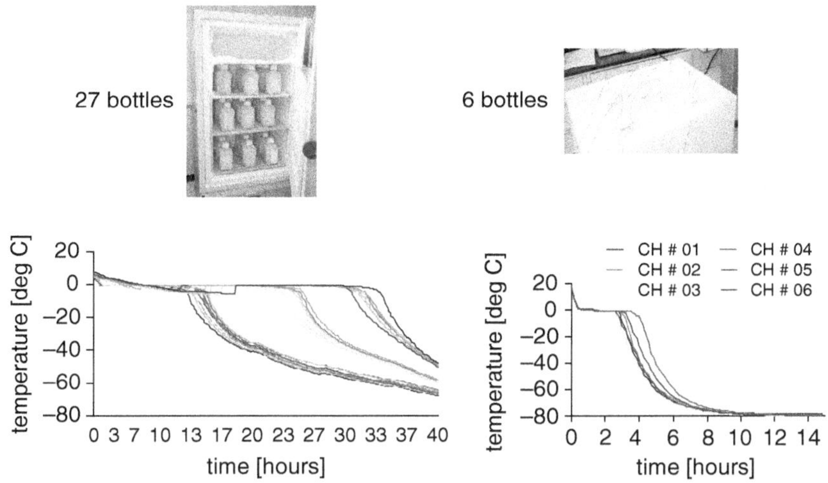

Figure 27.1. Freezing in ultra-low-temperature freezers or in dry ice. Product temperature profiles as measured by thermocouples as a function of location in the freezer. (Courtesy of Roland Schmidt.)

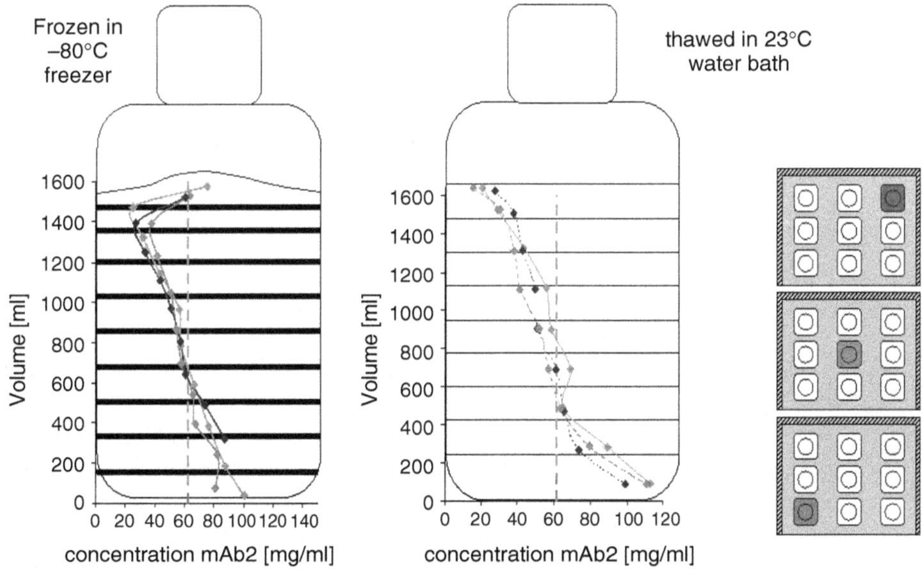

Figure 27.2. Gradient formation in frozen and thawed states. (Courtesy of Roland Schmidt.)

27.3.3.1.2. Process Characterization. One of the biggest challenges with uncontrolled freezing rate technology (e.g., in carboys) is the lack of interbatch consistency in the process. The variations stem from design of the freezers, space, cooling capabilities of the refrigeration system, and locations of the carboys within the freezer. Parameters such as the rate of air circulation, the number of carboys in the freezer and their fill volumes, the carboy design (e.g., heavy-duty vs. light-duty carboys), size, formulation composition, filling environment, and procedures can also play a role. It is important to study the effect of these variations on the process performance and on the product quality attributes before the formulation and the processes are locked down. The following are some factors that contribute to variations and need to be evaluated.

- *Effect of Freezer Design*. The design of the walk-in freezer impacts the freezing rate, which, in turn, can impact the degree of cryoconcentration and the morphology of ice structure. The refrigeration cooling capability of a freezer designed to cool to as low as -70°C would be different from the one designed to cool to -20°C, and the product would experience different freezing rates even when it is cooled to the same final temperature. Additionally, the freezer performances are typically quite variable; a temperature variation between identical models of $\pm 10^{\circ}$C is not unusual.
- *Number of Carboys (Batch Size) and Position of Carboys*. The performance of the freezing and thawing processes are also influenced by the number of carboys and their positions relative to others. The large heat of fusion of ice provides a demand for a large amount of heat removal to achieve the final frozen state, and the larger the liquid load, the more heat must be removed. If many carboys are used simultaneously, carboys that are surrounded by other carboys may be insulated by their neighbors, and thus experience slower cooling and freezing rates. In freezers, heat may be transferred via convection to circulating air, via conduction to freezer surfaces, and by thermal radiation. The air circulation pattern may be altered by the number and positioning of carboys placed in a freezer, and carboys can shield one another so as to also alter the pattern of radiative heat transfer to the freezer walls. The process and the product need to be characterized with respect to these parameters.
- *Fill Volume*. In addition to potential impacts on the integrity of the container, container fill volumes can also affect the performance of the process and product quality attributes. All these fill-volume-related contributions need to be factored in during the evaluation and determination of minimum and maximum working volumes. A concern at the maximum end is that volume expansion caused by formation of ice would exceed the tensile strength of the carboy and result in compromising its integrity, especially at the seams. We have also observed in our hands that the maximum fill volume defined solely on the basis of container integrity was not compatible or acceptable in terms of product integrity, and in fact a lower fill volume in the same-size container alleviated the damage

to the protein. On the other hand, underfilled containers are more susceptible to implosions because of pressure differentials arising from differences in atmospheric pressures and temperatures encountered during transportation or the freezing–thawing processes.

- *Effect of Multiple Freeze–Thaw*. Once the conditions for full-scale freeze–thaw processes are identified, the next step would be evaluation of effect of these process conditions on freeze concentration, specific ice surface area (ice–liquid interface), and low-temperature/cold denaturation temperatures and, in turn, their respective impacts on the stability of the drug substance. Since the availability of the material will be scarce during the early stages of product development, a scale-down model mimicking conditions that prevail at large scale needs to be developed and validated (see Section 27.3.3.3 on how to develop a scale-down model). It is recommended that, multiple freeze–thaw cycles (two or three) be performed on the scale-down model and that the samples removed at the end of each cycle be analyzed using stability-indicating assays. The justification for subjecting the drug substance to multiple freeze–thaw cycles comes from the possibility that during manufacturing the bulk drug substance may need to be returned to the freezer during drug product manufacturing process, because the bulk drug substance is not ready for further processing, the fill–finish line is malfunctioning, or there are other interrupting circumstances, and also because it is common practice to store remnants or intermediates frozen to be pooled at a later date. Since the subtle changes in the protein structure are seldom evident at time $t = 0$, the samples must be placed in long-term storage, and the stability and integrity of the protein must be evaluated. It is also important to study the effect of transferring from a -20°C or -30°C freezer to a -70°C freezer in order to simulate frozen transport conditions, where material may be shipped packed in dry ice. Particularly critical at this condition is the integrity of the container, which will be below the glass transition temperature of the polymeric material of most carboys.

27.3.3.2. Development and Characterization of "Controlled Rate" Freeze–Thaw Process. Integrated Biosystems Inc. (IBI), now called Sartorius-Stedim, has developed a controlled-rate freeze–thaw technology that is marketed in two different types of containers but that work on a similar principle. The "CryoFin" technology is based on a line of stainless-steel vessels, CryoVessels, when used in conjunction with CryoTrols freeze-thaw system, control the freezing and thawing rates thereby minimizing the denaturation at the ice–liquid interface and cryoconcentration. These CryoVessels are compartmentalized symmetrically through fins such that the distance between cooled surfaces and the LPTF is reduced and the heat transfer surface area is extended. Under optimal conditions, both factors can be combined to create an ice front velocity greater than the diffusion rates of solutes, thus minimizing the creation of large-scale concentration gradients. Additionally, the ice–liquid interface denaturation is minimized by preventing liquid supercooling by inducing ice nucleation at container surfaces. The vessels are also designed for dynamic mixing during the thawing process, which expedites thawing and removes any concentration gradients.

The Celsius-Pak technology is also based on concepts similar to those used in the CryoFin device, but it utilizes presterilized single-use plastic bags instead of stainless-steel containers. The bags are made from multilayer, coextruded high-gas barrier films containing ethylene vinyl acetate (EVA) copolymer as the fluid contact layer and EVA polymer as the barrier layer to water vapor. The bags are compressed between the heat exchange plates through which a silicone-based heat transfer fluid is circulated. One challenge with the freezing systems in general is the heat transfer limitation, however, the surface-to-volume ratio of the Celsius bags is designed in such a way that it provides high heat removal rate over a small freezing pathlength that enables controlled ice crystal growth and uniform product distribution of solutes across the frozen matrix. Obviously, other advantages are associated with disposability, including elimination of the need for clean-in-place and steam-in-place processing, minimizing the cost of operation and maintenance, and eliminating the risks associated with cross-contamination, especially in multiproduct facilities.

One important advantage, from a process perspective, of the controlled-rate technology is that one can achieve a high degree of uniformity and process reproducibility consistent with the expectations of validation and regulatory compliance. However, product stability is not ensured unless the process is well understood and optimal conditions are identified. Besides ensuring stability of the product during processing and on storage, the process should also be efficient and should not require major equipment modifications.

There are formulation-dependent interactions with the freezing process as well. The formulation determines the glass transition temperature of maximum freeze concentrate (T_g') of the product and also the maximum time and temperature for exposure of the product to freeze concentration without compromising the integrity of the protein. Generally, it is considered to be safe to store the product below the T_g', which minimizes mobility-driven reactions [30]. However, since the T_g' value for a given formulation also depends on the freezing rate, variations in the freezing process itself may result in different T_g' values [30].

Certainly, a goal during the development of a freezing process should be to avoid cycles that require deep-freezing storage conditions for the product, which, in turn, require specialized equipment and add to the cost of storage and shipment. Additionally, storing Celsius-Pak below the glass transition temperature (T_g) values of the plastic bags (around -20°C) can result in brittleness and risk the integrity of the bags. Brittleness is not determined by the T_g value alone; cold cracking temperatures or brittleness temperatures (thought to be around -80°C [31]) also need to be considered before finalizing the choice of controlled-rate technology.

27.3.3.2.1. Process Development. One of the current drawbacks with controlled-rate technologies is the lack of linear or direct scalability; that is, the process developed at small scale is not linearly transferable to large scale. The process first needs to be developed at large scale using a placebo solution, with the resulting product temperature profiles then used to develop freeze–thaw cycles at small scale to characterize the process and understand the impact of process variables on the stability of the drug substance. The set-point temperature profile of the heat transfer fluid for the scaled-down unit is then programmed and adjusted in such a way that the product temperature profiles observed at the LPTF in the small-scale system are superimposed over the ones observed at large scale. The multiple set-point profile adjustment on the scale-down unit must be done carefully, taking into account the differences in the cooling capacities and heat exchange efficiencies between the two scales. Supercooling is rarely observed with large-scale systems but is frequently seen in small-scale systems. Supercooling effects in small-scale units can be minimized by inducing ice nucleation by transiently dropping the setpoint temperature to a very low value (e.g., -70°C). The process performance equivalence between the two scales and performance consistency between large- and small-scale equipment can be assessed by comparing nominal freeze times (NFTs) and effective freeze times (EFTs), respectively. *Nominal freeze time* characterizes the ice crystal growth rate, which, in turn, determines the quality of the product as the rates influences both cryoconcentration and morphology of the ice. It is defined as the time it takes the last point to freeze to -5°C from an initial temperature of $+3^{\circ}$C. *Effective freeze time* is defined as the total time required to reduce the temperature at the thermal center of the sample to a sufficiently low temperature (e.g., -30°C) from an initial temperature (typically $\sim 10^{\circ}$C).

Since the controlled-rate technologies are designed and programmed to minimize the cryoconcentration effects through a controlled freezing rate (i.e., an ice front velocity of 20–25 mm/h and the creation of linear dendrites as opposed to branched dendrites), the cooling rates does not need to be monitored and assessed. However, the other critical parameters of the freezing process need to be evaluated to assess whether they need to be optimized to meet freezing requirements of the product. To minimize mobility-related interactions, it is advisable that the product be frozen to temperatures below T_g' and sufficient time be allowed for the product to equilibrate to that temperature to ensure complete conversion of all the free water to ice. The parameters that need to be evaluated include (1) the lowest temperature of solidification and (2) the time to reach complete solidification. These two parameters need to be determined by monitoring the thermocouple placed at the last point to freeze in the bag, which is usually at the top center of the bag.

The thawing process is normally characterized by the critical process parameters of time, temperature, and mixing speed. Their design is product-specific and is dictated by the stability of the protein. It is recommended that the parameters required for thawing to be completed as rapidly as possible, without the presence of ice and with minimal foam, be identified and optimized while maintaining the product temperature below the temperature limit imposed by the stability of the product. This can be done in two steps; initially a series of experiments can be performed where the temperature and hold times are varied while keeping the mixing speed constant to identify the optimum conditions that allow the thawing process to occur close to the maximum allowed product temperature. Next, various mixing speeds, typically ranging from 50 to 150 rpm, can be studied in an effort to identify an optimum mixing speed that results in a clear and homogenous solution with minimal generation of foam. If the formulation contains a crystallizable excipient such as mannitol, one may observe settling of fine mannitol crystals that formed during the freezing step at the bottom of the container. Increases in mixing speed will result in increased crystal dissolution rates; however, an increase in foaming tendency may also result.

27.3.3.2.2. Process Characterization. The operating parameters in the controlled-rate freezing–thawing method need to be characterized to understand the effect of various process variables on the performance of the process and the product quality attributes. The study should include the effects of fill volume, as this affects the process performance and, in turn, affects the product quality. Various fill volumes produce different product temperature profiles, and since it is not feasible to examine all possible freeze–thaw cycles and study their effects, a bracketed approach covering the extremes can be taken. The extremes consist of maximum load (100 L for Celsius-Pak and 200 or 300 L for CryoVessels) and a minimum load (4.2 L for Celsius-Pak and 30 L for a 200 or 300 L CryoVessels), and thus, only two cycles need to be developed at lab scale to study the effect of fill volumes. For commercial production, one freeze–thaw cycle covering all volumes and all denominations is preferred, as it will be easy to validate and use. In addition to fill volume, the effect of multiple freeze–thaw cycles, freezing rates (maximum and minimum), protein concentration, excipient weight ratios, formulation (liquid and lyophilized formulations), storage conditions (e.g., both relatively warmer temperatures such as −20°C or −30°C, and colder temperatures corresponding to dry ice after freezing to −50°C) should be studied.

27.3.3.3. Development of a Scale-Down Model. Because of the cost and limited availability of the drug substance during the early stages of product and process development, we recommend developing a scale-down model that mimics the conditions that prevail in the large-scale unit. Generally, these scale-down models are suitable for characterizing the process and/or studying the effect of various process variables on the integrity of the protein molecule in particular and overall quality attributes of the drug substance in general. These studies are key to identification of critical process parameters and to development of the "design space." However, in general, they are not suitable for developing a process or for identifying the process conditions for large-scale freeze–thaw processes, especially the thawing conditions. The development work for the large-scale freeze–thawing processes needs to be performed in a large-scale unit, typically using a placebo formulation in place of drug substance. A scale-down model can be said to mimic the large-scale unit when it demonstrates the equivalency in (1) product temperature–time profile (e.g., product temperature–time profiles on a scaled-down model must be closely superimposed on the profiles from a large-scale unit) and (2) freezing and thermal events (e.g., extent of cryoconcentration and specific surface area of ice). Both the temperature–time profile and the extent of thermal events depend on the freezing rate and ice front velocity, which, in turn, depends on the freezing pathlength. During the development of a scaled-down model, for both controlled- and uncontrolled-rate technology, if one keeps the freezing pathlength the same as that at large scale and manipulates the dimension of the container or insulates the container in such a way that the rate of freezing is determined predominantly by one-dimensional heat conduction, then the freeze–thaw conditions that prevail in the large-scale system can be easily mimicked. The flexibility to volume adjustments of the scaled-down model can be achieved by adjusting the other dimensions of the container, which may be challenging in the case of cylindrical carboys, but easily attainable in the case of rectangular containers. If this strategy is followed, freezer temperature is the only variable that determines the rate of freezing, and by making the temperature consistent between small- and large-scale systems, scale-independent freezing may be achieved.

27.3.3.3.1. CryoWedge. Sartorius-Stedim has designed a scaled-down system, called the CryoWedge, to accompany the large-scale CryoVessel that allows study of the impact of freezing and thawing conditions on the stability of the protein. The scaled-down system requires a minimum working volume of 350 mL, and samples as large as 4 L may be tested. Its wedge shape is designed to mimic one compartment of the symmetric compartments of the CryoVessel with identical configuration of the heat exchange surface angles, length, and material of construction [32].

The design of the CryoWedge models the compartments of the vessel, but it requires programming in order to generate product temperature–time profiles similar to those observed in the CryoVessel. A stepwise freezing–thawing program for the heat transfer fluid needs to be developed using the CryoTrol software associated with the CryoWedge unit so that it creates freezing and thawing conditions comparable to what the product would experience in the CryoVessel. The final optimized stepwise freeze–thaw conditions that resulted in superimposable product temperature profiles between CryoWedge and CryoVessel then can be used to study the effect of process variables on the stability of the product. Still, the CryoWedge connot be used to develop a large-scale freeze–thaw process; a large-scale CryoVessel is required for the development of the freeze–thaw process.

27.3.3.4. Considerations for Development of a Large-Scale Freeze–Thaw System. The objective of the design and development of a freeze–thaw system should be that it is easy to control, robust to operate, and linearly scalable from lab to manufacturing. Scalable techniques that can identify the necessary conditions to minimize such damage in manufacturing environment would be immensely valuable.

27.3.3.4.1. Design. Geometry of the container is a key design parameter. Computational fluid dynamics (CFD) studies can predict the impact of edges and other geometric considerations of the container on the temperature of the bulk drug substance stored in the container and provide useful insight into container design considerations. Dimensions of the container need to be designed such that the rate of freezing is determined predominantly by one-dimensional heat conduction, which is believed to be achieved more easily with rectangular as compared to cylindrical geometries [33]. In order to minimize cryoconcentration, the ice front velocity should be higher than the diffusion rate of solutes so that the protein molecules and solutes become entrapped by the moving freezing front. This can be achieved through the combination of shorter freezing pathlength and efficient external heat transfer. Computational studies suggest that a thickness of the heat transfer pathlength of 8.4 cm promotes linear dendrites, and the rate of growth of dendrites would be comparable to diffusional rate of solutes and the cryoconcentration phenomenon could be minimized [34]. One has to be careful while trying to design a container with small diameter and large height in an effort to increase the ice front velocity, because significant mechanical stress can be generated in the side walls from expansion of ice during freezing. It is safer to have the predominant heat sink located at the bottom of the vessel in order to prevent rapid freezing of the upper levels of the liquid and entrapment of unfrozen liquid phase under frozen surfaces.

For stainless steel containers besides heat transfer through conduction, heat transfer from the storage container to the atmosphere through radiation can be enhanced by polishing the heat transfer surface of the storage containers to 320 grit and electropolishing to provide a mirror finish [35]. Product transfer, storage conditions, recovery, handling, rating for full vacuum, ability to withstand Clean-in-place (CIP) and Sterilization-in-place (SIP), and compliance with ASME code are other considerations that need to be kept in mind during the design.

27.3.3.4.2. FREEZING. The limitations in heat transfer resulting in cryoconcentration in large-scale freeze–thaw systems arise mostly from their inability to quickly remove heat of fusion and cool the system as the ice front progresses. This challenge increases with the increase in thickness of ice due to poor heat conduction through the frozen mass. One way to address this problem is to increase the temperature differential between the heat transfer fluid and the product. Liquid nitrogen can be used, but thermal stress may be imparted to the container, and additional handling limitations may be needed. Use of cascade refrigeration design or cooling the heat transfer fluid using liquefied gases in combination with the use of extended heat transfer surfaces can be considered.

27.3.3.4.3. MIXING. Use of agitators in the form of an impeller or propeller may not be an efficient way of thawing the solutions as they can be effective only after significant amount is thawed statically, which takes considerable time. Gentle rocking of the container back and forth by placing it on a shaker or recirculation of the thawed material at a rate 2 times the average melting rate would be the ideal choices for ensuring production of homogenous solution.

27.3.3.4.4. REQUIREMENTS OF HEAT TRANSFER MEDIUM:. Heat transfer fluids should have low viscosities at low temperatures. Temperature-dependent viscosity changes may result in dramatic changes in heat transfer coefficients if the viscosity becomes large enough to cause transitions between turbulent and laminar flow. Heat transfer fluids ideally should be nontoxic, nonflammable, and high-boiling. Unfortunately, many low-viscosity fluids have relatively low flash point temperatures and thus may pose a flammability danger.

27.3.3.5. Practical Considerations.

27.3.3.5.1. FORMULATION CONSIDERATIONS. Purified proteins must be stored for an extended period of time before they can be processed for fill and finish operations. During this storage time, they must retain their native state conformation and avoid chemical degradation. Typically, the drug substance is expected to have a minimum of 2 years of storage stability. Although optimal conditions for storage are specific to each protein, an attempt is made here to provide some general considerations to keep in mind while designing the protein formulation and some guidelines for storage of bulk drug substance:

- *Storage in Liquid State.* Instability in aqueous solution can be minimized by optimizing solution conditions that increase thermodynamic stability of the native conformation of the protein. Important variables include solution pH, ionic strength, and protein concentration. In some situations, solution stability is enhanced as protein concentration increases, whereas in other cases aggregation follows second-order kinetics and rate of reaction is sensitive to protein concentration, and optimization of protein concentration helps to stabilize the protein solution [19]. Including some ligands that specifically bind to the protein and thereby increase the thermodynamic stability, such as calcium and other cofactors, can be considered. Solutes that are preferentially excluded from the surface of the protein (termed *excluded solutes*) increase the free energy of protein unfolding by destabilizing the unfolded state of the protein relative to the native conformation. Excluded solutes are effective stabilizers in aqueous solutions at relatively high molar concentrations (e.g., >500 mM) [36]. If the protein is known to contain sulfhydryl (–SH) groups, addition of a reducing agent such as dithiothreitol (DTT) to a final concentration of 1–5 mM may reduce unwanted disulfide formation [37].
- *Frozen Storage.* If the protein is found to be sensitive to freezing and thawing, and the mechanism of denaturation is suspected to be due to cold denaturation or interfacial denaturation, then including cryoprotectants and surfactants should be considered. The mechanism by which cryoprotectants protect the protein during freeze–thaw is the same as the mechanism that enables the exclusion solutes to protect the protein in solution by increasing the free energy of unfolding [36]. Since the mechanisms are the same, the same excluded solutes such as glycine, mannitol, disaccharides, and poly(ethylene glycols), PVP, certain amino acids, methyl amines, and salting-out-salts (ammonium sulfhate) typically can be used as cryoprotectants and are effective in molar concentrations in the range of 300–500 mM. If the mechanism of denaturation is through ice–liquid interfacial denaturation, then slow cooling combined with the use of a surfactant may alleviate denaturation at ice–liquid interfaces [38]. The addition of nonionic surfactants such as polysorbate 20 or 80, poloxamer 188, and pluronic F68 ranging from 0.01% to 0.1% protects the protein by competing with the protein molecules for ice–liquid interfaces [39]. Additionally, dilute solutions are more prone to inactivation and loss as a result of low-level binding to the storage vessel. In these situations the adsorption-related denaturation of the protein can be mitigated by increasing the initial protein concentration or by adding a carrier or filler protein such as purified hSA to 1–5 mg/mL (0.1%–0.5%).

Protein conformation and stability can be affected simply by exposure to low temperature without cryoconcentration effects [40], which means that formulation excipients and their concentrations should be selected such that they do not depress the freezing point substantially and require deep freezing for complete solidification of ice. Formulation excipients should be chosen to provide eutectic freezing and protein stability at warmer temperatures.

Storage of protein solutions in metal containers such as stainless steel can result in leaching of metals, and if the protein contains exposed methionine or cysteine or sulfhydryl groups, these groups may then be susceptible to metal-catalyzed oxidation. In this case, adding a metal chelator such as polyaminocarboxylate (PAC), citrate, or EDTA to a final concentration of 1–5 mM [41] should be considered. To prevent the proteolytic cleavage of proteins, inclusion of protease inhibitors can be considered and for any microbial growth, especially for liquid storage, including additives such as sodium azide (NaN_3) at a final concentration of 0.02–0.05% (w/v) or thimerosal at a final concentration of 0.01% (w/v) can be considered.

The use of pH indicators will help in identifying suitable environments for freezing protein solutions. Citrate, Tris, and histidine buffers are considered safe for freezing processes as they are less likely to undergo pH changes during freezing [42,43].

27.3.3.5.2. Storage Temperature Considerations. The physical characteristics of the containers with respect to their ability to withstand both freezing and autoclaving conditions and rigorous shipping and handling must be evaluated prior to their use; Table 27.1 [3] and Table 27.2 [44] summarize some of the important characteristics of plastic containers. In order to use plastic containers for freezing applications, their brittleness temperatures should be lower than their anticipated storage temperatures.

The storage temperature of biologics is dictated by the product stability; however, historically biologics have been stored in the range from −20°C to −80°C. It is critical to know the phase transition temperatures of product solutions, and storage

TABLE 27.1. Physical Properties of Plastic Materials

Type of Container	Clarity	Maximum Use Temperature, °C	Brittleness Temperature, °C	Autoclavable	Percentage of Water Absorption, %
High-density polyethylene	Translucent	120	−100	No	<0.01
Polycarbonate	Translucent	135	−135	Yes	0.35
Polypropylene	Translucent	135	0	Yes	<0.02
Polytetra fluoroethylene	Translucent	260	−267	Yes	0.01
Stedim Flexel® 2D bags	Translucent	—	−80	No, γ-irradiable	—
HyClone HyQ® CX5-14 Labtainer	Translucent	—	—	No	—

TABLE 27.2. Physical Properties of Plastic Materials

	Type of Plastic[a]						
	LDPE	HDPE	PP	MP	PC	PVC	PA
Maximum Use temperature, °C	80	120	135	175	135	70	121
Brittleness temperature, °C	100	100	0	20	135	30	40
Transparency	Translucent	Translucent	Translucent	Clear	Clear	Clear	Translucent
Autoclaving	No	No	Yes	Yes	Yes	No	Yes
Sterilization gas	Yes	Yes	Yes	Yes	Yes	Yes	Yes
Dry heat	No	No	No	Yes	No	No	No
Disinfectants	Yes	Yes	Yes	Yes	Yes	Yes	Yes
Specific gravity	0.92	0.95	0.90	0.83	1.20	1.34	0.90
Gravity flexibility	Excellent	Rigid	Rigid	Rigid	Rigid	Rigid	Moderate

[a]Resin codes: LDPE—low-density polyethylene; PP—polypropylene; PMP—polymethylpentane; PC—polycarbonate; PVC—poly(vinyl chloride); PA—polyallomer.

temperatures significantly away from phase transition temperature should be identified to avoid exposure to temperature-cycling effects that can inactivate the protein by exposing it to high salt concentrations, extreme pH, or other cryoconcentration effects. In general, storage at lower temperatures such as −40°C or −80°C is preferred over −20°C for a variety of reasons: (1) −20°C is close to the phase transition temperature (eutectic temperature) of salts that are commonly used in protein formulations, and cycling of the temperature by the freezers above and below −20°C can cause dissolution and recrystallization of salts, which could cause stress to the protein molecules in addition to the cryoconcentration effects as mentioned above; (2) the temperature of the freezer will not be the same −20°C at all locations but could differ by 5°C–10°C, which could potentially leave the product in a partially unfrozen state; and (3), on storage at lower temperatures such as −40°C or −80°C, the rates of adverse reactions arising from either cryoconcentration, cold denaturation, or interfacial denaturation can be retarded compared to those at −20°C. However, one must also realize that the lower the storage temperature, the higher will be the cost for energy, equipment, and operational logistics for storage and product. It may be possible to store some products at −20°C after they are frozen at lower temperatures if there are no phase transitions and if the subsequent annealing resulting in secondary crystallization would not impact product stability.

The adverse effects of repeated freeze–thaw can be mitigated by appropriate aliquoting so that containers are sized to fill and finish batch requirements. Alternatively, if the protein is sensitive to repeated freeze–thaws, then addition of glycerol or ethylene glycol will prevent solutions from freezing at −20°C, enabling repeated use from a single stock without warming (i.e., thawing). Ethylene glycol is a better choice than glycerol as it does not support microbial growth. To minimize cryoconcentration effects, an alternative is to use small containers that will reduce the freezing pathlength.

27.3.3.5.3. Directional Heat Flow. According to guidelines for the formulation and processing methods, the basic structural unit of ice formed could be (1) vitreous, (2) cubic, (3) dendritic (linear or branched), or (4) spherulitic (coarse or evanescent) [45]. The moving solid boundary during freezing could be flat or dendritic in nature. A moving dendritic ice front allows solutes to be entrapped in the interdendritic space, which promotes a more uniform macroscopic distribution of solutes in the frozen mass and minimizes freeze concentration. If the moving solid–liquid interface were flat, solutes could be more easily excluded from the frozen mass and become increasingly concentrated. One important strategy for controlling dendritic growth for freezing large volumes is to ensure directional heat flow. Convection or agitation in the liquid phase of the solution during freezing may cause an exclusion of solute molecules from the solution's solidifying mass and gradual cryoconcentration of solutes in the liquid phase. This is caused by the sweeping effect of liquid motion at the solid surface and the suppression of dendritic ice growth. It is important to avoid any mechanical agitation during the freezing.

27.3.3.5.4. Container–Closure Integrity (Carboys). A few practical considerations during the use of carboys as storage containers for freezing, shipping, and thawing of bulk drug substance are listed below:

- Avoid using polypropylene carboys (use polycarbonate carboys instead).
- Carboys may be autoclaved with a sterilization temperature of 121°C and a minimum time of 35 min.
- The autoclave cycle must be challenged with two bacterial aerosols.
- Define working volumes on the basis of integrity of the carboy and stability of the drug substance, generally recommended at 3–16 L for 20-L carboys and 1–8 L for 10-L carboys.
- Minimize the number of underfilled carboys to prevent implosions related to pressure differential when transported to global manufacturing sites.
- After filling of the bulk drug substance into carboys, the caps should be torqued to 125 in. · lb and retorqued to 125 in · lb after 10 min.
- Do not freeze carboys in upright or frost-free freezers.
- Do not freeze carboys on the floor of freezer; use vented stainless-steel wire racks of a walk-in freezer. Carboys on the floor will freeze atypically due to lack of heat transfer from the bottom, causing bulging and stress on the carboys underside. Leave a minimum of 2–3 in. of space between the carboys.
- Minimize any vibrations or agitation of carboys during freezing.
- Carboys may be wrapped in Mylar bags to avoid CO_2 ingress from dry ice used during shipment.
- Do not place carboys at an angle on gel packs during transportation to avoid bulging.
- Do not expose the frozen carboy to room temperature immediately; follow a gradual temperature exposure by exposing to 2°C–8°C, followed by exposure to room temperature.

- If stability data allow, consider dynamic thawing over static thaw at room temperature to reduce the thaw time, minimize residence time of protein in a concentrated state, and produce a homogeneous solution on the end of thaw.

27.3.3.6. Case Study. A systematic study was undertaken to understand the differences in the impact between controlled and uncontrolled freeze–thaw rates on product quality attributes. Conventional carboys were used for uncontrolled freeze–thawing rates study, while CryoVessels and Celsius-Pak technologies were explored for controlled-rate study. Two different types of biomolecules, a fusion protein and a peptibody, were involved in the study. The integrity of both types of molecules against multiple freeze–thaw cycles was evaluated using stability-indicating assays. The rationale behind studying multiple freeze–thawing cycles was (1) the impact is not clearly apparent from single cycle and (2) in practice, the products are frozen and thawed multiple times for various reasons. The case study results demonstrate that while some proteins may not be sensitive to the freezing-rate-dependent denaturation phenomenon, some other biomolecules that lose integrity on exposure to uncontrolled-rate freeze–thaw processes can be effectively protected using controlled-rate freeze–thaw technologies such as Celsius-Pak or CryoFin.

As stated earlier in this chapter, some proteins are cold-labile and undergo cold denaturation on exposure to subzero temperatures but others undergo denaturation during the freezing–thawing processes by two different mechanisms: cryoconcentration and/or ice–liquid surface denaturation. However, some proteins tend to survive against the uncontrolled freezing–thawing-related denaturation processes. These denaturation processes are believed to be influenced by the freezing rates, which, in turn, are influenced by the freezing and thawing methodologies.

27.3.3.6.1. Case Study 1: Fusion Protein. To mimic uncontrolled freezing and thawing rates, the bulk drug substance of the fusion protein solution was filled into carboys and frozen by placement in a walk-in freezer. Thawing was carried out by placing the frozen carboy in a cold room maintained at 2°C–8°C. A total of five freezing–thawing cycles were carried out, and samples were collected after each cycle. Size exclusion HPLC (SE-HPLC) was employed to resolve the higher-molecular-weight species (aggregates) from the main component and the results, as shown in Figure 27.3, indicate a slight increase in the percentage of aggregates, up to three freeze–thaw cycles, as compared to the pre-freeze–thaw control sample in a carboy. However, no further increase in the percentage of aggregates was observed beyond three consecutive freeze–thaw cycles. As no increasing trend in the % aggregates with increase in the number of freeze–thaw cycles was observed, it was concluded that no significant impact of freeze–thaw operation was observed on product quality of the fusion protein when frozen and thawed in carboys. All results from other stability indicating analytical tests indicated no significant alterations in the integrity of the protein. When this material was further processed for fill-finish operations, no change in the overall drug product quality attributes was observed, demonstrating no adverse impact of uncontrolled five freeze–thaw cycles.

CryoVessel technology was used as the controlled-rate freezing–thawing process, and the impact of the process conditions on the integrity of the fused protein was

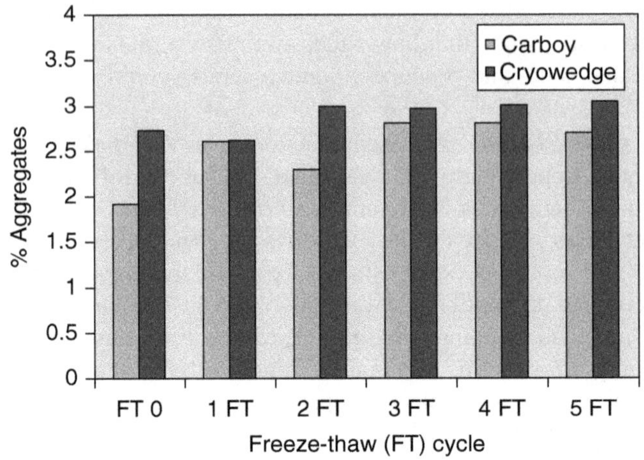

Figure 27.3. Results of stability-indicating assays of fusion protein as a function of number of freeze–thaw cycles in (■) carboy (uncontrolled freeze–thaw) and (■) cryowedge (controlled freeze–thaw).

studied in a CryoWedge, a scaled-down model that mimics the internal configuration of the CryoVessel and requires only a few liters of material. When the same fusion protein solution was then subjected to five consecutive freeze–thaw cycles in the CryoWedge using an optimized freeze–thaw cycle, sizing results indicated no change in the integrity of the protein structure, and the quality of fusion protein product compared to the control as shown in Figure 27.3. The side-by-side stability results of two technologies depicted in Figure 27.3 clearly demonstrate no impact of freezing methodology on the product quality, suggesting the robustness of the protein to freezing-induced denaturation processes.

27.3.3.6.2. Case Study 2: Peptibody. When the bulk drug substance of a peptibody was subjected to multiple freeze–thaw cycles under identical uncontrolled-rate conditions using 10-L carboys and the drug product quality attributes were evaluated, differences were observed against the control or prefreeze–thaw samples with the increase in number of freeze–thaw cycles as shown in Figure 27.4. Analysis of postfreeze–thaw samples using SE-HPLC indicates an increasing trend in the percentage of higher-order aggregates with increase in the number of freeze–thaw cycles. Differences in the percentage of main peak compared to control were also observed with scale from 2- to 10-L carboys (data not shown), in agreement with the fact that the freezing-induced denaturation is a scale-dependent phenomenon in the case of uncontrolled-rate freeze–thaw technology. On the other hand, product collected and analyzed from Celsius-Pak, a controlled-rate freeze–thaw technology, after each freeze–thaw cycle did not show any significant change in percentage aggregates even after five cycles, suggesting that the controlled-rate technology is better at preserving product quality during freeze–thaw than in a carboy.

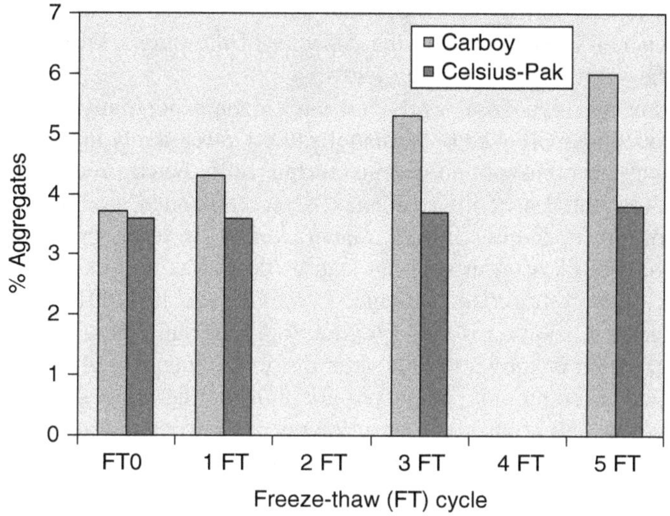

Figure 27.4. Results of stability-indicating assays of peptibody as a function of number of freeze–thaw cycles in (■) carboy (uncontrolled freeze–thaw) and (■) Celsius-Pak (controlled freeze–thaw).

The results of these case studies demonstrate that not all biomolecules are sensitive to freezing-induced denaturation phenomena, and their robustness against uncontrolled-rate freeze–thaw technology can be utilized advantageously by using walk-in freezers consisting of either plastic or stainless-steel containers, provided the material demand is not too high and initial capital investment in acquiring Celsius or CryoFin technologies is not excessive. However, for those biomolecules that are sensitive to freezing-induced denaturation phenomena and whose material demand is too high, use of controlled-rate technologies such as CryoFin or Celsius-Pak can prove to be quite beneficial in terms of efficiency and protection against denaturation.

27.3.4. Bulk Drying

Developing antibodies into therapeutic products involves administration of large doses for long periods of time, which translates into a challenge to the upstream/downstream scientist to produce high volumes of bulk. It becomes more challenging if the product is designed for subcutaneous administration where the maximum injection volume is typically 1.5–2.0 mL. One techniques that is commonly employed to concentrate the solution and is easily scalable is *tangential flow filtration* (TFF), which allows buffer exchange and concentration of protein to a certain level before it begins to build up higher concentrations at the boundaries of the membrane, causing decrease of flux and membrane fouling. Besides the impact of repeated circulation of protein solutions through tubing and pulsation of the pumps on the integrity of the protein caused by creation of cavities another factor that could potentially impact process performance and limit its use for concentrating protein solutions is the development of high backpressure during the process caused by increase in solution viscosity and increase in concentration exceeding the capacity of the pump. Furthermore, the high viscosity of the solution impacts the yield.

Considering the high dose levels and large patient populations, it is anticipated that some compounds will need to be manufactured in relatively high volume. To meet this requirement for high-volume manufacturing, bulk, batch processing and holding steps for pooling batches of intermediates is recommended, and should become an integral part of unit operations in bulk manufacturing. In terms of operational convenience, storage of bulk drug in solution state is desirable, and in most cases solution stability does allow refrigerated storage. However, the real difficulty or challenge arises when the bulk solution is subjected to vibrations and agitation during transport, creating foam, which is known to be deleterious to the integrity of the protein. Hence, cryopreservation is commonly employed for bulk storage and shipment. However, both uncontrolled- and controlled-rate cryopreservation technologies present logistical and operational challenges, as they are expensive to clean, maintain, and operate; require skilled personnel; and require a large facility space for storage of the in-process quarantined material. Moreover, shipping of the containers and frozen liquid is expensive.

In such situations, bulk drying can prove to be beneficial in addressing the limitation and challenges associated with UF/DF IV processes and cryopreservation. Various drying technologies such as freeze drying, spray–freeze drying, spray drying, and vacuum drying are available, and the selection of the appropriate drying system for a specific situation should be based on the product's characteristics and drying behavior as well as the requirements of the end product. Among all the available technologies, freeze drying is the most commonly used method for bulk drying of protein solutions, and among all types of freeze-drying trays, stainless-steel ones are perhaps the most commonly used. However, the selection of trays is crucial, as the material composition of the trays can impact heat transfer, which, in turn, can impact the nature of the cake structure and the efficiency and effectiveness of the process. The reader is referred to Chapters 23 and 28 for detailed information on process development and characterization of these technologies.

In addition to product quality, selection of the final bulk storage method also must factor in the business and logistical aspects, some of which are discussed in the next section.

27.4. BUSINESS AND LOGISTICS CONSIDERATIONS IN SELECTION OF BULK STORAGE/SHIPMENT TECHNOLOGIES

Business and logistic considerations in the selection of the bulk storage and shipment technology play a major role in determining the total cost of manufacturing, an important variable for any business. Value comparison analysis that includes risks, capital costs, depreciation value, implementation and operational costs, and other variables can be performed before finalizing the technology.

27.4.1. Capital Equipment Costs

These include

- *Technology-Specific Equipment*. A thorough evaluation of the technology in question should ensure that useful life-cycle assumptions are being made in the pursuit of this technology, determine the applicability of technology for multiple products, determine the ease and cost of implementation across multiple sites, and should check for robustness of the technology.
- *Facility Modification Requirements and Cost*. Issues such as the modifications necessary to install and operate the new equipment at the drug substance and drug product manufacturing sites, footprint space requirements, utility and equipment tie-ins, fit with current supply-chain configuration, and costs associated with reconfiguration of supply chain, if any, must be included in estimating the total cost of new-technology introduction.
- *Warehouse Requirements and Potential Expansion Costs*. Size and configuration of current coldrooms, freezers, pallet position requirements of both existing and new technologies, need for expansion of warehouse capacity, and associated costs should also be included as capital costs.

27.4.2. Operational Costs

These include

- *Full-time Employee (FTE)*. Comparison of ongoing staff costs to maintain status quo with the cost of implementing new technology, future benefits, and long-term savings after implementation, and the related underlying assumptions must be reviewed.
- *Transportation Costs*. Comparison of status quo transportation lane costs with assumptions of the new-technology transportation cost must be reviewed.
- *Disposable Cost*. Comparison of the cost of disposables for the old and new technologies must be reviewed.
- *Maintenance Costs*. Comparison of maintenance costs of old and new technologies must be reviewed.
- *Process Development Implementation Costs*. Prior to implementation of the new technology at the commercial site, a significant amount of process development work may be required to characterize the new technology and implement it at the various sites. Costs associated with the purchase of pilot-scale equipment and related FTE costs must be reviewed and included in the total operational cost.

ACKNOWLEDGMENT

The author would like thank Rene D' Inca, Amgen Inc., for useful insights on business and logistics considerations in the selection of bulk storage and shipment technologies.

REFERENCES

1. Akers, M. J. and Schmidt, D. J. (1997), Cryogranulation: A potential new final process for bulk drug substances, *BioPharm* **10**: 28–32.
2. Arford, J. R., Kendric, B. S., Carpenter, J. F., and Randolph, T. W. (2008), High concentration formulations of recombinant human interleukin-1 receptor antagonist: II. Aggregation kinetics, *Journal of Pharmaceutical Sciences* **97**(8): 3005–3021.
3. Hora, M. S. and Chen, B.-L. (1999), Stabilization of biopharmaceutical products and finished product formulation, in *Biopharmaceutical, an Industrial Perspective*, Walsh, G. and Murphy, B., eds., Kluwer Academic Publishers, pp. 217–248.
4. Banga, A. K. under print, 2010 *Therapeutic Peptides and Proteins—Formulation, Processing and Delivery Systems*, 2nd ed., Taylor & Francis, London.
5. Chen, B., Bautista, R., Yu, K., Zapata, G. A., Mulkerrin, M. G., and Chamow, S. M. (2003), Effect of histidine on stability and physical properties of an antibody, *Pharm. Res.* **20**(12): 1952–1960.
6. Lam, X. M., Yang, J. Y., and Cleland, J. L. (1997), Antioxidants for prevention of methionine oxidation in recombinant monoclonal antibody HER2, *J. Pharm Sci.* **86**: 1250–1255.
7. Waterman, K. C., Adami, R. C., Alsante, K. M., Hong, J., Landis, M. S., Lombardo, F., and Roberts, C. J. (2002), Stabilization of pharmaceuticals to oxidative degradation, *Pharm. Devel. Technol.* **7**(1): 1–32.
8. Lam, P., Keck, R., Lo, S., Wong, R., and Hsu, C. (2003), Effects of formulation components on the corrosion of 316L stainless steel, paper presented at American Chemical Society (ACS) New Orleans Conf. 2003.
9. Lentz, Y. K., Weck, S., and Lam, X. (2007), *Evaluation of Polymeric Bioprocess Bags for Frozen Storage and Shipping of Protein Drug Substance*, American Association of Pharmaceutical Scientists/National Biotechnology Conference.
10. CDER Guidance Document (1999), *Container Closure Systems for Packaging Human Drugs and Biologics*.
11. Jaenicke, R., (2000), Stability and stabilization of globular protein in solution, *J. Biotechnol.* **79**: 193–203.
12. Manning, M. C., Patel, K., and Borchardt, R. T. (1989), Stability of protein pharmaceuticals, *Pharm. Res.* **6**: 903–918.
13. Wang, W., Singh, S., Zeng, D. L., King, K., and Nema, S. (2006), Antibody structure, instability and formulation, *J. Pharm. Sci.* **96**(1).
14. Kerwin, B. and Remmele, R. L., Jr. (2007), Protect from light: Photodegradation and protein biologics, *J. Pharm. Sci.* **96**(6): 1468–1479.
15. Mahler, H. C., Müller, R., Friess, W., Delille, A., and Matheus, S. (2005), Induction and analysis of aggregates in a liquid IgG1-antibody formulation, *Eur. J. Pharm. Biopharm.* **59**(3): 407–417.
16. Cromwell, M. E. M., Hilario, E., and Jacobson, F. (2006), Protein aggregation and bioprocessing, *AAPS J.* **8**(3): E572–E579.
17. Maa, Y. F., and Hsu, C. C. (1998), Investigation on fouling mechanisms for recombinant human growth hormone sterile filtration, *J. Pharm. Sci.* **87**(7): 808–812.
18. Ahrer, K., Buchacher, A., Iberer, G., and Jungbauer, A. (2006), Effects of ultra-/diafiltration conditions on present aggregates in human immunoglobulin G preparations, *J. Membrane Sci.* **274**(1–2): 5 108–115.

19. Hanson, M. A. and Rouan, S. K. E. (1992), Introduction to formulation of protein pharmaceuticals, in *Stability of Protein Pharmaceuticals*, Ahern, T. J. and Manning, M. C., eds., Plenum Press, New York, pp. 209–233.

20. Arakawa, T. and Timasheff, S. N. (1985), The stabilization of proteins by osmolytes. *Biophys. J.* **47**: 411–414.

21. Franks, F. (1985), *Biophysics and Biochemistry at Low Temperatures*, Cambridge Univ. Press, New York.

22. Murase, N. and Franks, F. (1989), Salt precipitation during the freeze-concentration of phosphate buffer solutions, *Biophys. Chem.* **34**: 293–300.

23. van den Berg, L. and Rose, D. (1959), Effect of freezing on the pH and composition of sodium and potassium phosphate solutions: The reciprocal system KH_2PO_4-Na_2HPO_4-H_2O, *Arch. Biochem. Biophys.* **81**: 319–329.

24. Gomez, G., Pikal, M. J., and Rodriguez-Hornedo, N. (2001), Effect of initial buffer composition on pH changes during far- from-equilibrium freezing of sodium phosphate buffer solutions, *Pharm. Res.* **18**: 90–97.

25. Poulsen, K. P., and Lindelov, F. (1981), Acceleration of chemical reactions due to freezing in *Water Activity: Influences on Food Quality*, Rockand, L. B. and Stewart, G. F., eds., Academic Press, New York, pp. 651–677.

26. Ahmad, F., and Bigelow, C. (1986), Thermodynamic stability of proteins in salt solutions: A comparison of the effectiveness of protein stabilizers, *J. Protein Chem.* **5**: 355–367.

27. Hyclone Labs., Inc. (1992), Freezing and thawing serum and other biological materials: Optimal procedures minimize damage and maximize shelf-life, in *Art to Science in Tissue Culture*, Hyclone Labs, inc., Vol. 11, # 2 (spring 1992).

28. Webb, S. D., Golledge, S. L., Cleland, J. L., and Carpenter, J. F. (2002), Surface adsorption of recombinant human interferon-γ in lyophilized and spray-lyophilized formulations, *J. Pharm. Sci.* **91**: 1474–1487.

29. Pikal, M. J. (1999), Mechanisms of protein stabilization during freeze-drying and storage: The relative importance of thermodynamic stabilization and glassy state relaxation dynamics, in *Freeze-Drying/Lyophilization of Pharmaceutical and Biological Products*, Rey, L. and May, J. eds., Marcel Dekker, New York, Chapter 6.

30. Liesebach, J., Rades, T., and Mang, L. (2003), A new method for the determination of the unfrozen matrix concentration and the maximal freeze-concentration, *Thermochim. Acta* **401**(2): 159–168.

31. Barbaroux, M. and Voute, N., Characterization of Stedim 71 performance at sub-zero temperature, Sartorius-Stedim (unpublished).

32. Wilkins, J., Sesin, D., and Wisniewski, R. (2001), Large-scale cryopreservation of biotherapeutic products, *Innov. Pharm. Technol.* **1**(8): 174–180.

33. Lam, P. and Sane, S. (2007), Design and testing of a prototype large-scale bag freeze–thaw system, *BioPharm Int.* **20**: 6–16.

34. Wisniewski, R. (1996), Large-scale freezing and thawing of biopharmaceutical drug product, *Proc. 16th Annual Pharmaceutical Technology Conf.*, Sept. 16–19, 1996.

35. Incropera, F. P. and Dewitt, D. P. (1990), *Introduction to Heat Transfer*, Wiley, New York.

36. Carpenter, J. F. and Crowe, J. H. (1988), The mechanism of cryoprotection of proteins by solutes, *Cryobiology* **25**: 244–255.

37. Anderson, M. M., Breccia, J. D., and Hatti-Kaul, R. (2000), Stabilizing effect of chemical additives against oxidation of lactate dehydrogenase, *Biotechnol. Appl. Biochem.* **32**: 145–153.
38. Chang, B. S., Kendrick, B. S., and Carpenter, J. F. (1996), Surface-induced denaturation of proteins during freezing and its inhibition by surfactants, *J. Pharm. Sci.* **85**: 1325–1330.
39. Kerwin, B. A. et al. (1998), Effects of Tween 80 and sucrose on acute short-term stability and long-term storage at -20 degrees C of a recombinant hemoglobin, *J. Pharm. Sci.* **87**(9): 1062–1068.
40. Franks, F. (2003), Protein destabilization at low temperatures, *Adv. Protein. Chem.* **46**: 105–139.
41. Yin, J., Chu, J.-W., Ricci, M. S., Brems, D. N., Wang, D. I. C., and Trout, B. L. Effects of excipients on the hydrogen peroxide-induced oxidation of methionine residues in granulocyte colony-stimulating factor, *Pharm. Res.* **22**.
42. Hill, J. P. and Buckley, P. D. (1991), The use of pH indicators to identify suitable environments for freezing samples in aqueous and mixed aqueous/nonaqueous solutions, *Anal. Biochem.* **192**(2): 358–361.
43. Shalaev, E., Johnson-Elton, T., Chang, L., and Pikal, M. J. (2002), Thermophysical properties of pharmaceutically compatible buffers at sub-zero temperatures: Implications for freeze drying, *Pharm. Res.* **19**: 195–201.
44. MG Scientiifc Supplying Laboratories, http://www.mgscientific.com/service/plastics.asp.
45. Nail, S. and Akers, M. J., eds. (2002), *Fundamentals of Freeze Drying in Development and Manufacture of Protein Pharmaceuticals*, Springer, New York.

28

DRYING PROCESS METHODS FOR BIOPHARMACEUTICAL PRODUCTS: AN OVERVIEW

Ahmad M. Abdul-Fattah and Vu L. Truong

28.1. INTRODUCTION

The growing number of peptide, protein, and vaccine product candidates has greatly increased the number of drugs and drug candidates in which stabilization by drying and optimal dosage presentation are of increasing importance. The formulation, solid-state properties, and so-called thermal history to which the drug is subjected during downstream pharmaceutical processing are highly influential in determining key product characteristics such as shelf life. The thermal history of a dried product is defined by the drying process method and parameters used, including the heat and mass transfer mechanisms of the drying system, the drying time, the processing temperature, the chamber pressure (if drying under vacuum conditions), the moisture of the inlet air (if not drying under vacuum), and other factors. Therefore, it is important to understand the unit operations involved in a drying method and the key drying parameters that impact the solid-state properties, and ultimately the product's stability in storage. This chapter presents an overview of several conventional and unconventional drying methods used in the biopharmaceutical industry. Principles of biopharmaceutical stabilization, the unit operations of key pharmaceutical drying processes, thermal history

Formulation and Process Development Strategies for Manufacturing Biopharmaceuticals,
edited by Jameel and Hershenson

involved in each drying methods and their applications, and the challenges of each drying method are reviewed.

28.2. AN OVERVIEW OF THE MECHANISMS OF STABILIZATION IN FROZEN, LIQUID, AND DRIED STATES

More recent pharmaceutical drug development trends suggest that economical factors, marketing considerations, and life-cycle management have become major drivers in the selection of the final drug product form. While more biopharmaceutical products are manufactured in liquid dosage forms, dried forms continue to increase, facilitated by the advent of new drying process options.

Arguably the most important reason for drying a biological product is to stabilize and prolong its shelf life. An in-depth understanding of the mechanisms of pharmaceutical stabilization in the dry state requires an appreciation of the mechanisms that govern protein stability not only in the solid state but also in the liquid state, that is, the starting material prior to drying. The stabilizers that have been developed for liquid formulations including sugars (e.g., sorbitol, sucrose, trehalose), surfactants (e.g., polysorbates, pluronics), charge-shielding agents (e.g., small charged amino acids), specific protein ligands (e.g., heparin binding to FGF) also have a beneficial role in the dry state and will be briefly discussed here. An important mechanism of protein stabilization in the liquid state, particularly for sugar-based stabilizers, is believed to be "preferential exclusion." Essentially, the theory states that the stabilizing solutes exert their stabilization by being preferentially excluded from the protein surface [1–5]. Because the magnitude of solute exclusion is proportional to the surface area of the protein, the energy barrier is higher for the protein to unfold in the presence of these stabilizers. Therefore the system will favor the protein state with the smallest surface area; that is, a more ordered and native conformation of protein molecule is thermodynamically favored. In the absence of stabilizers, the free energy of unfolding (ΔG_{unf}) and total entropy of the system decrease, leading to destabilization. A second mechanism thought to be involved in protein stabilization in the liquid state is related to the colloidal stability as governed by the second virial coefficient [6,7]. However, the supportive data for this mechanism are still evolving and debated.

The use of dried formulations eliminates many problems associated with drug product in liquid or frozen formulations. In addition to a dramatic improvement in the shelf life of most biological materials [8], drying circumvents possible agitation or shear sensitivity, container closure integrity problems, freeze–thaw stresses, and manufacturability challenges at high concentration. Other benefits of drying generally include reduced product weight and bulkiness and hence ease of shipping and distribution, reduced cold chain storage requirements, and the possibility of increasing the reconstituted drug concentration range on reconstitution with a lower diluent volume. Drying a solution containing dissolved drug can be accomplished by unit operations as simple as the application of air drying, drying under gentle vacuum, or other heat-assisted evaporative processes [9]. In addition to freeze drying (lyophilization), other more complex drying methods that involve multiple unit operations such as spray drying, spray-freeze drying, and supercritical fluid drying have also been evaluated for biopharmceuticals and are the focus of this chapter.

The most commonly used stabilizers for pharmaceutical drying include nonreducing sugars (e.g., sucrose, trehalose, raffinose, stachyose), surfactants (e.g., polysorbates, pluronics), and some amino acids (such as arginine) and proteins (such as serum albumin). One common requirement for stabilization is that the stabilizer remain in the same amorphous phase as the biological material. Two generally accepted hypotheses to explain stabilization in the dried state for many of these stabilizers include the following [10–14]:

1. *Water Substitution Hypothesis*. Protein-bound water that is removed during the last stages of drying is believed to contribute to thermodynamic destabilization and activity loss. This dehydration or desiccation-induced loss is prevented by hydrogen bonding between the stabilizer and specific sites at the surface of the proteins, leading to preservation of native structure. In this theory, the stabilizer preserves the native protein conformation by maintaining high free energy of unfolding.
2. *Glass Dynamics Hypothesis*. Biological materials are kinetically immobilized in a rigid, glassy matrix in the dried form. Therefore, global motions (α relaxations due to translational and/or rotational motions) and local molecular motions (β relaxations) that may be relevant to protein instability are suppressed. Hence, protein conformation is stabilized regardless of the free energy of unfolding.

28.3. HEAT AND MASS TRANSFER DURING DRYING

The heat transfer to heat the solvent and mass transfer to remove the solvent from the product occur simultaneously during drying and can be described as three sequential processes [15]:

1. Heat transfer from an external source to the solvent containing the dissolved formulation and actives
2. Phase transformation and mass transfer of the solvent from a liquid or solid (e.g., ice) to vapor state
3. Mass transfer of vapor away from the dried material

28.3.1. Heat Transfer

During drying, energy is transferred in the form of heat by means of conduction, convection, and/or radiation. Heat transfer by conduction involves transfers of kinetic energy as temperature gradients across the solids from a region of higher temperature to a region of lower temperature. Conduction driers such as fluid-bed driers are also called "indirect" or "contact" driers [16]. Typically the heat is supplied through heated surfaces within the drier. The rate of conductive drying is proportional to the degree of contact between the wet material and the heated surface. Vaporized liquid is removed from the heating medium and carried away by vacuum or by a carrier gas. The rate of energy transferred by conduction (Q_{cond} in cal/s) is proportional to the magnitude of the temperature difference and opposite to its sign [15]:

$$Q_{\text{cond}} = -kA\frac{dT}{dX} \tag{28.1}$$

where A is heat transfer area (in cm^2) and k is a proportionality constant called *thermal conductivity* [in cal/(s·cm^2·°C)], a property that measures the ability of a material to conduct heat. Thermal resistance to heat transfer by conduction (R_{cond}), a value proportional to the reciprocal of k, reflects opposition to flow of heat energy by conduction and can be derived from the following equation [15]

$$Q_{\text{cond}} = \frac{\Delta T}{R_{\text{cond}}} \tag{28.2}$$

where ΔT is the temperature difference between the surface and the drying interface at product that drives heat transfer by conduction.

In many biopharmaceutical drying operations, heat is supplied externally to products by a drying medium such as heated dry gas to provide energy for solvent removal (mostly by evaporation). The mechanisms of heat supply in this case consist of convective with or without conductive heat transfer between the drying medium and the product [17]. Convective drying involves heat transfer through diffusion and by advection; that is, energy needed for solvent removal from the product is supplied by Brownian motion of molecules of the surrounding drying gas (e.g., air, nitrogen), as well as large-scale motion of currents of the drying gas. Thus, heat transfer occurs at the interface of a solid and fluid or gas as a result of the temperature gradient. Vaporized liquid is carried away by the drying medium [17,18]. Examples of convective dryiers, also called "direct" driers, are air suspension driers such as fluid-bed, flash, rotary, spray driers, and modifications of spray driers (e.g., spray coating) [16,18]. The rate of energy transfer by convection (Q_{conv}) can be given by [15]

$$Q_{\text{conv}} = hA\Delta T \tag{28.3}$$

where h is the convective heat transfer coefficient [in cal/(s · cm^2· °C)] a quantity that depends on conditions of fluid flow, fluid properties, and system geometry; ΔT is the temperature difference between drying medium and product. Thermal resistance to heat transfer by convection (R_{conv}) is proportional to the reciprocal of h and reflects opposition to heat flow by convection.

Radiation is the transfer of heat from a surface through electromagnetic radiation; that is, no medium is necessary for radiation to occur. The wavelengths of thermal radiation produced by emitting bodies range from 0.1 to 100 μm [15–17]. For a small spherical particle inside a large space, the rate of energy transfer by radiation (Q_{rad}) can be expressed as [15]

$$Q_{\text{rad}} = \frac{\Delta T}{R_{\text{rad}}} \tag{28.4}$$

where ΔT is the temperature difference between the radiating surface and the product. Thermal resistance to heat transfer by radiation (R_{rad}) reflects opposition to heat flow by radiation; R_{rad} is a function of emissivity, a dimensionless factor that expresses radiative emission from a surface as a fraction of emission from a blackbody. Heat needed for solvent removal may be generated volumetrically as in dielectric or microwave drying. This kind of heat supply eliminates the need to transport heat from the dry surface into the wet core. Hence, this eliminates most of the defects concerning external heat supply (to be discussed later). Despite numerous advantages, heat generated volumetrically is seldom frequently used in biopharmaceutical drying due to the difficulties associated with setup and high costs (17).

Pharmaceutical drying equipment typically uses more than one heat transfer mechanism for drying. For example, heat transfer in freeze drying and foam drying is accomplished mainly through gas convective (vapor in the space between the shelf and the bottom of the vial) and conduction from shelf to vial at points of contact. Additionally, there is some heat transfer via radiation (from the top, bottom, and sides to the vial) [19,20]. Under drying processes where all heat transfer mechanisms are involved, the total thermal resistance (R_T) can be described by [15]

$$R_T = R_{cond} + \left[\frac{R_{conv} \bullet R_{rad}}{(R_{conv} + R_{rad})}\right] \tag{28.5}$$

If $R_{rad} \gg R_{conv}$, then Equation (28.5) reduces to

$$R_T = R_{cond} + R_{conv} \tag{28.6}$$

and radiative heat transfer can be neglected. If R_{cond} is small (e.g., in the early stages of drying), then Equation (28.5) reduces to

$$R_T = \left[\frac{R_{conv} \bullet R_{rad}}{(R_{conv} + R_{rad})}\right] \tag{28.7}$$

Finally, if R_{rad} is large, then

$$R_T = R_{conv} \tag{28.8}$$

As drying proceeds, contribution of R_{cond} increases (through the growing dried layer).

28.3.2. Mass Transfer

The vapor generated during drying must migrate from the liquid–vapor interface through the dried material layer and be transported out of the drying equipment. Concentration difference is usually the driving force behind mass transfer. Diffusion of moisture in solids can occur by one or more of the following mechanisms [16,17, 19–21]:

1. Liquid and vapor flow due to differences in total pressure

2. Vapor diffusion due to differences in vapor concentration or partial vapor pressure gradients
3. Knudsen flow (effusion of vapor) if drying takes place at very low temperatures and pressures (e.g., freeze drying), where the mean free path of vapor molecules is of the order of diameter of the pores
4. Surface diffusion
5. Hydrodynamic flow (moisture movement due to differences in osmotic pressure); that is, hydrostatic pressure differences when internal vaporization rates exceed the rate of vapor transport through the solid to the surrounding
6. Capillary flow, referring to the liquid in capillaries

All of these phenomena can be combined into one, and the effective diffusivity can be defined by Fick's second law [15,21]. Thus, migration of vapor through the dried material layer can be expressed as

$$\frac{dm}{dt} = -DA\left(\frac{dc}{dx}\right) = -DA\left(\frac{M}{RT}\right)\left(\frac{dp}{dx}\right) = \left(\frac{p_{\mathrm{s}} - p_{\mathrm{i}}}{R_{\mathrm{D}}}\right) \tag{28.9}$$

where dc/dx is the concentration gradient in the direction of transfer, A is the transfer area, D is the diffusion coefficient (which includes combined effect from all mechanisms outlined above), M is the solvent molecular weight, R is the universal gas constant, p_{i} is vapor pressure at interface, p_{s} is partial pressure of solvent at solid surface, T is absolute temperature, and R_{D} is the effective mass transfer resistance of the dried layer.

Migration of vapor *away from* the dried material layer can be expressed as

$$\frac{dm}{dt} = \left(\frac{p_{\infty} - p_{\mathrm{S}}}{R_{\mathrm{C}}}\right) \tag{28.10}$$

where p_{∞} is partial pressure of solvent far away from solid surface and R_{C} is resistance to convective mass flow. Total mass flow can be expressed as

$$\frac{dm}{dt} = \left(\frac{p_{\mathrm{i}} - p_{\infty}}{R_{\mathrm{T}}}\right) = \left(\frac{p_{\mathrm{i}} - p_{\infty}}{R_{\mathrm{D}} + R_{\mathrm{C}}}\right) \tag{28.11}$$

where R_{T} is the total mass transfer resistance. At the beginning of a drying cycle from aqueous solution, there is no dried layer and hence $R_{\mathrm{T}} \cong R_{\mathrm{C}}$. As drying proceeds, a dried layer begins to form and its resistance increases with time. Thus, R_{D} significantly contributes to R_{T} and continues to increase with time, with a concurrent steady decrease in drying rate.

28.4. CONVENTIONAL AND UNCONVENTIONAL DRYING METHODS USED IN THE BIOPHARMACEUTICAL INDUSTRY

Freeze drying and spray drying are considered more conventional drying methods in the biopharmaceutical industry and have been applied to produce commercial biopharmaceutical products on large manufacturing scale. In contrast, drying methods such as spray-freeze drying, supercritical fluid drying, fluid-bed drying, foam drying, microwave-assisted foam drying, and Xerovac (Anhydro Ltd., North Lincolnshire, UK) are unconventional drying methods due to one or more limitations such as scalability and manufacturability. In addition to classifying drying methods as conventional or unconventional, the mechanism of heat transfer can be used as a basis for classification (drying by conduction, convection, radiation, or any combination thereof). Another approach to classify drying methods is by the mechanism of solvent removal (i.e., mass transfer). For example, solvent removal can be achieved via one or more of the following:

1. Evaporation in spray drying, spray coating, and foam drying and in the secondary drying stage of freeze drying and spray-freeze drying
2. Sublimation during the primary drying stage of freeze drying and spray-freeze drying
3. Precipitation in methods employing supercritical fluid technology.

Finally, the ability to affect particle attributes such as particle size and particle morphology has also been used to classify pharmaceutical drying processes. For example, spray drying and spray-freeze drying are considered drying methods that enable one to control particle attributes, whereas in freeze drying and foam drying, control of particle size is virtually impossible.

Since freeze drying is extensively covered elsewhere in this book, our focus is on other conventional and unconventional drying methods.

28.4.1. Spray Drying

Spray drying is a *bottom–up* processing method applied in the biopharmaceutical industry mainly to produce particles for pulmonary and respiratory drug delivery [22,23] or incorporation into complex systems such as sustained-release formulations. The key advantages of spray drying are shorter process cycle time (i.e., more batches per unit time), scalability (i.e., large batch size per unit operation, requiring fewer production units), processing of thermolabile materials at atmospheric pressure, and ability to control powder properties (e.g. particle size, particle morphology, powder density, surface composition) [22–24].

28.4.1.1. Unit Operations in Spray Drying

28.4.1.1.1. Atomization. Atomization is a process where the feedstock (solution, suspension, or colloidal dispersion) is sprayed in micrometer-sized droplets (1–200 μm) at high velocity. The surface area, surface tension, and surface free energy increase on atomization. The amount of work and factors that influence work (e.g., nozzle design, atomization pressure, physical properties of feedstock such as concentration, viscosity, surface tension, and density) control initial droplet sizes, and hence provide control over particle size distribution [22–24].

Atomization methods are classified according to the method of energy transfer to the liquid: centrifugal or rotary (e.g., spinning disk), kinetic (e.g., pneumatic or air-assisted), high-pressure, effervescent, ultrasonic, sonic, or electrostatic atomization [23]. Spray drying can also be classified according to the position of the atomizers inside a spray drier: concurrent spray drying (spray enters the drying chamber from the top in the same direction as the gas flow), countercurrent spray drying (the nozzle is located at the bottom of the chamber; hence droplets initially travel upward in a direction opposite to the gas flow leading to longer residence times), countercurrent, and mixed-flow spray drying (e.g., gas enters from the bottom and top of the drying chamber, while the atomizer is situated at the top) [23].

28.4.1.1.2. Drying. In the drying phase, the spray comes into thermal contact with the heated, dry gaseous stream (such as dry air, dry nitrogen). The atomized droplets have very large surface areas, resulting in rapid evaporation and a very short drying time when exposed to the hot gas in the drying chamber. Drying stages of a droplet inside the drying chamber can be divided into several periods [23–28]:

1. *Initial Period*. Droplet temperature increases to wet-bulb temperature T_{wb}, which is the lowest temperature attained by the edges or surface of the droplets during evaporation and results from evaporative cooling. It is also the temperature surrounding the droplet's vapor cloud during evaporation. The surface of the droplets would maintain 100% relative humidity at T_{wb}.
2. *Constant-Rate Period*. A concentration gradient builds up in the droplet and water activity at the surface decreases, causing the surface temperature to rise above that of T_{wb}. The temperature of the interior of the droplet is usually 10°C–15°C lower than T_{wb} [27,28]. Under such conditions, the evaporation rate is maximum at the surface of the droplet and decreases to a minimum just beyond the equator. Dissolved solute is transported through diffusion and convection to the droplet surface. At the droplet surface with the highest evaporation rate, the solute precipitates once its solubility limit is exceeded. This phase ends with crust formation.
3. *Falling-Rate Period*. After crust formation, and as moisture content in each particle drops, the diffusion coefficient of water to the surface becomes limited. A critical moisture level is reached below which the surface of the dried particle becomes impenetrable. Events during the falling-rate period depend on the material properties of the crust (e.g., porosity and rheology). Product temperature during this phase is usually lower than the outlet temperature. The outlet temperature depends on the inlet air temperature, airflow rate, product feed rate, and atomized droplet sizes [26].

Factors that affect drying are mainly the inlet temperature, outlet temperature, drying gas density, relative humidity in the drying chamber, drying gas flow rate,

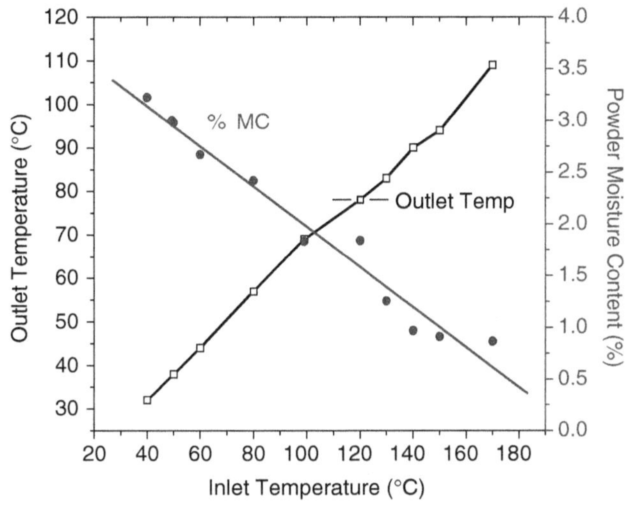

Figure 28.1. Effect of drying (inlet) temperature on outlet temperature and moisture content at fixed liquid feed rate of 1 mL/min (Unpublished data.)

residence time, nature of solute, solution concentration, solution feed rate, and solution boiling point [22,23]. Factors that one can practically control in a spray drying process include inlet temperature, drying gas medium, drying gas flow rate, feed rate, atomization pressure, and formulation. The rate of heat input must be sufficiently high to effectively evaporate the spray. Drying gas flow rate and inlet temperature determine the rate of heat input to the drying chamber, the minimum residence time of the droplets (τ) in the drying chamber, the evaporation rate, and the final moisture content of the powder. Outlet temperature and product moisture content are linearly dependent on inlet temperature [23] (Fig. 28.1) but vary inversely with the liquid feed rate (Fig. 28.2).

The drying behavior of single droplets differs according to the nature of the solute in solution to be dried [25]. For example, with mono- and disaccharides, droplet temperature tends to remain constant just after the start of drying (almost equal to T_{wb}). It is speculated that internal circulation of liquid may be involved at the early stage of drying. After the constant-rate period where drying rate decreases and internal circulation of the liquid is suppressed, droplet temperature will gradually rise to the outlet temperature. With high-molecular-weight sugars, constant-rate periods are much shorter, and droplets are deformed with progress of drying [25].

The fractional evaporation of water from a droplet per unit length of travel (dW/dl) at steady state can be represented by the Frossling equation:

$$\frac{dW}{dl} = \left[\frac{\Delta H}{V_f D^2}\right]^{(4.6e^{-7})[1+0.23(\mathrm{Re}^{0.5})]} \tag{28.12}$$

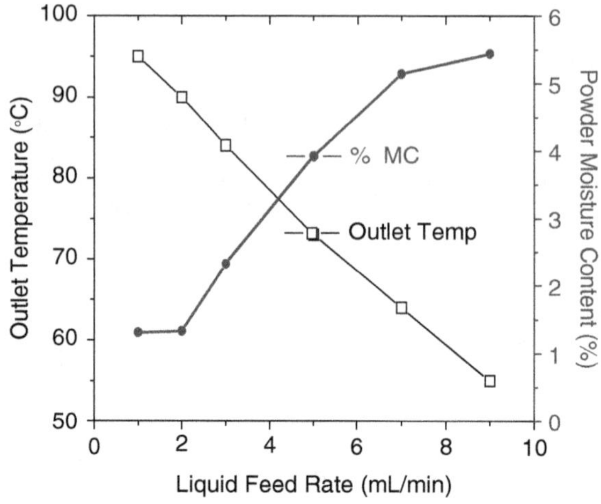

Figure 28.2. Effect of liquid feed rate on outlet temperature and moisture content at fixed inlet temperature of 150°C. (Unpublished data.)

where ΔH is humidity difference between the ambient condition in the drying chamber and that of the surface of the droplet, V_f is terminal velocity, and D is droplet diameter. It should be noted that the drying time is proportional to the square of the radius. For larger droplets, the drying time may become so long that the droplet reaches the wall of the drier while still wet. The latter problem is often encountered in small-scale driers. Heat and mass transfer can be improved by increasing the surface area-to-volume ratio of the droplets (i.e., generating smaller droplets). Smaller droplets can be generated by control of process variables such as increasing atomization pressure, using a drying gas with a higher density, increasing gas:liquid ratio (i.e., mass:flow ratio), and/or using a nozzle with a smaller diameter for atomization. More recently, high-pressure effervescent atomization, a technology borrowed from the diesel industry, has been successfully used in the spray-drying process of several solutions of vaccines and antibodies [29–32]. This atomization process generates an effervescent mixture of an insoluble gas with a liquid by colliding a gas such as nitrogen at high pressure [>1300psi (lb/in.^{2})] with a liquid feedstock at a small dead-volume "T" junction. As the atomizing gas–solution mixture is discharged from the atomizer orifice, the gas phase expands, breaking the liquid into small droplets of uniform droplet size distribution. This process generates an ultrafine droplet size distribution (mean size <7 μm) at lower mass:flow ratio [30] and a more uniform spray pattern than do conventional pneumatic two-fluid atomizers such as the Buchi two-fluid atomizer (Fig. 28.3). Consequently, a narrower particle size distribution of dried particles can be obtained (Fig. 28.4). Smaller droplets can also be achieved by control of feedstock physical properties such as decreasing viscosity and/or surface tension.

28.4.1.1.3. Powder Separation. The resulting powders, consisting of micrometer-sized particles, are separated from moist gas stream by means of a cyclone,

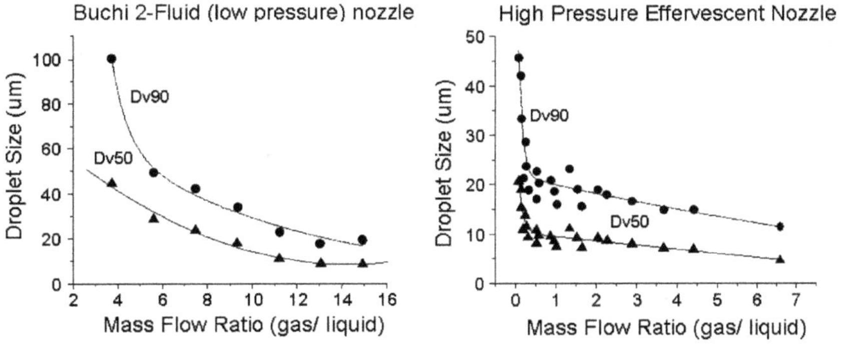

Figure 28.3. High-pressure effervescent atomization: generation of ultrafine spray droplets.

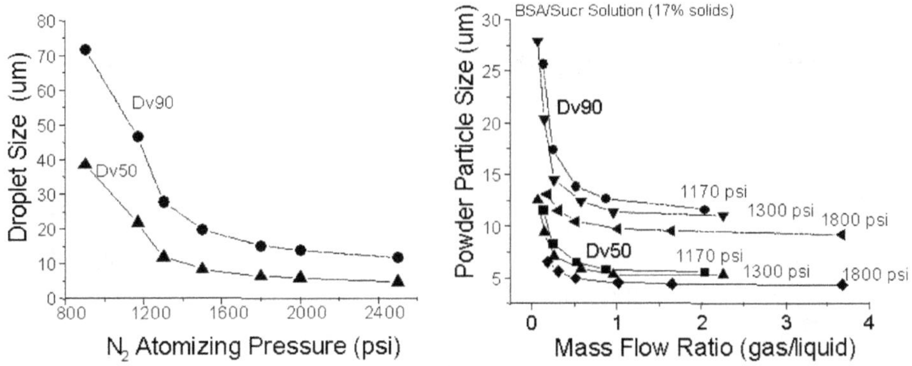

Figure 28.4. High-pressure effervescent atomization: controlling the droplet and dry powder size distribution and engineering powder size distribution by tuning atomizing pressure and flow rate.

electrostatic precipitator or bag filter and collected in a holding chamber (collector) [23,24]. The cyclone separator is the most widely used instrument for of powder collection. Collection efficiency from cyclones can be improved by increasing the gas inlet velocity and using higher initial solid content in solution to be spray-dried [23].

28.4.1.2. Morphology of Spray-Dried Particles. The morphology of the dried particle strongly influences the powder's behavior during reconstitution and in an air stream, and hence delivery characteristics from the delivery device. Spray drying mainly results in particles that may be spherical, dented, or raisin-like [33] (Fig. 28.5). Dynamics of evaporation, formulation, and processing conditions such as atomization conditions, and inlet temperature have been shown to affect spray dried

(a)

(b)

Figure 28.5. Scanning electron microscope scans showing (a) a typical hollow spray-dried particle, (b) a typical spherical spray-dried particle (with surfactant), (c) dented and hollow spray-dried particle, and (d) raisin-like spray-dried particles. (Unpublished data.)

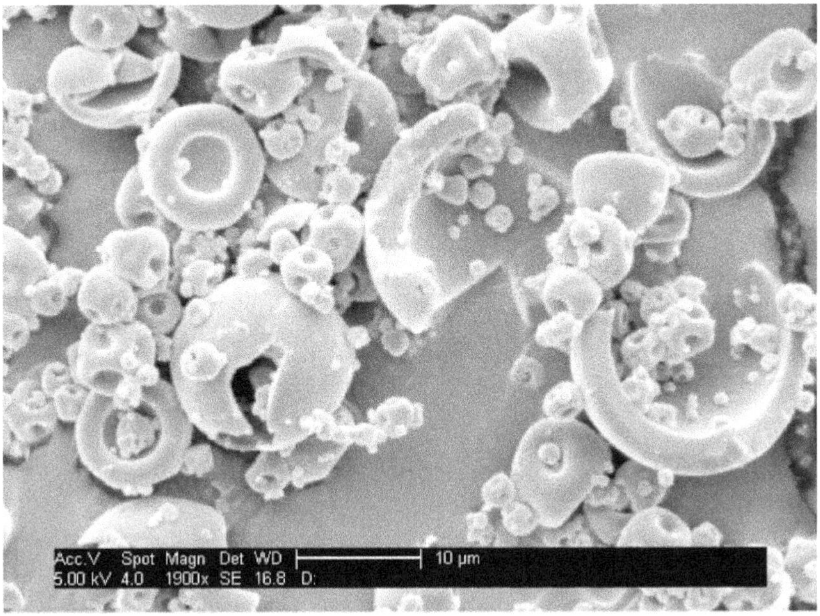

(c)

(d)

Figure 28.5. *(Continued)*

particle morphology [23,34]. Some of the formulation properties that may impact particle morphology include

1. *Nature of Solutes*. Pure lactose and formulations with a high surfactant concentration have been reported to dry into perfectly spherical particles. Formulations with higher-molecular-weight stabilizers, skim milk, caseinates, and other proteins shrink unevenly during drying, giving rise to surface folds [33,35].
2. *Reynolds Number*. This is a dimensionless parameter that reflects turbulent or laminar flow and is defined by the ratio of dynamic pressure and shearing stress.
3. *Boiling Point of Droplet Solution*. Depending on drying temperature relative to the boiling point of the droplet solution, spray drying can lead to multiple morphology and crust structures [23].

28.4.1.3. Challenges Associated with Spray Drying. While many proteins are not impacted by stresses associated with atomization and spray drying [36–38], those that exhibit high-surface-active properties, and thus tend to gravitate toward the air–water interface, are susceptible to atomization-related stress. This presents a challenge for spray drying because the atomization step involves not only shear stresses but also high surface tension as a result of surface area expansion. Exposing proteins to a large air–water interface can lead to protein unfolding followed by possibly aggregation of protein components. Human growth hormone (hGH) and bovine serum albumin (BSA) are examples of proteins that undergo extensive aggregation during atomization. It was found that the addition of surfactants and/or specific ions (such as zinc) helped alleviate this atomization-related [36–38] surface denaturation.

Outlet (or product) temperatures of typically >50°C are needed for efficient drying. While such temperature exposure can prove detrimental to some proteins, many biological materials are able to retain activity throughout the process. Several reasons for this include the short timescale of heat exposure (on the order of milliseconds), the evaporative cooling effect, and the rapid formation of an amorphous glass that encases the protein in a rigid matrix, thereby limiting mobility and decreasing chemical reactivity [22]. For more thermally labile materials, activity loss and poor shelf life are observed when processing at inlet temperatures higher than the thermal unfolding temperature. For example, the unfolding temperature (T_{unf}) of a monoclonal $\mathrm{IgG_1}$ antibody (MAb) corresponded to the observed onset of process induced MAb aggregation as measured by differential scanning calorimetry (DSC) (Fig. 28.6). It should be noted that depending on the protein structure and the process heat conditions, these changes may be reversible [26].

Under appropriate outlet temperature and formulation conditions, spray drying can be used to dry more complex biologicals, including live organisms. Generally for live bacteria, lower outlet processing temperature was associated with higher viability [26,39,40]. For example, gram-positive *Lactobacillus paracasei* bacteria were shown to be stabilized for long periods of time using a spray-drying process conducted at low outlet temperature conditions [41].

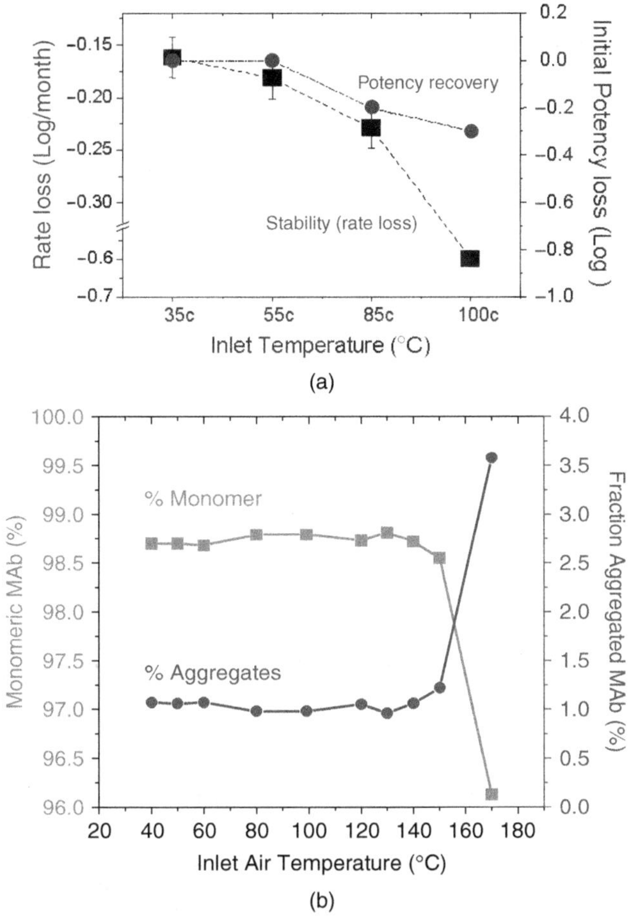

Figure 28.6. (a) Influence of drying temperature on activity recovery and stability of spray-dried live influenza virus; (b) spray drying above the $T_{\text{unfolding}}$ of a MAb adversely impacted product integrity. Below T_{outlet} of 90°C, MAb aggregation process loss was minimal. (Unpublished data.)

Other challenges associated with spray drying may be the need for an additional drying step to reduce the solvent content to pharmaceutically acceptable levels [23], difficulties in reconstitution, limited capacity for cGMP aseptic processing, yield problems (especially on smaller-scale spray driers), and electrostatic charges that may be problematic during powder filling.

28.4.2. Spray-Freeze Drying

Spray-freeze drying is a more recently developed drying process that involves elements of spray drying and freeze drying. This process has been used to produce free-flowing powder of porous, micrometer-sized, dense particles with a high specific surface area and improved bioavailability. Spray-freeze drying has potential application in pulmonary delivery of proteins, as well as intradermal delivery of vaccines from needle-free injections [42–45]. Spray-freeze drying can be used to dry a thermally labile and hygroscopic drug that requires a high outlet temperature and long residence time for drying [23]. Additional advantages of spray-freeze drying relative to spray drying may be greater efficiency in terms of product recovery, and the ability to produce highly dispersible porous particles [22,42].

28.4.2.1. Unit Operations in Spray-Freeze Drying

28.4.2.1.1. Atomization. As in spray drying, this process involves spraying or atomizing a feedstock (solution, suspension, or colloidal dispersion) using a variety of nozzles such as one-fluid, two-fluid, or ultrasonic nozzles. However, instead of atomizing into ambient gaseous medium, the feedstock is atomized directly into a cryogenic medium, where rapid freezing of the droplets takes place to form ice particles [22,45,46]. Where the cryogenic medium is a liquid such as liquid nitrogen, liquid helium, propane, or ethane, the process is referred to as *spray freezing into liquid* (SFL). If the cryogenic medium is a vapor, the process is referred to as *spray freezing into vapor* (SFV).

28.4.2.1.2. Rapid Freezing. Sprayed droplets freeze during time of flight through the cold vapor phase (in case of SFV) or inside the liquid cryogen (in case of SFL). The rate of freezing of the droplets in either case is rapid, and neither method allows for control of the rate of freezing. Freezing was estimated to be on the order of 0.84 and 3.2 ms for droplets of diameter 16 and 32 μm, respectively [36]. Rapid freezing results in a high degree of supercooling, and hence formation of very small ice crystals within the frozen droplets [44]. This, along with atomization, leads to high internal particle porosity and a very large specific surface area.

Liquid nitrogen is usually the preferred cryogen because of its inexpensive, environmentally friendly, and inert properties [45].

28.4.2.1.3. Harvest of Frozen Droplets. Suspended frozen droplets are collected by sieves or are collected after allowing the cryogen to boil off [45].

28.4.2.1.4. Primary Drying. The collected frozen droplets are transferred to prechilled shelves (typically $\leq -40°C$) of a lyophilizer for subsequent drying. The principles of drying by ice sublimation for this phase of the drying process are identical to those of primary drying in the conventional freeze-drying process. Specifically, the primary drying phase is performed at low pressure (typically ≤ 150 mTorr) and 2°C–5°C below the collapse temperature (T_c) to avoid product collapse and maintain elegant cake structure. The endpoint of primary drying is marked by a sharp decrease in the dew point (the temperature at which ice has an equilibrium vapor pressure equal to the measured partial pressure of water), an increase in product temperature to be similar or higher than shelf temperature, and a decrease in the pressure reading of a Pirani vacuum gauge [47,48]. By the end of this step, nearly 70%–80% of water has been removed in the form of ice, while the remaining water remains in the amorphous phase [13,48].

Primary drying at atmospheric pressure using cold dehumidified air for fluidization and drying of prefrozen droplets in a drying chamber was examined to determine whether fluidization of the prefrozen droplets could reduce the drying cycle time [49]. The results showed that this "atmospheric spray freeze drying" approach did not shorten the drying time even on comparison of primary drying in a conventional lyophilizer at 112, 225, and 750 mTorr.

An optional annealing step can be included before primary drying. *Annealing* is a process by which frozen samples are held at a temperature between the ice melt temperature and T_g' for a period of time. Annealing results in the growth of larger ice crystals and/or the crystallization of bulking agents (e.g., mannitol, glycine), which increases the collapse temperature, and allows for primary drying at higher temperature. The net effect is a decrease primary drying time and hence shorter process time. An additional benefit of annealing is a reduction in the specific surface area of the final product [31,32, 50–52].

28.4.2.1.5. Secondary Drying. As with primary drying, the spray-freeze drying process also involves a secondary drying step that is similar to the conventional freeze-drying process. A secondary drying step is a short step performed after primary drying at much higher shelf temperatures to remove the remaining water in the amorphous phase via evaporation. Ramping from primary drying to secondary drying final temperature is typically done gradually to avoid product collapse, which may occur in cases where the shelf temperature is higher than the glass transition temperature (T_g) of the product (since T_g increases very slowly relative to the shelf temperature depending on the rate of water evaporation).

28.4.2.2. Morphology of Spray-Freeze-Dried Particles. Unlike spray drying, spray-freeze drying results in nonhollow spherical particles with a highly porous inner structure [44]. Because of a high internal porous structure, the specific surface area of a spray-freeze-dried powder is reported to be as high as 40 times greater than that of a corresponding spray-dried powder [42]. The use of an annealing step before primary drying results in the loss of some of the porous structure in a spray-freeze-dried powder and hence a reduction in its specific surface area, as well as shrinking or shriveling of the surface of the particles [44,52].

28.4.2.3. Challenges Associated with Spray-Freeze-Drying. Aside from the challenges associated with the cost and the logistics of handling and developing a scaled-up process for liquid nitrogen, there are several process challenges in spray-freeze drying. For example, shear and surface tension are stress factors associated with the atomization step in the spray-freeze drying that can impact activity recovery for products [36,44,46,52,53]. Because of their amphiphilic nature, many biological materials readily adsorb onto the ice–water interface just as they do with the air–water interface. The hydrophobic ice–water interface may produce much damage if the interfacial area generated is large enough. In spray-freezing, the rapid and uncontrolled rate of freezing leads to generation of a very high interfacial surface area, causing destabilization equivalent in some proteins to 11 consecutive slow freeze–thaw cycles [54]. In the case of proteins, damage may occur at the ice–liquid interface, leading to unfolding and subsequently aggregation [54–56]. A common solution has been addition of a surfactant, which has been shown by several researchers to significantly increase recovery of many proteins, including trypsinogen [44], lysozyme [46], and recombinant human interferon-γ (rhIFNγ) [52]. Also, during rapid freezing, the generation of ice crystals may pose a challenge to biological materials that contain plasma membranes, including enveloped viruses and bacterial cells [57]. Other stresses are associated with freezing as a unit operation by itself, regardless of the freezing rate, include cryoconcentration effects, which may lead to salt concentration and/or large pH shifts due to buffer crystallization (as is the case with sodium phosphate buffer system), cold denaturation, and destabilizing dehydration effects [58–60].

Finally, the dehydration process, which removes hydrogen bonding from the protein surface, is known to be stressful. Therefore, the rational selection of stabilizers (lyoprotectants with or without surfactants) and/or process modifications such as annealing after spray-freezing will be critical to successful implementation of spray-freeze drying [36,44,46,52,53].

28.4.3. Vacuum (Foam) Drying

When processed without subjecting product to freezing conditions, vacuum drying is a convective form of drying in which moisture removal is achieved mainly by evaporation under reduced pressure. Vacuum drying, widely applied in the food industry, is an effective and convenient way to dry thermolabile materials, and has been used to dry and stabilize different biopharmaceutical formulations [61–64]. *Air drying*, on the other hand, refers to evaporative drying under normal atmospheric pressure. Vacuum drying process conditions usually result in a foamy sponge-like cake structure, and vacuum drying is sometimes referred to as *foam drying*. Foam drying can be conducted with transient freezing, in which case moisture removal is accomplished by both sublimation and evaporation [61–64]. Foam drying is characterized by relatively brief and minimal excursions to extreme temperature regimes, since drying is accomplished at ambient or near-ambient temperatures. Foam drying has lower energy requirements compared to a drying process such as freeze drying [61]. Hybrid drying technologies with vacuum drying are also emerging such as microwave vacuum drying, and vacuum drying with or without radiative heating [65–68]. Vacuum drying is a promising method for application to biopharmaceutical products when sterile process drying is required [69].

28.4.3.1. Unit Operations in Foam Drying. Foam drying was successfully used to dry and stabilize formulations of several biomolecules well over a half-century ago by several resourceful University of Cambridge microbiologists [31,32,70–73]. Among them was D. I. Annear, who described the foam-drying process with a good level of detail and demonstrated its use in stabilizing several bacteria. Annear used a simple setup starting with a high-concentration (40% w/v) solution of glutamate, a vacuum pump, and P_2O_5 moisture trap to cause boiling under vacuum conditions and simultaneous evaporation into a dry foamy cake with excellent preservation of product activity [70–72].

Foam drying can be done with better process control using conventional freeze driers [26,61–64,74]; however the basic unit operations remain the same; starting with the equilibration of the vials at ambient or near-ambient temperature conditions followed by rapid drop in chamber pressure, which decreases the boiling point of the solution, resulting in boiling (or foaming). As evaporation progresses, accompanied by a moisture trap, (e.g., a condenser in a freeze dryer) evaporative cooling may accelerate to cause ice nucleation and propagation to freeze the foam product. Depending on the shelf and hence product temperature setting, the solutions may or may not freeze, or may even become supercooled [68]. In cases where the pressure drop is slow, solutions may not foam and instead dry to a film. Drying in such case has been referred to as *film drying* [64,75,76], or Xerovac method of drying [77,78].

The foaming action is a cavitation process that involves cycles of solution bubbles percolating upward, bursting, and coalescence of neighboring bubbles. This process results in a variable foam structure with variable density, and likely variable moisture content within the cake. Viscous solutions and/or solutions with high solids content (>20% w/v) typically foam with less violent bubbling action, and hence lower likelihood of container closure integrity issues that can arise during boiling, such as splattering within the vial [22,31,32,73,74]. Aside from the use of different excipients to control solution viscosity, other strategies to control the foaming process include modifying the freeze drier to enable control vacuum rampdown rate, solution thickening by initial evaporation under low vacuum before rapid pressure drop to initiate the foaming, foaming from the partially frozen or supercooled state, the use of bubble seeding agents, and spinning the vial during foaming [79].

As with spray drying, any vacuum-drying process is characterized by two main drying periods: a constant-drying-rate period and a falling-rate period, the thermodynamics modeling of which has been described elsewhere [69]. *Primary drying* starts once pressure drops, during which evaporation initiates regardless of foaming and lasts for 1–2 h, when the bulk of heat and mass transfer occurs. Product temperature gradually rises until it reaches shelf temperature, marking the end of primary drying. The solution solidifies to form a foam or a film, and roughly 90% or more of the bulk water is removed by evaporation by the end of the primary drying step [31,32,73,80]. Drying proceeds under cake collapse conditions in cases where primary drying is initiated in the partially frozen state, that is, above the collapse temperature (Fig. 28.7). A *secondary drying* step follows, which further reduces the moisture content of the dried product. This step may last longer than the usual secondary drying step in freeze drying because of the tendency to form *closed-cell structure* (i.e., the roof of the foam cake is a continuous layer with no pore structures that could serve as a low-resistance path for moisture evaporation from the bottom layer of the foam). Besides convection, heat can also be supplied by conduction or radiation (e.g., IR and microwave) [109], either separately or in various combinations whereby the efficiency in moisture removal may be enhanced.

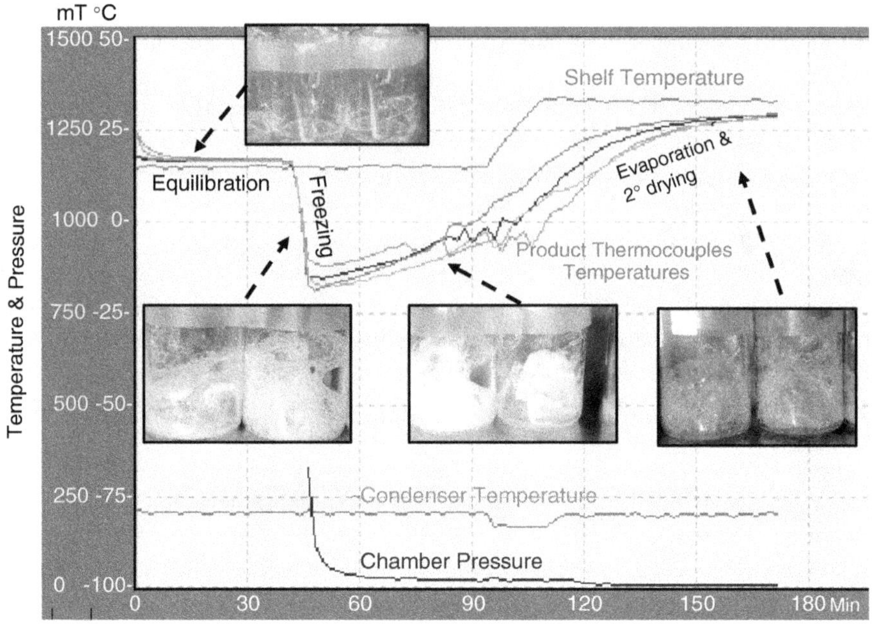

Figure 28.7. Product temperature profile and product foaming during a foam-drying cycle.

28.4.3.2. Morphology of Vacuum-Dried Products. The foam structure that is generated using sucrose- or trehalose-based excipients is amorphous with size and thickness, that can vary depending on formulation parameters such as the protein-to-stabilizer ratio and the nature of the stabilizer [31,64]. For example, with a low protein-to-stabilizer ratio (low-molecular-weight stabilizers such as sucrose, trehalose, or maltose), thicker glass foam is observed in film- or foam-dried products regardless of the presence or absence or surfactant, compared to freeze-dried particles (compare Fig. 28.8a and Fig. 28.8) [31,32,62,64]. The SEM data showed that the foam structures have no detectable pores (Fig. 28.8c) as compared to, for example, a spray-freeze-dried particle [64,69], which are porous and therfore exhibit larger specific surface area. A high protein-to-stabilizer ratio formulation or the use of special stabilizers such as dextran resulted in a more brittle product with a significantly higher specific surface area that can be broken into thin, foam glass (Fig. 28.8d) [31].

28.4.3.3. Challenges Associated with Foam Drying. Foam drying is an aggressive drying technique that involves several potential stresses on the product. The cavitation associated with the foaming action and the shear forces at the air–water interface could lead to protein denaturation and activity loss. However, under optimal formulation conditions, for several proteins and vaccines tested so far, shear stress does not seem to be damaging. Excellent recoveries have been observed after vacuum drying of the air–water interface sensitive protein human growth hormone [64], LDH

(a)

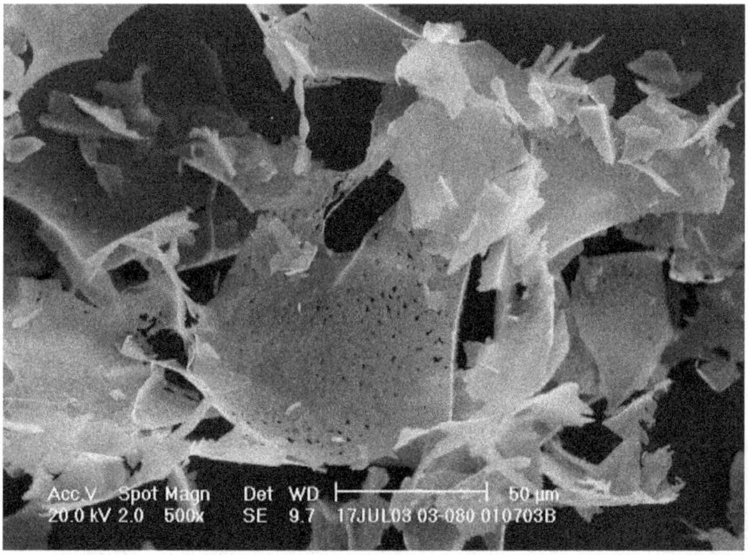

(b)

Figure 28.8. Scanning electron micrographs (a) taken after breaking a vacuum-dried product of human growth hormone and stachyose (1:3) showing large and thick plates, (b) of a freeze-dried product of human growth hormone and stachyose (1:3) showing very thin (and even some porous) leaflets or flakes, (c) showing magnified and side views of particles in (a) revealing a closed-pore structure in the plates, and (d) showing particles of a vacuum-dried product of a formulation of high IgG-to-stabilizer ratio consisting of small thin flakes. (Unpublished data.)

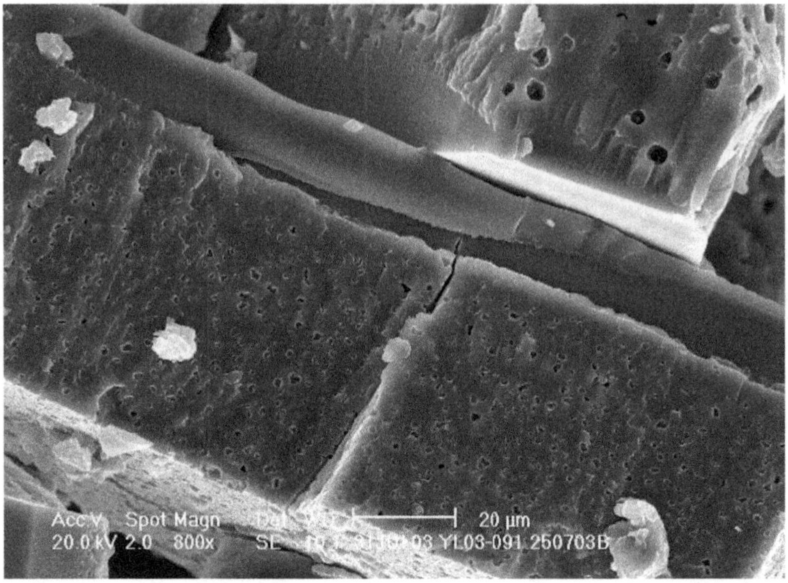

(c)

(d)

Figure 28.8. (*Continued*)

[61–63], immunoglobulin molecule [31], rhGCSF [62,63], a genetically engineered bivalent live attenuated virus vaccine for human parainfluenza virus 3 [32], and LaSota virus [74].

Another challenge to foam drying is optimizing the formulation and process parameters to achieve low (>4%) moisture contents. This may be important because increasing the shelf temperature during secondary drying is not always successful, especially once the closed-cell structure is formed. Strategies to achieve low moisture content include increasing the pressure rampdown rate to increase the foaming rate, increasing the surface-to-volume ratio and hence the surface area for evaporation, or adding a "crystallizable" component such as phenylalanine that improves the porous ultrastructure [66,67].

While it is feasible to scale up foam drying as a bulk process, implementing foam drying at commercial scale in single-unit-dose glass vials is very challenging because of the difficulty in controlling the foaming action well enough to avoid container–closure seal failure from splattering associated with bubble cavitation, while at the same time achieving good vial-to-vial consistency in cake structure, volume, and moisture content. The vial-to-vial variability in appearance also presents a challenge in establishing quality control acceptance criteria. Measures to reduce inconsistencies between vials are to control the foaming action by process and/or formulation control. Process controls include reducing the vacuum rampdown rate, operating at higher shelf temperatures, and using small fill volumes and large headspace (i.e., larger-volume vials). While percolating bubbles from the bottom of the container can be one solution, it is not considered a commercially feasible one. Formulation controls for foaming may be to increase solution viscosity (using initial high solids content or addition of viscosity-enhancing agents) and/or to control surfactant levels.

28.4.4. Supercritical Fluid (SCF) Drying

Supercritical fluid (SCF) technology, comprising methods used for either crystallization or drying (drying by precipitation); utilizes the high solvation power of supercritical fluids (SCFs) to produce particles at near ambient temperatures with narrow size distribution and controlled crystalline morphology suitable for dry powder inhalation delivery systems and needle-free dry powder injection systems [23,81–83]. This technology allows for drying of biological materials for which degradation is unavoidable by atomization and/or heat, and allows for control of solid state properties such as particle size. A unique aspect of SCF technology is the ability to rapidly vary solvent strength and the rate of supersaturation (and hence nucleation) of dissolved compounds [22,23,84]. The SCF method has been used for lab-scale production of powders of several biological materials such as insulin, catalase, trypsin, lysozyme, and human growth hormone [85–87].

A substance becomes a supercritical fluid (i.e., a highly compressed gas) when its pressure and temperature are above its critical pressure (P_C) and critical temperature (T_C), respectively (Fig. 28.9). In the supercritical regime, any lowering of temperature or pressure will bring the substance into liquid or vapor region, respectively. The SCF has unique thermophysical properties. A SCF has liquid-like density, large compressibilities, and viscosity values intermediate between those of a gas and a liquid.

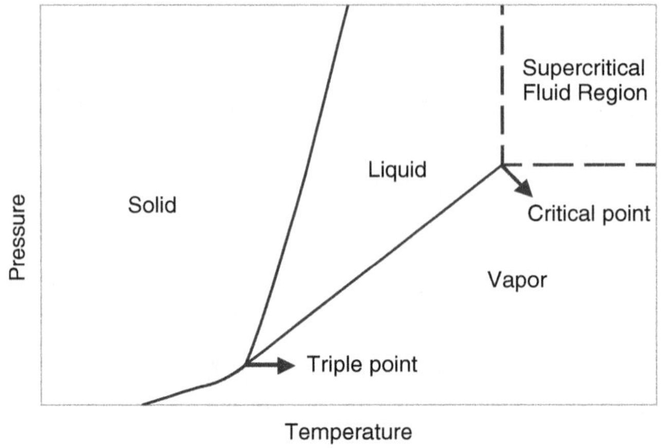

Figure 28.9. Pressure–temperature phase diagram for a pure component indicating the supercritical region.

Small changes in temperature or pressure can result in large changes in density and in solvent power. A SCF is able to penetrate substances such as a gas and dissolve materials such as a liquid [22,23,84].

Supercritical fluids are good solvents for nonpolar compounds such as cholesterol or α-tocopherol, but generally poor solvents for proteins [23,82,84]. Carbon dioxide is the most widely used SCF medium for pharmaceutical applications since it has a low critical temperature ($T_c = 31.1°C$) and pressure (75.8 bar). It is nontoxic, nonflammable, relatively inexpensive, recyclable, available in large quantity at high purity grade, and generally regarded as safe. As a nonpolar SCF, CO_2 is dissolvable in many low-molecular-weight (LMW) species with appreciable vapor pressures, but not with larger polymers that have low vapor pressures. It is used as an antisolvent for many lipophilic and hydrophilic compounds [23,82,84,88].

28.4.4.1. Methods Used in SCF Technology and Unit Operations. Supercritical fluid methods using CO_2 are typically divided into two groups according to its use as a solvent or an antisolvent [22,23].

28.4.4.1.1. Methods that Use Carbon Dioxide (CO_2) as a Solvent. Supercritical fluid methods that utilize CO_2 as a solvent are applied to materials soluble in or miscible with supercritical CO_2 to produce particles with controlled size range.

One SCF method that uses CO_2 as a solvent is rapid expansion of supercritical solutions (RESS), also known as *supercritical fluid nucleation* (SFN) [89]. The RESS method produces very fine particles in the submicrometer and micrometer size ranges and a solvent-free product. Recombinant human deoxyribonuclease (rhDNase), lysozyme, lactate dehydrogenase (LDH), ovalbumin, and trypsogen in formulations with less than 2% moisture were produced using RESS [82,84,86,90]. A typical RESS apparatus consists of a preheater, an extraction unit, and a precipitator, and the unit operations involved in RESS are [23,82,84,88]:

1. *Solubilization or Miscibility*. Compressed and preheated SCF (usually CO_2) is mixed with solute or solution containing the protein and excipients in the extraction unit at pressures between 80 and 100 bar and temperatures between 20°C and 50°C to achieve solubilization or miscibility. Heating at temperatures above the extraction temperature may be necessary to prevent condensation and blockage.
2. *Atomization*. The solution is expanded through a restriction such as a laser-drilled nozzle or capillary at supersonic velocities into atmospheric conditions in a precipitation unit. Rapid expansion of supercritical CO_2 in solution causes the breakup of drops into finer droplets. The SCF acts as a propellant and effervescent agent to enhance atomization and shorten the drying process.
3. *Precipitation*. The resulting pressure drop due to rapid expansion causes the density of the SCF to decrease rapidly, along with the solvent power, leading to supersaturation of the solute, homogeneous nucleation, and subsequent particle precipitation. The use of heated nitrogen may aid in drying.
4. *Powder Collection*. Particles generated from the nozzle are captured in a collection unit within a collection chamber. The collection chamber can operate at elevated or reduced pressure to control precipitation and allow more flexibility in controlling size and morphology of particles generated.

Particle morphology and particle size distribution of precipitated powder is a function of preexpansion conditions such as initial solute concentration, preexpansion composition, nature of solute, presence of cosolvent, preexpansion pressure, and temperature. The higher the preexpansion concentration, the smaller the particles will be and the narrower the particle size range will be [23,91]. Particle morphology and particle size distribution also depend on expansion/postexpansion conditions such as nature of the SCF, length:diameter ratio of expansion device, expansion trajectory from the preheater and the expansion device, particle agglomeration during free-jet expansion, postexpansion temperature and pressure, nozzle geometry, and solute properties [23,84,88,92].

For some RESS applications the high pressures and temperatures and the required high ratios of gas to liquid (especially in cases where solubility and miscibility are low) make it challenging to implement for labile proteins. Additionally, the separation of submicron particles from large volumes of gas can be difficult on an industrial scale. Limitations related to solubility can be partly overcome by using more polar cosolvents in the SCF [92].

The "particles from gas-saturated solutions" (PGSS) process is a unique SCF method based on the miscibility and solubility difference between CO_2 and other solutions (or melted compounds). The gas consumption in PGSS is typically lower than in RESS because of lower requirements for gas:liquid ratios (<0.1 kg gas/1 kg powder). This is perceived as a major advantage in cases where large-scale production of particles is involved [87,92,93]. The lower process pressure range of PGSS as compared to RESS means that the solvent can be recycled, the product is more likely solvent-free, and the process can be applied to thermolabile compounds since no hot streams are used in solvent evaporation [92].

In PGSS, a gas-saturated solution or gas-saturated melt is expanded by passing it through a nozzle or other expansion device. The combination of cooling and volume expansion on release leads to supersaturation, followed by particle formation and precipitation. The powder is separated from the gas by sedimentation in a spray tower or by centrifugal forces in a cyclone separator. In PGSS, particle size and size distribution are affected by temperature, nozzle geometry and dimensions, and SCF addition [92,93].

28.4.4.1.2. Methods that Use CO_2 as an Antisolvent. Supecritical fluid methods that utilize CO_2 as an antisolvent are known collectively as gas antisolvent (GAS) recrystallization methods or reverse crystallization. These methods are applied to formulations that are not soluble in the SCF. The same methods are also used for precipitation of biological materials from organic solvents, including insulin, lysozyme, and myoglobin [23,82,88]. The GAS method may be further classified based on application as batch or continuous modes [92].

In the batch GAS mode, SCF is added to a solution of the material. The rapid addition of the SCF causes a reduction in the liquid density, volumetric expansion of the liquid, followed by supersaturation and precipitation. Particle size is typically within the 0.5–500 μm range and can be controlled by the rate of addition of the antisolvent, the initial solute concentration, and temperature. Particle size distribution, on the other hand, is a function of volumetric liquid expansion. The latter depends on temperature, pressure, type of solvent, SCF, and stirring rate [92]. The GAS batch mode is limited by its application in batch operations, use of organic solvents and solvent removal.

Continuous GAS techniques include supercritical antisolvent (SAS) techniques and solution-enhanced dispersion by supercritical fluids (SEDS). The principles of precipitation are the same as in GAS batch techniques except that in continuous GAS techniques, the liquid and SCF are fed continuously into the particle formation vessel (precipitator/crystallizer) [92].

Supercritical antisolvent techniques are a subgroup of continuous GAS techniques used to generate particles in the size range of 1–250 μm [91]. Applications of SAS techniques include controlled-release formulations and protein powders. Advantages of SAS techniques include higher solute throughput and flexibility of solvent choice [84]. Biological materials applications of SAS use the supercritical fluids (usually CO_2) as an antisolvent for the solute and as an extraction medium for water or solvent. The process can be separated into several stages [82,88]:

- Aqueous or organic solution of the protein is introduced into the particle formation vessel by pumping through a capillary nozzle (typical orifice size is 20–70 μm). On mixing of SCF and solution of the formulation, rapid volume expansion occurs, and the solution becomes concentrated following extraction of solvent by the SCF.
- Solutes are precipitated out as a result of increasing concentration in the solution and the increasing miscibility between solvent and SCF.

- The precipitated solid is filtered from the solvent through the bottom of the pressure vessel and fresh gas or SCF can be introduced to remove trace solvent.

In SAS, the order of addition of solution and SCF in the particle formation vessel may differ. In normal addition SAS (NA-SAS), for example, SCF is injected into the solution. In NA-SAS, which is primarily a recrystallization technique, the rate of SCF addition affects particle size and particle morphology. However, NA-SAS lacks control of particle formation, and extensive particle agglomeration has been reported. Reverse addition SAS (RA-SAS) is primarily a precipitation technique and allows greater control of particle size distribution, particle shape, and physicochemical properties [82,84]. In RA-SAS, solution of the formulation is either injected [the method is also called *precipitation with compressed antisolvents* (PCA)] or aerosolized/sprayed [the method is called *aerosol spray extraction system* (ASES)] into the SCF. In general, SAS technology is noted for particle size control and morphology. For example, the use of SAS technology resulted in microspheres (particle size ≤ 1 μm) and needles (5 μm long) of insulin from an ethanol/water mixture at 90 bar and 35°C, and particles in the 2–4 μm range from a DMSO and *N,N*-dimethylformamide (DMFA) mixture. Using the same setup, a solution of catalase in ethanol and water yielded spherical and rectangular particles approximately 1 μm in size [23,81,94,95].

Factors affecting particle size in SAS include mass transfer rates between the droplet and antisolvent phases, mixing rate, rate at which solution is expanded, operating pressure, temperature and jet breakup. Jet breakup depends on nozzle configuration, spray velocity, and physical properties of different phases. The higher diffusion coefficient of water in supercritical CO_2 means that any increases in temperature and/or pressure can lead to particle size reduction, and favoring nucleation over crystal growth. An increase in solute concentration also results in particle size reduction. Finally, lowering the flow rate of supercritical CO_2 relative to the solvent and modifier can also result in smaller particles [82,84,96,97].

Particle morphology in SAS depends on formulation, droplet viscosity, and the concentration of solutes. The use of sucrose typically results in spherical particles with smooth surfaces. Mannitol tends to crystallize in SAS processes, resulting in less homogeneous particle distribution. High trehalose levels in lysozyme and alkaline phosphatase formulations have resulted in needle-shaped and spherical particles, whereas spherical particles are observed under lower trehalose concentrations [82,84,96,97].

Strategies to improve mass transfer during SAS operations include decreasing droplet size, increasing the drying medium:solvent ratio, and enhancing the solvent–SCF mixing during spraying process [82].

As compared to SAS, the solution-enhanced dispersion (SEDS) process offers the advantage of smaller droplet size and more intensified mixing that enhances mass transfer rates [92]. Biologics containing formulations that are not dissolvable in SCF or in an organic solvent can be processed by SEDS and maintained in an aqueous solution prior to particle formation. Three main stages are involved in SEDS [82,83,85,92]:

- Simultaneous introduction of aqueous and organic feeds into the particle formation vessel through a special coaxial nozzle to achieve dispersion and rapid

mixing. The coaxial nozzle is designed to use high-velocity supercritical CO_2 to disperse the aqueous feed and organic antisolvent into the vessel. Flow rates are tuned to a single equilibrium phase within the vessel.
- Rapid supersaturation followed by solute precipitation from the aqueous feed by the action of the antisolvent.
- Particles are retained in the vessel. The resultant supercritical mixture comprising water and organic solvent extracted into supercritical CO_2 is vented via the back pressure regulator outlet.

Strategies to minimize contact time between the proteins and organic solvents in SEDS are not always successful, and some studies have reported structural damage to proteins [85].

28.4.4.2. Challenges Associated with SCF Technology. Each SCF processing method has its own unique challenges with its use in the biopharmaceutical industry. Perhaps one common challenge among all is solvent and SCF selection. Ideal solvent requirements for use in SCF technology with biological materials include a high solvent power for and compatibility with the biological molecule, a solvent that is nondenaturing to biological materials and nontoxic. Solvents used in SCF seldom satisfy all these requirements. For example, dimethylsulfoxide (DMSO) is toxic and may cause significant perturbation of secondary structure of proteins (i.e., potential for protein denaturation). Other solvents result in considerable damage to antibody fragments and recombinant human growth hormone (hGH) on exposure to the solvent over long periods of time. Methanol, ethanol, and ethyl acetate have poor solvent power for proteins. Water, although the "friendliest" solvent to biomolecules, is poorly soluble in supercritical fluids such as CO_2. The addition of modifiers such as ethanol, acetone, or isopropanol to the aqueous solution or to the SCF may be necessary to enhance miscibility. Additionally, buffering agents can be added to aqueous solutions to prevent pH drop during drying or processing [82,83,85].

Other challenges associated with use of SCF technology with biomolecules include occasional high processing temperatures that may result in loss of biological activity, and occasional high initial moisture levels that necessitate an additional drying step to reduce initial moisture content to acceptable levels, in order to achieve appropriate shelf life [83,85].

28.4.5. Mixed Drying

Although not cost- or energy-efficient, there are some reports of the use of a combination of drying methods ("mixed" or "consecutive" drying) or drying methods with more complex unit operations. Spray coating, a form of fluid-bed drying, serves two consecutive purposes: (1) to coat carrier particles having a certain particle size range with the active ingredient and (2) to dry the coated particles [24]. Spray coating has promising application for producing vaccines for epidermal powder immunization [98]. In instances where the drying method of choice (i.e., "first option") was unsatisfactory in producing a final powder with low moisture content (to control yield and/or storage stability), mixed or consecutive drying systems were used. In other words, the moisture level in the powder produced after the main (first option) drying method was further reduced under "milder" temperature conditions by a second drying method (e.g., vacuum drying). For example, the use of a vacuum dryer as a second dryer is reported for removal of moisture from spray-dried lactic acid starter cultures [99], spray-dried protein formulations [31,64], and spray-dried vaccine formulations [32]. Also, a combination of fluid-bed drying and freeze drying is reported for the preparation of *Streptococcus thermophilus* [100].

28.5. CONCLUSIONS

An overview of several conventional and unconventional drying methods shows that a number of other drying methods are available for drying biopharmaceutical products in addition to freeze drying. The unit operations involved in each drying method, the thermal history, their applications, scalability and manufacturability, and the associated challenges are different for each process, and should be thoroughly evaluated in order to determine suitability. While freeze drying clearly has its place in pharmaceutical processing, under optimal formulation and processing conditions, alternative drying process options have the potential to enhance product performance beyond what can be achieved with the conventional freeze-drying process.

REFERENCES

1. Lee, J. and Timasheff, S. (1981), The stabilization of proteins by sucrose, *J. Biol. Chem.* **256**: 7193–7201.
2. Arakaw, T. and Timasheff, S. (1985), Mechanism of poly(ethylene glycol) interaction with proteins, *Biochemistry* **24**: 6756–6762.
3. Bhat, R. and Timasheff, S. N. (1992), Steric exclusion is the principal source of the preferential hydration of proteins in the presence of polyethylene glycols, *Protein Sci.* **1**: 1133–1143.
4. Kendrick, B., Chang, B., Arakawa, T., Peterson, B., Randolph, T., Manning, M., and Carpenter, J. (1997), Preferential exclusion of sucrose from recombinant interleukin-1 receptor antagonist: Role in restricted conformational mobility and compaction of native state, *Proc. Natl. Acad. Sci. USA* **94**: 11917–11922.
5. Xie, G. and Timasheff, S. (1997), The thermodynamic mechanism of protein stabilition by trehalose, *Biophys. Chem.* **64**: 25–43.
6. Chi, E., Krishnan, S., Kendrick, B., Chang, B., Carpenter, J., and Randolph, T. (2003), Roles of conformational stability and colloidal stability in the aggregation of recombinant human granulocyte colony-stimulating factor, *Protein Sci.* **12**: 903–913.
7. Bajaj, H., Sharma, V., Badkar, A., Zeng, D., Nema, S., and Kalonia, D. (2006), Protein structural conformation and not second virial coefficient relates to long-term irreversible aggregation of a monoclonal antibody and ovalbumin in solution, *Pharm. Res.* **23**: 1382–1394.
8. Rockland, L. B. (1987), in L. B. Rockland and L. R. and Beuchat (eds), *Water Activity: Theory and Applications to Food*, (Rockland, L. B. and Beuchat, L. R., eds., Marcel Dekker, New York.
9. Hinchley, J. (1922), Drying by pressure, *Trans. Inst. Chem. Eng.* **XLL**: 127–130.

10. Chang, L., Shepherd, D., Sun, J., Tang, X., and Pikal, M. J. (2005), Mechanism of protein stabilization by sugars during freeze-drying and storage: Native structure preservation, specific interaction, and/or immobilization in a glassy matrix? *J. Pharm. Sci.* **94**: 1427–1444.
11. Pikal, M. J. (2004), Mechanisms of protein stabilization during freeze-drying and storage: The relative importance of thermodynamic stabilization and glassy state relaxation dynamics, in *Drugs and the Pharmaceutical Sciences: Freeze-Drying/Lyophilization of Pharmaceutical and Biological Products*, Vol. 137, Koval, R. L. and Pikal, M. J., eds., Marcel Dekker, New York, pp. 63–107.
12. Carpenter, J., Chang, B., Rodriguez, W., Randolph, T., Carpenter, J. F. and Manning, M. C., eds., (2002), Vol. 13, *Rational Design of Stable lyophilized Protein Formulations: Theory and Practice*, Kluwer Academic/Plenum Publishers, New York, pp. 109–127.
13. Carpenter, J. F., Pikal, M. J., and Randolph, T. W. (1997), Rational design of stable lyophilized protein formulations: some practical advice, *Pharm. Res.* **14**: 969–975.
14. Franks, F., Hatley, R. H. M., and Mathias, S. F. (1991), Materials science and the production of shelf-stable biologicals, *BioPharm* **4**(9): 38–55, pages 38, 40–42, 55.
15. Hlinak, A. and Clark, B. (2002), Drying and dryers, in *Encyclopedia of Pharmaceutical Technology*, 2nd ed., Vol. 1, Swarbrick, J. and Boylan, J. C., eds., Marcel Dekker, New York, pp. 1018–1032.
16. Mujumdar, A. S., ed. (2007), Principles, classification and selection of dryers, *Handbook of Industrial Drying*, *3rd ed.*, CRC Press, Boca Raton, FL, pp. 3–32.
17. Kowalski, S. J., ed. (2003), Characterization of drying processes, *Thermomechanics of Drying Processes*, Vol. 8, Springer, Heidelberg, Germany, pp. 24–30.
18. Guerrero, M., Albet, C., Palomer, A., and Guglietta, A. (2003), Drying in pharmaceutical and biotechnological industries, *Food Sci. Technol. Int.* **9**: 237–243.
19. Rambhatala, S. and Pikal, M. J. (2004), Heat and mass transfer issues in freeze-drying process development, in *Lyophilization of Biopharmaceuticals*, Costantino, H. R. and Pikal, M. J., eds., AAPS Press, Arlington, VA, pp. 75–109.
20. Pikal, M. J., Shah, S., Senior, D., and Lang, J. E. (1983), Physical chemistry of freeze-drying: Measurement of sublimation rates for frozen aqueous solutions by a microbalance technique, *J. Pharm. Sci.* **72**: 635–650.
21. Marinos-Kouris, D. and Maroulis, Z. B. (2007), Transport properties in the drying of solids, in *Handbook of Industrial Drying*, *3rd ed.*, Mujumdar, A. S., ed., CRC Press, Boca Raton, FL, pp. 81–119.
22. Weers, J., Tarara, T., and Clark, A. (2007), Design of particles for pulmonary drug delivery, *Expert Opin. Drug Deliv.* **4**: 297–313.
23. Sacchetti, M. and Van Oort, M. (1996), Spray drying and supercritical fluid particle generation techniques, in *Physical and Biological Basis for Inhalation Aerosol Therapy*, Hickey, A., ed., Lung Physiology Series, Marcel Dekker, New York, pp. 337–383.
24. Jacquot, M. and Pernetti, M. (2004), Spray coating and drying processes, in *Fundamentals of Cell Immobilization Biotechnology*, Nedovic, V. and Willaert, R., eds., Kluwer Academic Publishers, Amsterdam, pp. 343–356.
25. Yamamoto, S. and Sano, Y. (1994), Drying of carbohydrate and protein solutions, *Drying Technol.* **12**: 1069–1080.
26. Santivarangkna, C., Kulozik, U., and Foerst, P. (2007), Alternative drying processes for the industrial preservation for lactic acid starter cultures, *Biotechnol. Progress* **23**: 302–315.

27. Masters, K. (1990), Understanding and applying spray dryers in chemical processing, *Powder Bulk Eng.* **4**: 36–44.
28. Maa, Y. and Hsu, C. (1997), Feasibility of protein spray coating using a fluid-bed Wurster processor, *Biotechnol. Bioeng.* **53**: 560–566.
29. Sovani, S. D., Chou, E., and Crofts, J. D. (2001), High pressure effervescent atomization: Effect of ambient pressure on spray cone angle, *Fuel* **80**: 427–435.
30. Truong-Le, V. and Scherer, T. (2005), *High Pressure Spray-Dry of Bioactive Materials*, Medimmune Vaccines, Inc., Application WO.
31. Abdul-Fattah, A. M., Truong-Le, V., Yee, L., Nguyen, L., Kalonia, D., Cicerone, M. T., and Pikal, M. J. (2007), Drying-induced variations in physico-chemical properties of amorphous pharmaceuticals and their impact on stability (I): Stability of a monoclonal antibody, *J. Pharm. Sci.* **96**: 1983–2008.
32. Abdul-Fattah, A. M., Truong-Le, V., Yee, L., Pan, E., Aso, Y., Kalonia, D. S., and Pikal, M. J. (2007), Drying-induced variations in physico-chemical properties of amorphous pharmaceuticals and their impact on stability II: Stability of a vaccine, *Pharm. Res.* **24**: 715–727.
33. Buma, T. J. (1971), Free fat in spray dried whole milk. IV. Significance of free fat for other properties of practical importance, *Neth. Milk Dairy J.* **25**: 88–106.
34. Charlesworth, D. H. and Marshal, W. R. J. (1960), Evaporation from drops containing dissolved solids, *Am. Inst. Chem. Eng. J.* **6**: 9–23.
35. Buma, T. J. and Henstra, S. (1971), Particle structure of spray-dried caseinate and spray-dried lactose as observed by a scanning electron microscope, *Neth. Milk Dairy J.* **25**: 278–281.
36. Maa, Y. and Prestrelski, S. J. (2000), Biopharmaceutical powders: Particle formulation and formulation considerations, *Curr. Pharm. Biotechnol.* **1**(3): 283–302.
37. Mumenthaler, M., Hsu, C. C., and Pearlman, R. (1994), Feasibility study on spray-drying protein pharmaceuticals: Recombinant human growth hormone and tissue-type plasminogen activator, *Pharm. Res.* **11**: 12–20.
38. Maa, Y., Nguyen, P. T., and Hsu, S. (2000), Spray-drying of air-liquid interface sensitive recombinant human growth hormone, *J. Pharm. Sci.* **87**: 152–159.
39. Kim, S. S. and Bhowmik, S. R. (1990), Survival of lactic acid bacteria during spray drying of plain yogurt, *J. Food Sci.* **55**: 1008–1011.
40. Prajapati, J. B., Shah, R. K., and Dave, J. M. (1987), Survival of Lactobacillus acidophilus in blended spray-dried acidophilus preparations, *Austral. J. Dairy Technol.* **42**: 17–21.
41. Desmond, C., Ross, R. P., O'Callaghan, E., Fitzgerald, G., and Stanton, C. (2002), Improved survival of Lactobacillus paracasei NFBC 338 in spray-dried powders containing gum acacia, *J. Appl. Microbiol.* **93**: 1003–1011.
42. Maa, Y., Nguyen, P., Sweeney, T., Shire, S. J., and Hsu, C. C. (1999), Protein inhalation powders: Spray drying vs spray freeze drying, *Pharm. Res.* **16**: 249–254.
43. Maa, Y., Ameri, M., Shu, C., Payne, L. G., and Chen, D. (2004), Influenza vaccine powder formulation development: Spray-freeze-drying and stability evaluation, *J. Pharm. Sci.* **93**: 1912–1923.
44. Sonner, C., Maa, Y., and Lee, G. (2002), Spray-freeze-drying for protein powder preparation: Particle characterization and a case study with trypsinogen stability, *J. Pharm. Sci.* **91**: 2122–2139.
45. Schiffter, H. (2007), Spray-freeze-drying in the manufacture of pharmaceuticals, *Eur. Pharm. Rev.* 67–71.

46. Yu, Z., Johnston, K., and Williams III, R. (2006), Spray freezing into liquid versus spray-freeze-drying: Influence of atomization on protein aggregation and biological activity, *Eur. J. Pharm. Sci.* **27**: 9–18.
47. Rambhatla, S., Ramot, R., and Pikal, M. J. (2004), Heat and mass transfer scale-up issues during freeze drying: II. Control and characterization of the degree of supercooling, *AAPS PharmSciTech.* **5** (article 58).
48. Tang, X. and Pikal, M. J. (2004), Design of freeze-drying processes for pharmaceuticals: Practical advice, *Pharm. Res.* **21**: 191–200.
49. Leuenberger, H., Plitzko, M., and Puchkov, M. (2006), Spray freeze drying in a fluidized bed at normal and low pressure, *Drying Technol.* **24**: 711–719.
50. Akers, M., Milton, N., Byrn, S. R., and Nail, S. L. (1995), Glycine crystallization during freezing: the effects of salt form, pH, and ionic strength, *Pharm. Res.* **12**: 1457–1461.
51. Lu, X. and Pikal, M. J. (2004), Freeze-drying of mannitol-trehalose-sodium chloride-based formulations: The impact of annealing on dry layer resistance to mass transfer and cake structure, *Pharm. Devel. Technol.* **9**: 85–95.
52. Webb, S. D., Cleland, J. L., Carpenter, J. F., and Randolph, T. W. (2003), Effects of annealing lyophilized and spray-lyophilized formulations of recombinant human interferon-gamma, *J. Pharm. Sci.* **92**: 715–729.
53. Nguyen, X. C., Herberger, J. D., and Burke, P. A. (2004), Protein powders for encapsulation: A comparison of spray-freeze drying and spray-drying of darbepoetin alfa, *Pharm. Res.* **21**: 507–514.
54. Chang, B. S. and Carpenter, J. F. (1996), Surface-induced denaturation of proteins during freezing and its inhibition by surfactants, *J. Pharm. Sci.* **85**: 1325–1330.
55. Strambini, G. B. and Gabellieri, E. (1996), Proteins in frozen solutions: Evidence of ice-induced partial unfolding, *Biophys. J.* **70**: 971–976.
56. Sarciaux, J. M., Mansour, S., Hageman, M. J., and Nail, S. L. (1999), Effects of buffer composition and processing conditions on aggregation of bovine IgG during freeze-drying, *J. Pharm. Sci.* **88**: 1354–1362.
57. Greiff, D. (1969), Comparative studies of the cryobiology of viruses classified according to their physicochemical characteristics in, *Freezing and Drying of Microorganisms*, T. Nei, ed., University Park Press, Baltimore, pp. 69–80.
58. Shalaev, E. Y., Johnson-Elton, T. D., Chang, L., and Pikal, M. J. (2002), Thermophysical properties of pharmaceutically compatible buffers at sub-zero temperatures: Implications for freeze-drying, *Pharm. Res.* **19**: 195–201.
59. Privalov, P. L., (1990), Cold denaturation of proteins, *Crit. Rev. Biochem. Mol. Biol.* **25**: 281–305.
60. Nail, S. L., Jiang, S., Chongprasert, S., and Knopp, S. A. (2002), Fundamentals of freeze-drying, *Pharm. Biotechnol.* **14**: 281–360.
61. Miller, D. P., Anderson, R. E., and De Pablo, J. J. (1998), Stabilization of lactate dehydrogenase following freeze thawing and vacuum-drying in the presence of trehalose and borate, *Pharm. Res.* **15**: 1215–1221.
62. Mattern, M., Winter, G., Rudolph, R., and Lee, G. (1997), Formulation of proteins in vacuum-dried glasses. I. Improved vacuum-drying of sugars using crystallizing amino acids, *Eur. J. Pharm. Biopharm.* **44**: 177–185.
63. Mattern, M., Winter, G., Kohnert, U., and Lee, G. (1999), Formulation of proteins in vacuum-dried glasses. II. Process and storage stability in sugar-free amino acid systems, *Pharm. Devel. Technol.* **4**: 199–208.

64. Abdul-Fattah, A. M., Lechuga-Ballesteros, D., Kalonia, D., and Pikal, M. (2008), The impact of drying method and formulation on the physical properties and stability of methionyl human growth hormone in the amorphous solid state, *Journal of Pharmaceutical Sciences* (2007). Volume Date 2008, **97**(1): 163–184.
65. Cuia, Z., Xu, S., and Sun, D. (2004), Microwave–vacuum drying kinetics of carrot slices, *J. Food Eng.* **65**: 157–164.
66. Yagi, S. (1999), *Microwave and Far Infrared Drying Under Reduced Pressure*, US Patent 5,859,412.
67. Mongpraneet, S., Abe, T., and Tsurusaki, T. (2002), Far infrared radiation drying of Welsh onion under reduced pressure and convection, *Proc. ASAE Annual Meeting*, The American Society of Agricultural and Biological Engineers, St. Joseph, MI, 2002, paper 026163.
68. Shibata, H. (2006), Vacuum drying of porous solids under supercooling, *Drying Technol.* **24**: 541–550.
69. Nastaj, J. F. (1994), Vacuum contact drying of selected biotechnology products, *Drying Technol.* **12**: 1145—1166.
70. Annear, D. I. (1956), The preservation of bacteria by drying in peptone plugs, *J. Hygiene* **54**: 487–508.
71. Annear, D. I. (1961), Recovery of Strigomonas oncopelti after drying from the liquid state, *Austral. J. Exp. Biol. Med. Sci.* **39**: 295–304.
72. Annear, D. I. (1964), Recoveries of bacteria after drying in glutamate and other substances, *Austral. J. Exp. Biol. Med. Sci.* **42**: 717–722.
73. Bronshtein, V. (2004), Preservation by foam formation, *Pharm. Technol.* **28**: 86–92.
74. Pisal, S., Wawde, G., Salvankar, S., Lade, S., and Kadam, S. (2006), Vacuum foam drying for preservation of LaSota virus: Effect of additives, *AAPS PharmSciTech.* **7** (article 60).
75. Franks, F., Hatley, R., and Mathias, S. (1991), Materials science and the production of shelf-stable biologicals, *BioPharm* **4**: 38–55.
76. Schroeder, A. L. and Schwarz, H. W. (1949), Low-temperature vacuum dehydration, *Chem. Eng. Progress* **45**: 370–376.
77. Litamoi, J., Ayelet, G., and Rweyemamu, M. (2005), Evaluation of the xerovac process for the preparation of heat tolerant contagious bovine pleuropneumonia (CBPP) vaccine, *Vaccine* **23**: 2573–2579.
78. Worrall, E., Litamoi, J., Seck, B., and Ayelet, G. (2001), Xerovac: An ultra rapid method for the dehydration and preservation of live attenuated Rinderpest and Peste des Petits ruminants vaccines, *Vaccine* **19**: 834–839.
79. Ratti, C. and Kudra, T. (2006), Drying of foamed biological materials: Opportunities and challenges, *Drying Technol.* **24**: 1101–1108.
80. Kramer, M., Sennhenn, B., and Lee, G. (2002), Freeze-drying using vacuum-induced surface freezing, *J. Pharm. Sci.* **91**: 433–443.
81. Debenedetti, P. G. and Sang, Y. (1993), *Formation of Biologically Active Protein Powders and Polymeric Micro-Fibers with Supercritical Anti-Solvents*, American Association of Aerosol Research, 1993, pp. 11D3.
82. Jovanovic, N., Bouchard, A., Hofland, G. W., Witkamp, G., Crommelin, D., and Jiskoot, W. (2004), Stabilization of proteins in dry powder formulations using supercritical fluid technology, *Pharm. Res.* **21**: 1955–1969.

83. Sarup, I., Servistas, M., Sloan, R., Hoare, M., and Humphreys, G. (2000), Investigation of supercritical fluid technology to produce dry particulate formulations of antibody fragments, *Trans. Inst. Chem. Eng.* **78**: 101–104.
84. Subramaniam, B., Rajewski, R., and Snavely, K. (1997), Pharmaceutical processing with supercritical carbon dioxide, *J. Pharm. Sci.* **86**: 885–890.
85. Velaga, S. and Carlfors, J. (2005), Supercritical fluids processing of recombinant human growth hormone, *Drug Devel. Industr. Pharm.* **31**: 135–149.
86. Yeo, S., Lim, G., Debendetti, P., and Bernstein, H. (2004), Formation of microparticulate protein powder using a supercritical fluid antisolvent, *Biotechnol. Bioeng.* **41**: 341–346.
87. Fages, J., Lochard, H., Letourneau, J., Sauceau, M., and Rodier, E. (2004), Particle generation for pharmaceutical applications using supercritical fluid technology, *Powder Technol.* **141**: 219–226.
88. Mayo, A. and Kompella, U. (2004), Supercritical fluid technology in pharmaceutical research, in *Encyclopedia of Pharmaceutical Technology*, Swarbrick, J., ed., Marcel Dekker, New York, pp. 1–17.
89. Matson, D. W., Fulton, J. L., Petersen, R. C., and Smith, R. D. (1987), Rapid expansion of supercritical fluid solutions; Solute formation of powders, thin films and fibers, *Industr. Eng. Chem Res.* **26**: 2298–2306.
90. Sellers, S., Clark, G., Sievers, R., and Carpenter, J. F. (2001), Dry powders of stable protein formulations from aqueous solutions prepared using supercritical CO_2-assisted aerosolization, *J. Pharm. Sci.* **90**: 785–797.
91. Vasukumar, K. and Bansal, A. K. (2003), Supercritical fluids technology in pharmaceutical research, *CRIPS* **4**: 8–12.
92. Rantakyla, M. (2004), Particle production by supercritical antisolvent processing techniques, Ph.D. dissertation in *Chemical Technology*, Helsinki Univ. Technology, Espoo, Finland, pp. 146.
93. Marentis, R. (2001), Processing pharmaceuticals with supercritical fluids, *Proc. 4th Brazilian Meeting on Supercritical Fluids EBFS 2001*, p. TN17.
94. Kiranand, E. Brennecke, J. F. (1993), in *Supercritical Fluid Engineering Science: Fundamentals and Applications, Symposium Series 514)*, Erdogan, K. and Brennecke, J. F. eds., *ACS Symp. Ser.* **514**.
95. Tom, J. W. and Debenedetti, P. B. (1991), Particle formation with supercritical fluids—a review, *J. Aerosol. Sci.* **22**: 555–584.
96. Bodmeier, R., Wang, H., Dixon, D. J., Mawson, S., and Johnston, K. P. (1995), Polymeric microspheres prepared by spraying into compressed carbon dioxide, *Pharm. Res.* **12**: 1211–1217.
97. Randolph, T., Randolph, A., Mebes, M., and Yeung, S. (1993), Sub-micrometer-sized biodegradable particles of poly(L-lactic acid) via the gas antisolvent spray precipitation process, *Biotechnol. Progress* **9**: 429–435.
98. Maa, Y., Ameri, M., Rigney, R., Payne, L. G., and Chen, D. (2004), Spray-coating for biopharmaceutical powder formulations: Beyond the conventional scale and its application, *Pharm. Res.* **21**: 515–523.
99. Stadhouders, J., Janson, L. A., and Hup, G. (1969), Preservation of starters and mass production of starter bacteria, *Neth. Milk Dairy J.* **23**: 182–199.
100. Wolff, E., Delisle, B., Corrieu, G., and Gibert, H. (1990), Freeze-drying of Streptococcus thermophilus: A comparison between the vacuum and the atmospheric method, *Cryobiology* **27**: 569–575.

29

SPRAY DRYING OF BIOPHARMACEUTICALS AND VACCINES

Jim Searles and Govindan Mohan

29.1. PROCESS DESCRIPTION

Spray drying is a continuous process that converts a solution or suspension into a powder by drying an aerosol spray of the solution with heated drying gas [1–3]. As the aerosol plume is sprayed into a drying chamber along with the drying gas, the mixture is allowed to flow out of the system through particle-capturing cyclones and/or filters. A wide range of solvents can be used, flows of the spray and drying gas can be in various directions and configurations, powder collection can be by any number of means, and the drying gas can be recycled or passed once-through. Spray drying is used extensively for food products and bulk chemicals, and is increasingly being used for small-molecule drugs, biopharmaceuticals, and vaccines.

A common scenario for biopharmaceuticals and vaccines is an aqueous solution feed (sometimes with ethanol as a cosolvent), top–down concurrent, coaxial flow of both the spray and the nitrogen drying gas, cyclone collection followed by a filter for fines removal, and recycling of drying gas using condensers to remove the water vapor. A diagram of such a setup is shown in Figure 29.1. Unless noted otherwise,

Formulation and Process Development Strategies for Manufacturing Biopharmaceuticals, edited by Jameel and Hershenson

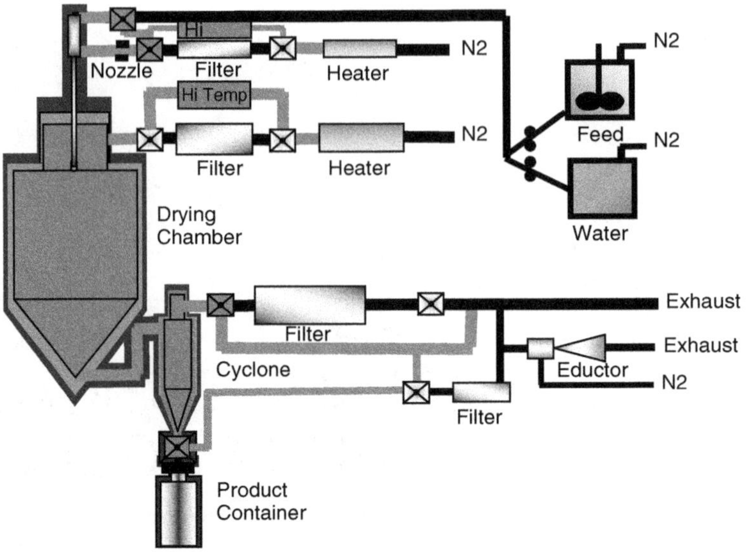

Figure 29.1. An open-mode spray-drying system with an in-place dry-heat sterilization arrangement. (Courtesy of Anhydro A/S, Denmark.)

in the remainder of this chapter all of the systems discussed use water as the solvent (e.g., aqueous solutions or suspensions).

To understand the process from the perspective of the product, one must envision the journey of a single sprayed droplet as it dries, becomes a particle, and is removed from the gas flow. Each particle making up the collected powder started out as one or more droplets emanating from the spray nozzle. Spray drying, like all drying processes, involves coupled heat and mass transfer. Heat to support vaporization must make its way from the bulk gas to the surface of the drying droplet, and vaporized solvent flows from the droplet surface to the bulk gas. During the later stages of drying a layer of dried material on the particle can impede these flows, adding resistance to both heat and mass transfer. Because these droplets or particles are usually very small, they travel *along with* the gas surrounding them, meaning that as the droplets dry into particles, the gas surrounding them becomes cooler and wetter.

When the spray and the drying gas are introduced co-currently, as is the case for biopharmaceuticals, droplets in the initial spray, which have just started the drying process, are exposed to the warmest, but driest, drying gas. Because these droplets do not yet have a dry layer on their surface, resistance to mass flow is low, and the temperature of each droplet rapidly equilibrates to the "wet bulb" temperature corresponding to the water vapor content of the gas surrounding it [4] and the droplets will be substantially cooler than the bulk drying gas. It is by this mechanism that spray drying can actually be a very thermally gentle process; one must keep in mind the evaporative cooling effect at the droplet level.

Biopharmaceuticals and vaccines are usually dried to improve their stability during shipping and storage. Lower molecular mobility in the dried state reduces the rate of chemical and physical degradation and aggregation processes. This same principle must be considered as one envisions an individual droplet or particle undergoing the spray-drying process. In a properly designed process the droplet or particle temperature will increase throughout the process, its temperature converging with that of the surrounding gas. As the droplets dry into particles, the temperature that they can safely withstand also increases. As a result of the coupled heat and mass transfer, the surrounding gas is decreasing in temperature and increasing in moisture while the droplet or particle is increasing in temperature and decreasing in moisture. Therefore the highest temperature of the droplet or particle is at the end of the process, where the gas is at its coolest. This is the exit temperature of the drier.

A terminal, or "additional" drying step is sometimes needed to achieve the desired final moisture content. This is because the relative humidity of the drying gas is at its highest at the exit, and the dried particles may require a lower ambient relative humidity to draw down particle moisture to the desired level without having to raise the product temperature to an objectionable level. This can be achieved by subsequent vacuum drying or pumping dry gas through the collected powder bed.

The fate of individual droplets or particles is also instructive to contemplate. They can either flow all the way through to be collected as product in the cyclone and/or filter, escape the system through the cyclone and/or filter, collide and merge with other droplets or particles, or deposit on the walls of the drying chamber. A major concern for someone considering spray-drying a product could be the product mass recovery fraction: How much of their product will deposit on the walls and not be collected as powder? The answer is that in laboratory-scale spray driers it is not uncommon to lose up to 50% of the product to wall deposition, whereas at full production scale (>1 m drying chamber diameter), this decreases to $\sim 5\%$. The composition of the product formulation can also have a pronounced effect on the mass recovery fraction. Materials with low glass transition temperatures will remain liquid-like and hence "stickier" throughout more of the process, and will have lower mass recovery fractions as a result. Sucrose, fructose, lactose, and glucose present in food products are difficult to spray dry and recover, so "drying aids" such as maltodextrins are used as formulation excipients [5].

29.2. SPRAY-DRYING EQUIPMENT

Spray-drier equipment consists of an atomizer nozzle located inside a cylindrical drying chamber, a cyclone particle separator, and sometimes a particulate filter as the core components. The atomizer nozzle is configured in such a way as to produce a fine mist of liquid droplets, which are carried through the drying chamber by a heated drying gas stream. Evaporation of the carrier solvent results in a cloud of solid particles, which are separated from the gas stream by the vortex action inside the cyclone and/or removed by membrane filtration. In systems using a cyclone a particulate filter is subsequently used to remove the smallest particles ("fines").

A spray-drying system can be operated in either an open-mode or closed-cycle design. In the open-mode operation, the process gas is discharged after a single pass through the dryer. This mode of operation is selected when air is used as a process gas and an aqueous feed solution is employed. However, a closed-cycle spray drier with inert gasses such as nitrogen or carbon dioxide provides emission-free drying and eliminates the hazards of explosion associated with the handling of organic solvents in the liquid feed. The solvent is recovered from the gas stream leaving the fines filter by a scrubber–condenser system in the closed-cycle mode, before returning the process gas to the drying chamber.

Various atomization techniques are applied in spray-drying technology. Five types of atomizers are used: rotary, conventional spray, ultrasonic, two-fluid low pressure, or two-fluid high pressure (see Table 29.1). All except the rotary type can be operated in either a concurrent mode or a fountain mode. In the concurrent mode, the nozzle is placed in the ceiling of the drying chamber near the process gas disperser, while in the fountain mode, the nozzle is placed in the cone section of the drying chamber, spraying upward into the hot process gas flowing out of the ceiling gas disperser. All the studies reported in this chapter used the concurrent configuration.

In rotary atomization, the feed solution is centrifugally accelerated to high velocity in the atomizer wheel, which typically spins at 10,000–30,000 revolutions per minute (rpm). The degree of atomization and particle morphology depends on wheel speed, feed rate, liquid properties, and the atomizer wheel design. Particle size is adjusted by changing the peripheral speed. Typically, a rotary atomizer design is selected for applications where the desired mean particle diameter is in the range of 20–50 μm, produced from non-heat-sensitive and shear-sensitive feed solutions.

With a conventional spray nozzle, atomization is the result of the conversion of pressure energy within the liquid feed into kinetic energy. Pressure applied to the liquid within the nozzle forces the liquid out of the orifice, creating the atomization. Particle size is adjusted by changing the feed pressure and nozzle orifice size. Pressure nozzles

TABLE 29.1. Types of Spray Drying Atomizers

Parameter	Rotary	Conventional Spray	Ultrasonic	Two-Fluid, Low Pressure	CAN-BD Two-Fluid, High Pressure, "Effervecent"
Nebulizing fluid	N/A	N/A	N/A	N_2 or air	CO_2 or N_2
Pressure of nebulizing fluid	N/A	N/A	N/A	2–10 atm	80–120 atm
Smallest particle size, μm	20	70	20	2	1
Biopharmaceuticals?	No	No	Yes	Yes	Yes
Vaccines other than live viruses?	No	No	No	Yes	Yes
Live virus vaccines?	No	No	No	No	Yes

will generally deliver a narrower particle size distribution but a larger particle than will other atomizer types. Typically, mean particle diameters in the range of 70–100 μm are achieved with a pressure nozzle. Ultrasonic vibration has also been used with this type of nozzle to achieve droplet diameters down to 20 μm (www.sono-tek.com).

Two-fluid low-pressure nozzle atomization is achieved pneumatically by a high-velocity gas stream, either internally or externally mixing with the liquid feed in the nozzle, thereby impacting atomization. The concurrent mode is selected when drying heat-sensitive biopharmaceutical products. Particle size is controlled by varying the nozzle flow ratio between atomizing gas and feed. Inlet gas pressures are 2–10 atm, and mean particle diameters in the range of 2–20 μm can be produced [6,7]. This nozzle type was used for most of the studies reported in this chapter.

A newer nozzle technology uses the two-fluid nozzle approach, but operates at much higher pressures of 80–120 atm. It has been called by different names. Beginning in 1996, it was termed "carbon dioxide–assisted nebulization with a bubble dryer" (CAN-BD) [8–27]; Truong-Le and co-workers have termed it "high-pressure spray drying" [28–34]; Reverchon uses the term "supercritical-assisted atomization" [35]; and more recent publications by Abdul-Fattah et al. have referred to it as "effervescent atomization" [36,37]. Room temperature is normally used, and both CO_2 and N_2 have yielded good results as the nebulizing gas. Particle sizes down to 1 μm can result. Two groups have used this type of nozzle successfully for live virus vaccines, which are especially thermolabile [26,27,37]. Their results are discussed in detail later in this chapter.

The particle characteristics of size distribution, residual solvent content, and chemical integrity are controlled by several variables, namely, the atomizer nozzle design, ratio of atomizer gas flow rate to liquid flow rate, inlet process gas temperature, process gas flow rate, and cyclone design. An integrated control system is necessary for achieving optimal process parameters to satisfy the required particle characteristics.

Equipment design considerations for *sterile* spray-dried particle manufacturing stipulate additional complexities required for ensuring sterility. In this case, aseptic boundaries such as sterilization-grade high-efficiency particulate air (HEPA) filters need to be introduced in the atomizer and process gas streams, and methods to render the liquid spray solution sterile need to be incorporated into the process. Furthermore, the spray-drier components should be presterilized and depyrogenated in preparation for the aseptic processing to render the product sterile and to conform to allowable endotoxin limits. The sterilization processes may encompass γ irradiation, steam sterilization, and dry-heat sterilization methods depending on the component material compatibility and the utilities available in the manufacturing facility. The sterilized process components are then introduced aseptically into a barrier isolator environment or a biosafety cabinet capable of being decontaminated. In all cases, a comprehensive validation program is instituted in order to comply with cGMP requirements and to assure reliability.

A more recent trend in the industry is moving towards an in-place sterilization capability for the spray drying system due to the complexity and staging of the presterilized components involved in the system assembly. A typical arrangement of a spray dryer system with a sterilization-in-place (SIP) scheme utilizing dry heat is shown in Figure 29.1. The entire system is assembled and then sterilized utilizing heated gas. Thermal stability and sealing integrity of the gaskets used in the component assembly often dictate the choice of the material of construction of the gaskets used in the assembly. A platinum-cured silicone gasket offers acceptable characteristics for high-temperature operations. The temperature limitation of the sterilization-grade HEPA filter element in the heated process gas stream can be overcome using a bypass arrangement as shown. The filter used in the bypass stream contains either a ceramic or stainless-steel filter media that can withstand the higher sterilization temperatures, while not requiring the postuse filter integrity testing as mandated by the FDA guidelines. On the product collection end, a barrier isolator environment is required to aseptically isolate the collection vessel.

The liquid spray solution for the atomizing nozzle should be presterilized using a sterilization-grade (0.22-μm presize) filter. Typically, the feedline is aseptically connected to the spray nozzle. It is highly recommended that the postuse integrity testing of the sterilizing filter be successfully demonstrated before the spray process commences. It is also a common practice to start the spray-drying system with sterile water and achieve thermal equilibration of the system before switching to the feed solution.

One major consideration in spray drying should be to optimize the process conditions so as to achieve the desired particle size distribution and the lowest possible moisture content. The process yield suffers if more finer particles are produced, which, in turn, are not captured by the cyclone separator; on the other hand, moisture-laden particles tend to stick to the walls of the piping and cyclone, significantly reducing the process yield. Therefore the atomizer nozzle design and the cyclone efficiency are of paramount importance. It is obvious that the residence time of the spray droplets in the drying chamber is primarily the controlling factor for achieving lower moisture content in the collected particles. The driving forces are process gas flow rate and the inlet temperature. In processing proteins and vaccines, there are limitations on the process temperatures due to thermal degradation considerations; on the other hand, decreasing the process gas flow rate with the intent of increasing the droplet residence time has a negative impact on the particle separation efficiency. Therefore, carefully designed process development studies are required to optimize the outcome. One way to decouple the complexities of drying from powder collection is to use filters to collect the powders. The filters are backpulsed at a continuous interval throughout the process to minimize flow resistance.

Conceptually the continuous nature of the spray drying process offers economy of scale in comparison to freeze drying; however, scale-up can be more technically complex for spray drying than for freeze drying.

29.3. GENERAL APPLICATIONS

Spray drying converts solutions or suspensions to powders. It can be carried out under full sterility and closed-system processing and configured in such a way as to manufacture particles anywhere in the size range of 1–100 μm mean particle diameter with a narrow size distribution. As a continuous process it is inherently more scalable than batch processes such as freeze drying. Furthermore, technology exists today to reproducibly fill powders in unit doses down to 1 mg aseptically.

Given these attributes, spray drying should be considered as an alternative for liquid products as well as those currently dried using lyophilization (freeze drying). This consideration should not be limited to final drug product. Bulk intermediates, sometimes requiring −70°C storage of hundreds of liters, could also be spray-dried and stored as stable bulk powders instead. If bulk intermediates are dried as powders, there may be no need for further wet processing. The powders, if formulated appropriately for final drug product, could be filled into primary containers as is or even admixed with other powders to achieve the desired final drug product. This would eliminate not only the bulk intermediate freezing, storage, and thawing but also the subsequent lyophilization step. Admixing of powders also reduces the complexity of designing liquid formulations for multiple active products, some of which may have conflicting formulation requirements (pH, excipients, etc.).

The use of spray drying for encapsulation of biopharmaceuticals or vaccines is beyond the scope of this review. The interested reader is referred to a number of recent articles on the subject [38–46]

29.4. SPECIFIC APPLICATIONS

Many papers in the scientific literature include the use of spray drying for biopharmaceuticals or vaccines. This section reviews this literature, highlighting insights gleaned about the advantages and disadvantages of spray drying for these applications. The interested reader is also referred to previous reviews [47–54].

29.4.1. Proteins, Antibodies, and Peptides

There are many examples of spray drying of proteins and peptides. Perhaps the best known is Exubera®, an insulin powder for inhalation developed by Pfizer and Nektar [55,56]. Exubera powder was manufactured using a spray drier with a proprietary design low-pressure two-fluid nozzle, and powder capture is with cyclones. An additional drying step was used to achieve the desired final moisture content of 2%. The formulation consists of 60% (w/w) insulin, nearly 30% sodium citrate dihydrate, and mannitol and glycine. The product is noncrystalline and has a glass transition temperature of 80°C. Exubera was discontinued by Pfizer because of poor sales in late 2007, and Nektar is currently seeking another marketing partner.

Eli Lilly and Alkermes are continuing to develop their inhaled dry powder insulin product, which is produced using the Alkermes AIR® spray-drying process [57–59]. Alkermes uses a synthetically manufactured natural lung surfactant DL-α-dipalmitoylphosphatidylcholine (DPPC) in their formulation. This surfactant has very low water solubility but good solubility in ethanol, which was used as a cosolvent for spray drying. They produce particles with 1–3 μm aerodynamic diameter for pulmonary delivery by inhalation, but the particles have a geometric diameter of 3–15 μm [57]. The very low density of their particles allows for this difference. The advantage of the large geometric size is that the AIR particles require less energy to achieve dispersion and deaggregation from the dry powder inhaler, and this allows the inhaler to be smaller, simpler, and less expensive than those required for powders with smaller geometric particle sizes. Emitted doses of up to 96% were achieved with a simple Spinhaler™ inhaler. Vanbever et al. [57] describe use of a benchtop spray drier with rotary atomizer processing a feed stream of product and excipients dissolved in 87% ethanol and 13% water with only 0.1% solute concentration. Experiments were run with inlet drying gas temperatures of 100°C, 110°C, 150°C, 175°C, or 200°C, and the outlet temperature ranged between 50°C and 130°C depending on the inlet temperature and other operating parameters. Yields varied between 3% and 80% with a trend toward higher yields at lower temperatures. Active ingredients were human albumin, insulin, albuterol, or estradiol. Formulation excipients DPPC and/or lactose were tested. One formulation that yielded good powder characteristics was albumin/DPPC/insulin in ratios of 38/57/5 (w/w/w) and an inlet temperature of 110°C.

Stahl et al. [60] successfully produced pure inhalable insulin powders using a benchtop spray drier. The formulation was simply 5 mg/mL insulin in water. Insulin degradation was avoided as long as the combination of spray rate, drying gas rate, and drying gas inlet temperature were operated at conditions that resulted in the outlet temperature of <120°C. In a 2007 study, full activity retention was obtained at an outlet temperature of 75°C for a range of insulin formulations as well [61].

The first published use of spray drying of a purified protein that we found was from 1989 when Labrude et al. reported on drying of human hemoglobin [62]. Their previous studies on lyophilization of hemoglobin found that sucrose was effective in preventing formation of methemoglobin (MetHb) during that desiccation process. In the spray-drying study a sucrose concentration of 0.25 M was found to be optimal for minimizing MetHb formation. Drying gas inlet temperatures greater than 120°C resulted in extensive degradation, and temperatures above 60°C also caused formation of low levels of MetHb.

Faldt and Bergenstahl [63] spray-dried sodium caseinate, bovine albumin, and glycine with a benchtop spray drier using an inlet drying gas temperature of 180°C, obtaining outlet temperatures of 80°C–90°C. The formulations also included lactose. They measured the surface compositions of the resulting powders using electron spectroscopy for chemical analysis (ESCA). Both proteins could be detected on the surface at liquid bulk concentrations as low as 0.01%. For both sodium caseinate and bovine albumin, the surface composition was enriched relative to what would be expected from the liquid bulk ratio of protein to lactose. This was not observed for powders in which the amino acid glycine was used instead of protein. Measurements of solution surface tensions revealed that the proteins were populating the air–liquid interface but that glycine was not. The authors attributed the enhanced presence of protein on the surface of dry particles solely to their surface activity in aqueous solution. Interested readers are referred to Vehring et al. [64] for an explanation of another hypothesis for surface enrichment—that the lower diffusivity of the higher-molecular-weight proteins relative to lactose and glycine caused or exacerbated surface enrichment during the rapid drying and concomitant shrinking of the droplets.

A study using β-galactosidase in 1994 revealed that residual enzymatic activity and product yield were significantly affected by process variables [65]. Among trehalose, mannitol, sucrose, and arginine hydrochloride, trehalose performed the best at yielding a fully active product stable over 1 year at 40°C. Examination by scanning electron microscope confirmed 1–5-μm-diameter particle diameters. A benchtop spray drier was used, and the best overall results were obtained using an inlet drying gas temperature of 140°C and resulting outlet temperature of 100°C. Activity losses correlated with formation of insoluble protein aggregates. Higher activity losses resulted from higher inlet drying air temperatures.

In 1999 Branchu et al. [66] reported on the effects of spray drying β-galactosidase alone or with either hydroxypropyl-β-cyclodextrin (HP-β-CD) and/or sucrose. The formulations were spray-dried at an outlet temperature of 61°C. Spray drying alone or with sucrose resulted in extensive inactivation of β-galactosidase. However, spray drying with HP-β-CD (with or without sucrose) in the formulation yielded fully active protein. No protein aggregation was detected. The authors attribute the damage to the air–water interface and suggest that HP-β-CD reduced the level of protein at the interface.

Tissue-type plasminogen activator (t-PA) was formulated in arginine phosphate with 0.004% polysorbate 80 and successfully dried using a benchtop spray drier with a two-fluid low-pressure nozzle [67]. The nozzle was operated a 5 mL/min liquid formulation and 600 L/h atomizing N_2 gas. Particle sizes were 3–8 μm. The protein was not damaged as long as the drying gas inlet temperature was maintained at 120°C or lower. Use of higher temperatures resulted in the formation of insoluble aggregates. The same paper reported that for recombinant human growth hormone (rhGH), spray drying caused extensive aggregation [67]. Testing atomized, but not dried, sample demonstrated that protein damage was caused by surface denaturation at the air–liquid interface of atomized droplets. Most of the aggregation could be avoided by adding 0.1% surfactant polysorbate 20. The mechanism at work here, and in examples cited later in this chapter, is that the protein structure is perturbed and/or protein molecules come into close contact with each other at the air–liquid interface, facilitating protein–protein aggregation. The addition of a surfactant reduces the concentration of protein at the interface, thereby maintaining more of it in the bulk liquid, where aggregation is less likely.

Maa et al. [68] followed up with more work on rhGH, further exploring the detrimental effect of protein degradation at the air–liquid interface. A benchtop spray drier was used with inlet drying gas temperature of 90°C and outlet temperature of 53°C, and a low-pressure two-fluid nozzle operated at 900 L/h atomizing N_2 gas and liquid formulation flow rate of 5 mLmin. They confirmed that rhGH insoluble aggregate formation could be prevented by using polysorbate 20 surfactant with no presence of sugar protectant. Addition of zinc suppressed the formation of soluble protein aggregates. The combination of the two yielded a spray-dried rhGH powder manifesting insignificant protein degradation. The authors speculated that the formation of soluble aggregates was eliminated by zinc ion formation of rhGH dimer complexes, which are resistant to such aggregation. Operating the spray nozzle at different conditions and using laser diffraction droplet sizes of the spray, they found that the percentage of aggregated protein in the sprayed liquid was inversely proportional to the mean droplet diameter. This confirmed spraying damage (without drying) related to the air–liquid interface, because smaller drops have a higher surface-to-volume ratio than do larger ones.

Recombinant human growth hormone with an added terminal NH_2 methionine is known as Met-rhGH. Abdul-Fattah et al. [69] used Met-rhGH in a comparative study of spray drying with a low-pressure two-fluid nozzle, lyophilization with and without an annealing step, and film drying. The latter involves very slow nonfreezing dehydration of a thin liquid layer on a surface. They also tested a number of different formulation excipients and surfactant at different levels. Spray drying was at benchtop scale with inlet drying gas temperature and outlet temperatures of 94°C and 58°C, respectively, and an atomization pressure of 40 psig (lb/in.^{2} gauge). Particle sizes of spray-dried powder were on the order of 2 μm. In results very similar to those from the studies above with rhGH, spray drying without surfactant produced high levels of protein aggregates, levels that were higher than those from the other drying methods. The inclusion of polysorbate 20 surfactant protected Met-rhGH against aggregation.

The first protein product approved for pulmonary delivery, recombinant human DNase (rhDNase), was tested with spray drying in an effort to develop a dry powder version (the original version was delivered as a nebulized liquid solution) [70]. A range of protein : NaCl formulation ratios were dried in a benchtop spray drier using an inlet temperature of 90°C and outlet temperature of 55°C. These were the only conditions reported, and inhalable, active particles were successfully produced.

In a subsequent paper members of the same research group state that spray drying rhDNase was impractical for large-scale production because it requires an additional drying step to achieve sufficiently low moisture content (3%) for adequate storage stability [71]. They tested a formulation consisting of rhDNase alone. Higher spray-drying temperatures, which would achieve the desired moisture, were found to denature the product. The authors also reported on the spray drying of a humanized anti-IgE monoclonal antibody in formulations with a range of mannitol contents [71]. They had similar findings of high moisture content, and also avoided antibody degradation through the process. They tested the anti-IgE antibody : mannitol 80 : 20 formulation ratio on storage stability for one year, and found that antibody protein aggregation could be avoided as long as the powder was retained in an environment of not more than 38% relative humidity at 2°C–8°C. Storage at 30°C or higher induced continuous increases in the amount of aggregated antibody over a year.

Maa et al. [72,73] present reports on performance of a benchtop spray drier for protein aerosols. They spray-dried anti-IgE monoclonal antibody in various formulations containing lactose or mannitol. The primary focus of the study was to optimize product recovery by modifications of operating conditions and the glass cyclone dimensions and interior surface coating for a benchtop system. Spray drying 80 : 20 anti-IgE Ab : mannitol at a solid concentration of 12.5 mg/mL over a dozen experiments with different conditions and cyclones yielded recoveries ranging from 73% to 78%, which is only a small improvement. Use of a cyclone with an electrically conductive inner coating did not affect recovery of Ab : mannitol formulations but did have a slight effect for rhDNase : mannitol formulations. Recovery of particles $< 2\ \mu m$ aerodynamic diameter was very poor, and the authors state that this as a fundamental limitation of using a cyclone. Along these lines, Sievers et al. have had success using 0.45-μm-pore-size membranes for powder recovery [26]. Cellulose acetate and nylon filters performed well, but PTFE membranes were found to impart a static electric charge to the powders (data not shown).

Costantino et al. [74] studied the same recombinant humanized anti-IgE monoclonal antibody, examining the effect of mannitol crystallization on powder aerodynamic diameter and formation of soluble antibody aggregates. Spray drying was at an inlet drying gas temperature of 105°C and outlet temperature of 55°C. Mannitol concentrations above 20% (w/w, dry basis) did not benefit protein stability. Applications of 10% and 20% mannitol powders were stable for particle size [fine-particle fraction (FPF)] during storage at 5°C and 30°C, but at 30% mannitol, crystallization occurred during stability testing, causing dramatic reductions of FPF and increases in protein aggregation rate.

In more work on recombinant humanized anti-IgE monoclonal antibody, Andya et al. [75] reported difficulties in meeting the competing needs of achieving powder particle size and dispersability with those of providing protein stabilization. As a reducing sugar, lactose was confirmed to cause glycation on storage. Mannitol had a positive effect on powder physical properties and protein stabilization, but above a molar ratio of 200 : 1 (excipient : protein) its crystallization during storage negated any benefit. Trehalose stabilized against aggregation, but powder physical properties were poor. Spray-drying conditions were the same as those reported in the previous paragraph.

Adler et al. published two papers [76,77] dealing with denaturation of proteins lactate dehydrogenase (LDH) and bovine serum albumin (BSA) during spray drying. In both cases the addition of sufficient quantities of the surfactant polysorbate 80 eliminated protein degradation during spray drying, but did not prevent degradation of LDH during storage [76].

Prinn et al. [78] used BSA as a model protein in a statistical study of the small-scale spray-drying effect of operating parameters on product quality results. Protein aggregation was measured using size exclusion chromatography (SEC). No straightforward relationship between outlet temperature (or any other parameters) and protein aggregation was found, except that high aggregate levels were not seen at lower temperatures. For example, at outlet temperatures of $\leq 60°C$, aggregate levels were always below 5%.

The 1–34 fragment of human parathyroid hormone was successfully spray-dried by Codrons et al. [79,80] for delivery by inhalation. The formulation included albumin, endogenous lung surfactant dipalmitoylphosphatidylcholine (DPPC), and trehalose or lactose. Spray drying was conducted on a benchtop system with an inlet drying gas temperature of 100°C and outlet temperature of 62°C. Rodent testing achieved good systemic delivery through pulmonary administration for formulations without albumin, which was found to absorb the peptide. Lactose yielded good results and no evidence of reducing; however, no stability testing was done.

Cetrorelix, a 10–amino acid peptide, was dissolved in 94% acetic acid and spray-dried, and the results compared to those for pearl milling [81]. Degradation products at a level of 1.3% were observed for 60°C outlet temperature and 0.7% at 50°C. There was no such degradation during pearl milling. Furthermore, the authors found that the pearl-milled powder dispersed well from lactose carrier particles while the spray-dried product did not.

In another study α-amylase was evaluated for residual enzymatic activity after spray drying [82]. At outlet temperatures up to 75°C small losses were seen, but activity was lost with higher temperatures, decreasing to 55% while remaining at 105°C.

Human polyclonal IgG was successfully spray-dried in histidine buffer, pH 6.0, at a concentration of 5 mg/mL with sucrose, trehalose, and 0.002% Tween 20 [83]. The benchtop spray drier was operated at only 70°C inlet drying gas temperature, and the outlet temperature was 40°C. Reconstitution with small water volumes to achieve concentrated reconstitution was successful up to a IgG concentration of 200 mg/mL without loss of monomer.

Jalalipour et al. [84] spray-dried recombinant human growth hormone (rhGH) in the development of an inhalable powder. Spray-drying inlet drying gas and outlet temperatures were 90°C and 56°C, respectively. Eight formulations were tested, each containing 2 mg/mL of protein in a ammonium bicarbonate buffer (pH 7.4) and various amounts of polysorbate 20 surfactant, zinc chloride, and lactose. The surfactant was effective at suppressing protein aggregation but had a detrimental effect on the fine-particle fraction (FPF). An intermediate level of polysorbate 20 (0.01%) was chosen as an optimum level for both FPF and elimination of aggregation. Zinc complexed with the protein and may have aided in its resisting aggregation as well.

Forbes et al. [85] studied thermal denaturation of trypsin, including that occurring during spray drying. Inlet drying gas and outlet temperatures for spray drying were 125°C and 65°C, respectively. Spray drying was compared to protein crystallization, but no stabilizing excipients were used. It is not surprising that pure trypsin retained its activity through the crystallization process but lost some activity during spray drying.

Spray drying of human antibody IgG1 with various levels of mannitol showed that high amounts of mannitol (50%–80%) exerted a destabilizing effect on the antibody and the aggregate level increased from 2.6% to 4.2%. In contrast, solutions with only 20%–40% mannitol showed the same amount of aggregates as did the pure antibody solution [86]. Using an extensive suite of analytical techniques, the authors showed that protein structure was retained, but did not use X-ray diffraction to determine whether the mannitol had crystallized, as had Costantino et al. earlier [74].

Antibody IgG1 was studied with high-pressure (CAN-BD) spray drying, freeze drying (with and without annealing during freezing), and foam drying with varying levels of sucrose and with or without surfactant [36]. The CAN-BD process was operated with nitrogen as the nebulizing fluid at 1800 psig and room temperature, at which conditions the nitrogen is a supercritical fluid. The drying gas inlet temperature was 87°C, and outlet temperature ranged from 50°C to 60°C. Additional drying after spray drying was conducted at 60°C for 30 min with dry nitrogen. All compositions and drying methods performed similarly with respect to protein aggregation across processing, with postprocessing aggregation levels ranging from 1.6% to 2.6%. Formulation and drying method both influenced the physical stability of IgG1. Stability of IgG1 improved as sucrose fraction increased, regardless of the drying method. The impact of drying method on stability varied depending on composition. In protein-rich formulations, freeze-dried powders showed the poorest storage stability, and the stability differences were correlated to differences in protein secondary structure. In sucrose-rich formulations without surfactant, CAN-BD spray-dried powders had the highest aggregation rate; however, inclusion of surfactant dramatically reduced the aggregation rate on storage. Foam drying performed well, but it would be expected that foam-dried materials would require long reconstitution times, and that it may be difficult to determine whether and when reconstitution was complete.

Hormones Met-rhGH and salmon calcitonin (sCT) were studied with trileucine, a novel excipient for aerosol powder formulations [87]. Spray drying was with a bench-top unit operated with an inlet drying gas temperature of 125°C and outlet temperature of 70°C. The study also included spray drying of netilmicin, gentamicin, albuterol, and cromolyn (individually). For sCT at nearly 80% degradation, 70% monomer loss was observed when a spray-dried powder of 100% sCT was stored at 25°C and 75% RH for 7 days. However, the initial sCT concentration and monomer content were nearly unchanged for a spray-dried powder of sCT formulated with 60% (w/w) trileucine. Mannitol, regardless of its concentration, failed to protect spray-dried sCT from elevated humidity. When formulated with trileucine, glycine, and mannitol, Met-hGH showed increased stability during the drying step, where 97% of the hGH was intact. This is a dramatic improvement over the same formulation without trileucine. The authors attribute this to the surface-active trileucine protecting Met-hGH from damage at the gas–liquid interface. Trileucine also improved storage stability. Trileucine has a high glass transition temperature, making it a good glass former and drying aid, and also improved the aerosol performance of everything tested in the study.

The CAN-BD high-pressure spraying method was also used for anti-CD4 antibody, α_1-antitrypsin and trypsinogen [88]. Carbon dioxide was used as the nebulizing fluid at 1200 psig and room temperature, at which conditions it is a liquid (below the critical point). Powders were stored in desiccated containers before testing. Anti-CD4 antibody was formulated in a proprietary buffered solution containing saccharide and surfactant. It was spray-dried at an outlet temperature of 50°C without any product degradation or reduction of ELISA binding activity. The particles were of respirable size with 2% (w/w) moisture. α_1-Antitrypsin was spray-dried with an outlet temperature of 40°C at a concentration of 4% (w/w) with 4% trehalose with 0.01% (w/w) Tween 20 in 0.1 M sodium phosphate, at pH 7.0. The resulting powder had 1.8% moisture, and the particles were of respirable size. The processed powder achieved 106% of its original enzymatic activity after reconstitution in water. The powders were free of protein aggregation as measured by size exclusion chromatography. Respirable sized particles were obtained. In the absence of a stabilizing sugar, the activity of the enzyme was only 48% of its original activity when redissolved in water after CAN-BD. The addition of either sucrose or trehalose in fractions constituting 80%–89% (w/w dry basis) was able to improve the retained activity of the trypsinogen processed by CAN-BD to nearly 100%. No protein aggregation was found, and the activity of the particles remained virtually unchanged after storage in a desiccated room temperature vacuum chamber for 6 months.

29.4.2. Vaccines

Spray drying has been used with vaccines as a way to switch from a liquid product to a dry product, as an alternative to lyophilization for injected vaccines, or as a method to manufacture powders for nasal or pulmonary delivery. Addressing the latter, direct immunization of the respiratory mucosa has the attraction that it offers needle-free delivery, that direct immunization of the mucosa may improve mucosal immunity, and that stronger mucosal immunity may, in turn, improve protection for pathogens that enter through the respiratory tract [89]. In fact, the respiratory system contains almost 80% of all immune system cells, and mucosal immunization can generate systemic immunity. Starting in 1952 live Newcastle disease vaccine was administered to chickens by inhalation. Other pulmonary immunization trials in animals have included fowlpox, infectious bronchitis, hog cholera, erysipelas, pseudorabies, gastroenteritis, pasteurellosis, and mycoplasmosis. Human vaccination by inhalation has been tested or practiced since 1957 in various countries for diseases such as measles, plague, tularemia, brucellosis, anthrax, and bacillus Calmette–Guérin (BCG) [89].

Live virus vaccines have a long history of creating longlasting immunity with broad protection; however, these vaccines are the most difficult type to manufacture. It is not unusual to lose 90% of vaccine potency between the initial cell culture production and product release. A key challenge has been the drying and stabilization of these vaccines. An example of this is Varivax® by Merck, which is a lyophilized varicella virus vaccine that, even though it has been freeze-dried, still must be maintained frozen during shipping and storage. It is not uncommon for live virus vaccines to lose half of their potency during lyophilization. However, lyophilized products are not suitable for pulmonary delivery unless the product is reconstituted and administered as a nebulized liquid, but reconstitution and nebulization can cause further potency losses, and nebulization requires electrical power. Conversion of lyophilized material to a powder with a particle size small enough for inhalation would require milling, which would add additional cost and incur potency losses due to fracturing of live virions (which can be on the order of inhalable particles in size) and frictional heating. Spray drying offers the convenience and simplicity of a single unit operation to convert a liquid formulation directly to a sterile, potent, inhalable dry powder. It is now possible to purchase filling machines for such powders that can reproducibly dispense single doses as small as 1 mg into foil blister packs, which prevent moisture ingress. The summaries presented below take examples of specific vaccines of several types from the most recent literature.

Alum-adjuvanted vaccines are of interest because international health authorities have identified that during cold chain distribution it is sometimes accidentally frozen, which causes coagulation of the alum-adjuvant particles and complete loss of vaccine potency [90]. Dry formulations would not be damaged by accidental freezing and may not even require refrigeration for at least part of their distribution. In a process comparison published in 2003 by Maa et al. [91], aluminum hydroxide-adjuvant hepatitis B surface antigen (Alum-HBsAg) and aluminum phosphate-adjuvant diphtheria and tetanus toxoids (Alum-DT) were dried by freeze drying, spray drying, air drying, or spray–freeze drying. Only spray–freeze drying yielded fully active vaccine. The ultrafast freezing of the droplets as they were sprayed into liquid nitrogen allowed insufficient time for the aluminum adjuvant particles to aggregate. With the other methods, including the conventional spray drying that they employed, the alum particles in the vaccines coagulated. In 2007 Sievers et al. [26] reported successful use of CAN-BD spray drying to create a dry powder vaccine from a commercial liquid Alum-HBsAg vaccine. Full preservation of ELISA activity was achieved for formulations containing sufficient amounts of stabilizing trehalose. The powders were stored for 43 days at either −20°C or +66°C without loss of potency, and testing in mice showed full retention of immunogenicity. The mechanisms of protection in this case may be steric hindrance of alum particle aggregation by sufficient trehalose, cooling imparted to the droplets from vaporizing CO_2 liquid during CAN-BD spraying; and/or the overall lower operating temperature of the CAN-BD system.

A 2003 patent application by Truong-Le discloses the use of "high-pressure gas spraying and/or near supercritical spraying" followed by drying [28], which seems to be virtually the same as the CAN-BD process described in a patent application filed 4 years earlier and issued as a patent in 2003 [14]. The Truong-Le application discloses successful spray drying of a live influenza vaccine B/Harbin in formulation AVO47A. The stability data shown indicate the loss of only 0.1 $\log_{10}$ of virus potency per month when stored at 25°C. The formulation consisted of 5% sucrose, 2% trehalose, 2% arginine, 0.2% pluronic F68, 50 mM KPO_4 buffer pH 7.2, 10 mM methionine, and 2 mM EDTA.

In 2004 Maa et al. described using a spray drier to coat 40–60-μm-diameter lactose particles with either diphtheria toxoid (dT) or alum-adjuvant hepatitis B surface antigen (Alum-HBsAg). The formulations also included trehalose, mannitol, dextran, and Tween 80. The particles were tested by epidermal powder immunization of mice and guinea pigs, and significant antibody responses were observed.

In 2004 US and international patent applications were published for "high pressure spraying" and drying disclose the drying of live influenza virus in formulation AV047r (5% sucrose, 2% trehalose, 10 mM methionine, 1% arginine, 0.2% 15 pluronic F68, 50 mM $KP0_4$, pH 7.2). [34,35] The spray nozzle was operated at 1300 psi with N_2, and drying gas was introduced at 55°C. Reconstitution of the dry powder showed no significant potency loss. The formulation lost only 1 $\log_{10}$ of potency over 23 days at 37°C storage.

In 2007 Corbaine et al. [92] published a paper on spray drying of an attenuated live virus vaccine with an ultrasonic nozzle, and a number of different formulations were tested. The drying gas inlet temperature was set at 90°C and the outlet temperature at 70°C. Particle aerodynamic diameters were in the vicinity of 30 μm, which is suitable for primary respiratory vaccination of chickens. Three of the formulations achieved full retention of activity through processing (trehalose, trehalose-BSA, and trehalose-PVP-BSA). The trehalose–BSA formulation was the most stable over 10 months at 25°C, with an average degradation rate of 0.2 $\log_{10}$ per month.

Abdul-Fattah et al. [37] reported using "effervescent atomization" (equivalent to one form of CAN-BD and high-pressure two-fluid nozzle spraying) to spray dry live attenuated parainfluenza virus vaccine in a sucrose-based formulation with and without pluronic F68 surfactant. This spray-drying method was compared with freeze drying (with and without annealing during freezing) and foam drying. Stability studies conducted at 25°C and 37°C showed that the vaccine was most stable in the foam-dried preparation with surfactant and least stable in spray-dried preparations without surfactant and in all freeze-dried preparations regardless of the presence of surfactant. The CAN-BD spray-drying method using surfactant performed best for retention of activity throughout the process; however, foam drying achieved greater stability over time. Surfactant improved thermal stability for all manufacturing types. The authors concluded that the presence of ice was a major destabilizing influence. They also introduced the concept of enthalpy recovery, and showed that better storage stability at both 25°C and 37°C corresponded with greater enthalpy recovery as measured during differential scanning calorimetry.

A vaccine strain of *Mycobacterium bovis* BCG and model organism *Mycobacterium smegmatis* were successfully spray-dried by removing salts, adding leucine and omitting glycerol from the bacterial suspension [93]. A benchtop spray drier was operated with an inlet drying gas temperature of 100°C–125°C and outlet temperature of 40°C. Retention of bacterial viability through the drying process was 48%, which is a 100-fold improvement over yields for the current freeze-drying process for commercial BCG vaccine against tuberculosis. Use of this new formulation for freeze drying increased yields 10-fold over those using the commercial formulation. The mass median aerodynamic diameter of particles containing viable mycobacteria was in the range of 3–5 μm, making the powder suitable for inhalation delivery. Storage stability at 25°C for 4 months was improved relative to that of the commercial product.

Sievers et al. [26,27] reported in 2007 and 2008 on stabilization and drying of live attenuated measles vaccine to be administered as a dry powder aerosol to populations in developing world countries. Starting with a reconstituted commercial lyophilized product and supplementing it with 4.9% trehalose, they showed 107% ± 23% retention of activity through CAN-BD (high-pressure two-fluid nozzle) using CO_2 at 1200 psig nozzle pressure, and a spray-drier outlet temperature of 50°C [26]. Flow rates of the vaccine formulation and CO_2 into the nozzle were 0.3 and 1.5 mL/min, respectively. Subsequent studies used bulk vaccine fluids as a starting material and introduced the use of myoinositol as a stabilizing sugar that has low hygroscopicity and imparts good aerosolization properties to the powder. Myoinositol was tested over a range of concentrations either alone or with mannitol, sorbitol, leucine, or gelatin. The CAN-BD spray-drying method was carried out with the same settings as the previous study. Retention of viral potency across spray drying ranged from 57% to 85%, and retention of activity through the World Health Organization (WHO) 7-day, 37°C stability test ranged from 5% to 21% (the WHO requirement is that at least 10% of potency must remain after this test). The formulation with the overall best performance contained 35 g/L of myoinositol and 15 g/L of sorbitol. It retained 60% through spray drying (equivalent to the freeze-drying process for the commercial vaccine) and 21% through the WHO stability test.

29.5. CONCLUSIONS

There is extensive documentation in the scientific literature of success with spray drying for the manufacture of proteins, peptides, and antibodies. Damage to these products during spray drying is usually either by high temperature or aggregation at the liquid–gas interface. In all cases where high temperature caused damage to the product, the process conditions were easily adjusted to eliminate the damage. The implication of running a spray dryer at lower temperatures is that it negatively impacts its throughput, which for spray driers is the evaporation rate (kg/h). For most cases where product aggregation was observed, the problem could be overcome by the addition of a surfactant such as polysorbate 80 up to 0.1%. Surfactant was not found to be damaging in any of the studies. Other stabilizing components such as trehalose, sucrose, sorbitol, myoinositol, leucine, trileucine, sodium citrate, and mannitol also proved to be beneficial.

The only reports of spray drying of live virus vaccines used CAN-BD (high-pressure two-fluid nozzle, effervescent) spraying. The exact mechanism that makes this method is more successful than other types of spray drying is not yet known. Both N_2 and CO_2 nebulizing fluids work well for this application. The phenomenon that is common to N_2 and CO_2, and which is absent with conventional spray drying, is the evaporative cooling provided by the N_2 or CO_2 as it flashes from liquid to gas. Much of this cooling would be imparted directly to the sprayed droplets. This cooling of the aerosol may protect the live virions from damage during the part of the process where they would be subjected to the hottest drying gas.

Overall, spray drying is a mature technology for spray drying of proteins, peptides, and antibodies at laboratory scale. It has been successfully scaled up, but these developments have not yet been shared in the scientific literature. Spray drying of vaccines is a less mature field. The mechanisms behind, and scalability of, high-pressure two-fluid nozzle spray drying should be examined further.

REFERENCES

1. Masters, K. (2002), *Spray Drying in Practice*, Danish Dairy Board, Aarhus, Denmark.
2. Mujumdar, A. S., ed. (2007), *Handbook of Industrial Drying*, 3rd ed., CRC Press, Boca Raton, FL.
3. APV Crepaco Inc., Lake Mills, WI (2000), Food dryers: Technology and engineering, in *Encyclopedia of Food Science and Technology*, Francis, F. J., ed., Wiley, New York, pp. 542–578.
4. Vehring, R., Tep, V., and Foss, W. (2003), Novel experimental method indicates proteins and peptides are protected from high gas temperatures during spray drying, *AAPS PharmSci* **5**(4): Abstract M1247 (available online at http://www.aapsj.org/abstracts/AM_2003/AAPS2003-002909.pdf).
5. Adhikari, B., Howes, T., Bhandavi, B. R., and Truong, V. (2004), Effect of addition of maltodextrin on drying kinetics and stickiness of sugar and acid-rich foods during convective drying: Experiments and modelling, *J. Food Eng.* **62**: 53–68.

6. Mizoe, T., Beppu, S., Ozeki, T., and Okada, H. (2007), One-step preparation of drug-containing microparticles to enhance the dissolution and absorption of poorly water-soluble drugs using a 4-fluid nozzle spray drier, *J. Control. Release* **120**(3): 205–210.
7. Chen, R., Tagawa, M., Hoshi, N., Ogura, T., Okamoto, H., and Danjo, K. (2004), Improved dissolution of an insoluble drug using a 4-fluid nozzle spray-drying technique, *Chem. Pharm. Bull.* (Tokyo) **52**(9): 1066–1070.
8. Sievers, R. E., Karst, U., Schaefer, J. D., Stoldt, C. R., and Watkins, B. A. (1996), Supercritical co_2-assisted nebulization for the production and administration of drugs, *J. Aerosol Sci.* **27**(Suppl. 1): S497–S498.
9. Sievers, R. E. and Karst, U. (1997), *Methods for Fine Particle Formation*, US Patent 5,639,441.
10. Sievers, R. E., Milewski, P. D., Sellers, S. P., Kusek, K. D., Kleutz, P. G., and Miles, B. A. (1998), Supercritical co_2-assisted methods for the production and pulmonary administration of pharmaceutical aerosols, *J. Aerosol Sci.* **29**(Suppl. 1): S1271–S1272.
11. Sievers, R. E., Karst, U., Milewski, P. D., Sellers, S. P., Miles, B. A., Schaefer, J. D., Stoldt, C. R., and Xu, C. Y. (1999), Formation of aqueous small droplet aerosols assisted by supercritical carbon dioxide, *Aerosol Sci. Technol.* **30**(1): 3–15.
12. Sievers, R. E. and Karst, U. (2000), *Methods and Apparatus for Fine Particle Formation*, US Patent 6,095,134.
13. Sievers, R. E., Milewski, P. D., Sellers, S. P., Miles, B. A., Korte, B. J., Kusek, K. D., Clark, G. S., Mioskowski, B., & Villa, J. A. (2000), Supercritical and near-critical carbon dioxide assisted low-temperature bubble drying, *Industr. Eng. Chem. Res.* **39**(12): 4831–4836.
14. Sievers, R. E., Sellers, S. P., and Carpenter, J. F. (2000), *Supercritical Fluid-Assisted Nebulization and Bubble Drying*, World International Property Organization Patent WO0075281.
15. Villa, J. A., Sievers, R. E., and Huang, E. T. S. (2000), Bubble drying to form fine particles from solutes in aqueous solutions, *Proc. 7th Meeting on Supercritical Fluids: Particle Design—Materials and Natural Products Processing*, Antibes/Juan-Les-Pins, France, Dec. 6–8, 2000, Perrut, M. and Reverchon, E., eds., International Society for Advancement of Supercritical Fluids, pp. 83–88.
16. Sievers, R. E., Sellers, S. P., and Carpenter, J. F. (2001), *Supercritical Fluid-Assisted Nebulization and Bubble Drying*, US Patent 6,630,121.
17. Sievers, R. E. and Karst, U. (2001), *Methods and Apparatus for Fine Particle Formation*, European Patent 1160018.
18. Sievers, R. E., Huang, E. T. S., Villa, J. A., Kawamoto, J. K., Evans, M. M., and Brauer, P. R. (2001), Low-temperature manufacturing of fine pharmaceutical powders with supercritical fluid aerosolization in a bubble dryer, *Pure Appl. Chem.* **73**(8): 1299–1303.
19. Sievers, R. E., Huang, E. T. S., and Villa, J. A. (2001), Generation of fine particles by a bubble dryer, *J. Aerosol Med.* **14**(3): 390.
20. Sellers, S. P., Clark, G. S., Sievers, R. E., and Carpenter, J. F. (2001), Dry powders of stable protein formulations from aqueous solutions prepared using supercritical CO_2-assisted aerosolization, *J. Pharm. Sci.* **90**(6): 785–797.
21. Sievers, R. E. and Karst, U. (2002), *Methods and Apparatus for Fine Particle Formation*, European Patent 0677332B1.
22. Sievers, R. E., Clark, G., Villa, J., Alargov, D., Rinner, L., Cape, S., and Huang, E. (2003), Micronization of inhalable drugs with liquid CO_2 at near ambient conditions, *J. Aerosol Med.* **16**(2): 213.

23. Sievers, R. E., Huang, E. T. S., Villa, J. A., Engling, G., and Brauer, P. R. (2003), Micronization of water-soluble or alcohol-soluble pharmaceuticals and model compounds with a low-temperature bubble dryer, *J. Supercrit. Fluids* **26**(1): 9–16.
24. Sievers, R. E., Sellers, S. P., and Carpenter, J. F. (2004), *Supercritical Fluid-Assisted Nebulization and Bubble Drying*, European Patent US2004067259.
25. Sievers, R. E., Sellers, S. P., and Carpenter, J. F. (2005), *Supercritical Fluid-Assisted Nebulization and Bubble Drying*, Chinese Patent 00811293.2.
26. Sievers, R. E., Quinn, B. P., Cape, S. P., Searles, J. A., Braun, C. S., Bhagwat, P., Rebits, L. G., McAdams, D. H., Burger, J. L., Best, J. A., Lindsay, L., Hernandez, M. T., Kisich, K. O., Iacovangelo, T., Kristensen, D., and Chen, D. (2007), Near-critical fluid micronization of stabilized vaccines, antibiotics, and anti-virals, *J. Supercrit. Fluids* **42**(3): 385–391.
27. Burger, J. L., Cape, S. P., Braun, C. S., McAdams, D. H., Best, J. A., Bhagwat, P., Pathak, P., Rebits, L. G., and Sievers, R. E. (2008), Stabilizing formulations for inhalable powders of live-attenuated measles virus vaccine, *J Aerosol Med.*
28. Truong-Le, V. and Pham, B. V. (2007), *Preservation of Bioactive Materials by Spray Drying*, US Patent Application 2003/0215515 A1 (Nov. 20, 2003); US Patent 7,258,873 (Aug. 21, 2007).
29. Truong-Le, V. and Pham, B. V. (2007), *Preservation of Bioactive Materials by Spray Drying*, US Patent Application 2007/0259334 (Nov. 8, 2007).
30. Truong-Le, V., Pham, B. V., and Carpenter, J. F. (2003), *Spray Freeze Dry of Compositions for Intranasal Administration*, International Patent Application WO 03/086443 A1.
31. Truong-Le, V., Pham, B. V., and Carpenter, J. F. (2003), *Spray Freeze Dry of Compositions for Pulmonary Administration*, International Patent Application WO03/087339A2.
32. Truong-Le, V. and Scherer, T. (2004), *High Pressure Spray-Dry of Bioactive Materials*, International Patent Application WO2004/058156A2.
33. Truong-Le, V. and Scherer, T. (2004), *High Pressure Spray-Dry of Bioactive Materials*, US Patent Application 2004/0185091A1 (Sept. 23, 2004).
34. Truong-Le, V. and Scherer, T. (2004), *High Pressure Spray-Dry of Bioactive Materials*, US Patent Application 2006/002862A1 (Jan. 5, 2006).
35. Reverchon, E. (2002), Supercritical-assisted atomization to produce micro- and/or nanoparticles of controlled size and distribution, *Industr. Eng. Chem. Res.* **41**(10): 2405–2411.
36. Abdul-Fattah, A. M., Truong-Le, V., Yee, L., Nguyen, L., Kalonia, D. S., Cicerone, M. T., and Pikal, M. J. (2007), Drying-induced variations in physico-chemical properties of amorphous pharmaceuticals and their impact on stability I: Stability of a monoclonal antibody, *J. Pharm. Sci.* **24**(4): 715–727.
37. Abdul-Fattah, A. M., Truong-Le, V., Yee, L., Pan, E., Aso, Y., Kalonia, D. S. & Pikal, M. J. (2007), Drying-induced variations in physico-chemical properties of amorphous pharmaceuticals and their impact on stability II: Stability of a vaccine, *Pharm. Res.* **24**(4): 715–727.
38. Dellamary, L., Smith, D. J., Bloom, A., Bot, S., Guo, G. R., Deshmuk, H., Costello, M., and Bot, A. (2004), Rational design of solid aerosols for immunoglobulin delivery by modulation of aerodynamic and release characteristics, *J. Control. Release* **95**(3): 489–500.
39. Burke, P. A., Klumb, L. A., Herberger, J. D., Nguyen, X. C., Harrell, R. A., and Zordich, M. (2004), Poly(lactide-co-glycolide) microsphere formulations of darbepoetin alfa: Spray drying is an alternative to encapsulation by spray-freeze drying, *Pharm. Res.* **21**(3): 500–506.

40. Alpar, H. O., Somavarapu, S., Atuah, K. N., and Bramwell, V. W. (2005), Biodegradable mucoadhesive particulates for nasal and pulmonary antigen and DNA delivery, *Adv. Drug Deliv. Rev.* **57**(3): 411–430.
41. Elversson, J. and Millqvist-Fureby, A. (2005), Aqueous two-phase systems as a formulation concept for spray-dried protein, *Int. J. Pharm.* **294**(1–2): 73–87.
42. Elversson, J. and Millqvist-Fureby, A. (2006), In situ coating—an approach for particle modification and encapsulation of proteins during spray-drying, *Int. J. Pharm.* **323**(1–2): 52–63.
43. Tewa-Tagne, P., Briancon, S., and Fessi, H. (2006), Spray-dried microparticles containing polymeric nanocapsules: Formulation aspects, liquid phase interactions and particles characteristics, *Int. J. Pharm.* **325**(1–2): 63–74.
44. Kwon, M. J., Bae, J. H., Kim, J. J., Na, K., and Lee, E. S. (2007), Long acting porous microparticle for pulmonary protein delivery, *Int. J. Pharm.* **333**(1–2): 5–9.
45. Takashima, Y., Saito, R., Nakajima, A., Oda, M., Kimura, A., Kanazawa, T., and Okada, H. (2007), Spray-drying preparation of microparticles containing cationic PLGA nanospheres as gene carriers for avoiding aggregation of nanospheres, *Int. J. Pharm.* **343**(1–2): 262–269.
46. Benchabane, S., Subirade, M., and Vandenberg, G. W. (2007), Production of bsa-loaded alginate microcapsules: Influence of spray dryer parameters on the microcapsule characteristics and bsa release, *J. Microencapsul.* **24**(6): 565–576.
47. Shoyele, S. A. and Cawthorne, S. (2006), Particle engineering techniques for inhaled biopharmaceuticals, *Adv. Drug Deliv. Rev.* **58**(9–10): 1009–1029.
48. Maa, Y. F. and Prestrelski, S. J. (2000), Biopharmaceutical powders: Particle formation and formulation considerations, *Curr. Pharm. Biotechnol.* **1**(3): 283–302.
49. Ameri, M. and Maa, Y.-F. (2006), Spray drying of biopharmaceuticals: Stability and process considerations, *Drying Technol.* **24**(6): 763–768.
50. Abdul-Fattah, A. M., Kalonia, D. S., and Pikal, M. J. (2007), The challenge of drying method selection for protein pharmaceuticals: Product quality implications, *J. Pharm. Sci.* **96**(8): 1886–1916.
51. Elversson, J. (2005), *Spray-Dried Powders for Inhalation*, Ph.D. thesis, Uppsala Univ.
52. Chow, A. H. L., Tong, H. H. Y., Chattopadhyay, P., and Shekunov, B. Y. (2007), Particle engineering for pulmonary drug delivery, *Pharm. Res.* **24**(3): 411–437.
53. Johnson, K. A. (1997), Preparation of peptide and protein powders for inhalation, *Adv. Drug Deliv. Rev.* **26**(1): 3–15.
54. Edwards, D. A., Ben-Jebria, A., and Langer, R. (1998), Recent advances in pulmonary drug delivery using large, porous inhaled particles, *J. Appl. Physiol.* **85**(2): 379–385.
55. White, S., Bennett, D. B., Cheu, S., Conley, P. W., Guzek, D. B., Gray, S., Howard, J., Malcolmson, R., Parker, J. M., Roberts, P., Sadrzadeh, N., Schumacher, J. D., Seshadri, S., Sluggett, G. W., Stevenson, C. L., and Harper, N. J. (2005), Exubera: Pharmaceutical development of a novel product for pulmonary delivery of insulin, *Diabet. Technol. Ther.* **7**(6): 896–906.
56. Harper, N. (2006), Pharmaceutical development of Exubera(r), paper presented at Massachusetts Biotechnology Council Drug Delivery Symp., May 24, 2006, Cambridge, MA. (available online at http://massbio.org/committees/event.php?eid=1820).

57. Vanbever, R., Mintzes, J. D., Wang, J., Nice, J., Chen, D., Batycky, R., Langer, R., and Edwards, D. A. (1999), Formulation and physical characterization of large porous particles for inhalation, *Pharm. Res.* **16**(11): 1735–1742.

58. Edwards, D. A. and Hrkach, J. S. (2005), *Stable Spray-Dried Formulations*, US Patent 6,956,021.

59. Muchmore, D. B., Silverman, B., De La Pena, A., and Tobian, J. (2007), The air inhaled insulin system: System components and pharmacokinetic/glucodynamic data, *Diabet. Technol. Ther.* **9**(Suppl 1): S41–S47.

60. Stahl, K., Claesson, M., Lilliehorn, P., Linden, H., and Backstrom, K. (2002), The effect of process variables on the degradation and physical properties of spray dried insulin intended for inhalation, *Int. J. Pharm.* **233**(1–2): 227–237.

61. You, Y., Zhao, M., Liu, G., and Tang, X. (2007), Physical characteristics and aerosolization performance of insulin dry powders for inhalation prepared by a spray drying method, *J. Pharm. Pharmacol.* **59**(7): 927–934.

62. Labrude, P., Rasolomanana, M., Vigneron, C., Thirion, C., and Chaillot, B. (1989), Protective effect of sucrose on spray drying of oxyhemoglobin, *J. Pharm. Sci.* **78**(3): 223–229.

63. Faldt, P. and Bergenstahl, B. (1994), The surface composition of spray-dried protein-lactose powders, *Colloids Surf. A Physicochem. Eng. Aspects* **90**(2): 183–190.

64. Vehring, R., Foss, W. R., and Lechuga-Ballesteros, D. (2007), Particle formation in spray drying, *J. Aerosol Sci.* **38**(7): 728–746.

65. Broadhead, J., Rouan, S. K., Hau, I., and Rhodes, C. T. (1994), The effect of process and formulation variables on the properties of spray-dried beta-galactosidase, *J. Pharm. Pharmacol.* **46**(6): 458–467.

66. Branchu, S., Forbes, R. T., York, P., Petren, S., Nyqvist, H., and Camber, O. (1999), Hydroxypropyl-beta-cyclodextrin inhibits spray-drying-induced inactivation of beta-galactosidase, *J. Pharm. Sci.* **88**(9): 905–911.

67. Mumenthaler, M., Hsu, C. C., and Pearlman, R. (1994), Feasibility study on spray-drying protein pharmaceuticals: Recombinant human growth hormone and tissue-type plasminogen activator, *Pharm. Res.* **11**(1): 12–20.

68. Maa, Y. F., Nguyen, P. A., and Hsu, S. W. (1998), Spray-drying of air-liquid interface sensitive recombinant human growth hormone, *J. Pharm. Sci.* **87**(2): 152–159.

69. Abdul-Fattah, A. M., Lechuga-Ballesteros, D., Kalonia, D. S., and Pikal, M. J. (2008), The impact of drying method and formulation on the physical properties and stability of methionyl human growth hormone in the amorphous solid state, *J. Pharm. Sci.* **97**(1): 163–184.

70. Chan, H. K., Clark, A., Gonda, I., Mumenthaler, M., and Hsu, C. (1997), Spray dried powders and powder blends of recombinant human deoxyribonuclease (rhdnase) for aerosol delivery, *Pharm. Res.* **14**(4): 431–437.

71. Maa, Y. F., Nguyen, P. A., Andya, J. D., Dasovich, N., Sweeney, T. D., Shire, S. J., and Hsu, C. C. (1998), Effect of spray drying and subsequent processing conditions on residual moisture content and physical/biochemical stability of protein inhalation powders, *Pharm. Res.* **15**(5): 768–775.

72. Maa, Y. F., Nguyen, P. A., Sit, K., and Hsu, C. C. (1998), Spray-drying performance of a bench-top spray dryer for protein aerosol powder preparation, *Biotechnol. Bioeng.* **60**(3): 301–309.

73. Maa, Y. F., Nguyen, P. A., Sweeney, T., Shire, S. J., and Hsu, C. C. (1999), Protein inhalation powders: Spray drying vs spray freeze drying, *Pharm. Res.* **16**(2): 249–254.

74. Costantino, H. R., Andya, J. D., Nguyen, P. A., Dasovich, N., Sweeney, T. D., Shire, S. J., Hsu, C. C., and Maa, Y. F. (1998), Effect of mannitol crystallization on the stability and aerosol performance of a spray-dried pharmaceutical protein, recombinant humanized anti-ige monoclonal antibody, *J. Pharm. Sci.* **87**(11): 1406–1411.
75. Andya, J. D., Maa, Y. F., Costantino, H. R., Nguyen, P. A., Dasovich, N., Sweeney, T. D., Hsu, C. C., and Shire, S. J. (1999), The effect of formulation excipients on protein stability and aerosol performance of spray-dried powders of a recombinant humanized anti-ige monoclonal antibody, *Pharm. Res.* **16**(3): 350–358.
76. Adler, M. and Lee, G. (1999), Stability and surface activity of lactate dehydrogenase in spray-dried trehalose, *J. Pharm. Sci.* **88**(2): 199–208.
77. Adler, M., Unger, M., and Lee, G. (2000), Surface composition of spray-dried particles of bovine serum albumin/trehalose/surfactant, *Pharm. Res.* **17**(7): 863–870.
78. Prinn, K. B., Costantino, H. R., and Tracy, M. (2002), Statistical modeling of protein spray drying at the lab scale, *AAPS PharmSciTech.* **3**(1): 1–8.
79. Codrons, V., Vanderbist, F., Ucakar, B., Preat, V., and Vanbever, R. (2004), Impact of formulation and methods of pulmonary delivery on absorption of parathyroid hormone (1–34) from rat lungs, *J. Pharm. Sci.* **93**(5): 1241–1252.
80. Codrons, V., Vanderbist, F., Verbeeck, R. K., Arras, M., Lison, D., Preat, V., and Vanbever, R. (2003), Systemic delivery of parathyroid hormone (1–34) using inhalation dry powders in rats, *J. Pharm. Sci.* **92**(5): 938–950.
81. Irngartinger, M., Camuglia, V., Damm, M., Goede, J., and Frijlink, H. W. (2004), Pulmonary delivery of therapeutic peptides via dry powder inhalation: Effects of micronisation and manufacturing, *Eur. J. Pharm. Biopharm.* **58**(1): 7–14.
82. Samborska, K., Witrowa-Rajchert, D., and Gonçalves, A. (2005), Spray-drying of α-amylase—the effect of process variables on the enzyme inactivation, *Drying Technol.* **23**(4): 941–953.
83. Dani, B., Platz, R., and Tzannis, S. T. (2007), High concentration formulation feasibility of human immunoglubulin g for subcutaneous administration, *J. Pharm. Sci.* **96**(6): 1504–1517.
84. Jalalipour, M., Gilani, K., Tajerzadeh, H., Najafabadi, A. R., and Barghi, M. (2007), Characterization and aerodynamic evaluation of spray dried recombinant human growth hormone using protein stabilizing agents, *Int. J. Pharm.* **352**(1-2): 209–216.
85. Forbes, R. T., Barrya, B. W., and Elkordy, A. A. (2007), Preparation and characterisation of spray-dried and crystallised trypsin: Ft-raman study to detect protein denaturation after thermal stress, *Eur. J. Pharm. Sci.* **30**(3–4): 315–323.
86. Schule, S., Friess, W., Bechtold-Peters, K., and Garidel, P. (2007), Conformational analysis of protein secondary structure during spray-drying of antibody/mannitol formulations, *Eur. J. Pharm. Biopharm.* **65**(1): 1–9.
87. Lechuga-Ballesteros, D., Charan, C., Stults, C. L., Stevenson, C. L., Miller, D. P., Vehring, R., Tep, V., and Kuo, M. C. (2008), Trileucine improves aerosol performance and stability of spray-dried powders for inhalation, *J. Pharm. Sci.* **97**(1): 287–302.
88. Cape, S. P., Villa, J. A., Huang, E. T., Yang, T.-H., Carpenter, J. F., and Sievers, R. E. (2008), Preparation of active protein fine powders using supercritical or near-critical carbon dioxide—protein formulation, micronization and delivery, *Pharm. Res.* **25**(9): 1967–1990.
89. Lu, D. and Hickey, A. J. (2007), Pulmonary vaccine delivery, *Future Drugs* **6**(2): 213–226.

90. Edstam, J. S., Dulmaa, N., Tsendjav, O., Dambasuren, B., and Densmaa, B. (2004), Exposure of hepatitis b vaccine to freezing temperatures during transport to rural health centers in Mongolia, *Prevent. Med.* **39**(2): 384–388.
91. Maa, Y. F., Zhao, L., Payne, L. G., and Chen, D. X. (2003), Stabilization of alum-adjuvanted vaccine dry powder formulations: Mechanism and application, *J. Pharm. Sci.* **92**(2): 319–332.
92. Corbanie, E. A., Remon, J. P., Van Reeth, K., Landman, W. J., van Eck, J. H., and Vervaet, C. (2007), Spray drying of an attenuated live newcastle disease vaccine virus intended for respiratory mass vaccination of poultry, *Vaccine* **25**(49): 8306–8317.
93. Wong, Y. L., Sampson, S., Germishuizen, W. A., Goonesekera, S., Caponetti, G., Sadoff, J., Bloom, B. R., and Edwards, D. (2007), Drying a tuberculosis vaccine without freezing, *Proc. Natl. Acad. Sci. USA* **104**(8): 2591–2595.

30

DEVELOPMENT AND OPTIMIZATION OF THE FREEZE-DRYING PROCESSES

Feroz Jameel and Jim Searles

30.1. INTRODUCTION

Lyophilization, which is also termed "freeze drying," is a dehydration process that is used to convert solutions of labile materials into solids to achieve enhanced stability during transportation and long-term storage. It is also being explored for its potential applicability as an alternate method to frozen storage and shipment of intermediates and bulk drug substances, and for diafiltration to concentrate protein solutions, beside its use in producing bulk crystalline material [1]. It comprises of three steps: (1) freezing, where the water of the solution is converted into solid ice; (2) primary drying, where the solid ice is sublimed under vacuum into vapor; and (3) secondary drying, where the water that did not form ice is removed through desorption at slightly elevated temperatures. Freeze drying is the most preferred drying technology among the other available drying methods for the simple reason that it is a low-temperature process and allows processing of biological solutions with less susceptibility to damage. However, development of a liquid dosage is preferred and prioritized over the lyophilized dosage form for cost (the process times are long and the cost of goods manufactured is high) and patient compliance (no reconstitution step is involved) reasons. Additionally, it is not an easy process; it is not uncommon to see lyophilized products that lack

Formulation and Process Development Strategies for Manufacturing Biopharmaceuticals, edited by Jameel and Hershenson

elegance, with high reconstitution time, freeze-drying cycles of several days, and lot-to-lot and vial-to-vial variability/heterogeneity in quality attributes; and most important is recommended storage temperature of 2°C–8°C (i.e., failure to achieve ambient storage stability). These issues arise from the lack of clear understanding of concepts relating to design of formulations and the freeze-drying process. Formulation and process are highly interdependent in lyophilization; what is in the formulation dictates the process design, and the process variations can completely change the physical state of the formulation. A poor formulation can be practically impossible to freeze-dry, and even with a well-designed formulation, a poorly designed process may take several days to produce material of suboptimal quality.

The dogma that the "process makes the product" is true. The process must be designed in such a way that the product looks pharmaceutically elegant without collapse and has low residual moisture content, short reconstitution time, in-process retention of activity, and adequate shelf-life. These quality attributes expectations require that the formulation be dried below the collapse temperature [2]. The commercial manufacturing requires that the process should be short (i.e ., economically viable), operative within the capabilities of the equipment with appropriate safety margins and efficient plant utilization. The above expectations require the design of lyophilization cycles to be such that the collapse temperatures are maximized and drying rates are as high as possible and robust enough to be implementable on typical production freeze dryers [3].

One also should be aware of the manufacturing challenges ahead, challenges relating to differences in environment (e.g., effect of a particle-free environment), differences in load size (scale-related issues), differences in equipment (dryer) design, and time and procedure differences [3].

This chapter discusses the scientific and engineering concepts required to develop a process for the manufacture of freeze-dried biopharmaceuticals to help address the challenges and issues stated above.

30.2. CHARACTERIZATION OF THE FORMULATION

As the physicochemical processes that occur with the excipients during the process of freezing and their influence on the values of critical freeze drying properties are not yet completely understood, we rely mostly on trial-and-error experiments to select formulation and process variables. The values of critical freeze-drying properties of the formulation such as eutectic melting temperature (T_e), the glass transition temperature of the maximum freeze concentrated solution (T_g'), crystallization and annealing temperatures and times, residual unfrozen water content at the T_g' temperature (W_g') and the glass transition temperature of the dry powder (T_g) dictate the selection of the excipients and the freeze-drying process conditions. The onset of crystallization and the annealing temperature dictate the freezing parameters and the protocol, with T_g' and/or T_e values setting the basis for the selection of primary drying conditions (shelf temperature, chamber pressure, and time). The W_g' value also forms the basis for selection of secondary drying conditions (ramping rate proceeding from primary to secondary drying temperature and time); T_g of the cake provides an assessment for the storage stability and forms the basis for the recommended transportation and storage conditions of the product [4]. The values of the abovementioned properties for the formulation depend heavily on the values of these properties for individual components present in the formulation and their mass ratios in the mixture. It is always desirable to have a simple formulation, and it is advisable to question the inclusion of any ingredient in the formulation; however, given the stability challenges associated with the labile protein coupled with end-use requirements, one may end up with a multicomponent salt system. Under those situations it is imperative that the formulator characterize the behavior of these components both individually and in a mixture and determine the values of the properties before locking in the formulation and selecting the process conditions. Most of the excipients that are used in the formulations either as buffers, stabilizers, or bulking agents behave differently because of the variations in their concentrations, temperature (cooling and warming), and presence and absence of other excipients [5].

30.2.1. Collapse Temperature

In order to meet the above mentioned desired product quality attributes requirements and to ensure an efficient process, the product needs to be dried at as high a temperature as possible without exceeding the collapse temperature for that product. The *collapse temperature* is defined as the maximum product temperature that allows drying to occur without the loss of porous "cake-like" structure with the dimensions equivalent to those of the frozen solid [6]. If all the excipients in the formulation form a single amorphous phase postfreezing and do not crystallize, then the T_g' value plus 2°C will be the collapse temperature of the formulation. Two degrees are added to the T_g' value because the system will not undergo viscous flow and loss of structure will not be observed until the product temperature exceeds the T_g' value by a couple of degrees. If all the excipients in the formulation crystallize on freezing, then the eutectic melting temperature (T_e) will be the collapse temperature of that formulation. A *eutectic mixture* is a physical mixture of two or more crystalline compounds that melt together at the same temperature as one compound. In a mixed system where crystalline phase constitutes the major component, the T_e would be the collapse temperature. Carrying out primary drying with the product temperature above T_g' but below T_e will dry the product with the collapse of the amorphous component on the surface of the crystalline phase, and the crystalline phase will render the necessary mechanical support to the cake structure, but the impact on the product stability must evaluated [5]. If the amorphous phase constitutes the major component and crystalline phase the minor component, then, under those situations drying above T_g' but below T_e will be risky and will depend on how far the product temperature is from the T_g' as the crystalline structure will not provide sufficient mechanical support.

Drying above the collapse temperature results in loss of pharmaceutical elegance, usually higher residual moisture content, and higher reconstitution time due to lack of porous structure, and with some proteins it is accompanied by loss of activity. The

TABLE 30.1. Collapse Temperature T_c and Glass Transition Temperature T_g' Data for Selected Excipients[a]

Material	T_g', °C	Reference	T_c, °C	Reference
BSA	−11	34	—	—
Dextran	−10	34,11	−10	37
Ficol	−19	11	−20	38
Gelatin	−9	11	−8	38
PVP (40k)	−20	11	−23	38
Dextrose	−44	11	—	—
Hydroxypropyl β-cyclodextrin	—	—	−18	37
Lactose	−28	34,11	−31	37
Mannitol	−35	34,11	—	—
Raffinose	−27	35	−26	38
Sorbitol	−46	11	−45	38
Sucrose	−32	34,36	−34,	37
Trehalose	−29	34,11	−34	37
β Alanine	−65	34	—	—
Glycine	(−62)	8	—	—
Histidine	−33	34	—	—
Acetate, potassium	−76	34	—	—
Acetate, sodium	−64	34	—	—
$CaCl_2$	−95	34		
Citric acid	−54	34		
Citrate, potassium	−62	34		
Citrate, sodium	−41	34		
HEPES	−63	34		
$NaHCO_3$	−52	34		
Phosphate, KH_2PO_4	−55	34		
Phosphate K_2HPO_4	−65	34		
Phosphate, NaH_2PO_4	−45	34		
Tris base	−51	34		
Tris HCl	−65	34		
Tris acetate	−54	34		
$ZnCl_2$	−88	34		

[a]Collapse temperature data were obtained by freeze-drying microscopy, and T_g' data were obtained using DSC at roughly 10° C/min heating rates and represent midponts of the glass transition region. Values in parentheses were estimated by extrapolation from noncrystallizing mixtures to the pure compound

collapse temperature, which is normally 2°–3° above the T_g' value, depends on on the chemical composition, hence, it is formulation-specific. The collapse temperatures of some excipients that are commonly used pharmaceutically are summarized in the Table 30.1. During selection of the excipients, care must be taken to avoid choosing those excipients that have low collapse temperatures. Some can be as low as −80°C and will have a significant impact on the overall formulation collapse temperature and product resistance (R_p), which, in turn, will impact drying rates and process times.

30.2.2. Characterization Techniques

Various techniques such as freeze-drying microscopy, modulated differential scanning calorimetry (MDSC), and electrical resistance are employed to determine the key freeze-drying properties. The T_g' of a multicomponent system can also be estimated from the T_g' values of the individual components using a Fox equation [7]

$$\frac{1}{\mathrm{T_g}} = \frac{W_1}{T_{g1}} + \frac{W_2}{T_{g2}} \tag{30.1}$$

where W_1 is the weight fraction of component 1 and T_{g1} is the glass transition temperature of pure component 1. Equation (30.1) can be applied to systems containing more than two components. The impact of a second solute component on T_g' of a formulation can be estimated from equation 1 if T_{gi} in Equation (30.1) is identified as T_g' of aqueous component i, and W_i are weight fractions of solute relative to the total mass of solute. For example, the effect of glycine on T_g' of aqueous sucrose systems [8] is fully consistent with Equation (30.1) (see Fig. 30.1). Here the T_g' value of aqueous glycine is obtained by fitting the data to the Fox equation as it is difficult to determine the value of T_g' of glycine as it readily crystallizes on freezing.

30.2.2.1. Freeze-Drying Microscopy. Freeze-drying cryostage, in combination with the light microscopy techniques such as polarized light, serves as an important tool in the screening and selection of excipients and in development and optimization of the freeze-drying process. Using freeze-drying microscopy, one can observe the behavior of the excipients during freezing and drying, and identify the

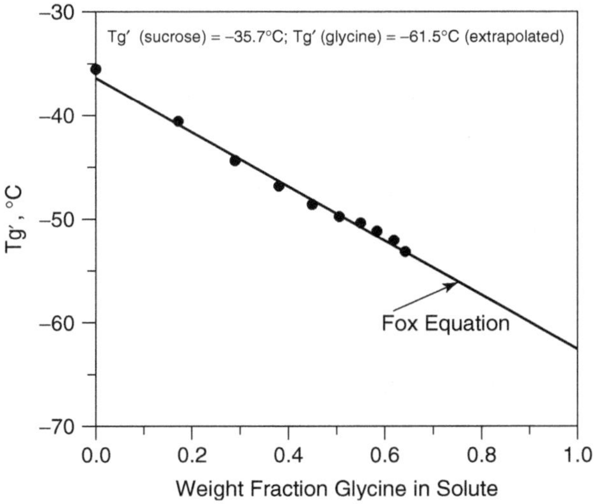

Figure 30.1. Effect of glycine on the T_g' of amorphous sucrose:glycine frozen systems. (Data adapted from Shalaev and Kanev [8].)

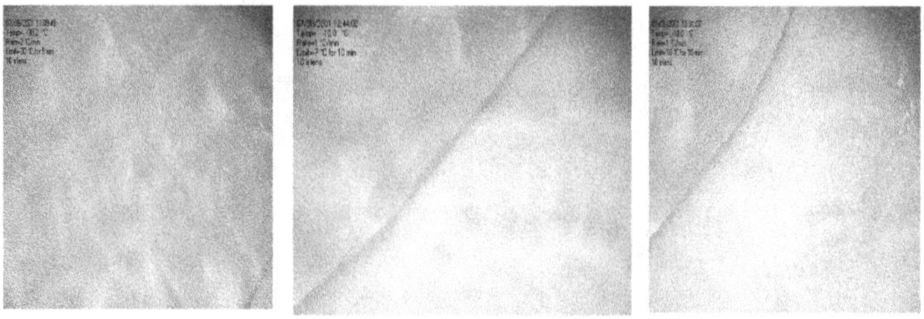

(a) (b) (c)

Figure 30.2. Depiction of phenomenon of ice dendrites, drying front with retention of structure and collapse: (a) dendritic ice formation after freezing and before the vacuum is applied; (b) drying with and (c) without retention of the structure.

temperatures at which various thermal events such as collapse, crystallization on set (annealing temperatures and time), and eutectic melt occur, and examine the morphology of the frozen solid and dried cake [9,10]. One may ask: "Do the events observed under freeze drying microscopy correlate with the events in the vial?" The answer is usually "Yes" in our experience; in most cases we have observed quite a good correlation. It is carried out by placing 4–5 μL of the sample between the two coverslips, cooling the samples to freezing temperatures, −40°C or −50°C, and holding them there for a few minutes. The temperature can be incrementally varied and the interesting thermal events during freezing such as the effect of formulation composition and freezing rates on the morphology of ice (dendritic or spherulitic, etc.), annealing conditions (temperature and time), and eutectic melt for crystallizable excipients can be observed and noted. Similarly, after applying vacuum, one can vary the temperature incrementally to observe the behavior of cake structure with the progress of drying, and can identify the collapse temperature. Figure 30.2 depicts the progress of freeze drying of a system containing 2% sucrose and 4% mannitol along with concentrated protein. Panel (a) shows the dendritic ice formation after freezing and before the vacuum is applied, and panels (b) and (c) depicts the drying with and without retention of structure, respectively. In this case we first observed the loss of amorphous phase on the crystalline matrix of mannitol followed by loss of crystalline phase, which resulted in complete loss of cake structure, that is, collapse of the cake. The temperature at which complete collapse was observed was −9°C.

One advantage of the latest freeze-drying microscope developed by Linkam is the ability to program and control the pressure and temperature to simulate vial freeze drying [10]. The temperature sensors need to be carefully calibrated. Inorganic salts such as NaCl (−21.5°C) and KCl (−11.1°C), whose eutectic melting temperatures are known and well documented, can be used as standards to calibrate the system.

30.2.2.2. Modulated DSC. Differential scanning calorimetry (DSC) has proved to be a very useful tool for characterizing the frozen formulation and acquiring most of the information one needs for the development of a process (e.g., see Table 30.1 for T_g' values found by DSC for various compounds). It measures

TABLE 30.2. Glass Transition Temperatures T_g of Selected Excipients Measured by DSC[a]

Compound	T_g, °C	Reference
Citric acid	11	54
Glycine	(~30)	13
Lactose	114	55
Maltose	100	55
Mannitol	13	21,56
Raffinose	114	55
Sorbitol	−1.6	56
Sucrose	75	55
Trehalose	118	55
Maltodextrin 860	169	55
PVP k90	176	55

[a] Consult the references for details on these techniques. Value in parentheses is extrapolated fom mixtures using Fox equation and is highly approximate.

the difference in heat flow to a sample compared to a reference and detects both (1) first-order irreversible/kinetic thermal events such as crystallization and eutectic melt (exotherms or endotherms) and (2) second-order reversible events such as glass transitions. However, one challenge with DSC is the detection of weak transitions due either to low concentration or low molecular weight of the excipient [11]. Sometimes it is also difficult to differentiate between a crooked baseline and a real transition. Such ambiguities can be overcome by rescanning or using modulated DSC, which separates the reversible components such as glass transitions from total heat flow. The sensitivity of the weak transitions can be improved by (1) increasing the sample size, which enhances the amplitude of the transition; (2) increasing the scanning rates; and (3) increasing the concentration but keeping the mass ratio of the components in the mixture constant. The caveats with these increasing sample size and scanning rates is that they move the transitions to higher temperatures and can also cause merging of the nearby thermal events. The dependence of glass transitions on heating rates needs to be corrected as heating rates employed in freeze drying are much lower and the real values will be lower by a few degrees, especially in cases where drying is carried out at product temperatures close to T_g'. Further resolution and sensitivity in the determination of T_g' can be obtained using the derivative of the power–time curve [12].

30.3. FREEZE-DRYING PROCESS

Once the formulation is completely characterized, its freeze-drying properties are understood, and their values are determined, the next step would be to identify optimal process conditions. Identification of optimal process conditions requires a good understanding of the various phases of freeze drying, the objectives of each phase, the process parameters, and the factors that affect both the independent and dependent variables. Freeze drying is carried out in three phases: freezing, primary drying, and secondary drying.

30.3.1. Freezing

Freezing is considered the most critical phase as the performance of the subsequent phases, and the quality attributes of the final drug product depends on the way in which the solution is cooled. The main objective of the freezing phase is to remove water from the system as much as possible by formation of ice and convert all solutes into solids, either crystalline (eutectic) or amorphous (a glass). As the formulation solution is cooled, two events occur in a fashion quite different from that normally observed with pure water: (1) the degree of undercooling will be less as the solutes will act as nuclei, so the nucleation temperature will be higher than that of pure water; and (2) the subsequent propagation of ice occurs below the equilibrium freezing point as the presence of solutes depresses the freezing point. As the water crystallizes into ice, the solutes concentrate into a *freeze concentrate*, and the continual increase in solute concentration further depresses the freezing point in this phase. If the solutes reach supersaturation, then crystallizable excipients such as mannitol or glycine will crystallize and precipitate. The other solutes will remain amorphous, and when the critical solute-dependent concentration is reached, the unfrozen amorphous freeze concentrate will exhibit restricted mobility and the physical state of the system will change from a viscoelastic liquid to an amorphous solid, called "glass" [10]. The temperature at which this occurs is called the *glass transition temperature of the maximally freeze concentrated system* (T_g'), and the corresponding unfrozen water and amorphous solutes

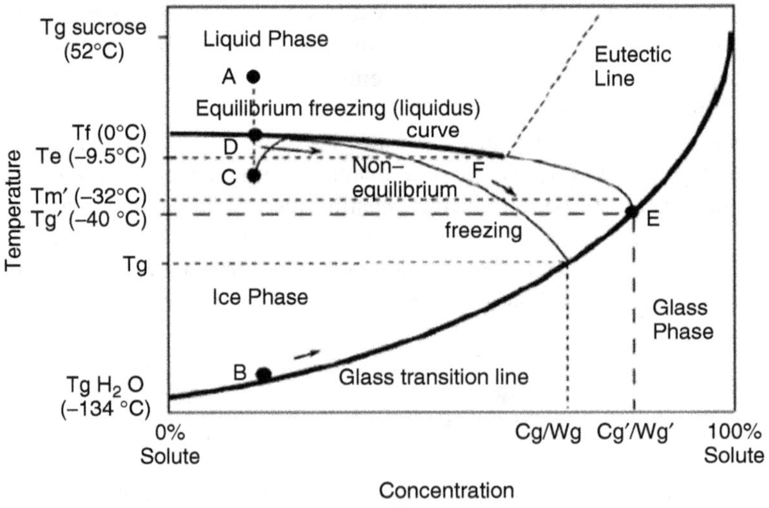

Figure 30.3. Equilibrium phase diagram and kinetically derived state diagram showing the solid–liquid coexistence boundaries and glass transition profile for a binary sucrose–water system.

concentrations are denoted as W_g' and C_g', respectively (see Fig. 30.3). The W_g' is significant during secondary drying as high values require that the product temperature be slowly ramped during secondary drying to ensure that the product temperature is below T_g/moisture curve. There are several methods for determining the W_g' values; they can be determined from the area under the ice-melting endotherm or by taking solutions of different concentrations, determining the amount of frozen water at each concentration, and then extrapolating to the solute/water ratio that has zero frozen water (W_g') [32]. The C_g' values can also be determined by methods proposed by Jens et al. and Bakul et al. [39,40].

30.3.1.1. Freeze Concentration. The time to reach the glassy state must be expedited in order to minimize the residence time that the protein spends in the freeze concentrate stage, where all the potential deleterious cryoconcetration effects can occur. The pH can shift because of selective crystallization of the buffer components, protein–protein interaction can result in colloidal instability, the ionic strength can increase several-fold [if a protein were formulated with 0.15 M NaCl, the NaCl concentration would increase to 6 M by the end of freezing (calculated data)], and an exposure of protein to such high ionic strength could destabilize the protein [13–15].

30.3.1.2. Crystallization of Excipients. Crystallization of an excipient depends on several factors, including their nature, solubility, and initial concentrations, and the presence of other formulation components and the details of the freezing protocol [16]. Depending the intended role of an excipient, crystallization may be desirable or undesirable, and it may be important when the crystallization occurs in the process. In terms of process performance and product quality, it is imperative that the intended role of each excipient be achieved during the freezing phase. If the intended role of an excipient is to function as a stabilizer, then it is important to ensure that it remains amorphous and the protein is molecularly dispersed in it [3,4]. Proteins do not crystallize on freezing but remain amorphous, and if the stabilizer does not remain amorphous and crystallizes out, then it will be a physical mixture and will not impart stability to the protein. If the intended role of an excipient is to function as a crystalline bulking agent and provide a crystalline matrix to the cake, then it is important to ensure that it totally crystallizes during the freezing phase. If it does not complete the crystallization process during the freezing phase, it will crystallize during primary drying and/or during storage. If it crystallizes during the primary drying phase, it will release water, and since the temperature during primary drying would still be below freezing point, the released water will form ice, which could break the vials. The classic example is the vial breakage of mannitol-based formulations [17]. If the partially crystallized bulking agent crystallizes during storage, then the water released on crystallization will depress the T_g of the dry amorphous component where protein is dispersed, as water acts as a plasticizer, and compromise the storage stability of the product [4]. A further, more amorphous component of the partially crystallized bulking agent will depress the T_g' in the frozen state and the T_g of the dried powder substantially; for example, amorphous mannitol has a T_g' of −34°C and T_g of 0°C. Additionally, longer reconstitution times are observed with partially crystallized products. On the other hand, if the intended role of an excipient is to act as a buffer, then it is important to ensure that none of its components crystallize out on freezing [18].

Because of the potential consequences of partially crystallizing the crystallizable excipients in the formulation during the freezing phase, as mentioned above, it is quite critical that optimal conditions be identified from both formulation and process aspects to ensure total crystallization during the freezing phase. With regard to formulation, crystallization can be enhanced by using a high mass ratio of the crystalline bulking agent (3 times the total amorphous solutes).

30.3.1.3. Annealing. From a process perspective, since few of the excipients complete their crystallization in normal freezing, a thermal treatment, called *annealing*, is required to ensure complete crystallization. The temperature and time of maximum crystallization of a particular excipient can be determined using frozen solutions by real-time X-ray diffraction techniques [19] and/or by the DSC and freeze-drying microscopy methods [16]. This can be confirmed or verified using any of several techniques: (1) T_g'–annealing temperature curves, (2) the area under the eutectic melting endotherm in a frozen system, (3) the area under the bulking agent melting endotherm in a dry powder system, or (4) by absence of an exotherm on heating the dry powder on DSC. Time of annealing can be determined by following the onset and offset of the crystallization peak.

The maximum rate of nucleation and the maximum rate of crystallization occur at two different temperatures (see Fig. 30.4). For crystallization to occur, nuclei need to form first. Nucleation typically occurs at very low temperatures when the critical mass is formed and the solution becomes supersaturated. However, at this temperature the growth of nuclei into crystals will be minimal to none because of the high viscosity and low mobility. It requires warmer temperatures to gain mobility and allow molecules to orient to form crystals; the rule of thumb is that you will see mobility when the temperature is 10°C above the glass transition. So, for total crystallization, the solution needs to be cooled below the T_g' then warmed to at least 10°C above the T_g' to start pinpointing the temperature at which maximum rate of crystallization occurs. Depending on the concentration of protein, mannitol-based formulations crystallize anywhere from −26°C to −10°C. For a mannitol-based formulation containing 25 mg/mL protein, we observed maximum crystal growth when annealed between −15°C and −12°C following −40°C cooling and, the T_g'–annealing temperature curve as illustrated in Fig. 30.5 was generated to confirm. Here the T_g' value has been used as an indirect measurement of degree of crystallinity.

It can be seen from Figure 30.5 that there was a good correlation between the T_g' value and the annealing temperature, with the increase in annealing temperature there has been an increase in T_g' value suggesting increase in degree of crystallinity, it plateaus between −15°C and −12°C, and then the values drop. The plateau was indicative of maximum crystallization, and the drop was due to the annealing temperature approaching the onset of eutectic melting temperature where ice melts and depresses T_g' values.

In addition to crystallization of crystallizable solutes, annealing enables secondary crystallization of ice and Ostwald ripening, where larger ice crystals grow at the

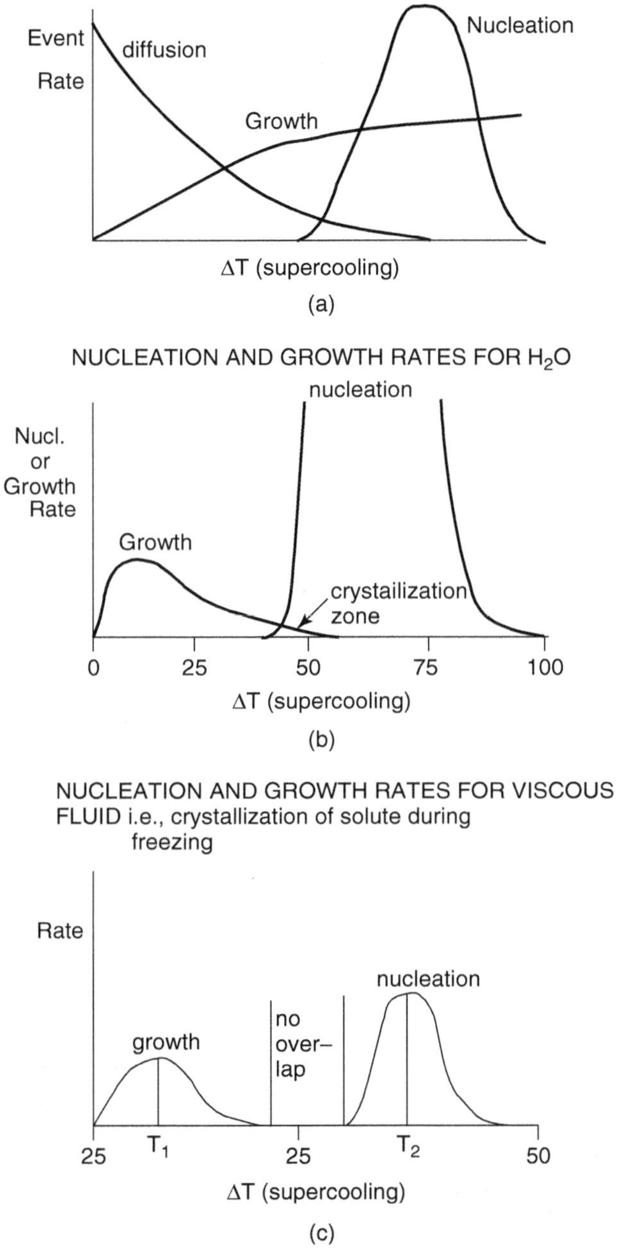

Figure 30.4. Effect of temperature on nucleation growth.

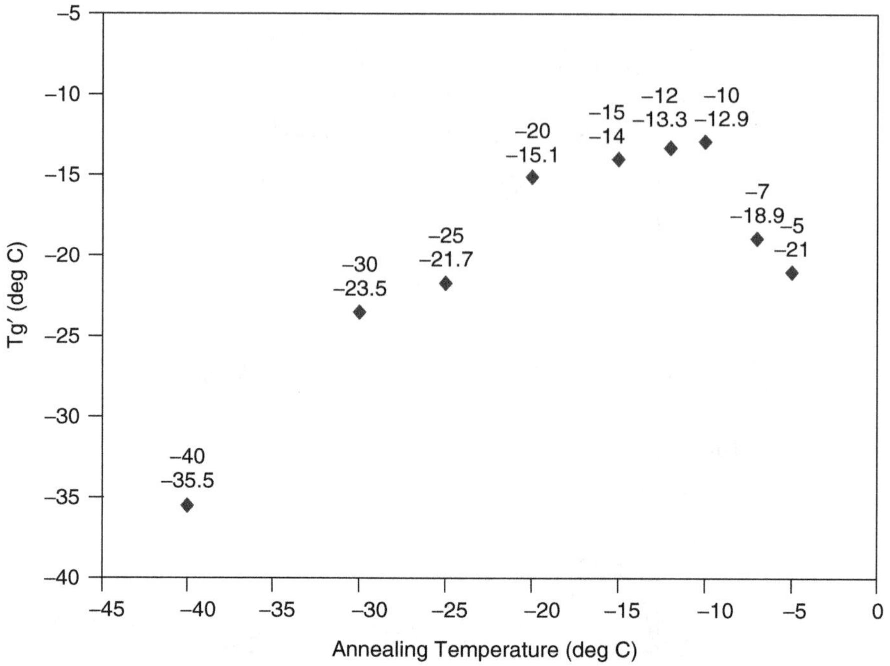

Figure 30.5. Glass transition temperatures T_g' of frozen formulation at various annealing conditions. The data points denote the annealing temperature and T_g', respectively. From A. Sethuraman and F. Jameel, IBC, 2006.

expense of smaller ones. Ostwald ripening is normally employed for two reasons: in situations where there is (1) heterogeneity in the product quality arising from heterogeneity in ice morphology or (2) mass transfer issue due to high product resistance. Ostwald ripening helps create larger ice crystals and decrease the product resistance to flow of water vapor by leaving larger channels in the dried cake, which results in a shorter primary drying time overall, as well as minimizing variation in primary drying time between vials [20].

As stated earlier, if both a bulking agent and a saccharide (stabilizer) are included in the formulation, the level of bulking agent must be either very low or very high relative to stabilizer. If the level of bulking agent in the amorphous stabilizer phase is very low, the impact on T_g' will be minimal. If the level of bulking agent is high, bulking agent can be induced to crystallize nearly completely, thereby minimizing the level of bulking agent in the amorphous phase and maximizing the T_g' of the formulation.

30.3.1.3.1. Freezing Parameters. The freezing parameters influence the ice morphology, which, in turn, influences the performance of the subsequent phases of freeze drying and also the quality attributes of the product. Additionally, the freezing protocol affects the integrity of the protein. Cryoconcentration and ice–liquid interface denaturation are the two mechanisms by which a protein tends to denature during freezing, and freezing rate affects both. Minimization of cryoconcentration effects requires that protein spend less time in the freeze concentrate stage, which means faster freezing. Establishing specifications against these effects and monitoring them during the process is key to minimizing process and product variability. Slow freezing is also deleterious to formulation systems that are prone to phase separation; slow freezing provides enough time for the phase separation process to occur. If the solution contains a stabilizer (cryoprotectant), the stabilizer is phase-separated.

Freezing rate determines the size and morphology of ice. The higher the freezing rate, the higher would be the *degree of undercooling*, which is defined as the difference between the equilibrium freezing point and the nucleation temperature at which ice crystals first form in the solution. Also, the higher the freezing rate, the smaller would be the size of ice crystals, which means larger ice-specific surface area. So, if one were to minimize the cryoconcetration effect by faster freezing, this would create larger ice-specific surface area and provide a larger interfacial area and more potential for denaturation, if the protein were to denature at the ice–aqueous interface. From the process perspective, smaller ice crystals means small "pore size" in the dried layer, which results in a high resistance to water vapor transport during primary drying. As a result, primary drying times can be long but secondary drying times could be shorter as high specific surface area in the dried product decreases secondary drying times [21–23].

The rate of ice crystal growth is a major factor in determining the solidification of ice and the residence time of the product in a "freeze-concentrated" fluid state. A rapid ice growth rate minimizes the residence time of solutes/protein, and minimizes the adverse conditions known to occur in the freeze concentrate. In commercial freeze-drying applications, heat transfer limits the rate of ice growth. Therefore, in vials and pans where heat removal is through the container bottom, rapid ice growth is facilitated by small fill volume : container area ratio (i.e., small fill depth) and "good contact" between the container bottom and the freeze-dryer shelf. A low shelf temperature also promotes rapid ice growth.

In freeze drying, various freezing methods are used. Depending on the product requirements to freeze the vials, this includes immersion of vial trays in liquid nitrogen, loading of vials or pans onto precooled shelves, or ramped cooling on the shelves. All these methods differ in freezing rates, and hence will differ in degree of undercooling. The highest degree of undercooling is normally observed with freezing small fill volumes using liquid nitrogen and minimum degree of undercooling is observed using the precooled shelf method.

One more concern before pulling down the vacuum and beginning the primary drying phase is to ensure the complete solidification of ice. A more uniform freeze and complete solidification can be achieved by lowering the final freezing shelf temperature to 5°C–10°C below the collapse temperature and hold for 1 h if the product is amorphous in nature and fill depth is ≤ 1 cm. Products with larger fill depths posses the potential for nonuniform freezing patterns. This can be eliminated or at least moderated by initially lowering the shelf temperature to an intermediate level, high enough to avoid crystallization of any ice (i.e., −5°C to −10°C) before the shelf temperature is lowered to the final temperature. This protocol provides an opportunity for a uniform freeze. For products that are crystalline in nature, the final freezing shelf temperature must be 10°C–20°C lower than the eutectic temperature since many eutectic systems can achieve high degrees of undercooling/supersaturation before crystallizing.

30.3.2. Primary Drying

Primary drying is the longest phase of freeze drying. In primary drying the crystalline ice is sublimed to vapor. In order to maintain product elegance and product stability, it is required that the primary drying be carried out at a product temperature below the collapse temperature. The details of determining the collapse temperature are mentioned above; however, it is emphasized here that for those systems that involve an annealing step in their freezing protocols, values determined that did not factor in annealing can be misleading.

Product temperature and sublimation rate are the two dependent variables that determine the performance of the process and the product quality attributes, which, in turn, depend on independent variables: shelf temperature and chamber pressure. The shelf temperature mainly determines the rate of heat transfer; the chamber pressure determines both heat and mass transfer as a significant amount of heat is also transferred through gas collisions; hence, product temperature is the result of shelf temperature and chamber pressure. The product temperature is also influenced by heat transfer characteristics of the vials and the mass transfer characteristics of the product and semi-stoppered vials [22,24,25].

30.3.2.1. Primary Drying Design Considerations.

30.3.2.1.1. TARGET PRODUCT TEMPERATURE. The most critical part of designing the primary drying phase is selection of the appropriate shelf temperature and chamber pressure profiles with time, which is very specific to formulation composition, fill volume, and type of vial. Depending on the composition of the formulation, the product temperature increases with drying time, due to resistance in mass transfer arising from dried layer thickness, which increases with time. In order to keep the product temperature below the collapse temperature to avoid collapse, the shelf temperature and chamber pressure need to be adjusted during primary drying. To avoid adjustments in process parameters during the process and to keep the process simple, it may be advisable to run primary drying at a constant shelf temperature–chamber pressure profile designed in such a way that product temperature is maintained at a safe level ranging from 2°C to 5°C below the collapse temperature or T_g' to accommodate increase in product temperature with time, and that will be considered as the target product temperature. In cases where the cycle time is short, a safety margin of up to 5°C can be used.

30.3.2.1.2. SELECTION OF CHAMBER PRESSURE. The rational selection of chamber pressure and shelf temperature requires knowledge and understanding of the coupled heat and mass transfer phenomena that combine to establish the product drying rate and temperature. Figure 30.6a shows drying rate and product temperature data for

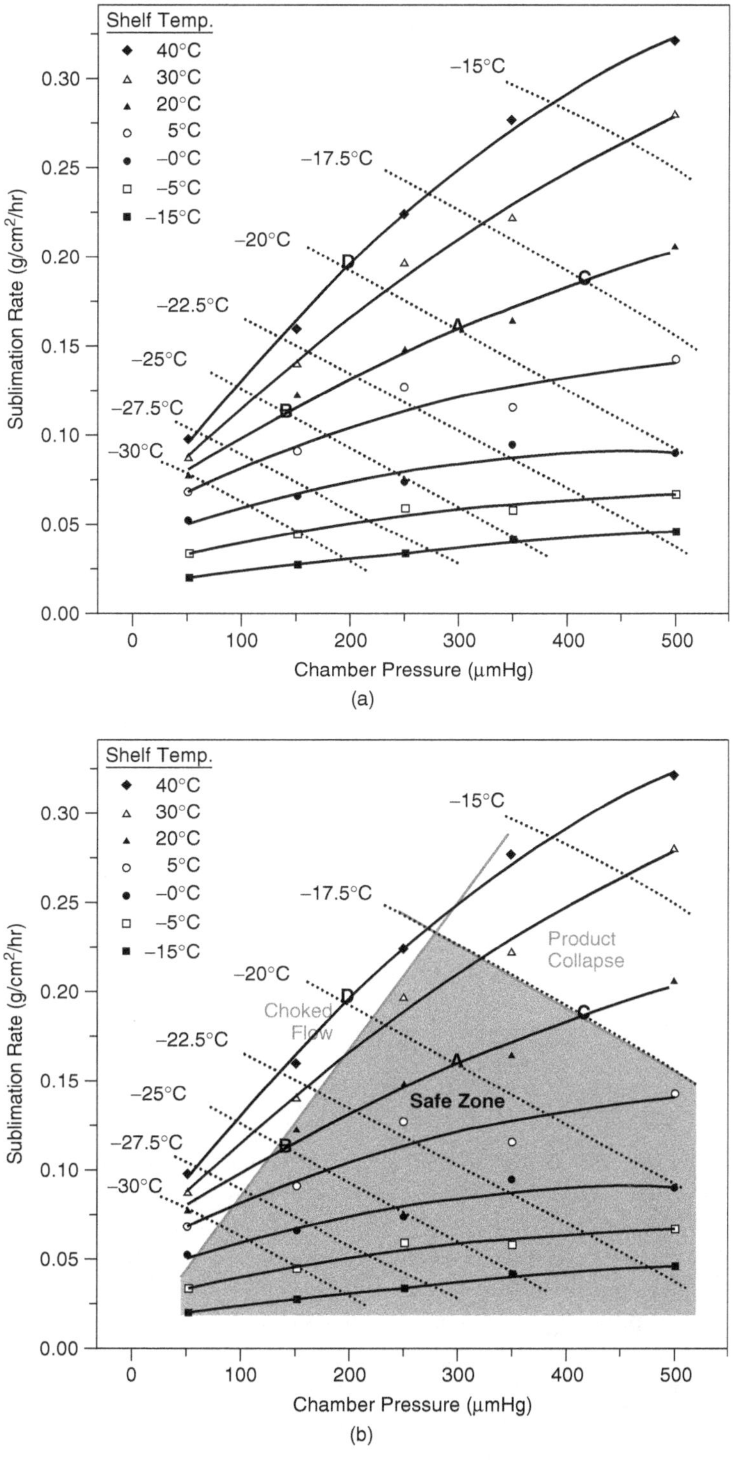

Figure 30.6. (Data from [28, 29]) Sublimation rate versus chamber pressure.

a typical product, across a range of chamber pressures and shelf temperatures [26]. Over this range of pressures, this is typical for pharmaceutical products in vials, increasing pressure leads to higher product temperature and drying rate. Increases in shelf temperature also lead to higher product temperature and drying rate. Also note that *at a given product temperature* the maximum drying rate can be achieved by using low pressure and high shelf temperature. However, the use of pressures below 50 mTorr increases heat transfer disparities among vials (especially edge vs. center shelf vials), exacerbating heterogeneity in both drying rates and product quality attributes. Such low pressures may also cause contamination of the product with volatile stopper components or pump oil. Another key consideration when setting these process conditions is to ensure that the drying rate capability of the freeze dryer is not exceeded (discussed in detail below).

The basic equations that govern these freeze-drying phenomena are as follows [22,24]:

$$\frac{dm}{dt} = \frac{P_{\text{ice}} - P_{\text{c}}}{R_{\text{p}} + R_{\text{s}}} \tag{30.2}$$

Here, dm/dt is the sublimation rate, P_{ice} is the vapor pressure of ice at the product temperature, P_{c} is the chamber pressure, R_{p} is the product resistance, and R_{s} is the stopper resistance. Equation (30.2) indicates that the driving force for mass transfer is the pressure gradient between the vapor pressure of ice and the chamber pressure; the higher the chamber pressure, the lower the driving force for mass transfer from the ice interface in the product to the freeze-dryer chamber. However, as discussed below, higher chamber pressure can improve heat transfer to the vial:

$$\frac{dQ}{dt} = \Delta H_{\text{s}} \frac{dm}{dt} \tag{30.3}$$

Equation (30.3) couples the heat and mass transfer phenomena, where (dQ/dt) is the rate of heat transfer, ΔH_{s} is the heat of sublimation, and dm/dt is the sublimation rate. Then

$$\frac{dQ}{dt} = A_{\text{v}} K_{\text{v}} (T_{\text{s}} - T_{\text{p}}) \tag{30.4}$$

Here, dQ/dt is the rate of heat transfer, A_{v} is the cross-sectional area of the vial, K_{v} is the vial heat transfer coefficient, T_{s} is the shelf temperature, and T_{p} is the product temperature at the bottom of the vial. The vial heat transfer coefficient is the sum of contributions arising from direct conduction from the shelf to the vial at the points of contact (K_{c}), the contribution from radiative heat transfer (K_{r}), and the contribution from conduction through the gas between the shelf and the bottom of the vial (K_{g}), represented as

$$K_{\text{v}} = K_{\text{c}} + K_{\text{r}} + K_{\text{g}} \tag{30.5}$$

Since radiation to the bottom of the vial accounts for a small fraction of the heat transfer, conduction through direct contact and through gas will be the major contributors. The coefficient of heat transfer through gas from the top of the shelf to the bottom of the vial (K_g) is in the range of pressures used for lyophilization, proportional to pressure [27]. Higher pressures increase the effective thermal conductivity of the gas in that narrow gap because the gap distance is on the order of the mean free path for molecular motions in the gas [27].

Returning to Figure 30.6a, we see that increasing pressure leads to an increase in product temperature and drying rate. Although the higher pressure decreases the driving force for mass transport, it increases the heat transfer coefficient for conduction through the gas between the shelf and the bottom of the vial (K_g).

Chang and Fischer reported a significant increase in drying rates with high shelf temperature and low chamber pressures with recombinant human interleukin-1 receptor antagonist formulation [26] but also found significant product temperature gradient across the cake, a source of heterogeneity. This could potentially become of significant concern with larger vials and fill volumes.

In summary, in most applications of practical interest, the chamber pressure is set between 50 and 50 to 200 mTorr. It is difficult to consistently maintain chamber pressure much below 50 mTorr, which will create heterogeneity problems, and there is no advantage in using pressures much higher than 200 mTorr. Hence, it is advisable to select a chamber pressure ranging from 10% to 30% of the vapor pressure of ice at the product temperature. For formulations with low collapse temperatures such as sucrose-based formulations, a chamber pressure near the upper limit of 30% would be considered optimal in terms of uniform heat transfer, and for formulations with high collapse temperatures such as -16°C one might consider the lower limit of 10%. The following equation or a readout from the graph in Figure 30.7 can be used to select the "optimal" chamber pressure at the known target product temperature (T_p).

$$P_c = 0.29 \times 10^{(0.019 \cdot T_p)} \tag{30.6}$$

Here, P_c is chamber pressure (in Torr) and T_p is the target product temperature (in $^{\circ}$C).

30.3.2.1.3. Selection of Shelf Temperature. After identifying the pressure conditions, the next question will be the appropriate shelf temperature–time profile in conjunction with the chamber pressure that would result in the target product temperature. This can be either through a trial-and-error approach, which requires many experiments, or through calculations using equations governing heat and mass transfer rates outlined above. In situations where the product temperature exceeds the target product temperature, the heat input needs to be reduced, and this can be done by decreasing either the shelf temperature or the chamber pressure.

For formulations with a low collapse temperature, use shelf temperature to lower the heat input rather than chamber pressure, as low collapse temperature formulations will have low pressures, and the flexibility to further lower pressure would be

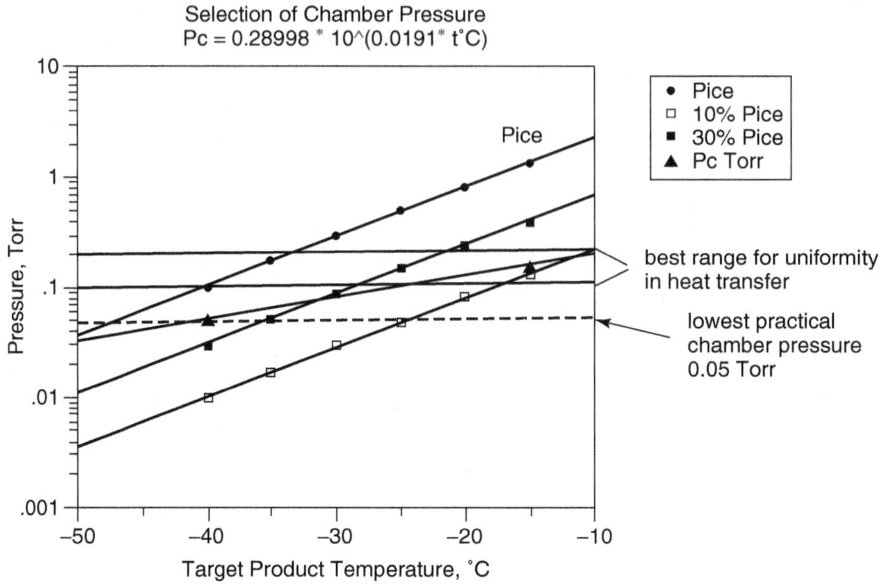

Figure 30.7. Initial chamber pressure estimation as a function of target product temperature. [*Note*: Low target product temperature forces compromises. Drying rate is small, $(P_{\mathrm{ice}} - P_{\mathrm{c}}) =$ "driving force," and to maintain positive driving force, chamber pressure is lower than optimal for best heat transfer uniformity!] Courtesy M. Pikal et al.

limited. For each 5°C change in shelf temperature, the product temperature changes by 1°C–2°C, and one also needs to keep in mind that there will be no added value in changing the settings after the process has progressed through two-thirds of the primary drying phase.

On the other hand, for formulations with a high collapse temperature, pressure can be used to lower the heat input as they are run at high pressure. Formulations with high collapse temperatures such as concentrated protein mannitol-based formulations allow primary drying at elevated shelf temperatures, but such drying conditions need to be questioned and reviewed carefully as they can overload the dryer. In general, shelf temperature and chamber pressure settings should be cautiously selected in such a way that the product temperature does not exceed −15°C; otherwise the heat and mass transfer capabilities of the commercial freeze dryer may be pushed beyond their limits, causing loss of pressure control, resulting in loss of product. The arrows in Figure 30.8 indicate the limits of the shelf temperatures above which the mass flow will exceed 1 kg $h^{-1}m^{-2}$ shelf area at $P_{\mathrm{c}} \sim 100$ mTorr [28,29] and can potentially cause choked flow. This topic is covered in more detail in section 30.4.

30.3.2.2. Determination of Product Temperatures. Product temperature is a critical parameter that needs accurate measurement as it is used to control and monitor the performance of the process. Temperature sensors in the form of thermocouples

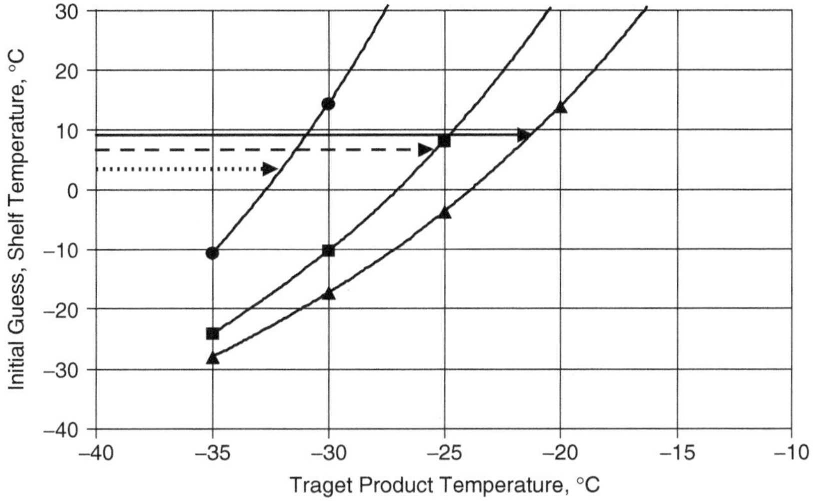

Figure 30.8. Initial shelf temperature estimation for a given target product temperatures at various solute concentrations, that is, different dry-layer resistance in primary drying. Circles represent low dry-layer resistance (solids ⩽ 1%), squares represent medium dry-layer resistance (10%> solids>1%), and triangles represent high dry-layer resistance (solids ⩾ 10%). The arrows specify the maximum shelf temperatures above which mass and heat transfer overload of a "typical" manufacturing dryer is likely. (Adapted with permission from Tang and Pikal [41].)

or resistance temperature device (RTD). RTD's of various gauges in the vials at selected locations are used to monitor the product temperature; however, it has two main drawbacks (1) the vials with probes will be biased and will not be representative of the whole batch, and will be misleading if the product temperature profile is used to identify the end of primary drying; and (2) it measures the product temperature at the bottom of the vial rather than at the sublimation interface. The precision and accuracy of the probes is important, and we recommend using 30-gauge thermocouples (Omega: 5SRTC-TT-T-30–36 or a similar model). Some of these limitations can be overcome by using the *manometric temperature measurement* (MTM) method, which is based on quickly isolating the freeze-drying chamber from the condenser for a short time during primary drying and subsequently analyzing the pressure rise data [33]. The advantage of the MTM method is that it determines the product temperature at the sublimation interface and would be representative of the whole batch. However, currently MTM is mostly used for developmental studies at the lab scale and has limited applicability for commercial purpose due to the following reasons. Firstly, MTM requires swift closure of the valve between the chamber and condenser and achieving that could be a challenge with production scale freeze dryers. Secondly, since pressure rise rate is dependent upon the product dry layer resistance, the ice sublimation area, and the chamber volume, large volume freeze dryers and small sublimation areas or small number of vials can significantly limit its applicability.

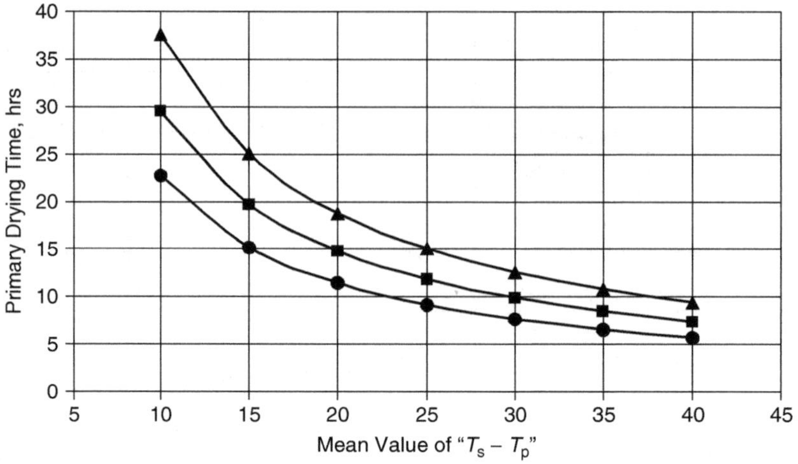

Figure 30.9. Estimated primary drying time as a function of temperature driving force for different solute concentrations for 1 cm fill depth. Circles represent low dry-layer resistance (solids $\leqslant$ 1%), squares represent medium dry-layer resistance (10%> solids>1%), and triangles represent high dry layer resistance (solids $\geqslant$ 10%). [Adapted with permission from Tang and Pikal [41].)

30.3.2.3. Primary Drying Endpoint. The primary drying time is directly related to the ice sublimation rate and is a dependent variable determined by independent variables-chamber pressure, shelf temperature, heat transfer coefficient of vials, fill volume, and product resistance. The primary drying time can be estimated using heat and mass transfer rates. Figure 30.9 depicts the primary drying time estimates for tubing vials as a function of $T_s - T_p$, where the product temperature (T_p) is the target product temperature in Figure 30.9 and the shelf temperature (T_s) is determined using Figure 30.8. Although these are only rough approximations, but can be used as an initial guess the method can be used to generate good initial guess.

30.3.2.4. Indicators of Primary Drying Endpoint. A method that determines precisely when the last piece of ice has sublimed would be very important not only for maximizing the throughput of a process but also from the product quality perspective, as advancing to secondary drying phase without completion of ice sublimation would carry the risk of product collapse. Several methods can be used to monitor the progress of primary drying and to indicate the endpoint of the phase; a few of these are summarized below.

30.3.2.4.1. Product Temperature Data. The product temperature data indicate the endpoint of primary drying when the product temperature profile converges to that of the shelf temperature. Normally, the product temperature, when plotted as a function of time, shows a steep increase in temperature at the end of primary drying followed by a plateau. Frequently, the vials containing thermocouples are not representative of the batch as a whole, because the vials that contain probes will freeze faster as probes act as nucleating sites and will have less undercooling, which means larger ice crystals. These larger ice crystals will form larger pores and offer less resistance to mass transfer and hence will dry faster, resulting in shorter primary drying time. Therefore, a "soak period" ranging from 10% to 15% of the primary drying time must be allotted following the time when all the thermocouple vials suggest the end of sublimation. Front-row vials complete primary drying far earlier than do the interior vials; this is due to atypical radiation heat transfer effects from the walls and door of the freeze dryer.

30.3.2.4.2. Dew Point. For water-based products moisture analyzers can be used to provide the data necessary to determine the end of primary drying phase. Moisture sensors have been considered too fragile for use in freeze drying, and temperature and pressure sensors have been the choice for monitoring the primary drying phase. But lately moisture sensors have demonstrated their ability to indicate the phase endpoint more reliably and have been designed to protect against steam sterilization. During the primary drying phase, water vapor continues to evolve from the product at a steady state and will be displayed on the moisture analyzer as a steady-state dew-point temperature, usually between −35°C and −65°C dew point. At the end of primary drying, the dew-point value drops sharply in response to the change in gas composition within the chamber, changing from almost 100% water vapor to essentially 100% nitrogen, indicating that the last piece of ice has sublimed.

30.3.2.4.3. Pirani Data. The design of the Pirani gauge is based on the transfer of heat from a hot wire, located inside the sensor, to the surrounding gases. Since gauge output depends on both thermal conductivity and pressure of the gases, all thermal conductivity gauges provide indirect, gas-dependent pressure readings. Traditional Pirani gauges provide useful pressure readings between 10^{-3} and 10 Torr. The thermal conductivity pressure gauges (Pirani pressure gauge) are generally calibrated against air or nitrogen, and since the thermal conductivity of water vapor is roughly 1.5 times that of air or nitrogen, it reads higher vapor pressure during primary drying. When the last piece of ice has sublimed and there is no water vapor but mostly nitrogen, the value drops and approaches the actual pressure determined by capacitance manometer (MKS Baratron gauge) indicating the end of the primary drying phase.

30.3.2.5. Drying Close to Collapse, Impact on Process Performance, and Product Quality

30.3.2.5.1. Relationship Between Resistance to Mass Transfer and Small-Scale Product Collapse in Protein Lyophilization. Conventional concepts suggest that resistance of the dried product to mass transfer is directly proportional to the thickness of the dried layer. Pikal et al. studied the effect of dried layer thickness on the mass transfer rate during the primary drying phase and reported that resistance to mass transfer increases with increase in the dried-layer thickness but in a nonlinear fashion, due to the existence of a surface barrier (i.e., R_p at $l = 0$ is not zero) [24]. They reported the dependence of the product resistance and dried-layer thickness profiles on the solute type and noted that their concentrations, in general, increase with the increase in the solute concentration. They also reported that the product resistance decreases sharply as the product temperature increases toward the collapse or eutectic temperature, and this was attributed to hydrodynamic surface flow of the water vapor through the pores. This was further investigated independently by Milton et al. and Overcashier et al. to better understand the mechanism behind it [33,42]. They all agreed that increase in resistance to mass flow correlates with the increase in thickness of the dried layer at lower temperatures but that resistance remains constant with the increase in the dried-layer thickness when the product is dried at a temperature close to the material's collapse temperature. They attributed this phenomenon to *microcollpase*, that is, development of additional pathways in the form fractures and/or holes in the dried material, which was further evidenced by SEM photomicrographs of the dried cake. Overcashier et al. also studied product resistance as a function of formulation at a constant solid mass, in protein product versus trehalose and sucrose, and reported that resistance was lower with trehalose than in the protein product and even lower in sucrose but that all resistance values were equivalent at lower shelf temperatures.

However, Chang and Fischer's studies with various combinations of shelf temperature and chamber pressure contradicted these findings and hypothesis [26]. They observed the resistance to vapor flow to be independent of product temperature over a wide range of temperatures from $-32°C$ to $-14°C$.

30.3.2.5.2. Impact on Product Stability. The relative impact of drying the product above and/or close to collapse temperature on the stability of the product depends on the nature of the protein, the formulation, and the mechanism of degradation. Various groups studied this effect and reported conflicting results. Lueckel and co-workers reported increased levels of higher-order aggregates of IL-6 in both glycine : sucrose and mannitol : sucrose formulations when the product was freeze-dried at collapse temperature [43]. Interestingly, they found more aggregates when annealed for glycine crystallization than when the annealing step was omitted, which they suspected to be due to extended exposure of the product above the T_g'. Contradictory to this observiation, Sarciaux and co-workers reported decreased levels of aggregates of human IgG when annealed, which they ascribed to decreased specific surface area of ice and hence decreased ice–liquid interfacial denaturation [44]. Adams and Irons studied the effect of structural collapse on the activity of L-asparaginase in 1% lactose formulation and found no correlation between the structural collapse and alteration of biological activity or the integrity of tetramer of asparaginase [45]. However, they reported good correlations between collapse and increased levels of residual moisture content and reconstitution time. Jiang and Nail studied the impact of drying at the collapse on the stability of three enzymes, catalase, β-galactosidase, and lactate dehydrogenase and reported loss of activity in all three [46]. Lactate dehydrogenase exhibited substantial loss compared to the other two, and this was attributed to a greater propensity of lactate dehydrogenase to denature on freeze–thaw. Most of these studies lacked evaluation of impact of collapse on long-term storage stability, which is of practical importance. Recognizing this as a gap and its practical significance, Wang and co-workers thoroughly investigated the impact of collapse on long-term stability of recombinant factor VIII and α-amylase using three different freeze-drying conditions; one produced a collapsed cake and the other two did not collapse the product but were either too aggressive or too conservative [47]. The results of 18-month storage stability studies indicated greater than 80% recovery of activity with both protein molecules, suggesting no adverse impact of drying with collapse on long-term storage stability. Additionally, the stability profile of products dried under aggressive conditions (i.e., very close to the collapse temperature) were indistinguishable from stability profiles of products dried under very conservation conditions.

Such studies involving the evaluation of impact of drying above and close to the collapse temperature on process performance and product stability are critical to (1) distinguishing between collapse merely as a cosmetic defect versus a product quality attribute that could lead to subtherapeutic dosing, (2) identifying maximum allowable product temperature and then developing process conditions to dry the product close to it and make the process efficient, and (3) collecting the storage stability data on the abused product to support the closure of nonconformances during commercial production.

30.3.3. Secondary Drying

At the end of the primary drying phase, depending on the composition of the formulation, there would be 5%–20% of water that did not form ice and remained bound to the product. If the formulation is predominantly amorphous in nature, then the moisture content after primary drying would be toward the higher end, and if it is crystalline in nature, it would be toward the lower end. Regardless, the storage stability of biological products depends on the residual moisture content, and the main objective of secondary drying is to remove that bound water by raising the temperature to a level that enhances the storage stability of the product. What is the optimal level is quite specific and varies from product to product, usually it is less than 1% w/w for proteins and varies between 2% and 3% for vaccines, but one key consideration that applies to all formulations is the ramp rate during transition from primary to secondary drying. As mentioned above, 5%–20% of residual water will remain at the end of primary drying and the T_g/collapse temperature of the product will be quite low, and fast ramping to the elevated temperature of secondary drying would carry the risk of product damage in terms of both quality and stability. It is advisable to ramp slowly until the set point is reached at a safe heating rate of 0.20°C/min for amorphous formulations. For crystalline formulations, it can be slightly faster (0.3°C or 0.4°C/min) as the collapse temperature would be higher and will not carry the potential for collapse during secondary drying; thus a higher ramp rate is suggested for such products. Since secondary drying is a desorption process and is independent of chamber pressure (at least up to 200 mTorr), the primary drying chamber pressure can be continued through secondary drying without change [22].

The shelf temperature–time profile for secondary drying varies from formulation to formulation, and protein stability as a function of the temperature profile must be considered while designing the secondary drying conditions for the product. The optimum secondary drying time can be determined by real-time residual moisture measurement during secondary drying by extracting samples from the freeze dryer without interrupting the freeze-drying cycle (i.e., using a "sample thief/extractor"). The moisture content can be measured using Karl Fischer titrimetry (KFT), thermal gravimetric analysis (TGA), or near-IR spectroscopy (NIR). For the uniformity of moisture across the cake and between the vials, it is advisable to dry at low temperatures for a longer time than at high temperatures for a shorter time. In general, for protein formulations, drying at 25°C–30°C for 3–4 h is adequate; however, for mannitol-based formulations and in situations where mannitol hemihydrates are formed and secondary drying is used as an opportunity to desolvate the mannitol hemihydrates, secondary drying at elevated temperatures of 50°C–55°C for 2–3 hours helps to eliminate mannitol hemihydrates [30].

30.4. CONSIDERATION OF SCALE-UP AND MANUFACTURING ISSUES DURING DEVELOPMENT

During development of the freeze-drying process one must be aware of the manufacturing challenges, challenges relating to differences in the environment, scale-up issues (differences in load size), differences in the design of the lyophilizer and associated equipment and, and time and procedure differences. A process can be assumed to be robust and easily transferable from one environment to another and from one scale to another if one can demonstrate equivalent product temperature profiles on different lyophilizers and if the lyophilizers can maintain the chamber pressure set point throughout the cycle. The shelf temperature is controlled as the temperature of the heat transfer fluid as it is pumped into the shelves. The fluid flow rate, internal shelf design, and fluid properties all contribute to determining the efficiency of heat transfer from the fluid to the shelf top surface. Shelf temperature adjustments can be made to the cycle recipe to compensate for slight differences in product temperatures. One must also ensure that the primary drying time is sufficient (but not excessive) for different lyophilizers, so the use of methods to monitor and adjust the duration is critical.

Thermal radiation and other heat transfer differences can cause significant heterogeneity in product temperatures and sublimation rates and hence variations in drying times [25,31]. Also, differences in the efficiency of the condenser or refrigeration system could limit the performance of one freeze dryer at a particular thermal load compared to another. Factors contributing to this are the surface area of the condenser coils and the cooling power provided through the refrigerant or heat transfer fluid pumped through those coils. Inadequacy of this part of the system is manifested by the product and condenser chamber pressures exceeding set point early in primary drying. This can be remedied by reducing the shelf temperature during at least that part of the cycle, but primary drying must be lengthened to account for the reduced sublimation rate.

It has been found that if the sublimation rate is too high and the pressure is too low, the water vapor can reach the speed of sound as it travels between the chamber and the condenser, causing "choked flow" [28,29]. This problem can be diagnosed by observing a very low condenser chamber pressure and by the product chamber pressure exceeding set point. It can be remedied by slowing the sublimation rate by reducing the shelf temperature, but this may extend the cycle duration unacceptably.

An alternative is to increase the chamber pressure and reduce the shelf temperature. Increasing the chamber pressure has the desired effect of increasing the density of gas, thereby reducing its velocity (for the same mass flow rate) and avoiding choked flow. However, we know from the section above that higher chamber pressure will also increase both the sublimation rate and the product temperature, so the shelf temperature must be reduced as well to the extent that these effects need to be reduced.

Most freeze dryer can support a sublimation rate of 1 kg/h per square meter of shelf area at 100 mTorr before reaching choked flow [28,29]. Searles [28] and Nail and Searles [29] suggest simple ways to measure this capability for every dryer. At twice the pressure, this capability is theoretically twice as high, but one must run tests at the desired pressures to be sure. In addition, product drying rate should be measured for the early part of primary drying (e.g., for the first 6 h) by stopping the run at that point and weighing vials. This can be done with a small batch in a laboratory dryer.

Figure 30.6b shows a way of matching product temperature requirements with cycle time benefits (by maximizing the drying rate) for a given dryer (see also Refs. 28 and 29). Here we have superimposed a red line for the maximum product temperature and a pink line for the dryer choked flow rate limit, each as functions of pressure. The area bounded by these lines represents the "safe zone" of operation for that particular product–dryer combination. One can see that increased drying rate, while it comes with the benefit of shorter cycle times, increases the risk that one will encounter either excessive product temperature or inability of the dryer to support that rate. Inability to support that rate will be manifested by pressures above set point which, unless counter balanced by an immediate shelf temperature decrease, will collapse the product through excessive temperature.

It has been shown that moving a lyophilization cycle from the lab bench to a cleanroom manufacturing environment can cause changes in product performance [23]. This can occur in the presence of higher levels of environmental particulates in a conventional laboratory setting, which can serve as ice nucleation sites, resulting in faster, cooler primary drying. On transfer to manufacturing in a cleanroom environment, ice nucleation temperatures will be lower and the product will dry more slowly and be warmer.

Considering these differences early on will help in minimizing the optimization of process parameters during technology transfer and scale-up in manufacturing.

30.5. CASE STUDY

Here we present a case study of a development of a freeze-drying process for a protein formulation that illustrates the importance of characterization and understanding of the thermal behavior of each excipient, both individually and in the mixture, in the design of a process (particularly the freezing protocol). It is a blood protein of molecular weight 240 kDa. When this protein was freeze dried in an unoptimized formulation using an unoptimized process, it was found to be unstable during the freezing phase and on storage. Several formulation candidates consisting of various combinations of stabilizers and bulking agents along with polysorbate 80, as described below, were tested, and on the basis of initial accelerated stability data and other business factors, three formulations were selected for further development. This case study relates to the formulation containing mannitol (as a bulking agent), trehalose (as a stabilizer), polysorbate 80, 0.1 mg protein, and residual salt (NaCl) at pH 7.0. The formulation was characterized using both modulated DSC (MDSC) and freeze-drying microscopy to determine the optimal annealing conditions for mannitol crystallization. The T_g' of the formulation was also determined before and after annealing. Mannitol was found to be easily crystallizable at any temperature above −26°C (see Fig. 30.10) and did not require much higher temperature for mobility, probably because of low viscosity and protein concentration.

However, the T_g' value of the formulation was found to be −53°C, surprisingly lower than expected. After freeze drying, the dried cake did not exhibit elegance and indicated signs of incomplete crystallization of mannitol. When the freeze-dried powder was scanned on MDSC, two interesting events were observed: (1) on the reversible component, which is based on heat capacity and captures only real transitions, a T_g of value 64°C was observed; and (2) on the nonreversible component, an exotherm value of ~ 60°C was observed (see Fig. 30.11).

When all the crystallizable excipients are completely crystallized and the intended role of each excipient is achieved, a trehalose-based formulation under ideal conditions should exhibit a T_g value close to 100°C. The T_g value of 64°C for the trehalose-based formulation was quite low, suggesting the presence of amorphous mannitol (amorphous mannitol has a T_g value of 0°C). This was further confirmed by the presence of an exotherm that was due to crystallization of amorphous mannitol. These pieces of information clearly indicated incomplete crystallization of mannitol despite annealing at optimal conditions. The incomplete crystallization of mannitol and low T_g' value of −53°C were found to be due to the presence of small amounts of NaCl. Sodium chloride was found to interfere with the crystallization of mannitol and depress the T_g' value of sucrose. Additionally, characterization of the physical state of the freeze-dried powder using X-ray powder diffraction (XRD) indicated the presence of mannitol hemihydrates (see Fig. 30.12). There were two options: (1) either eliminate NaCl completely from the formulation or (2) completely crystallize out NaCl before annealing mannitol and thereby prevent its inhibition of mannitol crystallization. Removing NaCl from the formulation was a difficult option as it would require a complete change of the downstream process, and it was too late to make any changes to the bulk manufacturing process. Crystallization of sodium chloride is not an easy process. As with other crystallizable excipients, crystallization of NaCl depends on the concentration, temperature, and time. So a thorough systematic characterization of sodium chloride, both individually and in the presence of other formulation excipients, was undertaken using both MDSC and freeze-drying microscopy to determine (1) minimum concentration required for crystallization, (2) nucleation temperature, and (3) optimal annealing temperature and time. It was determined that for crystallization of sodium chloride, the following were required: (1) a minimum concentration of 150 mM NaCl, (2) exposure to ≤ −50°C for a minimum of 1 h for nuclei to form prior to annealing, and (3) annealing at −35°C for 2 h for complete crystallization (see Fig. 30.13).

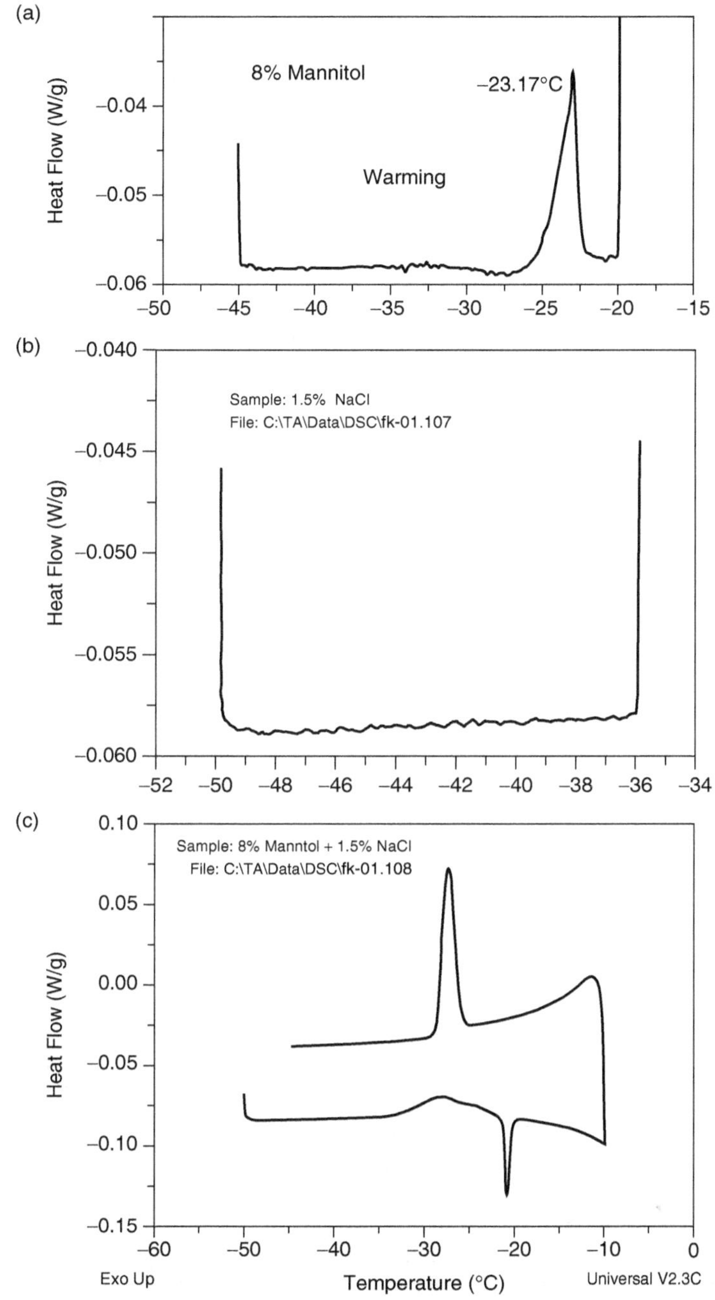

Figure 30.10. Modulated DSC thermograms exhibiting the crystallization of mannitol and sodium chloride: (a) crystallization temperature of mannitol on warming curve; (b) crystallization of mannitol and sodium chloride in a mixture solution followed by (c) a eutectic melt endotherm of sodium chloride and further crystallization of mannitol on freezing.

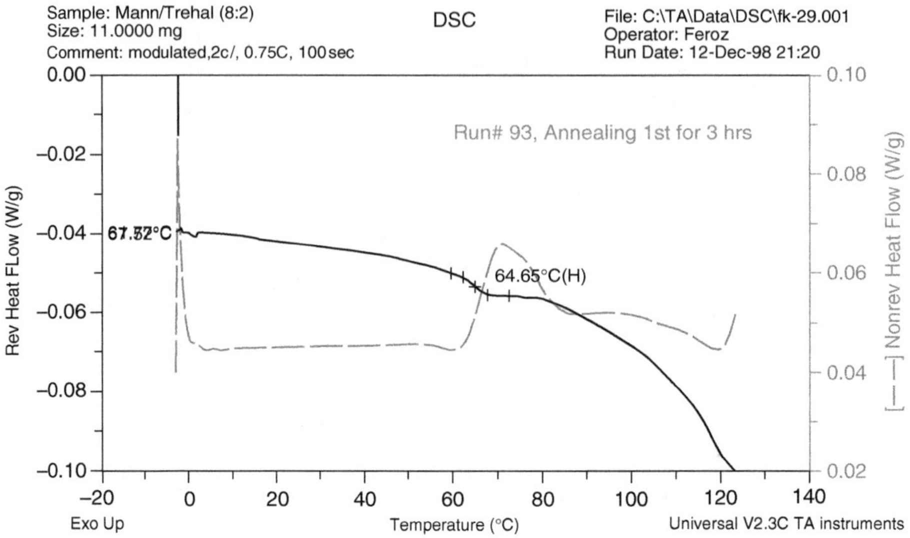

Figure 30.11. Modulated DSC thermogram of mannitol and trehalose freeze-dried powder depicting T_g on a reversible component and an exotherm of mannitol crystallization on a nonreversible component.

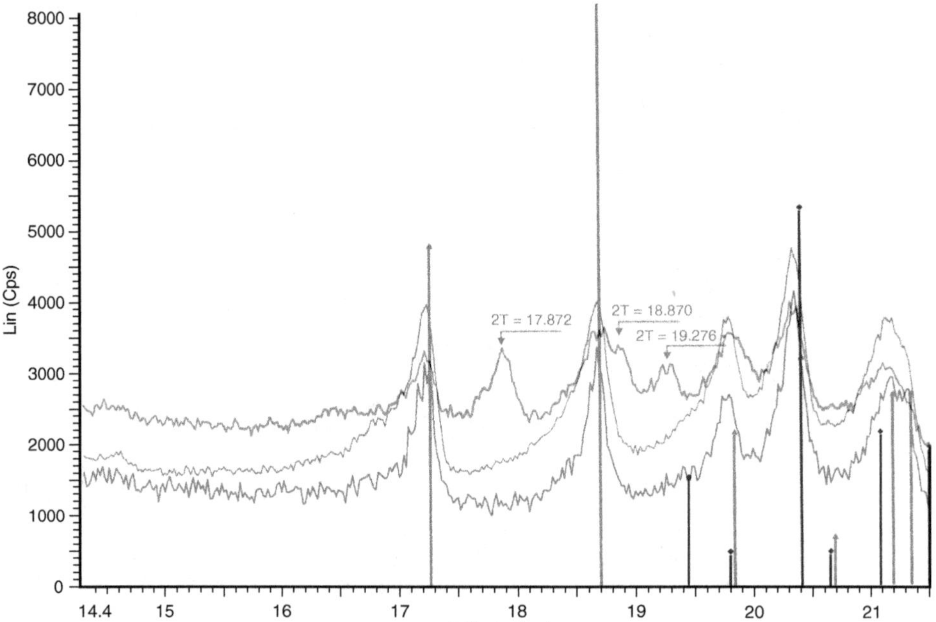

Figure 30.12. X-Ray powder diffraction of mannitol depicting the presence of hydrates, which are eliminated through secondary drying at elevated temperature.

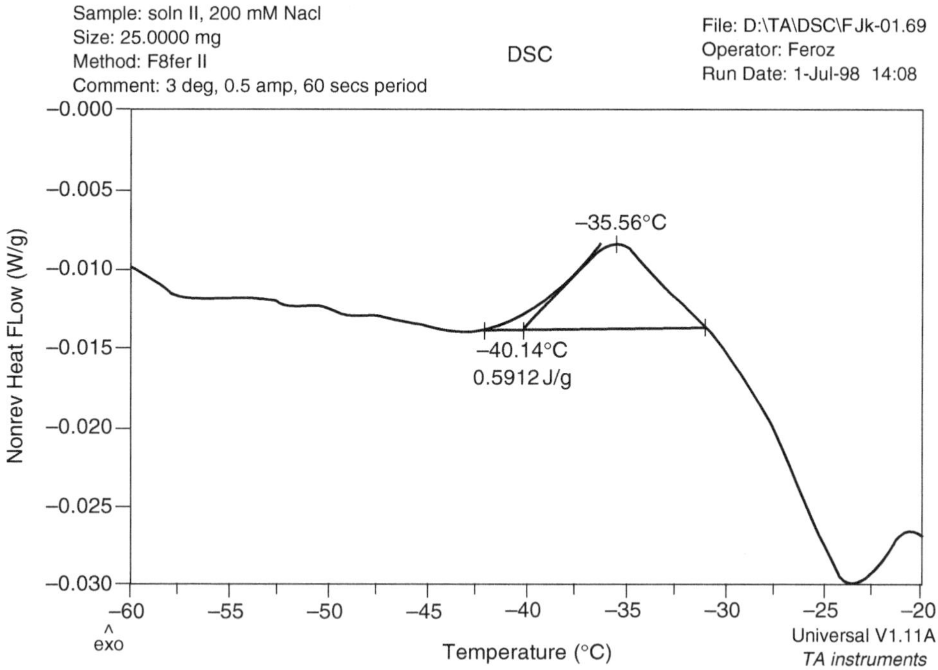

Figure 30.13. Thermogram exhibiting crystallization and eutectic temperature of sodium chloride.

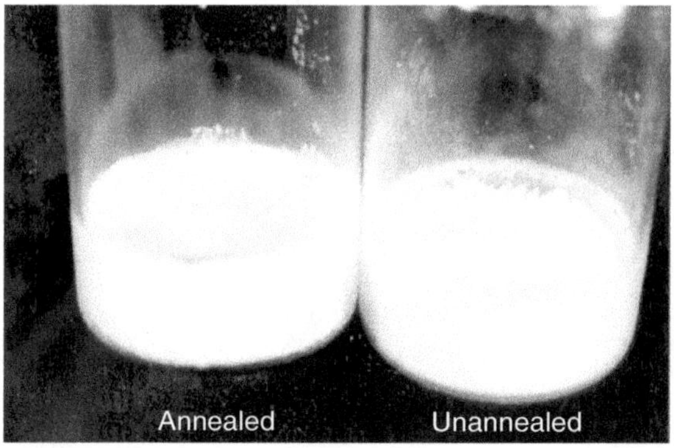

Figure 30.14. Appearance of cake before and after annealing of mannitol and sodium chloride.

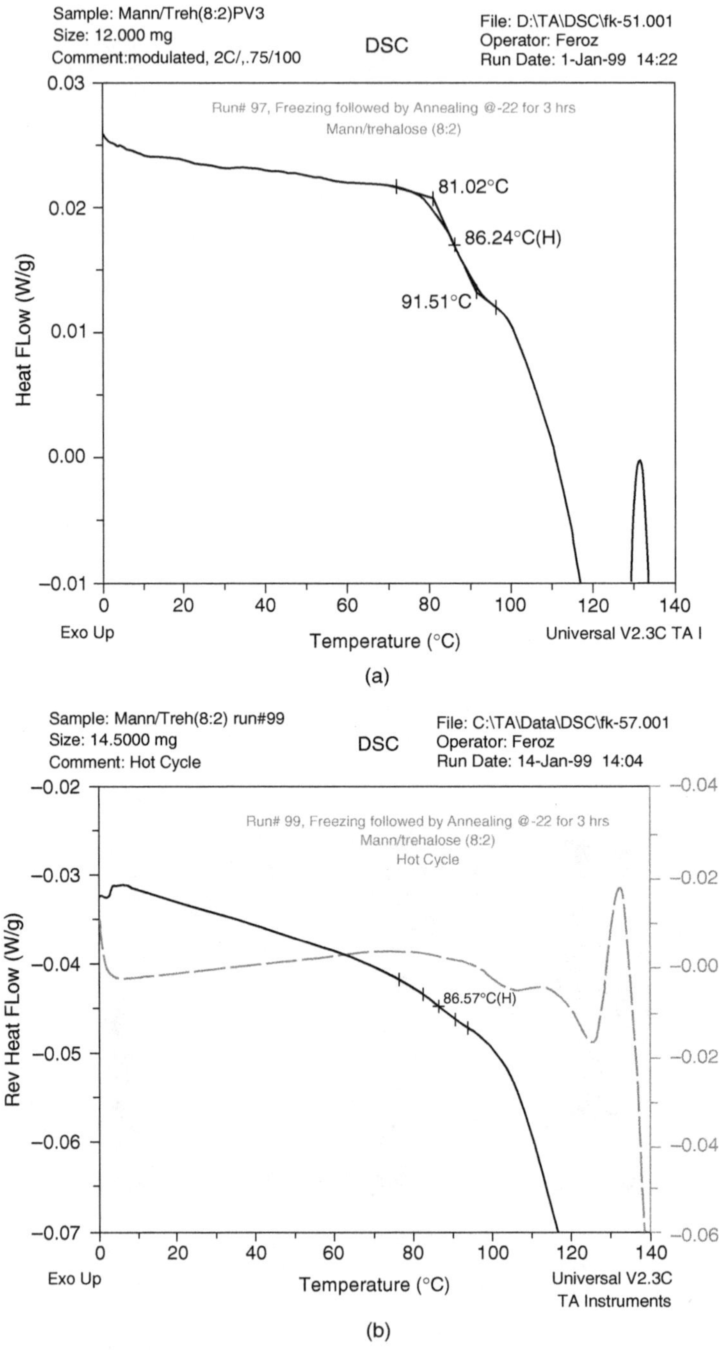

Figure 30.15. Thermograms depicting increase in T_g value and no crystallization peak of mannitol and sodium chloride after annealing.

The strategy undertaken was to first crystallize out sodium chloride before crystallizing mannitol. Accordingly, the freezing protocol was modified; the freezing temperature was changed from −40°C to −50°C to allow NaCl to nucleate, and an additional annealing step at −35°C for 2 h was added to facilitate crystallization of NaCl prior to annealing of mannitol at −26°C. This modified freezing protocol was first tested on MDSC complemented by freeze-drying microscopy before lyophilization in vials. An absence of any exothermic peaks on the final warming curve postannealing on the nonreversible component of the heat flow thermogram was observed, suggesting complete crystallization of all crystallizable excipients. As mentioned earlier in this chapter, the T_g' value can also be used to indicate the extent of crystallization. When the T_g' value closely matches that of the pure amorphous theoretical value, it is indicative of complete crystallization of all crystallizable excipients in the system. On the final warming curve a T_g' value of −41°C was observed, which was close to the theoretical value of −37°C. In addition to modification of the freezing protocol, the secondary drying conditions were modified such that the terminal temperature of secondary drying was changed to 50°C for 3 h to desolvate the mannitol hemihydrates. When the formulation was lyophilized with these altered parameters and the product was characterized for physicochemical properties, all the issues appeared to be completely resolved. The cake looked quite elegant as opposed to granular in the previous cycle (see Fig. 30.14).

Characterization of the powder on MDSC indicated an increase in the T_g value from 64°C to 84°C, and the absence of any thermal events on the nonreversible component confirmed near-complete crystallization of crystallizable excipients (Fig. 30.15). Physical-state characterization of the powder using XRD indicated not only an increase in degree of crystallinity of both NaCl and mannitol but also an absence of the characteristic peaks of mannitol hemihydrates.

These findings correlated very well with the improvement in physical properties (i.e., decrease in reconstitution time and the residual moisture content) and long-term storage stability of the product.

REFERENCES

1. Shire, S. J., Shahrokh, Z., and Liu, J. (2004), Challenges in the development of high protein concentration formulations, *J. Pharm. Sci.* **93**(6): 1390–1402.
2. Franks, F. (1982), The properties of aqueous solutions at subzero temperatures, *Water Compress. Treatises* **7**: 215–338.
3. Pikal, M. (2002), Lyophilization, in *Encyclopedia of Pharmaceutical Technology*, Vol. **6**, Swarbrick, J. and Boylan, J. C., eds., Marcel Dekker, New York, pp. 1299–1326.
4. Pikal, M. J., (1999), Mechanisms of protein stabilization during freeze-drying and storage: The relative importance of thermodynamic stabilization and glassy state relaxation dynamics, in *Freeze-Drying/Lyophilization of Pharmaceutical and Biological Products*, Louis Rey, L. and May, J., eds., Marcel Dekker, New York, Chapter 6.
5. Franks, F. (1990), Freeze-drying: From empericism to predictability, *Cryo-Letters* **11**: 93–110.

6. Pikal, M. J. and Shah, S. (1990), The collapse temperature in freeze drying: Dependence on measurement methodology and rate of water removal from the glassy phase, *Int. J. Pharm.* **62**: 165–186.
7. Hancock, B. C. and Zografi, G. (1994), The relationship between the glass transition temperature and the water content of amorphous pharmaceutical solids, *Pharm. Res.* **11**: 471–477.
8. Shalaev, E. Y. and Kanev, A. N. (1994), Study of the Solid-liquid state diagram of the water-glycine-sucrose system, *Cryobiology*, **31**: 374–382.
9. Mackenzie, A. P. (1964), Apparatus for microscopic observations during freeze drying, *Biodynamica* **9**: 213–222.
10. Nail, S., Schwegman, J. J., and Kamp, V. (2000), Analytical tools for characterization of frozen systems in the development of freeze dried pharmaceuticals, *Am. Pharm. Rev.* 17–25.
11. Her, L. M. and Nail, S. L. (1994), Measurement of glass transitions of the maximally freeze-concentrated solutes by differential scanning calorimetry, *Pharm. Res.* **11**: 54–59.
12. Levine, H. and Slade, L. (1988), Principles of cryostabilization technology from structure/property relationships of carbohydrate/water systems—a review, *Cryo-Letters* **9**: 21–63.
13. Ahmad, F. and Bigelow, C. (1986), Thermodynamic stability of proteins in salt solutions: A comparison of the effectiveness of protein stabilizers, *J. Protein Chem.* **5**: 355–367.
14. Koseki, T. et al. (1990), Freezing denaturation of ovalbumin at acid pH, *J. Biochem.* **107**: 389–394.
15. Murase, N. and Franks, F. (1989), Salt precipitation during the freeze-concentration of phosphate buffer solutions, *Biophys. Chem.* **34**: 293–300.
16. Kim, A. J., Akers, M. J., and Nail, S. L. (1998), The physical state of mannitol after freeze drying: Effects of mannitol concentration, freezing rate and a noncrystallizing scosolute, *J. Pharm. Sci.* **87**: 931–935.
17. Williams, N. A. and Dean, T. (1991), Vial breakage by frozen mannitol solutions: Correlation with thermal characteristics and effect of sterioisomerism, additives, and vial configuration, *J. Parenter. Sci. Technol.* **45**: 94–100.
18. Shalaev, E., Johnson-Elton, T., Chang, L., and Pikal, M. J. (2002), Thermophysical properties of pharmaceutically compatible buffers at sub-zero temperatures: Implications for freeze drying, *Pharm. Res.* **19**: 195–201.
19. Cavatur, R. K. and Suryanarayanan, R. (1998), Characterization of phase transitions during freeze-drying by in situ X-ray powder diffractometry, *Pharm. Devel. Technol.* **3**: 579–586.
20. Searles, J., Carpenter, J., and Randolph, T. (2001), Annealing to optimize the primary drying rate, reduce freezing-induced drying rate heterogeneity, and determine T_g' in pharmaceutical lyophilization, *J. Pharm. Sci.* **90**(7): 872–887.
21. Pikal, M. J., Shah, S., Roy, M. L., and Putman, R. (1990), The secondary stage of freeze-drying: Drying kinetics as a function of temperature and chamber pressure, *Int. J. Pharm.* **60**: 203–217.
22. Pikal, M. J., Roy, M. L., and Shah, S. (1984), Mass and heat transfer in vial freeze drying of pharmaceuticals: Role of the vial, *J. Pharm. Sci.* **73**: 1224–1237.
23. Searles, J., Carpenter, J., and Randolph, T. (2001), The ice nucleation temperature determines the primary drying rate of lyophilization for samples frozen on a temperature-controlled shelf, *J. Pharm. Sci.* **90**(7): 860–871.

24. Pikal, M. J. (1986), Use of laboratory data in freeze drying process design: Heat and mass transfer coefficients and the computer simulation of freeze drying, *J. Parenter. Sci. Technol.* **39**(3): 115–139.

25. Nail, S. L. (1980), The effect of chamber pressure on heat transfer in the freeze-drying of parenteral solutions, *J. Parenter. Drug Assoc.* **34**(5): 358–368.

26. Chang, B. S. and Fischer, N. L. (1995), Development of an efficient single-step freeze drying cycle for protein formulations, *Pharm. Res.* **12**(6): 831–837.

27. Searles, J. A. (2000), *Heterogeneity Phenomena in Pharmaceutical Lyophilization*, Ph.D. thesis, Univ. Colorado Chemical and Biological Engineering.

28. Searles, J. A. (2004), Observation and implications of sonic water vapor flow during freeze drying, *Am. Pharm. Rev.* **7**(2): 58.

29. Nail, S. L. and Searles, J. A. (2007), Quality by design in development and scale-up of freeze-dried parenterals, *BioPharm* **21**(1): 44.

30. Jameel, F., Bjornson, E., Hu, L., Kabingue, K., Besman, M., and Pikal, M. (2000), Effects of formulation and process variations on the stability of lyophilized recombinant human factor VIII, *Haemophilia* **6**(4): 343–368.

31. Rambhatla, S., Ramot, R., Bhugra, C., and Pikal, M. J. (2004), Heat and mass transfer scale-up Issues during freeze drying: II. Control and characterization of the degree of supercooling, *AAPS PharmSciTech.* **5**(4) (article 58).

32. Hatley, R. H. M. and Mant, A. (1993), Determination of the unfrozen water content of maximally freeze concentrated carbohydrate solutions, *Int. J. Biol. Macromol.* **15**: 227–232.

33. Milton, N., Pikal, M. J., Roy, M. L., and Nail, S. L. (1997), Evaluation of manometric temperature measurement as method of monitoring product temperature during lyophilization, *PDA J. Pharm. Sci. Technol.* **51**(1): 7–16.

34. Chang, B. S. and Randall, C. S. (1992), Use of subambient thermal analysis to optimize protein lyophilization, *Cryobiology* **29**: 632–656.

35. Slade, L. and Levine, H. (1995), Glass transitions and water-food. Structure interactions, *Crit. Rev. Food Sci. Nutr.* **30**: 115–360.

36. Pikal, M. J. and Chang, L. Q. (no date), unpublished data, Univ. Connecticut.

37. Pikal, M. J. and Shah, S. (no date), unpublished data, Eli Lilly & Co.,

38. Mackenzie, A. P. (1966), Basic principles of freeze drying for pharmaceuticals, *Bull. Parenter. Drug Assoc.* **26**: 101–129.

39. Liesebach, J., Rades, T., and Lim, M. (2003), A new method for the determination of the unfrozen matrix concentration and the maximal freeze-concentration, *Thermochim. Acta* **401**(2): 159–168.

40. Bhatnagar, B. S., Cardon, S., Pikal, M. J., and Bogner, R. H. (2005), Reliable determination of freeze-concentration using DSC, *Thermochim. Acta* **425**(1–2): 149–163.

41. Tang, X. and Pikal, M. J. (2004), Design of freeze-drying processes for pharmaceuticals: Practical advice, *Pharm. Res.* **21**(2): 191–200.

42. Overcashier, D. E., Patapoff, T. W., and Hsu, C. C. (1999), Lyophilization of protein formulations in vials: Investigation of the relationship between resistance to vapor flow during primary drying and small-scale product collapse, *J. Pharm. Sci.* **88**(7): 688–695.

43. Lueckel, B., Helk, B., Bodmer, D., and Leuenberger, H. (1998), Effects of formulation and process variables on the aggregation of freeze dried interleukin-6 after lyophilization and on storage, *Pharm. Devel. Technol.* **3**(3): 337–346.

44. Sarciaux, M., Hageman, M. J., and Nail, S. L. (1999), Effects of buffer composition and processing conditions on aggregation of bovine IgG during freeze drying, *J. Pharm. Sci.* **88**(12): 1354–1361.
45. Adams, G. D. J. and Irons, L. I. (1993), Some implications of structural collapse during freeze drying using erwinia caratovora L-asparaginase as a model, *J. Chem. Biotechnol.* **58**: 71–76.
46. Jiang, S. and Nail, S. L. (1998), Effect of process conditions on recovery of protein activity after freezing and freeze-drying, *Eur. J. Pharm. Biopharm.* **45**(3): 249–257.
47. Wang, D. Q., Hey, J. M., and Nail, S. L. (2004), Effect of collapse on the stability of freeze-dried recombinant factor VIII and α-amylase, *J. Pharm. Sci.* **93**(5): 1253–1263.

31

CONSIDERATIONS FOR SUCCESSFUL LYOPHILIZATION PROCESS SCALE-UP, TECHNOLOGY TRANSFER, AND ROUTINE PRODUCTION

Samir U. Sane and Chung C. Hsu

31.1. INTRODUCTION

The life cycle of a lyophilized pharmaceutical product can be classified into three main stages. The first stage, lyophilized product development, is where the formulation scientist defines the formulation, product container, and fill volume on the basis of the target product profile, and the process development engineer develops the lyophilization process on laboratory or pilot scale. Several excellent reports exist in the literature that discuss the science behind lyophilized product development along with case studies [1,2,20,22]. The second stage involves scaling up the process to production scale for clinical or commercial supply. Finally, the third stage involves continuously maintaining the clinical or commercial supply, including possible transfers between production facilities and process improvements. This chapter focuses on the last two stages of the life cycle for parenteral products contained in glass vials.

Throughout the life cycle, it is important for the critical product quality attributes of the lyophilized product to be identified; their acceptable ranges defined; and any changes in the process, scale, or facility evaluated for possible impact on those

Formulation and Process Development Strategies for Manufacturing Biopharmaceuticals,
edited by Jameel and Hershenson

attributes. The quality attributes that are unique to a lyophilized product are cake appearance, moisture content, and reconstitution time. These three attributes are generally the key indicators of process performance and can be correlated with biochemical attributes and long-term stability, which may be critical to the product quality [3–5]. Additionally, these three attributes are relatively easy to assess on a large number of samples. As a result, they are widely used during scale-up and routine manufacturing to assess process success.

The second unique aspect of lyophilization is that it is one of the few unit operations where the physicochemical treatment occurs after the product is aliquoted in the final container that is presented to the end user. However, except for cake appearance inspection, all the batch release testing is performed on only a very small number of product vials. Thus, batch uniformity is an important consideration during process scale-up, technology transfer, and routine production of a lyophilized product. Moisture content is especially suited for assessing and validating batch uniformity as will be evident during the course of the discussion in this chapter.

There are three main parts to this chapter. Section 31.2[1] discusses scale-up from lab/pilot scale to production scale, Section 31.3 describes process transfers between production sites, and Section 31.4 discusses supporting routine production. The scope of this chapter is limited primarily to considerations around lyophilization equipment differences during scale-up and process transfers. Indeed, there are numerous other considerations related to vial/stopper processing, filling equipment, product contacting materials, and other issues that may also impact the ultimate quality and stability of the lyophilized product. However, discussion of those factors is beyond the scope of this chapter.

31.2. SCALE-UP

Basic lyophilization equipment comprises of three major components:

1. A drying chamber, which typically consists of multiple shelves on which product containing vials are placed. Heat transfer fluid at a controlled temperature is circulated through the shelves to achieve the desired shelf temperature. The chamber can be sealed and evacuated to attain the low pressure levels required for lyophilization.
2. A condenser, which is generally another vessel connected to the drying chamber by a vapor tube. The condenser contains an array of coils that are cooled to a low temperature (typically below −50°C). These coils function as a water trap by condensing the water vapor from the drying chamber on the coil surface.
3. A refrigeration–heating–vacuum system that is used to control the temperature of the shelves and the condenser as well as to achieve low pressure levels during lyophilization.

[1]The scale-up section of this chapter evolved from a previously published article by Sane and Hsu [6].

These three components are responsible for controlling the overall heat transfer rate to and from the product vials and mass transfer of water vapor from the product vials. The rates of these heat and mass transport processes ultimately dictate the properties of the lyophilized product. Even though the user-specified lyophilization cycle parameters (viz. shelf temperature, chamber pressure and time) may be the same, the heat and mass transfer rates may differ depending on the design and performance of the lyophilizer. Therefore, during lyophilization scale-up it is important to understand the differences in the equipment at two different scales and their impact on lyophilization process dynamics [6].

31.2.1. What Could Change During Lyophilization Scale-Up

Figure 31.1 illustrates the various parameters that influence lyophilization process performance and the properties of a lyophilized product. Among the parameters listed, the "equipment variables" are the most difficult to match on different scales, and adjustments to the "cycle parameters" may be necessary to ensure comparable product quality. Some of those variables are discussed in greater detail below.

31.2.1.1. Environmental Conditions in the Processing Area. Production-scale machines are often operated in a cleanroom environment, while pilot-scale testing is generally performed in a lab environment, which typically has significantly higher particulate loading. Particles from the environment can be introduced in the bulk solution when it is filled and loaded in the lyophilizer. These particles can act as nucleation sites leading to higher ice nucleation temperature during freezing on pilot scale versus

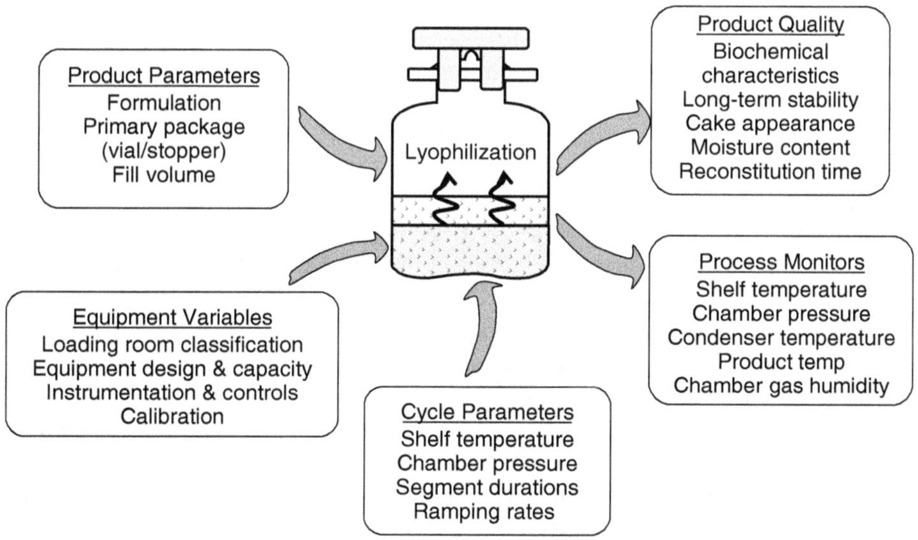

Figure 31.1. Lyophilization process variables.

production scale. It is widely reported in the literature that the nucleation temperature affects the ice crystal size—the lower the nucleation temperature is, the smaller is the ice crystal size and the greater is the ice interfacial area [7]. Small ice crystal size increases the mass transfer resistance offered by the dry cake during primary drying, and hence the material produced on production scale may experience slower ice sublimation and a higher product temperature. Additionally, size of the ice crystals could potentially impact biochemical stability of some proteins due to degradation at ice–liquid or ice–void interface [8]. To minimize the impact of environmental differences during scale-up, care should be taken while generating the pilot-scale data. The majors may include (1) use of vials and stoppers that have been washed using processes equivalent to the manufacturing scale; (2) handling of components and filling inside a laminar flow hood; (3) filtration of the bulk solution prior to filling using a filter equivalent to the manufacturing scale; (4) use of powder-free gloves, lab coat, and hair cover during material handling and filling; and (5) prompt transfer of filled and partially stoppered vials to the lyophilizer.

31.2.1.2. Shelf Surface Temperature. Shelf temperature during lyophilization is generally controlled according to the temperature of the heat transfer fluid (e.g., silicone oil) at the inlet of the shelf. On the other hand, the temperature experienced by the product is the shelf surface temperature. Depending on the differences in the equipment design and the shelf temperature control strategy, the shelf surface temperature may be different in different machines even though the shelf temperature set point is kept the same. Furthermore, production-scale machines tend to have wider temperature variation across shelves, especially under heavy heat load. A temperature difference of 2°C–5°C between the inlet and outlet of a shelf is frequently observed during the freezing ramp and the early stages of primary drying. This could lead to greater variability in the temperature ramping rates, extent of annealing, and sublimation rates among product vials from different regions of a shelf. Figure 31.2 shows an example of shelf inlet and outlet temperature profiles for a manufacturing-scale freeze drier. Clearly in this situation, the vials located near the shelf fluid outlet would experience somewhat slower freezing ramp and lower shelf temperature compared to the vials near the shelf fluid inlet. The lyophilization cycle on production scale has to be appropriately designed to yield uniform product quality despite this variability. Alternatively, worst-case locations need to be identified so that product from those locations can be monitored and tested as a part of validation and batch release analysis.

31.2.1.3. Radiation Effects. For a product vial undergoing lyophilization, the primary mode of heat transfer is gas-phase conduction to or from the shelf on which the vial is placed. However, there is some contribution via radiation from shelves and walls of the chamber. The differences in the shelf spacing, the wall temperature, and the proximity of the walls to the edge vials could lead to varying degree of radiation contributions on different scales. Figure 31.3 shows an example of product temperature profiles during primary drying of vials from different regions of a shelf. The vial near the edge completes primary drying sooner than do those near the center of the shelf. Additionally, the vials near the edge equilibrate to a somewhat higher

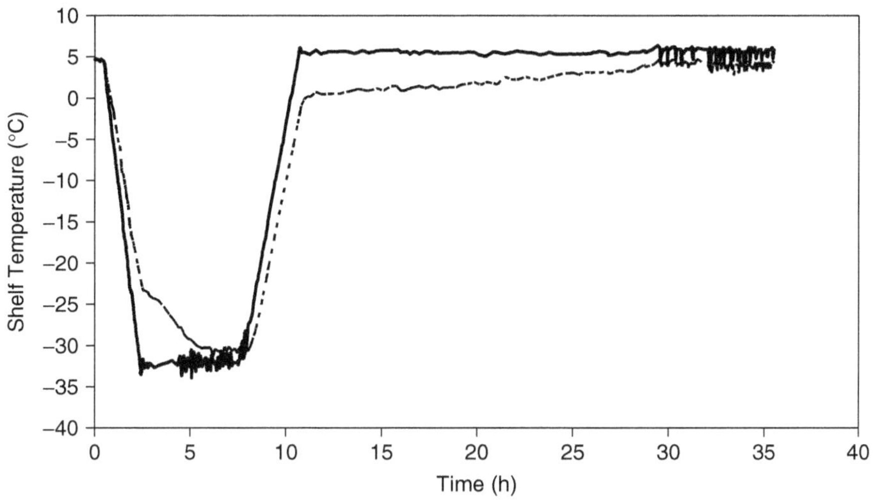

Figure 31.2. Example of shelf fluid inlet and outlet temperature profile during freeze drying in a 50-ft^2 lyophilizer with 45-kg water load (solid line—shelf fluid inlet; dashed line—shelf fluid outlet).

temperature at the end of primary drying. Both of these phenomena are a result of increased heat transfer to the edge vials via radiation. The extent of edge effect depends on the difference in the shelf temperature and the chamber wall temperature as well as the chamber pressure. Although lyophilizers with controlled wall temperature are commercially available, very few of them have such a capability. In lyophilizers without chamber wall temperature control, if the shelf temperature during primary drying is close to ambient temperature, the edge effect is reduced. On the other hand, if the shelf temperature is significantly below ambient (Fig. 31.3), a substantial difference is observed between the drying rate of the edge vials and the vials in the shelf interior. This effect is pronounced even further as chamber pressure decreases and radiation becomes a larger fraction of the overall heat transfer rate. Clearly in this case, if one relies on product temperature data from the edge vials, primary drying duration would be underestimated.

It can be argued that the edge vials may not necessarily experience higher product temperature during the sublimation phase of primary drying. The majority of the heat transfer due to radiation from chamber walls occurs in the radial direction through the walls of the product vial. If the product has a tendency to shrink, it can provide a low-resistance path for water sublimation via the gap between the vial wall and the cake. Effectively, the overall mass transfer resistance to water vapor flow decreases and the product temperature is relatively unaffected despite higher heat input to the product.

31.2.1.4. Chamber Pressure Control. The pressure control strategy could also have an influence on process dynamics because of its impact on chamber gas composition during drying. The majority of the commercial-scale lyophilizers today

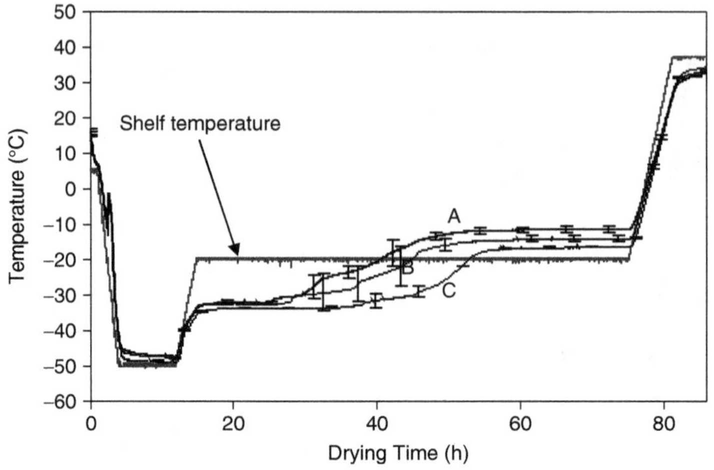

Figure 31.3. Product temperature profiles: radiation effect on edge vials. The data were gathered in a 20-ft^2 pilot-scale lyophilizer with transparent acrylic door and stainless-steel walls without wall temperature control. Shelf dimensions were 60 cm wide and 90 cm deep, while the chamber pressure during primary drying was 100 mTorr. Curve A—vials located in the front row facing lyophilizer door; curve B—vials located in the back row closest to the rear wall of the chamber; curve C—vials located in middle rows, roughly in the center of the shelf.

rely on introduction of nitrogen, at a controlled rate, to maintain the chamber pressure. However, some older lyophilizers may rely on the so-called on/off control where pressure in the chamber is controlled by opening and closing the vacuum valve without introducing any nitrogen. When nitrogen is used for chamber pressure control, the mole fraction of water vapor during drying is less than unity. This is especially true when working with small-vial loads relative to the capacity of the equipment or with freeze-drying cycles with low sublimation rates. On the other hand, with on/off control the drying chamber contains exclusively water vapor. Since primary mode of heat transfer from shelves to vials is gas phase conduction and since nitrogen has conductivity lower than that of water vapor [9], the differences in chamber gas composition could lead to differences in heat transfer rates and subsequently in primary drying duration. It is important to take these dependences into consideration while comparing data from different scales with different relative loadings and different pressure control strategies.

31.2.1.5. Type of Measuring Devices, Tolerances, and Calibrations.

The two most common types of gauges for chamber pressure (vacuum) measurement during lyophilization are capacitance manometer and Pirani. The latter is not recommended for chamber pressure control since its response is dependent on the composition of gas in the chamber. It is important to use the same type of gauge on pilot and production scales in order to ensure equivalent conditions for freeze drying. There could be differences in the shelf temperature and chamber pressure control tolerances and transducer calibrations on the two scales, which could lead to small differences in process dynamics.

Although any one of the factors discussed above may not be sufficient to significantly affect the performance of a lyophilization process, the combination of these factors, if not accounted for upfront, could lead to significantly different process performance upon scale-up.

31.2.2. Lyophilization Process Scale-Up Methodology

The main objective of lyophilization process scale-up is to specify a robust lyophilization cycle that can be executed reproducibly and yields uniform product attributes consistent with the material manufactured on pilot or clinical scale. To be successful in achieving this objective, one of the keys is to have a well-characterized formulation. The thermal characteristics of the formulation, in frozen state, such as the collapse temperature, should be evaluated to determine the critical product temperature that should not be exceeded during early stages of primary drying to avoid impact on product quality [10,11]. It is also important to characterize the lyophilized product for the target moisture range, cake appearance, reconstitution time, and any other critical attributes so that appropriate acceptance criteria can be developed for testing product from production scale.

Another key to successful scale-up is to perform pilot-scale process development while taking into account the capabilities of the production-scale machine. As was discussed in the previous section, the production scale equipment tends to have greater variability in shelf surface temperature and also some limitations in terms of maximum achievable ramping rates and condenser capacity. The pilot-scale process parameters need to be appropriately selected to account for these limitations (e.g., slower freezing ramp, longer primary drying duration). Additionally, vial and stopper components should be identical and be preprocessed in the same way as those for the production scale.

Once the lyophilization process parameters are established on pilot scale, one to four engineering runs should be performed on production scale to confirm that the equipment can execute the intended lyophilization cycle routinely and yields product of acceptable quality and uniformity. If the pilot-scale cycle has been properly designed and with adequate safety margin, no adjustments to the shelf temperature and chamber pressure set points and ramping rates for production-scale cycle should be necessary. Only adjustments to the drying durations may be required because of the scale-dependent factors described in the previous section. Often, active product is not available in sufficient quantities to perform these engineering runs. In these situations, a mixed load of active and surrogate vials can be used. To demonstrate that the product quality is uniform regardless of the location in the lyophilizer, groups of active vials can be dispersed throughout the lyophilizer and analyzed for cake appearance, moisture content, and any other relevant product quality attributes.

Engineering runs can also be utilized to demonstrate that the process on production scale is robust and to rationalize the choice of alarm limits. This can generally be accomplished with two runs. In one of the runs, the lyophilizer cycle can be set at the lower limits of primary and secondary drying shelf temperature and chamber

TABLE 31.1. Example Lyophilization Cycle Conditions for Robustness Testing

Parameter	Target	Run 1: Lower Process Limits	Run 2: Upper Process Limits
Primary drying			
Shelf temperature, °C	0	−3	3
Chamber pressure, mTorr	120	100	140
Secondary drying			
Shelf temperature, °C	20	17	23
Chamber pressure, mTorr	120	100	140
Time, h	8	7	9

pressure, and secondary drying duration. In the other run, the lyophilizer can be set at the upper limits of primary and secondary drying shelf temperature and chamber pressure, and secondary drying duration. Table 31.1 presents an example of target cycle conditions and the corresponding two runs that challenge the lower and upper limits of the process. It is well established that for the conditions typically used for vial lyophilization, the sublimation rate as well as the product temperature during early stages of primary drying increase with increasing shelf temperature and chamber pressure [12]. Additionally, higher temperature during secondary drying is expected to lead to lower moisture content. Thus, the choice of parameters presented in Table 31.1 (Run 1 and Run 2) brackets the entire range of product temperatures, sublimation rates, and residual moisture contents that may be observed within the alarm limits of the intended lyophilization cycle.

Once the lyophilization cycle on production scale is established and confirmed to be robust, it can be validated [3]. At this point, having comprehensive data from the engineering runs maximizes the probability of success for the validation batches. If the product uniformity has been well characterized during engineering runs, a reduced level of sampling for validation testing may be justifiable.

31.2.3. Key Considerations During Lyophilization Scale-Up

31.2.3.1. Equipment Capabilities. The production-scale machines generally deliver lower cooling and heating capacity per unit shelf area as compared to the

TABLE 31.2. Example Lyophilizer Performance With and Without Water Load

Test	No Load	With 1300 L Water
Shelf cooling rate (20°C to −35°C)	2.3°C/min	0.7°C/min
Shelf heating rate (−40°C to 20°C)	1.2°C/min	0.9°C/min
Dynamic condenser capacity	—	∼ 50 kg/h at 150 mTorr
Shelf temperature control at 50°C	± 0.5°C around set point	+1°C to −2.5°C around set point

Data from 56 m^2 lyophilizer designed for 1300 L maximum condenser capacity pilot-scale machines. One should be aware of the maximum ramping rates that can be executed on production scale with full product load. Lyophilizer manufacturers generally specify the maximum ramping rates for the empty chamber. However, the performance with product load tends to be significantly slower and should be determined during qualification of the machine (Table 31.2).

Another equipment parameter that could be potentially limiting is dynamic condenser capacity of the production-scale machine. This parameter dictates the maximum rate at which water can be captured (kg water/h) by the condenser while maintaining the process set points. Formulations with high collapse temperature (e.g., high-concentration protein formulations) can be freeze-dried using fairly aggressive conditions during primary drying. However, if the condenser is not able to capture the resulting water vapor at the same rate, then pressure control could be lost. Lyophilizer manufacturers often specify only the ultimate condenser capacity. The dynamic condenser capacity can be estimated experimentally by determining maximum achievable ice sublimation rate for water-filled vials or trays before the lyophilizer starts to lose chamber pressure control [13]. The dynamic condenser capacity is dictated by three characteristics that depend on the lyophilizer design: (1) capacity of the refrigeration system employed to cool the condenser; (2) overall resistance to vapor flow from chamber to the condenser coils, which is dictated by cross-sectional area, length, and shape of the flow path; and (3) Mach number for the water vapor flow from chamber to condenser. All three of these factors are directly or indirectly dependent on chamber pressure set point. The lower the chamber pressure set point, the lower is the dynamic condenser capacity. For instance, lowering the chamber pressure set point from 150 to 100 mTorr lowers the maximum allowable condenser temperature required for good pressure control. For a refrigeration system using piston compressors, this lower condenser temperature requirement could lower the overall cooling capacity by as much as 20% [14]. Therefore, the overall dynamic condenser capacity is expected to be lower by 20% as well. Figure 31.4 shows an example of a case where chamber pressure control was lost since the product sublimation rate during the early stages of primary drying exceeded the dynamic condenser capacity at the chamber pressure set point for the process. The same lyophilizer was able to maintain good pressure control under revised primary drying conditions, where the chamber pressure set point was increased by 20 mTorr and the shelf temperature was decreased by 5°C. The sublimation rate, and hence the condensation load, were roughly the same in both cases but the higher chamber pressure set point indirectly enhanced the ability of the condenser to capture water vapor.

31.2.3.2. Choice of Surrogate Formulation. As discussed in Section 31.2.2, during engineering runs on production scale, the lyophilizer can be loaded mostly with surrogate vials with groups of active vials dispersed across the lyophilizer. For this approach to be valid, the surrogate would have to mimic the freeze drying behavior of the active product. Accordingly, pilot-scale studies should be performed to demonstrate that the sublimation rates and product temperature profiles of the surrogate vials are comparable to that of the actives. Examples of sublimation rate measurements for an antibody formulation (active), the corresponding

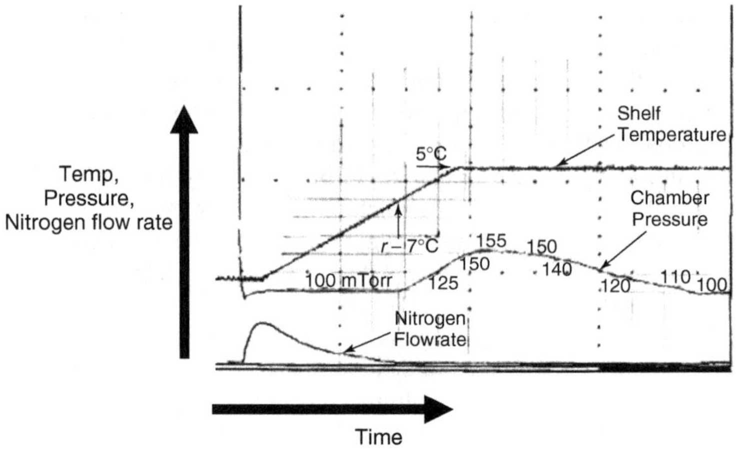

Figure 31.4. Loss of chamber pressure control during primary drying. For primary drying, the chamber was evacuated to 100 mTorr and the shelf temperature was ramped up from −35°C to 5°C. As the shelf temperature reached approximately −7°C, the process lost control on chamber pressure as indicated by increase in pressure to 155 mTorr. The corresponding condenser temperature was around −38°C, indicating that the pressure control was lost because of inadequate condenser cooling capacity. As the primary drying progressed, the overall sublimation rate decreased, the condenser temperature decreased to −40°C, and the process regained control on chamber pressure. (Condenser temperature data not shown.)

formulation buffer, and a surrogate formulation are shown in Figure 31.5. The protein to formulation: excipient weight ratio for this particular antibody formulation was about 1:1. Hence, when only the formulation buffer was freeze dried, the mass transfer resistance to sublimation was significantly lower than that for the active and the corresponding sublimation rate was significantly higher. On the other hand, the surrogate where the protein in the active formulation was replaced by a sugar with the same weight concentration yielded a sublimation rate that was comparable to that of the active. Use of such a surrogate during engineering runs ensures that the active vials experience the same conditions and the lyophilizer experiences the same thermal load as it would when the chamber is filled exclusively with active vials.

31.2.3.3. Lyophilization Cycle Monitoring. As part of the engineering runs, it is important to collect product temperature data during the lyophilization cycle. This allows one to confirm that the product is exposed only to intended temperatures (e.g., not exceeding the collapse temperature) during various stages of lyophilization. Additionally, from the data, one can confirm that the product temperatures have reached shelf temperature prior to start of the secondary drying ramp. If the product temperature information is available in real time, adjustments can also be made to the primary drying duration on the fly, in order to accommodate any ice sublimation rate differences that may be present between the pilot and production scales.

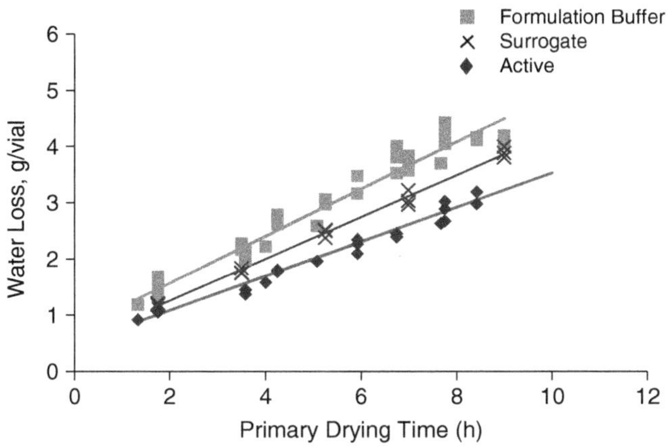

Figure 31.5. Comparison of sublimation rates for an antibody formulation. The protein constituted 50% of the total solids content of the formulation. The surrogate had the same solids concentration as the active material.

Often on production scale, especially in machines with automatic loading/unloading systems, use of product temperature probes on a routine basis is not possible. In such situations, one could rely on other measurements such as condenser temperature, chamber–condenser pressure differential, nitrogen bleed flow rate, and pressure rise measurement (PRM) to assess the progress of the primary drying phase. All these parameters except PRM can be monitored continuously and are indirect measures of sublimation rate. See Figure 31.6a for an example of these measurements along with product temperature data on production scale. During the early stages of primary drying when the sublimation rate is high, the condenser temperature is high because of large water load, the chamber–condenser pressure differential is high because of the too high volumetric flow rate of water vapor and the nitrogen bleed rate is low since the chamber pressure is being maintained mostly by the water vapor. As the primary drying is completed and the ice sublimation rate decreases, the condenser temperature and the chamber–condenser pressure differential decrease and the nitrogen bleed flow rate increases.

A PRM, on the other hand, is a discrete measurement where the rate of pressure increase in the chamber is measured while it is temporarily isolated by closing the valve between the chamber and the condenser. The rate of pressure rise measured in this manner is an indication of the amount of residual water in the product. Figure 31.6b shows an example of PRM results as a function of primary drying duration; as the primary drying progresses and the product temperatures approach the shelf temperature (Fig. 31.6a), the rate of pressure rise rapidly decreases. It should be noted that during the PRMs shown in Figure 31.6b, the chamber pressure was allowed to increase by only 20–30 μm Hg and the measurements lasted for only a few seconds. This is important because if the pressure is allowed to increase excessively while the ice sublimation phase is not complete, the product temperature could significantly increase and the cake could collapse.

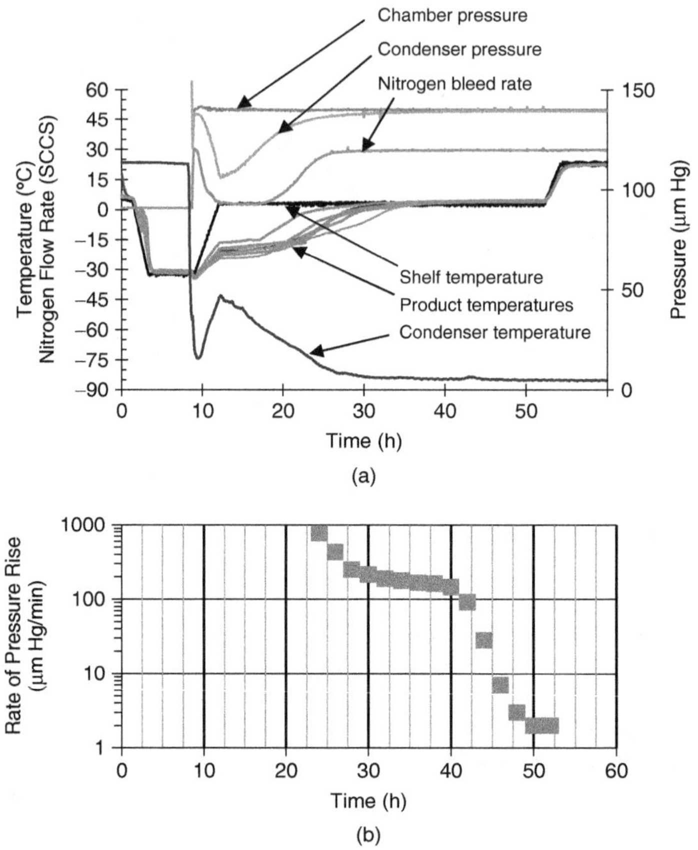

Figure 31.6. (a) Temperature and pressure profiles from an engineering run on production scale; (b) pressure rise measurement results from an engineering run on production scale. The corresponding temperature and pressure profiles from this run are presented in (a).

The condenser temperature, chamber pressure differential, nitrogen bleed rate, and PRM data in Figure 31.6 correlate well with the product temperature data and demonstrate that they can be useful in monitoring the progress of a lyophilization cycle during routine production. This type of correlation can be established during engineering runs on production scale so that their use during routine production can be justified.

31.3. LYOPHILIZATION PROCESS TRANSFER FROM ONE PRODUCTION SITE TO ANOTHER

Often, once a product is launched into market, it may become necessary to transfer the existing validated process from one production facility to another. These transfers are generally triggered by a need for additional capacity or dual sourcing to mitigate supply risks. The complexity and the amount of work required during these transfers is dependent on the extent of equipment and procedural differences between two sites. As a result, the first step toward successful interfacility transfer is to perform detailed gap analysis and identify risk areas that have the potential to impact process performance and/or product quality. If a contract manufacturing organization (CMO) is involved in the transfer, that could present some unique challenges as well. Table 31.3 lists some of the key technical areas that need to be included in the gap analysis during transfer of a lyophilized drug product. As part of the new CMO selection process, it is important to look at each of the technical areas listed in Table 31.3 in order to develop a realistic statement of work and risk management plans.

When it comes to the lyophilization part of the process transfer, the gap analysis should include a detailed assessment of the lyophilizer equipment specifications and performance parameters to establish anticipated comparability (or lack thereof) in process performance. Table 31.4 lists the lyophilizer characteristics that are relevant to comparability assessment. Essentially, any equipment parameter that may impact shelf surface temperature and chamber pressure during the course of the lyophilization cycle should be compared between facilities. Such data are generally available in the equipment's turnover package and qualification documents. Often, there may be significant mechanical differences between lyophilizers; however, the expected performance of the receiving site's equipment might still meet or exceed the process needs. Gap analysis may reveal some equipment limitations that can be addressed prior to proceeding with the process transfer. For example, if the dynamic condenser capacity of the lyophilizer at the receiving site is inadequate for the intended process, reduction in the batch size and/or a change to more conservative lyophilization cycle parameters may be necessary.

Unless significant prior data with similar product formulations exists at the receiving site, one cannot rely exclusively on comparison of equipment parameters on paper. Engineering runs with mixed loads of active and surrogate vials are often necessary to generate supporting data to establish the product comparability. These runs may include comparison of process trends (shelf temperature, chamber pressure, chamber Pirani gauge values, PRMs, product temperature, condenser temperature), evaluation of product manufactured at alarm limits, and moisture mapping to establish batch uniformity.

Facility change for lyophilized drug product manufacture generally requires process revalidation and establishment of product comparability. One may be able to employ a family or modular approach to validation when multiple products or configurations are involved in the transfer or when there is an existing body of data showing comparability for similar product formulations from the two sites.

31.4. SUPPORTING ROUTINE PRODUCTION

Lyophilization is often the last step in the manufacture of a pharmaceutical product. In some cases, a batch of lyophilized product could be worth millions of dollars.

TABLE 31.3. Key Technical Areas for Gap Analysis During Technology Transfer of a Lyophilized Drug Product

Facilities	
	Room classifications
	Environmental monitoring
	Storage areas for raw materials, bulk active pharmaceutical ingredient (API), intermediate, final product
	Temperature/humidity control
	Single use/multiple use
	Cleaning, carryover
Raw materials/accessories	
	Specifications
	Storage and handling
Manufacturing equipment (specifications, capacity, process control, cleaning/sterilization procedures)	
	API handling
	Final formulation
	Pooling/mixing
	Component preparation (e.g., vials, stoppers)
	Product sterilization
	Filling
	Lyophilization
	Inspection
	Packaging
	Product contacting materials (extractables/leachables, product compatibility)
Stress factors during processing	
	Temperature
	Hold/processing times
	Shear
	Air–liquid interface
	Humidity
Shipping of materials between sites	
	Shipping configuration
	Mode of transportation
	Temperature control

TABLE 31.4. Lyophilizer Characteristics for Comparability Assessment

Key Freeze-Dryer Performance Parameters	Lyophilizer Design Attributes that Could Influence Freeze-Dryer Performance
Heat Transfer	
Shelf surface temperature uniformity with and without load	Heat transfer fluid (HTF) path Shelf material of construction Shelf and shelf wall thickness *Shelf thermal mass*[a] HTF heat capacity, viscosity, and density HTF *circulation flow rate at various temperature set points* HTF distribution at the edges of the vial pack
Shelf temperature control with and without load Average deviation from set point over time Maximum controllable cooling and heating rates Ultimate low and high temperatures	*HTF cooling capacity* *HTF heating capacity* HTF temperature control strategy (based on shelf inlet vs. outlet temperature) Temperature measuring devices (accuracy, calibration)
Radiation effects	Shelf interdistance Shelf material of construction and surface finish Chamber walls (temperature control or not) Chamber wall material of construction and surface finish Proximity of chamber walls to product
Mass Transfer	
Pressure control Fluctuations/average deviation from set point over time Evacuation rate	*Vacuum system throughput* Control strategy (nitrogen inlet regulation vs. without nitrogen injection) Pressure transducer (capacitance manometer vs. Pirani type)
Dynamic condenser capacity *Maximum achievable water/condensation rate at various pressure set points*	*Refrigeration system design and capacity* *Condenser surface area and coil arrangement* Chamber–condenser flow path *Cross-sectional area of flow path* *Shape and length of flow path* Presence of any flow obstructions (e.g., radiation shield)
Facility	
Environment	Classification of filling area (particulate loading)

[a]The parameters listed in *italics* should be normalized on the basis of shelf area when comparing lyophilizers of different capacities.

Additionally, the loss of a batch during this final stage of manufacture could have far reaching impact on product's supply, company's production plan, and clinical trial timelines. Thus, maintaining high success rates during routine production of lyophilized products is essential. Keys to maximizing the success rate of a lyophilization operation are

1. Well-trained personnel with robust standard operating procedures
2. Well-engineered equipment and support systems with comprehensive maintenance program
3. Robust lyophilization process design
4. Vigilant engineering staff to collect and analyze performance trend data

In many facilities an empty-chamber plant test is performed prior to every batch to ensure that the equipment is ready for manufacture. Trending some of the plant test data (e.g., shelf cooling rate, evacuation rate, leak rate) may facilitate identification of potential issues before a failure occurs.

Even for a robustly designed lyophilization cycle and in a well-maintained facility, occasional deviations outside of the alert or alarm limits can result from mechanical failures or operator error. In such situations it is important to have quality systems and process or product experts who can quickly detect issues, troubleshoot, assess product impact, and implement corrective and preventive actions. Some of the more common failures and associated process deviations are listed below:

1. Extended freeze hold duration due to inability to reach target chamber evacuation level
2. Chamber pressure increase due to a variety of reasons (e.g., partial loss in condenser cooling capacity, nitrogen injection valve failures)
3. Shelf safety freeze due to increase in chamber pressure
4. Extension of secondary drying time due to stoppering issues

In these situations, depending on the type of deviation, supplemental testing may be required before the batch can be considered releasable to market. This testing may include additional product uniformity testing (e.g., cake appearance, moisture and reconstitution time mapping), characterization of the dried product (e.g., biochemical testing, crystallinity), and long-term stability (real-time and/or accelerated). Depending on the manufacturing license and the type of deviation, communication with health authorities may also be required. These activities not only are highly resource-intensive but also could have far reaching impact on the product's supply chain. Alternatively, one could prospectively define ranges for the various critical process parameters within which no product impact is expected, although the process may have deviated from normal equipment capability. In the current regulatory environment that encourages quality by design, this latter approach is greatly preferred if not expected.

31.4.1. Design Space for Lyophilization

In FDA's quality-by-design (QbD) framework, the multidimensional space that spans all possible combinations of the process inputs and parameters yielding acceptable product quality is termed *design space*. For the lyophilization cycle, shelf temperature, chamber pressure, and time are the three critical process control parameters. The interplay among these three parameters ultimately dictates the freeze-drying process performance and the product characteristics. The theory of lyophilization process dynamics is well established [15] and is highly conducive to rationally defining acceptable ranges of shelf temperature, chamber pressure, and time for a given process [21]. Tables 31.5–31.7 assess the potential impact of the critical control parameters on product quality for freezing, primary drying, and secondary drying. In each case, an example of a typical target and the associated acceptable range are also provided. The underlying principles in selection of these acceptable ranges, with the exception of freezing ramp, can be summarized as follows:

- Maintain product temperature below a critical temperature during every segment (e.g., below the phase transition temperature during freezing, below the collapse temperature during primary and secondary drying).
- During drying, maintain the product temperature high enough such that the drying processes are complete in the allotted segment durations.

31.4.1.1. Freezing. Two of the freezing segment parameters that are especially important are the freeze ramp rate and the final freezing temperature. The latter is generally chosen to ensure that the product is cooled below the phase transition temperature of the formulation (eutectic temperature or the glass transition temperature), and solidified prior to start of primary drying. Negative deviations in temperature (i.e., colder than the set point) are generally not an issue from a molecular stability perspective since the product is already solidified and there is limited mobility for the molecules to undergo degradative reactions. For some formulations, with high fill height : vial height ratio, vial breakage has been observed on cooling below a certain critical temperature (unpublished data).

Considerations around freezing ramp rate are more complex since there are multiple phenomena involved. The general understanding in the literature is that the faster the freezing ramp, the smaller are the ice crystals. This could result in higher resistance to mass transfer of water vapor through the partially dried cake, higher product temperature, and longer primary drying. From a molecular stability perspective, there are some reports in the literature suggesting that biomolecules (e.g., proteins) can degrade on exposure to ice–liquid surface [8,16]; smaller ice crystals resulting from faster freeze ramps would increase ice–liquid interfacial area. On the other hand, a slower freeze ramp increases the time for which the molecules experience potentially damaging cryoconcentrated state before solidifying below the phase transition temperature. Changes in freeze ramp rates could also impact the presence or extent of phase separation during freezing. This is especially of concern for formulations that

TABLE 31.5. Lyophilization Process Parameter Ranges and Rationale for their Selection: Freezing[a]

Lyophilization Cycle Parameter	Relevance	Example Settings: Target	Example Settings: Acceptable Limits (Lower limit–Upper limit)	Rationale for Lower and Upper Limit Selection
Initial equilibration time and shelf temperature	Ensures uniform starting point for product freezing	1 h	0.75–48 h	Lower limit on equilibration time is selected to ensure that entire content of all vials have same starting temperature
		5°C	2°C–8°C	Upper limit on equilibration time is dictated by the available stability data at the equilibration temperature
Freezing rate	May affect cake structure, morphology, drying resistance	20°C/h	10°C–30°C/h	Freezing rate may affect ice crystal size, mass transfer resistance, and product temperature during drying; also, appearance and crystallinity of the lyophilized drug product may be affected May need product quality and long-term stability data on lyophilized product at limiting freezing rates

Final freezing temperature	Ensures complete solidification	−40°C	NL[b] to −30°C	Generally no lower limit; some formulations may cause vial breakage if frozen below certain temperature Upper limit is generally the glass transition temp T'_g or the eutectic melting temperature (T_{eu}) to ensure complete solidification
Freeze hold duration	Ensures complete solidification of vial contents	5 h	3–NL h	Lower limit selected to ensure freezing of the entire content of each product vial in the batch Generally no upper limit as long as the product is maintained below T'_g or T_{eu}
Pressure during freeze hold	—	1 atm (except during evacuation)	NL–NL	Generally inconsequential; minimal drying is expected because of low product temperatures and essentially no driving force

[a] The information in this table is applicable in the majority of situations, but exceptions do exist.
[b] NL = no practically relevant limit from product quality perspective.

TABLE 31.6. Lyophilization Process Parameter Ranges and Rationale for their Selection: Primary Drying[a]

		Example Settings		
Lyophilization Cycle Parameter	Relevance	Target	Acceptable Limits (Lower limit–Upper limit)	Rationale for Lower and Upper Limit Selection
Primary drying shelf temperature ramp	Ensures gradual start to drying	10°C/h	NL[b]–15°C/h	Generally no lower limit Excessively fast ramp could cause a sublimation burst leading to loss of process control or product egress from the vials
Primary drying shelf temperature	Ensures fast drying while preventing cake collapse	5°C	0°C–15°C	Lower limit selected to ensure adequate sublimation rate so that primary drying is complete within the target primary drying duration Upper limit selected such that the resulting higher product temperature doesn't cause cake collapse and impact on critical product attributes; additionally, the resultant sublimation rate shouldn't exceed the dynamic condenser capacity

Primary drying chamber pressure	Ensures fast drying while maintaining acceptable cake appearance	120 mTorr	90–170 mTorr	Lower limit selected to ensure adequate sublimation rate so that primary drying is complete within the target primary drying duration Upper limit selected such that the resulting higher product temperature doesn't cause cake collapse and impact on critical product attributes
Primary drying duration	Ensures completeness of primary drying prior to start of secondary drying ramp	30 h	28–NL h	Lower limit selected such that the primary drying is complete at the target primary drying conditions Generally no upper limit; excessively long exposure to primary drying temperature may eventually cause product degradation

[a]The information in this tables is applicable in the majority of situations, but exceptions do exist.
[b]NL = no practically relevant limit from product quality perspective.

TABLE 31.7. Lyophilization Process Parameter Ranges and Rationale for their Selection: Secondary Drying[a]

Lyophilization Cycle Parameter	Relevance	Example Settings: Target	Example Settings: Acceptable Limits (Lower limit–Upper limit)	Rationale for Lower and Upper Limit Selection
Secondary drying ramp	Ensures gradual start to secondary drying while maintaining acceptable cake appearance	10°C/h	NL[b]–15°C/h	Generally no lower limit; slower ramp will only further ensure that there is no product collapse during ramp Upper limit selected such that the resulting higher product temperature doesn't cause cake collapse and impact on critical product attributes
Secondary drying shelf temperature	Ensures fast secondary drying without exposure of product to excessively high temperatures	25°C	20°C–35°C	Lower limit selected to ensure adequate drying rate so that target moisture content is achieved within the target secondary drying duration Upper limit selected to prevent exposure of product to temperatures that could impact critical product attributes within the timescale of secondary drying duration

Secondary drying chamber pressure	Ensures fast secondary drying	120 mTorr	60–250 mTorr	Generally doesn't significantly impact secondary drying kinetics; excessively low or high chamber pressures could lead to incomplete secondary drying within the allotted duration
Secondary drying duration	Ensures completion of secondary drying	8 h	6–24 h	Lower limit selected to ensure completion of secondary drying at the target secondary drying conditions Generally longer durations don't have a significant impact; in some cases, product may be unstable to extended exposure to secondary drying temperature

[a]The information in this table is applicable in the majority of situations, but exceptions do exist.
[b]NL = no practically relevant limit from product quality perspective.

contain crystallizing solutes. However, to see any impact on protein stability, the freeze ramp rates typically need to be varied much more drastically than what is feasible in manufacturing-scale lyophilizers. Hence, examples of product degradation due to changes in the freeze ramp rates on manufacturing scale are rare. A conservative approach would be to prepare lyophilized product at limiting shelf freezing rates and assess product quality and long-term stability as part of the lyophilized product development.

The freeze hold duration is generally inconsequential as long as a certain minimum freeze hold time is reached and the product is maintained below its phase transition temperature. Because of lack of mobility of the molecules at these temperatures, any degradation or morphological changes are highly unlikely.

The considerations are more complex when annealing is used as part of the freezing regimen. The annealing temperature as well as duration of hold at that temperature would have an impact on the type and extent of phase separation (e.g., crystallization), as well as the ice crystal size. As a result, there may be additional constraints on the temperature and durations of the various segments of the freezing regimen [17,18].

31.4.1.2. Primary Drying. During primary drying, the important parameters are shelf temperature ramp rate, hold temperature, hold duration, and chamber pressure.

As described in Table 31.6, a slower than target primary drying shelf temperature ramp is inconsequential. A faster than target ramp rate could temporarily overwhelm the condenser because of rapid sublimation of water at the top surface of the product combined with sublimation of any condensate that accumulated on the shelves during loading. For products that yield very low density and fluffy cakes, there is also a risk of product egressing from the vial because of an initial burst in sublimation.

Considerations involved in selecting lower and upper limits for shelf temperature and chamber pressure are very similar. For most lyophilization cycles and products, an increase in shelf temperature or pressure causes an increase in product temperature and sublimation rate [12]. Therefore, the upper limits for both these parameters are selected such that the product collapse does not occur. On the other hand, negative deviations (lower temperature or lower chamber pressure) generally cause a decrease in product temperature and sublimation rate. Hence the lower limit is selected such that the primary drying is still complete within the specified primary drying duration prior to the start of secondary drying.

The extension of primary drying duration is generally inconsequential since it would only further ensure that primary drying is completed prior to start of the secondary drying ramp. In these situations, the product would just equilibrate at the shelf temperature. Once again, there is limited mobility for the molecules for any degradative reactions or morphological changes to occur on timescales relevant to lyophilization, and any impact on product quality is unlikely.

31.4.1.3. Secondary Drying. During secondary drying, the important parameters are shelf temperature ramp rate, hold temperature, and hold duration.

It is well known that the secondary drying shelf temperature ramp rate should be slow enough such that the product temperature does not exceed the instantaneous collapse temperature of the partially dried cake. Slower than target ramp rate is inconsequential since it will only further ensure that the product is maintained below the collapse temperature.

The secondary drying kinetics is rarely impacted by the chamber pressures typically used during vial lyophilization [19]. Regardless of chamber pressure, the partial pressure of water in the drying chamber is extremely low, and hence there is sufficient driving force for moisture desorption. On the other hand, the secondary drying kinetics is generally a strong function of temperature. Hence the shelf temperature during secondary drying hold should be high enough such that target moisture content is achieved within the target duration. The upper limit of shelf temperature as well as the duration is dictated by the available stability data. In some cases the product may be unstable on excessively long exposure to secondary drying temperature.

31.4.1.4. Multidimensional Design Space. The information in Tables 31.5–31.7 and the discussion above assume deviations in a single process control parameter at a time. One can extend this analysis to encompass deviations in multiple parameters simultaneously. For the freeze hold segment, this is relatively straightforward since the chamber pressure has little influence on the freezing process, and the remaining two control parameters of shelf temperature and time do not interact (Fig. 31.7).

For primary drying, on the other hand, one has to take into account the delicate balance between the control parameters of shelf temperature and chamber pressure, and the key dependent variables of product temperature and sublimation rate. Typical relationships among these four variables are shown in a sketch in Figure 31.8a. The sketch was adapted from experimental data presented in the paper by Chang and Fischer [12], who measured sublimation rates and product temperatures for a protein

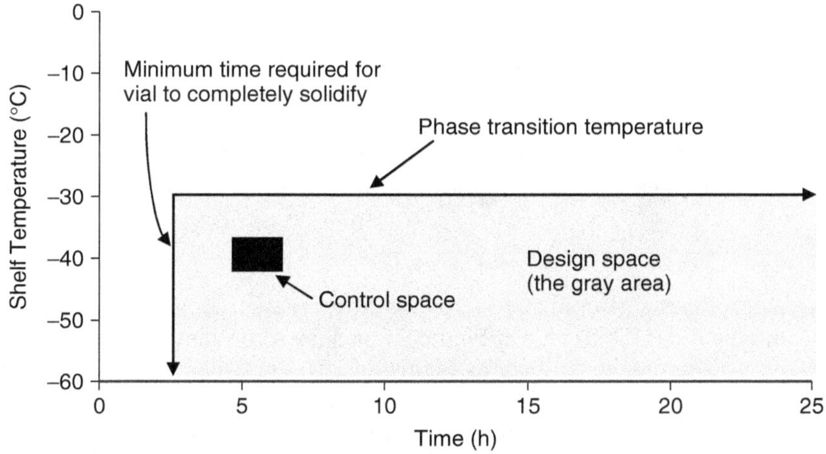

Figure 31.7. Design space for freeze hold. (Chamber pressure is generally inconsequential during freezing and hence not included in the illustration.)

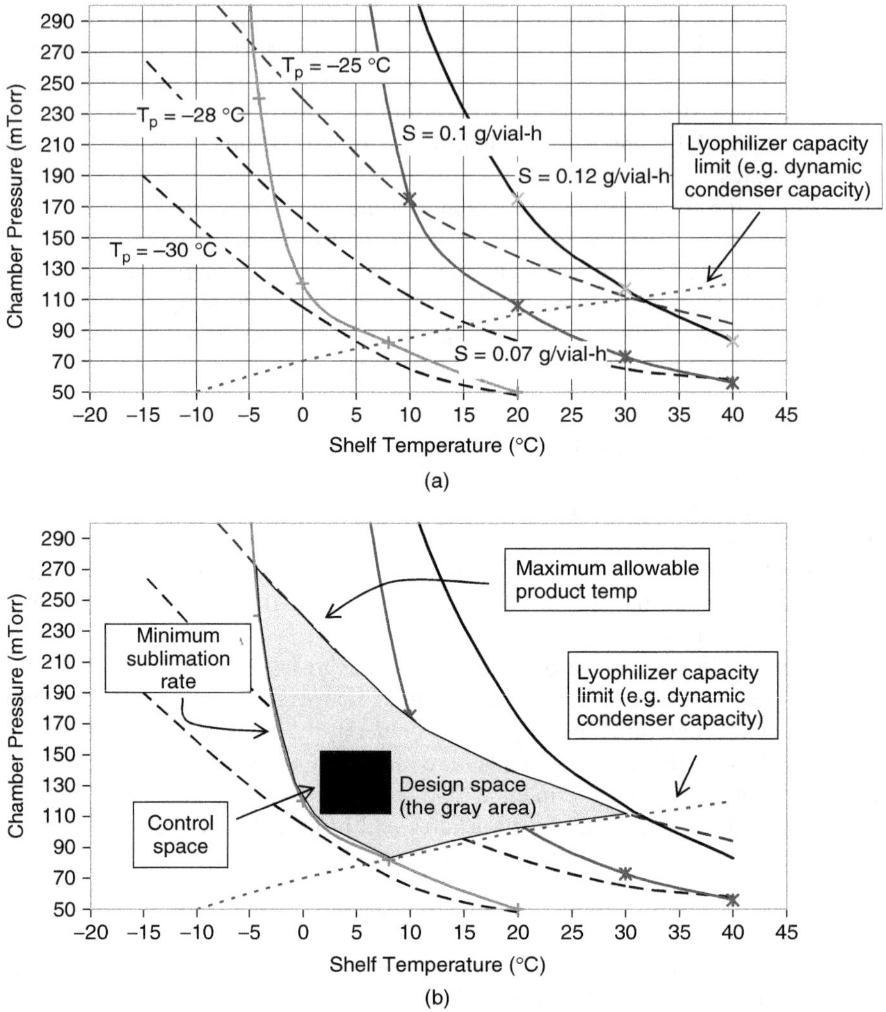

Figure 31.8. Design space for a fixed primary drying duration (solid lines—constant sublimation rate; dashed lines—constant product temperature; dotted lines—lyophilizer capacity limit).

formulation containing glycine and sucrose at different shelf temperature and chamber pressure combinations. The chart in Figure 31.8a shows constant sublimation rate and constant product temperature lines as a function of shelf temperature and chamber pressure. The chart also shows an example of the equipment's dynamic condenser capacity limit in terms of minimum achievable chamber pressure for a particular product configuration as a function of shelf temperature. In other words, the process would be out of control for any combination of shelf temperature and chamber pressure that causes the product's sublimation rate to exceed the dynamic condenser capacity.

Using the information summarized in Figure 31.8a, one can carve out the design space during primary drying of the product configuration in the corresponding lyophlizer (Fig. 31.8b). For illustration purposes, the collapse temperature for this product was selected to be −25°C. This sets one boundary of the design space for the primary drying segment. Another boundary can be set according to the equipment capacity limit (e.g., the lyophilizer's dynamic condenser capacity). The third and final boundary is set according to the allotted primary drying time, which dictates the minimum sublimation rate necessary to complete primary drying. In this illustration the minimum required sublimation rate was selected to be 0.07 g per vial per hour. Thus, the design space for primary drying hold segment for a fixed primary drying duration is fully defined. One could expand the design space even further if the primary drying duration is allowed to extend (by design) beyond the target. It is important to recognize that the design space for primary drying could also be impacted by the freezing regimen. If one introduces an annealing step, this will likely alter the cake resistance during primary drying, thereby altering the constant product temperature and constant sublimation rate lines in Figure 31.8.

For secondary drying, as discussed earlier, the chamber pressure is inconsequential. Hence only the shelf temperature and time need to be considered (Fig. 31.9). It is expected that the time required to achieve target residual moisture levels in the product will decrease as the shelf temperature is increased. This sets the lower limit on the secondary drying time. The upper limit would depend on the available stability data. With increasing temperature the allowable exposure time for which the product is stable, is expected to decrease.

The discussion here was limited to the main lyophilization process control parameters of shelf temperature, chamber pressure, and time. There are other factors that may influence the design space available during normal production. For instance, variations in the vial geometry and stopper placement could affect the heat and mass

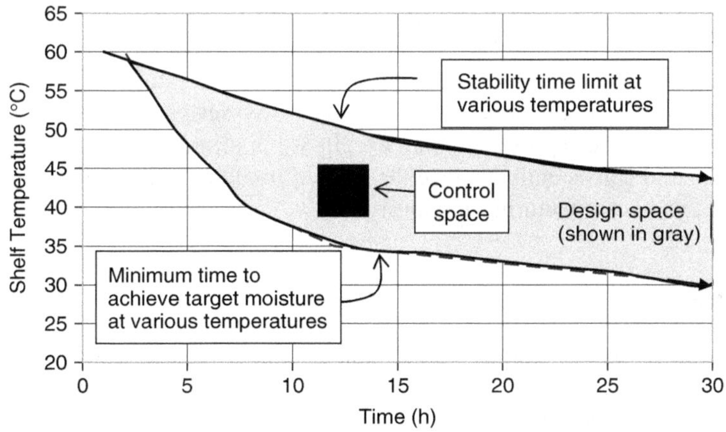

Figure 31.9. Design space for secondary drying hold. (Chamber pressure is generally inconsequential during secondary drying and hence not included in the illustration.)

transfer rates during the freeze-drying process. A drift in instrument calibration could be sufficient to alter the process failure limits over time, thereby altering the boundaries of the design space. But rather than including these parameters as additional dimensions during the development of design space, it is reasonable to limit the influence of those factors through control strategies such as raw-materials specifications, in-process monitoring, or calibration programs.

The knowledge gained from the information summarized in Tables 31.5–31.7 and Figures 31.7–31.9 is based on the physical understanding of the lyophilization process, and the illustrations serve as representative examples. The precise failure limits and quantitative relationships among the various parameters would vary with formulation and vial configuration. To define design space for lyophilization of a given product configuration, one can devise an experimental plan that maps the process dependencies and identifies the process failure limits. The plan may rely on design of experiment (DOE) or a rational selection of process conditions where bracketing strategy is used to study the effect of multiple factors simultaneously. These types of experiments for the most part can be performed only on small scale. Clearly, the applicability of the scaled-down model to the production scale system should be appropriately demonstrated.

31.5. CONCLUDING REMARKS

Lyophilization involves numerous input parameters pertaining to the process set points, equipment, manufacturing environment, and product configuration. It is important to fully understand how all these inputs influence the key output parameters related to process dynamics (e.g. product temperature), finished-product attributes (e.g. cake appearance, moisture content and reconstitution time), and ultimately the biochemical quality and long-term stability of the drug product. The underlying phenomena that govern these relationships during lyophilization are heat transfer to and from the product vial and mass transfer from the product vial. Mastering an understanding of these phenomena, coupled with comprehensive characterization of the product configuration and the equipment capabilities, is key to successful scale-up, technology transfer, and routine production. The FDA quality-by-design initiative and emphasis on characterizing the design space is an excellent opportunity to apply the science of lyophilization and consistently achieve the desired product quality, predictable process performance, and manufacturing flexibility.

REFERENCES

1. Chang, B. S., Patro, S. Y. (2004), Freeze-drying process development for protein pharmaceuticals, in Costantino, H. R. and Pikal, M. J., eds., *Lyophilization of Biopharmaceuticals*, AAPS Press, Arlington, VA: pp. 113–138.
2. Ma, X., Wang, D. Q., Bouffard, R., and Mackenzie, A. (2001), Characterization of murine monoclonal antibody to tumor necrosis factor (TNF-Mab) formulation for freeze-drying cycle development, *Pharm. Res.* **18**: 186–202.

3. Trappler, E. (2004), Validation of lyophilization: Equipment and process, in *Lyophilization of Biopharmaceuticals*, Costantino, H. R. and Pikal, M. J., eds., AAPS Press, Arlington, VA, pp. 113–138.
4. Breen, E. D., Curley, J. G., Overcashier, D. E., Hsu, C. C., and Shire, S. J. (2001), Effect of moisture on the stability of a lyophilized humanized monoclonal antibody formulation, *Pharm. Res.* **18**: 1345–1353.
5. Food and Drug Administration (1993), *Guide to Inspection of Lyophilization of Parenterals*.
6. Sane, S. U. and Hsu, C. C. (2007), Strategies for successful lyophilization process scale-up, *Am. Pharm. Rev.* **10**: 132–136.
7. Searles, J. A., Carpenter, J. F., and Randolph, T. W. (2001), The ice nucleation temperature determines the primary drying rate of lyophilization for samples frozen on a temperature controlled shelf, *J. Pharm. Sci.* **90**: 860–871.
8. Hsu, C. C., Nguyen, H. M., Yeung, D. A., Brooks, D. A., Koe, G. A., Bewley, T. A., and Pearlman, R. (1995), Surface denaturation at solid-void interface—a possible pathway by which opalescent particulates form during storage of lyophilized tissiu-type plasminogen activator at high temperature, *Pharm. Res.* **12**: 69–77.
9. Pikal, M. J., Roy, M. L., and Shah, S. (1984), The physical chemistry of freeze-drying: Measurement of sublimation rates for frozen aqueous solutions by a microbalance technique, *J. Pharm. Sci.* **73**: 1224–1237.
10. Nail, S. L., Jiang, S., Chongprasert, S., and Knopp, S. A. (2002), Fundamentals of freeze-drying, in Nail, S. L. and Akers, M. J., eds., *Development and Manufacture of Protein Pharmaceuticals*, Vol. 14 of Pharmaceutical Biotechnology series, Kluwer Academic/Plenum Publishers, New York, pp. 281–360.
11. MacKenzie, A. P. (1977), The physico-chemical basis for the freeze-drying process, *Devel. Biol. Stand.* **36**: 51–67.
12. Chang, B. S. and Fischer, N. L. (1995), Development of an efficient single-step freeze-drying cycle for protein formulations, *Pharm. Res.* **12**: 831–837.
13. Searles, J. A. (2004), Observation and implications of sonic water vapor flow during freeze-drying. *Am. Pharm. Rev.* **7**: 58–69.
14. Oetjen, G.-W. and Haseley, P. (2004), *Freeze-Drying*, Wiley-VCH, Weinheim, pp. 190–205.
15. Pikal, M. J., Cardon, S., Bhugra, C., Jameel, F., Rambhatla, S., Mascarenhas, W. J., and Akay, H. U. (2005), The nonsteady state modeling of freeze drying: In-process product temperature and moisture content mapping and pharmaceutical quality applications, *Pharm. Devel. Technol.* **1**: 17–32.
16. Chang, B. S., Kendrick, B. S., and Carpenter, J. F. (1996), Surface-induced denaturation of proteins during freezing and its inhibition by surfactants, *J. Pharm. Sci.* **85**: 1325–1330.
17. Hawe, A. and Friess, W. (2006), Impact of freezing procedure and annealing on the physico-chemical properties and the formation of mannitol hydrate in mannitol-sucrose-NaCl formulations, Eur. *J. Pharm. Biopharm.* **64**(3): 316–325.
18. Searles, J. A., Carpenter, J. F., Randolph, T. W. (2001), Use of annealing to optimize the primary drying rate, reduce freezing-induced drying rate homogeneity, and determine $T_{g'}$ in pharmaceutical lyophilization, *J. Pharm. Sci.* **90**: 872–887.
19. Pikal, M. J., Shah, S., Roy, M. L., and Putman, R. (1990), The secondary drying stage of freeze-drying: Drying kinetics as a function of temperature and chamber pressure, *Int. J. Pharm.* **60**: 203–217.

20. Colandene, J. D., Maldonado, L. M., Creagh, A. T., Vrettos, J. S., Goad, K. G., and Sptznagel, T. M. (2007), Lyophilization cycle development for a high concentration monoclonal antibody formulation lacking a crystalline bulking agent, *J. Pharm. Sci.* **96**: 1598–1608.
21. Nail, S. L. and Searles, J. A. (2008), Elements of quality by design in development of freeze-dried parenterals, *BioPharm Int.* **21**: 44–52.
22. Tang, X. L. and Pikal, M. J. (2004), Design of freeze-drying processes for pharmaceuticals: Practical advice, *Pharm. Res.* **21**: 191–200.

32

PROCESS ROBUSTNESS IN FREEZE DRYING OF BIOPHARMACEUTICALS

D. Q. Wang, D. MacLean, and X. Ma

32.1. INTRODUCTION

Recombinant coagulation factor VIII (rFVIII) is a therapeutic protein for the treatment of hemophilia A patients. It is the largest and most complex protein currently manufactured by the biopharmaceutical industry. It is a 300-kDa heterodimeric protein containing 2332 amino acids, 25 N-linked glycosylation sites, and 6 tyr-sulfation sites. The rFVIII protein is very unstable, and there are many challenges in its manufacture [1]. In brief, rFVIII drug substance is produced in baby hamster kidney (BHK) cells, and then purified by a series of chromatography steps. The drug product is formulated with stabilizers and freeze-dried in dose vials. For such a high-value product, process robustness in freeze drying is very critical to ensure process consistency and product quality in large-scale CGMP production. This chapter describes how we selected and tested the critical process steps and process parameter ranges in order to ensure a reliable large-scale freeze-drying production.

The rFVIII formulation used in our study is a crystalline-matrix-type formulation, consisting of 2.2% glycine, 1.0% sucrose, 0.02 M of histidine, 0.03 M of sodium chloride (NaCl), 0.0025 M of calcium chloride ($CaCl_2$), and 80 ppm Tween 80. The rFVIII as a therapeutic component is in the range of 50–200 μg/mL, which is relatively low compared with the inactive excipients. The critical consideration in the

Formulation and Process Development Strategies for Manufacturing Biopharmaceuticals, edited by Jameel and Hershenson

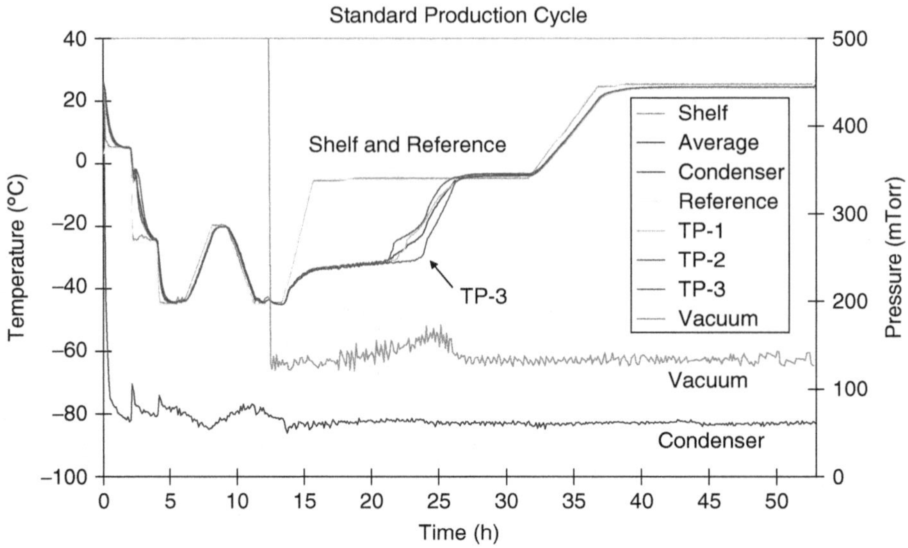

Figure 32.1. Freeze-drying cycle for rFVIII.

development of a freeze-drying cycle for a *crystalline-matrix*-type formulation is to use an annealing step to crystallize glycine to allow *microcollapse* of the amorphous components in between the glycine crystals. This allows primary drying to be performed at a relatively high product temperature without loss of the cake structure [2]. On the basis of our understanding of the formulation, we developed the freeze-drying cycle as given in Figure 32.1. The cycle begins with an annealing step in the freezing phase, then the product slowly warms from −45°C to −20°C ±3°C, is held there for one hour, then cools back to −45°C. Primary drying has a shelf temperature at −5°C ±3°C and a chamber pressure at 100 ± 50 mTorr for 16 h. An "overdrying" secondary drying has a ramped shelf temperature to 25°C ±3°C and a chamber pressure at 100 ± 50 mTorr for 16 h. Considering the robustness of this cycle, we must address three critical questions:

1. We need to know whether the microcollapse, which occurred when glycine crystallized, and even total collapse if glycine did not crystallize, would compromise product quality, particularly with product stability. In We discuss this issue in Section 32.2.
2. We need to provide a range for the process parameters in order to conduct large-scale CGMP production. Each process step of the freeze-drying cycle usually includes a set-point condition, defined as the *standard production condition*, and an allowable deviation from that set point, known as an *operating range* or *specification* of freeze-drying cycle parameters. Section 32.3 illustrates how we used a lab-scale freeze drier to investigate the impact of varying cycle parameters, such as shelf temperature, chamber pressure (vacuum), and phase

transition rate (ramp rate) on product quality consistency during freezing and the primary and secondary drying phases.

3. The capacity of a large scale production freeze drier could be up to 100 times that of the lab scale freeze drier that was used to develop a freeze-drying cycle. To demonstrate the robustness, we conducted a product mapping study in large production-scale freeze driers to ensure that the products at all locations within the freeze-drying chamber have the same quality (i.e., potency and moisture content). This is discussed in Section 32.4.

32.2. COLLAPSE STUDY

Details of the collapse study have been published previously [3] and therefore are not discussed further here. However, as this study is an important part of process robustness development, we briefly describe the concept here. To cause cake collapse, three different freeze-drying cycles, particularly the freezing step, were selected:

1. *No Collapse*. This was a very gentle cycle, in which the samples were annealed to allow crystallization of the glycine eutectic to occur, and the primary drying temperature was very low.
2. *Microcollapse*. This cycle also included an annealing step, but the primary drying temperature was much higher.
3. *Collapse*. This cycle did not include an annealing step, and the temperature during primary drying was also kept low to inhibit crystallization of the glycine eutectic.

Following freeze drying, the products were placed into stability chambers at 5°C, 25°C, and 40°C for up to 18 months in order to assess their long-term stability during storage. It was found that cake collapse is not necessarily detrimental to either the activity or stability of freeze-dried protein. In particular, data support the hypothesis that the more aggressive, shorter primary drying cycle permitted by the microcollapse cycle does not compromise the stability of the protein product. The similarity between stability results for the freeze-dried samples in groups 1 and 2 above (no collapse and microcollapse cycles) indicates that using the more aggressive cycle is not harmful to the protein. This is important because it permits a more efficient freeze-drying cycle to be used, without decreasing the stability of the protein.

32.3. RANGE-FINDING STUDY

As mentioned in Section 32.1, the goal of the second study is to validate the ranges of critical freeze-drying process parameters. A lab-scale freeze drier was used to check the impact of varying cycle parameters during freezing and primary and secondary drying phases, such as shelf temperature, chamber pressure (vacuum), and phase transition rate (ramp rate) on product quality consistency. Those process parameter conditions

TABLE 32.1. Summary of Cycle Parameter Conditions

Phase	Condition	Shelf Temperature, °C	Ramping, °C/min	Pressure, mTorr
Freezing and annealing	Upper	−42, −17, −42	0.3	Atmosphere
	Standard production	−45, −20, −45	0.2	Atmosphere
	Lower	−48, −23, −48	0.1	Atmosphere
Primary: drying	Upper	−2	0.3	150
	Standard production	−5	0.3	100
	Lower	−8	0.3	50
	Upper—higher ramping rate	−2	0.4	150
	Lower—lower ramping rate	−8	0.2	50
Secondary: drying	Upper	28	0.1	50
	Standard production	25	0.1	100
	Lower	22	0.1	150
	Upper—higher ramping rate	28	0.2	50

are summarized in Table 32.1. Combinations of the variables that generated extreme outcomes have been designed to determine whether a consistent rFVIII product is obtained. In total, 10 runs were conducted. Representative cycle charts are given in Figures 32.2–32.4, showing profiles of the freeze-drying cycles with shelf temperature, condenser temperature, vacuum pressure, and product temperatures under different conditions. The freeze-drying-related rFVIII properties, such as potency, reconstitution time, cake appearance, and moisture content from different freeze-drying cycle range limit conditions, were compared in this study. Test results are summarized in Table 32.2.

In the freezing phase, as shown in Table 32.1 and Figure 32.2, three different freezing and annealing cases were investigated. The standard production cycle specification states that the shelf temperature is lowered to −45°C ±3°C, held there for 2 h, then raised to −20°C ±3°C, held there for 1 h, then lowered to −45°C ±3°C for 2 h. The upper limit in the freezing phase, shown in Table 32.1, was set to determine whether the frozen product melted at the highest temperature within the freezing specification range. The lower limit evaluates whether the freezing phase is completed under the lowest temperature within the cycle specifications. Except for varying the freezing phase, primary and secondary drying phases followed the standard production parameters. The cycle profiles from the product temperature probes T_p indicated that the freezing phases were completed under each of the three freezing-phase conditions.

In the primary drying phase, as given in Table 32.1, the set point and specifications of standard production cycle for the primary drying phase are as follows: shelf temperature, −5°C ±3°C; chamber pressure, 100 ± 50 mTorr; and primary drying time, 16 h (not including the ramping time). Five different primary drying conditions were examined: a standard production set point, an upper limit, a lower limit, an

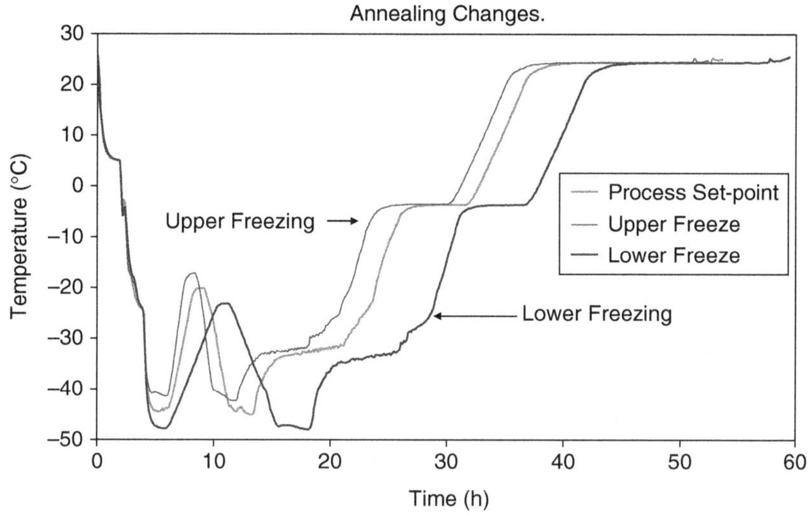

Figure 32.2. Freeze-drying cycle with process parameter changes in freezing phase.

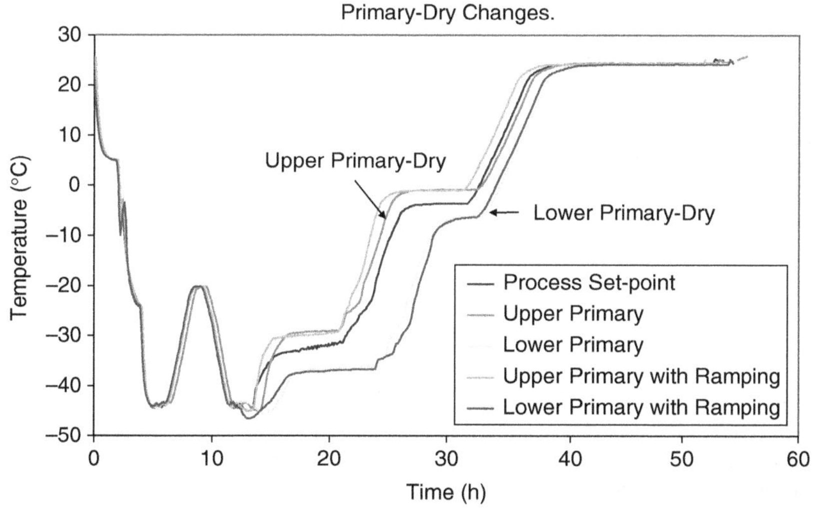

Figure 32.3. Freeze-drying cycles with process parameter changes in primary drying.

upper limit with increased ramping rate, and a lower limit with lower ramping rate (Table 32.1). The standard production set-point parameters were used in the freezing phase and secondary drying phase.

The upper-limit parameters included the highest specification values of shelf temperature and chamber pressure. This limit was to test if the highest product temperature

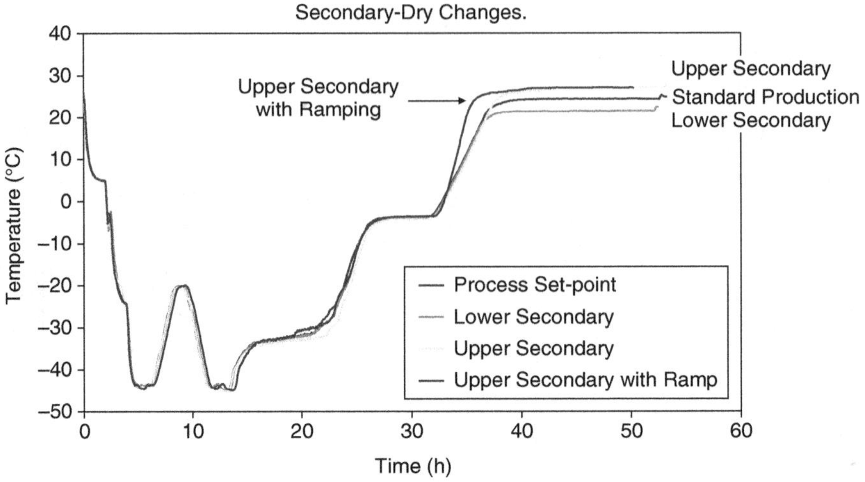

Figure 32.4. Freeze-drying cycles with process parameter changes in secondary drying.

TABLE 32.2. Summary of Results from 10 Freeze-Drying Cycles

Description Lot Comment	Lyophilized Potency $N = 6$, IU/mL	Liquid Potency $N = 6$, IU/mL	Appearance $N = 6$	Reconstitution Time $N = 6$, min	Moisture Content $N = 3$, %
Standard production	138.4 ± 7.2	131.7 ± 8.3	Acceptable	0.1 ± 0.0	1.18 ± 0.02
Upper freezing	137.1 ± 7.3	141.3 ± 7.4	Acceptable	0.1 ± 0.0	1.51 ± 0.09
Lower freezing	135.8 ± 7.4	133.6 ± 10.3	Acceptable	0.1 ± 0.1	1.00 ± 0.03
Upper primary: drying	136.3 ± 12.9	135.1 ± 8.7	Acceptable	0.1 ± 0.0	1.40 ± 0.03
Lower primary: drying	143.0 ± 8.9	132.3 ± 6.2	Acceptable	0.1 ± 0.0	1.14 ± 0.02
Upper primary: increase ramp	137.8 ± 4.3	132.5 ± 8.8	Acceptable	0.1 ± 0.0	1.18 ± 0.04
Lower primary: decrease ramp	134.0 ± 5.6	134.3 ± 9.5	Acceptable	0.1 ± 0.0	1.20 ± 0.03
Upper secondary	136.2 ± 7.3	129.3 ± 4.9	Acceptable	0.1 ± 0.0	1.04 ± 0.02
Lower secondary	133.4 ± 10.2	131.6 ± 6.9	Acceptable	0.1 ± 0.1	1.38 ± 0.02
Upper secondary: increase ramp	141.6 ± 10.1	132.0 ± 5.8	Acceptable	0.1 ± 0.0	1.22 ± 0.04

T_p, caused by both the higher shelf temperature and higher chamber pressure, would exceed the maximum allowable temperature in the primary drying phase. If the product temperature were above the eutectic or collapse temperatures, then melting or collapse could occur. In the standard production freeze-drying cycle, the product temperature T_p was approximately −33°C under the normal operation condition (Fig. 32.1). Under the upper-limit condition during the primary drying phase, the T_p was approximately −29°C (Fig. 32.3), and the cake appeared to be acceptable after being freeze-dried under the upper-limit condition.

The lower-limit parameters included lowest specification limit of shelf temperature and chamber pressure. In contrast to the previous experiment, this experiment was to see if the lower product temperature T_p, caused by both the lower shelf temperature and lower chamber pressure, would make the primary drying too slow to efficiently sublimate the product within the primary drying time stated in the standard production cycle. As expected, the product temperature T_p in the lower-limit cycle was approximately −37°C (Fig. 32.3), which was lower than the T_p in the normal cycle. The amount of time between the shelf temperature reaching −5°C and the product temperature T_p reaching shelf temperature was slightly longer than the standard production cycle, but was still completed within the given primary drying time in the standard production cycle (Fig. 32.1). The results suggest that product could be efficiently freeze-dried within a reasonable primary drying time even in the lower extreme primary drying conditions.

The upper and lower primary drying cycles were also tested using increased and decreased ramping rates from freezing to the primary drying phases. The primary drying time was slightly changed, but the phase was completed within the given primary drying time in the standard production cycle (Fig. 32.3).

Moisture content data indicated that freeze drying with extreme primary drying conditions did not significantly affect the moisture content in the final freeze-dried containers (Table 32.2) and the moisture contents were within the product release specification. There are no significant differences in the reconstitution times among the five cycles with different primary drying conditions.

Potency assay results, summarized in Table 32.2, were further analyzed by analysis of variance (ANOVA) statistical analysis. The statistical analysis of potency values post–freeze drying indicates that there were no statistical differences among the five cycles with different primary drying conditions with a 95% confidence level.

In the secondary drying phase, the specification ranges of process parameters, such as shelf temperature, chamber pressure, and cycle time, were also evaluated (Table 32.1). After primary drying was performed with the parameters determined above, secondary drying was evaluated in four tiers: an upper limit, an upper secondary limit with increased ramping rate, a lower limit, and a standard production set-point condition. The set point and specifications for the secondary drying phase are as follows: shelf temperature, 25°C ±3°C; chamber pressure, 100 ± 50 mTorr. The upper-limit parameters included the highest value of the shelf temperature and the lowest value of the chamber pressure specifications. The secondary drying, with highest temperature and lowest chamber pressure, provides the most efficient condition for water desorption. Therefore, such conditions were appropriate for testing whether the product could be excessively overdried within the extreme upper limit. The lower-limit parameters included lowest value of shelf temperature and highest value of chamber pressure in the specifications. In contrast, this experiment was to test whether product would be underdried with the lower-limit parameters.

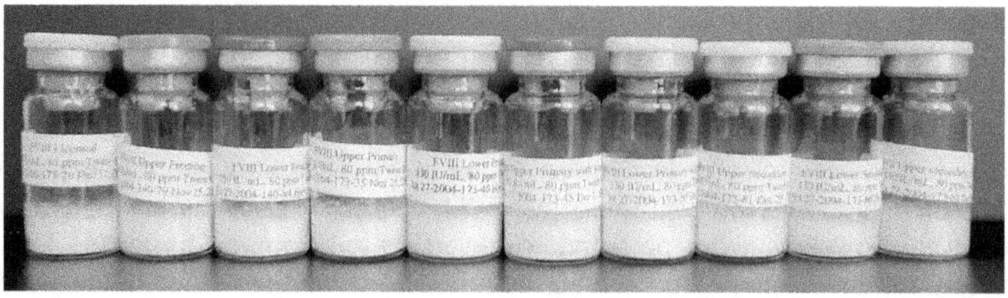

Figure 32.5. Photo of lyophilized product for all 10 cycles. Pictured from left to right: process set point, upper freezing, lower freezing, upper primary, lower primary, upper primary with increased ramping rate, lower primary with decreased ramping rate, upper secondary, lower secondary, and upper secondary with increased ramping rate.

The results shown in Table 32.2 indicated that the moisture contents in the freeze-dried product did not significantly change under the three different drying conditions and were within product release specification for moisture content. The cycle chart profile for the product temperature show all secondary drying phases were completed with the three range limit secondary drying conditions (Fig. 32.4). Potency, appearance, and reconstitution time were not affected by the secondary drying limit conditions. Potency assay results, summarized in Table 32.2, were further analyzed by ANOVA. Statistical analysis of potency values post–freeze drying indicates that there were no statistical differences among the four cycles with different secondary drying conditions with 95% confidence. In addition, all lyophilized products appeared as a white solid (Fig. 32.5). All reconstitution times were less than or equal to 0.1 min, and all vials appeared clear after reconstitution (Table 32.2).

In conclusion, this study showed that when rFVIII product is freeze-dried within the standard production cycle specifications, there are no significant differences in product performance.

32.4. PRODUCT-MAPPING STUDY

The third study focused on product mapping on large-scale production freeze-driers, which are 240-ft^2 STERIS Lyovac freeze driers. Product mapping is performed to evaluate the performance (uniformity and reproducibility) of the freeze-drying cycle and to confirm comparability of the scaled-up cycle. The study was designed to cover the minimum operating capacity and the maximum freeze-drier capacity. The product vials at the specific locations within the freeze-drier chamber, as shown in Figures 32.6 and 32.7 for the maximum operating capacity, were probed with thermocouples to monitor the product temperature during the freeze-drying process. Final container samples from specified sampling locations were tested according to data in Table 32.3.

Figures 32.6 and 32.7 provide tested product locations for the top five shelves and the bottom five shelves, respectively, in a freeze-drier chamber. The potency and

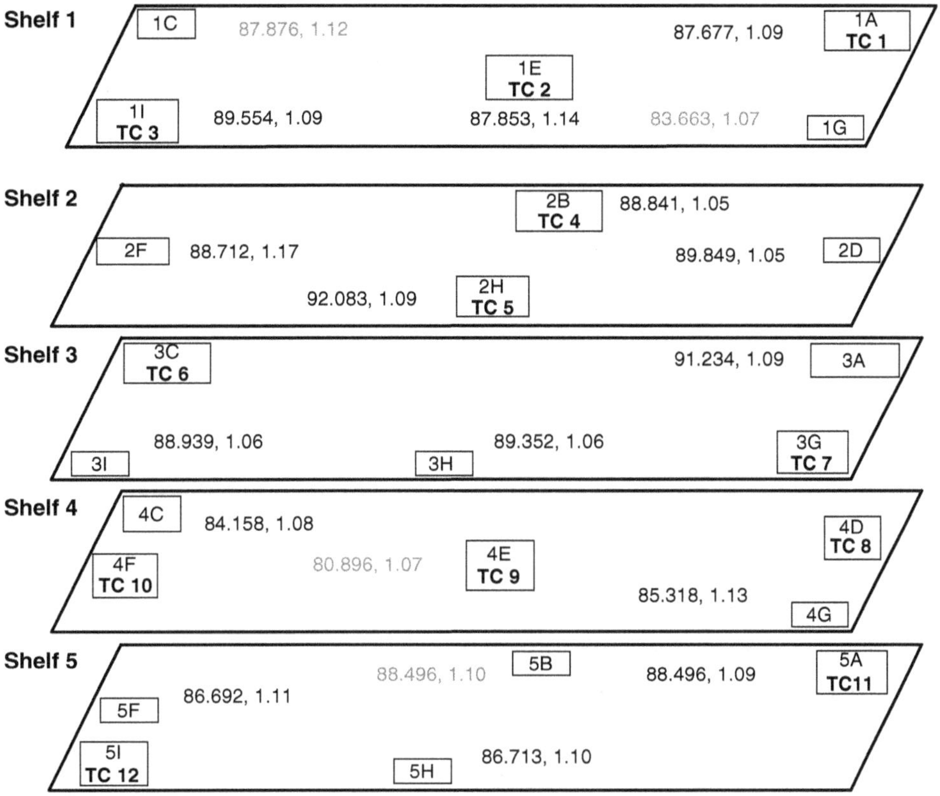

Figure 32.6. Product mapping sample locations, potency results, and moisture results for the top five shelves in a freeze-drier chamber.

moisture testing results of the product samples at the given locations are also presented in the figures. In total, four runs were conducted in the product mapping study in order to cover the two freeze-driers in the facility with the maximum (40,000 vials) and minimum (8000 vials) loads. Detailed results from all four runs are not provided here because of space limitations; however, no significant differences were observed in the four runs, demonstrating batch product quality uniformity and freeze-drying process consistency under large-scale cGMP production.

ACKNOWLEDGMENTS

The author wishes to thank Drs. Jeff Hey, Steve Nail, and the Bayer Berkeley New SFF (sterile filling and freeze-drying) project team for their work, which made this chapter possible.

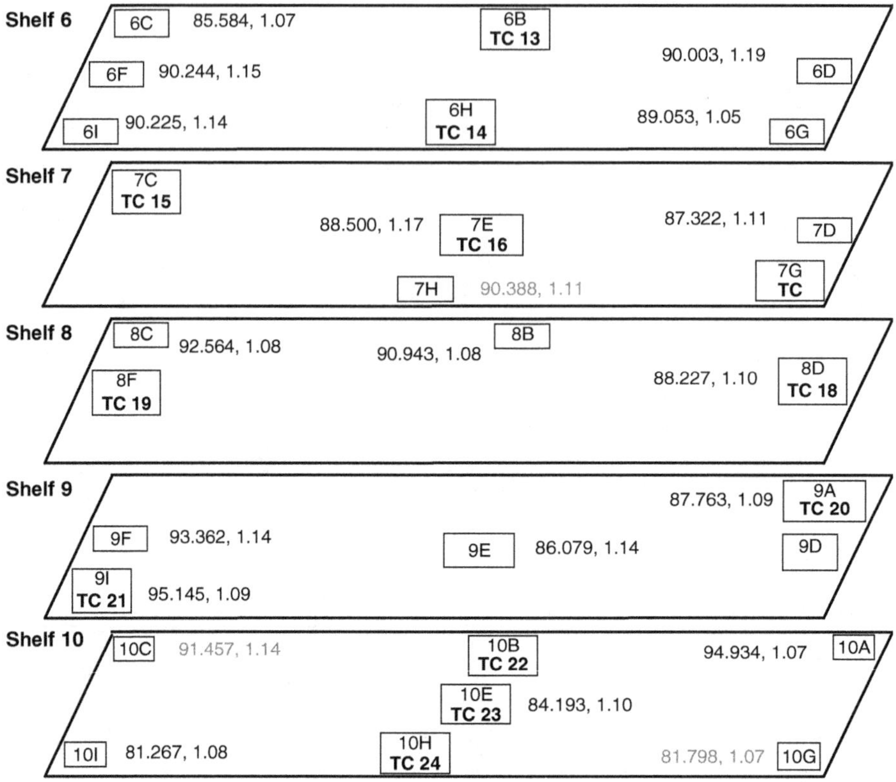

Figure 32.7. Product mapping sample locations, potency results, and moisture results for the bottom five shelves in a freeze-drier chamber.

TABLE 32.3. Tests Performed for Product Mapping Samples

Assay	Number of Vials per Location	Total Number of Vials per Freeze-Drying Run: Maximum Load (39 Sample Locations)	Minimum Load (12 Sample Locations)
Potency	3 samples/location	117 samples	36 samples
Solubility time			
Visual appearance before and after reconstitution			
Moisture by near-IR spectroscopy (NIR)	3 samples/location	117 samples	36 samples
Clarity	1 sample/location	39 samples	12 samples
pH			
Retains	3 samples/location	117 samples	36 samples

REFERENCES

1. Jiang, R., Monroe, T., McRoger, R., and Larson, P. J. (2002), Manufacturing challenges in the commercial production of recombinant coagulation factor VIII, *Haemophilia* **8**(Suppl. 2): 1–5.
2. Wang, D. Q. (2004), Formulation characterization, in *Freeze-drying/Lyophilization of Pharmaceuticals and Biological Products*, Rey, L. and Dr. Joan May, J., eds., Marcel Dekker, New York. pp. 213–238.
3. Wang, D. Q., Hey, J. M., and Nail, S. L. (2004), Effect of collapse on the stability of freeze-dried recombinant factor VIII and α-amylase, *J. Pharm. Sci*. **93**(5): 1253–1263.

33

FILLING PROCESSES AND TECHNOLOGIES FOR LIQUID BIOPHARMACEUTICALS

Ananth Sethuraman, Xiaogang Pan, Bhavya Mehta, and Vinay Radhakrishnan

33.1. INTRODUCTION

Challenges to a successful commercialization of a protein drug include maintaining the safety and efficacy during the drug product manufacturing process and storage until delivery to the patient. Purified protein substance is formulated and filled into typical primary containers such as glass vials, syringes, and dual-chamber cartridges. This review chapter is focused on filling process with specific emphasis on various filling technologies and process considerations. It is critical that the filling operation has minimal impact to product quality and provides a cosmetically acceptable drug product [1–3]. The filling technology must be evaluated for compatibility with the protein product and various process parameters, such as filling rates, pump size and volume, and needle size and configuration, which all must be optimized prior to manufacturing. Typically, filling technologies include positive-displacement pumps and time–pressure and mass flow pumps, which are discussed in the chapter. Filling evaluations need to be performed on early-stage as well as late-stage protein products to identify any potential process and product quality issues when products are processed through the filling equipment. Scale-down or benchtop equipment is often used to mimic

Formulation and Process Development Strategies for Manufacturing Biopharmaceuticals, edited by Jameel and Hershenson

manufacturing-scale equipment. The chapter discusses the various studies and also outlines process considerations for transfer to commercial manufacturing.

33.2. FILLING TECHNOLOGIES

The different types of filling technologies currently used in protein product manufacturing are positive displacement, time over pressure, peristaltic pumps, and mass flow systems. Selection of the type of system to be used for filling is based on solution properties, manufacturing needs, and availability of equipment at the manufacturing site. While positive-displacement pumps and peristaltic pumps have been widely used in the industry, time-over-pressure filling systems are becoming increasingly popular because of significant advantages that will be discussed later in this section.

33.2.1. Positive-Displacement Pumps

Positive-displacement pumps use a piston–pump mechanism to transfer liquid to a primary container through a needle. With these pumps, constant pressure is applied to a supply tank or manifold. The manifold is a supply tank that is connected to individual pumps and needles. The supply tank or manifold feeds solution to the piston chamber. The piston then volumetrically doses the required amount of solution via tubing and needle into a vial or syringe. There are different types of positive-displacement pumps: rolling-diaphragm, piston pumps, and rotary piston pumps, which are discussed further in this chapter.

33.2.1.1. Piston Pump Filler. A tabletop piston pump filling system is shown in Figure 33.1. This filling technology consists of a syringe or piston (a), a three-way valve (b), a cam (c), and a vernier volume adjustment (d). The vernier volume adjustment can be used to adjust the stroke travel length of the piston to control

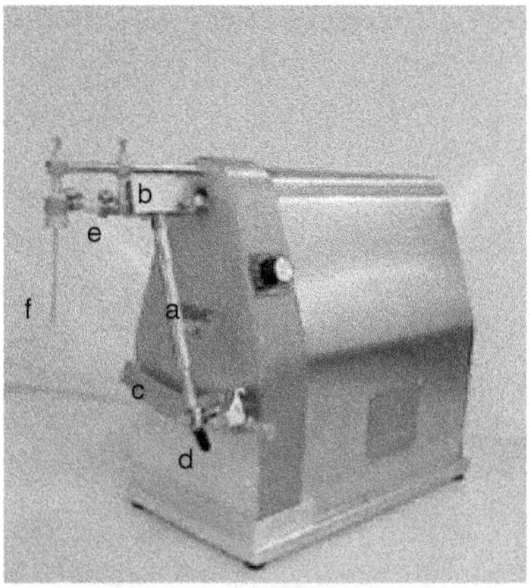

Figure 33.1. Tabletop piston pump filler.

the volume delivered. The three-way valve (b) is connected to the piston and the solution supply tank through a connection (e) and the filling needle (f). The piston (a) movement is governed by the cam (c). For the piston to move a predetermined distance, the lower piston position is set on the vernier (d).

The vernier then actuates the three-way valve (b) so as to move the volume of liquid product from the solution supply tank, through the valve into a syringe barrel and then back through the valve to the filling needle and into the primary container such as a vial or syringe.

Depending on the size of the batch to be filled and the line speed needed for filling, the size of the equipment will vary. However, the principle of operation is common to the basic mechanism described above. The benefits of this system include lower shear stress on the product and ability to deliver low fill volumes (<1 mL) accurately. The limitations are loss of product left over in the pumps, number of product contact parts, including seals, and the increased difficulty in cleaning and sterilizing the system.

33.2.1.2. Rotary Piston Pump Filler. The rotary piston filling system consists of a product tank that is connected to a manifold (Fig. 33.2). This manifold: (a) is connected to individual rotary piston pumps (b) using silicone tubing, (c) and each rotary piston pump is connected to an individual filling needle. The solution is filled into the product tank, which is then fed via the manifold and individual pumps into the filling needles.

The pump assembly and the different steps involved in the filling operation are illustrated in Figure 33.3. Each pump consists of a piston and cylinder (Fig. 33.3a), where the piston has a very low clearance when it is fitted into the cylinder (Fig. 33.3b). The piston is used to close and open the inlet and outlet ports by rotation as the pump

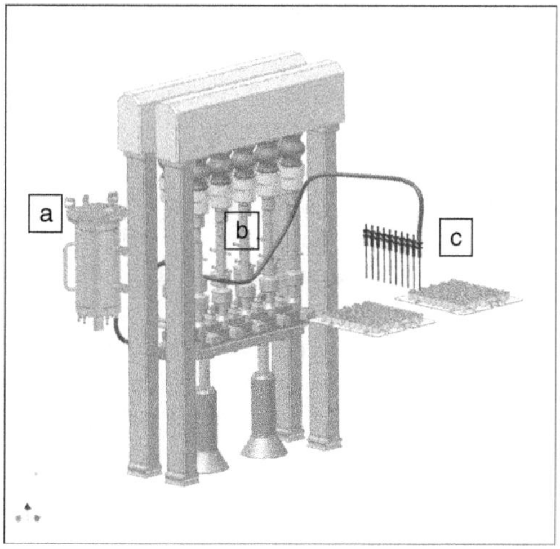

Figure 33.2. Schematic figure of a rotary piston pump filling system: (a) manifold; (b) fill pumps; (c) fill needles.

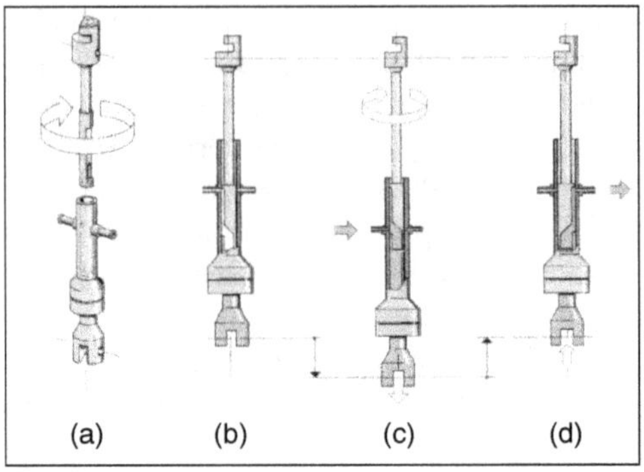

Figure 33.3. Rotary piston pump assembly.

cylinder moves vertically up and down. As the piston rotates and the cylinder moves in the downward direction, it provides a particular aperture for liquid to enter into the cylinder through the inlet port (Fig. 33.3c). The piston then rotates, and as the cylinder moves in the upward direction, the liquid is expelled out of the pump through the outlet port toward the filling needles (Fig. 33.3d). A control valve is integrated into the mechanism of the pump that controls the amount of liquid that flows in and out of the pump.

The benefits of this system include ability to accurately and repeatedly fill low fill volumes (<1 mL). In addition, the absence of seals makes the pumps easy to assemble and to clean and sterilize. The limitations are the need for a wetted surface between the piston and cylinder for use, the inability to use pharmaceutical solutions that crystallize, product wastage (volume of the pump), and need for handling the various parts with care.

33.2.1.3. Rolling-Diaphragm Filler. The rolling-diaphragm filler assembly is similar to that for the rotary piston assembly. The rolling-diaphragm pump and the different steps involved in the dosing are illustrated in Figure 33.4. As shown in Figure 33.4a, liquid is drawn into the dosing chamber (C) through port A when the piston (P) is moving in the downward direction and valves V_1 and V_2 are positioned toward the left. As the piston (P) reaches the bottom of the stroke and the liquid flow through port A ceases, the intake valve (V_1) is closed and the chamber (C) contains all the solution that will be dispensed in the subsequent pump cycle. Next the exhaust valve (V_2) is opened and the external force (F) causes the piston to travel upward (Fig. 33.4b) and liquid is dispensed through port B. The quantity of liquid to be dispensed in based on the amount of upward travel of the piston. As the piston reaches its preset upper position, the liquid ceases to flow through port B and the exhaust valve (V_2) is closed (Fig. 33.4c).

Figure 33.5 shows the disassembled (Fig. 33.5a) and assembled (Fig. 33.5b) rolling-diaphragm pump. The benefits of this system include ability to accurately and repeatedly fill low volumes (<1 mL), low shear on the product, ability to minimize

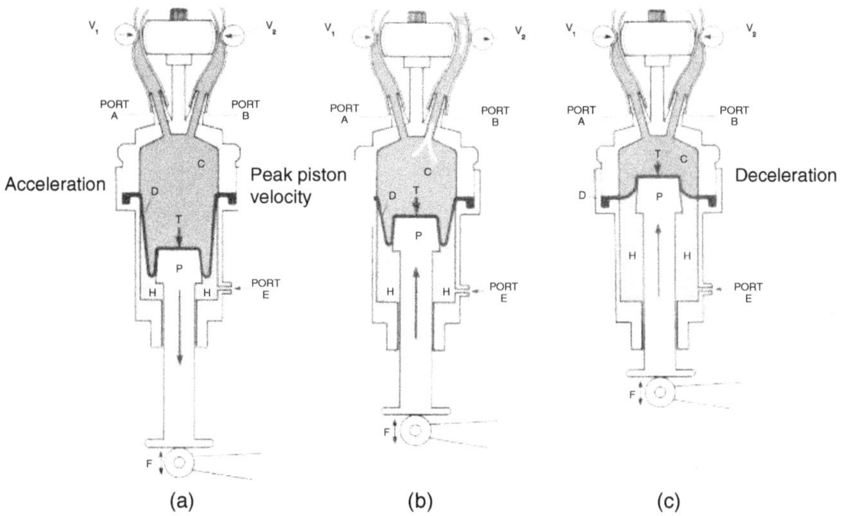

Figure 33.4. Operation of rolling-diaphragm pump. *Key:* V_1 = intake pinch valve; V_2 = exhaust pinch valve; D = diaphragm; A = inlet port; B = discharge port; E = vacuum port; P = piston; H = vacuum chamber; F = external force; T = internal differential thrust; C = liquid chamber.

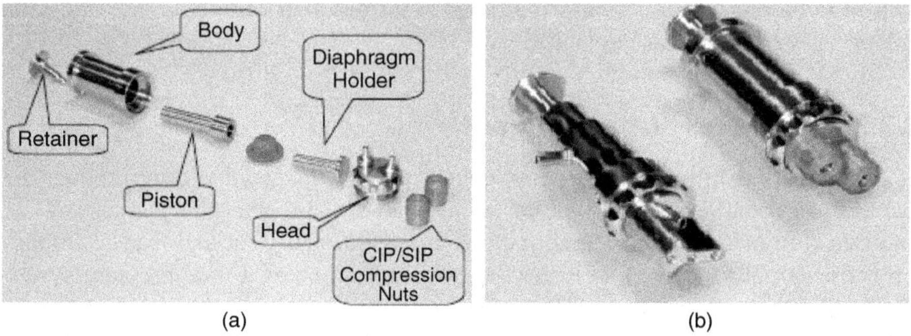

Figure 33.5. (a) Disassembled rolling-diaphragm pump; (b) assembled rolling-diaphragm pump.

particulate formation, and ease in cleaning and sterilizing. The limitations are the product wastage (volume of the pump), and need for assembly of number of parts.

33.2.1.4. Peristaltic Pump Filler. A peristaltic pump is a motor-driven mechanical roller applying pressure on flexible tubing, which performs both the suction and the compression of the liquid to be filled. The flexible tubing is squeezed along its length, positively displacing the fluid contained within the tube (as shown in Fig. 33.6a). The tube's restitution after the squeeze creates a vacuum, which draws

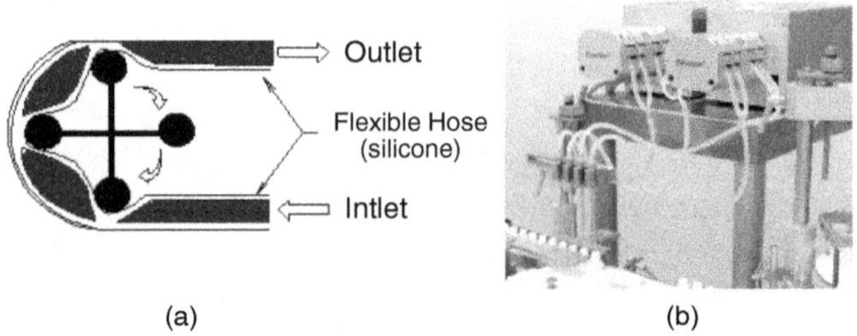

Figure 33.6. Peristaltic pump system: (a) mode of operation; (b) four-head peristaltic filling system.

more fluid into the tube, creating a gentle pumping action with minimal damage to the product inside the tube, particularly when compared to positive-displacement pump systems.

The peristaltic system can have a series of pumps that can provide a multiple-head filler (Fig. 33.6b). Each outlet can be connected to filling needles. The dosing can be controlled by a programmable logic controller (PLC) that can define the fill volume needed. The benefits of this system include ability to fill solutions that can produce crystals, ability to reduce product loss, use of disposable material for product flow path, and ability to fill chilled and viscous product. The limitations are the need to replace the tubing because of the additional wear and tear on the tubing, and tubing needs to be sterilized and autoclaved.

33.2.2. Time-over-Pressure Filler

The time pressure filling was first introduced into pharmaceutical drug product manufacturing in the 1980s. This type of filling does not involve any pumps and seals as needed for positive-displacement pumps and additional wear and tear on tubing as seen for peristaltic pumps. Figure 33.7 shows the outline of a time pressure system.

The product is transported through a pressurized surge or supply tank into the product manifold. The pressure in the tank is established and maintained at a constant value. The filling levels in the surge tank and manifold are controlled by separate level controls that constantly monitor the fill level. The pressure in the manifold is held constant at a preestablished value by a pressure sensor and a proportional pressure control valve. The product is supplied from the manifold to the filling needles using hoses. A pinch valve and an orifice are located on the hoses to control the flow of the liquid. This filling operation is sensitive to pressure and temperature changes. Hence there is a pressure sensor in the manifold that allows for constant and controlled pressure in the system. The temperature is measured with thermocouples located in the filling line, and electronic temperature compensation is possible to adjust for temperature fluctuation, especially during filling of cold liquids.

As mentioned earlier, a pinch valve and orifice are present in the hoses from the manifold to the filling needle (Fig. 33.8a). The pinch valves located in the hoses

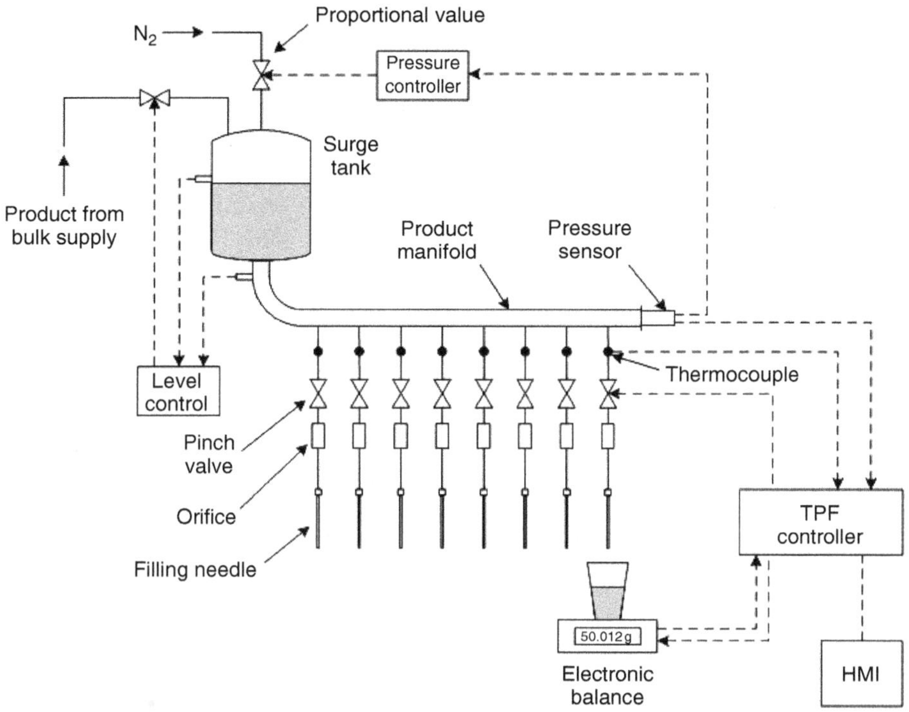

Figure 33.7. Time–pressure filling system outline.

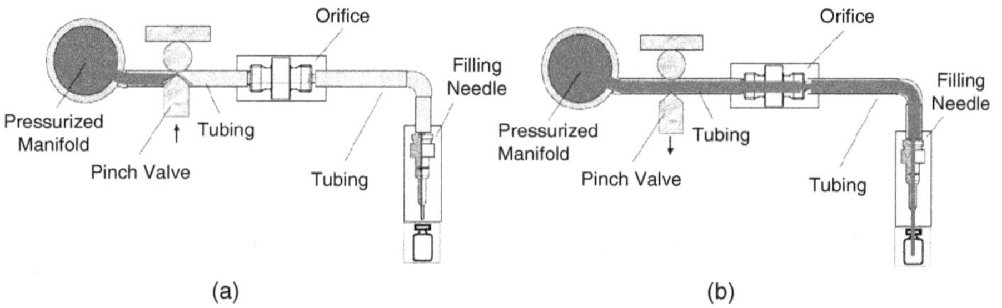

Figure 33.8. (a) Product flow through the pinch valve and orifice.

connected to the filling needles are driven by stepper motors. The opening of the pinch valve along with the flow through the orifice controls the filling process. Calculation of the filling time is based on the pressure and the temperature changes. As shown in Figure 33.8b, for the duration of filling, the pinch valves open and close, which leads to the volumetric filling of product. The design of the pinch valve and orifice is critical in ensuring a laminar flow for the liquid and minimizes shear stress on the product.

The fill accuracy is dependent on the temperature and viscosity of the product, manifold pressure, flow time, filling needle diameter, and flow orifice diameter. To limit temperature fluctuation impact to fill weight variations, temperature compensation function needs to be used on this system. The benefits of this system include decreased product contact parts, decreased change over time between batch manufacturing, ability to reduce product loss, ease in assembly, and ease for cleaning-in-place and steam-in-place. The limitations are increased shear stress impact on product and fill volume variability if product viscosity is sensitive to temperature.

33.2.3. Mass Flow Filler

Benefits of the mass flow system include low shear impact on the product, greater product recovery, ability to fill viscous and temperature-sensitive products, and ease in cleaning. The limitations are the infrequent use in pharmaceutical applications, limited accuracy range of the flow meters, slower fill speeds, and inability of flow meters to be easily changed.

In mass flow fillers, the product is filled into the manifold by pressure. The product is fed from the manifold into the mass flow meter (MFM). The programmable logic controller (PLC) uses an electronically controlled valve to regulate the volume of product in the manifold. The MFM measures and controls fill volume using an electronic transmitter that interfaces with the filling system PLC. The MFM measures fill weight using the Coriolis principle. The unit measures mass flow and density of the liquid being filled to convert mass flow (g/s) to fill volume (mL/unit).

33.2.4. Disposable Technologies

Disposable filling technology is based on liquid dosing principle that uses presterilized disposable components for all product contact surfaces. The product contact parts are presterilized or autoclaved. This technique offers an improved level of flexibility and ease of use with respect to quick changeover between batches and eliminates the need for cleaning and related validation.

One of the disposable filling techniques uses peristaltic pump systems. As mentioned earlier, tubing is the product contact part that needs to be replaced. Variability in tubing diameter, tubing wear, and flexibility can impact filling accuracy.

Another novel filling system has been introduced by the Millipore Corporation. The technology is based on the accurate volumetric filling driven by gravity [4]. The system consists of hardware and disposable reservoir, tubing, and needles (as shown in Fig. 33.9). The disposable components are presterilized.

The operation of the dispensing system is based on a gravimetric volumetric fill principle and works with a fill–discharge cycle as shown in Figure 33.10. The reservoir is filled to a set level with product solution from a hold tank after filtration and serves as a supply reservoir. The volume of liquid in this reservoir is controlled by a sensor located outside that detects and maintains the liquid level up to 0.5 L. This control is attained by using a feedback loop from the sensor to a peristaltic pump or pinch valve that feeds the solution from the hold tank to the reservoir. When the upper pinch valve opens, liquid is driven by a small head pressure from the reservoir

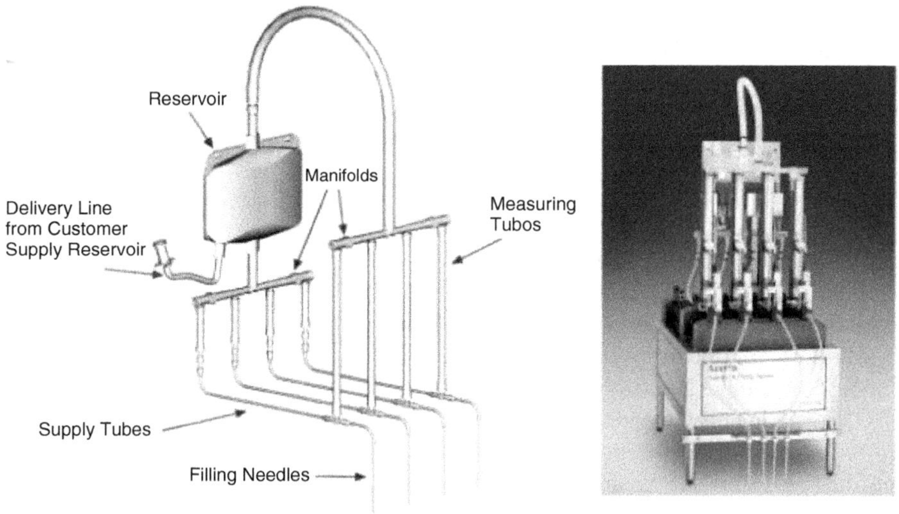

Figure 33.9. A disposable filling system showing the disposable reservoir, tubing, and needle assembly along with hardware for mounting on a filling line [4].

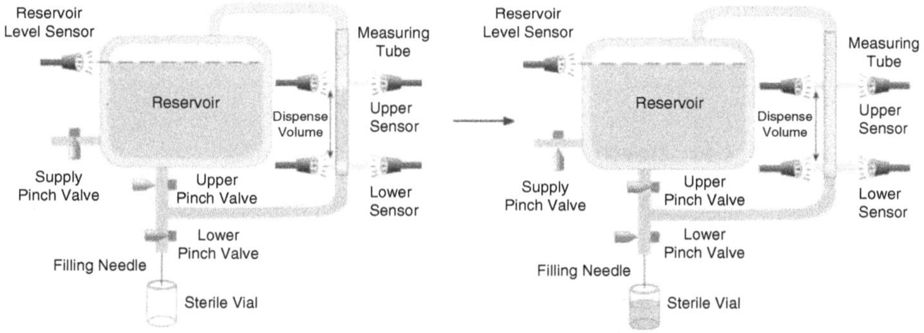

Figure 33.10. A disposable filling system operation; charging of the measuring tube and subsequent dispensing of the sterile product through the filling needles [4].

into the measuring tube. The upper sensor recognizes the meniscus and closes the upper pinch valve. The measuring tube is then filled with the volume of liquid. To allow for equilibration of pressure in the measuring tube, the top of the measuring tube is vented into the reservoir. The filled measuring tube is now ready for dispensing into the vial. When the vial is ready to be filled, the lower pinch valve opens. With the head pressure in the measuring tube, the solution flows into the vial. The liquid meniscus reaches a lower level; the lower sensor then triggers the lower pinch valve to close, thus ending the filling cycle. This cycle can be repeated when the next vial is ready to be filled. The reservoir can feed multiple measuring tubes in turn feeding filling tubes, thus increasing the number of filling heads on the line.

The advantages of this process are disposable product contact parts; accuracy in filling aqueous solutions; lower cost in utilities, parts, and maintenance; and quick changeover time between batch manufacturing. The disadvantage is the increase in costs of disposable material, inability to fill ready-to-use suspensions (solid–liquid dispersions), increased fill accuracy problems while filling liquids that foam, and relatively slow speed in filling viscous liquids.

33.3. PROCESS AND PRODUCT PARAMETERS

Currently marketed and developed protein therapeutics is available in multiple modalities. Antibody, globular protein, hormones, enzymes, conjugated protein, and vaccines are a few of them. These protein molecules typically have unique formulations. In this chapter, the protein and formulation combined are referred to as interchangeably *product* or *drug product*. These drug products are presented predominantly in three dosage forms. They are ready-to-use solutions, freeze-dried powders that are reconstituted into solutions immediately before the administration, and ready-to-use suspensions.

Each dosage form has important associated parameters and challenges with respect to the filling operation. These parameters and challenges can be broadly divided into two categories. The first category is associated with the product properties (mainly physical properties), and the second category is associated with the process parameters. In practical situation both sets of parameters and challenges are interrelated and should be approached and analyzed together.

33.3.1. Product Properties

33.3.1.1. Protein Concentration. Proteins are formulated in array of concentration ranging from μg/mL (low) to 10^2 mg/mL (high) levels. Both these extremes present unique challenges at filling operation. The higher fractional product loss due to absorption to surfaces will have to be considered for a low-concentration product. This challenge will require scrutiny to minimize surface interactions during the processing of the product to ensure that an adequate amount of protein is filled into the primary container. The high protein concentration will impact the physical properties of the drug product solution, specifically viscosity, density, and surface tension. The higher viscosity may result in dripping of product during the filling operation as well as variation in fill accuracy. The lower surface tension may result in higher foaming during filling. The complications around foaming and dripping are discussed below.

33.3.1.2. Density. The density of protein product increases with the concentration of protein in the formulation. The density measurement is important for accurate dosing in each primary container. All the filling technology discussed above employs density to determine appropriate dosage in each unit.

33.3.1.3. Viscosity. The viscosity of the drug product is a rheological property of the product. This parameter is useful in predicting the behavior of the product during the filling process. The viscosity of the protein products increases exponentially with increase in the protein concentration [5]. The product behaves as a Newtonian fluid for low-protein-concentration products. However, a solution with higher protein concentration may show non-Newtonian behavior. Typically, the product may show shear thinning behavior at higher protein concentration. Non-Newtonian behavior requires greater optimization of the filling process as dripping issues might be observed. Typically, viscosity is inversely proportional to the temperature of the solution. This may be useful if significant non-Newtonian behavior is observed at lower temperature. In such cases, it may be necessary to perform the filling operation when the liquid is at room temperature. However, it is necessary to ensure that there is no impact on product quality room temperature exposure for a prolonged period of time during the filling process.

33.3.1.4. Surface Tension. Protein products typically have surface-active agents in protein itself or surfactant as a formulation component. These agents reduce the surface tension of the product. This may result in process impact and product impact. The effect of lower surface tension may increase the dripping-, foaming-, and splashing-related challenges (process impact). The product impact includes higher product interactions with hydrophobic surfaces. Filling operation has hydrophobic surfaces in the form of air–liquid interfaces and tubing–liquid interfaces. These interfaces may absorb protein and result in surface denaturation of protein. This may lead in higher self-association or protein aggregation [6–8].

33.3.2. Impact on Product Quality

The product is exposed to high shear, multiple product contact surfaces, room temperature exposure, and photoexposure during the filling operation. These factors may have a product quality impact due to physical protein instability, which is less well characterized and more difficult to predict. This physical instability can result in protein association, protein aggregation, insolubility, and particle formation [5,9,10]. The product quality impact should be monitored by sizing, subvisible particle count, and visual inspection over the shelf life of the product. These analyses should be performed during the filling process evaluation at lab scale, characterization at pilot or commercial manufacturing scale, and shakedown and validation runs. The library of data generated during the filling operation–related process development should provide detailed insight into product behavior and might be useful in avoiding any challenges later on during product characterization.

33.3.2.1. Shear Impact. Generally, the filling operation constitutes the highest shear stress that can impact product quality among formulation, fill, and finish operations. The large shear generated by fluid flow could potentially yield sufficient torque to possibly perturb, partially unfold, or even denature protein drug products. Thus far a limited number of studies have investigated the impact of high shear during filling on the quality of biologics, and the results are inconsistent and contradictive. Several studies have shown that the loss of protein activity is proportional to the exposure time and shear rate, while others found no evidence that high shear could significantly destabilize proteins, but rather accelerate the turnover of interface interactions [7,11–13]. Ensuring exposure to high shear rates is critical when screening formulations, since it provides valuable information on the formulation's ability to protect the molecule from detrimental process stresses. In case of a shear-sensitive product, it might be necessary to optimize filler parameters to minimize the product impact. General approaches for filler parameter optimization will involve using lower fill speed or larger nozzle size. If the challenge is very severe, then low-shear filling technology might be preferred over high-shear filling technology.

33.3.3. Shear Evaluation for Different Filling Technologies

Shear rate in filling nozzle can be approximated as a pipe flow, or Hagen–Posieuille flow. The shear rate of a Newtonian fluid is simply represented as $\gamma = 4Q/\pi R^3$, where Q represents the flow rate and R represents the radius of the pipe.

However, for high-viscosity or non-Newton fluid, the dependence of viscosity on shear rate must be established. A model of viscosity will need to be developed to assess the effect on pipe flow equations.

The estimated shear rate for each filling technology is listed in the Table 33.1. To evaluate the impact of shear on product quality, a rheometer with precise shear gradient and temperature control could be used for medium-range shear rates. If higher shear rates are required (>8000 s^{-1}), they can be simulated with pressure-driven controlled flow through capillary tubing using a syringe pump.

33.3.4. Product-Specific Assays and Biophysical Characterization Techniques

Appropriate protein stability-indicating assays would be useful for measuring the protein structure, protein aggregation, or degradation during filling operation development

TABLE 33.1. Evaluation of Shear Impact on Product

Filling Technologies	Relative Flow Shear	Relative Product Contact Surface	Source of Product Quality Impact
Piston pump	Low	High	Flow through pump and filling nozzle
Peristaltic pump	Low	Low	Flow through filling nozzle
Rotary piston pump	Medium	High	Flow through pump (includes clearance between piston and cylinder) and needle
Rolling-diaphragm pump	Medium	Medium	Flow through pump and filling nozzle
Time over pressure	High	Low	Flow through orifice (in most cases) and filling nozzle
Mass flow	Low	Low	Flow through filling nozzle

and optimization. Besides typical chromatography and gel-based assay, peptide mapping for chemical modification, cell-based assays, and/or immunoassay for bioactivity or biophysical assays (FTIR, CD, DSC, DLS, AUC, fluorescence, Raman, etc.) could be applied for in-depth investigations. The investigator must develop a strategy for choosing analytical assays for each drug product. Additionally, methods for particulate characterization and turbidity should be included to determine any aggregation due to the filling operation.

33.3.5. Process Parameters

33.3.5.1. Filler Parameters. Filler parameters, such as pump size (for positive-displacement filling technologies), orifice size (for time-over-pressure filling technology), nozzle size, nozzle type, and fill speed should be appropriately optimized in light of product properties and challenges. These are the operational parameters of the fillers that can be controlled. The following general recommendations can be considered to address common challenges. A smaller pump and nozzle size is typically used for smaller fill volume to ensure tight fill tolerances. Typically, the largest orifice, nozzle, and tubing sizes are used for shear-sensitive products. Appropriate nozzle type should be selected to avoid product dripping, foaming, and splashing issues.

33.3.5.2. Drawback. *Drawback* is a common means of avoiding dripping challenges. The fill cycle is set in such a way that the nozzles are under slight vacuum from one filling cycle to the next. This results in a withdrawal of the product at the entrance of the nozzle, which substantially reduces the dripping challenges. The range of vacuum is product- and formulation-specific and is a parameter that should be characterized during the appropriate phase of product development.

33.3.5.3. Fill Weight. *Fill weight* is defined as amount of dose per container. The filling technologies discussed above employ product density and statistical weight measurement as a means of controlling appropriate dosage in each primary container. The statistical weight measurement is a common practice in industry. Generally 2%–3% of the containers are checked for appropriate dosage during the filling process. In most of the cases, the fill weights are fed back to the filler to ensure consistent operation during the process. The fill weight is typically greater than the delivered volume for the needed dose to account for container holdup volume and filler accuracy. More information on container holdup volume can be found in USP <698>. The filler accuracy is a process parameter that has to be optimized. Typically, for an early-stage clinical product, platform filler accuracy recommendation is provided and minimal optimization is performed. However, for commercial process development appropriate detailed evaluation should be performed using mimic solution and actual product to optimize the filler accuracy, and appropriate overfill should be recommended. It should be ensured that the overfill recommendations are within specification.

33.3.5.4. Determination of Equipment Tolerances. Fill weight checks are performed to monitor the filling accuracy during the filling process. Appropriate fill weight tolerance could ensure filled product to meet the label volume claim and a stable filling process without significant interventions. Tight tolerance and accurate filling capability are economically sound, especially for highly valuable biopharmaceutical product. Setting the appropriate fill weight depends on filling equipment type, product presentation (vial or syringe), and product physical properties, such as viscosity.

Fill weight tolerances are specified as an alert limit and an action limit. An *alert limit* is where an action is needed to adjust the fill volume, depending on the proximity of the net fill weight to the action alarm. The *action limit* is the maximum allowable acceptable fill weight from the net fill weight. A manual intervention is required once the action limit is reached.

Fill weight tolerance can be obtained by statistically analyzing historical fill weight data. The alert limit is normally set at 95% confidence interval, and action limit is calculated at 99% confidence interval. An alternate approach for determining fill targets is to use a formula. This approach requires additional testing to be performed with proper components.

33.3.6. Process-Associated Challenges

33.3.6.1. Dripping. *Dripping* means minor leakage of product on the filling line when the filler manifold is moving from one fill cycle to the next. The dripping issue can be due to multiple factors, including product viscosity and surface tension properties (discussed earlier) or damaged nozzle and inappropriate filler parameter selection. Product viscosity and surface tension are properties of the formulation, and it is seldom possible to modify at the stage of filler characterization. Modification of the fill speed, drawback, and nozzle type is recommended to overcome this challenge. Dripping could also be reduced by decreasing the fill speed. This will also lower the fluid velocity in the fill nozzle. The next option is to modify drawback and the nozzle type (drawback and nozzle type are discussed in Sections 33.3.5.3 and 33.3.5.4, respectively). Briefly, to resolve dripping issues, higher drawback should be set and an appropriate nozzle type should be selected. This parameter should be evaluated at the early-stage development with platform recommendations. However, a detailed characterization is necessary during commercial process development and technology transfer to commercial manufacturing sites.

33.3.6.2. Splashing. *Splashing* is a common challenge with high fill speeds. The cause of this phenomenon is higher solution velocity of product in the container, due to high fill speed. Splashing is also an undesirable phenomenon as it may result in container–closure integrity (CCI) implications, product spills on the fill line, and cosmetic defect implications for lyophilized products. The splashed product can be entrapped in the stopper placement or plunger placement, which can impact the CCI of the primary container. Splashing also increases the adsorption of product on the container surface. This may result in spots and streaks on the wall of the primary container for lyophilized products. This may also result in higher reject rates and decrease in yield, and will eventually affect the cost of goods. The factors affecting splashing are fill volume and fill speed (more specifically, cycle speed). Typically, the fill volume is determined as late as possible to provide appropriate dosage according to clinical requirements during product development. Therefore, splashing challenges have to be resolved by optimizing the fill speed and nozzle type. Splashing can be minimized by reducing the fill speed of the filler. The alternate option is to use a nozzle with large internal diameter, which will reduce the solution velocity in the container. The splashing can be further alleviated by moving the nozzle out of the container as filling progresses to avoid excessive upward velocity of product in the container.

33.3.6.3. Foaming. *Foaming* is a well-known phenomenon in which a foam layer is present on the surface of the solution. Protein product consists of a significant amount of surface-active agents in terms of protein proper and possible surfactants in the formulation. These surface-active agents lower the overall surface tension of the solution, thus rendering it more prone to significant air entrapment and foaming. Foaming is a highly undesirable phenomenon for dosage form manufacturing as it can impact container closure integrity, impact product quality, and lead to cosmetic defects for lyophilized products. Foaming increases the air–liquid interface significantly, thus facilitating higher protein absorption at the interface [7]. Finally, a persistent foam prior to lyophilization may result in significant spots and streaks on the primary container walls and product deposition on internal wall of stopper as well as lyophilized cake surface irregularities that may not be commercially acceptable. This may result in higher reject rates and affect the cost of goods. From this discussion, it is clear that foaming is a highly undesirable phenomenon. As discussed in the preceding sections, it is rarely possible to revise the formulation during process development. Therefore, any foaming-related challenges have to be mitigated during the filling process characterization.

33.3.6.4. Particle Shedding. Positive-displacement pumps, like piston pump and rotary piston pump filling technologies may shed particles as a result of the tight tolerance between the piston and the cylinder. Also, the product acts as a lubricant between piston and cylinder. Detailed product characterization is recommended to analyze for particles during filling operations.

33.4. PROCESS CHARACTERIZATION

33.4.1. Process Development, Scale-Up, and Scale-Down

After understanding the product physical properties (density, viscosity, etc.) and other parameters such as fill volume and vial or syringe size, the next step is to determine filling equipment to be used and how product may perform on equipment given the inherent physical properties of the product.

Because of high overhead cost, limited drug substance material, or unavailability of commercial filler, full-scale process characterization cannot be routinely performed during product development. As such, for the filling process to be effectively scaled down, several methods or instruments are typically available in the laboratory for scale-down evaluation. These studies seek to evaluate each candidate formulation for its behavior in scaled-down filling equipment representative of the filling equipment in the manufacturing setting for the desired product presentation.

Besides evaluating the impact of shear stress, additional filling-process-related issues need to be investigated, such as potential foaming, dripping, and material compatibility. For example, certain high-viscosity protein formulations tend to drip in the nozzle, which could compromise the fill accuracy. During filling process development, different nozzle sizes, nozzle shapes, and filling speeds should be evaluated to minimize the dripping problem.

To transfer the filling process to a clinical or commercial manufacturing setting, a machinability–shakedown run is recommended for evaluating product fill accuracy. Small-scale evaluation and characterization of the filling pump, filling speed, and nozzle size should be performed such that it results in shear rates comparable to those encountered at manufacturing scale. The engineering fill is typically used to confirm and set the validation acceptance criteria and process limits for the filling operation.

33.4.2. Mimic Solutions for Process Development Studies and Characterization

During the technology transfer/scale-up of a filling process, it is imperative that a few dry runs (machine or engineering runs) be performed on the commercial filler to evaluate the performance of filling line and process robustness. Generally, mimic solutions, which have viscosity, density, and other physical properties similar to those of the com solutions that will be used, are employed instead for purposes of cost-effectiveness or because of material unavailability. The commonly used mimic solution is the *placebo*, which contains all ingredients and excipients except the active pharmaceutical ingredient (API or therapeutic protein). However, if the drug product contains a large amount of API, such as high-concentration MAb formulation, a placebo alone would not simulate the drug product. Some other mimic solutions need to be developed. The common mimic solution could be made using polymeric solutions using poly(ethylene glycol) (PEG) or model protein solutions such as human serum albumin (HSA).

33.5. PROCESS TECHNOLOGY TRANSFER CONSIDERATIONS

If a product will be filled at a contract fill site, an assessment of the site's equipment and procedures is necessary. A product shakedown fill is recommended using the intended filling equipment at the contract site. Enough units must be filled to statistically determine the weight tolerances specific to this equipment and evaluate for the duration of a commercial batch fill.

The filling evaluation performed on a scaled-down model may provide accurate estimation of filling parameters such as pump size and nozzle size as well as alert and action limits during filling. This model cannot accurately predict possible issues that may be seen during filling in clinical as well as commercial manufacturing. It is important to note that a scaled-down model uses one or two pumps for filling evaluation, whereas manufacturing sites may use a higher number of pumps. Other variables such as filling line speed, vibration caused by increased line speed, and vial handling before-, after, and during filling are not taken into consideration during scale-down evaluation.

33.6. SUMMARY

The filling process is a critical unit operation during fill–finish manufacturing of protein drug products. An in-depth knowledge of protein product properties and optimized process parameters is required to successfully transfer and minimize any impact to product quality during filling. Various filling technologies, process considerations, and product compatibility studies were discussed in this chapter.

ACKNOWLEDGMENTS

The authors would like to thank Feroz Jameel, Narinder Singh, Cathryn Shaw-Reid, Lorena Barron, and Vick Wadhwa for technical review and comments on this chapter. The authors would like to acknowledge and thank Thilo E. Schaffert and Groninger & Co. GMBH for providing and granting permission to use Figures 33.2 and 33.3; Jeff Jackson and Bosch Packaging Technology for Figures 33.4–33.8; and Brett Belongia and Millipore Corporation, for Figures 33.9 and 33.10. The authors would also like to thank Matthias Poslowski, Optima Gp, Swabisch Hall, Germany, for providing supporting information on filler technologies.

REFERENCES

1. Patro, S. Y., Freund, E., and Chang, B. S. (2002), Protein formulation and fill-finish operations, *Biotechnol. Annu. Rev.* **8**: 55–84.
2. Rathore, N. and Rajan, R. S. (2008), Current perspectives on stability of protein drug products during formulation, fill and finish operations, *Biotechnol. Progress* **24**(3): 504–514.
3. Stockdale, D. (2004), Overview of aseptic fill/finish manufacturing, *Am. Pharm. Rev.*
4. Belongia, B., Blanck, R., and Tingley, S. (2003), Single-use disposable filling for sterile pharmaceuticals, *Pharm. Eng.* **23**(3): 126–134.
5. Shire, S., Shahrokh, Z., and Liu, J. (2004), Challenges in the development of high protein concentration formulations, *J. Pharm. Sci.*, **93**(6): 1390–1402.
6. Gabrielson, J., Bates, D. G., Williams, B. M., Hamel, J.-B., Carpenter, J. F., and Randolph, T. W. (2007), Silicone oil contamination of therapeutic protein formulations: Surfactant and protein effects, *Abstracts of Papers, 234th Natl. Meeting of the American Chemical Society*, Boston, American Chemical Society, Washington, DC.
7. Maa, Y. F. and Hsu, C. C. (1997), Protein denaturation by combined effect of shear and air-liquid interface, *Biotechnol. Bioeng.* **54**(6): 503–512.

8. Thurow, H. and Geisen, K. (2004), Stabilization of dissolved proteins against denaturation at hydrophobic interfaces, *Diabetologia* **27**(2): 212–218.
9. Cromwell, M. E. M., Hilario, E., and Jacobson, F. (2006), Protein aggregation and bioprocessing, *AAPS J.* **8**(3): E572–E579.
10. DeGrazio, F. L. (2006), Focus on fill and finish: Parenteral packaging concerns for biotech drugs—compatibility is key, *BioProcess Int.* **4**(2): 12–16.
11. Charm, S. E. and Wong, B. L. (1970), Enzyme inactivation with shearing, *Biotechnol Bioeng*, **12**(6): 1103–1109.
12. Jaspe, J. and Hagen, S. J. (2006), Do protein molecules unfold in a simple shear flow? *Biophys. J.* **91**(9): 3415–3424.
13. Oliva, A. et al. (2003), Effect of high shear rate on stability of proteins: Kinetic study, *J. Pharm. Biomed. Anal.* **33**(2): 145–155.

34

LEACHABLES AND EXTRACTABLES

Jim Castner, Pedro Benites, and Michael Bresnick

34.1. INTRODUCTION

Whenever a novel compound is identified as having potential pharmacological activity, whether it is therapeutic, palliative, prophylactic, or diagnostic, the primary focus of the research and development (R&D) organizations is to determine the chemical, physical, and toxicological properties of the identified molecular species. These studies are typically conducted in a sequence that conforms to topics listed in the *common technical document* (CTD). This strategy is designed to facilitate submission of reports for regulatory filings. Once all of the CTD topics have been addressed regarding the properties of the compound, and no chemical or toxicological impediments have been identified, efforts are then directed at selecting a clinical formulation matrix that effectively delivers the pharmaceutical agent to a patient, while maintaining the safety, stability and the effectiveness of the active ingredient. Once the clinical formulation has been identified and selected the drug development program finally addresses the issues of manufacturability. Following this linear mode of operation from drug discovery through its development, issues regarding process impurities, compatibility, and contaminates are addressed as an afterthought to the formulation selection process. The resulting drug development program would therefore impose the requirement that

Formulation and Process Development Strategies for Manufacturing Biopharmaceuticals, edited by Jameel and Hershenson

the manufacturing process and the primary packaging system have to conform to the selected clinical formulation. If an incompatibility does occur between the selected formulation and the manufacturing process, program delays will occur because of the narrow focus of the linear sequential strategy. Changing the composition or the process for preparing the clinical formulation, to address the manufacturability issues, comes at a significant expenditure of time and resources. These actions would involve amending the investigational new drug (IND) filing, which, in turn, would typically require submission of supporting safety data followed by active regulatory approval to implement the change(s). All of these actions relating to the filing of the amendment and its approval process usually result in significant delays in initiating the phase I or phase II clinical studies.

These delays can be alleviated if the formulation selection program adopts a wider scope earlier in the characterization studies than what occurs using the the linear model for product development. The additional studies would include, but not be limited to, compiling a database of the materials that would be exposed to the formulation. The compiled data would include the chemical composition and the physical properties of the primary packaging materials and the exposed surface materials of the manufacturing equipment. The data can be collected from a variety of sources, such as vendor certification sheets and published analytical data in reviewed journals, as well as test results from controlled extraction studies conducted as part of a qualification program for new materials. The data would then be used to expand the analytical scope of the drug substance impurity method(s) to include detection of not only related substance impurities but also extractable/leachable compounds that could contaminate the exposed formulation. In this way a strategy would be fashioned during the earlier stages of the drug development program to reduce the risk of proceeding to clinical trials with a product that could fail. A comparison of the two drug development programs is shown in Figure 34.1.

The discussion in this chapter will focus then on the development of analytical techniques using multiphase analyses designed to identify "at-risk-product" conditions early in the drug development program. Applying these analyses can provide tactical support early-on to address the contamination issues associated with exposure to the manufacturing equipment or the container–closure system. These topics will be addressed from both regulatory and technology perspective. Case studies will also be presented that demonstrate the application of these technologies.

34.2. REGULATORY CONSIDERATIONS

One of the hallmarks that collectively distinguish the pharmaceutical regulatory agencies (e.g., FDA, EMEA, Health Canada, MHLW Japan, ICH) is their overriding concern with ensuring the safety and purity of drug products that they approve and monitor for patient use. This concern, exemplified in the US Federal Food, Drug and Cosmetic Act[1], applies to both the materials and the processes that are used in the manufacturing of the pharmaceutical agent and includes the primary packaging. In this document an adulterated product is defined as being "a drug and the methods

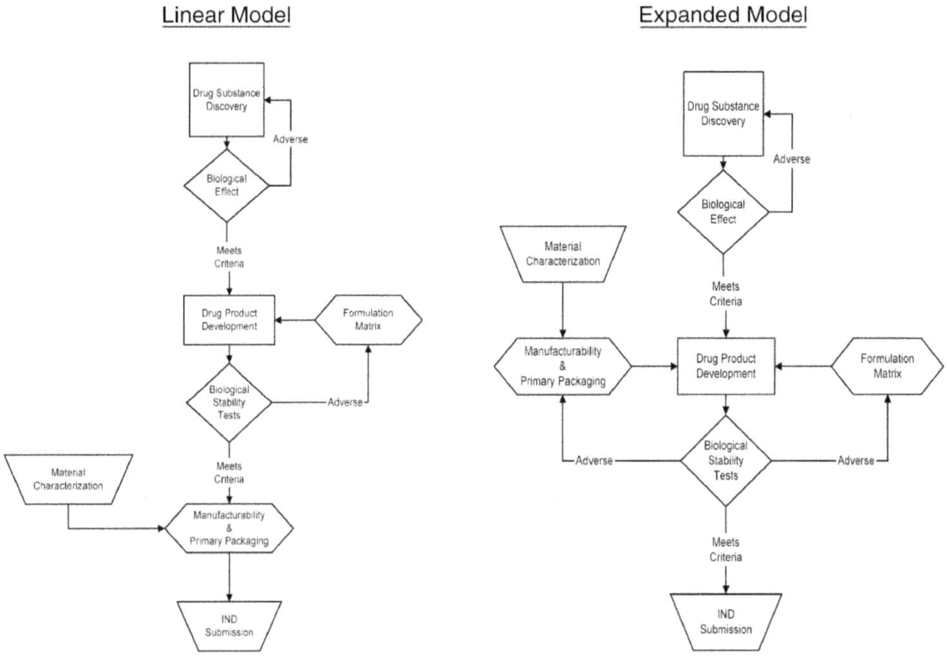

Figure 34.1. Models for drug product development programs.

used in or the facilities or controls used for, its manufacture, processing, packing, or holding do not conform to or are not operated or administered in conformity with current good manufacturing practice to assure that such drug meets the requirements of this Act as to safety and has the identity and strength, and meets the quality and purity characteristics, which it purports or is represented to possess."

The regulatory approval process thus not only applies for chemical or biochemical reagents but also includes how the pharmaceutical product is manufactured (i.e., equipment and process used) and how it is packaged. As a result of these concerns, the agencies have issued policies and guidelines instructing the industry as to the expectations needed to achieve approval of the product, process, and packaging. This is exemplified by the FDA guideline on current "good manufacturing practices" (cGMP) found in 21 CFR Part 210 and 211 and in particular for packaging of drug products in Section 211.194, the *Packaging Guidance* (May 1999). The EMEA has also issued *Guidelines on Plastic Immediate Packaging Material* (2005). These documents identify the requirements for approval, but they do not address the question of how to achieve the requirements. The national pharmacopoia or formularies (USP, EP, BP, JP, etc.) as well as the international organizations for standards (e.g., ISO for devices and materials) have attempted to take on this task. It is a daunting challenge to address the analytical testing issues needed to meet the regulatory requirements for a vast number of pharmaceutical agents with a variety of unique properties. As a result of this diversity of drug products, the published test procedures provided by these organizations have to be generic in design, technology, and application so that they will be readily available to the industry. Responsibility then shifts to the pharmaceutical companies to apply the test procedures for their products. Verbatim execution of the test procedure does not relieve the companies of their responsibility to demonstrate the applicability of the published compendial methods to evaluate the particular product, manufacturing process, or primary packaging material. In some cases monographs have been found to be unsuitable for testing a given product because of the uniqueness of the sample or the application of the technology. The option then is to either modify the published test procedure or find an alternate analytical technique. Modifying a test procedure carries with it the obligation of demonstrating that the change meets the regulatory requirements. An example of this process would be the application of the test procedures listed in USP <661> to determine the suitability of using a plastic container for a given drug product. While the USP tests are designed to provide test results regarding the reagents that can be extracted from the container and closure, the data may not have direct relevance about a given drug product matrix and its process manufacture or storage conditions. The reason then for the disconnect between direct application of the test procedures and obtaining relevant data for a particular drug product is in the design of the test methods. The chemical test procedures listed in USP <661> are designed to determine "extractables" and not "leachables." The difference between these two terms is critical to the appropriate use of the USP <661> test procedures.

In the glossary section of the FDA guidelines on container–closure systems, *extractables* are defined as "Compounds that can be extracted from elastomeric or plastic components of container closure system when in the presence of a solvent," while *leachables* are defined as "Compounds that leach into the formulation from elastomeric or plastic components of the drug product container closure system."

The critical difference between these two terms relates to the process in which a compound is removed from the container or closure. In the case of extractables, compounds are removed from the exposed surface under extractive or aggressive conditions that are not always representative of the drug product and how it is manufactured. These conditions would include the use of a variety of solvents ranging from strong organics, to acid/alkaline aqueous solutions, as well as operating at elevated temperature conditions. Leachables, on the other hand, are compounds that are removed from the test material by exposure to the formulation under conditions that are representative of the manufacturing process and product storage conditions.

The USP test procedures are designed to determine what compounds are capable of being extracted from either the container–closure system or the surfaces of the manufacturing equipment. However, the detected compounds using these procedures may not be found in the manufactured drug product. This disconnect between test results and the effect on product is due to the fact that the test conditions are not representative of the manufacturing process. In essence, detected extractables are potential leachables to the extent that the test conditions are representative of the manufacturing process and product. The sponsor of the drug product is therefore left with the problem of determining whether the USP test procedures are suitable analytical tests. If the test conditions are not suitable for determining leachable content, then a modified version or an alternate test is needed to address the regulatory issues. The issue is to demonstrate that the exposed surfaces will not adversely affect the final product by presenting a safety or efficacy concern.

To make this assessment about what analytical test to use, there first needs to be an understanding of the basic chemical property that is being examined. That property in this case is the partitioning of a compound between two phases. The two phases are the solid material of the container or the closure or the exposed manufacturing equipment vs. the formulation. The formulation can range from free-flowing to highly viscous liquid to solid or gases. In this chapter the discussion of partitioning effects will be limited to the interaction between a solid surface and a free-flowing liquid matrix, since this is one of the most common surface–product interactions that occur in the pharmaceutical industry.

34.3. TECHNOLOGY CONSIDERATIONS

Liquid chromatography, whether it is normal or reversed phase, is an appropriate model to use in describing the chemistry of the surface–product interaction that takes place during the formation of a leachable or extractable. This model is based on the similarity between the partitioning events in both cases. In chromatography, the compound of interest is initially retained on the column stationary phase and then desorbed as a function primarily by the hydrophobic or hydrophilic interaction with the liquid mobile phase. The same type of process occurs during the leaching or extracting of a compound from the exposed surfaces of the packaging materials or the manufacturing equipment into the liquid formulation. In both cases the partitioning between two phases is driven primarily by the hydrophobic or hydrophilic property of the respective interface. The partitioning driving force is commonly described in terms of a log P value. The log P value for a given compound is determined by the compound's preferential solubility in either a hydrophobic (organic) or a hydrophilic (aqueous) solvent. In practice, the organic solvent octanol is typically used as the hydrophobic phase, while purified water is used as the hydrophilic phase in determining the log P value for a given compound.

If the compound of interest also has an accessible pK_a in water, then the pH of the aqueous phase can be a driving force in determining the partitioning value. In this case the octanol–water partitioning value now becomes a series of ratio values that are reported in terms of log D (i.e., distribution ratio) for each pH condition. Application of the log P or log D ratio values provides an indication as to the hydrophobic (positive log ratio values) or the hydrophilic (negative log ratio values) property of the compound. Subsequently, knowing the octanol:water ratio values of a given compound can be used to estimate the likelihood that the compound will be partitioned or extracted into either a hydrophobic or hydrophilic solvent. For example, consider a compound that has a log P value of 3.0 and log D values that range from 3.5 to -1.5. The log D values for the compound rapidly transition from positive to negative at the pK_a value of 6.5. In this situation the compound would presumably be partitioned or extracted into an aqueous solution that was adjusted to a pH above 6.5. At more acidic pH conditions aqueous extraction of the compound would be less favorable. For many pharmaceutical reagents, log D values are more frequently reported than log P values, since they typically are used in formulation that contain a buffer to control the pH of the dosage form. The log D values of the active pharmaceutical ingredients (APIs) and the excipients can then aid in the selection of an appropriate pH buffer to ensure solubility of all the ingredients in the formulation. Additionally, log D values can be used to determine the need for a solubilizing reagent to ensure that the ingredients will stay in solution at a particular pH.

Besides their application in formulation development, log D values can also aid in the development of a liquid chromatographic (LC) method that is capable of identifying the related substance impurities and/or exogenous contaminates (i.e., leachables) in the drug product. This application is based on the assertion that the predominant factor that controls the chromatographic separation of these compounds is their respective hydrophilic or hydrophobic property. In short, this means that the retention time value for a given compound corresponds to its log P or log D value expressed in units of time. There are several citations in the literature [2] in which LC methods have been used to determine the partitioning or distribution value of an unknown compound. This determination is based on a comparison of retention time values for standards, which have known log P or log D values, and unknown test compounds. In these determinations of log P or log D values, the LC operational parameters of flow rate (for the mobile phase) and temperature (for the column) need be to be well controlled, since they directly affect retention time values independent of partitioning properties. Control of these parameters can be achieved by holding them constant during the analysis. Under these conditions, a partitioning value for a sample compound can be calculated from a regression equation derived from a fitting of the retention time data and the log P or log D values for the standard compounds. In practice, this process begins with a search of the literature for reported extractable compounds with known log D values. The list of these compounds should cover a range of at least one to two negative log D units of negative (hydrophilic) and three to four positive log D units (hydrophobic). An example of the type of compounds and the range of values that can be included in the list is shown in Table 34.1. The chemical structures of these compounds are shown in Figure 34.2.

The selection of which reference compounds to use for investigating potential leachables in a particular drug product can be based on the log D value for the API in the formulation. In the case where the API has a log D value of +2 in a formulation buffered at a specified pH, then a group of reference compounds that covers a range of $\pm$ 2 log D values at that pH would be appropriate, as shown in the chromatogram displayed in Figure 34.3. In this example a total of four reference compounds listed in Table 34.1 were selected with log D values that range from −0.34 (hydrophilic) to 5.32 (hydrophobic) at a pH of 5.0. The plotted data of the retention time and log D values for these compounds are then fitted to a linear regression equation as shown in Figure 34.4. In this fitted equation the $R^2 > 0.9$.

TABLE 34.1. Reference Extractable/Leachable Compounds

Compounds	CAS Number	log *D*	Source/Application[c]
2,4-Dichlorbenzoic acid (DCBA)	50-84-0	−0.34–2.7[a]	Silicone tubing peroxide catalyst
2-(2-butoxyethoxy)ethyl acetate (BEEA)	124-17-4	1.79[b]	Silicone tubing Pt catalyst binder
1,3,5-Triazine-2,4 -dithione (DBAT)	29529-99-5	0.3–2.11[a]	Crosslinking agent
4-Pentylphenol	87-26-3	3.88[b]	Lubricant, stabilizer, antioxidant
Phenol,2,6-bis(1,1-dimethylethyl)-4-methoxy (BHA)	489-01-0	4.69[b]	Antioxidant
Dibutylphthalate (DBP)	84-74-2	4.83[b]	Plasticizer
Butylated hydroxytoluene (BHT)	128-37-0	5.32[b]	Antioxidant
Dioctylphthalate (DOP)	117-81-7	8.7[b]	Plasticizer

[a]log *D* value over pH range 4.0–7.0.
[b]Nonionizable compounds over pH range 2–12 (log *D* = log *P*).
[c]*Source:* Jenke [11].

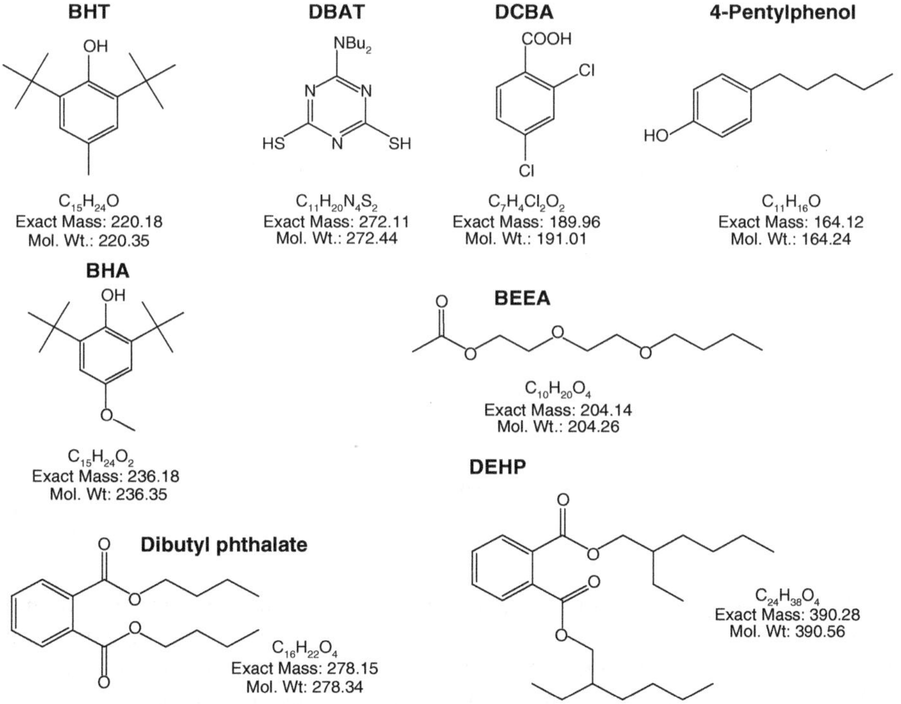

Figure 34.2. Chemical structures of extractable/leachable reference compounds.

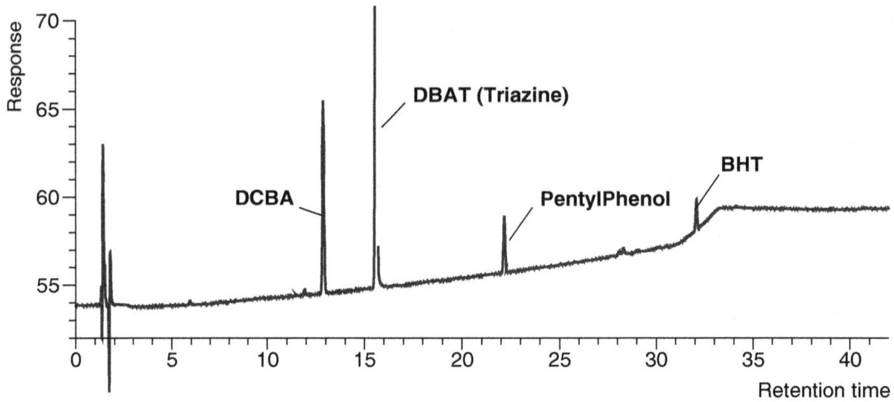

Figure 34.3. Chromatogram of reference compounds covering a log D values from −0.34 to 5.3.

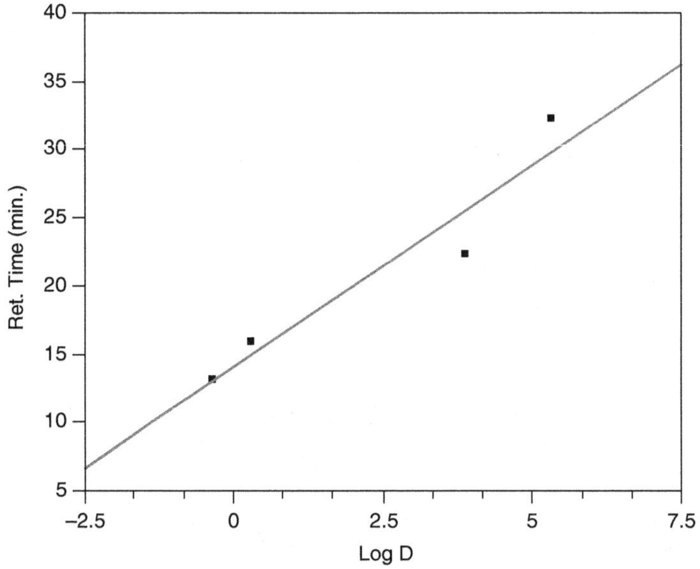

Figure 34.4. Comparison between log D and retention time values for reference compounds.

Changing the chromatographic operating parameters such as flow rate, composition of the mobile phase, or the shape of the mobile-phase gradient can, as mentioned before, result in modifying the type of function that will fit the data. For example, the fitted regression equation could change from a linear to a polynomial function. The log D values for unknown impurity peaks detected using this LC method can then be calculated using the fitted equation for the plotted data. Determination of the log D value for an unknown impurity peak can be helpful in identifying the unknown compounds, as will be shown in the case studies in Section 34.4.

The log D/retention time data can also provide an assessment of the range and number of compounds that the chromatographic method potentially can monitor. This assessment is based on a determination of the peak resolution (R_s) factor for the LC method as a function of log D. As shown in Equation (34.1) [3], the R_s factor between two chromatographic peaks is typically calculated in terms of the difference in retention time. In this equation the retention time of the first peak (t_1) is compared to that of the second peak (t_2); the difference between the two is then divided by the average baseline width (w_{avg}) for the two peak in units of time. The w_{avg} term in Equation (34.1) can be redefined in terms of retention time and theoretical plate number (N), as shown in Equation (34.2). The R_s factor can then be calculated on the basis of the retention time values and theoretical plate number, as shown in Equation (34.3):

$$R_s = \frac{t_2 - t_1}{w_{avg}} \tag{34.1}$$

$$w_{avg} = \frac{(w_1 + w_2)}{2} = \frac{(4t_1/\sqrt{N}) + (4t_2/\sqrt{N})}{2} = \frac{2(t_1 + t_2)}{\sqrt{N}} \tag{34.2}$$

$$R_s = \frac{(t_2 - t_1)\sqrt{N}}{2(t_2 + t_1)} \tag{34.3}$$

In Equation (34.4), the retention time value for a compound can also be described as a function of its hydrophobic or hydrophilic property in terms of log D. In the presented case in Figure 34.4 there is a linear relationship between the two terms, where the retention time (t_i) is a function of the slope of the regression line for the log D value (D_i) and the intercept of the regression line:

$$t_i = s \log D_i + a \tag{34.4}$$

After substituting Equation (34.4) for the respective retention time values in Equation (34.3), collecting terms, and simplifying the equation the; R_s value is now calculated (as shown in Equation 34.5) in terms of log D values, theoretical plate number N, and the intercept value for the regression equation.

$$R_s = \frac{\log(D_2/D_1)\sqrt{N}}{2[\log(D_2/D_1) + 2a]} \tag{34.5}$$

Redefining the R_s factor in terms of log D provides a means for assessing how many compounds with different partitioning values can be monitored for a given chromatographic method. As shown in Equation (34.6), the probability for resolving k

randomly spaced peaks between two reference peaks, using a single detection system, can be calculated in terms of the R_s factor,

$$P_1 = 100\left[\frac{R_s!}{R_s^k(R_s - k)!}\right] \tag{34.6}$$

where R_s values are rounded to the nearest integer and P_1 = percent probability of baseline resolution of "k" number of peaks. As an example, if this probability equation is applied to the chromatographic separation between two peaks (DCBA and BHT) as shown in Figure 34.4, the calculated probability of detecting four baseline resolved peaks between the two reference compounds is 70%, given R_s separation value of 21. The calculated probability value for resolving six peaks between these two compounds drops off exponentially to less than 50%. If the R_s factor is more than doubled to a value of 50 by expanding or lengthening the retention time window, the 70% probability level is extended out to six peaks as shown in Table 34.2. However in lengthening the retention time window by a factor of 2 will result in an exponential widening of the later eluting chromatographic peaks, correspondingly diminishing the signal intensity of the peaks.

Another approach that can be taken to improve the probability of detecting more peaks without altering the R_s factor is to change the single-detector LC system to a dual-detector system, in which each detector would operate independently of the other detector. The addition of a second independent detector results in a modification of the initial probability equation from a one- to a two-dimensional probability equation as shown in Equation 34.7

$$P_2 = 100[1 - (1 - [P_1/100])^2] \tag{34.7}$$

where P_2 = percent probability for a dual detection system. A typical example of a dual-independent-detector system coupled to an LC instrument would be the combined use of UV–visible and mass spectral detectors. In the dual-detector example the 70% probability is extended to near the six-peak resolution level, which is a similar result to

TABLE 34.2. Purity Probability

Number of Peaks	$R_s = 21$ (1),[a]%	$R_s = 50$ (1),[a]%	$R_s = 21$ (2),[b]%
2	95.2	98.0	99.8
3	86.2	94.1	98.1
4	73.9	88.4	93.2
5	59.8	81.4	83.8
6	45.6	73.2	70.4
7	32.5	64.4	54.5
8	21.7	55.4	38.7
9	13.4	46.5	25.1
10	7.7	38.2	14.8

[a] Single-detection system.
[b] Two independent detection systems.

the Rs value of 50 without incurring the problems of peak broadening. These results demonstrate that the use of two independent detectors provides greater analytical sensitivity than does lengthening the retention time window of the LC method, thus increasing the probability of resolving more compounds.

To recap, the technical issues that have been addressed in this section have focused on the use of the octanol–water partitioning model in both developing and analyzing pharmaceutical liquid drug products. In particular applying, the log P or Log *D* values to predict the solubility of a given compound in a formulation matrix, thus providing a tool to determine if modifications (i.e., change in buffer pH or addition of a solubilizing agent) are needed to assure solute stability in the drug products. Additionally, log *D* values can be used as a complementary analytical tool, coupled with other spectral techniques, in the identification of extractable or leachable compounds that can occur as a result of drug product exposure to the surfaces of primary packaging materials or manufacturing equipment. Practical examples of this synergy between partitioning (log *D*) and spectral (UV and MS detection) techniques aiding in the identification of leachable compounds are presented in the following three case studies.

34.4. CASE STUDIES

34.4.1. Polypropylene Extractable

During the development of a liquid drug product, a new impurity was discovered when the primary packaging material was changed from glass to plastic. In this case the plastic packaging material is high-density polypropylene (HDPP), a polymer that is commonly used to fabricate containers for a number of pharmaceutical liquid products. The formulation contained a drug substance with a log *D* value of 2.4, in an ethanol/water solution and a buffer targeted at a pH of 6.5 ± 0.5. Information provided by the manufacturer of the plastic container indicated that HDPP is chemically stable when exposed to an ethanol/water solution [4].

This leachable impurity appeared to be associated with the plastic container but did not exhibit a UV spectral response over a wavelength range of 200–500 nm, however it did have a mass selective ion peak at 282 amu scanning in positive mode, as shown in Figure 34.5. Scanning the chromatographic peak in negative ionization mode did not produce a spectral signal compared to background noise. This result suggests that the molecule contains a group favorable to protonation, such as $R–NH_2$ or R–NH–R. Further analysis of the mass spectral pattern indicates that the molecule could contain a maximum of 18 carbons, 1 nitrogen and 1 oxygen, based on a comparison of the signal intensities for the 282 (100%), 283 (20.2%), and 284 (2.1%) ion peaks shown in Figure 34.6 [5]. The retention time for the impurity peak is approximately 1.5 times greater than that for the API compound, indicating that the leachable is significantly more hydrophobic. Relative to the reference compounds list in Table 34.1, the estimated log *D* value for the 35-min retention time peak is > 6 to < 8 Log units.

In summary, the resulting profile from the compiled chromatographic and spectral data for the leachable compound is that it is a hydrophobic molecule, containing a linear chain of 16 or more carbons and has a molecular weight of 282 Daltons. The

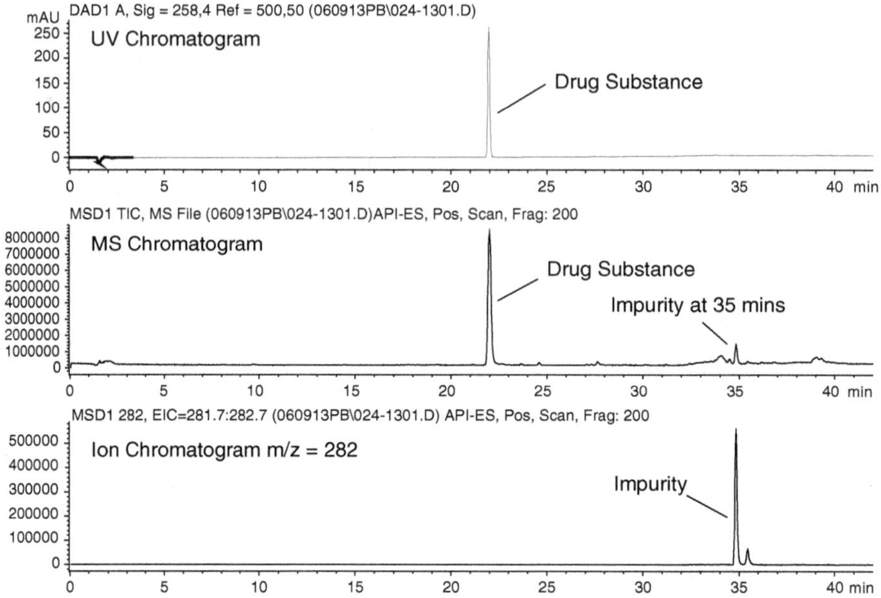

Figure 34.5. Chromatograms of the drug substance and impurity using multiple detection systems.

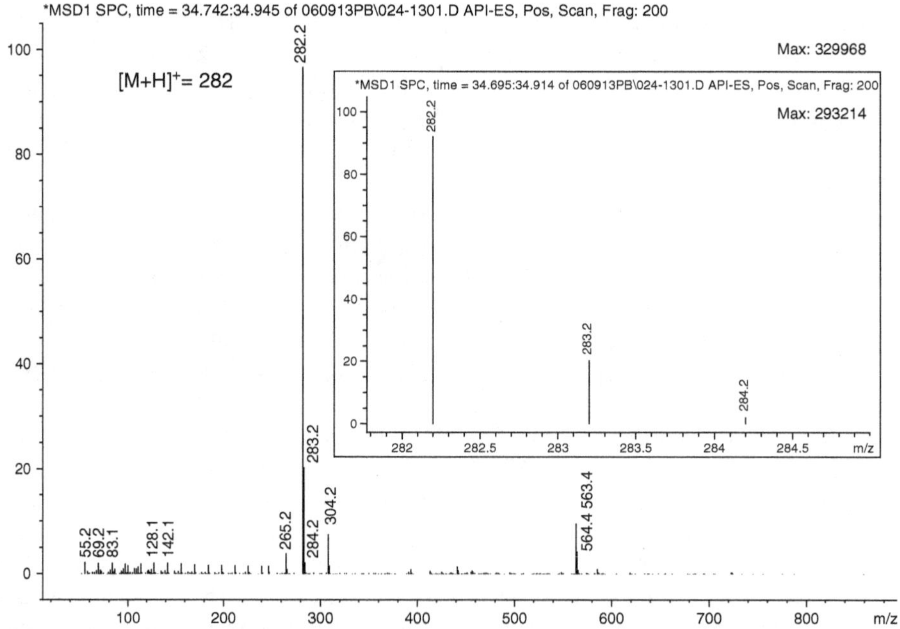

Figure 34.6. Mass spectral scan of leachable compound peak at 35 min retention time.

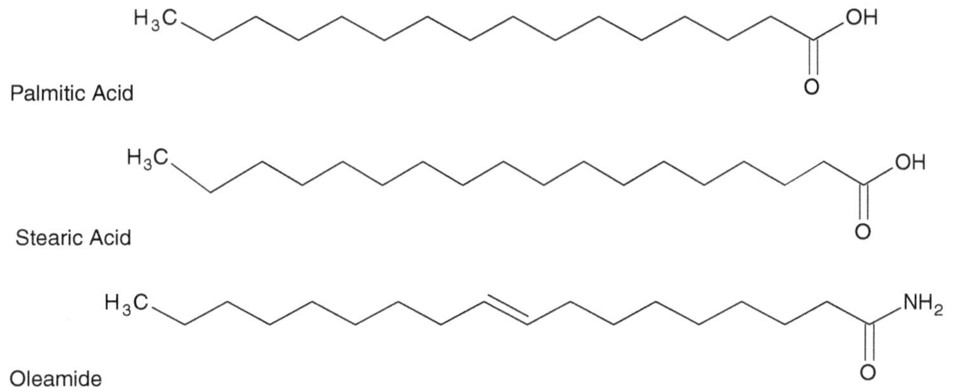

Figure 34.7. Molecular structure of palmitic acid, stearic acid and oleamide.

indication that this molecule is a linear chain is based on the absence of a UV spectral response in the 200–300 nm range, which is consistent for organic compounds that do not contain an aromatic or a conjugated substituent group. Also, linear chains or aliphatic molecules containing more than 10 carbons typically have hydrophobic log *D* values ≥6. A search of the literature for plastic additives [6] and extractables from plastic materials [7] with similar molecular properties to the unknown leachable compound revealed three possibilities: palmitic acid, stearic acid, and oleamide. All three of these compounds are linear chain organic molecules as shown in Figure 34.7. A comparison of their chemical properties along with the properties of the unknown impurity is shown in Table 34.3. Palmitic and stearic acids were eliminated from consideration as a match to the unknown compound even though they have similar molecular weights and log *D* values and no UV absorption. The mass spectral signal for these two molecules could be detected only during operation in negative ionization mode. The chemical properties of oleamide are a closer fit to those of the unknown compound since the signal is detected only during operation in positive ionization mode, plus the formula for oleamide is a closer match to the molecular formula of

TABLE 34.3. Comparison of Known and Unknown Extractable/Leachable Compounds

Compound	Formula	Molecular Weight	log *D*	Application
Palmitic	$C_{16}H_{32}O_2$	256.4	7.15	Lubricant and slip agent for plastics, etc.
Stearic	$C_{18}H_{36}O_2$	284.5	8.22	Lubricant and heat costabilizer, etc
Oleamide	$C_{18}H_{35}NO$	281.5	6.75	Lubricant and slip agent for plastics, antiblocking for extrusion of plastics, etc.
Unknown	$\geq C_{16}$	282(M + H)	> 6 to < 8	NA[a]

[a]NA = does not apply.

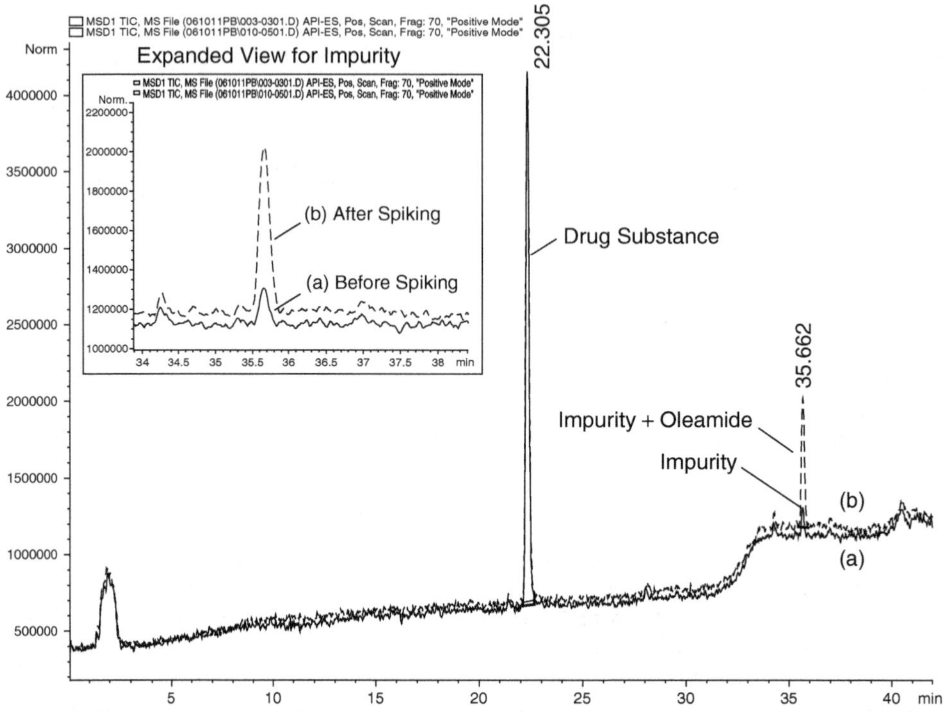

Figure 34.8. Mass spectral detection chromatograms operating in positive ionization mode for sample containing unknown leachable impurity before (curve a) and after (curve b) spiking with oleamide.

18 carbons, 1 nitrogen, and 1 oxygen. Confirmation as to the identity of the unknown leachable was verified by spiking the sample containing the leachable impurity with an authentic sample of oleamide, as shown in Figure 34.8. This action resulted in a measurable increase in peak response at the 35 min retention time, with no changes in the intensity of the other chromatographic peaks. Quantitation of the impurity in the product sample can then be determined using an oleamide standard.

34.4.2. Leachables in Sterile Water for Injection

The preparation of parenteral drug products frequently requires the use of sterile aqueous diluents to administer the dosage form. The two most commonly used sterile liquid diluents are normal saline and water for injection. The chemical and microbiological properties of these liquids are well defined in the respective national formularies. For example, in the USP <1231> monograph *Water for Pharmaceutical Purposes*, the quality of the starting material and the process conditions for manufacturing "sterile water for injection" (SWFI) are well defined. In addition, the monograph identifies the qualification tests (i.e., *Bacterial Endotoxin* <85>, *Sterility* <71>, *pH* <791>, *Total Organic Carbon* <643>, and *Conductivity* <645>) that certify the manufactured product as USP-grade SWFI. The requirements for the primary packaging material in the monograph are not as well defined. It is suggested that the container should be a glass bottle fabricated from either type I or type II pharmaceutical-grade glass. Plastic bottles or bags are not excluded as possible containers. The monograph also does not contain any specific requirement about the type of closures that can be used. The test procedures listed in the USP monographs <661> (*Glass and Plastic Containers*) and <381> (*Elastomeric Vial Closures of Injectable Products*) can provide guidelines on the general test procedures that can be used to assess the quality of primary packaging materials. These compendial test procedures, however, have limited capability to determine the suitability of using a given container or closure material for a specific product application. In particular, the tests are designed to address the general question about extractables as potential leachables.

One approach to address the issue of leachables in a specific drug product, as mentioned previously in the chapter, is to adopt the analytical test procedures that are developed for monitoring impurities in the specific drug product. This approach provides an economy of analysis for determining both related substance impurities and leachables in the final dosage. How this analysis could be conducted is demonstrated in the following example.

Sterile water for injection (SWFI) was purchased from a single vendor, but the diluent product was packaged in two types of glass containers. One was in a 1-L bottle with a large black elastomeric stopper, while the other was packaged in a 100-mL bottle with a smaller gray elastomeric stopper. Sterile diluent from each of the two containers was used to prepare the same drug product. The dosages prepared using the diluent from the 1-L bottle contained a series of five impurities detected using a modified version of the HPLC method that was developed to monitor the purity of the formulation. In this case the method modification was to alter the gradient profile of the mobile phase to provide a retention time window so that compounds with log D values ranging from -1 to 5 would be well retained. Selection of this range was based on the range of log D values for the ingredients in the formulation.

The impurities that were detected in the prepared product using the diluent from the 1-L bottle did not appear, by UV or mass spectral (MS) analyses, to be related to the drug substance or the other formulation ingredients. These impurities were not detected in the other dosages prepared using SWFI from the 100-mL bottle. Tests of aqueous extraction samples from the container and the closure of the 1-L bottle, using the modified HPLC method, attributed the source of the impurities to the black stopper. Representative chromatograms using UV and MS detectors of these impurities indicate that they are linked to the stopper as shown in Figure 34.9. The retention time values for the five impurity peaks suggest that the log D values for these compounds range from less than 1 to 5, based on a comparison of retention times to the reference compounds listed in Table 34.1. The estimated log D values, along with the UV and the MS spectral data for these five impurities, are summarized in Table 34.4.

As shown in Figure 34.9, three of the five impurities were detected in both the UV and the MS chromatograms, which provides a dual analytical approach for

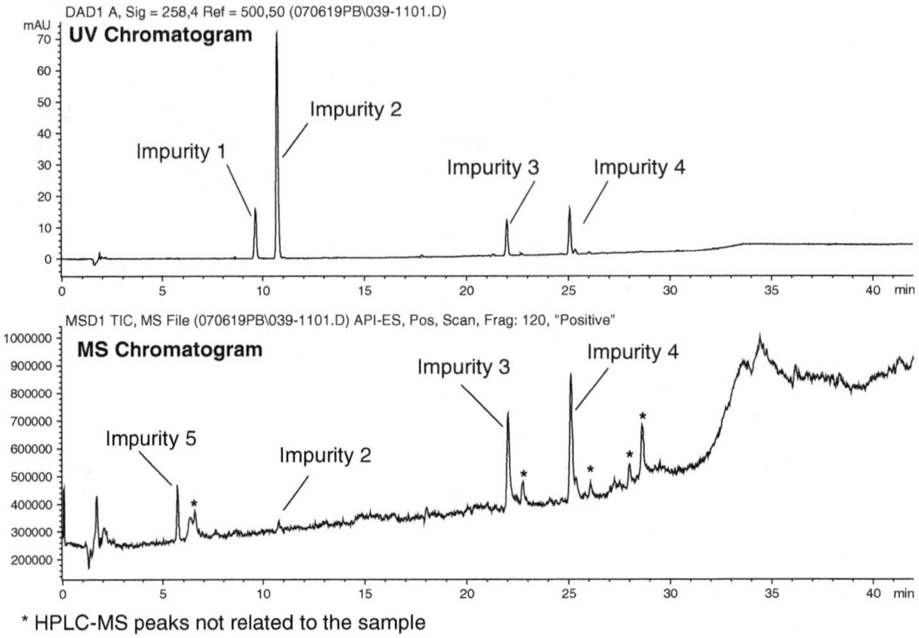

Figure 34.9. High-performance liquid chromatograms of leachable compounds from the stopper of the bottle containing SWFI.

TABLE 34.4. Summary of HPLC Data for Impurities Detected in SWFI Sample

Impurity	Retention Time, min	log D[a] (pH 5.8)	UV peaks, nm	MSI[b] (m/z[M + H])
1	9.67	<2.1	245, 286	NI[c]
2	10.76	~2.1	225, 252, 285	136
3	22.02	<3.9	285	170
4	25.13	<5.3	237, 282	210
5	5.72	0.3	220, 275	198

[a]Estimated by comparison to retention time versus log D values in Table 34.1.
[b]MSI = mass selective ion [positive mode (M+1).]
[c]NI = not ionizable in either positive or negative mode using an HPLC-electrospray ionization technique.

identifying the leachable compounds. The two early-eluting peaks, however, were not detected in both chromatograms. Impurity peak 1 was detected only in the UV chromatogram, while impurity peak 5 was detected only in the MS chromatogram. The ability, then, to identify these two compounds appears to be limited to a single analytical approach. In the case of impurity 5, at a retention time of the 5.7 min, the MS spectral signal provides information about the molecular ion, while the absence of a UV signal suggests that the compound may be aliphatic. This would be similar to the analysis conducted in identifying the leachable compound reported in Polypropylene Extractable Study. However, when the amount of sample injected onto the column was increased, a UV absorption peak at the 5.7 min retention time was detected. The UV spectral profile for this peak did contain two distinctive λ_{max} values, one at 220 nm and the other at 275 nm, indicating the presence of an aromatic or a conjugated moiety[8]. The initial absence of a UV signal response is attributed to sensitivity of the analytical method and not to chemical structure.

The UV spectral scan for impurity peak 1, at a retention time of 9.7 min, likewise produced two distinctive λ_{max} values at 245 and 286 nm. These two λ_{max} values distinguish between impurity peaks 1 and 5, but also indicate that peak 1 contains an aromatic or conjugated moiety.

Further determination as to the identity of this compound, based on the available HPLC data, is problematic because of the absence of an MS peak. No MS chromatographic peak was detected at the 9.7 min retention time, whether operating in either positive or negative ionization mode, as well as by increasing the amount of sample injected onto the column.

An alternate approach to determine the molecular mass of the impurity is the use of GC/MS analysis. The selection of the GC/MS technique is based on the assumption that the compound of interest is a volatile organic species. The aqueous sample was prepared for GC analysis using a solid-phase microextraction (SPME) technique to transfer the compounds of interest to a more suitable volatile matrix. A representative GC/MS chromatogram of the impurities linked to the stopper is shown in Figure 34.10. Three of the five impurities detected in the HPLC chromatograms, as shown in Figure 34.9, were identified in the GC chromatograms. The identification was based on a comparison of the MS scans of the peaks in the blank matrix to the peaks in the stopper extract samples. The molecular ion mass for impurity peak 1 was then determined by GC/MS to be 160 m/z. The chromatographic and mass spectral data for the SPME prepared samples are summarized in Table 34.5.

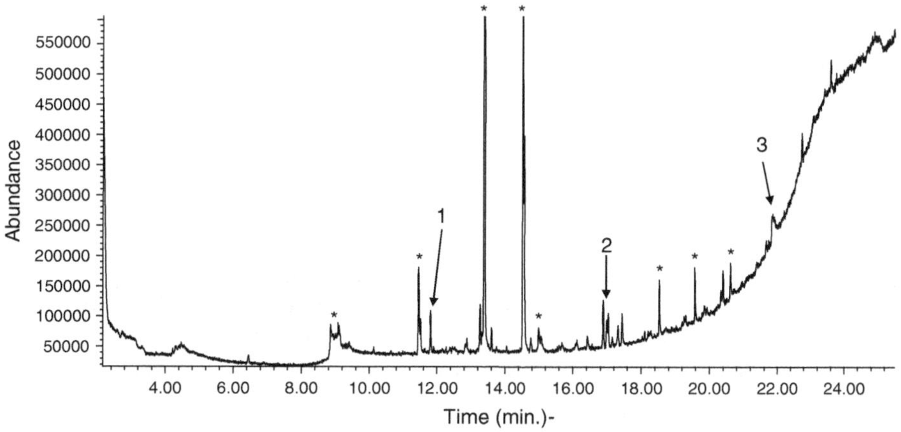

Figure 34.10. Gas chromatography/mass spectral chromatogram of leachable compounds from stopper of bottle containing SWFI.

TABLE 34.5. Summary of GC/MS Data for Detected Impurities

Impurity	Retention Time, min	MI[a] (m/z [M])
1	11.8	106
2	16.9	135
3	22.74	169

[a] MI = molecular ion.

The mass spectral patterns for the HPLC/MS and the GC/MS-detected impurity peaks could not be sufficiently resolved to determine the maximum number of carbon, nitrogen, and oxygen atoms in the respective molecules as presented Polypropylene Extractable Study. However, the molecular ion values do provide information about the atomic composition of these compounds using the "nitrogen exclusion rule"[9]. For example, in the case where the molecular ion value (m/z [M]) is an even number, the number of nitrogen atoms is expected to be equal to zero or some even number of atoms. When the molecular ion value is an odd number, the number of nitrogen atoms is expected to be equal to one or an odd number of atoms. Applying the rule to the MS data for the five impurities, impurity peak 1 would be expected to contain either no or an even number of nitrogen atoms, since it has an even number molecular ion (m/z [M]) value of 160. The four other impurities would theoretically contain at least one or an odd number of nitrogen atoms since the molecular ion (m/z [M]) values for these compounds have odd numbers: 135, 169, 209, and 197.

The combined UV and MS spectral data, along with the log D partitioning values, provide then a three-parameter assessment compounds. The identification is based on a comparison of data for aqueous extractable compounds from rubber stoppers listed in the literature. Extractable compounds from stoppers that have been reported in the literature are listed in Table 34.6.

The results of the data comparison identified the five compounds using the following criteria:

- Impurity 1: benzaldehyde
  1. There is exact match of UV and MS spectral profiles.
  2. The formula does not contain a nitrogen atom (nitrogen exclusion rule for even-number MI).
  3. The log P value is 1.64 compared to an estimated value of <2.1.
- Impurity 2: benzothiazole
  1. There is exact match of UV and MS spectral profiles.
  2. The formula contains at least one nitrogen atom (nitrogen exclusion rule for odd-number MI).
  3. The log D value (at pH >5) is 2.01 compared to an estimated value of 2.1.
- Impurity 3: diphenylamine
  1. There is exact match of UV and MS spectral profiles.
  2. The formula contains at least one nitrogen atom (nitrogen exclusion rule for odd-number MI).

TABLE 34.6. List of Aqueous Extracted Compounds Reported in the Literature Along with Molecular Formula and Mass as Well as Calculated log *D* Values

Compound	Formula	Molecular Weight	log *D* (pH >5)[a]
Benzaldehyde	C_7H_6O	106.12	1.64
Benzothiozole	C_7H_5NS	135.19	2.01
Cyclohexanone	$C_6H_{10}O$	98.14	0.76
Diphenylamine	$C_{12}H_{11}N$	169.22	2.97
9,10-Dihydro-9,9-dimethylacridine	$C_{15}H_{15}N$	209.29	4.19
2,6-Ditertbutylmethylphenol	$C_{15}H_{24}O$	220.35	5.32
Dibenzylamine	$C_{14}H_{15}N$	197.27	0.32
Ethylbenzene	C_8H_{10}	106.17	3.21
2-Butoxyethanol	$C_6H_{14}O_2$	118.17	0.80
Tetrachloroethane	$C_2H_2Cl_4$	167.85	2.17
o-Xylene	C_8H_{10}	106.17	3.14
N,*N*-Dibutylacetamide	$C_{10}H_{21}NO$	171.28	2.44
Dibutylformamide	$C_9H_{19}NO$	157.25	2.18
Acetophenone	C_8H_8O	120.15	1.66
2-Phenyl-2-propanol	$C_9H_{12}O$	136.19	1.73
2,2,5,5-Tetramethyltetrahydrofuran	$C_8H_{16}O$	128.21	2.39
Tetrachloroethylene	C_2Cl_4	165.83	2.95

[a]Values based on Advanced Chemistry Development (ACD/Labs) Software version 10.0 (2007).
Source: Jenke [12] and references cited therein.

3. The log *D* value (at pH >5) is 2.97 compared to an estimated value of <3.9.

- Impurity 4: dimethyldihydroacridine
  1. The UV and MS spectral profiles are similar to those of acridine (spectral profile for this analog not found in the literature).
  2. The formula contains at least one nitrogen atom (nitrogen exclusion rule for odd-number MI).
  3. The log *D* value (at pH >5) is 4.19 compared to an estimated value of <5.3.
- Impurity 5: dibenzylamine
  1. There is exact match of UV and MS spectral profiles.
  2. The formula contains at least one nitrogen atom (nitrogen exclusion rule for odd-number MI).
  3. The log *D* value (at pH >5) is 0.32 compared to an estimated value of 0.3.

The chemical structures of these compounds are shown in Figure 34.11. Confirmation of the identity of these impurities was verified by spiking the SWFI product with the five listed compounds, which produced chromatograms that matched those in Figure 34.9. Quantitation of these leachable impurities in the product sample was then referenced to the response factor for the respective standard compounds.

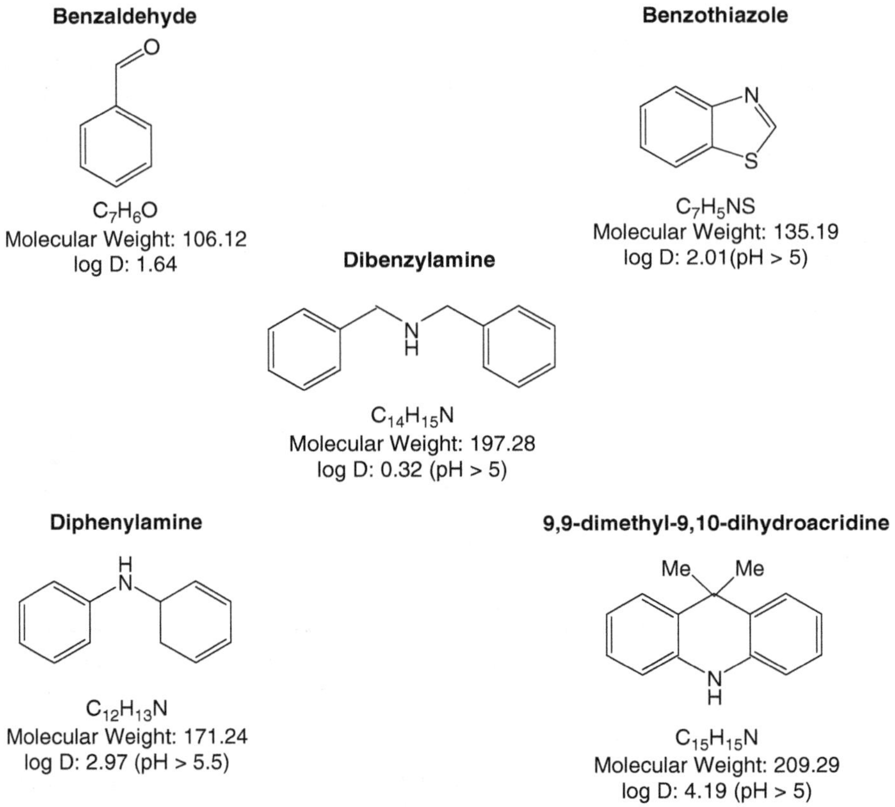

Figure 34.11. Match of impurity peaks to aqueous extractable compounds listed in the literature [11].

34.4.3. Degradent Leachable

One test that is typically conducted to qualify a piece of equipment for use in manufacturing more than one pharmaceutical agent is to expose the product contact surfaces to the solvents or solutions that are used in the respective procedures. After the surfaces have been exposed to these liquids, at the extremes of the operating conditions, the solvents and solutions are then analyzed for extractables. In these tests the type of liquids can range from pure organic solvents such as acetonitrile, to aqueous/organic mixtures such as water/ethanol. Additionally, pure aqueous solutions are frequently used, which can contain solutes such as acids, bases, or buffers at a specified ionic strength and pH. The results of these extraction tests can serve as predictors of potential leachable compounds that could occur in one of the manufactured drug products. Extraction test results, however, are not always a straightforward predictor of leachable compounds.

An example of this difficulty was the detection of an extractable compound in an aqueous buffered solution that was exposed to a polymeric O-ring. The O-ring served as a seal between two stainless-steel transfer lines. The UV spectral scan of the

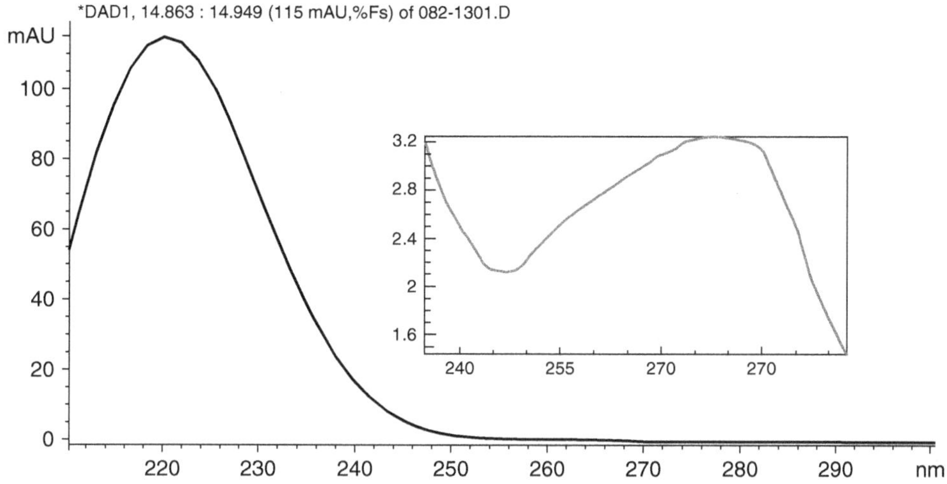

Figure 34.12. Ultraviolet spectral scan of detected extractable compound.

extractable compound, as shown in Figure 34.12, closely corresponds to the spectral pattern for a class of multisubstituted *tert*-butylphenol compounds. These compounds are frequently used as antioxidants to protect polymeric materials. In general, they have a characteristic UV absorption pattern with a primary λ_{max} peak between 222 and 224 nm and a secondary λ_{max} peak between 275 and 280 nm. Two examples of these compounds are BHT (butylated hydroxytoluene) and Irganox 1076 (octadecyl-di-*t*-butyl-hydroxhydrocinnamate). The molecular structures, formula weights, and log D values for these *tert*-butylphenolic antioxidants are shown in Figure 34.13.

The O-ring seal manufacturer confirmed that the bulk polymer contained both BHT and Irganox 1076. The rationale for using these two plasticizing reagents in the polymer is that the positive log D values for BHT at 5.3 and for Irganox 1076 at 13.9 (see Fig. 34.13) indicate that they will have a very low probability of partition out of the hydrophobic polymer matrix into an aqueous solution.

The HPLC retention time for the extractable peak indicates that it has a log D value estimated to be 3.8, when compared to the retention times of compounds with known log D values shown in Table 34.1. While a positive log D value of 3.8 indicates that the extractable is also a hydrophobic molecule, it is significantly less hydrophobic than either BHT or Irganox 1076. The mass spectral data for the extractable compound further distinguish the extractable from these two antioxidant reagents. The reported mass selective ion peak for the extractable compound is 279 m/z, when operating in positive ionization mode $(M + H)$. Compared to the formula weight for BHT at 220.35, the formula weight for the extractable compound at 278 is significantly larger, by 58 mass unites. This difference in mass units does not readily suggest a type of functional group that could be attached to the BHT molecule. Compared to the formula weight of Irganox 1076 at 530.86, it is 252 mass units lighter. This difference in formula weight, however, would correspond to the loss of the C_{18}

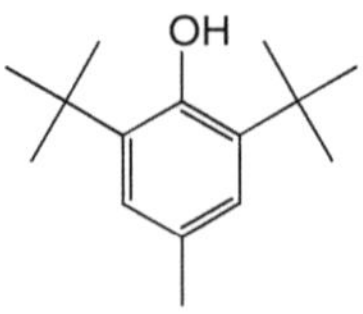

BHT
FW 220.35
Log D 5.3

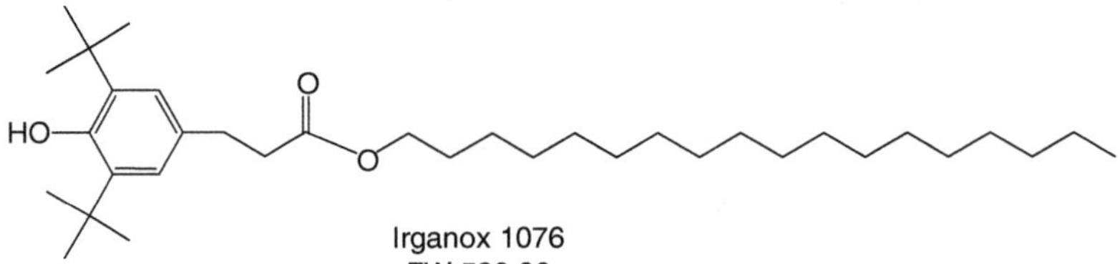

Irganox 1076
FW 530.86
Log D 13.9

Figure 34.13. Phenolic antioxidants BHT and Irganox 1078.

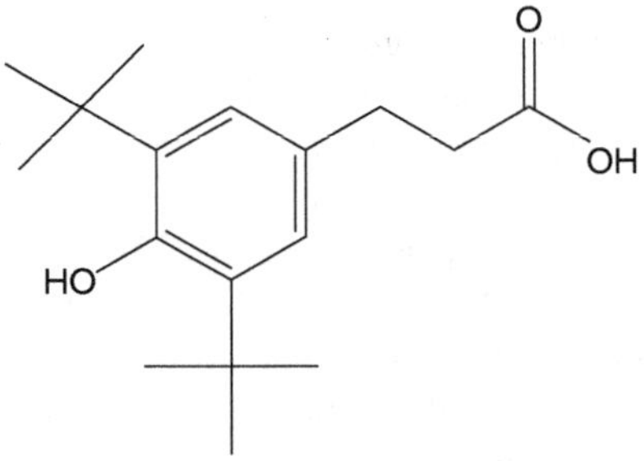

$C_{17}H_{26}O_3$
Exact Mass: 278.19
Mol. Wt.: 278.39
C, 73.34; H, 9.41; O, 17.24

Figure 34.14. Acid hydrolysis byproduct of Irganox 1076.

aliphatic chain on the esterified Irganox 1076 molecule, which would then produce a *tert*-butylhydroxybenzenepropanoic acid species as shown in Figure 34.14.

The calculated log *D* value[10] for the proposed aromatic carboxylic acid compound is 3.88 in the pH range of 4.5–5.0, which closely corresponds to the log *D* value for the unknown compound, based on an analysis of retention time data, as shown in Table 34.7. A tentative identification of the unknown extractable compound was based on the match of the molecular weight, consistency of the UV spectral scan, and the partitioning ratio value. Confirmation of the identified extractable compound was achieved by a direct comparison of the HPLC, UV, and mass spectral properties to those of an authentic sample.

TABLE 34.7. Determination of log *D* Value for Unknown Extractable Compound Based on Comparison to Reference Compounds

Compound	Retention Time, min	log D[a]
DCBA (2,4-dichlorbenzoic acid)	13.2	2.7
Unknown	14.8	3.8[b]
4-Pentylphenol	15.1	3.9
BHT	23.4	5.3
Irganox 1076	> 60	13.9

[a]Values obtained at pH 4.5–5.0.
[b]Extrapolated from retention times of reference compounds.

34.5. SUMMARY

In summary, improvements in the efficiency of conducting pharmaceutical development programs for liquid formulations can be achieved by expanding the analytical scope of the chromatographic methods that are initially designed to monitor impurities related to the formulation ingredients. The analytical focus for monitoring impurities thus needs to be inclusive not only of the formulation ingredients but also for potential leachable compounds, which result from drug product exposure to the surfaces of the processing equipment and/or primary packaging materials. The log *D* (or log *P*) partitioning values for the impurities due to ingredients and process can provide guidance for developing chromatographic methods that are capable of this expanded analytical role.

In addition to contributing to the design of chromatographic methods, log *D* (log *P*) partitioning data can also contribute to the identification of unknown impurities, as demonstrated in the case studies presented in this chapter (in Section 34.4). The combination of partitioning data with data obtained from two independent spectral detection systems (e.g., UV and MS) provides a three-parameter approach for identifying unknown compounds using readily available commercial instrumentation.

Adopting a broader analytical scope during the initial stages of the formulation selection program and incorporating partitioning values as an investigative tool for identifying unknown impurities can, therefore, provide early detection of leachables, avoiding later program delays and supporting product safety claims.

REFERENCES

1. *United States Code*, Chapter V, SubChapter A, Sec. 501(21 USC 351).
2. Carlson, R. M., Carlson, R. E., and Kopperman, H. L. (1975), *J. Chrom.* (1975) 107: 219–223; see also Mirrlees, M. S., Moulton, S. J., Murphy, C. T., and Taylor, P. J. (1976), *J. Med. Chem.* 19: 615–619, and references cited in both articles.
3. Snyder, L. R. and Kirkland, J. J. (1979), *Introduction to Modern Liquid Chromatography*, 2nd ed. Wiley-Interscience.

4. For a technical reference, see http://www.nalgenelabware.com/techdata/chemical/index2.asp.
5. Silverstein, R. M., Bassler, G. C., and Morrill, T. C. (1991), *Spectrometric Identification of Organic Compounds*, 5th ed., Wiley, New York.
6. Ash, M. and Ash, I. (2005), *Plastic and Rubber Additives*, 2005 ed., Synapse Information Resource Publisher.
7. Jenke, D. R. (2004, 2006), *J. Pharm. Sci. Technol.* 58: 24–31, 60: 191–207.
8. Pretsch, E. Buhlmann, B., and Affolter, C. (2000), *Structure Determination of Organic Compounds*, Springer-Verlag.
9. McLafferty, F. W. and Turecek, F. (1993), *Interpretation of Mass Spectra*, 4th ed., Univ. Science Books.
10. Calculated value based on Advanced Chemistry Development (ACD/Labs) software version 10.0 (2007).
11. Jenke, D. R. (2005), *J. Pharm. Sci. Technol.* 59: 265–281 (see also Ref. 7).
12. Jenke, D. R. (2002), *J. Pharm. Sci. Technol.* 56: 332–371.

35

PRIMARY CONTAINER AND CLOSURE SELECTION FOR BIOPHARMACEUTICALS

Olivia Henderson

35.1. INTRODUCTION

All drugs or medicinal products are shipped, stored, or marketed within a container–closure system, which is intended to contain and protect the contents. Whether this is a pill bottle, a tin, an inhaler, or the commonplace glass vial, this package is vital to maintain the quality attributes of the medicine contained during shipping, storage, and distribution. The packaging components that come in direct contact with the product are defined as the *primary package*. Examples of commonly used primary packaging are vials, stoppers, cartridges, plungers, prefilled syringes, and tip caps. Package components that do not directly contact the product such as carton, case, and dunnage are described as the *secondary package*. This chapter focuses on the primary package and what factors to consider when selecting the primary package components for use with a given product.

For purposes of this discussion, the *drug substance* is the final product of the cell culture, purification, and formulation processes, but that is not yet packaged for delivery to a patient. The *drug product* is defined as the formulated drug substance that has been filled or packaged into the appropriate final container–closure system for delivery to a patient, such as a vial, cartridge, or prefilled syringe. Container–closure

Formulation and Process Development Strategies for Manufacturing Biopharmaceuticals, edited by Jameel and Hershenson

systems for use with drug substance and drug product are discussed separately. Container–closure systems for use with pulmonary or buccal delivery systems are not discussed here.

35.2. DRUG SUBSTANCE CONTAINER–CLOSURE SYSTEMS

The biotechnology-derived drug substance provides some interesting challenges for both warehouse storage (due to the temperatures of −20°C or 2°C–8°C) and the container–closure system used during storage. The drug substance is typically composed of a protein or peptide formulated in a buffered solution and stored as a refrigerated liquid or frozen solution, depending on the inherent stability of the active biological molecule and any stabilization of the molecule provided by the formulation. The selection of a suitable container–closure system for the drug substance is dependent on the physical state (i.e., liquid or frozen), the batch volume, the anticipated storage time of the drug substance, and whether shipping will be necessary before the final manufacturing step of fill–finish into the final drug product.

Many biotechnology manufacturing processes are campaigned to optimize operational efficiency, and the resultant drug substance may be stored prior to further processing. This further processing into drug product may not take place for a period of time ranging from months to even years. If the drug substance must be stored for a significant period of time, it is imperative that the container–closure system be qualified to demonstrate that it is suitable for its intended use. A container or closure that is suitable for its intended use should

- Protect the product from degradation from sources such as light, loss of solvent, oxygen, and microbial contamination.
- Not interact with the contents to cause a change in product quality attributes such as identity, strength, safety, quality, purity, and potency. For example, this may be caused through mechanisms such as adsorption or absorption of the product to the container, leachate-induced degradation, pH change, or precipitation.
- Not release undesirable quantities of substances into the product solution that might affect the product stability or present a toxicity concern.
- Perform its function to protect the product during storage and shipping [1].

Therefore, the container–closure system must be relatively inert, rugged, cleanable or disposable, sterilizable, and last but not least, cost-effective. The most common materials currently in use for drug substance storage are glass, plastic, stainless steel, glass-lined metal, or epoxy-lined metal [1]. The plastic materials may be Teflon®, polypropylene, or polycarbonate. Finally, one of the newest options for drug substance storage is disposable bags.

Whichever material is selected for the main container, the materials must meet the appropriate standards described in the designated compendia for physical, chemical, and biological tests. The most commonly used compendia are the *United States*

Pharmacopoeia, the *European Pharmacopoeia*, and the *International Organization for Standardization*. These compendial tests may be conducted by the container manufacturer or by the user, but it is the user's responsibility to ensure the testing has been completed.

In addition to the physical, chemical, and biological testing, the materials must also be qualified for use within the manufacturing process. This qualification includes but is not limited to cleaning, sterilization, filling, and shipping. Each of the materials of construction listed below has advantages and disadvantages for use and must be evaluated for suitability of use according to the intended application and facilities available (i.e., autoclaves, stainless-steel tanks, piping, and clean-in-place or steam-in-place capability).

35.2.1. Glass

Glass containers have traditionally been used for commercial solution storage (i.e., solvent bottles, IV bottles) for decades. Glass is relatively inert, cleanable, compatible with moist heat sterilization, inexpensive, and compatible with smaller-scale shipping configurations. For parenteral products, USP Type I borosilicate glass (borosilicate glass is described further in Section 35.3) is highly recommended [2], although bottle sizes are usually restricted to about one liter. While this may be a very good option for smaller batch sizes, it may be inconvenient to subaliquot a 20-L batch into a series of 1-L containers. For an even larger batch size, the subaliquoting process becomes increasingly inconvenient. Larger glass bottles are typically made from soda-lime glass (soda-lime glass is described further in Section 35.3), which may be treated to meet USP type II criteria, in which case the user has the burden of proof of compatibility [2]. The additional risk of breakage for the larger bottle sizes makes the use of glass prohibitive because of weight and safety concerns. For batch volumes of several hundred liters, the use of glass containers becomes too risky to be feasible. For storage of a frozen solution, glass has the propensity to break as a result of ice formation, making it unsuitable for use.

Closures used for glass containers must be nonreactive to the contents, prevent leakage or evaporation, and protect the contents from bacterial incursion. Materials used for this purpose include plastic screwcaps (such as polypropylene), screwcaps with a plastic liner, or a rubber septum and overseal.

35.2.2. Stainless Steel

Stainless steel is another material that has been traditionally used in pharmaceutical manufacturing. Stainless steel can be used in either smaller cans (<20 L) or in larger tanks. The size of the container has few restrictions, and the tank design (inlet, outlet, and sampling ports) can be customized as needed. The most commonly used composition of stainless steel is 316 L, the low-carbon version of 316 stainless steel. Type 316 L stainless steel is more resistant to aqueous corrosion than is 316 or 304 stainless steel, two other common compositions. Because of the corrosion resistance, 316 L stainless steel is easily amenable to cleaning and moist heat sterilization (and is therefore reusable), and can be used for storage of refrigerated liquid and frozen drug substance. The strength of the steel makes this a very rugged container for storage and shipping. Stainless steel is also expensive to purchase as it is usually electro polished to remove surface irregularities, and requires routine maintenance in the form of passivation to prevent the occurrence of rouge, or rust. The cleaning and sterilization procedures must be validated, and its weight makes it relatively expensive to ship.

The closures used with stainless steel tanks or cans are also made of stainless steel, usually in combination with an elastomeric gasket to make a tight seal.

35.2.3. Teflon®

Teflon is the trademarked name for a series of fluoropolymers such as polytetrafluoroethylene (PTFE), polyfluoroalkoxy (PFA), fluorinated ethylene propylene (FEP), and tetrafluoroethylene (TFE). These materials are being used more frequently because of their desirable inertness. Teflon® is generally inert, can be washed and sterilized by moist heat, and can be used for refrigerated or frozen storage. It can be reused, but because of the high cost of cleaning validation as compared to the cost of the container, it is more typically used only once. Teflon® is lightweight, resistant to breakage, and easily shipped. Leachables must be evaluated since this is a polymeric material. It is visually slightly opaque, so a visual inspection of the contents will be impaired. Some researchers have reported increased protein aggregation in connection to the hydrophobic liquid–surface interface, but this should be evaluated by each user [3,4]. The size of the container may be a disadvantage; Teflon containers are commercially available only up to 2 L. For some applications (such as a low fill volume for the container size), the container wall thickness available in standard, off-the-shelf designs may need to be increased to prevent side wall collapse. This type of modification usually can be manufactured as a custom order. Teflon® closures are also commercially available for use.

35.2.4. Polypropylene and Polycarbonate

Polypropylene and polycarbonate are both commonly used materials for liquid storage. They both can be washed with standard solutions and sterilized using moist heat. As with Teflon®, polypropylene and polycarbonate can be reused, but because of the cost of cleaning validation as compared to the cost of the container, they are more typically used only once. Polycarbonate can be used for refrigerated or frozen storage, and is compatible with standard pharmaceutical solutions. Polypropylene can be used for refrigerated storage, but has a glass transition temperature between −5°C and −20°C, which increases its brittleness at temperatures below the glass transition point. Caution should be used if storing frozen drug substance in polypropylene because there is an increased possibility of breakage. Leachables must be evaluated for both materials. Polypropylene and polycarbonate may be purchased with polypropylene screwcap closures, which may also include a silicon rubber gasket. Other closure options are discussed in Section 35.2.6.

Polypropylene is visually slightly opaque, which impairs visual inspection of the contents. Polypropylene is available in sizes from 4 mL to 50 L in commercially available configurations. In contrast to polypropylene, polycarbonate is visually transparent, which is helpful for visual inspection of the contents. Polycarbonate is commercially available in sizes from 5 mL to 20 L. Again, as for Teflon® containers, the container wall thickness may need to be customized, depending on the application.

The end user must consider all the parameters described above when making a choice between Teflon®, polypropylene, and polycarbonate. The factors to consider include cost, volume, extractables, leachables, and availability. It is a safe statement to say that no one material or container meets all requirements for all users.

35.2.5. Disposable Bags

One of the newest container–closure systems to be used is disposable bags. These are typically manufactured using multilayered coextruded or laminated films that are heat welded into two-dimensional (pillow shaped) or three-dimensional (cube-shaped) bags as shown in Figures 35.1 and 35.2, respectively. The films are composed of multiple polymeric materials fused together in order to optimize the film characteristics. As an example, a film may have a product contact layer of low-density polyethylene (LDPE), but the next layers may include a gas barrier such as ethylene vinyl alcohol,

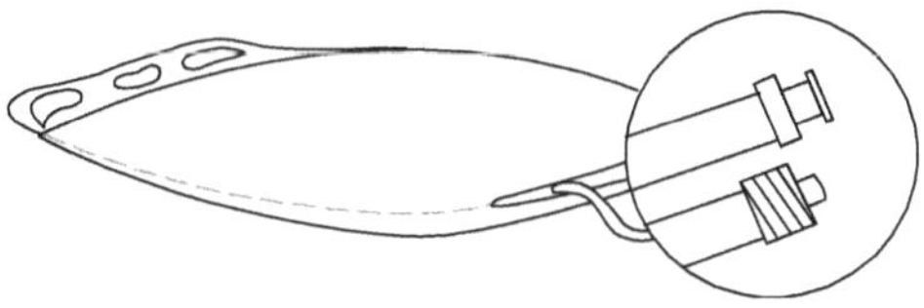

Figure 35.1. Diagram of a two-dimensional disposable bag. (Courtesy of Thermo Fisher Scientific, BioProcess Production, HyClone Products.)

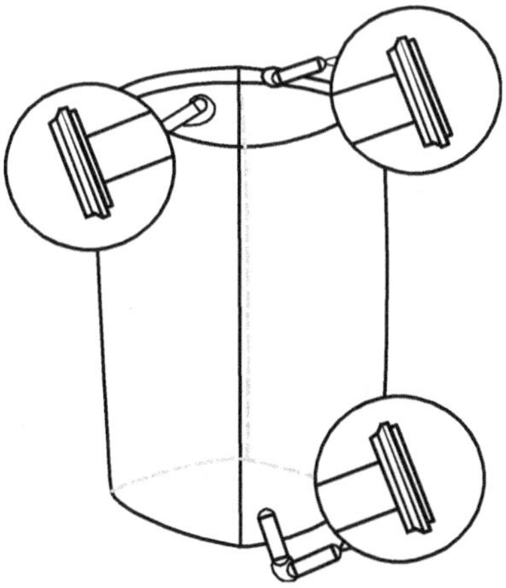

Figure 35.2. Diagram of a three-dimensional disposable bag. (Courtesy of Thermo Fisher Scientific, BioProcess Production, HyClone Products.)

and a polyester layer to add strength to the film. These bags have heat sealed ports that may be made of LDPE polyethylene or other suitable material to which inlet, outlet, and/or sample tubing is attached. There are several manufacturers of these disposable bag systems, each of whom maintains multiple proprietary films for various sizes and uses. These bags are manufactured in clean areas and sterilized by gamma irradiation so they are ready to use as purchased. The bags are available in standard sizes, but these sizes vary by manufacturer and film used. The bags are available (although not in all sizes, or even in all films) from 100 mL to 200 L, or even as much as 1000 L for some manufacturers. The advantages of bag use are that they are disposable, presterilized; are ready to use; can be easily filled, sampled, and emptied; have a short purchasing lead time; and are commercially available.

As may be imagined, these larger bags cannot be filled, stored, or transported without additional protection. Smaller bags that are filled with valuable drug substance may be handled with or without additional protection according to the user's preference. In either case, this additional protection may be provided by a hard plastic or stainless-steel secondary container of appropriate size. These secondary containers are non–product contact, but they are critical to the functionality of the bags. The secondary container must have a smooth interior surface (so as to not abrade or puncture the bags) and be sturdy, puncture-resistant, easy to move around (even when filled), and the bags must be readily accessible through the secondary container for inspection, sampling, and emptying. Finally, the secondary containers must be easily cleaned for reuse if needed and must also be cost-effective. These requirements may be met by several commercially available options; however, each user must qualify the container–closure system for use with the specific application.

The disadvantages in using disposable bags for drug substance storage include risks of leakage through bag punctures or tears and permeability of oxygen, carbon dioxide, and/or water vapor. The permeability of oxygen, carbon dioxide, and water vapor varies by film composition, surface area, and storage temperature. This sometimes low-level permeability may not be a concern for all applications, but it should be evaluated as part of the compatibility studies. An additional disadvantage is that there is a relatively short history of disposable bags usage. The cost of each bag is somewhat high, but this may be offset when one considers the costs of cleaning and validation for multiple-use containers. As for all container–closure systems, the extractable and leachable profile should be evaluated for the specific application.

In general, the bags are best used for refrigerated liquid drug substance storage; however, bag and storage systems adapted for frozen storage are commercially available. One such system provides specially designed equipment for freezing and thawing, storage, and transfer of several bags up to 16 L in size. This system has several advantages over the more traditional bottles for frozen storage. The configurations of the bag and the freezing unit provide tightly controlled freeze and thaw rates for the drug substance; the freeze–thaw, transfer, and storage units provide a compact design; and the storage racks used for the bags provide rather compact physical protection during storage for the frozen bag of drug substance. However, the costs versus benefits of this system must be evaluated by each user. Also, the bags for freezing are purchased as "ready to use," in contrast to other, more rigid containers such as polycarbonate carboys.

35.2.6. Container–Closure Design

Container–closure design is another consideration to be evaluated for each application. How materials are filled into and removed from the container–closure system must be determined in advance. If samples must be taken after the solution is filled and the container closed, there must be a means of pulling a sample without opening the container and risking microbial contamination. Steel tanks and disposable bags have a great deal of flexibility in design, as they are typically custom-designed to the user's specifications. Glass and plastic containers are more constrained since the container itself is usually of an essentially cylindrical or rectangular shape. For glass and plastic containers, one readily available closure is an appropriately sized screwcap; however, this may not provide the strongly preferred aseptic connection, especially if sampling is required. If there must be customization for glass and plastic container–closure systems, the simplest and most direct part to custom-design is the closure. There are commercially available silicone closures such as Topworks® (manufactured by Nalge Nunc, Intl.), with ports or welded tubing inserts that allow more sampling flexibility and a more sanitary connection than does the regular screwcap. In addition, if the commercially available Topworks does not exactly match a given application, modification may be done by specialty manufacturers to achieve the optimum design.

When a drug substance must be stored at frozen temperatures to optimize its shelf life, special care must be taken in selecting the container–closure system. Some factors to consider are the materials of construction, volume, shape, formulation, transportation, and storage capacity. The containers listed in the above sections are generally compatible with frozen storage, with the exceptions of glass and polypropylene. Glass containers can be used to store frozen materials, but glass is known to break during freezing, so care must be taken to ensure that the fill volume, formulation, freezing rate, and handling are compatible with the container volume. Even so, some breakage may be expected. As discussed earlier, polypropylene has a glass transition temperature between $-5^\circ C$ and $-20^\circ C$, which may increase brittleness at frozen temperatures, leading to an increase in breakage during handling.

35.2.7. Container–Closure Integrity

In order to protect the drug substance from microbial contamination, the container–closure system must be integral, or whole, not permitting leakage through which microbial contamination might enter the system. It is also critical that the contents not evaporate through the closure. Although container–closure integrity (CCI) is associated with sterile products, it is also important for biological drug substances due to their bacterial growth-promoting nature. Therefore CCI must also be established as part of the qualification exercise once the container/closure design has been established. Container–closure integrity tests have been thoroughly researched and the results reported by several authors [5–8]. Integrity can be established by a number of commonly performed leak tests, the most common of which test is microbial ingress. This test is easily performed for containers small enough to be submerged in a vat of microbial broth. This test also may be easily performed even when the entire container may be too large to be immersed as long as the critical seal portion is small enough to be tested. However, if the container and/or the critical seal portion are too large or cumbersome to be immersed in a laboratory-sized vat, then alternative tests must be evaluated to establish CCI. A report by the Parenteral Drug Association lists several test options for establishing CCI aside from the simplest immersion test mentioned above [9]. These tests may be performed by a contract laboratory; however, the sponsor is ultimately responsible for the quality of the data, so close review of the study design, execution, and results is highly recommended.

35.2.8. Stability

Stability studies for a drug substance are required for regulatory approval and generally are conducted in accordance with ICH Q1A and Q5C guidelines [1,10,11]. Stability studies can also be conducted to establish the compatibility with the container–closure system and evaluate the effects of leachates on drug substance over time. In addition, if the stability closure type (e.g., sanitary connection or screw thread) is comparable to the large-scale closure type, CCI is further supported by negative microbial tests over the course of the stability studies. With all the information that is gathered from one series of studies, and the time and resources needed to properly complete such studies, it is obvious that the stability containers' materials and design must be carefully considered.

The recommended practice is to scale down the size of the stability containers for conservation of precious drug substance [1,11]. However, in order to acquire the maximum amount of information from the studies, the stability container–closure system must be made of representative materials of construction, with a similar closure type as compared to the larger, bulk containers. When selecting the stability container it is important to consult closely with the manufacturers, as the different manufacturing processes used for smaller or larger containers may require slight modifications to the materials used. In this case the user must carefully evaluate any differences to determine whether the two materials may still be considered comparable for the purposes of the stability study.

35.2.9. Extractables and Leachables

Extractables are described as chemical compounds extracted from the container–closure system, typically with stronger solvents and/or a higher temperature than actual use would dictate. *Leachables* are described as chemical compounds that are extracted from the container–closure system under representative conditions of use, where the solvent and temperatures are similar to or representative of those found in actual use. The leachables may be considered as a subset of extractables. In actual use, leachables may be discovered during normal use or in stability study samples [1,12]. In a practical sense the end result of any extractable and leachables study should address two primary concerns: whether (1) the levels of leachables or extractables present in and of themselves pose a health risk to the end user and (2) the presence of leachables and extractables in any way alter the efficacy of the active pharmaceutical ingredient.

For most materials, the list of extractable compounds may be obtained from the supplier. In general, suppliers who sell to the pharmaceutical industry are aware of their customers' needs to satisfy regulatory requirements. These suppliers will often evaluate their materials for extractable compounds as well as compliance with compendial requirements such as USP <87>/<88> and/or the corresponding ISO standards tests, such as ISO 10993 [13–15]. This information should be evaluated by the user to ensure that the study performed was done appropriately and completely. This can be considered a starting point for the users' evaluation.

The user must then evaluate the contact materials for leachable substances under their conditions of use. The most direct situation would be to use the drug substance itself as the extracting medium at the expected storage conditions [1]. However, the drug molecule itself is usually present at concentrations much higher than that of the leachables, causing the analyst to search for the proverbial needle in a haystack. More often the formulation buffer or placebo vehicle will be used, which provides representative extraction of the leachables from the container under the expected conditions of use. The tests used to analyze the leachate must be appropriate to detect and quantify the expected leachates, given the list of extractable compounds and the conditions of use. Typical tests include but are not limited to nonvolatile residues, total organic carbon, reversed-phase high-pressure liquid chromatography (RP-HPLC), Fourier transform infrared (FTIR) spectroscopy, liquid chromatography/mass spectroscopy (LC/MS), gas chromatography/mass spectroscopy (GC/MS), and inductively coupled plasma mass spectroscopy, as appropriate. This testing may be performed in the user's analytical laboratories, but they are also frequently performed by contract laboratories. It is expected that studies performed in the user's analytical laboratories would be carefully planned and executed, but even when the tests are performed by a contract laboratory it is recommended that the study sponsor carefully review the contractor's protocol and results to ensure that the study and tests are representative of the sponsor's product and conditions of use.

35.3. DRUG PRODUCT CONTAINER–CLOSURE SYSTEMS

As defined above, drug product is a formulated drug substance that has been filled or packaged into the appropriate final container–closure system for delivery to a patient. The drug product container–closure selection may be the standard vial and stopper, a prefilled syringe, or a cartridge and plunger. Container–closure selection is frequently influenced by market demands as much as by compatibility or stability requirements. The traditional container–closure system for biotechnology derived parenterals is the glass vial and rubber stopper with an aluminum crimped seal. However, more recent trends in the therapeutic areas served by biotech drugs are to use prefilled syringes and/or autoinjector devices for patient convenience. The prefilled syringe is discussed in another chapter, and the complexity of an autoinjector device is outside the scope of this discussion.

35.3.1. Vials

By far the most commonly used material for drug product containers is glass. Glass vials used for parenteral administration are generally made from borosilicate glasses, which conform to USP/Ph.Eur. (*European Pharmacopoica*) Type I specifications. Borosilicate glass is described as highly chemically resistant, and is usually composed of approximately 70%–80% SiO_2, to which B_2O_3, Al_2O_3, and Na_2O are added to balance manufacturability with the desired physicochemical properties. Other inorganic oxides may be added in order to manipulate the physical properties of the glass or even the color (to create amber glass) during the manufacturing process [16]. These oxides can leach from the glass during storage, and may cause the pH of the contents to change as discussed below.

The USP also describes soda-lime glass, which, if treated (dealkalized) to meet Type II specifications, may be used if stability data are available to demonstrate suitability [1,2]. The soda-lime glass is also formulated primarily of SiO_2. However, the primary inorganic oxide modifiers are Na_2O and CaO rather than B_2O_3 and Al_2O_3. As a result of the different formulation, soda-lime glasses are less chemically resistant than are borosilicate glasses. When a soda-lime glass is treated with ammonium sulfate or SO_2 gas, the surface chemical resistance is improved, but does not reach the high level of inertness inherent in borosilicate glasses. In addition, soda-lime glasses, because of their higher thermal expansion coefficients, cannot endure sudden changes in temperature, which limits their use for lyophilized products, terminally sterilized products, or products stored as a frozen liquid. Given these characteristics of the soda-lime glasses, they are not commonly used for parenteral products.

There are two common glass vial manufacturing techniques in use: either shaping preformed tubing (for a tubing vial) or a molding process (for a molded vial). Tubing vials are currently the most commonly used vials for injectable pharmaceutical products. This type of vial is produced from a tube of glass cane, which is fed into a forming machine. The forming machine uses oxygen/gas flame jets to heat a specified area of the cane. When the specific area is softened by the heat, special tools push the glass tube together to form the crown, neck, and shoulder of the vial. After the top of the vial is formed, additional jets and tools repeat the softening and forming process to shape the bottom of the glass vial. As the vial is moved off the forming machine, a chemical treatment solution may be sprayed into the vial if this is desired. The vial is next moved through an annealing oven, or lehr, then cooled, inspected, and finally packaged for shipment. Syringes and cartridges are formed using a variation of this manufacturing technique, and further details can be found in specialized literature.

The molded vial may be used to create a specialized or unusual shape, a large volume, or for a screw-threaded closure (such as for a nonsterile product). In the manufacturing of the molded vial, glass is melted in a furnace, placed in a mold, and blown throughout the mold. The glass is then annealed, cooled, inspected, and packaged for shipment.

Vials and syringes are also available in polymeric materials such as cyclic olefin copolymer (COC) and Crystal Zenith® (CZ) (manufactured by Daikyo Seiko, Ltd.) resin. These materials are less commonly used as of the writing of this chapter, but they may provide a useful container for products that do not have optimum compatibility with more traditional, (i.e., glass) components. Both CZ resin and COC materials are transparent, much like glass, but are lightweight and break-resistant.

One area where these polymeric materials may be very useful is in resolving protein adsorption concerns [17]. Protein adsorption is frequently addressed by formulation excipients such as surfactants, amino acids, and carbohydrates [18–20]. In cases where the excipients do not adequately resolve the issue, polymeric containers or vials coated with pure silica (e.g., Schott Type I plus®, Schott, Germany) may offer a good alternative [21]. However, polymeric materials typically are more permeable to gases and water vapor than is glass, which may be of concern if the protein under development is sensitive to any of the following: oxygen; an increase in protein concentration; or an increase in excipient concentration due to water loss [22].

The advantage of using glass vials is the low cost, long history, and ease of use. Because of this long history, the problematic issues are generally known. Glass vials are generally fairly inert, although trace metals from the glass have been shown to leach into the solution. One example of such a leaching effect and the consequences occurs when a formulation is unbuffered or weakly buffered. In such a situation, Na^+ leaches from the surface of the interior glass wall into the solution, while H_3O^+ is removed from the solution, which in an unbuffered or weakly buffered solution will raise the pH of the contents [23,24]. To say the least, this can complicate adherence to a stability specification and may also impact product quality attributes. There are several options for resolving this situation: reformulating the drug product, changing to a treated glass vial, or changing to a vial made from a low-alkali forming process. If one reformulates the solution, extensive comparability studies must be conducted to establish suitable quality attributes of the new formulation. Changing to a treated vial may resolve the pH shift since dealkylizing treatment does remove surface alkali; however, it may lead to other issues such as delamination [23,24]. Alternatively, a low-alkali vial forming process may be selected; however, it will then be necessary to reestablish expiry (through comparability and/or stability studies), extractable or leachable profiles, and perhaps even container or closure integrity (if the vial dimensions change). All options are time- and cost-intensive, so selecting an appropriate combination of glass and formulation in the beginning of the development process is highly recommended.

Delamination, mentioned above, is the phenomenon where glass flakes peel off the interior surface of the vial after quantities of the surrounding glass have been etched or leached away. The glass flakes resulting from delamination are problematic in injectable solutions, where visible foreign particles are unacceptable. The basic cause of delamination has been attributed to complex interactions between vial surface alkalinity, vial surface defect structures, chemicals in solution (such as the drug and/or formulation), and heat stresses (such as depyrogenation). In an effort to mitigate the effects of surface alkalinity and defect structures, some researchers have suggested sulfate treatment, but this may actually exacerbate the surface defect structures that have been associated with delamination [23–25]. If delamination should occur during the development process, it may be necessary to redevelop the drug product formulation, change the manufacturing process (if terminal sterilization is involved), or change glass formulations or sulfate treatment options [23–25]. In such a case the CZ resin or COC vials may be an attractive alternative. It must be stated that the glass vial suppliers are extremely knowledgeable and helpful regarding the properties of their products and should be consulted with any questions regarding suitability of their product line for a given product as well as for assistance in solving any problems that may arise.

35.3.2. Closures

The purpose of the closure is to retain the contents and protect the contents from contamination. Closures used for vials, syringes, and cartridges are typically made from elastomeric materials, and are commercially available from several suppliers. These closures are manufactured with various formulations, each with differing properties. The different formulations include the base elastomer as well as some or all of the following: curing agents, accelerators, activators, antioxidants, plasticizers, fillers, lubricants, and pigments [16,26]. The most commonly used base elastomers are polyisoprene, butyl rubber, and halogenated-butyl rubber. Each of these materials has advantages and disadvantages for use. For example polyisoprene has good physical properties such as elasticity and resealability but has high extractables, is permeable to oxygen, and has poor stability, especially after autoclaving. Butyl and halogenated-butyl rubbers offer improved stability, decreased oxygen permeability, and fewer extractables, but have a poorer ability to reseal [16,26]. Similar to the vial suppliers, the stopper suppliers are also both knowledgeable and helpful regarding the properties of their products and here also, should be consulted with any questions regarding which item of their product line is most suitable for a given product.

The specific rubber formulation to be used with a given product must demonstrate suitable compatibility and leachable profiles when in contact with the product. To this end, a coating material either of Teflon or other fluoropolymer material can be advantageous in minimizing the number of chemical compounds that may leach from the stopper into the solution.

Another advantage of the coatings is improved lubricity of the stoppers. Most stoppers are siliconized prior to use in an automated filling machine during the fill–finish manufacturing process. The silicone oil used may be applied through a washing process, or poured on as the stoppers are tumbled (most stoppers are also available prewashed and presiliconized from the suppliers). However the silicone oil is applied, it is a necessity in order for the stoppers to move around the automated machines during the filling process. The use of silicone oil is nearly universal in the pharmaceutical industry; however, it has been implicated in a number of problems over the years. Silicone oil has been linked with haziness on reconstitution of lyophilized products as well as protein aggregation [27]. Therefore, it is best to minimize siliconization when possible, and stopper coatings may be advantageous in that regard. In addition, a secondary processing treatment that results in a crosslinked polymethylsiloxane layer has been shown to have quite good lubricity, and may be able to replace or minimize stopper siliconization [26].

Stopper configuration and size are dependent on the dosage form (liquid or lyophilized), the size of the vial, and the vial finish (commonly 13 or 20 mm). It is noteworthy that standard finish sizes may vary slightly between vials produced according to the Glass Packaging Institute and ISO standards, even though both bodies use the 13 and 20 mm designations. Liquid products are typically paired with a stopper with either a flat "button" or a concave-style plug insert, also known as a *serum stopper*. A lyophilized product requires a special stopper for use during the lyophilization cycle. These stoppers are available in single opening or "igloo" style, a double opening or "two-legged" style, or even in a "three-legged" format. Container–closure systems used for both liquid and lyophilized products may incorporate a "blowback" feature, which is a groove on the inside of the vial crown, very close to the top of the vial. Additionally, a stopper may be designed with a ridge that will rest in the groove or blowback to control the stopper position during lyophilization or before the crimp is applied (a "no pop" feature). The blowback and "no pop" features on stoppers may vary in size, shape, and location from one manufacturer to another, so compatability of the vial–stopper combination must be evaluated by the end user. Before selecting a particular vial or stopper, it is also advisable to consult with the proposed filling area to ensure machinability and physical compatibility with the proposed filling equipment.

One factor to consider for a lyophilization stopper is the interaction between the rubber formulation and moisture during the washing, sterilization, and drying processes. Several stopper formulations have been shown to increase their moisture content during sterilization, then later they have also been shown to desorb the moisture during storage—sometimes with a detrimental impact on the lyophilized product in the vial [28,29]. In addition to validating an appropriate drying cycle, a low-moisture-uptake rubber formulation is desired for a lyophilized product where low moisture is a critical parameter impacting product expiry.

35.3.3. Container–Closure Integrity (CCI)

One of the most critical purposes of the closure is to protect the contents from contamination and to retain the contents, so it is critical to demonstrate CCI. For sterile pharmaceutical container–closure systems, microbial contamination is the primary concern. The microbial ingress test is a direct measure of the system's ability to protect the product, and it is therefore the most common test used to demonstrate container–closure integrity. However, because of the shortcomings of the test (e.g., labor-intensive, long lag time from start to finish, probabilistic nature of microbial ingress through a small leakage path), there are other physical tests that have been correlated to the microbial ingress test for routine use [5–8]. These alternative physical tests include but are not limited to helium leak rate, vacuum decay, pressure decay, tracer gas leak detection, and liquid tracer [9].

35.3.4. Stability

Drug product stability is a requirement for regulatory filings, and is discussed in considerable detail in ICH Q1A and Q5C with regard to the storage temperatures and the testing frequency [10,11]. Given the thoroughness of the discussion in the ICH guidelines, only a few points are emphasized here. As discussed in Section 35.2, drug product stability studies also are used to demonstrate compatibility of the formulated drug with the container–closure system and its integrity throughout shelf life. Unlike the drug substance, the drug product is expected to be stored in the commercial container–closure system. For the drug product, a scaled-down version will not be sufficient unless a smaller-size container–closure system is one of the commercial presentations and a matrix approach is planned as described in ICH Q1A.

During the drug development process, it is important to start these stability studies as early in the process as possible using the planned commercial concentration, fill volume, and container–closure system to ensure that any physical or chemical incompatibility is discovered early. If incompatibilities are discovered early and the appropriate actions are taken to remediate the issue, the impact on development timelines is lessened. If the incompatibility is discovered late in the process, there may be costly delays in regulatory filings and/or approvals while the container–closure system is redeveloped and the stability studies are restarted.

35.3.5. Extractables and Leachables

Testing for extractables and leachables for drug product will generally follow the same procedure outlined for drug substance container–closure systems. Little more discussion is necessary except to add that for drug product, leachables testing is expected to be conducted on the drug product in the "actual" final container–closure system, rather than a "representative" one. It is noteworthy that for drug products the closure elastomers are the major source of leachables. These leachables in some cases may interact with the drug product and/or formulation to produce adverse events in patients. A recent example of this was reported where increased incidence of antibody-mediated pure red cell aplasia in Epoetin Alfa-treated patients was suggested to be associated with phenolic derivatives extracted from an uncoated syringe plunger in the presence of polysorbate 80 from the formulation [30]. Therefore, extractable and leachables should be evaluated during the initial drug product development, and also after a change in container closure component or drug product formulation.

35.4. CONCLUSIONS

Selecting the optimum container–closure system for drug substance or drug product is a critical activity during the drug development process. Given the volume of information about the container–closure system that is required for a regulatory filing, it is important that a great deal of thought be given to selecting the appropriate system for use. A well-designed container–closure system made of the appropriate materials will enable the formulation scientist to gather stability, compatibility, leachables, and CCI data to support regulatory filings. Additional information is available through many of the literature references listed at the end of this chapter.

ACKNOWLEDGMENTS

Few accomplishments are ever achieved through the efforts of only one person, and this chapter is no exception. The author would like to thank each of the following individuals for their keen insights and depth of technical knowledge, which contributed greatly to the clarity and accuracy of this chapter: Tia Estey, for asking such perceptive questions; Richard Strong, whose expertise in extractables and leachables testing is invaluable; Robert Swift, for lending his considerable experience with glass and the associated manufacturing processes; Hector Tamburini, for providing encouragement and editorial skills; and Tzung-Horng Yang, for his insightful discussions and willingness to proofread for the author.

REFERENCES

1. FDA (1999), Guidance for Industry: *Container/closure Systems for Packaging Human Drugs and Biologics: Chemistry, Manufacturing and Controls Documentation*, US Department of Health and Human Services, Food and Drug Administration, May 1999.
2. USP 30-NF25 (2007), *Second Supplement* <660> *Containers—Glass*, US Pharmacopoeial Convention, Rockville, MD, 2007.
3. Kueltzo, L. A., Wang, W., Randolph, T. W., and Carpenter, J. F. (2007), Effects of solution conditions, processing parameters, and container materials on aggregation of a monoclonal antibody during freeze-thawing, *J. Pharm. Sci.* (Epub ahead of print, 9/6/07, DOI: 10.1002/jps.21110).
4. Sluzky, V., Tamada, J. A., Klibanov, A. M., and Langer, R. (1991), Kinetics of insulin aggregation in aqueous solutions upon agitation in the presence of hydrophobic surfaces, *Proc. Natl. Acad. Sci. USA* **88**(21): 9377–9381.
5. Kirsch, L. E., Nguyen, L., and Moeckly, C. S. (1997), Pharmaceutical container/closure integrity I: Mass spectrometry-based helium leak rate detection for rubber-stoppered glass vials, *Parenter. Drug Assoc. (PDA) J. Pharm. Sci. Technol.* **51**(5): 187–194.
6. Kirsch, L. E., Nguyen, L., Moeckly, C. S., and Gerth, R. (1997), Pharmaceutical container/closure integrity II: The relationship between microbial ingress and helium leak rates in rubber-stoppered glass vials, *PDA J. Pharm. Sci. Technol.* **51**(5): 195–202.
7. Kirsch, L. E., Nguyen, L., Kirsch, A. M., Schmitt, G., Koch, M., Wertl, T., Lehmann, M., and Schramm, G. (1999), Pharmaceutical container/closure integrity V: An evaluation of the WILCO "LFC" method for leak testing pharmaceutical glass-stoppered vials, *PDA J. Pharm. Sci. Technol.* **53**(5): 235–239.
8. Morrical, B. D., Goverde, M., Grausse, J., Gerwig, T., Vorgrimler, L., Morgen, R., and Büttiker, J.-P. (2007), Leak testing in parenteral packaging: Establishment of direct correlation between helium leak rate measurements and microbial ingress for two different leak types, *PDA J. Pharm. Sci. Technol.* **61**(4): 226–236.
9. PDA (1998), *Pharmaceutical Package Integrity*, Technical Report 27, Parenteral Drug Association, Bethesda, MD.
10. ICH Q1A(R2) (2003), Stability testing of new drug substances and products, *Proc. Int. Conf. Harmonisation of Technical Requirements for Registration of Pharmaceuticals for Human Use*; ICH Harmonised Tripartite Guideline, 2003.
11. ICH Q5C (1995), Quality of biotechnological products: Stability testing of biotechnological/biological products, *Proc. Int. Conf. Harmonisation of Technical Requirements for Registration of Pharmaceuticals for Human Use*, ICH Harmonised Tripartite Guideline, 1995.

12. Paskiet, D. (2004), *Assessing Packaging & Processing Extractable/Leachables*, Parenteral Drug Association Course 190, Parenteral Drug Association Training and Research Institute, San Diego, CA, May 4–5, 2004.
13. USP (2007), *USP 30-NF 25 <87>: Biological Reactivity Tests, In Vitro*, United States Pharmacopoeial Convention, 2007, Rockville, MD.
14. USP (2007), *USP 30-NF 25 <88>: Biological Reactivity Tests, In Vivo*, United States Pharmacopoeial Convention, 2007, Rockville, MD.
15. ISO (1999), *ISO 10993-5: Biological Evaluation of Medical Devices—Part 5: Tests for In Vitro Cytotoxicity*, International Organization for Standardization, Geneva, 1999.
16. Wang, Y. J. and Chien, Y. W. (1984), *Sterile Pharmaceutical Packaging: Compatibility and Stability*, Parenteral Drug Association Technical Report 5, Philadelphia.
17. Qadry, S. S., Roshdy, T. H., Char, H., Del Terzo, S., Tarantino, R., and Moschera, J. (2003), Evaluation of CZ-resin vials for packaging protein-based parenteral formulations, *Int. J. Pharm.* **252**: 207–212.
18. Hawe, A. and Friess, W. (2007), Formulation development for hydrophobic therapeutic proteins, *Pharm. Devel. Technol.* **12**: 223–237.
19. Ruiz, L., Reyes, N., Aroche, K., Tolosa, V., Simanca, V., Rodriguez, T., and Hardy, E. (2005), Influence of packaging material on the liquid stability of interferon-α2b, *J. Pharm. Pharm. Sci.* **8**(2): 207–216.
20. Evans, R. K., Nawrocki, D. K., Isopi, L. A., Williams, D. M., Casimiro, D. R., Chin, S., Chen, M., Zhu, D.-M., Shiver, J. W., and Volkin, D. B. (2004), Development of stable liquid formulations for adenovirus-based vaccines, *J. Pharm. Sci.* **93**(10): 2458–2475.
21. Schwarzenbach, M. S., Reimann, P., Thommen, V., Hegner, M., Mumenthaler, M., Schwob, J., and Güntherodt, H.-J. (2002), *PDA J. Pharm. Sci. Technol.* **56**(2): 78–89.
22. Qadry, S. S., Roshdy, T. H., Knox, D. E., and Phillips, E. M. (1999), Model development for O_2 and N_2 permeation rates through CZ-resin vials, *Int. J. Pharm.* **188**: 173–179.
23. Ennis, R. D., Pritchard, R., Nakamura, C., Coulon, M., Yang, T., Visor, G. C., and Lee, W. A. (2001), Glass vials for small volume parenterals: Influence of drug and manufacturing processes on glass delamination, *Pharm. Devel. Technol.* **6**(3): 393–401.
24. Iacocca, R. G. and Allgeier, M. (2007), Corrosive attack of glass by a pharmaceutical compound, *J. Mater. Sci.* **42**: 801–811.
25. Adams, P. B. (1977), Surface properties of glass containers for parenteral solutions, *Bull. Parenter. Drug Assoc.* **31**: 213–226.
26. Brucker, B. A., Curry, W. T., Riter, J. L., Swift, R., Reynolds, G., and DeGrazio, F. (2005), *The Technology of Parenteral Packaging for Biotech Drug Products*, West the Source Educational Series, San Diego, CA, Oct. 11, 2005.
27. Jones, L. S., Kaufmann, A., and Middaugh, C. R. (2005), Silicone oil induced aggregation of proteins, *J. Pharm. Sci.* **94**(4): 918–927.
28. DeGrazio, F., and Flynn, K. (1992), Lyophilization closures for protein based drugs, *J. Parenter. Sci. Technol.* **46**(2): 54–61.
29. Duncan, P. D., Corvari, V., Burton, M. D., and Rajagopalan, N. (2007), Effect of stopper processing conditions on moisture content and ramifications for lyophilized products: Comparison of "low" and "high" moisture uptake stoppers, *PDA J. Pharm. Sci. Technol.* **61**(1): 51–58.
30. Sharma, B. (2007), Immunogenicity of therapeutic proteins. Part 2: Impact of container closures, *Biotechnol. Adv.* **25**: 318–324.

36

PREFILLED SYRINGES FOR BIOPHARMACEUTICALS

Robert Swift

36.1. INTRODUCTION

Traditionally, the most widely used container or closure system for biopharmaceuticals has been a glass vial, closed with an elastomeric stopper and secured by a crimped aluminum seal. However, in recent years, prefilled syringes have become increasingly popular [1]. The market trend toward prefilled syringes has been driven by advantages for several key stakeholders: ease of administration and convenience for patients and health-care providers, reduced risk of medication errors, market preference in a competitive environment, potential integration into automatic injection devices, and lower cost of goods for the manufacturer due to significantly lower overfill.

A *prefilled syringe* is a drug delivery syringe that has been prefilled with a formulated drug product by the manufacturer and is provided to the patient or health care provider in a convenient, "ready to use" format. As such, a prefilled syringe has a combined role as both a container–closure system and a drug delivery system. This dual role brings with it added complexity for the product development team as well as during the manufacturing and distribution of marketed products. This chapter begins with an overview of available prefillable syringe systems and compares the design features and materials of construction with those typically used in vial–stopper–seal

Formulation and Process Development Strategies for Manufacturing Biopharmaceuticals,
edited by Jameel and Hershenson

systems. The main body of the chapter then presents the challenges of developing and manufacturing a prefilled syringe presentation of a biopharmaceutical product focusing on the impact for formulation scientists, container engineers, manufacturing, and distribution. The chapter concludes with some special considerations when adapting a prefilled syringe for use in an automatic injection device and an overview of emerging prefilled syringe technologies seeking to address the specific requirements of biopharmaceutical products.

36.2. OVERVIEW OF PREFILLABLE SYRINGE SYSTEMS

The container portion of a prefilled syringe system is the syringe barrel. The vast majority of these are made from chemically resistant borosilicate glass. The barrel tip can be configured either with a preattached, or "staked," needle or a standard luer taper. A prefilled syringe system requires two closures. An elastomeric tip cap or needle shield is fitted to the tip prior to filling. After filling, the open end of the barrel is closed with an elastomeric piston or plunger–stopper.

Prefilled syringe components for therapeutic products are available in sizes ranging from 0.5 to 20 mL. For special applications, such as contrast media, systems as large as 150 mL can be made. However, most prefilled syringe systems for injectable therapeutic products fall into three standard formats: 0.5 mL, 1 mL "long" and 1–3 mL "standard." Examples of these three formats are shown in Figure 36.1. Note that the 0.5 mL and the 1-mL-long examples incorporate preattached (staked) needles while the 2.25-mL example incorporates a luer taper. Barrels for the so-called 1–3 mL standard sizes are of the same diameter. The desired capacity within the 1–3 mL range is achieved by selecting the appropriate barrel length. In this way, the same piston designs and molds can cover a wider range of products and fill volumes.

In addition to the broad range of prefilled syringe capacities, the components can be delivered to the biopharmaceutical manufacturer in a wide range of packaging

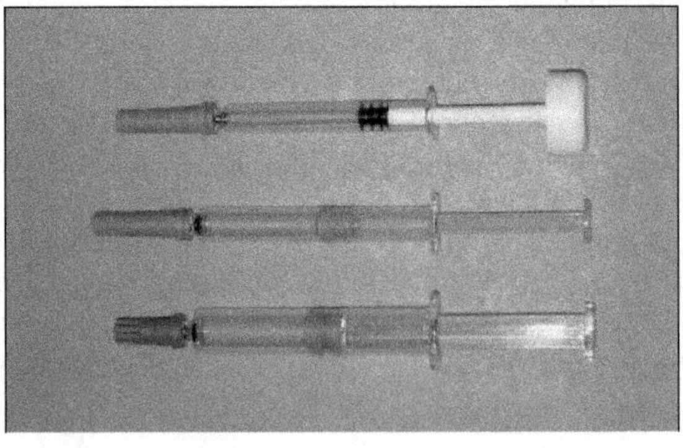

Figure 36.1. Photo of 0.5-mL staked needle (top), 1-mL-long staked needle (middle), and 2.25-mL luer tip syringe barrels.

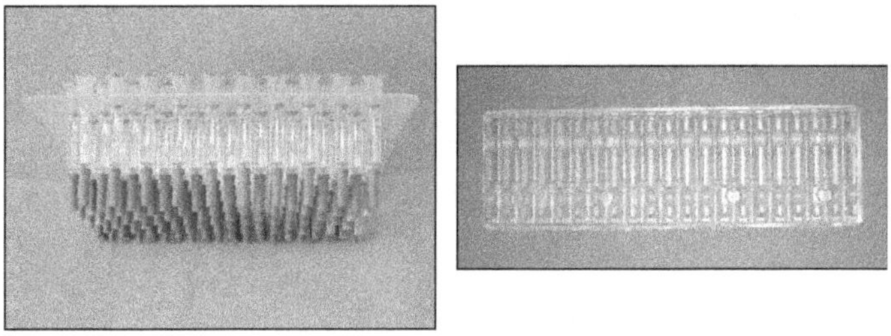

(a) (b)

Figure 36.2. Photos of 1-mL-long staked needle syringes in nested, ready-to-fill format (a) and unprocessed bulk format (b).

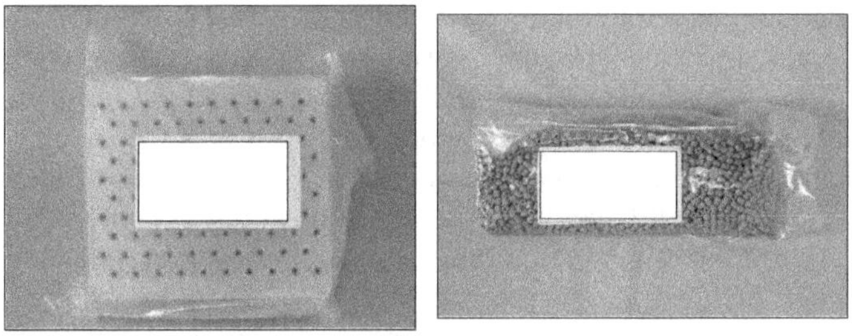

(a) (b)

Figure 36.3. Ready-to-use 1-mL-long pistons in nested (a) and bagged (b) formats.

formats to suit the specific design of the filling equipment. The barrels can be provided in bulk format, requiring rinsing, lubrication, and sterilization prior to filling or in presterilized, ready-to-fill format. Examples of bulk and ready-to-fill 1-mL-long syringes are shown in Figure 36.2. Similarly, pistons can be supplied in unprocessed bulk format, prewashed and packaged in sterilizable bags, or presterilized and ready to use. Even within the ready-to-use category, several packaging configurations are available as needed for specific piston placement equipment and filling line designs. Examples of two ready-to-use formats are shown in Figure 36.3. Pistons for prefilled syringes are available in a range of pharmaceutical-grade elastomer formulations. Barrier films or coatings are available to minimize potential interactions with the drug product formulation.

36.3. PREFILLED SYRINGE DESIGN FEATURES

A prefilled syringe system is similar to a vial–stopper–seal system in that both systems serve as the product storage container during shelf life. As such, both containers generally are formed using chemically resistant borosilicate glass and closed with high-purity pharmaceutical grade elastomers. However, a prefilled syringe minimizes user handling by serving as an integral part of the drug delivery device. This fundamental difference affects the number and nature of the materials of construction, imposes functional performance requirements on the container–closure interfaces, and influences the surface-to-volume ratio of the container system design. The remainder of this section addresses these points.

36.3.1. Materials of Construction

In its simplest form, the direct contact materials in a vial system during shelf storage are the glass vial and the elastomer stopper. Figure 36.4 shows the components and the limited number of direct product contact materials for a typical vial–stopper–seal system. These materials also are present in a typical prefilled syringe system as the glass barrel and the elastomer piston. However, a luer tip/luer lock prefilled syringe will include, at a minimum, a lubricant—usually polydimethylsiloxane (silicone fluid)—and a second elastomer formulation as the luer tip cap. A prefilled syringe with a preattached needle includes the adhesive and the stainless-steel needle as well as the elastomer needle shield. Figure 36.5 illustrates the added complexity of a typical staked needle prefilled syringe system. All components of the selected prefillable syringe system will have direct product contact throughout shelf life and have the potential to affect the biopharmaceutical product through interactions at the surface or as a result of substances leached into solution. These possible effects must be evaluated during formulation development and qualification of the container system. Some specific areas to consider are discussed later in this chapter.

36.3.2. Container–Closure Functional Requirements

As with the stopper on a vial, a prefilled syringe piston must provide a leak-free sterile barrier throughout product shelf life. At the same time, as a component of the drug

Figure 36.4. A 2-ml vial–stopper–seal system showing materials of construction: borosilicate glass vial, elastomer stopper (with or without barrier coating), and aluminum crimp overseal.

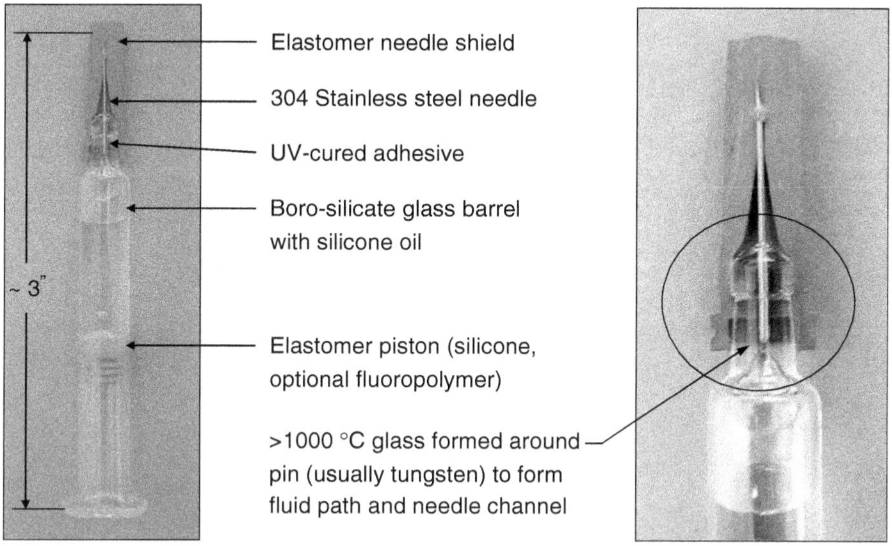

Figure 36.5. A 1-mL-long staked needle syringe system showing direct-contact materials.

delivery system, the piston must move easily and smoothly while requiring relatively low force to initiate and sustain the injection stroke at the time of use. The needle cover (for a staked needle prefilled syringe) or tip cap (for a luer tip or luer lock prefilled syringe) also must maintain a tight seal during storage yet be easy to remove at the time of use. Finally, luer tip or luer lock systems must be capable of forming a leak-free seal at the luer taper when assembled with a needle or other fitment at the time of use.

36.3.3. Geometry and Surface-to-Volume Ratio

For ergonomic ease of use, syringes tend to have a much greater length-to-diameter ratio than do vials of the same capacity. This means that a prefilled syringe may have more container surface area in contact with the drug product. On the other hand, the surface area of the air–liquid interface and the total headspace volume are likely to be much less than for a comparable vial–stopper–seal system. Figure 36.6 contrasts the wetted surface area, air–liquid interface area, and headspace for a 1 mL dose in a 1-mL-long prefilled syringe with that of a typical 2-ml vial with the same 1 mL fill volume. For the prefilled syringe, the wetted surface area of the glass is about 43% greater while the headspace volume may be as little as one-tenth and the air–liquid interface is about one-fifth that of the vial. Any of these characteristics of a prefilled syringe may affect product quality during transportation and storage differently than in a vial and may require formulation adjustments to compensate for the changes in geometry and contact area of the various surfaces.

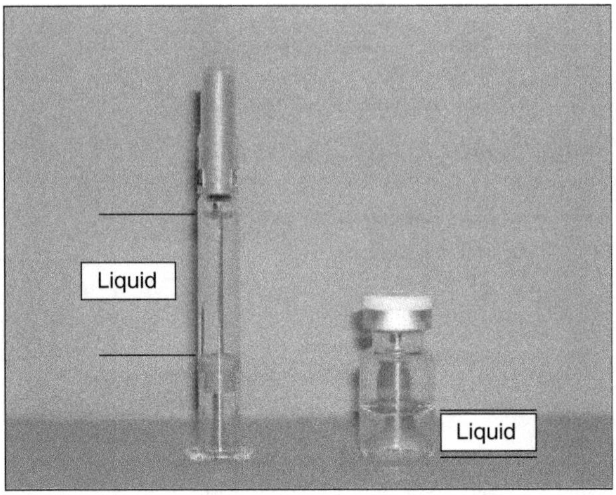

Figure 36.6. A 1-mL-long prefilled syringe and 2-ml vial—both with 1 mL fill showing differences in headspace volume and wetted container surface area.

36.4. FORMULATION DEVELOPMENT

A typical biopharmaceutical molecule may have been packaged for preclinical and early clinical studies as a frozen liquid in a vial and may have progressed to late-stage clinical studies and, possibly, commercial launch as a lyophilized powder or liquid formulation in a vial. To enable a biopharmaceutical product to gain the full benefit of a prefilled syringe presentation, the formulation scientist faces the challenge of developing a stable liquid formulation that is robust to the full range of direct and indirect product contact materials inherent in a prefilled syringe as well as the conditions imposed by the drug delivery functions of the syringe. In general, the development process follows the same path as for a vial, including, for example, an assessment of effect of the fill–finish process on the characteristics of the drug product. However, there are some materials and features that are unique to prefilled syringes. These deserve further discussion.

36.4.1. Direct- and Indirect-Contact Materials

36.4.1.1. Syringe System Lubricants. Syringes require some type of lubricant to moderate the friction forces that must be overcome to initiate and sustain the injection stroke at the time of use. Prefilled syringe systems are no exception. Medical-grade polydimethylsiloxane fluid, commonly called *silicone oil*, is the most widely used lubricant for prefilled syringes. In most applications, the silicone fluid is applied to both the inner diameter of the glass barrel and to the external surface of the piston. From the perspective of the formulation scientist, silicone oil is likely to be viewed as necessary but potentially problematic. In the literature, protein–silicone interactions have been implicated as causes of protein aggregation [2]. Such interactions may arise from silicone droplets in the solution, the hydrophobic surface of the siliconized glass or piston, a silicone film at the air–liquid interface or some combination of these effects. Moreover, any of these effects may be exacerbated during transportation [3].

Unfortunately for the formulation development team, it is challenging to anticipate or simulate at laboratory scale the range of silicone quantity and uniformity and potential transportation conditions to which a drug product may be exposed during commercial-scale manufacture and distribution. At a minimum, the team may consider testing candidate formulations against multiple syringe component lots and incorporating transportation studies into the formulation characterization protocols. One possible approach is to test formulation candidates using syringes to which excess silicone fluid has been added intentionally. As a cautionary note, silicone oil spiked directly into the drug product may react quite differently compared to silicone oil applied as an atomized mist to the inner surface of the glass. In addition, the protein therapeutic may respond to the combination of lubricant and surface in a way that differs from the response to either of the individual components. Therefore, it is important to test the drug product under actual storage conditions in the proposed prefilled syringe system.

36.4.1.2. "Staked" Needle Adhesives. In prefilled syringe systems with preattached, or "staked" needles, the cannula is preassembled into a small-diameter channel in the glass barrel at the time of syringe barrel manufacture. Typically, the cannula is bonded to the glass by means of a UV-cured adhesive. Testing for extractables should consider the adhesive as a potential contributor of extracted compounds during qualification of the syringe system. In addition, the formulation development team also must address the cured adhesive as a material in direct contact with the drug product throughout shelf life. Stability studies using staked needle syringes should include any effects of the adhesive on the drug product. However, the physical accessibility of the drug product to the adhesive may vary somewhat from one barrel lot to another. Therefore, the formulation development team may consider including syringe barrels from multiple supplier lots—even for small-scale, early-stage studies.

Beyond this, the team may consider additional, more comprehensive studies to evaluate potential interactions between candidate formulations and the adhesive in a proactive way. The adhesive contact area is very small relative to other direct-contact surfaces. Therefore, one could consider intentionally exposing candidate formulations to a larger adhesive surface area in order to assess the safety margin for adhesive contact. A similar approach could be considered to assess drug product sensitivity to volatile and semivolatile byproducts of the adhesive curing process—especially any compounds from the adhesive that are identified during extractables studies.

36.4.1.3. Tungsten from the Barrel-Tip-Forming Process. The hole in the syringe tip that becomes the fluid path or needle channel of a glass prefilled syringe is formed by pressing the flame-softened end of a glass tube against a metal forming pin. To be formed in this way, the glass temperature is about 1000°C. Very few of the materials that are available industrially as small-diameter wire are able to withstand repeated exposure to the mechanical and thermal stresses of this process. Traditionally, tungsten has been the material of choice for the forming pins. Despite its excellent high-temperature strength and oxidation resistance, under these conditions, even tungsten abrades and ablates slightly as each barrel is formed. Some of this material is deposited on the inner surface of the barrel tip, survives the rinsing process, and is present as a product contact material or leachate during product storage. The total quantity of tungsten to which the drug product may be exposed is quite small—in the range of 500 ng. However, the tungsten residues typically are at or near the tips of the syringes, which may lead to localized regions of higher concentration. Even at low concentration, leached tungsten ions may affect some proteins [4,5].

Moreover, the quantity of tungsten residue may vary significantly from barrel to barrel and lot to lot. The highest levels may be present in only a small proportion of the syringe population. Therefore, small-scale formulation development studies might not represent the range of tungsten exposure that may occur in commercial manufacture. In addition, the chemistry of soluble tungsten is quite complex and may involve a wide range of species that vary with pH and concentration [6,7]. Tungsten polyanions have been demonstrated to interact specifically with human serum albumin [8]. Protein precipitation in the presence of tungsten has been demonstrated to depend on the specific protein involved, the formulation pH, the concentration of both protein and tungsten, and the ratio of these two components [9]. Because of the complex chemistry of tungsten in aqueous solutions, unless the exact tungsten species from the prefilled syringe has been identified in the drug product, it can be very difficult to mimic these reactions by adding tungsten compounds. One approach to evaluating tungsten sensitivity is to perform spiking studies using high-concentration tungsten solutions extracted from used tungsten forming pins [9].

In response to the effect of tungsten on some biopharmaceutical products, suppliers have been working to develop alternative materials for the forming pins and new syringe concepts that are tungsten-free. Some of these are mentioned in Section 36.10, at the end of this chapter.

36.4.1.4. Stainless-Steel Cannula. Excipients, drug substance, and drug product have extensive contact with stainless steel, usually AISI 316L (American Iron and Steel Institute type 316L stainless steel), throughout many process steps. Therefore, the stainless-steel cannula in a prefilled syringe may be overlooked as a product contact material. However, it is worth noting that syringe cannulae typically are made from AISI 304 stainless steel. Although the compositions are similar, they are not identical—especially regarding the nickel and chromium content. As shown in Table 36.1, 316L stainless has greater nickel content and includes molybdenum. This combination improves the corrosion resistance of 316L stainless compared to 304 stainless. Therefore, development teams planning to use prefilled syringes with preattached cannulae may consider evaluating candidate formulations for potential interactions with 304 stainless steel.

36.4.1.5. Residual Ethylene Oxide from Sterilization of Ready-to-Fill Barrels. For those unfamiliar with the process, it is worth noting that presterilized, ready-to-fill glass syringe barrels are sterilized by ethylene oxide C_2H_4O exposure. Syringe barrel manufacturers typically certify that the residual C_2H_4O concentration

TABLE 36.1. Composition Comparison Between AISI 304 Stainless Steel and AISI 316L Stainless Steel

Element	AISI 304 Stainless Steel, %	AISI 316L Stainless Steel, %
Carbon	0.08	0.030
Manganese	2.00	2.00
Phosphorus	0.045	0.045
Chromium	18.0–20.0	16.0–18.0
Nickel	8.0–11.0	10.0–14.0
Molybdenum	None	2.00–3.00
Silicon	1.00	1.00
Sulfur	0.030	0.030
Iron	Balance	Balance

Source: Table compiled from composition data contained in ASTM A959-07, *Standard Guide for Specifying Harmonized Standard Grade Compositions for Wrought Stainless Steels (2007).*

in the syringes is less than 1 ppm. Nevertheless, formulation development teams should be alert to and consider evaluating possible effects of product exposure to trace quantities of C_2H_4O.

36.5. CONTAINER–CLOSURE SYSTEM QUALIFICATION

The general topics of selecting and qualifying final dosage form container–closure systems have been addressed in Chapter 33. This section highlights several points that are unique to prefilled syringe systems.

36.5.1. Container–Closure Integrity Testing

The principles of and some methods for container–closure integrity (CCI) testing, as described in Chapter 35, apply equally well to prefilled syringes. However, CCI test methods that are relatively straightforward for vials are more complicated to implement for prefilled syringes because of the functional requirements of prefilled syringes as a drug delivery device—namely, that the piston must be a movable seal and that the tip cap or needle shield must be removable.

Because of these functional design features, the dynamic stresses imposed during a prefilled syringe CCI challenge test must be managed to ensure that failures are not induced by the conditions of the test. For example, immersion in a microbial challenge bath coupled with an excessively aggressive dynamic pressure or vacuum challenge could cause the piston to move into a contaminated area of the barrel. In such a case, the loss of sterility may have been caused inadvertently by the exposure conditions, not by a container system failure. Conversely, incorporating external fixtures to prevent piston movement may mask a CCI problem by artificially altering the force exerted by the piston on the inner wall of the barrel. Similar challenges exist in evaluating CCI at the tip of the syringe in that dynamic pressure or vacuum challenges may induce movement of the tip caps or needle shields.

As of this writing, microbial aerosol, microbial immersion, and dynamic dye ingress are the primary methods used to establish CCI for prefilled syringe systems. An advantage of dye ingress is that it can be used with drug product and can be included as a test performed on stability. However, both dye-based and microbial methods lack the range and sensitivity to quantify seal performance. For example, when no dye ingress is observed, there is no way to distinguish between a robust system and one that is barely adequate. In addition, the physical manipulations of the syringes required for these methods can lead to false failures. Given these limitations, instrument manufacturers are working aggressively to adapt previously mentioned leakage-based methods, such as vacuum decay or tracer-gas detection, to prefilled syringes. Container engineering teams should monitor developments in this field in anticipation of sensitive quantitative methods that can be incorporated into syringe system qualification testing as well as stability programs for commercial products.

36.5.2. Evaluation of Extractables and Leachables

General concepts for evaluation of extractables and leachables from packaging components have been addressed in Chapters 34 and 35. Hojnicki has presented some strategies for selection and justification of solvents and extraction conditions for any type of container system [10]. This section addresses characterization of residual tungsten, a topic that is unique to glass syringe barrels, and a platform approach to assessment of extractables in prefilled syringe systems.

Residual tungsten in syringe barrels has been mentioned earlier in connection with formulation development. However, tungsten spiking studies to assess product sensitivity are not sufficient. In order to fully evaluate the risk of tungsten-induced protein degradation, the development team also should characterize the residual tungsten level in the proposed syringe barrels and explore which tungsten species may be present. Inductively coupled plasma/mass spectroscopy (ICP/MS) technology is well developed for analysis of trace metals in aqueous solutions and can be adopted readily to measure tungsten ion concentration in extract solutions as well as in drug product. Unfortunately, as is often the case with extraction of metal ions from glass containers, test container preparation and the exposure conditions of the extraction have been shown to have a strong influence on the efficiency of the extraction [5,11]. As of this writing, standardized extraction test methods for residual tungsten in prefilled syringes are not available. Therefore, it is recommended that any analytical method development studies for residual tungsten include successive extractions of the samples in order to assess the efficiency of the extraction procedure.

More generally, Nashed-Samuel et al. [12] have demonstrated a platform approach to evaluating potential leachables from a prefilled syringe system. In this approach, a series of model solvents was developed that encompasses the expected range of pH, buffer types, and solvation mechanisms likely to be used in formulating therapeutic protein products for prefilled syringes. Ionic strength of the model solvents is optimized to have minimal impact on the analytical techniques and instruments. Extraction studies were then conducted by filling the various model solvents into complete syringe systems using the planned commercial components. Parallel extractions, using the same model solvents, were performed on individual components (glass syringe barrels and elastomer needle shields and pistons) to establish the source of any compounds isolated from the complete system. Studies of this nature can yield a library of potentially leachable compounds and, where necessary, provide guidelines on the types of assays needed for leachable testing of drug products during formulation development and stability studies. The library also may be used proactively as the basis for spiking studies to assess sensitivity to potential leachables and to evaluate the impact of changes to processes or materials proposed by the supplier(s).

36.5.3. Graduation Marks and Delivered Volume

For drug products marketed in prefilled syringes, graduation marks often are needed—for example, when dosing is based on body weight. Graduation marks can be preprinted on the barrel by the barrel manufacturer or incorporated into the artwork of a preprinted label applied by the drug product manufacturer. For a given drug product concentration, the drug product volume in the syringe and, therefore, the available dose, is determined primarily by the precision and accuracy of the filling process. However, delivered volume accuracy, if based on graduation marks, includes the variability introduced by the tolerance of the barrel inside diameter as well as other factors such as print or label placement accuracy and user-induced variation.

Syringe barrel inner diameter tolerances, especially for the 1-mL-long format, are small compared to vials. The typical tolerance for a 1-mL-long syringe barrel is ±0.1 mm. By contrast, the typical inner diameter tolerance for tubular glass vials is ±0.25 mm. At the nominal 1-mL-long inner diameter (6.35 mm), a 1 mL dose fills 31.6 mm of the length of the barrel. Despite the relatively tight tolerance, the filled length of the 1 mL dose varies about ±1 mm over the inner-diameter tolerance range. This is illustrated in Figure 36.7, which shows two 1-mL-long barrels—one near the minimum barrel ID the other near the maximum. Each barrel contains 1.00 mL volume as determined gravimetrically.

Volumetric accuracy indicated by graduation marks also is affected by positional accuracy of the graduation marks. This variability can become a significant percentage of the total dose when the intended dose is less than the total capacity of the syringe—for example, ±0.5 mm variation in graduation mark position on a 1-mL-long barrel is ±1.6% of a 1 mL dose but ±5.3% of a 0.3 mL dose.

Both of these sources of variation in graduation mark accuracy as well as variation in the user's positioning of the piston in relation to the graduation mark need to be considered when developing artwork designs, manufacturing standard operating procedures (SOPs) for print or label placement and patient or physician insert information regarding dosing. These factors also need to be considered when the label design is required to include an indication of fill volume enabling users to verify that the syringe contains the correct volume.

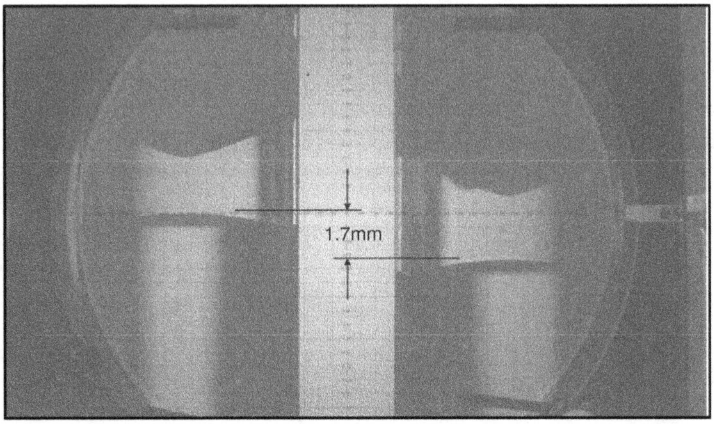

Figure 36.7. Optical comparator image (20×) of the liquid meniscus of 1-mL-long syringes with 1 mL fill volume showing the effect of barrel inner diameter on apparent fill level. The syringe on the left is near minimum specification for barrel inner diameter. The syringe on the right is near the maximum specification.

36.6. DRUG DELIVERY SYSTEM FUNCTIONAL PERFORMANCE

A number of characteristics affect the functional performance—or user experience—when a dose is delivered using a prefilled syringe. All of these need to be characterized during the qualification of the system and, for specific therapeutic settings, may need to be compared to the physical capabilities of the intended users.

Many of the functional characteristics are independent of the drug product and depend primarily on the design and materials of the components and the capability and variability of the suppliers' manufacturing processes; these include

- Force needed to remove the tip cap or needle shield
- Rotational torque resistance and removal force for a luer lock adapter
- Leak-free performance of a luer taper when assembled to a needle or device
- Needle pull out force for a preattached needle
- Needle penetration force and insertion force for a preattached needle

Each of these functional performance parameters can be characterized for the selected prefilled syringe system using straightforward mechanical test methods and compared to the physical abilities of the intended user population. It should be noted that some of the parameters may vary significantly from lot to lot or over time. Therefore, tests should be performed on multiple supplier lots and include the expected shelf life of the components and drug product.

One of the most important functional performance characteristics—the force that must be applied to deliver the dose—also is strongly influenced by the suppliers'

Figure 36.8. A 1-mL-long syringe barrel after breaking off the tip at the shoulder. The resulting hole in the barrel is large enough that product fluid flow resistance is no longer a factor when measuring glide force.

materials, designs, and process capabilities. The qualitative term *syringeability* sometimes is used to describe the perceived ease of use. Quantitatively, two aspects usually are assessed. The "break loose," "break out," or initial glide force is the force needed to initiate movement of the piston. The sustaining or extrusion or dynamic glide force is the force needed to maintain the forward motion of the piston. As one might expect, the dimensions of the components, the uniformity of the lubricating film, and the properties of the elastomer affect both parameters. These also may vary from lot to lot, may change over time, and may be characterized during the syringe system qualification independent of the drug product formulation.

However, in addition to performance characteristics attributable to the components, the characteristics of the drug product also may influence the forces, especially the extrusion force. Beyond the obvious effect of the viscosity, the formulation may tend to be self-lubricating or may interact with the barrel surface to inhibit the effectiveness of the lubricant. Some formulations may also have the potential to affect the physical properties of the piston. To isolate piston friction from the hydrodynamic resistance of fluid flow through a preattached needle, one may measure extrusion force after breaking off the tip of the barrel as shown in Figure 36.8. In this way, the backpressure caused by the flow of the product through the needle is eliminated and only the frictional forces are measured.

Overcashier et al. have reported the effect of the inner diameter of the cannula on extrusion force as a function of solution viscosity [13]. Their results show that the fluid flow forces conform to and may be predicted by the Hagen–Poiseulle equation for laminar flow in a tube [14]

$$Q = \frac{\pi R^4 (P_1 - P_2)}{8\ \mu L} \tag{36.1}$$

where Q = volumetric flow rate, μ = fluid viscosity, L = needle overall length, R = inner radius of needle, and P_1 and P_2 = pressure at inlet and outlet of needle. With respect to the cross-sectional area of the syringe inner diameter, the pressure P_1 at the needle inlet may be expressed as the force that must be applied to the piston to produce the desired flow rate. When testing in air, the pressure P_2 at the needle outlet is known to be atmospheric pressure. Thus, the equation can be solved for the force that must be applied to the piston to maintain a given flow rate.

It is especially important to recognize the fourth-power influence of the inner diameter of the needle. Because of this relationship, the effects of both the nominal needle inner diameter and the needle inner diameter tolerance should be considered, especially for viscous solutions. Note also that this equation models only the fluid flow force. The frictional force must be added in order to estimate the total extrusion force.

36.7. MANUFACTURING

Shifting the focus from developmental considerations to manufacturing, there are some aspects of filling, piston placement, inspection, and packaging operations for prefilled syringes that may be more challenging than similar processes for vials. Syringe filling may require smaller-diameter filling needles, and the filling equipment may use different pump technologies. The potential effects of these differences on the biopharmaceutical formulation will need to be characterized just as they would be for a vial filling process. In addition, the piston placement process may directly or indirectly affect the drug product and the container–closure system depending on the technique employed. The two main techniques available are vacuum placement and mechanical placement. With vacuum placement, the filled syringe is exposed to vacuum to remove most of the gas from the headspace. Then, while the headspace is at reduced pressure, the piston is partially inserted into the open end of the syringe. When the exterior of the system is restored to atmospheric pressure, the pressure differential between the inner and outer surfaces of the piston drives the piston down the barrel until the friction force balances the pressure difference. With mechanical placement, the piston is pushed through a thin-walled insertion tube that has been positioned inside the filled syringe. Both techniques have advantages and disadvantages with respect to equipment complexity and potential effects on the drug product, the piston, and the headspace volume. For this reason, some equipment manufacturers have developed hybrid systems in which the piston is placed under vacuum but a mechanical insertion rod also is used. As a further enhancement, some filling and piston placement machines can be adapted to evacuate the syringe prior to filling and then fill under vacuum. This technique, when combined with vacuum piston placement, can yield filled syringes with essentially zero headspace. Detailed discussion of the pros and cons of the various processing options is beyond the scope of this chapter. Readers are advised to consult with equipment suppliers and component manufacturers for information that is relevant to the specific syringe system and fill–finish equipment to be used. Similarly, product development teams are advised to work closely with their counterparts in manufacturing to assess the impact of the production filling and piston placement process on the drug product and the filled syringe system. This is especially important if the production process uses different technology than the laboratory-scale equipment used for formulation development.

When comparing the physical characteristics of a prefilled syringe to a vial of similar capacity, it should be obvious that the material handling equipment and the materials used for intermediate packaging and storage for syringes will be quite different from those used for vials. For example, vials can be manipulated and stored upright on their bases and packaged in dense, nested rows. By contrast, syringes typically are manipulated while hanging from the flange, which may expose the flange to forces that can lead to damage. In addition, the larger dimension of the flange relative to the syringe body means that syringes cannot be stored body to body. As such, more coldroom space may be needed for storage of prefilled syringes than would be needed for the same quantity of vials. In contrast to vials, which are normally intended to be stored upright, prefilled syringe packaging may be designed for many possible storage orientations, horizontally or vertically, and with the tip either up or down. Orientation of the syringes during storage may result in redistribution of lubricants or other effects. These possibilities should also be assessed during development.

In some manufacturing settings, packaging operations will integrate plunger rod insertion and labeling into the final packaging process. The potential effect of label placement on graduation mark or delivered dose accuracy and the need for robust SOPs and in-process control already has been discussed. Plunger rods typically are automatically threaded into the pistons. This process needs to be well controlled to ensure that pistons are not disturbed in a way that could compromise the integrity of the piston–barrel interface. Suitable equipment setup and maintenance procedures are needed to ensure adequate control of this process.

36.8. DISTRIBUTION

Ordinarily, one might not consider shipping and distribution to be part of the process of developing a biopharmaceutical product in a prefilled syringe. Nevertheless, there are some points to be considered that may reveal the need for significant developmental effort. First, a liquid biopharmaceutical product in a prefilled syringe is likely to require refrigeration during transportation and storage. Development, qualification, and validation of distribution packaging and cold chain monitoring systems are beyond the scope of this book. However, the work is not trivial. The time and resource requirements for these activities need to be included in the overall project plan. In addition, the formulation development team should evaluate the effects of temperature excursions—both above and below the intended storage temperature—in order to develop the user requirement specifications needed by the distribution packaging design team. On the package development side, the heat capacity and insulating properties of the dispensing carton and distribution package for a prefilled syringe presentation may be different from that used for vials. This may have an effect on the difference between temperature inside the shipper and actual product temperature.

In addition, environmental conditions during shipping may have direct effects on the container system and the drug product. For example, shipment may expose the syringes to reduced external pressure, which, in turn, may induce movement of the pistons. This increases the headspace volume while reducing the headspace pressure. The formulation development team may consider assessing these effects on candidate formulations. During transportation, packaged prefilled syringes also are exposed to shock, vibration, and other physical stresses. The combined effects of these stresses on the formulated drug product need to be tested as part of any transportation study, and transport validation protocols should span the expected range of shipping conditions. For marketed products, the team responsible for ongoing stability monitoring should consider whether samples for stability studies should be exposed to transportation (actual or simulated) prior to the start of the defined storage protocol.

36.9. AUTOMATIC INJECTION DEVICE INTEGRATION

Prefilled syringes originally were developed with the expectation that the users would be health-care professionals and that doses would be delivered by manual injection. However, as previously noted, prefilled syringes can bring much greater convenience and ease of use to an injection therapy compared to injection from a vial containing either a liquid or lyophilized powder. Partly for user convenience, the use of prefilled syringes increasingly has shifted to administration by the patient or caregiver. In this setting, user convenience takes on even greater importance—especially for patient populations with compromised manual dexterity. Market research indicates an increasing user preference for drug–device combination products in which a prefilled syringe is integrated into an automatic injection device [1]. Improved user convenience, enhanced ease of use, and reduced anxiety from "needle phobia," as well as integrated needle safety are among the possible benefits of these devices. A detailed discussion of the work required to develop, qualify, and commercialize an automatic injection device on the basis of a prefilled syringe is beyond the scope of this chapter. However, it is worthwhile to point out some topics where integration of a prefilled syringe into an injection device may affect drug product stability, the container–closure system, manufacturing operations, injection device function, and clinical performance.

36.9.1. Drug Product Stability

Automatic injection devices often are designed to mimic the injection rate typical of administration by health-care professionals. As such, the dose delivery is unlikely to have a different effect on the drug product than a manual injection. However, the device assembly and packaging processes may entail additional room temperature exposure and manipulations. In addition, the heat transfer characteristics of the assembled and packaged devices may differ from those of packaged prefilled syringes. These points need to be considered when assessing actual product temperature profile and total room temperature exposure during in-house operations as well as during transportation.

36.9.2. Container–Closure System Effects

Some automatic injection devices include design features that engage the outer surface of the needle during assembly of the device. These components of the device then become the means for removal of the needle shield by the user at the time of use. Since these features interact mechanically with the outer surface of the needle shield during assembly and also during storage and transportation of the assembled devices, it is necessary to demonstrate that these interactions do not alter the integrity of the container–closure interfaces at the inner surface of the needle shield. As is often the case in testing related to CCI, one challenging aspect is to develop a suitable positive control—in this case, an assembled device having a needle shield with compromised CCI. The time needed to develop and verify the concept for developing positive controls should be included in the assembly process development–validation plan.

36.9.3. Manufacturing Considerations

Several aspects of the assembly and packaging operations for an automatic injection device already have been mentioned in connection with possible effects on the drug product or the container-closure system. It should be obvious that SOPs for the corresponding equipment and processes also need to be integrated into manufacturing operations. In addition, the manufacturing and quality organizations should be prepared to conduct testing to monitor functional performance of assembled automatic injection devices. The quality unit also needs to develop test methods and procedures for evaluating components of the device for lot acceptance upon receipt. Finally, smooth assembly and proper operation of the assembled device may depend on tighter control of some parameters of the syringe components and the filled syringe. As such, suitable testing may be needed at incoming control and at appropriate intermediate points between filling and packaging. Some parameters that may need to be considered are discussed in the next section.

36.9.4. Injection Device Function

When a prefilled syringe, intended for manual injection, is assembled into an automatic injection device, some design features of the syringes may be subjected to mechanical contacts or forces for which they were not designed. For example, the thickness, diameter, contour, and strength of the flange of a glass syringe are relatively noncritical in a manual injection. However, in the assembly process and in the assembled automatic injection device, the flange may be required to fit within a precision assembly, may be used to orient the syringe in an assembly fixture, and may even transfer some of the energy of the injection sequence. Therefore, a thorough review of the device design and the assembly process should be conducted with the syringe supplier to assess whether any aspects of the syringe-forming process or dimensions need to be more tightly controlled to ensure reliable performance of the assembled automatic injection device.

Similarly, with many automatic injection devices, the speed of the injection sequence is a balance between the stored energy of the device and the friction and fluid resistance forces of the syringe during the injection. If the force needed to complete the injection varies significantly from lot to lot or over the shelf life of the product, the injection rate also may vary. Rapid injections may be more painful, while slow injections may seem uncomfortably long to the patient. Thus, the quantity and distribution of lubricant(s) on the syringe barrel and piston may need to be controlled more tightly or monitored differently for components that are intended for use in an automatic injection device.

Finally, the device designers may have designed the components and selected materials without considering that most biopharmaceuticals are stored and transported under refrigerated conditions. Durability of the device components and mechanical function of the device must be thoroughly tested with full consideration of the potential effects of reduced temperature, vibration, pressure change, and other stresses that can occur during shipment, as well as condensation that may form within the device on removal from the refrigerator just before use.

36.9.5. Clinical Performance

As with most aspects of injection device development, detailed discussion of the clinical performance of an automatic injection device is beyond the scope of this chapter. However, a few points can be mentioned. It may be necessary to demonstrate that the use of the syringe–device combination does not alter the pharmacologic attributes of the drug product when compared to a manual prefilled syringe or a vial with a disposable syringe. As mentioned earlier, the typical injection time for manual injection of the product should be assessed. The device parameters may be selected to mimic a manual injection or other injection parameters may be chosen to suit the specific application and user requirements.

36.10. EMERGING TECHNOLOGIES

To this point, several areas have been discussed where adapting a biopharmaceutical molecule to a prefilled syringe can be challenging. In this section, some current efforts to adapt prefilled syringe systems to special needs of biomolecules are mentioned.

Tungsten sensitivity is a concern for some molecules. Manufacturers are developing several ways to reduce or eliminate tungsten as a product contact material. For glass syringes, manufacturers have developed alternative materials to replace the use of tungsten forming pins. Manufacturers are likely to consider the details of the pin material to be proprietary. Therefore, development of drug products in tungsten-free glass syringes typically will be done through close cooperation between the syringe supplier and the drug product manufacturer. Potential interactions between the proposed new pin material and the formulated drug product need to be fully explored and approved prior to implementation. Many of the strategies mentioned for evaluating tungsten sensitivity may be applicable to alternative forming pin materials.

Another tungsten-free alternative is to use syringe barrels injection-molded from high-performance cycloolefin polymers (COPs). As with any new product contact material, these materials need to be thoroughly characterized and tested for drug product compatibility and safety.

Silicone sensitivity also has been mentioned as an area of concern for some molecules. Several syringe systems are being developed to either reduce silicone significantly or eliminate it altogether. The quality of silicone can be reduced either by "baking" the silicone after application or by using a reactive silicone system that can be applied as a liquid and then polymerized.

To completely eliminate silicone, one manufacturer has developed a system that combines a COP barrel with a piston that is fully protected by a fluoropolymer film. The fluoropolymer is sufficiently lubricious that the barrel does not need to be lubricated. This system also is tungsten-free.

Another approach to elimination of silicone is the use of chemical vapor deposition or plasma technology to generate nonsilicone lubricious layers on the barrel, piston, or both. Engineers and scientists are advised to monitor developments in these areas. As with any surface or material that comes into contact with the drug product, the effect of these coating materials on drug product quality during the shelf life of the product will need to be carefully assessed and approved before implementation into prefilled syringes used for marketed biopharmaceutical products.

ACKNOWLEDGMENTS

The author gratefully acknowledges the contributions of his co-workers and support of his managers in the preparation this manuscript. In particular, the author wishes to thank Linda Narhi, Margaret Ricci, and Yasser Nashed-Samuel for helpful discussions, especially in relation to the sections about product interactions with silicone and tungsten. Linda Narhi also made significant improvements in readability through her critical review of the manuscript. Mathias Romacker and Thomas Schoenknecht provided guidance with respect to development of the market for prefilled syringes and automatic injection devices. Finally, Ron Forster and Erwin Freund made many valuable suggestions that improved the organization and content of the chapter.

REFERENCES

1. Williams, D. and Thorpe, G. (2006), Marketing and end-user benefits (pre-filled syringes and self injection devices), presented at Parenteral Drug Assoc. conference, *The Universe of Pre-filled Syringes and Injection Devices*, Bethesda, MD, Oct. 2006 (unpublished data).
2. Jones, L. S., Kaufmann, A., and Middaugh, C. R. (2005), Silicone oil induced aggregation of proteins, *J. Pharm. Sci.* **94**(4): 918–927.
3. Thirumangalathu, R., Krishnan, S., Speed-Ricci, M., Brems, D., Randolph, T., and Carpenter, J. (2007), Silicone oil induced aggregation of a monoclonal antibody anti-streptavidin IgG1 in solution, *J. Pharm. Sci.*
4. Narhi, L. O., Wen, Z., and Jiang, Y. (2007), Tungsten and protein aggregation, presented at University of Colorado, *Protein Stability Conference.* Breckenridge, CO, 2007 (unpublished data).
5. Liu, W., Swift, R., Torraca, G., Nashed-Samuel, Y., Wen, Z., Jiang, Y., Vance, A., Mire-Sluis, A., Freund, E., Davis, J., and Narhi, L. O. (2010), Root cause analysis of Tungsten-Induced protein aggregation in pre-filled syringes, *PDA J. Pharm. Sci. and Tech.* **64**(1): 11–19.

6. Lassner, E. and Schubert, W. D. (1999), *Tungsten: Properties, Chemistry, Technology of the Element, Alloys, and Chemical Compounds*, Kluwer Academic/Plenum Publishers, New York.
7. Pope M. T. (1983), *Heteropoly and Isopoly Oxometalates*, Springer-Verlag, New York.
8. Zhang, G., Keita, B., Craescu, C. T., Miron, S., de Oliveira, P., and Nadjo, L. (2007), Polyoxometalate binding to human serum albumin: A thermodynamic and spectroscopic approach, *J. Phys. Chem. B* **111**(38): 11253–11259.
9. Jiang, Y., Nashed-Samuel, Y., Li, C., Liu, W., Pollastrini, J., Mallard, D., Wen, Z., Fujimori, K., Pallitto, M., Donahue, L., Chu, G., Torraca, G., Vance, A., Mire-Sluis, A., Freund, E., Davis, J., and Narhi, L. (2009), Tungsten-induced protein aggregation: Solution behavior, *J. Pharma. Sci.* **98**(12): 4695–4710.
10. Hojnicki, J. (2005), Chemical characterization of container closure systems: A systematic approach to the extraction process, presented at Parenteral Drug Assoc., *Extractables and Leachables Forum: The Extractables Puzzle: An Integrated Team Approach*, Bethesda, MD, May 2005 (unpublished data).
11. Swift, R. et al. (2006), Tungsten, Pre-filled syringes and protein aggregation, presented at Parenteral Drug Assoc. conference, *The Universe of Pre-filled Syringes and Injection Devices*, Bethesda, MD, Oct, 2006 (unpublished data).
12. Nashed-Samuel, Y., Silva Elipe, M., Swift, R., and Bordas-Nagy, J. (2005), Strategies for baseline extractables evaluation of a pre-filled syringe delivery system for waterbased biopharmaceutical formulations, presented at Parenteral Drug Assoc., *Extractables and Leachables Forum: The Extractables Puzzle: An Integrated Team Approach*, Bethesda, MD, May 2005 (unpublished data).
13. Overcashier, D., Chan, E., and Hsu, C. (2006), Technical considerations in the development of prefilled syringes for protein products, *Am. Pharm. Rev.* **9**(7): 77–83.
14. Whitaker, S. (1968), *Introduction to Fluid Mechanics*, International Series in the Physical and Chemical Engineering Sciences. Prentice-Hall, Englewood Cliffs, NJ, p. 457.

37

IMPACT OF MANUFACTURING PROCESSES ON DRUG PRODUCT STABILITY AND QUALITY

Nitin Rathore, Rahul S. Rajan, and Erwin Freund

37.1. INTRODUCTION

The development of manufacturing processes from drug substance to drug product is evolving from a "process is the product" practice to a more sophisticated approach that enables a proactive paradigm as an improvement over the traditional characterization approach. The latter is realized through improved molecular design allowing an upfront manufacturability assessment to aid in the selection of the optimal lead candidate. In addition, improved biophysical analysis and realistic scale down modeling allow more robust formulations. The arrival of platform approaches in product modalities such as monoclonal antibodies has fostered in the industry common approaches which shorten development time by the application of common process elements.

In addition to the advances in science, the regulatory agencies through the ICH (International Conference on Harmonization) sponsored the "a risk based approach to pharmaceutical cGMP of the 21st century" in 2003, which resulted in the issue of new ICH guidelines in 2005 and beyond. This stresses the need for a more rational design approach that will result in greater operational flexibility by the use of design of experiment (DOE) and process analytical technology (PAT) in achieving all three quality-by-design (QbD) concepts [1–4]: pharmaceutical development (Q8) and

Formulation and Process Development Strategies for Manufacturing Biopharmaceuticals, edited by Jameel and Hershenson

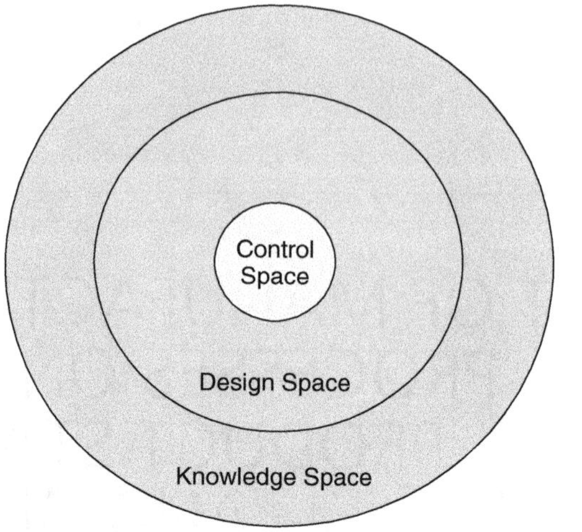

Figure 37.1. Schematic representation of the control, design, and knowledge spaces.

proactive risk management (Q9) within a well-defined quality system (Q10) supported by management. As a result, quality is not tested in the product and instead it is built in by the overall holistic design.

In addition to creating this advanced drug development model to improve the quality of pharmaceutical drugs, the agencies expect greater awareness of the possibility of immunogenicity as the result of protein degradation with aggregation as a leading but not the only cause [5]. This and the recent evolution toward doses with higher concentration proteins present greater challenges to scientist in the life-cycle development of the drug towards market approval.

A holistic approach has in place design elements that include the proactive activities stated above, a sophisticated mechanistic understanding of all drug product contact materials, process unit operations (Figure 37.1), and timely application of analytical tools for proactive characterization early in the commercial life cycle. A mechanistic understanding will lead to a more robust drug product and minimize the need to apply resources on investigations as a reaction to postapproval issues.

The regulatory agencies had a vision to see full implementation by 2007, but it is taking longer to put all guidelines in place and agree on case studies as examples. However, BLA-CMC (Biologic License Application-Chemistry, Manufacturing & Controls) submissions are expected to demonstrate increasing adherence to the principles outlined in the Q8, Q9, and Q10 guidelines, which cover pharmaceutical development and quality risk assessment.

In the traditional process, a univariate DOE resulted in "proven acceptable ranges" and subsequent process validation on three batches that served as proof of consistency and process control and evidence for a high degree of assurance that a specific process will consistently produce a drug product meeting its predetermined specification. Often this was achieved within a very narrow set of parameters as this took place without the benefit of multivariate DOE and risk assessment, which will permit the creation of "design space." Through the use of PAT-based automation controls, continuous validation by quality verification becomes achievable and could lead to real-time release.

A final benefit is continuous improvement being allowed within the established design space (Fig. 37.1) and pending regulatory acceptance, making postapproval changes less burdensome. The filing must convey a science-based complete mechanistic understanding of how the molecule stability, formulation, critical process unit operation steps, and container contact materials affect product performance and quality within the constraints of the drug target profile. Knowledge management defines knowledge space as consisting of data based on limited characterization. Within this area, a well-characterized design space allows for continuous improvement with fewer regulatory approval requirements as the specifications are all met. That is because working on unit operations within the design space does not represent a change. Note, however, that any move of values outside the design space constitutes a change and will be subject to the postapproval change process or scale-up postapproval changes (SUPACs). Regulators refer to exceeding the design space as the "edge of failure," prompting immediate action by the manufacturer. Finally, in order to issue tight manufacturing directions, the target values are centered within the control space using continuous monitoring verifying process control. When successful, QbD can substitute for the traditional statistical process control (SPC) applied during manufacturing based on traditional process development. In order to provide a path toward achieving the vision of the regulatory agencies, the following sections summarize the key expectations as outlined in the ICH guidelines.

37.1.1. Q8: Pharmaceutical Development

The first requirement is generation of the "drug target profile," stating upfront, for example, the desired drug indications; product attributes, including dose, container, injection aids, and desired rheology properties; intended use and practices; and dosage form (e.g., liquid vs lyophilized). This table of requirements is instructive to molecular engineer, formulator, PD scale-up scientist, and container engineers for design of experiment (DOE) and states clearly upfront expectations in the common technical document (CTD). It provides the context within which experimental protocols are generated. The product attributes form the landscape that must be covered by the design space in order to deliver process flexibility, paving the way for postapproval changes.

The relationship between the critical quality attributes (CQAs) and the [critical process parameter (CPP) as listed in Table 37.1] pertains to the specifications. The CQAs are properties of the molecule, output of a process, or contact materials that are critical to the process performance and are influenced by the CPPs. The relationships between these two are formalized in the "cause and effect" analysis, which is depicted in an Ishikawa diagram, also known as a *cause-and-effect diagram*, which resembles a fishbone; hence its colloquial name. It illustrates the relationship involved in meeting the design specifications as a function of the critical inputs. The design space can contain information beside unit operational parameters such as excipient properties, coatings within containers, and fill volumes.

TABLE 37.1. Process and Performance Parameters for Various Drug Product Processing Operations

Step	Unit Operation	Chapter Covered	Process Parameters (Examples)
1	Freeze–thaw	This chapter, Chapters 26 and 27	Temperature of thaw Freeze and thaw rates Agitation during thaw Storage container
2	Equipment cleaning	This chapter	Maximum allowable carryover Temperature of cleaning Cleaning agent concentration Agitation during cleaning
3	Formulation, mixing, and pumping	This chapter	Mixing speed Mixing shear Mixing time Batch size
4	Filtration	This chapter	Concentration of product Pressure Filtrate flow rate V_{max} Batch size Membrane adsorption
5	Component preparation		Wash cycle parameters Autoclave Drying cycle parameters
6	Filling	This chapter and Chapter 33	Fill speed Nozzle size Cam drawback Fill weight tolerances
7	Lyophilization	This chapter, Chapters 18–23, and Chapters 30–32	Primary drying temperature Ramping rate between primary and secondary drying Stopper moisture content Cake properties Vial bottom curvature
8	Inspection and light sensitivity	This chapter	Light exposure duration, wavelength of light exposure Machine settings (rpm, brakes) Sensitivity/false rejects
9	Storage container	This chapter	Storage temperature and time; Product degradation profile Material of construction of storage vessel
10	Transport	This chapter	Duration and temperature Vibrations and shock Ambient pressure Fill volume/headspace

The DOE requires the use of statistics for multifactorial and multivariate analysis to differentiate between independent and dependent variables combined with interaction analysis. Surface response analyses in more than one dimension exceed the information space compared to one-dimensional "proven acceptable ranges." Section 37.3.8 presents a case study demonstrating use of statistics for characterization of the design space of the cleaning process. Bench-scale characterization is performed to identify critical operating parameters for the process; such critical parameters should be further characterized using a DOE approach.

The history of the drug substance manufacturing steps has the potential to affect the CQA of the drug product. For instance, certain unit operations used for viral inactivation, such as heat, denaturants, lowering pH, or detergent, could result in irreversible denaturation of a small percentage of the drug product impacting quality and specifically aggregation, which could modify the appearance from a clear solution to a turbid solution. The application of a particle removal process by filtration does not always neutralize this impact to the drug substance as aggregates can break up and re-form or act as nuclei sponsoring visible aggregation.

37.1.2. Q9: Quality Risk Management

Risk analysis during development is a formal identification method to identify critical process variables and their relationship to the CQAs, such as yield, purity, solubility, and potency. Risk management has both the component of risk identification for which formal processes exists such as failure-mode effect and (criticality) analysis [FME(C)A] and assessment and control of the risk to the quality of the drug product through the product life cycle. Table 37.2 presents a hypothetical risk assessment listing potential impact of various unit operations on drug product CQAs.

Risk management is based on identifying and quantitating risks within the context of the defect categorization as function of risks to the manufacturer as well as to the consumer. General rules exist for classifying the implications by criticality on the CQAs (see Ref. 6, Table 15.2). Using these criteria, the risk can be understood and mitigated with respect to prioritization of potential failures through the use of risk priority numbers (RPNs). The risk to patients is managed by science-based development of the process (Q8) and robustness of the quality system (Q10). This approach allows establishing meaningful specifications that reflect the target profile and built-in expectations by the upfront molecular design.

Quality risk management is expected during the product life cycle and documented in the common technical document (CTD) in the form of tables mapping CQAs as a function of variables and unit operations. For additional clarification custom Ishikawa diagrams can be generated for each CQA. Another method is to link unit operations by incorporating CQA as an input allowing the elucidation of attributes on final product quality. This will demonstrate which CQA have wide versus narrow targets (i.e., a critical value), which goes into the *control strategy*.

A special case in risk assessment is on immunogenicity or an antigenic reaction of the host toward the administered protein. The response is a complex function involving the route of administration, type, size, adjuvant and repeat of antigenic epitope,

TABLE 37.2. A Hypothetical Risk Assessment to Identify Variables Potentially Impacting Drug Product (DP), Starting with Drug Substance (DS)

	Variables and Unit Operations									
DP CQA	Thawed DS	Formulation Mixing	Filtration UF/DF	Filling	Lyo[a1]	Inspection	Packaging	Storage	Transport	Device Delivery
Aggregation	Medium	Medium	High	Low	Low	Low	Low	Medium	High	Low
Potency	Low	Medium	Medium	Low	Low	Low	Low	Low	Low	Low
Impurities	Low	Low	Low	Medium	Low	Low	Medium	Low	Low	Low
Homogeneity	Medium	Medium	Medium	Low	Low	Low	Low	Low	Low	Low
Bioactivity	Low	Low	Medium	Low	Medium	Low	Low	Low	Low	Low
Bioburden	Medium	Low	Medium	Low	Low	Low	Low	Low	Low	Low
CCI	Low	Low	Low	Low	Low	Low	Low	Low	Medium	Low
Sterility	Low	Low	Low	Medium	Low	Low	Low	Low	Low	Low
Dose delivery	Low	Low	Low	Low	Low	Low	Low	Low	Low	High

[a]Lyophilization.

excipients, patient's immune system, and dose frequency. Models to predict immunogenicity in humans are lacking in the absolute sense, although sequence-based computer analysis, referred to as *in silico immunoinformatics*, exists as early algorithms. The surveillance of immune response during clinical trials is expected [7].

37.1.3. Q10: Quality Systems

This describes the comprehensive approach to an effective quality system, including CGMP guidelines, Q8, and Q9 by the application of business systems from development, technology transfer, manufacturing, and product discontinuation. Specifically it describes the corrective action–preventive action (CAPA) system. The QbD concept will through design space allow a more efficient CAPA system as well as change management provided appropriate risk management is conducted following the principles described in Q9. In summary, application of Q10 puts in place a planned set of controls also known as the *control strategy* that assures process performance and product quality. This is achieved by a multi-functional collaboration between the quality, regulatory, and process development units and written up as part of the CTD filing and very useful during the preapproval inspection (PAI) in order to demonstrate control of the DP CQAs. Now that the latest scientific trends and evolving expectations by the regulators have been described, the next section discusses the manufacturing process in more detail.

37.2. "OVERVIEW OF FORMULATION" AND FILL–FINISH OPERATIONS

The phrase "formulation, fill, and finish" refers to the series of processing steps that are required to convert the purified drug substance into the final dosage form, or the finished product, for the market [8]. The formulation step involves starting with the purified protein at the desired concentration and mixing it with the right excipients that can ensure product quality and integrity during the subsequent fill/finish steps, including filtration, filling, lyophilization, packaging, storage, transport, and delivery. A robust formulation would need to keep the biopharmaceuticals stable not only during shelf storage but also during these manufacturing steps. The problems for protein therapeutics are often very different from the traditional small-molecule pharmaceutical processing and may require special handling and storage conditions to ensure product quality [9,10].

An overview of proteins is provided in Chapter 1. As will be seen in many chapters of this book, the degradation pathways for protein therapeutics are numerous and often complex. Broadly speaking, these pathways can be divided into two categories: *physical degradations*, which do not involve covalent bond modifications, and *covalent modifications* (for reviews, see Refs. 8–12). Physical degradations are most commonly manifested by protein aggregation, while covalent degradation includes protein fragmentation, changes in glycosylation patterns and residue-specific modifications such as aspartate isomerization, protein oxidation, deamidation, pyroglutamic acid formation, and disulfide bond shuffling [12].

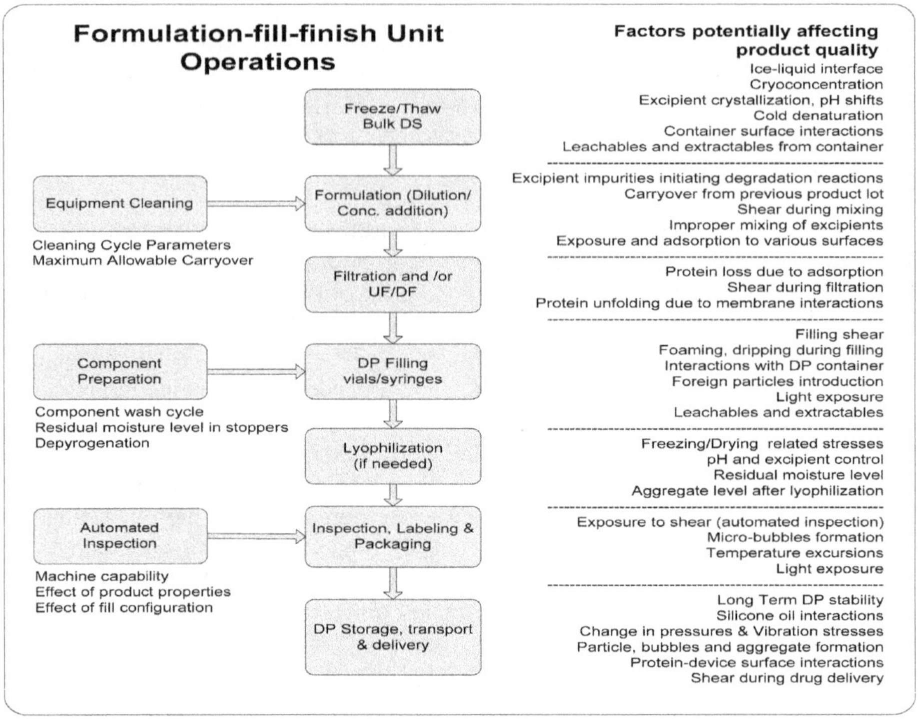

Figure 37.2. Overview of formulation, fill, and finish operations and process conditions that can impact product quality during manufacturing.

This chapter presents an overview of various formulation, fill, and finish operations. The key aspects of processing steps that can affect stability and integrity of a product are discussed. Figure 37.2 presents the series of operations and various factors that can impact product quality. In the sections that follow, each unit operation is described in detail. The stability consequences for some of the key operating inputs are included, and guidelines are is provided on scale-down studies needed to evaluate such destabilizing factors. For many fill and finish operations, such as filtration, holding, and formulation addition, the main concern would be physical stability of the protein. Nonetheless, orthogonal assays to determine the physical and covalent stability of the molecule should be carried out to determine the overall stability of the molecule to various bioprocessing stresses. In addition to formulation and other unit operations, protein instability issues may arise from interactions of drug product with packaging and components of the delivery device.

The current industry understanding of these challenges is included. Photodegradation of light-sensitive products is also discussed in a later section. Finally, one of the fill and finish operations, namely, cleaning, is considered in some detail with regard to identification of key operating parameters and how it relates to the concept of quality by design described in Section 37.1.

37.3. FORMULATION AND FILL–FINISH CONSIDERATIONS

37.3.1. Bulk Freeze–Thaw

Biopharmaceutical manufacturing often relies on bulk freeze–thaw in order to gain operational flexibility while maintaining product quality. Frozen drug substance provides several advantages, including increased product stability, reduced possibility of microbial growth, and alleviation of foaming issues during transportation. Freezing reduces the rates of degradation reactions and also immobilizes the protein molecule in a frozen matrix, thereby minimizing diffusive collisions that lead to aggregation. Free water content is reduced, which prevents several degradation reactions that are assisted by water, such as peptide bond hydrolysis and aspartic acid isomerization. The increase in protein stability, in turn, offers flexibility to schedule formulation–fill–finish operations according to the needs of the manufacturing facility. Bulk drug substance can be stored frozen and when needed, can be transported to the fill site where it is thawed, initiating a series of formulation and fill–finish steps. This chapter and Chapters 12 and 24 describe the freeze–thaw process in detail, and the reader is referred to these chapters for a more thorough understanding. In the following paragraphs we have provided a brief summary of these operations and the challenges that they face from the stresses associated with freeze and thaw operations.

37.3.1.1. Stability Challenges During Freezing. Bulk freeze–thaw, while offering numerous operational and product quality benefits, may also be detrimental to protein stability. Cryoconcentration of protein and excipient molecules is one of the most common mechanisms through which protein destabilization could occur during freezing [13–15]. Concentration buildup of excipients may result in pH change due to selective precipitation of buffer components, which can also result in protein destabilization [16]. An increase in protein concentration also increases the possibility of molecular collisions, thereby resulting in protein aggregation. Slow freezing rates, typically seen during uncontrolled rate freeze–thaw (e.g., in carboys), will maximize cryoconcentration. Smaller freezing pathlengths and higher heat transfer fluxes are recommended to minimize such freeze concentration effects. Mixing during the freezing process should, however, be avoided as it could suppress dendritic ice growth, making the ice–liquid interface more flat and therefore resulting in increased cryoconcentration. Additionally, cold denaturation [17–19] may also lead to spontaneous unfolding at cold temperatures, due to the weakening of the hydrophobic effect with decrease in temperature. Very high freezing rates can also be detrimental as they can result in large ice–liquid interfacial inducing protein molecules to concentrate and unfold [20–22] on the ice-water interface.

37.3.1.2. Stability Challenges During Thawing. Before a frozen drug substance can be formulated and processed, it needs to be thawed, which, in turn, can cause further stress and damage to the protein. Slow thawing rates can result in ice recrystallization with small ice crystals growing into larger ones. Proteins may get sheared at ice–liquid interfaces and lose their activity [21,22]. Concentration gradients created during freezing can further harm the protein during thawing. As are faster freezing rates, faster thawing rates are usually preferred for protein stability and are likewise achieved through improved heat transfer fluxes and smaller pathlengths. Controlled-rate technologies can be used to achieve such fast freeze–thaw rates [14]. Such technologies offer the additional advantage of process scalability as the heat transfer pathlengths are kept constant in both small- and large-scale equipments. Also, an appropriate level of mixing during thaw is the key to minimize recrystallization and cryoconcentration related effects. Very slow mixing would result in neither longer thaw times nor the ability to homogenize the solution. On the other hand, very high mixing rates could result in protein shearing, excessive foaming, and plausible protein denaturation on air–liquid interface.

37.3.2. Handling Stresses During Mixing and Filling

Protein products experience a variety of stresses during the series of formulation and fill operations. Table 37.1 summarizes some of the key factors contributing to these stresses. More generally, it includes temperature excursions during processing, prolonged exposure to various surfaces, and exposure to shear and air–liquid interfaces generated during formulation and filling. The mixing step during the formulation process may impact protein stability due to mechanical stresses created through stirring and shaking in presence of different contact surfaces [23–27]. Mixing and pumping process may also create air–liquid interfaces that can induce protein denaturation [48]. Pumping can also contribute to introduction of foreign particles that can further trigger protein aggregation. Similarly, during the filling operation, the protein solution may experience high shear force as well as interaction with different contact surfaces corresponding to containers (vials, syringes) or components (stoppers, plungers). Siliconized stoppers have been reported [29] to contribute to protein aggregation and particulate formation. Foreign-particle introduction during the filling process can not only adversely impact protein stability but also create immunogenicity risks [30,31]. High levels of protein aggregation and precipitation have been reported [26] for therapeutic antibodies exposed to high shear in presence of solid–liquid interfaces. In addition to these mechanical stresses, prolonged exposure to temperature and light outside the recommended window could also result in protein degradation. The impact of light on protein products is discussed in Section 37.3.7. Duration of formulation and fill processes, coatings on product contact surfaces, residual moisture levels in stoppers for lyophilized products; fill parameters, including fill speed and nozzle design; fill configuration, including the vial size; and the fill volume are some of the other process parameters that may have a direct bearing on the stability of the final drug product.

Small-scale characterization should therefore be conducted to evaluate the impact of these stresses on protein product prior to manufacturing. Hold-time studies under the worst case conditions of maximum exposure time and maximum contact surface area per unit volume of protein solution can be designed to assess the impact on protein stability. Scale-down mixing process design to generate representative levels of bulk and impeller shear as well as fluid turbulence can be used to mimic large-scale process conditions. Pilot-scale filler can be used to predict process performance as well as its impact on product quality. If such scaled-down models are not available, rheometers can be used to generate high shear under controlled conditions to evaluate the impact of mixing and filling shear on product. The impact of shear on product quality could, however, be coupled to the solid–liquid and air–liquid interfaces created during processing. It is therefore recommended that manufacturing process conditions be mimicked as closely as possible while performing process characterization.

37.3.3. Filtration

Drug product fill–finish operations invariably feature one or more filtration steps to ensure product sterility. An aspect of filtration that may impact product quality is adsorption of the protein on the membrane surface, resulting in either misfolding on the membrane surface or protein loss. This loss could be significant for low-concentration products, especially if the batch size is relatively small or if the bulk is not being pooled after filtration, prior to filling. Further, high transmembrane pressure could stress the protein while pushing it through the filter pores, leading to protein denaturation. It is therefore important to study the compatibility of the drug product to the membrane material.

Another factor to consider during filtration is the formulation composition. In particular, surfactants can become adsorbed on the membrane surface, causing the surfactant concentration in the solution to go below the target, which might result in product destabilization or deviation from the composition specification. Interestingly, a study by Mahler et al. [32] reported minimal loss of polysorbate 20 due to adsorption on filter membrane and also suggested that presence of protein in the formulation could influence surfactant concentration during the dialysis process. Although such losses may not be significant, it is advisable to test polysorbate concentrations under the process conditions at final scale. While a larger filter area reduces filtration time, it also maximizes protein and excipient losses associated with membrane adsorption.

It is therefore advisable for scale-down studies to be conducted prior to large-scale processing to assess the impact of filtration on product quality and to recommend the optimum filter size and the membrane type for the manufacturing process. Other process parameters to be monitored during filtration for their impact to product quality include the temperature of the bulk and the rate of liquid flow.

37.3.4. Lyophilization

In many cases, protein therapeutics are lyophilized to improve their stability toward physical and covalent degradation phenomena that may manifest in the liquid state (for expanded reviews, see Refs. 33–35). The process of freeze drying, also called *lyophilization*, removes most of the water in the drug product, except that presumably associated with the protein (which is typically less than 1%). Since the presence of water is needed for most covalent degradation phenomena, such as residue fragmentation and isomerization events, the process of lyophilization is often an effective way of overcoming these instabilities. Further, since diffusion is minimized in the solid compared to the liquid state, phenomena such as aggregation, which require diffusive encounters between monomers, can also be effectively mitigated.

However, the process of lyophilization, especially when not optimized, can also introduce instability in proteins, typically manifested by either an irreversible change in structure, or by greater levels of aggregation in the lyophilized product [36]. It is necessary to gain an understanding of the fundamentals of lyophilization to be able to design and optimize a lyophilization cycle. Part III of this book is devoted to describing development of lyophilized formulations. An overview of lyophilized formulation development is provided in Chapters 17, 19, and 20, while Chapter 18 and 21 discuss protein conformation and structural analysis methods in the freeze–dried state. Chapter 22 then combines these elements to discuss the impact of process and formulation parameters on the product profile, while in Part IV Chapters 26 and 27 discuss scale-up, technology transfer, and process development leading to large-scale production for lyophilized formulations.

37.3.5. Inspection, Labeling, and Packaging

Filled vials are crimped and inspected to ensure the absence of cosmetic defects, critical defects [such as use container–closure integrity (CCI)], and visible particulates. The latter previously focused only on foreign intrinsic or extrinsic matter, but with the use of biotech(nological) proteins, especially at elevated concentrations, protein particles can result in turbidity or visible particles. These can occur at such low levels that often the concentration of soluble protein does not decrease as detectable within the error of any assay. The use of forensic tools in identifying the nature of the particulates as pure proteins or as comingled forms is critical in establishing the associated risk and cause- and-effect analysis. Manual or automated inspection can be used at this stage to inspect the filled vials or syringes 100% for presence of visible particles. A confounding factor is the time-dependent kinetics that can cause the appearance of a newly filled unit to be "clear" at time of release but over time gradually change to "not clear" in all units at the expiry date. Note that most release criteria require testing for homogeneous attributes such as pH, osmolality, color, or a chromatography peak. In contrast, protein particles can impact a very small portion of the soluble fraction and be randomly distributed across a few or all filled units as a function of storage time and temperature. This heterogeneous property requires characterization studies with sufficient units per storage timepoint to be able to detect and quantitate this phenomenon. The product trend toward high-concentration doses with associated higher viscosities can result in entrained microbubbles masquerading as visible particles. Removal of these artifacts requires additional handling by manual or automated inspection machines using tilting, swirling, spinning, and braking manipulations that can impart additional shear-related stress that must be studied in advance.

When switching from manual to automated inspection machines (AIMs), the expectations are for the latter to be equal or greater in performance to human inspectors. This requires extensive evaluation of the manual process inspectors, including a probabilistic definition of what particle size is visible [37] using well-documented procedures, inspection tools, and certified inspectors.

The heterogeneous nature of particles can be a challenge in defining accept and reject criteria, even when obvious considerations such as proven clinical safety and maintenance of full potency are met. Proper characterization of aggregation, subvisible, and visible particles and forensic identification must be in place, combined with qualified methods to enable the manufacturer to detect and investigate trends when abnormal particulation occurs. Statistical uncertainty and nonquantitative analytical tools do impose limitations to measure the size and number of protein particles that behave in a nonideal way if they exhibit an amorphous shape, semitranslucency, and sometimes transient appearance. The latter is the result of the breakup of loosely associated aggregates that move detection in or out of the boundary of what is "visible," making quantitation a challenge. More recent application of flow image analysis using cameras and image analysis software is allowing additional characterization.

The labeling and packaging steps, combined with inspection (both manual and automated), would result in exposing the drug product to light for a certain duration of time. This photoexposure can result in affecting product quality and should be evaluated a priori for the worst-case conditions (maximum intensity and duration). In addition, the interactions between the drug product and the primary container during these steps can also impact product integrity. Light sensitivity and packaging-related issues are discussed further in the following sections.

37.3.6. Drug Product Storage, Transport, and Delivery

37.3.6.1. Drug Product Storage. Following filling into the primary container, inspection, labeling, and packaging, the drug product would be subjected to transportation to the clinical site, storage (at either the manufacturer site or the clinical site), and ultimately drug delivery to the patient. One of the main goals of formulation development is to ensure drug product stability during the abovementioned activities. Once the formulation is recommended, formal stability studies are conducted in a GMP environment (typically in a product quality laboratory) to establish shelf life and expiry dating for the product. For each lot of drug product manufactured, lot release assays would be conducted to ensure product purity, integrity, and safety. Some common problems that might be experienced during storage and transport would include nonconformances due to temperature excursions. A robust formulation development–product characterization program is therefore required to address these deviations.

It has been reported that product–surface interactions with the different contacting materials can significantly impact the stability and efficacy of protein products [24,38]. In the sections below, some common factors that are known to affect the product purity, activity, and shelf life during storage, delivery and transport are summarized.

37.3.6.2. Protein–Silicone Oil Interactions. Protein solutions come in contact with various surfaces during processing and prior to delivery to the patient. The surfaces of primary containers, such as syringe, vials, stoppers, and plungers, are treated with different coatings to minimize protein binding and facilitate their processing. As an example, rubber stoppers are lubricated with silicone to ensure appropriate closure fitting into the vial neck. To ensure smooth movement of the rubber plunger through the glass barrel, prefilled syringes are often coated with silicon oil. While silicone oil is considered relatively inert and its concentration in drug product container is significantly low, the interaction between proteins and silicone oil can induce protein aggregation or product inactivation around the oil droplets [26,39]. The molecular basis of silicone oil–induced aggregation is not completely understood, but in a 1997 study it was documented that no major protein structural changes were involved [39]. Glass surfaces that are commonly coated with silicone have been reported to increase protein adsorption [40]. To improve chemical inertness, alternative coatings in the form of fluoroelastomer have been introduced, in addition to the ability to lubricate [24].

37.3.6.3. Leachables and Extractables. Besides silicone oil, other constituents of the primary container have the potential to contribute to leachables and extractables, at concentrations that can affect protein stability and activity [24,41,42]. By definition, *leachables* are contaminants that have migrated from packaging components (plastic bottles, glass vials, prefilled syringes, rubber stoppers, etc.) or manufacturing equipment (tanks, tubing, etc.) into the product during drug product processing and over prolonged exposure. On the other hand, *extractables* are species that are released from the material of construction of the contacting surfaces following interactions with product over time and are therefore dependent on the solvent and temperature conditions. There are two main concerns with leachables and extractables: a potential contribution to toxicology and an impact on protein stability. Common examples of leachables and extractables include metal ions, peroxides, and aldehydes. Manufacturing equipment surfaces, unpassivated stainless-steel tanks, along with plastic bags, can be sources of metal ions. Peroxides can arise from packaging or impurities in excipients. Both metal ions and peroxides can induce protein degradation through oxidation via the formation of peroxy radicals, which is discussed further in Section 37.3.7. Aldehydes can react directly with the drug product or formulation excipients and are a greater concern for low-concentration products. Surfactants, which often used as stabilizers in liquid formulations, may potentially interact with stoppers and affect product quality.

Besides the material of the construction, the manufacturing history of the components can also contribute to product quality concerns. For example, residual tungsten [43,44] from the glass syringe manufacturing process can result in aggregation of therapeutic proteins. In this case study, the heat used during the molding process contributed to vaporizing the tungsten that settled in the syringe at the base of the needle. When the drug product was exposed to this residual tungsten over long perids of time, visible particle formation resulted. Unless systematically studied, such phenomena could be very difficult to resolve during drug product processing. Finally, mechanical failures in a delivery device can also affect the suitability of drug product for the patient, in addition to issues related to protein formulation, Needle breakage issues or other problems with plastic parts could easily result in dosing failure. Therefore, prior to the final selection of a delivery device, compatibility studies should be conducted to predict possible effects of the drug product container or device on the stability of drug product and the capability of the device to deliver it. As can be expected, higher viscosities associated with high-concentration formulations may cause problems while using delivery devices. Stability studies should be designed to monitor product quality over the shelf life to ensure that the components in contact with the drug product do not have any detrimental effects.

37.3.6.4. Transportation. Once fill and finish operations are complete, the drug product needs to be transported either to various storage sites or to the clinic. Especially for liquid protein drug products, transportation can introduce stresses related to temperature, pressure, vibration, and shock variations, which can pose a substantial risk for product degradation. Most commonly, transportation causes exposure of proteins to air–water interfaces, which can lead to protein misfolding and eventually aggregation, either soluble or insoluble. The *United States Pharmacopoeia* (USP) General Chapter 10 provides guidance on using specifications from the American Society for Testing and Materials document, *Standard Practice for Performance Testing of Shipping Containers and Systems* (ASTM D4169–98), and specifications from the International Safe Transit Association (ISTA) for the evaluation of shipping performance of drug products. A common practice for transportation development is to expose the drug product to laboratory simulation conditions that mimic shipping-induced stresses. To this end, vibrational tables are now available from various manufacturers to simulate transportation stresses, on which the stress profiles using ISTA or ASTM guidelines could be applied. These profiles could be generated using vibration, temperature, and pressure data from multiple shipments sent to different parts of the world. The idea would be to identify key operating parameters that affect the product stability and introduce monitoring and controls over them. It is important to realize that in certain cases simulated transportation studies may not adequately address transportation risks, especially for products that are found to be sensitive to transportation-induced stresses. In such instances, real-world transportation experiments must be conducted.

Transportation studies may need to be conducted at multiple times in the development life cycle of a molecule. It is generally a good idea to test for transportation-related stresses early during formulation development, to minimize risk to product quality at later stages. Adequate analytical testing in the form of visual and sub-visible particle assessments and soluble protein aggregation determination should be conducted. Transportation would not typically be expected to generate covalent modifications in proteins, although testing may be performed for these modifications for added quality assurance. Surfactants are often included in the formulation as they are known to minimize protein interactions with container surfaces, as well as unfolding at air–water interfaces [45]. In 1997 it was reported that protein aggregation during transportation, especially in prefilled syringes, may be induced by the presence of residual silicone oil in the barrel of the syringe [39].

Once the final formulation is determined, transportation validation activities are conducted for that specific product in later stages of its development. These activities are conducted to ensure that the transport process operates as designed when challenged using a simulated shipping protocol. Further, component qualification studies would be required to qualify packaging components as well as transport distribution lanes. Finally, performance qualification activities are conducted with a series of real-world shipments to demonstrate that the transport process performs according to plan.

37.3.6.5. Product Delivery. The most common routes of delivery of protein drugs to patients are subcutaneous, intravenous, or intramuscular [46]. However, depending on the physicochemical characteristics, pharmacology, and intended clinical application of a given protein drug, alternate delivery routes may be necessary [46,47]. Cleland et al. have outlined a useful decision tree for drug delivery, based on disease indication, duration of treatment, and pharmacology, depending in part on whether the patient will be dosed during hospitalizatioin (inpatient format) or in an outpatient format, which includes treatments either at home or in a physician's office. As would be expected, a key issue involved is the integrity of the protein during the manufacturing process for a given delivery route, as well as after delivery to the patient. Another consideration is the eventual protein yield if there are significant losses during delivery. For example, daily delivery of insulin in a vial would require the patient to withdraw it from the vial into a disposable syringe prior to injection. However, for greater convenience, prefilled syringes equipped with autoinjectors (automatic injection mechanisms) are now available. There are even devices with needleless injectors and pen devices that contain drug-loaded cartridges [48,49]. It is possible that during the entire fill–finish–storage–transportation process for the protein in the cartridge or syringe, the protein may be damaged by the processing (for, e.g., use of special pumping systems) or the use of new materials (e.g., syringe surfaces).

In addition, there would be an important need to develop detailed protocols for use of a drug product in a clinic, which would take into account product handling, exposure times at various temperatures (room, refrigerated), and impact of light exposure. Studies would have to be conducted to simulate the delivery of the drug to the patient (such as mixing with intravenous diluents and simulating the administration process via IV lines and pumps). The integrity and recovery of the product would then have to be examined to provide quality assurance to the clinic.

37.3.7. Photostability

During fill and finish operations, including visual inspections and packaging, the drug product is exposed to both natural and artificial light, of varying intensity and duration. There is also light exposure during clinical usage, such as dilution and administration via intravenous bags. Therefore, studies need to be designed and carried out to determine the maximum visible and UV light that can be tolerated by the biologic under consideration without affecting product quality.

In proteins, tryptophan, tyrosine, phenylalanine, and cysteine are residues that are more prone to modification due to light exposure than are other residues [50]. The peptide backbone may also be susceptible to photodegradation [50]. In combination with dissolved oxygen, peroxy radicals are often formed in the presence of light [51]. The following reactions are then possible by the peroxy species: converting methionine to methionine sulphoxide [50,52], cysteines to various sulfones and acids [50,52], and tryptophan to form kyneurenine and other hydroxylated derivatives [50,53–55].

Of all these degraded products, the kyneurenines are colored because of their capacity to absorb longer-wavelength light [56]. The residue phenylalanine is a special case: photolysis results in formation of the phenyl radical cation that can react with water to form ortho-, meta-, and paratyrosine byproducts [53–55]. The most common photodegradation pathway other than amino acid modification is protein aggregation. Photolytic aggregation can result, at least in part, from crosslinking of tyrosyl radicals, arising from tyrosine degradation, at the ortho positions of the tryrosine ring [50]. The dityrosine moiety has a characteristic fluorescence at 410–415 nm, following excitation at 315 nm [57].

Guidelines from the International Conference for Harmonization (ICH) recommends performing photostability testing in a controlled light chamber with one of the following two options for light sources:

1. A single lamp source with wavelengths below 320 nm filtered out, with an output similar to either D65 (international standard for outdoor light) or to ID65 (standard for equivalent window-filtered indoor direct daylight)
2. Two lamp sources, one for fluorescent lighting per ISO 10977 guideline and another for UV in the 320–400 nm range

The ICH Q1B outlines a "confirmatory approach," where an overall illumination level of at least 1.2 million lux hours is recommended, with integrated near-UV energy of $\geq 200\ (\mathrm{W} \cdot \mathrm{h})/\mathrm{m}^2$.

Because of the nature of protein-based therapeutics, it is likely that they may not be photostable under the levels recommended in ICH Q1B. As expected, product degradations represent a loss of quality, and many of them could lead to loss of efficacy of the drug product. Thus, in some cases photodegraded material can be deemed unacceptable in terms of patient consumption because of potential safety concerns. Therefore, there is a need for studies to be conducted, with appropriate analytical measurements to determine the extent of the product degradation, which, in turn, defines the light sensitivity of the product.

It is recommended that these light exposure studies be carried out under conditions that closely mimic a "worst-case scenario" for the light wavelengths, intensities, and total cumulative exposure that the product may encounter during fill and finish operations and patient dosing. The first step would be to determine the different sources of light exposure, such as the bulb types, and their spectral outputs as well.

The lighting conditions could be monitored using a light meter, and the extent of visible and UV light exposure would need to be determined. Once this is done, each process step, such as freeze–thaw, formulation, filling, inspection, and packaging, must be carefully examined to determine the maximum duration and light intensity borne by that step. For instance, while inspection is very brief in duration, the intensity of light exposure can be much greater than room light. Additional estimates are needed on clinical light exposure during end-user administration. These parameters, along with temperature of light exposure, can then be combined to design a light sensitivity study to capture the worst-case drug product manufacturing process.

It may be necessary to perform light-sensitivity studies during various stages of product development. For instance, a scaled-down version of the study, using smaller drug substance containers, may be performed at the time of deciding between the top formulation candidates. It is important to note that formulation components, such as antioxidants, may slow down the photodegradation profile of the protein therapeutic under consideration, while others may accelerate light-induced degradation. Once the final formulation is selected, a study to scale may be contemplated. Finally, additional studies may be required once the drug product manufacturing site is selected.

37.3.8. Cleaning

An effective and consistent cleaning process is mandatory for any biopharmaceutical manufacturing facility. Cleaning programs are designed to meet the regulatory expectations for equipment cleaning and maintenance [58]: "Equipment and utensils shall be cleaned, maintained and sanitized at appropriate intervals to prevent malfunctions or contaminations that would alter the safety, identity, strength, quality or purity of the drug product beyond the official or other established requirements." One of the main objectives of the cleaning operation is to ensure that the residuals from the previous product batch do not cross over to the next product. A robust cleaning process, with its consistency and effectiveness at removing the residual soils from the equipment surface, is critical to ensure product quality. In biotech industry, the cleaning process is designed to remove various process soils, including media, buffer, process intermediates, and purified bulk as well as final drug product.

Before each new-product introduction (NPI), the cleaning cycle is developed and challenged for its adequacy at removing process soils to a residual level that is below the established maximum allowable carryover (MAC) for that process step. Drug product processing is at the very end of the chain of manufacturing processes, and as a result the quality expectations are the highest. The US Food and Drug Administration (FDA) does not establish any specific analytical limit for the level of cleanliness that would be considered "clean enough." It is the responsibility of the manufacturer to establish MAC limits for their cleaning validations programs; these limits [59] should be scientifically justified, analytically verifiable, and achievable and most importantly ensure safety for the end users—the patients. Industry literature on cleaning validation has proposed various guidelines, such as detection limit of 10 ppm (no more than 10 ppm of any product will carry over to another product), a limit of 1 in 1000 of the dose for any biologically active component (no more than 0.01% of the daily therapeutic dose of any product, A, will carry over to the maximum daily dose of the next product, B) and visual verification (no residual amount should be visible on the equipment surfaces after the cleaning has been performed). In general, the cleaning process should be able to meet all these criteria or more specifically, the most stringent one. In addition to the carryover of the product residue, the cleaning process should also address the issue of microbial contamination. In biotech processes, any small remnant of microbial organisms in the equipment can quickly propagate and result in higher contamination levels in the following batch. A combination of different analytical methods such as total organic carbon (TOC) assay, product-specific assays,

HPLC or pH measurement, or bioburden (biological burden) assay can be employed to assess the cleaning status of the equipment.

As with any other unit operation, characterization of the cleaning process is essential to obtain a robust and consistent process. Laboratory-scale characterization work [59,61] can be used to evaluate various process conditions and their impact on the ability to clean process soils deposited over different equipment surfaces. Validation of cleaning process at large scale can benefit from such lab-scale and pilot-scale assessments. Characterizing the design space with respect to key parameters will help in developing an effective cleaning cycle for the commercial facility. The following case study provides an example of use of bench scale studies to characterize the simulated cleaning process.

37.3.8.1. Case Study for Cleaning Process: A Design Space Perspective. Process soils representing the final dosage form (protein drug products) were evaluated in this study [61]. Stainless-steel coupons were used as they represent the majority of the product contact equipment surfaces during drug product processing. The cleaning experiment involved spotting the products on a coupon and subjecting it to cleaning fluid in an agitated immersion bath. Key operating parameters, including temperature of the cleaning fluid (CIP-100), concentration of CIP-100, dirty hold time, and fluid velocity, were varied in a selected range, and the time required to clean the coupon (cleaning time) was measured. These studies were conducted by varying one parameter at a time, and their impact on the cleaning time was evaluated. Such studies can provide useful insights into which parameters have a significant impact on process performance.

It was observed that temperature and CIP-100 concentration were the critical parameters governing the performance of the cleaning cycle. Once the key parameters have been determined, a design of experiments (DOE) can then be performed to further understand the impact of individual parameters, as well as their cross-interaction, on cleanability. A design space can be constructed on the basis of these bench-scale data and may also include identification of potential failure modes. A scale-up analysis should then be performed to show the relevance of this design space generated on small scale to the proposed production scale. Based on the single-parameter screening study, a DOE was conducted to over a range of temperature and CIP-100 concentration values. Figure 37.3 shows the leverage plots analysis conducted to view the significance of the impact of the main effects and cross interaction terms on the final response, the cleaning time. P-value analysis shows strong correlation between the cleaning time and the cross interaction parameter between temperature and concentration. This suggests that the effect of a temperature change on cleaning time is strongly dependent on the value of CIP-100 concentration and vice-versa. Such interaction effects are more appropriately explained through interaction profiles and surface response plots as discussed later.

Figure 37.4 shows the cleaning time variation over the example design space created at lab scale by characterizing the two parameters: temperature and cleaning agent concentration. Such response profiles will have both generic and product specific behavior [61]. In general, longer cleaning times are observed in high temperature

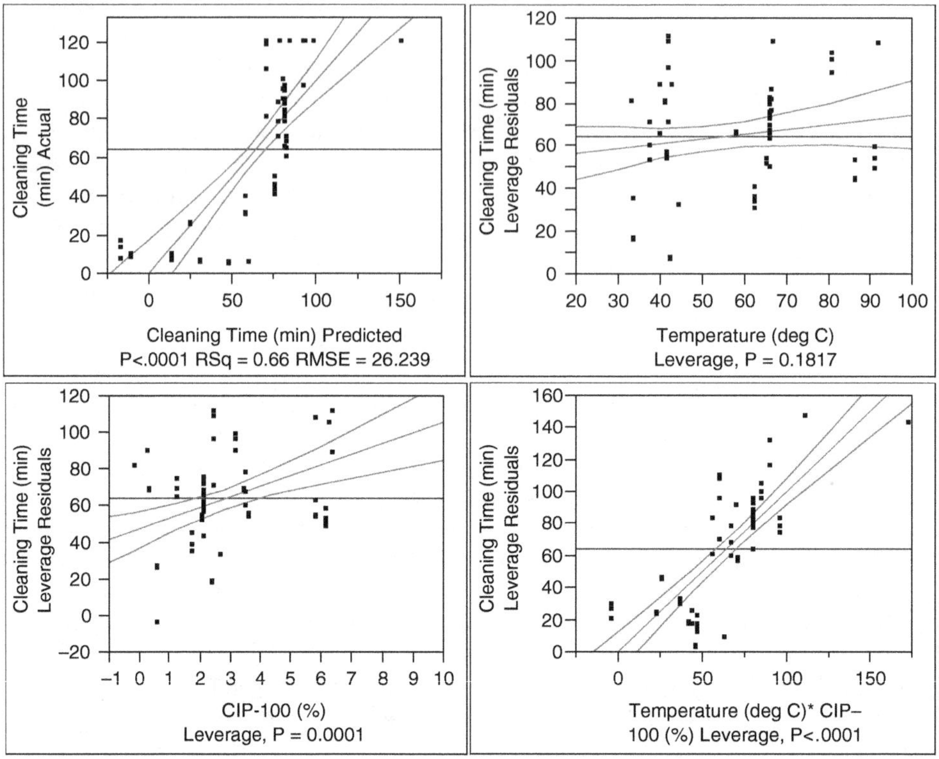

Figure 37.3. Leverage plots for two key parameters: Temperature and Concentration of cleaning solution. The two parameters are strongly coupled as shown by the significance of the cross interaction term.

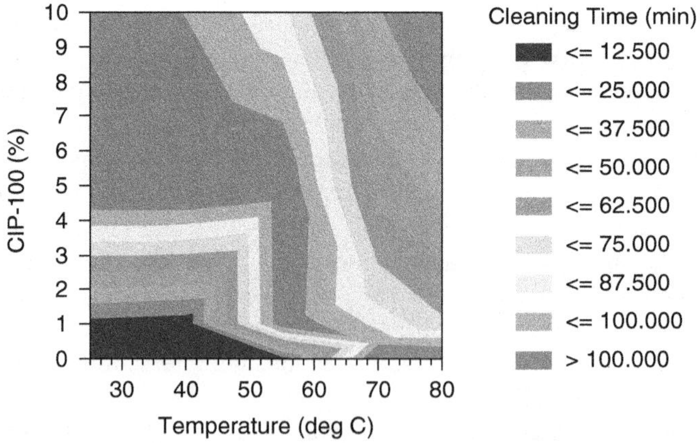

Figure 37.4. Design space characterization for cleaning time with respect to two parameters: Cleaning fluid Temperature and Concentration.

water (0% CIP-100) as well as low temperature and high CIP-100 concentration solutions. Improved performance is observed at low temperatures-low CIP-100 and high temperature-high CIP-100 values. The specific temperature and CIP-100 concentration values at which the response surface changes will be unique to each product depending on their solubility, degradability under cleaning conditions and other physical-chemical properties. It is therefore recommended to perform such bench scale characterization for each new product. These lab scale studies can be followed by appropriate scale-up studies to bridge the performance to large scale process and equipment.

37.4. SUMMARY

The bar has been raised in recent years as a result of advances in molecular design, analysis, quality levels, and evolving regulatory expectations. BLA submission is expected to include a holistic understanding of the risk-managed process based on the tripartite Q8, Q9, and Q10 harmonized guidelines. Benefits such as less cumbersome postapproval changes within the design space are reachable, provided a mechanistic understanding exists for all CQAs. This chapter seeks to capture our current understanding of various factors associated with formulation, fill, and finish processing that can impact the stability and quality of the drug product. Protein products can exhibit damage through a combination of covalent and physical degradations. The stability of these molecules should therefore be characterized over a range of process conditions achieved through well-designed scale-down and large-scale studies. Process operating conditions, quality of aseptic environment, quality of raw materials, and quality of various components, as well as the manufacturing history of the process, are pivotal in determining the purity, activity, and efficacy of the drug product and should therefore be included in the process characterization plan.

REFERENCES

1. ICH (2005), ICH Harmonised Tripartite Guideline: *Pharmaceutical Development*, Q8, version 4, Nov. 10, 2005.
2. ICH (2007), ICH Harmonised Tripartite Draft Consensus Guideline: *Pharmaceutical Development Annex to Q8*, version 2, Nov. 1, 2007.
3. ICH (2005), ICH Harmonized Tripartite Guideline: *Quality Risk Management*, version 4, Nov. 9, 2005.
4. ICH (2007), ICH Draft Consensus Guideline: *Pharmaceutical Quality system*, Q10, version 2, May 9, 2007.
5. Schellekens, H. (2004), *Nat. Biotechnol.* **22**: 1357–1359.
6. Gryna, F. M., Chua, R. C. H., and DeFeo, J. A., eds. (2007), *Juran's Quality Planning and Analysis*, 5th ed., McGraw-Hill, New York (see Section 15.11 and Table 15.2 on defect categorization).
7. Schellekens, H. and Jiskoot, W. (2003), *Nat. Biotechnol.* **21**: 959–957, **21**: 1265–1266.
8. Patro, S. Y., Freund, E., and Chang, B. S. (2002), Protein formulation and fill-finish operations, *Biotechnology Annu. Rev.* **8**: 55–84.

9. Wang, W. (1999), Instability, stabilization and formulation of liquid protein pharmaceuticals, *Int. J. Pharm.* **185**: 125–188.
10. Wang, W., Singh, S., Zeng, D. L., King, K., and Nema, S. (2007), Antibody structure, instability and formulation, *J. Pharm. Sci.* **96**(1): 1–26.
11. Dobson, C. M. (2004), Principles of protein folding, misfolding and aggregation, *Semin. Cell. Devel. Biol.* **15**(1): 3–16.
12. Wetzel, R. (2006), Kinetics and thermodynamics of amyloid fibril assembly, *Acc. Chem. Res.* **39**(9): 671–679.
13. Webb, S. D., Webb, J. N., Hughes, T. G., Sesin, D. F., and Kincaid, A. C. (2002), Freezing biopharmaceutical using common techniques and the magnitude of bulk scale freeze-concentration, *BioPharm* **15**(5): 22–34.
14. Lashmar, U. T., Vnanderburgh, M., and Little, S. J. (2007), Bulk freeze-thaw of macromolecules—effect of cryoconcentration on their formulation and stability, *BioProcess Int.* **5**(6): 44–54.
15. Glaser, V. (2005), Addressing stability of biological drugs, *Genet. Eng. Biotechnol. News* **25**(6).
16. Pikal-Cleland, K. A., Cleland, J. L., Anchordoquy T. J., and Carpenter, J. F. (2002), Effect of glycine on pH changes and protein stability during freeze-thawing in phosphate buffer systems, *J. Pharm. Sci.* **91**(9): 1969–1979.
17. Privalov, P. L. (1990), Cold denaturation of proteins, *Crit. Rev. Biochem. Mol. Biol.* **24**: 281–305.
18. Jayachithra, K., Kumar, T. K. S., Lu, T., Yu, C., and Chin, D. (2005), Cold instability of aponecarzinostatin and its stabilization by labile chromophore, *Biophys. J.* **88**: 4252–4261.
19. Antonino, L. C., Kautz, R. A., Nakano, T., Fox, R. O., and Fink, A. L. (1991), Cold denaturation and $2H_2O$ stabilization of a staphylococcal nuclease mutant, *Proc. Natl. Acad. Sci. USA* **88**: 7715–7718.
20. Cao, E., Chen, Y., Cui, Z., and Foster, P. R. (2003), Effect of freezing and thawing rates on denaturation of protein in aqueous solutions, *Biotechnol. Bioeng.* **82**(6): 684–690.
21. Strambini, G. B. and Gabellieri, E. (1996), Protein in frozen solutions: Evidence of ice-induced partial unfolding, *Biophys. J.* **70**(2): 971–976.
22. Hsu, C. C., Nguyen, H. M., Yeung, D. A., Brooks, D. A., Koe, G. S., Bewley, T. A., and Pearlman, R. (1995), Surface denaturation at solid-void interface—a possible pathway by which opalescent particulates form during the storage of lyophilized tissue-type plasminogen activator at high temperatures, *Pharm. Res.* **12**(1): 69–77.
23. Charm, S. E. and Wong, B. L. (1981), Shear effects on enzymes, *Enzyme Microb. Technol.* **3**: 111–118.
24. Colombie, S., Gaunand, A., and Lindet, B. (2001), Lysozyme inactivation under mechanical stirring: Effect of physical and molecular interfaces, *Enzyme Microb. Technol.* **28**: 820–826.
25. Mahler, H. C., Muller, R., Frie, W., Delille, A. and Matheus, S. (2005), Induction and analysis of aggregates in a liquid IgG1-antibody formulation, *Eur. J. Pharm. Biopharm.* **59**: 405–417.
26. Biddlecombe, J. G., Craig, A. V., Zhang, H., Uddin, S., Mulot, S., Fish, B. C., and Bracewell, D. G. (2007), Determining antibody stability: Creation of solid-liquid interfacial effects within a high shear environment, *Biotechnol. Progress* **23**: 1218–1222.

27. Bee, J., Stevenson, J., Chatterjee, K., Freund, E., Carpenter, J. F., and Randolph, T. W. (2007), Aggregation of a monoclonal antibody induced by adsorption to microparticle surfaces, *Abstracts for Papers, 234 ACS Natl. Meeting*, Boston, 2007.

28. Maa, Y. and Hsu, C. C. (1997), Protein denaturation by combined effect of shear and air-liquid interface, *Biotechnol. Bioeng.* **54**(16): 503–512.

29. Gabrielson, J., Bates, D. G., Williams, B. M., Hamel, J.-B., Carpenter, J. F., and Randolph, T. W. (2007), Silicone oil contamination of therapeutic protein formulations: Surfactant and protein effects, *Abstracts for Papers, 234 ACS Natl. Meeting*, Boston, 2007.

30. DeGrazio, F. L. (2006), Focus on fill and finish: Parenteral packaging concerns for biotech drugs—compatibility is key, *BioProcess Int.* **4**(2): 12–16.

31. Norde, W. and Giacomelli, C. E. (2000), BSA structural changes during homomolecular exchange between the adsorbed and the dissolved states, *J. Biotechnol.* **79**(2000): 259–268.

32. Mahler, H., Printz, M., Kopf, R., Schuller, R., and Müller, R. (2007), Behaviour of polysorbate 20 during dialysis, concentration and filtration using membrane separatin techniques, *J. Pharm. Sci.* **97**(2): 764–774.

33. Abdul-Fattah, A. M., Kalonia, D. S., and Pikal, M. J. (2007), The challenge of drying method selection for protein pharmaceuticals: product quality implications. *J. Pharm. Sci.* **96**(8): 1886–916.

34. MacKenzie, A. P. (1976), The physico-chemical basis for the freeze-drying process. *Devel. Biol. Stand.* **36**: 51–67.

35. Carpenter, J. F., Chang, B. S., Garzon-Rodriguez, W., and Randolph, T. W. (2002), Rational design of stable lyophilized protein formulations: Theory and practice, *Pharm. Biotechnol.* **13**: 109–133.

36. Carpenter, J. F., Prestrelski, S. J., and Dong, A. (1998), Application of infrared spectroscopy to development of stable lyophilized protein formulations, *Eur. J. Pharm. Biopharm.* **45**(3): 231–238.

37. Knapp, J. Z., Kushner, J. K., and Abramson, L. R. (1981), *J. Parenter. Sci. Technol.* **35**(176).

38. Lu, X., and Pikal, M. J. (2004), Freeze-drying of mannitol-trehalose-sodium chloride-based formulations: The impact of annealing on dry layer resistance to mass transfer and cake structure, *Pharm. Devel. Technol.* **9**: 85–95.

39. Carpenter, J. F., Pikal, M. J., Chang, B. S., and Randolph, T. W. (1997), Rational design of stable lyophilized protein formulations: Some practical advice, *Pharm. Res.* **14**(8): 969–975.

40. Carpenter, J. F., Izutsu, K.-I., and Randolph, T. W. (2004), Freezing- and drying-induced perturbations of protein structure and mechanisms of protein protection by stabilizing additives, *Drugs Pharm. Sci.* (special issue freeze-drying/lyophilization of pharmaceutical and biological products) **137**: 147–186.

41. Scott, C. (2006), Formulation development: Making the medicine, *BioProcess Int.* **4**(3): 42–47.

42. Pikal, M. J. (1994), Freeze-drying of proteins: Process, formulation, and stability, in *Formulation and Delivery of Proteins and Peptides*, Cleland, J. L. and Langer, R., eds., *ACS Symp. Ser.* **567**: 120–133.

43. Carpenter, J. F., Prestrelski, S. J., and Dong, A. (1998), Application of infrared spectroscopy to development of stable lyophilized protein formulations, *Eur. J. Pharm. Biopharm.* **45**(3): 231–238.

44. Schwegman, J. J., Carpenter, J. F., and Nail, S. L. (2007), Infrared microscopy for in situ measurement of protein secondary structure during freezing and freeze-drying, *J. Pharm. Sci.* **96**(1): 179–195.
45. Bai, S., Nayar, R., Carpenter, J. F., and Manning, M. C. (2005), Noninvasive determination of protein conformation in the solid state using near infrared (NIR) spectroscopy, *J. Pharm. Sci.* **94**(9): 2030–2038.
46. Cleland, J., Daugherty, A., and Mrsny, R. (2001), Emerging protein delivery methods, *Curr. Opin. Biotechnol.* **12**(2): 212–219.
47. Cleland, J. L. and Langer, R. (1994), Formulation and delivery of proteins and peptides: Design and development strategies, in *Protein Formulations and Drug Delivery*, Cleland, J. L. and Langer, R., eds., *ACS Symp. Ser.* **567**: 1–21.
48. Robertson, K. E., Glazer, N. B., and Campbell, R. K. (2000), The latest developments in insulin injection devices, *Diabet. Educ.* **26**: 135–152.
49. Arslanoglu, I., Saka, N., Bundak, R., Gunoz, H., and Darendeliler, F. (2000), A comparison of the use of premixed insulin in pen-injectors with conventional patient-mixed insulin treatment in children and adolescents with IDDM. Is there a decreased risk of night hypoglycemia? *J. Ped. Endocrin. Metabo.* **13**(3): 313–318.
50. Kerwin, B. A. and Remmele, R. L., Jr. (2007), Protect from light: Photodegradation and protein biologics, *J. Pharm. Sci.* **96**(6): 1468–1479.
51. Davies, M. J. and Dean, R. T. (1997), *Radical-Mediated Protein Oxidation*, Oxford Univ. Press, Oxford, UK.
52. Vogt, W. (1995), Oxidation of methionyl residues in proteins: Tools, targets, and reversal. *Free Radical Biol. Med.* **18**: 93–105.
53. Davies, M. J. and Truscott, R. J. (2001), Photo-oxidation of proteins and its role in cataractogenesis, *J. Photochem. Photobiol. B* **63**: 114–125.
54. Grossweiner, L. I. (1994), Photochemistry of proteins: A review, *Curr. Eye Res.* **3**: 137–144.
55. Laustriat, G. and Hasselmann, C. (1975), Photochemistry of proteins, *Photochem. Photobiol.* **22**: 295–298.
56. Dyer, J. M., Bringans, S. D., and Bryson, W. G. (2006), Characterisation of photo-oxidation products within photoyellowed wool proteins: tryptophan and tyrosine derived chromophores, *Photochem Photobiol Sci.* Jul **5**(7): 698–706. Epub 2006 Jun 7.
57. Amado, R., Aeschbach, R., and Neukom, H. (1984), Dityrosine: In vitro production and characterization, *Meth. Enzymol.* **107**: 377–388.
58. FDA (2004), *Food and Drug Administration Current Good Manufacturing Practice for Finished Pharmaceuticals*, CFR Title 21, Vol. **4**, Section 211.67, US Government Printing Office, Washington, DC.
59. Le Blanc, D. A. (2000), *Validated Cleaning Technologies for Pharmaceutical Manufacturing*, CRC Press, Boca Raton, FL.
60. Rathore, N., Qi, W., and Ji, W. (2008), Cleaning characterization of protein drug products using UV-Vis spectroscopy, *Biotechnol. Progress* **24**(3): 684–690.
61. Rathore, N., Qi, W., Chen, C., and Ji, W. (2009), Bench scale characterization of cleaning process design space for biopharmaceuticals, *BioPharm Intl* **22**(3): 32–44.

INDEX

A
Abatacept, 12
Abciximab, 28
Aβ peptide, 332, 334, 339–340
Activation energy, 461,485
Accelerated stability conditions, 140
Accretropin®, 8
Accuracy, volumetric, *see* Delivered volume
ACF, *see* Autocorrelation function
Actimmune®, 10, 15
Activase®, 11
Adam–Gibbs, 528, 531, 544–546
Adalimumab, 24, 28
ADCC, *see* Antibody-dependent cell-mediated cytotoxicity, 29
Adhesives, staked needle, 903
Adiamycin, 25
Administration:
    SC, 352
    IV, 352
    IM, 352
Advate®, 10, 18
Affilin, 32
Agalsidase β, 11
AGGRESCAN algorithm, 339
Aging, 526–528, 539–537, 548
Agitation, 681
Aggregates, 71, 74
Aggregates:
    irreversible, 429–430, 440, 442, 446
    reversible, 427–428, 431, 444, 447
Aggregation, 33, 258, 259, 260, 278, 280, 292, 294–297, 302, 568, 570–573, 578, 579, 581, 610, 612, 613
    high molar mass aggregate, 298–303
    dimer, 275, 276, 279, 281, 291–293, 299, 300, 302
    trimer, 273, 279, 290, 297
Aggregation-competent state, 340
Aggregation, covalent, 500
Aggregate detection, 232, 245, 247
Aggregation, nonreducible, 501
Aggregation prediction, 331–337, 341
Aggregation, protein, 87
Aggregation rates, 332, 333, 335–337
Aggregation, reducible, 500
Air-water interface, 610
Albumin, 17, 19
Albumin, bovine serum (BSA), 610, 611
Albumin, human serum (HSA), 614, 615
Aldesleukin, 9, 13
Aldurazyme®, 11
Alefacept, 12
Alemtuzumab, 27
Alferon N®, 9
Alglucosidase-α, 11
Alpha-1 proteinase inhibitor, 19
Alphanate, 19
Alteplase, 11, 18
Amevive®, 12, 150
Amorphous, 472
Ammonium sulfate, 682
Amyloid, *see* Fibril
Anakinra, 9, 16
Analytical methods, 358, 359
Analyzer, 526
    impedence, 526
    reflectometry frequency, 526
Analytical characteristics:
    physiochemical, 592,
    biological, 592
Analytical ultracentrifuge, 82

Formulation and Process Development Strategies for Manufacturing Biopharmaceuticals, edited by Jameel and Hershenson

Analytical ultracentrifugation (AUC), 230 –251, 360, 361, 362
Ancestim, 15
Ankyrin Repeat Proteins (ARPs), 31
Annealing, 520–521, 540–542, 544–545, 548, 574, 576–580, 721–722, 772, 641, 642, 660, 827
ANS:
  binding, 215–222
  fluorescence, 215–222
Antegren®, 27
Anti-IgE- Fc antibody, 350
Antibodies, 339, 606, 611, 612, 613, 614
Antibody products, approved, 384, 385
Antibody structure, 21
Antibody-dependent cell-mediated cytotoxicity (ADCC), 23
Antibody-dependent cell-mediated phagocytosis (ADCP), 23
Anticalins, 31
Antihemophilic factor, 10, 18
Antihemophilic factor VIII, 4
Antihemophilic factor, plasma/albumin free, 10
Antithrombin III, 11
Apidra®, 7, 8
Aralast, 19
Aranesp®, 9, 15
Aromatic-aromatic interactions, 330–331
Arrhenius, 524–525
Arrhenius kinetics, 139
Asp isomerization, 365
Aspartate proteolysis, 496
Aspartate isomerization, 44
Association constant, 431
Association:
  hetero, 281–284
  measurement of, 272, 278, 279–284
  self-, 280, 282, 284
Attenuated total reflectance (ATR), 551, 552
Atomization, 711–715, 720–721, 727, 729
Atomizer, 741
  rotary, 742
  ultrasonic, 742, 743
ATryn®, 11, 19
Autoantibodies, 107
Autocorrelation function, 286, 288, 296
Autoinjector(s), *see* Device(s)
Automated inspection, 926
Avastin®, 24, 27
Avimers, 31, 32
Avonex®, 10, 15
Azurin:
  denaturation, 222–226
  fluorescence, 223
  mutants, 219–222
  phosphorescence, 208–215

B

Basiliximab, 26
Becaplermin, 10, 17
Benefix®, 10
Benzaldehyde, 876
Benzothiazole, 876
2-(2-Butoxyethoxy) ethyl Acetate (BEEA), 863
Beta elimination, 47, 138, 501
Beta glucocerebrosidase, 19
Beta trefold, 16
Betaseron®, 10, 15
Bevacizumab, 24, 27
Bexxar ®, 24, 25, 26
β- Sheet, 331, 333, 341
B cell, 107
Biologically active drug-level assays, 117
Bioprocess containers, 679
Biophysical approaches, 130–131
Biophysical characterization of biopharmaceuticals, 175–178
Biosensor, 113
Blood-brain barrier, 34
BMP-7, 17
Brain-derived neurotrophic factor (BDNF), 16
Brittleness temperature, 678, 688
Bovine serum albumin, 275, 279, 294–303
Bubble dryer, 741
Buffer:
  (bi)carbonate, 510, 515
  CHES, 509
  citrate, 502, 509, 510, 513
  glycine, 515
  glycolate, 511
  L-histidine, 508, 515
  phosphate, 510, 511, 513, 514
  succinate, 508, 511
  TRIS, 509, 515
Bulking agent, 464
Bulk drying, 355, 699
Butylated hydroxytoluene (BHT), 863

C

Calcitonin-salmon, 9, 13
Calicheamicin, 25
Calorimetery:
  isothermal, 525
  scanning, 525
Cam, 840–841
Campath®, 27
Cannula(e), 904
CAN-BD, 742, 743

Capillary flow, 710
Capsomeres, 253–254
Carboys, 678
Celsius-Pak, 688
Cam path, 606
Ceprotin, 20
Cerezyme®, 11, 19
Certolizumab pegol, 28, 29
Cetuximab, 26
Challenges, 353
  analytical, 353
  manufacturing, 353
  stability, 353
Chamber pressure, 776
Characterization of an immune response, 114
Chemical degradation, 364, 365, 494
Chemical instabilities, 135–138
  analytical methods to characterize, 138
Chemical stability, 568, 570, 580
Chemical stability of lyophiles, role of:
  acid-base relationships, 508, 510, 512, 513
  buffer concentration, 511
  buffer crystallization, 510, 511
  media effect, 510, 512
  metal counterion, 511
  molecular mobility, 509, 510
  structure, 514
Chemical modification, 74
Chimeric antibodies, 24
Chimeric fab, 28
Choked flow, 780, 787
Chromatography, 605–606
Chromatography, size exclusion, 274–276, 287, 288, 295, 299–302
Choriogonadotropin α, 8
Chromatography, 441–442
Chromatographic peak resolution factor (Rs), 865–866
Cimzia®, 28, 29
Cinryze, 19
Circular dichroism (CD), 79, 175–176, 179–180, 189–194, 196–199
Closures, tip, 898
Colloidal stability, 72
Comparability, 593,
Component preparation, 597
Complement-dependent cytotoxicity (CDC), 23
Composition heterogeneity, 571, 575, 581
Confirmatory assay, 113, 115
Conformational change, 549, 555, 559
Conformation, molecular, 272, 273, 277, 295, 299, 301, 303
Conformational switches, 341
Conjugate analysis, 273, 277, 278
Container-closure system, 148
Container-closure:
  in devices, 912
  integrity of, 905
  testing, 905
Construction, materials of, 902, 903
Control space, lyophilization, 821, 822, 823
Correlation coefficient, spectral correlation coefficient, 567, 572 573, 581
Correlation Rate, 287, 289, 291, 292
Cosmetic defect(s), 852–853
Covalent characterization, 401
Covalent degradation, 396
Covalent stabilization, 406
Circular dichroism (CD) spectroscopy, 609
Cleaning, 934
Class switching of antibodies, 107
Clipping, 45
Closed cell structure, 723, 727
Cl esterase inhibitor, 19
Coagulation factor IX, 10
Coagulation factor VIIa, 10
Coefficient, 531–532
Collapse, 460,
Collapse study, 829
Collapse temperature, 765
Colloidal Stability, 706
Cold labile, 472
Cold denaturation, 462, 643, 643, 648–650
Compatibility studies, 166–167
Computational fluid dynamics, 691
Condensation, 498
Concentration polarization layer, 354
Condensation reactions, 55
Configuration, 522–524, 534
Configurational, 522
  entropy, 524–525
  mobility, 523
  motion, 527, 540
  rearrangements, 523
Conduction, 707–708
Container closure integrity, 887, 893
Control strategies, 164–167
Controlled rate freeze thaw, 687
Crosslinking, 500
Coupling, 523, 546
Convective drying, 708
Convective flow, 652
Critical quality attributes (CQA), 919, 921
Cryoconcentration, 461, 627, 628, 610, 633–634, 638–639, 641, 643, 645, 649–650, 656, 659–661, 665, 683
Cryopreservatives, 148, 683
Cryoprotection, 650–652

Cryoprotectants, 472
  glycerol, 213–214
  sucrose, 213–214
Cryo-transmission electron microscopy (cryo-TEM), 254, 263–265
CryoTrol, 687
Crystalline, 522
  non, 534
  phase, 523, 544
  solids, 523, 534
  state, 521, 523, 530
Crystallization, 464, 627, 630, 635, 639, 642, 644, 646, 647, 770
Crystalline-matrix-type, 827–828
Crystallization and annealing temperature, 764
Cryovessels, 687
Cryowedge, 691
Cumulants analysis, 289, 292
Cyclone, 714–715, 730, 739
Cystine knot, 16, 32
Cystine knot cytokines, 17
Cytokines, 13

D

Daclizumab, 24, 27
Darbepoetin alfa, 9, 15
Daunomycin, 25
Dead volumes, 352
Deamidation, 42, 149, 494
  effect on physical stability, 44
Debye equation, 437
Debye-Huckel theory, 127
Degradation, 529, 531–532, 549, 557, 589, 592
  chemical, 530, 531
  mechanism, 539
  pharmaceutical, 527
  processes, 521, 538, 539
  rate, 528, 531–532, 538, 546
  reaction, 528, 530
Degradation products, 73
Delivered volume
  accuracy of, 907
Delivery, product, 932
Deming, W. Edwards, 320
Denaturation, 460, 462
Denatured forms, 71
Dendritic cells, 106
Denileukin diftitox, 9, 20
Denosumab, 23
Depyrogenation, 892
Design of experiments (DOE), 162–163, 167–170, 313–314, 319–322, 326
Design space, 602, 918, 935
Design space, lyophilization, 813
Design variables, 323
Desiccation, 643, 648, 650
Detector:
  concentration, differential refractometer, 274, 275–277, 299, 300
  concentration, UV absorbance, 274, 277, 281, 284, 299
  dynamic light scattering, 285–288, 291, 292, 295
  multiple, 277
  single photon counting, 297
  static light scattering, 270, 274–278, 281, 283, 285, 299
Dewpoint, 783
D Free-radical oxidation, 50
Diabody, 22, 30
dibenzylamine, 876
dibutylphthalate (DBP), 863
2,4-dichlorbenzoic acid (DCBA), 863
9,9-dimethyl-9,10-dihydroacridine, 876
dioctylphthalate (DOP), 863
Diphenylamine, 876
Dibotermin-α, 10, 17
Diffusion, 531–532
  coefficient, 524, 529, 531–532
  controlled reaction, 531
  in food systems, 546
  jump, 524
  relaxation time, 524
  rotational, 535
  of water, 535
Diffusivity, 710
Diffusion coefficient, 233, 238, 243, 284, 285, 287, 293, 294
Diffuse reflectance infrared Fourier transform (DRIFT), 551
Differential scanning calorimetry, 82, 135, 141, 178, 179, 180, 196
Diketopiperazine formation, 497
Dimerization, 293
Disposable bags:
  ethylene vinyl alcohol, 885
  low-density polyethylene (LDPE), 885
  polyester, 885
Disposable technologies, 848
Dissociation constant, 431
Disulfide exchange, 138
Distribution, effects of, 911
Device(s), 912
  clinical performance, 914
  functionality in, 913
  manufacturing, 913
DNase, 18
DOE, 918, 935

Dornase alfa, 11, 18, 19
Dosage forms, 593, 594, 595, 596, 597
Drawback, 851–852
Dried matrices, 551, 554, 556, 559
Dripping, 848–849, 851, 852
Drotrecogin- α, 11
Drug product, 881, 889
Drug substance, 590, 591, 597, 881
Dynepo®, 9
Disaccharides,462
Drying, *see* Lyophilization
  additional, 741
  bubble, 743
  spray, 739
Drying method, 565, 572–574
Drying phase, 463
Drying techniques, 355
Dynamics, 521–523, 533, 535, 537–538, 543–544
  fast, 523, 538–539, 547–548, 543
  global, 523, 538
  local, 533, 535–536, 539, 541
  molecular, 521, 523, 526, 534, 538, 545, 547
Dynamic mechanical analysis, 527
Dynamic light scattering (DLS), 175, 178, 179, 180
Dynamic thawing, 684

E

Early immune response, 106
Eculizumab, 28
Efalizumab, 27
Effector functions, 23
Effective freeze time, 689
Electrostatic interactions, 121
Electron spectroscopy for chemical analysis, 571
Elaprase®, 11
Electrochemiluminescence assay (ECL), 112
Elitek®, 11
Elastomer(s)
  butyl rubber, 892
  butyl rubber, halogenated, 892
  polyisoprene, 892
EMEA, 859
Empirical phase diagram (EPD), 174, 178–180, 196–197
  dynamic, 195
  ionic strength versus pH, 193–194
  phase boundary, apparent, 179, 193, 194, 196
  temperature versus pH, 180, 189
  temperature versus concentration, 191–193
Enbrel®, 12, 50 20
Entanercept, 21
Enthalpy, 522, 526
  relaxation, 526, 530, 544–545, 548
Enzyme-linked immunosorbent assay (ELISA), 112
Epitope spreading, 107
Epoetin alfa, 9, 14
Epoetin beta, 15
Epoetin delta, 9
Epogen®, 9, 14
Eprex®, 9, 14
Erbitux ®, 26
Ergodic, 522, 543
  nonegodic, 522
Erythropoietin, 14
Etanercept, 12, 20
Ethylene oxide, 904
  residual, 905
Evithrom, 20
Extractable, 860, 906, 930,
Excipients, 464
Excipient robustness, 168–169
Excluded volume (crowding), 362, 363, 364, 367, 369
Eutectic, 625, 628–630, 640, 644
Eutectic mixture, 765
Eutectic temperature, 765
Excipient(s), 174, 197. S*ee also* Stabilizer(s)
Excluded solutes, 693
Experiments by design (EbD), 320–321, 326
Experimental trial, 323
Extractables, 679, 888, 894

F

Fab fragments, 22, 25, 29, 30
Fabrazyme®, 11
Factor VIII, 17, 18
Factorial designs, 167–169
Factor XIII, 607
FDA, 859
Fast dynamics, 569–571, 578–580
Fcγ receptors, 23
FGF-1, 16
FGF-2, 16
FGF-7/KGF (keratinocyte growth factor), 16
Fiblast, 16
Fibril packing, 336
Fibril structure, 336
Fibrillogenesis, 330, 334, 336
Fibroblast growth factor (FGF), 16, 610
Fibronectin, 31
Fibronectin type III domain (FN3), 31
Fick's Second Law, 710
Fictive temperature, 522, 527, 541
Filgrastim, 9, 13
Fill: accuracy, 846, 848, 854
  tolerances, 851
  weight, 851–852

Filler, *see also* Pump
  mass flow, 846
  needle(s), 841–847
  parameters, 851
  peristaltic pump, 843
  piston pump, 840, 850
  system(s), 840–847
  technology, 839
  time-over-pressure, 840, 844
Fill-finish, 922
Filling, 925
Filtration, 600, 926
Fluid bed, 707–708, 732–733
  azurin, 221–224
  ANS, 213–220
  fluorescence, 359, 361
  flushes, 355
  spectrum, 221, 214
  yield, 213–220
Field flow fractionation (FFF), 253–266
  accumulation wall, 255–256
  asymmetric flow (A-Fl), symmetric flow (Fl), 255
  channel flow, laminar, 255, 256
  channel height, 256
  crossflow, 255, 256
  diffusion coefficients, 255
  elution profile(s), absorbance, light scattering, 257, 259, 260
  method optimization, 258–261
  particle size, 256
  retention time, 255–256
  resolution, 256, 258–260
Fluorescence spectroscopy, 81, 175, 176–177, 178, 179, 196, 197
  8-anilino-1-naphthalenesulfonate (ANS), 176, 178, 189
  Congo red, 177
  propidium iodide, 177,
  red edge excitation, 195
  thioflavine T/S, 177
  tryptophan/intrinsic, 176–179, 189, 192, 196, 198
  YOYO-1, 177, 193
Fluorescence spectroscopy, solid state, 559
Foaming, 853–854
Follistim®, 9
Follitropin α, 8
Follitropin β, 9
Fourier transform infrared spectroscopy (FTIR), 551
Fill volume, 686
First-order irreversible/kinetic thermal events, 769
Foam-dried, 575–579
Foam drying, 709, 711, 722–723
Forced degradation studies, 162–166
Force, extrusion, 909. *See also* Performance, functional
  cannula diameter, effect of, 909
Formulation, 827–828
Formulation of biopharmaceuticals, 173–174
Formulation characterization, 161–170
Formulation development, 309, 311–313, 317, 322
Formulation, high concentration, 405
Formulation process, 412
Formulation robustness, 161–170
Formulation for freeze drying, 461
Forteo®, 9
Fortical®, 9
Four-helix bundle, 13
Fourier transform infrared (FTIR), 175, 178, 195, 196
Fox equation, 765
Fractionation, 274, 275, 278, 280, 284, 287, 295, 299, 300
Fractional evaporation, 713
Fragility, 524, 529, 531
Freezing, 461, 770, 775, 828–830
Freezing, lyophilization, 813, 814
Freezing:
  active technologies, 662–664
  alternate technologies, 664–665
  containers, 656–659
  controlled rate, 662
  large-scale, 656–658
  passive technologies, 662–664
  rate, 630, 639, 645
  scale-down model, 660–662
Freeze concentration, 607, 630, 631, 633, 639, 645
Freeze-drying, 555, 556, 829–830, 832, 834
Freeze-dried, 567, 568, 569–578, 580
Freeze concentrate, 461, 770
Freeze drying microscopy, 767
Freeze drying properties, 764
Freeze-thaw, 165–166, 924
Freeze-thaw cycling, 606–608
Freeze-thaw stress, 393
Freezing methods, 775
Freezing rate, 683, 775
Frictional coefficient, 233
FT Raman, or Fourier transform Raman, 80
Fourier transform infrared (FTIR), 80
Fusion protein, 20, 699

G

Galsulfase, 11
Gamimune, 17

Gatekeeper residues, 339–341
Gaucher's disease, 19
GEM 21S, 17
Gemtuzumab ozogamicin (calicheamicin), 25, 27
Genotropin®, 8
Gibbs–Helmholtz equation, 125
Glass, 472, 768
  borosilicate, 883, 889
  defect structures, 891
  delamination, 891
  soda-lime, 890
  sulfate treatment, 891
  surface alkali, 891
  type I, 883, 890
  type II, 883, 890
  vials, molded, 890
  vials, tubing, 890
Glass dynamics, 569, 572
Glassy matrix, 631–632
Glass transition, 630–633
Glass transition temperature (Tg'), 468, 608
  mixtures, 508
  buffers, 508
  pH pre-lyophilization solution, 509
  salt vs free acid, 508, 509
Gliding force, *see* Force, extrusion
Global:
  dynamics, 523, 541, 543, 707
  mobility, 530, 538, 539, 541, 544
  motions, 526, 539
  relaxation, 523
Global motions, 569, 578, 707
Glucagon®, 9, 13
Glutamate, 152
Goats, 19
Golimumab, 28
Gonal-f®, 8
Graduation marks, 907
  accuracy of, 907, 908
Granulocyte colony-stimulating factor (GCSF), 13
Granulocyte macrophage colony-stimulating factor (GMCSF), 13
Growth hormone, 3, 7
Growth hormone, human, 609

H

Hammett acidity function, 514
Handling stresses, 925
Heat, 163–164
Heat transfer rate, 707–709, 778
Hemoglobin, 608
Herceptin®, 24, 27, 150
hGH, *see* Growth hormone, human
High concentration, 447
  antibody solutions of, 447
  formulation, 350, 352, 364, 367
High concentration measurement, 278, 279
High clinical doses, 350
High density polypropylene (HDPP), 867
High molar mass aggregate, 298, 299
Hinge region, 23
Histidine, 149
Hofmeister series, 127
Hook effect, 112
Hydrodynamic analysis, 231, 244–246
Hydrogen-deuterium exchange, 556
Hydrodynamic flow, 708
Hydrodynamic volume, 357
Humalog®, 7, 8
Humate, 19
Humatrope®, 7, 8
Humira (U.S.) ®, 24, 28, 150, 350
Humulin®, 8
Hyaluronidase, 11
Hybridoma, 5
Hydrogen bonding, 122, 474
Hydrogen deuterium exchange, 83
Hydrophobic interaction chromatography (HIC), 87, 606
Hydrophobicity, 330, 332, 333
Hydrophobic effect, 123
Hydrolysis, 41, 149, 495
Hylenex®, 11
Hypersensitivity reactions, 116
  pH, 148
  polysorbate-80, 149
  Raptiva, 150
  Remicade, 150
  rhBMP-2, 153
  Rituxan, 150

I

Ibritumomab Tiuxetan $^{-111}$Inor^{-90}Y, 24, 26
Ice:
  dendritic, 637–639, 657, 665
  interface, 639, 641, 643, 645, 652–653, 660–661, 664
  morphology, 638
Ice-liquid interface, 462, 687
ICH Q1B, 164
Ice-morphology, 775
Ice-water interface, 607
Idursulfase, 11
IFN alfa- 2B, 15
IFN alpha-2, 15
IgG, *see* Antibodies
IgG subclass, 22
Il-2 aldesleukin, 13

Iletin, 3
IL-IRA, 16
Ion exchange chromatography, 606
Ionic strength, 148
Immunogenic reaction, 352
Immunogenicity, 613
Immune response maturation, 107
Immunoglobulin G, 288
Imiglucerase, 11, 19
Immunogenicity, 24, 32, 33
Impurities, 592
Increlex®, 8
Inflammatory response, 106
Infergen®, 9
Infliximab, 24, 26
Infrared microscopy, 556
Infuse, 10, 17
Injectability, 371, 374
Insect cell system, 4
Inspection, 927
Insulin, 5, 6, 8, 443–444
  aspart, 7, 8
  determir, 7, 8
  glargine, 7, 8
  glulisine, 7, 8
  lispro, 7, 8
Insulin deamidation, 495
Insulin dimerization, 501
Insulin-like growth factor, 497
Interactions, 430, 433–434, 436
  electrostatic, 443, 449
  fab–fab intermolecular, 449
  hydrophobic, 432–433, 446
  intermolecular, 449
  nonionic, 432–433
  protein–protein, 440, 448
  van der Waals, 433
Interaction profile, 167–168
Interferon, 15
  Alfa-2a, 9
  Alfa-2b, 9
  Alfacon-1, 9
  Alfa-n3, 9
  Beta-1a, 10, 15
  Beta-1b, 10, 15
  Gamma-1b, 10, 15
Intramuscular, 148
Intravenous, 148
Interfacial stabilization, 650
Interfacial stress, 394
Interference, 360
Interleukin-1 receptor antagonist, 607, 610
Interleukin-1, 16
Interleukin-2, 13
International Conference on Harmonization (ICH), 312
Intermediate forms, *see also* Misfolded species, 71
Intron A®, 9, 15
Intrinsically disordered proteins, 334
Ion Exchange Chromatography, 87
Irreversible aggregate, 71
Isoelectric point (pI), 330
Irradiation, 613
Isoaspartate, 42, 495
Isomerization, 137
Isotonicity, 483

K

Kepinvance®, 10
Keratinocyte growth factor (KGF), 16
Kineret ® (Palifermin), 9, 16
Kinetics, 482, 524
Kinetics, relaxation, 527, 529
Knottin family, 32
Knudsen flow, 710
Kogenate®, 10
Kogenate FS, 10, 18
Kohlrausch–Williams–Watt (KWW) , 525–526, 528–529
Kringle1, 18
Kynurenine, 46

L

Labeling, 927. *See also* Graduation marks
Lubricant(s), 902
  alternative, 912
Lamm equation analysis, 233, 245–246
Lantus®, 7, 8
Laronidase, 11
Last point to freeze, 684
Last point to thaw, 684
Lattice theory, 338
Leachables, 679, 680, 860, 888, 894, 906, 930
Leachables and extractables, 612
Lepirudin, 10
Leukine®, 9, 13
Levemir®, 7, 8
Ligand, 706
Ligand binding/stabilization, 129
Light chain variation, kappa and lambda, 21
Light exposure, 683
Light scattering, static, 436–441
Light scattering (LS), 83
  angular dependence, 257
  cumulants analysis, 288, 289, 291, 292, 296, 297
  dynamic, 269, 284–294, 295–298
  dynamic (DLS), batch-mode 254, 261, 263

limits of detection, 284, 294
molecular weight, 256–257
multiangle, 269, 271–284, 295, 299–304
static, 254, 256–258, 269, 275
particle size analysis, 256, 260
polydispersity index, 261–263, 291, 292, 296–298
Rayleigh-Gans-Debye optical model, 257, 258, 264
refractive index increment (dn/dc), 254
regularization analysis, 288–292, 294
root-mean-square (RMS) radius, 256, 257, 259–261
Z-average diameter, 254, 261–263
Zimm plot, format, 257
Limits of detection, 284, 294
Linker, 25, 30
Liquid dosage form, 351
Lipocalins, 31
Local molecular motion, 707
Log *D*, 851–862
Log *P*, 861
London dispersion forces, 122
Low affinity antibodies, 107, 113
Lucentis®, 28, 29
Lutropin α, 8
Luveris®, 8
Lyophilization, 149, 460,451, 549, 550, 554, 555, 557, 601, 926
annealing, 820
chamber pressure, 799, 802
collapse temperature, 803, 813
condenser, 798, 805
environmental conditions, 799
equipment, 798, 804
glass transition, 813
pirani, 802
pressure rise measurement, 807, 808
primary drying, 816, 820
radiation, 800
scale-up, 796
secondary drying, 818, 820
shelf temperature, 800
sublimation rate, 805
technology transfer, 808
Lyophilized formulation stability, 408
Lyophilization, high concentration, 410
Lyophilization induced stress, 394
Lyoprotectants, 357, 474

M

MAbs:
human, 24
humanized, 24
goats, 19
Maillard reaction, 55, 475, 479, 498, 614
Manifold, 840–841, 844–847
Manometric temperature measurement, 781
Manufacturing process, *See also* Production
cell culture, 590
drug product, 597, 598, 600, 602, 603
fermentation, 910
purification, 591
Mass flow ratio, 714
Mass transfer, 707–711, 723, 731
Mecasermin, 8
Metal-catalyzed oxidation (MCO), 49
Methionine oxidation, 50, 497
Methoxypolyethylene glycol epoetin beta, 9
Melt-back, 460
Membrane:
fouling, 368, 609–610
clogging, 354
Microbodies, 32
MIRCERA®, 9, 15
Microcollapse, 784, 827, 829
Misfolded species, 74. *See also,* Transitional conformation
Mixing, 692
Mobility, 521, 523, 527, 529–533, 535–539, 542, 546
alpha (α), 526–528, 531, 535
beta (β), 526, 533, 537, 539
configurational, 525
critical temperature, 534
global, 532, 540, 543
local, 538, 543
Molecular, 521–522, 524, 526, 527–534, 535, 538–547
NMR, 532
rotational, 533
Modulated differential scanning calorimetry, 767
Moisture content, 833–834
Moisture sorption, 566, 567, 574
Molar mass, 267, 271, 273–278, 281, 282, 284, 288, 294, 295, 297, 299–303
Molecular conformation, 277, 301, 303
Molecular interactions, 231, 234, 238, 241, 244–245
Molecular mobility, 565, 566, 568–573, 578–581
Molecular weight (or molar mass), 232, 236, 238, 241, 242
Monoclonal antibody degradation, 163–164
Mooney equation, 370, 371
Motions, 521, 523, 525–526, 533, 535, 537, 543
alpha (α), 526, 528, 531, 533, 535, 537
beta (β), 528, 535
cooperative, 523, 525

Motions (*Continued*)
- fast, 533
- molecular, 522, 523, 525, 526, 533, 535, 537, 543
- reorientation, 525
- rotational, 523, 526–526, 533
- thermal, 534
- translational, 522, 527

Multiple freeze-thaw, 687
Muromonab-CD3, 5, 26
Mutagenesis, 335, 338–339
Mylotarg ®, 25, 27
Myozyme®, 11

N

Naglazyme®, 11
Natalizumab, 27
Natrecor®, 7, 9
Near infrared spectroscopy (NIR), 554
Nebulizer, 19
Needle(s), *see* Cannula(e)
Nerve growth factor (NGF), 16
Nesiritide, 7, 9
Neulasta®, 9, 14
Neumega®, 9
Neupogen®, 9, 13
Neutron scattering, 533, 535, 537–538, 547
Neurotrophin 3 (NT3), 16
Neutralizing antibody, 114
Neutralizing antibody assay, 114–115
Nominal freeze time, 687
Nonideality, 362, 370, 436
Norditropin®, 8
NovoLog®, 7, 8
NovoSeven®, 10
Nozzle, 741, 850–854
Nozzle, two-fluid, 742
Nplate®, 12, 20
N-terminal cyclization, 46
Nucleation, 629–630, 634–638, 641, 643, 646, 661–662, 664–666, 772
Nucleic acid, 175, 176, 177, 194, 199
Nucleation temperature, 774
Nutropin®, 8

O

Oleamide, 869
Oligomerization, 429, 430–432, 437, 443–444
Omalizumab, 27
Omnitrope®, 8
Oncaspar®, 11
Ontak®, 9, 20
Opalescence, 155
Operating range, 827
Optical density (OD), 175, 177, 191, 192, 197–198
OP-1 Putty, 17
Oprelvekin, 9
Orencia®, 12
Orthoclone OKT-3®, 5, 26
Orthogonal approach, 360
Orifice, 844, 846, 851
Ostwald ripening, 772
Outlet temperature, 712–713, 718, 720
Ovalbumin, 315–316, 331
Overfill, 354
Ovidrel®, 8
Oxidation, 49, 137, 149, 497, 612
- control of, 53
- effect on physical stability, 54

P

Packaging, 927
- styles of, 898

Palifermin, 10
Palivizumab, 27
Palmitic acid, 869
Panitumumab, 24, 28
Packing density, 336
Particle formation, 387
Particle number density, 280, 295
Particle size distribution, 254, 258, 260–263
Particles from gas-saturated solutions (PGSS), 729–730
PASTA algorithm, 333
Pathogen inactivation heat treatment, 613–615
PEG, 14
Pegaspargase, 11
PEGasys®, 10
Pegfilgrastim, 9, 14
Peginterferon alfa-2a, 10
Peginterferon alfa-2b, 10
Pegintron®, 10, 15
PEGylatde Intron A, 15
Pegylation, 14, 15, 20, 29
4-Pentylphenol, 863
Peptibodies, 20, 699
Performance, functional, 908
Performance parameters, 920
Peroxides, 930
pH change, 646, 648
Phase diagram, 626–628, 630
- ice-water, 626
- sodium chloride-water, 627

Phase separation, 632, 642–643, 650
Phenol-2,6-bis(1,1-dimethylethyl)-4-methoxy (BHA), 861
Phosphorescence:
- azurin, 208–214

lifetime, 209–214
spectrum, 211
Phosphorescence spectroscopy, 81
Phosphorylation, 13
Photo acoustic, FTIR, PA-FTIR, 551, 552
Photooxidation, 51
Photo-stability, 51, 164–165, 932
Phylomers, 32
Physical characterization, 396
Physical stability, 124, 567–569, 571–573, 580
Physical stabilization, 403
π-stacking, 331, 332
Pinch valve, 843–847
Pirani gauge, 783
Plasticize, 530, 547
antiplasticizer, 537
Plasticizer, 475, 529, 537, 538, 547, 771
Platelet derived growth factor, 17
Polymers. *See also* Teflon
Crystal Zenith®, 890
cyclic olefin copolymer, 890
polycarbonate, 882, 884
polypropylene, 882, 884
Piston(s):
how supplied, 899
placement of, 910
Plunger(s), *see* Piston(s)
Polydispersity, 291, 292
Poly (ethylene glycol), 4
Polyglycine linker, 20
Polysorbate20, 926
Polyvinylpyrrolidone, 498
Potassium bromide pellet, 551
Precipitate, 71, 74
Preferential exclusion, 126, 473, 650, 682, 706
Preferential hydration, 73
Prediction profile, 168
Pre-formulation, 311, 460
Process parameters, 920
Process validation, 602, 603
Production, 589. *See also* Manufacturing
Product target profile (PTP), 313, 321, 323
Pre-formulation research, 119
Preservatives, 77, 392
Primary drying, 776, 828–830, 833–834
Principal properties, amino acids, 335
Procrit, 14
Process characterization, 686
Product heterogeneity, 412
Product mapping, 829, 834–836
Product resistance, 766, 784
Product temperature, 776, 828, 830, 832–833
Proinsulin, 7
Prokine®, 9, 13
Proleukin®, 9, 13
Protein:
crystallization, 357
denaturation, 222–226
excipients, 660
fluorescence, 223–226
interfacial, 649
kinetic , 648, 659
phosphorescence, 208–214
precipitation, 357
stabilization of, 650, 661
stabilizers, 632
tertiary structure, 208–222
thermodynamic, 650, 660
Protein A chromatography, 606
Protein adsorption, 611–612
Protein C, 20
Protein drug substance, 625
bulk storage, 652–656
Protein hydrolysis, 137
Protein stability, container-closure effect, 406
Protein stability, solution conditions, 391, 392
Protein–protein interaction, 362
Protein stabilization forces, 120
Protein structure, protein secondary structure, 565, 566, 567, 572, 573, 578, 580, 581
Protropin®, 7, 8
Protease, 679
Proteolysis, 44, 493
Pulmozyme®, 11, 19
Pump. *See also* Filler
positive-displacement pump, 840
rolling-diaphragm pump, 842–844
rotary piston pump, 841–842
Purfication, *see* Chromatography
Pyroglutamic acid formation, 46

Q

Quality, 603
Quality by Design (QbD), 320, 327, 813, 814, 917

R

Racemization, 47
Reconstitution, 409
Radiation, 708–709, 723
Radiation, lyophilization, 800
Radical scavengers, 53
Radius:
hydrodynamic, 284, 285, 287, 291, 293, 294, 296, 297, 299
root mean square of gyration, 271–273, 277–279, 290, 301
Rayleigh ratio, 271, 275, 278,
Raman spectroscopy, 551, 555–556

Raman microscopy, 559
Range-finding study, 829
Ranibizumab, 28, 29
Rapid expansion of supercritical solution (RESS), 728–729
Raptiva®, 27, 350
Rasburicase, 11
Rate constant(s), 431–432
Rayleigh ratio, 271
Reaction rate, 363
Reactive oxygen species, 50
Reactivity, structure and, 501
Rebif®, 10
Recombinate®, 10
Reconstitution time, 356
Recothrom®, 11
ReFacto®, 10, 18
Refludan®, 10
Registration, 603
Regranex®, 10, 17
Regularization analysis, 289, 292
Regulatory guidelines packaging, *see* EMEA and FDA.
Relaxation, 531, 533–534, 545–548
  alpha (α), 523–524, 530, 533, 538
  beta (β), 523, 527, 533, 535–537
  biexponential, 538
  cooperative, 528, 544, 546
  dielectric, 526–527, 544, 546
  diffusional, 522
  dynamics, 543
  enthalpy, 526, 530, 544, 547–548
  global, 523
  JG, 533
  kinetics, 525, 527
  NMR, 532
  relaxation time, 535, 546
  spin-lattice
  spin relaxation
  T_1, 533–537
  $T_{1\rho}$533– 535, 537–538, 541
  T_2, 531–533
  primary, 523, 527
  rate, 526, 530, 538
  secondary, 523, 527, 533–534
  structural, 523, 526, 530–531, 538, 541–542, 545
  time, 523–535, 538-541, 544, 547–548
Relaxin, 444–445, 496
Remicade ®, 24, 26
Reopro ®, 28
Response surface, 324–325
Response-surface design, 167
Retavase®, 11, 18
Reteplase, 11, 18
Reynold's number, 718
Reversible association, 359
Rheometry, 362
RhoGam [Rh (rhesus) factor], 17
RhuMAb VEGF, 445–446
Risk assessment, 928
Risk management, 921
Rituxan®, 24, 26
Rituximab, 24, 26
Robust formulation460
Roferon®, 9, 15
Romiplostim, 12, 20

S

Saizen/Serostim®, 8
Salting-out, 682
Sargramostim, 9, 13, 14
Scale-down model, 460
Scale-up, 353, 600, 601, 798
Schott Type I plus®, 891
scFv (single-chain variable fragments), 22, 30
SDS-PAGE, 86
SEC-HPLC, 440
  in conjunction with light scattering, 440
Second derivative UV, 79
Second viral coefficient, 73, 706
Secondary drying, 785, 828–831, 833–834
Second-order reversible thermal events, 769
Sedimentation:
  absorbance optics, 233–235, 242, 241
  concentration dependence, 233, 247
  data analysis, 233, 244–246
  data interpretation, 242
  experimental protocol(s), 242–244
  fluorescence optics, 237–239
  high concentration, 232, 233–234, 239, 240, 247
  interference optics, 236–237, 239, 240, 241, 242
  microfractionator, 234
  partial specific volume, 232, 233, 236, 242
  sample preparation, 241
  sample requirements, 240
Sedimentation coefficient, 233, 238, 243–246,
Sedimentation equilibrium, 361, 362, 433–436
Sedimentation velocity, 231–251
Self-association, 359, 360, 362, 363, 367
  discrete, 435
  indefinite, 431, 435
  mechanism, 430, 432
Sample viscosity, 241, 242

Serum albumin, 4
Set-point, 828, 830
Shear forces, 612–613
Shelf temperature, 780
Shipping. *See* Distribution.
Silicone, 902
alternatives to, 914
droplets, 903
interactions with, 903
variability, 903
Silicone oil, 365, 612, 892, 929
Simponi^{TM}, 28
Simulect ®, 26
Single chain variable fragments (scFv's), 29
Size(s), standard, 898
Shear, 849–850
Single-use containers, 679
Singular value decomposition, 132–134
Size exclusion chromatography, 85
Sodium chloride, 793
Sodium citrate, 149
Sodium phosphate, 149
Solid state nuclear magnetic resonance spectroscopy (ssNMR), 557
Solution enhanced dispersion, 730–731
Solubility, 72
Soluble aggregate, 71
Solute interaction, 76
Solute, partitioning of , 641
Solid-state acidity. *See* Hammett acidity function
Solid-state properties, 565, 564, 571, 572, 573, 574, 581
Solids, lyophilized, 501
Soliris®, 28
Somatostatin, 5
Somatrem, 7, 8
Somatrophin, 5, 7, 8, 499
Specific surface area, 566, 571, 575, 576, 577, 578
Spectroscopy, 132–135, 443
Splashing, 852–853
Spray coating, 732
Spray-dried, 569, 570, 573–581
Spray drying, 460, 711–719, 739
equipment, 741
proteins antibodies and peptides, 744
sterile, 743
vaccines, 752
Spray freeze drying, 719–721
Stability, 540–541, 543, 545–548, 598, 599
accelerated, 541
chemical, 541, 543, 546
pharmaceutical, 541
Stability indicating assay (SIA), 85, 314
Stabilizer(s), 180, 191, 194, 466, 476, 566, 567, 570, 572, 574, 576, 578, 580, 581
high-throughput screening of stabilizers, 174, 196–198
Stainless steel, 881
Stainless steel containers, 676
State diagram, 628–632
Stearic Acid, 869
Sterile water for injection (SWFI), 871–876
Sterilization, 904. *See also* Ethylene oxide
Steel-cup compatibility, 166
Stem cell factor, 15
Stemgen, 15
Storage and shipment, 680
Stokes–Einstein, 285, 524, 531
Stopper(s). *See* Piston(s).
Storage stability, 565–581
Stresses, 88, 681
cold denaturation, 91
dehydration, 91
freezing, 91
heat, 88
light, 90
pH, 92
pressure, 92
surface or shear, 89
Stress conditions, 162
Stress studies, 138–140
Structural analysis, 549, 554, 555
Supercooling, 626, 629, 636–641, 661–662
Sublimation. *See* Lyophilization.
Supercritical antisolvent, 730
Supercritical fluid drying, 727–728
Surface adsorption, 74, 78
Surface composition, 571, 575, 576, 577, 578, 579
Surface induced structural changes, 74, 77,89
Surface properties, 565, 566, 571, 578
Surface tension, 848–849
Surfactants, 89, 94, 392, 469
Structural changes, 70
Subcutaneous (SC), 148, 350
dose, 350
injection volume, 350
Sublimation rate, 778
Succinate, 152
Sucrose, 608, 614
Supercooled:
indomethacin, 544
liquid, 522–524, 529, 533, 540, 543
o-terphenyl, 544
sucrose, 525
Supercritical fluid drying, 707
Surfactant(s), 148, 607, 706, 707, 718, 722, 727

Sugars, 706–707, 713
Svedberg equation, 233, 244
Synagis®, 27, 350
Syringeability, 371, 373, 374
Syringes, prefilled, 612, 839–841

T

Tangential flow filtration (TFF), 354, 699
TANGO algorithm, 334
Technology transfer, 600, 601, 602
Technology transfer, lyophilization, 808
Teflon. *See also* Polymers
  fluorinated ethylenepropylene (FEP), 884
  polyfluoroalkoxy (PFA), 884
  polytetrafluoroethylene (PTFE), 884
  tetrafluoroethylene (TFE), 884
Temperature excursions, 163, 165–166
Tenecteplase, 11, 18
Teriparatide, 9
Tetranectin, 31, 32
Tev-Tropin®, 8
Thawing, 625, 643–646, 652, 656–658, 660–651, 664–665
  rate, 645, 664
Thermal cycling, 165–166
Thermodynamics, 124–126
Thermodynamic stability
  ΔG^0, 224–226
  freezing temperature, 226
  fraction of liquid water, 224
  guanidinium hydrochloride, 222–223
  m values, 222–224
Thermal conductivity, 708
Thermal gravimetric analysis, 786
Thermal history, 565, 572, 573, 581, 705, 733
Thermal resistance, 708–709
Thermally stimulated current (TSC), 526–527, 533
Thermocouples, 781
Thrombin, 20
Thrombin dermal, 11
Tryptophan hydrolysis, 46
Thiol-disulfide exchange, 500, 520
Tg', 478, 688, 770
Tissue Plasminogen Activator (r-tPA), 18
TNKase®, 11, 18
Topworks®, 887
Tositumomab (IgG & IgG-^{131}I), 24, 26
Transportation, 681, 931
Transforming growth factor (TGF β), 16
Trastuzumab, 24, 27
Trehalose, 608
1,3,5-Triazine-2,4-dithione (DBAT), 863
Triple point, 626
Tromethamine (TRIS), 152
Trudexa (EU), 28
Tungsten, 366, 903
  alternatives to, 914
  extractable, 902, 906
  test methods for, 906
Tysabri®, 27

U

Ubiquitin, 32
Ultrafiltration/diafiltration (UF/DF), 609–610
Ultraviolet absorption spectroscopy, 177
  second-derivative, 178, 179, 180, 187–189, 192
Uncontrolled rate freeze-thaw, 683
Under-cooling, 462, 775
Unit operation, 705, 707, 711,720, 722, 723, 729, 732
Urokinase, 614
USP <1231>, 870–871
USP <661>, 860

V

Vaccine, 593, 603
Vaccines, spray drying, 752
Validation, 682
Valtropin®, 8
Vapor pressure of ice, 778
Vascular endothelial growth factor (VEGF), 16
Vectibix^{TM}, 24, 28
Vial(s), 839–841
Vial breakage, 464, 641–642
Vial filling pumping issues, 610–611
Vibration motions, 551, 552, 555
Vial heat transfer coefficient, 778
Virial coefficients, 367
Virus-like particles (VLPs), 253–254, 257–265
Vogel–Tammann–Fulcher (VTF), 524–525, 527, 529
Viscosity, 367, 368, 406, 611, 848–850
Viscous formulations, 352
  intrinsic, 369
  kinematic, 372, 373

W

Water substitution, 567, 572, 707
Wheel of quality, 320–322
Wiener Skewing, 360
WLF633, 641
Wyman linkage function, 129

X
Xigris®, 11
Xolair ®, 27, 150, 350
X-ray powder diffraction, 788
Xyntha®, 10, 18

Y
Yeast, 7, 14

Z
Zemaira, 19
Zenapax®, 24, 27, 150
Zevalin ®, 24, 25, 26
Zorbtive®, 8
Zyggegator algorithm, 333

CPSIA information can be obtained
at www.ICGtesting.com
Printed in the USA
BVHW011459020522
635586BV00007B/1

9 780470 118122